REFERENCE

85

D0787501

GLENDALE
PUBLIC LIBRARY
_____

REFERENCE
_____

R   v.4
581.979   Abrams, L   c.1
A   Illustrated
            flora of the
            Pacific states

Not to be taken from   INV '76
      the Library
U-53   INV. '67 ●

17.50

ILLUSTRATED FLORA

ILLUSTRATED FLORA OF THE PACIFIC STATES
WASHINGTON, OREGON, AND CALIFORNIA

IN FOUR VOLUMES

VOL. I

OPHIOGLOSSACEAE TO ARISTOLOCHIACEAE
FERNS TO BIRTHWORTS

LEROY ABRAMS

VOL. II

POLYGONACEAE TO KRAMERIACEAE
BUCKWHEATS TO KRAMERIAS

LEROY ABRAMS

VOL. III

GERANIACEAE TO SCROPHULARIACEAE
GERANIUMS TO FIGWORTS

LEROY ABRAMS

VOL. IV

BIGNONIACEAE TO COMPOSITAE
BIGNONIAS TO SUNFLOWERS

ROXANA STINCHFIELD FERRIS

ILLUSTRATED FLORA

OF THE

PACIFIC STATES

WASHINGTON, OREGON, AND CALIFORNIA

BY

LEROY ABRAMS AND
ROXANA STINCHFIELD FERRIS

IN FOUR VOLUMES

VOL. IV

BIGNONIACEAE TO COMPOSITAE

BIGNONIAS TO SUNFLOWERS

BY

ROXANA STINCHFIELD FERRIS

STANFORD UNIVERSITY PRESS
STANFORD, CALIFORNIA

R
581.979
A
v.4
c.1

This ed. BIP 86-87 5000

STANFORD UNIVERSITY PRESS
STANFORD, CALIFORNIA

LONDON : OXFORD UNIVERSITY PRESS

© 1960 BY THE BOARD OF TRUSTEES OF THE
LELAND STANFORD JUNIOR UNIVERSITY

PRINTED AND BOUND IN THE UNITED STATES
OF AMERICA BY STANFORD UNIVERSITY PRESS

Library of Congress Catalog Card Number: 23-9934

APR 1 3 '60

PREFACE

On completion of this final volume of the *Illustrated Flora of the Pacific States*, there are many to whom I wish to express my sincere thanks. I am grateful to those who have contributed the text of genera and of families. Their willingness to prepare this material has shortened by years the time necessary to complete this volume for publication. My thanks are also given to the many others who have helped me with advice, material, and encouragement.

I wish especially to thank Sidney Fay Blake, though his name does not appear as author of any one group of the Compositae. Dr. Blake had originally planned to contribute the text of the Compositae but due to the pressure of other tasks was unable to do so. He generously made available to Dr. Abrams and later to me all the extensive notes and unfinished manuscript that he had assembled toward completion of that project. This material has been most helpful not only to me but to other contributors and much material included is based on his research. I wish to thank the following botanists for their contributions: Richard William Holm for the family Valerianaceae; Charles Bixler Heiser, Jr., for the genus *Helianthus*; David Daniels Keck for the genus *Eastwoodia*, the subtribe Madiinae, and the yellow-rayed genera of the tribe Astereae; Arthur John Cronquist for the genus *Erigeron* and the tribe Senecioneae; Malcolm Anthony Nobs for the genus *Achillaea*; George Henry Ward for the genus *Artemisia*; Carl William Sharsmith for the genus *Antennaria*; John Thomas Howell for the tribes Cynareae and Arctotideae; Kenton Lee Chambers for the genera *Nothocalais* and *Microseris*; and Quentin Jones for the genus *Agoseris*. Finally, I wish to express respect for and appreciation of LeRoy Abrams, to whom I owe my botanical training and who, long before his death in 1956, asked me to take the responsibility of preparing for publication this final volume of the *Illustrated Flora of the Pacific States*.

The illustrations in the book are for the most part the work of Jeanne Russell Janish, though many of the drawings of the Compositae are made by Doris Holmes Blake. The drawings of *Helianthus* and *Agoseris* are made by other illustrators under the direction of the respective authors of the text of those genera. Elisabeth F. Allen has ably assisted in reading proof. Barbara W. Law deserves especial credit for the arduous task of assembling the comprehensive index for all four volumes, in addition to the necessary editorial work of checking references.

Following the format of the preceding volumes, the description following each species heading, except in rare cases, refers to the name-bearing subspecies or variety where such segregations exist in literature. Deviations from this form are indicated in the text. The illustrations are reduced to one-half natural size except, of course, for the structural details. Any exception to this reduction is marked on the illustration to which it applies. Keys to the families of all four volumes are given in the Appendix, followed by a complete index of all scientific and common names that appear in the flora.

Roxana S. Ferris

Stanford University
November 1959

CONTENTS OF VOLUME IV

Sympetalae (concluded) 1–613

COMMON NAMES

Sympetalous Plants (concluded)

ILLUSTRATED FLORA

VOL. IV

Family 137. BIGNONIÀCEAE.

BIGNONIA FAMILY.

Trees, shrubs, or woody vines, or some exotic species herbs. Leaves opposite or rarely alternate, pinnately compound or simple, estipulate. Flowers showy, terminal or axillary, usually clustered. Corolla campanulate, funnelform, or tubular, 5-lobed and commonly bilabiate. Anther-bearing stamens 2 or 4, inserted on the corolla-tube and alternate with the lobes. Disk annular or cup-like. Ovary mostly 2-celled, with the placentae parietal or on the partitions of the ovary; style slender, with a terminal 2-lobed stigma. Fruit a 2-valved, sometimes woody, capsule. Seeds numerous, flat, winged, without endosperm; cotyledons broad, flat, entire or 2-lobed; radicle straight.

A family of about 100 genera and over 500 species, mainly tropical but a few in the warm temperate zones of both the northern and southern hemispheres.

1. CHILÓPSIS D. Don, Edinb. Phil. Journ. 9: 261. 1823.

Shrub with simple, usually alternate leaves. Flowers in terminal solitary racemes. Calyx inflated, deeply 2-lipped, upper lip 3-toothed, lower 2-toothed. Corolla funnelform, 5-lobed and obscurely 2-lipped. Anther-bearing stamens 4, the fifth represented by a rudimentary filament. Capsule linear and terete, somewhat woody. Seeds flat, with their wings dissected into long hairs. [Greek, meaning lip and resemblance.]

A monotypic genus of the arid southwestern United States and Mexico.

1. Chilopsis lineàris (Cav.) Sweet. Desert Willow. Fig. 4945.

Bignonia ? linearis Cav. Ic. **3**: 35. *pl. 269.* 1796.
Chilopsis saligna D. Don, Edinb. Phil. Journ. **9**: 261. 1823.
Chilopsis linearis Sweet, Hort. Brit. 283. 1827.
Chilopsis linearis var. *arcuata* Fosberg, Madroño **3**: 366. 1936.

A willow-like shrub, usually with several stems, 2–6 m. high. Leaves narrowly linear to linear-lanceolate, 5–15 cm. long, 2–6 mm. wide, long-attenuate at both ends, pale green, glabrous, and similar on both sides; racemes few- to many-flowered; calyx 6–8 mm. long, pubescent; corolla 2.5–3.5 cm. long, fragrant, white with pink lines or pink; capsule firm, 15–27 cm. long, 4–5 mm. in diameter, tapering at both ends.

Frequent along desert watercourses, Lower Sonoran Zone; Mojave and Colorado Deserts, California, and adjacent Nevada and Arizona east to western Texas and south to Lower California and the mainland of Mexico. Type locality: not stated. May–Aug. Sometimes cultivated as an ornamental.

Family 138. MARTYNIÀCEAE.

UNICORN-PLANT FAMILY.

Herbs with mostly opposite estipulate leaves and perfect irregular flowers. Calyx 4–5-cleft or 4–5-parted, or sometimes divided to the base on the lower side and spathaceous. Corolla sympetalous, irregular, the tube oblique, often decurved, the limb 5-lobed and slightly 2-lipped, the lobes nearly equal, upper 2 exterior in the bud. Anther-bearing stamens 2 or 4, didynamous or the posterior pair sometimes sterile; anthers 2-celled, longitudinally dehiscent. Ovary superior, 1-celled, with 2 parietal placentae expanded into broad surfaces, or appearing 2–4-celled by the intrusion of the placentae or by false partitions; ovules numerous or sometimes few, anatropous; style slender; stigma 2-lobed or 2-lamellate. Fruit various. Seeds often compressed; endosperm none; cotyledons large.

A family of 5 genera and about 15 species, natives of temperate and tropical regions of the western hemisphere.

1

Calyx of 5 free sepals; endocarp of the fruit strongly echinate.   1. *Ibicella.*
Calyx 4–5-dentate but split to the base on the lower side and more or less spathaceous.   2. *Proboscidea.*

1. **IBICÉLLA** Van Eseltine, N.Y. State Agr. Exp. Sta. Tech. Bull. No. 149: 31. 1929.

Stout, viscid-pubescent annuals and strongly scented. Leaves opposite or the upper sometimes alternate, ovate to suborbiculate. Inflorescence a dense terminal raceme. Calyx membranous, parted to the base into 5 sepals of unequal width. Corolla with a short tube, funnelform-campanulate throat, and 5-lobed limb, oblique. Fertile stamens 4, didynamous, the rudiment of the fifth stamen often present. Capsule 2-valved, with a curved beak, crested above, the exocarp rather fleshy, the endocarp densely echinate. Seeds compressed, rugose. [Name from the Latin, meaning chamois, because of the curved horns on the fruit.]

A genus of 2 species, natives of South America. Type species, *Martynia lutea* Lindl.

1. **Ibicella lùtea** (Lindl.) Van Eseltine. Yellow Unicorn-plant. Fig. 4946.

Martynia lutea Lindl. Bot. Reg. 11: *pl. 934.* 1825.
Proboscidea lutea Stapf in Engler & Prantl, Nat. Pflanzenf. 4[3b]: 269. 1895.
Ibicella lutea Van Eseltine, N.Y. State Agr. Exp. Sta. Tech. Bull. No. 149: 34. *pl. 14.* 1929.

Glandular-pubescent annual with stout spreading stems 3–6 dm. long. Leaves opposite or the upper sometimes alternate, suborbicular, mostly about 10 cm. broad, incised or subcordate at base, denticulate; inflorescence short-racemose, few-flowered and rather dense; sepals distinct, about 1.5 cm. long, the two lower much broader; corolla yellow and glandular on the outside, glabrous and orange or deep yellow within and often dotted with red, about 2.5 cm. long; stamens 4 with an obvious rudimentary fifth; fruit-body oblong-ovoid, crested above, about 5 cm. long; endocarp conspicuously echinate, with short ascending spines, the horns about twice as long as the body.

Sparingly introduced in California and mostly found in cultivated areas in the Sacramento and northern San Joaquin Valleys and in the central coastal area. Type locality: Brazil. June–Sept.

2. **PROBOSCÍDEA** Keller in Schmidel, Icon. (ed. Keller) 49. 1762.

Stout, annual or perennial herbs, glandular or viscid-pubescent and strongly scented. Leaves opposite or the upper sometimes alternate, long-petioled, and having broad blades. Flowers large in loose, open, terminal racemes. Calyx spathe-like, membranous, 5-lobed, split ventrally to the base, deciduous. Corolla funnelform-campanulate, declined, the limb oblique; lobes 5, nearly equal, spreading. Capsule 2-valved, loculicidal, crested below and also sometimes above, ending in a prominent incurved beak, becoming falsely 4-celled by the extension of the placentae; exocarp fleshy, separating in age from the woody sculptured endocarp. Seeds numerous, tuberculate. [From the Greek, meaning snout, in allusion to beak of fruit.]

A genus of about 9 species, native of central and southwestern United States, Mexico, and South America. Type species, *Martynia louisianica* Mill.

Fruit-body slender and subcylindrical, often crested both dorsally and ventrally; flowers yellowish or brownish; leaves 3–5 cm. wide.   1. *P. althaeifolia.*
Fruit-body stout, more or less ovoid, crested only on the dorsal side; flowers reddish purple or white dotted with purple, the throat often striped with yellow; leaves about 10 cm. wide.   2. *P. louisianica.*

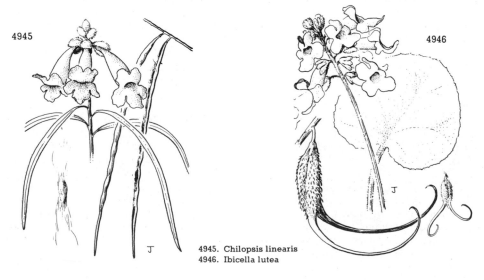

4945. Chilopsis linearis
4946. Ibicella lutea

1. Proboscidea altheaefòlia (Benth.) Decne. Desert Unicorn-plant. Fig. 4947.

Martynia altheaefolia Benth. Bot. Sulph. 37. 1844.
Proboscidea altheaefolia Decne. Ann. Sci. Nat. V. **3**: 324. 1865.
Martynia palmeri S. Wats. Proc. Amer. Acad. **24**: 66. 1889.

Spreading, viscid-pubescent perennial 3–4 dm. high with a long, yellow, fusiform root. Leaves 3–5 cm. broad, reniform to broadly ovate in outline, often broadly and shallowly 3–5-lobed; petioles 3–8 cm. long, densely glandular-pubescent; racemes several-flowered, often becoming 15–20 cm. long; calyx 1–1.5 cm. long; corolla 2.5–3.5 cm. long, buff to brownish yellow; body of the fruit slender, 5–6 cm. long, horns two to three times as long, the distal teeth of the crest on the dorsal suture often prolonged into a slender horn, the ventral suture not at all or only slightly crested.

Usually in sandy soils, Lower Sonoran Zone; Colorado Desert, California, and western Arizona southward into Sinaloa and southern Lower California. Rare within our range and limited to Imperial County, California. Type locality: Magdalena Bay, Lower California. May–Aug.

2. Proboscidea louisiánica (Mill.) Thell. Common Unicorn-plant. Fig. 4948.

Proboscidea jussieui Keller in Schmidel, Icon. (ed. Keller) 49. *pls. 12–13.* 1762. Not a binomial, Leaflets West. Bot. **1**: 80. 1934.
Martynia louisiana Mill. Gard. Dict. ed. 8. no. 3. 1768.
Martynia louisianica Mill. op. cit. in corrigenda.
Martynia proboscidea Gloxin, Obs. 14. 1785.
Martynia alternifolia Lam. Encycl. **2**: 112. 1786.
Proboscidea louisianica Thell. Mem. Soc. Sci. Cherbourg IV. **38**: 480. 1912.
Proboscidea louisiana Woot. & Standl. Contr. U.S. Nat. Herb. **19**: 602. 1915.
Martynia jussieui J. T. Howell, Leaflets West. Bot. **1**: 40. 1933.

Annual with prostrate or ascending branches 3–10 dm. long, slimy-viscid pubescent throughout. Leaves opposite or the upper subalternate, deeply cordate, broadly ovate to suborbicular in outline, 5–20 cm. broad, entire or sinuate; petioles 5–15 cm. long, stout, densely short-pubescent; racemes open, several-flowered, elongated in fruit; calyx 1.5–2 cm. long, the lobes acutish to obtuse; corolla 3.5–5 cm. long, yellow or dull white blotched with reddish purple or sometimes purple throughout; body of the fruit 4–6 (or rarely –10) cm. long, crested on the upper side only, the horns as long or three times as long as the body.

Usually in low moist ground, Sonoran Zones; Sacramento Valley and coastal valleys south to southern California, eastward to the southeastern United States, and southward into Mexico. Type locality: not definitely known; probably southeastern United States or eastern Mexico. May–Sept. Devils-claws.

Family 139. OROBANCHÀCEAE.

BROOMRAPE FAMILY.

Root-parasites without green foliage, the stems erect, simple or branched, usually yellowish or purplish. Leaves reduced to alternate appressed scales. Flowers perfect, irregular, solitary, or pedunculate or sessile in terminal bracteate racemes or spikes. Calyx free from the ovary, 4–5-toothed or 4–5-cleft, or split nearly or quite to the base on one or both sides. Corolla sympetalous, more or less irregular, the limb 2-lipped and 5-lobed, usually with a pair of bracteoles. Stamens 4, didynamous, slender, inserted on the corolla-tube, a rudimentary fifth one sometimes present; filaments slender; anthers 2-celled, the sacs parallel and equal. Ovary superior, 1-celled with 4 parietal placentae; ovules numerous; style slender; stigma discoid or broadly 2–4-lobed. Capsule 1-celled, 2- or 4-valved. Seeds numerous, reticulated, wrinkled or striate; embryo minute with the cotyledons scarcely differentiated.

About 14 genera and over 200 species of wide geographical distribution.

Capsule 2-valved with 2 placentae on each valve; filaments not with hairy tuft at base.   1. *Orobanche.*
Capsule 4-valved with 1 placenta on each valve; filaments with tuft of hairs at base.   2. *Boschniakia.*

1. OROBÁNCHE [Tourn.] L. Sp. Pl. 632. 1753.

Glandular-pubescent herbs, yellowish, purplish, or rarely white, parasitic on the roots of various plants with the leaves reduced to scattered scales. Flowers yellowish or purplish, solitary on long peduncles or more commonly spicate, racemose, or corymbose, with or without bracts. Calyx 4–5-cleft into acute or acuminate lobes. Corolla tubular, more or less 2-lipped, the upper lip erect, 2-lobed or entire, the lower lip spreading, 3-lobed. Stamens included; anther-sacs mostly mucronate at base. Ovary ovoid; placentae equidistant or approximate in pairs. Style curved outward at the apex; stigma peltate to funnelform, entire or bilobed. [Name Greek, meaning choke and vetch.]

About 100 species, native of North and South America and the Old World. Type species, *Orobanche major* L.

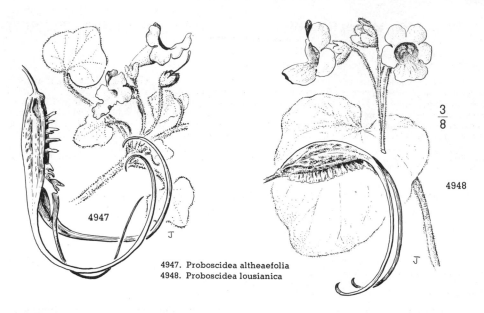

$\dfrac{3}{8}$

4948

4947

4947. Proboscidea altheaefolia
4948. Proboscidea lousianica

Flowers on long slender pedicels, solitary or several on the stem, without floral bracts.
Stems slender, very short, bearing 1 or sometimes 2–3 pedicels many times exceeding the length of the stem; calyx-lobes subulate-attenuate, longer than the tube.   1. *O. uniflora occidentalis.*
Stems stout, each bearing 3–12 pedicels equaling or shorter than the stem; calyx-lobes triangular or lanceolate, equaling or shorter than the tube.   2. *O. fasciculata.*
Inflorescence spicate, corymbose, or paniculate, the flowers sessile or pedicellate; floral bracts 2 (3 in *O. ramosa*).
Calyx not parted to the base before and behind.
Palatal folds obsolete or nearly so; inflorescence a compact, pyramidal, thyrsoid panicle of sessile flowers; base of stem greatly bulbose-thickened.   3. *O. bulbosa.*
Palatal folds present (small in *O. pinorum*); inflorescence various, the flowers pedicellate, or sessile in the spicate forms; base of stem not greatly bulbose-thickened (somewhat so in *O. corymbosa*).
Calyx 5-toothed or -divided, occasionally 6; native species.
Calyx short, the lobes equaling or a little surpassing the tube; basal attachment-body a con-spicuous solid mass 15–20 mm. in diameter.   4. *O. pinorum.*
Calyx 10–16 mm. long, the lobes usually much surpassing the tube; basal attachment, when evident, of loose thickened strands.
Flowers pedicellate, the lower pedicels 8–25 mm. long, sometimes shorter above; inflores-cence corymbose, subracemose, or paniculate.
Inflorescence openly or compactly corymbose; anthers woolly.
Inflorescence mostly open-corymbose; corolla-lips about 10–14 mm. long; lobes of the upper corolla-lip acute (rounded in var. *violacea*).   5. *O. grayana.*
Inflorescence compactly corymbose; corolla-lips about 7–8 mm. long; lobes of the upper corolla-lips obtuse or rounded.   7. *O. corymbosa.*
Inflorescence subracemose or paniculate; anthers glabrous or sparsely hairy.   6. *O. californica.*
Flowers sessile or the lower ones short-pedicellate; inflorescence essentially spicate, the axis much elongated in age.
Corolla-lobes, especially those of the lower lip, narrowly acute; calyx-lobes one and one-half times to twice the length of the tube.   8. *O. cooperi.*
Corolla-lobes obtuse or rounded; calyx-lobes twice or more the length of the tube.   9. *O. multiflora arenosa.*
Calyx 4-toothed; introduced species.   10. *O. ramosa.*
Calyx parted to the base before and behind, the divisions 2-toothed.   11. *O. minor.*

1. **Orobanche uniflòra** subsp. **occidentàlis** (Greene) Abrams ex Ferris. Naked Broomrape. Fig. 4949.

Aphyllon uniflorum var. *occidentale* Greene, Man. Bay Reg. 285. 1894.
Aphyllon sedi Suksd. Deutsch. Bot. Monatss. **18**: 155. 1900.
Aphyllon minutum Suksd. loc. cit.
Orobanche uniflora f. *sedi* Beck, Pflanzenreich 4[261]: 48. 1930.
Orobanche uniflora var. *minuta* Beck, op. cit. 49.
Orobanche uniflora var. *sedi* Achey, Bull. Torrey Club **60**: 446. 1933.
Orobanche uniflora subsp. *occidentalis* Abrams ex Ferris, Contr. Dudley Herb. **5**: 99. 1958.

Stems simple, 0.5–2 cm. long, subterranean, bearing several ovate-oblong scales. Pedicels usu-ally 1 (2 or rarely 3), erect, slender, scape-like, 3–8 cm. high, glandular-pubescent, mostly 1-flow-ered; calyx campanulate, 7–15 mm. high, lobes lanceolate-subulate, attenuate, much longer than the tube; corolla pale or dark purple to almost white, 12–22 mm. long, tube curved, throat 3–5 mm. wide, the lobes broadly ovate, subequal, scarcely spreading, 2–3.5 mm. long; anthers glabrous or sometimes pubescent.

Parasitic on various plants, principally on *Saxifragaceae* and on *Sedum*, open woods and mossy banks, Transition and lower Boreal Zones; Montana and British Columbia to southern California and east to the Rocky Mountains. Type locality: "Wooded stony hills," western central California. March–Aug.

Orobanche uniflora var. **purpùrea** (Heller) Achey, Bull. Torrey Club **60**: 445. 1933. (*Thalesia purpurea* Heller, Bull. Torrey Club **24**: 313. 1897; *Aphyllon inundatum* Suksd. Allg. Bot. Zeit. **12**: 27. 1906; *Orobanche uniflora* f. *inundata* Beck, Pflanzenreich **4**²⁶¹: 49. 1930; *O. porphyrantha* Beck, loc. cit.) Stems 2–7 cm. long, sometimes branched; pedicels usually 2, 5–15(20) cm. long, rather stout, glandular-villous; corolla 2.2–3 cm. long, pale to deep violet, somewhat streaked with yellow, tube somewhat curved, throat constricted below, about 6–8 mm. wide above; corolla-lobes rounded or obovate, spreading; anthers glabrous or hairy. Parasitic on various plants including the *Compositae*, open rocky ground; Idaho and British Columbia south through Washington and Oregon to Nevada County in the Sierra Nevada and Santa Cruz County in the Coast Ranges, California. Type locality: near Lewiston, Nez Perce County, Idaho.

Approaching the typical species, which is found mostly in the eastern United States, in growth habit but differing by having long-attenuate calyx-lobes instead of triangular-acute lobes that scarcely surpass the calyxtube, a characteristic of the eastern species.

2. **Orobanche fasciculàta** Nutt. Clustered Broomrape. Fig. 4950.

Orobanche fasciculata Nutt. Gen. **2**: 59. 1818.
Loxanthes fasciculatus Raf. Neogen. 3. 1825.
Anoplanthus fasciculatus Walp. Rep. **3**: 480. 1844–45.
Aphyllon fasciculatum Torr. & Gray ex A. Gray, Man. ed. 2. 281. 1856.
Thalesia fasciculata Britt. Mem. Torrey Club **5**: 298. 1894.

Stems fleshy, mostly subterranean but rising 2–6 cm. above ground, simple or compound with several erect branches bearing prominent scales and several to many 1-flowered pedicels, glandular-pubescent throughout. Calyx campanulate, 6–9 mm. long, the lobes broadly triangular-subulate to narrowly so, about equaling the tube; corolla usually purplish yellow or often strongly stramineous, 2–3 cm. long, the tube straight, its lobes usually 5–6 mm. long and usually rounded; capsule broadly ovoid, mostly 8–10 mm. long.

Parasitic on the roots of various plants, Transition Zone; British Columbia to southern California east to Michigan. In the Pacific States it is found in the Olympic, Cascade, and Blue Mountains of Washington and Oregon, and in the Coast Ranges and Sierra Nevada in California. Type locality: Fort Mandan, North Dakota. May–Aug. A variable species from which two intergrading varieties have been segregated.

Orobanche fasciculata var. **lùtea** (Parry) Achey, Bull. Torrey Club **60**: 449. figs. *12–13*. 1933. (*Phelipaea-lutea* Parry, Amer. Nat. **8**: 214. 1874; *Aphyllon fasciculatum* var. *luteum* A. Gray, Syn. Fl. N. Amer. **2**¹: 312. 1878; *Orobanche fasciculata* f. *lutea* Beck, Pflanzenreich **4**²⁶¹: 51. 1930.) Calyx-lobes equaling or shorter than the tube; corolla yellow and sometimes tinged with lavender, corolla-tube slightly constricted at base, corolla-lobes usually acute. Parasitic on grasses and other hosts; Alberta, Montana, and Idaho south in the Pacific States to northern California, east to North and South Dakota, and south to Chihuahua and Sonora. Type locality: Owl Creek, Wyoming.

Orobanche fasciculata var. **franciscàna** Achey, Bull. Torrey Club **60**: 450. figs. *14–15*. 1933. Pedicels 4–12; calyx-lobes somewhat longer than the tube; corolla straw-colored tinged with purple, corolla-tube not constricted, corolla-lobes rounded or truncate, sometimes acute. Parasitic on various hosts; southern Oregon south to San Diego County, California. Type locality: Mount Tamalpais, Marin County, California.

3. **Orobanche bulbòsa** Beck. Chaparral Broomrape. Fig. 4951.

Phelipaea tuberosa A. Gray, Proc. Amer. Acad. **7**: 371. 1868.
Aphyllon tuberosum A. Gray, Bot. Calif. **1**: 585. 1876.
Orobanche bulbosa Beck, Bibl. Bot. **4**: 83. *pl. 1, fig. 7.* 1890.
Orobanche tuberosa Heller, Cat. N. Amer. Pl. 7. 1898. Not Hook. Fl. Bor. Amer. **2**: 92. 1838.

Stems fleshy, dark purplish brown, arising from a tuber-like attachment to host, 1–2 dm. high, the stems much enlarged and sometimes bulbous at base. Scales at the base crowded, ovate, obtuse or rounded, appressed, the upper more scattered, triangular to lanceolate, acute, spreading, puberulent; inflorescence hoary-scabridulous, a dense and more or less pyramidal thyrsoid panicle; floral bracts lanceolate to lanceolate-subulate; flowers sessile; calyx 1 cm. long, unequally cleft, the lobes as long as or longer than the tube; corolla varying in color from yellow or brownish to purple or bluish, 10–15 mm. long, palatal folds obsolete or nearly so, the lobes 2 mm. or more long, acute, barely spreading; anthers white, glabrous or subglabrous.

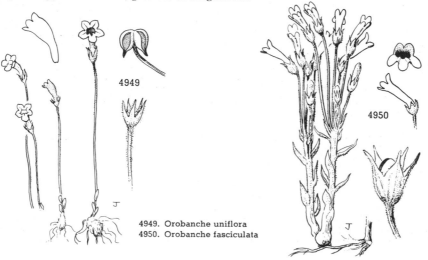

4949. Orobanche uniflora
4950. Orobanche fasciculata

Parasitic on various shrubs, usually in the dry chaparral, Upper Sonoran Zone; Solano County in the Coast Ranges and El Dorado County in the Sierra Nevada south to San Diego County, the southern California islands and northern Lower California. Type locality: "On a high and dry ridge of the Gavilan Mountains, in sandy soil," California. May–Oct.

4. Orobanche pinòrum Geyer ex Hook. Pine Broomrape. Fig. 4952.

Orobanche pinorum Geyer ex Hook. Kew Journ. Bot. **3**: 297. 1851.
Phelipaea pinetorum A. Gray, Proc. Amer. Acad. **7**: 371. 1868.
Aphyllon pinetorum A. Gray, Bot. Calif. **1**: 585. 1876.
Myzorriza pinorum Rydb. Bull. Torrey Club **36**: 695. 1909.

Stem arising from a large, tuber-like attachment to host, the base of the stem not greatly enlarged, rather slender and simple or branched above the middle into a few erect branches, growing stems lavender to purple becoming yellowish, minutely cinereous-pubescent and somewhat viscid. Scales of the stem lanceolate and mostly acute, 4–10 mm. long; flowers many, pedicellate, in a simple or branched raceme; pedicels 2–8 mm. long, strictly ascending; floral bracts linear-subulate, 3–5 mm. long; calyx 7–8 mm. long, cinereous-tomentulose, lobes subulate, about equaling the narrowly campanulate tube; corolla short-pubescent, yellowish or slightly purplish, 1.5–2 cm. long, the palatal folds small and obscure, tube constricted and curved outward at the base of the limb, obscurely 2-lipped, the lobes subequal, 3–4 mm. long, oblong; anthers glabrous or sparsely pubescent.

In coniferous forests, parasitic on the roots of various conifers and also reported from *Holodiscus,* mainly Transition Zone; British Columbia south in the Cascade Mountains to the Willamette Valley, Oregon, and the Siskiyou Mountains in northern Humboldt County, and in Santa Cruz County, California, and eastward to Idaho and the Blue Mountains, Oregon. Type locality: "Top of the high mountains near St. Joseph, Coeur d'Aleine country, growing on the roots of *Abies balsaminea,*" Idaho. June–Aug.

5. Orobanche grayàna Beck. Gray's Broomrape. Fig. 4953.

Orobanche comosa Hook. Fl. Bor. Amer. **2**: 92. *pl. 169.* 1838. Not Wallr. 1822.
Aphyllon comosum A. Gray, Bot. Calif. **1**: 584. 1876.
Orobanche grayana Beck, Bibl. Bot. **4**: 79. 1890.
Myzorrhiza grayana Rydb. Bull. Torrey Club **36**: 695. 1909.
Orobanche grayana var. *nelsonii* Munz, op. cit. **57**: 616. *pl. 38, fig. 7.* 1930.

Stems 5–20 cm. high, glandular-puberulent, simple below, the inflorescence corymbosely branched. Cauline scales lanceolate to ovate, 5–10 mm. long; inflorescence corymbose or corymbose-paniculate, 4–8 cm. long, the pedicels 5–25 mm. long; floral bracts linear-subulate, usually alternate on the pedicels; calyx 12–16 mm. long, the subulate lobes 10 mm. or more in length; corolla purplish with darker lines, occasionally white, 25–30 mm. long, the tube rather slender and palatal folds evident; corolla-lips 10–14 mm. long, the upper usually reflexed, the lobes acute, the lower with the narrow, sharply acute (rarely emarginate) lobes spreading; anthers with woolly hairs; stigma peltate.

Open ground or in woods, parasitic on *Grindelia* and other plants, Transition Zone; western Washington on the islands of Puget Sound and the adjacent mainland to the Willamette Valley, Oregon; also eastward along the Columbia River and southward through the Cascade Mountains to Lake Tahoe, California. The southern form varies somewhat in general appearance and has a less deeply incised corolla and usually wider corolla-lobes. Found growing with the closely related *O. corymbosa.* Type locality: "Banks of the Columbia. *Douglas, Dr. Scouler, Dr. Gairdner."* June–Oct.

The collections of Scouler and Gairdner were made west of the Cascade Mountains, principally near Fort Vancouver, a region in which Douglas also collected. Because of this and the close agreement of herbarium specimens from this region with the illustration in the original reference, the type locality is taken to be on the banks of the Columbia west of the Cascades.

Orobanche grayana var. *violàcea* (Eastw.) Munz, Bull. Torrey Club **57**: 616. *pl. 38, fig 6.* 1930. (*Aphyllon violaceum* Eastw. Zoe **5**: 85. 1900.) Inflorescence corymbose, the branches rather short; corolla deep purple or violet, 25–45 mm. long, the lobes of the upper lip rounded. Bluffs and sand dunes near the sea, parasitic on *Grindelia;* Humboldt County to San Luis Obispo County, California. Type locality: near Tomales Bay, Marin County.

Orobanche grayana var. *feùdgei* Munz, op. cit. 616. *pl. 38, fig 8.* Stems stout, 10–25 cm. high, simple, or additional stems arising from the base; inflorescence densely flowered, corymbosely branched or subpaniculate, 5–10 cm. long; calyx 14–18 mm. long; pedicels 5–18 mm. long; corolla purplish brown, marked with yellow at the palatal folds, the upper lip broad, 10–12 mm. long, the lower lip 9–12 mm. long. the narrow, usually lanceolate divisions about as long as the lip. Rocky slopes, Upper Sonoran and Transition Zones; Kern County, California, south in the mountains of southern California to northern Lower California, and on higher slopes of the Panamint Mountains, Inyo County, California. Apparently parasitic on *Artemisia.* Type locality: Baldwin Lake, San Bernardino Mountains, San Bernardino County, California. May–July.

Orobanche grayana var. *jepsònii* Munz, op. cit. 617. *pl. 38, fig. 10.* Stems 1–2 dm. high, corymbosely branched above, the divisions often long-racemose; pedicels longer than in related forms, 10–25 mm. high; corolla usually pinkish with darker veins, the lips 10–12 cm. long, the divisions acute. Open places, parasitic on *Baccharis, Grindelia,* and probably other composites, Upper Sonoran Zone; in the Sacramento Valley in Colusa County, California, south to Tulare County and in the adjacent foothills of the Coast Ranges from Sonoma County to Santa Clara County. Type locality: Princeton. Colusa County.

6. Orobanche califórnica Cham. & Sch. California Broomrape. Fig. 4954.

Orobanche californica Cham. & Sch. Linnaea **3**: 134. 1828.
Phelipaea californica G. Don, Gen. Hist. Pl. **4**: 632. 1838.
Aphyllon californicum A. Gray, Bot. Calif. **1**: 584. 1876.
Myzorrhiza californica Rydb. Bull. Torrey Club **36**: 695. 1909.
Orobanche comosa var. *vallicola* Jepson, Man. Fl. Pl. Calif. 952. 1925.

Plants 7–25(30) cm. high, glandular-puberulent, simple below or with axillary stems arising from the somewhat thickened base. Cauline scales narrowly or broadly ovate, acute, 8–10 mm. long; inflorescence subracemose or paniculate, densely flowered, 4–11 cm. long, the lower pedicels 10–20 mm. long, the upper often not more than 2 mm. long; floral bractlets linear, usually alternate, the uppermost subtending the calyx, 7–15 mm. long; calyx 10–15 mm. long, the linear lobes 7–10 mm. long; corolla with palatal folds, pale, the veins dark, 18–25 mm. long, straight, surpassing the calyx-lobes; upper and lower corolla-lips equal, 6 mm. long, the upper lip erect, cleft 3 mm. into 2 broadly acute or rounded lobes, the lower 3-cleft, the lobes 5 mm. long; anthers sparsely long-pilose or glabrous; mature capsule 9–10 mm. long, 6–7 mm. broad.

Open hillsides, mostly Upper Sonoran Zone; Shasta County, California, south in the Coast Ranges to the Santa Lucia Mountains, Monterey County. Parasitic on various shrubs; at the type locality growing on *Eriophyllum.* Type locality: San Francisco. July–Nov.

Orobanche californica var. **parishii** Jepson, Man. Fl. Pl. Calif. 952. 1925. Plants about 1 dm. high; inflorescence compact, subpaniculate to subracemose, the pedicels short; cauline bracts and calyx submembranaceous, many-nerved; calyx 10 mm. long; corolla yellowish with darker veins, 18–22 mm. long, the lips 5–7 mm. long, the lobes obtuse or rounded; anthers sparsely hairy to glabrate. Tehachapi Mountains (Fort Tejon) southward in the Transition Zone to the mountains of Lower California. Type locality: Bear Valley, San Bernardino Mountains, California.

Orobanche californica var. **claremonténsis** Munz, Bull. Torrey Club 57: 618. *pl. 39, fig. 11.* 1930. Plants 1–2 dm. tall; inflorescence corymbosely paniculate, 8–11 cm. long, densely flowered, the pedicels 5–10 mm. long; corolla pale, tinged with purple, 25 mm. long; corolla-lips 8 mm. long, the upper lip lobed to the base; anthers glabrate. Apparently parasitic on *Quercus agrifolia* and known only from the type locality, Claremont, Los Angeles County, California.

The characteristics of *Orobanche comosa* var. *vallicola* are to some degree intermediate between *O. californica* and *O. grayana.* The growth habit is that of *O. californica* with the inflorescence subracemose, densely flowered, and elongated in age, and the flowers with short pedicels. The main axis occasionally with 2 densely flowered branches, quite different from the type of *O. grayana* var. *jepsonii* Munz with corymbosely branched stems and long-pedicelled flowers. The divisions of the lower lip of the corolla, however, are rather narrowly lanceolate, a characteristic of the *grayana* complex.

7. **Orobanche corymbòsa** (Rydb.) Ferris. Rydberg's Broomrape. Fig. 4955.

Myzorrhiza corymbosa Rydb. Bull. Torrey Club **36**: 696. 1909.
Orobanche californica var. *corymbosa* Munz, op. cit. **57**: 618. 1930.
Orobanche corymbosa Ferris, Contr. Dudley Herb. **5**: 99. 1958.

Plants glandular-pubescent, 5–12 cm. high, the simple stems often bulbose-thickened at the

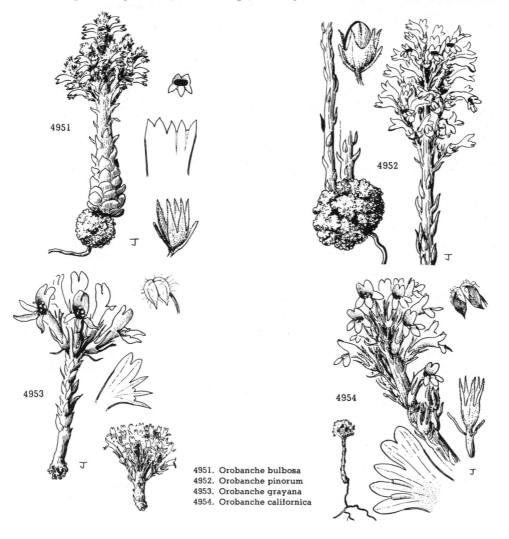

4951. Orobanche bulbosa
4952. Orobanche pinorum
4953. Orobanche grayana
4954. Orobanche californica

base. Cauline scales ovate or ovate-lanceolate; inflorescence corymbosely branched and very compact, 2.5–5 cm. long, the pedicels 3–10 mm. long; floral bracts linear-lanceolate, about 10 mm. long; calyx 13–15 mm. long, the lanceolate-subulate lobes about 10 mm. long; corolla purple, paler without, the upper lip darker than the lower, 25–30 mm. long, the corolla-lips 4–6(7) long, throat of the corolla curved toward the outer part of the inflorescence, the palatal folds evident; upper lip shallowly lobed or cleft, the lobes rounded or retuse; lower lip cleft to the base into lanceolate acutish divisions; anthers woolly; stigma peltate.

Sagebrush slopes and plains, Arid Transition Zone to Canadian and Hudsonian Zones; in the Pacific States east of the Cascade Mountains in Washington and Oregon and for the most part on the east face of the Sierra Nevada as far south as Tulare County, California; eastward to Montana, Wyoming, and Utah. Parasitic on *Artemisia* and associated shrubs. Type locality: Reynold's Creek, Idaho. June–Aug.

8. **Orobanche coòperi** (A. Gray) Heller. Cooper's or Desert Broomrape. Fig. 4956.

Aphyllon cooperi A. Gray, Proc. Amer. Acad. **20**: 307. 1885.
Orobanche ludoviciana var. *cooperi* Beck, Bibl. Bot. **4**: 81. 1890.
Orobanche cooperi Heller, Cat. N. Amer. Pl. 7. 1898.
Myzorrhiza cooperi Rydb. Bull. Torrey Club **36**: 695. 1909.
Orobanche ludoviciana var. *latiloba* Munz, op. cit. **57**: 621. 1930.

Plants fleshy and stout, 1–3 dm. tall, simple or with a few axillary stems arising from the base, viscid-puberulent. Cauline bracts obtuse, 5–10 mm. long; inflorescence densely flowered, 6–18 cm. long, spicate or the main axis branching, bearing 2–3 axillary spikes; the lowest flowers pedicellate, the rest sessile; calyx 5–9(12) mm. long, the tube 2–4 mm. long, the lobes lanceolate-attenuate; floral bracts shorter than the calyx; corolla purple within, with yellow markings at the palatal folds, grayish purple and puberulent without, 15–30 mm. long, the corolla-lips usually 4–8 mm. long; the upper lip erect, short-cleft at the apex, the narrowly or broadly acute lobes erect or somewhat reflexed; the lobes of the deeply cleft lower lip somewhat spreading, narrowly or broadly acute; anthers rather sparsely hairy; stigma peltate, with a crenulate reflexed margin, sometimes splitting and appearing bilobed.

Sandy desert flats and washes, Lower Sonoran Zone; southern Utah and Nevada and the Death Valley region, California, south through the Mojave and Colorado Deserts to Lower California and eastward to Sonora and New Mexico. Parasitic on *Franseria, Hymenoclea,* and other desert shrubs. Type locality: Fort Mohave, Arizona. April–June.

The flowers of some specimens from the southwestern edge of the Colorado Desert, California, resemble those of *O. multicaulis* Brandg. from Lower California. They lack, however, the bilobed stigma and other features of that species and are here considered to be variations of *O. cooperi.*

Orobanche válida Jepson, Madroño **1**: 255. 1929. (*Orobanche ludoviciana* var. *valida* Munz, Bull. Torrey Club **57**: 621. *pl. 39, fig. 16.* 1930.) Plants 1–2 dm. high, dark purplish brown throughout, rather slender; cauline bracts lanceolate; inflorescence spicate, the flowers separated, sessile except for the lowest; calyx 7–9 mm. long; corolla 12–14 mm. long, dark purple, the upper lip 3–4 mm. long, the lower a little shorter, yellowish, the lobes with dark purple veins; anthers glabrous. San Gabriel Mountains, at an altitude of 5,500–6,500 feet, Los Angeles County, California. Type locality: Rock Creek, San Gabriel Mountains. The relationship of this little-known species will doubtless be more clearly understood when further collections are made.

9. **Orobanche multiflòra** var. **arenòsa** (Suksd.) Munz. Suksdorf's Broomrape. Fig. 4957.

Aphyllon arenosum Suksd. Allg. Bot. Zeit. **17**: 27. 1906.
Orobanche multiflora var. *arenosa* Munz, Bull. Torrey Club **57**: 623. 1930.

Plants 5–18 cm. high with a grayish glandular pubescence. Cauline bracts ovate to ovate-lanceolate; inflorescence spicate, up to 10 cm. long, flowers except a few at the base sessile; floral bracts linear, shorter than the calyx; calyx rather irregularly cleft into linear-lanceolate lobes twice or more the length of the tube, 9–12 mm. long; corolla yellowish or pinkish, tinged with purple, glandular-pubescent, 15–20 mm. long, the upper lip about 5 mm. long, shallow-cleft with rounded lobes, the lower lip a little shorter, divided to the base into narrow, obtuse or rounded divisions; anthers glabrous.

Sandy places, Upper Sonoran and Arid Transition Zones; Idaho, Wyoming, Nevada, and Utah and south to Arizona; in the Pacific States in eastern Washington southward east of the Cascade Mountains through Oregon and occasionally collected east of the Sierra Nevada Divide in California; also reported from the Providence Mountains, San Bernardino County, California. Parasitic on the *Compositae.* Type locality: Bingen, Klickitat County, Washington. April–July.

This variety is far removed from the large-flowered, typical form with deeply cut corolla-lips which is found in the Rio Grande Valley, but until the relationship of the various forms is studied further it seems best to leave var. *arenosa* in its present status. A specimen collected south of Needles, California, near the Colorado River, having rounded corolla-lobes is doubtfully assigned to *O. multiflora* var. *pringlei* Munz (Bull. Torrey Club **57**: 623. 1930).

10. **Orobanche ramòsa** L. Branched Broomrape. Fig. 4958.

Orobanche ramosa L. Sp. Pl. 633. 1753.

Plants slender, yellowish, usually freely branching below into several to many elongated branches, puberulent below, glandular-tomentose above, 15–25 cm. high. Cauline bracts few, lanceolate, 7–8 mm. long; spikes loosely many-flowered, the lowest flowers short-pedicellate, 2–3 mm. long; floral bracts 3, one ovate subtending the flower, the others linear, about 3 mm. long; calyx including the 4 attenuate lobes 5 mm. long; corolla 10–15 mm. long, the tube yellowish, the lips bluish.

An adventive from Asia parasitic on tomatoes and other cultivated plants. Known from Alameda County, California.

11. **Orobanche mìnor** J. E. Smith. Clover Broomrape. Fig. 4959.

Orobanche minor J. E. Smith in Sowerby & Smith, Engl. Bot. **6**: *pl. 422.* 1797.
Orobanche columbiana St. John & English, Proc. Biol. Soc. Wash. **44**: 34. 1931.

Stem pubescent, rather stout, simple, 1–4.5 dm. high, pale yellowish brown. Lower scales numerous, oblong-ovate, the upper scattered, lanceolate, acute, 6–18 mm. long; spike rather loosely flowered, especially below, 8–20 cm. long; floral bracts lanceolate, about the length of the corolla-tube; calyx split almost to the base above and below, the lateral lobes 2-cleft into lanceolate-subulate segments; corolla 1–1.5 cm. long, the tube white or bluish, the limb white or bluish with violet markings, the lobes rounded and spreading, much shorter than the tube.

Parasitic on the roots of clover, naturalized from Europe; Clarke County, western Washington, and in Multnomah and Tillamook Counties. Oregon; also eastern United States. May–July.

2. **BOSCHNIÀKIA** C. A. Mey. ex Bong. Mém. Acad. St. Pétersb. VI **2**: 159. 1832.

Stems short and fleshy with a large tuber-like base at the point of attachment to host. Leaves closely imbricated, red or yellowish. Flowers short-pedicelled, ramose-spicate, closely subtended by foliaceous bracts. Calyx cupulate, truncate or 1–5-toothed. Corolla-tube curved, dilated at base, limb bilabiate; upper lip concave, entire or 2-toothed, lower lip 3-lobed. Stamens hairy, inserted near middle of tube; filaments hairy at base. Ovary 1-celled with 2–4 parietal placentae; stigma asymmetrical or distinctly lobed. Capsule 3–4-valved. Seeds large, alveolate. [Name in honor of Boschniaki, a Russian botanist.]

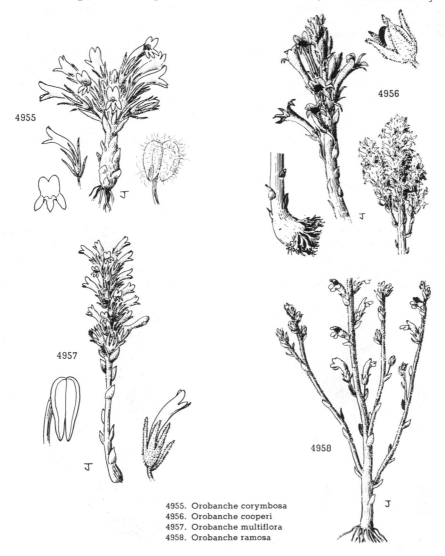

4955. Orobanche corymbosa
4956. Orobanche cooperi
4957. Orobanche multiflora
4958. Orobanche ramosa

A genus of 4 species inhabiting eastern Asia, the Aleutian Islands, and the Pacific Slope of North America. Type species, *Boschniakia glabra* C. A. Mey.

Bracts broadest at the middle or below; corolla 10–15 mm. long.   1. *B. hookeri.*
Bracts broadest above the middle; corolla 15–20 mm. long.   2. *B. strobilacea.*

1. Boschniakia hoòkeri Walp. Vancouver or Small Ground Cone. Fig. 4960.

Orobanche tuberosa Hook. Fl. Bor. Amer. **2**: 92. *pl. 168.* 1838. Not Vell. 1825.
Boschniakia hookeri Walp. Rep. **3**: 479. 1844–45.
Orobanche hookeri Beck, Bibl. Bot. **4**[19]: 85. 1890.
Boschniakia tuberosa Jepson, Man. Fl. Pl. Calif. 954. 1925.
Kopsiopsis tuberosa Beck, Pflanzenreich **4**[261]: 305. 1930.

Stems 1–2 dm. high, very thick and fleshy, dark reddish brown, simple and usually only one arising from the tuberous base. Scales dark red, closely imbricated, 8–12 mm. long, rhombic-dilated, more or less acute, usually broadest at the middle or below; spike usually longer than the sterile part of stems; bracts similar to the lower scales but spreading in flower; bractlets 2, subulate, rarely 1 or wanting; calyx cupulate, with 1–3 short slender teeth; corolla 12–15 mm. long, upper lip 3–5 mm. long, entire; lower lip 3-lobed, the lobes ovate or ovate-lanceolate, 3 mm. long, usually sparsely villous on the margins; anthers sparsely hairy, filaments conspicuously hairy at base; style asymmetrical or obscurely 3-lobed; placentae usually 2–3; seeds honey-colored.

Moist woods, Humid Transition Zone, parasitic on *Gaultheria shallon*; Vancouver Island south near the coast to Kitsap County, Washington, and Florence, Lane County, Oregon; Del Norte County to Marin County, California. Type locality: "N.W. Coast of America." May–Aug.

2. Boschniakia strobilàcea A. Gray. California Ground Cone. Fig. 4961.

Boschniakia strobilacea A. Gray, Pacif. R. Rep. **4**: 118. 1857.
Kopsiopsis strobilacea Beck, Pflanzenreich **4**[261]: 306. 1930.

Plants dark reddish brown, the stem simple and usually solitary, stout, 1–2 dm. high, the spike usually as long as or longer than the sterile part of stem and 3–5 cm. thick. Scales and bracts much imbricated, broadly obovate, broadest above the middle, obtuse, at least the upper 10–12 mm. long; bractlets subtending the flowers slender, surpassing the corolla-tube; calyx cupulate, sometimes truncate at apex but commonly with 2–4(5) lateral, lanceolate-attenuate teeth as long as or longer than the tube; corolla 15–20 mm. long, upper lip bifid or entire; style usually distinctly 4-lobed; placentae usually 4; seed-markings suggesting honeycomb.

Dry ground in chaparral or open woods, Upper Sonoran and Arid Transition Zones, parasitic on roots of *Arctostaphylos* and *Arbutus*; Josephine and Jackson Counties, southern Oregon, southward in the Coast Ranges and Sierra Nevada to southern California. Type locality: "Dry and rocky hills, South Yuba, California." May–July.

Family 140. LENTIBULARIÀCEAE.

BLADDERWORT FAMILY.

Small aquatic plants, or if terrestrial growing in moist places. Leaves sometimes wanting, when present in a basal rosette or along floating stems. Flowers solitary to several, borne on a scape, perfect and irregular. Calyx 2–5-lobed or -parted, persistent. Corolla sympetalous, 2-lipped, the upper lip entire or 2-lobed, the lower longer and entire or 3-lobed, usually with a prominent bearded palate and (in ours)

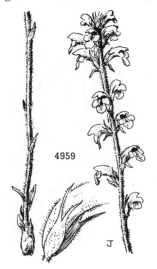

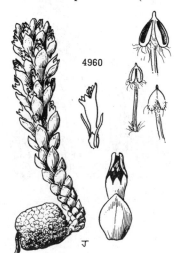

4960

4959

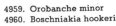

4959. Orobanche minor
4960. Boschniakia hookeri

with a nectariferous spur at base. Stamens 2, borne on the base of the corolla; anther-sacs confluent into one. Ovary superior, free from the calyx, 1-celled, with a free, cen-tral, subglobose placenta; ovules usually numerous; style very short or wanting; stigma 1–2-lipped. Capsule dehiscent by valves or often bursting irregularly. Seeds variously appendaged or sculptured; embryo in the axis, often imperfectly developed; endosperm none.

About 5 genera and about 350 species of wide geographical distribution.

Plant terrestrial; leaves entire, basal; calyx 5-lobed.   1. *Pinguicula.*
Plant aquatic; leaves dissected; calyx 2-lobed.   2. *Utricularia.*

1. PINGUÍCULA [Tourn.] L. Sp. Pl. 17. 1753.

Small, stemless, perennial herbs with fibrous roots, growing on damp rocks or in bogs. Leaves in a basal rosette, with the upper surface usually very viscid. Scapes naked, 1-flowered, circinnate. Calyx 2-lipped, the upper lip 3-lobed and the lower 2-lobed. Corolla 5-lobed, more or less 2-lipped, the upper lip 2-lobed, the lower 3-lobed, the lobes spreading; base of corolla saccate and contracted into a nectariferous spur. Capsule 2-valved. Seeds oblong, reticulate. [Name Latin, *pinguis,* meaning fat, in reference to the apparent greasi-ness of the leaves of some species.]

About 35 species of wide distribution in the northern hemisphere and southward along the Andes to Pata-gonia. Type species, *Pinguicula vulgaris* L.

1. Pinguicula vulgàris L. Common Butterwort. Fig. 4962.

Pinguicula vulgaris L. Sp. Pl. 17. 1753.

Scapes erect, glabrous or nearly so, 4–12 cm. high, only slightly elongated in fruit. Leaves ovate to elliptic, obtuse at apex, narrowed at base to a short winged petiole, the margins usually inrolled; calyx 3–5 mm. long, the lobes obtuse, the two lower more or less united; corolla violet-purple, 15–20 mm. long including the subulate spur, 2-lipped, the upper lip 2-lobed, the lower 3-lobed, both equally spreading; capsule ovoid, 6–8 mm. long, 5–6 mm. broad.

Growing on wet rocks or on sandy or gravelly banks; a circumpolar species, extending southward on the Pacific Coast from Alaska through western Washington to the Siskiyou Mountains, Oregon, and Smith River, Del Norte County, California, and across the continent to Minnesota, Michigan, and New York. Type locality: Europe. April–Aug.

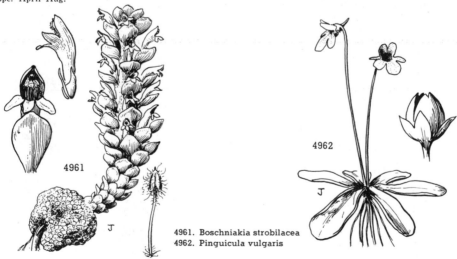

4961. Boschniakia strobilacea
4962. Pinguicula vulgaris

2. UTRICULÀRIA L. Sp. Pl. 18. 1753.

Aquatic or bog herbs with horizontal, submerged, branching stems. Leaves alternate, sometimes root-like, 2–8-parted at the very base and thus often appearing as if verticillate, the divisions dichotomously or pinnately finely dissected, at least some of them bladder-bear-ing. Bladders with a pair of bristles about the mouth. Flowers in racemes or sometimes solitary, the scapes naked or with a few scales basally attached or sometimes replaced by a whorl of inflated floats; pedicels from the axils of bracts, these sometimes auriculate. Calyx 2-lobed, the lobes concave, persistent. Corolla strongly 2-lipped, the palate at the base of the lower lip prominent, often 2-lobed. Anthers not lobed. Capsule few- to many-seeded. Seeds more or less peltate, flat-topped, and the margins sometimes winged. [Name Latin, *utriculus,* a little bag.]

About 300 species of wide geographic distribution in tropical and temperate regions. Type species, *Utricularia vulgaris* L.

Pedicels recurved in fruit.
 Leaves crowded, 3–4 cm. long; corolla-spur curved but little shorter than the lower lip.   1. *U. vulgaris.*
 Leaves scattered, mostly 1–2 cm. long; corolla-spur very short or sometimes almost rudimentary.
                                                                                                      2. *U. minor.*
Pedicels erect in fruit.
 Leaves once-parted; corolla-spur much shorter than lower lip.   3. *U. gibba.*
 Leaves 2–4-parted; corolla-spur about equaling lower lip.
   Corollas 10–18 mm. broad; ultimate leaf-divisions flat.   4. *U. intermedia.*
   Corollas 4–6 mm. broad; ultimate leaf-divisions terete.   5. *U. fibrosa.*

1. Utricularia vulgàris L. Greater or Common Bladderwort. Fig. 4963.

Utricularia vulgaris L. Sp. Pl. 18. 1753.
Utricularia vulgaris var. *americana* A. Gray, Man. ed. 5. 318. 1867.

Stems immersed, usually free-floating, rather stout and elongated, densely leafy. Winter buds 1–2 cm. long, often lobed in outline, leaf-divisions fringed with gray hairs; leaves 3–4 cm. long, 2–3-pinnately divided into filiform segments, midrib not evident, usually very bladdery, the bladders about 3–4 mm. long; scapes stout in comparison to other species, 10–30 cm. or more long, 5–16-flowered; pedicels reflexed in fruit; corolla yellow, 10–20 mm. broad, the sides of the lips reflexed and the upper nearly entire, barely longer than the palate, lower lip only slightly 3-lobed; spur 6–8 mm. long, somewhat curved, blunt or acutish, shorter than the lower lip; capsule globose.

Ponds and slow streams, Boreal, Transition, and sometimes Upper Sonoran Zones; British Columbia to southern California and eastward across the continent; also in Europe and Asia. Type locality: Europe. April–Aug.

2. Utricularia mìnor L. Lesser Bladderwort. Fig. 4964.

Utricularia minor L. Sp. Pl. 18. 1753.

Stems floating or creeping on the bottom in shallow water, very slender and thread-like, 1–3 dm. long, sparingly branched. Winter buds round, 1.5–5 mm. long, leaf-divisions not fringed; leaves about 1 cm. long, 2–4-parted, the divisions very slender but flattened, entire, of two types, the larger 25–30 mm. long and usually without bladders, the smaller 1–3 cm. long and bearing 1–5 bladders; scapes almost filiform, 5–15 cm. long, 3–6-flowered, and with 2–5 minute auriculate scales; pedicels capillary, 2–8 mm. long, recurved in fruit; calyx-lobes 1 mm. long; corolla pale yellow, upper lip 2–4 mm. long and about half as wide, lower lip 4–8 mm. long, palate obsolete, spur very short and saccate; capsule about 1 mm. in diameter.

In shallow water, mainly Boreal Zone; circumpolar in distribution, extending southward in North America from Alaska to Tulare County, California; also Utah, Colorado, Ohio, and Pennsylvania. Type locality: Europe. June–Aug.

Utricularia occidentàlis A. Gray, Proc. Amer. Acad. 19: 95. 1883. An imperfectly known species which differs from *U. minor* by the hairy-fringed leaf-segments of the winter buds. Known only from Falcon Valley, Klickitat County, Washington, the type locality.

3. Utricularia gíbba L. Swollen-spurred Bladderwort. Fig. 4965.

Utricularia gibba L. Sp. Pl. 18. 1753.
Utricularia fornicata Le Conte, Ann. Lyc. N.Y. 1: 76. 1824.

Stems slender, short, creeping over moss or mud in shallow water, the winter buds 1 mm. or less in diameter, the leaf-divisions thread-like, not fringed on the margins. Leaves once-parted, appearing root-like, divisions few, hair-like, without midribs; bladders small, scattered on leaf-divisions; scapes 2.5–10 cm. high, with short slender branches at base, 1–2-flowered, the peduncles erect in fruit; corolla yellow, 6–8 mm. broad, lips rounded, the lower reflexed and somewhat longer than the thick, blunt, gibbous spur; capsule globose.

In shallow water, Sonoran Zones; in the Pacific States occurring in California, in the San Joaquin Valley in San Joaquin County, and the San Francisco Bay area in Lake, Sonoma, and Contra Costa Counties and also in the northern Sierra Nevada; of wide distribution in the eastern states and also Mexico and Central America. Type locality: Virginia. July–Sept.

4. Utricularia intermèdia Hayne. Flat-leaved or Mountain Bladderwort. Fig. 4966.

Utricularia intermedia Hayne in Schrad. Journ. für de Bot. 1: 18. 1800.

Stems 1–2 dm. long, creeping on the bottom in shallow water and radiating from the base of the scape. Winter buds oval, 3–10 mm. long; the leaf-divisions fringed with gray hairs; leaves numerous and crowded, 2-ranked, rigid not flaccid, 4–5-forked, the segments linear, flat, bristly-serrulate, with evident midrib; bladders 2–5 mm. long, borne on separate leafless branches or sometimes on leafless portions of ordinary branches; scapes solitary, 5–20 cm. high, 1–4-flowered; pedicels slender, 8–20 mm. long, erect in fruit; corolla yellow, upper lip broadly triangular, 5–6 mm. long and about 1 mm. broader, the lower lip slightly 3-lobed, 10–12 mm. broad, palate prominent; spur conic at base, cylindric above, acute, a little shorter than the lower lip; capsule 3 mm. in diameter.

Shallow water in mountain bogs and ponds, Boreal and Transition Zones; cooler portions of the northern hemisphere in North America, ranging from British Columbia east to Newfoundland and south to California, Indiana, and New Jersey; also in Europe and Asia. In the Pacific States it is infrequent but ranges as far south as Plumas County, California. Type locality: Europe. June–Aug.

5. Utricularia fibròsa Walt. Fibrous Bladderwort. Fig. 4967.

Utricularia fibrosa Walt. Fl. Car. 64. 1788.

Stems creeping in shallow water, slender, radiating from base of the scape. Leaves dimorphic, those associated with the flowering scapes 2-forked and bearing bladders, the others 3-forked, about 15 mm. long and crowded, not bearing bladders; scapes 1.5–4 dm. long, with 2–6 pedicellate flowers; corolla yellow, upper and lower lips subequal, the palate conspicuous, the spur equaling or slightly exceeding the lower lip; capsule globose, the seeds winged and the body rough-tuberculate.

In shallow fresh water, Transition and Sonoran Zones; Mendocino County, California, and the Sacramento Valley; also eastern and central United States. Type locality: southeastern United States. June–Aug.

Family 141. **ACANTHÀCEAE.**

Acanthus Family.

Herbs, or some tropical genera shrubs or small trees, with opposite, simple, exstipulate leaves and regular or slightly irregular flowers. Calyx 4–5-parted or 4–5-cleft, the sepals or lobes imbricated in the bud, equal or unequal. Corolla sympetalous, 5-lobed and slightly irregular or distinctly 2-lipped, the lobes convolute in the bud. Perfect stamens 4 and didynamous, or only 2; anthers usually 2-celled, the sacs longitudinally dehiscent. Ovary superior, 2-celled; ovules 2–10 in each cell, anatropous or amphitropous; style simple, filiform; stigmas 1 or 2. Capsule 2-celled, loculicidally

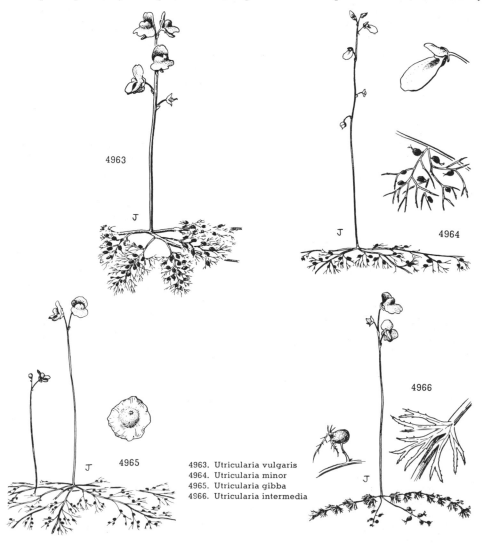

4963. Utricularia vulgaris
4964. Utricularia minor
4965. Utricularia gibba
4966. Utricularia intermedia

2-valved, the valves elastically dehiscent from the central column. Seeds globose or orbicular, borne on curved projections from the placentae; endosperm present or often wanting.

A family of about 200 genera and 2,000 species, natives of temperate and tropical regions of both the eastern and western hemispheres. Represented in the Pacific States by the single genus *Beloperone.*

1. **BELOPERÒNE** Nees in Wall. Pl. Asiat. Rar. **3**: 76, 102. 1832.

Shrubs or small trees with mostly opposite leaves. Flowers in the axils of bracts or small leaves, in ours showy and forming terminal bracteate racemes. Calyx 5-parted, bracteate. Corolla imbricate in bud, tubular, 2-lipped; throat narrow; lower lip spreading, 3-lobed, upper lip erect and more or less concave. Stamens 2, inserted in the throat, about equaling or shorter than the upper lip; anther-cells 2, disjoined and one inserted a little lower than the other on a broad connective. Capsule ovoid on an elongated clavate base; seeds sub-globose. [Name Greek, from *belos,* an arrow or dart, and *perone,* something pointed.]

About 100 species, inhabiting tropical and subtropical regions of the western hemisphere; our species is the only one within the borders of the United States. Type species, *Beloperone plumbaginifolia* Nees.

1. **Beloperone califórnica** Benth. Chuperosa. Fig. 4968.

Beloperone californica Benth. Bot. Sulph. 38. 1844.
Jacobinia californica Nees in A. DC. Prod. 11: 729. 1847.
Sericographis californica A. Gray in Torr. Bot. Mex. Bound. 125. 1859.

Low, much-branched shrub, the branches soon becoming leafless and somewhat rush-like, light gray-green with a close depressed puberulence. Leaves opposite, suborbicular to broadly oblong-ovate, 5–15 mm. broad, puberulent and more or less glandular on both sides, pale gray-green, more or less abruptly narrowed to a short petiole, most of them falling in the dry season; bracts and bractlets deciduous; calyx-lobes lanceolate-subulate, 4 mm. long; corollas dull red, 2.5–3.5 cm. long, the lips nearly as long as the tube, the lower 3-lobed, the upper emarginate and bearing a pair of longitudinal ridges; stamens about equaling the upper lip.

Dry rocky slopes and washes, Lower Sonoran Zone; western edge of the Colorado Desert, California, east to southwestern Arizona and south into Lower California and western Sonora. Type locality: Cape San Lucas, Lower California. Feb.–June.

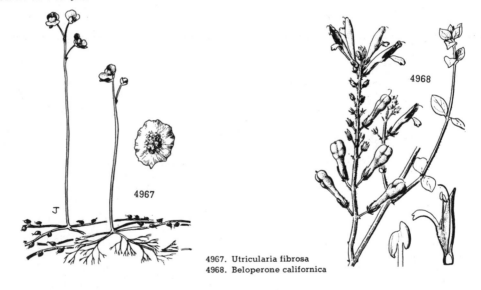

4967. Utricularia fibrosa
4968. Beloperone californica

Family 142. **PLANTAGINÀCEAE.**

PLANTAIN FAMILY.

Annual or perennial, acaulescent or short-stemmed, rarely stoloniferous herbs. Leaves basal, or in caulescent species opposite or alternate, estipulate, the venation seemingly parallel. Flowers hypogynous, small, perfect, polygamous or monoecious, bracteolate, in dense, terminal, long-scaped spikes or heads, or rarely solitary. Calyx 4-parted, persistent, the lobes imbricated and persistent. Corolla 4-lobed, scarious or membranous, usually marcescent. Stamens 4 or sometimes 2, or in one South American genus only 1, inserted on the tube or throat of the corolla; filaments filiform; anthers versatile, 2-celled, the sacs longitudinally dehiscent. Ovary superior, 1–2-celled, or falsely 3–4-celled. Stigma filiform, simple, mostly stigmatic longitudinally. Ovules 1 to several in each cell, peltate, amphitropous. Fruit a pyxis, circumscissile at or below the middle, or an indehiscent nutlet. Seeds 1 to several in each cavity of the fruit; endosperm fleshy; cotyledons narrow.

A family of 3 genera and about 250 species of wide geographical distribution.

1. PLANTÀGO [Tourn.] L. Sp. Pl. 112. 1753.

Acaulescent or rarely leafy-stemmed herbs, the scapes arising from the axils of the alternate leaves. Calyx-lobes equal or sometimes two of them larger. Corolla salverform, the tube cylindric or constricted at the throat, the limb spreading in anthesis. Fruit a pyxis, mostly 2-celled, the partitions falling away with the seeds; seeds various. [The Latin name.]

A genus of over 200 species of wide geographical distribution. Type species: *Plantago major* L.

Plants acaulescent; flowers spicate or capitate at apex of scapes.
  Leaves more or less laciniately divided (denticulate or often entire in variety); spikes nodding in bud.
                                                   1. *P. coronopus.*
  Leaves entire or denticulate; spikes erect in bud.
    Stout coarse perennials; stamens 4.
      Seeds 8–16; leaves ovate, distinctly slender-petioled; caudex not woolly-villous at leaf-bases.
                                       2. *P. major.*
      Seeds 2–4(6); leaves linear, linear-lanceolate, or if narrowly ovate these tapering to a short broad petiole; caudex usually woolly-villous at leaf-bases.
        Flowers polygamous, of 2 kinds; the pistillate corolla-lobes erect in anthesis, becoming conduplicate and beak-like over ripened capsule.   3. *P. hirtella galeottiana.*
        Flowers always perfect, alike, remaining spreading or reflexed over ripened capsule.
          Capsule 6–8 mm. long, splitting irregularly.   4. *P. macrocarpa.*
          Capsule 2–4.5 mm. long, circumscissile.
            Corolla pubescent without; leaves fleshy, glabrous; maritime species.
                               5. *P. juncoides.*
            Corolla glabrous without; leaves thin or somewhat coriaceous.
              Flower-bracts attenuate with hyaline tip, much surpassing the flower-buds in young inflorescence; seeds 2, conspicuously excavated on inner surface.
                                  6. *P. lanceolata.*
              Flower-bracts ovate, scarcely as long as flower-buds in young inflorescence; seeds usually 4, flattened on inner surface.   7. *P. eriopoda.*
  Slender annuals, or if biennial not coarse and stout; stamens 2 or 4.
    Stamens 4; corolla-lobes 2–3.5 mm. long.
      Flowers mostly polygamo-dioecious; the pistillate corolla-lobes conduplicate, forming a beak over ripened capsule.
        Seeds yellowish, the face concave; calyx-lobes rounded at apex.   8. *P. virginica.*
        Seeds wine red or brownish, the face plane.
          Calyx-lobes rounded and somewhat apiculate at apex; seeds brownish.
                              9. *P. truncata firma.*
          Calyx-lobes short-acuminate at apex; seeds wine red.   10. *P. rhodosperma.*
    Flowers perfect; corolla-lobes all spreading, or reflexed.
      Bracts linear or subulate, as long as to much surpassing the calyx, not scarious-margined or narrowly so at base.
        Bracts 6–35 mm. long, all much surpassing calyx; plant darkening on drying.
                              11. *P. aristata.*
        Bracts as long as or somewhat longer than calyx (lowest bracts in var. *oblonga* two to three times the length of calyces); plants grayish. 12. *P. purshii.*
      Bracts ovate with central midrib and wide scarious margins, about equaling to much shorter than calyces.
        Bracts resembling the sepals and about equaling them in length; mature seeds reddish yellow or reddish, shining.   13. *P. insularis.*
        Bracts not sepaloid, about one-half the length of the sepals; seeds dark brown, the surface dull and punctate   14. *P. erecta.*
    Stamens 2; corolla minute, the lobes less than 1 mm. long.
      Seeds (2)3–4(6); scapes and leaves erect.
        Capsule conoidal, twice the length of the calyx; corolla-lobes spreading or reflexed in age.
                              15. *P. bigelovii.*
        Capsule broadly ovoid to rotund, but little longer than the calyx; corolla-lobes mostly erect in age but not conduplicate.   16. *P. pusilla.*
      Seeds (6)8–14; scapes and leaves decumbent or spreading.   17. *P. heterophylla.*
Stems leafy; flowers capitate on axillary peduncles.   18. *P. indica.*

1. Plantago corónopus L. Cut-leaved Plantain. Fig. 4969.

Plantago coronopus L. Sp. Pl. 115. 1753.

Annual or short-lived perennial with a taproot, or in biennial or perennial plants the root-crown becoming lignescent, 2–2.5 cm. broad, somewhat lobed and sending out several tufts of leaves and scapes, grayish-pubescent throughout. Leaves spreading, 2.5–12 cm. long, pinnately divided into linear, acute or acuminate segments, or in robust plants some leaves bipinnatifid, the segments often remote; scapes more or less decumbent, 5–35 cm. long including the spike; spikes slender, 2.5–9 cm. long, dense, the flowers crowded and more or less imbricate; bracts ovate,

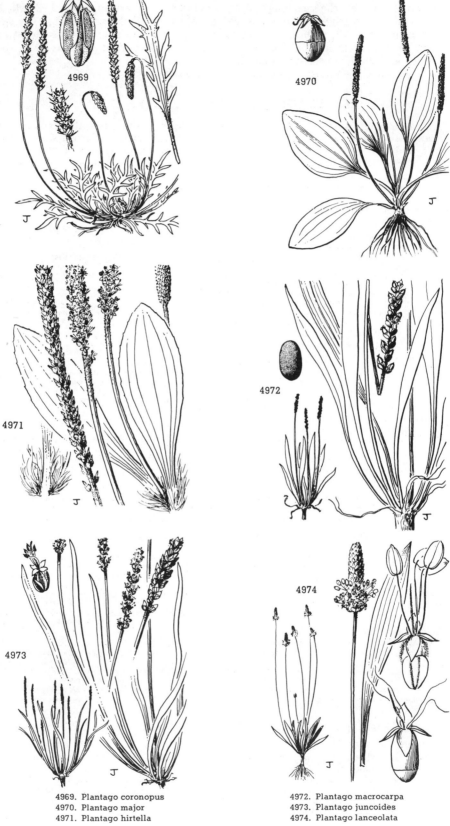

4969. Plantago coronopus
4970. Plantago major
4971. Plantago hirtella

4972. Plantago macrocarpa
4973. Plantago juncoides
4974. Plantago lanceolata

pubescent, with broad, hyaline, ciliate margins abruptly narrowed to a sharp point; flowers perfect; sepals pubescent, 3 mm. long with broad, white, scarious margins; corolla-lobes lanceolate, reflexed; capsule equaling the calyx, 2–4-seeded; seeds winged all around, or at least to below the middle.

Sparingly introduced along the coast, Whidby Island, Washington; Linnton (Portland) and Port Orford, Oregon; on the coast in California from southern Marin County to Monterey County, and on Santa Catalina Island, Los Angeles County, also adventive in Solano County. Type locality: Europe. April–Aug.

The annual specimens from Santa Catalina Island, Los Angeles County, are probably **Plantago coronopus** subsp. **commutàta** (Guss.) Pilger, Rep. Spec. Nov. **28**: 287. 1930. (*P. commutata* Guss. Prod. Sic. Suppl. 46. 1832; *P. parishii* J. F. Macbride, Contr. Gray Herb. No. 56: 61. 1918.) This Mediterranean form is characterized by the less acuminate and scarcely spreading bracts subtending the flower and the proportionately stouter peduncle. The leaves in depauperate plants are often entire. A complete list of varieties of this and the more common perennial form of the species is to be found in Pflanzenreich 4²⁶⁹: 126–155. 1937.

2. **Plantago màjor** L. Common Plantain. Fig. 4970.

Plantago major L. Sp. Pl. 112. 1753.

Perennial, glabrous or usually more or less sparsely short-pubescent. Leaves usually spreading horizontally, ovate, often subcordate, entire or coarsely and irregularly toothed, 5–20 cm. long, usually abruptly narrowed below to form a short wing on the apex of the petiole, this broad and channeled; scapes more or less curved, often decumbent, 8–40 cm. long including the more or less dense spike; sepals broadly ovate; corolla-lobes reflexed, 1–1.5 mm. long; capsule ovoid, about 3 mm. long, circumscissile near the middle, 8–18-seeded; seeds angled, reticulate.

A variable species of worldwide distribution. Occurring as a weed in Pacific States in a wide altitudinal range and in varied habitats. A complete list of the many segregate forms may be found in Pflanzenreich 4²⁶⁹: 41–56. 1937, written by Robert Pilger. Type locality: "*in Europa ad vias.*" May–Nov.

3. **Plantago hirtélla** var. **galeottiàna** (Decne.) Pilger. Mexican Plantain. Fig. 4971.

Plantago galeottiana Decne. in A. DC. Prod **13**¹: 726. 1852.
Plantago virginica var. *maxima* A. Gray, Bot. Calif. **2**: 611. 1880
Plantago subnuda Pilger, Notizblatt **5**: 260. 1912.
Plantago durvillei subsp. *subnuda* Pilger, Pflanzenreich 4²⁶⁹: 234. 1937.
Plantago hirtella var. *galeottiana* Pilger, op. cit. 257.

Perennials with short thick caudex clothed with persistent leaf-bases intermingled with long villous hairs, the stems erect, arcuate at base, 1.5–4 dm. long, with septate flattened hairs especially near the inflorescence. Leaves thin, elliptic to rather broadly obovate, 5–20 cm. long, 1.5–8 cm. broad, narrowed below to a broadly winged petiole less than one-half the length of the leaf-blades, the veins 5–7, prominent, margin sparingly denticulate or sometimes entire, more or less pilose especially on the upper surface; inflorescence polygamous, the spikes densely flowered except at base, 5–30 cm. long, 6–12 mm. broad; bracts pubescent, a little shorter than the calyx, carinate, ovate, acute at apex, the margin scarious and fringed with coarse cilia; sepals obtuse, scarious-margined; corolla-lobes of pistillate flowers inrolled and erect in anthesis; corolla-lobes of staminate or perfect flowers spreading, narrowly obovate, entire, 2 mm. long; stamens white, 4; capsule oblong, about as long as the calyx; seeds 3, flattened on one side.

Moist ground, Transition and Sonoran Zones; in the Pacific States mostly along the coast from Grays Harbor County, Washington, to San Diego and San Bernardino Counties, California; eastward to Arizona and Mexico. Type locality: "In Mexicanis Cordil. circa Real del Monte 2500 metr. (Galeotti n. 1427)." May–Oct.

4. **Plantago macrocárpa** Cham. & Sch. Alaska Plantain. Fig. 4972.

Plantago macrocarpa Cham. & Sch. Linnaea **1**: 166. 1826.

Glabrous perennial with a stout perpendicular root. Leaves several to many, erect, the blades thin and glabrous, broadly to narrowly lanceolate, 7–20 cm. long, 1–4 cm. wide, very acute or obtusish at apex, narrowed from the middle or above to a rather narrow petiole, equaling or usually exceeding the blade; scapes together with the spikes somewhat longer to a little shorter than the leaves, glabrous below, rather sparsely puberulent just below and on the spike; spikes loosely to rather densely flowered; bracts broadly oval, closely enfolding the capsule, 3 mm. long, obtuse, brownish and sparsely puberulent dorsally, the whitish hyaline margin very narrow; sepals elliptic, usually broadly so, glabrous or sparsely puberulent on the prominent, dark brown midrib; corolla-tube 3 mm. long, the lobes triangular-ovate, 2–2.5 mm. long; stamens much exserted; capsule ellipsoidal, 6–8 mm. long, not circumscissile but irregularly rupturing; seeds 4, black, 4–4.5 mm. long.

Edges of swamps, Humid Transition and Boreal Zones; known in the Pacific States from the Olympic Peninsula, Washington, and Lincoln County, Oregon (*Peck*), but ranging from Sitka, Alaska, and the Aleutian Islands to Vancouver Island, British Columbia. Type locality: Unalaska, Aleutian Islands. June–Aug.

5. **Plantago juncoìdes** Lam. Pacific Seaside Plantain. Fig. 4973.

Plantago juncoides Lam. Encycl. **1**: 342. 1783.
Plantago maritima var. *juncoides* A. Gray, Man. ed. 2. 268. 1856
Plantago maritima of many American authors, not L.

Perennial, the stout taproot with a simple or short-branched crown. Leaves linear to linear-lanceolate, only slightly fleshy, ascending, attenuate at apex, 3–20 cm. long, 2–5 mm. wide, entire or sparsely denticulate, glabrous or sparsely puberulent; scapes strongly ascending, about equaling to exceeding the leaves; spikes 1.5–7 cm. long, densely flowered, the flowers perfect; bracts fleshy-carinate, broadly ovate; sepals oblong-ovate, 2–3 mm. long, often ciliate on the margins at apex; corolla-tube pubescent, the lobes spreading, 1.5–2 mm. long; stamens and styles long-exserted; capsule 2–5-seeded, the seeds about 2 mm. long.

Banks and margins of marshes along the coast, Humid Transition Zone; southern Alaska to San Francisco Bay, California; also Patagonia. Type locality: probably Tierra del Fuego. Collected by Commerson. June–Oct.

Plantago juncoides var. califórnica Fernald, Rhodora 27: 100. 1925. (*Plantago maritima* var. *californica* Pilger, Pflanzenreich 4²⁶⁹: 187. 1937.) A more fleshy plant with caudex often branched, the leaves depressed or rosulate, broadly linear or linear-oblanceolate to subspatulate, obtuse; scapes stout, depressed or arcuate. Bluffs and beaches along the seashore, Clallam County, Washington, to Monterey County, California. Type locality: Montara Point, San Mateo County, California.

6. Plantago lanceolàta L. English Plantain or Ribwort. Fig. 4974.

Plantago lanceolata L. Sp. Pl. 113. 1753.

Perennial or biennial with short rootstock bearing conspicuous tufts of long brown hairs at the bases of leaves. Leaves narrowly oblong-lanceolate, mostly erect, acutish to acuminate at apex, narrowed below to the usually elongated, more or less winged petiole, 4–40 cm. long including both blade and petiole; blade 5–35 mm. broad, entire or irregularly and remotely denticulate, glabrous or nearly so above, more or less pubescent beneath especially in the rather prominent, 3–5, parallel rib-like veins; scapes rather slender, channeled, 1–6 dm. high; spikes dense, short and ovoid at first, becoming elongated and cylindric in age, 1.5–6(10) cm. long, 5–9 mm. thick; flowers perfect; bracts hyaline-margined, broad below with an abruptly long-acuminate hyaline tip, conspicuously surpassing the calyx; sepals ovate, with a narrow green midrib and broad scarious margins, the two lower ones usually united; corolla glabrous, brownish, its tube short, the lobes ovate, spreading, sometimes ascending in fruit; filaments white; capsule oblong, very obtuse at apex, slightly longer than the calyx, 2-seeded; seeds deeply excavated on the face.

Fields, roadsides, and waste places, frequent throughout the Pacific States and across the continent. Naturalized from Europe. April–Oct.

Highly variable as to length and pubescence of leaves, also as to length of mature spike, which may be head-like. (The tall robust form with stout spikes has been called *P. altissima* L.) A complete list of varieties, subvarieties, and forms is to be found in Pflanzenreich 4²⁶⁹: 313–327. 1937.

7. Plantago eriópoda Torr. Saline Plantain. Fig. 4975.

Plantago glabra Nutt. Gen. 1: 100. 1818. (Nomen dubium.)
Plantago attenuata James, Long Exped. 1: 445. 1823. Not Wallr. 1820.
Plantago eriopoda Torr. Ann. Lyc. N.Y. 2: 237. 1828.
Plantago virescens Barnéoud, Monogr. Plantag. 33. 1845.
Plantago oblongifolia Decne. in A. DC. Prod. 13¹: 700. 1852.
Plantago retrorsa Greene, Pl. Baker. 3: 32. 1901.
Plantago shastensis Greene, loc. cit.

Perennial; rootstock stout, sometimes elongate, the crown more or less long-villous with rust-colored hairs among the old leaf-bases. Leaves oblong-lanceolate to elliptic, acute at apex, 6–25 cm. long, 5–9-nerved, narrowed to a winged petiole about one-half the length of the blade, glabrous and somewhat leathery; spikes sparsely pubescent with septate hairs, loosly flowered, 8–18 cm. long; bracts broadly ovate to rounded, the narrow scarious margins sometimes ciliolate or erose, not keeled; sepals oval, scarious-margined, 2–2.5 cm. long; corolla-lobes ovate, often unsymmetrical; style much exserted, as long as or longer than the stamens; capsule broadly conical, about 3 mm. or more long, often tipped by persistent base of style; seeds 3–4, black, 2–2.5 mm. long.

In alkaline soil, Transition Zone; Nevada and Wyoming east to Nova Scotia and Quebec; on the Pacific Slope it is known from Windermere, British Columbia; Jordan Valley, Malheur County, Oregon; and eastern Siskiyou County, California, where it was first collected by Greene and later described by him as *P. shastensis*. More material is needed to evaluate this entity. Type locality: "Depressed and moist situations along the Platte." June–Aug.

8. Plantago virgínica L. Dwarf Plantain. Fig. 4976.

Plantago virginica L. Sp. Pl. 113. 1753.
Plantago caroliniana Walt. Fl. Car. 85. 1788.
Plantago missouriensis Steudel, Flora 32: 409. 1849.

Annual with a slender vertical taproot, glabrate or commonly more or less villous with septate hairs. Leaves spreading or ascending, 1.5–15 cm. long, the blades oblanceolate or spatulate to obovate or elliptic, entire or repand-dentate, 3–5-nerved, narrowed to slender or more or less margined petioles; scapes erect or ascending, 5–20 cm. long; spikes dense or sometimes interrupted below; flowers dioecious; bracts lanceolate to linear-lanceolate, 2 mm. long; calyx-lobes oblong to ovate, 2–2.5 mm. with prominent brownish midrib and broad, white, scarious margins; corolla of staminate flowers with spreading lobes, those of pistillate flowers erect and connivent in fruit; capsule ovoid or oblong-ovoid, about equaling the calyx-lobes; seeds 2–4, golden yellow.

Moist places, Transition and Sonoran Zones; Arizona, Texas, and across the United States to the Atlantic Coast and south into northern Mexico; introduced in the Pacific States in southwestern Oregon and Shasta County, California. Type locality: Virginia. May–July.

9. Plantago truncàta subsp. fírma (Kunze ex Walp.) Pilger. Chile Plantain. Fig. 4977.

Plantago firma Kunze ex Walp. Nov. Act. Nat. Cur. 19: Suppl. 1: 402. 1843.
Plantago virginica var. *firma* Reiche, Fl. Chile 6¹: 117. 1911.
Plantago truncata subsp. *firma* Pilger, Bot. Jahrb. 50: 223. 1913.

Annual with a slender taproot, herbage hirsute throughout the spreading whitish hairs. Leaves lanceolate, narrowed to a rather short petiole below, acute at apex, 2–4 cm. long or rarely longer and 3–5 mm. wide ; scapes slightly longer to shorter than the leaves ; spikes 1–2 cm. long, rarely longer ; bracts ovate-lanceolate, shorter than the calyx ; calyx-lobes broadly oval, with a broad hirsute midrib and broad scarious margins ; corolla-lobes 3 mm. long and closed over the fruit, ovate, acute ; capsule about equaling the calyx, 2-seeded ; seeds elliptical, plane on the inner face, brownish.

A native of Chile; sparingly introduced in Marin and Sonoma Counties, also in Butte and Calaveras Counties, California. Type locality: Chile. April–June.

10. **Plantago rhodospérma** Decne. Red-seeded Plantain. Fig. 4978.

Plantago rhodosperma Decne. in A. DC. Prod. **13**[1]: 722. 1852.
Plantago virginica var. *pectinata* Kuntze, Rev. Gen. Pl. **2**: 532. 1891.
Plantago rubra A. M. Cunningham, Proc. Indiana Acad. **1896**: 204. 1897.

Annual or winter annual with a slender elongated root. Leaves rosulate, several to many, narrowly elliptic to oblanceolate, narrowed to a winged petiole, acute or obtuse at apex, commonly 3–5 cm. or occasionally up to 15 cm. long, 4–25 mm. broad, entire or more or less pectinate-toothed, grayish green, pubescent on both surfaces, especially on the veins, with tapering multicellular hairs ; scapes 1 to several, longer or usually shorter than the spikes, hirsute-pubescent ; spikes 1.5–10 cm. long ; bracts lanceolate, sharply acute, scarious-margined, prominently keeled, ciliate on the margins and keel ; calyx-lobes 3 mm. long, elliptic-oblong, with broad scarious margins, and sparsely pubescent on the slightly keeled midrib, apiculate and often purplish at the apex ;

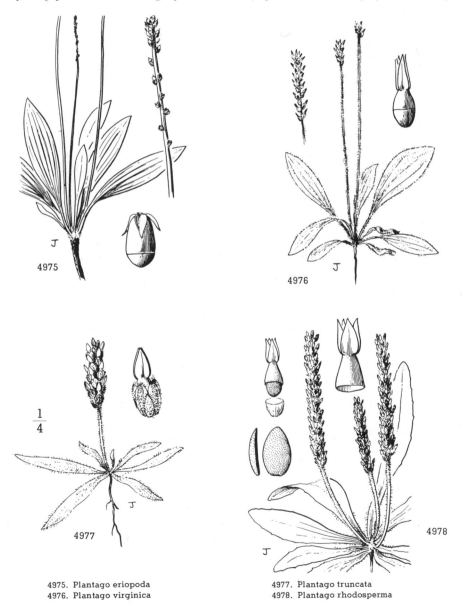

4975. Plantago eriopoda
4976. Plantago virginica

4977. Plantago truncata
4978. Plantago rhodosperma

corolla-lobes erect and closed over the capsule, 4–4.5 mm. long, acute and apiculate at apex, auriculate at base; corolla-tube puberulent on the outside; seeds 2, dark wine red, about 2.5 mm. long, narrowly elliptic-ovate, shallowly concave on the inner. surface, rounded on the outer, faintly reticulate-punctate.

Bluffs and dry mesas, Sonoran Zones; locally distributed in southern California from Corona Bluffs, Orange County, to San Diego; also, south of Tijuana to Rancho Cuevas, Lower California; east to Arizona, Texas, and adjacent northern Mexico. Type locality: "In Texas." April–May.

11. Plantago aristàta Michx. Bristly or Large-bracted Plantain. Fig. 4979.

Plantago aristata Michx. Fl. Bor. Amer. 1: 95. 1803.
Plantago nuttallii Rapin, Mem. Soc. Linn. Paris 6: 470. 1828.
Plantago patagonica var. *aristata* A. Gray, Man. ed. 2. 269. 1856.

Dark green annual 1–4 dm. high, loosely villous to glabrate, darkening on drying. Leaves ascending to erect, linear, acuminate, 0.5–1.5 cm. long, tapering to a petiole; spikes surpassing the leaves, pubescent with villous hairs, densely flowered, 3–10 cm. long; bracts linear, ascending, puberulent, five to seven times the length of the flower or longer at the base of the spike; sepals 3 mm. long, oblong, obtuse, villous, with a narrow scarious margin; corolla-lobes rotund-ovate, spreading; stamens small, scarcely exserted; capsule ellipsoid, 3 mm. long; seeds 2, the surface roughened, concave on the inner face.

Dry open ground, mainly Transition Zone; British Columbia and found as an adventive in the Pacific States at Seattle, Washington; Portland and in the Rogue River Valley, Oregon, and in Humboldt and Ventura Counties, California; extending eastward to the eastern seaboard and south to Texas. Type locality: Illinois. May–Aug.

12. Plantago púrshii Roem. & Sch. Pursh's Plantain. Fig. 4980.

Plantago purshii Roem. & Sch. Syst. Veg. 3: 120. 1818.
Plantago gnaphaloides Nutt. Gen. 1: 100. 1818.
Plantago patagonica var. *gnaphalioides* A. Gray, Man. ed. 2. 269. 1856.

Annual; scapes erect or ascending, slender, well exceeding the spikes, villous with ascending hairs and often also white-woolly. Leaves narrowly linear-oblanceolate, 1–4 mm. wide, 3–12 cm. long; spikes long-cylindric, 2–10 cm. long; bracts linear to linear-subulate, equaling the calyx-lobes or sometimes exceeding them below; calyx-lobes oval to obovate, 2–3 mm. long, densely long-villous; corolla-lobes spreading, broadly ovate, 1.5–2.5 mm. long, white with brownish spot at base; stamens small, equaling or slightly exceeding the corolla; capsule 3–4 mm. long, ellipsoid; seeds 2, dark brown, minutely pitted.

Dry hills, Arid Transition and Upper Sonoran Zones; British Columbia and eastern Washington from Okanogan County southward east of the Cascade Mountains to northern Arizona and adjacent California, and eastward to Minnesota, Texas, and Chihuahua. Much less abundant in California and Arizona and intergrading with *P. purshii* var. *oblonga*. Type locality: "In dry situations on the banks of the Missouri." Collected by Nuttall. May–Aug.

Plantago purshii var. oblónga (Morris) Shinners, Field & Lab. 18: 117. 1950. (*Plantago picta* Morris, Bull. Torrey Club 28: 118. 1901, not *P. picta* Colenso; *P. oblonga* Morris, op. cit. 119; *P. ignota* Morris, loc. cit.; *P. xerodea* Morris, op. cit. 36: 515. 1909; *P. spinulosa* var. *oblonga* Poe, op. cit. 55: 411. 1928; *P. purshii* var. *picta* Pilger, Pflanzenreich 4²⁶⁹: 369. 1937.) Plants commonly villous, sometimes glabrate; peduncles many (10–40), occasionally 2 or 3, 1–12 cm. long not including the inflorescence; spikes densely flowered, oblong to cylindric, 1.5–5(6) cm. long; bracts divergent, linear, broadly ovate to blunt at apex, sometimes scarious-margined at the base, 0.5–1 cm. long, the lower two to three times the length of the calyces, the upper but little surpassing them; flowers and seeds as in the species. Gravelly or sandy desert slopes, Sonoran Zones; Mojave and Colorado Deserts from San Bernardino County, California, to northern Lower California and eastward through southern and central Arizona to western New Mexico; also Sonora. Type locality: Colorado Desert. Collected by Orcutt. March–April.
The more common form in southeastern California and adjacent Arizona and Sonora, *P. ignota* Morris, represents the extreme of variation in length of bracts.

13. Plantago insulàris Eastw. Island Plantain. Fig. 4981.

Plantago insularis Eastw. Proc. Calif. Acad. III. 1: 112. 1898. Not *P. insularis* (Godr.) Nyman ex Briquet 1901.
Plantago brunnea Morris, Bull. Torrey Club 27: 115. 1900.

Very similar to *Plantago erecta* in general habit, stems slender, usually spreading, canescent with long silky hairs especially on the scapes below the spikes. Leaves narrowly to broadly linear-lanceolate, acuminate, often with a few callous teeth on the margins; spikes oblong-cylindric, 1–2 cm. long, 8–10 mm. wide; bracts broadly ovate or orbicular, about equaling the calyx, with a brownish or greenish midrib, broadly scarious-margined and closely resembling the calyx-lobes; corolla-lobes conspicuously brownish at base, often brown-striped above the middle; seeds oblong, reddish yellow or reddish, smooth and glossy, 2–3 mm. long, deeply excavated on the inner face.

Mostly sandy soils, Upper and Lower Sonoran Zones; along the coast of southern California from Point Concepcion, Santa Barbara County, to San Diego County and on the islands off the coast; also on the western side of Lower California to the Viscaino Desert, where it intergrades with the variety. Specimens from the San Joaquin Valley, except for corolla-markings, resemble the variety. Type locality: San Nicolas Island, "on sea-shore flats," Ventura County, California. March–May.

Plantago insularis var. fastigiàta (Morris) Jepson, Man. Fl. Pl. Calif. 956. 1925. (*Plantago minima* A. M. Cunningham, Proc. Indiana Acad. 1896: 202. 1897, not *P. minima* DC.; *P. fastigiata* Morris, Bull. Torrey Club 27: 116. 1900; *P. scariosa* Morris, op. cit. 117; *P. gooddingii* Nels. & Kenn. Muhlenbergia 3: 142. 1908; *P. insularis* var. *scariosa* Jepson, Man. Fl. Pl. Calif. 956. 1925.) Conspicuously white-silky annual or winter annual; bracts orbicular to broadly ovate, sepal-like, with greenish, usually hairy midribs and broad hyaline margins; calyx lobes faintly if at all brownish at base; seeds as in the species. Plants with the dense fastigiate habit represent the typical variety while those with the loose habit and more greenish appearance have been considered as a distinct entity (*P. scariosa*). Desert slopes and washes, Death Valley region, Inyo County, California, south through the desert to Lower California; eastward to Nevada and Utah and south to Arizona and Sonora. Type locality: Tucson, Arizona.

14. **Plantago erécta** Morris. California Plantain. Fig. 4982.

Plantago patagonica var. *californica* Greene, Man. Bay Reg. 236. 1894. Not *P. californica* Greene, Bull. Calif.
Acad. **1**: 123. 1885.
Plantago dura Morris, Bull. Torrey Club **27**: 113. 1900.
Plantago erecta Morris, op. cit. 118.
Plantago tetrantha Morris, op. cit. 119.
Plantago speciosa Morris, op. cit. **28**: 120. 1901.
Plantago obversa Morris, op. cit. 121.
Plantago hookeriana var. *californica* Poe, op. cit. **55**: 417. 1928.

Annual with 1 to several scapes commonly erect or ascending, 7–20 cm. high, thinly to rather densely villous-pubescent with appressed hairs. Leaves narrowly to rather broadly linear, narrowed above to an acutish apex and below to an elongated, somewhat winged petiole, 4–16 cm. long, 1–4 mm. wide, 3-nerved; spikes cylindric to ovoid-capitate, 1–3.5 cm. long, 6–8 cm. in diameter; bracts much shorter than the calyx, ovate, usually acute, with a broad, green or brownish, villous midrib, the scarious margins broad below, tapering toward the apex to the middle or above the middle of the bract; flowers perfect; calyx-lobes oblong, obtuse, 3 mm. long, brownish green, with broad scarious margins, villous; corolla-lobes 1.5–2.5 mm. long, strongly reflexed, brownish at base; seeds 2, dark brown, finely pitted, concave on inner surface.

Common on grassy hillsides and flats, Transition and Upper Sonoran Zones; Rogue River Valley, Oregon, southward west of the Cascade–Sierra Nevada Divide to northwestern Lower California. Type locality: "Abundant on grassy plains and hillsides," San Francisco Bay region. March–May.

This species shows a great variability of growth form, apparently due to habitat, and is expressed by height, number of scapes, and length of spikes.

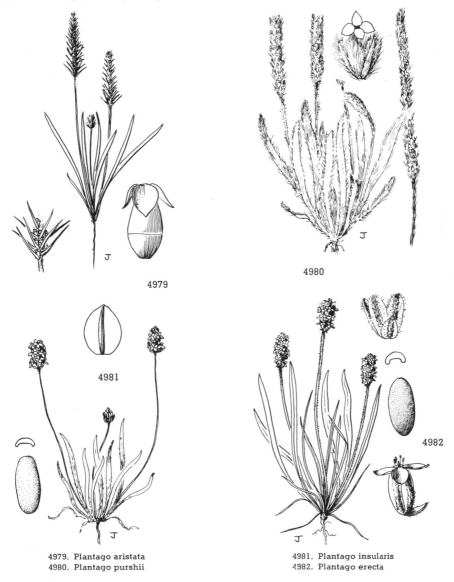

4979

4980

4981

4982

4979. Plantago aristata
4980. Plantago purshii

4981. Plantago insularis
4982. Plantago erecta

15. Plantago bigelòvii A. Gray. Annual Coast Plantain. Fig. 4983.

Plantago bigelovii A. Gray, Pacif. R. Rep. **4**: 117. 1857.

Annual, glabrous, sometimes puberulent on scapes below the spikes, scapes including spikes 2.5–20 cm. high, erect, 1 to many. Leaves erect, linear-lanceolate to linear-filiform, 2–7 cm. long, glabrous, the margins ciliate, rarely irregularly dentate; spikes densely flowered, 0.5–5 cm. long; bracts fleshy, ovate to nearly orbicular, carinate, scarious-margined, about the length of the sepals; sepals suborbicular, with broad scarious margins, 1.5–2 mm. long; corolla-lobes spreading to sharply deflexed in fruit, about 0.5 mm. long or a little longer, narrowly ovate, acute; stamens 2; capsule longer than the calyx, narrowly conoidal, truncate at apex, 2.5–3 mm. long, circumscissile a little below the middle; seeds 4–6, commonly 4, dull black, usually not coarsely pitted, winged at one end or narrowly so all around in occasional seeds, ellipsoid or oblong, about 1.5 mm. long.

Salt marshes along the coast and inland alkaline flats, Transition and Sonoran Zones; along the coast from British Columbia to Monterey County, California; also occurring in the Sacramento–San Joaquin Valley, California. Type locality: Benicia, Solano County, California. March–July.

16. Plantago pusílla Nutt. Slender Plantain. Fig. 4984.

Plantago pusilla Nutt. Gen. **1**: 100. 1818.
Plantago myosuroides Rydb. Mem. N.Y. Bot. Gard. **1**: 369. 1900.

Slender annual, puberulent with whitish hairs, scapes filiform, 5–12 cm. high including the spikes. Leaves narrowly linear, 0.5–2 mm. wide, entire, 1-nerved, long-villous on margin at base; spikes densely flowered, slender, 1–7 cm. long, imperfectly dioecious or polygamous; bracts triangular-ovate, 2 mm. long, shorter than the sepals, scarious-margined, slightly keeled and not subsaccate at base; corolla glabrous, the lobes triangular-ovate, 0.5 mm. or less long, erect but not connivent in fruit; stamens 2; capsule short-ovoid, equaling or slightly exceeding the calyx, circumscissile well below the middle; seeds 3–4, dark, finely pitted, sometimes winged at one end.

Moist places, Transition and Sonoran Zones; along the Columbia River in Oregon and Washington and east to New York, Virginia, Louisiana, and Texas. Type locality: "Arkansas." April–Aug.

Plantago elongàta Pursh (Fl. Amer. Sept. 729. 1814), a plant of northwestern United States and Canada, may be expected to occur in eastern Washington. The bracts in this species are as long as or longer than the sepals and are strongly keeled and subsaccate at base. The spikes are usually loosely flowered.

17. Plantago heterophýlla Nutt. Alkali Plantain. Fig. 4985.

Plantago heterophylla Nutt. Trans. Amer. Phil. Soc. II. **5**: 177. 1837.
Plantago perpusilla Decne. in A. DC. Prod. **13**[1]: 697. 1852.
Plantago californica Greene, Bull. Calif. Acad. **1**: 123. 1885.

Annual; scapes several to many, slender, arcuate, ascending or spreading, 5–10 cm. long, about twice the length of the leaves, sparingly puberulent to glabrate. Leaves spreading, linear, entire or often with a few teeth or linear lobes; spikes loosely flowered especially at the base, 0.5–4 cm. long; bracts fleshy, ovate to almost orbicular, broadly scarious-margined, strongly carinate, usually a little shorter than the sepals; sepals broadly oblong, obtuse, with a wide scarious margin, about 2 mm. long; corolla-lobes less than 1 mm. long, usually spreading in fruit; stamens 2; capsule broadly ovoid, somewhat rounded at apex, about twice the length of the calyx-lobes, (6)8–14-seeded, circumscissile below but near the middle; seeds blackish, irregularly and coarsely pitted, slightly angled, scarcely concave on the face, occasional seeds slightly winged at base.

Ours in alkaline soil, Lower Sonoran Zone; Sacramento Valley, California, from Colusa County south in the San Joaquin Valley to Tulare and Kern Counties and valleys in the adjacent ranges, and in cismontane southern California from Los Angeles County to San Diego County; also Arizona and Sonora eastward to Virginia and Florida. Type locality: Arkansas. Collected by Nuttall. April–May.

The plants described by Greene as *P. californica* usually have larger capsules and often fewer seeds than the eastern plants.

18. Plantago índica L. Sand Plantain. Fig. 4986.

Plantago indica L. Syst. Nat. ed. 10. 896. 1759.
Plantago arenaria Waldst. & Kit. Pl. Rar. Hung. **1**: 51. *pl. 51*. 1802.

Caulescent, pubescent and somewhat glandular annual with simple or much-branched leafy stems 8–40 cm. high. Leaves opposite, sometimes with shorter axillary leaves, sessile, hirsute or hirsute-villous, linear or linear-lanceolate, 6–8 cm. long, 2–4 mm. wide; inflorescence on axillary peduncles from the leaf-axils or umbellate at the apex of the stems, the heads oval or subglobose; lowest bracts ovate, abruptly long-acuminate, the upper ovate or oval; calyx-lobes obovate, hyaline-margined; corolla-lobes narrowly ovate, acute, shorter than the tube; seeds 2, about 2.5 mm. long, reddish brown, concave on the inner face.

Occurring sporadically in waste places in the vicinity of Seattle, Washington, Hood River, Oregon, and scattered localities in California. Native of Asia. For complete synonymy see Pflanzenreich 4[269]: 418–421. 1937.

Family 143. RUBIÀCEAE.

MADDER FAMILY.

Herbs, shrubs, or trees with simple, opposite or verticillate, stipulate leaves. Flowers regular and nearly symmetrical, perfect but often dimorphous. Calyx-tube adnate to the ovary, the limb various. Corolla sympetalous, 4–5-lobed, varying from rotate to campanulate or salverform or funnelform. Stamens as many as the lobes of the corolla and alternate with them, inserted on its tube or throat; anthers linear or oblong. Ovary 1–10-celled; style short or elongated, simple or lobed; ovules 1 to many in each cell. Fruit a capsule, berry, drupe, drupelet, or nutlet; seeds various.

Approximately 400 genera, and 6,000 species, of wide distribution but most abundant in tropical regions.

Large shrub; flowers in dense, globose, long-peduncled heads; corolla tubular-funnelform.	1. *Cephalanthus.*
Herbs or somewhat suffrutescent plants.	
Flowers in pedunculate involucrate heads.	2. *Sherardia.*
Flowers not in involucrate heads.	
Corolla funnelform.	
Leaves opposite; flowers in a loose forked cyme.	3. *Kelloggia.*
Leaves in whorls; flowers in clustered cymes.	4. *Asperula.*
Corolla rotate.	5. *Galium.*

1. CEPHALÁNTHUS L. Sp. Pl. 95. 1753.

Shrubs or small trees with opposite or whorled, short-petioled, entire leaves and terminal or axillary, densely capitate, small, bracteolate, white or yellowish flowers. Calyx-tube obpyramidal with 4 obtuse lobes. Corolla tubular-funnelform with 4 short, erect or spreading lobes. Stamens 4, inserted on the corolla-throat; filaments very short; anthers oblong, 2-cuspidate at base. Ovary 2-celled; ovules solitary in each cell, pendulous; style filiform,

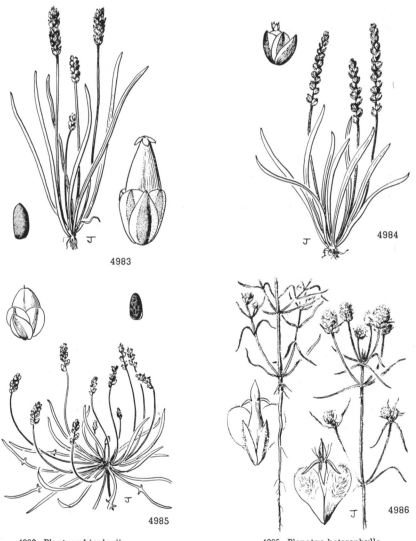

4983

4984

4985

4986

4983. Plantago bigelovii
4984. Plantago pusilla

4985. Planatgo heterophylla
4986. Plantago indica

exserted; stigma capitate. Fruit dry, obpyramidal, 1–2-seeded. Endosperm cartilaginous; cotyledons linear-oblong. [Name Greek, meaning head-flower.]

About 7 species, natives of America, Asia, and Africa. Type species, *Cephalanthus occidentalis* L.

1. **Cephalanthus occidentàlis** L. Button-bush. Fig. 4987.

Cephalanthus occidentalis L. Sp. Pl. 95. 1753.

Shrub or small tree 1–8 m. high with opposite or verticillate leaves and branches, glabrous or somewhat pubescent. Leaves petioled, ovate or oval, acuminate or acute at apex, truncate, rounded or narrowed at base, entire, 5–12 cm. long; petioles 8–10 mm. long; peduncles 3–7 cm. long; heads globose, 2–3 cm. broad, the receptacle pubescent; calyx greenish, sessile; corolla white, 8–12 mm. long, the lobes obtuse, often tipped with black; style very slender, about twice the length of the corolla; calyx persisting on the fruit, the fruit obpyramidal, about 4 mm. high; seed flattened, acutely margined.

Most soils on margins of streams and swamps, Upper and Lower Sonoran Zones; Lake and Napa Counties in the California Coast Ranges, and Sacramento River Canyon, Shasta County, south through the Great Valley and the Sierra Nevada foothills to Kern County, California; east to Arizona, Texas, and Florida; also northeast to eastern Ontario and New Brunswick. Type locality: eastern North America, but no definite locality given. June–Sept.

2. **SHERÁRDIA** [Dill.] L. Sp. Pl. 102. 1753.

Slender, procumbent or diffuse, annual herbs with verticillate, spine-tipped leaves and small, nearly sessile, pink or blue flowers in terminal and axillary, involucrate heads. Calyx-tube obovoid; lobes 4–6, lanceolate, persistent. Corolla funnelform, 4–5-lobed, the tube as long as the lobes or longer. Stamens 4 or 5, inserted on the corolla-tube; filaments slender; anthers small, oblong, exserted. Ovary 2-celled; style 2-cleft; ovules 1 in each cell. Fruit didymous, with indehiscent carpels. Seed erect. [Named in honor of Dr. William Sherard, patron of Dillenius.]

A monotypic genus of the Old World.

1. **Sherardia arvénsis** L. Blue Field Madder. Fig. 4988.

Sherardia arvensis L. Sp. Pl. 102. 1753.

Tufted annual with numerous, prostrate, decumbent or ascending stems 6–20 cm. long, herbage hispidulous. Leaves in whorls of 4–6, the lower often obovate and mucronate, the upper linear or lanceolate, acute and sharp-pointed, rough-ciliate on the margins, 6–10 mm. long, 2–4 mm. wide; flowers in few-flowered, slender-peduncled, involucrate heads; involucre deeply 6–8-lobed, the lobes lanceolate, sharp-pointed; corolla pink or bluish, 4–5 mm. long, the lobes spreading; fruit crowned with 4–6 lanceolate calyx-teeth.

Lawns, gardens, and pasture land; native of Europe and naturalized in the Pacific States mostly west of the Cascade Mountains and the Sierra Nevada; Washington to southern California. Type locality: Europe. April–July.

3. **KELLÓGGIA** Torr. Bot. Wilkes Exp. **17**: 332. *p. 6.* 1874.

Perennial herbs with opposite, entire, stipulate leaves and small flowers in a loose forking cyme terminating the simple or sparsely branched stems. Calyx-tube obovoid, somewhat flattened laterally, covered with short stiff bristles; teeth 4, very small, subulate-persistent. Corolla funnelform, with 4 (rarely 3) narrow ovate lobes, valvate in the bud. Stamens 4, inserted in the throat of the corolla; filaments short; anthers linear. Ovary 2-celled, with a solitary anatropous ovule attached at the base of each cell; style very slender; stigmas 2, filiform, papillose. Fruit small, oblong, dry and coriaceous, covered with hooked bristles, splitting at maturity into 2 closed carpels, to the walls of which the solitary seed adheres. [Named in honor of Dr. Albert Kellogg, an early California botanist.]

A monotypic genus of western North America.

1. **Kelloggia galioìdes** Torr. Kelloggia. Fig. 4989.

Kelloggia galioides Torr. Bot. Wilkes Exp. **17**: 332. *pl. 6.* 1874.

Plants with woody rootstocks, glabrous or nearly so, the stems several, simple or branched, 1–2.5 dm. high. Leaves lanceolate, 1–3.5 cm. long, about equaling or shorter than the internodes, darkened in dried specimens; inflorescence a loose, divergently branched cyme, the pedicels slender, divergent, 1.5–3 cm. long; corolla dull pink or lavender, 5–7 mm. long, the lobes about the length of the tube, acuminate; fruit oblong, 4–5 mm. long, densely covered with short uncinate bristles.

Dry ridges, Canadian Zone; in the Cascade Mountains of Washington south through Oregon to the Siskiyou Mountains, California, and the Sierra Nevada south to southern California; eastward to Idaho and Wyoming and south through Nevada to northern Arizona. Type locality: the "Walla-Walla River." May–July.

4. **ASPÉRULA** L. Sp. Pl. 103. 1753.

Perennial herbs with erect or ascending, 4-angled stems and small white, pink, or blue flowers in terminal or axillary, usually cymose clusters. Calyx somewhat didymous, the limb obsolete. Ovary 2-celled; ovules 1 in each cell; style 2-cleft. Fruit globose-didymous,

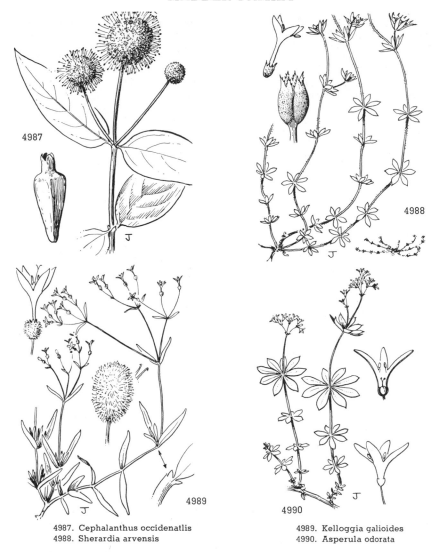

4987. Cephalanthus occidenatlis
4988. Sherardia arvensis

4989. Kelloggia galioides
4990. Asperula odorata

the carpels indehiscent. Seed adherent to the pericarp; endosperm fleshy; embryo curved. [Name Latin, diminutive of *asper*, rough, referring to the leaves.]

About 80 species, natives of the Old World. Type species, *Asperula odorata* L.

1. Asperula odoràta L. Sweet Woodruff. Fig. 4990.

Asperula odorata L. Sp. Pl. 103. 1753.

Stems erect, slender, smooth. Leaves usually in whorls of 8, varying from 6–9, thin, oblanceolate or oblong-lanceolate, acute or obtuse, mucronate, 1-nerved, roughish on the margins, 1–2.5 cm. long, the lower smaller and often obovate; flowers in several-flowered cymes on slender, terminal and axillary peduncles, white or pinkish, 3 mm. long; pedicels 2–4 mm. long; fruit very hispid, about 2 mm. broad.

Frequent escape from gardens, especially in open woods; western Washington and Oregon. Native of Europe. April–July.

Crucianélla angustifòlia L. Sp. Pl. 108. 1753. Slender annual with linear whorled leaves and narrow, terminal, and sometimes axillary spikes, the tubular flowers subtended by 2 scarious bracts with prominent mid-veins, the fruit glabrous. This European species has been collected in Tehama County, California.

5. GÀLIUM* L. Sp. Pl. 105. 1753.

Annual or perennial with slender 4-angled stems and branches, and opposite or appar-

---

* The following new names and combinations were made in the genus *Galium* by Lauramay Dempster (Brittonia **10**: 181–192. (Oct.) 1958) after manuscript was set up in type: *Galium nuttallii* var. *cliftonsmithii*, *G. nuttallii* var. *ovalifolium*, *G. nuttallii* var. *tenue*, *G. andrewsii* var. *gatense*, *G. mexicanum* var. *asperulum*, *G. catalinense* var. *buxifolium*, *G. angustifolium* var. *onycense*, *G. matthewsii* var. *magnifolium*, and *G. munzii* var. *kingstonense*. See also Dempster, Brittonia **11**: 105–22. 1959.

ently verticillate leaves (foliaceous stipules). Flowers perfect or sometimes dioecious, small, white, green, yellow, or purple, mostly in axillary or terminal cymes or panicles, the pedicels usually jointed with the calyx. Calyx-tube ovoid or globose, the limb minutely toothed or wanting. Corolla rotate or slightly campanulate, 4-lobed or rarely 3-lobed, the lobes involute and inflexed in the bud, often acuminate or mucronate at apex. Stamens 4, rarely 3; filaments short; anthers small, exserted. Ovary 2-celled with 1 ovule in each cell; styles 2, short; stigma capitate. Fruit didymous, dry or fleshy, smooth or often hispid, pubescent or villous, separating into 2 indehiscent carpels, sometimes only 1 of carpels maturing. Seed convex on the back and concave on the face, or spherical and hollow; endosperm horny; embryo curved; cotyledons foliaceous. [Name Greek, milk; one of the species, *G. verum*, once used to curdle milk.]

About 300 species of wide geographic distribution and especially well represented in the Pacific States. Type species, *Galium mollugo* L.

Plants annual.
    Carpels of fruit much longer than broad, curved outward on the inner face; weak-stemmed inconspicuous annuals.   1. *G. murale.*
    Carpels of fruit about as broad as long, rounded or nearly straight on the inner face; plants various.
        Fruit with coarse or slender, hooked bristles.
            Leaves in whorls of 4 or the upper stem-leaves opposite.
                Fruit solitary in leaf-axils on a slender reflexed pedicel; uppermost stem-leaves opposite, the lower in whorls of 3–4 of unequal length.   2. *G. bifolium.*
                Fruit sessile between leafy bracts on axillary branchlets; leaves in whorls of 4.
                                          3. *G. proliferum.*
            Leaves in whorls of 6–8.
                Fruit 1 mm. or less broad; plants slender, usually diffusely branched.
                                          4. *G. parisiense.*
                Fruit 2–5 mm. broad; plants coarse, with long, usually unbranched, reclining stems.
                                        6. *G. aparine.*
        Fruit smooth or tuberculate.
            Pedicels capillary; inflorescence much surpassing the leaves.   5. *G. divaricatum.*
            Pedicels stout, recurved; inflorescence axillary, not or scarcely longer than the leaves.
                                      7. *G. tricornutum.*
Plants perennial.
    Stems from slender creeping rootstocks, weak, delicate, seldom entirely erect, completely herbaceous.
        Fruit glabrous; corolla 3–4-parted.
            Flowers 2.5–3.5 mm. broad, rather numerous in small cymes on the upper branchlets; peduncles often curved but pedicels straight and divaricate in fruit.   8. *G. cymosum.*
            Flowers less than 2.5 mm. broad, 1–3 in upper leaf-axils or on branchlets; pedicels strongly arcuate in age (1-flowered and with pedicels usually straight in vars. *pusillum* and *pacificum*).
                                    9. *G. trifidum subbiflorum.*
        Fruit hispid or at least scabrous; corolla 4-parted.
            Leaves in whorls of (5–)6–8, veins not conspicuous.
                Cymes 3-flowered on axillary peduncles; carpels covered with slender hooked bristles about as long as width of carpel.   11. *G. triflorum.*
                Cymes terminal in a diffuse panicle; carpels with short hooked bristles or scabrous.
                                      10. *G. asperrimum.*
            Leaves in whorls of 4, conspicuously 3-nerved.   12. *G. oreganum.*
    Stems somewhat woody at least below, erect from woody root-crown or suffrutescent base, climbing, or tufted from branched underground stems and rootstocks; if herbaceous, plants from coarser rootstocks and without delicate stems.
        Leaves in whorls of 6–8; introduced species.
            Flowers bright yellow; leaves narrowly linear.   13. *G. verum.*
            Flowers white; leaves oblanceolate.   14. *G. mollugo.*
        Leaves in whorls of 4; native species; most species dioecious.
            Inflorescence a conspicuous, erect and compact, thyrsoid panicle of white flowers.
                                      15. *G. boreale.*
            Inflorescence various; if paniculate, not of conspicuous white flowers.
                Fruit glabrous or sparsely puberulent or pubescent, when mature dry or pulpy and berry-like.
                      Fruit dry, with short, curved or straight hairs or subglabrous.
                          Shrubby, 6–12 dm. high; inflorescence leafy.   16. *G. catalinense.*
                          Stems tufted, 1–2 dm. high; inflorescence much exceeding the leaves, subracemose.
                                      17. *G. jepsonii.*
                Fruit pulpy and berry-like when ripe, glabrous (pubescent to glabrate in *G. pubens* and sometimes in *G. californicum*).
                  Plants from slender rootstocks with slender creeping stems, diffuse or densely low-tufted.
                      Leaves linear or narrowly oblong, usually much longer than the internodes and often concealing the stems.
                          Leaves rigid, narrowly linear-subulate, acicular. 18. *G. andrewsii.*
                        Leaves firm but not rigid, 1–1.5 mm. wide, abruptly acute and cuspidate.
                            Longest leaves 9–12 mm. long; plants of northwestern California and adjacent Oregon.   19. *G. ambiguum.*
                            Longest leaves 4.5–6 mm. long; plants of the Santa Lucia Mountains.
                                  20. *G. clementis.*
                  Leaves ovate to ovate-lanceolate or elliptic or obovate, shorter than the internodes (nearly equaling them in *G. muricatum*).
                    Leaves dull, copiously covered on both surfaces with spreading hirsute hairs, ovate or ovate-lanceolate (more or less shining and hairy only on the margin in var. *miguelense*).   21. *G. californicum.*

Leaves shining, sparsely hispid or hirsute with curved or upwardly appressed hairs on the upper surface, margins, and sometimes on the midvein beneath, the lower surface glabrous or nearly so, broadly elliptic to oblanceolate. 22. *G. muricatum.*
Stems rather stout from a woody caudex and erect, or slender, long and vine-like.
Stems somewhat woody, long and slender, climbing or reclining; leaves ovate to oblong (linear-oblong in some intergrading forms). 23. *G. nuttallii.*
Stems erect from a woody caudex or rootstock, 1.5–3 dm. high.
Leaves firm and somewhat thickened, linear-oblong to ovate-oblong or ovate and large, the lower leaves 5–8 mm. wide, the marginal scaberulous hairs when present spreading.
Stems and leaves glabrous or nearly so, the leaves narrowly linear-oblong (young growth sometimes reclining). 24. *G. bolanderi.*
Stems and leaves pilose (less so in var. *scabridum*), the leaves ovate-oblong or ovate, narrower on much-branched plants. 25. *G. pubens.*
Leaves thin, mostly oval, the large lower leaves 8–10 mm. wide, the marginal cilia appressed; peduncles and pedicels of staminate inflorescence capillary. 26. *G. sparsiflorum.*
Fruit dry, conspicuously hirsute or villous, the spreading hairs about as long as or longer than the body of the fruit.
Leaves narrowly to broadly linear, acute.
Flowers red, with acuminate lobes, the staminate on capillary pedicels; plants polygamo-monoecious. 27. *G. wrightii rothrockii.*
Flowers yellowish green or yellowish white, with acute lobes, the staminate pedicels not markedly slender; plants dioecious.
Fruit 2.5–3 mm. wide including the stoutish spreading hairs; inflorescence paniculate.
Plants tall from a woody base, over 3 dm. tall; cismontane California. 28. *G. angustifolium.*
Plants low, about 1–2 dm. tall, tufted; San Gabriel Mountains. 29. *G. gabrielense.*
Fruit 5–8 mm. wide including the abundant, slender, spreading hairs; inflorescence narrow in cymose clusters on short axillary branchlets (including subsp. *puberulum*). 38. *G. watsonii.*
Leaves ovate, lanceolate-ovate, or orbicular and acute or rounded at the apex, or lanceolate-acuminate.
Lower stem-leaves closely clothing the older stems, persistent; plants low and tufted; inflorescence narrow and congested, the branches scarcely exceeding the leaves. 36. *G. parishii.*
Lower stem-leaves not approximate, not markedly persistent; inflorescence ample and widely spreading.
Fruit including the hairs 2.5–3 mm. wide; leaves rigid and acerose.
Intricately branched shrubs, hispidulous-puberulent; inflorescence congested-paniculate. 31. *G. stellatum eremicum.*
Plants glabrous, erect from a woody base; inflorescence open-paniculate. 32. *G. matthewsii.*
Fruit including the hairs 5–7 mm. or more wide; leaves thickish or thin but not rigid, sometimes apiculate but not acerose.
Fruiting branchlets as well as the fruiting pedicels recurved or pendent. 30. *G. hallii.*
Fruiting branchlets erect or widely spreading, the fruiting pedicels straight (somewhat curving in *G. hypotrichium*).
Inflorescence broad, many-flowered, the flowering branches divaricate; leaves of midstem one-third to one-sixth the length of the internodes.
Plants (in ours) more or less pubescent throughout with hispidulous hairs. 33. *G. munzii.*
Plants entirely glabrous (hirsute in f. *hirsutum*). 34. *G. multiflorum.*
Inflorescence narrow, few-flowered, the flowering branches erect or ascending; leaves of midstem usually one-half to nearly equaling the length of the internodes.
Leaves ovate to ovate-lanceolate; pubescence of hispidulous hairs, often scabridulous on midribs and margins of leaves; fruiting pedicels more or less curved. 35. *G. hypotrichium.*
Leaves broadly elliptic to oval; pubescence of dense hirsutulous hairs (becoming nearly glabrate in subsp. *glabrescens*); fruiting, pedicils straight. 37. *G. grayanum.*

1. Galium muràle All. Wall Bedstraw. Fig. 4991.

Galium murale All. Fl. Ped. 1: 8. *pl. 77, fig. 1.* 1785.

Diminutive annual with slender stems, simple or branching from the base, mostly about 2 cm. high, glabrous. Leaves in whorls of 4 or 5, broadly linear to oblanceolate, 2–4 mm. long, acute and mucronulate at apex, rather sparsely ciliate on the margins, glabrous or sparsely bristly-scabrous on lower surface; flowers usually 2 in each of the upper leaf-whorls, pedicels short, about 1 mm. long, spreading or somewhat recurved in fruit; corolla white, less than 1 mm. wide; fruit oblong, 1.5 mm. long and barely 1 mm. broad, the carpels divergent, body of fruit very sparsely clothed with short ascending bristles but densely tufted at apex.

An introduced plant becoming well established on open grassland in the Bay Region in Marin, Contra Costa, Alameda, San Mateo, and Santa Clara Counties, California. Native of southern Europe. March–May.

2. **Galium bifòlium** S. Wats. Low Mountain Bedstraw. Fig. 4992.

Galium bifolium S. Wats. Bot. King Expl. 134. *pl. 14.* 1871.

Annual, the stems erect, simple or with a few ascending branches, 5–15 cm. high, slender, glabrous even on the angles. Upper leaves 2, opposite, the lower usually in a whorl of 3 or 4, linear-lanceolate to oblong-lanceolate, 5–25 mm. long, usually of uneven length in some whorls, narrowed at base, acute or rounded at apex; flowers pedicellate, usually solitary in the axils; pedicels slender, 3–10 mm. long, widely spreading or reflexed, usually recurved at apex in fruit; corolla minute, white; fruit 2.5–3.5 mm. broad, covered with slender hooked prickles.

Open coniferous forests and edges of meadows, Transition and Canadian Zones; southern British Columbia, Washington, and Oregon southward in California in the Coast Ranges to Humboldt County, and from Modoc County through the Sierra Nevada to the San Bernardino Mountains, and east to Montana and Colorado. Type locality: "In the Trinity, Battle and East Humboldt Mountains, Nevada, and in the Wahsatch" Mountains, Utah. June–Aug.

3. **Galium prolíferum** A. Gray. Desert Bedstraw. Fig. 4993.

Galium virgatum var. *diffusum* A. Gray, Smiths. Contr. 3⁵: 80. 1852.
Galium proliferum A. Gray, op. cit. 5⁶: 67. 1853.
Galium proliferum var. *subnudum* Greenm. Proc. Amer. Acad. 33: 461. 1898.

Annuals, the stems simple or often branched, erect or somewhat decumbent at base, 1–3.5 dm. high, sparsely scabrous on the angles or glabrous. Cotyledons oblong, the blade about equaling the petiole, persisting; stem-leaves 1-veined, in whorls of 4, hispidulous to glabrate, oblong to narrowly ovate, 4–8 mm. long, short-petiolate on the lower stem to sessile above, the internodes much exceeding the leaves; flowers nearly sessile between 2 leaf-like bracts, borne on short pedunculate branchlets, these often proliferating; pedicels recurved, stout, much shorter than the fruit; flowers white or pale yellow, about 3 mm. wide; fruit 2.5–3 mm. wide, excluding the covering of fine uncinate bristles.

Rocky slopes, Upper Sonoran Zone; Kingston and Providence Mountains, San Bernardino County, California, east to western Texas and northern Mexico. Type locality: western Texas. Collected in 1851 by Wright. March–April.

4. **Galium parisiénse** L. Wall Bedstraw. Fig. 4994.

Galium parisiense L. Sp. Pl. 108. 1753.
Galium litigiosum DC. in Lam. & DC. Fl. Franc. ed. 3 4: 263. 1805.

Annual, the stems erect or ascending, 0.5–3.5 dm. high, very slender, much branched on larger plants from the base, rough on the angles, the branchlets flowering from the base to summit of the plants. Leaves usually in whorls of 6, varying from 4–7, linear to linear-lanceolate, often reflexed, 4–10 mm. long, cuspidate, minutely scabrous on the margins and midrib; cymes several-flowered, axillary and terminal on filiform peduncles, 4–6 mm. long, divaricately spreading; flowers greenish white, minute; fruit barely 1 mm. wide, densely covered with slender hooked bristles.

Hillsides and stream banks, Upper Sonoran and Transition Zones; California Coast Ranges from Humboldt County to Santa Cruz County, western slope of the Sierra Nevada from Butte County to Tuolumne County, California. Native of Europe. June–Aug.

5. **Galium divaricàtum** Lam. Lamarck's Bedstraw. Fig. 4995.

Galium divaricatum Lam. Encycl. 2: 580. 1788.
Galium anglicum var. *divaricatum* Koch, Syn. Fl. Germ. ed. 2. 1025. 1843–45.
Galium parisiense var. *leiocarpum* Tausch, Bot. Zeit. 18: 354. 1860.
Galium parisiense var. *anglicum* Huds. ex Jepson, Man. Fl. Pl. Calif. 958. 1925, as to California plants, not
    G. *anglicum* Huds.

Habit much as the preceding species with stems smooth or nearly so. Leaves usually in whorls of 6, linear, cuspidate, more or less scabrous on the margins, 3–10 mm. long; inflorescence as in the preceding species but the peduncles capillary, spreading and often curved, 8–15 mm. long; corolla white or tinged with pink; fruit granulate, devoid of hairs, less than 1 mm. wide.

Grassy hill slopes and fields, Upper Sonoran and Transition Zones; Willamette Valley, Oregon, south in the Coast Ranges to Marin County, California; also Yuba County. Native of Europe. June–July.

6. **Galium aparìne** L. Cleavers or Goose Grass. Fig. 4996.

Galium aparine L. Sp. Pl. 108. 1753.
Galium vaillantii DC. in Lam. & DC. Fl. Franc. ed. 3. 4: 263. 1805.
Galium agreste α *echinospermum* Wallr. Sched. 59. 1822.
Galium aparine β *minor* Hook. Fl. Bor. Amer. 1: 290. 1833.
Galium spurium var. *echinospermum* Hayak, Fl. Steierm. 2: 393. 1913.

Annual, from a slender root, the stems slender, weak, ascending or usually scrambling over other bushes, 1–15 dm. high, sometimes erect in dry open places, retrorsely hispid on the angles. Leaves in whorls of 6–8 or rarely 4, linear to narrowly lanceolate or oblanceolate, cuspidate at apex, tapering to the base, 2–5 cm. long, retrorsely hispid on the margins and midrib; flowers in 1–3-flowered cymes in the upper axils, becoming cymose-paniculate on older plants; corolla-lobes acute; fruiting pedicels straight, divaricately spreading, 5–20 mm. long; fruit 1.5–3.5 mm. broad, covered with hooked bristles shorter than the body of the fruit.

Growing in various habitats, especially edges of thickets and open woodland, and of wide distribution; common and widely distributed in the Pacific States and across the continent; also Europe and Asia. Type locality: Europe. March–Aug. Quite variable as to size of fruit and density of the covering of hooked bristles. Arizona.

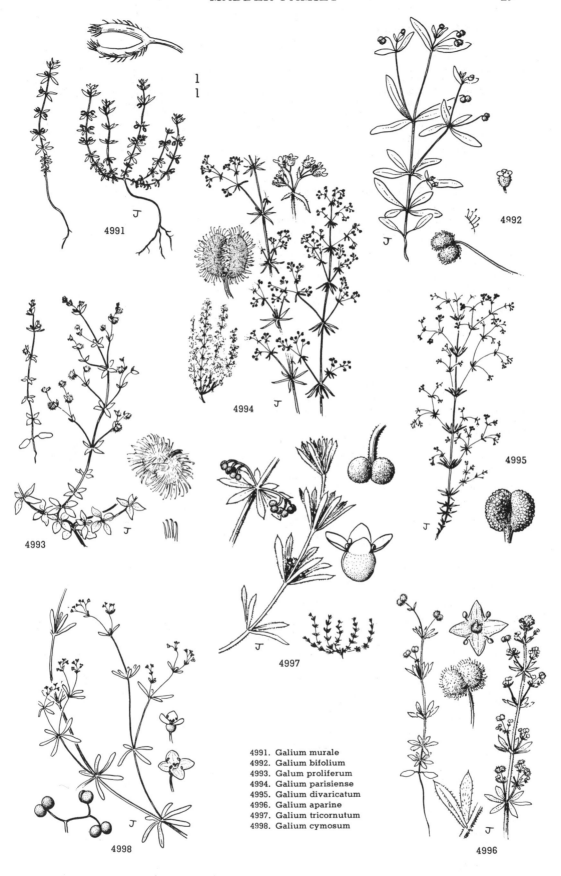

4991. Galium murale
4992. Galium bifolium
4993. Galum proliferum
4994. Galium parisiense
4995. Galium divaricatum
4996. Galium aparine
4997. Galium tricornutum
4998. Galium cymosum

7. Galium tricornùtum Dandy. Rough-fruited Corn Bedstraw. Fig. 4997.

Galium tricorne Stokes in With. Bot. Arr. Brit. Pl. ed. 2. 1: 153. 1787. (Illegitimate name.)
Galium tricornutum Dandy, Watsonia 4: 47. 1957.

Annual, the stems stout, decumbent or ascending, 15–45 cm. long, simple or little branched, retrorsely roughened on the angles, otherwise glabrous. Leaves in whorls of 6 or 8, linear to narrowly oblanceolate, 15–25 mm. long, 2–4 mm. wide, mucronate, densely and retrorsely barbed on the margins and on the midrib; inflorescence axillary, 3-flowered, the peduncles stout, about as long as or shorter than the leaves; pedicels stout,' strongly curved downward in fruit; corolla white, the lobes rounded; fruit 3–4 mm. broad, the carpels nearly spherical, tuberculate-roughened or granular.

Waste places and grain fields; sparingly introduced in the Willamette Valley and in Josephine County, western Oregon, and central California in the region of San Francisco Bay and the Sierra Nevada foothills and in San Luis Obispo County; also eastern United States. Native of Europe. June–Aug.

8. Galium cymòsum Wiegand. Pacific Bedstraw. Fig. 4998.

Galium cymosum Wiegand, Bull. Torrey Club 24: 401. 1897.

Perennial, the stems freely branching, slender and weak, 3–7 dm. long, more or less minutely scabrous on the angles. Leaves mostly in whorls of 6, oblanceolate to linear-oblanceolate, rounded at apex, unequal, 0.5–2 cm. long, 1-nerved, minutely scabrous on the margins and midveins, the lower stem-leaves linear and sometimes reflexed; flowers in small cymes at the ends of the numerous upper branches; pedicels rather short, straight, and divaricate in fruit; corolla white or tinged with rose, 2–4 mm. wide, 3–4-lobed, the lobes triangular-ovate; fruit glabrous, each carpel globose, about 1.5 mm. in diameter.

Moist meadows and sphagnum bogs, Humid Transition Zone; British Columbia south on the western side of the Cascade Mountains to Humboldt County, California. Type locality: Tacoma, Pierce County, Washington. June–Aug.

A variable species, perhaps conspecific with and at least much resembling the eastern *G. trifidum* var. *tinctorium* (L.) Torr. & Gray, Fl. N. Amer. 2: 22. 1841 (*G. tinctorium* L. Sp. Pl. 106. 1753; *G. claytonii* Michx. Fl. Bor. Amer. 1: 78. 1803; *G. trifidum* subsp. *tinctorium* Hara, Rhodora 41: 388. 1939). Some specimens are found in like ecological situations with few-flowered cymes and smaller flowers which are scarcely distinguishable from that species.

9. Galium trífidum var. subbiflòrum Wiegand. Trifid Bedstraw. Fig. 4999.

Galium trifidum var. *subbiflorum* Wiegand, Bull. Torrey Club 24: 399. 1897.
Galium tinctorium var. *submontanum* Wight, Zoe 5: 53. 1900.
Galium claytonii var. *subbiflorum* Wiegand, Rhodora 12: 229. 1911.
Galium tinctorium var. *subbiflorum* Fernald, op. cit. 39: 320. 1937.

Perennial with slender rootstock and slender, weak, erect or ascending stems 1.5–5 dm. high, sharply 4-angled, glabrous or somewhat scabrous on the angles. Leaves in whorls of 5–6, sometimes 4, the lower leaves often reflexed, somewhat unequal, linear-oblanceolate, obtuse, 7–12 mm. long, 0.5–1.5 mm. wide, somewhat flaccid, usually somewhat scabrous on midrib and margin; flowers pedicellate, 1–2 (sometimes 3) in the axils of the upper leaves or on short, spreading, axillary branchlets in bracteate cymes of 3 or 2, the pedicels slender, about equaling the leaves or a little longer, strongly deflexed-arcuate at least in fruit; corolla minute, about 0.5 mm. long, the lobes 3 (sometimes 4), obtuse; fruit glabrous, globose, each carpel about 1 mm. wide.

Moist ground or marshy meadows, Upper Sonoran and Transition Zones; British Columbia to southern California and east to the Rocky Mountains and across the northern United States and Canada. Type locality: Colorado. June–Sept.

The name-bearing taxon, a small plant with spreading slender branches, linear leaves in whorls of 4, and capillary, widely spreading pedicels, inhabits northeastern North America, Canada, and Alaska. Some specimens from Okanogan County, Washington, resemble the species and may on further study of more material prove to be typical. The complex is highly variable and the varieties in the Pacific States are not sharply defined; many intergrading forms are to be found, as has been pointed out by many authors.

Galium trifidum var. pacíficum Wiegand, Bull. Torrey Club 24: 400. 1897. (*Galium tinctorium* var. *diversifolium* Wight, Zoe 5: 54. 1900; *G. columbianum* Rydb. Fl. Rocky Mts. 808. 1917; *G. trifidum,* subsp. *columbianum* Hultén, Fl. Aleut. Isl. 307. 1937.) Perennial with slender rootstocks, the stems slender, diffuse and reclining, 2–6 dm. long, with creeping axillary branches sometimes arising from the lower nodes, glabrous except for the minutely retrorse-hispid angles; leaves oblong-spatulate to narrowly oblanceolate, obtuse, commonly in whorls of 4–5, more or less scabrous on the margins and midrib, 10–25 mm. long, 2–3.5 mm. wide; peduncles capillary, mostly 1-flowered or sometimes 2-flowered when terminal, divaricately spreading, often slightly arcuate, scabrous, longer than, equaling, or shorter than the leaves; corolla minute, about 0.5 mm. broad, 3–4-parted; fruit minute, glabrous.

Moist ground on edges of marshy meadows and streams, especially near the woods or thickets, Transition and Canadian Zones; Alaska and Vancouver Island south on the Pacific Slope and Cascade Mountains of Washington and Oregon to the Sierra Nevada and San Bernardino Mountains, California. Type locality: "Placer Co., Cal., Carpenter (1892), Bottom lands of the Columbia River, Suksdorf, no. 1661 (1893)." June–Nov.

In bogs along the coast of Washington and Oregon specimens occasionally have been collected that suggest *Galium tinctorium* var. *labradoricum* Wiegand (Bull. Torrey Club 24: 398. 1897; *G. labradoricum* Wiegand, Rhodora 6: 21. 1904). The plants are 1–2 dm. tall, branched; leaves in whorls of 4, sometimes 5 below, and reflexed, small, obtuse, and rather firm in texture; flowering material not seen.

Galium trifidum var. pusíllum A. Gray, Man. ed. 5. 209. 1867. Perennial with slender weak-branched stems forming mats, glabrous or nearly so, 5–12 cm. long; leaves in whorls of 4, glabrous, narrowly oblanceolate, unequal in length, 3–8 mm. long; pedicels solitary or sometimes in pairs, straight or somewhat curved, usually glabrous, rather slender, 2–5 mm. long; fruit as in the other varieties.

Wet mountain meadows, Canadian and Hudsonian Zones; Cascade Mountains in Washington and Oregon to the Sierra Nevada and San Bernardino Mountains, California, eastward to the Atlantic seaboard in mountainous areas. Type locality: not given in the original reference but said to be "in deep sphagnous swamps, northward." Intergrading with *Galium trifidum* var. *subbiflorum* and *G. trifidum* var. *pacificum*. Some of the specimens here included have been considered by western authors to be *G. brandegei* A. Gray (Proc. Amer. Acad. 12: 58. 1876). However, a specimen from the Gray Herbarium bearing the Synoptical Flora label, collected by Lemmon (no. 1217) at Webber Lake, Sierra County, California, and agreeing with the description given above, was identified by Gray as *G. trifidum* var. *pusillum*.

10. **Galium aspérrimum** A. Gray. Tall Rough Bedstraw. Fig. 5000.

Galium asperrimum A. Gray, Mem. Amer. Acad. II. **4**: 60. 1849.
Galium asperrimum var. *asperulum* A. Gray, Bot. Calif. **1**: 284. 1876.

Stems from a slender rootstock, erect or ascending and more or less supported, 3–8 dm. high, freely branching, minutely scabrous on the angles, otherwise glabrous. Leaves in whorls of 6–8 or rarely 4, linear-lanceolate to oblong-lanceolate, cuspidate, conspicuously retrorse-scabrous on the margins and midribs, otherwise glabrous; flowers on rather long capillary pedicels, mostly in terminal, many-flowered, leafy, cymose panicles; corolla white, 2.5–4 mm. broad, the lobes acuminate; mature fruit 2.5 mm. broad, granulate-scabrous or minutely hispidulous with hooked prickles; pedicels 5–10 mm. long, capillary but slightly clavate just below the fruit.

Usually in moist thickets, Arid Transition Zone; Kittitas County, Washington, southward east of the Cascade Mountains to northern California, and south to the central Sierra Nevada, Mariposa County, California, and east to Idaho, Utah, New Mexico, western Colorado, western Texas, and Sinaloa, Mexico. Type locality: "wet places, near irrigating ditches, Santa Fe," New Mexico. June–Aug.

11. **Galium triflòrum** Michx. Fragrant Bedstraw. Fig. 5001.

Galium triflorum Michx. Fl. Bor. Amer. **1**: 80. 1803.
Galium triflorum f. *glabrum* Leyendecker, Iowa State Coll. Journ. Sci. **15**: 180. 1941.
Galium triflorum f. *hispidum* Leyendecker, loc. cit.

Sweet-scented perennials, several-stemmed from a clustered root, the simple weak stems 1–4 dm. long, occasionally remotely branched, reclining or erect, and rather narrow-leaved in

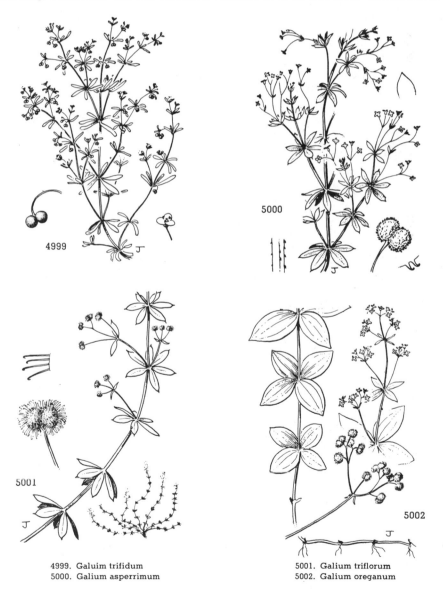

4999. Galuim trifidum
5000. Galium asperrimum

5001. Galium triflorum
5002. Galium oreganum

exposed situations, glabrous or somewhat barbed or hirsute on the angles. Leaves in whorls of 6, usually but little reduced above, elliptic-lanceolate to obovate, 1-nerved, somewhat scabrous on margins and midvein, cuspidate at apex, 1.5–4.5 cm. long, 3–12 mm. wide; inflorescence a simple 3-flowered cyme, often forked and each branch 1–3-flowered; peduncles axillary, slender, surpassing the leaves, bracteate at base of cyme; corolla greenish, 2–3 mm. wide, the lobes acuminate; fruit 3–4 mm. broad, densely hispid with rather slender, long, hooked hairs.

In woods, Upper Sonoran Zone to Canadian Zone; Alaska to southern California and across the continent; also in Japan, the Himalaya Mountains, and Europe. Type locality: Canada. April–Aug.

12. Galium oregànum Britton. Oregon Bedstraw. Fig. 5002.

Galium oreganum Britton, Bull. Torrey Club 21: 31. 1894.
Galium kamtschaticum subsp. *oreganum* Piper, Contr. U.S. Nat. Herb. 11: 526. 1906.

Perennial with very slender, creeping rootstock producing slender rootlets at the nodes; stems slender, 2–4 dm. high, erect or decumbent at base, glabrous or slightly scabrous on the angles. Leaves in whorls of 4, ovate to ovate-oblong or ovate-elliptic, acute or acutish at apex, the lower stem-leaves smaller than the upper, 1.5–3.5 cm. long, thin, 3-nerved, ciliate on the margins and often on the nerves above, glabrous below or sparsely hairy, especially toward the apex, the leaves of the occasional lower branchlets obtuse; inflorescence from the upper axils and terminal, the terminating, ascending, peduncle-like branches more or less ternately branched, each branch bearing 6–15 flowers; corolla greenish yellow, 3.5–4 mm. wide, the lobes 3-nerved, about 1 mm. wide, acute at apex; fruit sparsely to densely covered with hooked bristles, about 4 mm. broad.

Moist woods, Transition and Canadian Zones; Olympic Mountains and western slopes of the Cascade Mountains in Washington and Oregon, and in Jackson and Josephine Counties, Oregon, and the adjacent Siskiyou Mountain area, California (*Bacigalupi*). Type locality: Oregon. Collected by Howell. June–Aug.
The related *G. kamtschaticum* Steller differs by the terminal fewer-flowered inflorescence bearing 1–4 flowers and also by the oval to orbicular leaves which are occasionally to be seen on specimens of *G. oreganum* only on axillary basal branchlets.

13. Galium vèrum L. Yellow Bedstraw. Fig. 5003.

Galium verum L. Sp. Pl. 107. 1753.

Perennial, from a creeping rootstock, with a slightly woody base, the stems slender, erect or ascending, 1.5–9 dm. high, simple or with many short axillary branchlets, smooth or minutely pubescent; herbage darkening on drying. Leaves in whorls of 6–8, narrowly linear, 6–20 mm. long, shorter than the internodes, about 0.5 mm. wide, roughened on the inrolled margins, usually becoming deflexed in age; flowers yellow, cymose, the cymes in dense narrow panicles; fruit glabrous, about 1 mm. broad.

Ballast and waste places at Linnton and sparingly distributed in lawns of the Willamette Valley, Oregon, and Seattle, Washington; also reported from San Mateo County, California. Type locality: Europe. July–Oct.

14. Galium mollùgo L. Wild Madder. Fig. 5004.

Galium mollugo L. Sp. Pl. 107. 1753.

Perennial, glabrous throughout, the stems smooth or sometimes with spreading pubescence, diffusely branched or erect, 3–10 dm. long. Leaves in whorls of 6, rarely 8, linear to oblanceolate, 1–3 cm. long, 1.5–5 mm. wide, cuspidate at apex, sometimes slightly rough on the margins; flowers in terminal, many-flowered, panicled cymes; corolla white, veined with purple, 2–2.5 mm. broad; pedicels filiform, mostly divaricate; fruit smooth and glabrous.

Waste places and lawns, sparingly introduced in British Columbia, and around Portland and in the Willamette Valley, Oregon; along the coast in Humboldt County and occurring sporadically in San Mateo and Santa Clara Counties, California; also in eastern United States; native of Europe. May–Sept.
Galium saxátile L. Sp. Pl. 106. 1753. A matted perennial with leaves usually in whorls of 6 and trailing stems, the flowering stems ascending, bearing leafy panicles of white flowers. Reported from lawns in the Willamette Valley, Oregon.
Galium sylváticum L. Sp. Pl. ed. 2. 155. 1762. Perennial with many stems, erect or nearly so, up to 6 dm. tall; leaves in whorls of 6–8, glabrous, lanceolate, acuminate, mostly about 3–5 cm. long; inflorescence a loose, terminal, bracteate panicle with ascending capillary pedicels; the flowers numerous, white; fruit smooth. A European plant occasionally escaping from cultivation in Salem, Oregon, and Pullman, Washington. July–Sept. Scotchmist or Baby's Breath.

15. Galium boreàle L. Northern Bedstraw. Fig. 5005.

Galium boreale L. Sp. Pl. 108. 1753.
Galium septentrionale Roem. & Sch. Syst. 3: 253. 1818.

Erect, leafy, perennial herb, smooth and glabrous, simple or branched, often strict, 1.5–8 dm. high. Leaves in whorls of 4, lanceolate or sometimes nearly linear, 1-nerved or the lower broader leaves 3-nerved, obtuse or acute, 1.5–6 cm. long, 2–8 mm. wide, margins sometimes ciliate; flowers in compact terminal cymes forming many-flowered panicles; corolla white, 3–4 mm. broad; fruit about 1 mm. broad, hispid with short incurved hairs, sometimes glabrate.

Borders of meadows and open woods, mainly Canadian Zone; Alaska south to Humboldt and Shasta Counties, California, east across the continent; also in Europe and Asia. Type locality: Europe. May–Aug.
A circumboreal species quite variable in width of leave and pubescence of fruit. *G. boreale* var. *hyssopifolium* (Hoffm.) DC., with glabrous fruit, and *G. boreale* var. *intermedium* DC., with fruit covered with short, appressed or incurved hairs, are segregate forms that occur within our limits, as well as the species with its short, villoushirsute fruit. There appears to be no geographical significance in the distribution of the variants and they are here considered to be minor variations of the species.

16. **Galium catalinénse** A. Gray. Catalina Bedstraw. Fig. 5006.

Galium catalinense A. Gray, Syn. Fl. N. Amer. ed. 2. 1²: 445. (Jan.) 1886.

Stems woody, forming an erect, rather compact, branching shrub 4–12 dm. or more high, scars of leaf-bases prominent, the younger branches sharply 4-angled, the ultimate ones numerous and very leafy, sparsely and microscopically scabrous. Leaves in whorls of 4 or merely opposite on the branchlets, persistent, always longer than the internodes, midvein and usually the lateral veins prominent, linear-oblong, mucronulate, those of the main branches 15–25 mm. long, 2–5 mm. wide, of the ultimate branches often only 5 mm. in length, thinly hispidulous with stout curved hairs or glabrous; inflorescence leafy, cymosely branching, the flowers on short pedicels; flowers white, 3–4 mm. broad, the lobes ovate; fruit dry, glabrous or with a few short straight hairs.

Rocky ledges and slopes, Upper Sonoran Zone; an insular species growing on Santa Catalina and San Clemente Islands off the coast of southern California. Type locality: Santa Catalina Island, Los Angeles County. March–Aug.

Galium buxifólium Greene, Bull. Calif. Acad. **2**: 150. (Nov.) 1886. Identical in habit with *G. catalinense*; leaves rather firm and shining, broadly oblong to obovate-oblong, 15–25 mm. long, 4–9 mm. broad, glabrous or sparsely hispidulous; inflorescence as in *G. catalinense*; fruits glabrous or with short, stout, curved hairs. Rocky slopes, Santa Cruz and San Miguel Islands, Santa Barbara County, California. Type locality: Santa Cruz Island. Perhaps only subspecifically distinct from *G. catalinense*.

17. **Galium jepsònii** Hilend & Howell. Jepson's Bedstraw. Fig. 5007.

Galium angustifolium var. *subglabrum* Jepson, Man. Fl. Pl. Calif. 962. 1925.
Galium jepsonii Hilend & Howell, Leaflets West. Bot. **1**: 135. 1934.

Stems cespitose at the ends of the slender branching rootstocks, erect or decumbent at base, 10–20 cm. high, smooth and glabrous. Leaves in whorls of 4, longer than the internodes, midvein prominent, linear-oblong or the lower varying to lanceolate or lanceolate-ovate, obscurely cus-

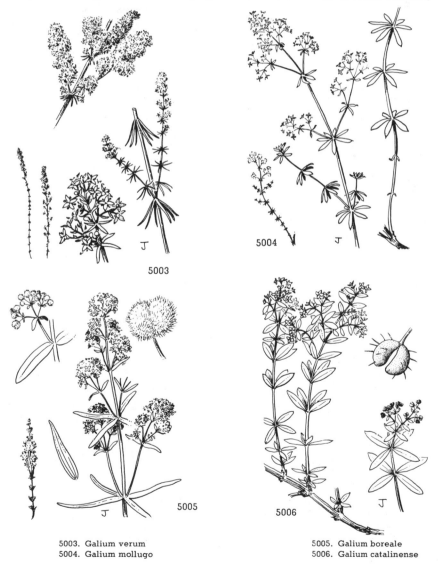

5003. Galium verum
5004. Galium mollugo

5005. Galium boreale
5006. Galium catalinense

pidate, setose-ciliate on the margins, otherwise glabrous and shining; flowers few, in the axils of the upper bract-like leaves; inflorescence narrow, the lower ascending, bearing a bract and 1–2 flowers on very short pedicels; corolla pale, about 2 mm. broad, glabrous; fruit 2.5–3 mm. wide, rather thinly pilose with ascending or appressed hairs.

Gravelly or rocky ridges, near the borders of the Arid Transition and Canadian Zones; San Bernardino Mountains, southern California. Type locality: Whitewater Basin, San Bernardino Mountains. June–Aug.

18. **Galium andrèwsii** A. Gray. Phlox-leaved Bedstraw. Fig. 5008.

Galium andrewsii A. Gray, Proc. Amer. Acad. **6**: 537. 1865.

Plants cespitose and usually densely matted, grayish green, with branching, slender, woody roots, the stems much branched and densely leafy from tangled mats or tufts 4–12 cm. high, glabrous or often sparsely scabrous on the angles. Leaves persistent, in whorls of 4, crowded, 4–10 mm. long, linear-subulate, rigid and sharply pungent, grayish green, sometimes spinulose-ciliate on the thickened margins; staminate flowers borne on the upper branchlets in small clusters or singly, on capillary pedicels as long as or somewhat surpassing the leaves; pistillate flowers solitary, axillary on short pedicels; corolla about 2.5 mm. wide, greenish white, the lobes acute; fruit 2–3 mm. broad, glabrous, on short, spreading or slightly recurved pedicels, berry-like when ripe and blackish in dried specimens.

Dry stony ridges, especially in chaparral and yellow pine belts, Upper Sonoran and Arid Transition Zones; in the Sierra Nevada foothills and the Californa Coast Ranges from Glenn County to San Diego County, and adjacent Lower California. Type locality: "California, Dr. Andrews . . ." May–Aug.

19. **Galium ambíguum** Wight. Yolla Bolly Bedstraw. Fig. 5009.

Galium ambiguum Wight, Zoe **5**: 55. 1900.

Low tufted perennial from a branched woody root, with branching underground stems from which arise tufted, leafy, lax stems 5–10 cm. long, rarely branched, the whole plant grayish with a spreading hirsute pubescence. Leaves 5–12 mm. long, 1–2 mm. wide, linear, acute, ascending, much surpassing the internodes and concealing the stem except on young growth, firm but not at all rigid, midrib prominent beneath and the margins somewhat inrolled, more or less bristly-hirsute on both surfaces, short axillary branchlets often present at the internodes; inflorescence polygamo-dioecious (?), flowers few, solitary in the axils of the upper leaves or terminal in small, short-peduncled cymes; corolla greenish white, about 3 mm. broad, the lobes sometimes sparsely hairy without; fruit fleshy, glabrous or somewhat hairy, blackish in dried specimens.

Dry open forests or rocky places, Arid Transition Zone; Siskiyou Mountains in Josephine County, Oregon, south to Trinity County, California; also Bartlett Mountain, Lake County, California. Type locality: "Yolo Bolo, Yolo Co., Cal., T. S. Brandegee, Sept. 20, 1892." March–Sept.

Galium ambiguum var. **siskiyouénse** Ferris, Contr. Dudley Herb. **4**: 337. *fig. 1a*. 1955. Resembling the name-bearing taxon in habit but often more diffuse (except in exposed situations), the stems lax, up to 16 cm. long, the longer stems sometimes branched above; leaves 6–10 mm. long, like the species in shape and texture, dark green, shining, glabrous or with a few scattered, bristly-ciliate hairs. Siskiyou Mountains of Curry and Josephine Counties, Oregon, south to Del Norte and Siskiyou Counties and the mountains of Trinity and Humboldt Counties, California. Type locality: near Gasquet, north side of Middle Fork of Smith River, Del Norte County.

20. **Galium cleméntis** Eastw. Santa Lucia Bedstraw. Fig. 5010.

Galium clementis Eastw. Leaflets West. Bot. **1**: 56. 1933.

Low, tufted or densely matted perennial with stems about 3.5–8 cm. long, the stems and leaves quite densely beset with bent or straight, hirsute hairs. Leaves in whorls of 4, surpassing the internodes except on young shoots, narrowly elliptic or oblong, all appearing linear because of the strongly inrolled margins, 3–6 mm. long, 1–2 mm. wide; flowers yellowish, few in the upper axils or in small, short-peduncled cymes, the lobes sparsely pubescent without; fruit fleshy, somewhat hairy, blackish in dried specimens.

Rocky outcrops, Arid Transition Zone; known from Cone Peak and Santa Lucia Peak, Santa Lucia Mountains, Monterey County, California. Type locality: Santa Lucia Peak. March–May.

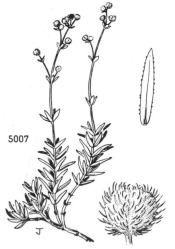

5007

5007. Galium jepsonii

5008

5008. Galium andrewsii

21. Galium califórnicum Hook. & Arn. California Bedstraw. Fig. 5011.

Galium californicum Hook. & Arn. Bot. Beechey 349. 1838.
Galium californicum β *crebrifolium* Nutt. ex Torr. & Gray, Fl. N. Amer. **2**: 20. 1841.
Galium flaccidum Greene, Pittonia **1**: 34. 1887.
Galium occidentale McClatchie, Erythea **2**: 124. 1894.

Perennial with slender, branching, woody rootstock, the stems erect, forming low tufts 8–15 cm. high or more diffuse and often 30 cm. high, slender, grayish-hirsute throughout. Leaves in whorls of 4, the whorls distant except in the more tufted plants, oblong-lanceolate to oval or ovate, rather abruptly apiculate-acuminate, 5–12 mm. long, thin, hirsute with spreading hairs on both surfaces or rarely only on the margins; flowers dioecious, solitary and axillary or the staminate ones in twos or threes on short axillary pedicels; corolla yellowish, about 2–2.5 mm. broad; fruit small, fleshy, whitish-translucent when ripe, dark when dry, glabrous or hairy.

Open woods or shaded rocky slopes, Upper Sonoran and Transition Zones; southward in California through the Coast Ranges from Humboldt County to the San Gabriel Mountains, Los Angeles County, and on Santa Cruz Island, Santa Barbara County, and locally in the central Sierra Nevada in Amador County; much less common from Santa Barbara County southward. Type locality: California. Collected by Douglas probably in the vicinity of Monterey. May–July.

Occasional forms are found having the typical leaf-shape and habit of *G. californicum* but having the stout scabrous hairs on margin and midrib typcal of *G. nuttallii* instead of the spreading-hirsute pubescence of *G. californicum*. Other specimens from rocky ridges of San Francisco, San Mateo, and Monterey Counties and southward have the crowded and thickened leaves of var. *miguelense* but, in common with the specimens mentioned above, the scabrous marginal trichomes of *G. nuttallii*. They appear to be transitional between *G. californicum* and *G. nuttallii*, as has been suggested by Hilend and Howell (Leaflets West. Bot. **1**: 167. 1935).

Galium californicum var. **miguelénse** (Greene) Jepson, Man. Fl. Pl. Calif. 961. 1925. (*Galium miguelense* Greene, Pittonia **1**: 34. 1887.) Stems matted or sometimes elongated; leaves crowded, oblong to round-ovate, chartaceous in texture, glabrous or with few hairs on the margins; fruit white. San Miguel and Santa Rosa Islands, Santa Barbara County, California. Type locality: San Miguel Island.

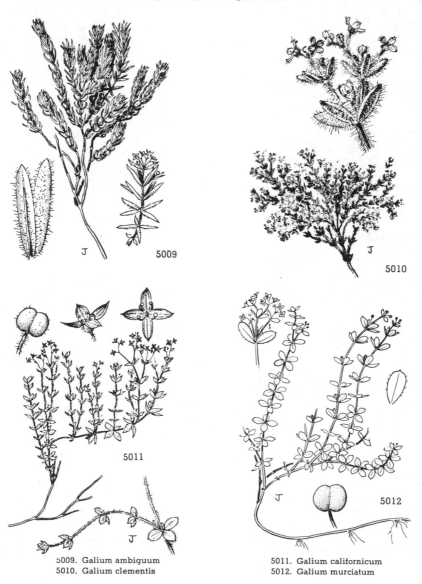

5009. Galium ambiguum
5010. Galium clementis

5011. Galium californicum
5012. Galium murciatum

22. Galium muricàtum Wight. Humboldt Bedstraw. Fig. 5012.

Galium muricatum Wight, Zoe **5**: 56. 1900.
?Galium chartaceum Wight, loc. cit.

Diffuse or somewhat tufted, dioecious perennial with creeping stems 3–12 cm. long, the stems glabrous and shining. Leaves in whorls of 4, spreading, shorter than the internodes or equaling them in the more tufted plants, 5–9 mm. long, firm in texture, dark green, shining, narrowly to broadly elliptic and sometimes obovate, short-cuspidate, the margins hispid-ciliate with curved ascending hairs, the upper surface sparsely short-hispid with appressed hairs to glabrous, the lower surface glabrous or with occasional hispid hairs on the midvein; inflorescence axillary or terminal with slender pedicels, in small clusters of 2 or 3 flowers, little surpassing the leaves; corolla 4-parted, white, 1.5–2 mm. wide; fruit glabrous, black in herbarium specimens, fully mature fruit not seen.

Wooded or brushy slopes, Arid Transition Zone; Agness, Curry County, Oregon (*Peck, Henderson*), and Humboldt and Mendocino Counties, California. Type locality: "Westport, Mendocino County." June–July. Much resembling the more tufted forms of *G. californicum* and apparently some forms intergrading with that species.

23. Galium nuttállii A. Gray. Climbing Bedstraw. Fig. 5013.

Galium suffruticosum Nutt. in Torr. & Gray, Fl. N. Amer. **2**: 21. 1841. Not Hook. & Arn. 1833.
Galium nuttallii A. Gray, Smiths. Contr. **3**⁵: 80. 1852.

Dioecious perennial with 1 or more slender stems arising from a woody root, suffrutescent below and elongate with long internodes, reclining or climbing on shrubs and rocks, 1.5–2 m. in length, compactly much branched above and leafy, the seasonal stems and branches herbaceous, more or less retrorse-scaberulous. Leaves in whorls of 4, usually rather thin, becoming thicker in age and somewhat shining, 2.5–8 mm. long, ovate-oblong to broadly linear-oblong (narrowly linear-oblong or even linear in some forms), acute at apex or rarely obtuse, the upper surface very sparsely hispid to glabrous, the margins inrolled with age, sparsely to densely scabrous with usually retrorse hairs; staminate inflorescence in few-flowered cymules in leafy branchlets: pistillate inflorescence solitary from leaf-axils of branchlets, the pedicels 2–4 mm. long; corolla usually greenish yellow, 2–2.5 mm. broad, the lobes narrowly ovate-acuminate; fruit pulpy, 2–2.5 mm. in diameter, often but 1 carpel developing, pearly white when mature, often tinged with lavender, turning black when dry.

Shrubby hillsides, Sonoran Zones; Jackson and Josephine Counties, Oregon, south through the Inner and Outer Coast Ranges to northern Lower California and from Siskiyou and Shasta Counties south along the Sierra Nevada foothills. Type locality: San Diego, California. March–Aug. Intergrades with *G. bolanderi* are frequently found in its northern limits and along the Sierra Nevada foothills.

Galium nuttallii subsp. **insulàre** Ferris, Contr. Dudley Herb. **4**: 338. *figs. 2a–c.* 1955. Habit as in the species but having the main stems rather stout, secondary stems and branches glabrous or more rarely very sparsely scaberulous, the branches conspicuously ridged on the angles, rather short and densely leafy; leaves 5–12 mm. long, linear, cuspidate-acute, thick and rather coriaceous in texture, the leaf-margin inrolled, entire or with occasional scabrous hairs, the leaves of the young shoots often linear-oblong as in typical specimens of *G. nuttallii* but thick in texture and bearing a cusp. Santa Rosa, Santa Cruz, and Santa Catalina Islands, California. Type locality: Santa Catlina Island, Los Angeles County.

24. Galium bolánderi A. Gray. Bolander's Bedstraw. Fig. 5014.

Galium bolanderi A. Gray, Proc. Amer. Acad. **7**: 350. 1868.
Galium margaricoccum A. Gray, op. cit. **13**: 371. 1878.
Galium arcuatum Wiegand, Bull. Torrey Club **24**: 395. 1897.
Galium trifidum var. *cuspidulatum* A. Gray, loc. cit., as a synonym.

Stems loosely branched above, several to many arising from woody roots, strongly angled and smooth, more rarely scaberulous, 1.5–4 dm. high, erect or the longer stems somewhat reclining, the new growth with long internodes, the stems sometimes reaching a length of 7 dm. Leaves in whorls of 4, sometimes in pairs above, linear to linear-elliptic, acute at the apex and often apiculate, 8–22 mm. long, those of the main stems the longer, thin in texture and thickening in age and darkened in dried specimens, completely glabrous above or sparsely pubescent, the inrolled margins entire or with 3–10 scabrous hairs on each side; staminate inflorescence leafy with divaricate branchlets, many-flowered; pistillate flowers axillary in the upper branchlets, the pedicels 2–7 mm. long, spreading or arcuate; corolla purple or greenish purple, 3–4 mm. broad, the lobes ovate-acuminate; fruit juicy or pulpy, pearly white when ripe, sometimes but one carpel developing, dark in dried specimens.

Dry, open, wooded slopes or in thickets, Transition and Canadian Zones; southern Oregon in the Siskiyou Mountains region south through the Coast Ranges of California to the mountains of Glenn and Lake Counties and Mount Hood. Sonoma County, and in the Sierra Nevada to Madera County. Type locality: Mono Trail, Yosemite National Park. Collected by Bolander. April–June.

25. Galium pùbens A. Gray. Gray's Bedstraw. Fig. 5015.

Galium pubens A. Gray, Proc. Amer. Acad. **7**: 350. 1868.
Galium culbertsonii Greene, Leaflets Bot. Obs. **1**: 80. 1904.

Stems usually several from a woody caudex, usually erect but often diffuse and much branched, 3–7 dm. high, more or less hirsutulous with spreading grayish hairs and also scaberulous, rather sparsely leafy on the main stem with internodes 3–5 cm. long or the much-branched plants with short internodes. Leaves in whorls of 4, firm, oblong-ovate, acute, or linear-oblong and acute on much-branched plants, hirsutulous on both surfaces and scabrous, sometimes obscurely so on the often inrolled margins, 10–12 mm. long and barely one-half as long on the ultimate branches:

flowers dioecious, the staminate inflorescence paniculate, the cymules few-flowered on short pedicels, the pistillate solitary or in pairs in the upper leaf-axils; corolla purplish or greenish; fruit large, fleshy, puberulent, becoming glabrate, dark in dried specimens.

Rocky ridges and loose talus slopes, Arid Transition Zone; in California on the western slopes of the Sierra Nevada from Butte County to Tulare County and in the Inner Coast Range in eastern Humboldt, Lake, and Napa Counties; also from southern Jackson and Josephine Counties, Oregon. Type locality: "Yosemite Valley and adjacent mountains." May–Aug.

Galium pubens, G. pubens var. *scabridum* A. Gray, *G. subscabridum* Wight (which is here considered to be a synonym of var. *scabridum*), *G. bolanderi* and its synonym *G. margaricoccum* A. Gray all have Yosemite Valley or the area immediately adjacent to it as their type locality. *Galium sparsiflorum* Wight is also common in the same region. With the occurrence of so many entities from this region, it is not surprising that variants are found which are difficult to assign precisely to their proper places.

Galium pubens var. **scàbridum** A. Gray, Bot. Calif. 1: 285. 1876. (*Galium subscabridum* Wight, Zoe 5: 56. 1900.) Habit as of *G. pubens*, the spreading hirsutulous pubescence sparse or completely lacking; leaf-margins, midribs, and stem-angles always scabrous; fruit glabrous. Growing with the species from the type locality in Yosemite Valley to the Tehachapi Mountains (*Dudley, 377*).

Galium pubens var. **gránde** (McClatchie) Jepson, Man. Fl. Pl. Calif. 961. 1925. (*Galium grande* McClatchie, Erythea 2: 124. 1894.) Plants rather strict, 0.5–2 m. high with spreading branches above, densely cinereous-pubescent; leaves mostly less than 10 mm. long, revolute on the margins, narrowly ovate; mature fruit sparsely puberulent, the pedicel shorter than the width of the mature berry. Upper Sonoran and Transition Zones; southern Sierra Nevada in the Tule River drainage south in the mountains of southern California to northern Lower California. Intergrading with the species and with var. *scabridum* at its northern limit. Type locality: Mount Wilson, Los Angeles County.

26. **Galium sparsiflòrum** Wight. Sequoia Bedstraw. Fig. 5016.

Galium sparsiflorum Wight, Zoe 5: 55. 1900.

Perennial from branching elongated rootstocks, stems several, erect and branching above, 10–30 cm. high, shining, glabrous or thinly and minutely hispidulous. Leaves in whorls of 4,

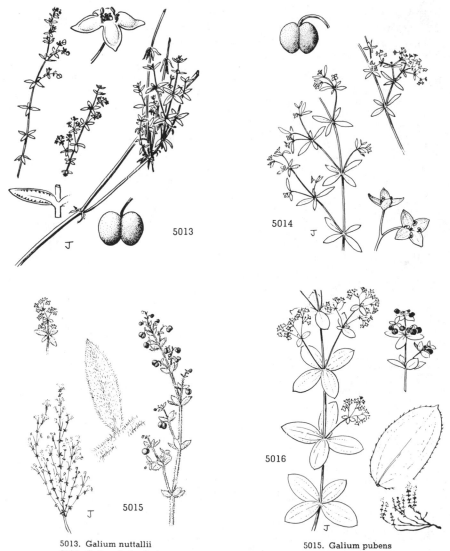

5013

5014

5015

5016

5013. Galium nuttallii
5014. Galium bolanderi

5015. Galium pubens
5016. Galium sparsiflorum

unequal, oval to broadly ovate, mostly 1.5–2.25 cm. long and 6–10 mm. broad or sometimes reduced above, obtuse at apex and apiculate, thin, 3-veined, glabrous to thinly pubescent above and sometimes below with short ascending hairs, the margins with upwardly appressed cilia; inflorescence axillary or terminal, the pistillate flowers solitary or 2–3 in the upper leaf-axils, the pedicels slender, 3–6 mm. long, the staminate flowers several in bracteate, axillary and terminal cymes, the peduncles and pedicels capillary, up to 10 mm. long; corolla greenish, about 2.5 mm. broad, the lobes attenuate; fruit blackish in dried specimens, glabrous, about 3 mm. broad.

Rocky wooded slopes, Arid Transition Zone; Sierra Nevada from Plumas County to Tulare County, California, where it is more abundant. Type locality: Big Meadows, Plumas County, as designated by Hilend & Howell (Leaflets West. Bot. 1: 164. 1935). June–Aug.

27. Galium wrìghtii var. rothróckii (A. Gray) Ehrendorfer. Slender-branched or Rothrock's Bedstraw. Fig. 5017.

Galium rothrockii A. Gray, Proc. Amer. Acad. 17: 203. 1882.
Galium wrightii var. rothrockii Ehrendorfer in Ferris, Contr. Dudley Herb. 5: 99. 1958.

Perennial from a woody base, stems several, slender, mostly erect, glabrous or nearly so. Leaves in whorls of 4, rigid, narrowly linear, acute, about 0.5–1 cm. long; inflorescence a few-flowered, diffusely branched panicle terminating the stems; flowers polygamo-monoccious, dark red or purplish brown, 1–2 mm. broad, the lobes acuminate; style and style-branches slender; fruit about 1.5–2 mm. wide including the bristles, dark-colored, the bristles white, straight, spreading, not very dense.

Rocky slopes and canyons, Arid Transition Zone; known only in California from Clark Mountain, San Bernardino County; ranging eastward through Arizona to southern New Mexico and adjacent Mexico. Type locality: Santa Cruz County, Arizona. July–Aug.

28. Galium angustifòlium Nutt. Narrow-leaved Bedstraw. Fig. 5018.

Galium angustifolium Nutt. ex Torr. & Gray, Fl. N. Amer. 2: 22. 1841, as a synonym.
Galium angustifolium Nutt. ex A. Gray, Bot. Calif. 1: 285. 1876.

Erect shrubby plant 3–10 dm. high with slender stiff branches, glabrous or puberulent. Leaves mostly 3 or 4, narrowly linear, 5–15 mm. long, those of the branchlets and inflorescence much shorter than those of the main stem, scabrous or ciliate on the margins, or glabrous, or minutely puberulent; flowers many, dioecious, on slender pedicels, forming narrow or diffuse panicles; corolla greenish white, 1.5–2.5 mm. broad; fruit small, densely covered with conspicuous white bristles about equaling the width of the body.

Dry, sandy or gravelly washes, mesas, and hillsides, Upper and Lower Sonoran Zones; San Benito, Monterey, and Kern Counties, California, south to northern Lower California. Type locality: San Diego, San Diego County. April–Aug.

The typical form is characterized as follows: numerous pistillate flowers generally congested at the ends of the branches of the inflorescence; internodes of the stems glabrous; leaves scabrous on the margins; fruit 1.5–2 mm. broad. A species highly variable and from which the following segregates have been keyed out by Hilend & Howell (Leaflets West. Bot. 1: 154. 1935). As stated by them much intergradation occurs between neighboring forms, but their treatment of the group is most helpful in distinguishing the extremes of variation in this group.

Galium angustifolium var. siccàtum (Wight) Hilend & Howell, Leaflets West. Bot. 1: 135. 1934. Galium siccatum Wight, Zoe 5: 54. 1900.) Closely resembling the typical form in virgate habit and inflorescence but cinereous-puberulent on stems and leaves. Outer Coast Ranges from Santa Barbara County to San Diego County and in the Inner Coast Ranges from San Benito and Monterey Counties south to Los Angeles County, California. Type locality: Del Mar, San Diego County, the first cited specimen.

Galium angustifolium var. pinetòrum Munz & Jtn. Bull. Torrey Club 49: 357. 1923. Loose open habit, the stems lax, branched at the base and slightly woody, 1.5–3 dm. high, glabrous throughout; leaves 1–3 cm. long; panicles few-flowered, lax, 3–10 cm. long. Dry mountain ridges, Arid Transition Zone; Santa Barbara County to Riverside County, California. Type locality: Sierra Madre Mountains, Los Angeles County.

Galium angustifolium var. diffùsum Hilend & Howell, Leaflets West. Bot. 1: 134. 1934. Stems lax and diffusely branched; staminate flowers in broad diffuse panicles, flowers of pistillate inflorescence scattered, solitary on slender pedicels. Dry desert slopes and ridges, Greenhorn Mountains, Kern County, and Holcomb Valley on desert slopes of the San Bernardino Mountains south to the western edge of the Colorado Desert, southern California. Type locality: upper Holcomb Valley, San Bernardino County.

Galium angustifolium var. foliòsum Hilend & Howell, loc. cit. Compact leafy shrub 1–3 dm. high; leaves 4–10 mm. long, mucronate, scabrous on the margins; fruit 2 5 mm. broad; staminate flowers in a narrow panicle. Santa Rosa, Santa Cruz, and Anacapa Islands off the coast of southern California and occasional on the adjacent mainland. Type locality: exposed rocky slopes, Anacapa Island, Ventura County.

Galium angustifolium var. bernardìnum Hilend & Howell, loc. cit. Low and compact, 1.5–3 dm. high; leaves 0.5–1 cm. long; staminate inflorescence narrow and elongated, the pistillate pyramidal with divaricate branches and numerous congested flowers; fruit 1.5 mm. broad. Dry, rocky or gravelly places, Arid Transition Zone; Bear Valley and Cactus Flats, San Bernardino Mountains, to the Laguna Mountains, southern California. Type locality: near Cactus Flats, San Bernardino Mountains.

29. Galium gabrielénse Munz & Jtn. Old Baldy Bedstraw. Fig. 5019.

Galium gabrielense Munz & Jtn. Bull. Torrey Club 51: 229. 1924.
Galium siccatum var. anotinum Jepson, Man. Fl. Pl. Calif. 962. 1925.

Dioecious perennial herb with a woody base, stems crowded, much branched and tufted, somewhat woody at base, 5–15 cm. high, canescent throughout with a short, spreading, hirsute pubescence. Leaves in whorls of 4, linear, 5–10 mm. long, acute but not cuspidate or only obscurely so, crowded; flowers in short cymes in the axils of the upper leaves; corolla yellowish, 3 mm. broad, the lobes hairy on the outside; fruit including the hairs about 3 mm. wide.

Dark rocky ridges and slopes, Arid Transition and Canadian Zones; mountains of southern California in Los Angeles, Riverside, and San Bernardino Counties. Intergrades with Galium angustifolium. Type locality: "ridges east of Ontario Peak, alt. 8400 ft.," San Bernardino County. June–Aug.

30. **Galium hállii** Munz & Jtn. Hall's Bedstraw. Fig. 5020.

Galium hallii Munz & Jtn. Bull. Torrey Club **49**: 358. 1923.

Dioecious suffrutescent perennial, woody at base, with relatively few decumbent or ascending, quadrangular stems 3–4 dm. long, leafy above, the younger stems short-hirsute with hispidulous angles, older stems glabrate with thin, white, exfoliating bark. Leaves in whorls of 4, much shorter than the internodes on the lower stems, two to three times shorter above, light green, ovate-elliptic to elliptic, 5–11 mm. long, with revolute margins and prominent midrib, lateral veins more obscure, both surfaces hirsute; inflorescence rather narrowly pyramidal, especially in the pistillate inflorescence, very leafy, the pistillate flower-clusters on nodding branches 2–3 cm. long, the fruiting pedicels about 5 mm. long, usually curved to one side; corolla yellowish, about 2–3 mm. broad, sparsely hirsute on outer surface; fruit black, slightly juicy, 3 mm. long, densely covered with spreading white hairs about 2 mm. long.

Dry flats and slopes, Arid Transition Zone; southern Sierra Nevada, California, Kern County, adjacent Tulare County, Tehachapi Mountains, and the Mount Pinos region southward to the San Gabriel and San Bernardino Mountains. Type locality: "Coldwater Fork of Lytle Creek, San Gabriel Mountains, in gravelly ground at 5200–5700 feet altitude." June–Aug.

31. **Galium stellàtum** subsp. **eremicum** (Hilend & Howell) Ehrendorfer.
Desert Bedstraw. Fig. 5021.

Galium stellatum var. *eremicum* Hilend & Howell, Leaflets West. Bot. **1**: 137. 1934.
Galium stellatum subsp. *eremicum* Ehrendorfer, Contr. Dudley Herb. **5**: 7. 1956.

Dioecious low shrub 2–4 dm. high, intricately branched except sometimes on young growth, woody toward the base, the stems light gray and shreddy, stramineous above with exfoliating

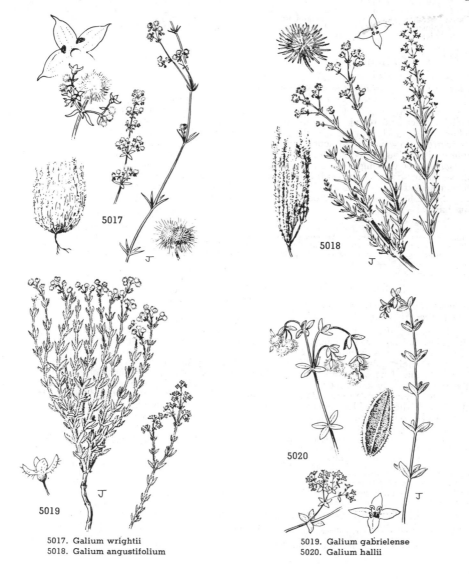

5017. Galium wrightii
5018. Galium angustifolium
5019. Galium gabrielense
5020. Galium hallii

bark, the ultimate branches strongly 4-angled, the angles white, more or less scabridulous and densely puberulent with hispidulous hairs. Leaves 4–7 mm. long, in whorls of 4, those of vigorous shoots longer, ovate-lanceolate to linear-lanceolate, rigid, divergent, slightly revolute, acuminate-cuspidate, pale gray-green with a scabrous puberulence, midvein white, prominent, the upper reduced leaves sometimes 2 or 3 at a node; inflorscence, both staminate and pistillate, many-flowered, of crowded diffuse panicles, leafy, the leaves somewhat reduced in size, the fruiting pedicels straight, usually 2–3 mm. long; corolla greenish yellow, 2.5–3.5 mm. broad, the lobes ovate, acute, the outer surface with scattered hirsute hairs; fruit including the copious white hairs 2.5–3 mm. wide.

Dry desert slopes, Sonoran Zones; common in Inyo County, California, and southern Nevada and Utah southward through the Mojave and Colorado Deserts to western Arizona and Sonora. Type locality: western end of the Sheep Hole Mountains, San Bernardino County, California. March–May.

32. Galium matthèwsii A. Gray. Matthews' Bedstraw. Fig. 5022.

Galium matthewsii A. Gray, Proc. Amer. Acad. **19**: 80. 1883.

Suffrutescent, glabrous, much-branched perennial 2.5–3.5 dm. high, the stems rather slender and persisting more than one season, the bark whitish and shining, exfoliating in age. Leaves in whorls of 4, much shorter than the internodes, lanceolate-ovate, 4–8 mm. long, cuspidate-acute, 1-nerved, rigid, bright green, the upper smaller; inflorescence ample and open, many-flowered, the branches of the flower-clusters divaricate, rather long, and the subtending leaves much reduced, pedicels always straight in fruit; corolla greenish white, 1–2 mm. wide, the lobes acute to acuminate, the outer surface sparsely covered with spreading hirsute hairs; fruit small, 2–3 mm. broad including the dense covering of short white hairs.

Rocky slopes of desert ranges, Upper Sonoran and Arid Transition Zones; eastern face of the Sierra Nevada, California, in Inyo County and the eastern border of Kern County eastward to the Inyo Mountains and the Death Valley region and also the Kingston Mountains, San Bernardino County. Type locality: "Arid district in Inyo Co., California," but according to label on type specimen in the Gray Herbarium collected "among sagebrush near Camp Independence." May–July.

33. Galium múnzii Hilend & Howell. Munz's Bedstraw. Fig. 5023.

Galium munzii Hilend & Howell, Leaflets West. Bot. **1**: 135. 1934.
Galium munzii var. *carneum* Hilend & Howell, op. cit. 136.
Galium munzii f. *glabrum* Ehrendorfer, Contr. Dudley Herb. **5**: 9. 1956.

Stems many from a branching, somewhat woody base, erect or decumbent, 1.5–3.5(5) dm. high, slender and pallid, short-hispid with spreading hairs or sometimes glabrous. Leaves in whorls of 4, much shorter than the internodes, usually broadly ovate-lanceolate, 5–12 mm. long, dark green, tapering to an acute, more or less apiculate and mucronulate apex, rather thin, and thinly hispidulous on both sides, midvein prominent, white, the 2 lateral veins short and slender or sometimes wanting, the margin often sparsely scaberulous; inflorescence pyramidal, many-flowered and rather open with ascending or spreading branches, the leaves reduced in size, the flowering branchlets divaricate; fruiting pedicels 4–6 mm. long, straight and divaricate; corolla 1.5–2 mm. broad, greenish or occasionally reddish, hispid exteriorly; fruit including the spreading hairs 3–4.5 mm. wide.

Dry, rocky or gravelly slopes, Upper Sonoran and Arid Transition Zones; desert ranges of Inyo County including the west face of the southern Sierra Nevada, and adjacent Nevada, south to the San Bernardino Mountains and east to northwestern Arizona and southern Utah. Type locality: "Bonanza King Mine, 3000 feet, east slope of the Providence Mountains, San Bernardino County, California." June–Sept.

Galium munzii is quite variable as to leaf-shape, and especially as to pubescence in its more easterly range where f. *glabrum* is sometimes collected.

34. Galium multiflòrum Kell. Many-flowered Bedstraw. Fig. 5024.

Galium multiflorum Kell. Proc. Calif. Acad. **2**: 96. *fig. 27.* 1863.
Galium bloomeri A. Gray, Proc. Amer. Acad. **6**: 538. 1865.

Perennial, stems erect from a woody rootstock, often somewhat suffrutescent below with the epidermal layer exfoliating in thin papery shreds, the leaf-bearing branches numerous, rather slender, erect or ascending, 1.5–2.5(3) dm. high, green and glabrous, the sharp whitish angles rarely minutely scabrous. Leaves in whorls of 4, widely spreading, broadly ovate to ovate-lanceolate, 5–10 mm. long, apiculate at the apex, pale green, shining, and rather firm, midvein prominent, the lateral more obscure, margin somewhat thickened, slightly revolute and paler; staminate and pistillate inflorescences pyramidal, many-flowered, the branches spreading, definitely longer than the leaves, the branchlets divaricate, the fruiting pedicels 3–8 mm. long, divaricate, straight; corolla glabrous, greenish yellow, 2–2.5 mm. wide; fruit 4–6 mm. wide including the long, spreading, white hairs.

Dry rocky ridges and talus slopes, mainly Arid Transition Zone; Nevada west to Modoc County, California, south to Inyo County and east to Utah. Type locality: "vicinity of Virginia City, Washoe," Nevada. June–Aug.

Galium multiflorum f. hirsùtum (A. Gray) Ehrendorfer, Contr. Dudley Herb. **5**: 10. 1956. (*Galium bloomeri* var. *hirsutum* A. Gray, Bot. Calif. **1**: 285. 1876; *G. matthewsii* var. *scabridum* Jepson, Man. Fl. Pl. Calif. 963. 1925.) Differing from the preceding only by the presence of longish, spreading, hirsute hairs on stems and leaves. From the type locality south to Inyo County and also adjacent Nevada. Type locality: Sierra Valley, Sierra County, California. Collected by Lemmon.

35. Galium hypotríchium A. Gray. Alpine Bedstraw. Fig. 5025.

Galium hypotrichium A. Gray, Proc. Amer. Acad. **6**: 538. 1865.

Herbaceous perennial, the slender woody rootstocks becoming much branched, the stems and branches numerous, forming low tufts or mats 5–10 cm. high, very leafy, the internodes mostly shorter than the leaves, slender, scabrous-puberulent, with very short divergent hairs. Leaves 3–6(10- mm. long, 2–4 mm. broad, rather thin, ovate to ovate-lanceolate, acute, 1-nerved, green but with a short, scabrid, spreading puberulence, markedly so on the margins; inflorescence narrow and rather few-flowered, flowering branchlets erect or ascending, not much exceeding the leaves, the pedicels 1.5–2 mm. long, curved; corollas of staminate and pistillate plants greenish yellow, sometimes reddish, the lobes about 1.5 mm. long; fruit densely covered with whitish or very faintly tawny bristles longer than the body.

Rocky alpine summits, Hudsonian and Arctic-Alpine Zones; central Sierra Nevada from the type locality in Tuolumne County, California, south to Inyo County where it intergrades with *G. hypotrichium* subsp. *subalpinum* and eastward to the White Mountains of California and Nevada and to the Wassuk and Toyabe Ranges of central Nevada where intergrades are found with other subspecies of the *hypotrichium* complex. Type locality: "Sonora Pass, in the Sierra Nevada, alt. 8000 to 9000 feet in dry and rocky places." July–Aug.

The taxon *G. hypotrichium* (including all of the subspecies) is one of the more variable units of the *G. multiflorum* complex. It is probably most closely related to *G. munzii* from which it differs in the narrower inflorescence and the fruiting pedicels which, though short, tend to be nodding. Its distribution occupies a somewhat intermediate position between that of *G. multiflorum* and of *G. munzii* though it is found at higher elevations than either of those taxa.

Galium hypotrichium subsp. **subalpìnum** (Hilend & Howell) Ehrendorfer, Contr. Dudley Herb. **5**: 12. 1956. (*Galium munzii* var. *subalpinum* Hilend & Howell, Leaflets West. Bot. **1**: 136. 1934.) Habit as in the preceding but differing by the ovate leaves; the pubescence of stems and leaves a little longer than that of *G. hypotrichium* subsp. *hypotrichium*. At high elevations in the Sierra Nevada, California, from the southern border of Mono County south to Tulare County; also in the White Mountains in Nevada. Type locality: Sawmill Canyon, Inyo County.

Galium hypotrichium subsp. **tomentéllum** Ehrendorfer, Contr. Dudley Herb. **5**: 12. 1956. Stems about 5–8 cm. high, more tufted than subsp. *hypotrichium* or subsp. *subalpinum*, the herbage densely white-tomentulose; leaves smaller than the preceding taxa. Known in its typical form from the type locality, Telescope Peak, Panamint Mountains, Inyo County, California, but collections of subsp. *subalpinum* from Boundary Peak in the White Mountains show some characteristics of this taxon.

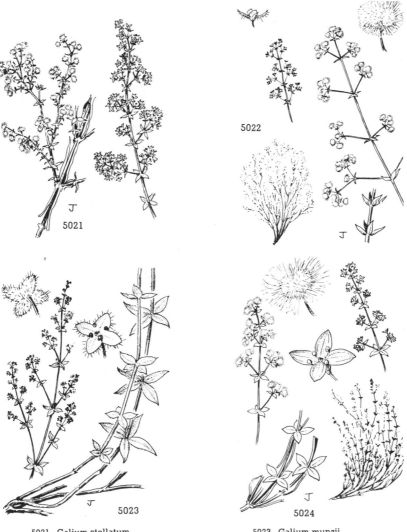

5021. Galium stellatum
5022. Galum matthewsii
5023. Galium munzii
5024. Galium multiflorum

36. **Galium paríshii** Hilend & Howell. Parish's Bedstraw. Fig. 5026.

Galium multiflorum parvifolium Parish, Zoe **5**: 75. 1900.
Galium parvifolium Jepson, Man. Fl. Pl. Calif. 963. 1925. Not Gaud. 1800.
Galium parishii Hilend & Howell, Leaflets West. Bot. **1**: 136. 1934.

Perennial, with woody branching rootstocks, the stems slender, usually many, tufted or matted, 5–20 cm. high, ascending or somewhat decumbent at base, simple or with ascending branchlets, light gray-green, scabrous on the angles and minutely rougish-puberulent between. Leaves in whorls of 4, often approximate near the base and clothing the stems, tardily deciduous and smaller than the more distant ones on the upper stem, ovate, acute, often broadly so, abruptly mucronate at apex, about 2–5 mm. long, about 4 mm. wide, minutely scabrous, puberulent on both sides; inflorescence narrow and elongate, in small close cymes in the leaf-axils on upper part of stems, the flowers monoecious; corolla 1–2 mm. wide, reddish, rarely greenish, sparsely puberulent exteriorly; body of fruit about 1 mm. broad, densely covered with spreading hairs about 1.5 mm. long.

Rocky slopes and flats, Arid Transition and Canadian Zones; San Antonio, San Bernardino, and San Jacinto Mountains, southern California, east to the New York Mountains, California, and the Charleston Mountains, southern Nevada. Type locality: Bear Valley, San Bernardino Mountains. July–Aug.

37. **Galium grayànum** Ehrendorfer. Gray's Bedstraw. Fig. 5027.

Galium grayanum Ehrendorfer, Contr. Dudley Herb. **5**: 15. 1956.

Herbaceous cespitose perennial about 1–2 dm. high with several erect or ascending stems arising from a woody rootstock, stems greenish, inconspicuously angled, and clothed as are the leaves with a dense, soft, short, even puberulence. Leaves 8–15 mm. long, 5–10 mm. wide, a little less than one-half the length of the internodes to nearly equaling them, 3-nerved, thickish and dark grayish green, ovate to nearly orbicular, obtuse to rounded at the apex; inflorescence narrow and elongate, the flowering branches ascending, leafy, usually little exceeding the leaves, the flowers rather congested, the fruiting pedicels straight; flowers 3–4 mm. broad, whitish, often tinged with purple; fruit including the copious white hairs 7–9 mm. broad.

Rocky slopes, Canadian and Hudsonian Zones; in the Coast Ranges of California from Siskiyou and Trinity Counties south to Lake County and to the east from Butte County to Placer County. Type locality: Jonesville, Butte County. July–Aug.

Galium grayanum subsp. **glabréscens** Ehrendorfer, Contr. Dudley Herb. **5**: 15. 1956. Like *G. grayanum* subsp. *grayanum* in habit and shape of leaves but often up to 2.5 dm. in height; nearly glabrous or sparsely pubescent with somewhat longer hairs (hispidulous hairs rarely present) to nearly glabrous; inflorescence as in the preceding but fruiting pedicels often 5 mm. or more long and usually somewhat curved downward. Crater Lake, Oregon, south to Shasta and northern Trinity Counties, California, eastward to the Warner Mountains in Lake County, Oregon; also Modoc County, California, where intergrading forms with *G. multiflorum* and *G. watsonii* are to be found. Type locality: Castle Lake near Sisson, Trinity Mountains, Siskiyou County, California.

38. **Galium watsònii** (A. Gray) Heller. Watson's Galium. Fig. 5028.

Galium multiflorum var. *watsonii* A. Gray, Syn. Fl. N. Amer. **1²**: 40. 1884.
Galium watsonii Heller, Bull. Torrey Club **25**: 627. 1898.
Galium watsonii f. *scabridum* Ehrendorfer, Contr. Dudley Herb. **5**: 16. 1956.

Herbaceous, somewhat cespitose perennial (sometimes woody at base), with several erect stems, usually 1–2.5 dm. high from a woody rootstock, the herbage glabrous throughout (long scabrous hairs in f. *scabridum*) and usually brownish green in dried specimens. Leaves in whorls of 4, rather thin, 5–15 mm. long, 1–3 mm. wide, linear-oblong or linear-oblanceolate, the midvein prominent; branches of the inflorescence usually not much surpassing the leaves, erect or ascending, flowers 3–4 mm. wide, greenish yellow; fruit densely covered with long spreading hairs, 6–8 mm. wide.

Dry slopes, Arid Transition and Canadian Zones; east of the Cascade Mountains from Chelan County, Washington, south to the Warner Mountains of Modoc County and the Scott Mountains of Siskiyou County, California, where it is represented by f. *scabridum* only, east to central Idaho and northern Nevada. Type locality: not given but lectotype designated by Ehrendorfer as Union County, Oregon, *Cusick*. May–July.

Galium watsonii subsp. **pubérulum** Ehrendorfer, Contr. Dudley Herb. **5**: 16. 1956. (*Galium multiflorum* subsp. *puberulum* Piper, Contr. U.S. Nat. Herb. **11**: 527. 1906; *G. multiflorum* var. *puberulum* St. John, Northwest Sci. **2³**: 90. 1928.) Very similar to *G. watsonii* subsp. *watsonii* in habit, but the plants more or less densely hirtellous throughout. Dry slopes on serpentine, Arid Transition and Canadan Zones; common in Chelan and Kittitas Counties, Washington. Type locality: Ellensburg, Kittitas County.

Family 144. **CAPRIFOLIÀCEAE.**

Honeysuckle Family.

Shrubs, trees, vines, or perennial herbs with opposite, simple or pinnate, estipulate leaves (rarely stipulate in *Sambucus* or stipules reduced to nectiferous glands in some genera) and perfect, regular or irregular, mostly cymose flowers. Calyx adnate to the ovary, its limb 3–5-toothed or 3–5-lobed. Corolla 5-lobed, sometimes 2-lipped. Stamens 5 or rarely 4, inserted on the corolla-tube and alternate with the lobes; anthers versatile, 2-celled, longitudinally dehiscent. Ovary inferior, 1–6-celled; style slender; stigma capitate or 2–5-lobed; ovules 1 to several in each cavity. Fruit a berry, drupe, or capsule.

A family of 11 genera and about 350 species, mostly in the northern hemisphere, a few in South America and Australia.

Corolla rotate or urn-shaped; flowers in compound cymes; shrubs or trees.
    Leaves pinnate; drupe 3–5-seeded.   1. *Sambucus.*
    Leaves simple; drupe 1-seeded.   2. *Viburnum.*
Corolla tubular or campanulate, often 2-lipped.
    Creeping evergreen herb; flowers geminate, on long slender peduncles.   3. *Linnaea.*
    Shrubs or woody twining vines; flowers if geminate not on long slender peduncles.
        Corolla campanulate, short, regular or nearly so; berry white.   4. *Symphoricarpus.*
        Corolla tubular or campanulate, more or less irregular.   5. *Lonicera.*

1. **SAMBUCUS** [Tourn.] L. Sp. Pl. 269. 1753.

Shrubs or trees with opposite pinnate leaves, serrate or laciniate leaflets, and small, white, yellowish, or pinkish flowers in compound, depressed or thyrsoid cymes. Calyx-tube ovoid or turbinate, 3–5-toothed or 3–5-lobed. Corolla rotate or slightly campanulate, regular, usually 5-lobed, the lobes imbricate or valvate. Stamens 5, inserted at the base of the corolla; filaments slender. Ovary 3–5-celled; style 3-parted; ovules 1 in each cell, pendulous. Fruit a berry-like drupe, with 3–5 one-seeded nutlets. [Latin name of the elder.]

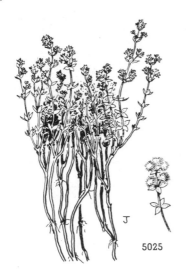

5025

5026

5027

5028

5025. Galium hypotrichium
5026. Galium parishii

5027. Galium grayanum
5028. Galium watsonii

About 20 species of wide geographical distribution, about half of them occurring in North America. Type species, *Sambucus nigra* L.

Inflorescence a flat-topped, compound cyme; ripe berries with white evanescent bloom.
Many-stemmed shrubs; cymes 12–30 cm. broad; leaflets usually gradually acuminate and markedly asymmetrical at base.   1. *S. caerulea.*
Usually arborescent; cymes 7–15 cm. broad; leaflets typically abruptly acuminate and usually slightly to not at all asymmetrical at base.   2. *S. mexicana.*
Inflorescence pyramidal or dome-shaped in outline; berries without bloom.
Ripe berries black.   3. *S. melanocarpa.*
Ripe berries bright red, rarely chestnut or yellow.
Leaflets closely serrate to the apex, more or less pubescent beneath.   4. *S. pubens arborescens.*
Leaflets coarsely and sharply serrate except at the apex, mostly glabrous beneath.   5. *S. microbotrys.*

1. **Sambucus caerùlea** Raf. Blue Elderberry. Fig. 5029.

Sambucus caerulea (cerulea) Raf. Alsog. Amer. 48. 1838.
Sambucus glauca Nutt. in Torr. & Gray, Fl. N. Amer. **2**: 13. 1841.
?Sambucus californica K. Koch, Dendr. **2¹**: 72. 1872.
Sambucus cuerulea var. *glauca* Schwerin, Deuts. Dendr. Ges. **1909**: 37, 328. 1909.
Sambucus decipiens M. E. Jones, Bull. Univ. Mont. Biol. Ser. **15**: 46. 1910.
Sambucus ferax A. Nels. Bot. Gaz. **53**: 225. 1912.
Sambucus tomentella Heller ex Schwerin, Deuts. Dendr. Ges. **1920**: 218. 1920, in synonymy.

Shrub of clustered erect stems 3–7 m. high sprouting freely from the base, the principal stems usually a few centimeters in diameter, the young stems and branches brown and often glaucous. Leaves 1.5–3.5 dm. long, the petioles 4–8 cm. long and often tomentulose (leaves of vigorous stems up to 6 dm. long and not infrequently bipinnate); leaflets 5–9, green above and paler beneath, rather thick, 6–16 cm. long, often long-petiolulate, broadly lanceolate to oblong-lanceolate and tapering to the acuminate apex, rounded at base and typically strongly asymmetrical, the margin serrulate or serrate, glabrous or sparingly to densely tomentose; compound, flat-topped cymes 12–30 cm. broad, the flowers pale yellow or creamy white; ripe fruit about 5 mm. in diameter, dark blue or blackish, and when fully ripe covered with a dense white bloom which may disappear under moist conditions.

Woods and thickets, Transition and Canadian Zones; southern British Columbia southward west and east of the Cascade Mountains to the Siskiyou region of northwestern California, and to northeastern California southward through the Sierra Nevada to the mountains of San Bernardino County and the Mount Pinos region; also Alberta and Montana south to northern Arizona and New Mexico. In the Sierra Nevada and the desert ranges eastward in Nevada, many specimens are found with a close velvety tomentum on the petioles and undersurfaces of the leaves. Glabrous but similar plants occur in the same localities. Type locality: "near Oregon Mts." Collected on the Lewis and Clark Expedition. May–Sept.

Sambucus caerulea var. **neomexicàna** Rehd. Deuts. Dendr. Ges. **1915**: 228. 1915. (*Sambucus neomexicana* Wooton, Bull. Torrey Club **25**: 309. 1898; *S. intermedia* var. *neomexicana* Schwerin, Deuts. Dendr. Ges. **1909**: 38, 328. 1909; *S. glauca neomexicana* A. Nels. in Coult. & Nels. New Man. Bot. Rocky Mts. 469. 1909; *S. caerulea* 3. *trifida* Schwerin, Deuts. Dendr. Ges. **1920**: 218. 1920; *S. trifida* Heller ex Schwerin, loc. cit. in synonymy.) Habit of *S. caerulea* var. *caerulea*; leaflets 3–5(7), about 3–8 cm. long, 2–3 mm. wide, in ours oblong-lanceolate, serrate with rather small teeth; thickish, pale green; inflorescence rather open.

In canyons or among boulders, Arid Transition Zone; Masonic Peak, Mono County, and the White Mountains of Inyo County, California, and in southern Nevada, New Mexico, and Arizona and also Sonora. Type locality: Ruidoso Crossing, White Mountains, New Mexico.

2. **Sambucus mexicàna** Presl ex DC. Southwestern or Desert Elderberry. Fig. 5030.

Sambucus mexicana Presl ex DC. Prod. **4**: 322. 1830.
Sambucus velutina Dur. & Hilg. Journ. Acad. Phila. II. **3**: 39. 1855.
Sambucus canadensis var. *mexicana* Sarg. Silva **5**: 88. *pl. 221.* 1893.
Sambucus caerulea var. *glauca* Schwerin, Deuts. Dendr. Ges. **1909**: 37, 328. 1909, in part.
Sambucus caerulea var. *velutina* Schwerin, loc. cit., in part.
Sambucus orbiculata Greene, Leaflets Bot. Obs. **2**: 99. 1910.
Sambucus coriacea Greene, loc. cit.
?Sambucus fimbriata Greene, loc. cit.
Sambucus caerulea f. *semperflorens* Schwerin, Deuts. Dendr. Ges. **1920**: 218. 1920.
Sambucus orbiculata var. *puberula* Schwerin, op. cit. 219.
Sambucus orbiculata var. *glabra* Schwerin, loc. cit.
Sambucus caerulea var. *arizonica* Sarg. Man. Trees N. Amer. ed. 2. 885. 1922.
Sambucus glauca var. *arizonica* Sarg. ex Jepson, Man. Fl. Pl. Calif. 965. 1925.
Sambucus glauca of authors, not Nutt.
Sambucus caerulea var. *mexicana* L. Benson, Amer. Journ. Bot. **30**: 240. 1943.

Trees or shrubs 3–10 m. high with 1 or more trunks having furrowed bark (these 15–45 cm. in diameter), and also several smaller stems arising from the base, the smaller branches glabrous or tomentulose. Leaflets 5–7, less often 3, oval, ovate-lanceolate, or oblong, abruptly acuminate, sometimes cuspidate, and rounded or cuneate at base, usually not markedly asymmetrical, petiolulate to nearly sessile, 2–6 cm. long, coarsely or finely serrate except at base and apex, firm, equally green on both surfaces, glabrous to densely tomentulose including the petioles and petiolules; inflorescence a compound, flat-topped cyme 5–15 cm. broad, the flowers pale yellow to creamy white; ripe fruit 5–6 mm. in diameter, dark blue or blackish, and when fully ripe covered with a dense white bloom.

Stream banks, open flats, and hill slopes, mostly confined to the Sonoran Zones; Sacramento–San Joaquin

Valley, California, and coastal and interior valleys on either side to cismontane southern California and northern Lower California; also Arizona, New Mexico, western Texas, and Mexico. Type locality: Mexico. April-Aug.

As in the preceding species, *S. mexicana* shows much variation throughout its extensive range, and local forms are commonly found. *Sambucus orbiculata* Greene, under which Schwerin described two varieties, *puberula* and *glabra*, representing tomentose and glabrous phases, is characterized by orbicular to oval leaflets that are abruptly cuspidate, and are dentate or serrate. This form with broad leaflets occurs rather frequently from the San Francisco Bay region southward to Santa Barbara County and the Channel Islands. Plants with ovate-lanceolate leaflets, often cuneate-based and sparsely pubescent to glabrous, which are considered to be typical of *S. mexicana*, occur just as frequently in this area.

The locality at which *S. velutina* was originally found is given as Posé Creek [Poso Creek]. According to the itinerary given in the fifth volume of the report on the exploration for a Pacific railroad route, the Ocoya or Posé Creek depot camp seven miles north of the Kern River was made in a flat along the stream in the lower foothills where the vegetation contrasted with "the barren and parched surface of the surrounding hills, which were without trees or any green vegetation." This velvety tomentose-leaved form of *S. mexicana* occurs rather commonly in the San Joaquin Valley, California, and is not to be confused with the tomentose-leaved forms of *S. caerulea* found at higher elevations in the adjacent Sierra Nevada.

3. Sambucus melanocárpa A. Gray. Black Elderberry. Fig. 5031.

Sambucus melanocarpa A. Gray, Proc. Amer. Acad. **19**: 76. 1883.
Sambucus melanocarpa communis Schwerin, Deuts. Dendr. Ges. **1909**: 43, 328. 1909.
Sambucus melanocarpa fürstenbergii Schwerin, loc. cit.
Sambucus seminata Vilmorin ex Schwerin, op. cit. 53.
Sambucus racemosa var. *melanocarpa* McMinn, Ill. Man. Calif. Shrubs 529. 1939.

Shrub 1–3 m. high with smooth, erect, glabrous stems. Leaflets usually 5–7, dark green above

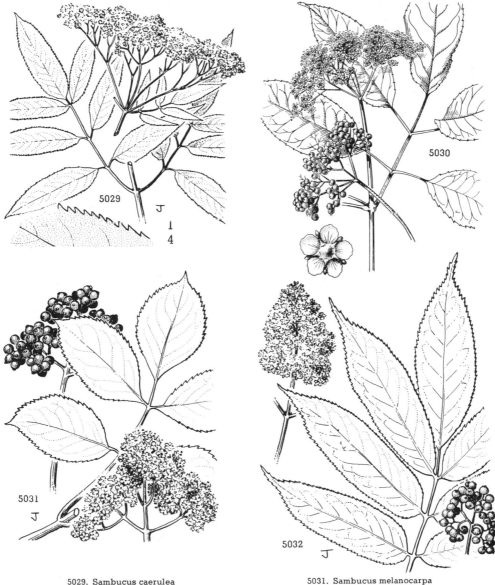

5029. Sambucus caerulea
5030. Sambucus mexicana

5031. Sambucus melanocarpa
5032. Sambucus pubens

and paler beneath, rather thin in texture, the younger leaflets sparsely short-pubescent becoming glabrate in age, 8–15 cm. long, one-half to one-third as broad, rather narrowly ovate-lanceolate, tapering abruptly to the acuminate apex, rounded or cuneate at base and usually oblique, the margin coarsely and rather unevenly serrate; inflorescence glabrous, about 5–7 cm. high, 6–10 cm. broad, ovoid in outline, rather openly branched; flowers creamy white, darkening on drying; corolla-lobes oval to oblong; ripe fruit globose, about 6 mm. in diameter, black, without bloom.

By lakes and streams, mainly in the Canadian Zone; British Columbia south in the Cascade and Blue Mountains of Oregon and Washington and occurring rarely in the Siskiyou region and the Sierra Nevada of California as far south as Olancha Peak, Tulare County; also Idaho and Montana south to northern Arizona and New Mexico. Type locality: New Mexico. Collected by Fendler. June–Aug.

4. Sambucus pùbens var. arboréscens Torr. & Gray. Coast Red Elderberry. Fig. 5032.

Sambucus pubens γ arborescens Torr. & Gray, Fl. N. Amer. **2**: 13. 1841.
Sambucus racemosa var. *arborescens* Torr. & Gray ex A. Gray, Syn. Fl. N. Amer. **1²**: 8. 1884.
Sambucus callicarpa Greene, Fl. Fran. 342. 1892.
Sambucus maritima Greene, Pittonia **2**: 297. 1892.
Sambucus leiosperma Leiberg, Proc. Biol. Soc. Wash. **11**: 40. 1897.
Sambucus racemosa var. *maritima* Jepson, School Fl. 85. 1902.
Sambucus racemosa var. *callicarpa* Jepson, Fl. W. Mid. Calif. 471. 1901.

Shrub 2–5 m. high with smooth bark. Leaves 5–7(9)-foliolate, the leaflets lanceolate to oblong-ovate or sometimes oblanceolate, acuminate, 6–15 cm. long, the margin finely and sharply serrate to the apex, the teeth usually incurved, paler beneath and more or less pubescent; inflorescence commonly broadly ovoid and dense, the flowers cream-colored; berries bright red or rarely chestnut brown or yellowish, about 4 mm. in diameter, without bloom.

Stream banks and mud flats, mostly near the coast, Humid Transition and Canadian Zones; southern Alaska southward west of the Cascade Mountains to southern Oregon, where it also occurs in the Crater Lake region, and along coastal California to San Mateo County. Type locality: not given. April–June.

5. Sambucus microbótrys Rydb. Mountain Red Elderberry. Fig. 5033.

Sambucus microbotrys Rydb. Bull. Torrey Club **28**: 503. 1901.
Sambucus acuminata Greene, Leaflets Bot. Obs. **2**: 100. 1910.
Sambucus racemosa var. *microbotrys* Kearney & Peeb. Journ. Wash. Acad. **29**: 492. 1939.

Low shrub 0.5–1.5(2) m. high, the stems and herbage glabrous throughout. Leaves 5–7-foliolate, the leaflets 6–12 cm. long, ovate to ovate-lanceolate, acute or acuminate at apex, rounded at base and equal or slightly asymmetrical, sessile or short-petiolulate, coarsely serrate except at the rather short-acuminate tip, pale green; inflorescence ovoid, 4–7 cm. broad at the base and about as high, dense, the flowers cream-colored to pale yellow; fruit bright red, 4–5 mm. in diameter, without bloom.

Rocky slopes, mostly Canadian Zone; Sierra Nevada and San Bernardino Mountains, California, east to Colorado, Utah, and Arizona. Type locality: Pikes Peak, Colorado. Collected by Bessey. May–July.

2. VIBÚRNUM [Tourn.] L. Sp. Pl. 267. 1753.

Shrubs or trees with toothed or lobed leaves, and white or pink flowers in compound cymes, the outer flowers sometimes radiate and sterile. Calyx with an ovoid or turbinate tube and short, 5-toothed limb. Corolla rotate or deeply saucer-shaped, regular or the margins slightly irregular, 5-lobed. Stamens 5, inserted on the corolla-tube; anthers oblong, exserted. Ovary 1–3-celled; style short, 3-lobed or 3-parted; ovules solitary in each cell, pendulous. Fruit a drupe, with an ovoid, globose, or sometimes flattened stone; endosperm fleshy; embryo minute. [The ancient Latin name.]

About 100 species of wide geographic distribution. Besides the following, about 20 other species occur in North America. Type species, *Viburnum tinus* L.

Fruit red; leaves more or less 3-lobed, sharply toothed.   1. *V. edule.*
Fruit black; leaves crenate-dentate, not lobed.   2. *V. ellipticum.*

1. Viburnum édule (Michx.) Raf. High Bush-cranberry. Fig. 5034.

Viburnum opulus γ edule Michx. Fl. Bor. Amer. **1**: 180. 1803.
Viburnum edule Raf. Med. Repos. N.Y. II. **5**: 354. 1808.
Viburnum pauciflorum La Pylaie ex Torr. & Gray, Fl. N. Amer. **2**: 17. 1841.

Shrub 1–3 m. high; branches glabrous. Leaves 4–9 cm. long, 3–5-nerved from the base, broadly oval to orbicular or broadly obovate, truncate or often subcordate at base, usually rather shallowly 3-lobed above the middle, coarsely and unevenly dentate and usually with an acumination at apex, glabrous or sparsely pilose beneath; cymes peduncled, 1.5–3 cm. broad, short-rayed; flowers small, all perfect; drupes light red, globose to ovoid, 8–10 mm. long; stone orbicular, flat, obscurely grooved.

Mountain woods, Canadian and Hudsonian Zones; Alaska to Washington and south in the Cascade Mountains to central Oregon east to Colorado, Pennsylvania, and Newfoundland. Type locality: Canada. June–July.

2. **Viburnum ellípticum** Hook. Oval-leaved Viburnum. Fig. 5035.

Viburnum ellipticum Hook. Fl. Bor. Amer. 1: 280. 1833.

Deciduous slender shrub 1–4 m. high, the young branches cinnamon-brown. Leaves opposite, thick in texture, 3–5-nerved from the base, 2.5–7 cm. long, the blade 2–6 cm. long, 1.5–5.5 cm. wide, elliptic to orbicular or ovate, the margin rather coarsely crenate-dentate or sometimes dentate above and entire at base, dark green and glabrous to sparsely pubescent above, paler and more pubescent beneath, the petioles often conspicuously so; inflorescence cymose on the ends of the branches, the branches and branchlets of the cyme bracteate and more or less glandular; calyx-tube glandular, 5–7 mm. long; corolla white, the lobes rounded, rotately spreading, 7–9 mm. broad; stamens exserted; fruit a black ellipsoid drupe about 1 cm. long, containing a grooved stone.

Wooded slopes, Transition Zones; west of the Cascade Mountains in Washington and Oregon and occurring less commonly in the north Coast Ranges of California as far south as Contra Costa County; also in the Sierra Nevada in Eldorado County and Fresno County (*Buckalew*). Type locality: "branches of the Columbia, near its confluence with the Pacific." April–June.

Viburnum ópulus L. Sp. Pl. 268. 1753. The much-cultivated snowball or guelder-rose has been collected in a very few localities in Washington but is evidently not established as an introduction.

3. **LINNAÈA** [Gronov.] L. Sp. Pl. 631. 1753.

Creeping evergreen herbs with opposite leaves and perfect flowers borne in pairs at the end of elongated terminal peduncles. Calyx ovoid, 5-lobed. Corolla campanulate or funnelform, 5-lobed, the lobes imbricate. Stamens 4, didynamous, inserted near the base of the corolla-tube and included. Ovary 3-celled, two of the cells with several abortive ovules, the other with 1 perfect pendulous ovule. Fruit nearly globose, 3-celled, two of

5033. Sambucus microbotrys
5034. Viburnum edule
5035. Viburnum ellipticum

which are empty, the other with a solitary oblong seed. Endosperm fleshy; embryo cylindric. [Named for Linnaeus by Gronovius.]

A monotypic genus inhabiting the north temperate regions.

1. Linnaea boreàlis var. longiflòra Torr. Western Twin-flower. Fig. 5036.

Linnaea borealis var. *longiflora* Torr. Bot. Wilkes Exp. 327. 1874.
Linnaea longiflora Howell, Fl. N.W. Amer. 280. 1900.
Linnaea borealis subsp. *longiflora* Hultén, Fl. Aleut. Isl. 310. 1937.

Stems creeping, slender, somewhat woody, 3–10 dm. long with erect, somewhat pubescent branches 5–20 cm. high. Leaves suborbicular to obovate, 8–20 mm. long, usually toothed above the middle, short-petioled, glabrate or rather thinly pilose; peduncles solitary, elongated; pedicels 2, from the axils of a pair of small, linear or narrowly oblanceolate bracts, each bearing at the apex a pair of minute bractlets and a short-pediceled flower; calyx-lobes linear-subulate, 3 mm. long, pubescent; the inferior ovary closely subtended by a pair of broad, glandular-hairy bractlets; corolla funnelform, 12–15 mm. long, the tube exceeding the calyx, rose-colored, pubescent within; fruit ovoid, glandular-pubescent.

Forests, Boreal Zone; the longer-flowered form ranges along the Pacific Coast region from Alaska to Humboldt County, northern California, east to Idaho. Type locality: Not definitely stated. May–July.

This larger, long-tubed form is varietally distinct from the Rocky Mountain and eastern North American plants (*L. borealis* var. *americana* Rehd. Rhodora **6**: 56. 1904).

4. SYMPHORICÁRPOS [Dill.] Duhamel, Traité Arb. **2**: 295. *pl. 82.* 1755.

Shrubs with opposite, simple, short-petioled, deciduous leaves and small perfect flowers in axillary or terminal clusters. Calyx-tube nearly globular, the limb 4–5-toothed. Corolla campanulate or salverform, regular or sometimes gibbous at base, 4–5-lobed, white or pink, glabrous or pilose in the throat. Stamens 4–5, inserted on the corolla. Ovary 4-celled, two of the cells containing several abortive ovules, the other two each with a solitary suspended ovule; style filiform; stigma capitate or 2-lobed. Fruit a 4-celled, 2-seeded, berry-like drupe. Seeds oblong; endosperm fleshy; embryo minute. [Name Greek, meaning to bear together, and fruit, from the clustered berries.]

A genus of about 17 species inhabiting northern North America and the mountains of Mexico, and central and southwestern China. Type species, *Lonicera symphoricarpos* L. (*Symphoricarpos orbiculatus* Moench).

Corolla short-campanulate with a broad, often ventricose tube; corolla-lobes nearly equaling or a little exceeding the tube.
  Corolla 3–5 mm. long; style shorter than the corolla.
    Young twigs glabrous; corolla-throat densely white-villous within.   1. *S. albus laevigatus.*
    Young twigs tomentulose-puberulent or sparsely to densely pilose; corolla sparsely villous within.
                                         2. *S. mollis.*
  Corolla 6–9 mm. long; style exserted.   3. *S. occidentalis.*
Corolla cylindric-campanulate or salverform, the tube never ventricose; corolla-lobes shorter than the tube.
  Corolla 6–9 mm. long, cylindric-campanulate.
    Shrubs erect; corolla 7–9 mm. long.   4. *S. vaccinioides.*
    Shrubs low and spreading; corolla 6–7 mm. long.   5. *S. parishii.*
  Corolla 10–15 mm. long, salverform with a narrow tube.   6. *S. longiflorus.*

1. Symphoricarpos álbus var. laevigàtus (Fernald) Blake. Common Snowberry. Fig. 5037.

Symphoricarpos racemosus var. *laevigatus* Fernald, Rhodora **7**: 167. 1905.
Symphoricarpos albus var. *laevigatus* Blake, op. cit. **16**: 119. 1914.
Symphoricarpos rivularis Suksd. Werdenda **1**: 41. 1927.

Erect shrub 0.5–2.5 m. tall, young branches slender, glabrous or nearly so, bark of older stems gray, smooth, little or not at all shreddy. Leaves oval, 1.5–3 cm. long, acute or obtusish at apex, acutish at base, entire or on vigorous shoots often few-toothed or -lobed, dark green and glabrous above, lower surface a little paler and glabrous or with a few scattered hairs, glabrous or sparsely ciliate on the margins; petioles 2–4 mm. long; flowers often numerous in terminal clusters, 1–2.5 cm. long; bracts and calyx glabrous; corolla 5–7 mm. long, rosy pink to white, densely white-villous within, the lobes about as long as the strongly ventricose tube; stamens 2.5–3 mm. long, included; style 2 mm. long; fruit usually several in terminal clusters, subglobose, 10–15 mm. in diameter; nutlets oval, planoconvex, obtuse at both ends, 4–6 mm. long.

Shaded canyons and slopes, Canadian, Transition, and Upper Sonoran Zones; southern Alaska to southern California east to Alberta, Montana, and Wyoming. Frequently cultivated and commonly escaping in eastern United States. Type locality: not designated. May–July.

2. Symphoricarpos móllis Nutt. Creeping Snowberry. Fig. 5038.

Symphoricarpos mollis Nutt. in Torr. & Gray, Fl. N. Amer. **2**: 4. 1841.
Symphoricarpos ciliatus Nutt. in Torr. & Gray, loc. cit.
Symphoricarpos nanus Greene, Fl. Fran. 345. 1892. (Provisional name for *S. ciliatus*)
Symphoricarpos albus var. *mollis* Keck, Bull. S. Calif. Acad. **25**: 72. 1926.

Low, diffusely branched shrub 3–10 dm. high, sometimes with trailing stems, the older branches gray and shreddy, the young twigs finely tomentulose-puberulent. Leaves rather firm, oval to orbicular, obtuse at apex, entire, ciliate on the margin, 1–4 cm. long, rather thinly short-pilose above, often becoming glabrate, paler beneath and reticulate and more densely pilose, especially along the veins, the leaves of the young growth often sinuate; flowers in small clusters on the upper part of the stems; calyx-lobes deltoid, ciliate; corolla pinkish or pink, campanulate, the tube somewhat swollen on the lower side, 3–5 mm. long, the lobes 2–3 mm. long, sparsely hairy within; stamens about equaling the lobes; style glabrous, about as long as the corolla-tube; fruit white, 5–6 mm. in diameter.

Open woods and brush land, Upper Sonoran and Transition Zones; Oregon Caves National Monument, Josephine County, Oregon, and from Mendocino County, California, southward in the Outer Coast Ranges to southern California; also in the San Bernardino and Cuyamaca Mountains. Of more common occurrence from the San Francisco Bay region southward. Type locality: Santa Barbara, California. Collected by Nuttall. April–May.

Symphoricarpos mollis subsp. **hesperius** (G. N. Jones) Abrams ex Ferris, Contr. Dudley Herb. **5**: 99. 1958. (*Symphoricarpos hesperius* G. N. Jones, Journ. Arnold Arb. **21**: 220. 1940; *S. pauciflorus* of western authors, not Robbins. 1867.) A trailing shrub with twigs sparsely pilose to glabrate; leaves rather thin, oval, 1–3 cm. long, acute at both ends, margins ciliate, entire or sometimes toothed on young shoots, nearly glabrous above, pilose beneath, especially along the veins; corolla 3–8 mm. long, the tube slightly gibbous, sparsely pilose within. British Columbia southward mostly on the western slope of the Cascade Mountains to the mountains of Siskiyou, Trinity, and Humboldt Counties, California. Type locality: upper valley of the Nisqually River, Pierce County, Washington.

Symphoricarpos mollis var. **acùtus** A. Gray, Syn. Fl. N. Amer. 1²: 14. 1884, in part. (*Symphoricarpos acutus* Dieck, Hamb. Gart. Blumenzeit **44**: 562. 1888; *S. racemosus* var. *trilobus* Durand, Journ. Acad. Phila. II. **3**: 89. 1855.) Low diffuse shrub with arching or trailing branches often rooting at the nodes, the twigs short spreading villous, mostly densely so; leaves mostly densely short-villous on both surfaces, oval to ovate, acute but varying to suborbicular, entire but frequently sinuate-toothed or -lobed, especially on new growth. Southern Oregon in Jackson, Klamath, and Lake Counties to adjacent California through the Sierra Nevada to the Tehachapi Mountains, Kern County; also Nevada in the Lake Tahoe region and in the Santa Lucia Mountains, Monterey County, California. Type locality: Lassen Peak, Shasta County, California.

3. **Symphoricarpos occidentàlis** Hook. Wolfberry or Northern Snowberry. Fig. 5039.

Symphoria occidentalis R. Brown ex Richards. in Frankl. Journ. Bot. App. 733. 1823. (Nomen nudum)
Symphoricarpos occidentalis Hook. Fl. Bor. Amer. **1**: 285. 1833.
Symphoricarpos heyeri Dippel, Handb. Laubh. **1**: 281. *fig. 187.* 1889.

Erect shrub 3–10 dm. high, often forming dense colonies; branches reddish, the twigs rather stout, puberulent to glabrate. Leaves rather thick, oval or oblong, usually acute at apex, 4–6 cm.

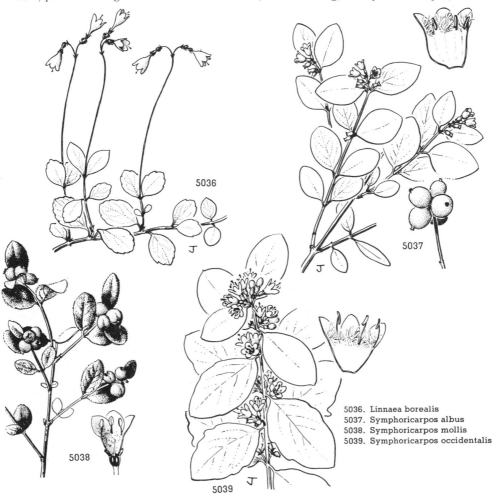

5036. Linnaea borealis
5037. Symphoricarpos albus
5038. Symphoricarpos mollis
5039. Symphoricarpos occidentalis

long, glabrous above, paler and sparsely pilose beneath, entire or sinuate-dentate on young growth; flowers many in axillary clusters on the ends of the branches; calyx-teeth ovate, ciliate; corolla broadly campanulate, pinkish, 6–9 mm. long, the obtuse lobes 3–4 mm. long, densely villous-hirsute within; stamens somewhat exceeding the corolla-lobes; style pilose or glabrous, exserted; fruit 6–8 mm. in diameter, greenish white, becoming discolored in age.

Mountains, Transition and Canadian Zones; British Columbia and Okanogan County, Washington, eastward to Michigan and Illinois and south to northern New Mexico. Type locality: "between lat. 54° and 64°,". Collected by Richardson. July–Aug.

4. Symphoricarpos vaccinioìdes Rydb. Mountain Snowberry. Fig. 5040.

Symphoricarpos mollis var. *acutus* A. Gray, Syn. Fl. N. Amer. 1²: 14. 1884, in part.
Symphoricarpos vaccinioides Rydb. Mem. N.Y. Bot. Gard. 1: 371. 1900.
Symphoricarpos austiniae Eastw. Bull. Torrey Club 30: 499. 1903.
Symphoricarpos rotundifolius vaccinioides A. Nels. in Coult. & Nels. New Man. Bot. Rocky Mts. 471. 1909.
Symphoricarpos rotundifolius acutus Frye & Rigg, Northwest Fl. 366. 1912.

A low erect shrub 8–15 dm. high with dark brown branches and smooth or shreddy bark; young branchlets light brown, finely grayish-puberulent with short curved hairs. Leaves oval, acute or acutish at both ends, 1–2 cm. long, entire or slightly dentate, especially on young growth, puberulent; bracts, pedicels, and calyces puberulent; calyx-lobes triangular, about 1 mm. long; corolla 7–8 mm., cream-colored or pinkish, cylindric-campanulate, not at all ventricose, tube sparsely pilose within, the lobes rounded and about one-third the length of the tube; anthers 1.5 mm. long, about equaling the filaments, included; style glabrous, included; fruit about 10 mm. long and 6–8 mm. wide, ellipsoid.

Dry slopes and ridges, Arid Transition and Canadian Zones; British Columbia south through the Cascade Mountains, mostly the eastern slopes, and the Siskiyou Mountains area, Oregon and California; also the Sierra Nevada to Tulare County, California, and extending eastward to Montana, Colorado, Nevada, and Utah. Type locality: "Forks of the Madison," Montana. June-Aug.

5. Symphoricarpos paríshii Rydb. Parish's Snowberry. Fig. 5041.

Symphoricarpus parishii Rydb. Bull. Torrey Club 26: 545. 1899.
Symphoricarpos glaucus Eastw. op. cit. 30: 497. 1903.
Symphoricarpos parvifolius Eastw. op. cit. 498.

Low spreading shrub, the branches often declined and sometimes rooting at the tip, 5–10 dm. long, bark on older branches thin and shreddy, young branches thinly pilose or rarely glabrate. Leaves glaucous, oval to linear-elliptic, acute, 8–25 mm. long, 3–13 mm. wide, firm, grayish green, sparsely short-pilose on both sides, rarely glabrate; petioles 1–3 mm. long; flowers commonly in pairs in the upper axils and in small terminal bracteate racemes; calyx campanulate, glaucous, the lobes scarious-margined, 1 mm. long; corolla pink, elongate-campanulate, 6–7 mm. long, the tube pilose within, the lobes 2–3 mm. long; stamens included; style glabrous, short; fruit ellipsoid to subglobose, about 6–8 mm. in diameter.

Dry mountain slopes and ridges, Transition and Canadian Zones; southern Sierra Nevada, Tulare County, and Tehachapi Mountains and the Mount Pinos region south to the San Jacinto and Santa Rosa Mountains, southern California, east to southwestern Nevada and northwestern Arizona. Type locality: San Bernardino Mountains, California. June–July.

6. Symphoricarpos longiflòrus A. Gray. Desert Snowberry. Fig. 5042.

Symphoricarpos longiflorus A. Gray, Journ. Linn. Soc. 14: 12. 1873.
Symphoricarpos fragrans Nels. & Kenn. Muhlenbergia 3: 143. 1908.

Low, intricately branched shrub with divaricately spreading or even declined branches 3–10 dm. high, glabrous throughout, the older branches with light gray, shredded bark, young branches glabrous or sparsely puberulent, light reddish brown and more or less glaucous. Leaves lanceolate or oblanceolate to elliptic or oval, 8–15 mm. long, 3–6 mm. wide, gray-green and glaucous, glabrous or sparsely puberulent toward the base; petioles 1–3 mm. long; flowers fragrant, in the axils of the upper leaves or racemosely disposed on young branchlets; calyx 1.5–2 mm. long; corolla salverform, 10–15 mm. long, white to pink, glabrous within and without, the lobes 2–3 mm. long; style usually hairy below; berries ellipsoid, white or pinkish, 8–10 mm. long.

Rocky canyons and slopes in desert mountain ranges, Upper Sonoran Zone; southeastern Oregon and Nevada, and in the White, Panamint, and Cottonwood Mountais, Inyo County, and Kingston and Providence Mountains, San Bernardino County, California, east to Utah, Arizona, and western Texas. Type locality: "Pahranagat Mountains, in the south-eastern part of Nevada." May–June.

5. LONÍCERA L. Sp. Pl. 173. 1753.

Erect shrubs or woody climbers with opposite, mostly entire leaves and spicate, capitate, or geminate flowers subtended by bracts and bractlets, the bractlets distinct, connate, or sometimes wanting. Calyx-tube ovoid to nearly globulose, the limb shallowly 5-toothed. Corolla usually irregular, tubular, funnelform, or campanulate, often gibbous at base, the limb 5-lobed and more or less oblique or 2-lobed. Stamens 5, inserted on the corolla-tube; anthers linear or oblong. Ovary 2–3-celled; ovules many in each cell, pendulous; style slender; stigma capitate. Fruit a fleshy berry, 2–3-celled or rarely 1-celled, few-seeded. Seeds oval or oblong, with fleshy endosperm and a terete embryo. [Named in honor of Adam Lonitzer, a sixteenth century German botanist.]

A genus of about 150 species, natives of the north temperate and a few of the tropical regions. Type species, *Lonicera caprifolium* L.

Flowers borne in pairs on a common peduncle in the leaf-axils; erect shrubs with deciduous leaves.
    Bractlets becoming fleshy and juicy, enclosing the 2 ovaries and berries, and the whole suggesting a single
        berry.   1. *L. cauriana.*
    Bractlets not becoming fleshy and sac-like.
        Native shrubs; flowers 1–2 cm. long.
            Bractlets large and involucre-like, becoming reddish in fruit.   2. *L. involucrata.*
            Bractlets minute or sometimes wanting.
                Flowers black-purple, the 2 ovaries partially or completely united.   3. *L. conjugialis.*
                Flowers white or pale yellow, the 2 ovaries entirely distinct.   4. *L. utahensis.*
        Introduced vine; flowers 3–4 cm. long.   5. *L. japonica.*
Flowers borne in whorls and sessile, the whorls solitary or more commonly in terminal spikes or panicles.
    Corollas with a short, nearly regular limb; leaves deciduous.   6. *L. ciliosa.*
    Corollas conspicuously 2-lipped, the limb revolute, more or less equaling the tube; leaves mostly evergreen.
        Corollas 3–5 cm. long, the slender tube not at all gibbous.   7. *L. etrusca.*
        Corollas 2–2.5 cm. long, tube gibbous.
            Upper leaves connate; stems glaucous.
                Inflorescence glandular-pubescent; corolla glandular-pubescent without. 8. *L. hispidula.*
                Inflorescence glabrous; corolla glabrous throughout.   9. *L. interrupta.*
            Upper leaves not connate; inflorescence glandular-pubescent.   10. *L. subspicata.*

1. **Lonicera cauriàna** Fernald. Blue Fly Honeysuckle. Fig. 5043.

Lonicera cauriana Fernald, Rhodora **27**: 10. 1925.
Lonicera coerulea of American authors, not L.

Erect shrub 0.5–1 m. high, the branches strongly ascending, young branches pruinose or

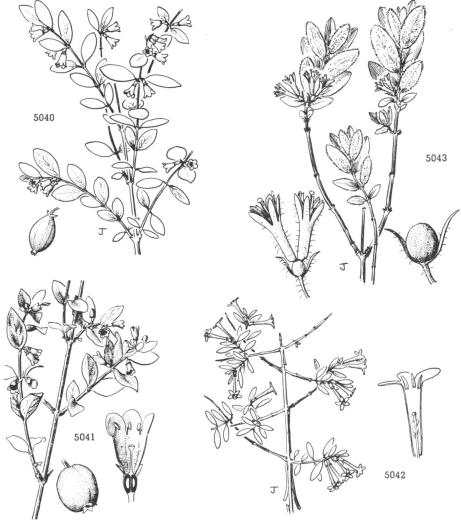

5040. Symphoricarpos vaccinioides   5042. Symphoricarpos longiflorus
5041. Symphoricarpos parishii   5043. Lonicera cauriana

puberulent, and sparsely hirsute. Leaves thin, narrowly obovate or oblong, 2–9 cm. long, 1–4 cm. broad, glabrous above, villous-ciliate on the margins, and more or less villous beneath especially on the veins; bracts linear-setaceous, about twice as long as the ovary; peduncles 3–5 mm. long; calyx-lobes very short, sparsely ciliate; corolla yellow or tinged with red, narrowly funnelform, nearly regular, 12–18 mm. long, the tube much longer than the lobes; ovaries of the two flowers completely united into a blue ellipsoid fruit.

Stream banks, Canadian Zone; Washington to Wyoming and south in the Cascade Mountains, Oregon, and occurring more rarely in the Sierra Nevada as far as Tulare County, California; also in western Nevada. Type locality: "Alpine meadows, Mt. Paddo [Adams]," Washington. June–July.

2. Lonicera involucràta (Richards.) Banks. Black Twin-berry. Fig. 5044.

Xylosteum involucratum Richards. in Frankl. 1st Journ. Bot. App. 733. 1823.
Lonicera involucrata Banks ex Spreng. Syst. 1: 759. 1825.
Lonicera ledebourii Eschsch. Mém. Acad. St.-Pétersb. 10: 284. 1826.
Lonicera intermedia Kell. Proc. Calif. Acad. 2: 154. *fig. 47.* 1862.
Caprifolium involucratum Kuntze, Rev. Gen. Pl. 1: 274. 1891.
Lonicera involucrata var. *ledebourii* Jepson, Man. Fl. Pl. Calif. 968. 1925.

Erect shrub, with freely exfoliating bark, 1–3 m. high, with principal branches mostly erect, the branches leafy, light brown or purplish, glabrous or sparsely pubescent. Leaves lanceolate to oblong, oblong-ovate, or obovate, acute to rounded and usually with an acumination at apex, caudate to cuneate at base, 3–10 cm. long, 2–6 cm. wide, glabrous or sparsely pubescent especially on the veins above, ciliate on the margins, and more or less pilose beneath, petioles short; peduncles axillary, 1–3 cm. long; involucral bracts 2, separate or more or less united, dark reddish purple, glabrous or often glandular-pubescent and ciliate; flowers 2 in the axils of the bractlets; corolla yellow, 12–15 mm. long, narrowly funnelform, glandular on the outer surface, the lobes nearly equal, short and rounded; berries separate, black.

Moist ground, Upper Sonoran Zone, and Transition Zones; southern Alaska south along the coast to Santa Barbara County, California, and in the interior parts of the Pacific States to the southern Sierra Nevada, ranging eastward to Quebec, Lake Superior, Colorado, and Arizona. Type locality: "wooded country from latitude 54° to 64° north." Canada. March–July.

3. Lonicera conjugiàlis Kell. Double Honeysuckle. Fig. 5045.

Lonicera conjugialis Kell. Proc. Calif. Acad. 2: 67. *fig. 15.* 1863.
Lonicera breweri A. Gray, Proc. Amer. Acad. 6: 537. 1865.
Caprifolium conjugiale Kuntze, Rev. Gen. Pl. 1: 274. 1891.
Xylosteon conjugialis Howell, Fl. N.W. Amer. 282. 1900.
Lonicera sororia Piper, Bull. Torrey Club 29: 644. 1902.

Erect shrub 6–15 dm. high with more or less shredded bark and many persistent bud-scales, young branches glabrous or with scattering subsessile glands, usually purplish. Leaves mostly obovate, varying to elliptic or ovate, 2–6 cm. long, sparsely pubescent at least beneath; petioles 2–4 mm. long; peduncles slender, 1–3 cm. long; bracts very small; corolla very dark purple, 6–8 mm. long, bilabiate, the upper lip 4-toothed, the lower strongly deflexed, entire, the throat and also the filaments and style conspicuously white-hairy below; the pairs of ovaries joined about two-thirds their length, blackish red when ripe.

Margins of streams and lakes, Canadian Zone; Washington and Idaho south along the Cascade and Siskiyou Mountains, and the Sierra Nevada to Tulare County, California, and western Nevada. Type locality: "Washoe," Nevada. Collected by Dr. J. A. Veatch. June–July.

4. Lonicera utahénsis S. Wats. Utah Honeysuckle. Fig. 5046.

Lonicera utahensis S. Wats. Bot. King Expl. 133. 1871.
Lonicera ebractulata Rydb. Mem. N.Y. Bot. Gard. 1: 372. 1900.
Xylosteon utahensis Howell, Fl. N.W. Amer. 282. 1900.

Low shrub with slender spreading branches 6–15 dm. high. Leaves oblong or elliptic to oblong-ovate or ovate, obtuse or rounded at apex, obtuse to blunt at base, 2–5 cm. long, entire, thin, glabrous on both sides or more or less sparsely pubescent beneath; petioles 3–4 mm. long; flowers in pairs on a slender peduncle 1–1.5 cm. long; bracts 2, very narrow, about 1.5 mm. long; corolla funnelform, 1.5–2 cm. long, white or pale yellow, fading to salmon-yellow, the base gibbous on one side and pinkish, the lobes about equal in size and shape but arranged to form a 2-lipped corolla-limb; berry red, globose, about 6 mm. in diameter.

Stream banks and bogs, Boreal Zone; British Columbia and Olympic Mountains, Washington, south to the Siskiyou Mountains near Yreka, California, in the Cascade Mountains to Crater Lake, Oregon, and in the Blue Mountains, northeastern Oregon; east to Montana and Utah. Type locality: "Wahsatch Mountains, Utah, in Cottonwood Cañon; 9,000 feet altitude." May–July.

5. Lonicera japónica Thunb. Japanese Honeysuckle. Fig. 5047.

Lonicera japonica Thunb. Fl. Japonica 89. 1784.

Vigorous, half-evergreen, trailing or climbing vines with stems 2–4 m. in length. Leaves 3–8 cm. long, 1.5–3 cm. wide, ovate to oblong, rounded or sometimes subcordate at base, pubescent, becoming glabrate above; petioles short, 5–8 mm. long; peduncles axillary, about the length of, or exceeding the petioles; bracts subtending the flowers green, leafy, petiolate, shorter than the leaves; flowers 3–4 cm. long, the tube slender, about equaling the limb, pubescent without with long hairs and glandular, white fading to dull yellow, fragrant; style and stamens exceeding the limb; berries black.

A native of Japan, much cultivated in gardens and occasionally becoming established in the west and reported from Sonoma, Marin, Santa Barbara, and San Bernardino Counties, California; common in eastern United States. May–Sept.

For a complete list of synonyms and segregates of *L. japonica* see Rehder (Rep. Mo. Bot. Gard. **14**: 159–162. 1903).

6. **Lonicera cilìòsa** (Pursh) Poir. Northwest or Orange Honeysuckle. Fig. 5048.

Caprifolium ciliosum Pursh, Fl. Amer. Sept. **1**: 160. 1814.
Lonicera ciliosa Poir. Encycl. Supp. **5**: 612. 1817.
Caprifolium occidentale Lindl. Bot. Reg. **17**: *pl. 1457*. 1831.
Lonicera ciliosa a *youngii* Dippel, Handb. Laubh. **1**: 215. 1889.

Stems woody, widely branching and usually climbing and twining, often 5–6 m. long, glabrous, younger parts glaucous. Leaves mostly elliptic, occasionally ovate or obovate, 6–10 cm. long, the upper 1–3 pairs on flowering branches connate, green and glabrous above, glaucous beneath and sparsely pilose, ciliate on the margins; flowers sessile, in 1–3 terminal whorls or clusters; calyx-teeth short-triangular to almost obsolete; corolla reddish yellow or orange, tubular-funnel-

5044. Lonicera involucrata
5045. Lonicera conjugialis

5046. Lonicera utahensis
5047. Lonicera japonica

form, 2.5–3 cm. long, rather obscurely bilabiate, the lobes short, broadly oblong-ovate; stamens and style slightly exserted, hairy below; berries red, about 6 mm. in diameter.

Open woods, often on hillsides and ridges, Transition Zones; British Columbia south on both sides of the Cascade Mountains to the Siskiyou and Trinity Mountains and Butte County, California, and east to Montana and Arizona. Type locality: "On the banks of the Kooskoosky. *M. Lewis.*" May–Aug.

7. **Lonicera etrúsca** Santi. Etruscan Honeysuckle. Fig. 5049.

Lonicera etrusca Santi, Viaggio Montam. **1:** 113. *pl. 1.* 1795.

Vigorous, evergreen or half-evergreen vine 2–4 m. long, ours glabrous to the inflorescence. Leaves 4–9 cm. long, broadly oblong-ovate to oblong or oval, obtuse, sessile and not connate except just below the inflorescence, glaucous and veiny beneath; inflorescence a terminal pedunculate head or sometimes with 2 or 3 axillary heads, glandular-stipitate and sometimes pubescent, the bracts subtending the flowers orbicular, nearly as long as the ovary; flowers 3–5 cm. long, yellowish white tinged with purplish red, glandular without, the tube and limb nearly equal; fruit red.

A European vine sparingly introduced along the coast from Douglas County to Curry County, Oregon, which also has been collected in Del Norte and Humboldt Counties, California. July–Aug.
For a complete list of synonyms and segregates see Rehder (Rep. Mo. Bot. Gard. **14:** 194–197. 1903).

8. **Lonicera hispídula** Dougl. Hairy Honeysuckle. Fig. 5050.

Caprifolium hispidulum Lindl. Bot. Reg. **21:** *pl. 1761.* 1836.
Lonicera hispidula Dougl. ex Lindl. loc. cit., as a synonym.
Lonicera californica Torr. & Gray, Fl. N. Amer. **2:** 7. 1841.
Lonicera hispidula Dougl. ex Torr. & Gray, op. cit. 8.
Lonicera hispidula var. *vacillans* A. Gray, Pros. Amer. Acad. **8:** 628. 1873.
Lonicera hispidula var. *douglasii* A. Gray, loc. cit.
Lonicera hispidula var. *californica* Rehd. in Bailey, Cyclop. Hort. **3:** 943. 1900.
Lonicera catalinensis Millsp. Field Mus. Bot. Ser. **5:** 252. 1923.

Stems slender, trailing or more frequently climbing, freely branched, 1–3 m. long, villous-hirsute with long spreading hairs interspersed with shorter ones and also with stalked glands, varying to glabrous. Leaves ovate to ovate-oblong, 3–6 cm. long, often subcordate at base, glaucous beneath and more or less pubescent to glabrate, the lower leaves with stipule-like appendages at the base of the short petiole, the upper 1–3 pairs connate; spikes usually glandular-pubescent, pedunculate, terminal and from the upper axils, sometimes becoming paniculate; corolla 2–2.5 cm. long, reddish purple without, yellow within, strongly bilabiate, the tube slightly gibbous near the base, upper lip 4-lobed, filaments and style long-exserted, hairy below; berries red.

Thickets in open woods and clearings. Humid Transition Zone; Vancouver Island and western British Columbia south, west of the Cascade Mountains, to southern Oregon and southern California; also on Santa Cruz and Santa Catalina Islands. Type locality: "woods of North West America." June–July.

9. **Lonicera interrúpta** Benth. Chaparral Honeysuckle. Fig. 5051.

Lonicera interrupta Benth. Pl. Hartw. 313. 1849.
Lonicera hispidula var. *interrupta* A. Gray, Proc. Amer. Acad. **8:** 628. 1873.
Caprifolium interruptum Greene, Fl. Fran. 347. 1892.

Stem stout and woody below, becoming widely branched above, often climbing or reclining on other shrubs; young branches glabrous and decidedly glaucous. Leaves elliptic to broadly ovate or suborbicular, 2–3.5 cm. long, the first and often the second pair below the inflorescence connate at the base or rarely distinct, the others short-petioled, coriaceous, glabrous, glaucous on both sides, but more so on the lower side; inflorescence glabrous, 3–12 cm. long; calyx glabrous; corolla 8–10 mm. long, yellowish, often with a reddish tinge, the tube obscurely gibbous at base, shorter than the limb, glabrous without and nearly so within; filaments sparsely hairy; berries red.

Dry slopes and ridges, Upper Sonoran and Transition Zones; Jackson County, southern Oregon, south through the Coast Ranges and the western slopes of the Sierra Nevada to the San Bernardino Mountains, southern California; also in Arizona. Type locality: "Juxta fl. Carmel prope Monterey." May–July.

10. **Lonicera subspicàta** Hook. & Arn. Southern Honeysuckle. Fig. 5052.

Lonicera (?) *subspicata* Hook. & Arn. Bot. Beechey 349. 1838.
Lonicera hispidula var. *subspicata* A. Gray, Proc. Amer. Acad. **8:** 628. 1873.
Caprifolium subspicatum Greene, Fl. Fran. 348. 1892.
Lonicera interrupta var. *subspicata* Jepson, Fl. W. Mid. Calif. 474. 1901.

Stems woody and stout below, loosely clambering over shrubs, young twigs light brown, puberulent and glandular. Leaves coriaceous, linear-oblong, 1–3 cm. long, margin revolute, dark green and glabrate or with a few scattering hairs and subsessile glands on the upper side, cinereous-puberulent beneath; petioles 1–5 mm. long; spikes 2–4 cm. long or sometimes longer, with several interrupted, compact, few-flowered whorls; corolla yellowish to cream-colored, 8–10 mm. long, short-pubescent, the limb 2-lipped and somewhat recurved; stamens about equaling the lobes; filaments thinly pilose below, also the style; berries 5–7 mm. long, seeds 2.5–3 mm. long, light brown, shallowly pitted.

Dry slopes, Upper Sonoran Zone; Santa Ynez Mountains, Santa Barbara County, California. Type locality: California. Collected by Douglas, probably at Santa Barbara. June–Aug.

Lonicera subspicata var. **johnstònii** Keck, Bull. S. Calif. Acad. **25**: 67. 1926. (*Lonicera johnstonii* McMinn, Ill. Man. Calif. Shrubs 541. 1939.) Leaves broader, varying from oblong-ovate to broadly orbicular, whitish beneath; corolla 10–14 mm. long. Dry slopes, California Coast Ranges, San Benito County, and Tehachapi Mountains south to the mountains of southern Lower California. Type locality: Poppet Flat, Idyllwild Road, Riverside County. Southern Honeysuckle.

Lonicera subspicata var. **denudàta** Rehd. Rep. Mo. Bot. Gard. **14**: 176. 1903. (*Lonicera denudata* Davids. & Moxley, Fl. S. Calif. 344. 1923.) Similar to the preceding variety but leaves mostly oblong-oval or oblong-ovate, 5–15 mm. broad, very thinly puberulent, shining and green, not whitish beneath. Southern San Diego County and adjacent Lower California. Type locality: San Diego. San Diego Honeysuckle.

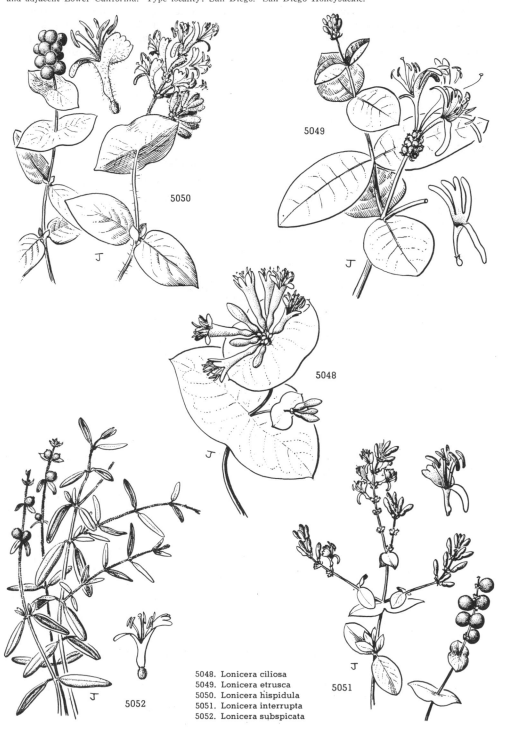

5049
5050
5048

5048. Lonicera ciliosa
5049. Lonicera etrusca
5050. Lonicera hispidula
5051. Lonicera interrupta
5052. Lonicera subspicata

5052
5051

Family 145. **VALERIANÀCEAE.***

VALERIAN FAMILY.

Annual or perennial, odoriferous herbs with basal or opposite, simple or pinnately divided, estipulate leaves and usually small, more or less irregular, perfect or polygamous flowers in corymbose, paniculiform, or capitate cymes (sometimes monochasial). Calyx with the limb inconspicuous or absent in the flower, sometimes developing into an awned or plumose pappus in the fruit. Corolla epigynous, 5-lobed and usually more or less zygomorphic, the tube narrow, often gibbous or spurred at the base. Stamens 1–4, epipetalous. Ovary inferior, 1–3-celled with a solitary pendulous ovule. Seed without endosperm, the embryo straight and with oblong cotyledons.

A family of about 10 genera and 370 species of wide geographic distribution, but most abundant in the northern hemisphere.

Stamens 3; corolla gibbous or spurred, white, pink, or bluish, less than ½ cm. long (except *Valeriana columbiana*).
Calyx-limb composed of plumose bristles in the fruit; native perennials with rhizomes or taproots.
1. *Valeriana.*

Calyx-limb absent; annual species.
Flowers in cymose clusters forming a more or less flat-topped inflorescence; stem dichotomously branched; annuals, adventive from Europe. 2. *Valerianella.*
Flowers densely clustered in capitate or interrupted, spike-like inflorescences; native annual species.
3. *Plectritis.*

Stamens 1; corolla with a long spur, usually magenta or red (rarely white), more than 1 cm. long; perennial, adventive from Europe. 4. *Kentranthus.*

1. **VALERIÀNA**† [Tourn.] L. Sp. Pl. 31. 1753.

Perennial, strong-smelling herbs from rhizomes or taproots. Leaves decussate, basal and cauline, spatulate and undivided to deeply pinnatifid. Inflorescences thyrsiform. Flowers perfect, gynodioecious or polygamodioecious. Calyx initially involute, later spreading, its lobes much divided, setose-plumose. Corolla infundibuliform, subcampanulate or rotate, the tube gibbous or straight, the 5 lobes equal or subequal. Stamens 3, rarely 4, epipetalous. Ovary inferior, 3-carpellate with 1 fertile adaxial cell at maturity. Fruit a cypsela with 3 abaxial and 3 adaxial veins. [Named for the Roman emperor Valerianus.]

A genus of about 200 species in both hemispheres, chiefly in the boreal and temperate regions but also in the tropics in the mountains. Lectotype species, *Valeriana officinalis* L.

Leaves mostly pinnatifid, generally petiolate; plants from rhizomes or stolons.
Leaves usually ovate to spatulate in outline; corolla funnelform to subsalverform, 3–19 mm. long, the tube obviously gibbous.
Stamens exserted, longer than the corolla-lobes; corolla-lobes less than half the length of the tube.
Leaves ascending-ciliate, the terminal lobe linear to elliptic or oblanceolate, the lateral lobes of the basal leaves in 8–12 pairs; adventive from Europe. 1. *V. officinalis.*
Leaves glabrous or with spreading pubescence.
Plants relatively robust, 3–10 dm. tall; at least the lower cauline leaves petioled, glabrous or essentially so. 2. *V. sitchensis.*
Plants relatively slender, 1–6 dm. tall; cauline leaves essentially sessile and glabrous or the stem and/or the leaves puberulent. 3. *V. capitata californica.*
Stamens included, shorter than the corolla-lobes; corolla-lobes 3–6 mm. long, about half the length of the tube; Wenatchee Mountains of Washington. 4. *V. columbiana.*
Leaves mostly oblong in outline; corolla subrotate to rotate, 2.0–3.5 mm. long, the tube short, straight or indistinctly gibbous.
Plants relatively robust and leafy, 3–9 dm. tall; fruits sparsely to densely pilosulose, rarely glabrous; extreme northern California through Oregon. 5. *V. occidentalis.*
Plants relatively slender and less leafy, 1.5–4.5 dm. tall; fruits glabrous; northern Washington.
6. *V. dioica sylvatica.*
Leaves mostly ligulate-spatulate, gradually decurrent to the clasping leaf-bases (the cauline often pinnatifid but more or less decurrent); plants from vertical, usually forked taproots.
7. *V. edulis.*

1. **Valeriana officinàlis** L. Valerian or All-heal. Fig. 5053.

Valeriana officinalis L. Sp. Pl. 31. 1753.

Stems from a short perennial rhizome, pilosulose to short-pilose toward the base, glabrescent above. Leaves mostly cauline, petiolate below, sessile above, oblong to oblong-ovate, pinnate to pinnatifid, 9–36 cm. long, glabrous to short-pilose, particularly on the veins beneath, glabrous above, ascending-ciliate, the lateral lobes 4–8 pairs, linear to oblanceolate, acute, more or less falcate, 2.0–7.5 cm. long, 0.5–3.0 cm. wide, the terminal lobe two-thirds as long but as wide as the lateral lobes; basal leaves 15–30 cm. long; inflorescenses thyrsiform with perfect flowers;

---

* Text contributed by Richard William Holm.
† This treatment is based largely on the monograph of F. G. Meyer, *Ann. Mo. Bot. Gard.* **38**: 377–503. 1951.

corolla infundibuliform, 3–5 mm. long, white, the lobes half as long as the tube; stamens and style exserted; fruits lanceolate to ovate-oblong, 2.5–3.0 mm. long, glabrous or pilosulose, tawny or rubiginous. $n = 14, 28$.

Introduced into gardens in the United States and Canada, and established as a garden escape in parts of Washington. May–July.

2. Valeriana sitchénsis Bong. Mountain Heliotrope. Fig. 5054.

Valeriana sitchensis Bong. Mém. Acad. St. Pétersb. VI. 2: 145. 1832.
Valeriana hookeri Shuttlw. Flora 20²: 450. 1837.
Valeriana suksdorfii Gandoger, Bull. Soc. Bot. Fr. 65: 36. 1918.
Valeriana frigidorum Gandoger, op. cit. 37.
Valeriana anomala Eastw. Leaflets West. Bot. 3: 22. 1941.

Stems robust from a rather stout rhizome, 3.5–12.0 dm. tall, glabrescent. Basal leaves often wanting, petiolate, undivided or pinnate to pinnatifid, ovate-elliptic to obovate, 10–40 cm. long, dentate to repand or entire, the terminal lobe 5–10 cm. long, 1.5–6.5 cm. wide, glabrous; cauline leaves predominating, 2–5 pairs, pinnate to pinnatifid, 10–20 cm. long, the lateral lobes 1–4 pairs, crenate to irregularly repand-dentate or essentially entire, the terminal lobe obovate, ovate-rhombic, to suborbicular, 2.5–4.5 cm. wide, acute or obtuse; inflorescences thyrsiform with perfect or gynodioecious flowers; corolla infundibuliform, 4.5–7.0 cm. long, white to pinkish, the limb usually less than half as long as the gibbous tube; stamens and style exserted; fruits ovate to oblong-ovate, 3–6 mm. long, 2.0 mm. wide, tawny or purpurascent, frequently purple-maculate.

Subalpine meadows and woodlands, Canadian and Hudsonian Zones; Kenai Peninsula southward on the Alaska coastal mountains to east-central and southern British Columbia eastward to south-central Idaho and western Montana; Washington and Oregon to the Siskiyou Mountains, California. Type locality: "Habitat in montanis, l'Isle Sitka" [now Baranof Island], Alaska. June-Sept. Northern Valerian.

Valeriana capitata subsp. pubicárpa (Rydb.) F. G. Mey. Ann. Mo. Bot. Gard. 38: 406. 1951. (*Valeriana scouleri* Rydb. Mem. N. Y. Bot. Gard. 1: 377. 1900; *V. adamsiana* Eastw. Leaflets West. Bot. 2: 196. 1939; *V. follettiana* Eastw. op. cit 197; *V. humboldtiana* Eastw. op. cit. 198.) Differing from *V. sitchensis* subsp. *sitchensis* in the following characteristics: plants slender, to 7 dm. tall; leaves mostly basal, essentially entire, glabrous, the terminal lobe 0.5–4.0 cm. wide; fruits mostly oblong-linear, 5–6 mm. long, 1–2 mm. wide, tawny to rubiginous, rarely purpurascent. Lower elevations on bluffs along forest streams west of the Cascade Mountains from southwestern British Columbia to Mendocino County, California. Type locality: "On moist rock and islands of the Columbia River near Oak Point." March–July.

3. Valeriana capitàta subsp. califórnica (Heller) F. G. Mey. California Valerian. Fig. 5055.

Valeriana capitata Pall. ex Link, Jahrb. d. Gewächsk. 1³: 66. 1820, in part.
Valeriana californica Heller, Muhlenbergia 1: 60. 1904.
Valeriana puberula Piper, Smiths. Misc. Coll. 50: 202. 1907.
Valeriana seminuda Piper, Proc. Biol. Soc. Wash. 37: 95. 1924.
Valeriana sylvatica var. *glabra* Jepson, Man. Fl. Pl. Calif. 970. 1925.
Valeriana whiltonae Eastw. Leaflets West. Bot. 3: 24. 1941.
Valeriana capitata subsp. *californica* F. G. Mey. Ann. Mo. Bot. Gard. 38: 404. 1941.

Stems with mostly basal leaves from stout rhizomes, 2–6 dm. tall, glabrescent above, puberulent to pilosulose toward the base. Basal leaves mostly undivided or pinnate to pinnatifid, oblanceolate- to obovate-spatulate, 4.5–15.0 cm. long, 0.5–2.5 cm. wide, truncate or retuse at the apex, 3–7-toothed or essentially entire, puberulent, ciliate, or glabrous, the lateral lobes of the divided leaves 1–4 pairs; cauline leaves petiolate, pinnate to pinnatifid, 2.5–8.0 cm. long, rarely simple, the petioles 2–4 cm. long; inflorescences thyrsiform with perfect flowers; corolla infundibuliform, 3.0–5.5 cm. long, white to pinkish, the lobes half as long as the gibbous tube; stamens and style longer than the corolla-limb; fruits elliptic to ovate-oblong, 4.0–6.5 mm. long, 2.0–2.8 mm. wide, glabrous or pilosulose.

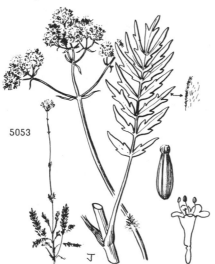

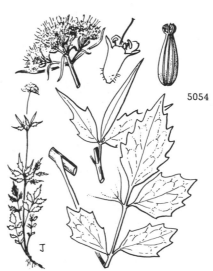

5054

5053. Valeriana officinalis

5054. Valeriana sitchensis

Subalpine slopes, meadows, along creek banks, and also in coniferous woods, Boreal Zone; southern Oregon and the Sierra Nevada of California to Tulare County, and in adjoining Nevada. Type locality: ridge south of Donner Pass at 8,500 feet, Nevada County, California. June–Aug.

Valeriana capitata subsp. **pubicárpa** (Rydb.) F. G. Mey. Ann. Mo. Bot. Gard. 38: 406. 1951. (*Valeriana pubicarpa* Rydb. Bull. Torrey Club 36: 697. 1909; *V. puberulenta* Rydb. loc. cit.; *V. cusickii* Gandoger, Bull. Soc. Bot. Fr. 65: 36. 1918; *V. utahensis* Gandoger, op. cit. 37; *V. maculata* Eastw. Leaflets West. Bot. 3: 23. 1941.) This subspecies differs from the above in its uniformly puberulent stems and mostly entire and glabrous leaves. On stony sagebrush slopes or beneath pines and on talus slopes in the mountains; southwestern Montana, western Wyoming, and central Idaho to northern Utah; southwestern Oregon to the Charleston Mountains of Nevada. Type locality: Mount Nebo, Juab County, Utah. June–Aug.

4. **Valeriana columbiàna** Piper. Wenatchee Valerian. Fig. 5056.

Valeriana columbiana Piper, Bot. Gaz. 22: 489. 1896.

Stems leafy from stout rhizomes, 0.5–3.0 dm. tall, glabrescent above, uniformly puberulent below. Basal leaves forming a loose rosette, petiolate, undivided, broadly ovate to ovate-oblong or sometimes suborbicular, 6–15 cm. long, 1.0–3.5 cm. wide, irregularly dentate to entire or nearly so, glabrous or the veins sometimes pilosulose; cauline leaves 1–4 pairs, 4–12 cm. long, the lowermost petiolate, the uppermost sessile and much reduced, pinnate to pinnatifid, acute, irregularly repand-dentate or essentially entire, the terminal lobe oblong to oblanceolate or obovate, 2.0–6.5 cm. long, 1.0–2.7 cm. wide, occasionally 3 lobed, the lateral lobes 1–2 pairs, sometimes equaling the terminal lobe in length or grading smaller; inflorescences thyrsiform with perfect flowers; corolla infundibuliform to subsalverform, 11–18 mm. long, white, the tube gibbous; stamens and style included, shorter than the corolla-lobes; fruits oblong-linear, 5–7 mm. long, 1.5–2.0 mm. wide, tawny, smooth, sometimes purple-maculate.

Open to forested rocky slopes in the mountains, Arid Transition and lower Canadian Zones; north-central Washington in the Wenatchee Mountains. Type locality: "side hill above Farwell's house west of Wenatchee," Chelan County. May–July.

5. **Valeriana occidentàlis** Heller. Western Valerian. Fig. 5057.

Valeriana occidentalis Heller, Bull. Torrey Club 25: 269. 1898.
Valeriana micrantha E. Nels. Erythea 7: 166. 1899.

Stems more or less leafy from the rhizomes, 4.5–9.0 dm. tall, glabrous or glabrescent. Basal leaves petiolate, undivided or pinnate to pinnatifid, oblong to narrowly ovate or more or less spatulate, rarely suborbicular, 12–30 cm. long, essentially entire, the blades and terminal lobe of the divided leaves 2–10 cm. long, 1.3–6.0 cm. wide, short-acuminate or obtuse, the lateral lobes 1–2 pairs, grading smaller; cauline leaves 2–4 pairs, the lowermost short-petiolate, pinnate to pinnatifid or sometimes undivided, 4.5–14.5 cm. long, the uppermost sessile and much reduced, the terminal lobe oblong-linear, ovate to obovate, 2.0–6.8 cm. long, 1–4 cm. wide, acute or obtuse, the lateral lobes 1–6 pairs, grading smaller; inflorescences thyrsiform with gynodioecious flowers; corolla rotate to subrotate, 3.0–3.5 mm. long, white, the tube slightly gibbous; stamens and style exserted; fruits linear– to ovate-oblong, 3–5 mm. long, 1–2 mm. wide, sparsely to densely pilosulose or glabrous, tawny.

Aspen glens and yellow pine woods, wet meadows or grassy places among willows, in the mountains, Boreal Zone; in the Pacific States in eastern Oregon from Deschutes and Grant Counties south to Modoc County, California, and eastward to Montana, Nevada, and Utah; also Wyoming and Colorado. Type locality: "near the western end of the Craig Mountain plateau, above Lake Waha, Nez Perces county, Idaho." May–Sept.

6. **Valeriana dioìca** subsp. **sylvática** (Sol. ex Richards.) F. G. Mey. Wood Valerian. Fig. 5058.

Valeriana sylvatica Sol. ex Richards. in Frankl. 1st Journ. Bot. App. 730. 1823.
Valeriana wyomingensis E. Nels. Erythea 7: 167. 1899.
Valeriana septentrionalis Rydb. Mem. N.Y. Bot. Gard. 1: 376. 1900.
Valeriana psilodes Gandoger, Bull. Soc. Bot. Fr. 65: 37. 1918.
Valeriana dioica subsp. *sylvatica* F. G. Mey. Ann. Mo. Bot. Gard. 38: 417. 1951.

Stems more or less leafy from slender rhizomes, 1.5–3.0 dm. tall, glabrescent to glabrous. Basal leaves petiolate, undivided, oblong, ovate-oblong, or spatulate, rarely suborbicular, 3–27 cm. long, essentially entire, the blades 1.5–8.0 cm. long, 0.5–3.0 cm. wide, short-acuminate, acute, or obtuse; cauline leaves 3–4 pairs, the lowermost short-petiolate, pinnate to pinnatifid, rarely undivided, the uppermost sessile and much reduced above, oblong, ovate, to ovate-oblong, 2.5–11.5 cm. long, glabrous, the terminal lobe linear– to ovate-oblong, acute or short-acuminate, 1–5 cm. long, 0.2–2.5 cm. wide, the lateral lobes 1–7 pairs, grading smaller; inflorescences thyrsiform with gynodioecious flowers; corolla rotate to subrotate, 2–3 mm. long, white, the tube slightly gibbous; stamens and style exserted; fruits ovate to ovate-oblong, 3–5 mm. long, glabrous, tawny to rubiginous.

Meadowlands, moist wooded hillsides in rocky duff-covered clay soil, or on talus slopes, Arid Transition Zone; Newfoundland northwesterly across Canada to the Rocky Mountains, thence southward to central Idaho and eastern Washington. Type locality: "on the Clear-water River." May–Aug.

7. **Valeriana édulis** Nutt. ex Torr. & Gray. Tobacco Root. Fig. 5059.

Patrinia ceratophylla Hook. Fl. Bor. Amer. 1: 290. 1834.
Valeriana edulis Nutt. ex Torr. & Gray, Fl. N. Amer. 2: 48. 1841.
Valeriana furfurascens A. Nels. Bull. Torrey Club 28: 232. 1901.
Valeriana trachycarpa Rydb. op. cit. 31: 645. 1904.
Valeriana ceratophylla Piper and western authors, not H.B.K.

Stems robust, 1–12 dm. tall, usually glabrous, with the leaves mostly basal, from vertical, often forked taproots, 0.5–3.0 cm. thick. Basal leaves linear or oblong- to obovate-spatulate, undivided or pinnate to pinnatifid, subacute to obtuse, 6–40 cm. long, 0.3–6.5 cm. wide, gradually tapering to the base, more or less repand to entire, spreading-ciliate, pilosulose to appressed white-puberulent or glabrous, or the veins often puberulent, the lateral lobes of the divided leaves 1–4 pairs, mostly distinct or sometimes narrowly decurrent, the terminal lobe 4.5–9.0 cm. long, 0.5–2.0 cm. wide; cauline leaves 2–3 well-developed pairs, pinnate to pinnatifid, rarely undivided, short-petiolate or sessile below, much reduced above, 3–22 cm. long, elliptic- to obovate-spatulate; inflorescences thyrsiform, large, with polygamodioecious flowers; corolla rotate, the perfect and staminate flowers 3.0–3.5 mm. long, the pistillate flowers 0.5 mm. long, white, the tube slightly gibbous; stamens and style exserted; fruits 2.5–4.5 mm. long, glabrous to densely hirsutulous.

In moist pastures, creek bottoms, yellow pine and aspen woods, sagebrush plains, cliffs, and subalpine parks, Arid Transition and lower Boreal Zones; southern British Columbia and western United States except California to northwestern Mexico. Type locality: Walla Walla, Walla Walla County, Washington. May–Sept.

2. **VALERIANÉLLA** [Tourn.] Mill. Gard. Dict. abr. ed. 4. 1754.

Dichotomously branched annuals with basal leaves tufted and entire, the stem-leaves sessile and often dentate. Flowers in terminal and usually corymbosely arranged cymose clusters. Calyx-limb short or obsolete in the flower, short-tooth when present; in the fruit various but not divided into plumose segments. Corolla bluish, white, or pink, small, nearly regular, funnelform or salverform, the limb 5-lobed. Stamens 3; style minutely 3-lobed at the summit. Ovary 3-celled, one cell fertile, the others empty but one usually about as large as the fertile one. [Name diminutive of *Valeriana*.]

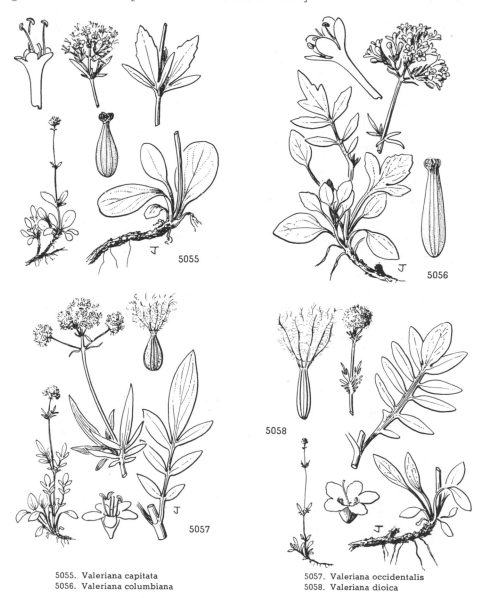

5055. Valeriana capitata
5056. Valeriana columbiana

5057. Valeriana occidentalis
5058. Valeriana dioica

About 50 species, native of the northern hemisphere, and most abundant in the Mediterranean region. Type species, *Valeriana locusta* L.

Fruits elliptic in cross section, with a prominent corky mass on the dorsal side of the fertile cell, margins of the empty cells usually not produced, separated by a narrow shallow groove.   1. *V. locusta.*
Fruits square in cross section, without a corky mass on the dorsal side of the flattened fertile cell, margins of the empty cells incurved, forming a broad deep groove.   2. *V. carinata.*

1. **Valerianella locústa** (L.) Betcke. Lamb's Lettuce. Fig. 5060.

Valeriana locusta a olitoria L. Sp. Pl. 33. 1753.
Valerianella olitoria Poll. Hist. Pl. Palat. **1**: 30. 1776.
Valerianella locusta Betcke. Anim. Val. 10. 1826.

Stems simple or usually dichotomously branched, 15–30 cm. tall, glabrous to more or less puberulent, at least at the nodes, the hairs somewhat reflexed. Basal leaves spatulate to oblanceolate, rounded or obtuse at the apex, 3–5 cm. long, 6–15 mm. wide, entire; upper stem-leaves broadly linear to oblong, 1.5–2.0 cm. long, entire or sometimes sparingly toothed; peduncles short; cymes 6–12 mm. broad, subcapitate; bracts linear to narrowly oblong; corolla bluish or pink, about 2 mm. long; fruit 2 mm. long, about as broad as long, laterally compressed, finely foveolate, glabrous, the empty cells separated by a narrow shallow groove, the fertile cell with a thick corky mass on the dorsal side. $n = 7$.

Occasional in waste places and cultivated fields; Washington, Oregon, and central California and across the continent to the eastern and southern states; naturalized from Europe and sometimes cultivated as a salad plant. April–June.

5059
5060
5061
5062

5059. Valeriana edulis
5060. Valerianella locusta
5061. Valerianella carinata
5062. Plectritis congesta

2. **Valerianella carinàta** Loisel. Corn Salad. Fig. 5061.

Valerianella carinata Loisel. Not. Pl. Fr. 149. 1810.
Valerianella praecox Willk. Linnaea **30**: 104. 1859.

Slender erect annual, 1–2 dm. high, simple below, once to thrice dichotomously forked above, glabrous to more or less puberulent, especially on the lower nodes and leaf-bases, the hairs somewhat reflexed. Basal leaves spatulate, 1–2 cm. long, 3–5 mm. wide; upper stem-leaves linear-oblong, 1–2 cm. long, 2–3 mm. wide, sessile, entire or sparsely and obscurely denticulate; peduncles slender, 1.0–2.5 cm. long; cymes subcapitate, 6–8 mm. broad; bracts narrowly oblanceolate to linear, 3–4 mm. long, ciliate on the margins; corolla light blue to pink, 1.0–1.5 mm. long; fruit about 2 mm. long and barely half as broad, pubescent, not foveolate, the fertile cell flattened, the margins of the empty cells incurved, forming a broad deep groove. $n = 9$.

Locally adventive on shady banks and slopes; Amador and Eldorado Counties, California, and Marion and Clackamas Counties, Oregon. April–May.

3. **PLECTRÌTIS*** DC. Prod. **4**: 631. 1830.

Usually slender, annual, glabrous herbs with opposite, oblong to ovate leaves. Inflorescences thyrsiform, densely to loosely capitate or verticillate to spiciform; bracts linear, fused into a multifid structure. Calyx reduced to a narrow ring. Corolla subcampanulate to narrowly infundibuliform, white to deep pink, occasionally with spots of dark red at the base of the midventral lobe, strongly bilabiate to essentially regular, the tube spurred or merely gibbous. Stamens 3, exserted to included. Pistil inferior, 1-celled, the stigma bifid, rarely trifid. Fruit pale straw-yellow to dark red-brown, vernicose to dull, usually more or less pubescent, dorsally keeled, the ventral wall with glands either in 2 longitudinal rows near the base of the wings or as 2 prominent mammillae; wings connivent or spreading, often obsolete, occasionally with glands on the dorsal surface opposite those on the ventral surface. Cotyledons accumbent or incumbent. [Name from the Greek, *plectron*, meaning a spur.]

A genus of 3 species in the Pacific States, formerly distributed into many species because of the extreme variation of the fruits. Type species, *Valerianella congesta* Lindley.

Corolla pale to medium pink, bilabilate, funnelform, spur usually less than one-third the length of the corolla or obsolete; fruit keeled, the keel rarely grooved, acutely angled (with prominent veins) or smoothly rounded (with pronounced vacuities between the walls of the fruit and the embryo), winged or wingless, the wings thin and thin-margined, pubescent, the trichomes flexible, evenly tapered, and more or less obtuse; cotyledons accumbent, dull green, with only the midvein prominent on germination.
                                                                     1. *P. congesta.*
Corolla white and essentially regular to light red and strongly bilabiate, tubular or somewhat funnelform, spurred, the spur usually more than one-third the length of the corolla; fruit keeled, the keel often with a dorsal groove, obtusely angled, rarely with visible dorsal veins, winged or essentially wingless, the wings stiff and often thick, the wing-margins thickened, often grooved, pubescent, the trichomes stiffly clavate, cylindrical, glandular- cylindrical or long and curly; cotyledons incumbent, silvery green, with secondary and tertiary veins prominent on germination.
   Corolla essentially regular, white or light pink, usually without red spots at the base of the ventral lip, stout and with a stout clavate spur; fruit with the trichomes stout and clavate or more or less narrowly cylindrical and glandular-tipped, sometimes long and curly, the keel without 2 brush-like rows of hairs, the ventral surface often with a multiseriate row of hairs along the midrib.
                                                                     2. *P. macrocera.*
   Corolla strongly bilabiate, pink to light red, usually with 2 red spots at the base of the ventral lip, slender and with a slender spur; fruit with the trichomes stout, cylindrical or rarely subclavate, never long and curly, the keel often with 2 brush-like rows of hairs, the ventral surface usually lacking a multiseriate row of hairs along the midrib.
                                                                     3. *P. ciliosa.*

1. **Plectritis congésta** (Lindl.) DC. Pink Plectritis. Fig. 5062.

Valerianella congesta Lindl. Bot. Reg. **13**: *pl. 1094.* 1827.
Plectritis congesta DC. Prod. **4**: 631. 1830.
Plectritis congesta β *minor* Hook. Fl. Bor. Amer. **1**: 291. 1833.
Plectritis microptera Suksd. Deutsch. Bot. Monatss. **14**: 119. 1896.
Plectritis racemulosa Gandoger, Bull. Soc. Bot. Fr. **65**: 35. 1918.
Plectritis suksdòrfii Gandoger, loc. cit.

Plants 1.5–6.0 dm. tall, erect. Leaves obovate to ovate, petiolate below to sessile or amplexicaule above, entire to submucronate or emarginate; inflorescence densely capitate to densely verticillate-spiciform; corolla pale pink to pink, subcampanulate to infundibuliform, 4.5–9.5 mm. long, strongly bilabiate, the dorsal lobes fused and extended, the ventral free and patent, the spur short to obsolescent, the tip usually conspicuously expanded; fruit pale yellow to brown, 2.0–4.5 mm. long, variously pubescent, the hairs long, gradually and evenly attenuated to a thin point, the walls thin and readily collapsed on drying; fertile cell strongly trigonous, deeper than broad; keel sharply angled, acute, very slightly indented dorsally, smooth, or with 1–3 prominent veins, variously pubescent or glabrous, occasionally with 1 to several glands at the wing-roots; ventral wall planar to convex, glands rarely present, few, above the middle and in 2 longitudinal rows near the lateral limits; wings broad to obsolescent, margins thin, subscarious, sharp, strongly connivent basally, usually strongly spreading above; cotyledons accumbent, veins above the first order inconspicuous when expanded.

---

*This treatment is based largely on the recently published work of Dennison Harlow Morey, Jr.

Among shrubs and bushes in shady places (usually with marine influences) at low altitudes, Upper Sonoran and Humid Transition Zones; British Columbia, Washington, Oregon, and northern California south along the coast to Monterey and San Luis Obispo Counties. Type locality: "north-west coast of North America." Collected by Douglas. April–May.

Plectritis congesta subsp. **nítida** (Heller) Morey, Contr. Dudley Herb. **5**: 120. 1959. (*Plectritis nitida* Heller, Muhlenbergia **2**: 328. 1907.) Differing from the typical subspecies in the vernicose or nitid fruits with the keel smoothly rounded and the wings when present appressed and connivent top and bottom; walls of the fruit usually more or less separate from the embryo. Slender decumbent plants of more or less shady, wet places, usually on the coastal lowlands from Monterey to Fort Bragg, California; rarely elsewhere. Type locality: Crystal Springs, San Mateo County.

Plectritis congesta subsp. **brachystèmon** (Fisch. & Mey.) Morey, Contr. Dudley Herb. **5**: 119. 1959. (*Plectritis brachystemon* Fisch. & Mey. Ind. Sem. Hort. Petrop. **2**: 47. 1835; *Betckea major* Fisch. & Mey. Ann. Sci. Nat. II. **5**: 189. 1836; *Plectritis samolifolia* Hoeck, Bot. Jahrb. **3**: 37. 1882, in part; *Valerianella anomala* A. Gray, Proc. Amer. Acad. **19**: 83. 1883; *V. aphanoptera* A. Gray, loc. cit.; *V. magna* Greene, Proc. Acad. Phila. **1895**: 548. 1895; *Plectritis involuta* Suksd. Deutsch. Bot. Monatss. **15**: 144. 1897; *P. gibbosa* Suksd. Werdenda **1**: 43. 1927.) Differing from the typical subspecies in the smaller flowers (1.0–3.5 mm. long), which may lack a spur, and which have a tube 0.5 mm. long or longer, and in the typically tubular-funnelform corolla. In brushy but relatively open, montane habitats from British Columbia to Monterey County, California. Type locality: "prope coloniam Ross in California," Sonoma County.

2. **Plectritis macrócera** Torr. & Gray. White Plectritis. Fig. 5063.

Plectritis macrocera Torr. & Gray, Fl. N. Amer. **2**: 50. 1841.
Aligera eichleriana Suksd. Deutsch. Bot. Monatss. **15**: 147. 1897.
Aligera ostiolatata Suksd. loc. cit.
Aligera jepsonii Suksd. Erythea **6**: 24. 1898.
Plectritis glabra Jepson, Fl. W. Mid. Calif. 475. 1901.
Plectritis collina Heller, Muhlenbergia **2**: 329. 1907.

Plants slender, 1.0–6.5 dm. tall. Leaves obovate to ovate-lanceolate, diminishing in size upward, petiolate below, sessile above, entire to coarsely serrate; inflorescence verticillate (sometimes capitate in small specimens); corolla white to pale pink, lacking pigment spots at the base of the ventral lobes, 2.0–3.5 mm. long, broadly tubular to subcampanulate, stout, essentially regular to weakly bilabiate, the spur stout, about twice as long as broad, about one and one-half to two times as long as the tube, saccate, often rather abruptly narrowed at about the middle, tip slightly or not at all expanded; fruit pale straw-yellow to red-brown, 2–4 mm. long, more or less pubescent, the hairs stout, clavate to glandular-cylindric, rarely obtusely pointed, the larger often collapsed on drying, or long and curly; keel rounded to rather angular, obtusely angled, often with a dorsal groove; ventral wall usually planar without a median longitudinal ridge, sometimes with a pair of equatorial mammillae; wings thick, expanded, or essentially obsolete, bounded with a broad, cylindrical, usually grooved, marginal thickening, connivent basally and apically to an equal degree and presenting an orifice that is broader than tall; cotyledons incumbent, very conspicuously veined to the third order when expanded.

In usually heavily shaded, but not brushy habitats, Upper Sonoran and Humid Transition Zones; San Francisco Bay region of California from sea level to 4,000 feet, with occasional isolated stations to southern Washington and southern California. Type locality: "California." Collected by Douglas. April–May.

Plectritis macrocera subsp. **gràyii** (Suksd.) Morey, Contr. Dudley Herb. **5**: 120. 1959. (*Aligera grayi* Suksd. Deutsch. Bot. Monatss. **15**: 147. 1897; *A. mamillata* Suksd. loc. cit.; *Plectritis macroptera* Rydb. Fl. Rocky Mts. 818. 1917, not Suksd.) Differing from the typical subspecies in that the wings when expanded are thin and have a narrow, but usually obvious marginal thickening which is only rarely or inconspicuously grooved; wings are connivent top and bottom to present a large circular orifice; and there is usually a median longitudinal ridge on the ventral surface of the fruit which usually bears a multiseriate row of stiff, clavate or cylindrical bristles. In open, moderately shaded, but not brushy habitats in piedmont and moist, low, montane localities; north-central California northward to the Columbia River and its tributaries, eastward to southwestern Montana, and southward to southern California, Nevada, and Utah. Type locality: California.

3. **Plectritis ciliòsa** (Greene) Jepson. Long-spurred Plectritis. Fig. 5064.

Valerianella ciliosa Greene, Proc. Acad. Phila. **1895**: 548. 1896.
Aligera macroptera Suksd. Deutsch. Bot. Monatss. **15**: 146. 1897.
Aligera macrocera Suksd. op. cit. 147. Not Torr. & Gray.
Plectritis ciliosa Jepson, Man. Fl. Pl. Calif. 971. 1925.
Aligera californica Suksd. Werdenda **1**: 44. 1927.

Plants slender, 1.0–5.5 dm. tall. Leaves obovate to oblong-fusiform, petiolate below, sessile above; inflorescence capitate to verticillate-spiciform; corolla deep pink, with dark spots at the base of the ventral lobes, 5.5–8.5 mm. long, slender and narrowly infundibuliform, bilabiate, the spur slender, usually exceeding the ovary, the tip not conspicuously expanded; fruit pale straw-yellow to brown, 3–4 mm. long, more or less pubescent, the hairs stout, stiff, cylindrical, rounded or bluntly pointed, rarely somewhat attenuate, thick-walled and not readily collapsed on drying; fertile cell essentially trigonous, broader than deep; keel rounded, obtusely angled, with a deep dorsal groove and lateral, multiseriate, brush-like rows of hairs; ventral wall usually with a thin, prominent, longitudinal ridge, occasionally with the epidermis split from the underlying tissues and stretched from ridge to inner wing-margins, often with a pair of equatorial mammillae; wings variously expanded, rarely obsolescent, bounded with a cylindrical, usually grooved, marginal thickening, connivent basally and connivent to more or less spreading above; cotyledons incumbent, more or less conspicuously veined to the second and third order when expanded.

In open sunny habitats, grassy hillsides, and stream margins, Upper Sonoran and lower Humid Transition Zones; central California at higher altitudes in the interior from San Benito County north to Mendocino County and from Tuolumne County west to Monterey County; rarely elsewhere. Type locality: Oakville, Napa County. April–May.

Plectritis ciliosa subsp. **insígnis** (Suksd.) Morey, Contr. Dudley Herb. **5**: 121. 1959. (*Aligera insignis* Suksd. Deutsch. Bot. Monatss. **15**: 146. 1897; *A. rubens* Suksd. loc. cit.; *A. patelliformis* Suksd. W. Amer. Sci. **12**: 54. 1901; *Plectritis davyana* Jepson, Fl. W. Mid. Calif. 475. 1901; *Aligera intermedia* Suksd. Werdenda 1: 44. 1927; *A. glabrior* Suksd. loc. cit.; *A. barbata* Suksd. loc. cit.) Differing from the typical subspecies principally in the smaller flowers (1.5–3.5 mm. long). In hotter and drier situations, often in dappled shade in the southern part of the range; southern California and the inland coastal valleys north to Napa County, occasional to the Columbia Gorge, and in Arizona. Type locality: northern Lower California. Collected by Orcutt.

4. KENTRÁNTHUS Neck. Elem. 1: 122. 1790.

Perennial or annual, glabrous herbs with simple or pinnatifid leaves and large, terminal, thyrsiform inflorescences. Calyx inrolled in flower, becoming a large plumose pappus in fruit. Corolla-tube usually with a long spur, the limb somewhat zygomorphic. Stamen 1. Fruit somewhat compressed, crowned by the persistent plumose calyx. [Name from the Greek, meaning spurred flower.]

A genus of about 12 species in the Mediterranean region. Type species, *Valeriana rubra* L.

1. **Kentranthus rùber** (L.) DC. Jupiter's Beard or Red Valerian. Fig. 5065.

Kentranthus ruber DC. in Lam. & DC. Fl. Franc. ed. 3. 4: 238. 1805.

Glabrous and somewhat glaucous, perennial herb to 1 m. high. Leaves simple, entire or rarely toothed, ovate to lanceolate, to 10 cm. long, sessile; inflorescence a large paniculiform cyme elongating in fruit; calyx-limb with 5–15 linear plumose lobes 6–8 mm. long; corolla red, magenta, or white, with a prominent spur at the base, about 16 mm. long including the spur. *n* = 7.

A garden plant occasionally escaped from cultivation, in moist areas, particularly along the coast, in California. Jan.–Dec.

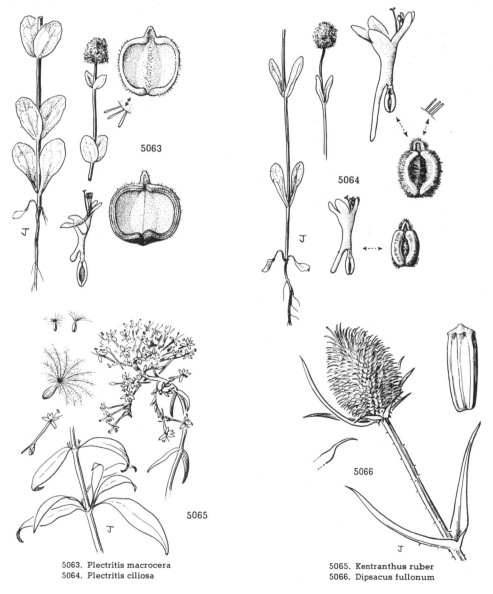

5063. Plectritis macrocera
5064. Plectritis ciliosa

5065. Kentranthus ruber
5066. Dipsacus fullonum

Family 146. DIPSACÀCEAE.

TEASEL FAMILY

Annual or perennial herbs with opposite or verticillate, exstipulate, simple or pinnatifid leaves. Flowers perfect, borne on an elongated or globose receptacle, bracted and involucrate. Calyx-tube adnate to the ovary, the limb cup-shaped or disk-shaped, or divided into spreading bristles. Corolla epigynous, the limb 4–5-lobed. Stamens 2–4, inserted on the tube of the corolla and alternate with its lobes; filaments distinct. Ovary inferior, 1-celled; style filiform; stigma undivided, terminal or lateral; ovule 1, pendulous. Fruit an achene, its apex crowned with the persistent calyx-lobes.

A family of about 10 genera and 180 species, natives of the Old World.

Bracts of the involucre prickly-pointed, some exceeding the heads. 1. *Dipsacus.*
Bracts of the involucre herbaceous, shorter than the heads. 2. *Scabiosa.*

1. DÍPSACUS [Tourn.] L. Sp. Pl. 97. 1753.

Tall, erect, rough-hairy or prickly, biennial or perennial herbs with usually large leaves and bluish or white flowers in dense, more or less elongated heads terminating stout peduncles. Bracts of the involucre and scales of the receptacle rigid or spiny-pointed. Flowers each surrounded by an involucel of 4 bractlets. Limb of calyx cup-shaped, 4-toothed or -lobed. Corolla oblique or 2-lipped, 4-lobed. Stamens 4. Stigma oblique or lateral. Achene free from or adnate to the involucre. [Greek name for the teasel.]

An Old World genus of about 15 species. Lectotype, *Dipsacus fullonum* L.

Bracts of the receptacle hooked at the apex; flowers white. 1. *D. fullonum.*
Bracts of the receptacle straight; flowers lavender. 2. *D. sylvestris.*

1. Dipsacus fullònum L. Fuller's Teasel. Fig. 5066.

Dipsacus fullonum L. Sp. Pl. 97. 1753.
Dipsacus fullonum β *sativus* L. loc. cit.
Dipsacus fullonum of Hudson, 1762; Miller, 1768.
Dipsacus sativus Honck. Vollst. Syst. Verz. 16. 1782.

Stout biennial, 1 m. or more high, with scattered short prickles on the stems, midribs of the leaves, and involucre, otherwise nearly or quite glabrous. Leaves sessile or the upper usually connate-perfoliate, lanceolate or oblong, entire, the upper acute, the lower obtuse and crenate; foliaceous bracts of the involucre spreading or reflexed, some shorter than the head; heads ovoid becoming cylindric, 6–10 cm. long; bracts of the receptacle with hooked tips, about equaling the flowers; corolla white or tinged with lavender, 8–12 mm. long; achenes 6– 8 mm. long, strongly 4-angled.

Waste places especially in low damp ground; frequent in central and southern California. Native of Europe. March–Oct.

2. Dipsacus sylvéstris Huds. Wild Teasel. Fig. 5067.

Dipsacus sylvestris Huds. Fl. Anglica 49. 1762.

Stout biennial, 1–2 m. high, with numerous short prickles on the stems especially below the inflorescence, the midrib of the leaves and the involucral bracts glabrous or nearly so. Lower leaves obtuse, crenate or sometimes pinnatifid, often 2–3 dm. long; upper stem-leaves sessile or the uppermost slightly connate, acuminate at apex, and usually entire; foliaceous bracts of the involucre as long as the heads or longer, linear and curved upward; bracts of the receptacle ovate, tipped by a long, straight, subulate, forked awn usually exceeding the flowers; corolla lavender, 9–12 mm. long; achene 6–8 mm. long, angled.

Waste places, common in Washington and Oregon and occurring in northwestern California as far south as San Francisco County; also eastern United States. Native of the Old World. June–Sept.

2. SCABIÒSA [Tourn.] L. Sp. Pl. 98. 1753.

Glabrous or pubescent herbs with leaves opposite and no prickles. Bracts of the involucre herbaceous, distinct or slightly united at the base. Receptacle pubescent, without scales. Involucels compressed, their margins often minutely 4-toothed. Calyx-limb 5–10-awned. Corolla-limb 4–5-cleft, oblique or 2-lipped. Stamens 4 or rarely 2. Stigma oblique or lateral. Achene more or less adnate to the involucel, crowned with the persistent calyx-limb. [Name Latin, meaning scale, being a reputed remedy for scaly eruptions.]

An Old World genus of about 75 species. Lectotype, *Scabiosa columbaria* L.

1. Scabiosa atropurpùrea L. Mourning Bride. Fig. 5068.

Scabiosa atropurpurea L. Sp. Pl. 100. 1753.

Stems branching, 6–10 dm. high, glabrous or sparsely and retrorsely hispidulous. Lowest leaves lyrate, the upper pinnately divided or the uppermost coarsely serrate, or the smaller ones narrower and entire; peduncles 2-3 dm. long; heads 3-4 cm. broad, the flowers very dark purple varying to pale lavender or white; corolla 10–12 mm. long; awns of fruiting calyx 6–10 mm. long, scabrous.

Escaped from gardens, and naturalized in the Sacramento Valley and San Francisco Bay region; also Santa Barbara, San Bernardino, and San Diego Counties, California. Native of Europe. May–Nov. Old Lady's Pin-cushion.

Family 147. **CUCURBITÀCEAE.**

Gourd Family.

Mostly herbaceous annual or perennial vines, trailing or climbing by means of tendrils, with alternate petioled leaves and solitary, racemose, paniculate or cymose, monoecious or dioecious flowers. Calyx-tube adnate to the ovary, its limb usually 5-lobed. Petals usually 5, separate or united, inserted on the limb of the calyx. Stamens mostly 5, the androecia highly modified throughout the family. Ovary 1–3-celled; style simple or lobed; ovules few or numerous. Fruit a pepo, indehiscent, or rarely dehiscent at the summit, or bursting irregularly. Seeds commonly flat; endosperms none.

Flowers yellow, large; pistillate and staminate flowers solitary.   1. *Cucurbita.*
Flowers white or greenish, small; staminate flowers racemose or paniculate.
    Fruit symmetrical, oblong, oval, or globose, 4–15 cm. long; ovary 2–4-celled.
        Seeds flat; roots slender and fibrous.   2. *Echinocystis.*
        Seeds turgid; roots very large and tuberous.   3. *Marah.*
    Fruit oblique, 1 cm. or less long; ovary 1-celled.   4. *Brandegea.*

1. **CUCÚRBITA** [Tourn.] L. Sp. Pl. 1010. 1753.

Rough prostrate vines, rooting at the nodes, with branched tendrils, usually lobed leaves which are often cordate at the base, and large, yellow, axillary, monoecious flowers. Calyx-tube campanulate, usually 5-lobed. Corolla campanulate, 5-lobed to about the middle, the lobes recurving. Staminate flowers with 3 stamens, the anthers linear, contorted, more or less united. Pistillate flowers with 1 pistil; ovary oblong, with 3–5 many-ovuled placentae; style short, thick; stigmas 3–5, each 2-lobed, papillose; staminodia 3. Fruit large, fleshy, with a thick rind, indehiscent; seeds horizontal, many. [Latin name for the gourd.]

An American genus of about 15 species, mainly tropical. Type species: *Cucurbita pepo* L.

Leaves triangular-ovate, longer than broad, scarcely lobed and often angled at base; tendrils long-petiolate.
                                                                  1. *C. foetidissima.*
Leaves conspicuously lobed or divided to the base, as broad as or broader than long; tendrils sessile or short-petiolate.
    Leaves 5-cleft to the base with narrowly lanceolate or linear lobes, these often sublobed toward the base.
                                                            2. *C. digitata.*
    Leaves 5-lobed to the middle or a little below, the lobes broadly lanceolate or triangular.   3. *C. palmata.*

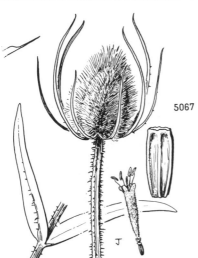

5067

5067. Dipsacus sylvestris

5068

5068. Scabiosa atropurpurea

1. Cucurbita foetidíssima H.B.K. Calabazilla or Mock-orange. Fig. 5069.

Cucurbita foetidissima H.B.K. Nov. Gen. & Sp. 2: 123. 1817.
Cucumis perennis James in Long, Exped. 2: 20. 1820.
Cucurbita perennis A. Gray, Bost. Journ. Nat. Hist. 6: 193. 1850.

Stems stout, rough, hirsute, trailing to a length of 2–5 m.; root very large, carrot-shaped. Leaves ovate-triangular, cordate or truncate at the base, acute at the apex, 1–2 dm. long, denticulate, usually slightly 3–5-lobed, rough above, canescent beneath, on stout petioles 8–15 cm. long; peduncles 2.5–5 cm. long; flowers mostly solitary; corolla 7–10 cm. long; pepo globose, 5–10 cm. in diameter, smooth.

Wastelands, in dry, usually sandy soil, Upper and Lower Sonoran Zones; San Joaquin Valley and adjacent foothills of the Sierra Nevada south to Lower California, Arizona, and Sonora; also Nebraska and Missouri south to Texas and northern Mexico. Type locality: Guanajuato, Mexico. April–Sept.

2. Cucurbita digitàta A. Gray. Finger-leaved Gourd. Fig. 5070.

Cucurbita digitata A. Gray, Smiths. Contr. 5⁶: 60. 1853.

Trailing vine with deep-seated, fleshy, fusiform root and slender stems pustulate on the angles. Tendrils sessile or nearly so, 3–5-parted, the divisions gland-tipped; leaves 8–12 cm. broad, more green and less scabrous-pubescent above than below except on midveins, pedately 5-lobed to the base or nearly so, the lobes 4.5–10 cm. long, linear-lanceolate, sometimes attenuate at apex, each division sublobed near the base or bearing 1–2 distinct lobes or teeth; petiole ribbed and scabrous, 4–7.5 cm. long; flowers sparingly pilose, 4–6 cm. long; staminate calyx cylindric or slightly flared below the short subulate lobes; pedicels enlarged at base of fruit; fruit depressed-globose, dark green, somewhat mottled, and marked with 10 narrow stripes; seeds thickened on the margin, ovate, 10–11 mm. long.

In sand, desert valleys and washes, Lower Sonoran Zone; Riverside and Imperial Counties, California, to New Mexico and northwestern Sonora. Type locality: New Mexico. Collected by Charles Wright. July–Oct.

3. Cucurbita palmàta S. Wats. Palmate-leaved Gourd or Coyote Melon. Fig. 5071.

Cucurbita palmata S. Wats. Proc. Amer. Acad. 11: 137. 1876.
Cucurbita californica Torr. ex S. Wats. op. cit. 138.

Gray trailing vine from a deep-seated, large, sometimes divided root with stems strongly ribbed and muriculate on the angles. Tendrils sessile or subsessile, weak, short, mostly 3-divided; leaf-blades more or less cordate at base, thick, gray, palmately 3–5-parted, the two basal lobes shorter than the upper three, these triangular or lanceolate, sometimes sublobed, 3–10(15) cm. long and about as wide, pustulate and densely appressed-hispid above especially along the veins, less so beneath; petioles stout, muricate, shorter than the blades; flowers 6–8 cm. long; staminate calyx 2–2.5 cm. long, narrowly campanulate, with short subulate lobes; fruit globose, mottled, and marked with 10 faint lines, 8–9 cm. broad and long; seeds 10–14 mm. long, without a raised margin.

In arid situations, Lower Sonoran Zone; lower Sacramento Valley southward through the San Joaquin Valley, California, to northern Lower California and eastward in the Mojave and Colorado Deserts to Mohave and Yuma Counties, Arizona. Type locality: Cajon Valley, San Diego County, California. July–Sept.
Cucurbita californica Torr. ex S. Wats., which was collected by Pickering on the Wilkes Expedition, presumably in the Sacramento Valley, is still imperfectly known as to floral and fruit characters. As interpreted by recent authors, it differs from *C. palmata* by having the upper lobe of the leaf triangular and less than twice as long as broad and is found in the Sacramento and San Joaquin Valleys south to San Diego County, California, east to the Death Valley region and adjacent Nevada and Arizona, and in the Colorado Desert, California.

2. ECHINOCÝSTIS (Michx.) Torr. & Gray, Fl. N. Amer. 1: 542. 1840.

Climbing herbs with branched tendrils and slender stems. Leaves alternate, palmately lobed or angled. Flowers monoecious. Staminate inflorescence paniculate. Pistillate flowers usually solitary. Stamens 3; anthers more or less coalesced. Ovary 2-celled; ovules 2 in each cell. Fruit fleshy, becoming dry at maturity and dehiscing from the summit. Seeds flat, dark-colored. [Name Greek, meaning hedgehog and bladder.]

A monotypic genus of eastern and central United States and adjacent Canada, occurring sporadically in western United States. Type species, *Sicyos lobata* Michx.

1. Echinocystis lobàta (Michx.) Torr. & Gray. Wild Balsam-apple. Fig. 5072.

Momordica echinata Muhl. Trans. Amer. Phil. Soc. 3: 180. 1793. (Nomen nudum)
Sicyos lobata Michx. Fl. Bor. Amer. 2: 217. 1803.
Echinocystis lobata Torr. & Gray, Fl. N. Amer. 1: 542. 1840.
Micrampelis lobata Greene, Pittonia 2: 128. 1890.

Plants nearly glabrous, roots slender, fibrous, stems slender, climbing, 3–8 m. long. Leaves 7–8 cm. long including petiole, the blade with broad open sinus, 5-(rarely 7-)lobed to about the middle, the lobes narrowly triangular, acute or acuminate, sparsely serrulate on the margin; staminate inflorescence axillary, narrowly paniculate, densely flowered, the corolla-lobes rotately spreading, slender, attenuate; pistillate flowers similar, solitary, or occasionally more than one in the leaf-axils; fruit about 5 cm. long, ovoid, clothed with weak prickles; seeds 4, with rough thickish coat.

Thickets and waste places, eastern United States and Canada westward to Montana but evidently escaping from cultivation and appearing sporadically in the Willamette Valley, Oregon, and eastern Washington and western Idaho. Type locality: Pennsylvania. Aug.–Sept.

3. **MÁRAH** Kell. Proc. Calif. Acad. **1** : 38. 1855.

Climbing or trailing, herbaceous, monoecious vines from a deep-seated, greatly enlarged, tuberous root with stems arising annually, these reaching several meters in length. Tendrils usually branched. Leaves palmately angled, lobed or divided. Staminate flowers axillary, in racemes or panicles. Pistillate flowers solitary, axillary, staminodea sometimes present. Calyx vestigial or obsolete. Corolla campanulate or subrotate, 5–6-parted, the lobes often papillate on the inner surface. Stamens 3, the anthers more or less coherent. Pistillate flowers somewhat larger than the staminate, the ovary 2–4-celled, tapering, or abruptly contracted into a beak. Style very short, the stigma hemispheric or lobed. Fruit dry at maturity, 2–4-celled, fibrous within, dehiscing irregularly from the apical portion. Seeds turgid, often margined with a darker line ; germination hypogeous. [Name Hebrew, meaning bitter.]

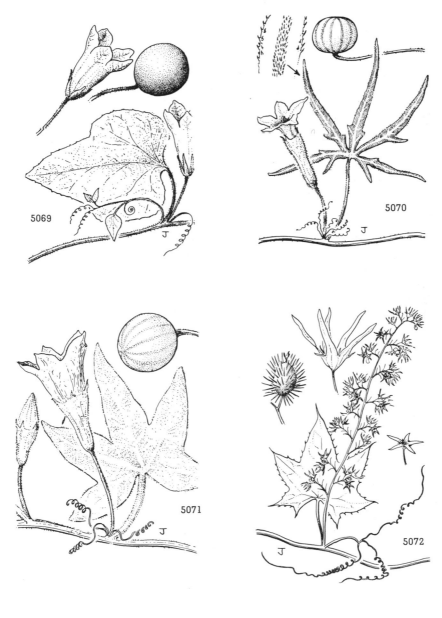

5069. Cucurbita foetidissima
5070. Cucurbita digitata

5071. Cucurbita palmata
5072. Echinocystis lobata

An American genus of about 10 species, confined to western North America and adjacent Mexico. Type species, *Marah muricatus* Kell.

Mature fruit completely smooth or bearing scattered weak, fleshy spines; seeds spherical or orbicular in outline and strongly compressed.
    Ovary of pistillate flowers tapering to a beak; fruit ovoid, usually attenuate at the apex.
                                                                                        1. *M. oreganus.*
    Ovary of pistillate flowers spherical with the beak arising abruptly from the ovary; fruit spherical.
                                                                                        2. *M. watsonii.*
Mature fruit with crowded, slender, acicular prickles or stout spines; seeds neither spherical nor orbicular in outline and strongly compressed (compressed in *M. guadalupensis*).
    Staminate flowers markedly campanulate, the lobes erect and recurved distally; fruit 10–15 cm. long.
                                                                                        3. *M. horridus.*
    Staminate flowers with rotate-spreading lobes, the tube shallowly basin-shaped; fruit 2.5–10 cm. long.
        Mature fruit ovoid to oblong.
            Fruit 8–14-seeded, the seeds not over 2 cm. long, not compressed; leaves usually lobed to below the middle.   4. *M. macrocarpus.*
            Fruit 2–4-seeded, the seeds about 3 cm. long, compressed; leaves mostly not lobed to the middle.
                                                                                        5. *M. guadalupensis.*
        Mature fruit globose, except for the short beak.
            Staminate inflorescence mostly few-flowered, the flowers bright white; leaves mostly palmately parted to the middle.   6. *M. inermis.*
            Staminate inflorescence many-flowered, the flowers greenish; leaves typically with shallow, broad, triangular lobes.   7. *M. fabaceus.*

1. **Marah oregànus** (Torr. & Gray) Howell. Western Wild Cucumber or Coast Man-root. Fig. 5073.

Sicyos angulatus Hook. Fl. Bor. Amer. **1**: 220. 1833. Not L.
Sicyos oreganus Torr. & Gray, Fl. N. Amer. **1**: 542. 1840.
Marah muricatus Kell. Proc. Calif. Acad. **1**: 38. 1855. (Nomen confusum, not *Echinocystis muricatus* Kell.)
Megarrhiza oregana Torr. Pacif. R. Rep. **6**: 74. 1858. (Nomen nudum)
Megarrhiza marah S. Wats. Proc. Amer. Acad. **11**: 138. 1876.
Echinocystis oregana Cogn. Mem. Acad. Sci. Belg. **28**: 87. 1878.
Marah oreganus Howell. Fl. N.W. Amer. 239. 1898.

Stems glabrous or somewhat scabrous, reaching several meters in length. Leaves suborbicular to rounded-cordate, uusally scabrous above and often pubescent beneath especially along the veins, 8–20 cm. wide, mostly longer than broad, 5–7-angled or some leaves 5–8-lobed to about the middle, the lobes sometimes dentate on the margin, basal sinus deep and usually closed, blade much exceeding the petiole in length; staminate flowers always in long slender racemes, including the peduncle 20–35 cm. long, the pedicels slender, 5–20 mm. long; corolla white, 8–15 mm. broad, the corolla-tube basin-shaped, the lanceolate-oblong lobes about equaling or a little shorter than the tube, rotately spreading and the margin sometimes recurved; pistillate flowers white, the lobes rotately spreading, the pedicels 6–20 mm. long; ovary ovoid, tapering to a glabrous beak, the young spines sometimes bearing soft septate hairs; fruit ovoid, glabrous, 4–7 cm. long, rounded and sometimes somewhat narrowed at the base, attenuate and prominently beaked at the apex, completely glabrous or sparsely muricate with weak fleshy spines up to 1 cm. in length; seeds 5–6, orbicular in outline, decidedly flattened and rather undulate, 18–22 mm. wide, 4–7 mm. thick, often dark brown.

    Open hillsides and edges of woods, mainly Humid Transition Zone; southern British Columbia southward west of the Cascade Mountains in Washington and Oregon along the Coast Ranges in California to San Mateo County; also east of the Cascade Mountains through the Columbia River gap. Type locality: "On the Oregon [Columbia River] from near its mouth to Kettle Falls." April–Aug.

2. **Marah watsònii** (Cogn.) Greene. Taw Man-root. Fig. 5074.

Echinocystis muricatus Kell. Proc. Calif. Acad. **1**: 57. 1855. Not *Marah muricatus* Kell.
Megarrhiza muricata S. Wats. Proc. Amer. Acad. **11**: 139. 1876.
Echinocystis watsonii Cogn. in DC. Monogr. Phan. **3**: 819. 1881.
Marah watsonii Greene, Leaflets Bot. Obs. **2**: 36. 1910.

Stems slender, 2–3 m. long, climbing by simple or branched tendrils. Leaves glabrous, paler beneath, suborbicular to reniform in outline, broader than long, the blades 7–11 cm. wide, the basal sinus open, rounded, the petioles shorter than or equaling the blades in length, deeply 3–5-lobed, sometimes to within 1 cm. of the base of the blade, the sinuses between the lobes acute or rounded; lobes of the leaf narrowed below, widened above and 3-angled or -lobed, the secondary lobes sometimes angled; staminate inflorescence racemose, 4–20 cm. long, few-flowered (2–10), the flowers campanulate, white, 4.5–6.5 mm. long, the tube rounded at base, much surpassing the obtuse or triangular, spreading, papillate lobes; pistillate flowers axillary, the pedicel 2–2.5 cm. long, similar to the staminate flowers or a little larger, the ovary spherical, the beak about 4 mm. long, arising abruptly from the ovary; fruit globose or depressed-globose, smooth or sparsely muricate with broad-based weak spines, pale green and marked with darker veins; seeds 2 or 4, spherical, 11–14 mm. in diameter, not at all compressed.

    Climbing over shrubs in the digger pine belt, Upper Sonoran Zone; Shasta County south in the Inner Coast Ranges to Solano County and in the Sierra Nevada to Mariposa County, California. Type locality: Placerville, Eldorado County. March–April.

3. **Marah hórridus** (Congdon) Dunn. Sierra Man-root. Fig. 5075.

Echinocystis horrida Congdon, Erythea **7**: 184. 1900.
Marah horridus Dunn, Kew Bull. **1913**: 151. 1913.

Stems striate, glabrous, long, and branching. Leaves glabrous or scaberulous, 4–12 cm. long, the width greater than the length, the sinus broadly rounded at the base of the blade, open or closed, the petiole equaling or a little longer than the blade, this 3–7-lobed, usually 5, cleft to well below the middle, to a lesser degree on leaves late in the season, the sinuses between the lobes acute; lobes of the leaf 1–2 cm. broad at base, narrowed to an acute, bristle-tipped apex, the margins of the lobes with 1–4 broad, triangular, bristle-tipped lobes on each side, sometimes merely sinuate-toothed; staminate inflorescence 10–25 cm. long, racemose, rarely branched, loosely flowered, the flowering portion shorter than the peduncle; staminate flowers white, 1–1.5 cm. long, the tube 5–8 mm. long, markedly campanulate with a truncate base, the lobes revolute, oblong, the pedicels as long as or longer than the flowers; pistillate flower axillary, tube more shallow than the staminate, the lobes spreading; ovary pubescent, narrowly elliptic, 5–7 mm. long; fruit 10–16(20) cm. long, usually 4-celled, the surface strongly echinate with many long spines of unequal length, these stout or slender on the same fruit, and sometimes pubescent with soft hairs even in age; seeds 8–10, oblong, 26–32 mm. long, 15–18 mm. wide, one end slightly compressed, the other terete.

Climbing over shrubbery in the digger pine belt and lower edge of the ponderosa pines, Upper Sonoran Zone; western slopes of the Sierra Nevada from Tuolumne County to Kern County, California. Type locality: Mariposa County. Collected by Congdon. March–April.

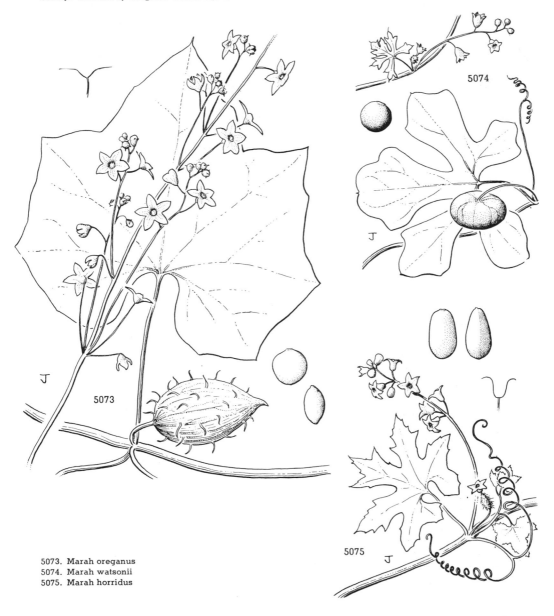

5073. Marah oreganus
5074. Marah watsonii
5075. Marah horridus

4. Marah macrocárpus (Greene) Greene. Chilicothe. Fig. 5076.

Echinocystis macrocarpa Greene, Bull. Calif. Acad. **1**: 188. 1885.
Micrampelis leptocarpa Greene, Pittonia **2**: 282. 1892.
Marah macrocarpa Greene, Leaflets Bot. Obs. **2**: 36. 1910.

Stems reaching a few meters in length, striate, glabrous or slightly scaberulous or the new growth pubescent with septate hairs and becoming glabrate. Leaves broader than long, 5–9(11) cm. broad, the sinus obtuse, usually open, typically 5–7(8)-lobed to the middle or below, rounded between. the lobes, these 7–15 mm. broad, broadly acute or obtuse or sometimes rounded at the apex and usually mucronulate, the margins entire or shallowly lobed or toothed; staminate inflorescence with flowers more or less densely arranged, pubescent, paniculate or subpaniculate, 4–12 cm. long, in new growth up to 20 cm. with elongate branches; staminate flowers white, 7–13 mm. broad, the lobes triangular-acute, rotately spreading; pistillate flowers white, 16–22 mm. broad, the ovary ovoid, sometimes narrowly so, and beaked, the ovary and beak pubescent, the pedicel 1–2 cm. long; fruit 6–10 cm. long, ovoid or oblong, densely echinate with basally broadened spines of different lengths, some up to 25 mm. long, vestiges of pubescence remaining on body of fruit and bases of spines; seeds 6–14(24), obovoid, 13–20 mm. long, 8–10 mm. wide.

Hillsides and canyons, mainly Upper Sonoran Zone; from Santa Barbara County south to San Diego County, California, and Lower California, including the islands off the coast, where it occurs with *M. guadalupensis*, eastward to the desert slopes of the mountains of southern California. Type locality: Cuacamonga, San Bernardino County, California. Feb.–April.

Specimens from the islands off Los Angeles and San Diego Counties appear to be *M. macrocarpus* var. *micranthus* Stocking, Madroño **13**: 134. 1955 (*M. micranthus* Dunn, Kew Bull. **1913**: 150. 1913), a species with small flowers and seeds 10–13 mm. long, originally described from Cedros Island, Lower California.

5. Marah guadalupénsis (S. Wats.) Greene. Island Man-root. Fig. 5077.

Megarrhiza guadalupensis S. Wats. Proc. Amer. Acad. **11**: 115, 138. 1876.
Marah guadalupensis Greene, Leaflets Bot. Obs. **2**: 36. 1910.
Marah major Dunn, Kew Bull. **1913**: 151. 1913.
Marah macrocarpus var. *major* Stocking, Madroño **13**: 134. 1955.

Stems striate, glabrous or nearly so, reaching several meters in length. Leaves orbicular in outline, more or less scaberulous above, the new growth sparsely pubescent and becoming glabrate, 12–32 cm. broad, the sinus obtuse or rounded, usually closed in the larger leaves, mostly 5-lobed (occasionally 3- or 7-) to above or near the middle, the lobes broadly triangular, mucronulate, the margins usually triangular-toothed, sometimes sinuate-lobed, more rarely entire; staminate inflorescence loosely flowered, pubescent, often becoming glabrate, racemose or paniculate, 2.5–4 cm. long, often shorter late in the season; staminate flowers white, pubescent without, 18–30 mm. wide, the lobes broadly or narrowly lanceolate, the pedicels 10–20 mm. long; pistillate flowers white, equaling the staminate in size, pubescent without, ovary and beak pubescent; fruit ovoid, 5–7 cm. long, echinate with stout spines of varying lengths, the longest about 1.5 cm. in length, the body of the fruit and spines for the most part pubescent; seeds 2–4, oval and somewhat compressed, 28–34 mm. long, 22–28 mm. wide.

Moist ravines and canyons, Upper Sonoran Zone; islands off the coast of southern California from Santa Barbara County south to Guadalupe Island, Lower California; more abundant on Santa Catalina, San Clemente, and San Nicolas Islands. Type locality: Guadalupe Island. Feb.–Aug.

6. Marah inérmis (Congdon) Dunn. Foothill Man-root. Fig. 5078.

Echinocystis inermis Congdon, Zoe **5**: 134. 1901.
Echinocystis scabrida Eastw. Bull. Torrey Club **30**: 500. 1903.
Marah inermis Dunn, Kew Bull. **1913**: 153. 1913.
Echinocystis fabacea var. *inermis* Jepson, Fl. Calif. **2**: 554. 1936.

Stems striate, glabrous, long, and branching, young growth beset with soft septate hairs which persist to some extent about the inflorescence and fruit. Leaves glabrous or scaberulous, 5–10(13) cm. broad, the width greater than the length, 5–7-lobed to about the middle, the lower pair of lobes less deeply divided, the angle between the lobes typically narrow, the margin of the lobes entire or more commonly toothed, the apex attenuate and usually bristle-tipped; staminate inflorescence long-pedunculate, 10–30 cm. long, narrowly paniculate at the base, sometimes racemose on shorter inflorescences, 11–35-flowered; staminate flowers bright white, 1–1.5(2) cm. broad, the broadly lanceolate corolla-lobes longer than the shallow, basin-shaped tube, rotately spreading and plane, papillose on the inner surfaces; pistillate flowers similar, somewhat larger, the pedicel 1–3 cm. long; ovary globose, beak abruptly attenuate, the young prickles and the beak with soft hairs; mature fruit globose, 2–4-seeded but mostly 4, a short, cone-shaped beak sometimes evident in the growing fruit, 2.5–4 cm. broad, covered with acicular prickles 5–9 mm. long, these becoming glabrate in age; seeds about 1.5 cm. long, oblong-ovoid, scarcely compressed.

Climbing on shrubs or often prostrate on grassy slopes and in fields, Upper and Lower Sonoran Zones; lower foothills of the Sierra Nevada from Butte County south to Kern County and the Inner Coast Ranges from Yolo County to Kern County, California. Type locality: Sherlocks, along the canyon of the Merced River, Mariposa County. March–May.

Close to *M. fabaceus* var. *agrestis* (Greene) Stocking, and the fruit of the two scarcely distinguishable. The ranges of the two forms overlap. The white flowers and usually more deeply lobed leaves distinguish it from *M. fabaceus* var. *agrestis,* and the vesture of the fruit as well as the white flowers and more deeply lobed leaves distinguish it from *M. fabaceus* var. *fabaceus.*

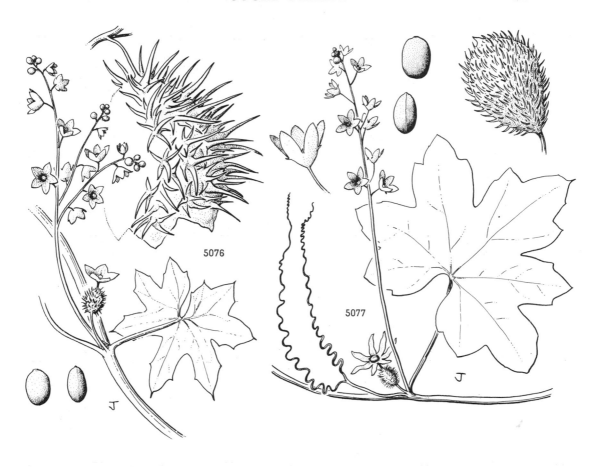

5076. Marah macrocarpus

5077. Marah guadalupensis

7. Marah fabàceus (Naudin) Greene. California or Valley Man-root. Fig. 5079.

Megarrhiza californica Torr. Pacif. R. Rep. **6**: 74. 1858. (Nomen nudum)
Echinocystis fabacea Naudin, Ann. Sci. Nat. IV. **12**: 154. *pl. 9*. 1859.
Megarrhiza californica Torr. ex S. Wats. Proc. Amer. Acad. **11**: 138. 1876.
Marah fabacea Greene, Leaflets Bot. Obs. **2**: 36. 1910.

Stems reaching several meters in length, nearly glabrous to densely scabrous, the young growth densely beset with soft septate hairs which are more or less deciduous in age except about the inflorescence and leaf-bases. Leaves 4–8(10) cm. broad, about as broad as long, typically 5–7-angled, ivy-like in form, or 5–7-lobed to above the middle, the lobes entire or sometimes dentate; staminate inflorescence many-flowered, paniculate, 8–25 cm. long, the panicle-branches on early growth long and ascending, much shorter and sometimes lacking on later growth, the flowers green or greenish-white, the pedicels 10 mm. or less long; staminate corollas 6–8 mm. broad, corolla-tube very shallowly basin-shaped, margin of the corolla-lobes recurved, the rotately spreading lobes thus appearing linear-acute; pistillate flowers similar but larger; ovary clothed with septate hairs, these on the spines as well as surface of the ovary, the beak also pubescent, arising abruptly from the ovary; fruit globose, with stout, stiff, usually pubescent spines 6–12 mm. in length; seeds 2–4, oblong-ovoid, bordered lengthwise with a shallow groove or line, somewhat flattened at one end.

Slopes and washes, Upper Sonoran Zone; Outer Coast Ranges from Northern Sonoma County to Santa Barbara County, California; also reported from the Channel Islands. Type locality: described from plants cultivated at Paris, the material presumably collected from the San Francisco Bay region. Feb.–May.

Marah fabaceus var. **agréstis** (Greene) Stocking, Madroño **13**: 130. 1955. (*Micrampelis fabaceus* var. *agrestis* Greene, Fl. Fran. 236. 1891; *Echinocystis fabaceus* var. *agrestis* Congdon, Zoe **5**: 133. 1901.) Leaves as in the species, mostly 3.5–5 cm. broad; flowers greenish; fruit 1.5–2.5 cm. broad, globose, the spines slender and acicular, 2–5 mm. long; seeds 1–4 developing, usually 2. Lower slopes of Inner Coast Ranges and the adjacent valley from Glenn County to San Luis Obispo County and apparently less common in the Sierra Nevada foothills from Butte County to Fresno County, and also on the coastal slopes in Santa Barbara County, California. Without mature fruit not distinguishable from the species. Type locality: "grain field of the valley of the San Joaquin."

4. BRANDÈGEA Cogn. Proc. Calif. Acad. II. **3**: 58. 1890.

Perennial monoecious vines with simple tendrils. Leaves 3–5-parted or -angled. Staminate flowers in racemes; pistillate flowers solitary, often axillary to the staminate racemes. Style short; stigma hemispheric. Fruit asymmetrical, long-beaked, tardily and irregularly dehiscent, 1-seeded. Seed irregularly obovate, warty or echinate. [Named in honor of T. S. Brandegee, a pioneer western botanist.]

A monotypic genus of the Sonoran Desert region. Type species, *Elaterium bigelovii* S. Wats.

1. Brandegea bigelòvii (S. Wats.) Cogn. Brandegea. Fig. 5080.

Elaterium bigelovii S. Wats. Proc. Amer. Acad. **12**: 252. 1877.
Echinocystis parviflora S. Wats. op. cit. **17**: 373. 1882.
Cyclanthera monosperma Brandg. Proc. Calif. Acad. II. **2**: 159. 1889.
Brandegea bigelovii Cogn. op. cit. **3**: 58. 1890.
Brandegea monosperma Cogn. op. cit. 59.

Slender-stemmed perennial vines of dense growth from large thick roots. Leaves variable, hastate to quadrate in outline with deep narrow sinus, 3–5-cleft or merely angled, 1–5.5 cm. long, the upper lobe the longer, sharply triangular-acute, smaller leaves often cleft nearly to the base with narrow, linear-acute divisions, the upper surface conspicuously pustulate; petioles shorter than the leaves; pistillate and staminate flowers similar, minute, whitish, rotate and deeply lobed, about 1.5 mm. broad; fruit straw-colored, 8–12 mm. long including the beak, body of fruit more or less echinate, about equaling the beak in length; seeds about 5 mm. long.

Sandy arroyos and canyons, Lower Sonoran Zone; Colorado Desert, California, east to the western parts of Maricopa, Pima, and Yuma Counties, Arizona, south to Lower California and Sonora. Type locality: "the Lower Colorado Valley." March–April.

Family 148. CAMPANULÀCEAE.

BLUEBELL FAMILY.

Herbs, shrubs, or trees with alternate, simple (rarely pinnatisect), exstipulate leaves. Flowers perfect, epigynous. Calyx 5-parted. Corolla gamopetalous, regular, or zygomorphic. Stamens free or united. Style 1. Ovary 2–10-celled, rarely 1-celled, the ovules on axile placentae. Fruit a capsule or rarely a berry; seeds numerous; embryo straight.

A family of about 60 genera and 1,600 species, widely distributed in subtropical and temperate regions.

Corolla regular; anthers and filaments free.   Subfamily 1. *CAMPANULOIDEAE.*
Corolla irregular; anthers and filaments united into a tube (anthers free in *Nemacladus* and *Parishella*).
         Subfamily 2. *LOBELIOIDEAE.*

Subfamily 1. CAMPANULOÌDEAE.

BELLFLOWER SUBFAMILY.

Ours herbs, with acrid or milky juice. Flowers racemose, spicate, paniculate, or solitary, perfect. Calyx-tube adnate to the ovary, its limb mostly 5-lobed or 5-parted, the lobes equal or slightly unequal, valvate or imbricate in the bud, commonly persistent. Corolla gamopetalous, regular, usually 5-lobed or -parted, blue or white, inserted on the calyx-tube where that becomes free from the ovary. Stamens 5, inserted with the corolla, alternate with the corolla-lobes; anthers 2-celled, introrse, separate. Ovary usually 2–5-celled. Style 1; stigmas 2–5, usually distinct. Fruit a capsule or berry, usually opening by clefts or pores; seeds minute and numerous.

Capsule dehiscent by small valves or circular perforations below the calyx-lobes.
    Plants (in the Pacific States) perennial (except *C. exigua* and *C. angustiflora*); earliest flowers not cleistogamous.
       Corolla shallowly lobed above or to the middle, rarely a little below, the lobes triangular to oblong-lanceolate.   1. *Campanula.*
       Corolla deeply divided, the lobes linear-lanceolate.   3. *Asyneuma.*
    Plants annual; earliest flowers on the lower part of the stem cleistogamous.
       Capsule oblong-cylindric to narrowly turbinate; corolla divided below the middle.   2. *Triodanus.*
       Capsule short and broad, nearly hemispheric; corolla not divided to the middle.   4. *Heterocodon.*
Capsule dehiscent at the apex of the calyx-limb within the lobes.   5. *Githopsis.*

1. CAMPÁNULA [Tourn.] L. Sp. Pl. 163. 1753.

Perennial or annual herbs with alternate or basal leaves. Flowers large or small, solitary, racemose or paniculate, regular, blue, violet or white. Calyx-tube adnate to the ovary, hemispheric to turbinate, prismatic, the limb 5-lobed or -parted or rarely 3–4-parted. Corolla campanulate or rotate, more or less deeply 5-lobed. Stamens 5, free from the corolla; filaments usually dilated at base; anthers distinct. Ovary inferior, 3–5-celled; stigma 3–5-lobed. Capsule partly or wholly inferior, crowned by the persistent calyx-lobes, opening by 3–5 small valves or perforations situated on the sides near the top, the middle, or the base. [Name diminutive of *campana*, the Latin word for bell.]

A genus of about 230 species, mostly in the boreal and north temperate regions of both the Old and New Worlds. Type species, *Campanula latifolia* L.

Perennials.
 Style longer than the corolla; reflexed lobes of the corolla nearly equaling the tube.
                       1. *C. scouleri.*
 Style shorter than to about the length of the corolla; corolla-lobes shorter than the tube, erect or nearly so.
  Leaves crenate or crenate-serrate, the stems and leaves sparsely retrorse-ciliate; weak-stemmed, swamp
   or bog plants.   2. *C. californica.*
  Leaves entire or toothed, not retrorse-ciliate; not weak-stemmed, swamp or bog plants.
   Leaves dimorphic, the upper linear and sessile, the basal orbicular to ovat and petiolate.
                    3. *C. rotundifolia.*
   Leaves essentially alike.
    Leaves coarsely few-toothed.
     Calyx-lobes entire; plants completely glabrous; natives of northern California.
                  4. *C. wilkinsiana.*
     Calyx-lobes usually toothed; plants variously pubescent, especially around the inflorescence
      and leaf-bases; plants of Washington and northward.
      Calyx-tube more or less pubescent with long hairs; corolla 15–25 mm. long.
                  5. *C. lasiocarpa.*
      Calyx and pedicels with rather sparse bristly puberulence; corolla 10–12 mm. long.
                  6. *C. piperi.*
   Leaves and calyx-lobes entire.
    Plants uniformly short-bristly pubescent throughout.   7. *C. scabrella.*
    Plants nearly glabrous but some or all of the lower leaf-bases long-ciliate; calyx-tube usu-
     ally somewhat puberulent.   8. *C. parryi idahoensis.*
Annuals.
 Corolla twice the length of the calyx-lobes; plants branching above the base. 9. *C. exigua.*
 Corolla about the length of the calyx-lobes; plants branching from the base. 10. *C. angustiflora.*

5078. Marah inermis   5079. Marah fabaceus

1. Campanula scoùleri Hook. Scouler's Harebell or Campanula. Fig. 5081.

Campanula scouleri Hook. ex A. DC. Monog. Camp. 312. 1830.
Campanula scouleri α hirsutula Hook. Fl. Bor. Amer. **2**: 28. *pl. 125.* 1834.
Campanula scouleri β glabra Hook. loc. cit.

Stems from slender rootstocks, erect or decumbent at base, 1–3 dm. high, slender, simple or rarely branched below, herbage glabrous throughout or nearly so. Leaf-blades broadly ovate to lanceolate, 1.5–4.5 cm. long, sharply serrate, often remotely so, acute or acuminate at apex, narrowed at base to a margined petiole often as long as the blade; flowers nodding, in a terminal, usually few-flowered, raceme-like panicle; the pedicels very slender, 0.5–4 cm. long, bearing a subulate bractlet near the middle; calyx-tube campanulate, 2–3 mm. long, the lobes subulate, about twice as long, erect or spreading; corolla light blue to white, the lobes reflexed, as long as the tube, acute; capsule broadly ovoid.

Open woods and slopes, mainly Humid Transition Zone; southern Alaska southward west of the Cascade Mountains to Humboldt County in the Coast Range, California, and also to Sierra County in the Sierra Nevada. Type locality: Fort Vancouver on the Columbia River, Washington. June–Aug.

2. Campanula califórnica (Kell.) Heller. Swamp Harebell. Fig. 5082.

Wahlenbergia californica Kell. Proc. Calif. Acad. **2**: 158. *fig. 49.* 1862.
Campanula linnaeifolia A. Gray, Proc. Amer. Acad. **7**: 366. 1868.
Campanula californica Heller, Muhlenbergia **1**: 46. 1904.

Perennial, from slender creeping rhizomes, with slender weak stems simple or sparingly branched above, 1–4 dm. long, retrorsely hispid on the angles or ribs. Leaves oblong-ovate to obovate, sessile or subsessile, narrowed below to a petiolar base, commonly broadly and shallowly crenate-serrate above the middle, entire below, 0.5–2 cm. long, the margins more or less retrorsely hispid; flowers solitary, terminating the branches or on slender, elongated, axillary peduncles; calyx-tube turbinate-campanulate, the lobes subulate, 4–5 mm. long; corolla pale blue, 10–15 mm. long; style included, 6–7 mm. long; capsule broadly obovoid to globular, 10-ribbed with thin-walled interspaces.

Swamps and bogs along the coast, Humid Transition Zone; Mendocino County to Point Reyes, Marin County, California. Type locality: "coast range at some point north of San Francisco." June–Sept.

3. Campanula rotundifòlia L. Scotch Bluebell. Fig. 5083.

Campanula rotundifolia L. Sp. Pl. 163. 1753.
Campanula petiolata A. DC. Monog. Camp. 278. 1830.
Campanula sacajaweana M. E. Peck, Proc. Biol. Soc. Wash. **50**: 123. 1937.

Perennial, with few to many underground stems arising from root-crown with slender simple stems 1–10 dm. high, glabrous throughout or more rarely sparsely puberulent. Basal leaves often from offshoots, less commonly appearing on the flowering stems, soon withering, on petioles much surpassing the blade, ovate to nearly orbicular, cordate to subcordate at base, irregularly sinuate-dentate to sinuate; lower stem-leaves short-petioled, spatulate, oblanceolate or lanceolate and often with a sinuate margin, merging into the sessile, entire, linear or linear-filiform upper leaves; flowers in loose panicles with slender branches, sometimes solitary; calyx-lobes narrowly subulate to filiform, spreading, longer than the turbinate tube; corolla violet-blue, campanulate, 1.5–2.5(3) cm. long, the tube much longer than the lobes; capsule nodding, opening by valves near the base.

Meadows or rocky slopes, Transition and Boreal Zones; Alaska to the mountains of northern California and eastward across the continent; also northern Europe and Asia. Type locality; Europe. July–Sept.

A highly variable and widespread polymorphic species with many segregates. High-mountain forms are low in stature, have lanceolate leaves, and are often 1-flowered. The taller plants of lower elevations are usually characterized by more flowers in the panicles and by linear to linear-filiform leaves. Some large-flowered forms along the Washington coast and in the Columbia Gorge with long, filiform, divergent calyx-lobes resemble the Alaskan form of the species.

4. Campanula wilkinsiàna Greene. Wilkins' Harebell. Fig. 5084.

Campanula wilkinsiana Greene, Pittonia **4**: 38. 1899.
Campanula baileyi Eastw. Bull. Torrey Club **29**: 525. 1902.

Plants from slender creeping rootstocks, the leafy stems slender, 3–14 cm. high, glabrous throughout. Leaves sessile, 8–15 mm. long, cuneate to obovate, coarsely few-toothed on the upper part of leaf, the upper leaves oblanceolate and somewhat reduced in size; flowers mostly solitary, terminating the stems; calyx-tube obconic, the lobes entire, erect, triangular-subulate, about equaling to a little longer than the tube; corolla blue, 12–14 mm. long, cleft to about the middle or a little below, the lobes ovate-lanceolate and spreading; style about the length of the corolla; capsule turbinate, about 7 mm. long, the calyx-lobes shorter than the capsule and erect.

In meadows and along streambanks, Canadian and Hudsonian Zones; Mount Shasta, Siskiyou County, and the Trinity Mountains, Trinity County, California. Type locality: head of Squaw Creek, Mount Shasta. Aug.

5. Campanula lasiocárpa Cham. Common Alaska Bluebell or Harebell. Fig. 5085.

Campanula lasiocarpa Cham. Linnaea **4**: 39. 1829.

Perennial, the stems 4–18 cm. high, arising from a creeping rootstock. Leaves 2–7 cm. long, spatulate to narrowly oblanceolate, tapering to the base to a winged petiole, the uppermost sessile and bract-like, the margin laciniate-denticulate to laciniate with rather few teeth, glabrous except for marginal white hairs along the petioles, these more abundant on the bracts and upper leaves;

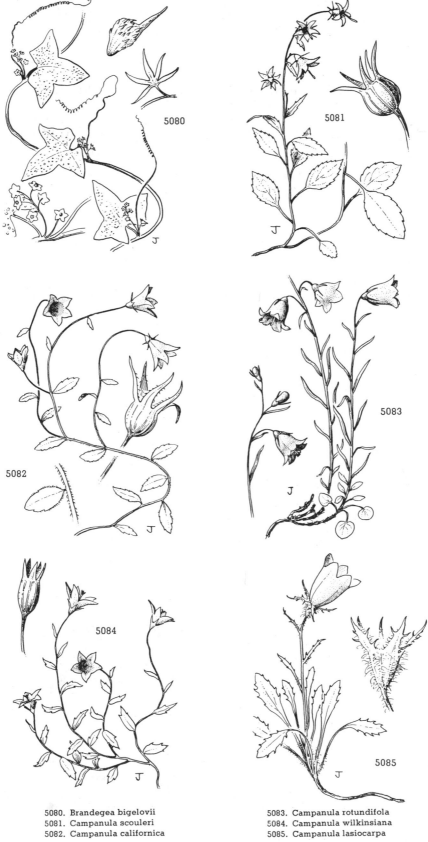

5080. Brandegea bigelovii
5081. Campanula scouleri
5082. Campanula californica

5083. Campanula rotundifola
5084. Campanula wilkinsiana
5085. Campanula lasiocarpa

flowers usually solitary, terminating the stem; calyx-tube broadly obconic, 4–6 mm. long, densely beset with long villous hairs; calyx-lobes 7–15 mm. long, becoming divergent, laciniate, more or less ciliate on the margin; corolla blue, 15–28 mm. long, the lobes triangular-ovate, 5–10 mm. long; capsule turbinate.

Grassy banks and uplands, Boreal Zone; northwestern Asia and Alaska south to western Alberta and British Columbia and in Snohomish County, Washington (*Thompson*). Type locality: Unalaska. July–Sept.

6. Campanula pìperi Howell. Olympic Harebell. Fig. 5086.

Campanula piperi Howell, Fl. N.W. Amer. 1: 409. 1901.

Stems clustered from a much-branched rootstock, 3–10 cm. high, leafy to the tops, the leaves scarcely reduced in size, herbage glabrous throughout, or more or less finely scabrous above. Leaves cuneate to spatulate, with broad petioles, 8–25 mm. long, coarsely and sharply serrate to irregularly dentate, the basal withering, persistent; flowers solitary or commonly in few-flowered, terminal, bracteate racemes; calyx-tube obconic, short, the lobes subulate to linear-lanceolate, 5–15 mm. long, sparsely short-hirsutulous; corolla blue, open-campanulate, about 1 cm. long; style slightly exceeding to shorter than the corolla; stigma-lobes strongly recurved; fruiting capsule subglobose.

Rocky crevices of cliffs, Boreal Zones; Olympic Mountains, Washington. Type locality: "On cliffs, Mount Steele, Olympic Mountains, Washington." July–Aug.

7. Campanula scabrélla Engelm. Rough Harebell. Fig. 5087.

Campanula scabrella Engelm. Bot. Gaz. 6: 237. 1881.
Campanula uniflora Jepson, Man. Fl. Pl. Calif. 973. 1925. Not *C. uniflora* L.

Plants 3–12 cm. high; stems usually tufted, arising from the much-branched underground stems, densely short-bristly pubescent throughout. Leaves clustered at base of stems, entire, lanceolate to oblanceolate, acute or the lower usually obtuse, 1–5 cm. long, 2–5 mm. broad, thickish; flowers solitary and terminal, or in a few-flowered raceme; sepals subulate, 2–8 mm. long; corolla light blue, 6–9 mm. long; capsule ellipsoidal to obconic, 6–7 mm. long, opening near the summit.

Rocky alpine slopes, Boreal Zones; Wenatchee Mountains and Mt. Adams, Washington, south on high peaks of the Cascade Mountains to Mount Shasta and Mount Eddy, Siskiyou County, California. Type locality: "On bleak rocky ridges of Scott Mountain, west of Mount Shasta." June–Aug.

Intermediates between this and the following taxon are occasionally found in Chelan and Kittitas Counties, Washington.

8. Campanula párryi var. idahoénsis McVaugh. Parry's Northern Harebell. Fig. 5088.

Campanula parryi var. *idahoensis* McVaugh, Bull. Torrey Club 69: 241. 1942.
?*Campanula rentoniae* Senior, Rhodora 51: 302. 1949.

Plants perennial from much-branched, slender rootstocks, the sparsely leafy, glabrous, simple stems 6–35 cm. high arising from leaf-clusters at the apex of the rootstocks. Leaves sessile, lanceolate, elliptic-lanceolate or oblanceolate, acute at apex, tapering to the base, the lower sometimes rounded, 6–28 mm. long, the uppermost leaves reduced and sometimes bract-like, glabrous except for sparse ciliae mostly on the basal margin, these usually not present on the upper leaves; flowers solitary, terminating the stem, sometimes 1–3 more on axillary branchlets; calyx-tube obconic or campanulate, often with a few minute hairs; calyx-lobes narrowly subulate to linear, erect, shorter than the tube, lengthening somewhat and often spreading in age; corolla blue, campanulate, 9–15 mm. long, the lobes about one-third the length of the corolla; filaments with an expanded ciliate base; capsule 7–11 mm. long, the 3 valvular openings near the summit of the capsule.

Open rocky slopes, Boreal Zones; Montana and Idaho west to Chelan and Kittitas Counties, Washington. Type locality: near Friday's Pass, Idaho County, Idaho. July–Aug.

9. Campanula exígua Rattan. Chaparral Campanula. Fig. 5089.

Campanula exigua Rattan, Bot. Gaz. 11: 339. 1886.

Annual, 7–15 cm. high, simple below and divergently branched above, or rarely branching from the base, hispidulous, especially below, with spreading or somewhat reflexed hairs. Basal leaves obovate or oblong-obovate, entire, those of the lower stem mostly oblong, toothed or entire, the upper most often subulate; flowers axillary and terminal; calyx-lobes subulate, 4–5 mm. long, twice as long as the turbinate tube, glabrate or appressed-hispidulous; corolla blue, 8–12 mm. long, the lobes ovate or oblong-ovate, acutish; stamens 4–6 mm. long; filaments dilated below the middle to a ciliolate base; style about equaling the rim of the throat, linear-clavate; capsule somewhat urceolate, opening by 3 valves just above the middle.

Rocky ridges and talus slopes, Arid Transition Zone; summit of the Coast Ranges, mainly Inner Coast Ranges; Mount St. Helena, Mount Tamalpais, and the Mount Diablo and Mount Hamilton Ranges, central California. Type locality: Mount Diablo. May–June.

10. Campanula angustiflòra Eastw. Eastwood's Campanula. Fig. 5090.

Campanula angustiflora Eastw. Proc. Calif. Acad. III. 1: 132. *pl. 11, figs. 2a–2d.* 1898.
Campanula angustiflora var. *exilis* J. T. Howell, Leaflets West. Bot. 2: 102. 1938.

Annual, 18–25 cm. high, usually diffusely branched from the base, the upper branches erect to spreading and the stems with short hispid hairs on the angles. Leaves glabrous, sessile, 3–10 mm. long, narrowly to broadly ovate, coarsely and deeply dentate below, less deeply to subentire on the upper reduced leaves; flowers solitary, lateral or terminal on the branches; pedicels 3–10 mm. long; calyx 4–6 mm. long, the subulate lobes exceeding the tube; corolla blue, tubular or slightly flaring above, inconspicuous, shorter than to equaling the calyx-lobes; stamens 2 mm. long, about equaling the filaments; style tapering from the base; capsule strongly ribbed, about 6 mm. long.

Rocky chaparral slopes, Arid Transition Zone; Marin County to Santa Cruz County and at The Pinnacles, San Benito County, California. Type locality: Mount Tamalpais, Marin County. May–June. The more slender plants with longer internodes represent the variety.

2. TRIODÀNUS Raf. New Fl. 4:67. 1838.

Annual herbs with erect or reclining stems, the branches if any basal. Leaves alternate, toothed or entire. Inflorescence spiciform, the axillary flowers 2-bracted, sessile or nearly so. Early flowers small and cleistogamous, the later with a blue or purple, nearly rotate corolla. Calyx-tube narrow, the lobes in the earlier flowers 3–4, in the later ones 4–5. Corolla 5-lobed or -parted, the lobes imbricated in the bud. Filaments dilated to a ciliate base; anthers separate, linear. Ovary 3-celled, or rarely 2-celled or 4-celled; stigma usually 3-lobed. Capsule opening at the apex or near the middle. [Name Greek, meaning three unequal teeth.]

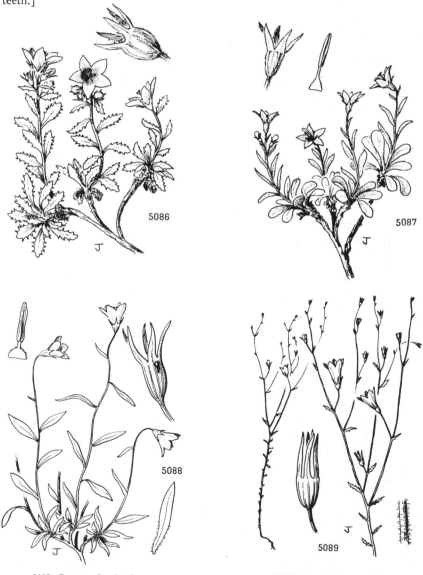

5086. Campanula piperi
5087. Campanula scabrella

5088. Campanula parryi
5089. Campanula exigua

A genus of 8 species, native of Europe, the Mediterranean region, and North and South America. Type species: *Campanula perfoliata* L.

Leaves sessile, not cordate-clasping; pores near the top of the capsule.   1. *T. biflora.*
Leaves mostly cordate-clasping; pores near the middle of the capsule.   2. *T. perfoliata.*

1. **Triodanus biflòra** (Ruiz & Pav.) Greene. Small Venus' Looking-glass. Fig. 5091.

Campanula biflora Ruiz & Pav. Fl. Peruv. **2**: 55. *pl. 200.* 1799.
Specularia biflora Fisch. & Mey. Ind. Sem. Hort. Petrop. **1**: 17. 1835.
Dycmicodon californicum Nutt. Trans. Amer. Phil. Soc. II. **8**: 256. 1843.
Dycmicodon ovatum Nutt. loc. cit.
Triodanus biflora Greene, Man. Bay Reg. 230. (Feb.) 1894.
Legouzia biflora Britt. Mem. Torrey Club **5**: 309. (Oct.) 1894.

Plants scabrous on the angles, otherwise glabrous or nearly so throughout; stems simple or branched, slender and often weak, 1.5–6 dm. high. Leaves ovate to oblong or the uppermost lanceolate, sessile, 6–20 mm. long, entire, or inconspicuously crenate-serrate with very low broad teeth; earlier flowers with 3–4 ovate to lanceolate calyx-lobes, the later ones with 4–5 longer, lanceolate-subulate calyx-lobes; capsule 6–10 mm. long, oblong-cylindric, opening by valves close under the calyx-lobes.

Dry soils, mostly in disturbed places, Transition and Sonoran Zones; southwestern Oregon southward to Lower California and eastward across the southern United States to Virginia and Florida; also Mexico and South America. Type locality: Peru. April–July.

2. **Triodanus perfoliàta** (L.) Nieuwl. Venus' Looking-glass. Fig. 5092.

Campanula perfoliata L. Sp. Pl. 169. 1753.
Specularia perfoliata A. DC. Monogr. Camp. 351. 1830.
Triodanus rupestris Raf. New Fl. **4**: 67. 1838.
Legouzia perfoliata Britt. Mem. Torrey Club **5**: 309. 1894.
Triodanus perfoliata Nieuwl. Amer. Midl. Nat. **3**: 192. 1914.

Stems simple or branched, rather stiffly erect or sometimes decumbent at base and more or less spreading, hirsute especially on the angles with spreading, bristle-like hairs of uneven length, 1.5–6 dm. long. Leaves 1–2 cm. long, broadly ovate to suborbicular, strongly cordate-clasping or the lower merely sessile, acutish or rounded at apex, rather shallowly crenate-dentate or sometimes entire, short-hispid on the veins and margins; flowers 1–3 in the axils, sessile; upper flowers (the later ones) with 5 (rarely 4) rigid, triangular-lanceolate, acuminate calyx-lobes and a rotate, blue or white corolla 8–20 mm. broad, the earlier flowers smaller with 3–4 shorter calyx-lobes exceeding the rudimentary corolla; capsule oblong to narrowly turbinate, 4–6 mm. long, tardily opening near the middle; seeds lenticular.

Dry woods and rocky banks, Transition and Upper Sonoran Zones; British Columbia southward through Washington and Oregon to Siskiyou and Humboldt Counties, northwestern California; eastward across the continent. Type locality: Virginia. April–July.

3. **ASYNEÙMA** Griseb. & Schenk in Wiegm. Archiv Naturgesch. **18**[1]: 335. 1852.

Perennial or sometimes biennial herbs. Inflorescence an interrupted panicle or sometimes a raceme, the short-pedicelled or subsessile flowers clustered or solitary. Flower-buds cylindric. Calyx 5-cleft. Corolla deeply 5-parted, the linear or linear-lanceolate lobes rotate-spreading to reflexed. Style pilose, the stigma trifid. Capsule 3-celled, dehiscing by pores which are near the apex, the base, or at the middle of the capsule wall. [Name derived from the Greek, the application obscure.]

A genus of about 40 species, mostly natives of southeastern Europe and Asia Minor but occurring also in eastern Asia and the Pacific States. Type species, *Asyneuma canescens* Griseb. and Schenk.

1. **Asyneuma prenanthoìdes** (Durand) McVaugh. California Harebell. Fig. 5093.

Campanula prenanthoides Durand, Journ. Acad. Phila. II. **3**: 93. 1855.
Campanula filiflora Kell. Proc. Calif. Acad. **2**: 5. 1863.
Campanula roezlii Regel, Gartenfl. **21**: 239. 1872.
Asyneuma prenanthoides McVaugh, Bartonia No. 23: 36. 1945.

Stems arising from creeping rootstocks, 2–8 dm. high, erect or sometimes decumbent in shade forms, simple or much branched, longitudinally ribbed or angled, glabrate above, rather sparingly hispidulous with divergent or somewhat reflexed hairs. Leaves sessile or short-petioled, ovate-lanceolate to narrowly lanceolate, acute or acuminate at apex, cuneate at base, 1–4 cm. long, sharply and conspicuously serrate, minutely hispidulous below, especially on the veins and margins; flowers in small clusters forming an interrupted, terminal, racemose panicle or simple, few-flowered racemes, the lower subtended by small leaves, the upper by small acuminate bracts; calyx campanulate, the lobes subulate, as long as or longer than the tube; corolla bright blue, 10–12 mm. long, the lobes linear, two to three times as long as the tube; style long-exserted, often recurved; capsule hemispheric to short-oblong, subcordate at base and thin-walled between the nerves.

Dry open woods, Transition Zones; British Columbia south, mostly west of the Cascade Mountains, to Tulare County in the Sierra Nevada and to Monterey County in the Coast Ranges, California. Type locality: Nevada County, California. June–Aug.

4. HETEROCÒDON Nutt. Trans. Amer. Phil. Soc. II. **8**: 255. 1843.

A small delicate annual with weak slender stems and sessile or partly clasping, sub-orbicular leaves. Flowers of two forms, the earlier ones with rudimentary corollas and self-fertilized in the bud. Calyx with the tube adnate to the ovary, the lobes broad and foliaceous, much longer than the tube, 3–4 in the early cleistogamous flowers, 5 in the later open ones. Corolla blue, broadly campanulate, 5-lobed. Filaments of the stamens ciliate at base. Styles as in *Campanula*. Capsule 3-celled, 3-angled, dehiscent at base by a valve-like opening in the capsule-wall over adjacent locules. Seeds numerous. [Name from two Greek words meaning different and bell, for the two kinds of bell-shaped flowers.]

A monotypic genus of western North America.

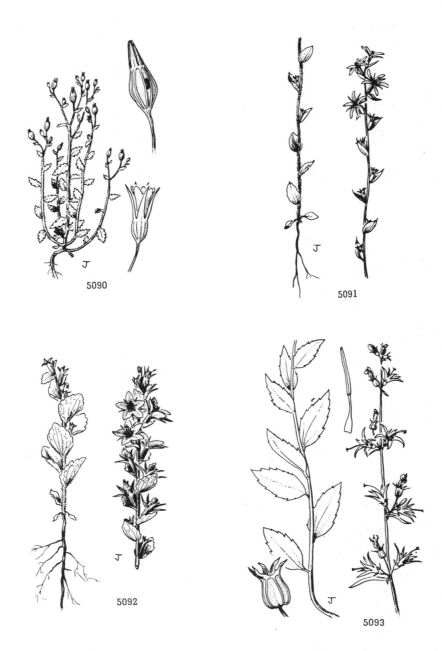

5090. Campanula angustiflora
5091. Triodanus biflora

5092. Triodanus perfoliata
5093. Asyneuma prenanthoides

1. **Heterocodon rariflòrum** Nutt. Heterocodon. Fig. 5094.

Heterocodon rariflorum Nutt. Trans. Amer. Phil. Soc. II. **8**: 255. 1843.
Specularia rariflorum McVaugh, Leaflets West. Bot. **3**: 48. 1941.

Delicate annual with almost simple to diffusely branched, almost filiform stems 1–2 dm. long, sparsely hirsute. Leaves sessile, cordate or rarely clasping, orbicular or nearly so, 2–8 mm. broad, dentate-serrate with broadly triangular teeth; flowers subsessile; calyx-lobes of the later flowers triangular-ovate, 3–5 mm. long and foliaceous in fruit, often with a few broad, bristle-tipped teeth; calyx-tube broadly obpyramidal; corolla of the upper flowers light blue or the lobes often darker, about equaling the calyx-lobes; capsule short and broad.

Moist banks and boggy meadows, mostly Transition Zones; British Columbia and Idaho southward through the Pacific States and Nevada to the mountains of San Diego County, California. Type locality: "Grassy plains of the Wahlamet and Oregon." May–July.

5. **GITHÓPSIS** Nutt. Trans. Amer. Phil. Soc. II. **8**: 258. 1843.

Small annual herbs with angled stems and cuneate-obovate, sessile, usually inconspicuous leaves. Flowers all alike, solitary in the axils. Calyx with the tube 10-ribbed and the 5 lobes long, narrow and foliaceous. Corolla tubular-campanulate, 5-lobed. Stamens with short filaments, dilated at base; anthers long and narrow. Ovary 3-celled; stigma 3-lobed. Capsule clavate or obconic, strongly striate-ribbed, crowned with rigid calyx-lobes, somewhat shorter to longer than the tube, opening in the summit by a circular orifice left by the falling away of the style. [Name Greek, from the generic name Githago, on account of the similarity of the calyces.]

A genus of probably 3 or 4 variable species, natives of western North America. Type species: *Githopsis specularioides* Nutt.

Capsule narrowly obconic to clavate, 2.5–4 mm. wide at summit; calyx-lobes as long as and usually longer in age than the calyx-tube.
    Corolla shorter than the calyx-lobes or rarely equaling them.   1. *G. specularioides.*
    Corolla half again as long as the calyx-lobes.   2. *G. pulchella.*
Capsule but little enlarged at the summit; calyx-lobes in fruit shorter than the tube.   3. *G. diffusa.*

1. **Githopsis specularioìdes** Nutt. Common Blue-cup. Fig. 5095.

Githopsis specularioides Nutt. Trans. Amer. Phil. Soc. II. **8**: 258. 1843.
Githopsis specularioides β *hirsuta* Nutt. loc. cit.
Githopsis calycina Benth. Pl. Hartw. 321. 1849.
Githopsis calycina var. *hirsuta* Benth. loc. cit. (Nomen nudum)
Githopsis latifolius Eastw. Proc. Calif. Acad. IV. **20**: 154. 1931.
Githopsis specularioides subsp. *candida* Ewan, Rhodora **41**: 308. 1939.

Annual with stems profusely branched, erect or the branches ascending or sometimes somewhat decumbent, simple or nearly so particularly in dwarfed plants, 3–15 cm. long, light green or grayish green (glabrate and hirsute plants in same colony), glabrate or more or less densely hirsute and hispid with spreading or deflexed hairs, especially on the ribbed angles of the stem. Leaves ovate to oblong or broadly linear, sessile, more or less toothed, 3–20 mm. long, abundant on younger plants, mostly deciduous in age; calyx-tube 7–12 mm. long, prominently ribbed longitudinally in fruit and 2.5–4 mm. wide at the apex; calyx-lobes linear, mostly as long as the tube or longer in fruit and more or less spreading; corolla blue or sometimes white, 3–10 mm. long, the lobes shorter than the flaring tube, usually shorter than the calyx-lobes or equaling them.

Mostly in clay or adobe soil on dry slopes, often in burned areas in the chaparral, Sonoran and Transition Zones; western Washington (Chehalis County) and the Blue Mountains and western Oregon southward through the Coast Ranges and Sierra Nevada foothills of California to the southern border of the state. Type locality: "plains of the Oregon, near the outlet of the Wahlamet." April–June.

2. **Githopsis pulchélla** Vatke. Large-flowered Blue-cup. Fig. 5096.

Githopsis pulchella Vatke, Linnea **38**: 714. 1874.
Githopsis specularioides var. *glabra* Jepson, Man. Fl. Pl. Calif. 974. 1925.

Annual, 5–15 cm. high, widely divaricate-branching from the base and above, in vigorous plants as wide as high, stems angled, glabrous or hirsute. Leaves sessile, 2–10 cm. long, lanceolate to linear-lanceolate, few-toothed, rarely entire, the lower leaves withering at anthesis; flowers sessile or short-peduncled in the leaf-axils; flowering calyx 11–16 mm. long; fruiting calyx-tube strongly ribbed, glabrous or with retrorse hirsute hairs; fruiting calyx-lobes linear, acicular, glabrous or hirsute, widely spreading in age, straight or curved; corolla blue, 15–20 mm. long.

Dry slopes, Upper Sonoran and Arid Transition Zones; foothills of the Sierra Nevada from Butte County to Mariposa County, California. Type locality: "California," *Bridges 153*, possibly near the Calaveras Sequoia Grove which Bridges is known to have visited. May–June.

3. **Githopsis diffùsa** A. Gray. Southern Blue-cup. Fig. 5097.

Githopsis diffusa A. Gray, Proc. Amer. Acad. **17**: 221. 1882.

Annual with slender stems, simple or branched above and below with ascending branches, herbage glabrous except for occasional hairs. Leaves narrowly obovate, serrulate; flowers scat-

5094. Heterocodon rariflorum
5095. Githopsis specularioides

5096. Githopsis pulchella
5097. Githopsis diffusa

tered, sessile or short-pedicellate on the branches; flowering calyx 8–10 mm. long, the lobes one-half or less the length of the tube; fruiting calyx 9–11 mm. long, essentially linear, little enlarged at the apex, conspicuously ribbed; fruiting calyx-lobes linear, ascending or spreading, shorter than the tube in fruit; corolla "light purple," 3–5 mm. long, shorter than the calyx-lobes.

In moist places and shaded banks, Arid Transition Zone; San Gabriel Mountains, Los Angeles and San Bernardino Counties, California. Type locality: Mount Cucamonga, San Gabriel Mountains. May–June.

Githopsis gilioides Ewan, Rhodora 41: 311. 1939. Low plant much branched from the base with persistent, hirsutulose or glabrate leaves; corolla 4–5 mm. long, equaling the calyx. Known only from the San Gabriel Mountains, southern California. Type locality: forks of the San Gabriel River. Possibly a luxuriant hirsutulous form of *G. diffusa*.

Githopsis filicaulis Ewan, op. cit. 312. Slender lax annual with few widely spreading branches and minute flowers 3–6 mm. long and calyces 3–5 mm. long. Known from Mission Canyon, San Diego County, California, the type locality, and from Vallecito (*Orcutt*), Lower California.

Subfamily 2. LOBELIOÌDEAE.*

LOBELIA SUBFAMILY.

Ours herbs with acid juice, alternate or basal, exstipulate, simple leaves and spicate or racemose, often leafy-bracted flowers. Calyx-lobes 5, equal; calyx-tube adnate to or partly free from the ovary, turbinate, ovoid or hemispheric. Corolla gamopetalous, usually persistent on withering, often split down one side, 2-lipped, with 2 lobes

---

* Keys and taxonomic treatment are based largely on the works of Rogers McVaugh.

on the upper lip and 3 on the lower, these often reversed by the inversion of the flower. Stamens 5, free from the corolla, or lower part of the filament sometimes adnate; filaments flattened, coherent into a tube. Anthers zygomorphic, introrse, forming an anther-tube, often with tuft of hairs or cusped, the longer anthers usually curved, the opening of the anther-tube thus lateral. Ovary inferior, at least partly so, usually 2–5-celled. Style 1; stigma 1–2-lobed; ovules numerous. Fruit a 1–5-celled capsule. Seeds numerous, smooth or variously marked with transverse and longitudinal striae.

Filaments free or united distally; anthers distinct; flowers never blue.
  Flowers solitary on filiform pedicels borne on branches of erect stems.   1. *Nemacladus.*
  Flowers subsessile in capitate clusters on diffuse or prostrate stems.   2. *Parishella.*
Filaments united; anthers united into a tube, two shorter, the opening of the tube thus oblique; flowers variously colored.
  Flowers pedicellate; capsule obconic to hemispheric or narrowly clavate.
    Corolla including lobes 2–10 mm. long or lacking; more or less succulent annuals.
      Plants immersed aquatics; corolla when present less than 3 mm. long.   3. *Howellia.*
      Plants terrestrial; corolla 3.5–10 mm. long.
        Corolla 3.5–4 mm. long; capsule clavate oblong, 1.5–2 mm. wide.   4. *Legenere.*
        Corolla about 10 mm. long; capsule obconic, 3–4 mm. wide.   6. *Porterella.*
    Corolla including lobes (10)12–50 mm. long; coarse perennials (except *L. dortmanna* and *L. kalmii*).   7. *Lobelia.*
  Flowers sessile; capsule very narrowly linear, simulating a flower-pedicel.   5. *Downingia.*

1. NEMÁCLADUS Nutt. Trans. Amer. Phil. Soc. II. **8**: 254. 1843.

Small, diffusely branched annuals with numerous slender branches. Leaves mostly basal, the cauline leaves minute, sessile, subtending the dichotomous branches. Inflorescence racemose. Flowers minute on proportionately long, usually capillary pedicels, these bracteate at base. Calyx partly or entirely free, the lobes triangular to ovate. Corolla bilabiate, the upper lip 2-lobed or -parted, the lower 3-lobed or-parted. Filaments monadelphous to near the base; anthers oval, glabrous, distinct and stellately spreading. Style incurved at the tip; stigma capitate, 2-lobed. Capsule 2-celled, 2-valved from the apex, few- to many-seeded. [Name from two Greek words meaning thread and branch, in reference to the very slender stems and branches.]

    A genus of about 12 species, natives of Oregon and California, and also adjacent Arizona, Nevada, and Mexico. Type species: *Nemacladus ramosissimus* Nutt.

Calyx free from the ovary, 5-parted to the base; capsule two to three times as long as the calyx; corolla-tube well exceeding the calyx.   1. *N. longiflorus.*
Calyx-tube adnate to the ovary; capsule not much exceeding the calyx; corolla-tube not exceeding the calyx-lobes (slightly exceeding in *N. secundiflorus*).
  Seeds with undulate or zigzag, longitudinal ridges, separating the well- or ill-defined pits.
    Corolla-tube evident, about equaling the lobes, not exceeding the calyx-lobes except in *N. secundiflorus.*
      Axes of the racemes zigzag and to a less degree the stems; seeds distinctly longer than broad.
        Mature capsule obscurely pointed and very thin at apex; mature pedicels widely spreading with a double curve (this less marked in *N. secundiflorus*).
          Basal leaves rhombic-ovate to elliptic, entire or irregularly crenate-dentate; plants (the larger ones) intricately branched; cismontane southern California and deserts.   2. *N. sigmoideus.*
          Basal leaves oblong-lanceolate, irregularly crenate-dentate to subpinnatifid; Inner South Coast Ranges and Greenhorn Range.
            Corolla-tube mostly included within the calyx; racemes distichous, secund.   4. *N. secundiflorus.*
            Corolla-tube mostly included within the calyx; racemes distichous.   3. *N. gracilis.*
        Mature capsule sharply pointed and distinctly firm at apex; mature pedicels ascending and curved upward only toward the tip.   5. *N. pinnatifidus.*
      Axes of the racemes straight and also the stems; seeds nearly globose.   6. *N. ramosissimus.*
    Corolla divided almost to the base, the tube thus extremely short.
      Anthers 0.5–0.8 mm. long.
        Branches diffuse and spreading; pedicels mostly spreading; corolla-lobes ciliate; Mojave and Colorado Deserts and adjacent Nevada and Arizona.   7. *N. rubescens.*
        Branches and pedicels more or less stiffly ascending; corolla-lobes not ciliate; western slopes of Sierra Nevada from Butte County to Kern County.   8. *N. interior.*
      Anthers 0.2–0.3 mm. long.
        Plants stoutish; capsule 3–4 mm. long; base of fruiting calyx oblique to rounded.   11. *N. rigidus.*
        Plants delicate; capsule 1.5–2.7 mm. long; base of calyx long-turbinate.   12. *N. capillaris.*
  Seeds with longitudinal impressed lines, the flat ridges between marked transversely with fine lines.
    Anthers 0.1–0.3 mm. long; calyx-tube broadly rounded to hemispheric in fruit.   9. *N. glanduliferus.*
    Anthers 0.6–0.7 mm. long; calyx-tube turbinate to oblique in fruit.   10. *N. montanus.*

1. Nemacladus longiflòrus A. Gray. Long-flowered Nemacladus. Fig. 5098.

Nemacladus longiflorus A. Gray, Proc. Amer. Acad. **12**: 60. 1876.

Plants mostly 5–18 cm. high, stems 1 to several, branching from the base and above, minutely pubescent at least below or glabrous throughout, brownish or purplish, often shiny. Basal leaves few to many, 3–12 mm. long, 1.5–4 mm. wide, oblanceolate to obovate or spatulate, mostly pubescent on both sides; racemes with a zigzag axis; pedicels mostly 1–2 cm. long in fruit, wide-spreading, ascending at base in an upward curve and abruptly turned again upward at tip, glabrous; bracts 2–4 mm. long, lanceolate to ovate or elliptic, folded around the base of the pedicel, shortly and rather stiffly ciliate-pubescent on the margins and sometimes on one or both surfaces, occasionally entirely glabrous; calyx divided nearly to the base, the lobes linear-lanceolate; corolla 5–8 mm. long, the tube much exceeding the calyx-lobes, white or slightly tinged with pink, upper lip bearded at the base within, with a yellowish spot at base of each lobe; filament-tube 3.5–7.5 mm. long; capsule 3–5 mm. long, fusiform, entirely free from the calyx or nearly so; seeds broadly ellipsoid, with obscure, undulate, longitudinal ridges and ill-defined pits.

Dry, sandy or gravelly plains or slopes, Sonoran Zones; cismontane southern California from Los Angeles County and western San Bernardino County south to northern Lower California. Type locality: "S. E. California, Wallace, Lemmon." The Wallace collection probably was made between Los Angeles and San Bernardino. May–June. Long-flowered Thread-plant.

Nemacladus longiflorus var. **breviflòrus** McVaugh, Amer. Midl. Nat. **22**: 526. 1939. Plant pubescent, the leaves especially more densely pubescent, as well as the stem-pedicels and calyx-lobes; corolla 3–3.5 mm. long; filament-tube 2.2–2.8 mm. long, included in the corolla. Desert slopes of the mountains bordering the western edges of the Mojave and Colorado Deserts, southern California, and adjacent Lower California. Type locality: "Roadside Mine, between Tucson and Sells, Pima Co., Arizona."

2. **Nemacladus sigmoìdeus** G. T. Robbins. Small-flowered Nemacladus. Fig. 5099.

Nemacladus sigmoideus G. T. Robbins, Aliso **4**: 144. 1958.

Plants 4–12 cm. high, the stems often many and intricately much-branched, more or less short-pubescent at base and glabrous above, purplish brown. Basal leaves 1.5–7(10) mm. long, ovate to elliptic, thickish, short-pubescent, entire or irregularly crenate-dentate; racemes markedly zigzag; pedicels 10–18 mm. long, finely capillary, wide-spreading in a double curve, this more conspicuous in fruiting plants, subtended by minute, ovate, conduplicate bracts; flowers erect; corolla 2 mm. long, campanulate, the tube included, white; capsule about one-half inferior, about as broad as long and scarcely exceeding the calyx-lobes; seeds broadly ellipsoid, marked with obscure, undulate or zigzag, longitudinal lines and shallow pits.

Slopes and open desert, Sonoran Zones; Ormsby County, Nevada, southward along the eastern face of the Sierra Nevada, California, to the Mojave and Colorado Deserts and western ranges adjacent to the deserts; also Mohave County, Arizona. Type locality: south fork of Little Rock Creek Canyon, Los Angeles County, California. May–July. Small-flowered Thread-plant.

3. **Nemacladus grácilis** Eastw. Slender Nemacladus. Fig. 5100.

Nemacladus gracilis Eastw. Bull. Torrey Club **30**: 500. 1903.
Nemacladus ramosissimus var. *gracilis* Munz, Amer. Journ. Bot. **11**: 240. 1924.

Plants 10 cm. high, the stems simple or branched, zigzag, pubescent below (rarely throughout) or wholly glabrous, brownish or purplish. Basal leaves 2.5–6 mm. long, rather narrow, oblong-lanceolate, more or less crenate-dentate to subpinnatifid, usually pubescent below and often densely so; inflorescence 2-ranked, zigzag; pedicels finely capillary, wide-spreading in a double curve with flower erect or nearly so; bracts 2 mm. or more long, glabrous or ciliate on the margins and sometimes below, blunt at apex, conduplicate at base concealing the pedicel; corolla 1 mm. long, campanulate, white, the lobes oblong, usually a little longer than the tube, this included or barely surpassing the calyx; filament-tube abruptly curved near the tip; capsule 1.5–2.5 mm. long,

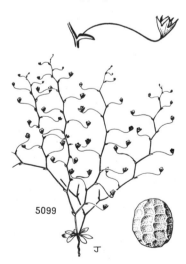

5098. Nemacladus longiflorus

5099. Nemacladus sigmoideus

almost as wide, rounded above, about half-adnate to the calyx-tube; seeds broadly ellipsoid, with 10–12 rows of shallow pits alternating with as many undulating or zigzag longitudinal lines.

Dry, sandy or gravelly soils, Sonoran Zones; Inner Coast Ranges from Merced County south to Kern County, California. Type locality: "Alcalde, Fresno County, California." April–June.

4. Nemacladus secundiflòrus G. T. Robbins. Secund Nemacladus. Fig. 5101.

Nemacladus secundiflorus G. T. Robbins, Aliso **4**: 142. 1958.

Plants about 2.5–10 cm. high, the stems simple or somewhat branched, glabrous, essentially straight, purplish brown. Basal leaves as in *N. gracilis*; racemes mostly secund, not conspicuously zigzag; pedicels about 9–12 mm. long, finely capillary, horizontally spreading and more or less double-curved, the subtending bracts about 2 mm. long, basally clasping the pedicels; corolla about 3 mm. long, with spreading lobes, the tube usually surpassing the calyx-lobes; stamens about 4 mm. long, the filament-tube curved above; capsule and seeds (some approaching globose) as in *N. gracilis*.

In sand and gravel, dry streambeds or slopes, Upper Sonoran Zone; Inner Coast Ranges in southern San Benito County to San Luis Obispo County, California, and also the Greenhorn Range, Kern County. Type locality: east fork of Huerhuero Creek, 3 miles north of Creston, San Luis Obispo County. April–June.

5. Nemacladus pinnatífidus Greene. Comb-leaved Nemacladus. Fig. 5102.

Nemacladus pinnatifidus Greene, Bull. Calif. Acad. **1**: 197. 1885.
Nemacladus ramosissimus var. *pinnatifidus* A. Gray, Syn. Fl. N. Amer. ed. 2. 2¹: 393. 1886.

Plants 6–15 or rarely 20 cm. high, the stems several from the base, intricately branched, with the principal branches and branchlets strongly erect or ascending, entirely or essentially glabrous, greenish or more commonly brownish or purplish. Basal leaves 1.5–4 cm. long, oblanceolate, rounded to acutish at apex, deeply pinnatifid with toothed lobes, or entire, or some merely deeply toothed; pedicels 5–15 mm. long in fruit, capillary, ascending, curved upward, usually bent above the middle, the flowers and fruit erect; bracts elliptic to linear, 2–5 mm. long; calyx short-campanulate in flower, ellipsoid to campanulate in fruit; corolla 1.5–2 mm. long, white or tinged with rose-purple; filament-tube 1.3–2 mm. long, glabrous; capsule 3–4 mm. high, acutely pointed at apex; seeds ellipsoid, with 8–10 longitudinal rows of 8–10 broad rounded pits each, the rows separated by narrow ridges.

Dry open washes or chaparral slopes, Sonoran Zones; coastal region of southern California from the vicinity of Los Angeles and San Bernardino south to northern Lower California. Type locality: "All Saints Bay, Lower California." May–July.

6. Nemacladus ramosíssimus Nutt. Nuttall's Nemacladus. Fig. 5103.

Nemacladus ramosissimus Nutt. Trans. Amer. Phil. Soc. II. **8**: 254. 1843.
Nemacladus tenuissimus Greene, Bull. Calif. Acad. **1**: 198. 1885.

Plants 5–20 cm. high, intricately branched, the stems and branches ascending and, including the more or less secund inflorescence, straight and not at all zigzag, glabrous throughout or pubescent below, greenish above, brownish purple toward the base. Basal leaves 5–15 mm. long, oblanceolate, narrowed to a winged petiole at base, toothed or pinnatifid, glabrous or sparsely pilose especially on the margins toward the base; pedicels very finely capillary, spreading nearly horizontally but curved upward near the apex, the flowers erect; calyx campanulate in flower, broadly conic in fruit; corolla 1.5–2.5 mm. long; filament-tube 1.3–2 mm. long, smooth and glabrous, usually curved at apex; capsule 1.6–2.5 mm. high, broad and rounded at apex; seeds about 0.5 mm. long, almost spherical, with 10 rows of usually 6 rounded pits.

Dry, sandy or gravelly soil, Sonoran Zones; Tassajara Hot Springs, Monterey County, California, to northern Lower California; also inland in western Inyo and in San Bernardino Counties. Type locality: "San Diego." May–July.

7. Nemacladus rubéscens Greene. Desert Nemacladus. Fig. 5104.

Nemacladus rubescens Greene, Bull. Calif. Acad. **1**: 197. 1885.
Nemacladus adenophorus Parish, Bull. S. Calif. Acad. **2**: 28. 1903.
Nemacladus rigidus var. *rubescens* Munz, Amer. Journ. Bot. **11**: 245. 1924.

Plants 5–20 cm. high, repeatedly forked and becoming diffuse and bushy in well-developed plants, glabrous or the stems sparsely puberulent below, usually shiny gray-green. Basal leaves 8–15 mm. long, usually elliptic, obtuse or rounded at apex, mostly entire, glabrous or sparsely and rather coarsely puberulent especially on the margins; pedicels 8–15 mm. long, capillary, slightly ascending, straight or somewhat curved upward at the tip; calyx broadly rounded in flower, mostly hemispheric in fruit; corolla with a short tube and spreading lobes, yellow, with purplish brown margins on the lobes, these at least the upper ciliate on the margins; filament-tube straight or slightly curved above, glabrous and smooth, conspicuously exserted; seeds broadly ellipsoid, with 8–10 undulate or zigzag longitudinal ridges and poorly defined pits between the ridges.

Dry, sandy or gravelly soils, Sonoran Zones; from Inyo and Kern Counties, California, southwestern Nevada, and adjacent Arizona south through the Mojave and Colorado Deserts to northern Lower California. Type locality: Mojave Desert, California. April–June.

Nemacladus rubescens var. **ténuis** McVaugh, Amer. Midl. Nat. **22**: 536. 1939. Lower part of stem silvery gray, and shining as in the typical species; leaves oblanceolate, pinnatifid-toothed; pedicels very finely capillary. Dry sandy soils, Lower Sonoran Zone; Mojave Desert, California, in southern Inyo County southward, and most common in the Colorado Desert at least as far south as Borego Valley, San Diego County. Type locality: Indio Mountain, Riverside County.

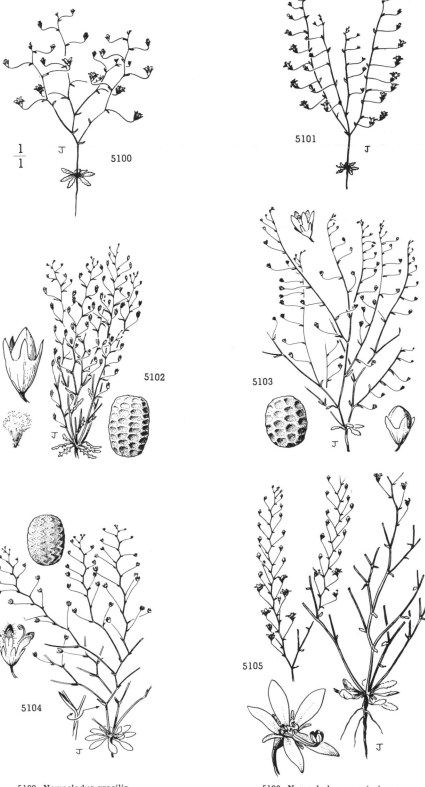

$\frac{1}{1}$

5100

5101

5102

5103

5104

5105

5100. Nemacladus gracilis
5101. Nemacladus secundiflorus
5102. Nemacladus pinnatifidus

5103. Nemacladus ramosissimus
5104. Nemacladus rubescens
5105. Nemacladus interior

8. Nemacladus intèrior (Munz) G. T. Robbins. Sierra Nemacladus. Fig. 5105.

Nemacladus rigidus var. *interior* Munz, Amer. Journ. Bot. 11: 243. 1924.
Nemacladus rubescens var. *interior* McVaugh, Amer. Midl. Nat. 22: 537. 1939.
Nemacladus interior G. T. Robbins, Aliso 4: 146. 1958.

Plants (7)15–25 cm. or more high, the stems and branches stiffly ascending, zigzag, brownish or purplish, rather dull. Basal leaves 1–2 cm. long, oblanceolate to elliptic, sometimes with a distinct petiole, the margins irregularly serrate; racemes strongly zigzag; pedicels 7–13 mm. long, slender, ascending, straight except at the extreme tip beneath the erect flowers; bracts 1–3 mm. long, linear or lanceolate, scarcely or not at all enfolding the pedicel; corolla 2.5–5 mm. long, the tube very short, thĕ lobes spreading, eciliate; calyx-tube turbinate or obconic in fruit; capsule 2–3.5 mm. long, rounded or obscurely pointed at apex; seeds ellipsoid, with undulate or zigzag ridges and well-defined pits.

Dry, gravelly or rocky slopes, Arid Transition Zone; along the western face of the Sierra Nevada from Butte County to Kern County, California. Type locality: "North Fork, Peckinpah, Fresno Co. [Madera]." May–July.

9. Nemacladus glandulíferus Jepson. Glandular Nemacladus. Fig. 5106.

Nemacladus glanduliferus Jepson, Man. Fl. Pl. Calif. 975. 1925.

Plants 7–20 cm. high, the stems several to many from the base, with rather stiff and ascending branches, or sometimes loose and flexuous, pubescent below or sometimes glabrous or nearly so, brownish or purplish at least below. Basal leaves mostly 0.5–1.5 mm. long, oblanceolate or rarely elliptic, obtuse at apex, and narrowed to a broadly winged base, often toothed or pinnatifid, green to purplish brown, pubescent near the base on the margins and upper side; pedicels very slender, spreading horizontally at base and usually curved upward near the apex; bracts 2–5 mm. long, linear to lanceolate, spreading; corolla about 2 mm. long, with acute spreading lobes, the tube very short; filament-tube glabrous, curved at tip; calyx-lobes linear to narrowly triangular, the tube hemispheric in fruit, about as broad as long; capsule about half-inferior, acute at apex; seeds varying from cylindric with truncate ends to narrowly ellipsoidal, with 6–8 longitudinal ridges separated by sharply impressed lines and divided by fine transverse lines into 15–20 narrow cross-ridges.

Dry desert slopes, Lower Sonoran Zone; southern Mojave Desert and the Colorado Desert (mainly in western portions) from Sheep Hole Mountains, San Bernardino County, southward to northern Lower California. Type locality: "Wagon Wash, near Sentenac Canon," Colorado Desert, San Diego County, California. March–May.

Nemacladus glanduliferus var. orientàlis McVaugh, Amer. Midl. Nat. 22: 540. 1939. Very similar to the typical species but pedicels stiffly spreading-ascending, straight or very slightly curved upward at tip; calyx-lobes 0.8–1.5 mm. long as compared with mostly 1.5–2.5 mm. long in the typical species. Mojave Desert region of Inyo and Los Angeles Counties south through the Colorado Desert to northern Lower California and east to southern Nevada, southwestern Utah, western Arizona, and Sonora. Type locality: shore of Lake Mead near Hoover Dam, Clark County, Nevada.

10. Nemacladus montànus Greene. Mountain Nemacladus. Fig. 5107.

Nemacladus montanus Greene, Bull. Calif. Acad. 1: 197. 1885.
Nemacladus ramosissimus var. *montanus* A. Gray, Syn. Fl. N. Amer. ed. 2. 2.2^1: 393. 1886.
Nemacladus rigidus var. *montanus* Munz, Amer. Journ. Bot. 11: 243. 1924.

Plants 8–18 cm. high, the stems erect, few from the base, several times forked above with ascending branches, glabrous or nearly so, dull or shining, brownish or purplish. Leaves 5–18 mm. long, 2–8 mm. broad, oblanceolate to spatulate-elliptic, narrowed to a short broad petiole, glabrous or obscurely and very sparsely puberulent; pedicels 10–15 mm. long, strongly ascending, not capillary, usually with a few ciliate hairs on the upper side near the base, otherwise glabrous; bracts 1–3 mm. long, linear to narrowly lanceolate, scarcely or not at all enfolding the base of the pedicel; calyx turbinate; corolla 1.5–2.5 mm. long, white to purplish, the lobes free almost to the base; filament-tube 2–2.5 mm. long, glabrous; capsule 2.5–3 mm. high, about half-inferior, blunt at apex; seeds ellipsoid, with 10–12 broad longitudinal ridges separated by sharply impressed longitudinal lines, ridges marked by about 30 fine cross-ridges.

Usually on serpentine outcrops, Upper Sonoran and Arid Transition Zones; Inner Coast Ranges, Lake and Napa Counties, California. Type locality: Allen's Springs, Lake County. May–July.

11. Nemacladus rígidus Curran. Rigid Nemacladus. Fig. 5108.

Nemacladus rigidus Curran, Bull. Calif. Acad. 1: 154. 1885.

Plants 4–9 cm. high, the stems simple or branching from the base, stoutish, glabrous or sparsely pubescent, the branches often decumbent. Leaves 7–9 mm. long, 2–5 mm. wide, elliptic, narrowed to a broad base, flowering from near the base of stems, strongly zigzag; pedicels 8–11 mm. long, stiff and comparatively short, spreading horizontally or widely ascending; bracts 2–3 mm. long, spreading from the base of pedicel, elliptic, usually about one-half as wide as long; corolla 1–1.5 mm. long, white or purplish, with short tube; filament-tube 1.2–1.6 mm. long, the anthers with minute reflexed appendages; capsule 3–4 mm. long, 2–2.5 mm. broad, about half-enclosed in hypanthium, oblique at base; seeds ellipsoid, with 8–10 longitudinal ridges, with narrow pits between the ridges.

Sandy hillsides, Upper Sonoran and Arid Transition Zones; southeastern Oregon and adjacent California south to Washoe and Storey Counties, Nevada. Type locality: Geiger Grade near Virginia City, Storey County. May–July.

12. Nemacladus capillàris Greene. Common Nemacladus. Fig. 5109.

Nemacladus capillaris Greene, Bull. Calif. Acad. 1: 196. 1885.
Nemacladus rigidus var. *capillaris* Munz, Amer. Journ. Bot. 11: 244. 1924.

Plants 7–18 cm. high, the stems glabrous or minutely puberulent, brownish or purplish, the branches usually several times forked, rather stiffly ascending, the axis of the raceme more or less zigzag especially in fruit. Basal leaves 5–15 mm. long, usually ovate, narrowed abruptly at base to a short petiole, entire or obscurely crenate, usually glabrous; pedicels 8–12 mm. long, spreading or ascending, capillary but usually straight, the flower then usually horizontal and not strongly upturned, glabrous; bracts 1–3 mm. long, narrowly oblong; corolla about 1 mm. long, white, lobed almost to the base; filament-tube 0.8–1.2 mm. long; calyx-tube turbinate in flower and in fruit, the lobes 0.6–1.2 mm. long, elliptic or obtusely triangular; capsule 1.5–2.5 mm. long, rounded at apex, about half-inferior; seeds few, broadly ellipsoid, with 8–10 narrow longitudinal ridges alternating with rows of 9–12 shallow pits.

Dry gravelly slopes usually of igneous formations, mainly Arid Transition Zone; southern Oregon in Jackson and Klamath Counties southward in the North Coast Ranges and Sierra Nevada to the northern parts of the Mojave Desert, California. Type locality: "Mohave Desert." Collected by Curran. June–Aug.

2. PARISHÉLLA A. Gray, Bot. Gaz. 7: 94. 1882.

Low, diffuse, annual plants with the leaves and flowers in subcapitate tufts at the base of the plant and at the ends of the branches. Calyx-tube campanulate, shorter than the spatulate lobes. Corolla rotate, almost equally 5-parted. Stamens with minute appendages and ovary as in *Nemacladus*; filaments united and anthers distinct. Capsule circumscissile at the rim of the adnate calyx-tube; placentae axile on a central septum which divides the inferior part of ovary and capsule into 2 cells, the lid of the capsule without septum at maturity. Seeds pitted. [Name in honor of S. B. and W. F. Parish, botanical collectors in southern California.]

A monotypic genus of the Mojave Desert.

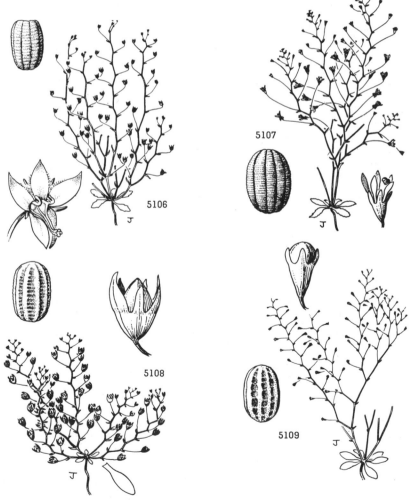

5106. Nemacladus glanduliferus
5107. Nemacladus montanus

5108. Nemacladus rigidus
5109. Nemacladus capillaris

1. **Parishella califòrnica** A. Gray. Parishella. Fig. 5110.

Parishella californica A. Gray, Bot. Gaz. **7**: 94. 1882.

Low annuals with 1 to several stems from the base, diffusely branched, glabrous or nearly so and usually purple. Leaves in a basal rosette or rarely the floral bracts foliaceous, oblanceolate, obtuse at apex and about 5 mm. or less wide, gradually tapering into a margined petiole, 8–12 mm. long, or sometimes ovate and abruptly narrowed to a slender petiole; flower-clusters few- to 20-flowered; corolla white, 3–4.5 mm. high, tubular-campanulate, the lobes subequal, about 1 mm. wide; seeds ellipsoid to oblong, pitted with 8–10 longitudinal rows of 10–12 pits each.

Gravelly slopes, deserts and plains, Sonoran Zones; eastern San Luis Obispo County and in the western and central parts of the Mojave Desert, California. Type locality: Rabbit Springs, Mojave Desert, San Bernardino County. May–June.

3. HOWÉLLIA A. Gray, Proc. Amer. Acad. **15**: 43. 1879.

Delicate, immersed, aquatic, annual plants with flaccid stems and narrow leaves. Flowers both apetalous and petal-bearing. Calyx-tube adnate throughout to the ovary, the 5 lobes narrow. Corolla-tube short, cleft nearly to the base on one side, 3 of the lobes united farther than the other 2; anthers unequal, 2 of them smaller than the other 3. Ovary 1-celled, with 2 parietal placentae; ovules few. Fruit a capsule, irregularly dehiscent by the rupture of the very thin lateral walls; seeds large, smooth. [Name in honor of Thomas Howell, a pioneer botanist of northwest America.]

A monotypic genus of northwestern United States.

1. **Howellia aquátilis** A. Gray. Howellia. Fig. 5111.

Howellia aquatilis A. Gray, Proc. Amer. Acad. **15**: 43. 1879.

Stems flaccid and somewhat fistulous, sparingly branched, 4–7 dm. long. Leaves narrowly linear-subulate, 2–5 cm. long, entire or with a few slender teeth; earlier flowers cleistogamous in the axils of ordinary leaves, the later on branches with more or less verticillate leaves, some with and others without petals; corolla 2–2.7 mm. long, whitish or pale lavender, about equaling the linear, acute, unequal calyx-lobes; ovary wholly inferior, in fruit with a depressed-conic summit capped by the persistent base of the style, narrowly clavate, 8–10 mm. long; seeds 1–5 maturing, smooth and shining, 2–4 mm. long, cylindric, rounded at one end, pointed at the other.

In stagnant ponds, mainly Humid Transition Zone; west-central Washington and northern Idaho to the Willamette Valley, Oregon. Type locality: Sauvies Island, near the mouth of Willamette River, Multnomah County, Oregon. May–Aug.

4. LEGENÉRE McVaugh, N. Amer. Fl. **32A**: 13. 1943.

Annual herbs growing in moist or wet ground or the base of the plants often immersed, rooting at the nodes. Flowers loosely racemose, both with and without corollas. Corollas of the petaliferous flowers cleft dorsally. Stamens with the filaments and also the anthers connate; 2 of the anthers shorter than the other 3. Fruit a 1-celled capsule with parietal placentae, dehiscent at the apex. [Name an anagram of E. L. Greene, the discoverer of the only known species.]

A monotypic genus of central California.

1. **Legenere limòsa** (Greene) McVaugh. Legenere. Fig. 5112.

Howellia limosa Greene, Pittonia **2**: 81. 1890.
Legenere limosa McVaugh, N. Amer. Fl. **32A**: 13. 1943.

Stems weakly erect or somewhat decumbent and rooting at the nodes, simple or often with few to many lateral branchlets, the whole plant green, smooth and glabrous. Leaves few, the cauline entire and sessile, the lower acute and early deciduous; floral bracts foliaceous, elliptic, mostly obtuse at apex, somewhat rounded and sessile at base, 2–3 mm. wide, 6–12 mm. long; corollas 3.5–4 mm. long, white, the lobes about 2 mm. long, the 2 lower ones distinctly narrower than the 3 upper lobes; capsule wholly inferior, dehiscent at apex by short, thin-walled valves; calyx-lobes in fruit subulate to broadly deltoid, usually 4 in apetalous flowers, 5 in petaloid ones; seeds smooth and shining.

Low, moist or wet ground, especially vernal pools, Sonoran Zones; lower Sacramento and San Joaquin Valleys, also ponds on Coal Mine Ridge, Santa Cruz Mountains, California. Type locality: lower Sacramento Valley near Elmira, Solano County, according to type specimens in Greene Herbarium. May.

5. DOWNÍNGIA Torr. Pacif. R. Rep. **4**⁵: 116. 1857. Nomen conservandum.

Low, soft-stemmed, annual, spring-flowering herbs, erect or decumbent and sometimes rooting at the nodes when partially immersed, glabrous or essentially so. Basal and lower stem-leaves narrowly linear and entire or with a few narrow teeth or sometimes pinnatifid, usually not functioning at flowering time, those of the flowering branches thicker and firmer, narrowly linear to broadly lanceolate or oblong. Flowers 5-merous, perfect, inverted, solitary in axils of upper leaves or foliaceous bracts, sessile, the elongate, stalk-like, inferior ovary resembling a stout pedicel. Calyx-lobes 5, usually with the three upper lobes longer than the other two, normally entire and linear. Corolla violet varying to blue, pink, or white, usually with a white or yellow blotch at the base of the lower lip; tube entire, the limb abruptly bilabiate; the 2 lobes of the upper lip usually narrower and smaller than the 3 broader but partially fused lobes of the lower lip; the 2 smaller anthers with a terminal tuft of bristles and also often with an apical, horn-like process. Ovary and capsule 1- or 2-celled; ovules and seeds numerous, when mature fusiform, usually shining, smooth or with faint striae. [Name in honor of A. J. Downing, an American horticulturalist.]

A western North American genus of about 12 species. Type species, *Clintonia elegans* Dougl.

Anther-tube straight or only slightly incurved.
    Lower lip of corolla conspicuously reflexed, forming a sharp angle with the tube; corolla 7–20 mm. long, usually much exceeding the calyx-lobes.
        Corolla-tube densely white-bearded within on the lower side; tip of anther-tube with reflexed and intertwined, bristle-like processes; capsule 2-celled.   4. *D. bicornuta.*
        Corolla-tube glabrous within (sparsely pilose in *D. ornatissima*); tip of anther-tube with or without processes, these never bristle-like and intertwined.
            Ovary and capsule 2-celled, the ovules attached to the longitudinal septum.
                Lateral sinuses of corolla deeply cut, the lower lip thus appearing hinged, the dorsal sinuses equally cut; corolla-tube sparsely pilose within on the lower side.
                    3. *D. ornatissima.*
                Lateral sinuses of the corolla scarcely or not at all deeply cut; corolla-tube glabrous within.
                    Filament-tube shorter than corolla-tube, the anther-tube thus at least partially included; anther-tube not tapering distally, the apex rounded or obtuse; corolla-lobes not in one plane.
                        Seeds smooth or lightly marked longitudinally with straight lines.
                          Lobes of the upper corolla-lip marginally ciliate (under hand-lens); lower corolla-lip with broad purple blotch.   1. *D. concolor.*
                          Lobes of the upper corolla-lip eciliate; lower corolla-lip with 3 purple spots.
                            2. *D. bella.*
                        Seeds appearing twisted, marked obliquely with fine lines on the long axis of the seed.
                          5. *D. cuspidata.*
                    Filament-tube longer than the corolla-tube, the anther-tube thus exserted; anther-tube inclined to taper distally to the acute or slightly rounded apex; corolla-lobes in one plane.   7. *D. pulchella.*
            Ovary and capsule 1-celled, the ovules attached to ovary-wall.
                Dorsal anthers of anther-tube pilose at apex; seeds dull, lightly marked with longitudinal lines.   10. *D. montana.*
                Dorsal anthers of anther-tube glabrous or sparsely pilose; seeds smooth and somewhat shining, without longitudinal markings.   11. *D. yina.*
    Lower lip of corolla ascending; corolla 2–7 mm. long, not exceeding the calyx-lobes or merely equaling them.
        Corolla 2–4 mm. long; seeds appearing twisted, lightly marked with oblique lines.   6. *D. pusilla.*
        Corolla 4–7 mm. long; seeds without markings.   9. *D. laeta.*
Anther-tube strongly incurved, usually at right angles to the filament-tube.
    Ovary and capsule 2-celled, the ovules attached to the longitudinal septum.   8. *D. insignis.*
    Ovary and capsule 1-celled, the ovules attached to the ovary-wall.   12. *D. elegans.*

5111

5110

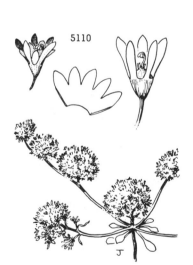

5112

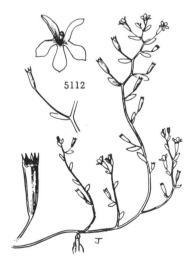

5110. Parishella californica   5111. Howellia aquatilis   5112. Legenere limosa

1. **Downingia cóncolor** Greene. Maroon-spotted Downingia. Fig. 5113.

Downingia concolor Greene, Bull. Calif. Acad. **2**: 153. 1886.
Downingia tricolor Greene, Pittonia **2**: 79. 1890.
Bolelia concolor Greene, op. cit. 127.
Bolelia tricolor Greene, loc. cit.
Bolelia concolor var. *tricolor* Jepson, Fl. W. Mid. Calif. 481. 1901.

Stems branched from the base, few to many or sometimes simple, 4–20 cm. high, glabrous except for the calyx-tube which is sometimes minutely puberulent. Leaves linear, 0.5–2 mm. wide, 5–20 mm. long; inflorescence loosely few-flowered, the floral bracts elliptic to ovate; calyx-lobes linear-elliptic or oblanceolate, ascending to rotate, 3–8 mm. long, nearly equal with the two lower often shorter; corolla 7–13 mm. long, glabrous except for the two upper lobes, these ciliate-scabrous on the margins near the apex, blue or sometimes deep blue, the lateral sinuses more deeply cleft than the upper and cut below the surface of the lower lip; lower lip with a quadrate or 2-lobed, purple or reddish purple spot at base of the central white area, the white sometimes lacking and the whole lip purple except for the spot at base, this bearing 2 ridges or low, nipple-like processes; corolla-tube 3–5 mm. long, narrowly funnelform; anther-tube slightly exserted or sometimes included in the corolla-tube, glabrous or somewhat pubescent on the back, minutely tufted at apex, the 2 shorter anthers with short, horn-like, apical processes; capsule narrowly fusiform, 1–2 mm. broad and 3–5 cm. long; seeds not twisted, the markings parallel to long axis of seed.

Low moist depressions, Upper Sonoran Zone; valleys of the North Coast Ranges of California from Lake County to Monterey County, and also in the Sacramento Valley; collected at Deer Creek, Tulare County, by Congdon. Type locality: wheat field near Suisun, Solano County. March–June.

Downingia concolor var. **brévior** McVaugh, Mem. Torrey Club 19⁴: 20. 1941. Mature capsule mostly 12–25 mm. long, early dehiscent, the valves separated by lines of delicate hyaline tissue. Cuyamaca Lake, San Diego County, California. Known only from the type locality.

2. **Downingia bélla** Hoover. Hoover's Downingia. Fig. 5114.

Downingia bella Hoover, Leaflets West. Bot. **2**: 2. 1937.

Plants up to 17 cm. high, glabrous throughout or the calyx-tube slightly scabrous, the stems somewhat fistulous. Leaves 5–12 mm. long, 1–1.5 mm. wide; inflorescence loosely 3–7-flowered; bracts oblong to elliptic, acute or obtuse, 7–18 mm. long, 1–2.5 mm. wide; calyx-lobes ascending or rotately spreading, 3–6 mm. long; corolla 10–12 mm. long, bright blue, lateral sinuses more deeply cleft than upper sinuses and cut below the surface of lower lip; lobes of the upper lip lanceolate or nearly ovate, erect or slightly recurved, parallel with each other and with the tube; lower lip with central white area with a yellow center and 2 yellow ridges at base, these alternating with 3 small purple spots or the central purple spot wanting; corolla-tube 3–4 mm. long, funnelform; anther-tube more or less scaberulous, the 2 lower anthers bearing short, horn-like processes at apex; capsule nearly linear, 3.5–5 cm. long; seeds with longitudinal markings.

Vernal pools on alkaline plains, Sonoran Zones; Sacramento and San Joaquin Valleys from Colusa County to Tulare County, California. Type locality: near San Joaquin River southwest of Modesto, Stanislaus County. April–Aug.

3. **Downingia ornatíssima** Greene. Solano Downingia. Fig. 5115.

Downingia ornatissima Greene, Pittonia **2**: 80. 1890.

Stems simple to few-branched above, or in more vigorous plants branched below and above, 6–20 cm. high, glabrous throughout or the calyx-tube sometimes minutely scabrous. Leaves 5–12 mm. long, 1–2.5 mm. wide; inflorescence loosely 1–20-flowered; bracts lanceolate to ovate or elliptic, 6–15 mm. long, obtuse or acute; calyx-lobes linear to elliptic, subequal or the 2 lower shorter, 2–6(rarely –9) mm. long, erect or somewhat spreading; corolla 8–13 mm. long, the tube 2–3 mm. long, narrowly funnelform, sparsely bearded within on the lower side, the lateral and and upper sinuses about equally cleft, the lower lip appearing hinged, the margin of the sinus usually turned backward into a horn-like projection; lobes of the upper corolla-lip lanceolate, spreading, the tips curving backward into a ring; lower corolla-lip plane or concave, deep blue to nearly whitish with a white patch bearing 2 yellow or yellowish green spots, the corolla-lobes rounded or truncate, mucronulate, the base of the lip with 2 prominent folds; filaments pubescent near base; anther-tube wholly exserted; anthers white-tufted at apex, especially the two shorter ones, each also with a sharp, horn-like process, glabrous or pubescent dorsally; capsule linear or narrowly subulate, 2.5–6.5 cm. long, 1–1.5 mm. thick, the outer walls tough; seeds with fine longitudinal lines.

Along ditches and in the beds of vernal pools, Sonoran Zones; Butte County to Merced and Solano Counties, California. Type locality: lower Sacramento Valley near Elmira, Solano County. April–May.

Downingia ornatissima var. **exímia** (Hoover) McVaugh, Mem. Torrey Club 19⁴: 24. 1941. (*Downingia mirabilis* J. T. Howell, Leaflets West. Bot. **1**: 221. 1936; *D. mirabilis* var. *eximia* Hoover, op. cit. **2**: 6. 1937.) Two upper corolla-lobes minutely pubescent within near the apex, divergent but not curled into a ring. Moist depressions, San Joaquin Valley, California. Type locality: vicinity of Orange Cove, Fresno County.

4. **Downingia bicornùta** A. Gray. Double-horned Downingia. Fig. 5116.

Downingia bicornuta A. Gray, Syn. Fl. N. Amer. ed 2. 2¹: 395. 1886.
Bolelia bicornuta Greene, Pittonia **2**: 127. 1890.
Downingia sikota Applegate, Contr. Dudley Herb. **1**: 97. *pl. 6.* 1929.

Plant 6–30 cm. high, sometimes more in partially submerged plants, glabrous. Leaves linear-lanceolate, 0.5–2 cm. long; flowers usually few; calyx-lobes widely spreading, linear to narrowly elliptic, 3–10 mm. long, the upper surpassing the lower in length; corolla 9–19 mm. long, glabrous without, densely bearded within on the lower side, the lateral sinuses more deeply cleft than the upper and cut below the surface of the lower lip, the corolla-tube broadly funnelform; upper corolla-lobes 4–7 mm. long, ovate or lanceolate, erect or somewhat divergent, the tips usually turned backward; lower lip folded or reflexed, about 10 mm. long, much wider than long, the 3 lobes truncate or rounded, mucronulate, purplish blue with white patch marked with yellow or greenish yellow and deep purple at base of lip, this bearing 2 prominent, nipple-like protrusions centrally and 2 less prominent ones laterally; stamineal column included in the corolla-tube; anther-tube puberulent apically with 2 apical bristles arising from the 2 shorter anthers reflexed and twisted together; ovary 2-celled; seeds faintly marked longitudinally.

Low moist depressions, usually in adobe soil, Upper Sonoran and Transition Zones; southeastern Oregon and adjacent Idaho and Nevada southward in California to Sierra and Merced Counties. Type locality: Chico, Butte County, California. May–Aug.

Downingia bicornuta var. **pícta** Hoover, Leaflets West. Bot. **2**: 4. 1937. Flowers and fruit slightly smaller than in the typical species; corolla-tube with a brownish yellow spot on its upper side; lower lip strongly concave, with the 2 upper lobes white or pale blue often tipped with darker blue, not divergent but directed toward each other so their tips cross, strongly reflexed and appressed to the tube; horn-like anther-processes often longer than the anthers themselves. Low moist depressions, Upper Sonoran Zone; San Joaquin Valley, California. Type locality: near Le Grand, Merced County.

5. **Downingia cuspidàta** Greene. Cuspidate Downingia. Fig. 5117.

Bolelia cuspidata Greene, Erythea **3**: 101. 1895.
Downingia cuspidata Greene ex Jepson, Fl. W. Mid. Calif. ed. 2. 403. 1911.
Downingia pulchella var. *arcana* Jepson, Madroño **1**: 100. 1922.
Downingia immaculata Munz & Jtn. Bull. Torrey Club **51**: 300. 1924.
Downingia pallida Hoover, Leaflets West. Bot. **2**: 1. 1937.

Stems simple or branched, 6–25 cm. high, glabrous throughout or the calyx-tube sparsely and

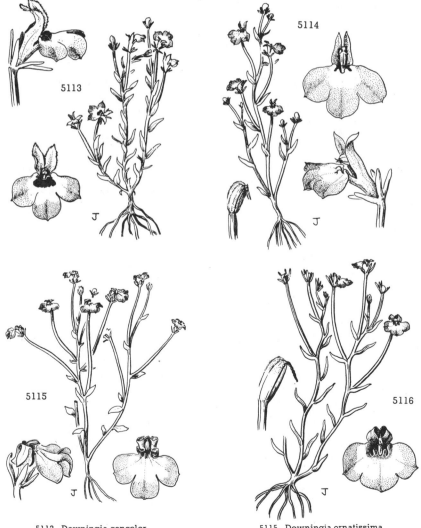

5113. Downingia concolor
5114. Downingia bella
5115. Downingia ornatissima
5116. Downingia bicornuta

minutely scabrous. Leaves narrowly linear to linear, 0.2–2 mm. wide, 3–13 mm. long; inflorescence loosely 1–20-flowered; bracts linear to broadly elliptic, 4–12 mm. long, 1–4 mm. wide; calyx-lobes elliptic to oblanceolate, ascending, 3–10 mm. long; corolla 7–15 mm. long, glabrous, bright or pale blue or lavender, rarely white, the tube narrowly funnelform to almost cylindric, the dorsal and lateral sinuses about equally cleft, the lateral slightly cut below the surface of lower lip; upper lobes ovate, acute, slightly divergent or recurved and overlapping in age; lower lip plane or nearly so, with a central white area bearing 1 or 2 more or less confluent spots at base and with 2 low yellow ridges at base, the lobes broadly ovate or oblong, rounded to retuse at apex; upper lip darker than the lower, purple-veined; filament-tube glabrous, included; anther-tube glabrous or sparsely pubescent dorsally, all minutely white-tufted at apex, the 3 shorter also bearing a short, horn-like process; calyx-tube linear or slightly fusiform in fruit, 2–4 cm. long in fruit; seeds shining, the fine lines running obliquely to the low axis of the seed.

Clay soils of desiccated vernal pools and flats, Sonoran Zones; Humboldt and Shasta Counties south to San Diego County, California. Type locality: "grain field west of Yountville, Napa Co., Calif." April–June.

6. **Downingia pusílla** (G. Don) Torr. South American Downingia. Fig. 5118.

Lobelia pusilla Poepp. ex Cham. Linnaea **8**: 217. 1833. (Nomen nudum)
Clintonia pusilla G. Don, Gen. Hist. Pl. **3**: 718. 1834.
Downingia pusilla Torr. Bot. Wilkes Exp. **17**: 375. 1874.
Bolelia humilis Greene, Pittonia **2**: 226. 1892.
Downingia humilis Greene, Leaflets Bot. Obs. **2**: 45. 1910.

Plants glabrous throughout or the calyx-lobes minutely scabrous, 2–12 cm. high. Leaves 0.5–1 mm. wide, 4–7 mm. long; inflorescence 1–7-flowered; floral bracts elliptic or lanceolate, 2–8 mm. long; calyx-lobes 3–8 mm. long, erect or appressed, surpassing the corolla; corolla 2.5–4 mm. long, glabrous, white or the lower lip blue-tipped, the upper and lateral sinuses about equally cleft; tube narrowly campanulate, 1.5–2 mm. long, the 2 upper lobes deltoid-lanceolate, about 1.5 mm. long, erect; lower lip spreading, with white area centrally bearing a yellow patch near base, the 3 lobes deltoid, 1.2 mm. long; stamen-tube 2–3 mm. long; anthers white-apiculate, the 2 shorter with a blunt, horn-like process and a few bristles at apex; ovary 2-celled; capsule linear, 2–3 cm. long and subulate; seeds marked with oblique lines, appearing twisted.

Moist flats, Sonoran Zones; sparingly introduced or possibly native in Napa, Sonoma, Stanislaus, and Merced Counties, California; also Chile and Argentina. Type locality: Chile. April–June.

7. *Downingia pulchélla* (Lindl.) Torr. Flat-faced Downingia. Fig. 5119.

Clintonia pulchella Lindl. Bot. Reg. **22**: *pl. 1909.* 1836.
Downingia pulchella Torr. Pacif. R. Rep. **4⁴**: 116. 1857.
Bolelia pulchella Greene, Pittonia **2**: 126. 1890.

Plants up to 25–35 cm. high, glabrous throughout or the calyx-tube sometimes sparsely scabrous. Leaves narrowly linear, 1–2 mm. wide, 4–12 mm. long; inflorescences few-flowered up to 15–20-flowered; floral bracts varying from lanceolate to ovate, usually elliptic; calyx-lobes usually rotately spreading, 3–10 mm. long; corolla 8–13 mm. long, glabrous, deep bright blue varying to pink or white; tube purple, 2–3 mm. long, broadly funnelform; upper lip deeply 2-cleft, the lobes elliptic to oblanceolate, spreading, 6–8 mm. long, acute at apex; lower lip spreading in same plane as upper, the lateral sinuses somewhat deeper than upper, with a central white area bearing 2 yellow spots that extend down the low yellow folds at base of lip, 3 dark purple spots alternating with the yellow folds, the lobes oblong, acute, or mucronate; filament-tube longer than corolla-tube, 3–4.5 mm. long; anther-tube attenuate, the anthers glabrous or minutely ciliate-tufted apically, the 2 shorter anthers with slender, horn-like processes; capsule linear, 3–7 cm. long, narrowly subulate or fusiform, the walls tough in texture; seeds shining, without fine lines.

Adobe or alkaline soil of vernal pools and ditches, Sonoran Zones; California from Lassen and Colusa Counties south to Merced and Monterey Counties. Type locality: California. Collected by Douglas. April–June.

8. **Downingia insígnis** Greene. Cupped Downingia. Fig. 5120.

Downingia insignis Greene, Pittonia **2**: 80. 1890.
Bolelia insignis Greene, op. cit. 126.

Stems very slender, somewhat zigzag, 1–3 dm. high, glabrous throughout or the calyx-tube sometimes somewhat scabrous. Leaves narrowly linear, 5–15 mm. long, 1–2 mm. wide; inflorescence 4–20 cm. long, usually about 5-flowered but varying to 16-flowered; floral bracts elliptic to ovate, 6–20 mm. long, 1–5 mm. broad; calyx-lobes ascending, narrowly elliptic, obtuse at apex, 3–8 mm. long; corolla glabrous, sky-blue with darker veins, 9–15 mm. long, lateral sinuses much more deeply cleft than upper; corolla-tube 3.5–5 mm. long, broadly funnelform; upper corolla-lobes ascending and parallel, elliptic, 6–10 mm. long, 2–3 mm. wide; lower lip with central white area bearing 2 oblong, parallel, green spots and 2 golden yellow folds at base of lip in a dark purple patch, this sometimes reduced to 3 purple spots, the lip concave, not reflexed, and usually shorter than the upper lip; filament-tube glabrous; anther-tube usually strongly incurved at right angles to the tube; anthers and connectives whitish in color, granular-roughened throughout, the 2 shorter ones white-tufted at apex; ovary 2-celled; capsule 3–8 cm. long, the lateral walls tough and breaking with difficulty; seeds not twisted.

Low fields and depressions, Sonoran Zones; California in northern Sacramento Valley south to eastern Contra Costa County and Stanislaus County and occurring in Lassen and Modoc Counties; also Washoe County, Nevada. Type locality: lower Sacramento Valley near Elmira, Solano County, California. April–June.

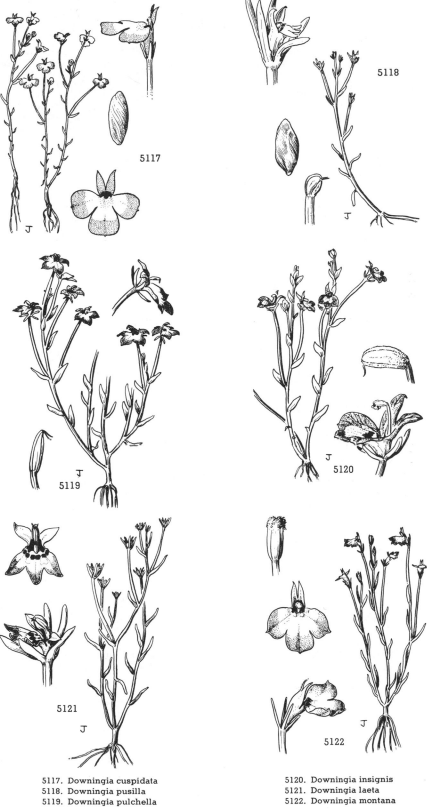

5117. Downingia cuspidata
5118. Downingia pusilla
5119. Downingia pulchella
5120. Downingia insignis
5121. Downingia laeta
5122. Downingia montana

9. **Downingia laèta** (Greene) Greene. Great Basin Downingia. Fig. 5121.

Bolelia laeta Greene, Erythea **1**: 238. 1893.
Bolelia brachyantha Rydb. Mem. N.Y. Bot. Gard. **1**: 483. 1900.
Downingia laeta Greene, Leaflets Bot. Obs. **2**: 45. 1910.

Plants 2–3 dm. high, sometimes shorter, glabrous throughout. Leaves lanceolate to elliptic, 5–18 mm. long; inflorescence rather loosely 1–10-flowered; floral bracts elliptic or lanceolate to ovate, 7–20 mm. long; calyx-lobes elliptic, somewhat unequal, 3–7 mm. long; corolla light blue or purplish, 4–7 mm. long; corolla-tube funnelform, 1–2 mm. long, yellow below on the lower side beneath the purple area; upper corolla-lobes lanceolate or triangular, acute; lower lip with central area white or yellow with the transverse band of purple at base sometimes reduced to 2 or 3 purple spots, the lip spreading, its lobes oblong, acute, 2–3.5 mm. long; anther-tube little or not at all incurved; anthers glabrous or ciliate on the margins, the 2 shorter ones with a tuft of white hairs; ovary 2-celled; capsule 2–4 cm. long, the walls thin but tough; seeds slightly or not at all appearing twisted.

In depressions, Upper Sonoran and Arid Transition Zones; southern Oregon east of the Cascade Mountains and northeastern California eastward to Nevada, southwestern Wyoming, and southern Saskatchewan. Type locality: Humboldt Wells, Elko County, Nevada. June–Aug.

10. **Downingia montàna** Greene. Sierra Downingia. Fig. 5122.

Downingia montana Greene, Pittonia **2**: 104. 1890.
Bolelia montana Greene, op. cit. 127.
Downingia bicornuta var. *montana* Jepson, Madroño **1**: 102. 1922.

Plants up to 15 cm. high, glabrous throughout but calyx-lobes sometimes minutely scabrous. Leaves linear to subulate, 1–5 mm. wide; inflorescence 1–10-flowered; floral bracts subulate to narrowly linear, similar to the leaves but slightly larger; calyx-lobes linear-subulate, acute, sometimes minutely scabrous on the margins, the 2 lower distinctly shorter than the others; corolla 9–12 mm. long, dark blue or violet, glabrous, the lateral sinuses cut below the surface of the lower lip; corolla-tube 3.5–5 mm. long, narrowly funnelform and somewhat gibbous, exceeding the calyx; upper corolla-lobes erect, narrowly triangular; lower lip of corolla with central white patch, the base bluish purple with 2 prominent purple folds at the angle at mouth of tube; filament-tube completely included in corolla-tube, glabrous or minutely puberulent at base; all of the anthers bearded at apex, the 2 shorter ones tufted with short bristles and a horn-like process; capsule subulate, 2–4 cm. long; seeds not shining, marked longitudinally with fine lines.

Moist mountain meadows, Arid Transition and Canadian Zones; northern California in Siskiyou County and south in the Sierra Nevada to Tuolumne County. Type locality: Lake Eleanor, Tuolumne County. July–Aug.

11. **Downingia yína** Applegate. Cascade Downingia. Fig. 5123.

Downingia yina Applegate, Contr. Dudley Herb. **1**: 97. *pl. 5, fig. 2.* 1929.

Plants small and delicate, glabrous or the calyx-tube slightly scabrous, the stems simple and 1-flowered or branched and few to several-flowered, 2.5–7 cm. high. Leaves 10–30 mm. long, linear-lanceolate, acute; calyx-lobes shorter than the corolla-tube; corolla 8–10 mm. long, dark blue with a yellow throat, usually about 7 mm. broad, the lateral sinuses more deeply cleft than the upper, the tube narrowly funnelform; upper lip with lobes linear-lanceolate, acute, erect, and usually approximate, about 4 mm. long and 1 mm. wide; lower lip plane, spreading, bluish purple, the central yellowish area surrounded by white except for the 2 yellow ridges at base of lip, the lobes oblong, acute, about 3 mm. long and 2 mm. wide; ovary slender, 8–10 mm. long, 1-celled; stamen-column erect, included; anther-tube somewhat pubescent dorsally; valves of the capsule separated by hyaline lines, extending the length of the capsule and visible as impressed lines, mature capsule broadest near the middle; seeds dullish or somewhat shining, without longitudinal lines.

Moist or boggy meadows, Canadian Zone; Cascade Mountains of southern Oregon to Siskiyou County, California. Type locality: Four Mile Lake, Klamath County, Oregon. July–Aug.

Downingia yina var. màjor McVaugh, N. Amer. Fl. **32A**: 24. 1942. (*Downingia willamettensis* M. E. Peck, Proc. Biol. Soc. Wash. **47**: 187. 1934; *D. pulcherrima* M. E. Peck, op. cit. **50**: 94. 1937.) Plants more robust, up to 35 cm. high; corolla 7–12 mm. long; mature capsule broadest near the base; valves of capsule usually invisible before splitting, with no impressed lines on the surface; seeds shining and without lines. Central and western Washington southward mostly west of the Cascade Mountains to Humboldt County, California. Type locality: near Aumsville, Marion County, Oregon.

12. **Downingia élegans** (Dougl.) Torr. Common Downingia. Fig. 5124.

Clintonia elegans Dougl. ex Lindl. Bot. Reg. **15**: *pl. 1241.* 1829.
Downingia elegans Torr. Bot. Wilkes Exp. **17**: 375. 1874.

Plants mostly 1–4 dm. high, glabrous except for the scabrous calyx-tube, the stems usually rather stout, few- to many-flowered. Leaves linear-lanceolate to broadly lanceolate, 2–25 mm. long, 1–6 mm. wide; floral bracts lanceolate to ovate, 8–25 mm. long; calyx-lobes 4–12 mm. long, narrowly to broadly linear; corolla 8–18 mm. long, minutely puberulent within at base, otherwise glabrous, bright blue varying to lavender-pink or white, the lateral sinuses more deeply cleft than the upper; tube 2–3.2 mm. long, broadly funnelform, lighter blue with purple veins and often with 3 oblong purple blotches at base on the lower side; upper lip somewhat longer than the lower, the lobes acute, erect, or somewhat divergent; lower lip with a central white spot and beneath this 2 low yellow ridges distinctly white-margined, concave, not reflexed, the 3 lobes oblong to deltoid, acute, 2–4 mm. long; filament-tube glabrous; anther-tube strongly incurved usually at right angles to the filament-tube; anthers bluish gray with white connectives, smooth and glabrous or with a few cilia, the 2 shorter ones tufted apically and with short, recurved, horn-like processes; capsule 2.5–4.5 cm. long, subulate, and broadest near the base, the walls papery, easily ruptured; ovary 1-celled; seeds shining, without surface lines.

Moist meadows and borders of ponds, Transition and Canadian Zones; northeastern Washington and northern Idaho south to Mendocino, Stanislaus, and Plumas Counties, California, and Elko County, Nevada. Type locality: "plains of the Columbia, near Wallawallah river." June–Aug.

Downingia elegans var. **brachypétala** (Gandoger) McVaugh, Mem. Torrey Club **19**⁴: 55. 1941. (*Downingia brachypetala* Gandoger, Bull. Soc. Bot. Fr. **65**: 55. 1918.) Flowers generally smaller; corolla 5–9 mm. long; filament-tube 3–4 mm. long; anther-tube 2–2.5 mm. long. Southern Washington and adjacent Idaho south through the Cascade Mountains to northeastern California. Type locality: Falcon Valley, Klickitat County, Washington.

6. **PORTERÉLLA** Torr. in Porter, Hayden Geol. Rep. Montana **1871**: 488. 1872.

Slender annuals with lanceolate or linear-subulate leaves. Flowers on slender, axillary, inverted pedicels. Calyx-tube united to the wholly inferior ovary, the lobes narrow. Corolla-tube entire, not cleft dorsally, the limb bilabiate, with the upper lip 2-cleft and the lower 3-lobed and spreading. Filaments and anthers connate, 2 of the anthers shorter than the others. Fruit a 2-celled capsule, many-seeded, apically dehiscent. Seeds numerous, smooth and apiculate. [Name in honor of Thomas C. Porter, one of the authors of the *Flora of Colorado Territory*.]

A monotypic genus of western North America.

1. **Porterella carnósula** (Hook. & Arn.) Torr. Porterella. Fig. 5125.

Lobelia carnosula Hook. & Arn. Bot. Beechey 362. 1838.
Porterella carnosula Torr. in Porter, Hayden Geol. Rep. Montana **1871**: 488. 1872.
Laurentia carnosula Benth. ex A. Gray, Bot. Calif. **1**: 444. 1876.
Porterella eximia A. Nels. Bull. Torrey Club **27**: 270. 1900.

Stems simple or sometimes branching below, somewhat succulent, erect, 1–3 dm. high, glabrous. Leaves lanceolate to linear-subulate, 5–20 mm. long; flowers in a simple terminal raceme, ebracteolate; pedicels slender, the lower usually exceeding the foliaceous bracts; calyx-lobes linear, obtuse, 3–7 mm. long, divergent; corolla excluding hypanthium about 8–10 mm. long, blue, narrowly funnelform, the tube equaling the limb; lower lip deeply 3-lobed, the lobes obovate, divergent, with a long yellow area at base; upper lip 2-lobed, the lobes with acuminate upturned tips; stamens completely united, minutely tufted apically, the somewhat curved tip of the anther-tube barely exserted; capsule obconic, 5–7 mm. long.

Muddy edges of ponds and pools, Arid Transition and Canadian Zones, Idaho and northwestern Wyoming southwest to the Cascade Mountains of Oregon and the southern Sierra Nevada (Tulare County), California, and south to Utah and northern Arizona. Type locality: "Blackfoot River, Snake Country." (Southeastern Idaho.) June–Aug.

7. **LOBÈLIA** [Plumier] L. Sp. Pl. 929. 1753.

Herbs, or some tropical species shrubs, with alternate leaves and spicate, racemose, or paniculate, leafy-bracted flowers. Calyx adnate to the ovary, turbinate, hemispheric, or ovoid. Corolla red, yellow, blue, or white; corolla-tube straight, oblique, or incurved, mostly divided to the base on one side, usually 2-lipped, the lobe on each side of the cleft usually turned away from the other three which are somewhat united, the sinuses sometimes splitting to the base at maturity dividing the corolla into 5 petals. Stamens free from

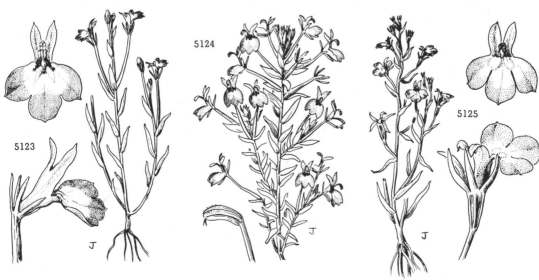

5123. Downingia yina   5124. Downingia elegans   5125. Porterella carnosula

the corolla-tube, united at least above; anthers united into a tube, 2 or all 5 with a tuft of hairs at the tips, three usually larger than the other two. Ovary 2-celled, the 2 placentae many-ovuled; stigma 2-lobed or -cleft. Capsule loculicidally 2-valved. [Name in honor of Matthias de L'Obel, a Flemish botanist of the Sixteenth Century.]

A genus of about 250 species of wide geographic distribution. Type species, *Lobelia dortmanna* L.

Corolla red, 25–50 mm. long.   1. *L. cardinalis graminea.*
Corolla blue, violet or whitish, 10–20 mm. long.
    Plant aquatic; leaves all in a basal rosette.   2. *L. dortmanna.*
    Plant not aquatic; stems leafy.
        Corolla-tube 2.5–4 mm. long, conspicuously cleft dorsally; slender perennials; Canada and adjacent
           United States.   3. *L. kalmii.*
        Corolla-tube 10–18 mm. long, not cleft dorsally at the apex; coarse perennials; southern California and
           adjacent Lower California.   4. *L. dunnii serrata.*

1. Lobelia cardinàlis subsp. gramínea (Lam.) McVaugh. Western Cardinal Flower. Fig. 5126.

Lobelia graminea Lam. Encycl. **3**: 583. 1791.
Lobelia splendens Willd. Hort. Berol. *pl. 86.* 1809.
Lobelia ignea Paxton, Paxton's Mag. Bot. **6**: 247. 1839.
Lobelia cardinalis subsp. *graminea* McVaugh in Woodson & Schery, Ann. Mo. Bot. Gard. **27**: 347. 1940.

Perennial by offsets, with fibrous roots and stout, simple, erect stems often purplish below, glabrous or sparsely hirsutulous. Lower leaves oblanceolate, tapering to a winged petiole, 5–10 cm. long, entire and somewhat wavy on the margin or broadly crenate; stem-leaves linear-lanceolate to linear, mostly acute at apex and narrowed below the middle to the subsessile base, 5–15 cm. long, 5–12 mm. wide, subentire to rather distantly and finely pectinately toothed; racemes few- to many-flowered, 5–15 cm. long; bracts broadly to narrowly subulate, the lower often simulating the upper leaves; pedicels ascending, the lower usually exceeding the calyx, the upper gradually shorter; calyx-lobes lanceolate-subulate, 6–8 mm. long, much longer than the tube; corolla crimson, 2.5–4 cm. long, the tube split to near the base dorsally, fenestrate; anthers ex-serted, 4–5 mm. long; capsule cup-shaped, strongly ribbed, broader than high.

Moist shady slopes and creek banks, Upper Sonoran and Arid Transition Zones; Death Valley region, Inyo County, to the San Gabriel and San Bernardino Mountains, California, south to Lower California and east to southern Utah and western Texas and in Chihuahua. Type locality: probably Panama. July–Oct.

Two varieties of this variable subspecies occur in the Pacific States: *Lobelia cardinalis* subsp. *graminea* var. *pseudosplendens* McVaugh (Ann. Mo. Bot. Gard. **27**: 348. 1940) and *L. cardinalis* subsp. *graminea* var. *multiflora* (Paxton) McVaugh (op. cit. 349). For a complete account of varieties, see N. Amer. Fl. **32A**: 80–82. 1942. The variety *pseudosplendens* is an essentially glabrous plant having narrow, finely serrulate leaves and a rather short inflorescence. It occurs around springs and streams in the mountains of the Death Valley region south to Los Angeles and San Diego Counties, California, and in Lower California, eastward to Nevada, western Texas, and Chihuahua; also Oaxaca. The second variety, *multiflora,* characterized by greater stature, wider leaves, an ample, many-flowered inflorescence, and a short pubescence throughout, seems to occur more sparingly. It has been collected in San Bernardino and San Diego Counties, California, and ranges through western Texas and Mexico to Guatemala and Honduras.

2. Lobelia dortmánna L. Water Lobelia or Water Gladiole. Fig. 5127.

Lobelia dortmanna L. Sp. Pl. 929. 1753.
Dortmanna lacustris G. Don, Gen. Hist. Pl. **3**: 715. 1834.
Rapuntium dortmanna Presl, Prodr. Mon. Lob. 18. 1836.

Glabrous, simple, rarely branched, aquatic herbs 20–35 cm. high, immersed except for the inflorescence, with fibrous roots and erect stems bearing scattered fleshy bracts about 5 mm. long. Leaves several to many in a basal rosette, fleshy, linear, mostly obtuse at apex, 2–5 cm. long; in-florescence loosely racemose, usually 5–6-flowered, the pedicels more or less horizontal in anthesis and recurved in fruit, about 10 mm. long, each subtended by a fleshy bract 2–3 mm. long; calyx-lobes entire, 1.5–2.5 mm. long; corolla pale violet to white, glabrous except for lightly bearded base of lower lip; corolla-tube 6–7 mm. long, split dorsally nearly to the base; lower lip little or not at all reflexed, with 3 narrowly ovate lobes longer than the recurved, linear, upper lobes; anther-tube gray or black, the 3 larger anthers more densely bearded than the other 2; capsule pendant, obconic, 6–12 mm. long.

Ponds and lake margins, Humid Transition and Boreal Zones; British Columbia south to Jefferson County, Oregon, and eastward to Newfoundland and northeastern Pennsylvania; also in Europe. Type locality: Europe. July–Sept.

3. Lobelia kálmii L. Brook or Kalm's Lobelia. Fig. 5128.

Lobelia kalmii L. Sp. Pl. 930. 1753.
Rapuntium kalmii Presl, Prodr. Mon. Lob. 23. 1836.
Lobelia falcata Raf. New Fl. **2**: 18. 1837.
Lobelia kalmii var. *strictiflora* Rydb. Mem. N.Y. Bot. Gard. **1**: 378. 1900.

Plants perennial, slender, simple or diffusely branched, glabrous or slightly pubescent at base, 1.5–3.5 dm. high or sometimes tufted from a rosette-like mat. Basal leaves when present spatu-late, 0.5 cm. wide, 1.5 cm. long; stem-leaves few, thin, subentire to denticulate, narrowly to broadly linear, acute or obtuse at apex, 0.5–4 cm. long; inflorescence about one-half the height of plant, in a loose raceme, bracts linear; pedicels 8–18 mm. long with 2 bracteoles about midway of the pedicel; calyx-lobes equaling the tube or longer; corolla blue marked with white, or white, the tube dorsally cleft, 2.5–4 mm. long; lobes of the upper lip lanceolate, curved upward, about equaling the tube in length; lower lip longer than the upper, the 3 lobes ovate, apiculate; anther-tube about 2 mm. long, bluish gray, the 2 smaller anthers white-tufted, the 3 larger smooth or pubescent; capsule not completely inferior, campanulate or subglobose, 4–9 mm. long.

In moist places, Canadian Zone; in the Pacific States this species has been collected only at Priest Rapids on the Columbia River, Kittitas County, Washington (*Cotton 1382*); however, it is found in the Columbia drainage in British Columbia; northern Canada south to Pennsylvania, the Great Lakes region, and west to Montana and British Columbia. Type locality: Canada. Collected by Kalm. July–Sept.

4. **Lobelia dúnnii** var. **serràta** (A. Gray) McVaugh. Rothrock's Lobelia. Fig. 5129.

Palmerella debilis var. *serrata* A. Gray, Bot. Calif. 1: 620. 1876.
Lobelia rothrockii Greene, Pittonia 1: 297. 1889.
Laurentia debilis var. *serrata* McVaugh, Bull. Torrey Club 67: 144. 1940.
Lobelia dunnii var. *serrata* McVaugh, op. cit. 795.

Plants perennial, stems simple or branched, mostly decumbent at base, 3–5 dm. high, glabrous or more or less pilose below, glabrate or often puberulent or hispidulous above. Basal leaves spatulate or broadly obovate to oblanceolate, gradually narrowed at base to a short winged petiole or the very lowest broadly ovate to suborbicular and abruptly narrowed to a short winged petiole; stem-leaves many, oblanceolate or linear-lanceolate, usually broadest near the middle, sessile or tapering to a winged petiole, acute or acuminate at apex, 5–25 mm. wide, 3–15 cm. long, more or less conspicuously and sharply serrate, thin, glabrate or sometimes rather sparsely puberulent or appressed-pilose; raceme 2–3 cm. long, often appearing subcapitate, 3–15-flowered; lower pedicels 3–10 mm. long; floral bracts subulate, longer than the pedicels, the lower exceeding the calyx; calyx-tube 4–8 mm. long; calyx-lobes subulate, 8–14 mm. long, bristly-pubescent or glabrous; corolla blue, with whitish tube, slender, 12–18 mm. long, the tube split from near the base to about half its length; upper lip erect, 3–4 mm. long; lower lip widely spreading, 6–9 mm. long, deeply

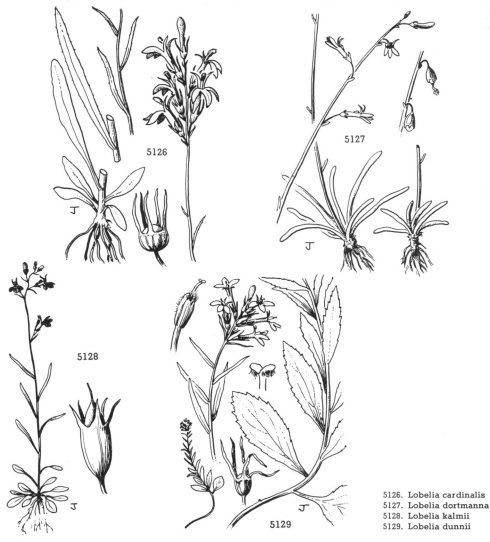

5126
5127
5128
5129

5126. Lobelia cardinalis
5127. Lobelia dortmanna
5128. Lobelia kalmii
5129. Lobelia dunnii

3-lobed; capsule 6–7(12) mm. long, glabrous or sparsely bristly, the seeds about 0.5 mm. long, smooth, shining, light brown.

Shaded canyons and stream banks, Upper Sonoran Zone; California Coast Ranges, Monterey County to San Diego County, south to Lower California, where it occurs and merges with the typical species in the Sierra Juarez and Sierra San Pedro Martir. Type locality: valley of Ojai Creek, Ventura County, California. Collected by Rothrock. June–Nov.

Family 149. COMPÓSITAE.

Sunflower Family.

Annual or perennial herbs, subshrubs, or shrubs, sometimes scandent. Leaves exstipulate, opposite or alternate or in some herbs basal only, entire or more or less dissected. Flowers usually many, borne in a head* (rarely with 1 flower only) on the enlarged summit of the peduncle (receptacle), surrounded by an involucre of variously modified bracts (phyllaries), these in 1 to several series and often imbricate. Heads solitary or in an inflorescence of several to many heads. Flowers of the heads epigynous, gamopetalous, with or without subtending bracts (receptacular bracts), all ligulate with strap-shaped corollas, or all discoid with tubular or expanding, equally toothed or lobed corollas (homogamous), or the ligulate and discoid corollas combined in 1 head (heterogamous), the ligulate marginal and called rays, or the corollas rarely lacking; flowers hermaphrodite (functionally perfect), staminate, pistillate, or neutral with vestigial parts. Stamens 5, the anthers united into a tube (rarely distinct or nearly so in some genera) around the lengthening style, rounded, sagittate or caudate at base, sometimes appendaged above. Style 1, usually with 2 branches, these bearing a stigmatic surface and often appendaged; functionally staminate flowers often with an undivided style. Fruit an achene, with or without a pappus of scales, teeth, awns, or bristles.

A family of cosmopolitan distribution, probably the largest family of plants with 900 or more genera and over 15,000 species.

Key to the Tribes of the Compositae
(The characters used in this key apply to the species in the Pacific States)

Corollas of the heads all with 3-toothed marginal ligules (rays) and the remainder tubular, or all flowers of the heads tubular and regularly toothed or divided, or all flowers completely or in part bilabiate; herbage without latex (except 146, *Gazania*).
    All corollas deeply cleft, either bilabiate or regularly cleft, with the lobes linear or nearly so, commonly one-third or more the length of the corolla.
        Corollas bilabiate (except 149, *Hecastocleis*, with 1 flower in each involucre); styles not thickened nor with a ring of hairs below the truncate style-branches. Tribe 11. *MUTISEAE* (genera 147–149, p. 549).
        Corollas regularly cleft (some *Centaurea* spp. irregular and falsely radiate); styles thickened below or with a ring of hairs below the style-branches, these in the perfect flowers short-branched in relation to the length of exserted style.
            Herbage, at least the phyllaries (except 137, *Saussurea*, and some spp. of *Centaurea* with fringed phyllaries), prickly or spiny; receptacle long-fimbrillose or setose (except 139, *Onopordum*). Tribe 9. *CYNAREAE* (genera 135–145, p. 506).
            Herbage (in ours) not at all prickly or spiny; phyllaries fused basally; receptacle naked; garden ruderal. Tribe 10. *ARCTOTIDEAE* (genus 146, *Gazania*, p. 548).
    Disk-corollas mostly toothed or short-lobed, the lobes not markedly narrow, if rather long not associated with short-lobed style-branches.
        Anthers distinct or nearly so; pistillate corollas lacking or rudimentary; monoecious, with the unisexual heads separated on the inflorescence in genera 22–25. Subtribe *AMBROSIINAE*, Tribe 1. *HELIANTHEAE* (genera 19–25, p. 140).
        Anthers connate into a tube; pistillate corollas present; heads either homogamous or heterogamous (dioecious in 94, *Baccharis*).
            Receptacle with long (at least equaling the achenes) chaffy, hyaline or sometimes bristle-like bracts subtending all the flowers of the heads or only those either toward the margin of the receptacle or toward its center.
                Heads discoid; anthers caudate at base; woolly, mostly dwarfed annuals with phyllaries and hyaline receptacular bracts often much modified. Tribe 7. *INULEAE*, in part (genera 125–129, p. 465).
                Heads radiate,† the rays sometimes inconspicuous; anthers sagittate or rounded at base; perennials or shrubs, or if annuals not densely woolly and dwarfed.
                    Style-branches truncate or penicillate; phyllaries with wide scarious or hyaline margins. Tribe 4. *ANTHEMIDEAE*, in part (genera 95–97, p. 387).
                    Style-branches subulate, the appendages hairy on both sides; phyllaries without wide scarious margins.
                        Receptacle chaffy throughout; receptacular bracts long, folded about the achene (either flat or bristle-like in 11, *Eclipta*; 12, *Bidens*; 13, *Coreopsis*; 14, *Galinsoga*). Tribe 1. *HELIANTHEAE*, in part (genera 1–18, p. 99).

---

* The central portion of the flowering head, i.e., exclusive of the rays, known as the disk.

† Heads discoid in 15, *Bebbia*; 16, *Eastwoodia*; 97, *Santolina*; and some spp. in 1, *Wyethia*; 5, *Rudbeckia*; 8, *Encelia*; 9, *Geraea*; 12, *Bidens*.

Receptacular bracts mostly in 1 or 2 series, variously modified. Subtribe *MADI-INEAE,* Tribe 1. HELIANTHEAE (genera 26–35, p. 154).

Receptacle naked (shallowly or deeply pitted), sometimes inconspicuously short-fimbrillate or short-hairy (densely short-hairy in 38, *Whitneya;* with long setae in 46, *Gaillardia*). *

Style-branches subterete and clavate (except 130, *Trichocoronis,* with style-branches short and somewhat flattened); heads homogamous, discoid; corollas white or pink. Tribe 8. *EUPATORIEAE* (genera 130–134, p. 496).

Style-branches not subterete nor clavate; heads heterogamous, discoid or radiate; disk-corollas commonly yellow, the rays when present yellow, purple, blue or white.

Phyllaries dry or scarious or with a wide scarious margin (concealed by wool in some *Inuleae;* scarious margin not conspicuous in 103, *Soliva*).

Pappus a short crown or paleaceous or none; heads radiate or eradiate. Tribe 4. *AN-THEMIDEAE,* in part (genera 98–105, p. 387).

Pappus of capillary bristles; heads eradiate (except 120, *Inula*). Tribe 7. *INULEAE,* in part (genera 120–124, p. 465).

Phyllaries herbaceous or firm (sometimes partially cartilagenous), the scarious margins very narrow or lacking (scarious margin conspicuous and colored in 63, *Blennosperma*).

Phyllaries few- to many-seriate and well imbricated (except 80, *Bellis;* 82, *Monoptilon;* 91, *Erigeron;* 93, *Conyza;* 94, *Baccharis*); style-branches (perfect flowers) flattened, the stigmatic lines conspicuous and usually with a hairy appendage glabrous on the inner face. Tribe 3. *ASTEREAE* (genera 68–94, pp. 253–254).

Phyllaries mostly in 1 or 2 series, sometimes more, little or not at all imbricate (except 39, *Venegasia;* 40, *Jaumea;* 118, *Lepidospartum*), subequal (short calyculate phyllaries sometimes present).

Pappus of usually copious capillary bristles (epappose in 106, *Adenocaulon;* plumose in 109, *Raillardella;* pappus scanty in some spp. of 108, *Arnica*); style-branches (perfect flowers) truncate and penicillate at apex (short-appendaged in 111, *Crocidium*). Tribe 5. *SENECIONEAE* (genera 106–118, p. 416).

Pappus awn-like, bristly, paleaceous or none; style-branches (perfect flowers) various.

Disk-flowers perfect and fertile (sterile in 38, *Whitneya;* 63, *Blennosperma*); anthers sagittate; achenes various. Tribe 2. *HELENIEAE* (genera 36–67, pp. 193–194).

Disk-flowers sterile; anthers caudate; rare ruderal with curved warty achenes. Tribe 6. *CALENDULEAE* (genus 119, *Calendula,* p. 464).

Corollas of the heads all with 5-toothed ligules and perfect; herbage with latex. Tribe 12. *CICHORIEAE* (genera 150–176, pp. 551–552).

Tribe 1. HELIANTHEAE

Receptacular bracts folded about the achene or at least strongly concave.
Pappus various or lacking, never plumose.
Inner phyllaries plane or nearly so, never completely enclosing the ray-achene; coarse herbs or shrubs.
Achenes sharply or obscurely angled or lenticular, not noticeably margined.
Shrubs or distinctly shrubby-based plants.
Heads rayed; leaves opposite at least below; deserts and coastal San Diego County.
3. *Viguiera.*
Heads rayless; leaves alternate; California Inner Coast Ranges. 16. *Eastwoodia.*
Annual or perennial herbs.
Rays 2–4 mm. long, firm and persisting on the achene; small annual of eastern San Bernardino County. 18. *Sanvitalia.*
Rays if present 1.5–5 cm. long, deciduous or withering; coarse annuals or perennials.
Plants scapose or subscapose from a heavy aromatic root. 2. *Balsamorhiza.*
Plants evidently leafy-stemmed, the basal leaves of some species conspicuous and longer than the stem-leaves.
Disk flat or low-convex.
Pappus-paleae persistent; ray-flowers fertile. 1. *Wyethia.*
Pappus-paleae deciduous; ray-flowers sterile. 4. *Helianthus.*
Disk conic, 2–6 cm. high (low-conic in *R. hirta*). 5. *Rudbeckia.*
Achenes flat (rather plump but thin-margined in *Helianthella castanea*) with a white wing or margin, this sometimes ciliate.
Scapose perennial herbs; heads very large. 7. *Enceliopsis.*
Leafy-stemmed herbs or shrubs; heads of medium size.
Basal leaves many, persistent, the stem-leaves few; pappus of 2 persistent awns and also squamellae (lacking in *H. californica*). 6. *Helianthella.*
Basal leaves not many, not persistent; pappus various.
Achenes conspicuously long-ciliate on the margins (less conspicuous in *Geraea viscida*).
Shrubs. 8. *Encelia.*
Annuals or short-lived perennials. 9. *Geraea.*
Achenes not ciliate on margin but conspicuously winged. 10. *Verbesina.*
Inner phyllaries completely enclosing the ray-achenes; introduced annual in southern California.
17. *Melampodium.*
Pappus of 15–20 plumose bristles; intricately branched desert shrub. 15. *Bebbia.*
Receptacular bracts flat or bristle-like.
Phyllaries in 2 distinctly unlike series; rays yellow, usually conspicuous (lacking or whitish in *B. frondosa* and *B. pilosa*).
Pappus of retrorsely hispid awns. 12. *Bidens.*
Pappus various or lacking, not of retrorsely hispid awns. 13. *Coreopsis.*
Phyllaries with the series essentially alike; rays white or pinkish, inconspicuous.
Receptacular bracts bristle-like; achenes epappose or essentially so. 11. *Eclipta.*
Receptacular bracts flat and narrow; achenes with scarious fimbriate or aristate paleae.
14. *Galinsoga.*

---

* A few longer setae or bracts sometimes present in some spp. of *Chaenactis* and *Eriophyllum.*

1. WYÈTHIA Nutt. Journ. Acad. Phila. 7 : 39. *pl. 5.* 1834.

Leafy-stemmed, balsam-scented, perennial herbs, glabrous, pubescent, or tomentose, resinous, usually from a thick, vertical, fusiform or subcylindric caudex, sometimes branching below the ground to produce cylindrical underground stems, those above the ground simple or branched above, erect or ascending. Basal leaves often large, the stem-leaves alternate, linear to oblong or suborbicular or ovate-deltoid, usually petioled, entire or toothed. Heads radiate (discoid in one species), yellow (or rays white or whitish in one species), medium-sized or large, solitary or rather few, terminal and axillary. Involucre campanulate or hemispheric, the phyllaries subequal or sometimes graduate, about 2–4-seriate, mostly lanceolate to obovate, herbaceous throughout or with indurate base, the outer sometimes foliaceous. Receptacle flattish or rounded, the receptacular bracts firm, persistent, folded, half-embracing the achenes. Ray-flowers pistillate, fertile, with an oblong to oval, usually 2–3-denticulate ligule. Disk-flowers perfect, fertile, their corollas with short tube and subcylindric or cylindric-funnelform throat, 5-toothed. Anthers subentire or shortly sagittate at base, the appendages ovate. Style-branches hirsutulous nearly to base, with slenderly subulate terminal appendages. Achenes oblong or linear-oblong, more or less regularly quadrangular (or those of rays trigonous), plump, with intermediate nerves, glabrous or puberulous. Pappus a crown of basally united, more or less lacerate squamellae and often 1–4 slender awns or these reduced to teeth or lacking. [Named for Captain Nathaniel J. Wyeth, the collector of the original species, with whom Nuttall subsequently crossed the continent.]

A genus of 14 species, of the western United States and British Columbia. Type species, *Wyethia helianthoides* Nutt.

Natural hybrids have been reported by W. A. Weber (Amer. Midl. Nat. **35**: 400–452. 1946) as occurring not infrequently in the sections *Alarconia* and *Euwyethia*.

Leaves essentially uniform, ovate to orbicular, broad at base and all on definite petioles, without conspicuous tufts of basal leaves.
  Plants low, 0.5–3.5 dm. high; involucres campanulate.
    Heads axillary, much surpassed by the subtending leaves; rays few (5–8), about 1.5 cm. long.
      1. *W. ovata.*
    Heads terminal, the inflorescence surpassing the leaves; rays more numerous (8–12), about 3 cm. long.
      2. *W. bolanderi.*
  Plants tall, 4–10 dm. high; involucres hemispheric.
    Rays present, 2–5 cm. long.
      Leaves green beneath, merely sparsely pilose chiefly on the nerves; plant resinous-viscid; achenes with minute, coroniform, awnless pappus.
        3. *W. reticulata.*
      Leaves cinereously pilose-subtomentose beneath; plant glandular but not resinous-viscid; achenes with conspicuous pappus bearing short stout awns.
        4. *W. elata.*
    Rays absent or occasionally with 1–3 short rays.
      5. *W. invenusta.*
Leaves not uniform, mostly oblong, elliptic, or lanceolate, the basal tufted, usually much larger than the often sessile cauline leaves.
  Involucre very large, 2.5–8.5 cm. high, foliaceous, usually equaling or surpassing the rays, the phyllaries numerous, oblong or lanceolate to broadly ovate.
    Plants stipitate-glandular throughout, otherwise glabrous; achenes obscurely glandular; receptacular bracts merely glandular.
      6. *W. glabra.*
    Plants gray-tomentose throughout, sometimes glabrescent in age; achenes puberulous above; receptacular bracts densely pubescent above.
      7. *W. helenioides.*
  Involucre smaller, 1.3–4 cm. high, usually much shorter than the rays.
    Plants densely and somewhat floccosely gray-tomentose throughout at least when young, somewhat glabrescent in age but with persistent tomentum at least on the petioles and apex of peduncles.
      8. *W. mollis.*
    Plants glabrous or pubescent, never tomentose.
      Plants definitely pilose or hirsute; achenes pubescent above.
        Rays pale yellow to white; leaves conspicuously ciliate or ciliolate.
          9. *W. helianthoides.*
        Rays bright yellow; leaves not obviously ciliate.
          10. *W. angustifolia.*
      Plants glabrous or nearly so throughout or slightly puberulous on leaves or hirsute on involucre; achenes glabrous.
        Upper stem-leaves narrowed to base; plant not strongly resinous.
          11. *W. longicaulis.*
        Upper stem-leaves conspicuously clasping; plant usually strongly resinous.
          12. *W. amplexicaulis.*

1. Wyethia ovàta Torr. & Gray. Southern Wyethia. Fig. 5130.

Wyethia ovata Torr. & Gray ex Torr. in Emory, Notes Mil. Rec. 143. 1848.
Wyethia coriacea A. Gray, Proc. Amer. Acad. 11 : 77. 1876.
Wyethia ovata var. *funerea* Jepson, Man. Fl. Pl. Calif. 1170. 1935. As to type, *Aster tortifolius* var. *funereus* Jepson.

Low perennial from a thick rootstock; stems usually several, 5–30 cm. high, stout, densely and loosely canescent-pilose or subtomentose, becoming glabrate, simple or little-branched, bearing a few scales at base, few-leaved. Leaves much surpassing the stems and flowers, the blades suborbicular to oval or broadly ovate, 5–20 cm. long, about equaling the petioles, obtuse, rounded to subcordate at base, entire, coriaceous, feather-veined or obscurely triplinerved and strongly reticulate, subcanescently subappressed-pilose or subtomentose, in age green and often glabrescent; heads usually solitary in the axils and at tip of stem, short-peduncled; involucre rather narrowly campanulate, 2–5.8 cm. high, the outer phyllaries 4–6, lanceolate to obovate, usually acuminate, usually equaling or surpassing the rays, coriaceous-herbaceous, veiny, pubescent like the leaves, the inner much shorter; rays 5–9, 1–2 cm. long; achenes about 1 cm. long, glabrous or very sparsely pubescent; pappus a variously toothed or lobed crown 1.5 mm. or less high, not awned.

Grassy hills and open pine woods, Arid Transition Zone; southern Sierra Nevada in Tulare County, California, and the mountains of southern California and adjacent Lower California. Type locality: "Abundant on the western side of the Cordilleras of California" between Warner's Ranch and San Isabel, San Diego County, California. May–Aug.

2. **Wyethia bolánderi** (A. Gray) W. A. Weber. Bolander's Wyethia. Fig. 5131.

Balsamorhiza bolánderi A. Gray, Proc. Amer. Acad. **7**: 356. 1868.
Wyethia bolanderi W. A. Weber, Amer. Midl. Nat. **35**: 419. 1946.

Perennial, 20–30 cm. high, essentially glabrous throughout, glutinous, the stem stoutish, 2–3-leaved, bearing at base a few scales about 2.5 cm. long. Basal and stem-leaves similar, alternate, ovate to suborbicular, 6–12 cm. long and nearly as wide, rounded to apiculate, at base rounded to subcordate and unequal, entire, coriaceous, feather-veined and strongly reticulate, the lower pair of veins conspicuous, thus appearing to be triplinerved; head solitary, short-peduncled; involucre in 2 series; outer phyllaries 4–6, obovate to ovate or oblong, acute or apiculate, coriaceous, reticulate, resinous, ciliate, leaf-like, thinly pilose, 2–2.5 cm. long, 5–12 mm. wide; inner phyllaries linear-lanceolate, acuminate, pilose, about 1.5 cm. long; rays 7–12, yellow, 2.5–3.5 cm. long; achenes glabrous, brownish, about 7.5 mm. long, crowned with a shallow, sometimes 4-toothed ring, those of the ray obcompressed, those of the disk clavate-quadrangular.

Mostly growing in chaparral, lower Arid Transition Zone; foothills of the Sierra Nevada, California, from Butte County to Mariposa County. Type locality: Auburn, Placer County. March–May.

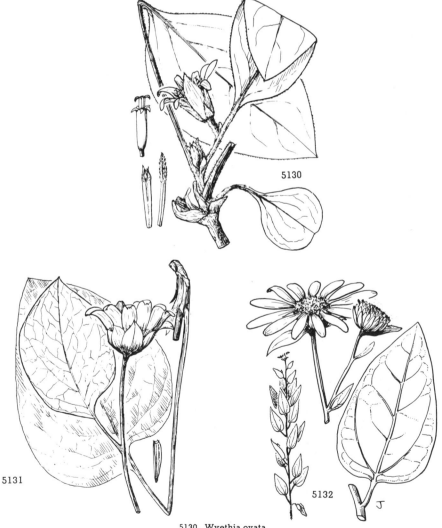

5130. Wyethia ovata
5131. Wyethia bolanderi
5132. Wyethia reticulata

3. **Wyethia reticulàta** Greene. Eldorado Wyethia. Fig. 5132.

Wyethia reticulata Greene, Bull. Calif. Acad. **1**: 9. 1884.

Perennial, resinous-viscid and very sparsely hispid, becoming scabrous on older herbage, the stem reddish, 4–7 dm. high, simple, viscid-glandular and leafy. Leaves all similar, alternate; petioles stout, naked, similarly pubescent, 1.5–3.5 cm. long; blades triangular-ovate, 5.5–14.5 cm. long, 2.5–8 cm. wide, acute, truncate to subcordate at base, crenate, serrulate or subentire, obscurely triplinerved, glossy, reticulate-veiny, becoming coriaceous; heads solitary or few in the axils of the somewhat reduced upper leaves, the peduncles about 4 cm. long; involucres hemispheric, 1–2 cm. broad; outer phyllaries longer than the inner, linear-lanceolate, acute, with spreading tips, becoming reticulate and thickened, the inner rigid, erect; rays 10–16, 20–25 cm. long; achenes about 6 mm. long, glabrous; pappus coroniform and lacking awns.

Growing under live oaks or in chaparral, Arid Transition Zone; foothills of the Sierra Nevada in Eldorado County, California. Type locality: banks of Sweetwater Creek, Eldorado. County. Collected by Mrs. Curran. May–July.

4. **Wyethia elàta** H. M. Hall. Hall's Wyethia. Fig. 5133.

Wyethia ovata A. Gray, Proc. Amer. Acad. **7**: 357. 1868. Not Torr. & Gray, 1848.
Wyethia elata H. M. Hall, Univ. Calif. Publ. Bot. **4**: 208. 1912.

Perennial, the stem simple or branched above, densely spreading-pilose, glandular above, 0.6–1 m. high. Leaves alternate, somewhat reduced toward the apex of the stem, on stout naked petioles 1–6 cm. long, the blades ovate, 8–21 cm. long, 4.5–12 cm. wide, acute, at base broadly rounded to shallowly cordate, entire to crenate, triplinerved or feather-veined, rather thin but firm, above rather densely incurved-pilose but green, in age sometimes scabrous, beneath softly and cinereously pilose-subtomentose and densely gland-dotted; heads solitary at tips of stem and branches, 2.5–4 cm. wide, the peduncles 2–4 cm. long; involucre hemispheric, 1.5–2 cm. high, the phyllaries about 3-seriate, slightly graduate or subequal, or the outer enlarged and surpassing the disk, linear-lanceolate to obovate, acute or acuminate, pubescent like the leaves, herbaceous with indurate base; rays about 10–20, 3–5 cm. long; achenes about 8–12 mm. long, glabrous or sparsely strigillose; pappus a toothed crown 1–2 mm. high, two of the teeth usually larger than the others.

Open pine woods, Arid Transition Zone; slopes of the Sierra Nevada from Mariposa County to Tulare County, California. Type locality: Clark's Ranch (now Wawona), Mariposa County. Collected by Bolander. May–Aug.

5. **Wyethia invenústa** (Greene) W. A. Weber. Coville's Wyethia. Fig. 5134.

Helianthus (?) invenustus Greene, Pittonia **1**: 284. 1889.
Balsamorhiza invenusta Coville, Contr. U.S. Nat. Herb. **4**: 130. 1893.
Wyethia invenusta W. A. Weber, Amer. Midl. Nat. **35**: 421. 1946.

Resinous-viscid perennial 4–10 dm. high, the stem leafy, simple or branched above, densely and finely glandular and sparsely spreading-hispid. Leaves alternate; petioles 1–6 cm. long; blades oblong- to triangular-ovate, 8–17 cm. long, acuminate to obtuse, at base shallowly cordate to cuneate, hispid or hirsute with tuberculate-based hairs especially along the veins and somewhat glandular, feather-veined or obscurely triplinerved; heads about 2 cm. thick, solitary or paired at tip of stems and branches, on peduncles 3–15 cm. long; outer phyllaries about 8, oblong-lanceolate to obovate, 1.5–2 cm. long, 3–7 mm. wide, acute, strongly veined, herbaceous above the short indurated base, glandular and hispid especially on margin; inner phyllaries about 2-seriate, shorter, with short herbaceous tip, pilose; ray-flowers none or a few occasionally present; disk-flowers orange; achenes dull black, subquadrangular, striate, about 8 mm. long, the pappus lacking.

Open woodland in pine and oak forest, Arid Transition Zone; Sierra Nevada and the Greenhorn Range, California, from Fresno County to Kern County. Type locality: Greenhorn Mountains 10 or 12 miles west of Kernville, Kern County. June–July.

6. **Wyethia glàbra** A. Gray. Mule-ears. Fig. 5135.

Wyethia glabra A. Gray, Proc. Amer. Acad. **6**: 543. 1865.

Perennial from a vertical, subcylindric, resiniferous caudex, stipitate-glandular throughout and balsam-scented, the stems often several, 2–6 dm. high, stout, simple, few-leaved, bearing a few membranous scales at base. Basal leaves on petioles 4–11 cm. long, the blades oblong or oval, 18–38 cm. long, 7.5–12 cm. wide, obtuse or acute, at base cuneate to rounded, entire to crenate-toothed, rather thin, feather-veined; stem-leaves similar but smaller, sometimes elliptic or ovate, short-petioled; heads solitary, very large, peduncled; involucre broadly hemispheric, 4–7 cm. long, of numerous oblong to obovate or broadly ovate, foliaceous phyllaries, subequal or the outer usually larger, glandular; rays about 12–27, usually equaled or surpassed by the involucre, 2–3.5 cm. long; receptacular bracts stipitate-glandular; achenes 9–11 mm. long, obscurely glandular; pappus 2.5–5 mm. long, cleft into about 2–8 unequal, lanceolate or ovate teeth.

Dry foothills, Upper Sonoran Zone; in the Coast Ranges from Mendocino and Lake Counties to San Luis Obispo County, California. Type locality: Marin County, California. March–June.

7. **Wyethia helenioìdes** (DC.) Nutt. Gray Mule-ears. Fig. 5136.

Alarconia helenioides DC. Prod. **5**: 537. 1836.
Wyethia helenioides Nutt. Trans. Amer. Phil. Soc. II. **7**: 353. 1840.
Melarhiza inuloides Kell. Proc. Calif. Acad. **1**: 37. 1855.

Stout perennial 3–6 dm. high with the habit of *W. glabra* but densely and somewhat floccosely gray-tomentose throughout at least when young, in age often glabrescent, somewhat resinous but not obviously glandular. Leaves essentially as in *W. glabra* but almost always entire; heads 1–3, very large; involucre 2.5–8.5 cm. high, sometimes leafy-bracted, mostly subequal, the phyllaries numerous, oblong or lanceolate to broadly ovate, herbaceous, gray-tomentose; rays about 13–20, usually equaled or exceeded by the involucre, 2–3.5 cm. long; receptacular bracts densely pubescent above; achenes 12–15 mm. long, sparsely puberulous above; pappus 2.5–5 mm. long, cleft into 4–8 often unequal, mostly triangular, ciliolate teeth.

Grassy slopes or oak woodland, Upper Sonoran Zone; in the Sierra Nevada foothills, California, from Eldorado County to Mariposa County and in the Coast Ranges from Mendocino, Tehama, and Colusa Counties south to San Luis Obispo County. Type locality: California. March–July.

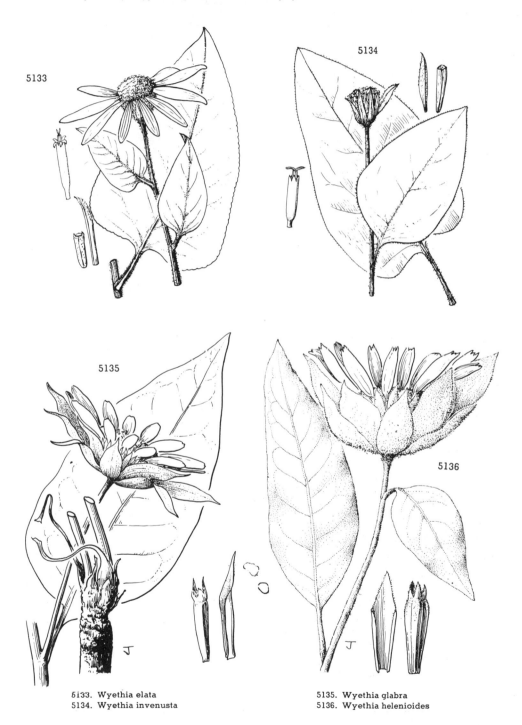

5133. Wyethia elata
5134. Wyethia invenusta
5135. Wyethia glabra
5136. Wyethia helenioides

8. Wyethia móllis A. Gray. Mountain Mule-ears. Fig. 5137.

Wyethia mollis A. Gray, Proc. Amer. Acad. **6**: 544. 1865.

Stout perennial from a thick vertical caudex, densely and somewhat floccosely canescent-tomentose throughout at least when young, in age greener and somewhat glabrescent; stems simple, 0.4–1 m. high. Basal leaves on thick naked petioles 4–20 cm. long, their blades elliptic to oblong-ovate, 9–50 cm. long, acute or obtuse, at base cuneate to subcordate, firm, entire, feather-veined and reticulate; stem-leaves few, much smaller, elliptic, oblong, or ovate, short-petioled; heads 2–4, solitary at apex of stem and in the upper axils, peduncled, medium-sized; involucre campanulate, 1.3–3 cm. high, the principal phyllaries rather few, elongate-triangular to oblong, acute or acuminate, herbaceous, erect, usually equaling or exceeding the disk but shorter than the rays; rays about 5–9, 1.5–3.5 cm. long; receptacular bracts pubescent above; disk-achenes appressed-pubescent above, 8–11 mm. long; pappus a short, lacerate, pubescent crown and usually also 2–4 slender, unequal, linear-lanceolate awns 7 mm. long or less, sometimes reduced to short teeth or wanting.

Open ridges and dry woods, Arid Transition and Canadian Zones; southern Oregon in Klamath and Lake Counties south through northeastern California and the Sierra Nevada and adjacent Nevada to Fresno County, California. Type locality: Mono Lake and summit of Sonora Pass in the Sierra Nevada, California. May–Aug. Woolly Dwarf Sunflower.

9. Wyethia helianthoìdes Nutt. White-rayed Wyethia or Mule-ears. Fig. 5138.

Wyethia helianthoides Nutt. Journ. Acad. Phila. **7**: 40. *pl. 5.* 1834.

Perennial, from a thick, fusiform or carrot-shaped caudex and root; stems ascending, 15–40 cm. high, stipitate-glandular especially above and loosely spreading-pilose, stout, leafy. Basal leaves on petioles 3.5–5 cm. long, these often margined above, the blades elliptic to oblong or oval, 8–30 cm. long, 3–9 cm. wide, obtuse to acute, at base cuneate to subcordate, thinnish, feather-veined, pilose or hirsute especially toward margin, entire or slightly serrate, often wavy-margined; stem-leaves much smaller, about 7–12, elliptic to lanceolate or lance-ovate, narrowed into a short, usually margined petiole, not at all clasping, pilose-ciliate, on surface loosely pilose or subglabrous, entire; heads 1(–4), short-peduncled, 6–9 cm. wide; involucre about 4-seriate, usually subequal, 1.8–2.5 cm. high, hemispheric, sometimes foliaceous-bracted, the phyllaries numerous, linear to lance-ovate, acuminate, densely pilose-ciliate, essentially glabrous dorsally; rays about 16–20, "sulphur yellow" to cream or white, 2.5–4 cm. long; disk-flowers yellow; receptacular bracts glandular and ciliate above, sparsely pubescent on keel; achenes pubescent above; pappus coroniform, denticulate, 0.5 mm. long or less.

Wet hillsides and meadows, Arid Transition and Canadian Zones; eastern Oregon to southwestern Montana, northern Wyoming, and northern Nevada. Type locality: "Camas Plain, Flathead River." Collected by Wyeth, probably in eastern Idaho. April–July.

10. Wyethia angustifòlia (DC.) Nutt. Narrow-leaved Mule-ears. Fig. 5139.

Helianthus longifolius Hook. Fl. Bor. Amer. **1**: 313. 1834. Not Pursh, 1814.
Alarconia ? angustifolia DC. Prod. **5**: 537. 1836.
Helianthus hookerianus DC. op. cit. 590.
Wyethia angustifolia Nutt. Trans. Amer. Phil. Soc. II. **7**: 352. 1840.
Wyethia robusta Nutt. loc. cit.
Wyethia angustifolia var. *solanensis* Jepson, Man. Fl. Pl. Calif. 1080. 1925.

Perennial herb from a cylindric or fusiform vertical caudex, pilose or hirsute and somewhat resinous; stems ascending, 1.5–3 dm. high, usually simple, stout, sparsely to densely pilose or hirsute, the hairs usually spreading at least below, sometimes appressed above. Basal leaves on petioles 2–11 cm. long (often margined above), the blades linear-lanceolate to oblong, 8–50 cm. long, 1–8 cm. wide, acuminate, long-cuneate at base, papery, entire to serrate, sparsely to densely hirsute or hirsutulous on both sides with spreading to subappressed hairs; stem-leaves about 5–10, the lower similar to the basal but smaller, the upper gradually reduced, lanceolate or elliptic, short-petioled or sessile, not clasping; heads 1(–4), about 5–9 cm. wide; involucre broad, 1.5–3 cm. high, about 3-seriate, subequal or somewhat graduated, the phyllaries numerous, lanceolate to lance-ovate or obovate, acuminate to obtuse, herbaceous, strongly hirsute-ciliate, sparsely hirsute or subglabrous dorsally; rays 10–18, 1.5–3.5 cm. long; achenes puberulous at apex, about 8 mm. long; pappus a ciliolate-denticulate or lacerate crown of squamellae 2 mm. long or less and 1–4 slender hispidulous awns 9 mm. long or less, sometimes reduced to teeth.

Dry hills, wet meadows, and open pine woods, Upper Sonoran and Transition Zones; southern Washington (Klickitat County) and western Oregon south to Monterey County, California, and sparingly in the Sierra Nevada. Type locality: Monterey, California. Collected by Douglas. April–Aug.

Wyethia angustifolia var. foliòsa (Congdon) H. M. Hall, Univ. Calif. Publ. Bot. **4**: 207. 1912. (*Wyethia foliosa* Congdon, Erythea **7**: 186. 1900.) Stems 4–5.5 dm. high; upper leaves ovate to lanceolate with clasping base; heads 1–4, usually smaller; phyllaries mostly lance-linear or lanceolate. Mainly in northern California and adjacent Oregon southward in the Sierra Nevada from Shasta County to Fresno County and in the North Coast Ranges in Humboldt and Mendocino Counties, California. Type locality: "wooded slopes of the Sierras, ranging from 3,000 to 8,000 ft."

Not strongly differentiated from *W. angustifolia* var. *angustifolia* because of existing intermediate forms but recognizable by the characters outlined above.

11. Wyethia longicaùlis A. Gray. Humboldt Wyethia. Fig. 5140.

Wyethia longicaulis A. Gray, Proc. Amer. Acad. **19**: 4. 1883.

Perennial herb, essentially glabrous throughout except sometimes on involucre, resinous and shining; stem essentially simple, ascending, about 4 dm. high. Lower leaves oblanceolate, about 20 cm. long (including the narrowly margined petiole), 3 cm. wide, obtuse, long-cuneate at base, serrulate except toward base, hispidulous-ciliolate, rather firm, feather-veined; stem-leaves few, similar or lanceolate, 7–15 cm. long, the upper entire, all narrowed at base; heads solitary or few, long-peduncled, medium-sized; involucre broadly campanulate, 2 cm. high, subequal, the phyllaries rather few, oblong to lance-oblong, obtuse or acute, surpassing the disk but much shorter than the rays, thick-herbaceous, glabrous or hirsute-ciliate and somewhat hirsute dorsally; rays about 6–15, 2.5 cm. long; receptacular bracts ciliolate above, otherwise sparsely pubescent or nearly glabrous; achenes about 7 mm. long, glabrous; pappus a ciliolate-denticulate crown 0.5 mm. long.

Ridges and prairies, Transition Zones; North Coast Ranges of California in eastern Humboldt, adjacent Trinity, and northeastern Mendocino Counties. Type locality: prairies of eastern Humboldt County. May–July.

12. **Wyethia amplexicaùlis** Nutt. Northern Mule-ears. Fig. 5141.

Espeletia amplexicaulis Nutt. Journ. Acad. Phila. **7**: 38. 1834.
Wyethia amplexicaulis Nutt. Trans. Amer. Phil. Soc. II. **7**: 352. 1840.
Silphium ? laeve Hook. Lond. Journ. Bot. **6**: 244. 1847.
Wyethia lanceolata Howell, Fl. N.W. Amer. 341. 1900.
Wyethia amplexicaulis subsp. *major* Piper, Proc. Biol. Soc. Wash. **27**: 98. 1914.
Wyethia amplexicaulis subsp. *subresinosa* Piper, loc. cit.

Stout perennial, glabrous and more or less balsamic-resinous throughout, often vernicose, from a usually fusiform caudex, the stems ascending, 3–8 dm. high, essentially simple, leafy. Basal leaves on short petioles, these usually margined above, the blades oblong, elliptic-oblong or obovate,

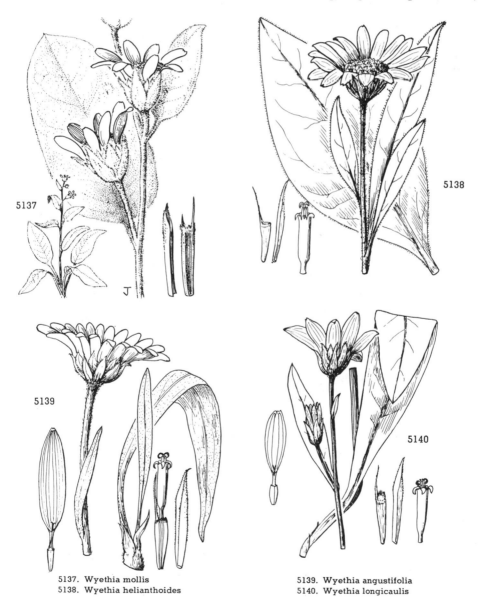

5137. Wyethia mollis
5138. Wyethia helianthoides
5139. Wyethia angustifolia
5140. Wyethia longicaulis

15–50 cm. long, acute or acuminate, cuneate to rounded at base, entire or serrate, feather-veined, rather firm; stem-leaves several, gradually reduced above, elliptic, oblong, or oblanceolate to ovate, the lower short-petioled, the upper sessile and clasping; heads 1–8, terminal and axillary, peduncled, 4–10 cm. wide; involucre hemispheric, 1.8–4 cm. high, subequal or somewhat graduate, the phyllaries numerous or rather few, lanceolate to oblong or occasionally obovate, acuminate to obtuse, mainly herbaceous, glabrous, not ciliate; rays 8–18, bright yellow, 2–5 cm. long; receptacular bracts glabrous; achenes glabrous, 7–10 mm. long; pappus a lacerate crown of squamellae 2 mm. long or less and often 1–2 slender, essentially glabrous awns 7 mm. long or less or these reduced to teeth.

Wet or dry open places, occasionally in woods, Arid Transition Zone; British Columbia, eastern Washington, and eastern Oregon to Montana, Wyoming, Colorado, Utah, and Nevada. Type locality: about Flathead River, Montana. May–July. Smooth Dwarf Sunflower.

Wyethia cusickii Piper, Proc. Biol. Soc. Wash. 27: 98. 1914. Intermediate between *W. amplexicaulis* and *W. helianthoides*, with both of which it grows, and evidently a hybrid between the two species. Type locality: Blue Mountains, Union County, Oregon.

2. **BALSAMORHÌZA** Hook. ex Nutt. Trans. Amer. Phil. Soc. II. **7**: 349. 1840.

Perennial herbs, scapose, subscapose, or sparingly leafy-stemmed, from a thick, vertical, fusiform or subcylindric, terebrinthine root and simple or branched caudex. Leaves tufted and basal, the reduced stem-leaves alternate or subopposite and those of the subscapose stems nearly basal and opposite, entire to bipinnatifid, petioled. Heads radiate, usually large, solitary or few-peduncled. Involucre broad, 2–4-seriate, graduate or subequal, the phyllaries mostly herbaceous, the outer sometimes enlarged and much surpassing the inner. Receptacle flattish, the receptacular bracts firm, conduplicate. Ray-flowers pistillate, fertile, the ray 2–3-denticulate; disk-flowers perfect, fertile, with a short tube and cylindric or cylindric-campanulate throat, 5-toothed. Style-branches slender, hispid or hispidulous with slender, subulate, terminal appendages. Anthers subentire or shortly sagittate. Achenes oblong, epappose, glabrose or pubescent, the ray-achenes obcompressed and trigonal, those of the disk quadrangular, usually with intermediate nerves. [Name from the Greek, referring to the resinous roots.]

A genus of about 12 species, natives of southwestern Canada and northwestern United States. Type species, *Heliopsis ? balsamorhiza* Hook.

The taxa of *Balsamorhiza* hybridize freely (Ownbey & Weber, Amer. Journ. Bot. **30**: 179–187. 1943) and certain of the named entities have been proven to be of hybrid origin: **Balsamorhiza × terebinthàcea** (Hook.) Nutt. Trans. Amer. Phil. Soc. II. **7**: 349. 1840 (*Heliopsis ? terebinthacea* Hook. Fl. Bor. Amer. 1: 310. 1833), derived from *B. deltoidea* and *B. hookeri* or *B. sagittata* and *B. hookeri*, and **Balsamorhiza × bónseri** St. John, Fl. S. E. Wash. 429. 1937, derived from *B. deltoidea* and *B. rosea*.

Plants with woody root and multicipital caudex; basal leaf-blades broadly deltoid or sagittate, entire or coarsely crenate or toothed, the stem-leaves borne near the inflorescence.
  Plants densely canescent-tomentose or -tomentulose at least when young.   1. *B. sagittata.*
  Plants not canescent-tomentose or -tomentulose.
    Achenes glabrous; rays deciduous.   2. *B. deltoidea.*
    Achenes pubescent (except in var. *intermedia*); rays mostly persistent and becoming papery.
         3. *B. careyana.*
Plants with a thickened woody taproot (up to 3 cm. in diameter) and a typically simple short caudex; basal leaf-blades lanceolate or oblong in outline, pinnately or bipinnately divided (except sometimes in *B. serrata* and *B. rosea*), the stem-leaves if present borne near the base.
Leaves pinnately or bipinnately divided, not noticeably reticulate-veined and subcoriaceous.
  Outer phyllaries 2.5–3.5 cm. long, much surpassing the inner phyllaries and the disk, often toothed or incised distally.   5. *B. macrolepis.*
  Outer phyllaries 1–1.5(2) cm. long, about the length of the inner phyllaries and equaling or shorter than the disk, entire.
    Phyllaries linear-lanceolate or lanceolate.   4. *B. hookeri.*
    Phyllaries triangular or ovate, broadly attenuate or abruptly attenuate with a caudate tip.
      Leaves greenish, hirsute, bipinnately divided or bipinnatifid with many narrow divisions.
         6. *B. hirsuta.*
      Leaves whitish or silvery, not at all hirsute, pinnatifid, the divisions entire or shallowly incised.
        Rays 3–6 cm. long; pubescence densely lanate with long, soft, tangled hairs.
           7. *B. incana.*
        Rays 2–3.5 cm. long; pubescence densely appressed-sericeous or short-sericeous.
           5. *B. macrolepis platylepis.*
Leaves crenate or sharply serrate, if at all incised not parted to the midrib, reticulate-veined and subcoriaceous.
  Leaves sharply serrate; rays yellow, the width of ray about one-half the length.
         8. *B. serrata.*
  Leaves crenate; rays becoming roseate, the width of ray about four-fifths the length.
         9. *B. rosea.*

1. **Balsamorhiza sagittàta** (Pursh) Nutt. Arrow-leaved Balsamroot. Fig. 5142.

Buphthalmum sagittatum Pursh, Fl. Amer. Sept. 2: 564. 1814.
Espeletia helianthoìdes Nutt. Journ. Acad. Phila. 7: 39. 1834.
Balsamorhiza sagittata Nutt. Trans. Amer. Phil. Soc. 7: 350. 1840.

Perennials 2–6.5 dm. high from a deep-seated, woody taproot and multicipital caudex clothed with fibrous leaf-bases, the sparsely leafy, white-tomentose stems bearing 1–3 heads. Basal leaves 2–5 dm. long, up to 1.5 dm. wide, the petioles exceeding to nearly equaling the blades, these cordate-sagittate or -hastate at base, triangular or triangular-ovate, entire, densely white-canescent or cinereous-tomentulose and gland-dotted, usually becoming more or less glabrate in age on the upper surface; stem-leaves much reduced and bract-like, spatulate or lanceolate; involucres 1–2 cm. wide, densely white-lanate-tomentose; outer phyllaries about 2.5 cm. long, lanceolate-acuminate, longer than the inner; rays 2–4 cm. long, 13–21, withering, soon deciduous; achenes 7–8 mm. long, glabrous.

Open hillsides and valleys, mostly in the Arid Transition Zone; southern British Columbia and Idaho south through eastern Washington and Oregon to northeastern California and adjacent Nevada and the eastern slope of the Sierra Nevada to the Tahoe region; Saskatchewan and Montana south to Utah and Colorado and east to the Black Hills of South Dakota. Type locality: "On dry barren hills, in the Rocky-mountains." Collected by Lewis. May–Aug.

2. **Balsamorhiza deltoìdea** Nutt. Deltoid Balsamroot. Fig. 5143.

Balsamorhiza deltoidea Nutt. Trans. Amer. Phil. Soc. II. **7**: 351. 1840.
Balsamorhiza glabrescens Benth. Pl. Hartw. 317. 1849.

Perennials with the habit of *B. sagittata,* the stems 2–9 dm. high, bearing 1–4 heads, sparingly leafy, green, glandular, pilose or pilosulose. Basal leaves 1–5 dm. long, up to 2 dm. broad, the petiole longer than the blade, elongate-triangular to deltoid-ovate, cordate or sagittate with blunt lobes at the base, the margin entire or sometimes partially coarsely dentate or crenate, green, glandular, inconspicuously hirsute or hispidulous in age, subcoriaceous and prominently veiny beneath; stem-leaves few, much reduced, lanceolate to linear-lanceolate; involucres 1.5–3 cm. wide, the terminal one the larger, more or less short-pilose; outer phyllaries 3–6 cm. long, lanceolate to oblong, much surpassing the inner and longer than the disk; rays 2–5 cm. long, about 12–20, not papery in age; achenes 7–8 mm. long, glabrous.

Open rolling hill slopes, Transition Zones; British Columbia south on the west side of the Cascade Mountains to California, thence to Los Angeles County in the Coast Ranges and the western slope of the Sierra Nevada to Kern County. Type locality: Willamette Valley, Oregon. March–July. Northwest Balsamroot.

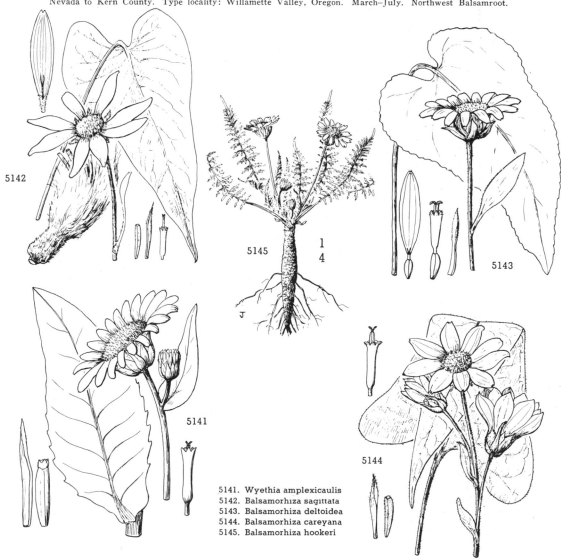

5142

5145   1
        4

5143

5141

5144

5141. Wyethia amplexicaulis
5142. Balsamorhiza sagittata
5143. Balsamorhiza deltoidea
5144. Balsamorhiza careyana
5145. Balsamorhiza hookeri

3. Balsamorhiza careyàna A. Gray. Carey's Balsamroot. Fig. 5144.

Balsamorhiza careyana A. Gray, Mem. Amer. Acad. II. **4**: 81. 1849.

Perennials 2–6 dm. high from a deep-seated woody taproot and multicipital caudex, clothed with fibrous leaf-bases, the sparsely leaved stems bearing 3 or more closely racemose heads. Basal leaves 1.5–5 dm. long, up to 1.5 dm. wide, the petioles nearly equaling to exceeding the blades, these cordate at base with a broad sinus, mostly triangular-hastate, entire or nearly so, subcoriaceous and veiny, green, glandular, and hispidulous on both sides; cauline leaves few, subopposite, becoming bracteate and sessile above; involucres 0.8–2.5 cm. broad; outer phyllaries 1.5–2 cm. long, herbaceous, triangular-lanceolate, exceeding the inner, more or less hirsute to glabrate; rays 2–3 cm. long, 8–13, persisting and becoming somewhat papery; achenes 6–7 mm. long, hairy.

Plains and scablands, Arid Transition Zone; central Washington in Grant and Lincoln Counties south to northern Umatilla and Morrow Counties, Oregon. Type locality: given as "Clearwater" but collected by Spalding probably west of the Clearwater River. March–July.

Balsamorhiza careyana var. **intermèdia** Cronquist, Vasc. Pl. Pacif. Northw. **5**: 101. 1955. Differing from *B. careyana* var. *careyana* in having the leaves often crenate, the terminal head of the inflorescence enlarged, the outer phyllaries more enlarged, and the achenes glabrous. East of the Cascade Mountains from the British Columbia border south to Deschutes and Wheeler Counties, Oregon. Type locality: northeast of Clarno, Wheeler County, Oregon.

4. Balsamorhiza hoòkeri (Hook.) Nutt. Hooker's Balsamroot. Fig. 5145.

Heliopsis ? balsamorhiza Hook. Fl. Bor. Amer. **1**: 310. 1833.
Balsamorhiza hookeri Nutt. Trans. Amer. Phil. Soc. II. **7**: 349. 1840.
Balsamorhiza balsamorhiza Heller, Cat. N. Amer. Pl. 7. 1898.

Perennials from thick woody taproot with black rugose bark and short unbranched caudex, stems 1–3 dm. high, scapose or with a pair of nearly basal stem-leaves, sericeous with some longer hairs near the solitary heads. Basal leaves 10–20 cm. long, lanceolate in outline, the blade much longer than the petioles, pinnately parted to the midrib, the divisions cleft into spreading lobes, thinly sericeous, more densely so on the smaller leaves, the petioles often pilose as well; involucres about 1.5–2.5 cm. wide; phyllaries 1–1.5 cm. long, 2–4 mm. wide, lanceolate, attenuate at apex, the outer phyllaries not exceeding the disk, glandular, lanate on the margins and densely so toward the base, rather thinly so on the surface and appearing greenish; rays 1.5–2.5 cm. long, about 10–16; achenes 4–5 mm. long, glabrous.

Rocky slopes, Arid Transition Zone; western Klickitat County, Washington, to Clark County. Type locality: Fort Vancouver, Clark County. April–June.
Intergrading with *B. hookeri* var. *lagocephala* in eastern Klickitat and Yakima Counties.

Balsamorhiza hookeri var. **lanàta** Sharp, Ann. Mo. Bot. Gard. **22**: 130. 1935. Habit as in *B. hookeri* var. *hookeri*, the plants low, densely lanate throughout with long silky hairs; leaves bipinnately divided; phyllaries as in *B. hookeri* var. *hookeri* but with a longer, more dense pubescence. Known only on dry hills near Yreka, Siskiyou County, California, the type locality.

Balsamorhiza hookeri var. **lagocéphala** (Sharp) Cronquist, Vasc. Pl. Pacif. Northw. **5**: 103. 1955. (*Balsamorhiza hirsuta* var. *lagocephala* Sharp, Ann. Mo. Bot. Gard. **22**: 139, 1935.) Habit as in *B. hookeri* var. *hookeri*; leaves pinnately divided, the lobes cleft into a few acute broad lobes or teeth, green, hirsute or hirsutulous with appressed hairs and often glandular; phyllaries seriate, the outer rarely surpassing the inner, lanceolate to linear-lanceolate, more or less villous-tomentose especially at the base of the outer phyllaries. Rocky soil and sagebrush, east of the Cascade Mountains, Washington. Type locality: Ellensburg, Kittitas County. A highly variable taxon interbreeding with *B. hookeri* var. *hookeri* at its southern limit and with other taxa in *Balsamorhiza* where the ranges overlap.

Balsamorhiza hookeri var. **neglécta** (Sharp) Cronquist, Vasc. Pl. Pacif. Northw. **5**: 113. 1955. (*Balsamorhiza hirsuta* var. *neglecta* Sharp, Ann. Mo. Bot. Gard. **22**: 139. 1935.) Habit as in *B. hookeri* var. *hookeri*, the plants 2.5–3.5 dm. high; leaves lanceolate-attenuate in outline, pinnately divided, the lobes narrow, shallowly to deeply cleft with blunt divisions, usually grayish green, with an appressed-hirsute pubescence; phyllaries more or less seriate, the outer seldom surpassing the inner series, typically lanceolate to linear-lanceolate, more or less villous-tomentose, densely so on the margins. Rocky soil, central Utah westward through northern and central Nevada to northeastern California. Type locality: Truckee Pass, Washoe County, Nevada. Many aberrant forms probably of hybrid origin are found in Washoe County and in adjacent California which may have ovate-lanceolate or ovate-attenuate, granular-glandular phyllaries from which the villous-tomentose pubescence may be lacking. *Balsamorhiza hookeri* var. *lanata*, *B. macrolepis*, and *B. macrolepis* var. *platylepis* occur in this area.

5. Balsamorhiza macrolèpis Sharp. California Balsamroot. Fig. 5146.

Balsamorhiza macrolepis Sharp, Ann. Mo. Bot. Gard. **22**: 132. 1935.

Perennials from a coarse woody root and short caudex, the several stems 2–6 dm. high, scapose or with a pair of reduced pinnate leaves near the base, appressed-pubescent, appressed or spreading above. Basal leaves 2–4.5 dm. long, 5–10 cm. wide, blade much longer than the petioles, lanceolate in outline, pinnate, the divisions lanceolate, usually deeply lobed, the lobes rounded or acute at apex, very thinly sericeous or appressed-tomentose and obscurely glandular; involucres 2–3 cm. broad, appressed short-sericeous and obscurely glandular; outer phyllaries 2–3.5 cm. long, much surpassing the inner and the disk, oblong or oblong-lanceolate, usually incised or toothed at the apex, the inner lanceolate; rays about 10–15 and 2.5–3.5 cm. long; achenes 5–7 mm. long, glabrous.

Fields and rocky hill slopes, Upper Sonoran Zone; Butte County, California, south to Mariposa County and in the Inner Coast Ranges from Sonoma County south to Santa Clara County. Type locality: Clear Creek, Butte County. April–May.

Balsamorhiza macrolepis var. **platylèpis** (Sharp) Ferris. Contr. Dudley Herb. **5**: 99. 1958. (*Balsamorhiza platylepis* Sharp, Ann. Mo. Bot. Gard. **22**: 131. 1935.) Plants 1–3.5 dm. high; pubescence usually more dense and silvery than *B. macrolepis* var. *macrolepis*; pinnate divisions of the leaves shallowly to not at all lobed, rounded to broadly acute; outer phyllaries shorter than to equaling the disk, the outer and the second series triangular or ovate, broadly or abruptly attenuate and often spreading, sericeous or short-sericeous and sometimes velutinous particularly in plants of the Siskiyou Mountains area. Siskiyou Mountains in Josephine County, Oregon, to Siskiyou County, California, where it grows on serpentine; also Modoc County, California, south to Nevada County and adjacent Washoe County, Nevada. Type locality: Marmol Station, Washoe County. Silvery Balsamroot.

6. **Balsamorhiza hirsùta** Nutt. Hairy Balsamroot. Fig. 5147.

Balsamorhiza hirsuta Nutt. Trans. Amer. Phil. Soc. II. **7**: 349. 1840.
Balsamorhiza hookeri var. *hirsuta* A. Nels. in Coult. & Nels. New Man. Bot. Rocky Mts. 546. 1909.

Perennials with the habit of *B. hookeri,* the stems 2–5 dm. high, scapose or with a pair of reduced, pinnate or bipinnatifid leaves near the base, spreading-hirsute or -hirsutulous and glandular. Basal leaves 2.5 dm. long, 1.5–15 cm. wide, the blade much longer than the petiole, lanceolate to oblong-oblanceolate in outline, the midveins prominent, pinnate with bipinnatifid divisions to bipinnatifid, the ultimate divisions many and narrow, green, more or less hirsute; involucres 2–3 cm. broad, more or less hirsute or hispid and pilose-ciliate; outer phyllaries 1–2 cm. long, not surpassing the inner, ovate or ovate-lanceolate with an abruptly attenuate, spreading tip; rays 2.5–4.5 cm. long, 10–18; achenes 3–6 mm. long, glabrous.

Rocky soil, Arid Transition Zone; local in eastern Oregon. Type locality: "Dry plains east of Walla-Walla, near the Blue Mountains, and in the Grand Ronde prairie." Probably collected at the second locality as the species is not known from Walla Walla. Collected by Nuttall. May–July.

7. **Balsamorhiza incàna** Nutt. Woolly Balsamroot. Fig. 5148.

Balsamorhiza incana Nutt. Trans. Amer. Phil. Soc. II. **7**: 350. 1840.
Balsamorhiza hookeri var. *incana* A. Gray, Syn. Fl. N. Amer. 1²: 266. 1884.
Balsamorhiza floccosa Rydb. Bull. Torrey Club **27**: 629. 1900.

Perennials with the habit of *B. hookeri,* the stems 2–7 dm. high, scapose except for a pair of

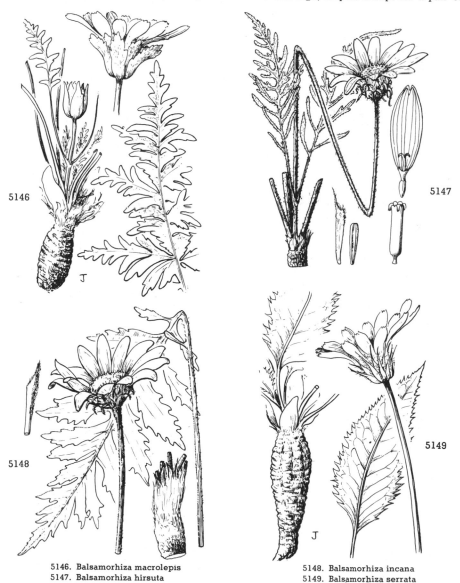

5146. Balsamorhiza macrolepis
5147. Balsamorhiza hirsuta

5148. Balsamorhiza incana
5149. Balsamorhiza serrata

reduced pinnatifid leaves near the base, loosely lanate. Basal leaves 1–4.5 dm. long, 3–10 cm. wide, lanceolate to oblong in outline, the blade much exceeding the petiole, pinnate, the lanceolate or ovate divisions large, coarsely toothed or more deeply incised, pubescence loosely lanate and usually densely so; involucres mostly 2–2.5 cm. wide, densely canescent-tomentose with loosely spreading hairs; phyllaries 1.5–2 cm. long, the outer not surpassing the inner, ovate to ovate-lanceolate with an abruptly attenuate, spreading tip; rays 3–5 cm. long, 13 or more; achenes 3–6 mm. long, glabrous.

Meadows and grassy slopes, Arid Transition and Canadian Zones; southeastern Montana west to southeastern Washington and south to central Oregon east of the Cascade Mountains. Type locality: "Rocky Mountains." Collected by Nuttall. May–July.

8. Balsamorhiza serràta Nels. & Macbr. Serrated Balsamroot. Fig. 5149.

Balsamorhiza serrata Nels. & Macbr. Bot. Gaz. **56**: 479. 1913.

Perennial from a coarse woody taproot and short simple caudex, the several stems 1–3 dm. high, scapose or with 2 reduced leaves near the base, sparsely pilose and usually reddish. Basal leaves 6.5–25 cm. long, 2–10 cm. wide, the petioles shorter than the blades, lanceolate to ovate-lanceolate, the margin shallowly serrate to incised or divided often on the same plant, subtruncate or cuneate at base, green, coriaceous, conspicuously reticulate-veined, scabrous; involucres about 1.5–2.5 cm. wide, nearly glabrous to rather densely white-pilose; phyllaries linear-lanceolate, the outer not surpassing the inner; rays 2–4 cm. long, mostly 10–16; achenes 6–7 mm. long, glabrous.

Dry rocky soil in open forest or sagebrush, Arid Transition Zone; southeastern Washington south to Lake and Harney Counties, Oregon, and adjacent Modoc County, California; also northwestern Nevada. Type locality: near Rock Creek, Morrow County, Oregon. May–June.

9. Balsamorhiza ròsea Nels. & Macbr. Rosy Balsamroot. Fig. 5150.

Balsamorhiza rosea Nels. & Macbr. Bot. Gaz. **56**: 478. 1913.
Balsamorhiza hookeri var. *rosea* Sharp, Ann. Mo. Bot. Gard. **22**: 130. 1935.

Perennial with the habit of *B. serrata*, the stems 0.6–3 dm. high, purplish, scapose or with a pair of reduced, bract-like leaves near the base, strigose. Basal leaves 3–20 cm. long, 1–10 mm. wide, the blade longer than the petiole, oblong to deltoid, crenate or lobed partway to the midrib with ovate or oblong-ovate, crenate divisions, more or less reticulate-veined, short-strigose and glandular; involucre about 2 cm. broad, densely white-pilose; phyllaries about 1 cm. long, linear-lanceolate, acute, the outer not surpassing the inner; rays 1–2.5 cm. long, up to 1 cm. wide, pale yellow(?) becoming roseate in age; achenes 5–6 mm. long, strigose.

Rocky soil, Arid Transition Zone; known from three localities, Yakima County, Walla Walla County, and Spokane County, Washington. Type locality: Rattlesnake Mountains, Yakima County. April–May.

3. VIGUÍERA H.B.K. Nov. Gen. & Sp. **4**: 224. 1820.

Herbs or shrubs, more or less pubescent. Leaves usually opposite at least below, linear to ovate or orbicular, entire, toothed, or pinnate-lobed. Heads small to large, radiate and yellow (in ours), solitary, cymose or panicled. Involucre usually campanulate to hemispheric, 2–7-seriate, graduate or subequal, the phyllaries linear to oval or oblong, herbaceous throughout or usually indurated and ribbed at base. Receptacle flattish to low-conical, the receptacular bracts folded, firm, embracing the achenes. Rays neutral, ligulate, spreading, oval or elliptic, 2–3-denticulate. Disk-flowers perfect, fertile, their corollas tubular, with short tube and cylindric throat, 5-toothed. Anthers cordate-sagittate at base. Style-branches hispid above, with acutish to acuminate, usually rather short, sterile appendages. Achenes of ray sterile, of disk fertile, more or less thickened, oblong or obovoid, pubescent or glabrous; pappus (disk-achenes) of 2 usually persistent awns and several shorter, free or basally united, persistent squamellae, or entirely wanting. [Name in honor of L. G. A. Viguier of Montpelier, author of a work on *Papaver*.]

A genus of about 145 species, native of North and South America. Besides the following, six others occur from Montana to Arizona and Texas, and another species occurs in Georgia. Type species, *Viguiera helianthoides* H.B.K. (= *Viguiera dentata* var. *helianthoides* (H.B.K.) Blake).

Pappus present; achenes pubescent; leaves ovate or triangular.
Leaves narrowly triangular, laciniate-toothed and usually hastate-lobed at base; plant resinous.
1. *V. laciniata.*
Leaves ovate or deltoid-ovate, serrate or entire; plant not resinous.
Leaves densely and closely and rather prominently reticulate beneath, canescently pilose to subtomentose above.
2. *V. reticulata.*
Leaves somewhat veiny but not closely nor prominently reticulate beneath, green and tuberculate-hispidulous above.
3. *V. deltoidea parishii.*
Pappus absent; achenes glabrous; leaves linear or lance-linear.
4. *V. multiflora nevadensis.*

1. Viguiera laciniàta A. Gray. Laciniate Viguiera. Fig. 5151.

Viguiera laciniata A. Gray in Torr. Bot. Mex. Bound. 89. 1859.

Shrubby, much-branched, leafy, resinous, scabrous-hispidulous plants up to 1.3 m. high, the branches slender, hispidulous with mostly incurved hairs. Leaves alternate, short-petioled, often with axillary fascicles, the blades narrowly triangular, 2–3.5 cm. long, 7–20 mm. wide, acute or acuminate, at base shortly cuneate, laciniate-toothed and hastate-lobed toward base or rarely subentire, firm, glandular-punctate, triplinerved and veiny, harshly and rather sparsely hispidulous, the hairs especially of upper face with tuberculate bases; heads 3–13 in a terminal cyme or cymose panicle, 1.7–3 cm. wide; involucre 3-seriate, slightly graduate, 5–7 mm. high, the phyllaries with ovate indurate base and abrupt, shorter to longer, acuminate, linear to triangular, loose, herbaceous tip, hispidulous; rays 8–12, 0.8–1.5 cm. long; achenes sparsely strigose, 3 mm. long; pappus of 2 broad, somewhat deciduous, paleaceous awns and about 8 short, persistent, lacerate squamellae.

Dry slopes and mesas, Sonoran Zones; southwestern San Diego County, California, and northern Lower California. Type locality: "Rancho Gamacha, east of San Diego, California, September, 1855; *Schott*." April–Nov.

2. **Viguiera reticulàta** S. Wats. Leather-leaved Viguiera. Fig. 5152.

Viguiera reticulata S. Wats. Amer. Nat. **7**: 301. 1873.

Suffrutescent branching perennial 0.6–1.3 m. high, the stem and branches slender, whitish, appressed- or incurved-pilose, glabrescent above. Leaves opposite or alternate, short-petioled, the blades ovate or deltoid-ovate, 2.5–8 cm. long, 1.5–7 cm. wide, acute, broadly rounded to subcordate at base, entire or rarely serrate, coriaceous, triplinerved and strongly reticulate especially beneath, canescently pilose-subtomentose above with mostly antrorse hairs, less conspicuously so beneath with looser, more spreading hairs and gland-dotted; heads 1.8–2.5 cm. wide, in small, rather close, usually long-peduncled clusters at tips of stem and branches; involucres about 3-seriate, graduate, about 5 mm. high, the phyllaries with ovate, indurate, ribbed body and usually shorter, narrower, oblong or triangular, usually appressed, herbaceous tip, cinereous-pilosulose especially toward margin; rays about 8–15, about 8–12 mm. long; achenes rather densely appressed-pilose, 3 mm. long; pappus-awns 2, paleaceous, 2 mm. long or less, about twice as long as the fimbriate squamellae.

Dry rocky slopes and canyons, Sonoran Zones; desert ranges of Inyo County, California. Type locality: Telescope Peak, Panamint Mountains. April–Sept. Death Valley Golden-eye.

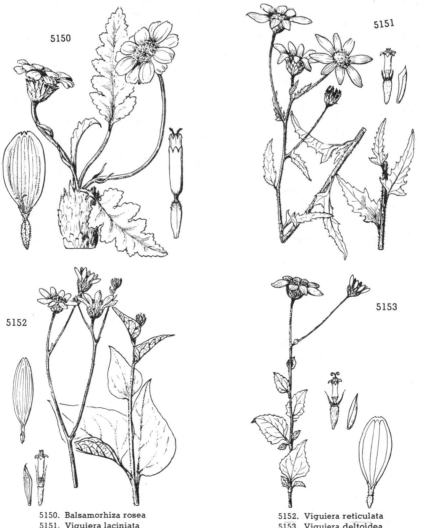

5150. Balsamorhiza rosea
5151. Viguiera laciniata

5152. Viguiera reticulata
5153. Viguiera deltoïdea

112 COMPOSITAE

3. **Viguiera deltoìdea** var. **paríshii** (Greene) Vasey & Rose. Parish's Viguiera.
Fig. 5153.

Viguiera parishii Greene, Bull. Torrey Club **9**: 15. 1882.
Viguiera deltoidea var. *parishii* Vasey & Rose, Contr. U.S. Nat. Herb. **1**: 72. 1890.

Branching shrub 0.8 m. high or less with gray-barked stems, the branches slender, densely cinereous-hispidulous, often sparsely hispid. Leaves chiefly opposite, short-petioled, the blades deltoid-ovate, 1.5–3.5 cm. long, 0.8–2.5 cm. wide, acute or obtuse, rounded to subcordate at base, serrate to subentire, triplinerved and veiny, tuberculate-hispidulous and gland-dotted; heads 1–6 toward tips of stems and branches, mostly long-peduncled, 1.5–5 cm. wide; involucre 5–9 mm. high, 2–3-seriate, somewhat graduate, the phyllaries with ovate to lance-ovate, indurate, ribbed body and usually longer, linear to subulate, herbaceous tip, hispid-hirsute to strigillose; rays about 8–10, 0.8–1.5 cm. long; achenes appressed-pilose, 3 mm. long; pappus-awns 2, about 3 mm. long, paleaceous below, about four times as long as the fimbriate squamellae.

Mesas and rocky canyons, Sonoran Zones; southern Nevada south through the Mojave and Colorado Deserts, California, to northern Lower California, east to Arizona and northern Sonora; also coastal San Diego County, California. Type locality: San Luis Rey, San Diego County. April–Sept.

4. **Viguiera multiflòra** var. **nevadénsis** (Λ. Nels.) Blake. Nevada Viguiera.
Fig. 5154.

Gymnolomia nevadensis A. Nels. Bot. Gaz. **37**: 271. 1904.
Gymnolomia linearis Rydb. Bull. Torrey Club **37**: 327. 1910.
Viguiera multiflora var. *nevadensis* Blake, Contr. Gray. Herb. No. 54: 110. 1918.

Slender, several-stemmed perennial from a short rootstock with stems 25–90 cm. high, simple or branched, strigose or strigillose. Leaves opposite below or nearly throughout on petioles 2–5 mm. long, the blades narrowly linear or lance-linear, 3–5.5 cm. long, 2–5 mm. wide, usually revolute-margined, entire or obscurely serrulate, usually acute or acuminate at each end, feather-veined or obscurely triplinerved, tuberculate-strigillose or -strigose or antrorse-hirsutulous, sometimes densely so beneath; heads few to rather numerous and in the upper axils terminal, slender-peduncled, 1.5–5 cm. wide; involucre 2-seriate, subequal, 5–7 mm. high, the phyllaries linear-lanceolate to lanceolate, herbaceous, acuminate or acute, densely strigose and strigillose; rays 10–14, 8–25 mm. long; achenes 1.8–3 mm. long, glabrous, epappose.

Mesas and canyons, Upper Sonoran Zone; southern Utah and southern Nevada south to Inyo County, California (Inyo, Argus, and Panamint Ranges), and San Bernardino County in the Kingston Range and Clark Mountain and eastward to Arizona. Type locality: Meadow Valley Wash, Lincoln County, Nevada. May–Sept. Nevada Showy Golden-eye.

Viguiera ciliàta (Robins. & Greenm.) Blake, Contr. Gray Herb. No 54: 113. 1918. (*Gymnolomia longifolia* Robins. & Greenm. Proc. Bost. Soc. Nat. Hist. **29**: 92. 1899.) An annual with hispid-ciliate leaves and the phyllaries merely hispid-ciliate or also sparsely hispid dorsally. This native of southwestern United States and adjacent Mexico has been found as an adventive plant at Santa Monica and Los Angeles, California.

4. **HELIÁNTHUS*** L. Sp. Pl. 904. 1753.

Usually tall, coarse, annual or perennial herbs. Leaves simple, opposite below, linear-lanceolate to ovate, usually petiolate. Heads solitary or corymbed. Involucre 2-seriate or more. Receptacle flat or low-convex. Heads radiate; ray-flowers yellow, neutral; disk-flowers perfect, the corollas yellow throughout or with reddish purple lobes. Chaff (receptacular bracts) 3-cuspidate to entire. Pappus of 2 chaffy scales, rarely more, caducous. Achenes obovate or oblong-linear, 4-angled to lenticular, slightly laterally compressed. Basic chromosome number, $\times = 17$. [Name Greek, meaning sunflower.]

A genus of about 75 species, in North and South America. Type species, *Helianthus annuus* L.

Lobes of disk-corollas red or purple.
  Annuals.
    Phyllaries ovate or ovate-lanceolate, abruptly attenuate; lower leaves usually ovate or lance-ovate, cordate
      at base.   1. *H. annuus.*
    Phyllaries lanceolate or ovate-lanceolate, gradually attenuate; lower leaves lanceolate or lance-ovate,
      cuneate to truncate at base.
      Phyllaries hirsute; receptacular bracts generally exceeding disk-flowers in length, glabrous at tips.
         2. *H. bolanderi.*
      Phyllaries appressed-strigulose; receptacular bracts generally equaling disk-flowers, tips of center-
        most white-hispid or hirsute.   3. *H. petiolaris.*
  Perennials.
    Leaves whitish or greenish, usually petiolate; rhizomes lacking.
      Leaves, stems, and phyllaries densely white-sericeous-villous; achenes 7–8 mm. long.
         4. *H. tephrodes.*
      Leaves, stems, and phyllaries hispid to glabrous; achenes 4 mm. long.   5. *H. gracilentus.*
    Leaves bluish green, subsessile; rhizomes present.   6. *H. ciliaris.*
Lobes of the disk-corollas yellow.
  Annuals.
    Chaff exceeding disk-flowers in length; disks 1.5–2.5 cm. in diameter.   2. *H. bolanderi.*
    Chaff not exceeding disk-flowers; disks 3.0 cm. or more in diameter.   1. *H. annuus.*

---

* Text contributed by Charles Bixler Heiser, Jr.

Perennials.
    Phyllaries shorter than or equaling disk in length, not attenuate; leaves 5–11 cm. in length.
                                      5. *H. gracilentus.*

    Phyllaries usually exceeding disk in length, somewhat attenuate; leaves 6–20 cm. in length.
        Plants to 1.5 m. tall; stems usually hispid, with 1 to few heads.   7. *H. cusickii.*
        Plants over 1.5 m. tall; stems usually glabrous, with numerous heads.
            Phyllaries conspicuously dilated near base, 3–4 mm. broad, long-attenuate and reflexed.
                                      9. *H. californicus.*
            Phyllaries not conspicuously dilated near base, 2–3 mm. broad, short-attenuate, not reflexed.
                                      8. *H. nuttallii.*

1. **Helianthus ánnuus** L. Common Sunflower. Fig. 5155.

Helianthus annuus L. Sp. Pl. 904. 1753.

    Annual, 0.5–4 m. tall, branched or unbranched; stems usually rough-hispid, green or purple-mottled. Leaves alternate above, ovate to ovate-lanceolate, usually cordate at base, serrate, 7.5–40 cm. long, 3.2–40 cm. wide, long-petiolate; disks 2 cm. or more in diameter, flat or low-convex; phyllaries 3 mm. or more broad, ovate or ovate-lanceolate, abruptly attenuate, ciliate on margins, glabrous or pubescent on backs; middle cusp of chaff usually hispid at apex; lobes of disk-flowers reddish purple, rarely yellow; achene glabrous or obscurely pubescent, 4 mm. or more long. $n = 17$.

    A widespread weed throughout much of North America. Type locality: eastern North America. July–Oct.

    Helianthus annuus var. **macrocárpus** (DC.) Cockerell, Science **40**: 709. 1914. (*Helianthus macrocarpus* DC. Pl. Rar. Jard. Gen. Vme Not. 8. 1826.) Stems unbranched; disks over 5.5 cm. in diameter; rays 30–70; achenes 6.5–15 mm. long. Based on the cultivated plant. This variety is cultivated throughout the world for its oily seeds.

    Helianthus annuus subsp. **lenticuláris** (Dougl.) Cockerell, Bot. Gaz. **45**: 338. 1908. (*Helianthus lenticularis* Dougl. ex. Lindl. Bot. Reg. **15**: *pl. 1265.* 1829; *H. aridus* Rydb. Bull. Torrey Club **32**: 127. 1905, in part.) Branched; stems rough-hispid; leaves ovate-lanceolate to ovate, hispid, prominently serrate; disks 2–3.5 cm. in diameter; phyllaries 4–7 mm. broad; rays 17–26; achenes 4–5.5 mm. long. Western North America. Type locality: California. July–Nov. This taxon hybridizes with the preceding variety and with both *H. bolanderi* and *H. petiolaris.*

    Helianthus annuus subsp. **jàegeri** (Heiser) Heiser, Contr. Dudley Herb. **4**: 317. 1955. (*Helianthus jaegeri* Heiser, Bull. Torrey Club **75**: 513. 1948.) Branched; stems sparsely hispid to glabrous; leaves lanceolate to lance-ovate, sparingly hispid, shallowly serrate to entire; disk 1.5–2 cm. in diameter; phyllaries 3–4 mm. wide; rays 10–15; achenes 4–5 mm. long. In dry places, San Bernardino and Inyo Counties, California, and Clark County, Nevada. Type locality: Soda Dry Lake, San Bernardino County.

    Helianthus maximiliànii Schrader, Ind. Sem. Hort. Götting. 1835. Perennial from short rhizome; stems 0.5–2.5 m. tall, scabrous-hispidulous; leaves mostly alternate, lanceolate, scabrous, often infolded, 14–30 cm. long, 2.0–5.5 cm. wide, subsessile; inflorescence spiciform or racemiform; phyllaries linear-lanceolate, exceeding disk in length, canescent; disk yellow, 1.5–2.5 cm. broad. $n = 17$. Minnesota and Saskatchewan southward to Texas. Cultivated as an ornamental and frequently escaping. This plant has been collected in Fresno County, California. Maximilian's Sunflower.

    Helianthus laetiflòrus Pers. Syn. Pl. **2**: 476. 1807. This central and eastern North American plant is sometimes cultivated in California as an ornamental and may persist in old gardens. This species is similar to *H. tuberosus* but lacks tubers and has shorter, more tightly appressed phyllaries. The disk may be either yellow or brown.

2. **Helianthus bolánderi** A. Gray. Bolander's Sunflower. Fig. 5156.

Helianthus scaberrimus Benth. Bot. Sulph. 28. 1844. Not Ell. 1824.
Helianthus bolanderi A. Gray, Proc. Amer. Acad. **6**: 544. 1865.
Helianthus exilis A. Gray, op. cit. 545.

    Annual, 0.3–1.3 m. tall, branched; stems rough-scabrous to hirsute-villous. Leaves alternate above, ovate to linear-lanceolate, entire to irregularly serrate, largest blades 6.0–15.0 cm. long, 3.0–12.0 cm. wide, petiolate, hispid to hirsute-villous on both surfaces, more densely so below, petiolate; disk 1.5–2.5 cm. in diameter; phyllaries oblong to lanceolate, 3.0–4.5 mm. broad,

5154

5155

5154. Viguiera multiflora

5155. Helianthus annuus

hirsute to hirsute-villous; middle cusp of chaff usually conspicuously exceeding disk-flowers in length, glabrous at apex; rays 10–17; lobes of disk-flowers reddish purple or yellow; achenes 3.0–4.5 mm. long, minutely villous. $n = 17$.

Dry ground, Upper Sonoran and Transition Zones; Josephine and Jackson Counties, Oregon, south to Tulare and San Luis Obispo Counties, California. Type locality: geysers in Lake County, California. July–Nov.

3. Helianthus petiolàris Nutt. Prairie Sunflower. Fig. 5157.

Helianthus petiolaris Nutt. Journ. Acad. Phila. **2**: 115. 1821.
Helianthus patens Lehm. Ind. Sem. Hort. Hamb. 8. 1828.
Helianthus petiolaris var. *humilis* Nees in Neuwied. Reise N.-Amer. **2**: 441. 1841.
Helianthus integrifolius Nutt. Trans. Amer. Phil. Soc. II. **7**: 366. 1841.
Helianthus integrifolius var. *gracilis* Nutt. loc. cit.
Helianthus petiolaris patens Rydb. Mem. Torrey Club **5**: 334. 1894.
Helianthus petiolaris var. *phenax* Cockerell, Nature **66**: 174. 1902.
Helianthus aridus Rydb. Bull. Torrey Club **32**: 127. 1905, in part.

Annual, 0.4–2.0 m. tall, branched; stems strigose to nearly glabrous. Leaves mostly alternate, oblong-lanceolate to deltoid-ovate, entire or obscurely serrate, cuneate to truncate at base, blade 4.0–15.0 cm. long, 1.0–10.0 cm. wide, strigose on both surfaces, frequently bluish green in color, petiolate; disk 1.0–2.5 cm. in diameter; phyllaries lanceolate to ovate-lanceolate, hispidulous on backs, short-ciliate on margins, 2.0–4.0 mm. broad; middle cusp of chaff not conspicuously exceeding disk-flowers, those in center densely white-hispid or hirsute at apex; disk-flowers reddish purple; achenes lightly villous, 3.5–4.5 mm. long; pappus of 2 awns. $n = 17$.

Sandy soils and waste places, Manitoba and Minnesota to Washington south to Texas and northern Mexico. Type locality: Upper Missouri River. June–Nov.

Helianthus petiolaris var. **canéscens** A. Gray, Smiths. Contr. **3**⁵: 108. 1852. (*Gymnolomia encelioides* A. Gray, Proc. Amer. Acad. **19**: 4. 1883; *Helianthus petiolaris canus* Britt. Mem. Torrey Club **5**: 334. 1894; *H. canus* Woot. & Standl. Contr. U.S. Nat. Herb. **16**: 190. 1913.) Leaves, stems, and involucres densely strigulose or canescent; middle cusp of chaff merely hispid to densely ciliate at tip. Open sandy soils, Lower Sonoran Zone; Texas to southern California and northern Mexico. Type locality: southwestern Texas.

4. Helianthus tephròdes A. Gray. Desert Sunflower. Fig. 5158.

Helianthus tephrodes A. Gray in Torr. Bot. Mex. Bound. 90. 1859.
Viguiera nivea A. Gray, Bot. Calif. **1**: 354. 1876.
Viguiera tephrodes A. Gray, Proc. Amer. Acad. **17**: 218. 1882.
Viguiera sonorae Rose & Standl. Contr. N.S. Nat. Herb. **16**: 20. *pl. 16.* 1912.

Perennial from stout taproot, erect or decumbent, 1.0–1.5 m. tall; stem appressed sericeous-villous. Leaves alternate above, deltoid or deltoid-ovate, 3.0–7.0 cm. long, 2.0–4.0 cm. broad, appressed white sericeous-villous, petiolate; disks 1.4–2.8 cm. in diameter; phyllaries lanceolate, evenly appressed sericeous-villous, 2.5–4.0 mm. wide; lobes of the corolla reddish purple; achenes 7–8 mm. long, densely villous with long hairs; pappus of 2(–3) long awns and several squamellae.

Sand hills, Lower Sonoran Zone; Imperial County, California, and northwestern Sonora. Type locality: Colorado Desert of California. Feb.–May.
This species has previously been confused with *H. niveus* (Benth.) Brandg. of Lower California.

5. Helianthus graciléntus A. Gray. Slender Sunflower. Fig. 5159.

Helianthus gracilentus A. Gray, Proc. Amer. Acad. **11**: 77. 1876.

Perennial from thickened root, 1–2 m. tall; stems strigose-hispid or rarely glabrous. Leaves opposite to near inflorescence, lanceolate to lance-ovate; blades 5–11 cm. long, 2–3.5 cm. wide, hispid both surfaces, subsessile or with petioles to 3 cm. long; disks 2.0–2.5 cm. wide; phyllaries lanceolate to lance-ovate, equal to or shorter than disks, puberulent, 2.5–3.5 mm. broad; lobes of disk-corolla yellow or reddish purple; achene about 4 mm. long, glabrous. $n = 17$.

Dry slopes, Upper Sonoran Zone; Contra Costa County, California, south to northern Lower California. Sea level to 6,000 feet. Type locality: San Diego County, California. May–Oct.

6. Helianthus ciliàris DC. Blueweed. Fig. 5160.

Helianthus ciliaris DC. Prod. **5**: 587. 1836.
Helianthus laciniatus A. Gray, Mem. Amer. Acad. II. **4**: 84. 1849.

Perennial from long stout rhizome, 0.5–1.0 m. high, glaucous, glabrous or rarely hispid. Leaves mostly opposite, linear-lanceolate to lanceolate, entire to laciniate, usually somewhat undulate on margins, glaucous, bluish green in color, subsessile; stems monocephalic or with few heads; disks 0.9–2.0 cm. wide; phyllaries ovate, obtuse, shorter than disk; rays short; lobes of disk-flowers reddish purple; achene glabrous, 3–4 mm. long. $n = 51$.

Sandy soil and waste places, Texas to Arizona south to central Mexico. This species is an extremely troublesome weed and at one time was established in Tehama, San Luis Obispo, Santa Barbara, Ventura, Orange, and Los Angeles Counties, California, but appears to have been largely eradicated. Type locality: near Reynosa de Tamaulipas, Mexico.

7. Helianthus cusíckii A. Gray. Cusick's Sunflower. Fig. 5161.

Helianthus cusickii A. Gray, Proc. Amer. Acad. **21**: 413. 1886.

Perennial from thick fleshy root; stems glabrous or with scattered long hairs. Leaves mostly alternate, linear-lanceolate to lanceolate, frequently strongly 3-nerved, entire, scabrous-hispid

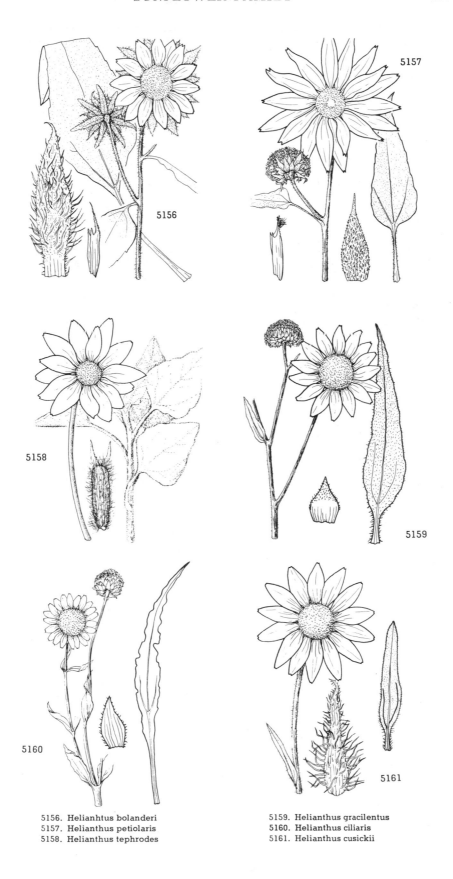

5156. Helianhtus bolanderi
5157. Helianthus petiolaris
5158. Helianthus tephrodes
5159. Helianthus gracilentus
5160. Helianthus ciliaris
5161. Helianthus cusickii

to glabrous, 6.0–15.0 cm. long, 1.0–2.5 cm. broad, sessile or with very short petiole; stems monocephalic or with few heads; disks 2.0–3.0 cm. in diameter; phyllaries lanceolate or oblong-lanceolate, generally exceeding disk in length, 3.4 mm. broad, appressed-pubescent to long-villous; achene glabrous, 4–5 mm. long. $n = 17$.

Dry slopes and open woods, Upper Sonoran and Arid Transition Zones; central Washington south to Owyhee County, Idaho, Storey County, Nevada, and Shasta County, California; 2,000–6,500 feet. Type locality: Malheur River, southeastern Oregon. April–July.

8. Helianthus nuttállii Torr. & Gray. Nuttall's Sunflower. Fig. 5162.

Helianthus nuttallii Torr. & Gray, Fl. N. Amer. **2**: 324. 1842.
Helianthus giganteus var. *utahensis* D. C. Eaton, Bot. King. Expl. 169. 1871.
Helianthus californicus var. *utahensis* A. Gray, Syn. Fl. N. Amer. **1²**: 277. 1884.
Helianthus fascicularis Greene, Pl. Baker. **3**: 28. 1901.
Helianthus utahensis A. Nels. Bull. Torrey Club **29**: 405. 1902.
Helianthus coloradensis Cockerell, Proc. Biol. Soc. Wash. **27**: 6. 1914.
Helianthus parishii coloradensis Cockerell, Torreya **18**: 181. 1918.
Helianthus nuttallii subsp. *coloradensis* Long, Rhodora **56**: 199. 1954.

Perennial from thick fascicled roots; stems 1.0–3.0 m. tall, glabrous to slightly hispid, occasionally glaucous. Leaves opposite to top or alternate above, lanceolate, entire to serrate, 10–15 cm. long, 1.0–3.0 cm. wide, nearly glabrous to hispid above, hispid below, subsessile to short-petiolate; peduncles glabrous or sparingly pubescent; disk 1.5–2.0 cm. in diameter; phyllaries linear-lanceolate, 2–3 mm. broad, equaling or slightly exceeding disks in length, appressed-pubescent, rarely ciliate on margins, not reflexed; lobes of the disk-corollas yellow; achene glabrous, about 4 mm. long. $n = 17$.

Dry plains, Upper Sonoran Zone to Canadian Zone; Saskatchewan to Alberta to eastern Oregon south to New Mexico and San Bernardino County, California; 2,300–9,000 ft. Type locality: plains of the Lewis River, Wyoming. Aug.–Nov.

Helianthus nuttallii subsp. paríshii (A. Gray) Heiser, Contr. Dudley Herb. **4**: 316. 1955. (*Helianthus parishii* A. Gray, Proc. Amer. Acad. **19**: 7. 1883; *H. oliveri* A. Gray, op. cit. **20**: 299. 1885; *H. parishii* f. *oliveri* Cockerell, Torreya **18**: 181. 1918; *H. californicus* var. *parishii* Jepson, Man. Fl. Pl. Calif. 1077. 1925; *H. californicus* var. *oliveri* Blake in Munz, Man. S. Calif. 549. 1935.) Leaves mostly alternate, 10–20 cm. long, 1.5–25 cm. wide, hispid to densely tomentose above, tomentulose to densely villous-tomentose below, subsessile to short-petiolate; peduncles pubescent near summit; heads rather numerous; phyllaries tomentulose to tomentose. In swampy or wet places, Upper Sonoran Zone; Los Angeles, Orange, and San Bernardino Counties, California; 1,000–1,500 feet. Type locality: San Bernardino County. Aug.–Oct.

9. Helianthus califórnicus DC. California Sunflower. Fig. 5163.

Helianthus californicus DC. Prod. **5**: 589. 1836.
Helianthus giganteus var. *insulus* Kellogg, Proc. Calif. Acad. **5**: 17. 1873.
Helianthus californicus var. *mariposianus* A. Gray, Syn. Fl. N. Amer. **1²**: 277. 1884.

Perennial from thickened root; stem 1.5–3.0 m. tall, glabrous, glaucous, sulcate. Leaves mostly alternate, lanceolate, rarely ovate-lanceolate, entire or remotely serrate, hispid above, hispid to hirsute beneath, 10–20 cm. long, 2.5–6.0 cm. wide, sessile or with petioles to 3 cm. long; peduncles glabrous or hispid near summit; disks 1.5–2.5 cm. in diameter; phyllaries broadly lanceolate, 3–5 mm. wide, hispidulous to glabrous on margins, conspicuously exceeding disk in length and reflexed at maturity; lobes of disk-corollas yellow; achene glabrous, about 5 mm. long. $n = 51$.

In rather wet soil, Upper Sonoran and Transition Zones; Napa County to Mariposa and Santa Clara Counties, California, and Los Angeles County south to northern Lower California; sea level to 5,500 feet. Type locality: California. June–Oct.

Helianthus tuberòsus L. Sp. Pl. 905. 1753. (*Helianthus tuberosus* var. *subcanescens* A. Gray, Syn. Fl. N. Amer. **1²**: 280. 1884; *H. besseyi* J. M. Bates, Amer. Botanist **20**: 16. 1914; *H. subcanescens* E. E. Wats. Papers Mich. Acad. **9**: 430. 1929; *H. mollissimus* E. E. Wats. op. cit. 432. *pl. 68; H. formosus* E. E. Wats. op. cit. 445. *pl. 72*, in part.) Tuber-bearing perennial; stems 1.5–2.5 m. tall, scabrous-hispid to glabrous; leaves mostly alternate, ovate or ovate-lanceolate, serrate, upper surface scabrous-setose, lower surface whitish-puberulent, rarely glaucous, 10–25 cm. long, 7–15 cm. wide; petioles 3–7 cm. long; phyllaries linear-lanceolate, equaling or slightly exceeding disk in length, appressed-pubescent; disk 1.5–2.5 cm. in diameter. $n = 51$. Northeastern North America. The "Jerusalem artichoke" is sometimes cultivated for its edible tubers. The plant has been collected as a weed or escape in Asotin County, Washington.

5. RUDBÉCKIA L. Sp. Pl. 906. 1753.

Annual, biennial, or perennial herbs. Leaves alternate, entire to bipinnatifid, the lower usually long-petioled, the upper usually sessile. Heads rather large and usually showy, radiate or rarely discoid, solitary or rather few; rays yellow or with purple-brown base, rarely crimson; disk hemispheric to conic, purple-brown, fulvous, or greenish yellow. Involucre usually about 2-seriate, of unequal, usually linear or linear-oblong, herbaceous phyllaries, squarrose or reflexed. Receptacle conic to slenderly subulate; receptacular bracts boat-shaped, firm, sometimes deciduous with the achenes. Rays neutral, usually linear-oblong; disk-corollas with short tube and cylindric or subcylindric throat, 5-toothed. Anthers subentire at base, with ovate terminal appendages. Style-branches linear or linear-oblong, with short and obtuse or longer and subulate, hispid or hispidulous appendages. Achenes quadrangular (in one species obscurely so), oblong, glabrous in ours; pappus none or a firm, often 4-toothed crown or reduced to 2–3 teeth. [Name in honor of Prof. Olof Rudbeck, 1630–1702, and his son of the same name, 1660–1740, predecessors of Linnaeus in the chair of botany at Uppsala.]

A genus of about 25 species, all North American. Type species, *Rudbeckia laciniata* L.

Rays present.
    Disk purple-brown; stem densely spreading-hispid or -hirsute throughout; pappus none.
                                                                              1. *R. hirta pulcherrima*.
    Disk greenish yellow; stem glabrous except toward apex; pappus present.   2. *R. californica*.
Rays wanting.
    Disk greenish yellow; leaves, except the uppermost, pinnately 3–9-lobed or -parted.
                                                                    3. *R. alpicola*.
    Disk purple-brown; leaves entire or merely serrate.   4. *R. occidentalis*.

1. **Rudbeckia hírta** var. **pulchérrima** Farwell. Black-eyed Susan. Fig. 5164.

Rudbeckia hirta L. Sp. Pl. 907. 1753, in part.
Rudbeckia serotina Nutt. Journ. Acad. Phila. **7**: 80. 1834.
Rudbeckia hirta var. *pulcherrima* Farwell, Rep. Mich. Acad. **6**: 209. 1904.

Biennial or short-lived perennial, 3–10 dm. high, often several-stemmed, the stems erect, simple or branched, reddish at the base and spotted with red above, rather slender, spreading-hirsute and also hispid. Lower leaves ovate, obovate, or oblanceolate, acute or obtuse, contracted into margined petioles, entire or serrate, triplinerved, hirsute or hispid, rather thin, 8–25 cm. long including the petiole; stem-leaves obovate to lanceolate, short-petioled or sessile, sometimes somewhat clasping, gradually reduced above; heads 1 to several, 4–7 cm. wide, borne on long peduncles; involucre 8–25 mm. high; phyllaries linear or lanceolate, unequal, rather densely hirsute, soon reflexed, the outer sometimes enlarged; rays about 14, golden yellow, 2–4 cm. long; disk about 1.5 cm. thick, purple-brown, subglobose, becoming low-conic; receptacle bluntly

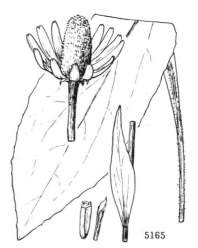

5162. Helianthus nuttallii
5163. Helianthus californicus

5164. Rudbeckia hirta
5165. Rudbeckia californica

conic, the receptacular bracts narrow, acute, purplish above, hispid at apex; style-branches with subulate hispidulous appendages; achenes 1.8–2 mm. long, epappose.

In meadows or disturbed areas, native of eastern and central United States; introduced in the Pacific States in Washington, and in California in the central Sierra Nevada and Mendocino County in the North Coast Ranges. Type locality: Virginia; Canada. June–Sept. For additional synonymy, see Perdue, Rhodora 59: 296. 1957.

2. Rudbeckia califórnica A. Gray. California Rudbeckia. Fig. 5165.

Rudbeckia californica A. Gray, Proc. Amer. Acad. **7**: 357. 1868.

Stout perennial, about 1 m. high, stem usually simple, sparsely short-hirsute above, foliage green. Lower leaves oval or elliptic-oblong, obtuse or acute, cuneate at base, subentire to irregularly dentate, feather-veined, pubescent, and usually roughish above, hirsute to pilose beneath with sometimes tuberculate-based hairs, the blade 8–18 cm. long, shorter than or equaling the hairy petiole; middle leaves similar but shorter-petioled or subsessile, often coarsely toothed; upper leaves smaller, ovate to lanceolate, sessile, and sometimes clasping, entire, coarsely toothed or deeply 3-lobed; heads 1–2, long-peduncled; involucre 1–2 cm. long, of lanceolate, unequal, dorsally pubescent phyllaries; rays about 12, yellow, usually 2–4 cm. long; disk greenish yellow, subglobose, soon becoming conic or oblong-ellipsoid, 2.8–5.5 cm. long, about 1.8 cm. thick; receptacle elongate-conic, the receptacular bracts acute or obtusish, canescent-puberulous toward the tip; style-appendages triangular-ovate, obtusish, hispidulous; achenes oblique at base, 4–5 mm. long; pappus a firm, few-toothed crown 1 mm. long.

Moist places, Canadian Zone; Sierra Nevada, California, from Eldorado County to Kern County. Type locality: Mariposa Grove, Yosemite National Park. July–Aug.

Rudbeckia californica var. **glaùca** Blake, Journ. Wash. Acad. **21**: 330. 1931. (*Rudbeckia glaucescens* Eastw. Leaflets West. Bot. **2**: 55. 1937.) Leaves glaucous, tuberculate-hispidulous on margin, otherwise glabrous, entire or obscurely denticulate; phyllaries hispidulous on margin, glabrous or nearly so on back. Douglas County, Oregon, to Del Norte and Siskiyou Counties, California. Type locality: on road to Grants Pass, 20 miles northeast of Crescent City, Del Norte County.

Rudbeckia californica var. **intermèdia** Perdue, Rhodora **59**: 289. 1957. Similar to *R. californica* var. *glauca;* leaf-blades of the basal leaves 1.5–2 dm. long, the margins irregular, either crenate, dentate, or serrate; phyllaries ciliate, otherwise glabrous. Mountains of Siskiyou County, California, at 3,500–5,000 feet. Type locality: Mount Eddy, Siskiyou County.

3. Rudbeckia alpícola Piper. Washington Rudbeckia. Fig. 5166.

Rudbeckia alpicola Piper, Erythea **7**: 173. 1899.
Rudbeckia occidentalis var. *alpicola* Cronquist, Vasc. Pl. Pacif. Northw. **5**: 280. 1955.

Stout perennial, about 1 m. high, stem usually simple, somewhat pilose at least toward the apex. Upper leaves ovate, entire or merely toothed, the others ovate or deltoid in outline, pinnately 3–9-lobed or -parted, with ovate or oblong, acuminate, and entire, toothed, or sometimes basally pinnatifid lobes, roughish above, spreading-pilose beneath especially along the nerves and on the undersurface, rather glabrate in age, the lower long-petioled, the upper sessile, the larger blades 13–20 cm. long; involucre about 2–3 cm. long, of linear to lanceolate, acuminate, soon reflexed phyllaries; mature disk dark purplish brown, subcylindric, 2.5–6 cm. long, 1.8–3 cm. thick; receptacular bracts obtuse, cinereous-pilosulose toward the tip; style-branches clavately dilated toward apex, with triangular-ovate, obtusish, hispidulous appendages; achenes oblique at base, 3.8–4 mm. long; pappus 0.5–0.7 mm. high, coroniform and irregularly toothed or reduced essentially to 2–3 teeth.

Mountain slopes, Transition and Canadian Zones; Cascade Range, Chelan and Kittitas Counties, Washington. Type locality: Mount Stuart, Kittitas County. July–Aug.

4. Rudbeckia occidentàlis Nutt. Niggerheads. Fig. 5167.

Rudbeckia occidentalis Nutt. Trans. Amer. Phil. Soc. II. **7**: 355. 1840.

Stout perennial, 0.5–2 m. high, stem striate, glabrous or sparsely appressed-pubescent above, usually simple. Lower leaves broadly ovate to lanceolate, long-petioled, cuneate to unequally cordate at base, entire to serrate; stem-leaves usually broadly ovate, 7–20 cm. long, acuminate, cuneate to broadly rounded or subcordate at base, sessile or short-petioled, coarsely repand-serrate to entire, very rarely deeply 3–5-lobed, triplinerved, usually roughish-hirsutulous at least beneath, with some pilose hairs when young, rarely glabrous; heads 1–4, on elongate peduncles; involucre about 2-seriate, 1–3 cm. long, the phyllaries very unequal, linear to oblong-ovate, herbaceous, usually reflexed in age; receptacle becoming subcylindric, the receptacular bracts obtuse or acutish, pilosulose toward tip; disk purple-brown or fuscous, at first subglobose, becoming 1.8–5 cm. long, about 1.5 cm. thick; style-branches with short, obtuse, ovate, hispidulous appendages; achenes very oblique at base, 4–5 mm. long; pappus coroniform, irregularly and shallowly lobed or dentate, about 0.7 mm. long.

Mountain slopes and along woodland streams, Arid Transition Zone; eastern Washington to Montana and Utah and south in the Sierra Nevada, California, to Placer County. Type locality: "Rocky Mountains and woods of the Oregon, particularly in the Blue Mountain range, by small streams." July–Sept.

6. HELIANTHÉLLA Torr. & Gray, Fl. N. Amer. **2**: 333. 1842.

Herbaceous, more or less leafy-stemmed perennials with stout taproots and branching caudex. Leaves alternate or opposite, oblong to lanceolate or linear, entire, usually triplinerved, the basal and lower usually larger and petioled, the others often sessile. Heads solitary or few (in ours), medium or large, radiate, the rays yellow, the disk yellow, purple, or brown. Involucre hemispheric or broader, 2–4-seriate, subequal or graduate, often subtended by a few leafy bracts, the phyllaries mostly lanceolate, herbaceous throughout or somewhat indurate at base. Receptacle flat, the receptacular bracts folded, firm or soft and scarious, partially enclosing the achenes. Rays ligulate, spreading, oblong or oval, 2–3-denticulate; disk-flowers perfect, fertile, with slender tube and cylindric throat, 5-toothed. Anthers minutely cordate-sagittate at base. Style-branches slender, hispidulous, with deltoid or short-triangular, obtusish or acutish appendages. Achenes obovate to obcordate in outline, notched at apex, flatly compressed or in one species decidedly thickened medially, narrowly whitish-margined (except in one species), glabrous or pubescent. Pappus of 2 slender persistent awns and a crown of thin, more or less united, scarious squammellae, rarely reduced to 2 awns or teeth or wanting. [Name Latin, little sunflower, from the likeness to *Helianthus*.]

A genus of 8 species, natives of British Columbia, western United States, and northern Mexico. Type species, *Helianthus uniflorus* Nutt.

Heads reflexed or turned horizontally at anthesis; receptacular bracts thin, scarious.　　　2. *H. quinquenervis.*
Heads erect at anthesis; receptacular bracts firm, chartaceous.
　　Middle cauline leaves largest, nearly all sessile or very short-petioled; achenes appressed-pilose on sides, more or less ciliate.　　　1. *H. uniflora.*
　　Lower cauline leaves largest, narrowed into definite petioles; achenes glabrous or sparsely hispidulous at apex.
　　　Phyllaries not curved over the disk at maturity, the outer rarely enlarged; achenes very flatly compressed.
　　　　3. *H. californica.*
　　　Phyllaries curved over the disk at maturity, some of the outer enlarged and leaf-like; achenes distinctly thickened medially.　　　4. *H. castanea.*

1. **Helianthella uniflòra** (Nutt.) Torr. & Gray. Rocky Mountain Helianthella. Fig. 5168.

Helianthus uniflorus Nutt. Journ. Acad. Phila. 7: 37. 1834.
Leighia lanceolata Nutt. Trans. Amer. Phil. Soc. II. 7: 365. 1841.
Helianthella uniflora Torr. & Gray, Fl. N. Amer. 2: 334. 1842.
Helianthella multicaulis D. C. Eaton, Bot. King Expl. 170. 1871.

Perennial from a short, thick, branching caudex; stems several, 2.5–12 dm. high, simple or erect-branched, erect or ascending, more or less appressed-pubescent with a few spreading hirsute hairs, becoming glabrate below. Basal and lowermost leaves oblanceolate, reduced, petioles 2–5 cm. long, the middle cauline leaves larger, lanceolate or elliptic, 3-nerved near the middle, 8–20 cm. long, opposite or alternate, the upper leaves reduced and often sessile; heads erect, terminal, long-peduncled, the disk mostly 1.5–2 cm. wide; involucre sometimes subtended by leafy bracts up to 5 cm. long; phyllaries lanceolate-attenuate, rather uniformly pubescent; rays 11–20, about 2–3 cm. long; receptacular bracts firm, hispidulous at the blunt apex; achenes obovate, flatly compressed, about 9 mm. long, narrowly wing-margined, appressed-pilose, more or less ciliate; pappus normally of 2 slender, upwardly pubescent, unequal awns 5 mm. long or less and several scarious fimbriate squamellae about 1 mm. long.

Dry hillsides and woods, Arid Transition Zone; Idaho and Montana to Utah, Colorado, and Nevada; known in the Pacific States from Harney County, Oregon. Type locality: "On the borders of the upper branches of the Columbia." Collected by Wyeth. According to Weber (Amer. Midl. Nat. **48**: 15. 1952), the type was collected on the western border of the lava beds along the boundary between Jefferson and Butte Counties, Idaho. May–Aug.

Helianthella uniflora var. **douglásii** (Torr. & Gray) W. A. Weber, Madroño **9**: 186. 1948. (*Helianthella douglasii* Torr. & Gray, Fl. N. Amer. **2**: 334. 1842.) Stems spreading-hirsute; disk 2–2.5 or rarely 3 cm. wide; phyllaries ciliate-hirsute only, the outermost rarely enlarged; rays 3–4 cm. long. Southeastern British Columbia south to central and eastern Washington and central Oregon, and adjacent Idaho. Type locality: subalpine ranges of the Blue Mountains, Oregon. Collected by Douglas. Douglas' False Sunflower.

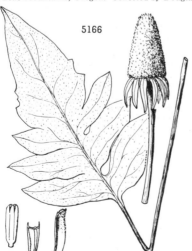

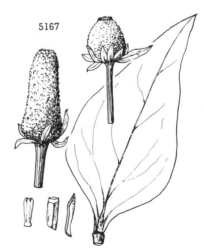

5166. Rudbeckia alpicola

5167. Rudbeckia occidentalis

2. Helianthella quinquenérvis (Hook.) A. Gray. Nodding Helianthella. Fig. 5169.

Helianthus quinquenervis Hook. Lond. Journ. Bot. **6**: 247. 1847.
Helianthella quinquenervis A. Gray, Proc. Amer. Acad. **19**: 10. 1883.
Helianthella quinquenervis var. *arizonica* A. Gray. Syn. Fl. N. Amer. **1²**: 284. 1884.
Helianthella majuscula Greene, Leaflets Bot. Obs. **1**: 148. 1905.

Perennial from a stout branching caudex, the stems stout, erect, 5–15 dm. high, glabrous or becoming glabrate below, conspicuously but rather sparsely hirsute above. Leaves opposite, sparsely hairy, up to 50 cm. long, with 2 pairs of prominent nerves, elliptic-lanceolate, acuminate at each end, the basal and lower leaves petiolate, the upper 2 pairs sessile and subclasping; heads long-peduncled, mostly solitary, some axillary and short-peduncled, nodding; phyllaries mostly ovate-lanceolate, acuminate, 1.5–2 cm. long, ciliate-margined; disk 2.5–4 cm. wide; rays about 20, 2.5–4 cm. long; receptacular bracts thin, soft and scarious; achenes 8–10 mm. long, ciliate-margined, appressed-hairy, the broad pappus of 2 slender awns 4–5 mm. long with well-developed squamellae.

Mountain meadows and boggy areas in the forests, Canadian and Hudsonian Zones; known in the Pacific States from the Warner Mountains, Lake County, Oregon, and probably occurring at higher elevations in the mountains of southeastern Oregon; southern Idaho, Montana, and the Black Hills of South Dakota south to Chihuahua and Nuevo Leon. Type locality: northwestern Natrona County, Wyoming. (See Weber, Amer. Midl. Nat. **48**: 20. 1952.) Collected by Geyer. July–Aug.

3. Helianthella califórnica A. Gray. California Helianthella. Fig. 5170.

Helianthella californica A. Gray, Pacif. R. Rep. **4**: 103. 1857.

Perennial from a branching caudex; stems several, ascending, slender, simple or rarely erect-branched, sparsely pilose with ascending to spreading hairs, 1.5–6 dm. high. Leaves mostly basal or subbasal and opposite, those of stem only 1–2 pairs and 1–3 alternate ones above; petioles slender, sparsely pilose, 3–10 cm. long; blades linear to lanceolate or oblanceolate, 5–20 cm. long, acuminate at each end, entire, feather-veined or triplinerved near middle, in age firm, sparsely appressed-hirsutulous with subappressed to spreading hairs especially beneath and on margin, becoming glabrate or scaberulous in age; heads solitary, long-peduncled, 3.5–5.5 cm. wide; involucre often subtended by 2–3 lanceolate foliaceous bracts 2–6 cm. long, subequal, 1–2 cm. high, the phyllaries lanceolate, acuminate, hirsute-ciliate, more or less hirsute dorsally, erect or squarrose; rays about 12–16, 1–2.5 cm. long; disk yellow, the receptacular bracts firm, hispidulous at tip; achenes cuneate-obovate, deeply notched at apex, glabrous; pappus none. $n = 15$.

Open places on ridges of the Coast Ranges, Upper Sonoran Zone; Mendocino and Glenn Counties, California, south through Santa Clara County. Type locality: Napa Valley, Napa County. Collected by Bigelow. March–June.

Helianthella californica var. **nevadénsis** (Greene) Jepson, Man. Fl. Pl. Calif. 1081. 1925. (*Helianthella nevadensis* Greene, Bull. Calif. Acad. **1**: 89. 1885.) Leaves scabrous-puberulent; achenes obovate-oblong, often puberulent toward the apex, marginal wings vestigial, awns 2, slender, persistent, 1–2 mm. long and often having between them a border of much shorter lacerate squamellae. Mostly in open forests in the ponderosa pine belt, Jackson and Klamath Counties, Oregon, to Modoc and Shasta Counties, California, extending southward through the Sierra Nevada to Tulare and Kern Counties, and in adjacent Washoe County, Nevada; also in the inner North Coast Ranges from Trinity County to Lake County. Type locality: "in the higher Sierra," California.

Helianthella californica var. **shasténsis** W. A. Weber, Amer. Midl. Nat. **48**: 30. 1952. Leaves densely appressed-pubescent; achenes cuneate-obovate, marginal wings narrow, broader at the apex, awns 2, about 1 mm. long. Open forests in the Siskiyou and Trinity Mountains, Siskiyou and Trinity Counties, Californa. Type locality: Mount Eddy, Siskiyou County. Intermediate forms between this variety and the preceding have been collected in adjacent Shasta County.

4. Helianthella castànea Greene. Diablo Helianthella. Fig. 5171.

Helianthella castanea Greene, Erythea **1**: 127. 1893.
Helianthella cannonae Eastw. Zoe **5**: 82. 1900.
Helianthella castanea var. *cannonae* Jepson, Man. Fl. Pl. Calif. 1081. 1925.

Perennial, from much-branched caudices, 1.5–4.5 dm. high, similar to *H. californica* but with larger leaves and more leafy-bracted involucre; stems erect or ascending, simple or rarely erect-branched, 14–45 cm. high, stoutish, hirsute with spreading hairs. Leaves chiefly basal or sub-basal, those of stem usually 3–8 and alternate (rarely with a pair below), on hirsute petioles 3–11 cm. long, the blades elliptic to lance-oblong or oblanceolate, 6–16 cm. long, 2–4.5 cm. wide, obtuse or acute, long-cuneate at base, entire, firm at maturity, feather-veined or usually triplinerved near middle, sparsely hirsute chiefly beneath and on margin; heads solitary, long-peduncled; involucre subtended by about 4 usually elliptic foliaceous bracts about 4 cm. long, about 3-seriate, subequal, about 2 cm. high, the phyllaries lanceolate or lance-oblong, acuminate, hispid-ciliate and hispid or subglabrous dorsally, inflexed over disk after anthesis; rays 12–20, 1.8–2.5 cm. long; disk yellow; achenes broadly obovate or suborbicular, with small terminal notch, glabrous, not wing-margined, 8–9 mm. long, 5–6 mm. wide, distinctly thickened medially when mature; pappus none or 2 minute teeth, without squamellae. $n = 15$.

Grassy hillsides, Upper Sonoran Zone; Alameda and Contra Costa Counties, California; San Francisco County, *Congdon.* Type locality: summit of Mount Diablo, Contra Costa County. April–May.

7. ENCELIÓPSIS (A. Gray) A. Nels. Bot. Gaz. **47**: 432. 1909.

Scapose xerophytic perennials with stout root and branched caudex, the short branches bearing tufts of leaves and monocephalous, essentially naked scapes. Leaves densely cinereous- or canescent-velutinous or hispid-canescent, all basal or subbasal, the bases of leaves of the preceding season persisting on the stout caudex, ovate to rhombic or orbicular, entire, 3–5-nerved, petioled. Heads large, radiate or discoid, yellow, sometimes nodding. Involucre hemispheric, 2–3-seriate, subequal or graduated, the phyllaries herbaceous, lanceolate to lance-ovate. Receptacle somewhat convex, the receptacular bracts softly scarious, folded, embracing the achenes and falling with them. Rays when present neutral, linear-elliptic, tridenticulate or entire, more or less puberulent on the back; disk-flowers perfect, fertile, their corollas tubular with slender tube and cylindric throat, 5-toothed. Anthers cordate-sagittate at base. Style-branches slender, hispidulous, with obtuse appendages. Achenes of ray triquetrous, sterile; of disk broadly cuneate, flatly compressed, silky-villous particularly on margin (glabrate or merely puberulous in one species), with blackish body and whitish corky border, the pappus of 2 short subulate awns and an erose crown of united squamellae, these in age thickened and corky, continuous with the margins of the achenes. [Name Greek, meaning Encelia-like.]

A genus of 4 species, natives of desert regions from Idaho through the Great Basin to Inyo County, California, and northwestern Arizona. Type species, *Tithonia argophylla* D. C. Eaton.

Petioles broadly winged, shorter than the blade or rarely equaling it; face of achene puberulent to glabrate even when immature. 1. *E. covillei.*
Petioles narrowly to scarcely winged, one and one-half to over two or more times as long as the blade, sometimes equaling it in young leaves; face of achene partially or completely covered with straight, silky-villous hairs. 2. *E. nudicaulis.*

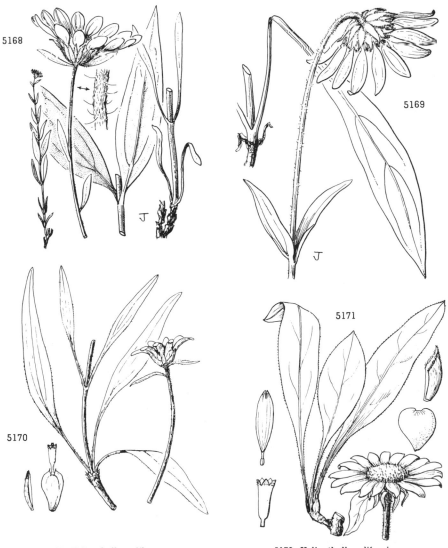

5168. Helianthella uniflora
5169. Helianthella quinquenervis

5170. Helianthella californica
5171. Helianthella castanea

1. **Enceliopsis covíllei** (A. Nels.) Blake. Panamint Daisy. Fig. 5172.

Encelia grandiflora M. E. Jones, Proc. Calif. Acad. II. **5**: 702. 1895. Not *E. grandiflora* (Benth.) Hemsl. 1881.
Helianthella covillei A. Nels. Bot. Gaz. **37**: 273. 1904.
Enceliopsis grandiflora A. Nels. op. cit. **47**: 433. 1909.
Enceliopsis argophylla var. *grandiflora* Jepson, Man. Fl. Pl. Calif. 1081. 1925.
Enceliopsis covillei Blake, Journ. Wash. Acad. **21**: 334. 1931.

Scapose perennial with numerous scapes and tufts of leaves from a branched caudex. Leaves basal or subbasal, numerous, crowded, on broad margined petioles about as long as the blade or shorter, the blades rhombic-oval to suborbicular, 4.5–9.5 cm. long, 3–6 cm. wide, obtuse or acutish, cuneate or cuneate-rounded at base, entire, 3-nerved, finely and densely silvery-velutinous with mostly appressed hairs, thick; scapes 4–6 cm. high, naked or essentially so, cinereous-puberulous; heads 9–13 cm. wide including the rays; involucre 3-seriate, graduate, about 1.8–3 cm. high, the phyllaries lance-ovate, acuminate to an obtuse apex, densely cinereous-puberulous or subvelutinous; rays 3–5 cm. long, about 20–34; achenes broadly obovate-cuneate, 10 mm. long, 6.5 mm. wide, with blackish, finely puberulent or glabrate broad and broad, yellowish white, corky margin and crown, the awns 2, nearly smooth, subulate, 1 mm. long.

Clay banks and cliffs and gravelly slopes, Lower Sonoran Zone; along the west face of Panamint Mountains, Inyo County, California. Type locality: Hall Canyon, Panamint Mountains. April–June. Large-flowered Sunray.

Enceliopsis argophýlla (D. C. Eaton) A. Nels. Bot. Gaz. **47**: 433. 1909. (*Tithonia argophylla* D. C. Eaton, Bot. King Expl. *423*. 1871; *Encelia argophylla* A. Gray, Proc. Amer. Acad. **8**: 657. 1873; *Helianthella argophylla* A. Gray, op. cit. **19**: 9. 1883.) Growth habit like the preceding taxon; leaves oblong-obovate to rhombic-ovate, acute at apex, silvery-velutinous with appressed hairs, tapering below to a broadly margined petiole distinctly shorter than the blade; disk large; rays about 2 cm. long, 3–4.5 mm. wide; achenes oblong, about 10 mm. long, 3.5 mm. wide, silky-villous on the body of the achene and the margins, awnless or with 2 subulate awns. Clayey cliffs and sandy washes, southern Utah, southern Nevada, and northwestern Arizona. A specimen collected by Billings in 1932 in the "Mohave Desert near the Nevada line" agrees with the diagnostic characters given above. Type locality: St. George, Utah. Collected by Palmer.

2. **Enceliopsis nudicaùlis** (A. Gray) A. Nels. Naked-stemmed Sunray. Fig. 5173.

Encelia nudicaulis A. Gray, Proc. Amer. Acad. **8**: 656. 1873.
Helianthella nudicaulis A. Gray, op. cit. **19**: 9. 1883.
Enceliopsis nudicaulis A. Nels. Bot. Gaz. **47**: 433. 1909.
Enceliopsis tuta A. Nels. loc. cit.

Scapose perennial with numerous scapes 1.5–4.5 dm. high arising from a short branched caudex. Leaves basal or subbasal, abruptly narrowed to a wingless or nearly wingless petiole one to three times the length of the blade, the blades broadly ovate to oval or nearly orbicular, 3-nerved, abruptly acute at apex or rounded, 2–6 cm. long, 1.5–6 cm. wide, mostly densely puberulent with often spreading hairs, dull not silvery; scapes naked or essentially so, less puberulent than the leaves and heads; heads 4–9 cm. wide including the rays; involucre 3-seriate, the phyllaries subulate-lanceolate from an ovate base, bluntish at apex; rays 20–21, 2–4 cm. long; achenes cuneate, 9 mm. long, about 3.5 mm. wide, the body and margins silky-villous except for the narrow, white, marginal border; awns short, nearly concealed by the silky-villous hairs, the squamellae nearly fused into a crown.

Clayey soil and gravel, Sonoran Zones; Custer and Lemhi Counties, Idaho, to Nevada and southern Utah, and the Death Valley National Monument in Inyo County, California. Type locality: St. Thomas or St. George, Utah. Collected by Bishop. May–Aug.

8. **ENCÈLIA** Andans. Fam. Pl. **2**: 128. 1763 (Hyponym); Lam. Encycl. **2**: 356. 1786.

Perennial herbs or shrubs, more or less pubescent, sometimes tomentose. Leaves alternate, linear to ovate, entire to laciniate-lobed. Heads small or medium, radiate or discoid, the rays yellow, the disk yellow or purple. Involucre hemispheric, 2–3-seriate, graduate or subequal, the phyllaries linear to ovate, herbaceous, sometimes with indurate base. Receptacle flattish, the receptacular bracts softly scarious, folded, embracing the achenes and falling with them. Ray-flowers when present sterile, ligulate, spreading, oval or oblong, entire or 2–3-toothed; disk-corollas with short tube and usually cylindric-funnelform throat, 5-toothed. Anthers minutely cordate-sagittate at base. Style-branches linear-oblong or clavate-oblong, hispidulous, with nearly glabrous, obtusish, deltoid appendages. Achenes of ray sterile, of disk obovate or oblong, flatly compressed, notched at the apex, with narrow, white, villous-ciliate, callous margin, glabrous or pubescent on the sides. Pappus wanting or of 1–2 slender pubescent awns. [Named for Christopher Encel, who published a work on oak galls in 1577.]

A genus of about 14 species, natives of southwestern United States to Mexico; in Peru, Chile, and in the Galapagos Islands. Type species, *Encelia canescens* Lam.

Heads cymose or panicled; inflorescence glabrous or nearly so; leaves densely tomentulose.   1. *E. farinosa.*
Heads solitary at tips of stem and branches; peduncles pubescent; leaves not tomentulose.

    Disks yellow; involucre hispid or hirsute; deserts and interior ranges.

        Rays normally none; leaves scabrous with scattered pustulate-based hairs; peduncles 6–12 cm. long.
                2. *E. frutescens.*

        Rays present, 1–3 cm. long; leaves finely appressed-pubescent, sometimes with slender scabrous hairs between; peduncles 15–30 cm. long.   3. *E. virginensis.*

    Disks purple; coastal.   4. *E. californica.*

1. **Encelia farinòsa** A. Gray. Brittlebush or Incienso. Fig. 5174.

Encelia farinosa A. Gray in Torr. in Emory, Notes Mil. Rec. 143. 1848.

Woody below, much branched, forming a rounded bush 1–1.6 m. high, exuding a fragrant resin, the branches finely canescent-tomentose, glabrescent. Leaves on slender petioles 1–4 cm. long, the blades ovate to lanceolate, 3–10 cm. long, 2–5 cm. wide, obtuse or acute, cuneate at base, entire but often undulate-margined or rarely repand-dentate, 3-nerved, densely and canescently farinose-tomentulose, often glabrescent; branches essentially naked and glabrous or nearly so above, bearing several cymose or panicled heads 1.8–4 cm. wide, these nodding in fruit; involucre 3–4-seriate, 3.5–7 mm. high, graduate, the phyllaries linear to ovate, obtuse or acutish, loosely short-pilose, more or less glabrate; rays about 8–18, yellow, 7–15 mm. long; disk yellow; achenes about 4 mm. long, villous-ciliate and villous on center of faces; pappus wanting.

Dry slopes and washes, especially in the desert, Sonoran Zones; Inyo County, California, south through the Mojave and Colorado Deserts to central Lower California and extending westward in California to the San Bernardino Valley and about Lake Elsinore, Riverside County; also southern Nevada south to Arizona, Sonora, and Sinaloa. Type locality: vicinity of Carrizo Creek, San Diego and Imperial Counties, California. March–June.

Encelia farinosa var. **phenicodónta** (Blake) I. M. Johnston, Proc. Calif. Acad. IV. **12**: 1198. 1924. (*Encelia farinosa* f. *phenicodonta* Blake, Proc. Amer. Acad. **49**: 362. 1913.) A color form which differs from *E. farinosa* var. *farinosa* only in having the disk-flowers purple above. Occurring with *E. farinosa* var. *farinosa* but only in the southern part of its range.

5172. Enceliopsis covillei
5173. Enceliopsis nudicaulis

5174. Encelia farinosa
5175. Encelia frutescens

2. Encelia frutéscens A. Gray. Bush Encelia. Fig. 5175.

Simsia frutescens A. Gray in Torr. Bot. Mex. Bound. 89. 1859.
Encelia frutescens A. Gray, Proc. Amer. Acad. **8**: 657. 1873.
Encelia frutescens f. *ovata* H. M. Hall, Univ. Calif. Pub. Bot. **3**: 135. 1907.
Encelia frutescens f. *radiata* H. M. Hall, loc. cit.

Much-branched shrub up to 1.6 m. high, white-barked, the branches usually scabrous with short, spreading to appressed hairs. Leaves short-petioled, the blades 1–3 cm. long, 6–16 mm. wide, oblong or elliptic to ovate, obtuse to acute, entire or rarely repand-toothed, triplinerved, green, scabrous with scattered, short, conical, tuberculate-based, white hairs; heads solitary at tips of branches, on peduncles usually 6–12 cm. long, discoid or rarely radiate, the disk 1–2.5 cm. wide; involucre 3-seriate, 6–10 mm. high, graduate, the phyllaries with usually ovate or lance-ovate body and rather abrupt, narrowly triangular-acute, glandular, herbaceous tip, hispid-ciliate and hispid; rays when present about 12, yellow, about 9 mm. long; disk yellow; achenes about 7 mm. long, villous-ciliate, sparsely pubescent on sides; pappus none or 1–2 weak villous awns.

Rocky desert slopes or washes, mostly Lower Sonoran Zone; southern Nevada and Utah south to Arizona and Sonora, and in California in Inyo County south through the Mojave and Colorado Deserts. Type locality: Agua Caliente, Maricopa County, Arizona. March–Nov. Rayless Encelia.

3. Encelia virginénsis A. Nels. Virgin River Encelia. Fig. 5176.

Encelia virginensis A. Nels. Bot. Gaz. **37**: 272. 1904.
Encelia frutescens f. *virginensis* H. M. Hall, Univ. Calif. Pub. Bot. **3**: 135. 1907.
Encelia frutescens var. *virginensis* Blake, Proc. Amer. Acad. **49**: 364. 1913.

Much-branched, shrubby-based perennial up to 1 m. or more high, the pubescence of the stems spreading-hispidulose and also glandular, becoming glabrous on the older stems. Leaves short-petioled, the blades 1–2.5 cm. long, deltoid-ovate to ovate, triplinerved, entire, somewhat cinereous with sparse or sometimes more abundant appressed hairs and mixed on both surfaces with ascending, tuberculate-based hairs, glandular; heads radiate, solitary on elongate peduncles up to 3 dm. long, pubescence like the stems; involucre 3-seriate, grayish, the phyllaries mostly ovate with an abruptly acuminate tip or lance-ovate, spreading-hirsute and usually densely glandular; rays 12–20, about 1.5–2 cm. long; disk yellow; achenes 4–5 mm. long, glabrous on the sides, long villous-ciliate on the margins.

Gravelly slopes and washes, Sonoran Zones; southwestern Utah and southern Nevada to northern Arizona and occurring rather sparingly in the Death Valley region, Inyo County, California, and more abundant southward in eastern San Bernardino County. Type locality: "The Pockets," Virgin River, southern Nevada. Collected by Goodding. April–June.

Encelia virginensis subsp. actònii (Elmer) Keck, Aliso **4**: 101. 1958. (*Encelia actonii* Elmer, Bot. Gaz. **39**: 47. 1905; *E. frutescens* f. *actonii* H. M. Hall, Univ. Calif. Pub. Bot. **3**: 135. 1907; *E. frutescens* var. *actonii* Blake, Proc. Amer. Acad. **49**: 365. 1913.) Like the preceding in habit; leaves 1.5–4.5 cm. long, ovate or sometimes elliptic, white-cinereous with dense appressed hairs, the tuberculate-based, hispid hairs lacking except in intermediate forms in the Death Valley region; heads long-peduncled as in *E. virginensis* subsp. *virginensis* but mostly larger, the phyllaries more densely pubescent and scarcely glandular; rays 1.5–3 cm. long. Dry or gravelly slopes and washes, Kern County, California, in the southernmost Sierra Nevada south to eastern San Diego County and the desert ranges east of the Sierra Nevada in Mono and Inyo Counties and adjacent Nevada south to northwestern Arizona. Type locality: Acton, Los Angeles County, California. Intermediate forms between this taxon and *E. virginensis* subsp. *virginensis* are not infrequently found and in the area about Palm Springs, Riverside County, some have been seen that approach *E. frutescens*.

4. Encelia califórnica Nutt. California Encelia. Fig. 5177.

Encelia californica Nutt. Trans. Amer. Phil. Soc. II. **7**: 357. 1841.

Much-branched, shrubby-based perennial up to 1 m. or occasionally 3.6 m. high, strong-scented, the stem incurved-puberulous to cinereous-pilose, glabrescent. Leaves on slender petioles 0.5–3 cm. long, the blades lanceolate to ovate, 3–6 cm. long, 1–3 cm. wide, acuminate to obtuse, at base cuneate to rounded-cuneate, entire or repand-toothed, 3-nerved, sparsely or densely incurved- or appressed-pubescent with soft hairs; heads about 3.5–7.5 cm. wide, solitary at tips of stems and branches, long-peduncled; involucre 2–3-seriate, 1–1.3 cm. high, the phyllaries lanceolate to lance-ovate, acutish, subtomentose to densely pilose; rays 14–25, yellow, 1–3 cm. long; disk purple; achenes 5–6 mm. long, villous-ciliate, villous down middle of each side; pappus wanting.

Ocean bluffs and dry slopes and canyons. Upper Sonoran Zone; coastal ranges from Santa Barbara County, California, south to northwestern Lower California and on the islands off the coast; also eastward to Riverside, Riverside County. Type locality: near Santa Barbara, Santa Barbara County. Jan.–Aug.

9. GERAÈA Torr. & Gray ex A. Gray, Proc. Amer. Acad. **1**: 48. (Jan.?) 1847.

Annual or perennial herbs, glandular and spreading-pubescent. Leaves alternate, obovate to ovate or the upper linear, sessile or narrowed into margined petioles, toothed or entire. Heads medium or rather large, radiate or discoid, yellow. Involucre usually hemispheric, about 3-seriate, subequal or graduated, the phyllaries linear to oblong or ovate, herbaceous, appressed. Receptacle flattish, the receptacular bracts softly scarious, conduplicate, embracing and falling with the achenes. Rays when present neutral, wedge-shaped; disk-flowers perfect, fertile, their corollas tubular, with slender tube and cylindric-funnelform or cylindric throat, 5-toothed. Anthers cordate-sagittate or subentire at base. Style-branches slender, essentially hispid or hispidulous throughout, with obtuse sterile appendages. Achenes cuneate, flatly compressed, villous on sides and villous-ciliate, with blackish body and narrow, whitish, crustaceous margin produced into 2 strong persistent awns connected at base by a conspicuous, entire, crustaceous crown. [From the Greek word meaning old, referring to the canescent-villous achenes.]

A genus of the following 2 species. Type species, *Geraea canescens* Torr. & Gray.

Heads radiate; phyllaries linear to linear-lanceolate, densely white-ciliate.	1. *G. canescens.*
Heads discoid; phyllaries oblong to ovate, densely glandular, obscurely ciliate.	2. *G. viscida.*

1. **Geraea canéscens** Torr. & Gray. Desert Sunflower. Fig. 5178.

Geraea canescens Torr. & Gray ex A. Gray, Proc. Amer. Acad. 1: 49. (Jan.?) 1847.
Simsia canescens A. Gray, Mem. Amer. Acad. II. **4**: 85. 1849.
Encelia eriocephala A. Gray, Proc. Amer. Acad. **8**: 657. 1873.

Erect annual 1–6 dm. high; stem simple or branched, glandular and spreading-hirsute or -hispid. Lower leaves cuneate-obovate or obovate, 1.5–13 cm. long, 0.5–4 cm. wide, usually acute, cuneately narrowed into a margined base, remotely and usually sharply toothed above the middle or entire, triplinerved, hirsute with tuberculate-based hairs, the upper gradually reduced, lanceolate to linear and sessile; heads 2.5–6 cm. wide, solitary, cymose, or numerous in nearly naked panicles; involucre 2–3-seriate, subequal or graduate, 8–12 mm. high, the phyllaries linear or lance-linear, acuminate, green and usually glandular on back above, densely hispid-ciliate except

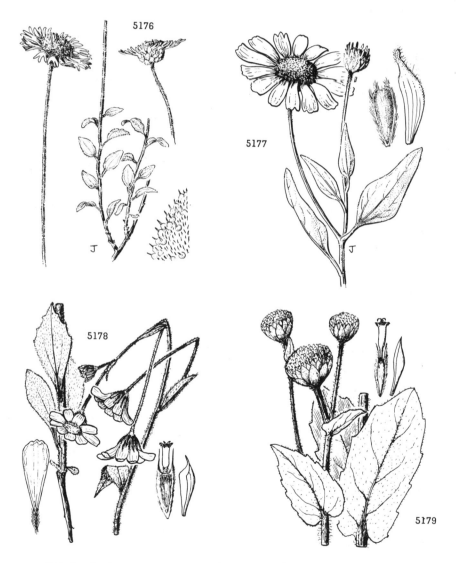

5176. Encelia virginensis
5177. Encelia californica

5178. Geraea canescens
5179. Geraea viscida

at tip with long white hairs; rays about 10–21, golden yellow, 1–2 cm. long, wedge-shaped; achenes about 6 mm. long, silky-villous except toward margin and villous-ciliate; pappus-awns about 2.5 mm. long, lance-linear, subentire, villous at base, connected by a thick yellowish crown.

Sandy desert places, mostly Lower Sonoran Zone; southern Utah and southern Nevada southward through the northern Mojave Desert and Death Valley region to the Colorado Desert and northeastern Lower California; eastward to southern Arizona and adjacent Sonora. Type locality: California. Jan.–May. Hairy-headed Sunflower.

2. Geraea víscida (A. Gray) Blake. Sticky Geraea. Fig. 5179.

Encelia viscida A. Gray, Proc. Amer. Acad. **11**: 78. 1876.
Geraea viscida Blake, op. cit. **49**: 357. 1913.

Stout perennial, the stems numerous from the crown of a thick, conical, vertical root and up to 8 dm. high, simple or branched, viscid-glandular and spreading-villous, very leafy. Lowest leaves scale-like or small and spatulate or obovate, narrowed into a petioliform base, the others ovate, oval, or oblong, 3–10 cm. long, 1.5–5 cm. wide, acute or obtuse, sessile and cordate-clasping, coarsely dentate or serrate to entire, somewhat triplinnerved well above base, glandular and villous especially on costa beneath; heads few, discoid, 1–3 cm. thick, pedunculate; involucre about 3-seriate, graduate or subequal, 11–15 mm. high, the phyllaries ovate to elliptic or oblong-ovate, obtuse or acutish, densely glandular, sparsely ciliate; achenes about 8 mm. long, awns about 5 mm. long, linear-lanceolate, villous below, hirsutulous above.

Dry hills, Upper Sonoran Zone; southern San Diego County, California, and adjacent northern Lower California. Type locality; near Larkin's Station, 80 miles east of San Diego. Collected by Palmer. May–June.

10. VERBESÌNA L. Sp. Pl. 901. 1753.

Herbs, shrubs, or rarely trees, usually pubescent. Leaves usually opposite at least below, entire, toothed, or pinnatifid. Heads small to rather large, radiate or discoid, usually yellow. Involucre 2- to several-seriate, usually graduated, at least the outer phyllaries usually subherbaceous. Receptacle convex or conical, the receptacular bracts folded, embracing the achenes. Ray-flowers usually present, pistillate or rarely neutral, their corollas ligulate, usually oblong or oval, 2–3-toothed. Disk-flowers perfect, fertile (sterile in one species), their corollas tubular, 5-toothed. Anthers subentire at base. Style-branches slender with short or elongate, acute, hispidulous appendages. Achenes strongly compressed, 2-winged (very rarely wingless), the wings usually whitish. Pappus of 2 slender awns, rarely wanting. [Derivation uncertain, possibly from *Verbena*.]

A genus of about 200 species, all American. Besides the following, about 9 others occur in the southern United States. Type species, *Verbesina alata* L.

Leaves green on both sides, sessile or subsessile.   1. *V. dissita*.
Leaves canescent-strigose beneath, slender-petioled.   2. *V. encelioides exauriculata*.

1. Verbesina díssita A. Gray. Big-leaved Crownbeard. Fig. 5180.

Verbesina dissita A. Gray, Proc. Amer. Acad. **20**: 299. 1885.

Stout perennial, suffrutescent at base, 1 m. high or more, erect-branched above, the stems striate, glabrous below, hispidulous toward the inflorescence. Leaves opposite below, alternate above, remote; principal leaves ovate, on short petioles 2–8 mm. long and winged to base, the blades 4–8 cm. long, acute or acuminate, cuneate or rounded-cuneate at base, remotely repand-serrate or -dentate, subtriplinerved, veiny, firm, bright green on both sides and sparsely tuberculate-hispidulous; upper leaves ovate, sessile, clasping, subentire; heads several, yellow, large, 4.5–6 cm. wide, irregularly cymose-panicled, on peduncles 1.5–12 cm. long; involucre about 3-seriate, graduate, 7–11 mm. high, the phyllaries obovate or spatulate to oblong, blunt or the inner acute, appressed, minutely hispidulous, the outer subherbaceous; rays about 12, 1.5–2.5 cm. long; achenes glabrous, the body about 9 mm. long, broadly 2-winged, the awns slender, about 3 mm. long.

Bluffs, Upper Sonoran Zone; coastal southern California (also reported from the San Bernardino Mountains) to northern Lower California. Type locality: near All Saints Bay, Lower California. Collected by Orcutt. May.

2. Verbesina encelioìdes var. exauriculàta Robins. & Greenm. Crownbeard. Fig. 5181.

Verbesina encelioides var. *exauriculata* Robins. & Greenm. Proc. Amer. Acad. **34**: 544. 1899.
Verbesina exauriculata Cockerell, Nature **66**: 607. 1902.
Ximenesia exauriculata Rydb. Bull. Torrey Club **33**: 154. 1906.

Annual, usually branched at least above, 1 m. high or less, the stem cinereous-strigose. Lowest leaves opposite, the others alternate; petioles of lower leaves slender, unappendaged, those of the upper leaves short, often with a pair of linear to half-ovate foliaceous, stipuliform appendages at base; blades ovate to (upper) lanceolate, 3–10 cm. long, acute or acuminate, at base cuneate to subcordate, sharply and irregularly toothed, rather thin, green above, densely canescent-strigose beneath; heads several, yellow or orange, cymose-panicled, peduncled, 2.5–4.5 cm. wide; involucre 7–12 mm. high, 2-seriate, subequal or subgraduate, the phyllaries lance-ovate to lance-linear, acuminate, subherbaceous, strigose, appressed or loose-tipped; rays 10–15, 3-toothed or -lobed, 1–2 cm. long; achenes broadly 2-winged, pubescent, the body 4–6 mm. long, the pappus of 2 short slender awns.

Plains, foothills, mountains, and beds of winter streams, Sonoran Zones; Kansas to Montana, Texas, and Arizona south into Mexico, and also in the warmer regions of the Old World; introduced in California scatteringly from the Salinas River in Monterey County southward to Los Angeles and Ventura Counties; also occurring in Fresno and Kern Counties. Type locality: not definitely stated. Aug.–Dec.

11. **ECLÍPTA** L. Mant. 157. 1771.

Annual or perennial, weedy plants, diffuse or erect, strigose or hirsute. Leaves opposite, lanceolate to oval, toothed or entire. Heads small, radiate, white or yellow, usually solitary in the forks of the stem and the axils, pedunculate. Involucre hemispheric or campanulate, 2-seriate, usually subequal; phyllaries lanceolate to oval, herbaceous above, somewhat indurate below. Receptacle flat or somewhat convex, the receptacular bracts very narrow, mostly filiform or aristiform, subtending but not embracing the achenes. Rayflowers fertile, numerous, more than 1-seriate, the ray small, narrow, entire or 2-toothed; disk-corollas funnelform, 4–5-toothed. Anthers cordate at base, with ovate, obtuse, terminal appendages. Style-branches oblong, hispidulous, with short obtuse appendages. Achenes obovate-oblong, plump, truncate at apex, those of the ray trigonous, obcompressed, of the disk bluntly quadrangular, compressed in age, often corky-margined and corky-bullate. Pappus a short, thick, ciliolate crown, sometimes produced into 1–3 teeth. [The Greek word meaning deficient, referring to the absence of pappus.]

A genus of about 4 species, in the warmer regions of the world. Type species, *Eclipta erecta* L. (= *Verbesina alba* L.).

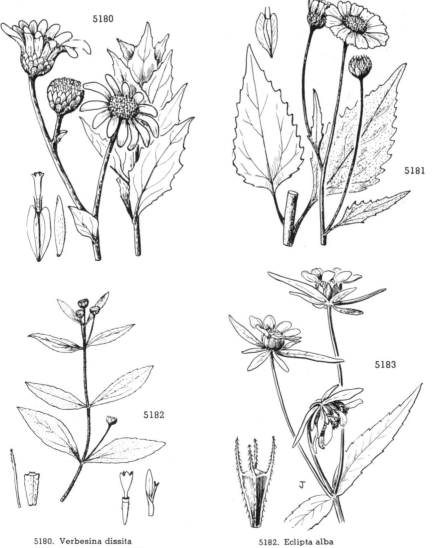

5180. *Verbesina dissita*
5181. *Verbesina encelioides*

5182. *Eclipta alba*
5183. *Bidens cernua*

1. Eclipta àlba (L.) Hassk. False Daisy. Fig. 5182.

Verbesina alba L. Sp. Pl. 902. 1753.
Verbesina conyzoides Trew, Pl. Rar. 8. *pl. 8.* 1763.
Eclipta erecta L. Mant. 286. 1771.
Eclipta alba Hassk. Pl. Jav. Rar. 528. 1848.

Diffuse or erect, branching, annual or biennial herbs 2–8 dm. high, strigose throughout. Leaves opposite, lanceolate to oblong, 2–10 cm. long, sessile and often clasping, remotely serrulate, tripli-nerved; peduncles mostly solitary in the forks and in the upper axils, 0.5–4 cm. long; heads white-rayed, about 4–8 mm. thick; involucre 2-seriate, 3–8 mm. high, the phyllaries subequal, lanceolate to broadly ovate, acuminate, strigose, herbaceous above; ray-corollas numerous, not surpassing involucre, the ray linear, about 2 mm. long; disk-corollas 4-toothed, greenish white; achenes ovate-oblong, plump, about 2 mm. long, the ray-achenes trigonous, corky-margined and corky-bullate, hirsutulous on the truncate apex; pappus a minute ciliolate crown, often 1- or 2-toothed.

Waste places and along streams and irrigation ditches, Lower Sonoran Zone; in central California along the lower Sacramento and San Joaquin Rivers and in southern California, eastward to the Atlantic seaboard and southward through Mexico and Central America to South America; introduced in its northern range in Nebraska and Massachusetts. Type locality: Virginia; Surinam. March–Nov. Yerba-de-tago.

12. BÌDENS L. Sp. Pl. 831. 1753.

Ours annual or perennial herbs, glabrous or somewhat pubescent. Leaves opposite, serrate to pinnately divided or dissected. Heads radiate or discoid, yellow (in ours), solitary or cymose at apex of stem and branches. Involucre double, the outer of 5–16 usually narrow, essentially 1-seriate, herbaceous phyllaries, shorter or longer than the inner; inner of more numerous 2-seriate, equal, membranous, oblong or ovate, several-nerved phyllaries, usually yellow-margined and brown-nerved. Receptacle flat, the receptacular bracts flat, narrow, membranous or subscarious, usually yellow, subtending the achenes. Rays neutral, oval or oblong, subentire to tridenticulate, shorter or longer than the involucre, yellow or yellowish white in ours; disk-flowers perfect, fertile, tubular, 4–5-toothed. Anthers minutely bidentate at base; style-branches with triangular or subulate, hispid tips. Achenes linear or fusiform to obovate, not beaked, strongly obcompressed and flat to quadrangular; pappus of 2–6 persistent, stiff, retrorsely hispid awns (in ours). [Latin for two-toothed, from the bi-aristate achene.]

A genus of perhaps 200 species, nearly throughout the world. Type species, *Bidens tripartita* L.

Stems not floating; leaves not divided into filiform or capillary segments; awns 5 mm. long or less.
  Leaves merely serrate to subentire; achenes retrorse-hispidulous on margin, very narrowly cuneate.
    Rays wanting or not over 1.5 cm. long; receptacular bracts yellow, conspicuously brown-nerved; achenes often somewhat bent, with a strong rib on each face; awns usually 4.   1. *B. cernua.*
    Rays 1.5–3 cm. long; receptacular bracts orange-tipped; achenes straight and flat, usually without facial ribs; awns usually 2.   2. *B. laevis.*
  Leaves pinnately 3–5-parted; achenes antrorse-ciliate or -hispidulous.
  Achenes cuneate to broadly obovate, very flat, 2–4 mm. wide.
    Outer phyllaries 5–8, remotely ciliate; achenes usually blackish, narrowly cuneate, 2–3.3 mm. wide.   3. *B. frondosa.*
    Outer phyllaries 10–16, densely hirsute-ciliate; achenes yellowish or olive-brown, broadly obovate or cuneate-obovate, 2.5–4 mm. wide.   4. *B. vulgata.*
  Achenes linear to linear-fusiform, more or less quadrangular, 1 mm. wide or less.   5. *B. pilosa.*
Stems floating; submerged leaves divided into filiform or capillary segments; awns 1.5–2.3 cm. long.
    6. *B. beckii.*

1. Bidens cérnua L. Nodding Bur-marigold. Fig. 5183.

Bidens cernua L. Sp. Pl. 832. 1753.
Bidens minima Huds. Fl. Angl. 310. 1762.
Bidens cernua elliptica Wiegand, Bull. Torrey Club **26**: 417. 1899.
Bidens cusickii, B. glaucescens, B. kelloggii, B. lonchophylla, B. prionophylla Greene, Pittonia **4**: 256–267. 1901.

Annual, simple or branched, usually under 1 m. high, stems stoutish or slender, glabrous or sometimes hirsute or hispid. Leaves lanceolate to oblong-lanceolate, 4–20 cm. long, 0.4–5 cm. wide, acuminate, sessile or the lowest contracted into a petiole, connate at base, sharply serrate to subentire, glabrous or sparsely hirsute along midrib beneath; heads solitary to several at apex of stem and branches, usually short-peduncled, radiate or discoid, soon nodding, the disk in fruit 2.5 cm. thick or less; outer phyllaries 6–8, linear or spatulate to narrow-oblong, 2.5 cm. long or less, hispidulous-ciliolate, soon reflexed; rays often wanting, when present about 8, golden-yellow, 1.5 cm. long or less; receptacular bracts golden-yellow, brown-nerved, the apex rarely orange; achenes narrowly cuneate, 5–7 mm. long, often somewhat curved, usually strongly 1-ribbed on each face and quadrangular at least above, retrorse-hispidulous on the margins with usually tuberculate-based hairs; awns (2–)4, retrorse-hispid, subequal or 2 usually shorter, 2–3 mm. long.

Marshy places, sloughs, and ponds; a native of Europe and Asia occurring in eastern Canada south in eastern United States and west to British Columbia and New Mexico; in the Pacific States it is found occasionally in Washington and Oregon south to Del Norte and Plumas Counties, California, and also in Sonoma and San Francisco Counties. Type locality: Europe. June–Oct. Stick-tight.

2. Bidens laèvis (L.) B.S.P. Bur-marigold. Fig. 5184.

Helianthus laevis L. Sp. Pl. 906. 1753.
Bidens chrysanthemoides Michx. Fl. Bor. Amer. **2**: 136. 1803.
Bidens laevis B.S.P. Prel. Cat. N.Y. 29. 1888.
Bidens speciosa Parish, Zoe **5**: 75. 1900. Not Gardn. 1845.
Bidens expansa Greene, Pittonia **4**: 266. 1901.

Usually stout annual (or perennial by the persistent base), up to 1 m. high, glabrous throughout; stems usually decumbent and rooting at base, simple or branched above. Leaves lanceolate to elliptic- or obovate-oblong, 4–22 cm. long, 1–6 cm. wide, usually acuminate, sessile, slightly or scarcely connate at base, serrate or serrulate, thickish, glabrous; heads showy, 4–7 cm. wide, mostly solitary or in threes at apex of stem and branches, often long-peduncled, at length nodding; outer phyllaries 6–9, usually narrowly oblong, obtuse, hispidulous-ciliolate, 2.5 cm. long or usually much less, rarely equaling the rays, tending to reflex; disk golden-yellow; rays usually 8, oval-oblong, 2–3 cm. long, golden-yellow, often deeper-colored at base; receptacular bracts usually orange-brown at tip; achenes narrowly cuneate, 5–8 mm. long, usually straight and flat, sometimes weakly 1-ribbed on each face, harshly retrorse-hispidulous on margin, often also on faces; awns 2(–4), retrorse-hispid, 2.2–3 mm. long.

Streambanks, swamps, and sloughs; in the Pacific States in the lower Sacramento Valley, California, southward in the San Joaquin Valley and South Coast Ranges to southern California, Arizona, and from Mexico to South America; on the eastern seaboard from New England south to southern United States. Type locality: Virginia. July–Dec. Often distinguishable with difficulty from *B. cernua*.

Bidens tripartita L. Sp. Pl. 831. 1753. A cosmopolitan annual weed with petioled, tripartite or deeply serrate leaves, with rayless or short-rayed heads, and outer enlarged foliaceous phyllaries has been collected on ballast at Portland, Oregon, and at Bingen, Klickitat County, Washington, where it was collected by Suksdorf. Type locality: described from a cultivated specimen in Holland.

Bidens amplíssima Greene, Pittonia **4**: 268. 1901. (*Bidens cernua* β *elata* Torr. & Gray, Fl. N. Amer. **2**: 352. 1842.) Tall, branched annual with large, sessile, pinnately tripartite leaves, the uppermost entire; outer phyllaries foliaceous, unequal, and coarsely incised. Known only from Vancouver Island, British Columbia. May be expected to occur in adjacent Washington. Type locality: Lomas River, Vancouver Island.

3. Bidens frondòsa L. Stick-tight. Fig. 5185.

Bidens frondosa L. Sp. Pl. 832. 1753.
Bidens melanocarpa Wiegand, Bull. Torrey Club **26**: 405. 1899.

Annual, up to about 1 m. high, stem slender, branching, sparsely pilose to essentially glabrous. Leaves opposite, slender-petioled, pinnately 3(–5)-parted, the segments lanceolate to oblong-ovate, 9 cm. long or less, acuminate or caudate, cuneate at base, sharply serrate, thin, sparsely pubescent to nearly glabrous, the lateral short-stiped or subsessile, the terminal long-stiped; heads discoid or inconspicuously radiate, solitary or few and irregularly cymose at apex of stem and branches; disk in fruit 8–12 mm. high and about as thick; outer phyllaries 5–8, spatulate or oblanceolate, 0.4–2.5 cm. long, acute, remotely ciliate; inner phyllaries ovate-oblong, about equaling the disk in young flower; rays golden-yellow, the rays when present 1–5, not surpassing the disk; achenes cuneate or narrowly cuneate-obovate, 5–10 mm. long, 2–3.3 mm. wide, antrorse-ciliate at least below the apex (sometimes retrorse-ciliate above), usually sparsely pilose on the face, blackish; awns 2, retrorse-hispid, 4 mm. long or less.

Moist open places, frequently occurring as a weed; eastern Canada west to British Columbia and south to Colorado, Louisiana, and Florida; occasional in Washington, Oregon, and California as far south as the San Bernardino Mountains; also introduced in Europe. Type locality: "America." July–Oct.

5184. Bidens laevis

5185. Bidens frondosa

4. **Bidens vulgàta** Greene. Western Stick-tight. Fig. 5186.

Bidens vulgata Greene, Pittonia 4: 72. (July) 1899.
Bidens frondosa var. *puberula* Wiegand, Bull. Torrey Club 26: 408. (Aug.) 1899.
Bidens vulgata var. *puberula* Greene, Pittonia 4: 250. 1901.

Annual, similar to *B. frondosa* but coarser and taller; terminal division of leaves often short-stiped; heads larger, the disk in fruit 1.3–2.5 cm. thick; outer phyllaries 10–16, densely hirsute-ciliate or merely pubescent or puberulous, up to 5.5 cm. long; corolla yellow; achenes broadly cuneate or obovate-cuneate, 6–12 mm. long, 2.5–4 mm. wide, usually merely antrorse-ciliolate, on faces less pubescent or nearly glabrous, yellowish or olive-brown; awns 2, retrorse-hispid, 5 mm. long or less.

Moist places and roadsides; Quebec southward to North Carolina and Missouri and westward to Washington, Oregon, and in California in the Sacramento Valley, where it apparently is introduced; also adventive in Europe. Type locality: not given. June–Oct.

5. **Bidens pilòsa** L. Beggars-ticks. Fig. 5187.

Bidens pilosa L. Sp. Pl. 832. 1753.
Bidens californica DC. Prod. 5: 599. 1836.

Slender annual, branched chiefly above, up to 1.3 m. high, stem striate-angled, sparsely hispid or hirsute to nearly glabrous, with long internodes. Leaves slender-petioled, pinnately 3–5 parted, the divisions lance-oblong to ovate or rhombic-ovate, crenate-serrate to incised, acuminate to obtuse, sparsely pubescent, usually thin, the lateral ones subsessile or short-stiped, the terminal long-stiped; heads many-flowered, solitary or few and irregularly cymose at apex of stem and branches, 6–8 mm. thick in flower, 1–2 cm. thick in fruit; outer phyllaries 6–8, linear or spatulate, about 3.5 mm. long, ciliate especially above, shorter than the inner; disk yellow; rays none or few, small, yellowish white; achenes linear or linear-fusiform, 5–13 mm. long, obcompressed or sub-equally quadrangular, few-striate, tuberculate-hispidulous at least above with antrorse hairs, the outer much shorter than the inner; awns 2–4, erect or divergent, about 2 mm. long, naked below, retrorse-hispid for about the upper half of their length.

Introduced weed in waste places and cultivated ground; native of the American tropics, of which many forms have been segregated; occurring in California from Monterey and Tulare Counties south into Arizona and Mexico. Type locality: "America." March–Oct.

6. **Bidens béckii** Torr. Water Marigold. Fig. 5188.

Bidens beckii Torr. ex Spreng. Neue Entdeck. 2: 135. 1821.
Megalodonta beckii Greene, Pittonia 4: 271. 1901.
Megalodonta remota Greene, op. cit. 272.
Megalodonta beckii var. *oregonensis* Sherff, Bot. Gaz. 97: 609. 1936.
Megalodonta beckii var. *hendersonii* Sherff, Amer. Journ. Bot. 25: 589. 1938.

Aquatic perennial, glabrous throughout, the stems floating or rarely emersed, simple or few-branched above, up to 2 m. long or more. Submersed leaves subsessile, 1.5–4 cm. long, 3-parted almost to base, the lobes many times forked into narrowly linear to capillary, acute segments; emersed leaves few (usually 2–5 pairs, rarely wanting), lanceolate to ovate, 1–3.5 cm. long, sessile, acute or acuminate, narrowed at base, sharply serrate, pectinate, or laciniate; peduncles solitary at tips of stems and branches, 1–3.5 cm. long; heads in flower 1.5–3.5 cm. wide; outer phyllaries about 5–6, oblong-ovate, thick-herbaceous, the inner somewhat longer, submembranous, narrowly yellow-margined; rays about 8, golden yellow, oblong-oval, 1–1.5 cm. long; achenes linear, 1 cm. long, obcompressed-quadrangular, striate-ribbed, glabrous, yellowish or greenish brown; awns 3–6, slender-subulate, 1.5–2.3 cm. long, divergent, rigid, somewhat unequal, sulcate inside, naked below, densely retrorse-hispidulous all around at tip for about one-fourth their length.

In lakes, rivers, and marshes; Quebec to New Jersey west to Manitoba, Minnesota, and Missouri; in the Pacific States in Washington and Oregon, where perhaps introduced by aquatic birds. Type locality: near Schenectady, New York. July–Sept.

The garden cosmos (*Cosmos bipinnatus* Cav. Ic. 1: 10. 1791), a plant with leaves divided into linear, more or less filiform divisions and with white-, pink-, or crimson-rayed flowers, may be found occasionally as an adventive around habitations.

13. **COREÓPSIS** L. Sp. Pl. 907. 1753.

Herbs or shrubs, usually glabrous or nearly so. Leaves opposite or alternate, entire to dissected. Heads small to medium, radiate, usually yellow, the rays sometimes with purple-brown spot at base, the disk sometimes purple-brown. Involucre hemispheric, double; outer phyllaries 1-seriate, usually few (6–10), herbaceous, usually much shorter than the inner, free or united at base; inner more numerous, 2-seriate, equal, membranous, free, usually brown or yellow. Receptacle flattish, the receptacular bracts membranous or sub-scarious, flat. Ray-flowers neutral or pistillate, their corollas ligulate, usually wedge-shaped, often toothed or 3–5-lobed. Disk-flowers fertile, their corollas tubular, with slender tube and 5-toothed limb. Anthers bidentate or subentire at base. Style-branches tipped with a conical appendage or subtruncate. Achenes obcompressed, usually oblong or oval, flat or meniscoid, usually winged. Pappus of 2 teeth or awns (never downwardly barbed), or sometimes a small whitish cup, or wanting. [Name Greek, bug-like, from the achenes of the original species.]

A genus of over 100 species, mainly American but found also in Africa and the Hawaiian Islands; numerous others occur in the United States. Type species, *Coreopsis lanceolata* L.

Perennials; comparatively tall and stout.
    Leaves deeply bi- to tripinnately divided; ray-flowers fertile.
        Heads solitary or few, the peduncles usually 2–5 dm. long.   1. *C. maritima.*
        Heads cymose-clustered, the peduncles usually 1–1.5 dm. long.   2. *C. gigantea.*
    Leaves entire, sometimes divided, with few lanceolate lobes; ray-flowers sterile.   9. *C. lanceolata.*
Annuals; slender and usually low.
    Rays yellow throughout (or white-tipped); disk yellow; ray-flowers fertile.
        Disk- and ray-achenes dissimilar; pappus of disk-achenes of 2 conspicuous awns.
           Outer phyllaries linear.
               Rays horizontal at anthesis; pappus-paleae of disk-flowers 2–3 mm. long; Monterey County, California, south and east to the Mojave Desert.   3. *C. bigelovii.*
               Rays strongly reflexed at anthesis; pappus-paleae of disk-flowers 1 mm. long; Mount Hamilton Range.   4. *C. hamiltonii.*
           Outer phyllaries broadly ovate or deltoid; pappus-paleae of disk-flowers mostly 4–5 mm. long.   5. *C. calliopsidea.*
        Disk- and ray-achenes alike; pappus a small white cup.
           Leaves entire or with 1–2 linear pinnae, the terminal portion not broadened; annulus of disk-corollas bearded.
               Achenes pubescent on both faces with clavellate or capitate hairs, the corky wings irregularly much thickened; plants of the deserts and cismontane southern California.   6. *C. californica.*
               Achenes glabrous on both faces or nearly so, the corky wings thin and flat; plants of the Inner Coast Ranges from Mount Hamilton Range to San Rafael Range.   7. *C. douglasii.*
           Leaves pinnate or bipinnate (sometimes entire in depauperate specimens); annulus of disk-corollas glabrous or nearly so.   8. *C. stillmanii.*
    Rays purple-brown at base; disk purple-brown; ray-flowers sterile.
        Achenes wingless.   10. *C. tinctoria.*
        Achenes winged.   11. *C. atkinsoniana.*

1. **Coreopsis marítima** (Nutt.) Hook. f. Sea-dahlia. Fig. 5189.

Tuckermannia maritima Nutt. Trans. Amer. Phil. Soc. II. 7: 363. 1841.
Leptosyne maritima A. Gray, Proc. Amer. Acad. 7: 358. 1868.
Coreopsis maritima Hook. f. Bot. Mag. 102: *pl. 6241.* 1876.

Perennial, glabrous throughout, the stems fleshy-herbaceous, spreading, 3–8 dm. high, from a thick woody base. Leaves alternate, deltoid in outline, up to 2 dm. long including petiole, fleshy, bi- to tripinnately dissected into narrowly linear, obtuse, divergent lobes 1–3 mm. wide; peduncles solitary or few at tips of stem and branches, naked, 1-headed, usually 2–5 dm. long; heads 6–10 cm. wide, golden-yellow; outer phyllaries 6–10, lanceolate to ovate, mostly obtuse, 8–16 mm. long, the inner lance-oblong to ovate, usually acute or acuminate, up to 2 cm. long; rays 16–20, 2–4 cm. long, disk-corollas with puberulous annulus; receptacular bracts 8–12 mm. long, linear to oblanceolate, midrib somewhat callous-thickened, falling separately from the achenes; achenes narrowly oblong-obovate, 6–7 mm. long, glabrous, narrowly winged, the body of the achene and the narrow wings brown, epappose or rarely with 2 teeth or short awns.

Beaches and coastal bluffs and hillsides, Upper Sonoran Zone; San Diego County, California, to northern Lower California and adjacent islands. Type locality: San Diego, San Diego County. Collected by Nuttall. March–June.

5186. Bidens vulgata

5187. Bidens pilosa

2. **Coreopsis gigantèa** (Kell.) H. M. Hall. Giant Coreopsis. Fig. 5190.

Leptosyne gigantea Kell. Prod. Calif. Acad. **4**: 198. 1872.
Coreopsis gigantea H. M. Hall, Univ. Calif. Pub. Bot. **3**: 142. 1907.
Tuckermannia gigantea M. E. Jones, Contr. West. Bot. No. 15: 74. 1929.

Stout, fleshy-woody, glabrous perennial 3–30 dm. high, the stem up to 1 dm. thick or more, branched only above, the branches spreading or ascending, leafy only toward the ends. Leaves alternate, deltoid or ovate in outline, up to 3 dm. long, bi- to tripinnately dissected into filiform or narrowly linear, fleshy, obtuse lobes 0.5–1.5 mm. wide; heads 4–6 cm. wide, golden-yellow, cymosely clustered toward tips of branches on peduncles usually 1–1.5 dm. long; involucre essentially as in *C. maritima*; rays about 13, 1.5–2 cm. long; disk-corollas with puberulent annulus; receptacular bracts 2–10 mm. long, linear, midrib somewhat callous-thickened below, falling separately from the achenes; achenes oval to oblong, 5–7 mm. long, glabrous, body of the achene and the narrow wings brown, epappose.

Sand dunes and sea cliffs, Upper Sonoran and Humid Transition Zones; coast of southern California from San Luis Obispo and Santa Barbara Counties to Los Angeles County and adjacent islands (except San Clemente Island); also Guadalupe Island, Lower California. Type locality: near Cuyler Harbor, San Miguel Island, Santa Barbara County. Collected by Harford. March–May.

3. **Coreopsis bigelòvii** (A. Gray) H. M. Hall. Bigelow's Coreopsis. Fig. 5191.

Pugiopappus bigelovii A. Gray, Pacif. R. Rep. **4**: 104. 1857.
Leptosyne bigelovii A. Gray, Syn. Fl. N. Amer. **1²**: 300. 1884.
Coreopsis bigelovii H. M. Hall, Univ. Calif. Pub. Bot. **3**: 141. 1907.

Erect, essentially glabrous, scapose or subscapose annual, 3–60 cm. high, the scapose or scapiform stems mostly 5–10, naked or very rarely sparsely leafy for one-third their length. Leaves chiefly or entirely in a basal tuft, ovate or deltoid in outline, 2–12 cm. long including petiole, fleshy, pinnately or bipinnately dissected into linear or linear-filiform, obtuse segments 2 cm. long or less or the lowest sometimes linear-filiform and unlobed; heads golden-yellow, 2–4 cm. wide; outer phyllaries 5–7, linear or lance-linear, obtuse, 4–9 mm. long, the inner about 8, ovate, somewhat longer, brown with narrow yellow margin; rays about 8–10, obovate or wedge-obovate, 1–2.5 cm. long; disk-corollas with puberulent annulus; receptacular bracts 4–10 mm. long, lanceolate or oblanceolate, hyaline, falling with the disk-achenes; ray-achenes about 5 mm. long, obovate-oblong, glabrous or essentially so, epappose, often becoming corky-winged and corky-ribbed; disk-achenes linear-oblong, about 6 mm. long, densely villous-ciliate, glabrous to sparsely villous on the faces; pappus of 2 flattened, lanceolate, persistent paleae 2–3 mm. long.

Desert places and hillsides, Sonoran Zones; west of the Sierra Nevada in California from Fresno and Monterey Counties south to San Diego County and in the Mojave and Colorado Deserts from Inyo County to Riverside County. Type locality: "On the Mohave Creek, in the desert east of the Colorado." Feb.–June.
Plants of the southern Sierra Nevada and the Tehachapi Mountains are often rather coarse and leafy-stemmed but have the characteristic linear outer phyllaries of *C. bigelovii*.

4. **Coreopsis hamiltònii** (Elmer) H. K. Sharsmith. Mount Hamilton Coreopsis. Fig. 5192.

Leptosyne hamiltonii Elmer, Bot. Gaz. **41**: 323. 1906.
Coreopsis hamiltonii H. K. Sharsmith. Madroño **4**: 214. 1938.

Erect, essentially glabrous annual 10–15 cm. high, scapose, sometimes branching at the base, the scapes mostly 10–15. Leaves all basal or nearly basal, 1–5 cm. long, spreading, bipinnate into short, linear, obtuse lobes; heads 1–2 cm. broad, golden-yellow, the rays reflexed; outer phyllaries 4–7, broadly linear, obtuse, 3–6 mm. long, glabrous at the truncate base, the inner 6–8, lanceolate to narrowly ovate, 5–8 mm. long, not constricted medially in age; rays 5–8, oblong to obovate, mostly truncate at apex; disk-flowers with a glandular pubescent annulus; receptacular bracts 5–6 mm. long, narrowly linear, obtuse, falling separately from the disk-achenes; ray-achenes 5 mm. long, meniscoidal, obovate, brown or mottled with brown, both the body and marginal wings smooth and glabrous, eppapose; disk-achenes 5–6 mm. long, narrowly obovate, both faces more or less villous, the margins densely villous-ciliate; pappus of 2 persistent paleae 1 mm. long.

Open rocky slopes, Upper Sonoran Zone; Mount Hamilton Range, Santa Clara County, California. Type locality: Mount Hamilton. March–April.

5. **Coreopsis calliopsídea** (DC.) A. Gray. Leafy-stemmed Coreopsis. Fig. 5193.

Agarista calliopsidea DC. Prod. **5**: 569. 1836.
Coreopsis calliopsidea A. Gray in Torr. Bot. Mex. Bound. 90. 1859.
Leptosyne calliopsidea A. Gray, Syn. Fl. N. Amer. **1²**: 300. 1884.

Erect, essentially glabrous annual 1–4.5 dm. high, stems 4–7, stout, usually sparsely leafy one-third to one-half their length and with less pronounced basal tufts of leaves. Leaves deltoid in outline, 3–8 cm. long including petiole, once or twice ternately or quinately parted into linear, obtuse, fleshy divisions mostly 1–1.5 mm. wide; heads normally golden-yellow, 2.5–6 cm. wide; outer phyllaries 5, sparsely pilose on their venose, turbinate, united, lower part, their free parts deltoid- or suborbicular-ovate, obtuse or acutish, 3.5–5 mm. long and about as wide; inner phyllaries about 7–8, about twice as long as the outer, oblong-ovate, brown with narrow yellow margin; rays about 8, sometimes white on upper third, broadly wedge-obovate, 1–2.5 cm. long; disk-corollas with bearded annulus; receptacular bracts 5–6 mm. long, broadly linear, obtuse, hyaline, falling separately from the achenes; ray-achenes 5–6 mm. long, broadly oval, tan or brown, winged, glabrous, smooth or with a thick corky covering, epappose; disk-achenes 6–7 mm. long,

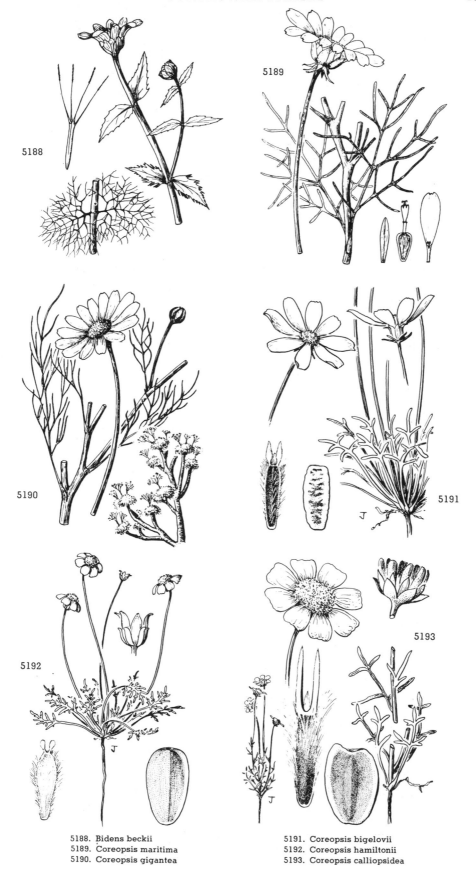

5188. Bidens beckii
5189. Coreopsis maritima
5190. Coreopsis gigantea

5191. Coreopsis bigelovii
5192. Coreopsis hamiltonii
5193. Coreopsis calliopsidea

narrowly cuneate-oblong, both faces more or less villous, the margins densely villous-ciliate with persistent, long, white cilia, the pappus of 2 linear-lanceolate, persistent paleae, about 4 mm. long.

Moist hillsides and plains, Sonoran Zones; eastern slopes of the Inner Coast Ranges, Alameda County, California, and the adjacent San Joaquin Valley south to the Cuyama Valley, Santa Barbara County, and east to the western edge of the Mojave Desert in Kern, San Bernardino, and Los Angeles Counties; also Saugus, Los Angeles County. Type locality: California. Collected by Douglas. March–May.

6. Coreopsis califórnica (Nutt.) H. K. Sharsmith. California Coreopsis. Fig. 5194.

Leptosyne californica Nutt. Trans. Amer. Phil. Soc. II. 7: 363. 1841.
Leptosyne newberryi A. Gray, Proc. Amer. Acad. 7: 358. 1868.
Coreopsis californica H. K. Sharsmith, Madroño 4: 217. 1938.
Coreopsis douglasii of authors. Not *Leptosyne douglasii* DC.

Erect, scapose, essentially glabrous annual usually 2–3.5 dm. high, the scapes about 5–15, slender. Leaves basal, erect, 2–15 cm. long, linear-filiform, entire or with a few short slender pinnae; heads 1–3.5 cm. broad, golden-yellow; outer phyllaries 2–8, linear, the obtuse apex with a red callous tip, 3–8 (rarely more) mm. long, papillate at the gibbous base; inner phyllaries 5–8, broadly lanceolate to narrowly obovate, in age yellow and shiny and constricted, the tufted apices spreading; rays 5–12, narrowly to broadly obovate, 5–16 mm. long, often pale at the tip; disk-corollas with a bearded annulus, the central corollas often sterile; receptacular bracts usually 4–5.5 mm. long, linear, obtuse at apex, not falling with the disk-achenes; achenes 2.5–5 mm. long, dull and finely pubescent with clavate hairs on both surfaces, ventrally ribbed with a corky ridge, tan or light brown, the wings of the same color, these strongly and irregularly corky-thickened, the ventral side of the wings with few to several reddish spots or blotches; pappus a small white cup.

Plains and sandy places. Sonoran Zones; southern California ranges from Santa Barbara County to northern Lower California eastward through the Mojave and Colorado Deserts to Inyo County, California, and western Arizona. Type locality: near San Diego, San Diego County. Collected by Nuttall. March–May.

7. Coreopsis douglásii (DC.) H. M. Hall. Douglas' Coreopsis. Fig. 5195.

Leptosyne douglasii DC. Prod. 5: 531. 1836.
Coreopsis douglasii H. M. Hall, Univ. Calif. Pub. Bot. 3: 140. 1907, as to name only.
Coreopsis stillmanii var. *jonesii* Sherff, Bot. Gaz. 97: 605. 1936.

Scapose or subscapose annual 5–25 cm. high or more, essentially glabrous throughout except for the short-pilose base of involucre, the scapes mostly 2–5, slender, erect or ascending. Leaves basal or subbasal, spreading when young, later becoming suberect, 2–12 cm. long including petiole, narrowly linear, entire or with 1–2 filiform pinnae 3–10 mm. long; heads solitary, 1.5–3.5 cm. wide, golden-yellow; outer phyllaries 4–7, linear, obtuse, 3–7 mm. long, papillate at the gibbous base; inner phyllaries more numerous, obovate or oval, 6–10 mm. long, constricted at the middle in age, the tufted apex spreading; rays about 8–10, cuneate, 0.8–1.5 cm. long; disk-corollas with bearded annulus; receptacular bracts 4–5 mm. long, linear, obtuse at apex, hyaline, falling separately from disk-achenes; achenes often meniscoidal, 2.5–5 mm. long, the body narrowly obovate, glabrous or nearly so, broadly 2-winged, the wings stramineous, sometimes tan or brownish, slightly corky but scarcely thickened, unspotted on the inner face, the body of the achene shining, dark brown, glabrous or occasionally with a few inconspicuous papillae; pappus a small whitish cup.

Dry rocky slopes, Upper Sonoran Zone; Inner Coast Ranges, California, from Santa Clara County to Santa Barbara County and also the inner face of the southern Santa Lucia and the San Rafael Mountains. Type locality: California. Collected by Douglas. March–May (Aug.).

8. Coreopsis stillmánii (A. Gray) Blake. Stillman's Coreopsis. Fig. 5196.

Leptosyne stillmanii A. Gray in Durand, Journ. Acad. Phila. II. 3: 91. 1855.
Coreopsis stillmanii Blake, Proc. Amer. Acad. 49: 342. 1913.

Essentially glabrous annual, sparsely leafy to the middle and with stems occasionally branched below or subscapose. Leaves 2–10 cm. long, pinnately or sometimes bipinnately parted into 2–7 linear spatulate lobes 0.5–3 mm. wide, the terminal pinna usually broader than the lateral ones, entire on depauperate plants; heads 1–3.5 cm. broad, orange-yellow; outer phyllaries 4–8, linear to subspatulate, obtuse, 3–10 cm. long, bearded at the nongibbous base; inner phyllaries 5–10, ovate, constricted at flowering, the tufted tips spreading; rays 5–15 mm. long, obovate and deeply 3-toothed at the apex; ring of disk-corollas merely puberulous or the corollas glabrous; receptacular bracts 5–6 mm. long, linear-lanceolate, obtusish, hyaline, not falling with the achenes; achenes 2.5–5 mm. long, obcompressed, the body of the achene dark brown, rounded on the back, essentially glabrous, the ventral surface usually with only a central row of callous papillae, the wings narrow, stramineous, corky-thickened; pappus a whitish cup, occasionally with 1–2 short awns from the rim.

Arid slopes, often on serpentine, Sonoran Zones; foothills of the Sierra Nevada, California, from Butte County to Tulare County and those of the Inner Coast Ranges from Contra Costa County to Stanislaus County. Type locality: upper Sacramento Valley. March–April.

9. Coreopsis lanceolàta L. Garden Coreopsis. Fig. 5197.

Coreopsis lanceolata L. Sp. Pl. 908. 1753.

Perennial herb 2–6 dm. high, mostly glabrous except for the lower leaves, with several to many erect or ascending stems arising from a woody rootstock, the stems branching, leafy at or near the base, rather naked above. Lower leaves 10–15 cm. long, long-petiolate, spatulate to oblanceolate, entire or with 2 or 3 lateral lobes at the base, usually hirsute, the upper leaves reduced in size, entire or sometimes lobed, glabrous or nearly so; heads many, 3–5 cm. wide at anthesis on peduncles 15–30 cm. long; outer phyllaries usually narrower and shorter than the inner, lanceolate to oblong-ovate, the inner 8–12 mm. long; rays about 8, yellow, 1.5–3 cm. long, obovate or cuneate, 2–3-toothed at the apex; receptacular bracts 4–6 mm. long; achenes 2.5–3 mm. long, obcompressed, the body black, orbicular in outline, the wings somewhat incurved; pappus of 2 small fimbriolate teeth.

A native of the United States often planted in gardens in the Pacific States and sometimes becoming locally established (Santa Cruz County, California, *Thomas*; Santa Barbara County, California, *C. F. Smith*); occurring naturally from Michigan to Florida and to Texas and New Mexico. Type locality: unknown. May–June.
For complete synonymy see N. Amer. Fl. II. 2: 18–19. 1955.

10. **Coreopsis tinctòria** Nutt. Calliopsis. Fig. 5198.

Coreopsis tinctoria Nutt. Journ. Acad. Phila. **2**: 114. 1821.
Calliopsis bicolor Reichb. Ic. Pl. Cult. **2**: *pl. 70*. 1823.

Erect glabrous annual 6–12 dm. high, usually single-stemmed, the stem leafy, much branched, the branches angulate. Leaves opposite, 5–10 cm. long, subsessile or short-petiolate, pinnately

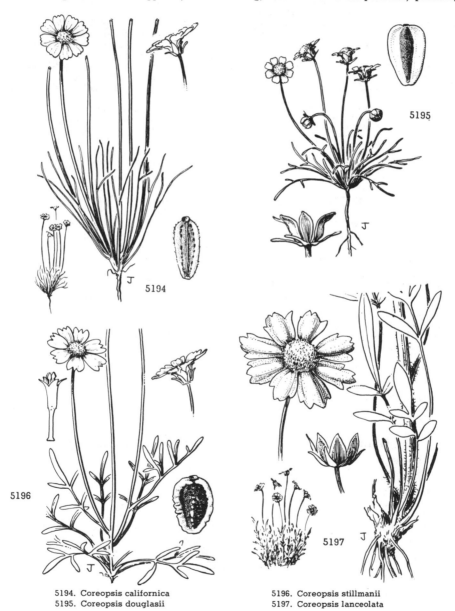

5195

5194

5196

5197

5194. Coreopsis californica
5195. Coreopsis douglasii

5196. Coreopsis stillmanii
5197. Coreopsis lanceolata

or bipinnately divided into linear or linear-lanceolate divisions, the upper leaves usually un-divided; heads many, 2–3 cm. wide, more or less corymbosely arranged, the peduncles 4–10 cm. long; outer phyllaries about 8, linear-oblong or triangular, about 2 mm. long; inner phyllaries deltoid- to oblong-ovate, 5–6 mm. long; ray-flowers sterile, rays 7–8, obovate and apically 3-lobed, 7–15 mm. long, yellow with a brownish red base; disk-flowers dark red; receptacular bracts more or less filiform, reddish; disk-achenes of variable length, 1.2–4 mm. long, linear-oblong, com-pressed, wingless, black, glabrous or nearly so, the pappus obsolete.

A native of central United States much resembling *C. atkinsoniana,* occurring in gardens and occasionally becoming locally established; Washington and in Shasta, Madera, Stanislaus, and Tulare Counties, California. Type locality: Red River, Arkansas. Collected by Nuttall. June–Sept. For complete synonymy of the various forms see N. Amer. Fl. II. 2: 30. 1955.

Coreopsis básalis var. **wrìghtii** *(A. Gray)* Blake (Proc. Amer. Acad. **51**: 526. 1916) has been collected in San Diego County, California. It differs from *C. tinctoria,* which it much resembles, in having the outer phyllaries linear to linear-lanceolate and spreading, 5–8 mm. long, and nearly equaling in length the ovate inner phyllaries.

11. **Coreopsis atkinsonìana** Dougl. ex Lindl. Columbia Coreopsis. Fig. 5199.

Coreopsis atkinsoniana Dougl. ex Lindl. Bot. Reg. **16**: *pl. 1376.* 1830.
Calliopsis atkinsoniana Hook. Fl. Bor. Amer. **1**: 311. 1833.

Annual (or biennial?), usually single-stemmed, 0.4–1.3 m. high, glabrous throughout, the stems slender, erect-branched, striate-angulate, rather naked above. Leaves opposite, the lower 5–16 cm. long including petiole, pinnately or bipinnately divided into elliptic to linear segments 1–6 mm. wide, the middle and upper pinnately divided into 3–5 linear segments 0.5–3 mm. wide, or the uppermost entire; heads 2–4 cm. wide, several to numerous, cymose-panicled, the peduncles 5–15 cm. long; outer phyllaries 7–8, narrowly triangular, obtuse, 2–3 mm. long; inner phyllaries 6–8 mm. long, ovate, brown with yellow margin; ray-flowers sterile, the rays about 8, wedge-obovate, 1–2 cm. long, golden-yellow with purple-brown basal spot; disk purple-brown; recep-tacular bracts linear-oblong, orange-red, with a scarious margin; disk-achenes narrowly oblong, about 2.5 mm. long, glabrous, often papillose, 2-winged, the thin wings half as wide as the body or less; pappus none or of 2 small teeth.

River banks and lake shores, Transition Zones; British Columbia south to northern Oregon (along the Colum-bia River) and eastern Washington east to Idaho and Montana; also north-central United States and introduced elsewhere in the United States. Type locality: Mewries [Menzies?] Island, in the Columbia River. Collected by Douglas. May–Sept. Probably included in mixed wildflower seed-packets with *C. tinctoria* and becoming locally established around dwellings elsewhere in the Pacific States.

Guizòtia abyssínica (L. f.) Cass. Dict. Sci. Nat. **59**: 248. 1829. An annual species which resembles *Bidens* but has epappose achenes has been found growing spontaneously in San Francisco and Santa Barbara Counties, California.

14. **GALINSÒGA** Ruiz & Pav. Fl. Peruv. 110. *pl. 24.* 1794.

Annual herbs, erect or diffuse, more or less pubescent to glabrate. Leaves opposite, simple, thin, triplinerved, the margins entire or toothed. Heads rather small, radiate, in leafy cymes. Involucres hemispheric or campanulate, the phyllaries in 2 series, the outer mostly incomplete, few, thin, green, and several-nerved. Receptacle conic, covered with narrow membranous bracts. Marginal flowers pistillate, fertile, few, with short, white or pink rays. Disk-flowers perfect, fertile, yellow, tubular, 5-toothed. Anthers minutely or scarcely sagittate. Style-branches flattened, the minute appendages minutely hairy on the outer face. Achenes dark, obovoid or slightly angled, glabrous or hairy, those of the ray-flowers somewhat flattened parallel to the phyllaries. Pappus-paleae several to many, thickish, the margins fimbriate and often awned above, sometimes reduced or lacking on the ray-achenes. [Name in honor of Mariano Martinez de Galinsoga, a Spanish botanist of the eighteenth century.]

A genus of about 6 species, natives of southern United States and Central and South America. Type species, *Galinsoga parviflora* Cav.

Pappus of disk-flowers tapering to an awn; pappus of ray-flowers present.   1. *G. ciliata.*
Pappus of disk-flowers truncate or rounded, the margins densely and finely fimbriate; pappus of ray-flowers want-ing or vestigial.   2. *G. parviflora.*

1. **Galinsoga cilìata** (Raf.) Blake. Ciliate Galinsoga. Fig. 5200.

Adventina ciliata Raf. New Fl. Pt. 1: 67. 1836.
Galinsoga parviflora hispida DC. Prod. **5**: 677. 1836.
Galinsoga aristulata Bickn. Bull. Torrey Club **43**: 270. 1916.
Galinsoga ciliata Blake, Rhodora **24**: 35. 1922.

Freely branching, annual herb 1.5–6 dm. high, the stems, particularly the upper part, and the peduncles rather thickly beset with spreading hirsute hairs and often glandular. Leaves deltoid-ovate, the petioles 1–3 cm. long, the blade 2–5.5 cm. long, 1–3 cm. wide and somewhat decurrent on the petiole, the margin typically coarsely toothed, sparsely hirsute; involucre 2–3 mm. high, the phyllaries ovate; ray-flowers 4–5, the rays white, 1–2 mm. long; disk-flowers yellow, but little surpassing the pappus-paleae; achenes black, 1.5–2 mm. long, hispidulous, obscurely angled, the disk-achenes compressed, hispidulous on the inner face; pappus-paleae of ray- and disk-flowers 9–14, more or less fimbriate-ciliate and tapering into an awn.

Waste places and gardens; occurring as a weed in British Columbia and western Washington, and widespread in central and northeastern United States and adjacent Canada. Naturalized from Mexico. June–Oct.

2. **Galinsoga parviflòra** Cav. Small-flowered Galinsoga. Fig. 5201.

Galinsoga parviflora Cav. Ic. **3**: 41. *pl. 281.* 1794.

Branching, fibrous-rooted, annual herb 1.5–6 dm. high with long internodes, the stems and peduncles with short appressed pubescence and some spreading glandular hairs or nearly naked. Leaves narrowly ovate to lanceolate-ovate, the petioles 5–12 mm. long, the blades 2.5–4 cm. long, 0.5–2.5 cm. wide, the margins crenate-denticulate, rarely entire, mostly glabrous; involucres 2–3 mm. high, the phyllaries ovate; ray-flowers 4–5, white-rayed, the ray about 1 mm. long, the tube villous; disk-flowers yellow, shorter than the pappus-paleae; ray-achenes black, compressed, sparsely hispidulous, the pappus vestigial or lacking; disk-achenes about 2 mm. long, black, obovoid or truncate or rounded, slightly angled, sparsely hispidulous, the pappus-paleae 9–18, puberulent, the margine finely and densely fimbriate.

Waste places and orchards or gardens; occurring in the Pacific States in Inyo County, California, and cismontane southern California; also central, eastern, and southern United States. Introduced from South America. Aug.–Feb.

15. **BÉBBIA** Greene. Bull. Calif. Acad. **1**: 179. 1885.

Suffrutescent xerophyte, intricately branched, nearly leafless, strongly scented. Leaves opposite below, alternate above, linear to triangular-ovate, entire or hastately lobed. Heads medium-sized, light yellow, solitary or few at tips of branches, discoid, many-flowered.

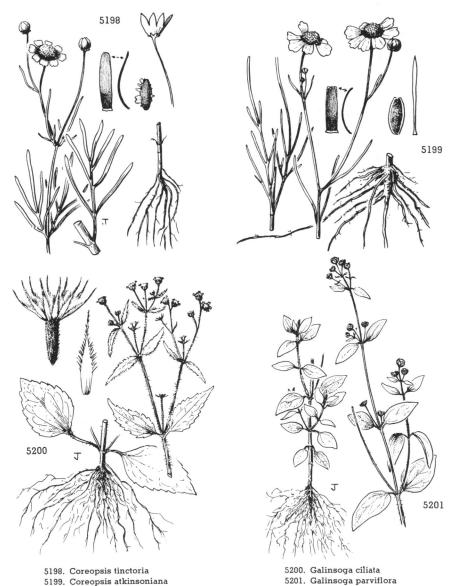

5198. Coreopsis tinctoria
5199. Coreopsis atkinsoniana
5200. Galinsoga ciliata
5201. Galinsoga parviflora

Involucre hemispheric, 4–5-seriate, strongly graduate, the phyllaries ovate or oval to lanceolate or lance-linear, the outer subherbaceous, the inner thinner, dry, usually whitish. Receptacle flattish, the receptacular bracts subscarious, folded, embracing the achenes. Corollas all tubular with short tube and elongate subcylindric throat, 5-toothed. Anthers subentire at base with ovate terminal appendages. Style-branches recurved, with subulate hispidulous appendages. Achenes slenderly obconic, somewhat compressed, weakly angled, densely silky. Pappus of about 20 slender, equal, 1-seriate, plumose awns longer than the achene. [Named for Michael Schuck Bebb, 1833–1895, student of American willows.]

A monotypic genus of the Sonoran region.

1. Bebbia júncea (Benth.) Greene. Sweetbush. Fig. 5202.

Carphephorus junceus Benth. Bot. Sulph. 21. 1844.
Bebbia juncea Greene, Bull. Calif. Acad. 1: 180. 1885.

Suffrutescent, much-branched shrub 1 m. high or less; stem and branches junciform, slender, glabrous, pale green, the stem in age whitish-barked. Leaves remote, linear or lance-linear and entire or the larger lanceolate and with one or two pairs of hastate lobes 3 cm. long or less, the upper usually reduced and squamiform; heads about 1 cm. high, usually long-peduncled, the peduncles hispidulous toward apex; involucre 4–7 mm. high, the phyllaries ovate to (inner) lanceolate or linear-lanceolate, mostly acute or acuminate, the outer subherbaceous, grayish-hispidulous, the inner whitish or brownish, dry, ciliate and dorsally pubescent toward apex; achenes about 3 mm. long, pappus whitish, 4–6 mm. long.

Desert stream beds, washes, and dry hills, Sonoran Zones; Orange and San Bernardino Counties, California, east to Nevada and Arizona, south to Lower California and Sonora. Type locality: Magdalena Bay, Lower California. April–Oct. Chuckwalla's Delight.

Bebbia juncea var. **áspera** Greene, Bull. Calif. Acad. 1: 180. 1885. (*Bebbia aspera* A. Nels. Bot. Gaz. 37: 273. 1904.) Plant more or less densely hispidulous with usually tuberculate-based, sometimes deciduous hairs. Similar range but extending north to Inyo County, California, and southern Nevada and east to New Mexico (Dona Ana County).

16. EASTWOÓDIA* Brandg. Zoe 4: 397. *pl. 30*. 1894.

Xerophytic shrub, somewhat glutinous. Stem white-barked, striate, glabrous, brittle. Leaves alternate, essentially linear, entire. Heads discoid, many-flowered, yellow, solitary at ends of the often cymosely arranged branches. Involucre hemispheric, about 4-seriate, graduate, the phyllaries appressed, narrow, firm, 1-nerved, smooth, narrowly scarious-margined. Receptacle broad, paleaceous throughout, its bracts firm, boat-shaped, caducous. Corollas funnelform, the 5 lobes long. Style-branches linear, the acuminate appendages equaling the stigmatic portion. Achenes more or less quadrangular, narrowly obpyramidal, silky-pilose especially on the ribs. Pappus of 5–6 free, persistent, linear, acute or acuminate, firm paleae about two-thirds as long as the corolla. [Named in honor of Alice Eastwood, California botanist.]

A monotypic California genus.

1. Eastwoodia élegans Brandg. Eastwoodia. Fig. 5203.

Eastwoodia elegans Brandg. Zoe 4: 397. *pl. 30*. 1894.

Rounded desert shrub 3–10 dm. high, erect-branched, the herbage minutely and sparingly hispidulous, rather glutinous, pallid. Leaves linear to linear-oblanceolate or the upper subulate, up to 2.5 cm. long, to 3 mm. wide; heads depressed-subglobose, 1–1.5 cm. thick; involucre impressed at base, 5–7 mm. high; corollas about 5.5 mm. long; achenes about 2 mm. long; pappus about 4 mm. long.

Hot dry hillsides, up to 2,500 feet, Upper Sonoran Zone; foothills on the west and south sides of San Joaquin Valley, California, from Alameda County to Santa Barbara County east to Tehachapi Mountains, Kern County. Type locality: not definitely stated. April–July.

17. MELAMPÒDIUM L. Sp. Pl. 921. 1753.

Herbs, rarely suffrutescent, glabrous or pubescent, the stems usually dichotomously branched. Leaves opposite, entire, dentate or pinnatifid. Heads small or medium, usually solitary in the forks and terminal, heterogamous, radiate, the rays pistillate, fertile, the disk perfect, sterile; corollas yellow or white. Involucre 2-seriate; outer phyllaries 3–5, broad, herbaceous, sometimes connate at base, often accrescent; inner phyllaries as many as the ray-flowers, each closely enveloping an ovary, coriaceous and often appendaged, deciduous with the enclosed achene at maturity. Receptacle convex, the receptacular bracts membranous, enfolding the disk-flowers. Ray-corollas few or several, 1-seriate, oblong or broader, entire or denticulate. Disk-corollas regular, tubular, 5-toothed, sterile. Anthers bluntly cordate at base, with ovate terminal appendages. Style in ray-flowers

---

* Text contributed by David Daniels Keck.

2-parted, in disk-flowers undivided, cylindric, hispidulous. Achenes of ray obovoid, covered by the variously muricate and often hooded or beaked inner phyllaries which are deciduous with the achenes. Achenes of disk sterile, stipitiform. Pappus none. [An ancient name of black hellebore from the Greek, arbitrarily transferred to this genus.]

A genus of about 12 species of subtropical America but mostly Mexico; one or two introduced into the Old World. Type species, *Melampodium americanum* L.

1. Melampodium perfoliàtum (Cav.) H.B.K. Melampodium. Fig. 5204.

Alcina perfoliata Cav. Ic. 1: 11. *pl. 15.* 1791.
Melampodium perfoliatum H.B.K. Nov. Gen. & Sp. 4: 274. 1820.

Coarse annual, sparsely hispid especially toward the nodes, the peduncles puberulous, the stems about 1 m. high or less, dichotomously branched. Leaves opposite, in rather few pairs, the internodes long; petioles broadly wing-margined, dilated at base and usually connate; blades deltoid-ovate or rhombic-ovate, often hastately lobed at base, acute, serrate or serrulate, 5–20 cm. long and wide, rough above, hispidulous and gland-dotted beneath, green on both sides, thin; heads solitary in the forks and terminal, on peduncles 2–11 cm. long; phyllaries 5, ovate or oval, 7–18 mm. long, obtuse, ciliolate, united at extreme base; rays about 8-13, yellow, much shorter than involucre; fruit rounded on back, muricate toward the tip, about 5 mm. long.

Occurring as a weed in swampy and waste places, vicinity of Los Angeles, Los Angeles County, California; introduced from Mexico; occurring also in Guatemala. Type locality: Mexico. May–Nov.

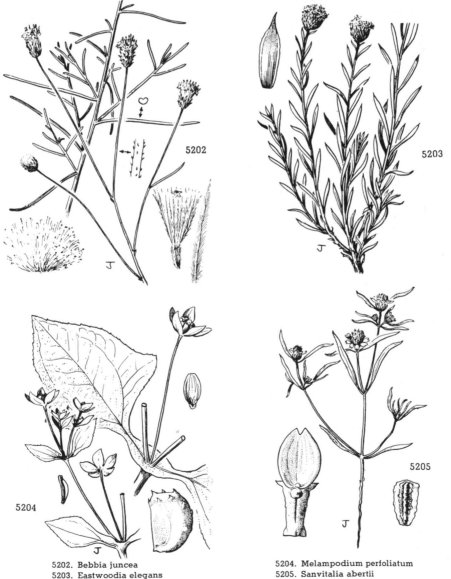

5202. Bebbia juncea
5203. Eastwoodia elegans

5204. Melampodium perfoliatum
5205. Sanvitalia abertii

18. SANVITÁLIA Lam. Journ. Hist. Nat. 2: 176. *pl. 33.* 1792.

Simple or branching, annual or perennial herbs. Leaves opposite, entire, petioled or sessile. Heads heterogamous, radiate, terminal on the branches or sessile in the upper leaf-axils. Involucre hemispheric or campanulate. Phyllaries in 2 or 3 series, obscurely imbricate, dry or partly herbaceous, the outer spreading in fruit. Receptacle hemispheric to narrowly conical, the receptacular bracts scarious and rather firm, subtending and folded around the disk-flowers, tardily deciduous. Ray-flowers pistillate, fertile, the rays white or yellow, sessile and persistent on the achene. Disk-flowers perfect, purple or greenish white, tubular or nearly so, 5-toothed, commonly shorter than the receptacular bracts. Anthers dark with an acute pale appendage, nearly entire below. Style-branches truncate and shortly appendaged. Ray-achenes thick, irregularly 3–5-angled and bearing rigid awns or horns. Disk-achenes various, often 4-angled and warty or tuberculate, the pappus none or of 1–2 slender awns or teeth. [Name in honor of a noble Italian family, Sanvitali.]

A genus of about 4 species, natives of southwestern United States and Mexico. Type species, *Sanvitalia procumbens* Lam.

1. Sanvitalia abértii A. Gray. Abert's Sanvitalia. Fig. 5205.

Sanvitalia abertii A. Gray, Mem. Amer. Acad. II. 4: 87. 1849.

Slender annual from a taproot, 1–3 dm. high, simple or with ascending opposite branches, the stems hispidulous with curved ascending hairs, usually becoming glabrate. Leaves linear-lanceolate to lanceolate, shorter than the internodes, the petioles 2–12 mm. long, the blades 2.5–5 cm. long, cuneate at base and somewhat decurrent on the petiole, the leaf-margins and surfaces hispidulous with pustulate-based hairs; phyllaries few, subequal in 2 series; receptacle narrowly conical to nearly subulate, the receptacular bracts much surpassing the disk-flowers; ray-flowers 6–10, the ray 2–4 mm. long, emarginate (rarely lobed) and many-veined; disk-flowers about 1.5 mm. long, greenish yellow; ray-achenes 3–3.5 mm. long, thick and irregular, white, 5-grooved, smooth or obscurely tuberculate and bearing 3 short, stout, horny awns; disk-achenes 2.5–3 mm. long, 4-angled, the angles often compressed and wing-like, conspicuously and irregularly warty, blackish or sometimes with a chalky surface, epappose.

Arid mesas and slopes, Sonoran Zones; in California on Clark Mountain, eastern San Bernardino County (*Roos 4954, 4893*), eastward in northern and central Arizona to Texas; also Chihuahua and Sonora. Type locality: "between Bent's Fort and Santa Fé," New Mexico. July–Oct.

Subtribe AMBROSIINAE

Heads heterogamous (a few unisexual staminate heads present on inflorescence of *Dicoria*); pistillate flowers and achenes not enclosed in a bur-like or nut-like involucre.

Achenes more or less turgid, wingless; phyllaries in 1 or 2 series, the inner when present not noticeably larger and not becoming accrescent.

Achenes densely villous; leaves or their lobes linear-filiform. 19. *Oxytenia.*

Achenes not densely villous; leaves entire, lobed, or pinnately cleft. 20. *Iva.*

Achenes flattened and with pectinate or toothed wings; inner phyllaries larger than the outer, becoming accrescent. 21. *Dicoria.*

Heads unisexual, the staminate borne above the pistillate; pistillate flowers and achenes in a nut-like or bur-like involucre.

Phyllaries of the staminate heads united; fruiting involucres winged, tuberculate, or if hooked spines present these few.

Fruiting involucres with conspicuous transverse wings; leaf divisions or leaves linear-filiform.
22. *Hymenoclea.*

Fruiting involucres without transverse wings; leaves toothed or pinnatifid, the lobes not linear-filiform.

Fruiting involucres unarmed or with a single row of tubercles or teeth. 23. *Ambrosia.*

Fruiting involucres with several spines in more than 1 row. 24. *Franseria.*

Phyllaries of the staminate heads free; fruiting involucres conspicuously covered with hooked spines or prickles. 25. *Xanthium.*

19. OXYTÈNIA Nutt. Journ. Acad. Phila. II. 1: 172. 1848.

Strigillose perennial, shrubby and soft-woody, with striate stems. Leaves alternate, petioled, pinnately divided nearly to midrib or the upper entire, the divisions linear-filiform, revolute-margined. Heads heterogamous, disciform, numerous, spicate-panicled, sessile or short-pedicelled. Outer flowers 5, pistillate. Inner flowers 10–20, perfect, sterile. Corollas yellowish white. Involucre of about 5 free, imbricated, subequal, subcoriaceous, broadly ovate phyllaries with somewhat rigid, subherbaceous, abruptly acuminate tips. Receptacle small, the receptacular bracts membranous, cuneate-spatulate, villous. Corollas of the pistillate flowers reduced to a thick fleshy ring, of the perfect ones funnelform, 5-toothed, stipitate-glandular, and sparsely pilose above. Filaments free; anthers lightly connate, at maturity disjunct, with subentire bases and inflexed, ovate, terminal appendages. Style in pistillate flowers with 2 short, oblong, obtuse, dorsally glabrous branches; in perfect flowers undivided, at apex slightly dilated and fringed. Achenes obovoid, convex dorsally, flattish ventrally, 1-ridged on each face, densely long-villous; pappus none.

[Name Greek, meaning sharp, referring to the rigid, subulate-pointed divisions of the leaves.]

A monotypic genus of the southwestern United States.

1. Oxytenia aceròsa Nutt. Oxytenia. Fig. 5206.

Oxytenia acerosa Nutt. Journ. Acad. Phila. II. 1: 172. 1848.

Strigillose shrubby perennial 1–2 m. high, the stems conspicuously striate, greenish, leafy, erect branched above, sometimes leafless and rush-like. Leaves petioled, 3–14 cm. long, pinnately 3–7-lobed or the upper entire and linear-filiform, the lowest lobes sometimes with 1–3 similar lobes, the rachis, lobes, and usually the narrowly winged petiole all strongly revolute-margined, 0.5–1.5 mm. wide, the lobes callous-pointed; heads 3–4 mm. thick, usually not nodding; achenes 2 mm. long.

Dry plains and along creeks, Sonoran Zones; southwestern Colorado to New Mexico, southern Utah, Nevada, northern Arizona, and Inyo County, California, in the Death Valley region. Type locality: "Rocky Mountains, near Upper California." July–Sept.

20. ÌVA L. Sp. Pl. 988. 1753.

Annual or perennial herbs or shrubs, glabrous or pubescent, sometimes tomentose. Leaves usually opposite at least below, linear to broadly ovate, entire, toothed, or dissected, sessile or petioled. Heads heterogamous, disciform, nodding, racemose, spicate, spicate-panicled, or scattered on the branches, sometimes ebracteate; outer flowers (1–8) pistillate, inner (3–20) perfect, sterile; corollas whitish. Involucre nearly hemispheric, usually double, the outer of 3–9 free or rarely connate, sometimes imbricated, mostly oval or ovate, subequal, subherbaceous phyllaries; the inner usually present, of free membranous phyllaries, opposite the outer phyllaries, each subtending and sometimes half-enwrapping a pistillate flower, linear-spatulate to broadly obovate. Receptacle small, sometimes naked toward the center, the receptacular bracts linear to spatulate, flat, membranous, glandular. Corollas of pistillate flowers tubular or campanulate, short, sometimes truncate, sometimes 4–5-toothed, in some species reduced to a fleshy ring or obsolete; corollas of perfect flowers tubular, campanulate or broadly funnelform from a slender base, 5-toothed. Filaments free or somewhat connate; anthers lightly connate, at maturity disjunct and usually exserted, with subentire to sagittate bases and inflexed terminal appendages. Style in the perfect flowers undivided, thickened at apex and there fringed; in the pistillate flowers with oblong to linear, dorsally glabrous branches. Achenes of the pistillate flowers obovoid, thickened, somewhat obcompressed, glabrous, glandular, or sparsely pilose toward apex; of the disk rudimentary. Pappus none. [Named because of its similarity in smell to *Ajuga iva;* the name of the latter derived from an old Latin word signifying an abortifacient.]

A genus of about 15 species, all North American. Type species, *Iva annua* L.

Heads racemose, mostly solitary in the axils of reduced leaves; inner phyllaries much narrower than the ovaries (and achenes); corolla of pistillate flowers evident, tubular, glandular; leaves obovate to linear, entire; plants perennial.
    Phyllaries united at least to middle; herb, from running rootstocks.   1. *I. axillaris.*
    Phyllaries free; plant suffrutescent.   2. *I. hayesiana.*
Heads scattered on the upper branches or spicate-panicled, mostly ebracteate; inner phyllaries broader than the ovaries (and achenes) and infolding them on the edges; corolla of pistillate flowers villous or glabrous, sometimes obsolescent; leaves serrate to bipinnatifid; plants annual.
    Plant 15 cm. high or less; leaves bipinnatifid, small; corolla of pistillate flowers tubular, enlarged below, villous.   3. *I. nevadensis.*
    Plant up to 2 m. high; leaves sharply serrate, large; corolla of pistillate flowers short, tubular-campanulate, glabrous.   4. *I. xanthifolia.*

1. Iva axillàris Pursh. Poverty Weed. Fig. 5207.

Iva axillaris Pursh, Fl. Amer. Sept. 743. 1814.
Iva axillaris β robustior Hook. Fl. Bor. Amer. 1: 309. 1833.
Iva foliolosa Nutt. Trans. Amer. Phil. Soc. II. 7: 346. 1840.
Iva axillaris var. *pubescens* A. Gray in Torr. Bot. Wilkes Exp. 17: 350. 1874.

Perennial herb from running rootstocks; stems several, scattered, simple or simply branched, 18–60 cm. high, strigose, incurved- or rarely spreading-pubescent, somewhat glandular. Leaves opposite below, alternate above, obovate to lance-elliptic, 1.5–4.5 cm. long, 4–15 mm. wide, obtuse or acute, narrowed to the sessile or subsessile base, entire, thick, triplinerved, pubescent like the stem; heads nodding, 4–7 mm. thick, solitary in the axils of the usually not much reduced upper leaves, pedicellate; involucre gamophyllous, shallowly or to middle about 5-lobed; pistillate flowers 5–8, perfect, about 6–18; achenes sessile-glandular, 2–3 mm. long.

Dry or moist, open, alkaline places, often becoming a troublesome weed, Transition and Sonoran Zones; Manitoba to British Columbia south to southern California and Nevada, Utah, New Mexico, and Oklahoma. Type locality: "In Upper Louisiana." May–Sept.

2. **Iva hayesiàna** A. Gray. San Diego Poverty Weed. Fig. 5208.

Iva hayesiana A. Gray, Proc. Amer. Acad. 11: 78. 1876.

Suffrutescent, about 1 m. high, sparsely strigose or ascending-pubescent, especially on the leaves, and sessile-glandular; branches erectish. Leaves opposite below, alternate above, obovate to lanceolate or the upper nearly linear, 2–6 cm. long, 5–16 mm. wide, obtuse or acute, the larger narrowed at base into a short petiole, thick, entire or rarely with a single tooth on each side, triplinerved; heads mostly solitary in the axils of the strongly reduced upper leaves, short-pedicelled, 3–7 mm. thick; pistillate flowers about 5, the perfect up to 20; involucre of about 5 free, obovate or oval phyllaries; achenes 2 mm. long.

On brackish alkaline flats, Upper Sonoran Zone; San Diego County, California, south to central Lower California; also on Cedros Island. Type locality: Jamul Valley southeast of San Diego, California. April–Sept.

3. **Iva nevadénsis** M. E. Jones. Nevada Poverty Weed. Fig. 5209.

Iva nevadensis M. E. Jones, Amer. Nat. 17: 973. 1883.
Chorisiva nevadensis Rydb. N. Amer. Fl. 33: 9. 1922.

Diffusely branched, single-stemmed annual from a taproot, 7–15 cm. high, cinereous with mostly incurved hairs and somewhat sessile-glandular. Leaves alternate, petioled, the blades ovate or deltoid in outline, 8–18 mm. long, ternately bipinnatifid, the ultimate segments rather few, obtuse, thick, the bracts of inflorescence much smaller, mostly cuneate and 3-lobed; heads 2–3 mm. thick, scattered on the upper branches, mostly not subtended by bracts, sessile or subsessile; phyllaries 3, free, ovate, with short, recurved-spreading, obtuse, herbaceous tips; receptacular bracts subtending the pistillate flowers broadly cuneate, villous, partly enfolding the achenes; pistillate flowers usually 3, the perfect 8–10; achenes obovoid, planoconvex, 2 mm. long, at length crustaceous-muricate on margins and both faces.

In desert places, Lower Sonoran Zone; western Nevada and adjacent California from Mono County to the Death Valley region, Inyo County. Type locality: Hawthorne, Mineral County, Nevada. June–Oct.

4. **Iva xanthifòlia** Nutt. Marsh-elder. Fig. 5210.

Iva xanthifolia Nutt. Gen. 2: 185. 1818.
Cyclachaena xanthifolia Fresen. Ind. Sem. Hort. Frankf. 4. 1836.

Coarse annual up to 2 m. high, simple or branched; stem glaucescent, essentially glabrous below, pilose and glandular in the inflorescence. Leaves opposite at least below, slender-petioled, the blades ovate or deltoid-ovate, 4–15 cm. long or more, about as wide, acuminate, the larger truncate or subtruncate at base, sharply and unequally serrate and sometimes shallowly laciniate-lobed, above green, beneath canescent-strigillose at least when young; heads 2–4 mm. thick, not bracted, sessile or subsessile, numerous and crowded in axillary, simple or branched spikes and in a terminal panicle; phyllaries 5, broadly ovate, free, abruptly and shortly herbaceous-acuminate; pistillate flowers 5, the perfect about 15; achenes obovoid, biconvex, sparsely pilose toward apex, minutely papillose in lines, about 2 mm. long.

Along creeks and in waste places, Arid Transition Zone; Alberta and Saskatchewan to Nebraska, and south through eastern Washington and Idaho to Arizona and New Mexico; also as an occasional weed east to the Atlantic. Type locality: Fort Mandan [North Dakota]. Aug.–Oct.

21. **DICÒRIA** Torr. & Gray ex A. Gray in Torr. Bot. Mex. Bound. 86. 1859.

Coarse, much-branched, weedy, autumn annuals, aromatic; pubescence of the stems more or less strigose interspersed with hispid hairs and becoming scabrous in age. Leaves mostly alternate, petioled, entire or toothed, densely strigose when young especially on the upper surface, hispid hairs also present on the lower surface; juvenile leaves linear to lanceolate-attenuate and cuneate at base. Inflorescence of open leafy panicles, the heads heterogamous on short, slender, usually nodding peduncles, the terminal ones usually stami-nate only. Outer phyllaries about 5, spreading, herbaceous, the inner (1–3, subtending the pistillate flowers) larger, broad, more or less scarious, becoming strongly accrescent and hooded and (in ours) continuing to enlarge after the seeds have matured. Receptacle small, flat, with narrow, scarious, receptacular bracts. Pistillate flowers 1–5, without corollas; staminate corollas about 6–15, regular, funnelform. Stamens obtuse at apex, free, the filaments united. Style-branches linear. Achenes dorsoventrally flattened, dark, sometimes with a mottled corky coat on outer face, with a broad or narrow, toothed, pecti-nate, or lobed, wing-like margin. Pappus none or rudimentary (a tuft of short white bristles). [Name from the Greek, meaning twice and bug.]

A genus of possibly 4 closely related species, natives of southwestern United States and adjacent Mexico. Type species, *Dicoria canescens* A. Gray.

Achenes 3.5–4.5 mm. long, the margin one-fourth or more the width of the body of the achene, the teeth or toothed lobes distinct; stems with many pustulate-based, spreading, hispid hairs and appressed finer pubescence.
1. *D. canescens hispidula.*
Achenes 4.5–6 mm. long, the margin one-half to one-third the width of the achene, its toothed or laciniate lobes contiguous; stems with appressed pubescence, the spreading hispid hairs sparse, rarely completely absent.
Leaves oval to orbicular; margin one-half or more the width of the body of the achene; lower Colorado Basin and adjacent Mexico.
1. *D. canescens.*
Leaves mostly ovate; margin about one-third the width of the body of the achene; southern Nevada and adjacent California.
2. *D. clarkiae.*

5206. Oxytenia acerosa
5207. Iva axillaris
5208. Iva hayesiana

5209. Iva nevadensis
5210. Iva xanthifolia
5211. Dicoria canescens

1. **Dicoria canéscens** A. Gray. Desert Dicoria. Fig. 5211.

Dicoria canescens A. Gray in Torr. Bot. Mex. Bound. 87. *pl. 30.* 1859.
Dicoria calliptera Rose & Standl. Contr. U.S. Nat. Herb. **16**: 18. *pl. 12.* 1912.

Coarse annual 5–10 dm. high, the stems spreading from the base, divaricately much branched, striate, clothed with upward-curved, appressed hairs, these tardily deciduous, and with a few pustulate-based, hispid hairs. Stem-leaves 1.5–5 cm. long, reduced above, rather thick, the petioles about one-third the length of the blades to equaling them, the blades suborbicular or oval, the margin denticulate with broad teeth, crisped, densely white-hirsute with appressed hairs, becoming scabridulous in age; inflorescence many-flowered, paniculately branched, the ultimate branches slender; outer phyllaries about 3 mm. long, oval to elliptic, the inner 7–8.5 mm. long, usually 2, glandular-puberulent, nearly orbicular, erose-margined, accressed in age ultimately to 10 mm. or more long and becoming hooded; body of the achene oblong, 5–6(8) mm. long, 1.75–2 mm. wide, dark brown, dotted with yellow glands and sometimes sparsely hirsute, medially ridged on both surfaces, the inner often with a distinct crest or tooth at the summit and sometimes partially winged on the ridge, the outer often partly corky, the winged margin one-half or more the width of the body of the achene, with broad, laciniate or pectinate lobes, usually straw-colored, thickened toward the base.

In sand, Lower Sonoran Zone; Riverside and Imperial Counties, California, southwestern Arizona, and northwestern Sonora. Type locality: "In the sandy desert of the Gila and of the Colorado." Collected by Emory. Oct.–Jan.

Dicoria canescens subsp. **hispídula** Keck, Aliso **4**: 101. 1958. (*Dicoria hispidula* Rydb. N. Amer. Fl. **33**: 12. 1922.) Habit of *D. canescens* subsp. *canescens* and *D. clarkiae*, the pubescence of the stem much the same except for the more abrupt, spreading, pustulate-based, hispid hairs, becoming scabrid in age; leaves and inflorescence as in the preceding; outer phyllaries about 2 mm. long, the inner 4–7 mm. long, 2–3, elongating in age to 8–10 mm. and concave and oval when young, becoming hooded in age, glandular-puberulent and erose on the margin; body of the achene 3.5–5 mm. long, about 1.5(2) mm. wide, oblong or slightly narrowed toward the base, dark brown, the inner face flat, scarcely ridged, the outer rounded, with a distinct ridge, both surfaces glandular-dotted and sparsely hirsute, the margins toothed or double-toothed and irregularly so, usually about one-fourth or more the width of the body of the achene, usually straw-colored and somewhat thickened at attachment with the body of the achene. In sand, Lower Sonoran Zone; southern Mojave Desert, California, south through the Coachella Valley. Type locality: Whitewater Desert, Riverside County. Differing principally by the more distinctly toothed and narrower margins of the achenes. Growing with *D. canescens* in the Coachella Valley.

Dicoria oblongifolia Rydb. N. Amer. Fl. **33**: 12. 1922, is said by Rydberg to grow in southern California and Lower California. There is some question about the area in which the type of *D. oblongifolia* was collected. The data on the type specimen indicate that it was collected in southern California by Palmer (no. *636*) in 1875. According to Dr. Rogers McVaugh, the collections made by Palmer in 1875 included 462 numbers only, but there were some supplementary series from St. George, Utah, in that year. The following year Palmer collected 660 numbers, many of them from the Mojave Desert region. The type locality, then, could be St. George, Utah, if the year given on the label is correct. If it is not, the type locality could be the Mojave Desert. The achenes of the type are about 5 mm. long, the margined lobes about 0.5 mm. wide, distinct, double-toothed, darkened at the base, somewhat hairy or glandular-dotted.

2. **Dicoria clárkiae** Kennedy. Clark's Dicoria. Fig. 5212.

Dicoria clarkiae Kennedy, Muhlenbergia **4**: 2. *fig.* 1908.
Dicoria canescens subsp. *clarkiae* Keck, Aliso **4**: 101. 1958.

Coarse annuals up to 7 dm. high, stems spreading from the base and branching above, striate, puberulent with upward-curved hairs and some longer appressed hairs, becoming scabrous in age. Upper leaves 1.5–5 cm. long, the petiole about one-half or less the length of the blade, broadly ovate, repand-dentate, white-stigose above and below, the lower surface becoming scabrous; heads many in paniculate inflorescence, rather leafy; outer phyllaries 3 mm. or more long, the inner 3 (2 or 4), 8–13 mm. long, strongly accrescent and hooded, glandular-pubescent, erose-margined; achenes 5–6(7) mm. long, the body of the achene mostly 2 mm. wide, oblong, more often more or less oblanceolate, inner surface brown, flat, 1-nerved or ridged, rarely partially winged, often glandular-dotted and sparsely hirsute, the outer rounded, ridged, brown, and often corky-mottled, more or less gland-dotted and sometimes hirsute, the margin 0.4–0.5 mm. or more wide, more narrow at base of achene, either evenly or unevenly lobed and toothed, pale, usually darkened where attached to the body of the achene.

In sand, Upper Sonoran Zone; Churchill County, Nevada, to Inyo County, California, and adjacent San Bernardino County. Type locality: Soda Lake, Churchill County, Nevada. Oct.–Nov.

Seemingly merging with *Dicoria canescens* subsp. *hispidula* in its southern extension. When the life-cycles of all the species of *Dicoria* are thoroughly studied, particularly with the possibility in mind of variation of the winged margins of the achenes, more nomenclatural change may become necessary.

22. **HYMENOCLÈA** Torr. & Gray ex A. Gray, Mem. Amer. Acad. II. **4**: 79. 1849.

Monoecious or subdioecious shrubs with white or straw-colored bark and greenish, striate, glabrous or sparsely hispidulous, more or less resinous branches. Leaves alternate, filiform or linear-filiform, entire or pinnately few-lobed with filiform divisions, green beneath, above densely canescent-hispidulous, sulcate, and often strongly involute-margined. Heads unisexual, more or less nodding, sessile or subsessile, intermixed or the staminate above the pistillate, clustered or solitary in the axils of the leaves, often forming spike-like panicles. Staminate heads about 5–15-flowered; involucre saucer-shaped or turbinate, of about 5–6 mostly ovate, herbaceous phyllaries united about to middle. Receptacle small, bearing few spatulate, glandular, and ciliate, membranous bracts. Pistillate involucre 1-flowered, subtended by 1–3 usually ovate and subherbaceous or mostly scarious phyllaries, turbinate-fusiform, angled below, terete and tubular-beaked above, indurated, bearing about 5–12 cuneate or oblanceolate to flabellate, horizontal, scarious wings either in a single whorl at middle or spirally from base to middle or rarely above it. Staminate corollas funnelform, 5-toothed, somewhat glandular; pistillate flowers without corolla. Stamens 5; filaments usually lightly connate to middle or nearly to apex; anthers lightly coherent, with sagittate bases and deltoid, inflexed, terminal appendages. Style in staminate flowers undivided, dilated at apex and there fimbriate; in pistillate flowers with 2 linear, obtuse, exserted, stigmatic branches. Achene permanently enclosed by the fruiting involucre, obovoid-fusiform, blackish brown, glabrous, epappose. [Name Greek, meaning enclosing membrane.]

A genus of 3 species, natives of the southwestern United States and northern Mexico. Type species, *Hymenoclea salsola* Torr. & Gray.

Wings of fruiting involucre spirally arranged, 5–8 mm. wide.   1. *H. salsola.*
Wings of fruiting involucre in a single whorl, or with 1 or 2 additional wings above or below middle, 1–5 mm. wide.
    Wings 5–7, flabellate or reniform-orbicular, 2.5–5 mm. wide.   2. *H. pentalepis.*
    Wings 7–12, mostly cuneate or obovate, 1–2 mm. wide.   3. *H. monogyra.*

1. **Hymenoclea salsòla** Torr. & Gray. White Burrobush. Fig. 5213.

Hymenoclea salsola Torr. & Gray ex A. Gray, Mem. Amer. Acad. II. 4: 79. 1849.
Hymenoclea fasciculata A. Nels. Bot. Gaz. 37: 270. 1904.
Hymenoclea fasciculata var. *patula* A. Nels. op. cit. 47: 431. 1909.

Shrub 1–2 m. high, erectish or spreading-branched; branches sparsely hispidulous when young. Leaves mostly deciduous except from the younger branches, linear or linear-filiform, 2–6 cm. long, 0.5–1.5 mm. wide, entire or pinnately few-lobed, obtuse, green and often sparsely hispidulous beneath, densely canescent-hispidulous and often strongly involute-margined above; heads intermixed or the staminate often above the pistillate, clustered or solitary in the axils of reduced leaves; staminate involucre about 2.5 mm. wide, 6–15-flowered; pistillate involucre turbinate-fusiform, about 6 mm. long, glandular, bearing about 12 spirally arranged, suborbicular or flabellate, often mucronate, scarious wings 5–8 mm. wide, their broad, short, stipe-like bases usually deeply pitted; fruit cone-like when dry, 12–14 mm. in diameter when wings are opened.

Desert washes and alkaline soil, Lower Sonoran Zone and lower part of Upper Sonoran Zone; San Joaquin Valley in Kern County, California, and eastern Santa Barbara and Ventura Counties (Cuyama Valley; Mount Pinos region) southward along the desert slopes of the mountains to San Diego County and eastward across the Mojave Desert and ranges of the Death Valley region to southern Nevada, northern Arizona, and southern Utah. Type locality: uplands of the Mojave River, San Bernardino County, California. Collected by Fremont. March–July. Desert Pearl.

2. **Hymenoclea pentalèpis** Rydb. Southern Burrobush. Fig. 5214.

Hymenoclea pentalepis Rydb. N. Amer. Fl. 33: 14. 1922.
Hymenoclea hemidioica A. Nels. Amer. Journ. Bot. 25: 117. 1938.
Hymenoclea salsola var. *pentalepis* L. Benson, op. cit. 30: 631. 1943.

Intricately branched shrub about 1 m. high, plants often dioecious or essentially so, branches slender, often glandular-granular and more or less resinous, sparsely hispidulous when young. Leaves mostly deciduous, filiform, entire or pinnately few-lobed, 0.5–5 cm. long, about 0.5 mm.

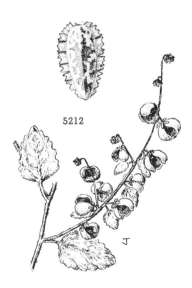

5212. Dicoria clarkiae

5213. Hymenoclea salsola

wide, obtuse, green and glabrous beneath, densely canescent-hispidulous above but strongly involute-margined so as to conceal the hairs; heads intermixed, thickly clustered in glomerules along the branches in the axils of reduced leaves, the wings remaining expanded in age; staminate involucres about 2.5 mm. broad, about 7-flowered; pistillate involucres turbinate-fusiform, about 5 mm. long, glandular, bearing at or near the base 2 or 3 suborbicular scarious bractlets and wings in a single whorl or shortened spiral at the middle, these 5–7, flabellate or reniform-orbicular, with a short broad plane or concave base, usually obtuse and often mucronate at apex, pinkish or straw-colored, 3–4 mm. wide when mature, additional wings sometimes present, the whorl about 7–8 mm. in diameter when spread.

Desert washes and alluvial fans, Lower Sonoran Zone; northern border of the Colorado Desert and in the vicinity of Needles, San Bernardino County, California, southward to the eastern side of Lower California as far south as Santa Rosalia eastward to southern Arizona and adjacent Sonora. Type locality: Pima Canyon, Pima County, Arizona. Feb.–June.

The western specimens are usually more densely flowered and the flowers are a little larger than the more typical material in Arizona and Sonora.

3. Hymenoclea monogỳra Torr. & Gray. Leafy Burrobush. Fig. 5215.

Hymenoclea monogyra Torr. & Gray ex A. Gray, Mem. Amer. Acad. II. 4: 79. 1849.

Slender shrub 1–4 m. high with numerous erectish branches, usually very leafy at least above, branches glandular-granular. Leaves filiform, 1.5–8 cm. long, 0.3–1 mm. wide, entire or the lower pinnately divided, callous-pointed, green and glabrous or glandular-granular beneath, densely canescent-hispidulous above and strongly involute-margined, often with axillary fascicles; heads many in axillary clusters on the branches, forming long, leafy, spike-like panicles, the staminate and pistillate usually intermixed; staminate involucre about 2.5 mm. broad, 8–12-flowered; pistillate involucre about 4 mm. long, turbinate-fusiform, glandular, bearing at or near base 2 or 3 scarious suborbicular bractlets and at middle a single whorl of 7–12 mostly cuneate or obovate, acute or obtuse, scarious wings 1–2 mm. wide, the whorl about 5 mm. in diameter when spread.

Sandy washes and desert grassland, mostly Upper Sonoran Zone; San Bernardino County (Rialto), California, and eastern San Diego County southward throughout Lower California and eastward to Texas; also northern Mexico. Type locality: "Along the valley of the Gila," Arizona. Collected by Emory. Aug.–Nov.

23. AMBRÒSIA L. Sp. Pl. 987. 1753.

Annual or perennial, normally monoecious, pubescent herbs. Leaves opposite at least below, entire or usually lobed or dissected. Heads unisexual, the staminate nodding, in ebracteate terminal racemes or spikes, the pistillate sessile, usually clustered in the upper axils or at base of the staminate racemes, subtended by reduced leaves and a few small herbaceous bracts. Staminate heads usually many-flowered; involucre gamophyllous, turbinate or saucer-shaped, more or less oblique, herbaceous, the phyllaries connate nearly to apex, the margin crenate or irregularly toothed; receptacle small, the receptacular bracts mostly linear or filiform. Pistillate involucre 1-flowered, completely enclosing the achene, in fruit indurated and nut-like, obovoid, usually with a circle of small conic tubercles or spines around apex, short-beaked, the beak conic, denticulate. Staminate corollas funnelform or campanulate, 5-toothed; corolla wanting in pistillate flowers. Stamens 5 (rarely 4); filaments lightly connate; anthers lightly coherent, subentire at base, with deltoid terminal appendages tipped with an inflexed cusp about as long as the appendage. Style in staminate flowers undivided, at apex slightly dilated and fimbriate; in pistillate flowers with 2 linear, exserted, stigmatic branches. Achene obovoid, tightly enclosed by the fruiting involucre, glabrous, epappose. [Greek, meaning the food of the gods, a classical name of various plants, its application to these weeds obscure.]

A genus of about 20 species, nearly all American, one or two in the Mediterranean region. Type species, *Ambrosia maritima* L.

Leaves entire or palmately cleft.	1. *A. trifida.*
Leaves pinnately lobed to bi- or tripinnatifid.	
Leaves green, coarsely lobed or divided.	
Annual; leaves thin, at least the lower usually bipinnatifid.	2. *A. artemisiifolia.*
Perennial; leaves thickish, mostly only once-pinnatifid.	3. *A. psilostachya.*
Leaves canescently silky-pilose, finely dissected, the ultimate divisions about 1 mm. wide.	4. *A. pumila.*

1. Ambrosia trífida L. Giant Ragweed. Fig. 5216.

Ambrosia trifida L. Sp. Pl. 987. 1753.
Ambrosia integrifolia Muhl. ex Willd. Sp. Pl. 4: 375. 1805.
Ambrosia variabilis Rydb. Brittonia 1: 97. 1931.

Coarse annual 0.5–6 m. high, the stems rough-hispid or scabrous above, often glabrate below. Leaves opposite, broadly elliptic to ovate, palmately 3–5-cleft into serrate lobes, the lobes more rarely lacking; petioles 1.5–3 cm. long, more or less wing-margined; blades 4.5–15 cm. long, sparsely scabrous on both surfaces, pale beneath and dark green above, prominently palmate-veined; staminate racemes 1 to several, many-flowered, terminal and from the upper leaf-axils; pistillate heads few, at the base of staminate racemes and in leaf-axils; staminate involucre sparsely hirsute, regular, with short slender pedicel, strongly 3-nerved on one side; fruiting pistillate involucre 6–12 mm. long with an acute beak, several-ribbed, the ribs terminating in acute tubercles.

Moist rich soil, roadsides and waste places; in the Pacific States occurring occasionally as a weed in Washington and California, and widely distributed throughout the United States and Canada. Type locality: "Virginia, Canada." July–Oct.

2. Ambrosia artemisiifòlia L. Annual Ragwood. Fig. 5217.

Ambrosia elatior L. Sp. Pl. 987. 1753.
Ambrosia artemisiifolia L. op. cit. 988.
Ambrosia artemisiifolia diversifolia Piper, Contr. U.S. Nat. Herb. 11: 551. 1906.
Ambrosia elatior var. *artemisiifolia* Farwell, Rep. Mich. Acad. 15: 190. 1913.

Branching annual 3–10 dm. high, the stem slender, usually strigose or incurved-pubescent, sometimes spreading-hirsute. Leaves opposite below, alternate above, triangular or ovate in outline, on narrowly margined petioles 1–5 cm. long, the blades 1–2-pinnatifid with often toothed or incised, mostly acute divisions, rather thin, pubescent like the stem, paler and more densely pubescent beneath, the uppermost leaves often linear and entire; staminate heads in terminal, often panicled racemes, about 2–3 mm. in diameter, about 12–32-flowered; pistillate heads clustered in the upper axils or at the base of staminate racemes, subtended by reduced leaves and a few small herbaceous bracts; fruit turgid-obovoid, 3–3.8 mm. long (including beak, this 0.7–1.2 mm. long), pubescent and glandular, bearing around apex of body 1–7 usually acute, short, conic tubercles, rarely unarmed.

An abundant weed nearly throughout the United States and southern Canada south into tropical America, occasional in our range (Washington to California); also naturalized in Europe. Type locality: "Virginia, Pennsylvania." June–Oct.

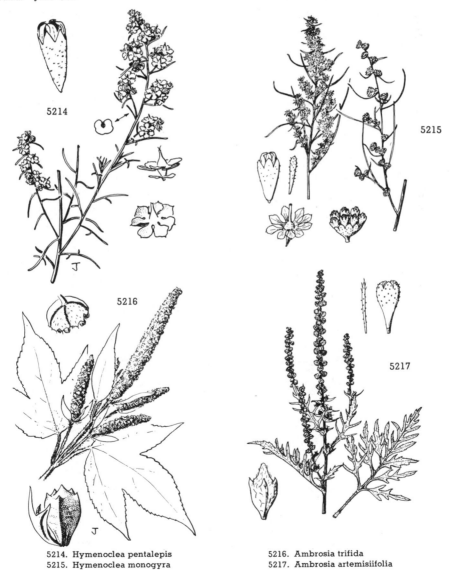

5214. Hymenoclea pentalepis
5215. Hymenoclea monogyra

5216. Ambrosia trifida
5217. Ambrosia artemisiifolia

3. Ambrosia psilostáchya DC. Common Ragweed. Fig. 5218.

Ambrosia psilostachya DC. Prod. **5**: 526. 1836.
Ambrosia coronopifolia Torr. & Gray, Fl. N. Amer. **2**: 291. 1842.
Ambrosia californica Rydb. N. Amer. Fl. **33**: 20. 1922.
Ambrosia psilostachya var. *californica* Blake in Tidestrom, Contr. U.S. Nat. Herb. **25**: 581. 1925.

Perennial, the stems mostly scattered, 0.3–1 m. high, from running rootstocks, strigous or hirsute with ascending or spreading hairs, usually branched at least above. Leaves opposite below, ovate or triangular in outline, 4–10 cm. long, on usually short-winged petioles, thickish, shallowly or deeply pinnatifid with lanceolate, acute, often toothed or lobed, forward-pointing divisions, strigose or ascending-hirsute or -hispid and glandular, the upper narrower and less divided, often subentire; staminate heads 1–5 mm. in diameter, in usually elongate and often-panicled racemes, about 12–25-flowered, the involucre hirsutulous with usually tuberculate-based hairs; pistillate heads solitary or clustered in the upper axils; fruit obovoid, about 3 mm. long (the short blunt beak about 0.6 mm. long), glandular, hirsute, often rugose, unarmed or with 1–6 short, obtuse or acute, conic tubercles around apex of body.

Dry or moist places, Sonoran and Transition Zones; common weed in California, Arizona, and northern Mexico and in Washington, where probably introduced; also Illinois to Saskatchewan. Type locality: between San Fernando and Matamoros, Tamaulipas. Collected by Berlandier. April–Oct.

4. Ambrosia pùmila (Nutt.) A. Gray. San Diego Ragweed. Fig. 5219.

Franseria pumila Nutt. Trans. Amer. Phil. Soc. II. **7**: 344. 1840.
Hemiambrosia heterocephala Delpino, Studi Lign. Anem. 61. 1871.
Ambrosia pumila A. Gray, Proc. Amer. Acad. **17**: 217. 1882.

Perennial herb 1–2.5 dm. high with running rootstocks; stems erect, canescent-pilose, at least when young, with mostly appressed hairs and sparsely hispid. Leaves mostly alternate, ovate or oblong in outline, bi- or tripinnatifid, 5–10 cm. long including petiole, densely and canescently silky-pilose with mostly appressed hairs, the ultimate divisions mostly linear-oblong, obtusish, crowded, 1–1.5 mm. wide; staminate heads about 25-flowered, short-pediceled in a short raceme, the involucre crenate-lobed, pilose; pistillate heads clustered in the axils of reduced upper leaves; fruit obovoid, 2–2.5 mm. long, pubescent, longitudinally ribbed below, obscurely reticulate above, unarmed.

Alkaline soil, Sonoran Zones; southern San Diego County, California, southward to central Lower California. Type locality: near San Diego. June–Sept.

24. FRANSÉRIA Cav. Ic. 2: 78. 1793.

Monoecious pubescent herbs or shrubs. Leaves mostly alternate, toothed to tripinnatifid. Heads unisexual, the staminate nodding, in ebracteate terminal racemes, the pistillate sessile at base of staminate racemes and clustered in the upper axils or rarely intermixed with the staminate, subtended by reduced leaves and a few small herbaceous bracts. Staminate heads many-flowered, the involucre gamophyllous, saucer-shaped or turbinate, crenate, toothed or deeply lobed, herbaceous; receptacle small, the receptacular bracts membranous, narrowly linear or spatulate. Pistillate involucre 1–4(5)-flowered, 1–4(5)-celled, 1–4(6)-beaked, completely enclosing the achenes, usually ovoid or fusiform, in fruit coriaceous, nut-like or bur-like, bearing few to many straight or hooked, rigid spines in 2 or more series, the beaks conic, opening obliquely, often unequally 2-toothed. Staminate corollas mostly funnelform, 5-toothed; corolla wanting in pistillate flowers. Stamens 5; filaments connate or lightly coherent; anthers lightly coherent, subentire at base, with ovate or deltoid, terminal appendages often tipped with an inflexed cusp. Style in staminate flowers undivided, at apex slightly dilated and fimbriate; in pistillate flowers with 2 linear, exserted, stigmatic branches. Achenes usually obovoid, tightly enclosed by the fruiting involucre, glabrous, epappose. [Name in honor of Antonio Franser, physician and botanist of Madrid.]

A genus of about 30 species, all American, mostly North American but a few on the western coast of South America. Type species, *Franseria ambrosioides* Cav.

Herbs.
    Annual; staminate involucre with 3 of the nerves thickened and blackish below; fruit armed with flattened straight spines.   1. *F. acanthicarpa.*
    Perennial; nerves of the staminate involucre not thickened and blackish below.
        Erect; fruit 2–4 mm. long, obovoid, bearing 10–20 hooked spines 0.8 mm. long or less; leaves usually greenish, more or less sparsely strigose.   2. *F. confertiflora.*
        Procumbent; fruit ovoid or fusiform, 6–9 mm. long, the spines not hooked, 2–4 mm. long; leaves densely canescent-strigose.
            Leaves mostly cuneate or obovate, crenate-serrate.   3. *F. chamissonis.*
            Leaves bi- to tripinnatifid.   3. *F. chamissonis bipinnatisecta.*
Shrubby or suffrutescent.
    Leaves canescent at least beneath, the lower stem-leaves not more than 4 or 5 cm. long.
        Leaves densely canescent-strigilose, bi- to tripinnately divided into small, mostly ovate or obovate divisions; fruit glandular and sparsely pilose.   4. *F. dumosa.*

Leaves greenish above, densely whitish- or canescent-tomentulose beneath, from crenate-serrate to pin-
   natifid; fruit densely villous or tomentose.
   Leaves subsessile, sinuate-toothed or pinnatifid; fruit fusiform, 8–10 mm. long, densely long-villous,
      bearing about 20 or fewer straight spines.   5. *F. eriocentra.*
   Leaves slender-petioled, crenate-serrate; fruit turbinate-subglobose, about 5 mm. long, the body
      densely lanate-tomentose, thickly covered with mostly hooked spines.
                                                                    6. *F. chenopodiifolia.*
Leaves green on both sides, lower stem-leaves 6–18 cm. long.
   Leaves merely toothed, acuminate, petiolate.   7. *F. ambrosioides.*
   Leaves coarsely spiny-toothed, not acuminate, sessile and clasping.   8. *F. ilicifolia.*

1. **Franseria acanthicárpa** (Hook.) Coville. Annual Burweed. Fig. 5220.

Ambrosia acanthicarpa Hook. Fl. Bor. Amer. **1**: 309. 1833.
Franseria montana Nutt. Trans. Amer. Phil. Soc. II. **7**: 345. 1840.
Franseria hookeriana Nutt. loc. cit.
Franseria acanthicarpa Coville, Contr. U.S. Nat. Herb. **4**: 129. 1893.
Gaertneria acanthicarpa Britt. Mem. Torrey Club **5**: 332. 1894.
Franseria palmeri Rydb. N. Amer. Fl. **33**: 25. 1922.

Weedy branching annual, 2–7 dm. high, stem and branches strigose or hispidulous and often
spreading-hispid. Leaves slender-petioled, the blades ovate or deltoid in outline, 3–10 cm. long,
shallowly or deeply and usually ternately bi- to tripinnatifid, with obovate to linear, obtuse or acute
divisions, strigillose and often hirsute, green above, paler beneath, sometimes cinereous; staminate
racemes usually panicled, the pistillate flowers at their base and in the upper axils; staminate heads
slender-pediceled, 2–4 mm. thick, about 15-flowered, the involucre about 6-lobed, 3 upper lobes

5218. Ambrosia psilostachya   5220. Franseria acanthicarpa
5219. Ambrosia pumila   5221. Franseria confertiflora

larger than the others and with the midrib thickened and blackish below; fruit very variable, 1-flowered, 1-beaked, obovoid to fusiform, 4–8 mm. long (the beak usually subulate or acicular and 2–3 mm. long), usually rugose, glandular and sparsely pilose or essentially glabrous, the body armed with 6–30 strongly flattened, lanceolate, acerose-tipped, straight, spreading spines 2–5 mm. long, these rarely on the same plant reduced to minute incurved hooks or wanting.

Dry or moist, sandy soil, Arid Transition and Sonoran Zones; Minnesota to Alberta south to Arizona, New Mexico, and western Texas; Washington south to southern California, not common in our range. Type locality: "Banks of the Saskatchewan and Red River." Collected by Douglas and by Drummond. June–Dec.

2. Franseria confertiflòra (DC.) Rydb. Weak-leaved Burweed. Fig. 5221.

Ambrosia confertiflora DC. Prod. **5**: 526. 1836.
Franseria tenuifolia Harv. & Gray ex A. Gray, Mem. Amer. Acad. II. **4**: 80. 1849.
Franseria confertiflora Rydb. N. Amer. Fl. **33**: 28. 1922.
Franseria strigulosa Rydb. loc. cit.
Franseria incana Rydb. op. cit. 30.

Slender perennial, 0.2–1.5 m. high, usually simple below the inflorescence, the stem strigose or strigillose and often spreading-hispid or -hirsute. Leaves slender-petioled, the blades 1.5–12 cm. long, ovate in outline, interruptedly bi- to tripinnatified with mostly linear to lanceolate, acute divisions (those of the lower leaves often broader and obtuse), the terminal segments usually elongated, all green above, paler beneath, strigose or strigillose and often hirsute, rarely canescent; staminate racemes slender, usually 8 cm. long or less and panicled at apex of stem, the pistillate heads at their base and in often dense clusters in the upper axils; staminate heads 2–4 mm. thick, about 10–20-flowered, the involucre about 6–10-lobed, some of the lobes often acute; fruit obovoid, 2–4 mm. long, usually 1-flowered, 1-beaked, with turbinate or stipitiform base and short conic beak, glandular and sparsely pilosulose, the body deeply areolate-reticulate, bearing about 10–20 uncinate-tipped, usually incurved, flattish-based, subulate spines 0.8 mm. long or less.

Plains, valleys, and waste places; occurring as an occasional weed in San Francisco, Alameda, and Mariposa Counties, California, and more commonly in southern California and Lower California; Texas and Oklahoma south to Mexico. Type locality: Matamoros, Tamaulipas, Mexico. May–Nov.

3. Franseria chamissònis Less. Beach-bur. Fig. 5222.

Franseria chamissonis Less. Linnaea **6**: 507. 1831.
Franseria cuneifolia Nutt. Trans. Amer. Phil. Soc. II. **7**: 345. 1840.
Gaertnera chamissonis Kuntze, Rev. Gen. Pl. **1**: 339. 1891.

Perennial herb, the stems numerous, up to 1.6 m. long, decumbent from a deep root, stout, mostly simple, pubescent and often cinereous with mostly appressed or sometimes spreading hairs. Leaf-blades ovate or oblong to cuneate or obovate, 3–6 cm. long, on petioles about as long, usually obtuse, cuneate at base, coarsely crenate-serrate or obscurely lobed, thick, densely silky-strigose on both sides; staminate flowers in a dense terminal raceme or spike, usually about 8 cm. long, the pistillate flowers crowded at its base and in the upper axils; staminate heads about 4–6 mm. thick, about 25–50-flowered, the involucre shallowly about 9-toothed; fruit ovoid, 1-flowered, 1-beaked, 7–9 mm. long, glandular and pilose, covered with about 20–30 stout, rigid, triangular-subulate spines 2.5–4 mm. long, these flattened and grooved on the upper surface, often with deep axillary pits.

Coastal sands, Humid Transition and Upper Sonoran Zones; Vancouver Island, British Columbia, to Monterey County, California, and on the Santa Barbara Islands, Ventura County. Type locality: California. July–Oct.

Franseria chamissonis subsp. **bipinnatisécta** (Less.) Wiggins & Stockwell, Madroño **4**: 120. 1937. (*Franseria chamissoni* var. *bipinnatisecta* Less. Linnaea **6**: 508. 1831; *F. bipinnatifida* Nutt. Trans. Amer. Phil. Soc. II. **7**: 344. 1840; *F. lessingii* Meyen & Walp. ex Walp. Nova-Acta Acad. Leop.-Carol. **19**: suppl. 268. 1843; *F. bipinnatifida* var. *dubia* Eastw. Proc. Calif. Acad. III. **1**: 117. 1898; *F. bipinnatifida* var. *insularis* Reiche, Fl. Chile **4**: 80. 1905; *F. villosa* Eastw. ex Rydb. N. Amer. Fl. **33**: 26. 1922.) With the habit, pubescence, and general characters of *F. chamissonis* subsp. *chamissonis*; leaves ovate or triangular in outline, slender-petioled, the blades 3–7 cm. long, often as wide, pinnatifid to bi- or tripinnatifid, the principal ultimate divisions mostly ovate or oblong and obtuse; staminate heads usually on evident peduncles; fruit ovoid or fusiform, 6–8 mm. long, glandular and usually pilose, bearing about 9–25 slender or usually stout, subulate or triangular-subulate, mostly spreading, straight or slightly curved spines 2–4 mm. long, usually the lower or sometimes all more or less flattened and grooved on the upper surface at least toward the base. Coastal sands, Vancouver Island, British Columbia, south to northern Lower California; also on the coast of Chile. Type locality: Santa Barbara, Santa Barbara County, California. May–Nov. Hybrids with *F. chamissonis* subsp. *chamissonis* occur.

4. Franseria dumòsa A. Gray ex Torr. White Bur-sage. Fig. 5223.

Franseria dumosa A. Gray ex Torr. in Frem. Second Rep. 316. 1845.
Franseria albicaulis Torr. Smiths. Contr. **6**²: 16. 1854.
Gaertnera dumosa Kuntze, Rev. Gen. Pl. **1**: 339. 1891.

Shrub 2–6 dm. high, white-barked, rigidly branched, the branches spinescent, young growth densely canescent-strigillose. Leaves 1–3 cm. long including the usually short but slender petiole, oblong to ovate or deltoid in outline, bi- to tripinnately divided into usually few, small, mostly ovate or obovate, obtuse divisions, densely canescent-strigillose; racemes or racemiform panicles terminal, usually 5 cm. long or less, the staminate and pistillate flowers often intermixed; staminate heads 3–5 mm. thick, about 20–30-flowered, the involucre bluntly 6–9-lobed about to middle; fruit ovoid or fusiform, 4–6 mm. long, 2-flowered, 2-beaked, glandular and sparsely pilose, the beaks slenderly conic or subulate from a thickened base, the body covered with about 25–40 triangular-subulate, rigid, flattened, acerose-tipped, straight spines 2.5–4 mm. long.

Desert washes and plains, Sonoran Zones; southeastern California (Colorado and Mojave Deserts) to southern Utah, southern Arizona, northern Lower California, and Sonora. Type locality: sandy uplands of the Mojave River, California. April–Nov. Burrobush.

5. **Franseria eriocéntra** A. Gray. Woolly-fruited Burbush. Fig. 5224.

Franseria eriocentra A. Gray, Proc. Amer. Acad. **7**: 355. 1868.
Gaertnera eriocentra Kuntze, Rev. Gen. Pl. **1**: 339. 1891.

Branching shrub, 3–9 dm. high, white-barked, the stem glabrate, the branches puberulent and spreading-pilose. Leaves ovate, triangular, or oblong, 2–5.5 cm. long, narrowed into a short petiole or subsessile, sinuate-toothed to pinnatifid with laciniate lobes, tomentulose but green above, densely canescent-tomentulose and sparsely hirsute along the veins beneath; heads in short racemes or racemiform panicles toward tips of branches and branchlets, the staminate above; staminate heads 4–6 mm. in diameter, about 30-flowered, the involucre acutely about 7–10-lobed to middle, the corollas pilose above; fruit 1-flowered, fusiform, 8–10 mm. long, stipitate-glandular and (especially on the spines) densely long-villous, bearing mostly near middle about 20 or fewer rigid, straight, subulate, acuminate, crowded spines 3–4 mm. long, flattish or grooved above.

Desert washes and canyons, Upper Sonoran Zone; southeastern California in eastern San Bernardino County to southern Utah and southern Nevada to northwestern Arizona. Type locality: Providence Mountains, San Bernardino County, California. April–July.

6. **Franseria chenopodiifòlia** Benth. San Diego Burbush. Fig. 5225.

Franseria chenopodiifolia Benth. Bot. Sulph. 26. 1844.
Gaertnera chenopodiifolia Abrams, Bull. N.Y. Bot. Gard. **6**: 461. 1910.
Franseria lancifolia Rydb. N. Amer. Fl. **33**: 36. 1922.

Suffrutescent, leafy, about 0.5 m. high, at length whitish-barked, branching, the branches tomentose and glutinous, glabrescent. Leaves ovate or deltoid-ovate, 2–3.5 cm. long, nearly as wide, slender-petioled, acute or obtuse, broadly cuneate or rounded at base, crenate-serrate or obscurely lobed, subcoriaceous, tomentulose but greenish above, densely whitish-tomentulose and

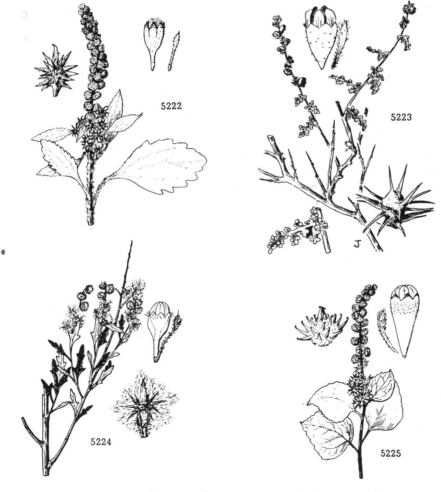

5222. Franseria chamissonis
5223. Franseria dumosa

5224. Franseria eriocentra
5225. Franseria chenopodiifolia

reticulate beneath; racemes simple or branched below, the pistillate heads at their base and clustered in the upper axils; staminate heads about 3–5 mm. thick, 16–30-flowered, the involucre shallowly and bluntly 6–10-toothed; fruit turbinate-subglobose, about 5 mm. long, 2–3-flowered, 2–3-beaked, the body densely lanate-tomentose, thickly covered with numerous rigid, shallowly grooved, mostly uncinate-tipped spines 3–5 mm. long, the lower mostly linear-lanceolate and flattened, the upper subulate, subterete.

Dry slopes, Lower Sonoran Zone; southwestern San Diego County, California, and Lower California. Type locality: Magdalena Bay, Lower California. April–May.

7. Franseria ambrosioìdes Cav. Ambrosia-leaved Burbush. Fig. 5226.

Franseria ambrosioides Cav. Ic. **2**: 79. 1793.
Gaertnera ambrosioides Kuntze, Rev. Gen. Pl. **1**: 339. 1891.

Suffrutescent perennial, 1–2 m. high, the ascending stems several from the base, leafy, mostly unbranched below the inflorescence, glandular-puberulent and grayish-hirsute. Leaves 6–18 cm. long with petioles 1–4 cm. long. elongate-triangular to lanceolate, acuminate, cordate to broadly cuneate at base, margin irregularly and coarsely toothed, entire at the acuminate apex, thick, reticulate-veined beneath, dull green and grayish hirsutulous; inflorescence branched or less commonly simple, the terminal staminate racemes about 10 cm. long, the branches mostly pistillate, the heads leafy-bracted; staminate involucres many-flowered, pedunculate, about 5–8 mm. broad, hirsutulous and glandular; fruit globose-fusiform, 1–2 cm. long, 1–2-beaked, the beaks 3–4 mm. long, spines many, spreading, slender, slightly flattened, glandular-granuliferous, hooked at the tip.

Desert slopes and sandy washes, Lower Sonoran Zone; near San Diego, San Diego County, California (where it probably was introduced), south to Lower California and from southern Arizona south to Sinaloa and Durango. Type locality: Mexico. Feb.–May.

8. Franseria ilicifòlia A. Gray. Holly-leaved Burbush. Fig. 5227.

Franseria ilicifolia A. Gray, Proc. Amer. Acad. **11**: 77. 1876.
Gaertnera ilicifolia Kuntze, Rev. Gen. Pl. **1**: 339. 1891.

Suffrutescent, very leafy, at length whitish-barked, about 1 m. high, the stem and branches densely stipitate-glandular and spreading-hirsute. Leaves ovate, 3–10 cm. long, sessile and cordate-clasping, coarsely and irregularly sinuate-toothed with spinescent teeth, rigidly coriaceous, green and glandular on both sides, pubescent on the veins, more or less reticulate especially beneath; staminate racemes about 8 cm. long, the pistillate heads at their base and mostly solitary in the upper axils; staminate heads 7–10 mm. wide, about 60-flowered, the involucres considerably exceeding the flowers, of about 14 lance-ovate or lanceolate, spinescent-tipped phyllaries, connate half their length or less, the corolla stipitate-glandular; fruit fusiform or globose-fusiform, 1–2.3 cm. long, 1–2-beaked, usually 2-flowered, densely stipitate-glandular, bearing below the beaks numerous rigid, subulate, uncinate-tipped, flattish or grooved spines 4–6 mm. long.

Desert washes and canyons, Lower Sonoran Zone; Riverside and Imperial Counties, California, to northern Lower California and southwestern Arizona. Type locality: Great Canyon of the Cantillas ["Tantillas"] Mountains, Lower California. March–May.

25. XÁNTHIUM L. Sp. Pl. 987. 1753.

Coarse monoecious annuals with branching stems. Leaves alternate, petiolate, dentate to nearly entire, or lobed. Staminate heads borne above the pistillate; involucres many-flowered, the phyllaries free in 1–3 series; receptacle cylindric, paleaceous. Pistillate involucre fused, enclosing the 2 flowers, prickly, indurated. Staminate flowers tubular, 5-toothed above; anthers free, slender, the filaments monodelphous. Pistillate flowers without corolla, the style-branches exserted from the 2 beaks of the indurated involucre. Achenes linear or sometimes ovate, compressed, epappose. [From the Greek word meaning yellow, the ancient name of some plant producing a dye that color.]

A genetically unstable genus of uncertain origin with probably about 3 species. Many species have been described on variation of the bur. Type species, *Xanthium strumarium* L.

Spines lacking in the leaf-axils; leaves deltoid-ovate to subtriangulate, long-petioled.   1. *X. strumarium.*
Conspicuous 3-forked spines at axils of leaves; leaves lanceolate, short-petioled.   2. *X. spinosum.*

1. Xanthium strumàrium L. Cocklebur. Fig. 5228.

Xanthium strumarium L. Sp. Pl. 987. 1753.

Erect, simple or branching annual 2.5–8 dm. high with stout, striate, scabridulous stems. Leaf-blades 6–12 cm. long and usually equally as wide, the petioles as long or longer, deltoid-ovate or -triangular, sometimes obscurely 3-lobed, truncate or subcordate at base and also glandular, the margin dentate or serrate, scabrid above and below; pistillate involucres crowded, 1–3 in leaf-axils or on short lateral branches; burs about 1–1.5 cm. long, ellipsoid or fusiform, pale brown to greenish, the body with a fine puberulence, the prickles up to 2 mm. long, straight except for the hooked apex, glabrous above and puberulent and glandular at base, the beaks stouter, about the length of the prickles, straight or incurved.

In damp areas; known in the Pacific States only from the Colorado Desert, California (Cameron Lake, *Brandegee*; Fort Yuma, *Parish*), but widespread in the American tropics and in Europe. Type locality: not known.

Many forms of this cosmopolitan weed are to be found and innumerable races have developed locally which have received specific names. For the complete and involved synonymy see Widder, Rep. Spec. Nov. Beihefte **20**: 1–221. 1923. The synonyms listed with the following taxa are those more frequently found in western floras.

Xanthium strumarium var. **canadénse** (Mill.) Torr. & Gray, Fl. N. Amer. **2**: 294. 1838. (*Xanthium canadense* Mill. Gard. Dict. ed. 8. 1768; *X. italicum* Moretti, Brugnatelli Giorn. fis. Chim. Dec. II. **5**: 326. 1822; *X. saccharatum* Wallr. Beitr. Bot. 1²: 238. 1844; *X. oviforme* Wallr. op. cit. 240; *X. campestre* Greene, Pittonia **4**: 61. 1899; *X. palustre* Greene, op. cit. 63.) Burs usually 2–3(3.5) cm. long, variable in shape and density of prickles, the beaks exceeding the prickles in length, the prickles more or less densely hirsute toward the base as well as glandular. A cosmopolitan weed growing in moist valleys and flood plains throughout the Pacific States. Type locality: not known.

Xanthium strumarium var. **glabràtum** (DC.) Cronquist, Rhodora **47**: 403. 1945. (*Xanthium macrocarpum* β *glabratum* DC. Prod. **5**: 523. 1836; *X. pennsylvanicum* Wallr. Beitr. Bot. 1²: 236. 1844; *? X. commune wootoni* Cockerell, Proc. Biol. Soc. Wash. **16**: 9. 1903; *X. calvum* Millsp. & Sherff, Field Mus. Bot. Ser. **4**: 35. *pl. 12*. 1919.) Burs usually 2–3 cm. long, sometimes shorter, variable in shape and density of prickles, the beaks exceeding the prickles in length, the prickles glandular toward the base and mostly lacking the hirsute hairs of *X. strumarium* var. *canadense*. A cosmopolitan weed growing in moist valleys and flood plains throughout the Pacific States but more common from central California south. Type locality: "in Carolina."

2. **Xanthium spinòsum** L. Spiny Clotbur. Fig. 5229.

Xanthium spinosum L. Sp. Pl. 987. 1753.

Leafy, erect, branching annual up to 1 m. high, the leaf-axils armed with paired, tripartite, yellowish spines 3 cm. long or less, the stem whitish, strigillose or erectish pubescent. Leaves mostly lanceolate or lance-ovate, 5–12 cm. long including petiole, acuminate, cuneate at base, usually laciniately few-lobed toward base, green above, densely canescent-strigillose beneath; pistillate involucres solitary or clustered in the axils and on short lateral branches but not crowded, in fruit

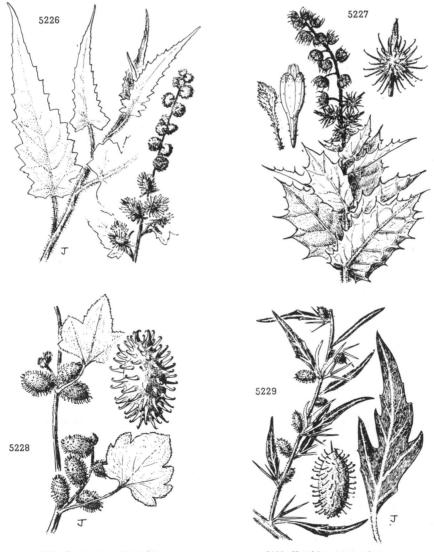

5226. Franseria ambrosioides
5227. Franseria ilicifolia
5228. Xanthium strumarium
5229. Xanthium spinosum

the burs yellowish green or brownish white, oblong, about 10–14 mm. long, 6–9 mm. thick (including prickles), usually somewhat compressed, ribbed, loosely pilosulose and glandular, not stipitate, at apex bearing a single slender straight beak 2 mm. long or less or 2 unequal or short and subequal beaks, the body rather densely armed with slender, yellow-brown (often blackish-based), hooked prickles 2–3 mm. long.

In waste places, usually pasture lands and around habitations. A cosmopolitan weed rather common on the Pacific Slope and nearly throughout the United States; a native of South America but originally described from Portugal. June–Nov. Spanish Thistle.

Subtribe MADIINAE*

Ray-achenes obcompressed, each hidden within the obcompressed phyllary, the abruptly infolded lateral margins of which more or less overlap.

Disk-flowers 6 or more.
   Achenes 10-costate and tuberculate-scabrous; pappus oblong, obtuse.   26. *Achyrachaena.*
   Achenes not costate or scabrous; pappus when present at least attenuate to apex.
     Disk-achenes sterile, undeveloped; heads closing during midday.
      Pappus none; annuals.   27. *Lagophylla.*
      Pappus present; perennials.   28. *Holozonia.*
     Disk-achenes fertile; heads open continuously.   29. *Layia.*
Disk-flower 1.   30. *Madia.*
Ray-achenes not obcompressed or each hidden by the enveloping lateral margins of the subtending phyllary.
   Style of disk-flowers glabrous below the subulate branches.
     Disk-achenes not prominently ribbed; ray-achenes usually glabrous and epappose.
      Ray-ligules 3-toothed or -lobed, the lobes subparallel; herbage without tack-shaped glands; leaves not narrowly linear or grass-like.
       Ray-achenes usually laterally compressed and finely longitudinally striated, each completely enfolded by the deeply sulcate subtending phyllary; basal leaves subentire.   30. *Madia.*
       Ray-achenes broader, not laterally compressed or striated, each only partially enfolded by the subtending phyllary; basal leaves pinnatifid or toothed, rarely subentire.
        Upper leaves and phyllaries not terminated by open pit glands.   31. *Hemizonia.*
        Upper leaves and phyllaries terminated by open pit glands; ligules yellow; each disk-flower subtended by a bract; pappus none.   32. *Holocarpha.*
      Ray-ligules 3-cleft or -parted into palmately spreading lobes; tack-shaped glands present (excepting *C. tenella*); leaves narrowly linear and grass-like.   33. *Calycadenia.*
     Disk-achenes prominently ribbed; ray-achenes pubescent and pappose; pappus plumose.   34. *Blepharizonia.*
   Style of disk-flowers hairy below the 2-cleft tip; pappus plumose.   35. *Blepharipappus.*

26. **ACHYRACHAÈNA** Schauer, Del. Sem. Hort. Vratisl. 3. 1837.

Vernal mesophytic annuals. Leaves opposite and clasping below, alternate above, linear. Heads solitary, long-peduncled, terminating the stem and few ascending branches, heterogamous, radiate, all flowers fertile. Involucre oblong-campanulate, nearly equaling the inconspicuous flowers. Ray-flowers 1-seriate; ligules yellow, turning crimson-red. Receptacular bracts in a single row between ray and disk, free. Achenes 10-costate, tuberculate-scabrous; pappus of silvery scales, spreading on ripe achenes to form a globose head. [Name Greek, meaning chaff and achene, in reference to the chaffy pappus.]

A monotypic genus of the Pacific Coast.

1. **Achyrachaena móllis** Schauer. Blow-wives. Fig. 5230.

Achyrachaena mollis Schauer. Del. Sem. Hort. Vratisl. 3. 1837.
Lepidostephanus madioides Bartl. Ind. Sem. Hort. Gotting. 1837, ex Linnaea **12**: Litt. 82. 1838.

Stem simple or few-branched, 1–4 dm. high, the herbage villous, becoming moderately glandular above. Leaves entire or remotely serrulate, up to 13 cm. long and 7 mm. wide, mostly much less; heads in flower 1.5–2 cm. high, in fruit forming a globose cluster 3 cm. across; phyllaries completely investing the ray-achenes; ray-flowers about 8, the ligules inconspicuous; disk-flowers about 15–35; achenes clavate, black, about 5 mm. long, those of the ray epappose and smooth-ribbed, those of the disk pappose and scabrous with brown teeth; pappus of 10 oblong blunt scales, the 5 outer half as long as and alternate with the inner, the inner 7–9 mm. long.

Frequent in moist grassy fields with heavy soils, Sonoran and Transition Zones; cismontane valleys from Douglas County, Oregon, south to northern Lower California. Type locality: California. April–May.

27. **LAGOPHÝLLA** Nutt. Trans. Amer. Phil. Soc. II. **7**: 390. 1841.

Mesophytic or xerophytic annuals. Basal leaves serrate-dentate to subentire; cauline leaves entire, readily caducous. Involucre turbinate to hemispheric, the phyllaries completely enfolding the obcompressed ray-achenes and caducous with them. Flowers yellow, the heads opening toward evening and closing in the morning. Receptacle penicillate-pubescent centrally, its bracts in a single row between ray and disk, slightly united. Ray-flowers 5, fertile. Disk-flowers 6, sterile. Pappus none. [Name from two Greek words

---

* Text contributed by David Daniels Keck.

meaning hare and leaf, in allusion to the copious sericeous pubescence of the upper leaves of the original species.]

A genus of 5 species, natives of the western United States. Type species, *Lagophylla ramosissima* Nutt.

Ligules 8–13 mm. long, conspicuous, bright yellow.
    Stem dichotomously branched; involucre hemispheric; chiefly Coast Ranges.
        Ray-achene broadly oblanceolate, glossy, smooth, midnerve evident; phyllaries pilose-ciliate; low, slender, flexuous habit; vernal.   1. *L. minor.*
        Ray-achene obovate, dull, striate, mid-nerve obscure; phyllaries barbellate-ciliate; taller and stricter habit; late vernal and early aestival.   2. *L. dichotoma.*
    Stem simple or paniculately branched above; involucre turbinate; late vernal to autumnal; Sierra Nevada.
                                                3. *L. glandulosa.*
Ligules 3–5.5 mm. long, inconspicuous, pale lemon-yellow; vernal to autumnal.
    Heads scattered or at least not densely glomerate.   4. *L. ramosissima.*
    Heads densely congested in compact glomerules.   5. *L. congesta.*

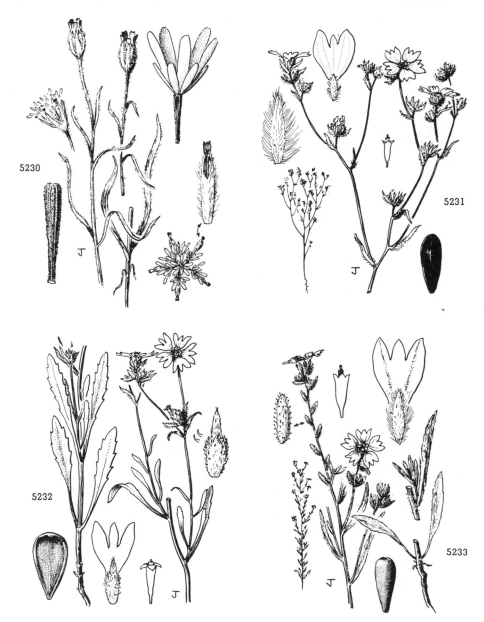

5230. Achyrachaena mollis   5232. Lagophylla dichotoma
5231. Lagophylla minor   5233. Lagophylla glandulosa

1. Lagophylla mìnor (Keck) Keck. Lesser Hareleaf. Fig. 5231.

Lagophylla dichotoma subsp. *minor* Keck, Madroño **3**: 16. 1935.
Lagophylla minor Keck, Aliso **4**: 105. 1958.

Slender herb 1–3 dm. high, the stem repeatedly forking, at least in upper half, to form an open corymbose crown, the ultimate twigs very dark, strigose. Leaves linear, some tapering to base, up to 5.5 cm. long and 3 mm. wide, entire except for the basal, narrowly involute, soon deciduous, strigulose-puberulent, becoming hirsute-ciliate toward the heads; inflorescence bearing few or no stalked glands, if present, not extending down the peduncles; involucre 4–5 mm. high, pilose, the phyllaries long-ciliate on the angles with hairs almost as long as bract is wide, acuminate; ray-achenes smooth and glossy black, 0.8–1.3 mm. wide.

Dry serpentine slopes of the foothills, Upper Sonoran Zone; Inner North Coast Range from Glenn County to Napa County and on the Sierra foothills of Eldorado County, California. Type locality: Pope Creek, just south of Walters Spring, Napa County. April–May.

2. Lagophylla dichótoma Benth. Forking Hareleaf. Fig. 5232.

Lagophylla dichotoma Benth. Pl. Hartw. 317. 1849.

Stouter than *L. minor*, 1.5–6 dm. high. Inflorescence bearing few to many evident stalked glands, these extending down the peduncles; involucre 4.5–5.8 mm. high, hispid-hirsute, the hairs much shorter than the width of the acute bracts; ray-achenes blackish or lightly mottled, often with cellular surface, with 20–30 striae on each face, 1.4–1.9 mm. wide.

Rather rare, in grassy places in the valleys and foothills, Upper Sonoran Zone; east side of the Central Valley from Butte County to Tulare County and in the Inner South Coast Range from San Benito County to Monterey County, California. Type locality: Sacramento Valley (probably in what is now Butte County). April–May.

3. Lagophylla glandulòsa A. Gray. Glandular Hareleaf. Fig. 5233.

Lagophylla glandulosa A. Gray, Proc. Amer. Acad. **17**: 219. 1882.

Stem 1–10 or 15 dm. high, simple or paniculately branched, the plant slender or bushy with uneven crown; herbage densely appressed-hirtellous and distinctly glandular. Lower leaves linear-oblanceolate to spatulate, lost before flowering, 3–12 cm. long, 5–12 mm. wide, cauline leaves bearing fascicles or short sterile shoots in their axils, readily caducous, reduced to bracts upward; heads on short peduncles; involucre turbinate, 5–7 mm. high; ligules 8–13 mm. long, 6–9 mm. wide, obcordate, the lateral lobes rounded, much wider than the central one; ray-achenes clavate-obovate, smooth, 2.7–3.9 mm. long.

Hot dry slopes, Upper Sonoran Zone; common in the Sierran foothills from Shasta County to Kern County and occasional in the upper Sacramento Valley and in Mendocino County, California. Type locality: "California, in the Sierra Nevada from Auburn to near the Yosemite." July–Oct.

Lagophylla glandulosa subsp. **serràta** (Greene) Keck, Aliso **4**: 105. 1958. (*Lagophylla serrata* Greene, Bull. Calif. Acad. **1**: 280. 1886.) Inconspicuous herb 1–6 dm. high, simple or with few ascending branches, almost glandless below the crown; leaves without fascicles or sterile shoots in their axils, the lower often persistent after time of flowering; heads few, on short or elongated, very slender peduncles; involucre 4.75–6 mm. high; ray-achenes 2.5–3.2 mm. long. A seasonal ecotype growing with the species. Type locality: Grass Valley, Nevada County. May–July.

4. Lagophylla ramosíssima Nutt. Common Hareleaf. Fig. 5234.

Lagophylla ramosissima Nutt. Trans. Amer. Phil. Soc. II. **7**: 391. 1841.
Lagophylla minima Kell. Proc. Calif. Acad. **5**: 1873.
Lagophylla hillmani A. Nels. Proc. Biol. Soc. Wash. **17**: 98. 1904.

Stiffly erect, 2–10(–15) dm. high, racemosely or paniculately branched above, sometimes from the base, rarely simple; herbage grayish or dull green, densely white-hirtellous or white-sericeous, the prominent, yellow, stipitate glands confined to upper leaves and heads. Leaves as in *L. glandulosa*; heads short-peduncled or subsessile, racemosely disposed along the branchlets and in small clusters at their ends; involucre 4.4–6.7 mm. high, the lanceolate phyllaries densely villous-ciliate on the angles; ray-achenes clavate and slightly arcuate, 2.5–4 mm. long.

Abundant in open places in hard dry soils, Sonoran and Transition Zones; eastern Washington and western Idaho south to San Diego County, California, and northern and western Nevada. Type locality: "In the prairies near Walla-Walla, in Oregon" (possibly in what is now Washington). May–Oct.

5. Lagophylla congésta Greene. Rabbit-foot. Fig. 5235.

Lagophylla congesta Greene, Bull. Torrey Club **10**: 87. 1883.
Lagophylla ramosissima var. *congesta* Jepson, Fl. W. Mid. Calif. 539. 1901.
Lagophylla ramosissima subsp. *congesta* Keck, Madroño **3**: 16. 1935.

Similar to *L. ramosissima* except more robust, the heads densely congested in compact glomerules 1.5–6 cm. thick, the glomerules simply terminal or in interrupted spikes or more rarely on short lateral branches racemosely arranged, the stems usually simple; involucre 5–7.5 mm. high.

Hot, dry, open slopes with *L. ramosissima*, Upper Sonoran Zone; from Humboldt County to Santa Cruz County and from Eldorado County to Stanislaus County, California. Type locality: Mendocino County. May–Sept.

28. HOLOZÒNIA Greene, Bull. Torrey Club **9**: 122, 145. 1882.

Perennial herb, hemicryptophyte, with elongated fleshy rootstocks. Stems erect, usually simple and densely leafy below, diffusely branched at the nearly leafless apex, frequently with short leafy spurs or peduncles in the leaf-axils. Leaves opposite and connate below, alternate above, ligulate, reduced to small ovate bracts on peduncles. Heads scattered, solitary on long slender peduncles. Involucre broadly turbinate. Ray-flowers 1-seriate, fertile; ligules white, purplish dorsally. Receptacular bracts united into a cup surrounding the sterile disk-flowers. Ray-achenes clavate, finely striate, the cupulate areola entire or apiculate-margined. Disk-achenes with or without a fragile pappus of 1–5 caducous paleaceous bristles nearly as long as the corolla. [From the Greek words for whole and girdle, from the wholly enclosed ray-achenes.]

A monotypic genus of California.

1. **Holozonia fílipes** (Hook. & Arn.) Greene. Holozonia. Fig. 5236.

Hemizonia filipes Hook. & Arn. Bot. Beechey 356. 1838.
Lagophylla filipes A. Gray, Pacif. R. Rep. **4**: 109. 1857.
Holozonia filipes Greene, Bull. Torrey Club **9**: 122, 145. 1882.

Stems 3–10 dm. high, stramineous, the mature herbage cinereous and villous, beset with stalked glands toward the inflorescence. Lower leaves remotely serrate to subentire, 3–10 cm. long, 4–8 mm. wide, progressively smaller upward; involucre 3.5–5 mm. high, as broad, the linear-lanceolate (as folded) phyllaries prominently hirsute; ray-flowers 4–8, the ligule 3.5–4.5 mm. long, deeply 3-cleft into linear lobes; disk-flowers 10–20(–28).

In dry alkaline or rocky gulches or beds of intermittent streams, Sonoran and Transition Zones; Shasta County to Mariposa County and from Napa County to Marin, Santa Clara, and Monterey Counties, California. Type locality: California. June–Oct.

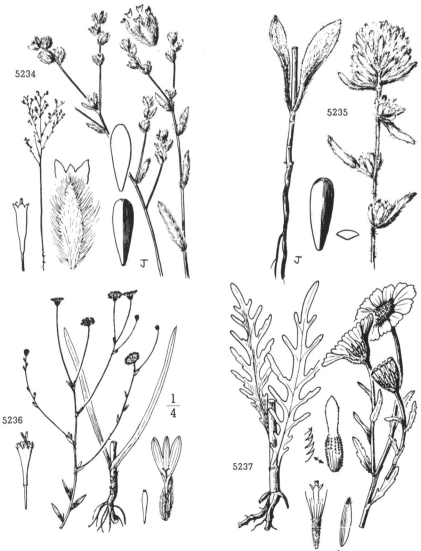

5234. Lagophylla ramosissima
5235. Lagophylla congesta
5236. Holozonia filipes
5237. Layia chrysanthemoides

29. **LÀYIA** Hook. & Arn. Bot. Beechey 148. 1833, nomen provisorum; DC. Prod. **7**: 294. 1838, nomen conservandum.

Vernal annuals with chiefly alternative, subentire or toothed to pinnatifid, narrow leaves. Heads many-flavored, on usually naked terminal peduncles, heterogamous and radiate (except in *L. discoidea*), both ray- and disk-flowers fertile. Involucre campanulate to broadly hemispheric, the thin, dilated, lower margins of the phyllaries abruptly infolded and enclosing the ray-achenes, the tip flat (except in *L. discoidea*). Receptacle broad, chaffy marginally or throughout. Ray-flowers 8–24; ligules 3-lobed or -toothed, white, yellow, or yellow with white tip. Disk-flowers numerous, yellow; anthers black or yellow. Ray-achenes obcompressed, commonly smooth and glabrous, with prominent terminal areola, epappose. Disk-achenes pubescent and pappose, rarely glabrous and epappose; pappus of numerous bristles, awns, or paleae, the bristles often plumose below. [Name in honor of G. Tradescant Lay, botanist with Captain Beechey in the *Blossom*, which visited California in 1827.]

A genus of 15 species, all of which occur in, and 13 of which are restricted to, California. Type species, *Tridax* (?) *galardioides* Hook. & Arn.

Bracts of receptacle subtending each disk-flower; involucre (and herbage) not glandular, the phyllaries with prominent pustulate processes at base of hairs.

Pappus of rigid subulate awns; disk-achenes obviously obcompressed, cuneate; phyllaries hispid-ciliate or pectinate along the folded edge, otherwise nearly smooth; Coast Ranges.   1. *L. chrysanthemoides.*

Pappus of ovate-lanceolate thin paleae; disk-achenes plump, clavate; phyllaries uniformly papillate-hispid dorsally; Great Valley.   2. *L. fremontii.*

Bracts of receptacle limited to a ring between ray- and disk-flowers; involucre (and herbage) glandular, the phyllaries pubescent but not as above.

Pappus paleaceous, glabrous except for capillary hairs radiating from very base.

Ray-flowers in 1 series, their ligules obovate to flabelliform, their achenes dull, glabrous or pubescent, flattened at least ventrally; involucre hemispheric, green, the bracts not inflated, with prominent tips; pappus 2–3.5 mm. long.

Ligules white; anthers yellow; ray-achenes sericeous; leaves mostly entire. 3. *L. leucopappa.*

Ligules yellow with long white tips; anthers black; ray-achenes glabrous, or lightly pubescent toward apex; leaves mostly pinnately lobed.   4. *L. munzii.*

Ray-flowers usually in 2 series, their ligules oblong-spatulate, their achenes polished, glabrous, plump; involucre broadly urceolate, dark-dotted, the bracts inflated, with inconspicuous tips; pappus 1–2 mm. long.   5. *L. jonesii.*

Pappus setaceous or if palaceous, plumose.

Ligule yellow with white tip (rarely yellow throughout); pappus commonly merely scabrid, if plumose, also with inner wool, of 18–30 bristles; anthers black; tips of phyllaries longer than basal portion.   6. *L. platyglossa.*

Ligule yellow or white (none in *L. discoidea*) but not as above (except sometimes in *L. gaillardioides*); pappus plumose, with inner woolly hairs only in *L. septentrionalis* and *L. glandulosa;* tips of phyllaries shorter than the basal portion (except in *L. heterotricha*).

Pappus persistent; stem not prominently fistulous, pubescent below (glabrate in *L. carnosa*); disk-achenes strigose.

Ligules more than 5 mm. long, conspicuous (none in *L. discoidea*); self-sterile.

Anthers yellow; involucre not urceolate; stems not dark-dotted; pappus white.

Involucre ovoid, 7–12 mm. high; ligules golden-yellow; disk-corollas 5.5–8 mm. long; ray-achenes 3.8–5.2 mm. long; pappus 3.5–6.2 mm. long, the slender bristles 16–21, densely plumose, with inner wool; Lake and Colusa Counties.   7. *L. septentrionalis.*

Involucre turbinate or campanulate to hemispheric, 5–9.5 mm. high; disk-corollas 3–6 mm. long.

Heads discoid; pappus palaceous, 1–1.5 mm. long, fulvous; San Benito County.   8. *L. discoidea.*

Heads radiate; pappus setaceous, 2–5 mm. long, white.

Stems hispid; basal leaves dentate or lobed, the cauline entire; ray-achenes 3–4 mm. long; pappus 3–5 mm. long, the bristles 10(–12), flattened, linear-attenuate, densely plumose, usually with inner wool.   9. *L. glandulosa.*

Stems pilose or hirsute; basal leaves pinnatifid to bipinnatifid; ray-achenes 2.5–4 mm. long; pappus 2–4 mm. long, the bristles 10–18 (rarely less), filiform, moderately plumose, without wool. 10. *L. pentachaeta.*

Anthers black; involucre urceolate; stems dark-dotted; pappus rufous.   11. *L. gaillardioides.*

Ligules 2–4 mm. long, inconspicuous; self-fertile; pappus rufous to whitish; anthers black.

Ligules yellow; pappus-bristles 11–15; stem rigidly erect, 20–100 cm. tall, dark-dotted; pungently odorous; ray-flowers 8–16.

Lower leaves laciniate-dentate, broadly oblanceolate or oblong; stems mostly stout; San Francisco Bay to Santa Cruz.   12. *L. hieracioides.*

Lower leaves pinnatifid, if merely dentate narrowly linear, narrower; stems mostly slender; Mount Diablo to Santa Barbara.   13. *L. paniculata.*

Ligules white; pappus-bristles 25–32; stem freely branched, to 15 cm. tall, without dots; not odorous; ray-flowers 5–7(–9); coastal dunes.   14. *L. carnosa.*

Pappus deciduous; stem prominently fistulous, glabrate below; disk-achenes sericeous; ligules creamy white to pale yellow.   15. *L. heterotricha.*

1. **Layia chrysanthemoìdes** (DC.) A. Gray. Smooth Layia. Fig. 5237.

Oxyura chrysanthemoides DC. Prod. **5**: 693. 1836.
Hartmannia ciliata DC. op. cit. 694.
Calliglossa douglasii Hook. & Arn. Bot. Beechey Suppl. 356. 1838.

Callichroa douglasii Torr. & Gray, Fl. N. Amer. **2**: 396. 1843.
Layia calliglossa A. Gray, Mem. Amer. Acad. II. **4**: 103. 1849.
Layia chrysanthemoides A. Gray, Proc. Amer. Acad. **7**: 360. 1868.
Layia calliglossa var. *oligochaeta* A. Gray, Bot. Calif. **1**: 370. 1876.
Blepharipappus chrysanthemoides Greene, Pittonia **2**: 247. 1892.
Blepharipappus douglasii Greene, loc. cit.
Blepharipappus douglasii var. *oligochaeta* Greene, Man. Bay Reg. 201. 1894.

Stem erect, corymbosely branched, 1–4 dm. high. Leaves scabrociliate, otherwise smooth, the basal pinnately parted into linear or oblong, blunt lobes; involucre 6–11 mm. high; ligules yellow with white lobes, 7–15 mm. long, 5–9 mm. wide; disk-flowers 30–100, the corolla 3–5 mm. long; ray-achenes 2.7–3.9 mm. long, strongly compressed, glabrous; disk-achenes 2.5–4 mm. long, densely strigose, sometimes lacking pubescence and pappus; pappus of 9–17 tawny, rigid, subulate, scabrous awns, the lateral ones 1.5–3.5 mm. long, the intervening ones usually shorter or even reduced to rudiments.

Moist heavy soils of valley floors, Upper Sonoran Zone; Mendocino and Glenn Counties to Monterey County, California. Type locality: "California." March–May.

Layia chrysanthemoides subsp. **marítima** Keck, Aliso **4**: 108. 1958. Stem prostrate, even the central peduncle horizontal; leaves succulent, the lower with broad rounded lobes; tips of phyllaries dilated, rounded; late-flowering. Rare; on the immediate coast of Mendocino and Sonoma Counties, California. Type locality: south of Jenner, Sonoma County. May.

2. **Layia fremóntii** (Torr. & Gray) A. Gray. Fremont's Layia. Fig. 5238.

Calliachyris fremontii Torr. & Gray, Bost. Journ. Nat. Hist. **5**: 110. 1845.
Layia fremontii A. Gray, Mem. Amer. Acad. II. **4**: 103. 1849.
Blepharipappus fremontii Greene, Pittonia **2**: 246. 1892.

Stem 1–4 dm. high. Leaves scabrous or short-hispid, not at all viscid, the basal pinnately parted; involucre 6–11 mm. high; ligules yellow, the outer half white, 9–18 mm. long, 7–12 mm. wide; disk-flowers 40–100, the corolla 3.6–5.3 mm. long; ray-achenes 2.5–3.7 mm. long, strongly compressed, glabrous; disk-achenes 2–3.5 mm. long, densely strigose and with a row of long capillary hairs on the areola; pappus of 10 tawny, white, ovate-lanceolate paleae with attenuate tips, 2–4.5 mm. long.

Grassy fields, Upper Sonoran Zone; Great Valley and Sierran foothills, Tehama County to Tulare County, California. Type locality: California. Collected by Fremont. March–May.

3. **Layia leucopáppa** Keck. Comanche Layia. Fig. 5239.

Layia leucopappa Keck, Madroño **3**: 17. 1935.

Stem branched from base upward, to 5 dm. high; herbage more or less glaucescent. Leaves hispidulous-ciliate, villous on upper surface, viscidulous and with scattered black glands, the basal pinnately toothed or lobed, the cauline mostly entire; involucre 6–9 mm. high, the bracts papillate, short-hirsute and black-glandular on the back, the infolded margin lanuginous-ciliate; ligules white, 8–12 mm. long, 8–14 mm. wide; disk-flowers 50–100, the corolla 3.7–5.2 mm. long; ray-achenes 2.5–2.9 mm. long; disk-achenes 2.5–4 mm. long, densely white-sericeous and with a row of long capillary hairs on the areola; pappus of 10(–13) bright, white, lanceolate, acuminate paleae 2–3.5 mm. long.

Forming colonies on moist benches, Upper Sonoran Zone; very local at the head of the San Joaquin Valley, Kern County, California. Type locality: Comanche Point. March–April.

5238. Layia fremontii

5239. Layia leucopappa

4. **Layia múnzii** Keck. Munz's Layia. Fig. 5240.

Layia munzii Keck, Madroño **3**: 16. 1935.

Stem branched from base upward, 2–3.5 dm. high. Leaves mostly glabrous except for scabrous or ciliate margin, the upper sparsely villous and viscid, the basal and lower cauline pinnately lobed to parted; involucre 7–8.5 mm. high, the bracts glabrate to sparsely hispid; ligules yellow tipped with white for one-third to two-thirds their length, 9–12 mm. long, 6–9 mm. wide; disk-flowers 50–100, the corolla 3.6–5 mm. long; ray-achenes 2.8–3.5 mm. long; disk-achenes 2.5–4 mm. long, densely strigose and with a row of long capillary hairs on the areola; pappus of 9–12 dirty white, linear-lanceolate, attenuate paleae, usually with a spot of anthocyanin at base, 2.3–3.4 mm. long.

Alkaline flats of heavy clay, Upper Sonoran Zone; west side of San Joaquin Valley and adjacent Inner Coast Range from Merced County to Kern and San Luis Obispo Counties, California. Type locality: near Cholame, San Luis Obispo County. March–April.

5. **Layia jònesii** A. Gray. Jones's Layia. Fig. 5241.

Layia jonesii A. Gray, Proc. Amer. Acad. **19**: 18. 1883.
Callichroa jonesii Greene, Pittonia **2**: 228. 1892.
Blepharipappus jonesii Greene, op. cit. 247.

Stem usually corymbosely branched, up to 3.5 dm. high. Leaves glandular-puberulent and moderately short-hispid, more or less pustulate, the basal and lower cauline pinnately parted; involucre 6–8 mm. high; ligules yellow, with short white tips, 6–9 mm. long, 4–5.5 mm. wide; disk-flowers 35–75(–100), the corolla 3.1–5.2 mm. long; ray-achenes 2.3–3.3 mm. long; disk-achenes 2.4–4 mm. long, moderately strigose and with a row of capillary hairs on the areola; pappus of 10–14 dirty white, ovate or oblong-ovate, imbricate, acuminate paleae 1–2 mm. long.

Pastures and grassy slopes, Upper Sonoran Zone; local along the coast of San Luis Obispo County, California. Type locality: near San Luis Obispo. March–May.

6. **Layia platyglóssa** (Fisch. & Mey.) A. Gray. Tidy Tips. Fig. 5242.

Callichroa platyglossa Fisch. & Mey. Ind. Sem. Hort. Petrop. **2**: 31. 1835.
Madaroglossa hirsuta Nutt. Trans. Amer. Phil. Soc. II. **7**: 394. 1841.
Layia platyglossa A. Gray, Mem. Amer. Acad. II. **4**: 103. 1849.
Blepharipappus platyglossus Greene, Pittonia **2**: 246. 1892.

Stems prostrate or decumbent, stout, succulent, corymbosely branched, 1–3 dm. long. Leaves short-hirsute or pilose, the basal and lower cauline dentate to pinnatifid with salient, rotund, short lobes; involucre 6–12 mm. high, the bract-tips rounded; ligules 6–15 mm. long, 5–10 mm. wide; disk-flowers 30–100, the corolla 4–6 mm. long; ray-achenes 3–3.8 mm. long; disk-achenes 2.8–5 mm. long, strigose; pappus of 18–32 white (or tawny) scabrid bristles.

Grassy flats, Humid Transition Zone; along the immediate coast from Mendocino County to Point Concepcion and Santa Cruz Island, Santa Barbara County, California. Type locality: Russian colony at Fort Ross, Sonoma County. May–June.

Layia platyglossa subsp. **campéstris** Keck, Aliso **4**: 106. 1958. (*Madaroglossa elegans* Nutt. Trans. Amer. Phil. Soc. II. **7**: 393. 1841; *Layia elegans* Torr. & Gray, Fl. N. Amer. **2**: 394. 1843; *L. platyglossa* var. *breviseta* A. Gray, Proc. Amer. Acad. **9**: 193. 1874; *Blepharipappus elegans* Greene, Pittonia **2**: 246. 1892.) Stem erect, the side branches ascending to erect, slender, less succulent, to 5 dm. high, the herbage a grayer green; leaf-lobes narrower and longer; phyllary-tips not dilated; pappus as in the species or the bristles densely plumose and interlaced with woolly hairs within. Grassy plains and foothills, Upper Sonoran Zone; the widespread form of the species, mostly from the Bay Region, California, southward to central Lower California, occasional northward to Mendocino and Butte Counties; abundant on the coastal plain, infrequent in the Great Valley. Type locality: Pacheco Creek, Santa Clara County. March–May.

7. **Layia septentrionàlis** Keck. Colusa Layia. Fig. 5243.

Layia septentrionalis Keck, Aliso **4**: 106. 1958.

Stem often corymbosely branched, 1.5–4 dm. high. Leaves hirsute, glandular, the lower pinnately toothed to parted; involucre 7–12 mm. high, rather densely pilose and glandular; ligules golden yellow, 8–16 mm. long, 7–9 mm. wide; disk-flowers 35–60, the corolla 5.5–8 mm. long; ray-achenes 3.8–5.2 mm. long; disk-achenes 4.5–7.5 mm. long; pappus of 16–21 glistening white bristles, not appreciably flattened, plumose to above middle with capillary hairs outside and tangled woolly hairs inside.

Grassy fields or slopes, in sandy or serpentine soils, Upper Sonoran Zone; Inner North Coast Ranges from Colusa County to Sonoma County, California; Sutter [Marysville] Buttes, Sutter County. Type locality: near Lakeport, Lake County. April–May.

8. **Layia discoìdea** Keck. Rayless Layia. Fig. 5244.

Layia discoidea Keck, Aliso **4**: 106. 1958.

Stem simple or corymbosely few-branched, 5–12(–15) cm. high. Leaves white-hispid, all but those of the rosette moderately black-glandular, the basal pinnately short-lobed, the cauline few-toothed or mostly entire; involucre turbinate-campanulate, anthocyanous, 5–6 mm. high, moderately villous and rather densely beset with stalked black glands; flowers 10–25, the corolla 3.5–4 mm. long; achenes 3.5–4.5 mm. long; pappus of (8–)11–15 lanceolate, acuminate, entire or deeply lacerate (and then often truncate), fulvous paleae 1(–1.5) mm. long, plumose with short capillary hairs.

Serpentine slopes in chaparral openings, Upper Sonoran Zone; rare endemic of southern San Benito County, California. Type locality: New Idria. May.

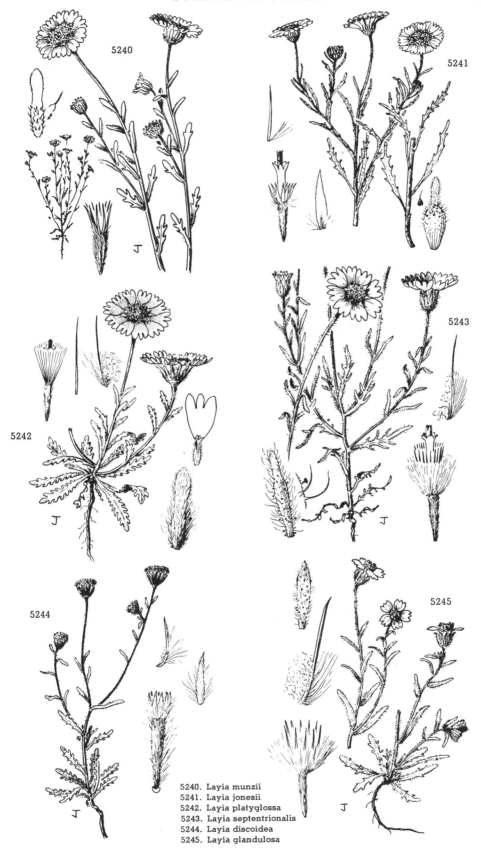

5240. Layia munzii
5241. Layia jonesii
5242. Layia platyglossa
5243. Layia septentrionalis
5244. Layia discoidea
5245. Layia glandulosa

9. Layia glandulòsa (Hook.) Hook. & Arn. White Layia. Fig. 5245.

Blepharipappus glandulosus Hook. Fl. Bor. Amer. **1**: 316. 1833.
Eriopappus glandulosus Arn. in Lindl. Nat. Syst. ed. 2. 443. 1836.
Madaroglossa angustifolia DC. Prod. **5**: 694. 1836.
Layia glandulosa Hook. & Arn. Bot. Beechey Suppl. 358. 1838.
Layia douglasii Hook. & Arn. loc. cit.
Madaroglossa douglasii Walp. Rep. **2**: 632. 1843.
Layia glandulosa var. *rosea* A. Gray, Bot. Calif. **1**: 368. 1876.
Layia hispida Greene, Pittonia **2**: 20. 1889.
Blepharipappus hispidus Greene, op. cit. 246.
Blepharipappus oreganus Greene, loc. cit.
Layia glandulosa var. *hispida* Jepson, Man. Fl. Pl. Calif. 1101. 1925.

Stem commonly corymbosely branched, 1–4(–6) dm. high, often anthocyanous. Leaves hispid, often densely strigose on upper surface, the basal dentate or lobed, the cauline mostly entire; involucre 6.5–9.5 mm. high, short-hispid and usually stipitate-glandular; ligules white, often fading rose-purple, 6–15 mm. long, 5–9(–12) mm. wide; disk-flowers 25–50(–100), the corolla 4–6 mm. long; ray-achenes 3–4 mm. long; disk-achenes 3.5–6 mm. long; pappus of 10(–12) glistening, white, flattened, linear-attenuate paleae, plumose to above middle with capillary hairs outside and tangled woolly hairs inside.

Open sandy places, Sonoran and Transition Zones; central Washington and Idaho to Lower California and New Mexico, extending to the coast in central and southern California. Type locality: "plains of the Columbia." March–June.

Layia glandulosa subsp. lùtea Keck, Madroño **3**: 18. 1935. Ligules golden yellow. Sandy soils, Upper Sonoran Zone; inner South Coast Ranges, San Benito County to Monterey County, California, and locally in the Santa Lucia Mountains and the mountains of Ventura County. Type locality: Bear Valley, north of The Pinnacles, San Benito County. April–May.

10. Layia pentachaèta A. Gray. Sierran Layia. Fig. 5246.

Layia pentachaeta A. Gray, Pacif. R. Rep. **4**: 108. 1857.
Blepharipappus pentachaetus Greene, Pittonia **2**: 246. 1892.
Layia pentachaeta var. *hanseni* Jepson, Madroño **1**: 201. 1929.

Stem branching above, 2–8 dm. high; herbage with acrid odor. Leaves hirsute and finely glandular, pinnatifid to bipinnatifid; involucre broad, 6–9 mm. high, pustulate-hirsute and moderately glandular; ligules 5–18 mm. long, 4–9 mm. wide; disk-flowers 20–125, the corolla 3–5.5 mm. long; ray-achenes 2.5–4 mm. long; disk-achenes 2.5–4(–4.7) mm. long; pappus of 10–18 (rarely 5 or less) filiform bristles, moderately plumose at base with rather short, straight, capillary hairs.

Grassy flats and slopes, Upper Sonoran Zone; Sierran foothills of California from Placer County to Kern County, and in a more robust form on the plains of Tulare County. Type locality: Knight's Ferry, Stanislaus County. March–May.

Layia pentachaeta subsp. álbida Keck, Aliso **4**: 107. 1958. Ligules white. Open plains and foothills, Upper Sonoran Zone; west side and head of the San Joaquin Valley and adjacent hills from western Fresno County to Kern County, California. Type locality: Polonio Pass, east of Cholame, San Luis Obispo County. March–May. Readily separable from *L. glandulosa* by the pappus, odor, leaf-cut, etc.

11. Layia gaillardioìdes (Hook. & Arn.) DC. Woodland Layia. Fig. 5247.

Tridax (?) galardioides Hook. & Arn. Bot. Beechey 148. 1833.
Layia gaillardioides DC. Prod. **7**: 294. 1838.
Blepharipappus gaillardioides Greene, Pittonia **2**: 246. 1892.
Blepharipappus nemorosus Greene, Man. Bay Reg. 200. 1894.
Layia nemorosa Jepson, Fl. W. Mid. Calif. ed. 2. 448. 1911.
Layia gaillardioides var. *nemorosa* Jepson, Man. Fl. Pl. Calif. 1102. 1925.

Stem simple or freely branched, 2–6 dm. high; herbage pungently odorous. Leaves hirsute to hispid, pustulate, somewhat viscid or stipitate-glandular or both, remotely serrate or dentate to pinnatifid (or bipinnatifid); involucre 5–8.5 mm. high, the bracts ventricose, villous and stipitate-glandular; ligules golden yellow throughout to pale yellow with whitish tip, 7–12 mm. long, 4–7 mm. wide; disk-flowers (20–)40–100, the corolla 3.5–5.2 mm. long; ray-achenes 3–4.4 mm. long; disk-achenes 3.2–5 mm. long; pappus of 17–21 slender, reddish brown to white bristles 1–4 mm. long, plumose to above middle with straight capillary hairs.

Open grassy places, Humid Transition and Upper Sonoran Zones; both Outer and Inner Coast Ranges from Humboldt and Colusa Counties to northern Monterey County, California. Type locality: not stated. April–June. Especially variable in flower-color.

12. Layia hieracioìdes (DC.) Hook. & Arn. Tall Layia. Fig. 5248.

Madaroglossa hieracioides DC. Prod. **5**: 694. 1836.
Layia hieracioides Hook. & Arn. Bot. Beechey Suppl. 358. 1838.
Blepharipappus hieracioides Greene, Pittonia **2**: 246. 1892.

Stem stout, rigidly erect, corymbosely branched toward apex, 2–8(–10) dm. high; herbage pungently odorous. Leaves hispid, usually pustulate, viscid, laciniate-dentate (or pinnatifid); heads terminating short leafy branches; involucre 5–8 mm. high, villous and stipitate-glandular; ligules 2–4 mm. long, 1.5–2.5 mm. wide; disk-flowers 15–75, the corolla 2.3–3.3 mm. long; ray-achenes 3–3.3 mm. long; disk-achenes 3–4 mm. long; pappus of 11–15 slender, purplish to white bristles 2–4 mm. long, plumose to middle with straight capillary hairs.

Grassy openings in wooded hills, mostly Humid Transition Zone; Berkeley Hills, Mount Diablo, Mount Hamilton, and Santa Cruz Mountains, California. Type locality: California. April–June.

13. **Layia paniculàta** Keck. Slender Layia. Fig. 5249.

Layia paniculata Keck, Aliso **4**: 107. 1958.

Similar to the preceding except stems more slender. Leaves narrower and more cut; disk-corolla 3–4.3 mm. long; ray-achenes 3.5–4 mm. long; disk-achenes 3.8–5 mm. long. This species is tetraploid, with 16 pairs of chromosomes, in contrast with *L. hieracioides*, a diploid with 8 pairs.

In sandy soils on wooded or brushy slopes, Upper Sonoran Zone; both Coast Ranges from San Benito County and Monterey Bay to Santa Barbara, Santa Barbara County, California. Type locality: Jolon Grade, Monterey County. April–May.

14. **Layia carnòsa** (Nutt.) Torr. & Gray. Beach Layia. Fig. 5250.

Madaroglossa carnosa Nutt. Trans. Amer. Phil. Soc. II. **7**: 393. 1841.
Layia carnosa Torr. & Gray, Fl. N. Amer. **2**: 394. 1843.
Blepharipappus carnosus Greene, Pittonia **2**: 246. 1892.

Stem freely branched, erect, the branches divergent to decumbent, up to 15 cm. high and 4 dm. across, usually much less. Leaves scabrociliate, glabrous beneath, villous above, the uppermost sparingly black-glandular, the basal sinuate-pinnatifid, the cauline mostly entire; involucre 5–7.5 mm. high; ligules white (or fading pink), 2–2.4 mm. long, 1.5 mm. wide; disk-flowers 12–26, the corolla 2.5–3.5 mm. long; ray-achenes 3.7–4.5 mm. long; disk-achenes 3.8–5.5 mm. long; pappus of 25–32 slender bristles 2.5–3.7 mm. long, plumose for two-thirds their length with rather short, straight, capillary hairs.

Widely scattered stations on coastal sand dunes, Humid Transition Zone; Humboldt County to near Point Concepcion, Santa Barbara County, California. Type locality: "St. Diego, Upper California" but probably from farther north. April–June.

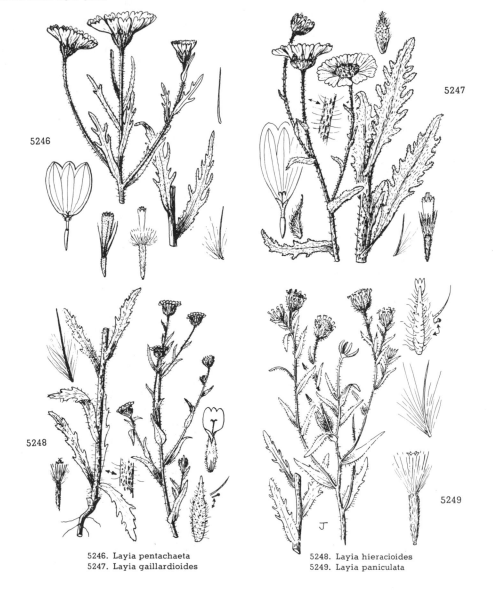

5246. Layia pentachaeta
5247. Layia gaillardioides
5248. Layia hieracioides
5249. Layia paniculata

15. Layia heterotrìcha (DC.) Hook. & Arn. Pale-yellow Layia. Fig. 5251.

Madaroglossa heterotricha DC. Prod. **5**: 694. 1836.
Layia heterotricha Hook. & Arn. Bot. Beechey Suppl. 358. 1838.
Layia graveolens Greene, Bull. Calif. Acad. **1**: 92. 1885.
Blepharipappus heterotrichus Greene, Pittonia **2**: 245. 1892.
Blepharipappus graveolens Greene, op. cit. 246.
Blepharipappus glandulosus var. *heterotrichus* Jepson, Fl. W. Mid. Calif, 536. 1901.
Layia glandulosa var. *heterotricha* H. M. Hall, Univ. Calif. Pub. Bot. **3**: 157. 1907.

Stem stout, simple or with few sharply ascending branches, 2–9 dm. high; herbage pallid, more or less succulent, heavily odorous. Leaves short-hispid, strigulose and glandular, crenate or short-dentate, the upper entire; involucre 8–12 mm. high, densely hispid and glandular; ligules creamy white to pale yellow, not tipped with white, often fading rose-purple, 6–15(–22) mm. long, 4.5–12(–15) mm. wide; disk-flowers 35–85, the corolla 4.5–7 mm. long; ray-achenes 4.2–5 mm. long; disk-achenes 4.3–6.3 mm. long; pappus of 14–19 very slender, glistening white, deciduous bristles 3.5–6 mm. long, plumose to above middle with straight capillary hairs equaling the bristles.

Grassy places, in mud flats or heavy adobe soils, Sonoran and Transition Zones; west side of San Joaquin Valley and adjacent Inner Coast Range (also Santa Lucia Mountains) to Mount Pinos and Tehachapi Pass, California. Type locality: California. March–June.

30. MÀDIA Molina, Sagg. Chile ed. 1. 136. 1782.

Mesophytic or xerophytic herbs, usually very glandular and heavy-scented. Leaves linear or oblong, the basal entire to remotely denticulate. Involucre more or less angled by its deeply sulcate phyllaries which completely enclose the ray-achenes. Receptacle bearing between ray and disk a single row of chaffy bracts, these usually more or less united into an often persistent prismatic cup. Ray-flowers few to many, fertile. Disk-flowers fertile or sterile. Ray-achenes usually laterally compressed, with flat sides, narrow back, and sharp ventral angle (see also *M. minima*), longitudinally striate. Pappus usually none in ray-achenes, sometimes present in disk-achenes. [*Madi*, the Chilean name of the original species.]

A genus of 18 species of Pacific North and South America. Type species, *Madia sativa* Molina.

Pappus present; heads not closing (Section *Anisocarpus*).
  Tall perennials of dense woods; leaves opposite well up stem, those of the rosette wider; pappus of soft lanceolate paleae.
    Disk-achenes fertile; ray-achenes without beak; involucre 10–12 mm. high; anthers black.
        1. *M. bolanderi.*
    Disk-achenes sterile; ray-achenes beaked; involucre 5–6 mm. high; anthers yellow. 2. *M. madioides.*
  Low vernal annuals; leaves alternate above first basal pairs, those of the rosette narrow.
    Disk-achenes fertile; pappus of wider paleae; anthers yellow; ligules flabelliform; plants 5–25 cm. tall.
      Branches flexuous; buds and fruiting heads nodding; pappus only on disk-achenes, lance-attenuate, fimbriate, 2.3–3.7 mm. long. 3. *M. nutans.*
      Branches strict, subumbellate; heads strictly erect; pappus of ray and disk similar, oblong or quadrate, erose-fimbriate, 0.2–0.3 mm. long. 4. *M. hallii.*
    Disk-achenes sterile; pappus of long flexuous awns; anthers black; ligules obovate.
      Heads medium; ray-flowers 7–10, ligules 5–9 mm. long; disk-flowers 20–65; plants 20–60 cm. high.
        5. *M. rammii.*
      Heads small; ray-flowers 4–7, ligules 2.5–3 mm. long; disk-flowers 2–7; slender plants 7–20 cm. high. 6. *M. yosemitana.*
Pappus none; annuals (Section *Eumadia*).
  Disk-flowers more than 1 (except rarely in *M. glomerata*), pubescent; phyllary with empty flat tip.
    Ray-achenes prominently beaked; receptacular bracts deciduous; heads not closing; the ligules golden yellow; disk-achenes fertile. 7. *M. radiata.*
    Ray-achenes not prominently beaked; receptacular bracts more or less persistent; heads mostly closing at midday.
      Disk-achenes sterile; receptacle pubescent; ray-flowers conspicuous (except in *M. citriodora*).
        Ray-achenes compressed, with narrow backs (except in subsp. *wheeleri*); ligules bright yellow, conspicuous; herbage not lemon-scented. 8. *M. elegans.*
        Ray-achenes scarcely compressed, with broad backs; ligules greenish yellow, inconspicuous; herbage lemon-scented. 9. *M. citriodora.*
      Disk-achenes fertile; receptacle glabrous; ray-flowers inconspicuous.
        Ray-flowers 6–15; ray-achenes broadest toward top with small areola.
          Achenes obovoid, black and shining. 10. *M. anomala.*
          Achenes strongly laterally compressed, dull.
            Glandular above the middle, slender and flexuous; heads small (to 9 mm. high); early; mostly interior.
              Heads in a spike-like raceme; herbage yellow-green; foothills, Butte County to Mariposa County, California. 11. *M. subspicata.*
              Heads paniculate (or openly racemose); herbage grayer or deeper green.
                Side branches not surpassing main axis; cauline leaves narrow at base; herbage coarsely pilose with spreading hairs. 12. *M. gracilis.*
                Side branches surpassing main axis; cauline leaves broad at base; herbage densely villous with rather appressed hairs. 13. *M. citrigracilis.*
            Glandular well down toward base; stem stout; heads usually large; late; mostly near the coast; often a ruderal.
              Heads short-peduncled, scattered or subglomerate, the subtending bracts narrowly lanceolate, short; involucre 7–12 mm. high; stems rather slender but tall, the branches, if any, usually ascending. 14. *M. sativa.*

Heads congested in terminal glomerules, very large, the subtending (mature)
bracts deltoid-lanceolate and usually exceeding them; involucre 8–15 mm.
high; stems stout, not tall, the branches, if any, divaricate.
15. *M. capitata.*
Ray-flowers 0–3; ray-achenes broadest at middle, truncate at both ends with broad areola.
16. *M. glomerata.*
Disk-flowers 1 (or 2), glabrous, phyllary embracing the achene and sulcate to tip; tiny herbs.
Ray-achenes laterally compressed, with sharp ventral angle, glabrous; involucre strongly stipitate-
glandular, the glands yellow.   17. *M. exigua.*
Ray-achenes obcompressed, not angled, pubescent; involucre minutely stipitate-glandular, the glands
black.   18. *M. minima.*

1. **Madia bolánderi** (A. Gray) A. Gray. Bolander's Madia. Fig. 5252.

Anisocarpus bolanderi A. Gray, Proc. Amer. Acad. **7**: 360. 1868.
Madia bolanderi A. Gray, op. cit. **8**: 391. 1872.

Perennial herb with woody rootstocks; stem simple, 5–12 dm. high, densely hirsute to glabrate
below, strongly stipitate-glandular above. Leaves entire, hirsute, linear, attenuate, the basal crowded
into an erect tuft, the lower pairs connate, 10–30 cm. long, 4–12 mm. wide, the cauline often abruptly
reduced in size and frequency; heads few, openly corymbose, terminating peduncles up to 25 cm.
long; involucre campanulate to hemispheric; ray-flowers 8–12; disk-flowers 30–65; ray-achenes
5.7–6.5 mm. long, broadly linear, obscurely 5-nerved, glabrous (when pappose somewhat hispid);
disk-achenes 6–8 mm. long, 5-angled, tawny-hispid; pappus of ray none or rudimentary, of disk
of 5–10 lance-attenuate to lance-ovate, ciliate, stramineous paleae 1.8–5 mm. long.

Damp mountain meadows or along streamways, Transition Zone; Cascade-Sierran axis from Lane County,
Oregon, to Tulare County, California; also in Scott and Marble Mountains and Trinity Summit, California. Type
locality: Mariposa Big-tree Grove, Mariposa County, California. July–Sept.

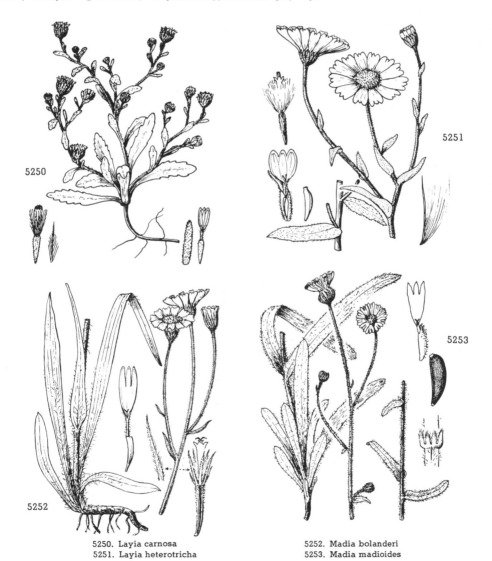

5250. Layia carnosa
5251. Layia heterotricha

5252. Madia bolanderi
5253. Madia madioides

2. Madia madioìdes (Nutt.) Greene. Woodland Madia. Fig. 5253.

Anisocarpus madioides Nutt. Trans. Amer. Phil. Soc. II. 7 : 388. 1841.
Madia nuttallii A. Gray, Proc. Amer. Acad. **8** : 391. 1872.
Madia madioides Greene, Man. Bay Reg. 193. 1894.

Perennial herb with a short, usually simple rootstock; stem simple, 3–6.5 mm. high, hirsute below, glandular-pubescent above. Leaves coarsely strigose, linear to linear-oblong, the basal horizontal in the rosette, then erect, 6–12 cm. long, 5–15 mm. wide; heads few, racemose or cymose; involucre globose-urceolate; ray-flowers 8–15; disk-flowers 10–30; ray-achenes 3.5–5 mm. long, semilunar; disk-achenes 3–4 mm. long, undeveloped, pubescent; ray-pappus none or vestigial, the disk-pappus or 5–8 lanceolate to quadrate, discrete or somewhat united, unequal, fimbriate paleae 1 mm. or less long.

Moist coniferous woods, Humid Transition Zone; Vancouver Island south to Monterey County, California. Type locality: "On the banks of the Oregon, among rocks, in shady forests, at the outlet of the Wahlamet." May–Sept.

3. Madia nùtans (Greene) Keck. Nodding Madia. Fig. 5254.

Callichroa nutans Greene, Pittonia **2** : 227. 1892.
Blepharipappus nutans Greene, op. cit. 247.
Layia nutans Jepson, Fl. W. Mid. Calif. ed. 2. 449. 1911.
Madia nutans Keck, Madroño **3** : 5. 1935.

Stem erect, commonly with few ascending or decumbent branches from the base, 1–2.5 dm. high, hirsute below, puberulent and beset with minute stipitate glands above. Leaves mostly entire, linear, the lower 2–3 cm. long, up to 2.5 mm. wide; heads solitary or loosely paniculate; involucre 4–7 mm. high; ray-flowers 5–8; disk-flowers 7–30; ray-achenes 2.7–4.1 mm. long, narrowly clavate, slightly arcuate, without prominent nerve; disk-achenes 2.3–4.5 mm. long, tapering to a slender base, the stipe dilated.

Woodland slopes, in volcanic ash, Upper Sonoran Zone; Napa and Sonoma Counties, California, from Mount St. Helena to Napa Range and Hoods Peak. Type locality: "mountains of Sonoma County." April–May.

4. Madia hállii Keck. Hall's Madia. Fig. 5255.

Madia hallii Keck, Madroño **3** : 5. 1935.

Principal stem short, very leafy, hirsute, topped by a dense rosette of leaves, then 2–5-ramified into assurgent, leafless, glandular-pubescent branches terminating in solitary heads or in leafy rosettes from which similar peduncles arise, 5–18 cm. high. Leaves hirsute, linear, the margins thickened, 0.5–3 cm. long, 0.7–1.7 mm. wide; heads solitary on black-glandular peduncles 1–5 cm. long, or more or less congested in corymbs terminating peduncles 5–9 cm. long; involucre obovoid, 4.5–5.2 mm. high; ray-flowers 3–6; disk-flowers 8–20; ray-achenes 2.7–3.1 mm. long, narrowly clavate, arcuate, slightly laterally compressed; disk-achenes 2.8–3.2 mm. long, prismatic, the stipe scarcely dilated.

Dry, stony, serpentine ridges, Upper Sonoran Zone; a rare plant of southern Lake and northern Napa Counties, California. Type locality: near Knoxville, Napa County. May.

5. Madia rámmii Greene. Ramm's Madia. Fig. 5256.

Madia rammii Greene, Bull. Calif. Acad. **1** : 90. 1885.
Anisocarpus rammii Greene, Fl. Fran. 415. 1897.

Stem usually simple below, openly branching above, slender, pilose below, glandular above, 2–6 dm. high. Leaves scattered, essentially entire, linear, strigose to hirsute, 1.5–6(–10) cm. long, 1–4 mm. wide; heads cymose, terminating long naked peduncles; involucre broadly urceolate, 3.8–4.8 mm. high; ray-achenes 2.4–3 mm. long, laterally flattened, ventral angle more or less sharp, slightly concave, the back rounded, with about 12 longitudinal striae per face, the stipe none, the areola on a short, stout, eccentric beak; ray-pappus a microscopic ring of fimbriate rudimentary paleae less than 0.3 mm. long, the disk-pappus equaling or exceeding the disk-flowers, the exposed portions anthocyanous, of 5–7 ciliate flexuous awns.

Moist grassy slopes, Sonoran and Transition Zones; Sierran foothills from Butte County to Calaveras County, California. Type locality: Nevada City, Nevada County. May–July.

6. Madia yosemitàna Parry ex. A. Gray. Yosemite Madia. Fig. 5257.

Madia yosemitana Parry ex A. Gray, Proc. Amer. Acad. **17** : 219. 1882.
Anisocarpus yosemitanus Greene, Fl. Fran. 416. 1897.

Stem usually simple below, di- or trichotomously branching above, very slender, 6–15(–24) cm. high; habit as in *M. rammii*; involucre broadly turbinate, 3.2–4.2 mm. high; ray-achenes 2.7–3.8 mm. long, the beak slender, otherwise these and the pappus similar to *M. rammii*.

Grassy places, Arid Transition Zone; Sierra Nevada from Tuolumne County to Tulare County and near Fallen Leaf Lake, Eldorado County, California. Type locality: "in damp moss at the foot of the Upper Yosemite Fall." May–July.

7. Madia radiàta Kell. Golden Madia. Fig. 5258.

Madia radiata Kell. Proc. Calif. Acad. **4** : 190. 1870.
Anisocarpus radiatus Greene, Fl. Fran. 416. 1897.

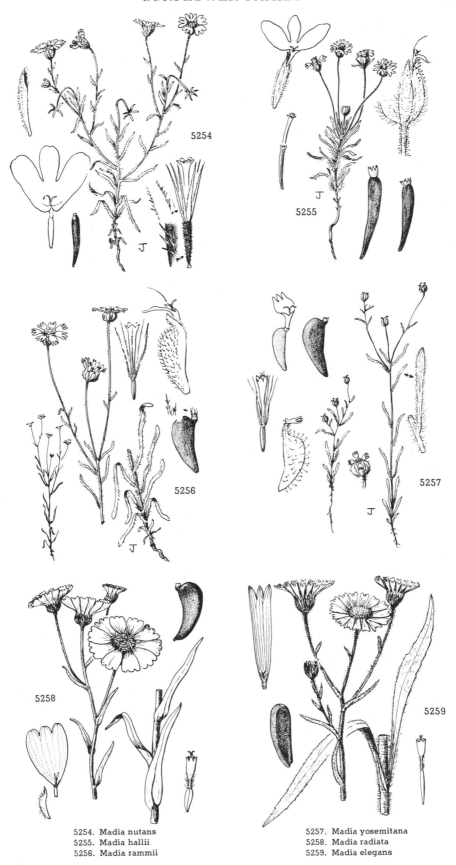

5254. Madia nutans
5255. Madia hallii
5256. Madia rammii

5257. Madia yosemitana
5258. Madia radiata
5259. Madia elegans

Stem commonly branched above or throughout, 1.5–9 dm. high, fistulous, yellowish, glandular-pubescent. Leaves linear-lanceolate, acuminate, glandular and often hirsute, the lower 4–10 cm. long, 4–15 mm. wide, the upper gradually reduced in size and broadest at the base; heads many, in an irregular compound cyme; involucre depressed-hemispheric, 4.5–6.5 mm. high, bearing glands on pustulate stipes and additionally crisped-pubescent or pilose; ray-flowers 8–16, the obovate ligules 6–16 mm. long; disk-flowers 20–65, the anthers yellow; ray-achenes 3.3–4.2 mm. long, flattened, semilunar; disk-achenes 3.2–4.4 mm. long.

Grassy foothills, Upper Sonoran Zone; Inner South Coast Range from eastern Contra Costa County to western Kern County, California. Type locality: mouth of the San Joaquin River. March–May.

8. Madia élegans D. Don ex Lindl. Common Madia. Fig. 5259.

Madia elegans D. Don ex Lindl. Bot. Reg. **17**: *pl. 1458*. 1831.
Madaria elegans DC. Prod. **5**: 692. 1836.
Madaria corymbosa β? hispida DC. loc. cit.
Madaria racemosa Nutt. Trans. Amer. Phil. Soc. II. **7**: 386. 1841.
Madaria elegans var. *depauperata* A. Gray, Proc. Amer. Acad. **7**: 361. 1868.
Madia hispida Greene, Pittonia **2**: 217. 1891.
Madia polycarpa Greene, op. cit. **3**: 167. 1897.
Madaria polycarpa Greene, Fl. Fran. 418. 1897.
Madia villosa Eastw. Proc. Calif. Acad. III. **2**: 293. 1902.
Madia elegans var. *hispida* H. M. Hall, Univ. Calif. Pub. Bot. **3**: 147. 1907.

Stem commonly corymbosely branching above, 2–8 dm. high, villous below, inconspicuously to densely glandular above. Leaves linear to broadly lanceolate, the basal few or in a small rosette, the lower cauline often crowded, the upper well spaced; involucre campanulate to hemispheric, to 10 mm. high, hirsute and stipitate-glandular (sometimes hispid-hirsute and eglandular), the attenuate tips of the phyllaries equaling the basal portion; ray-flowers 8–16, the ligules 6–15 mm. long, yellow or with maroon blotch at base; disk-flowers 25 or more, yellow or maroon; anthers purple-black.

Dry slopes, meadow borders, Transition and Canadian Zones; northern Oregon to Lower California, largely west of the Cascade-Sierran axis. The summer-flowering montane ecotype. Type locality: "North-west coast of North America." June–Aug.

Madia elegans subsp. **vernális** Keck, Aliso **4**: 108. 1958. Stem simple to openly branching throughout, 3–8 dm. high, the herbage sparingly pubescent to densely hispid or pilose, sparingly glandular below the inflorescence; basal rosette scarcely developed, cauline leaves scattered; heads large, remaining open longer through the day than the other subspecies; spring-flowering; flowers all yellow, rarely with maroon blotch at base of ligules; anthers black. Valley floors and foothills, Upper Sonoran and Transition Zones; northern Oregon south to Kern and San Luis Obispo Counties, California. The spring-flowering lowland ecotype. Type locality: east of Clarksville, Eldorado County, California. March–June.

Madia elegans subsp. **densifòlia** (Greene) Keck, Madroño **3**: 4. 1935. (*Madaria corymbosa* DC. Prod. **5**: 692. 1836; *Madia corymbosa* Greene, Pittonia **2**: 218. 1891; *M. densifolia* Greene, Fl. Fran. 417. 1897; *M. elegans* var. *densifolia* Jepson, Fl. W. Mid. Calif. 528. 1901.) Stem stout, usually branching well above the middle to form a corymbose panicle, to 25 dm. high, usually strongly glandular-pubescent above; basal leaves forming a large rosette, the lower cauline closely imbricated, densely villous or hirsute, the upper scattered, strongly glandular and pubescent; involucre broad, to 12 mm. high, the tips of the phyllaries often exceeding the basal portion; ray-flowers 12–20, the ligules 10–20 mm. long, all yellow or usually with maroon blotch at base; disk-flowers always yellow; anthers purple-black. Valley floors and foothills, Upper Sonoran and Transition Zones; western Oregon to San Diego County, California. The fall-flowering lowland ecotype. Type locality: "Western California." Aug.–Nov.

Madia elegans subsp. **wheèleri** (A. Gray) Keck, Aliso **4**: 108. 1958. (*Hemizonia wheeleri* A. Gray, Bot. Calif. **1**: 617. 1880; *Madia tenella* Greene, Pittonia **3**: 167. 1897; *M. wheeleri* Keck, Madroño **3**: 4. 1935.) Stem slender, simple or commonly divaricately branching from the middle, 1–4.5 dm. high, moderately villous, sparingly glandular; basal rosette small or none, cauline leaves scattered; heads small, scattered and solitary on leafy peduncles or in leafy cymules, not closing; involucre 4.5–5.5 mm. high, the phyllaries three-fourths enclosing ray-achenes, villous, inconspicuously yellow-glandular, the tip shorter than the basal portion; ligules 4–5 mm. long, flabelliform, yellow; disk-flowers yellow; anthers yellow, with rounded ovate appendages; ray-achenes much less laterally compressed, triquetrous. Gravelly mountainsides. Arid Transition Zone to Hudsonian Zone; southern Sierra Nevada and Mount Pinos to San Jacinto Mountains, California; also Sierra Juarez, Lower California. Type locality: Monache Meadows, Tulare County. June–Aug.

9. Madia citriodòra Greene. Lemon-scented Madia. Fig. 5260.

Madia citriodora Greene, Bul. Torrey Club **9**: 63. 1882.
Hemizonia citriodora A. Gray, Syn. Fl. N. Amer. **1²**: 307. 1884.

Stem usually simple below, corymbosely or subumbellately branching above, rather slender, 2–7 dm. high, villous-hirsute, especially below, with black stipitate glands above. Leaves not crowded, linear, the lower 4–9 cm. long, 3–7 mm. wide; heads in a corymbose panicle; involucre broadly turbinate to hemispheric-urceolate, 6–8 mm. high, villous-hirsute, scarcely glandular; ray-flowers 5–12; disk-flowers 15–50; anthers black; ray-achenes 3.3–4.4 mm. long, the lateral angles rounded.

Dry open slopes, Upper Sonoran and Arid Transition Zones; central Washington and Oregon south to Amador and Napa Counties, California. Type locality: hills about Yreka, Siskiyou County, California. May–July.

10. Madia anómala Greene. Plump-seeded Madia. Fig. 5261.

Madia anomala Greene, Bull. Calif. Acad. **1**: 91. 1885.
Hemizonia anomala A. Gray, Syn. Fl. N. Amer. ed. 2. **1²**: 451. 1886.
Madia dissitiflora var. *anomala* Jepson, Fl. W. Mid. Calif. ed. 2. 441. 1911.

Stem slender, usually flexuously branching from the middle, 2–5 dm. high, glandular only in upper half; herbage rather anthocyanous. Leaves not very crowded, villous, the upper ones somewhat glandular, linear, up to 7 cm. long and 7 mm. wide; heads paniculate to subracemose, not congested; involucre globose, about 6–7 mm. high, the phyllaries dilated but usually entirely enclosing the achene, thickly beset with stout, gland-tipped hairs and viscid-puberulent; bracts of the receptacle united only toward base, ciliate; ray-flowers 3–7; disk-flowers 3–6; ray- and disk-achenes similar.

Grassy slopes at low elevations, Upper Sonoran and Transition Zones; North Coast Ranges from Humboldt County to Mount Diablo, Contra Costa County, California; also Sacramento County. Type locality: probably Sacramento County. May–June.

11. **Madia subspicàta** Keck. Slender Tarweed. Fig. 5262.

Madia subspicata Keck, Carnegie Inst. Wash. Pub. No. 564: 45. 1945.

Stem strict, slender, usually simple, or sparingly short-branching above, 1–1.5 dm. high, the herbage yellow-green, pilose and viscid-puberulent with prominent stipitate glands. Leaves linear, 2–7 cm. long, to 3 mm. wide; heads subspicate on very short peduncles, overtopped by the subtending leaves; involucre ovate, 6–7 mm. high; disk-flowers 5–15, the anthers black; ray-achenes about 3 mm. long, minutely striate, purple-spotted, the disk-achenes similar.

Grassy slopes, Upper Sonoran Zone; Sierra foothills from Butte County to Mariposa County, California. Type locality: near Knights Ferry, Stanislaus County. May–June.

12. **Madia grácilis** (Smith) Keck. Gum-weed or Slender Tarweed. Fig. 5263.

Sclerocarpus gracilis Smith in Rees, Cycl. **31** : 1815.
Madorella dissitiflora Nutt. Trans. Amer. Phil. Soc. II. **7**: 387. 1841.
Madorella racemosa Nutt. loc. cit.

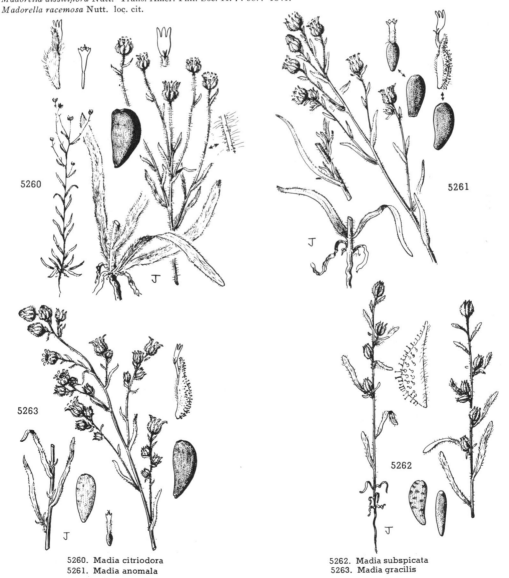

5260. Madia citriodora
5261. Madia anomala

5262. Madia subspicata
5263. Madia gracilis

Madia dissitiflora Torr. & Gray, Fl. N. Amer. **2**: 405. 1843.
Madia racemosa Torr. & Gray, loc. cit.
Madia sativa var. *dissitiflora* A. Gray, Proc. Amer. Acad. **9**: 189. 1874.
Madia sativa var. *racemosa* A. Gray, loc. cit.
Lagophylla hillmani A. Nels. Proc. Biol. Soc. Wash. **17**: 98. 1904.
Madia gracilis Keck, Madroño **5**: 169. 1940.
Madia sativa subsp. *dissitiflora* Keck, Madroño **3**: 4. 1935.

Stem usually slender, simple or flexuously branching from the middle, the branches not over-topping the main stem, 1–10 dm. high; herbage not as viscid as *M. sativa*, resinously fragrant. Leaves not very crowded even at base, mostly linear, sessile by a narrow base, to 10 cm. long and 5 mm. wide; heads paniculate to racemose, not congested, the leafy bracts rarely prominent; involucre ovoid to depressed-globose, 6–9 mm. high; phyllaries thickly beset with stout, gland-tipped hairs, the acuminate tips short; ray-flowers 8–12; ligules 3–8 mm. long; disk-flowers 15–35; anthers included, black; ray-achenes 2.8–5 mm. long, gibbously obovate, often mottled; disk-achenes similar but straighter.

Open or wooded areas, sometimes in disturbed soil, Sonoran Zones to Canadian Zone; British Columbia to Lower California east to Montana and Utah; Chile and adjacent Argentina. Type locality: west coast of North America. April–Aug.

Madia gracilis subsp. **collina** Keck, Aliso **4**: 108. 1958. More robust, usually branching near the top, the heads spicate or glomerulate, the inflorescence very viscid, the stem not glandular below; heads large, the globose-urceolate involucres 8–10 mm. high, the phyllaries with elongated tips. Upper Sonoran Zone: Sierran foothills of Amador and Calaveras Counties, California. Type locality: Vallecito, Calaveras County. May–June.

Madia gracilis subsp. **pilòsa** Keck, Aliso **4**: 108. 1958. Like the species but moderately pilose throughout, especially on peduncles and involucres, rather sparsely glandular but usually so well down toward the base; leaves few, large, the upper ones rather broad; heads solitary or racemose on elongated peduncles, large, the involucres depressed-globose, contracted above the achenes, the phyllaries somewhat dilated, lightly holding the achenes, the erect tips elongated. Humid Transition Zone; Humboldt County, California. Type locality: Buck Mountain.

13. **Madia citrigrácilis** Keck. Shasta Tarweed. Fig. 5264.

Madia citrigracilis Keck, Carnegie Inst. Wash. Pub. No. 564: 44. 1945.

Stem sparsely branched throughout, the branches strict, slender, often exceeding the main stem, 2.5–5 dm. high, hispid-hirsute below, villous and viscid-puberulent above with prominent stipitate glands. Leaves linear-oblong, often subamplexicaule, the lower 4–8 cm. long, to 6 mm. wide; heads racemose or terminating leafy peduncles; involucre obovate, 6–8 mm. high; phyllaries somewhat hirsute and densely stipitate-glandular, villous-ciliate over the face of the achene; ray-flowers 5(–14); disk-flowers 3–10(–30); ray-achenes about 4 mm. long, broadly lanceolate in cross section.

Forest openings, Arid Transition Zone; Modoc, Shasta, and Lassen Counties, California. Type locality: Burney Spring, Shasta County. July–Aug.

14. **Madia satìva** Molina. Chile Tarweed or Coast Tarweed. Fig. 5265.

Madia sativa Molina, Sagg. Chile ed. 1. 136. 1782.

Stem usually stout, often rigidly branched above, glandular well down toward base, 5–10(–20) dm. high; herbage strongly odorous. Leaves rather crowded, sessile by a broad base, up to 18 cm. long and 12 mm. wide; heads paniculate, racemose or subspicate, often approximate along the branchlets, not foliose-bracted; ray-achenes falcate-oblanceolate; disk-achenes oblanceolate, the sides sometimes 1-nerved.

Grasslands, roadsides, and waste places, Transition and Upper Sonoran Zones; behaving as a ruderal; along the coast from southern Alaska to Los Angeles County, California; Chile; Argentina. Type locality: Chile. May–Oct.

15. **Madia capitàta** Nutt. Headland Tarweed. Fig. 5266.

Madia capitata Nutt. Trans. Amer. Phil. Soc. II. **7**: 386. 1841.
Madia sativa var. *congesta* Torr. & Gray, Fl. N. Amer. **2**: 404. 1843.
Madia sativa subsp. *capitata* Piper, Contr. U.S. Nat. Herb. **11**: 576. 1906.

Stem stout, simple or rigidly corymbosely branched above, glandular to base, more viscid than *M. sativa*, 3–6 dm. high; herbage strongly odorous. Leaves sessile by a broad base; heads in congested spikes or in terminal or lateral glomerules, often with foliose bracts exceeding the heads; ray-achenes narrower and longer than in *M. sativa*, the sides usually 1-nerved.

Low fields and coastal headlands, Transition and Upper Sonoran Zones; sometimes weedy; British Columbia south near the coast to Santa Barbara County, California. Type locality: "Wappatoo Island, at the outlet of the Wahlamet." May–Oct. but usually earlier than *M. sativa*.

16. **Madia glomeràta** Hook. Mountain Tarweed. Fig. 5267.

Madia glomerata Hook. Fl. Bor. Amer. **2**: 24. 1834.
Amida gracilis Nutt. Trans. Amer. Phil. Soc. II. **7**: 390. 1841.
Amida hirsuta Nutt. loc. cit.
Madia ramosa Piper, Bull. Torrey Club **29**: 222. 1902.
Madia glomerata var. *ramosa* Jepson, Man. Fl. Pl. Calif. 1098. 1925.

Stem rigid, very leafy, simple or with assurgent or fastigiate branches, 1.5–8 dm. high, villous to hispid, with yellow stipitate glands above; herbage strongly and unpleasantly odorous. Leaves narrowly linear, often with fascicles in their axils, the lower 3–9 cm. long, 2–7 mm. wide;

heads in dense terminal glomerules of 5–30, or in more open cymes and panicles; involucre narrowly ovoid, 5.5–9 mm. high; ligules inconspicuous, greenish yellow to purplish, 1.5–2.5 mm. long; disk-flowers 1–10; achenes 4–6 mm. long, 5-nerved.

Grassy forest-openings, Transition and Canadian Zones; southern Alaska south to the Sierra Nevada and San Bernardino Mountains, California, and through the Rocky Mountains to South Dakota and New Mexico. Type locality: plains of the Saskatchewan. July–Sept.

17. **Madia exígua** (Smith) A. Gray. Small Tarweed. Fig. 5268.

Sclerocarpus exigua Smith in Rees, Cycl. **31**: 1815.
Harpaecarpus madarioides Nutt. Trans. Amer. Phil. Soc. II. **7**: 389. 1841.
Harpaecarpus exiguus A. Gray in Torr. Bot. Mex. Bound. 101. 1859.
Madia exigua A. Gray, Proc. Amer. Acad. **8**: 391. 1872.
Madia filipes A. Gray, loc. cit.
Madia filipes var. *macrocephala* Suksd. Deuts. Bot. Monatss. **18**: 97. 1900.
Harpaecarpus exiguus var. *macrocephalus* Suksd. loc. cit.
Madia exigua subsp. *macrocephala* Piper, Contr. U.S. Nat. Herb. **11**: 576. 1906.
Harpaecarpus californicus Gandoger, Bull. Soc. Bot. Fr. **65**: 43. 1918.
Harpaecarpus longipes Gandoger, loc. cit.
Harpaecarpus suksdorfii Gandoger, loc. cit.

Stem simple, corymbosely branched above or paniculately branched throughout, very slender, 0.5–3(–4) dm. high, hirsute, prominently stipitate-glandular above; herbage aromatic. Leaves

5264. Madia citrigracilis
5265. Madia sativa

5266. Madia capitata
5267. Madia glomerata

strigose, linear, 1–4 cm. long, 2 mm. or less wide; heads on filiform, divaricate, often elongated peduncles in corymbose panicles; involucre depressed-globose, 2.5–4.8 mm. high; phyllaries early-deciduous with the mature fruit, linear dorsally, lunate in outline laterally, covered with prominent glands on thick, often pustulate stalks; ray-flowers 5–8, the ligule 1 mm. long; ray achenes 1.8–2.8 mm. long, crescentic, stoutly beaked; disk-achenes fertile.

Dry grasslands and woodlands, Sonoran and Transition Zones; southern British Columbia and western Montana to northern Lower California and northern Nevada. Type locality: west coast of North America. May–July.

18. **Madia mínima** (A. Gray) Keck. Hemizonella. Fig. 5269.

Hemizonia minima A. Gray, Proc. Amer. Acad. **6**: 548. 1865.
Hemizonia parvula A. Gray, op. cit. 549.
Hemizonia durandii A. Gray, loc. cit.
Hemizonella minima A. Gray, op. cit. **9**: 189. 1874.
Hemizonella parvula A. Gray, loc. cit.
Hemizonella durandii A. Gray, loc. cit.
Harpaecarpus parvulus Greene, Fl. Fran. 416. 1897.
Harpaecarpus minimus Greene, op. cit. 417.
Hemizonella minima var. *parvula* H. M. Hall, Univ. Calif. Pub. Bot. **3**: 148. 1907.
Melampodium minimum M. E. Jones, Contr. West. Bot. No. 15: 156. 1929.
Melampodium durandii M. E. Jones, loc. cit.
Madia minima Keck, Madroño **10**: 22. 1949.

Stems 1 to several from near the base, divaricately branched to form often a hemispheric plant, villous below, glandular-puberulent above, 2–15 cm. high. Leaves often in little clusters at the nodes, otherwise usually scattered, linear-oblong, 1–2 cm. long; heads solitary or in small terminal glomerules, napiform, 2–3 mm. high; phyllaries loosely appressed, becoming arcuate with the ripening achenes, rounded on the back, their glands tiny, on prominent pustulate processes; ray-achenes incurved, more or less beaked.

Gravelly slopes, usually in coniferous woods, Transition and Canadian Zones; British Columbia and northern Idaho south to San Diego County, California. Type locality: "Dry soil, near Soda Springs, alt. 8,680 feet" (Tuolumne Meadows, Yosemite National Park). May–July.

31. **HEMIZÒNIA** DC. Prod. **5**: 692. 1836.

Annual or perennial herbs or shrubs, usually very glandular and aromatic, mostly fall-flowering xerophytes, very few spring-flowering. Basal leaves variously lobed, rarely subentire; upper leaves and bracts not terminated by open pit-glands. Phyllaries half-enclosing the ray-achenes. Receptacular bracts in a single row (and often more or less united) outside the disk-flowers or scattered. Flowers yellow or white. Ray-achenes beaked (except in Section *Hemizonia*), triquetrous, the odd angle posterior, epappose, fertile. Disk-achenes usually bearing a paleaceous pappus. [Name Greek, *hemi*, half, and *zone*, girdle, the phyllaries but half-enclosing the ray-achenes.]

A genus of 31 species, essentially confined to California and northern Lower California. Type species, *Hemizonia congesta* DC.

Shrubs; leaves small, crowded (Section *Zonamra*).
　　Flocs present in axils of older leaves; phyllaries scarcely keeled; anthers black; insular.　1. *H. clementina*.
　　Flocs absent; phyllaries strongly keeled; anthers yellow; Santa Susana Mountains.　　2. *H. minthornii*.
Herbs; leaves larger.
　　Ray-achenes obviously beaked; ligules yellow; inner receptacular bracts when present not adnate nor deliquescent.
　　　Leaves not spine-tipped; receptacular bracts confined to a row surrounding the outer disk-flowers and united into a cup; pappus of quadrate or oblong paleae (Section *Deinandra*).
　　　　Disk-achenes sterile.
　　　　　Phyllaries not keeled; anthers yellow (except in *H. corymbosa*).
　　　　　　Heads not glomerate.
　　　　　　　Ray-flowers 5; disk-flowers 6; Inner Coast Ranges to southern California.
　　　　　　　　　　　　　　　　　　　　　　　　　　　　　　　　　3. *H. kelloggii*.
　　　　　　　Ray-flowers 8 or more; disk-flowers 10 or more.
　　　　　　　　Leaves pubescent throughout; stems not strongly fistulous.
　　　　　　　　　Ray-flowers 8–12; disk-flowers 10–20; herbage pallid; inland.
　　　　　　　　　　Pappus obvious; radical leaves lobed; herbage villous below; ligules pale yellow; cismontane.　　　　　　　　　　　　4. *H. pallida*.
　　　　　　　　　　Pappus none or vestigial; radical leaves entire or obscurely toothed; herbage hispid-hirsute below; ligules deep yellow; transmontane.　　　　　　　　　　　　　　　　　　　　　5. *H. arida*.
　　　　　　　　　Ray-flowers 18–32; disk-flowers 28–48; herbage bright green; radical leaves pinnatifid or bipinnatifid; coastal.　　　　　　6. *H. corymbosa*.
　　　　　　　Leaves merely ciliolate, mostly entire; stems strongly fistulous; pappus usually none; Inner Coast Range.　　　　　　　　　7. *H. halliana*.
　　　　　　Heads glomerate, small; ray-flowers 5; disk-flowers 6; Mojave Desert.
　　　　　　　　　　　　　　　　　　　　　　　　　　　　　　　　　8. *H. mohavensis*.
　　　　　Phyllaries keeled; anthers black; Coast Ranges.
　　　　　　Ray-flowers 3; disk-flowers 3; herbage hispid and gray-green; plants low and divaricate; Bay Region to southern Monterey County.　　　　　9. *H. lobbii*.

Ray- and disk-flowers more than 3.
  Ray-flowers 5; disk-flowers 6.
    Herbage bearing pustulate hairs, gray-green; plants low, intricately branched; southern Monterey County to San Luis Obispo County.
        10. *H. pentactis.*
    Herbage bearing simple filamentous hairs, yellow-green; plants taller, less intricately branched; southern California.
      Heads paniculate; stems tall, scarcely hispid, flexuous and much branched; San Luis Obispo County to San Diego County.   11. *H. ramosissima.*
      Heads in dense glomerules; stems lower, rather hispid above, with strict divaricate branches; coastal plain to Lower California.
        12. *H. fasciculata.*
  Ray-flowers 8; disk-flowers 13–21; southwestern San Diego County.
        13. *H. conjugens.*
Disk-achenes mostly fertile.
  Radical leaves pinnatifid or bipinnatifid; herbage grayish-hirsute; ray-flowers 8–13, ligules as broad as long; common; central California to Lower California.   14. *H. paniculata.*
  Radical leaves merely lobed; herbage bright green, soft-pubescent; ray-flowers 13–20, ligules half as broad as long; rare; southern San Diego County.   15. *H. floribunda.*
Leaves tipped with a rigid spine or apiculation; receptacular bract subtending each disk-flower free, persistent; pappus none or of very narrow paleae (Section *Centromadia*).
  Pappus none; anthers yellow.
    Receptacular bracts pungent; northern and central California.   16. *H. pungens.*
    Receptacular bracts obtuse or more or less acute, not cuspidate; southern California.
        17. *H. laevis.*
  Pappus present.
    Receptacular bracts fleshy at tip, not long-villous; pappus-paleae 3(–5), sparsely ciliolate; herbage bearing small yellowish glands, if any, of mild odor.
      Anthers yellow; central California.   18. *H. parryi.*
      Anthers black; inflorescence very glandular; southern California.   19. *H. australis.*
    Receptacular bracts long-villous at tip; pappus-paleae 8–12, densely fimbriate at tip; herbage bearing large, stalked, black glands, of rank odor; anthers black.   20. *H. fitchii.*
Ray achenes not obviously beaked; ligules white or yellow; inner receptacular bracts adnate, forming a cell about each disk-flower, deliquescent (Section *Hemizonia*).
  Ligules yellow, dorsally veined with purple.
    Spring-flowering (May–June); leaves prominent at time of flowering, elongated, not crowded into a basal rosette; herbage glandular only above.   21. *H. multicaulis.*
    Fall-flowering (Aug.–Oct.), taller; leaves inconspicuous at time of flowering, very short, crowded into a basal rosette; densely glandular throughout.   22. *H. lutescens.*
  Ligules white, dorsally veined with purple.
    Heads paniculate or terminal; herbage puberulent to sericeous or villous with soft hairs; lower leaves broad.
      Peduncular bracts short, not overtopping the heads.
        Herbage green, merely puberulent; heads scarcely glandular; spring-flowering.
          23. *H. tracyi.*
        Herbage gray or silvery, pubescent; heads obviously glandular; spring- to fall-flowering.
          Inflorescence widely paniculate, the heads scattered; herbage villous and copiously dark-glandular, or sericeous; phyllaries 3.5–6 mm. long; mostly inland, northern and central California.   24. *H. luzulaefolia.*
          Inflorescence corymbosely paniculate, the heads usually glomerate; herbage shaggy and with few pale glands; phyllaries 6–9 mm. long; coastal valleys, north of San Francisco Bay.   25. *H. congesta.*
      Peduncular bracts long, overtopping the heads, calyculate; herbage prominently glandular; Mendocino and Lake Counties.   26. *H. calyculata.*
    Heads spicate, at least on the fully developed side-branches, obviously glandular; herbage pilose; lower leaves narrow; basal rosette not obvious; Oregon to northern California.
        27. *H. clevelandii.*

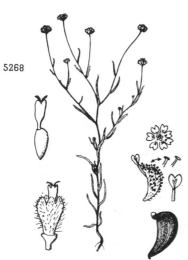

5268

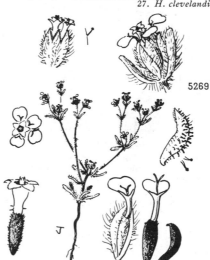

5269

5268. Madia exigua   5269. Madia minima

1. Hemizonia clementìna Brandg. Catalina Tarweed. Fig. 5270.

Hemizonia clementina Brandg. Erythea **7**: 70. 1899.
Zonanthemis clementina Davids. & Moxley, Fl. S. Calif. 401. 1923.

Shrub 3–8 dm. high with many erect, ascending or decumbent branches from base, densely leafy above, the older leaves deciduous; herbage sparsely hirsute, viscid. Lower leaves opposite, narrowly linear, 3–8 cm. long, 1.5–6 mm. wide, remotely dentate; upper leaves alternate, entire, with fascicles or leafy shoots in their axils; heads in compound cymes; involucre broadly campanulate to hemispheric, 5–8 mm. high; ray-flowers 13–14, the ligules yellow, 4.5–6.5 mm. long; receptacular bracts in 2 series; disk-flowers 18–30, the central ones without bracts; ray-achenes 2–3 mm. long, transversely rugose, the prominent recurved beak one and one-half to three times as long as thick; disk-achenes sterile; pappus of 7–10(–15) linear-lanceolate, attenuate, fimbriate, unequal paleae 1–3 mm. long.

In heavy soils, Upper Sonoran Zone; Anacapa, San Nicholas, and Santa Barbara Islands, Ventura County, and on San Clemente and Santa Catalina Islands, Los Angeles County, California. Type locality: San Clemente Island. May–Aug.

2. Hemizonia minthórnii Jepson. Santa Susana Tarweed. Fig. 5271.

Hemizonia minthornii Jepson, Man. Fl. Pl. Calif. 1092. 1925.

Shrub 6–10 dm. high, 10–30 dm. wide, with up to 500 stiff, woody, ascending stems from base; herbage short-hirsute, viscid, fragrantly resinous. Leaves alternate, somewhat thickened, like those of *clementina*; heads mostly solitary on long peduncles, racemosely or corymbosely paniculate; involucre 5.5–6 mm. high, 4–6 mm. broad; ray-flowers 8; the ligules yellow, 5.5–6.5 mm. long; receptacular bracts in about 3 series; disk-flowers 18–23, each subtended by a bract; ray-achenes 2.5–3 mm. long, smooth, the beak scarcely longer than thick; disk-achenes developed but sterile; pappus of 8–12 linear, more or less fimbriate, subequal paleae 1.2–2.6 mm. long.

Dry chaparral slopes, Upper Sonoran Zone; Santa Susana Mountains, Ventura County, California, the type locality. July–Oct.

3. Hemizonia kellóggii Greene. Kellogg's Tarweed. Fig. 5272.

Hemizonia kelloggii Greene, Bull. Torrey Club **10**: 41. (Apr.) 1883.
Hemizonia wrightii A. Gray, Proc. Amer. Acad. **19**: 17. (Oct.) 1883.
Deinandra kelloggii Greene, Fl. Fran. 424. 1897.
Deinandra wrightii Greene, loc. cit.
Hemizonia wrightii var. *kelloggii* Jepson, Man. Fl. Pl. Calif. 1090. 1925.

Stem 2–10 dm. high, corymbosely branching above or intricately branching throughout, soft-villous at base, hispid-hirsute and densely glandular above. Lower leaves narrowly oblong, remotely sharp-toothed or pinnatifid, 3–8 cm. long, 3–10 mm. wide; upper leaves entire; heads pedunculate, in an open panicle or thyrse; involucre 4–5.5 mm. long, 3.5–5 mm. wide, densely stipitate-glandular, usually hirsute; ligules 4–7 mm. long, 3.5–5.5 mm. wide; pappus of 6–12 white, linear to oblong paleae.

Dry plains and hillsides, Upper Sonoran Zone; San Francisco Bay region south through the San Joaquin and Coast Range valleys to San Diego County, California; adventive in Mendocino, Colusa, and Imperial Counties. Type locality: near Antioch, Contra Costa County. April–July.

4. Hemizonia pállida Keck. Kern Tarweed. Fig. 5273.

Hemizonia pallida Keck, Madroño **3**: 8. 1935.

Stem 2–8 dm. high, branching above or throughout with ascending or divergent branches, whitish or reddish; herbage villous-hirsute and hispidulous, lightly glandular and mildly odorous. Lower leaves linear to oblanceolate, remotely sharp-toothed or cleft, 5–10 cm. long, 3–6(–10) mm. wide, the upper ones becoming entire; heads numerous, cymose; involucre 4.5–6.5 mm. high, 5–8 mm. wide, hispid-hirsute and minutely stipitate-glandular; ligules 6–10 mm. long, 3–5 mm. wide; pappus of 4–8 narrow distinct paleae 0.8 mm. or less long.

Dry plains, Upper Sonoran Zone; head of San Joaquin Valley, California. Type locality: 5.3 miles north of Grapevine, Kern County. April–May.

5. Hemizonia árida Keck. Red Rock Tarweed. Fig. 5274.

Hemizonia arida Keck, Aliso **4**: 109. 1958.

Stem 2–4 dm. high, intricately corymbosely branching throughout; herbage hispid-hirsute and hirsutulous, lightly glandular and mildly odorous. Leaves about as in *pallida*; heads numerous, cymose-paniculate; involucre 4 mm. high, 5 mm. wide, hispid-hirsute and stipitate-glandular; ligules 5–6 mm. long, 3–4 mm. wide; pappus vestigial or none.

Dry sandy canyon-bottom, Lower Sonoran Zone; Red Rock Canyon, Kern County, California, the type locality. May–Nov.

6. Hemizonia corymbòsa (DC.) Torr. & Gray. Coast Tarweed. Fig. 5275.

Hemizonia angustifolia DC. Prod. **5**: 692. 1836.
Hartmannia corymbosa DC. op. cit. 694.
Hemizonia corymbosa Torr. & Gray, Fl. N. Amer. **2**: 398. 1843.
Hemizonia decumbens Nutt. Journ. Acad. Phila. II. **1**: 175. 1848.

Hemizonia balsamifera Kell. Proc. Calif. Acad. **2**: 64. 1860.
Zonanthemis angustifolia Greene, Fl. Fran. 425. 1897.
Zonanthemis corymbosa Greene, loc. cit.
Hemizonia corymbosa var. *angustifolia* Jepson, Man. Fl. Pl. Calif. 1091. 1925.

Stem erect or decumbent, 2–10 dm. long, branching from base upward or from above the middle, the branches ascending or spreading; herbage densely villous and glandular, strongly and pleasantly odorous. Lower leaves linear to oblanceolate, deeply pinnatifid, 3–10 cm. long, 3–20 mm. wide, the upper leaves becoming entire in older plants; heads cymose or paniculate; involucre 5–9 mm. high, 6–12 mm. wide; ligules 4–7 mm. long, 2.5–4 mm. wide; pappus of narrow, distinct, unequal, entire or laciniate paleae mostly 1 mm. or less long, or wanting.

Grassy slopes, Humid Transition Zone; along the coast, Mendocino County to Monterey County, California. Type locality: California. Collected by Douglas. May–Oct.

Hemizonia corymbosa subsp. **macrocéphala** (Nutt.) Keck, Aliso **4**: 109. 1958. (*Hemizonia macrocephala* Nutt. Journ. Acad. Phila. II. 1: 175. 1848; *H. angustifolia* subsp. *macrocephala* Keck, Madroño **3**: 12. 1935.) Stouter, denser; heads densely glomerate at the apices of the branches, cymose, larger; ray-flowers 20–35; disk-flowers 40–70. Coastal southern Monterey and northern San Luis Obispo Counties, California. Type locality: "St. Simeon," San Luis Obispo County.

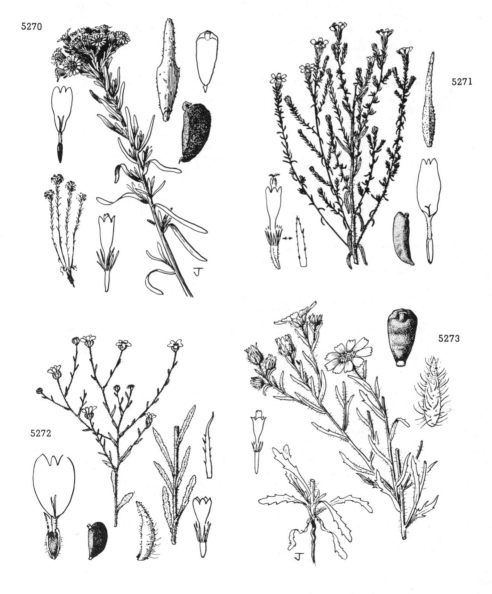

5270. Hemizonia clementina
5271. Hemizonia minthornii

5272. Hemizonia kelloggii
5273. Hemizonia pallida

7. **Hemizonia halliàna** Keck. Hall's Tarweed. Fig. 5276.

Hemizonia halliana Keck, Madroño **3**: 12. 1935.

Stem 2–12 dm. high, up to 16 mm. thick, simple and glabrous below, corymbosely branching (only near summit or to below the middle) and becoming pilose and viscid-puberulent above; herbage pungently odorous. Leaves linear-lanceolate, sessile, glabrous except for the scabrous margin, the lower remotely short-dentate, 5–8 cm. long, 3–9 mm. wide; involucre 5–7 mm. high, 8–10 mm. wide; ray-flowers 10–14, the ligule 5.5 mm. long; disk-flowers 30–60; ray-achenes 3.5–4 mm. long; disk-achenes rarely bearing a rudimentary paleaceous pappus.

Heavy adobe or serpentine soils, Upper Sonoran Zone; rare, in the Inner South Coast Range, San Benito County to San Luis Obispo County, California. Type locality: 1.5 miles east of Cholame, San Luis Obispo County, April–May.

8. **Hemizonia mohavénsis** Keck. Mojave Tarweed. Fig. 5277.

Hemizonia mohavensis Keck, Madroño **3**: 9. 1935.

Stem 1.5–3 dm. high, subsimple or divaricately branched above; herbage very soft-pubescent and viscid, pleasantly odorous. Lower leaves oblanceolate, subentire, the upper leaves oblong-lanceolate, entire, amplexicaul, much reduced toward the expanded-corymbose inflorescence; heads sessile in glomerules at the ends of the branches; involucre 5–6 mm. high, 5 mm. wide; ligules 5–6 mm. long; pappus of 6–8 quadrate, more or less connate paleae 0.5 mm. long.

Dry ground, Lower and Upper Sonoran Zones; known from but two stations in California, Mojave River at its confluence with Deep Creek, Mojave Desert (the type locality), and Banning-Idyllwild road, Mount San Jacinto, Riverside County. July–Sept.

9. **Hemizonia lóbbii** Greene. Three-rayed Tarweed. Fig. 5278.

Hemizonia lobbii Greene, Bull. Torrey Club **9**: 109. 1882.
Hemizonia fasciculata var. *lobbii* A. Gray, Syn. Fl. N. Amer. 1²: 310. 1884.
Deinandra lobbii Greene, Fl. Fran. 425. 1897.

Stem erect, slender, 2–5 dm. high, rigidly branched above or throughout, the branches divaricate and again very freely branching toward tips; herbage glabrate to sparingly hispid-hirsute, often beset with sessile glands above, mildly odorous. Leaves small, linear and entire at anthesis, those of peduncles often crowded, the deciduous lower leaves laciniately sharp-toothed or pinnatisect; heads very numerous, subsessile to short-peduncled; involucre 4–6 mm. long, 2.5–4 mm. wide, beset with sessile glands and sometimes more or less hispid with pustulate hairs; ray- and disk-flowers 3 each; pappus of 6-8 lanceolate or oblong, entire or toothed paleae equaling the corolla-tube.

Dry, barren, interior hills, Upper Sonoran Zone; Solano County to southern Monterey County, California. Type locality: "probably near Monterey" but more likely in the Salinas Valley. May–Nov.

10. **Hemizonia pentáctis** (Keck) Keck. Salinas River Tarweed. Fig. 5279.

Hemizonia lobbii subsp. *pentactis* Keck, Madroño **3**: 8. 1935.
Hemizonia pentactis Keck, Aliso **4**: 109. 1958.

Having the low, intricately branched habit, grayish herbage, and pinnatisect basal leaves of *lobbii*, the stems and heads more definitely pustulate-hirsute and hispid, the leaves whiter-pilose; ray-flowers 5; disk-flowers 6; otherwise similar to *lobbii*.

Dry barren hills, Upper Sonoran Zone; southern Monterey and San Luis Obispo Counties and introduced about Stanford University, California. Type locality: San Miguel, San Luis Obispo County. May–Oct.

11. **Hemizonia ramosíssima** Benth. Slender Tarweed. Fig. 5280.

Hemizonia ramosissima Benth. Bot. Sulph. 30. 1844.
Hemizonia fasciculata var. *ramosissima* A. Gray, Syn. Fl. N. Amer. 1²: 310. 1884.
Deinandra fasciculata var. *ramosissima* Davids. & Moxley, Fl. S. Calif. 401. 1923.
Hemizonia fasciculata subsp. *ramosissima* Keck, Madroño **3**: 8. 1935.

Stem erect, 2–10 dm. high, simple below and corymbosely branching above or with ascending branches from the base upward; herbage moderately hirsuite to glabrate, the glands (most frequent on involucres) sessile, yellow. Lower leaves (mostly missing at anthesis) linear-oblanceolate, remotely dentate, 3–15 cm. long, 3–30 mm. wide, upper leaves becoming entire, small, and bract-like; heads pedunculate, numerous, more or less remote and solitary, sometimes in twos at the ends of branches; ray-flowers 5; disk-flowers 6; pappus as in *lobbii*.

Dry fields and hillsides, Upper Sonoran Zone; Santa Barbara County and the Channel Islands to Orange County, largely replaced by the next in San Diego County, California, and Lower California. Type locality: probably San Pedro, Los Angeles County, California. May–Aug.

12. **Hemizonia fasciculàta** (DC.) Torr. & Gray. Fascicled Tarweed. Fig. 5281.

Hartmannia fasciculata DC. Prod. **5**: 693. 1836.
Hartmannia glomerata Nutt. Trans. Amer. Phil. Soc. II. **7**: 391. 1841.
Hemizonia fasciculata Torr. & Gray, Fl. N. Amer. **2**: 397. 1843.
Hemizonia glomerata A. Gray, Syn. Fl. N. Amer. 1²: 309. 1884.
Deinandra fasciculata Greene, Fl. Fran. 424. 1897.
Deinandra simplex Elmer, Bot. Gaz. **39**: 48. 1905.

Stem 1–10 dm. high, mostly branching from above the middle, the branches rigid, sharply ascending and with comparatively few twigs; heads subsessile in glomerules of 3 to many, the

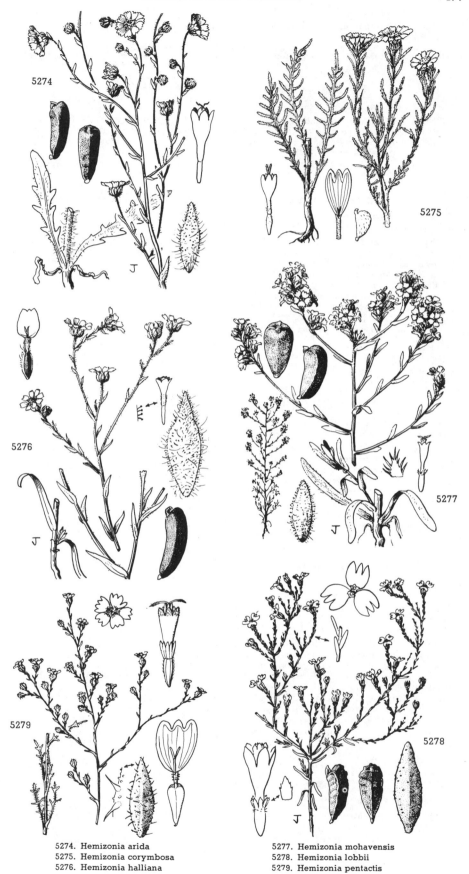

5274. Hemizonia arida
5275. Hemizonia corymbosa
5276. Hemizonia halliana

5277. Hemizonia mohavensis
5278. Hemizonia lobbii
5279. Hemizonia pentactis

glomerules terminating short leafy branches and sometimes a few solitary heads below; otherwise similar to *ramosissima*.

Dry coastal plains, Upper Sonoran Zone; frequent in California from San Luis Obispo County to Riverside County, abundant thence to Lower California. Type locality: "Nova-California." Collected by Douglas, probably near Santa Barbara. May–Sept.

13. **Hemizonia cónjugens** Keck. Otay Tarweed. Fig. 5282.

Hemizonia conjugens Keck, Aliso **4**: 109. 1958.

Stem 2–5 dm. high, branching as in *fasciculata*; foliage and involucres soft-hirsute and sometimes hispidulous, especially the involucre bearing large, flat and small, capitate, sessile or subsessile glands; heads solitary on short peduncles or subsessile in few-headed glomerules; ray-flowers 8–10; disk-flowers 13–21.

Dry mesas, Upper Sonoran Zone; southwestern San Diego County, California. Type locality: river bottom near Otay. May–June.

14. **Hemizonia paniculàta** A. Gray. San Diego Tarweed. Fig. 5283.

Hemizonia paniculata A. Gray, Proc. Amer. Acad. **19**: 17. 1883.
Deinandra paniculata Davids. & Moxley, Fl. S. Calif. 401. 1923.

Stem 2–8 dm. high, the central shaft displaced about midway by numerous ascending branches which ramify to form a twiggy inflorescence; herbage moderately hispid-hirsute (especially below) and stipitate-glandular (especially above), fragrant. Lower leaves persistent until anthesis, linear to oblanceolate, sharp-toothed or pinnatifid with entire or dentate lobes, 1.5–8 cm. long, up to 25 mm. wide; heads obviously peduncled; involucre 5–7 mm. high, 5–7 mm. wide, the phyllaries densely glandular-pubescent; ray-flowers 8; disk-flowers 8–13; disk-achenes sparsely pubescent, some fertile and dark-colored, the majority sterile and pale; pappus of 6–12 white, oblong, fimbriate paleae.

Dry hills and mesas, Upper Sonoran Zone; western Riverside County to San Diego County, California, and northern Lower California. Type locality: near Temecula, Riverside County (lectotype). May–Nov.

Hemizonia paniculata subsp. **incrèscens** Hall ex Keck, Madroño **3**: 11. 1935. Plants erect or somewhat spreading, deep green, even the ultimate twigs rather rigid (never capillary); ray-flowers 8–13; disk-flowers 14–30. Abundant in fields near the coast, Upper Sonoran Zone; Monterey County to Santa Barbara County and on Santa Rosa Island, Santa Barbara County, California. Type locality: 7.5 miles southwest of Arroyo Grande, San Luis Obispo County. May–Nov.

15. **Hemizonia floribúnda** A. Gray. Tecate Tarweed. Fig. 5284.

Hemizonia floribunda A. Gray, Proc. Amer. Acad. **11**: 79. 1876.
Deinandra floribunda Davids. & Moxley, Fl. S. Calif. 401. 1923.

Stem 3–10 dm. high, corymbosely branched above or throughout with strict ascending branches; herbage moderately pilose and densely stipitate-glandular throughout, mildly odorous. Leaves mostly entire, the lower oblanceolate, the middle cauline linear-lanceolate, 1.5–3 cm. long, 2–3.5 mm. wide; heads short-peduncled, in racemosely compound cymes or panicles, not glomerate; involucre 5–6 mm. high, 6–7.5 mm. wide, soft-pubescent and very glandular; ray-achenes 13–20; disk-flowers about 28, their achenes all or mostly fertile; pappus of 6–9 oblong or elliptic, closely fimbriate, rufous and flecked paleae.

Dry hills and montane valleys, Upper Sonoran Zone; southern San Diego County, California, and adjacent Lower California. Type locality: California, near the southern boundary, on the Fort Yuma road, 80 miles east of San Diego. Aug.–Oct.

16. **Hemizonia púngens** (Hook. & Arn.) Torr. & Gray. Common Spikeweed. Fig. 5285.

Hartmannia ? pungens Hook. & Arn. Bot. Beechey Suppl. 357. 1838.
Hemizonia pungens Torr. & Gray, Fl. N. Amer. **2**: 399. 1843.
Centromadia pungens Greene, Man. Bay Reg. 196. 1894.
Hemizonia pungens subsp. *interior* Keck, Madroño **3**: 14. 1935.

Stem 1–12 dm. high, divergently and rigidly branched above or throughout, leafy, sparsely to copiously hirsute, not glandular; herbage yellowish green, inodorous. Basal leaves linear-lanceolate in outline, 5–15 cm. long, 10–40 mm. wide, bipinnately parted into segments; upper cauline leaves with fascicles in their axils, mostly entire, spine-tipped, stiffly ciliate and scabrous-puberulent; heads subsessile in upper axils and terminal; involucre hemispheric, 3–6 mm. high, usually overtopped by bracts; phyllaries keeled, pungent, scabrous, persistent; ray- and disk-flowers numerous, the ligules only 2-toothed; ray-achenes angular, often rugose; disk-achenes in part sterile, the fertile ones more or less fusiform.

Dry valley floors, Upper Sonoran Zone; San Joaquin Valley from San Joaquin County to Kern County and Salinas Valley, California; introduced in Los Angeles, San Bernardino, and San Diego Counties and in northern Oregon. Type locality: California. April–Oct.

Hemizonia pungens subsp. **marítima** (Greene) Keck, Aliso **4**: 110. 1958. (*Centromadia maritima* Greene, Man. Bay Reg. 196. 1894.) Plants fairly low, the branches usually divaricate and strict; leaves and bracts not scabrous; heads in glomerules of 5–40 members, large. Salt marshes and valley land around San Francisco Bay south to northern Monterey County and northeast to San Joaquin and Yolo Counties, California. Type locality: Alameda, Alameda County.

Hemizonia pungens subsp. **septentrionàlis** Keck, Aliso **4**: 110. 1958. Plants large, the branches elongated and often lax; leaves and bracts not scabrous; heads scattered, small. Dry ground, Upper Sonoran Zone; Sacramento Valley and Shasta Valley, California; introduced in northern Oregon and Washington. Type locality: Shasta Valley, Siskiyou County.

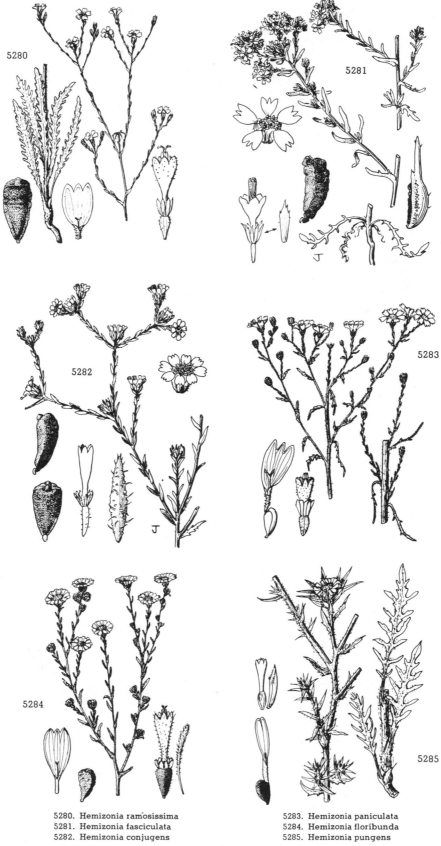

5280. Hemizonia ramosissima
5281. Hemizonia fasciculata
5282. Hemizonia conjugens

5283. Hemizonia paniculata
5284. Hemizonia floribunda
5285. Hemizonia pungens

17. Hemizonia laèvis (Keck) Keck. Smooth Spikeweed. Fig. 5286.

Hemizonia pungens subsp. *laevis* Keck, Madroño **3**: 14. 1935.
Hemizonia laevis Keck, Aliso **4**: 110. 1958.

Similar to *pungens* but of somewhat lower habit. Upper leaves and bracts sparsely setose-ciliate, otherwise very glabrous; heads small, scattered, or approximate in loose glomerules; receptacular bracts obtuse or slightly acute, sometimes minutely but weakly mucronate, not cuspidate.

Dry fields, Upper Sonoran Zone; cismontane southern California away from the immediate coast from Los Angeles County south; also southwestern Kern County. Type locality: San Bernardino Valley. April–Sept.

18. Hemizonia párryi Greene. Pappose Spikeweed. Fig. 5287.

Hemizonia parryi Greene, Bull. Torrey Club **9**: 16. 1882.
Centromadia parryi Greene, Man. Bay Reg. 197. 1894.
Centromadia pungens var. *parryi* Jepson, Fl. W. Mid. Calif. 532. 1901.
Hemizonia pungens var. *parryi* H. M. Hall, Univ. Calif. Pub. Bot. **3**: 155. 1907.

Stem 1–7 dm. high, erect, corymbosely branching above, the branches rigid or lax, or rather prostrate, more or less leafy, sparsely to copiously hirsute. Basal leaves linear- to lance-oblong, 5–20 cm. long, 10–30 mm. wide, pinnatifid to bipinnatisect, the lower cauline more or less hirsute, their lobes more or less spine-tipped, the upper mostly entire, linear-subulate, spine-tipped, with fascicles (incipient shoots or heads) in their axils; heads sessile or short-peduncled along the branches and terminal, or in small clusters; peduncular bracts often exceeding the involucre; involucre hemispheric, 5–10 mm. high, the phyllaries glandular-pubescent, copiously ciliate, the glands subsessile; ray-flowers 9–20, the ligules 3–6 mm. long, yellow, not fading saffron, 2-lobed; receptacular bracts with resinous-thickened, soft, obtuse to acute (not cuspidate) tip; disk-flowers 40–60 or more; disk-achenes in part sterile; pappus of 3–5 linear-subulate, usually fimbriate paleae about equaling the corolla, often united at base and contorted.

Low, often alkaline fields, Upper Sonoran Zone; Colusa and Napa Counties to San Mateo and Monterey Counties, California. Type locality: Calistoga Springs, Napa County. June–Oct.

Hemizonia parryi subsp. **rùdis** (Greene) Keck, Madroño **3**: 15. 1935. (*Centromadia rudis* Greene, Man. Bay Reg. 197. 1894.) Stem erect, corymbosely or diffusely branched, not prostrate, herbage pale green; leaves and bracts scabrid-puberulent; inflorescence obscurely if at all glandular; peduncular bracts not exceeding the involucre; heads small; ray-flowers often fading saffron. Common in low fields, Upper Sonoran Zone; Great Valley of California from Glenn County to Merced County. Type locality: Vacaville, Solano County. June–Oct.

Hemizonia parryi subsp. **congdónii** (Robins. & Greenm.) Keck, Madroño **3**: 15. 1935. (*Hemizonia congdonii* Robins. & Greenm. Bot. Gaz. **22**: 169. 1896; *Centromadia congdonii* C. P. Smith, Muhlenbergia **4**: 73. 1908; *C. pungens* var. *congdonii* Jepson, Man. Fl. Pl. Calif. 1087. 1925.) Stem erect with ascending or horizontal branches, or all branches closely prostrate; herbage yellow-green, sparsely to copiously hirsute but smooth between the long hairs; peduncular bracts exceeding the involucre; ray-flowers not fading saffron. Low ground, Upper Sonoran Zone; San Ramon Valley, Contra Costa County; lower end of San Francisco Bay, mostly Alameda County; and lower Salinas Valley, Monterey County, California. Type locality: Salinas, Monterey County. June–Oct.

19. Hemizonia austràlis (Keck) Keck. Southern Spikeweed. Fig. 5288.

Hemizonia parryi subsp. *australis* Keck, Madroño **3**: 15. 1935.
Hemizonia australis Keck, Aliso **4**: 110. 1958.

Similar to *parryi*; stem erect, the central leader of medium length and with long and lax, divaricate branches, not prostrate, the twiggery above dense; herbage dark green. Leaves and bracts densely glandular-puberulent and villous; peduncular bracts not exceeding the involucre; heads small; phyllaries 4–5.5 mm. long; ray-flowers often fading saffron; anthers black.

Low alkaline fields near the coast, Upper Sonoran Zone; Santa Barbara County to San Diego County, California, and adjacent Lower California. Type locality: Seal Beach, Orange County. June–Sept.

20. Hemizonia fítchii A. Gray. Fitch's Spikeweed. Fig. 5289.

Hemizonia fitchii A. Gray, Pacific R. Rep. **4**: 109. 1857.
Centromadia fitchii Greene, Man. Bay Reg. 197. 1894.

Stem 2–8 dm. high, rigid, erect, above diffusely branched; herbage dark, densely villous and beset with prominent stalked glands, unpleasantly heavy-scented. Leaves crowded, at least in the rosette, very slender, the lower pinnatisect, up to 15 cm. long, the upper entire, rigid, linear-subulate, acerose-tipped, often with fascicles in their axils, the uppermost involucroid, spreading, radiating around the head which they overtop; involucre hemispheric, densely villous, with few prominent glands, the phyllaries rigidly 1-nerved, with subulate tip slightly surpassing the rays; receptacular bracts soft, pointless, hairy, enfolding the disk-flower; ray-flowers 10–20, the short, light yellow ligules bifid; pappus of 8–12 glistening-white, soft, oblong paleae united at base and long-fimbriate above, equaling the disk-corolla.

Dry hills and plains away from the immediate coast, Upper Sonoran Zone; southwestern Oregon to San Luis Obispo County, California; Sierran foothills from Siskiyou County to Fresno County, California; Central Valley and north base of the San Bernardino Mountains, California. Type locality: "Plains of the Sacramento, California." May–Nov.

21. Hemizonia multicaùlis Hook. & Arn. Seaside Tarweed. Fig. 5290.

Hemizonia multicaulis Hook. & Arn. Bot. Beechey Suppl. 355. 1838.

Stem erect, ascending or decumbent, 1–3(–5) dm. high, subsimple or openly branching above or throughout, pilose and glandular-pubescent. Leaves not crowded into a basal rosette, the lower remotely serrulate, linear-lanceolate, 6–15 cm. long, 6–12 mm. wide, deep green; heads corymbose; involucre hemispheric, 5–6.5 mm. high, the tips of the phyllaries about equaling the body; ray-flowers 8–13, showy, clear yellow (sometimes drying greenish).

Seaward bluffs along the immediate coast, Sonoma and Marin Counties, California. Type locality: not given. May–July.

Hemizonia multicaulis subsp. **vernàlis** Keck, Aliso **4**: 110. 1958. (*Hemizonia citrina* Greene, Man. Bay Reg. 194. 1894; *H. luzulaefolia* var. *citrina* Jepson, Fl. W. Mid. Calif. 530. 1901, probably belongs in this taxon.) Stem erect, corymbosely branching above, slender; leaves grass-like, narrower, usually silvery-sericeous. Mostly away from the immediate coast, Sonoma and Marin Counties, California. Type locality: Tiburon, Marin County. April–June.

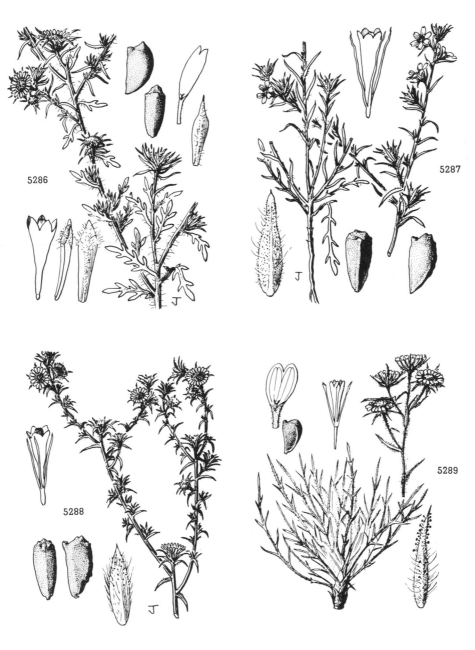

5286. Hemizonia laevis
5287. Hemizonia parryi
5288. Hemizonia australis
5289. Hemizonia fitchii

22. Hemizonia lutéscens (Greene) Keck. Sonoma Tarweed. Fig. 5291.

Hemizonia luzulaefolia var. *lutescens* Greene, Bull. Torrey Club **9**: 16. 1882.
Hemizonia congesta subsp. *lutescens* Babc. & Hall, Univ. Calif. Pub. Bot. **13**: 38. 1924.
Hemizonia congesta var. *lutescens* Jepson, Man. Fl. Pl. Calif. 1089. 1925.
Hemizonia lutescens Keck, Aliso **4**: 110. 1958.

Stem erect, robust, 2–7 dm. high, usually much branched from the base upward; herbage dark-glandular within inflorescence or throughout, heavily scented; basal rosette at first prominent. Leaves short and narrow (up to 12 cm. long and 5 mm. wide), often silvery-villous with long cobwebby hairs; heads small; involucre 4.5–5 mm. high, the tips of the phyllaries shorter than the body; ray-flowers deep yellow.

Grassy valleys and hillsides away from the immediate coast, Upper Sonoran Zone; southern Mendocino County to Marin and Alameda Counties, California. Type locality: San Pablo, Contra Costa County. July–Nov.

23. Hemizonia tràcyi (Babc. & Hall) Keck. Tracy's Tarweed. Fig. 5292.

Hemizonia congesta subsp. *tracyi* Babc. & Hall, Univ. Calif. Pub. Bot. **13**: 46. 1924.
Hemizonia congesta var. *tracyi* Jepson, Man. Fl. Pl. Calif. 1089. 1925.
Hemizonia tracyi Keck, Aliso **4**: 111. 1958.

Stem slender, 1.5–6 dm. high, subsimple or openly few-branched above, somewhat villous or merely densely puberulent; herbage soft-pubescent, glandular only around the heads. Leaves not crowded into a basal rosette, grass-like, very narrow, the lower up to 15 cm. long and 5 mm. wide, bright green; heads corymbose, relatively few; involucre 5–7 mm. high, the tips of the phyllaries shorter than the body; ray-achenes one-half to three-fifths as broad as long.

Bald hills and forest openings, Humid Transition Zone; southern Humboldt and northern Mendocino Counties, California. Type locality: valley of the Van Duzen River opposite Buck Mountain, Humboldt County. May–July.

24. Hemizonia luzulaefòlia DC. Hayfield Tarweed. Fig. 5293.

Hemizonia luzulaefolia DC. Prod. **5**: 692. 1836.
Hemizonia sericea Hook. & Arn. Bot. Beechey Suppl. 356. 1838.
Hemizonia luzulaefolia var. *fragarioides* Kell. Proc. Calif. Acad. **2**: 69. 1860.
Hemizonia grandiflora Abrams, Torreya **2**: 122. 1902.
Hemizonia congesta subsp. *luzulaefolia* Babc. & Hall, Univ. Calif. Pub. Bot. **13**: 43. 1924.
Hemizonia congesta var. *luzulaefolia* Jepson, Man. Fl. Pl. Calif. 1089. 1925.

Stem slender, 2–5 dm. high, openly branching above or throughout, mostly viscid-pubescent well down toward base, often also more or less villous; basal rosette none or obscure. Lower leaves prominent at anthesis, subentire, up to 15 or more cm. long and to 13 mm. wide, green and chiefly puberulent, or silky or silvery with cobweb-tomentum, not glandular; heads showy; involucre 5–6 mm. high, the tips of the phyllaries shorter than the body; ray-achenes about one-half as broad as long.

Open ground, coastal valleys and foothills, Upper Sonoran Zone; Lake County to San Luis Obispo County, California. Type locality: California. April–June.

Hemizonia luzulaefolia subsp. **rùdis** (Benth.) Keck, Aliso **4**: 111. 1958. (*Hemizonia rudis* Benth. Bot. Sulph. 31. 1844.) More robust, 2–8 dm. high, branching and densely glandular-viscid throughout; basal rosette well developed, the shorter and narrower leaves often silvery-villous; heads smaller but very numerous; involucre 3.5–5 mm. high. The most widespread and abundant member of the section, in stubblefields and pastures, Upper Sonoran Zone; Sacramento Valley from Tehama County to San Joaquin County, Coast Range valleys from Lake County to Santa Barbara County, very rare north of Sonoma County, and in Monterey County, California. Type locality: Santa Clara, Santa Clara County. July–Nov.

25. Hemizonia congésta DC. Hayfield Tarweed. Fig. 5294.

Hemizonia congesta DC. Prod. **5**: 692. 1836.

Stem virgate, 1–5 dm. high, with rather few strict or lax branches only above or from the base upward; herbage more or less densely villous throughout, the hairs sometimes appressed but never silky, somewhat viscid toward heads but the glands usually obscure; basal rosette none or obscure. Lower leaves usually prominent at anthesis, remotely denticulate; heads in small terminal clusters or solitary at the ends of the short branches and often a few scattered lateral heads; involucre 6–9 mm. high, the tips of the phyllaries equaling to much exceeding the body; ray-achenes mostly three-fifths as broad as long.

Fields near the coast, Humid Transition Zone; Del Norte County to San Mateo County, California, but frequently only in Sonoma and Marin Counties. Type locality: California. May–Oct.

26. Hemizonia calyculàta (Babc. & Hall) Keck. Mendocino Tarweed. Fig. 5295.

Hemizonia congesta subsp. *calyculata* Babc. & Hall, Univ. Calif. Pub. Bot. **13**: 42. 1924.
Hemizonia congesta var. *calyculata* Jepson, Man. Fl. Pl. Calif. 1089. 1925.
Hemizonia calyculata Keck, Aliso **4**: 111. 1958.

Stem 2–8 dm. high, divaricately branching above; herbage moderately villous and densely glandular-pubescent, the dark stalked glands abundant in the inflorescence. Lower leaves not prominent at anthesis, linear-lanceolate, 4–10 cm. long, 2–5 mm. wide; heads terminating short stiff peduncles, the floral bracts crowded, tending to surpass the heads; involucre 6–12 mm. high, the tips of the phyllaries exceeding the body; ray-achenes one-half to three-fifths as broad as long.

Fields and open hillsides, mostly Upper Sonoran Zone; Lake and southern Mendocino Counties, California. Type locality: east of Orr's Springs, Mendocino County. July–Oct.

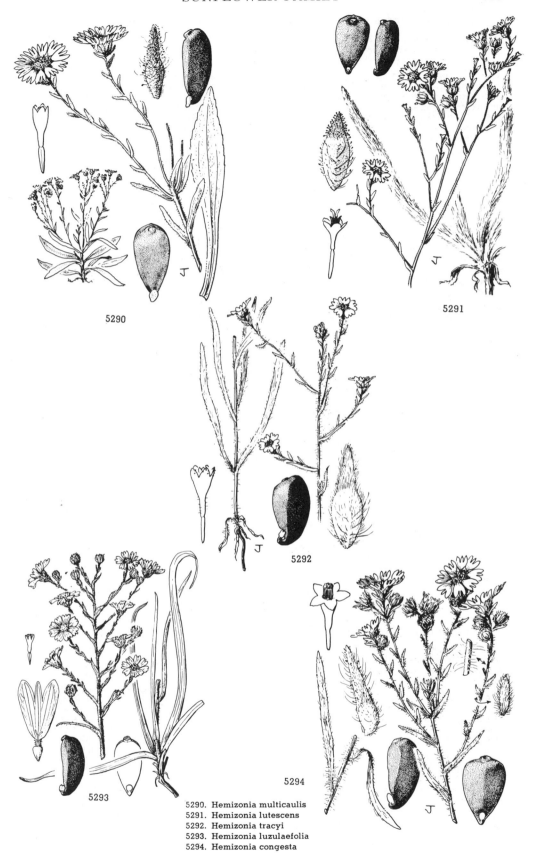

5290

5291

5292

5293

5294

5290. Hemizonia multicaulis
5291. Hemizonia lutescens
5292. Hemizonia tracyi
5293. Hemizonia luzulaefolia
5294. Hemizonia congesta

27. Hemizonia clevelándii Greene. Cleveland's Tarweed. Fig. 5296.

Hemizonia clevelandii Greene, Bull. Torrey Club **9** : 109. 1882.
Hemizonia congesta subsp. *clevelandii* Babc. & Hall, Univ. Calif. Pub. Bot. **13** : 48. 1924.

Stem 2–6 dm. high, simple to divaricately branched above or throughout, or with several ascending branches from base; herbage of early-flowering forms usually densely villous especially toward base of stem and rosette and moderately glandular above, of late-flowering forms moderately villous or merely tomentose but strongly glandular-pubescent especially toward the inflorescence; basal rosette small or none. Basal leaves narrow, to 15 cm. long, to 7 mm. wide; heads terminal on the branches in early forms, also spicate-paniculate in late forms; involucre 4–7 mm. high, the tips of the phyllaries usually shorter than the body; ray-achenes about two-thirds as broad as long.

Dry grassy hills of the interior, Upper Sonoran Zone; southern Oregon south through both Coast Ranges to Sonoma and Napa Counties, California. Type locality: Allen Springs, Lake County, California. June–Oct.

32. HOLOCÁRPHA (DC.) Greene, Fl. Fran. 426. 1897.

Annual herbs, mostly very glandular and aromatic. Leaves linear, the basal remotely serrulate to dentate with slender teeth, the upper reduced to entire bracts usually bearing fascicles or peduncles in their axils, truncated at apex by a crateriform gland. Phyllaries half-enclosing the ray-achenes, bearing on back and at apex stoutly stalked pit-glands. Receptacular bracts subtending each disk-flower, enclosing the achene and corolla-tube, free, persistent. Flowers yellow, the rays 3-lobed. Ray-achenes laterally beaked, obcompressed, triquetrous, the odd angle anterior. Disk-achenes sterile or outermost fertile. Pappus none. [Greek, *holos*, whole, and *karphos*, chaff, the entire receptacle chaffy.]

A genus of 4 species, all endemic to California. Type species, *Holocarpha macradenia* DC.

Anthers black; herbage not puberulent above.
    Disk-flowers 40–90; ray-flowers 8–16; heads very large, densely glomerate; low, with robust divaricate
      branches; rare, near the coast.   1. *H. macradenia.*
    Disk-flowers 10–25; ray-flowers 3–7; heads small, fusiform to subglobose, racemose; tall, with virgate ascending branches; interior.   2. *H. virgata.*
Anthers yellow; medium tall; interior.
    Herbage not densely puberulent above; inflorescence corymbose-paniculate; involucre obconical, the phyllaries
      bearing 5–15(–20) stout, gland-tipped processes; receptacle broad.   3. *H. obconica.*
    Herbage densely puberulent above; inflorescence racemose; involucre subglobose, the phyllaries bearing 25–50
      very slender, gland-tipped processes; receptacle narrow.   4. *H. heermannii.*

1. Holocarpha macradènia (DC.) Greene. Santa Cruz Tarweed. Fig. 5297.

Hemizonia macradenia DC. Prod. **5** : 693. 1836.
Holocarpha macradenia Greene, Fl. Fran. 426. 1897.

Stem 1–5 dm. high, rigid, divaricately branched above the base, the few branches arising near together and usually again similarly branched, rather densely leafy. Basal leaves usually lost before anthesis, broadly linear, up to 12 cm. long and 8 mm. wide including the slender remote teeth; upper leaves becoming bract-like with revolute margin, conspicuously to obscurely glandular-pubescent and often hirsute, the apical gland sometimes occurring only on the uppermost; heads mostly in terminal glomerules; involucre subglobose, 5.5–8 mm. high, the phyllaries bearing about 25 stout, terete, gland-tipped processes; ray-achenes 2.5–3 mm. long.

Colonial in heavy soils in grassy flats along the coast, Humid Transition Zone; Marin, Alameda, and Santa Cruz Counties, California; now possibly extinct. Type locality: California. June–Oct.

2. Holocarpha virgàta (A. Gray) Keck. Virgate Tarweed. Fig. 5298.

Hemizonia virgata A. Gray in Torr. Bot. Mex. Bound. 100. 1859.
Deinandra virgata Greene, Fl. Fran. 425. 1897.
Holocarpha virgata Keck, Aliso **4** : 111. 1958.

Stem 2–12 dm. high, rigid, usually simple below, with few to several strict, sharply ascending branches above; herbage sparingly pilose below and glandular-puberulent, canescent or hispidulous above and strongly resinous and odorous. Basal leaves 6–15 cm. long, 3–10 mm. wide, mostly lost before anthesis, the upper reduced to bracts, often fasciculate, those of the peduncles crowded, heath-like, spreading or recurved, the apex terete, prominently gland-tipped; heads short-peduncled, spicately or racemosely disposed along the virgate branches; involucre 5–6 mm. high, the phyllaries bearing 5–20 stout, terete, ascending, gland-tipped processes, otherwise glabrous; ray-achenes 2.4–3.5 mm. long.

Abundant in hard-baked, valley and foothill soils, Upper Sonoran Zone; Central Valley from Glenn County to Fresno County, the Sierran foothills, and the Inner Coast Ranges (Lake, Napa, Santa Clara, and San Benito Counties), California. Type locality: "California, probably on the Sacramento." Collected by Fremont. June–Nov.

Holocarpha virgata subsp. **elongàta** Keck, Aliso **4** : 111. 1958. Stems slender, 5–12 dm. high, multibranched, the ultimate branchlets gracefully curving, elongated, the heads more scattered on less leafy peduncles up to 15 cm. long. Southwestern San Diego County, California. Type locality: San Diego. July–Nov.

3. **Holocarpha obcónica** (Clausen & Keck) Keck. San Joaquin Tarweed.
Fig. 5299.

Hemizonia obconica Clausen & Keck, Madroño **3**: 7. 1935.
Hemizonia vernalis Keck, loc. cit.
Holocarpha obconica Keck, Aliso **4**: 111. 1958.

Stem 1.5–8(–12) dm. high, the central shaft replaced by several divaricate branches below the middle, the branches paniculately compound to form a dense twiggery; herbage strongly resinous and odorous. Leaves like *virgata* except the basal more hispid and the peduncular often well spaced; heads paniculate on long spreading peduncles; involucre 4–5 mm. high, the phyllaries bearing relatively few gland-tipped processes, the terminal one less prominent than in *virgata*, the surface otherwise glabrous or minutely glandular, rarely more or less hispid but never perubulent, often failing to cover completely the ripe achenes.

In open fields and hillsides, Upper Sonoran Zone; Inner South Coast Ranges from Alameda County to western Fresno County and Sierran foothills of Fresno and Tulare Counties, California. Type locality: Tesla, Alameda County. April–Oct.

Holocarpha obconica subsp. **autumnàlis** Keck, Aliso **4**: 112. 1958. Central shaft quite concealed by short sterile shoots densely clothed with bract-like leaves and progressively shorter from the base upward, so that before anthesis the plants form tapering green columns; principal leaves exceptionally narrow; flowering branches relatively short. On arid plains, Upper Sonoran Zone; north and east of Mount Diablo, Contra Costa County, California. Type locality: west of Byron. Sept.–Nov.

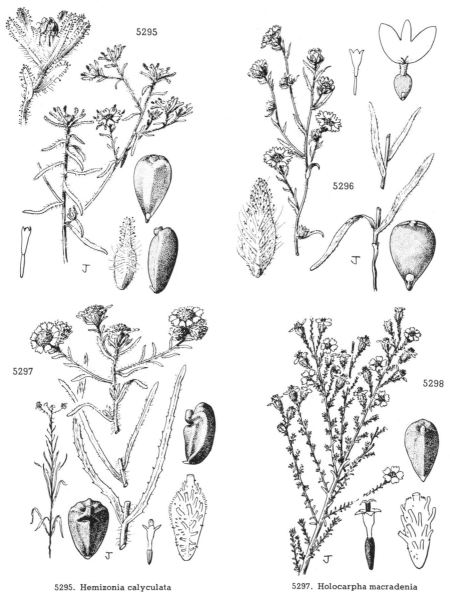

5295. Hemizonia calyculata
5296. Hemizonia clevelandii
5297. Holocarpha macradenia
5298. Holocarpha virgata

4. Holocarpha heermánnii (Greene) Keck. Heermann's Tarweed. Fig. 5300.

Hemizonia heermannii Greene, Bull. Torrey Club **9**: 15. 1882.
Deinandra heermannii Greene, Fl. Fran. 425. 1897.
Hemizonia virgata var. *heermannii* Jepson, Man. Fl. Pl. Calif. 1090. 1925.
Holocarpha heermannii Keck, Aliso **4**: 112. 1958.

Stem 2–12 dm. high, simple below and virgately few-branched above to diffusely branched throughout, pilose or hispid below, cinereous above, with interspersed glandular pubescence throughout; herbage strongly resinous and odorous. Leaves canescent, the upper also densely stipitate-glandular, otherwise like *virgata*, the peduncular bracts usually overlapping each other but not the base of the involucre, appressed to moderately recurved, flattened to apex, 2–5 mm. long; heads paniculate or openly racemose; involucre 5–6 mm. high, puberulent and viscid in addition to bearing many gland-tipped processes.

In hard-baked soils up to 4,000 feet, Upper Sonoran Zone; foothills surrounding the San Joaquin Valley from Contra Costa County to the Mount Hamilton Range and from Calaveras County to Tehachapi Pass and Ventura County, local in Monterey and San Luis Obispo Counties, California. Type locality: Tehachapi Pass, Kern County, near Keene Station. May–Oct.

33. CALYCADÈNIA DC. Prod. **5**: 695. 1836.

Xerophytic annuals with linear to filiform, entire, revolute, grass-like leaves, those at the base of the rigid stem crowded into an erect rosette, more scattered above and often fasciculate, those of the fascicles and the uppermost usually bearing a prominent, tack-shaped or saucer-shaped gland terminally and often similar glands along their margins, these bract-like leaves frequently bearing also pectinate cilia. Ray-flowers few, 1–5(–8), with very broad, palmately, 3-lobed or -parted ligules. Receptacular bracts united into a cup surrounding the few disk-flowers. Ray-achenes with nearly central terminal areola. Disk-achenes likewise fertile, angular, usually bearing a paleaceous pappus. [Greek, *kalux*, cup, and *adenos*, gland, in reference to the cup-shaped glands of the inflorescence.]

A genus of 11 species, nearly confined to California. Type species, *Calycadenia truncata* DC.

Tack-shaped or saucer-shaped glands none; lobes or ray-flowers linear; ray-achenes distinctly beaked; southern California (Section *Osmadenia*).
    One species; pappus-paleae 10, alternating long and short.   1. *C. tenella.*
Tack-shaped or saucer-shaped glands present on bracts; lobes of ray-flowers oblanceolate or broader; ray-achenes not distinctly beaked but areola sometimes elevated; southern Oregon to central California.
    Ray-achenes rugose (Section *Eucalycadenia*).
        Stems glabrous throughout or scabrid toward summit; pappus of about 10–12 broad, blunt, fimbriate paleae 0.5–1.8 mm. long or none.   2. *C. truncata.*
        Stems softly pubescent, never scabrid; pappus of 5 long-attenuate paleae 4–5 mm. long, sometimes alternating with 1–3 short blunt ones.   3. *C. mollis.*
    Ray-achenes smooth; stems pubescent, at least above; herbage scabrous.
        Bracts subtending the heads nerved quite to apex, i.e., more or less flattened throughout, the rounded or truncate apex not impressed, irregularly shorter or longer than the involucre.
            Tack-shaped glands confined to the one terminating most of the peduncular bracts; ray-achenes villous.   4. *C. villosa.*
            Tack-shaped glands more numerous and on the heads (see *C. fremontii*); ray-achenes glabrous to moderately hairy.
                Stem simple or openly branched; heads medium large, crowded and often glomerate; ray-flowers 2–8.
                    Ligules with very broad lateral lobes; late-flowering.
                      Lobes of ligule equal; bracts much shorter than the involucre, densely and finely hispidulous; pappus much shorter than the disk-achene.   5. *C. ciliosa.*
                      Lobes of ligule unequal, the central one only half as wide as the laterals; bracts equaling or exceeding involucre, coarsely hispid; pappus equaling or exceeding disk-achene.
                        Pappus-paleae 2.7–3.5 mm. long, alternating long and short; peduncular bracts nearly straight, rigidly ascending or erect, rarely longer than involucre; inflorescence more or less anthocyanous.   6. *C. multiglandulosa.*
                        Pappus-paleae 4–6.5 mm. long, all long; peduncular bracts recurved from heads, usually much longer than involucre (except in subsp. *reducta*); herbage not anthocyanous, more finely and densely pubescent.
                            7. *C. hispida.*
                Ligules equally 3-parted to base with linear-oblanceolate divaricate lobes; early-flowering.
                  Leaves all opposite, the uppermost three times as long as the involucres; heads in verticillate glomerules.   8. *C. oppositifolia.*
                  Leaves alternate above, the uppermost but slightly exceeding the involucres; heads racemose.   9. *C. fremontii.*
             Stem densely and often dichotomously branched above; heads small, never glomerate; ray-flowers usually 1, white.   10. *C. pauciflora.*
        Bracts subtending the heads not nerved to apex, the outer third terete, the apex truncate and impressed, strongly imbricated, regularly equaling the involucre.   11. *C. spicata.*

1. Calycadenia tenélla (Nutt.) Torr. & Gray. Osmadenia. Fig. 5301.

Osmadenia tenella Nutt. Trans. Amer. Phil. Soc. II. **7**: 392. 1841.
Calycadenia tenella Torr. & Gray, Fl. N. Amer. **2**: 402. 1843.
Hemizonia tenella A. Gray, Proc. Amer. Acad. **9**: 191. 1874.

Stem 1.5–5 dm. high, divaricately branched, often intricately twiggy, the ultimate branchlets subcapillary, leafy and sparsely villous, becoming somewhat viscid-puberulent within the inflorescence. Leaves 1.5–5 cm. long, scabrous and somewhat hirsute, the floral bracts white-ciliate; heads scattered; involucre 3.5–4.8 mm. high, ovoid; ray-flowers 3–5, the ligules white, occasionally with crimson blotch at center of each narrow lobe, the lateral lobes opposite; disk-flowers 3–10, white, like the rays fading roseate, the lobes medially streaked with crimson; ray-achenes 1.8–2.7 mm. long, the central areola strongly beaked; pappus of 4 or 5 rufous-flecked, aristoform-tipped paleae 3.5 mm. long, alternating with acute paleae 1 mm. long.

In light soils, Upper Sonoran Zone; cismontane California from Los Angeles County south to Lower California. Type locality: "St. Diego, Upper California." May–July.

2. **Calycadenia truncàta** DC. Rosin Weed. Fig. 5302.

Calycadenia truncata DC. Prod. **5**: 695. 1836.
Hemizonia truncata A. Gray, Proc. Amer. Acad. **9**: 192. 1874.

Stem 3–11 dm. high, arcuately few-branched above, reddish, smooth and glabrous, or minutely scabrous toward summit; herbage glaucescent, odorous. Leaves sessile by a widened base, firm, 2–9 cm. long, 1–4.5 mm. wide, minutely scabrous, the lower sparsely ciliate, the upper tipped with a broad, saucer-shaped gland, the floral bracts scabrous and pectinate-ciliate; heads spicately

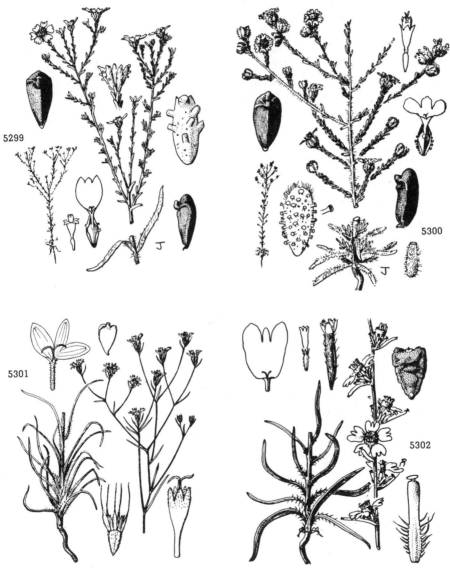

5299. Holocarpha obconica
5300. Holocarpha heermannii

5301. Calycadenia tenella
5302. Calycadenia truncata

scattered and terminal, involucre 4–7 mm. high, campanulate; ray-flowers 3–8, the ligules light to deep yellow, sometimes with small red "eye" at base or suffused with red dorsally, flabelliform-orbicular, 7–12 mm. long; disk-flowers 10–25, bearing short, quadrate, brownish pappus.

Dry sunny slopes, Upper Sonoran Zone; Sierran foothills from Plumas County to Mariposa County, the Inner Coast Ranges from Mendocino County to Santa Clara County, and in the Santa Lucia Mountains, Monterey County, California. Type locality: California. Collected by Douglas. June–Oct.

Calycadenia truncata subsp. scabrélla (E. Drew) Keck, Leaflets West. Bot. 4: 260. 1946. (*Hemizonia scabrella* E. Drew, Bull. Torrey Club 16: 151. 1889; *Calycadenia scabrella* Greene, Fl. Fran. 420. 1897; *Hemizonia speciosa* Congdon, Zoe 5: 135. 1901; *Calycadenia truncata* var. *scabrella* Jepson, Man. Fl. Pl. Calif. 1093. 1925; *Layophylla scabrella* M. E. Jones, Contr. West. Bot. No. 18: 79. 1933.) Stem more slender, usually less branched or even simple, nearly always glabrous throughout; floral bracts smooth, neither glandular nor ciliate; heads smaller, fewer-flowered, terminal, neither scabrous nor glandular; epappose. In dry ground, Upper Sonoran and Arid Transition Zones; Douglas County, Oregon, south to Placer and Lake Counties, California. Type locality: "Hillsides near Grouse Creek," Humboldt County, California. June–Oct.

Calycadenia truncata subsp. microcéphala H. M. Hall ex Keck. Leaflets West. Bot. 4: 259. 1946. Slender habit of subsp. *scabrella*, the branching stem glabrous; bracts smooth or sparsely scabrous, some with a few cilia; heads 5 mm. high, spicate or nearly all terminal, the phyllaries smooth or with a few stiff bristles dorsally; receptacular bracts 3(–6); ray-flowers 3, the ligules 2–3 mm. long; disk-flowers 3–4, epappose. Dry hillsides, Upper Sonoran and Transition Zones; Inner North Coast Range from Trinity County to Lake County and local in the Santa Lucia Mountains, southern Monterey County, California. Type locality: Mill Creek Canyon east of Ukiah, Mendocino County. May–Aug.

3. Calycadenia móllis A. Gray. Soft Calycadenia. Fig. 5303.

Calycadenia mollis A. Gray, Proc. Amer. Acad. 7: 360. 1868.
Hemizonia mollis A. Gray, op. cit. 9: 191. 1874.

Stem 2.5–7 dm. high, rigidly erect, virgate, usually simple or with 2–6 main stems from base and few short divergent branches, arachnoid-villous at base, villous and viscid above; herbage gray-green, the odor acidulous. Leaves 2–7 cm. long, not rigid, appressed-pilose and beset with granular glands, the floral bracts also white-ciliate and bearing 1–10 saucer-shaped glands; heads congested in axillary and terminal glomerules of 5–50; involucre 4–5.5 mm. high, the phyllaries often reduced to 1; ray-flowers 1–3, the ligules deep yellow, white, or deep rose, with or without crimson "eye," 5–6 mm. long, 8–10 mm. wide; disk-flowers 5–7(–10).

Drier borders of meadows, Upper Sonoran and Arid Transition Zones; Sierra Nevada from Tuolumne County to Tulare County, California. Type locality: "At Clark's [Meadows], Mariposa Co.," California. Collected by Bolander. June–Sept. The type was white-flowered. Colonies of it extend over the range of the species. Colonies with golden-yellow flowers (f. *aurea* Keck, Aliso 4: 112. 1958) occur in pure stands from Mariposa County south, most abundantly in Madera County (type locality: Raymond), and colonies with pure rose to deep claret flowers (f. *rosea* Keck, Aliso 4: 112. 1958) are best developed in Tuolumne County, (type locality: Mather).

4. Calycadenia villòsa DC. Dwarf Calycadenia. Fig. 5304.

Calycadenia villosa DC. Prod. 5: 695. 1836.
Hemizonia douglasii A. Gray, Proc. Amer. Acad. 9: 192. 1874.

Stem 1–3 dm. high, commonly with several ascending branches, often somewhat zigzag above, densely leafy at base and moderately so to summit, scabrid-puberulent and more or less hirsute, sometimes viscidulous. Leaves rigid, 2–5 cm. long, 1 mm. wide, scabrous, long-ciliate with stiff bristles; peduncular bracts about as long as and closely investing the involucre, terete toward apex, truncate, tipped by a large, solitary, T-gland, scabrous, densely ciliate and glutinous from pellucid glands; heads 1(–3) at a node along the interrupted spike; involucre 5–6 mm. high; ray-flowers 1–4, the ligues white (or pinkish), 6 mm. long, 8–9 mm. wide; disk-flowers 6–14; pappus of 10–11 linear-lanceolate, awn-pointed paleae 4–6 mm. long.

In hard stony soil of open hillsides, Upper Sonoran Zone; away from the coast, Monterey County to Santa Barbara County, California. Type locality: California. June–Oct.

5. Calycadenia ciliòsa Greene. Klamath Calycadenia. Fig. 5305.

Calycadenia ciliosa Greene, Fl. Fran. 421. 1897.

Stem 2–8 dm. high, usually corymbosely few-branched above, densely leafy toward base, hirsute and viscidulous-puberulent. Leaves rigid, 2–10 cm. long, up to 2 mm. wide, scabrous, sparsely bristly ciliate; peduncular bracts not crowded, tipped with a T-gland and often bearing a subapical pair or so, scabrous, copiously white-ciliate and viscid with pellucid glands; heads spicate; involucre 5–6 mm. high; ray-flowers (1–)3–6, the ligules yellow to white, with or without red "eye" at base, 6–7 mm. long; disk-flowers 7–18; pappus of 5 awn-tipped paleae 2.2–3.2 mm. long alternating with much shorter, blunt, erose paleae, variable.

Stony barren soils, Upper Sonoran and Arid Transition Zones; Rogue River region, Oregon, south to Butte and Lake Counties, California. Type locality: Humboldt County, California. June–Sept. Most frequently found with yellow flowers as originally described, but white-flowered colonies (f. *alba* Keck, Aliso 4: 112. 1958) occur throughout the range of the species (type locality: Burney, Shasta County).

6. Calycadenia multiglandulòsa DC. Sticky Calycadenia. Fig. 5306.

Calycadenia multiglandulosa DC. Prod. 5: 695. 1836.
Hemizonia multiglandulosa A. Gray, Proc. Amer. Acad. 9: 192. 1874.

Stem 2–7 dm. high, strict, usually simple, leafy throughout, glabrous or puberulent below, puberulent to villous and often more or less crisp-pubescent above, scarcely viscid; herbage yellowish green to anthocyanous, pungently odorous. Leaves rigid, 3–9 cm. long, to 2 mm. wide, scabrous, hispid-ciliate toward base; heads spicate or terminally glomerate; involucre 5–6.5 mm. high, prominently glandular; ray-flowers 2–5, the ligules white, often tinged with rose, with or without a red eye-spot at base, 5–7 mm. long, 9–12 mm. wide; disk-flowers 4–12.

A variable species separable into three geographic subspecies, but the type ("in California legit. cl. Douglas") cannot with surety be included in any one of them, so nomenclaturally it stands alone. It probably is closest to subsp. *cephalotes.*

Calycadenia multiglandulosa subsp. **bicolor** (Greene) Keck, Aliso **4**: 113. 1958. (*Calycadenia bicolor* Greene, Fl. Fran. 421. 1897.) Stems slender; upper leaves usually short; heads dark-anthocyanous and black-glandular, usually in spicate racemes well down the stem, sometimes glomerate, especially in Eldorado County, but in this case the glomerules rarely limited to the apex of the stem. In dry stony soils, Upper Sonoran Zone; foothills of the Sierra Nevada from southern Butte County to Tulare County, California. Type locality: "Calaveras County." May–Aug.

Calycadenia multiglandulosa subsp. **cephalòtes** (DC.) Keck, Aliso **4**: 113. 1958. (*Calycadenia cephalotes* DC. Prod. **5**: 695. 1836; *Hemizonia cephalotes* Greene, Bull. Torrey Club **9**: 110. 1882; *H. multiglandulosa* var. *cephalotes* A. Gray, Syn. Fl. N. Amer. 1²: 312. 1884; *Calyacadenia multiglandulosa* var. *cephalotes* Jepson, Man. Fl. Pl. Calif. 1095. 1925.) Stems rather slender, mostly 1–3 dm. high; leaves long, the uppermost usually well exceeding the heads, bright green; heads congested at apex of stem in a terminal glomerule, their bracts anthocyanous apically and black-glandular. Grassy valleys and ridges, Humid Transition Zone; Coast Ranges from southern Mendocino County to Napa and San Mateo Counties, California. Type locality: California. Collected by Douglas. May–Oct.

Calycadenia multiglandulosa subsp. **robústa** Keck, Aliso **4**: 113. 1958. Stems stout, often tall; leaves long, the floral ones tending to exceed the heads and recurve, yellowish green; heads aggregated into apical and axillary or subaxillary glomerules, their bracts yellow-green and bearing honey-colored glands. Dry chaparral openings and open savannah, Upper Sonoran Zone; Outer and Inner Coast Ranges, Santa Clara County, California. Type locality: near Isabel Creek, Mount Hamilton. June–Sept.

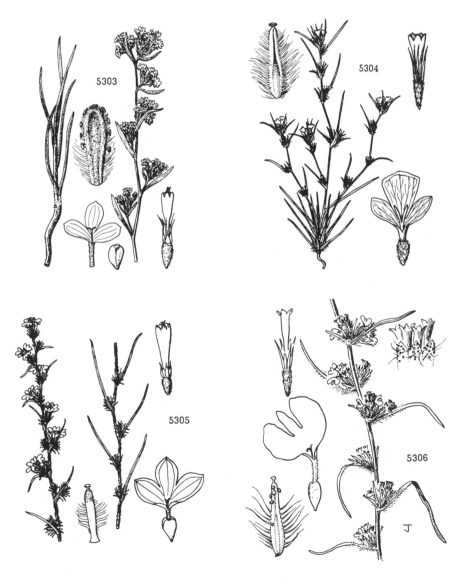

5303. Calycadenia mollis
5304. Calycadenia villosa

5305. Calycadenia ciliosa
5306. Calycadenia multiglandulosa

7. **Calycadenia híspida** (Greene) Greene. Hispid Calycadenia. Fig. 5307.

Hemizonia hispida Greene, Bull. Torrey Club **9**: 63. 1882.
Calycadenia hispida Greene, Fl. Fran. 421. 1897.
Calycadenia campestris Greene, loc. cit.

Stem 2–7 dm. high, usually robust, simple or with several short branches above, leafy throughout and with conspicuous basal rosette, puberulent and long-villous or hispid, usually viscidulous; herbage yellowish green, with fragrant acetic odor. Leaves 2–7 cm. long, mostly 1 mm. wide, considerably reduced in length upward, almost all tipped with a small T-gland, the upper ones also margined with several, viscid with pellucid glands, scabrous, long-ciliate; heads spicate, racemose or paniculate, not glomerate; involucre 6–7 mm. high; ray-flowers 3–8, the ligules deep yellow, pale yellow, or white, without red "eye," 7–8 mm. long, 9–10 mm. wide; disk-flowers 10–20, their corollas 7–9 mm. long.

Dry ground, Upper Sonoran Zone; foothills of the Sierra Nevada and adjacent plains from Placer County to Kern County and on the Diablo Range, southern Monterey County, California. Type locality: near Atwater, Merced County. July–Sept. The type form was yellow; the form with white rays that turn pink in age (*f. albiflora* Keck, Aliso **4**: 113. 1958) occurs on Parkfield Grade, southeastern Monterey County.

Calycadenia hispida subsp. **redúcta** Keck, Aliso **4**: 114. 1958. More slender, without conspicuous rosette; heads exceeding their subtending bracts; ligules short, 1–4, white; disk-flowers 4–8, their corollas 6–6.8 mm. long; pappus 4–4.8 mm. long. Grassy slopes, below 3,000 feet, Upper Sonoran Zone; from Antioch, Contra Costa County, to the east side of Mount Hamilton Range, Santa Clara County, California. Type locality: Colorado Creek, Mount Hamilton Range. July–Sept.

8. **Calycadenia oppositifòlia** (Greene) Greene. Butte Calycadenia. Fig. 5308.

Hemizonia oppositifolia Greene, Bull. Torrey Club **9**: 110. 1882.
Calycadenia oppositifolia Greene, Fl. Fran. 423. 1897.

Stem 1–3 dm. high, simple or corymbosely branched, cinereous-strigose, equally leafy throughout. Leaves opposite, 1–5 cm. long, up to 2 mm. wide, increasing in length up the stem, scabrous, covered with pellucid punctate glands, rigidly ciliate near base; peduncular bracts short, crowded, with terminal and often a few lateral T-glands; heads in verticillastrate glomerules from about midway up stem to apex; involucre 5–6 mm. high; ray-flowers 2–4, the ligules white, fading deep rose, 6–9 mm. long; disk-flowers 6–20; pappus-paleae brownish, alternating awn-pointed to 3 mm. long and shorter blunt ones.

Dry valley floors and hillsides, Upper Sonoran Zone; Butte County, California. Type locality: Chico. April–July.

9. **Calycadenia fremóntii** A. Gray. Fremont's Calycadenia. Fig. 5309.

Calycadenia fremontii A. Gray in Torr. Bot. Mex. Bound. 100. 1859.
Hemizonia fremontii A. Gray, Proc. Amer. Acad. **9**: 191. 1874.
Hemizonia multiglandulosa var. *sparsa* A. Gray, Syn. Fl. N. Amer. **1²**: 312. 1884.

Stem 1.5–2.5 dm. high, usually with several ascending branches above, strigose, not viscid. Leaves alternate above, 2–4 cm. long, up to 2 mm. wide, scabrous, sparsely bristly-ciliate near base; peduncular bracts with or without a T-gland, not otherwise glandular; heads paniculate, few; involucre 5–6 mm. high; ray-flowers 3–6, the ligules white tinged with rose, 8 mm. long; disk-flowers 10–12; pappus-paleae subequal or alternating short and medium long.

Dry plains, Upper Sonoran Zone; known only from the vicinity of Chico, Butte County, California. May.

10. **Calycadenia pauciflòra** A. Gray. Small-flowered Calycadenia. Fig. 5310.

Calycadenia pauciflora A. Gray in Torr. Bot. Mex. Bound. 100. 1859.
Hemizonia pauciflora A. Gray, Proc. Amer. Acad. **9**: 191. 1874.
Calycadenia elegans Greene, Fl. Fran. 423. 1897.
Calycadenia ramulosa Greene, loc. cit.
Calycadenia pauciflora var. *elegans* Jepson, Man. Fl. Pl. Calif. 1093. 1925.

Stem 1.5–5 dm. high, corymbosely branched above or divaricately branched throughout with dichotomously forking twigs ascending in a more or less zigzag manner, villous at base, often glabrate below but always strigose-puberulent above, scarcely at all viscid. Leaves 1–5 cm. long, 1–2 mm. wide, scabrous, sparsely ciliate, the upper often gland-tipped; peduncular bracts more ciliate, with terminal and often several axial T-glands; heads spicately scattered, never clustered, very small; ray-flowers 1 (or 2), the ligules white, fading pink, 4.5–6.5 mm. long, 10–13 mm. wide; disk-flowers 3–6; pappus-paleae unequal, up to 2.3 mm. long.

Exposed, dry, rocky slopes in the foothills, Upper Sonoran Zone; Inner North Coast Range from Glenn County to Sonoma County and on Sierran foothills of Calaveras and Stanislaus Counties, California. Type locality: California. Collected by Fremont. May–Sept.

11. **Calycadenia spicàta** (Greene) Greene. Spicate Calycadenia. Fig. 5311.

Hemizonia spicata Greene, Bull. Torrey Club **9**: 16. 1882.
Calycadenia spicata Greene, Fl. Fran. 422. 1897.

Stem 1–6 dm. high, often unbranched or with several rigidly ascending branches, leafless at base at anthesis, densely leafy-bracteate above, densely puberulent and more or less strigose; herbage gray-green, with strong but pleasant odor. Leaves 2–6 cm. long, 1.5 mm. or less wide; peduncular bracts closely investing and concealing involucre, glutinous from pellucid glands and tipped with a T-gland; ray-flowers 1–5, the ligules white, fading old rose, 6-9 mm. long, 7–9 mm.

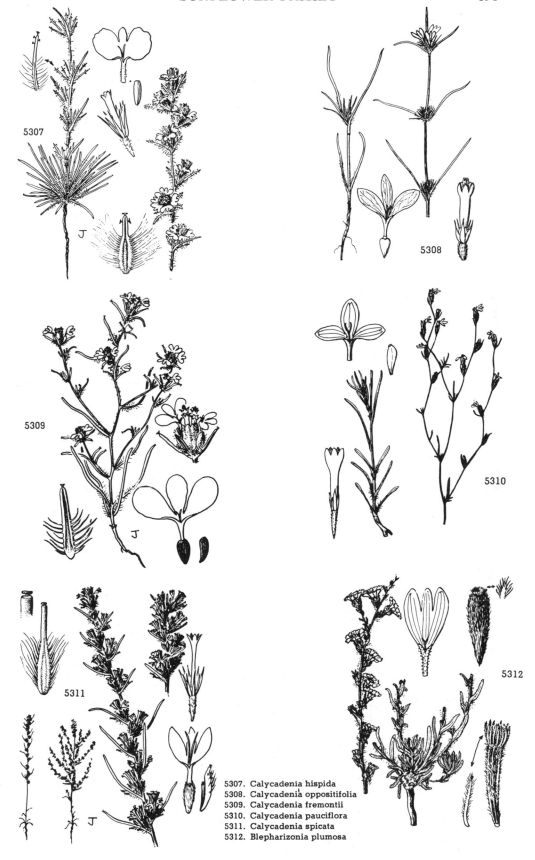

5307. Calycadenia hispida
5308. Calycadenia oppositifolia
5309. Calycadenia fremontii
5310. Calycadenia pauciflora
5311. Calycadenia spicata
5312. Blepharizonia plumosa

wide; disk-flowers 5–9(–12); pappus-paleae rather unequal, all awn-tipped, the longer 4–5.5 mm. long.

Rolling dry hills below 2,000 feet, Upper Sonoran Zone; foothills of the Sierra Nevada from Butte County to Tulare County, California. Type locality: Milton, Calaveras County. June–Sept.

34. BLEPHARIZÒNIA Greene, Bull. Calif. Acad. 1: 279. 1886.

Stout, fall-flowering, xerophytic, odorous annuals with white, densely hirsute stems and pallid herbage. Uppermost leaves copiously covered with prominent, yellow, tack-shaped glands. Heads large, scattered. Phyllaries concave, not surrounding the subtended ray-achenes. Ray-flowers 5–11, white, dorsally veined with rose. Ray- and disk-achenes similar, 10-ribbed, densely rufous-hirtellous, all fertile. Pappus of ray very minute, of disk of 15–20 plumose narrow paleae united at base. [Greek, *blepharis*, eyelash, and *zone*, girdle, from the short plumose pappus-paleae of the typical form.]

A monotypic genus of California.

1. Blepharizonia plumòsa (Kell.) Greene. Big Tarweed. Fig. 5312.

Calycadenia plumosa Kell. Proc. Calif. Acad. 5: 49. 1873.
Hemizonia plumosa A. Gray, Proc. Amer. Acad. 9: 192. 1874.
Hemizonia plumosa var. *subplumosa* A. Gray, Syn. Fl. N. Amer. 1²: 312. 1884.
Blepharizonia laxa Greene, Bull. Calif. Acad. 1: 279. 1886.
Blepharizonia plumosa Greene, loc. cit.
Blepharizonia plumosa var. *subplumosa* Jepson, Man. Fl. Pl. Calif. 1095. 1925.

Stem 3–18 dm. high, paniculately branching above, not rigid; herbage gray-green, densely pilose, sparingly glandular below the heads. Leaves crowded at base, less congested above, the basal linear, 4–15 cm. long, 5–8 mm. wide, sharply incised-serrate to subentire, the cauline entire, broadest at base, mostly with fascicles in their axils, the fascicular leaflets tipped with several prominent T-glands; heads subracemose to loosely paniculate; involucre 5.5–7.5 mm. high, subglobose, canescent, with large T-glands and sometimes also small granular glands; disk-flowers 10–35; pappus half as long as the disk-achenes or longer.

Dry hills and plains, Upper Sonoran Zone; Solano and San Joaquin Counties to Alameda and Stanislaus Counties, California. Type locality: Stockton, San Joaquin County. July–Oct.

Blepharizonia plumosa subsp. **víscida** Keck, Aliso 4: 114. 1958. Herbage yellowish green, moderately to densely hispid and very conspicuously glandular; involucre hispid; pappus reduced to less than 1 mm. in length and frequently obsolete. Dry open ground below 3,000 feet, Upper Sonoran Zone; Inner South Coast Range from San Benito County to Kern County and also in the Santa Lucia Mountains, Monterey County, California. Type locality: 3.8 miles south of Tres Pinos, San Benito County. July–Oct.

35. BLEPHARIPÁPPUS Hook. Fl. Bor. Amer. 1: 316. 1833.

Late, vernal, xerophytic annuals corymbosely branched with alternate, narrowly linear, entire leaves. Heads many-flowered, terminating the branchlets, heterogamous, radiate, all flowers fertile. Involucre 1–2-seriate, the outer phyllaries concave, lightly holding the ray-achenes. Ray-flowers 1-seriate; ligules broad, 3-lobed. Disk-flowers each subtended by a distinct, more or less scarious bract, the style linear, hairy, bearing glabrous, extremely short, truncate branches; anthers black. Achenes all alike and fertile, turbinate, silky-villous; pappus present or absent. [Name from two Greek words meaning eyelash and pappus.]

A monotypic genus of the northwestern United States.

1. Blepharipappus scàber Hook. Blepharipappus. Fig. 5313.

Blepharipappus scaber Hook. Fl. Bor. Amer. 1: 316. 1833.
Ptilonella scabra Nutt. Trans. Amer. Phil. Soc. II. 7: 386. 1841.
Blepharipappus scaber var. *subcalvus* A. Gray, Proc. Amer. Acad. 9: 193. 1874.
Ptilonella scabra var. *subcalva* Greene, Fl. Fran. 433. 1897.

Stem erect, corymbosely branched from about the middle or throughout, sometimes proliferous, 1–3 dm. high, puberulent, becoming glandular upward. Leaves spreading or ascending, up to 3 cm. long, rarely more than 1 mm. wide, rapidly reduced to small bracts on the peduncles, scabrous, margins becoming revolute; involucre turbinate to hemispheric, 4–6 mm. high, glandular-puberulent and somewhat hispid; ray-flowers 3–6, the ligules white with 3 purple dorsal veins; disk-flowers 8–20; pappus of 12–18 linear, hyaline, densely fimbriate scales up to 2 mm. long or frequently very short, rarely lacking.

Arid plains and slopes, Upper Sonoran and Arid Transition Zones; southeastern Washington south through eastern Oregon and adjacent Idaho to Sierra County, California, and northern Nevada. Type locality: "Sandy plains of the Columbia." May–Aug.

Blepharipappus scaber subsp. **laèvis** (A. Gray) Keck, Aliso 4: 114. 1958. (*Blepharipappus scaber* var. *laevis* A. Gray, Bot. Calif. 1: 358. 1876; *B. laevis* A. Gray, Bot. Gaz. 13: 73. 1888; *Ptilonella laevis* Greene, Fl. Fran. 433. 1897.) Branching more subumbellate; leaves all small, becoming very small, closely appressed bracts on the peduncles; peduncles sharply ascending, wiry, nearly filiform, surmounted by unusually small heads. The usual absence of pappus in Oregon material does not hold for California material. Dry ground, often with yellow pine, Transition Zone; west side of the Cascade Mountains from Lane County, Oregon, south to Siskiyou and Scott Mountains, northern California. Type locality: not indicated. Collected by Bridges. June–Aug.

Tribe 2. **HELENIEAE**

Involucres woolly, hairy, or glabrous, often glandular-pubescent or minutely glandular-dotted, lacking conspicuous impressed oil-glands; herbage sometimes aromatic, not ill-smelling.

Phyllaries herbaceous or more or less chartaceous at base, the margins narrowly or not at all scarious, un-
colored; achenes glabrous or pubescent, not conspicuously papillate on both faces (except *Chaenactis
nevii*).

Ray-flowers persistent, becoming papery.
Leaves opposite; receptacle short-villous throughout; plants of the Sierra Nevada.
38. *Whitneya.*
Leaves alternate (basal only, in *Hymenoxys acaulis arizonica*); receptacle naked; plants mostly of
desert areas.
Many-stemmed subshrubs; leaves linear, entire; phyllaries 1-seriate. 36. *Psilostrophe.*
Annuals or perennials; leaves pinnatifid or parted into linear divisions (entire in *Hymenoxys
acaulis arizonica* but plants scapose); phyllaries 2–3-seriate.
Achenes epappose; leaves pinnatifid with broad divisions. 37. *Baileya.*
Achenes with pappus-paleae; leaves with linear divisions or entire. 47. *Hymenoxys.*
Ray-flowers when present deciduous or, if persisting, withering and not becoming papery.
Receptacle naked or if hairs or bracts present these few and inconspicuous.
Phyllaries imbricate, broad and rounded at apex.
Leaves alternate; rays 1.5–2 cm. long; erect perennials of southern California canyons.
39. *Venegasia.*
Leaves opposite; rays scarcely exceeding disk; widespread creeping perennials of saline
habitats. 40. *Jaumea.*
Phyllaries uniseriate, if 2-seriate or more, subequal (somewhat imbricate in *Hulsea*), linear
or linear-lanceolate, acute at apex (broader and obtuse in *Amblyopappus*).
Achenes strongly compressed, the margin fringed with hairs (see also *Crockeria*) or with
a callous or thickened margin.
Phyllaries flat, in 2 subequal series, not prominently ribbed; plants white-floccose.
41. *Eatonella.*
Phyllaries more or less boat-shaped, uniseriate or in 2 subequal series, strongly single
or double-ribbed; plants never white-floccose.
Leaves hastate-triangular and caudate-acuminate, 5–10 cm. long.
44. *Pericome.*
Leaves not hastate-triangular and caudate-acuminate.
Annuals; heads with white rays; lower leaves 5 or more cm. long.
42. *Perityle.*
Perennials; heads rayless; leaves 1–3 cm. long. 43. *Laphamia.*
Achenes terete, linear-clavate, turbinate or obpyramidal, obscurely or sharply angled, often
ribbed, rarely somewhat compressed (see *Pseudobahia*).
Rays and phyllaries reflexed during anthesis; disk large, globose or dome-shaped.
48. *Helenium.*
Rays if present not reflexed during anthesis; phyllaries erect or spreading, some-
times reflexed when achenes are mature; disk small to medium, if large not
globose or markedly dome-shaped.
Heads radiate, the rays sometimes small and inconspicuous (lacking in some desert
annual species of *Eriophyllum*).
Phyllaries 20 or more, somewhat imbricate in 2–3 series, not subtending
the rays. 45. *Hulsea*
Phyllaries 5–13, subequal in 2 series, more or less subtending the ray-
flowers.
Rays obvious or showy; annuals or perennials.
All leaves opposite; spring annuals.
Margins of the achenes naked.
Phyllaries distinct or slightly united at the base.
49. *Baeria.*
Phyllaries partially united, forming a cup.
50. *Lasthenia.*
Margins of the achenes prominently fringed with clavate
glandular hairs. 51. *Crockeria.*
Leaves alternate, sometimes the lower opposite; annuals or peren-
nials.
Herbage greenish; inflorescence with gland-tipped hairs; south-
ern California ranges. 52. *Bahia.*
Herbage more or less white-floccose (leaves sometimes becom-
ing greenish above) or dense white-tomentose.
Perennials or woody-based subshrubs.
54. *Eriophyllum.*
Annuals.
Leaves pinnatifid or bipinnatifid; glandular hairs in
a ring at base of corolla-throat and upper tube.
55. *Pseudobahia.*
Leaves entire, lobed or toothed; glandular hairs on
the corolla scattered.
Achenes epappose; small lobe opposite the ray on
summit of tube; medium-sized plants of
Sacramento-San Joaquin Valley and central and southern California Coast Ranges.
53. *Monolopia.*
Achenes with pappus paleaceous or a ring of
bristles or reduced to a crown (epappose
in *Syntrichopappus lemmonii*); mostly
small annuals.
Pappus paleaceous or reduced to a crown;
central and southern California ranges
and deserts.
54. *Eriophyllum.*

Pappus of barbellate bristles or if none the
rays white or pinkish; desert annuals.
56. *Syntrichopappus.*
Rays filiform, obscure, scarcely surpassing the disk; wiry-stemmed
annual.   57. *Rigiopappus.*
Heads discoid.
Some or all of the leaves incised, or pinnately dissected; annuals or perennials.
Pappus-paleae deeply parted into numerous bristles; achenes obpyramidal.
58. *Trichoptilium.*
Pappus-paleae not deeply parted into numerous bristles; achenes linear-
clavate, terete.   59. *Chaenactis.*
All leaves entire (see also *Chaenactis cusickii*); annuals.
Involucres 3–7 mm. high; corollas yellowish.
Pappus-paleae not deciduous in a ring, entire; plants subglabrous
but gummy; coastal southern California south.
62. *Amblyopappus.*
Pappus-paleae deciduous in a ring, lacerate-dissected; plants glandular-hairy; southern Sierra Nevada.
60. *Oreochaenactis.*
Involucre 13–20 mm. high; corollas pinkish or whitish; desert plants.
61. *Palafoxia.*
Receptacle setose-fimbriate throughout, the hairs exceeding the achenes.   46. *Gaillardia.*
Phyllaries with conspicuous, thin, scarious, colored margins; achenes papillate on both faces; vernal annuals.
63. *Blennosperma.*
Involucres glabrous or nearly so, the phyllaries and usually the leaves dotted or striped with conspicuous im-
pressed oil-glands; herbage ill-smelling; deserts.
Heads radiate, pink or yellow, the disk-corollas yellow; pappus of paleae or of paleae and bristles, if of
bristles only, the plants annual.
Annuals; leaves all opposite, entire and with a few coarse setae at base.   64. *Pectis.*
Perennials or shrubby perennials; leaves mostly alternate, toothed or lobed, lacking coarse setae.
Rays pinkish; phyllaries in 1 series with subtending calyculate bractlets.   65. *Nicolletia.*
Rays yellow; phyllaries in 2 series, also with calyculate bractlets.   66. *Dyssodia.*
Heads discoid, the corollas white with purple lines; pappus of numerous scabrous bristles only; perennial.
67. *Porophyllum.*

36. PSILOSTRÒPHE DC. Prod. 7: 261. 1838.

Perennial herbs or small shrubs, more or less woolly-tomentose. Leaves alternate,
linear, the margin entire or slightly lobed. Inflorescence corymbose, the heads loosely or
densely arranged. Involucre cylindric, campanulate, or rarely turbinate; the phyllaries
in 1 series, subequal, the inner usually with membranous margins. Receptacle small, naked.
Ray-flowers pistillate, the ligule spreading and becoming papery, conspicuously nerved,
toothed at the apex. Disk-flowers perfect, somewhat glandular, with short tube, cylindric
throat wider than the tube, 5-lobed above. Anthers obtuse at base. Style-branches truncate and thickened. Achenes slender, obtusely angled, glabrous or villous. Pappus of 4–6
hyaline paleae, their margins entire or laciniate. [Name Greek, meaning bare and to turn.]

A genus of 7 or 8 species, natives of southwestern United States and Mexico. Type species, *Psilostrophe
gnaphalodes* DC.

1. Psilostrophe coòperi (A. Gray) Greene. Paper-flower. Fig. 5314.

Riddellia cooperi A. Gray, Proc. Amer. Acad. 7: 358. 1868.
Psilostrophe cooperi Greene, Pittonia 2: 176. 1891.

Plants shrubby, 3–6 dm. high, stems many from the woody base with ascending branches,
densely white-lanate especially on the young growth. Leaves mostly linear, narrowed to a short
petiole, 3–7 cm. long, tomentulose, becoming glabrate in age; heads loosely corymbose on the
upper branches, the peduncles 4–10 cm. long; involucre subcylindric or somewhat spreading,
6–9 mm. high; phyllaries 10–20, oblong, the outer lanate and firm, the inner thinner; pistillate
flowers 4–8, the ligules 8–20 mm. long, pale yellow, papery and persistent, quadratish, nerved,
toothed at the apex; disk-flowers somewhat granular and glandular, much exserted at anthesis;
achenes sparsely hairy when young, becoming glabrate, 5–7 mm. long, the pappus-paleae shorter
or nearly equaling the achenes, oblong to spatulate, entire or erose-laciniate.

Gravelly desert mesas and washes, Sonoran Zones; southern Utah, southern Nevada, and eastern Mojave
Desert in eastern San Bernardino County, California, to the Colorado Desert in Riverside and Imperial Counties
and also in Lower California, eastward to Arizona and New Mexico. Type locality: Fort Mojave. Collected by
Cooper. May–Sept.

37. BÀILEYA Harv. & Gray ex Torr. in Emory, Notes Mil. Rec. 144. 1848.

Annual or perennial, densely floccose-woolly herbs, the stems branching from or above
the base. Leaves alternate, pinnatifid or bipinnatifid, the upper entire or merely lobed.
Heads pedunculate, terminal on the branches, loosely cymose or solitary, radiate; the rays
yellow, persistent and becoming papery and reflexed in age. Involucre hemispheric, the
phyllaries subequal, in 2 series. Receptacle naked, flat or convex. Ray-flowers 5–50 or
more, pistillate, the ray nerved, broadly rounded at the apex or toothed. Disk-flowers many,

tube short, throat funnelform, shortly 5-lobed at the summit and glandular. Anthers minutely sagittate at base. Style-branches with truncate thickened tips. Achenes oblong, linear, with many striae and usually obtusely ribbed, truncate at apex, glandular-hispidulous and granular, becoming glabrate. Pappus none. [Name in honor of J. W. Bailey, early American microscopist.]

A genus of 4 or 5 species, natives of southwestern United States and Mexico. Type species, *Baileya multiradiata* Harv. & Gray.

Ray-flowers 5–7; heads relatively small, the disk about 6 mm. broad.	1. *B. pauciradiata.*
Ray-flowers 20–50 or more; heads large, the disk about 10 mm. or more broad.	
Stems leafy to the middle or above; peduncles 10 cm. long or less.	2. *B. pleniradiata.*
Stems leafy only below; peduncles 10–20 cm. long.	3. *B. multiradiata.*

1. Baileya pauciradiàta Harv. & Gray. Colorado Desert Marigold. Fig. 5315.

Baileya pauciradiata Harv. & Gray ex A. Gray, Mem. Amer. Acad. II. 4: 105. 1849.

Densely floccose-woolly annual herbs, sometimes persisting, 1–6 dm. high, with erect stems and ascending branches, leafy throughout with rather long internodes. Leaves linear-oblanceolate to lanceolate, sessile, 3–10 cm. long, mostly entire, the soon-withering basal leaves and lower stem-leaves irregularly pinnatifid with 2–5 short linear lobes; heads many, cymose-paniculate,

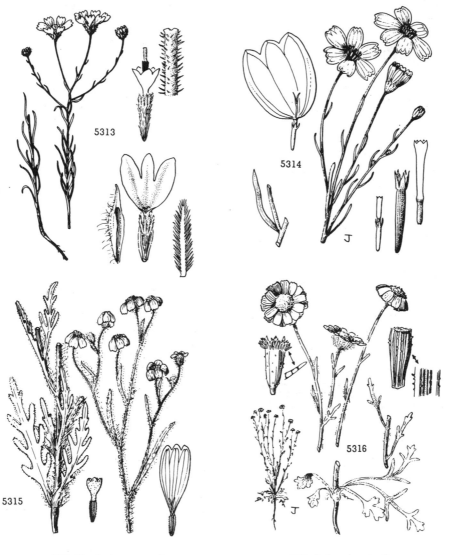

5313. Blepharipappus scaber
5314. Psilostrophe cooperi

5315. Baileya pauciradiata
5316. Baileya pleniradiata

the peduncles 2–5 cm. long; involucres 5–6 mm. high, a little broader than long; ray-flowers 5–7, the ligules greenish yellow, 5.5–8 mm. long, oval, shallowly crenate or crenate-dentate at apex; disk-flowers 10–20; achenes straw-colored, about 4 mm. long, nearly linear, strongly striate-ribbed, scabrous.

In sand, desert washes and dunes, Lower Sonoran Zone; southeastern Mojave Desert in San Bernardino County, California, south through the Colorado Desert to Lower California and east to southern Arizona and adjacent Sonora. Type locality: California. Collected by Coulter. March–Oct. Laxflower.

2. Baileya pleniradiàta Harv. & Gray. Woolly Marigold. Fig. 5316.

Baileya pleniradiata Harv. & Gray ex A. Gray, Mem. Amer. Acad. II. 4: 105. 1849.
Baileya multiradiata pleniradiata Coville, Contr. U.S. Nat. Herb. 4: 133. 1893.
Baileya nervosa M. E. Jones, Contr. West. Bot. No. 8: 34. 1898.

Annuals or winter annuals, branching from the base and above, 2–5 dm. high, more or less floccose, more conspicuously so about the upper stems. Leaves spatulate to oblanceolate in outline, basal leaves soon withering, 4–7 cm. long, pinnately obtuse-lobed, tapering into a petiole much longer than the blade; stem-leaves reduced upward, pinnatifid, those of the upper stems sessile and often entire; heads solitary at the ends of the branches, the peduncles 3–10 cm. long; involucre hemispheric, 6–8 mm. high, 8–14 mm. broad; phyllaries many, linear; ray-flowers bright yellow, 20–40, the ligules 8–10 mm. long, oval or obovate, crenate at the apex or shallowly 3–4-toothed; achenes straw-colored or somewhat chalky, 4–5-ribbed with striae in the intervals, smooth or occasionally with a few glandular dots.

In sand, on desert slopes and mesas, Sonoran Zones; Death Valley region and the Mojave and eastern Colorado Deserts, San Bernardino County, California; southern Nevada and Utah to Arizona, New Mexico, and western Texas and south to northern Mexico. Type locality: California. Collected by Coulter. Feb.–Nov. Closely allied to *B. multiradiata*.

3. Baileya multiradiàta Harv. & Gray. Wild Marigold. Fig. 5317.

Baileya multiradiata Harv. & Gray ex Torr. in Emory, Notes Mil. Rec. 144. *pl. 6.* 1848.
Baileya multiradiata var. *nudicaulis* A. Gray, Syn. Fl. N. Amer. 1²: 318. 1884.

Floccose-woolly stems 1 to many, erect, 3–4 dm. high, arising from a perennial or biennial base, unbranched, the heads terminating the stems. Leaves much as in the preceding species but the basal ones persisting and stem-leaves few and near the base of the stems; peduncles 1–3 dm. long; involucres hemispheric, 7–8 mm. high, 11–16 mm. broad; phyllaries many, linear; ray-flowers bright yellow, 25–50, the ligules oblong, 11–15 mm. long, shallowly 3–5-toothed at the apex; achenes sublinear, truncate at apex, chalky white, sparsely covered with glandular dots.

Desert slopes, mostly Upper Sonoran Zone; eastern Mojave Desert in San Bernardino County, California, and adjacent Nevada and from southern Utah to Arizona and western Texas; also Sonora and Chihuahua. Type locality: "along the Del Norte [Rio Grande] and in the dividing region between the waters of the Del Norte and those of the Gila." Collected by Emory. April–Oct.

38. WHÍTNEYA A. Gray, Proc. Amer. Acad. 6: 549. 1865.

Herbaceous perennials clothed with a very short, dense tomentum, longer and looser on the inflorescence and also glandular. Stems few or solitary from a rather slender, branching, woody rootstock. Leaves opposite, entire or rarely somewhat denticulate, the basal and subbasal leaves several, obovate to suborbicular, apiculate at the apex, 3-nerved, tapering into the petiole, these scarious-winged at the extreme base and somewhat united; stem-leaves few, reduced above, oblanceolate. Flower-heads few, cymose, long-pedunculate from the upper leaf-axils. Involucre in 1 series. Receptacle short-villous throughout. Ray-flowers 6–9, the ligules many-veined, 2–3-toothed at apex, becoming papery. Disk-flowers many, with short slender tube and expanded throat, the limb of 5 short spreading lobes. Anthers obtuse at base. Style-branches hirsute, acute. Achenes obovoid, somewhat compressed, strongly nerved, without pappus. [Name in honor of J. D. Whitney of the California Geological Survey.]

A monotypic genus.

1. Whitneya dealbàta A. Gray. Whitneya. Fig. 5318.

Whitneya dealbata A. Gray, Proc. Amer. Acad. 6: 550. 1865.

Plants 18–33 cm. high with long internodes. Lower leaves including petiole about 9–10 cm. long, the upper shorter and sessile or nearly so; peduncles 5–15 cm. long, rarely with 2–3 short alternate bracts, more or less glandular below the heads; phyllaries erect, narrowly ovate or broadly lanceolate, thin, loosely pubescent and more or less glandular, glabrous within except at the tip; ray-flowers about 2 cm. long, spreading, the disk-flowers about 8 mm. long, both pubescent with spreading hairs; achenes dark brown when mature, hairy, strongly angled, epappose.

Found infrequently in light soil at edges of forests, Canadian Zone; Sierra Nevada, California, from Shasta County to Fresno County. Type locality: Yosemite, Mariposa County. June–Aug.

39. VENEGÀSIA DC. Prod. 6: 43. 1837.

Stout, leafy, branching perennial. Leaves alternate. Heads showy, yellow, arising from the upper leaf-axils, pedunculate. Involucre hemispheric, broad, the round-ovate bracts imbricated in 3 series, the outer somewhat foliaceous, the innermost narrow and scarious. Receptacle plane or slightly rounded, naked. Ray-flowers pistillate, many, long, narrow, entire or 3-toothed. Disk-flowers glandular-bearded especially at the base of the tube. Anthers subentire at base. Style-branches thickened, obtuse or truncate, at apex. Achenes papillose, angulate, about 10-nerved, without pappus. [Name in honor of Michael Venegas, a Jesuit missionary.]

A monotypic genus.

1. **Venegasia carpesioìdes** DC. Venegasia. Fig. 5319.

Venegasia carpesioides DC. Prod. **6**: 43. 1837.
Parthenopsis maritimus Kell. Proc. Calif. Acad. **5**: 101. 1873.
Venegasia deltoidea Rydb. N. Amer. Fl. **34**: 5. 1914.

Coarse herbs with stems widely branched from a somewhat lignescent base, 1.5 m. high or less, reddish, glabrous except above. Leaves thin, deltoid-ovate or ovate-cordate, somewhat attenuate at the apex, the margin crenate or crenate-dentate, resinous-dotted beneath, 7–15 cm.

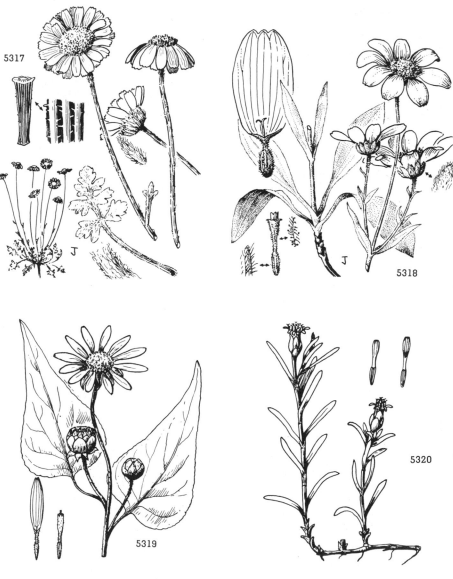

5317. Baileya multiradiata
5318. Whitneya dealbata

5319. Venegasia carpesioides
5320. Jaumea carnosa

long, 4–8 cm. wide, the petioles 3–5 cm. long and sparsely pubescent; flower-heads yellow, scattered in the upper axils, usually solitary, the peduncles 2–3.5 cm. long, sparsely glandular-pubescent, the outer phyllaries glandular-puberulent toward the base; ray-flowers 13–20, about 2 cm. long, entire, acute at the apex, sometimes toothed at the tip; achenes about 6 mm. long.

Shaded moist canyons, Upper Sonoran Zone; southern border of Monterey County southward in the Coast Ranges and the Channel Islands to San Diego County and adjacent Lower California. Type locality: California. Collected by Douglas, presumably at Santa Barbara. March–Sept.

40. JAÚMEA Pers. Syn. Pl. 2: 397. 1807.

Shrubby or herbaceous (in ours) perennial. Leaves opposite, linear or linear-lanceolate, entire. Heads terminal at the ends of the branches or axillary on long peduncles. Phyllaries mostly rounded above, graduate, in 3–4 series. Ray-flowers pistillate, lacking in some species; disk-flowers perfect. Athers obtuse at base. Style-branches papillose or hairy, the apices obtuse or truncate. Receptacle conical or plane. Achenes angled, attenuate at base, the pappus-paleae various, lacking in ours. [Name in honor of the French botanist, J. H. Jaume St. Hilaire.]

A genus of about 9 species, one species native in the Pacific States, the rest natives of Mexico, South America, and southern Africa. Type species, *Jaumea linearis* Pers.

1. Jaumea carnòsa (Less.) Á. Gray. Fleshy Jaumea. Fig. 5320.

Coinogyne carnosa Less. Linnaea 6: 521. 1831.
Jaumea carnosa A. Gray in Torr. Bot. Wilkes Exp. 17: 360. 1874.

Glabrous, somewhat succulent, perennial herbs with several, mostly simple stems 1–2.5 dm. long arising from a thickened root crown, the stems prostrate and rooting at the nodes or ascending. Leaves entire, subterete, slightly connate at the bases, 1.5–2.5 cm. long; heads solitary, terminating the ascending branches or axillary, short-peduncled; involucres 1 cm. or more high, the phyllaries in 3 series, broad, rounded above; ray-flowers pistillate, the rays yellow, 6–10, but little surpassing the yellow disk-flowers; receptacle conical; achenes epappose, glabrous, 10-nerved, about 3 mm. long.

In salt marshes and mud flats along the coast, Transition and Sonoran Zones; the Puget Sound area on the Washington coast southward to southern California and adjacent Lower California. Type locality: California. Collected by Chamisso. June–Oct.

41. EATONÉLLA A. Gray, Proc. Amer. Acad. 19: 19. 1883.

Annual, more or less floccose herbs, subacaulescent or loosely branching from the base. Leaves alternate or subrosulate. Heads in the upper leaf-axils or clustered at the ends of the branchlets, sessile or short-peduncled, about 20-flowered, discoid or rayed, the marginal or ray-flowers pistillate, the disk-flowers perfect. Receptacle flat, naked. Phyllaries in 1 series, few, equal, oval or oblong. Marginal flowers with or without rays, the tube glandular-granuliferous, the throat much expanded-campanulate. Disk-flowers nearly like the marginal, 4–5-toothed. Anthers subentire. Style-branches short, obtuse or truncate. Achenes flattened, the faces often somewhat ridged, those of the marginal flowers subtriquetrous, the lightly callous margin densely hirsute. Pappus-paleae few, erose and sometimes awn-tipped. [Name in honor of Daniel Cady Eaton, an American botanist.]

A genus of 2 species, natives of the Great Basin area of western United States and the San Joaquin Valley, California. Type species, *Burrielia nivea* D.C. Eaton.

Plants 2.5–4.5 cm. high; rays present; plants of the Great Basin.   1. *E. nivea.*
Plants 1–2.5 dm. high; rays absent; plants of the southern San Joaquin Valley, California.
                                                                                                     2. *E. congdonii.*

1. Eatonella nívea (D. C. Eaton) A. Gray. White Eatonella. Fig. 5321.

Burrielia nivea D.C. Eaton, Bot. King Expl. 174, *pl. 18, figs. 6–14.* 1871.
Actinolepis nivea A. Gray, Bot. Calif. 1: 379. 1876.
Eatonella nivea A. Gray, Proc. Amer. Acad. 19: 19. 1883.

Depressed tufted annual 2.5–4.5 cm. high, much branched from the base, loosely white-woolly throughout. Leaves many and persisting, linear-spatulate to linear-oblanceolate, 8–15 mm. long; heads sessile or on filiform axillary peduncles; involucre campanulate, about 5–6 mm. high, the phyllaries about 8, oval to oblong; ray-flowers yellow or purplish, short, scarcely exceeding the disk-flowers; achenes linear-oblanceolate, compressed, black except for the white ciliate margins, about 3 mm. long; pappus-paleae 2, usually quadratish, erose and shortly awn-tipped.

Dry sandy deserts, Arid Transition Zone; southwestern Idaho and southeastern Oregon to Nevada and the desert slopes of the Sierra Nevada as far south as Inyo County, California. Type locality: Virginia Mountains, western Nevada. Collected by Watson. May–June.

2. Eatonella congdònii A. Gray. Congdon's Eatonella. Fig. 5322.

Eatonella congdonii A. Gray, Proc. Amer. Acad. 19: 20. 1883.
Lembertia congdonii Greene, Fl. Fran. 441. 1897.

Annuals 1–2.5 dm. high, more or less floccose-woolly throughout, more densely so on the upper stems and inflorescence, several stems from the base, these often branched above. Leaves alternate, mostly longer than the internodes, thinly floccose-woolly, sparsely sinuate-dentate, more rarely entire, 2–4.5 cm. long, not much reduced above; heads from the leaf-axils of the upper branchlets or clustered on more vigorous plants, subsessile or on peduncles up to 2 cm. long; involucres hemispheric, about 4 mm. high and 6 mm. wide; phyllaries oval, 5 or 6; marginal flowers without rays; disk-flowers yellow, glandular-granuliferous, the throat campanulate, 4–5-lobed; achenes flattened, about 3 mm. long, with a dense, marginal, hirsute fringe longer than the width of the achene, the marginal achenes more hairy than the disk-achenes, the ridge on inner face also more prominent, disk-achenes black: usually with a pubescent line on each face; pappus-paleae usually 2, erose, mostly shorter than the marginal hairs.

Alkaline or loamy plains, Lower Sonoran Zone; in the San Joaquin Valley, California, from Fresno County to Kern County; also on the Carrizo Plain, San Luis Obispo County. Type locality: Deer Creek, Tulare County. March–April.

42. PERÍTYLE Benth. Bot. Sulph. 23. *pl. 15.* 1844.

Annual or perennial, usually branched herbs, sometimes shrubby at least at the base. Leaves petioled, opposite at least below or alternate, the margin entire to deeply lobed or toothed. Heads numerous, radiate or discoid, corymbose at the ends of the branches. Involucre hemispheric or turbinate, the phyllaries more or less boat-shaped, subequal in 1 or 2 series. Receptacle naked, flat. Ray-flowers pistillate, glandular-puberulent, the rays when present white or yellow, 3-toothed. Disk-flowers perfect, many, the tube short and glandular-puberulent and, in some species, also the funnelform or campanulate throat, the limb 4-toothed. Anthers subentire at base. Style-branches linear, obtuse at apex. Achenes flat, oblong to elliptic in outline, with a callus at base and a ciliate or villous, more or less cartilaginous margin. Pappus of scarious scales, these sometimes much reduced, and 1–2 awns, though sometimes awnless. [From the Greek words meaning around and callus.]

A genus of 25 species, natives of Mexico and southwestern United States. Type species, *Perityle californica* Benth.

1. Perityle emòryi Torr. Emory Rock-daisy. Fig. 5323.

Perityle emoryi Torr. in Emory, Notes Mil. Rec. 142. 1848.
Perityle nuda Torr. Pacif. R. Rep. **4**: 100. 1857.
Perityle californica var. *nuda* A. Gray, Syn. Fl. N. Amer. 1²: 321. 1884.
Perityle fitchii var. *palmeri* A. Gray, loc. cit.
Perityle rothrockii Rose, Bot. Gaz. **15**: 114. 1890.
Perityle emoryi var. *orcuttii* Rose, op. cit. 117.
Perityle greenei Rose, loc. cit.
Perityle grayi Rose, op. cit. 118.

Winter annual 1–4 dm. high, much branched, the branches arched, widely spreading, sparsely hirsute and glandular-puberulent to nearly glabrous. Leaves mostly alternate, 1.5–8 cm. long, the petiole about equaling the cordate, suborbicular to ovate blade, shorter above, irregularly palmately lobed, the lobes laciniate-toothed; involucres 5–6 mm. high, broader than high; phyllaries in 2 subequal series, the outer 2-veined, ciliate at the tip, the inner narrower and scarious-margined, reflexed in age; ray-flowers about 10, white, 2–3 mm. long; tube of the disk-flowers about equaling the throat, glandular-puberulent; achenes 2–3 mm. long, black, the flattened faces shining and nearly glabrous, the margin inconspicuously thickened and thickly beset with short hirsute hairs; pappus of very short scales and a single awn about as long as the disk-corolla, this sometimes absent.

In desert ranges among rocks and cliffs, Sonoran Zones; Channel Islands and adjacent Ventura County, California, and southern Nevada and the Death Valley region in Inyo County, California, south through the Mojave and Colorado Deserts to Lower California and east to southwestern Arizona and in Sonora and Sinaloa. Type locality: not stated. Collected by Emory. Dec.–June.

5321

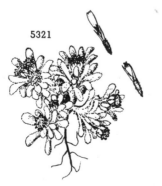

5322

5321. Eatonella nivea
5322. Eatonella congdonii

43. **LAPHÁMIA** A. Gray, Smiths. Contr. 3⁵: 99. 1852.

Low, suffruticose, mostly glandular perennials or subshrubs with scabrous-puberulent to villous pubescence, the stems many, simple or corymbosely branched above. Leaves mostly opposite below, alternate and often somewhat reduced above, entire, toothed, lobed or parted. Heads small to medium, radiate or discoid, borne at the ends of the stems and branches. Involucres hemispheric, the phyllaries in 2 subequal series, 2–3-ribbed, somewhat boat-shaped, the outer achenes resting in the groove thus formed or merely concave and flattened above, in ours with a villous tip. Receptacle flat or slightly convex, naked. Marginal pistillate flowers eradiate or with a small, white or yellow ray (ours rayless). Disk-flowers 15–50, perfect, the tube short, the throat more or less ampliate, the limb lobed, the lobes spreading. Anthers auriculate at base. Style-branches slender, subulate, and hispidulous. Achenes flattened, dark when mature, the face with appressed hirsute hairs, the margins with a thickened rib. Pappus wanting or of 1 or 2 awn-like bristles. [Name in honor of I. A. Lapham, a Wisconsin botanist.]

A genus of about 18 species, natives of southwestern United States and of Mexico. Type species, *Laphamia halimifolia* A. Gray.

Herbage hispidulous or granular-puberulent; leaves thickish.
    Leaf-blades about as broad as long, ovate to suborbicular.   1. *L. megalocephala.*
    Leaf-blades definitely longer than broad, linear, oblanceolate, or spatulate.   2. *L. intricata.*
Herbage villous with lax spreading hairs; leaves thinnish.
    Leaf-blades entire, more or less tapering to the petiole; villous hairs up to 0.6 mm. long.   3. *L. villosa.*
    Leaf-blades triangular-toothed, mostly truncate at base; villous hairs up to 1–1.5 mm. long.
                                                                4. *L. inyoensis.*

1. **Laphamia megalocéphala** S. Wats. Large-headed Laphamia. Fig. 5324.

Laphamia megalocephala S. Wats. Amer. Nat. **7**: 301. 1873.
Monothrix megacephala Rydb. N. Amer. Fl. **34**: 20. 1914 (*megalocephala*).

Aromatic perennial 2–3.5(5) dm. high, forming rounded subshrubs, the stems many, arising from a suffrutescent base and woody root, leafy with long internodes, simple or branched, the branches usually spreading and arcuate, herbage granular to scabridulous throughout, hispidulous on new growth. Leaves 3–10 mm. long including the short petiole, the blade 2.5–6 mm. long and usually as wide as long (uppermost leaves sometimes longer than wide), broadly ovate, sometimes deltoid or suborbicular, thickish, entire or rarely with 1 or 2 triangular teeth; heads bright yellow-flowered, terminal and 1 to few on each stem; involucre about 6 mm. high, the phyllaries linear or linear-elliptic; achenes 2.75–3.25 mm. long, epappose.

Rock crevices or among loose rocks in canyons, Arid Transition Zone; desert ranges of Mineral, Esmeralda, and Nye Counties, Nevada, to the White Mountains, Mono County, California. Type locality: Nevada. Collected by Wheeler. July–Sept. Some southern forms approach *L. intricata* in leaf-shape.

2. **Laphamia intricàta** Brandg. Narrow-leaved Laphamia. Fig. 5325.

Laphamia intricata Brandg. Bot. Gaz. **27**: 450. 1899.
Monothrix intricata Rydb. N. Amer. Fl. **34**: 20. 1914.

Aromatic subshrub 1.5–3.5 dm. high from a stout woody root and short, suffrutescent, branching caudex, the striate stems herbaceous, many, erect as are the many branches, young growth and protected stems sometimes lax, herbage densely hispidulous throughout, the hairs usually upward-curved, becoming scabrid in age, also with sessile glands. Lower leaves 7–9 mm. long, oblanceolate or narrowly spatulate, narrowed to a petiole, upper leaves much reduced, 2–5 mm. long, linear to narrowly oblanceolate, upper and lower leaves thickish; heads yellow-flowered, terminating the branches, several on older stems; involucres 3–3.5 mm. high, the phyllaries linear-oblong; achenes 2.75–3 mm. long, epappose.

Rock crevices and canyon walls, Upper Sonoran and Arid Transition Zones; desert ranges of western Nye and Clark Counties, Nevada, west to the Inyo Mountains and ranges surrounding Death Valley, Inyo County, California. Type locality: "Pahrump and on Sheep mountain, Nevada." Collected by Purpus. July–Sept. The specimen collected at Sheep Mountain is designated as the type by Brandegee.

3. **Laphamia villòsa** Blake. Hanaupah Laphamia. Fig. 5326.

Laphamia villosa Blake, Proc. Biol. Soc. Wash. **45**: 142. 1932.

Aromatic perennial 1–2.5 dm. high from a woody root and branching suffrutescent base, the stems slender, flexuous or ascending, striate, glandular-dotted or the glands short-stalked, and loosely covered with spreading, jointed, villous hairs. Leaves shorter to about equaling the internodes, 9–15 mm. long including the petiole (this shorter on upper leaves), 2–8 mm. wide, ovate to elliptic, rounded to cuneate at base, entire or rarely with a small lobe, thin, 3-nerved, pubescence as that of the stems; heads solitary or few on the stems, bright yellow or flowers turning reddish at the tips; involucres about 5 mm. high, the phyllaries thin; achenes 2 to nearly 3 mm. long, epappose.

Shaded rock crevices, Arid Transition Zone; known only from the type locality, Hanaupah Canyon, Panamint Mountains, Inyo County, California. Aug.–Sept.

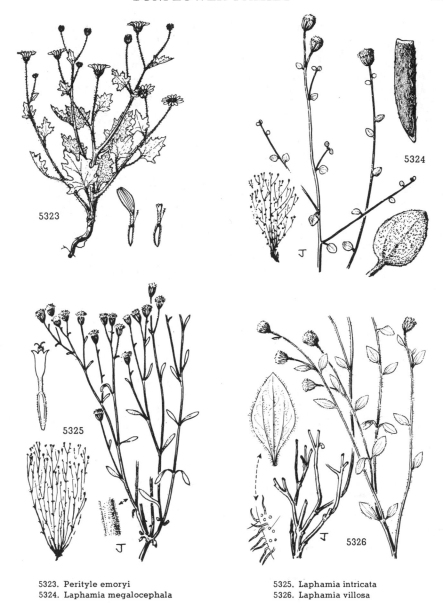

5323. Perityle emoryi
5324. Laphamia megalocephala

5325. Laphamia intricata
5326. Laphamia villosa

4. **Laphamia inyoénsis** Ferris. Inyo Laphamia. Fig. 5327.

Laphamia inyoensis Ferris, Contr. Dudley Herb. **5**: 104. 1958.

Aromatic perennials, the stems several, leafy, 2–2.5(3) dm. high from a suffrutescent base and woody root, erect to laxly spreading, more or less branched, the branches ascending or spreading, often arcuate, pubescent throughout with long, villous, spreading hairs and also glandular. Leaves 5–15 mm. long and as broad as long, somewhat reduced above, subsessile or the petiole about 2 mm. long, deltoid or ovate-deltoid, thin, the veins more or less prominent beneath, irregularly triangular-toothed, the teeth 1–8; heads terminating the branches, yellow-flowered (?); involucres about 6 mm. high, the phyllaries linear, acute; achenes about 3 mm. long, epappose.

Rock crevices, Transition Zone; Cerro Gordo Peak and Talc Canyon, Inyo Mountains, California. Type locality: east face of Cerro Gordo Peak. July–Aug.

44. **PERÍCOME** A. Gray, Smiths. Contr. **5**⁶: 81. 1853.

Perennial herbs, suffrutescent at base, with many widely spreading branches, glabrous or puberulent. Leaves opposite, petiolate, hastate or deltoid, long caudate-acuminate. Inflorescence many-flowered, cymose. Heads discoid, yellow, the flowers much exserted. Involucre turbinate-campanulate. Phyllaries many, carinate, more or less loosely united, the tips free and densely tomentulose within. Receptacle small, low-conic. Flowers all perfect, the corolla glandular-hairy, short-tubed, the abruptly expanded throat cylindric, 4-toothed with spreading lobes. Anthers subsagittate. Style-branches slender, scarcely flattened, minutely hirtellous without and at the tip. Achenes flattened, black, the callous margin densely hirsute-fringed. Pappus-crown of fimbriate, lacerate paleae, sometimes with 1 or 2 marginal awns. [From the Greek meaning around and tuft of hair, in reference to the achene.]

A genus of 2 species, natives of southwestern United States and Mexico. Type species, *Pericome caudata* A. Gray.

1. **Pericome caudàta** A. Gray. Tailed Pericome. Fig. 5328.

Pericome caudata A. Gray, Smiths. Contr. 5[6]: 82. 1853.

Large, somewhat aromatic, much branched perennial, suffrutescent at base, up to 1.3 m. high and widely spreading. Leaves sparsely hirsutulous to glabrate, palmately 3-veined, the petioles 1–5 cm. long, the long-caudate blades 3–5 cm. long, 1.5–2.5 cm. wide at the base, the larger often with 1–2 pairs of marginal lobes; involucre 5–6 mm. high, nearly as broad; phyllaries about 20, linear, hirsutulous, lightly connate and mostly separating in age; corollas 4–4.5 mm. long; achenes dark when mature, glabrous, with a hirsute, ciliate, white, callous margin and narrowed callous base, 5 mm. long, about 0.5 mm. broad; pappus-paleae on the truncate apex about 1 mm. long.

Canyon slopes and washes, Upper Sonoran and Arid Transition Zones; in the Pacific States occurring only in California in Inyo and Tulare Counties in the higher desert ranges and eastern face of the Sierra Nevada; extending eastward to Colorado and south to Arizona, New Mexico, western Texas, and adjacent Chihuahua. Type locality: "copper mines [Santa Rita del Cobre], New Mexico." Aug.–Oct.

45. **HÚLSEA** Torr. & Gray ex A. Gray in Torr. Pacif. R. Rep. **6**: 77. 1857.

Annual, biennial, or perennial and aromatic herbs, erect and leafy-stemmed, scapose or subscapose. Leaves alternate, entire or pinnatifid. Heads rather large, many-flowered, solitary or paniculate. Involucre hemispheric; phyllaries herbaceous, in 2 or 3 series, finally reflexed in age. Receptacle flat, naked except for minute horny teeth. Ray-flowers pistillate and fertile, yellow or purplish red, the rays narrow. Disk-flowers shades of yellow, perfect and fertile; tube slender; throat cylindric; limb 5-lobed. Anthers minutely sagittate; style-branches flattened, the minute hairy appendages obscure. Achenes sublinear, somewhat compressed, more or less villous with straight hairs; pappus-paleae 4, hyaline, lacerate or erose. [Name in honor of Dr. G. W. Hulse, who collected in California.]

A genus of 8 or 9 species, all natives of western North America. Type species, *Hulsea californica* Torr. & Gray ex A. Gray.

Leafy-stemmed annuals and biennials (perennial in *H. brevifolia*); basal leaves few, not persisting after anthesis.
    Basal and lower stem-leaves (as well as upper leaves) green and glandular.
        Rays 50–70, mostly purplish red, 1 mm. or less wide; stems stout and fistulous.   1. *H. heterochroma.*
        Rays 10–17, yellow, 1.5–2 mm. wide; stems slender.   2. *H. brevifolia.*
    Basal and lower stem-leaves white, woolly-villous, the upper green and glandular.   3. *H. californica.*
Scapose or subscapose perennials (branched above in *H. callicarpha*); basal leaves many, persisting after anthesis.
    Herbage, especially the leaves, white with woolly-villous hairs (glandular hairs when present not obvious); rays yellow on inner surface, reddish orange on the outer.
        Flowering stems branched, 15–40 cm. high.   4. *H. callicarpha.*
        Flowering stems unbranched, 4–15 cm. high.
            Leaves strongly dentate or lobed, rather sparsely white-villous; plants of southern California mountains.   5. *H. parryi.*
            Leaves entire or somewhat crenate, densely white-villous; plants of the Sierra Nevada.   6. *H. vestita.*
    Herbage green and glandular, the woolly-villous hairs when present not conspicuous; rays yellow on both surfaces.
        Flowering stems naked; phyllaries lanceolate-oblong.   7. *H. nana.*
        Flowering stems with some leaves; phyllaries linear-lanceolate.   8. *H. algida.*

1. **Hulsea heterochròma** A. Gray. Red-rayed Hulsea. Fig. 5329.

Hulsea heterochoma A. Gray, Proc. Amer. Acad. **7**: 359. 1868.

Stout aromatic biennials or perennials, 4–15 dm. high; stems 1 to few, leafy, fistulous, simple below, branching above at the inflorescence; herbage green, more or less viscid-villous throughout. Leaves 3.5–10 cm. long, 1–3 cm. wide, oblong, saliently toothed, the basal leaves with a broad petioliform base, soon withering, the stem-leaves sessile at least at midstem, longer than the internodes; heads many, with reddish purple or orange rays, corymbosely or paniculately arranged, the peduncles 2–10 cm. long; phyllaries 11–12 mm. long, 2–3 mm. at the broadest point, lanceolate, attenuate at the apex; rays 50–70, about 9 mm. long, linear, 1 mm. or less wide, more or less laciniate-toothed at apex, hirsute and glandular; achenes 5–6.5 mm. long; pappus-paleae usually 1–2.5 mm. long, the pairs unequal, oblong or truncate, laciniate.

In openings in the yellow pine or juniper forest or in chaparral, Arid Transition and Canadian Zones; in the Sierra Nevada, California, from Alpine County to Tulare County and in the Coast Ranges from Santa Clara and Monterey Counties south to the San Jacinto and San Gabriel Mountains; also in the Panamint Mountains, Inyo County. Type locality: Yosemite Valley, Mariposa County. Collected by Bolander. June–Aug.

2. **Hulsea brevifòlia** A. Gray. Short-leaved Hulsea. Fig. 5330.

Hulsea brevifolia A. Gray, Proc. Amer. Acad. 7: 359. 1868.

Perennial, 3–5 dm. high, with 1 to several slender, leafy, viscid-villous, branching stems from the woody root-crown. Leaves 1.5–5 cm. long, 5–12 mm. wide, shorter than to somewhat exceeding the internodes, sessile, spatulate, evenly repand-dentate above the base, green, the midvein white, viscid-villous; heads yellow-flowered, solitary on the branches, the peduncles 3–7 cm. long; phyllaries 9–11 mm. long, mostly lanceolate to oblong-lanceolate, often reddish at the tip, viscid-villous; rays 10–20, yellow, 9–13 mm. long, 1–2 mm. wide, toothed at the apex; achenes about 7 mm. long; pappus-paleae 1–1.5 mm. long, the pairs somewhat unequal, more or less truncate and fimbriate-margined.

Open places in the forest, Canadian and Hudsonian Zones; in the Sierra Nevada, California, from Tuolumne County to Fresno County. Type locality: Yosemite Valley, Mariposa County. Collected by Bolander. July–Aug.

3. **Hulsea califórnica** Torr. & Gray ex A. Gray. San Diego Hulsea. Fig. 5331.

Hulsea californica Torr. & Gray ex A. Gray in Torr. Pacif. R. Rep. 6: 77. 1857.

Robust annual 5.5–10 dm. high, stems single or 2 or 3 from the base, unbranched below the inflorescence, more or less leafy, loosely but abundantly white-woolly below and glandular-villous

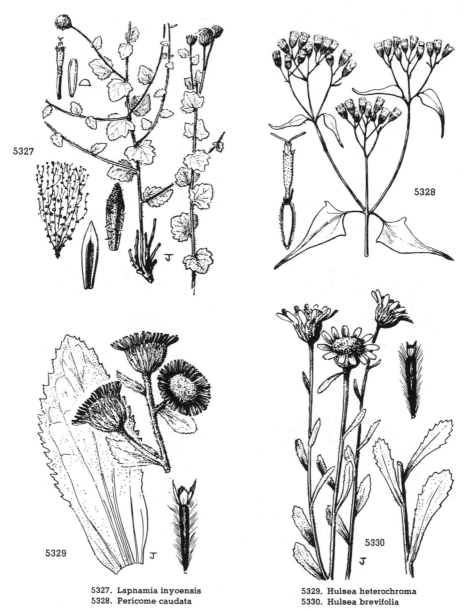

5327. Laphamia inyoensis
5328. Pericome caudata
5329. Hulsea heterochroma
5330. Hulsea brevifolia

above. Basal leaves 7–10 cm. long, not persisting, the blades spatulate, about 2 cm. long, narrowed to a slender petiole, white-woolly; leaves of the lower stem all about equal in length, spatulate or lingulate with a broad petioliform base, entire, white-woolly; uppermost leaves becoming reduced and distant, ovate-lanceolate, greenish and glandular-villous; heads 3–8, large, yellow-flowered, paniculately arranged on sparsely bracteate peduncles 9–12 cm. long; phyllaries 9–12 mm. long, lanceolate, acuminate, glandular-villous; rays 20–30, bright yellow, 12–15 mm. long, 1.5–3 mm. wide, rounded or obscurely toothed at apex; disk-flowers 7–8 mm. long; achenes about 5 mm. long; pappus-paleae about 2 mm. long, the pairs nearly equal, quadratish, laciniate.

On brushy slopes, Upper Sonoran Zone; mountains of San Diego County and adjacent southern California. Type locality: mountains east of San Diego, San Diego County. Collected by Parry. May–July.

Hulsea californica subsp. **inyoénsis** Keck, Aliso **4**: 101. 1958. Lower as well as upper leaves green and glandular, irregularly and broadly toothed to entire; rays yellow; pairs of pappus-paleae equal, about 1 mm. long, quadrate, laciniate. Growing in the pinyon and juniper belt, Inyo and Panamint Mountains, Inyo County, California. Type locality: Mazourka Canyon, Inyo Mountains.

Hulsea mexicàna Rydb. N. Amer. Fl. **34**: 41. 1914. Anual, leafy-stemmed, branching at the inflorescence, resembling *H. californica* in habit and size of heads; basal and lower stem-leaves green, viscid-villous, the lowest leaves with broad petioliform bases, saliently toothed, the upper leaves shallowly toothed to entire; phyllaries more slender and attenuate than *H. californica,* purple-tipped; rays yellow; pairs of pappus-paleae unequal. Laguna Mountains, San Diego County, California, and the Sierra San Pedro Martir in Lower California. Type locality: La Grulla in the San Pedro Martir.

4. **Hulsea callicárpha** (H. M. Hall) S. Wats. ex Rydb. El Chaparossa. Fig. 5332.

Hulsea callicarpha S. Wats. in A. Gray, Syn. Fl. N. Amer. **1²**: 342. 1884, in synonymy.
Hulsea vestita var. *callicarpha* H. M. Hall, Univ. Calif. Pub. Bot. **1**: 129. 1902.
Hulsea callicarpha S. Wats. ex Rydb. N. Amer. Fl. **34**: 39. 1914.

Biennial or short-lived perennial from a stout root, stems several, unbranched or with 2 or 3 long erect branches terminating in heads of medium size. Leaves persistent, 3–10 cm. long, to 35 mm. wide, mostly basal or near the base, obovate or spatulate with a broad petiolar base, the few upper leaves sessile and becoming bract-like on the elongate erect peduncles, these up to 20 or more cm. long; heads medium; phyllaries 9–11 mm. long, lanceolate; rays 6–7 mm. long, orange-red at least on the outer surface; achenes 5–7 mm. long; pappus-paleae up to 2 mm. long, the pairs nearly equal, broadly to more narrowly oblong, lacerate.

Occ: ional on open gravelly benches and ridges, Upper Sonoran and lower Arid Transition Zones; mountains of Los Angeles and Riverside Counties, California, south to the Cuyamaca Mountains, San Diego County. Type locality: Thomas Ranch [Hemet Valley], San Jacinto Mountains, Riverside County. Collected by Parish. May–July.

5. **Hulsea párryi** A. Gray. Parry's Hulsea. Fig. 5333.

Hulsea parryi A. Gray, Proc. Amer. Acad. **12**: 59. 1876.
Hulsea vestita var. *pygmaea* A. Gray, Syn. Fl. N. Amer. **1²**: 343. 1884.

Cespitose, scapose or subscapose perennial from a stout branched caudex, the stems few to several, 2.5–18(–20) cm. long, leafless or with a few narrowly spatulate to linear-oblong, reduced leaves on the longer stems, reddish and viscid-villous. Leaves persistent, densely clustered at or near the base, 1–7 cm. long, 7–10 mm. wide, spatulate, narrowed to a petiolar base, the margins rather evenly toothed or shortly lobed, loosely woolly-villous, the indument rather thin and upper surface of the leaves often greenish; heads medium, the phyllaries 8–10 mm. long, lanceolate, glandular-pubescent and strongly reddish-tinged; rays 5–6 mm. long, orange-red or red on both surfaces; achenes 5–7 mm. long; pappus-paleae 2–3 mm. long, the pairs more or less unequal in length, broadly or narrowly oblong, the margins lacerate.

Gravelly and rocky summits and under pines, Upper Transition and Boreal Zones; upper slopes and summits of the San Bernardino Mountains, San Bernardino County, California, and the Mount Pinos region. Type locality: Bear Valley, San Bernardino Mountains, headwaters of the Mojave River. Collected by Parry. June–Aug.

6. **Hulsea vestìta** A. Gray. Pumice Hulsea. Fig. 5334.

Hulsea vestita A. Gray, Proc. Amer. Acad. **6**: 547. 1865.

Cespitose, scapose or subscapose perennial from a woody taproot and branching subterranean caudex, stems several, 4–25 cm. high, leafless or with few narrowly spatulate to linear-oblong, reduced leaves, rather sparsely viscid-villous. Leaves persistent, densely clustered at the base, 2–5 cm. long, about 1.5–2 cm. broad, spatulate to obovate, narrowed to a petiolar base, the margin entire or obscurely undulate, densely white-woolly-villous; heads medium, the phyllaries 8–10 mm. long, lanceolate, glandular and villous; rays 8–12 mm. long, toothed at the apex, golden-yellow on the inner surface, orange-red or red on the outer; achenes 5–7 mm. long; pappus-paleae 1.5 mm. or more long, the pairs nearly equal, more or less quadrate, the margins lacerate.

On pumice flats, Canadian Zone; in the Sierra Nevada, California, from Mono County to Tulare County; also western Nevada. Type locality: near Mono Lake, Mono County. June–Sept.
This and the two preceding taxa are closely related.

7. **Hulsea nàna** A. Gray. Dwarf Hulsea. Fig. 5335.

Hulsea nana A. Gray in Torr. Pacif. R. Rep. **6**: 76. *pl. 13.* 1857.
Hulsea nana var. *larseni* A. Gray, Bot. Calif. **1**: 387. 1876.
Hulsea larseni Rydb. N. Amer. Fl. **34**: 40. 1914.
Hulsea vulcanica Gandoger, Bull. Soc. Bot. Fr. **65**: 45. 1918.

Cespitose subscapose perennials 4–10(–15) cm. high, with a stout woody root and rather slender, multicipital caudex. Leaves 2–6 cm. long, many on the short stout stem, green, thickish,

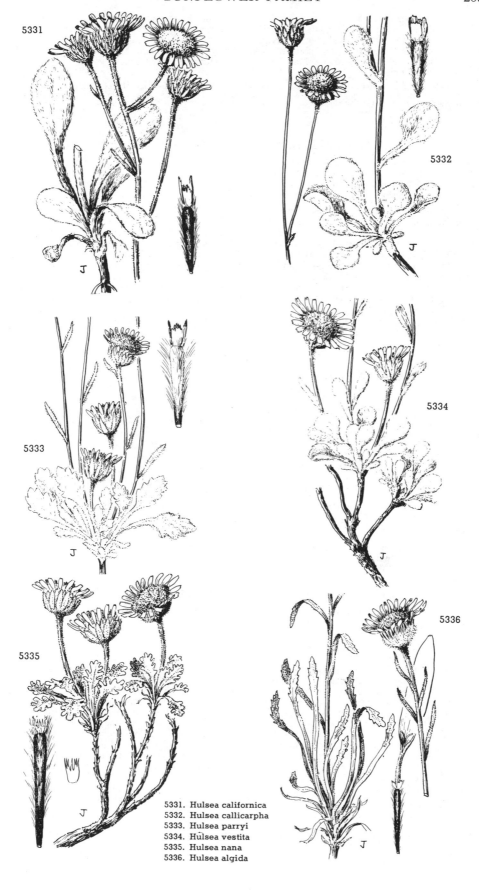

5331. Hulsea californica
5332. Hulsea callicarpha
5333. Hulsea parryi
5334. Hulsea vestita
5335. Hulsea nana
5336. Hulsea algida

oblanceolate, tapering to a petioliform base, the blade lobed with short blunt divisions one-half to one-third the width of the blade, glandular and sparsely to rather densely woolly-villous; heads yellow-flowered, on naked stems; phyllaries 10–12 mm. long, some 2–4 mm. wide, oblong-lanceolate, acute at the apex, glandular and somewhat woolly-villous at the base; rays about 20 and 8–10 mm. long; achenes 6.5–8 mm. long; pappus-paleae 1–1.5 mm. long, the pairs nearly equal, deeply fimbriate.

On cinder cones and pumice flats, Boreal Zone; in the Cascade Mountains from Mount Rainier, Washington, south through Oregon to Lassen Peak, Shasta County, California; also at high elevations in the Wallowa Mountains, Oregon. Type locality: "Crater pass, Cascade mountains, lat. 44° 10'," Oregon. Collected by Newberry. July–Sept.

8. Hulsea álgida A. Gray. Alpine Hulsea. Fig. 5336.

Hulsea algida A. Gray, Proc. Amer. Acad. **6**: 547. 1865.
Hulsea carnosa Rydb. Mem. N.Y. Bot. Gard. **1**: 423. 1900.
Hulsea caespitosa Nels. & Kenn. Proc. Biol. Soc. Wash. **19**: 38. 1906.
Hulsea nevadensis Gandoger, Bull. Soc. Bot. Fr. **65**: 44. 1918.

Cespitose perennial with sparsely leafy stems 1.5–3.5 dm. high, each bearing a single head, arising from a heavy taproot and stout forked caudex, more or less clothed with old scarious leaf-bases, herbage viscid-pubescent and green throughout except for the white-villous involucre. Leaves many, 5–10 cm. long, 3–11 mm. wide, sessile, mostly basal and on the lower stem, the upper stem-leaves few, longer than the internodes and somewhat reduced, linear, with 1–3 white nerves, the margin nearly entire to dentate with short, narrowly lanceolate teeth; heads large, yellow-flowered; phyllaries 11–14 mm. long, linear-lanceolate, attenuate, woolly-villous and also viscid-pubescent; rays 25–55, naked, 7–15 mm. long, the apex entire or shallowly toothed; achenes 6–8 mm. long, the pappus-paleae short, deeply fimbriate, the shorter pair less than 1 mm. long.

Talus slopes in granitic sand or gravel, usually among boulders, mostly Arctic-Alpine Zone; Idaho and southwestern Montana and higher peaks in Nevada, in the Wallowa Mountains, Oregon, and the central Sierra Nevada, California, south to Tulare County, and also the White Mountains, Mono County. Type locality: Mount Dana, Yosemite National Park. July–Aug.

46. GAILLÁRDIA Fouq. Mém. Acad. Paris **1786**: 5. 1788.

Annual, biennial, or perennial herbs, simple-stemmed or branched, scapose in some species. Leaves alternate or entirely basal, entire, toothed, or pinnatifid. Heads large, radiate or sometimes discoid, usually long-peduncled. Involucre shallowly basin-shaped. Phyllaries 2–3-seriate, subequal, ovate, oblong, or lanceolate, thickened at base and herbaceous above, spreading and reflexed in age. Receptacle convex to subglobose, bristly or scaly. Ray-flowers pistillate and fertile or neutral; rays when present cuneate to flabelliform, 3-cleft, yellow, reddish purple, or more often bicolored. Disk-flowers perfect, fertile, yellow or purplish, the tube very short, the throat nearly cylindric or flaring, the lobes short, glandular. Anthers auricled at the base. Style-branches flattened, appendaged, the appendages hispidulous or glabrous. Achenes broadly pyramidal, covered at the base or completely with long ascending hairs. [Name honoring Gaillard de Marentonneau, a French botanist.]

A genus of 12 species or more, natives of North America, one species occurring in South America. Type species, *Gaillardia pulchella* Fouq.

1. Gaillardia aristàta Pursh.* Blanketflower. Fig. 5337.

Virgilia grandiflora Nutt. in Fraser's Cat. 1813. Not *Gaillardia grandiflora* Hort. ex Lemaire, Ill. Hort. **4**: *pl. 139*, 1857.
Gaillardia bicolor Simms, Bot. Mag. **39**: *pl. 1602*. 1813. Not *G. bicolor* Lam. 1788.
Gaillardia aristata Pursh, Fl. Amer. Sept. 573. 1814.
Gaillardia hallii Rydb. N. Amer. Fl. **34**: 135. 1915.

Perennial with 1 to several usually reddish, striate, leafy stems 2–6 dm. high arising from a woody taproot, herbage more or less hirsute throughout with jointed hairs. Leaves oblanceolate to linear-oblong in outline, entire or with a few incised teeth or lobes, 5–12(15) cm. long, 8–15 mm. wide, equalling or longer than the internodes, the basal and lower stem-leaves tapering to a petiolar base, the upper stem-leaves but little reduced in length, sessile, dull green, the midrib white, rather prominent; 1 or sometimes 2 heads on each stem, the peduncles 8–20 cm. long; phyllaries about 10 mm. long, loose and spreading, lanceolate, attenuate, densely villous-tomentose; rays 6–16, usually about 2.5–3 cm. long, purple at the base, the remainder yellow; disk-flowers 7–9 mm. long, purple or brownish purple at the tips; setae of receptacle longer than the achenes; achenes about 4 mm. long or less, densely hairy; pappus-paleae about 6 mm. long, the awn surpassing the palea in length.

Dry meadows and open places, Arid Transition Zone; British Columbia east of the Cascade Mountains south through Washington to central Oregon and eastward to Saskatchewan and North Dakota, Colorado, and northern Utah. Type locality: Rocky Mountains. May–Sept. Great-flowered Gaillardia.

---

* For a discussion of the validity of the name used, see Rhodora **57**: 290–293. 1955; op. cit. **58**: 23–24, 281–289. 1956; op. cit. **59**: 100. 1957.

47. HYMENÓXYS Cass. Dict. Sci. Nat. **55**: 278. 1828.

Annual or perennial, aromatic herbs from a simple taproot or branched root-crown, simple-stemmed or branched, some scapose, variously pubescent. Leaves alternate or basal only, entire or pinnately or ternately divided, more or less glandular-punctuate. Heads radiate or discoid, pedunculate, yellow-flowered. Involucres mostly hemispheric. Phyllaries in 2–3 series, the outer distinct or connate at base, subequal to noticeably different in length, herbaceous or cartilaginous in age and erect. Receptacle mostly naked, convex or low-conic or hemispheric. Ray-flowers pistillate, fertile, their rays prominently veined, 3-toothed at apex in ours, reflexed after flowering, becoming papery. Disk-flowers perfect, fertile, the tube very short, the throat cylindric to funnelform, the lobes short. Anthers entire or sagittate at base, the lobes short, the tips triangular or rounded. Style-branches flattened, truncate, minutely penicillate. Achenes obpyramidal, more or less 5-angled, densely appressed-hairy. Pappus-paleae usually 5 (in ours), hyaline, obscurely nerved or nerveless, the nerve often produced into an awn. [From the Greek words meaning membrane and sharp, referring to the paleae.]

A genus of 20 species or less, natives of western North America and of South America. Type species, *Hymenopappus anthemoides* Juss.

Leaves entire, all basal; heads solitary on the naked scapes.   1. *H. acaulis arizonica.*
Leaves pinnately divided into narrow divisions (at least all the lowermost); heads numerous **on the** branching inflorescence.
     Biennials or perennials, branching only at the inflorescence.
         Upper and lower leaves with linear-filiform divisions, 1.5 mm. or less wide.
                                                      2. *H. cooperi.*
         Uppermost leaves usually entire, the divisions of the basal and lower stem-leaves 2–3 mm. wide.
                                                        3. *H. lemmonii.*
     Annuals, usually branching throughout.   4. *H. odorata.*

1. **Hymenoxys acaùlis** var. **arizònica** (Greene) Parker. Arizona Hymenoxys. Fig. 5338.

Tetraneuris arizonica Greene, Pittonia **3**: 266. 1898.
Actinea acaulis var. *arizonica* Blake ex Munz, Man. S. Calif. 570. 1935.
Hymenoxys acaulis var. *arizonica* Parker, Madroño **10**: 159. 1950.

Densely cespitose perennial with woody taproot and compact, stout, much-branched caudex densely covered with the long-silky, pubescent leaf-bases, the scapose stems 1–3 dm. high, striate, thinly appressed-sericeous. Leaves 2–5 cm. long, 2–5 mm. wide, basal, crowded, thick, linear-oblanceolate, appressed-sericeous, nearly glabrate toward the apex, obscurely glandular; heads solitary, the involucres hemispheric, 6–9 mm. high, about 10–15 mm. wide; phyllaries distinct, oblong to oval, thin and scarious at least on the margins, villous; receptacle hemispheric; rays 10–15 mm. long, broad, often glandular-dotted at the broad apex, veiny, reflexed in age and somewhat papery; achenes 3 mm. long, prominently nerved; pappus-paleae shorter than the achenes, hyaline, ovate, the midnerve extending above the body into an awn.

Rocky slopes in pinyon and juniper belt, upper part of the Upper Sonoran and the Arid Transition Zones; western Colorado and southern Utah to central and southern Nevada south to the mountains of San Bernardino County, California, and northern Arizona. Type locality: Mount Trumbull, Mohave County, Arizona. Collected by Palmer. May–June.

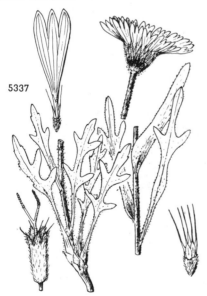

5337

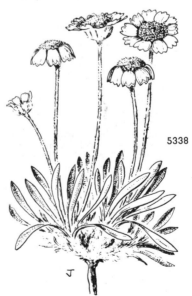

5338

5337. Gaillardia aristata   5338. Hymenoxys acaulis

2. **Hymenoxys coòperi** (A. Gray) Cockerell.   Cooper's Goldflower.   Fig. 5339.

Actinella cooperi A. Gray, Proc. Amer. Acad. **7**: 359. 1868.
Actinella biennis A. Gray, op. cit. **13**: 373. 1878.
Actinea cooperi Kuntze, Rev. Gen. Pl. **1**: 303. 1891.
Actinea biennis, Kuntze, loc. cit.
Picradenia cooperi Greene, Pittonia **3**: 272. 1898.
Picradenia biennis Greene, loc. cit.
Hymenoxys canescens subsp. *biennis* Cockerell, Bull. Torrey Club **31**: 482. 1904.
Hymenoxys cooperi Cockerell, op. cit. 494.
Hymenoxys biennis H. M. Hall, Univ. Calif. Pub. Bot. **3**: 204. 1907.

Stout taprooted biennial (short-lived perennial ?), the stems 2.5–8 dm. high, leafy, essentially solitary, unbranched below the open cymose inflorescence, usually reddish, sparsely scurfy-canescent. Leaves 1.5–5 cm. long, pinnately 3–5-divided, the linear-filiform divisions often again parted, 1.5 mm. or less wide, spreading, sparsely scurfy-canescent to glabrate, the stem-leaves longer than the internodes, the basal and lowest leaves usually withering at anthesis; heads 3–20, the peduncles about 6–10 cm. long; involucres 5–5.5 mm. high, hemispheric, much broader than high; phyllaries strongly indurated, more or less pubescent and also glandular-punctate, the outer lanceolate, united at base, midrib prominently ridged, the inner oblong-ovate; rays 8–14 mm. long; receptacle subconical, short-villosulous; achenes about 2.5 mm. long; pappus-paleae variable, 1–2 mm. long, oblong-ovate, acute, sometimes abruptly cuspidate, usually one-half or less the length of the disk-flowers.

Brushy flats and juniper-covered slopes, mostly Arid Transition Zone; Little San Bernardino, New York, Providence, and Clark Mountains, San Bernardino County, California; also southern Nevada, Utah, and northern Arizona. Type locality: Providence Mountains. Collected by Cooper. May–Sept.

Hymenoxys cooperi var. **canéscens** (D. C. Eaton) Parker, Madroño **10**: 159. 1950. (*Actinella richardsonii* var. *canescens* D. C. Eaton, Bot. King Expl. 175. 1871; *Picradenia canescens* Greene, Pittonia **3**: 271. 1898; *Hymenoxys canescens* Cockerell, Bull. Torrey Club **31**: 484. 1904.) Habit much like the preceding but usually perennial, 1–3 dm. high, and usually with more than 1 stem from the base; herbage densely scurfy-canescent; heads larger, the involucres 5–6 mm. high, 10–15 mm. broad. Grassy slopes at high elevations; southern Idaho southward through Nevada and in California from Modoc County south to the White Mountains, Mono County; also in Arizona. Type locality: east Humboldt Mountains, Nevada.

3. **Hymenoxys lemmònii** (Greene) Cockerell.   Lemmon's Goldflower.   Fig. 5340.

Picradenia lemmonii Greene, Pittonia **3**: 272. 1898.
Hymenoxys lemmonii Cockerell, Bull. Torrey Club **31**: 477. 1904.
Hymenoxys lemmonii subsp. *greenei* Cockerell, op. cit. 479.
Actinea lemmonii Blake, Contr. U.S. Nat. Herb. **25**: 596. 1925.

Perennial, from a woody taproot and short branching caudex, the stems striate, leafy, 2–6 dm. high, 2 or more, unbranched below the cymose inflorescence, glabrous. Leaves 5–15(20) cm. long, pinnately parted, the divisions linear, 2–3 mm. wide or less, mostly ascending, glabrous and glandular-punctate, the stem-leaves longer than the internodes, the uppermost mostly entire; heads 5–12, the peduncles about 4–8 mm. long; involucres 6 mm. or more high, hemispheric, broader than high; phyllaries more or less indurate, sparsely pubescent to glabrous, glandular-punctate, the outer lanceolate, subcarinate, united at base, the inner oblong-ovate; rays 8–13 mm. long; receptacle conic, more or less finely villosulous, becoming glabrate; achenes about 3 mm. long; pappus-paleae one-half or more the length of the disk-corollas, lanceolate or ovate-lanceolate, attenuate into an awn.

Damp alkaline flats and borders of meadows, Arid Transition Zone; Siskiyou and Shasta Counties, California, and central Nevada southward. Type locality: not definitely stated. July–Aug.

4. **Hymenoxys odoràta** DC.   Fragrant Bitter-weed.   Fig. 5341.

Hymenoxys odorata DC. Prod. **5**: 661. 1836.
Actinella odorata A. Gray, Mem. Amer. Acad. II. **4**: 101. 1849.
Actinea odorata Kuntze, Rev. Gen. Pl. **1**: 303. 1891.
Picradenia odorata Britton ex Britt. & Brown, Ill. Fl. **3**: 449. 1898.
Picradenia davidsonii Greene, Pittonia **4**: 240. 1901.
Hymenoxys chrysanthemoides var. *excurrens* Cockerell, Bull. Torrey Club. **31**: 501. *pl. 22 fig. 3.* 1904.

Leafy annual, 1–5(6) dm. high, branching from the base and above, the heads terminating the many branches, stems striate, more or less short-pubescent. Leaves 2–5 cm. long, 3–5-parted (rarely more) into filiform divisions, obscurely glandular-punctate, nearly glabrous, the basal leaves early-deciduous; peduncles 4–10 cm. long, enlarged below the head; involucres 4–5 mm. high; phyllaries lanceolate, both series strongly indurated, more or less pubescent, the outer shorter than the inner, united at the base, the tip herbaceous, midvein not prominent; rays 5–8 mm. long; achenes about 2 mm. long; pappus-paleae a little shorter than the achene, about two-thirds the length of the disk-corolla, ovate, with a cuspidate apex.

Weedy plants in disturbed areas and bottom lands. Lower Sonoran Zone; adjacent to the Colorado River from Riverside County to Imperial County, California, and eastward to Kansas and Texas; also in Mexico. Type locality: Mexico. Feb.–May. Bell Goldflower.

48. **HELÉNIUM** L. Sp. Pl. 886. 1753.

Annual or perennial herbs, the stems simple or branched, striate (in ours). Leaves alternate, glandular-punctate, glabrous or pubescent, often decurrent. Heads very many-flowered, solitary or cymose, pedunculate, usually radiate. Phyllaries in 2–3 series, sub-equal, herbaceous, soon deflexed or merely loosely spreading in age. Receptacle naked, globose or ovoid, sometimes convex. Ray-flowers when present pistillate (more rarely sterile), mostly yellow, 3–5-lobed or -toothed, reflexed or spreading. Disk-flowers many, perfect, yellow or purplish brown especially on the lobes, tube very short, throat cylindric or narrowly campanulate, the lobes with glandular hairs. Anthers minutely auricled or sagittate at base. Style-branches flattened, more or less truncate, penicillate. Achenes turbinate, angled and ribbed, in ours hirsute with ascending hairs, at least on the angles and ribs; pappus-paleae several, hyaline in ours, awn-tipped and often fimbriate on the margins. [A Greek word for some unknown plant, said to be named for Helen of Troy.]

A genus of about 40 species, all natives of North and South America. Type species, *Helenium autumnale* L.

Leaves crowded and mostly fascicled, filiform.   1. *H. amarum.*
Leaves not fascicled or crowded, not filiform.
  Rays linear or narrowly oval, much longer than wide, spreading.   2. *H. hoopesii.*
  Rays flabellate or cuneate, one-half or more as wide as long distally, always reflexed.
    Coarse annuals or biennials; rays small and often inconspicuous, much shorter than the disk; stems paniculately and openly branched.   4. *H. puberulum.*
    Perennials; rays conspicuous, nearly equaling to much surpassing the disk; stems simple or corymbosely branched above.
      Heads several to many on the main stems, the peduncles mostly short; leaves not much reduced in size on the upper stems, at least some denticulate.   3. *H. autumnale.*
      Heads solitary or the stems (sometimes 2 or 3) long-peduncled; leaves typically reduced on the upper stems, always entire.
        Pappus-paleae one-half (sometimes two-thirds, rarely longer) as long as the disk-corollas; peduncles rather slender and not conspicuously enlarged below the disk; plants of foothills and mountains, California and southern Oregon.   5. *H. bigelovii.*
        Pappus-paleae nearly or quite as long as the disk-corollas; peduncles stout and noticeably enlarged below the disk; coastal plants of southwestern Oregon and northwestern California.
                                                           6. *H. bolanderi.*

5339. Hymenoxys cooperi
5340. Hymenoxys lemmonii
5341. Hymenoxys odorata

1. Helenium amàrum (Raf.) H. Rock. Fine-leaved Sneezeweed. Fig. 5342.

Galardia amara Raf. Fl. Ludov. 69. 1817.
Helenium tenuifolium Nutt. Journ. Acad. Phila. **7**: 66. 1834.
Helenium amarum H. Rock, Rhodora **59**: 131. 1957.

Slender glabrous annual from a taproot, 2–7 dm. high, fastigiately branching above the base, densely leafy. Leaves 2–4 cm. long, sessile, not decurrent, linear-filiform, mostly fascicled, bearing impressed glands; heads several, corymbiform, on slender peduncles; disk about 1 cm. wide, depressed-globose; phyllaries 5–10 mm. long, lanceolate, reflexed; ray-flowers about 8, gland-dotted, 9–13 mm. long, reflexed; achenes 1 mm. or more long, hirsute; pappus-paleae ovate to orbiculate, tipped with awn as long as the body of the paleae.

A native of eastern United States, becoming established as a weed in Marin and Stanislaus Counties, California. Type locality: Alexandria, Louisiana. Collected by Ball (neotype). July–Nov.

2. Helenium hoòpesii A. Gray. Tall Mountain Helenium. Fig. 5343.

Helenium hoopesii A. Gray, Proc. Acad. Phila. **1863**: 65. 1864.
Heleniastrum hoopesii Kuntze, Rev. Gen. Pl. **1**: 342. 1891.
Dugaldea hoopesii Rydb. Mem. N.Y. Bot. Gard. **1**: 425. 1900.

Perennials 5–10 dm. high with 1 or more leafy stems arising from a stout rhizome or caudex, the herbage yellowish green, more or less loosely villous-tomentose throughout and becoming glabrate. Basal leaves 1–3 dm. long, veiny, oblanceolate, attenuate into a broad petioliform base, the upper leaves somewhat reduced and mostly clasping; heads yellow, hemispheric, large, 3–8 on peduncles 4–15 cm. long; phyllaries thin, loosely imbricate, 8–11 mm. long, linear-lanceolate to -attenuate, more or less reflexed in age; receptacle broadly dome-shaped; ray-flowers pistillate, fertile, the ray 1.5–3 cm. long, very narrow, subentire; disk-flowers about 5 mm. long; achenes about 3 mm. long, densely silky-hirsute; pappus hyaline, lanceolate-acuminate, the margins somewhat lacerate, a little shorter than the corolla.

Meadows and edges of forests, Canadian and Hudsonian Zones; Steen Mountains, Harney County, and southern Klamath and Lake Counties, Oregon, southward in adjacent California to Tulare County; eastward through the mountains of Nevada and to Wyoming, Colorado, Arizona, and New Mexico. Type locality: South Park, Colorado. July–Sept. Orange Sneezeweed; Owlsclaw.

3. Helenium autumnàle var. montànum Fernald. Western Sneezeweed. Fig. 5344.

Helenium montanum Nutt. Trans. Amer. Phil. Soc. II. **7**: 384. 1841.
Helenium autumnale var. *montanum* Fernald, Rhodora **45**: 491. 1943.

Fibrous-rooted perennials 1.5–8 dm. high, herbage more or less densely puberulent with upwardly curved hairs, stems 1–3, mostly simple below, branched above, very leafy, the internodes one-half or less than one-half the length of the leaves, the basal leaves usually withering and deciduous at anthesis. Leaves 3–8 cm. long, ovate-lanceolate, tapering at the base and conspicuously decurrent on the stem, thick, dull green with straw-colored veins, the margin obscurely serrulate or sometimes serrate, occasionally completely entire, not conspicuously reduced in size on the upper stems; heads yellow, 1–1.5 mm. broad, 3 to many in a corymbiform inflorescence, the peduncles 3–6 cm. long; phyllaries 6–8 mm. long, linear or linear-lanceolate, attenuate, reflexed; receptacle globose; rays 10–20, about 1 cm. long (rarely 1.5), 3-lobed at apex, the lobes round; disk-flowers about 3 mm. long, the lobes pubescent; achenes about 1 mm. long, appressed-hirsute on the angles; pappus about one-half the length of the disk-corolla, broadened at base and attenuate, somewhat lacerate on the margin.

Bottom lands and edges of marshes, Arid Transition Zone; southeastern British Columbia and Washington east of the Cascade Mountains to eastern Oregon and Siskiyou and Modoc Counties, California; eastward through Idaho and Nevada to the Rocky Mountains. Type locality: "In the Rocky Mountain range, on the borders of Lewis' [Snake] River." July–Aug.

Helenium autumnale var. **grandiflòrum** (Nutt.) Torr. & Gray, Fl. N. Amer. **2**: 384. 1842. (*Helenium grandiflorum* Nutt. Trans. Amer. Phil. Soc. II **7**: 384. 1841; *H. macranthum* Rydb. N. Amer. Fl. **34**: 127. 1915.) Stout, leafy-stemmed perennials 4–12 dm. high; leaves broadly lanceolate to ovate-lanceolate, obscurely toothed, thinner than the preceding taxon and sparsely puberulent; peduncles nearly glabrous; rays 1.5–2.5 cm. long, narrowed to a slender base. Damp places west of the Cascade Mountains in the Puget Sound area, Washington, and northwestern Oregon west of the Columbia Gorge to Clatsop County; also Del Norte County, California (*Wolf 836*). Type locality: "Banks of the Oregon and Wahlamet."

4. Helenium pubérulum DC. Rosilla. Fig. 5345.

Cephalophora decurrens Less. Linnaea **6**: 517. 1831.
Helenium pubescens Hook. & Arn. Bot. Beechey 149. 1833; 355. 1838. Not Ait. 1789.
Helenium puberulum DC. Prod. **5**: 667. 1836.
Helenium californicum Link, Ind. Sem. Hort. Berol. 21. 1840.
?Helenium rosilla Turcz. Bull. Soc. Nat. Mosc. **24**[1]: 186. 1841.
Helenium decurrens Vatke, Ind. Sem. Hort. Berol. **1875**: App. 1875. Not *Helenia decurrens* Moench. 1794.

Coarse annual or biennial, 3–15 dm. high from a fibrous root, paniculately branched, the branches ascending, more or less puberulent throughout. Basal leaves narrowed to a petioliform base, soon withering; stem-leaves sessile, linear-lanceolate to linear, 3–15 cm. long, 0.5–2.5 cm. wide, reduced above, conspicuously and strongly decurrent, spreading and often sinuate-margined; heads globose, borne on the ends of the paniculate branches on long peduncles, the disks 1–2 cm. broad, brownish yellow to purplish brown; phyllaries 8–10 mm. long, loose and reflexed, hairy; receptacle globose or depressed-globose; ray-flowers 2–8(–10) mm. long, usually gland-dotted on the outer surface and sparsely short-villous; disk-flowers about 2 mm. long, glabrous except for the puberulent lobes, often gland-dotted; achenes 2 mm. or less long, sparsely hirsute along the ribs; pappus-paleae 5–6, ovate or narrowly ovate-lanceolate, 0.5–1 mm. long, awn-pointed.

Stream beds, wet meadows, and springy areas, sometimes in alkaline areas, mostly in the Sonoran Zones; occurring sparingly in Oregon in the Willamette Valley and at Klamath Lake, Klamath County, southward through California where it becomes more common, to Lower California. Type locality: California. Collected by Douglas. April–Sept.

Apparent intergrades with *Helenium bigelovii* var. *bigelovii* which may have slender leaves, longer rays and longer awns on the paleae are found occasionally in the Inner Coast Ranges, often growing on serpentine, from Lake County, California, southward.

5. **Helenium bigelòvii** A. Gray. Bigelow's Sneezeweed. Fig. 5346.

Helenium bigelovii A. Gray in Torr. Pacif. R. Rep. **4**: 107. 1857.
Heleniastrum bigelovii Kuntze, Rev. Gen. Pl. **1**: 342. 1891.
Heleniastrum occidentale Greene, Man. Bay Reg. 202. 1894.
Heleniastrum rivulare Greene, Fl. Fran. 435. 1897.
Helenium rivulare Rydb. Fl. N. Amer. **34**: 126. 1915.

Perennial, 3–9 dm. high, stems 2–3 or arising singly from a stout, erect, woody caudex and fibrous roots, reddish at the base and sometimes above, glabrate but with villous hairs below and

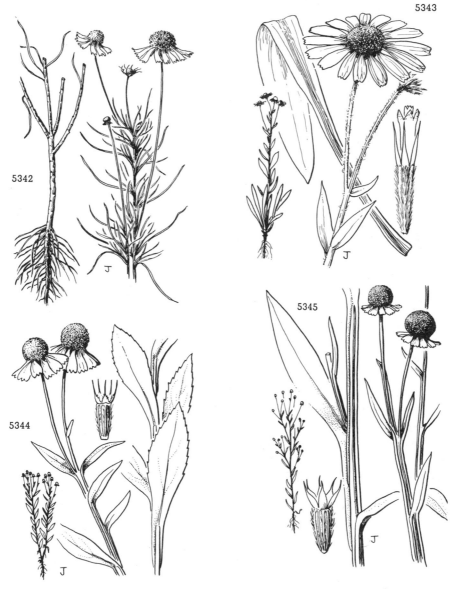

5342. Helenium amarum
5343. Helenium hoopesii

5344. Helenuim autumnale
5345. Helenium puberulum

on the involucres, rather sparingly leafy. Leaves about 8–15 cm. long, somewhat reduced in length above, pale green, linear-lanceolate or lanceolate, entire, thickish in texture, glabrous or sparsely puberulent with curved hairs, the basal and lowest stem-leaves narrowed to a long petioliform base, the upper narrowed at the base but sessile, clasping, decurrent, often narrowly so; heads, not including the rays, 1.5–2 cm. broad on long, rather slender peduncles up to 30 cm. long, solitary or sometimes 2–3, globose; receptacle globose; phyllaries linear-lanceolate, attenuate, loosely villous, reflexed; rays 10–18 mm. long, lobed, villous on the back; disk-flowers 3.5–4 mm. long, glabrous except for the hairy lobes, the hairs somewhat purplish brown; achenes 2 mm. long, hirsute on the ribs; pappus-paleae 6–8, tapering into attenuate awns, the paleae 1.5–2 mm. long, shorter than one-half to about one-half the length of the corolla.

Open marshy meadows or along streams in the foothills and lower mountain slopes, Upper Sonoran and Transition Zones; southern Josephine and Curry Counties, Oregon, south through the Coast Ranges to San Luis Obispo County and occurring in the mountains of southern California; also in the Sierra Nevada where the typical form is occasionally found at higher elevations within the range of the variant noted below with which it appears to merge completely. Type locality: Santa Rosa, Sonoma County, California. June–Aug.

Helenium rivulare (Greene) Rydb. appears to be a tall mesophytic form of *H. bigelovii*. The leaves are thin and rather large, 8–15 cm. long, 2.5–4.5 cm. wide, lanceolate or oblanceolate, the basal tapering to a petiolar base, the upper leaves usually more conspicuously decurrent than in *H. bigelovii;* the rays are 10–15 mm. long; and the disk is usually dark brownish purple, sometimes yellow. In wet places throughout the range of *H. bigelovii* but for the most part occurring at higher elevations; frequent in the mountains of southern Oregon and northern California, the Sierra Nevada particularly in the southern part, and occasional in the mountains of southern California. The type locality was not designated.

6. **Helenium bolánderi** A. Gray. Coast Sneezeweed. Fig. 5347.

Helenium bolanderi A. Gray, Proc. Amer. Acad. **7**: 358. 1868.
Heleniastrum bolanderi Kuntze, Rev. Gen. Pl. **1**: 342. 1891.
Dugaldia grandiflora Rydb. N. Amer. Fl. **34**: 120. 1915.
Helenium bigelovii var. *festivum* Jepson, Man. Fl. Pl. Calif. 1132. 1925.

Perennial 3–7 dm. high with stout to very stout stems, these arising from a heavy caudex and usually horizontal rootstock, herbage reddish below and often above, villous-tomentose, sometimes densely so especially about the heads, and often becoming glabrate in age. Basal leaves 7–15 cm. long, 1.5–4 cm. wide at the broadest, narrowly obovate to spatulate, tapering to a broad, clasping, petioliform base, usually withering at anthesis; stem-leaves 3–20 cm. long, ovate and acute, oblanceolate or broadly lanceolate, reduced upward, sessile with a broad clasping base, not at all or but little decurrent; heads yellow-rayed with a purplish disk, 2 or 3 but more commonly 1 per stem borne on thick peduncles 15–35 cm. long which are enlarged below the heads and densely villous-tomentose; disk 2–3 cm. broad, wider than high, the receptacle hemispheric; phyllaries lanceolate, 7–11 mm. long, loosely reflexed in age, villous-tomentose; ray-flowers pistillate, many, 18–25 mm. long, deeply 3(2–6)-lobed and more or less villous on the outer surface; disk-flowers 4.5–5 mm. long, sparsely hairy and densely penicillate on the short lobes with broad, often purplish hairs; achenes about 2 mm. long, angled, strongly hirsute on the ribs and at the base; pappus-paleae subulate or lanceolate, 6 or 7, often of unequal length, slightly more than 2 to a little over 3 mm. long, laciniate below to nearly entire.

Bogs and wet areas in the coastal terrace, Humid Transition Zone; western Coos and Curry Counties, Oregon, south to Mendocino County, California. Type locality: Mendocino County. Collected by Bolander. June–Aug.

Helenium bigelovii var. *festivum* Jepson is a form growing with *H. bolanderi* and resembling it in the large hemispheric heads and long pappus-paleae which are two-thirds to nearly as long as the disk-corollas. It differs in the greater stature, the sometimes more slender peduncles, and the long petioliform bases of the lower leaves. Its variability suggests that it may be of hybrid origin as it often shows characteristics of the mesophytic form of *H. bigelovii* as well as *H. bolanderi*. Its type locality is designated as Humboldt Bay, Humboldt County, California.

49. **BAÈRIA** Fisch. & Mey. Ind. Sem. Hort. Petrop. **2**: 29. (Jan.) 1836.

Spring annuals (often perennial or biennial in coastal forms), glabrous or variously pubescent, the pubescence sometimes of septate or basally jointed hairs. Leaves opposite, sessile, often somewhat connate at the base, entire or laciniately pinnatifid. Heads pedunculate, terminal on the stems and branches, mostly many-flowered. Involucre cylindric to hemispheric; phyllaries in 1–2 series, subequal, distinct or somewhat united toward the base, some species with a prominent midvein, nearly plane and slightly carinate in age at the base to more or less strongly carinate above the base with the ray-achenes resting in the groove so formed, deciduous after achenes are shed, variously pubescent on the outer face and usually pubescent within at the tip. Receptacle subulate, conic or hemispheric, coarsely muricate to scrobiculate, the depressions conforming to the bases of the achenes. Ray-flowers yellow, often deeper yellow toward the base (very rarely white or white-tipped), pistillate, fertile. Disk-flowers perfect, fertile, with a slender glandular tube, campanulate or broadly funnelform throat, and spreading, 5-lobed limb. Anthers entire or subentire at base, appendaged apically with a deltoid tip or the appendages subfiliform in one species. Style-branches ovate or most often capitate, with or without an apiculate tip. Achenes usually obscurely 4-angled, somewhat compressed, narrowly clavate to nearly linear. Pappus of awns or scales or both or often wanting. [Name in honor of the Russian zoologist, Karl Ernst von Baer.]

A genus of 10 variable species, all natives of the Pacific Slope. Type species, *Baeria chrysostoma* Fisch. & Mey.

Quite often different taxa of specific rank grow in the same locality and even in the same habitat. In addition many aberrant forms occur, some of which may be derived from hybridization between otherwise distinct species.

Leaves entire.
  Plants perennial.
    Leaves oblong to broadly linear, 2–10 mm. wide.   1. *B. macrantha.*
    Leaves linear-filiform, 0.5–2 mm. wide.   2. *B. bakeri.*
  Plants annual.
    Receptacle subulate.
      Involucres nearly cylindric; rays wanting or vestigial.   5. *B. microglossa.*
      Involucres campanulate or turbinate; rays 2–5 mm. long.
        Stems wiry, erect, scarcely branched; anther-appendages subfiliform.
                        4. *B. leptalea.*
        Stems weak, freely branching; anther-appendages deltoid, acute.
                        6. *B. debilis.*
  Receptacle broadly or narrowly cone-shaped or dome-shaped.
    Receptacle dome-shaped (except in *B. fremontii heterochaeta*), muriculate and often finely hirsute;
      achenes 1 mm. long (leaves entire only on depauperate plants).
                        9. *B. fremontii.*
    Receptacle conical, muricate, often coarsely so; achenes 2.5–3.5 mm. long.
      Herbage spreading-hirsute or strigose or both.
        Herbage more or less hirsute with spreading hairs (with some finer, strigose or spreading
          hairs particularly about the inflorescence), and a few coarsely ciliate hairs on lower
          leaf-margins; achenes subclavate, the pappus-paleae 2–7 when present, clear or
          brownish, of subulate bristle-like awns tapering gradually from the base.
          Leaves linear, acute; stems slender, usually with erect or ascending branches.
                        3. *B. chrysostoma chrysostoma.*
          Leaves linear-oblong to oblong-spatulate; stems stoutish and widely branching.
                        3. *B. chrysostoma hirsutula.*
        Herbage mostly strigose only, with very few or no ciliate hairs on the lower leaf-margins;
          achenes nearly linear, the pappus-paleae 2–5 when present, whitish, ovate-lanceolate
          basally and tapering into a long awn.   3. *B. chrysostoma gracilis.*
      Herbage villous-tomentose or woolly to glabrous or nearly so.
        Phyllaries carinate in age only at the base, midvein not prominent, scarcely evident except
          at the base; plants villous-tomentose to woolly, scarcely glabrate.
                        7. *B. minor.*
        Phyllaries strongly carinate in age, the midvein prominent for nearly the whole length;
          plants glabrous to slightly villous-pubescent around the heads.
                        8. *B. platycarpha.*
Leaves pinnately parted or lobed, the lobes sometimes minute.
  Herbage variously pubescent but not glandular.
    Receptacle subulate; leaves sometimes with 2 or 3 short lobes.   6. *B. debilis.*
    Receptacle conical or dome-shaped.
      Pubescence hirsute and also strigose, ciliate on the leaf-bases.
        Perennial; 1–2 short lobes rarely present on some leaves of occasional plants.
                        1. *B. macrantha.*
        Annual; 1–2 minute lobes rarely present on some leaves of occasional plants.
                        3. *B. chrysostoma hirsutula.*
      Pubescence when present never hirsute or ciliate on the leaf-bases, finely pubescent or loosely vil-
        lous-tomentose or even woolly.
        Leaves broadly linear (2–11 mm. wide), the salient lobes 1–10 mm. long.
                        7. *B. minor.*
        Leaves linear-filiform, the divisions filiform-spreading, mostly 0.5–1 cm. long.
          Pappus when present alike, all paleae ovate, tapering to an awn; achenes 3–3.5 mm. long;
          involucre broadly turbinate.   8. *B. platycarpha.*
          Pappus when present dissimilar, of awns and intervening scales; achenes 1–1.5 mm. long;
         involucre broadly hemispheric.   9. *B. fremontii.*
  Herbage variously pubescent and also glandular; leaves nearly entire on depauperate plants.
                        10. *B. californica.*

5346

5347

5346. Helenium bigelovii   5347. Helenium bolanderi

1. Baeria macrántha (A. Gray) A. Gray. Perennial Baeria. Fig. 5348.

Burrielia chrysostoma var. *macrantha* A. Gray in Torr. Pacif. R. Rep. **4**: 106. 1857.
Baeria chrysostoma var. *macrantha* A. Gray, Proc. Amer. Acad. **9**: 196. 1874.
Baeria macrantha A. Gray, op. cit. **19**: 21. 1883.
Lasthenia macrantha Greene, Man. Bay Reg. 205. 1894.
Baeria macrantha var. *littoralis* Jepson, Man. Fl. Pl. Calif. 1112. 1925. (Nomen confusum)

Perennial, often subcespitose, the stems erect or ascending, leafy below, simple or branched, 12–40 cm. high, more or less pubescent with ascending or appressed hairs, arising from a stout taproot and several thickened, decumbent, underground stems. Leaves linear, slightly broadened at the somewhat clasping bases, 2–9 cm. long, the upper shorter than the lower, 2–4(5) mm. wide, acute to obtuse at apex, margins more or less ciliate, more conspicuously so at the clasping bases, occasionally with some cilia on leaf-surfaces, these sparsely to rather densely pubescent with ascending or appressed hairs; heads solitary on peduncles 5–12 cm. long; involucre broadly hemispheric, the phyllaries in 2 series 9–11.5 mm. long, ovate, acute, rather thin, the veins not prominent except in age; receptacle about 5 mm. high, conical, muricate; rays in the typical form 16–19 mm. high, elliptic, emarginate at apex; achenes 3 mm. long, clavate, somewhat flattened, surface papillate with minute, yellow, pointed hairs, the outer cellular coat easily separating from the black inner surface, completely epappose or with 1–3 brownish awns broadened at the base and nearly as long as the disk-corollas.

Along beaches, mostly in sand, Humid Transition Zone; limited to southern Sonoma County and Marin County, California. Type locality: Point Reyes, Marin County. Collected by Bigelow. March–June.

Baeria macrantha var. **thalassóphila** J. T. Howell, Leaflets West. Bot. **5**: 108. 1948. Low plants spreading by stout underground stems, the stems above the surface short and leafy, the leaves 3–5 cm. long, 7–10(15) mm. wide, usually densely ciliate and also pubescent, oblong or obovate; heads characteristically as large as *B. macrantha* var. *macrantha*, the rays usually shorter than those of var. *macrantha*; achenes epappose or sometimes with 1 or 2 brownish awns, these about as long as the disk-flowers or sometimes much reduced. Ocean bluffs, often in seepage areas, Mendocino County to San Mateo County and also in northern San Luis Obispo County, California. Type locality: ocean bluffs at Dillons Beach, Marin County. Intergrading freely with *B. macrantha* var. *pauciaristata* in Mendocino County.

Baeria macrantha var. **pauciaristàta** A. Gray, Proc. Amer. Acad. **19**: 21. 1883. Plants more or less cespitose from perennial roots, the stems ascending, 10–20 cm. long, leafy at the base, the leaves on the upper stems shorter than the internodes, mostly rounded at the apex, pubescence as in *B. macrantha* var. *macrantha*; rays about equaling the phyllaries in length or slightly shorter, scarcely emarginate; achenes as in the preceding taxa, the pappus of 1, 2, or 3 awns or none, the awns sometimes very short. Bluffs or coastal terraces, Curry County, Oregon, south to Sonoma County, California. Intergrading with var. *thalassophila* and also with *B. bakeri*. The specimens from Curry County often are linear-leaved. Type locality: Mendocino County, California. Collected by Bolander and by Pringle.

The taxa *B. macrantha*, *B. bakeri*, *B. chrysostoma* subsp. *hirsutula*, and *B. chrysostoma* subsp. *chrysostoma* occur on the coastal strip from Curry County, Oregon, to San Luis Obispo County and the Channel Islands, California. These highly variable taxa are closely related morphologically, all having to a greater or lesser degree broad-based, tapering, marginal, ciliate hairs at least on the leaf-bases, as well as the finer, appressed, or ascending hairs that are found on leaves, upper stems, and inflorescence. The pappus, if present at all, consists of 1–5 pale brownish awns slightly widened at the base. The achenes are subclavate in outline and somewhat compressed, glabrous or more or less covered with yellowish, pointed papillae. Occasional individuals also are found that appear to be hybrids between one or another of the preceding taxa and the maritime representatives of *B. minor* that have a similar distribution.

2. Baeria bàkeri J. T. Howell. Baker's Baeria. Fig. 5349.

Baeria bakeri J. T. Howell, Leaflets West. Bot. **1**: 7. 1932.
Baeria macrantha var. *bakeri* Keck, Aliso **4**: 101. 1958.

Erect perennial from fascicled thickened roots, stems erect, solitary or sometimes 2, sometimes branched, 25–40 cm. high, more or less pubescent with asecnding or appressed hairs. Basal leaves 6–10 cm. long, linear-filiform, ciliate-margined, usually not persisting; stem-leaves connate at the base, as long as or longer than the basal but reduced in length above, linear, acute at apex, up to 2 mm. wide, nearly glabrous except for marginal cilia; heads solitary on peduncles 6–15 cm. long; involucre hemispheric, the phyllaries thin, 5–8 mm. long, ovate or elliptic, pubescent, carinate at the base in age; rays 8–9 mm. long, elliptic, usually emarginate at apex; disk-flowers 2.5 mm. long; receptacle about 3 mm. high, conic, muricate; achenes 2.5–3 mm. long, narrowly subclavate, somewhat compressed, more or less scabrous with minute, yellowish, pointed hairs; pappus lacking or of 1 or 2 short awns.

Open grassy slopes and swales in forested or shrubby areas adjacent to the ocean, Humid Transition Zone; Mendocino and Sonoma Counties, California. Type locality: 6 miles south of Point Arena, Mendocino County.

Intergrading with *B. macrantha* the resulting forms, though of different growth habit, showing to some degree characteristics of *B. bakeri*, such as narrow leaves or a tendency toward thickened clustered roots.

3. Baeria chrysóstoma Fisch. & Mey. Coast Goldfields. Fig. 5350.

Baeria chrysostoma Fisch. & Mey. Ind. Sem. Hort. Petrop. **2**: 29. (Jan.) 1836.
Baeria gracilis var. *aristosa* A. Gray, Proc. Amer. Acad. **19**: 21. 1883.
Baeria aristosa Howell, Fl. N.W. Amer. 354. 1900.

Rather slender, usually reddish-stemmed annual 1–2.5 dm. high, branching from the base and above, the branches ascending and rather sparsely pubescent with spreading and appressed hairs, more densely so below the heads. Leaves greenish, (1.5)2.5–6 cm. long, 1–3 mm. wide, linear, acute at the apex, sparsely hirsute and hirsute-ciliate on the margins at the slightly widened, clasping base; heads solitary on the branches, the peduncles 3–10 cm. long; involucres broadly hemispheric; phyllaries 8–12, ovate-lanceolate or oblong-ovate, 5–9 mm. long, rather thin, midvein not prominent but phyllaries carinate below in age, more or less hirsute; receptacle 2.5–3 mm. high, conical, muricate; rays (7)8–10 mm. long, rather narrowly oblong, sometimes reddish-veined; disk-flowers 2–2.5 mm. long; achenes 2 mm. or more long, linear-clavate and somewhat rounded above, smooth or with minute yellowish papillae or coarse hairs; pappus completely lacking or 2–7 clear or brownish, slender, subulate awns of variable lengths, often as long as the achenes.

Fields and open slopes, Upper Sonoran and Humid Transition Zones; Curry County, Oregon, southward mostly on the coast and in the Outer Coast Ranges to Santa Barbara County, California, and extending into the Inner Coast Ranges where it merges with *B. chrysostoma* subsp. *gracilis* and possibly with other named forms growing in that area. Type locality: vicinity of Bodega Bay, Sonoma County. March–May.

Part of an unstable complex which is related to the perennial species of the genus.

Baeria chrysostoma subsp. **hirsùtula** (Greene) Ferris, Contr. Dudley Herb. **5**: 99. 1958. (*Lasthenia hirsutula* Greene, Man. Bay Reg. 206. 1894; *Baeria hirsutula* Greene, Fl. Fran. 438. 1897.) In its characteristic form 2.5–10 cm. tall, rather stout, much branched from the base and above, the width of the plants sometimes greater than the height, more rarely simple-stemmed, short-hirsute throughout and also more or less strigose; leaves 10–20 mm. long, oblong-spatulate or linear-oblong, rounded at the apex, very rarely with 1 or 2 minute spreading lobes; phyllaries about 5 mm. long, usually broader than those of the preceding taxon, strongly carinate in age, the midvein evident below; rays about 5 mm. long, typically broadly oblong; achenes as in *B. chrysostoma* subsp. *chrysostoma*, smooth or with minute, coarse, yellowish hairs, epappose or with 2–4 clear or brownish, slender, subulate awns as long as the disk-flowers. Coastal terraces and dunes or among rocks, from Mendocino County, California, to San Luis Obispo County and the Channel Islands, Santa Barbara County. Type locality: not given, but said to grow "from Marin Co. southward."

Possibly interbreeding with *B. chrysostoma* subsp. *chrysostoma* with which it grows. Forms occur frequently which variously combine the characteristics of both taxa. On the coast of San Mateo County specimens having large heads and broad leaves but annual roots appear to be intergrades with the *B. macrantha* complex.

Baeria chrysostoma subsp. **grácilis** (DC.) Ferris, Contr. Dudley Herb. **5**: 100. 1958. (*Burrielia gracilis* DC. Prod. **5**: 664. 1836; *B. tenerrima* DC., loc. cit.; *B. longifolia* Nutt. Trans. Amer. Phil. Soc. II. **7**: 380. 1841; *B. parviflora* Nutt. op. cit. 381; *B. hirsuta* Nutt. loc. cit.; *Baeria gracilis* A. Gray, Proc. Amer. Acad. **9**: 196. 1874; *B. palmeri* A. Gray, Bot. Calif. **1**: 376. 1876; *B. gracilis* var. *paleacea* A. Gray, Proc. Amer. Acad. **19**: 21. 1883; *B. clevelandii* A. Gray, op. cit. 22; *B. gracilis* var. *tenerrima* A. Gray, Syn. Fl. N. Amer. **1²**: 326. 1884; *B. palmeri* var. *clementina* A. Gray, op. cit. ed. 2. **1²**: 452. 1886; *B. chrysostoma* var. *gracilis* H. M. Hall, Univ. Calif. Pub. Bot. **3**: 170. 1907; *B. chrysostoma* var. *gracilis* f. *nuda* H. M. Hall, loc. cit.; *B. chrysostoma* var. *gracilis* f. *tenerrima* H. M. Hall, op. cit. 171; *B. chrysostoma* var. *gracilis* f. *paleacea* H. M. Hall, loc. cit.; *B. chrysostoma* var. *gracilis* f. *clementina* H. M. Hall, loc. cit.; *B. chrysostoma* var. *gracilis* f. *crassa* H. M. Hall, op. cit. 172; *B. chrysostoma* var. *gracilis* f. *curta* H. M. Hall, loc. cit.). Slender annuals 1–2 dm. high, in the typical form strigose throughout with some spreading or ascending hairs present, more conspicuously pubescent on the peduncles, heads, and leaves; leaves 8–15(20) mm. long, usually about 2 mm. wide toward the base, somewhat ciliate on the lower margins of the leaves but rarely with coarser cilia; involucre hemispheric, the phyllaries (4)8–12, ovate-lanceolate, usually about 5 mm. long, strongly carinate in age, the midvein prominent for one-half or more the length of the phyllary; receptacle as in the other two related taxa; achenes 2 mm. long, nearly linear, apically truncate (sometimes rounded in epappose heads), glabrous or minutely strigulose with whitish hairs; pappus-paleae when present 2–5, opaque, whitish (somewhat brownish in older herbarium specimens), about the length of the corollas, typically a small ovate-lanceolate scale attenuate into a long slender awn. Open slopes and plains, Jackson County, Oregon, south through the Sacramento and San Joaquin Valleys and adjacent foothills to southern California and northern Lower California and eastward through the deserts to central Arizona. Type locality: California. Collected by Douglas, presumably near Monterey. Goldfields.

Very abundant and more widespread than the other members of the *B. chrysostoma* complex. Variable but characterized principally by the more or less dense strigose pubescence, the usually small heads, and the well-defined pappus-scales terminated by attenuate awns. The pappus-paleae vary considerably in different areas. For a discussion of this variation see H. M. Hall (Univ. Calif. Pub. Bot. **3**: 170–172. 1907).

4. **Baeria leptàlea** (A. Gray) A. Gray. Salinas Valley Baeria. Fig. 5351.

Burrielia leptalea A. Gray, Proc. Amer. Acad. **6**: 546. 1865.
Baeria leptalea A. Gray, Syn. Fl. N. Amer. **1²**: 325. 1884.

Very slender, erect annual 7–15 cm. high, mostly simple-stemmed, glabrous or somewhat pubescent on peduncles and involucres. Leaves 3–10 mm. long, entire, filiform or nearly so; involucres campanulate to nearly turbinate; phyllaries 4–6, rather broadly oblong, 4–5 mm. long, thin, ciiiate on the margins and nearly glabrous, obscurely carinate at the base; rays 3–4 mm. long, oblong, notched at the apex; lobes of the disk-corollas narrowly triangular; anther-appendages 0.5 mm. or more long, filiform or nearly so; receptacle subulate, shorter than the phyllaries;

5348

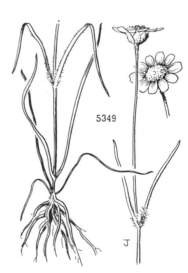

5348. Baeria macrantha

5349. Baeria bakeri

achenes 2 mm. long, brown, glabrous or with minute hairs toward the apex; pappus of 1–4 pale or brownish bristles tapering from a somewhat broadened base, the bristles about equaling the disk-corollas.

Fields and slopes, Upper Sonoran Zone; of rather rare occurrence in interior Monterey and San Luis Obispo Counties, California (*Brewer, Brandegee, Abrams, Ornduff*). Type locality: Nacimiento River, Monterey County. April.

Under adverse growing conditions depauperate specimens of *B. chrysostoma* subsp. *gracilis* have few-flowered heads and nearly subulate receptacles. The plants are strigose and have phyllaries that are strongly carinate and lack the subfiliform anther-appendages of *B. leptalea*. More difficult to separate from *B. leptalea* are aberrant, dwarfed, nearly glabrous specimens from the Inner North Coast Ranges of the awned forms of *B. chrysostoma* subsp. *chrysostoma*. None of those examined, however, has the diagnostic characters of subulate receptacles and subfiliform anther-appendages.

5. Baeria microglóssa (DC.) Greene. Small-rayed Baeria. Fig. 5352.

Burrielia microglossa DC. Prod. **5**: 664. 1836.
Lasthenia microglossa Greene, Man. Bay Reg. 205. 1894.
Baeria microglossa Greene, Fl. Fran. 438. 1897.
Pentachaeta laxa Elmer, Bot. Gaz. **41**: 318. 1906.

Weak, slender-stemmed, erect or reclining annuals 5–20 cm. high, the stems arising singly or several from the base, usually branched above, sparsely pubescent throughout with ascending hairs. Leaves thin, linear, entire, 1.5–4 cm. long, usually about 2 mm. wide; peduncles 1–3 cm. long; involucres cylindric, the phyllaries 3 or 4, oblong or narrowly ovate and acute at apex, 6.5–8 mm. long, ciliate-margined, thin and leafy, not carinate in age; receptacle subulate; ray-flowers 1–3, the rays minute or wanting; disk-flowers usually less than 10, about 2 mm. long, the throat broadly funnelform; achenes black, nearly linear, 3.5–4 mm. long, minutely and sparsely hispidulous; pappus-paleae 2–4, surpassing the disk-flowers, narrowly lanceolate and attenuate into an awn, or pappus-paleae lacking in some flowers of the heads.

Grassy shaded slopes or among rocks and boulders, Sonoran Zones; Tehama County, California, southward in the Coast Ranges to San Diego County; also in the Death Valley region in Inyo County and the western edges of the deserts from Kern County to Riverside County. Type locality: California. Collected by Douglas. March–April.

6. Baeria débilis Greene ex A. Gray. Greene's Baeria. Fig. 5353.

Baeria debilis Greene ex A. Gray, Syn. Fl. N. Amer. 1²: 325. 1884.

Lax-stemmed annual, erect or reclining, 5–30 cm. high, the stems arising singly or several from the base, simple or with few ascending branches, sparsely soft-villous with microscopically septate ascending hairs, more densely so below the heads. Leaves flaccid, linear, 1–6(8) cm. long, 1–5 mm. wide, entire, more rarely with 1–3 lobes; peduncles 2–7 cm. long; involucre campanulate to broadly turbinate, the phyllaries 5, about 5 mm. long, broadly elliptic to narrowly ovate, thin, veins not prominent in age, keeled at the extreme base; receptacle subulate, 2.5–4 mm. high, about 1 mm. wide at base; ray-flowers yellow or white, the rays 3–4(5) mm. long, narrowly elliptic and usually toothed; disk-flowers 15–20, about 2 mm. long, tube slender, throat campanulate; achenes narrow, 2.5–3 mm. long, hispidulous; pappus-paleae 2–4, white, ovate-lanceolate and tapering to a slender awn, a little shorter than the disk-flowers, or pappus-paleae lacking.

Grassy woodlands or among boulders, Upper Sonoran Zone; lower foothills of the Sierra Nevada and adjacent plains, Mariposa County, California, south to Kern County. Type locality: plains of Fresno County. March–April.

7. Baeria mìnor (DC.) Ferris. Woolly Baeria or Goldfields. Fig. 5354.

Monolopia minor DC. Prod. **6**: 74. 1837.
Dichaeta tenella Nutt. Trans. Amer. Phil. Soc. II. **7**: 383. 1841.
Dichaeta uliginosa Nutt. loc. cit.
Baeria uliginosa A. Gray, Proc. Amer. Acad. **9**: 197. 1874.
Baeria uliginosa var. *tenella*, A. Gray, loc. cit.
Baeria uliginosa var. *tenera* A. Gray, op. cit. **19**: 22. 1883.
Baeria tenella Greene, Fl. Fran. 439. 1897.
Eriophyllum minus Rydb. N. Amer. Fl. **34**: 86. 1915.
Baeria minor Ferris, Contr. Dudley Herb. **4**: 332. 1955.

Annuals 1–3 dm. high, erect and simple-stemmed below or diffusely branched from the base or even prostrate; stems weak and somewhat succulent or firm and strictly erect in drier situations; herbage very sparsely villous to densely lanate-villous with long, septate, often twisted hairs. Leaves broadly or narrowly ligulate or linear, 2–6 cm. long, the undivided portion of the leaves 2–11 mm. broad, completely entire or with (1)2–5 salient linear lobes 2–10 mm. long; peduncles of various lengths, usually about 1–5 cm. long but as much as 9 cm. long; involucres broadly hemispheric, the phyllaries 8–15, thin, ovate, acute at apex, somewhat carinate toward the base in age; receptacle glabrous, conic or sometimes in large heads merely convex, coarsely muricate when mature, the papillae bearing the achenes scarcely fused toward the apex of the receptacle and appearing as distinct pedicels; ligules of ray-flowers usually oblong and emarginate or toothed at the apex, 4–8 mm. long; disk-flowers many, tube glandular, throat campanulate; achenes somewhat angled, about 2–2.5 mm. long, hispidulous or smooth; pappus completely lacking on ray- and disk-achenes or of 2–4 awns nearly as long as the corolla and (2)4–6 truncate fimbriate scales, these variable and often completely cleft, particularly on the ray-achenes where they may be vestigial and awns only present.

In damp soil near streams, seepage areas, or depressions in fields and by roadsides, Sonoran and Transition Zones; common along the coast from Sonoma County to Santa Barbara County, California, and eastward through

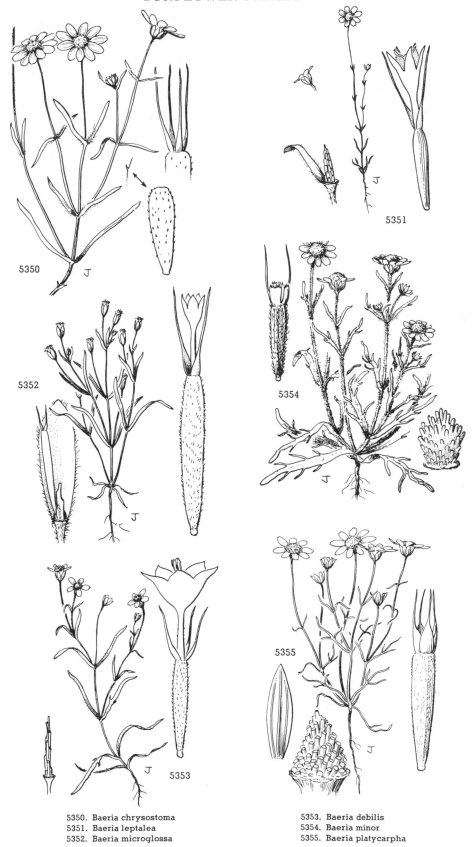

5350. Baeria chrysostoma
5351. Baeria leptalea
5352. Baeria microglossa

5353. Baeria debilis
5354. Baeria minor
5355. Baeria platycarpha

the Coast Ranges and Sacramento–San Joaquin Valley to the lower Sierra foothills, where it occurs from Amador County south to Kern County. Type locality: California. Collected by Douglas. March–June.

Baeria minor subsp. **marítima** (A. Gray) Ferris, Contr. Dudley Herb. **4**: 334. 1955. (*Burrielia maritima* A. Gray, Proc. Amer. Acad. **7**: 358. 1868; *Baeria maritima* A. Gray, op. cit. **9**: 196. 1874.) Diffuse succulent annuals, sparsely soft-villous as in the preceding taxon to completely glabrous except for the inflorescence; leaves broadly linear, entire to ligulate, and laciniately lobed, the rachis up to 11 mm. wide; peduncles usually short; involucres and receptacles as in the preceding; ray-flowers few, 3–8, the ligule minute, about 2 mm. long, oval or oblong, often emarginate at apex; achenes glabrous and epappose, or hispidulous and bearing 3–5 awns alternating with the deeply cleft, truncate paleae. Occasional along the coast in seepage areas, from British Columbia to Monterey County, California. Type locality: Farallone Islands, San Francisco County, California.

This taxon differs from *B. minor* subsp. *minor* in the few and minute rays and the tendency to have very short peduncles on the spring plants.

8. Baeria platycárpha (A. Gray) A. Gray. Alkali Goldfields. Fig. 5355.

Burrielia platycarpha A. Gray in Torr. Bot. Mex. Bound. 97. 1859.
Baeria platycarpha A. Gray, Proc. Amer. Acad. **9**: 196. 1874.
Baeria carnosa Greene, Bull. Torrey Club **10**: 86. 1883.
Lasthenia platycarpha Greene, Man. Bay Reg. 205. 1894.
Lasthenia carnosa Greene, loc. cit.

Erect annuals 10–30 cm. high with rather wiry, often reddish stems branching from the base (single-stemmed in small plants), also somewhat branched above with ascending branches, glabrous or with sparse, soft, short-villous pubescence below the heads. Leaves 1.5–7 cm. long, about 1.5 mm. wide, somewhat succulent, linear, usually with 2(3) short linear lobes or entire, glabrous or sparsely short-villous about the leaf-bases; peduncles 2–8 cm. long; involucres broadly turbinate, the phyllaries 8–9, about 7 mm. long, elliptic to ovate, rather firm in texture and somewhat reticulate in age, keeled in age, usually 3-nerved, the midvein conspicuous, extending almost to the tardily reflexed tip; receptacle conical, strongly muricate, especially toward the summit of the receptacle; rays 7–8 mm. long, elliptic; disk-flowers 3 mm. or more long; achenes 3–3.5 mm. long, somewhat angular, hispidulous; pappus-paleae 4–7, when present as long as or longer than the corolla-tube, rather firm, ovate, abruptly tapering to an awn, the margin microscopically erose.

Alkaline or saline soil, Sonoran Zones; Sacramento Valley, California, from Tehama County to the lower San Joaquin Valley in Merced and Fresno Counties. Type locality: valley of upper Sacramento. March–May.

Plants from Tehama County differ in having longer pappus-paleae and more abundant pubescence.

9. Baeria fremóntii (Torr. ex A. Gray) A. Gray. Fremont's Baeria. Fig. 5356.

Dichaeta fremontii Torr. ex A. Gray, Mem. Amer. Acad. II. **4**: 102. 1849.
Burrielia fremontii Benth. Pl. Hartw. 317. 1849.
Baeria fremontii A. Gray, Proc. Amer. Acad. **9**: 196. 1874.
Baeria burkei Greene, Bull. Calif. Acad. **2**: 151. 1886.

Somewhat succulent annuals 10–30 cm. high, stems arising singly or several from the base, usually branching below the middle with erect branches, herbage pubescent (often sparsely so) with fine, soft, appressed or ascending hairs, mostly of equal length, densely so on the upper part of the peduncles. Leaves 1–3.5(5.5) cm. long, mostly linear-filiform, with 2–4 divergent filiform lobes 0.5–1 cm. long; heads many-flowered, the involucre broadly hemispheric, rarely obscurely glandular, the erect peduncles 5–15 cm. long; phyllaries 4–5 mm. long, thin, spreading in age and 1–3 keeled at the base, the margins finely ciliate; receptacle 2 mm. or more high, dome-shaped, glabrous or hairy, very finely muriculate to nearly scrobiculate; rays broadly elliptic, emarginate at apex; disk-flowers about 2 mm. long, the tube conspicuously glandular, the throat broadly funnelform; achenes 1 mm. long or slightly longer, angled, scarcely compressed, more or less finely scabridulous, rarely glabrous; pappus of 4 (sometimes 1–3) awns as long as or somewhat longer than the corolla-tube, with usually 2 minute, broadly acute, lacerate scales between the awns, these rather variable in length and sometimes completely divided; pappus rarely absent.

Mostly in vernal pools or ditches, Lower Sonoran Zone; Tehama and Butte Counties, California, south to Kern County in the Sacramento–San Joaquin Valley and in adjacent valleys in the Inner Coast Ranges as far south as the eastern edge of Santa Barbara County and the Sierra Nevada foothills. Type locality: California. April–May.

Baeria burkei Greene is an aberrant form characterized only by the pappus, which consists of 10–12 very minute, acute scales and a single long slender awn about as long as the disk-corolla. Known only from the type collection from Ukiah, Mendocino County, California, and northwest of Windsor, Sonoma County (*Jepson 9303*).

Baeria fremontii var. **heterochaèta** Hoover, Leaflets West. Bot. **1**: 229. 1936. Like the preceding taxon in growth-form; differing in the broadly conical, glabrous or hairy, muriculate receptacle, achenes to 1.5 mm. or more in length, and pappus-scales between the awns much dissected and of unequal length, with some awn-like. Growing with *B. fremontii* var. *fremontii* and occasional from Sacramento and San Joaquin Counties southward to Kern County, California. Type locality: Semitropic, Kern County.

The longer achenes and the conical receptacle suggest relationship to the *B. californica* complex of southern California.

Baeria fremontii var. **cónjugens** (Greene) Ferris, Contr. Dudley Herb. **5**: 100. 1958. (*Lasthenia conjugens* Greene, Pittonia **1**: 221. 1888. A less pubescent plant in which the phyllaries are plainly united for less than half their length and the achenes (about 1.5 mm. long) are smooth and without pappus. Found only in Solano and Contra Costa Counties, California, where it occurs with *B. fremontii* var. *fremontii*. Type locality: Antioch, Contra Costa County.

10. Baeria califórnica (Hook.) Chambers. Glandular or Southern Baeria. Fig. 5357.

Hymenoxys californica Hook. Bot. Mag. **67**: *pl. 3828*. 1840.
Ptilomeris aristata Nutt. Trans. Amer. Phil. Soc. II. **7**: 382. 1841.

Ptilomeris coronaria Nutt. loc. cit.
Ptilomeris mutica Nutt. loc. cit.
Ptilomeris anthemoides Nutt. loc. cit.
Hymenoxys calva Torr. & Gray, Fl. N. Amer. **2**: 381. 1842.
Ptilomeris affinis Nutt. Proc. Acad. Phila. **4**: 20. 1848.
Ptilomeris tenella Nutt. loc. cit.
Baeria tenella A. Gray, Proc. Amer. Acad. **19**: 23. 1883. Not *Dichaeta tenella* Nutt. 1841.
Baeria aristata Coville, Proc. Biol. Soc. Wash. **13**: 121. 1889.
Baeria parishii S. Wats. Proc. Amer. Acad. **24**: 83. 1889.
Baeria aristata f. *mutica* H. M. Hall, Univ. Calif. Pub. Bot. **3**: 173. 1907.
Baeria aristata f. *anthemoides* H. M. Hall, op. cit. 174.
Baeria aristata var. *affinis* H. M. Hall, loc. cit.
Baeria aristata var. *affinis* f. *truncata* H. M. Hall, loc. cit.
Baeria aristata var. *parishii* H. M. Hall, op. cit. 175.
Baeria aristata var. *parishii* f. *varia* H. M. Hall, loc. cit.
Baeria aristata var. *parishii* f. *quadrata* H. M. Hall, loc. cit.
Baeria californica Chambers, Contr. Dudley Herb. **5**: 68. 1957.

Sweet-scented annual 1–3 dm. high, stems simple or branched above, with ascending or spreading branches, more rarely with several stems from the base, sparsely (more densely so below the heads) pubescent throughout with slender, spreading, jointed hairs of unequal length and also glandular-pubescent. Leaves 1.5–5 cm. long, typically shorter than the internodes, linear or linear-filiform, pinnately parted into long filiform lobes, sometimes bipinnate on vigorous plants and rarely nearly entire on depauperate specimens; peduncles 1–10 cm. long, the involucre hemispheric; phyllaries 4–5 mm. long, ovate-lanceolate or narrowly ovate, thin, the midrib prominent, strongly carinate in fruit and partly enclosing the ray-achenes; ray-flowers (6)8–15, the rays 3–10 mm. long, oblong to linear-oblong, emarginate or toothed at the apex, rarely entire; disk-flowers about 2 mm. long, tube slender and densely glandular, throat campanulate; receptacle about 2 mm. high, conical, glabrous or hirsute, mostly scrobiculate; achenes 2 mm. or more long, slender, linear, scabridulous, slightly compressed and obscurely 2-4-nerved; pappus-paleae 8–12, often of variable shapes in the same head, margins erose or fimbriate especially above, lanceolate or oblong and tapering to awns as long as the disk-corollas or shorter, or truncate or obtuse and either awnless or abruptly awned, or completely lacking; paleae of the ray-flowers usually differing from those of the disk-flowers.

Open places, Sonoran Zones; cismontane southern California in Ventura, Los Angeles, and San Bernardino Counties south to northern Lower California. Type locality: San Diego, San Diego County. Collected by Nuttall. April–May.

Extremely variable as to pappus and pubescence. *Baeria aristata* var. *affinis* and the other synonyms associated with it (*B. tenella* and *B. parishii*) differ somewhat in having fewer rays (6–8), these 3–4 mm. long, which are often entire at the tip. The plants are sometimes less glandular. These forms are more commonly found away from the coast but scarcely reaching as far as the desert. The original collction was made at Los Angeles.

50. **LASTHÈNIA** Cass. Opusc. **3**: 88. 1834.

Somewhat succulent spring annuals, glabrous to sparsely puberulent especially about the inflorescence. Leaves opposite, entire, sessile and slightly connate at base. Heads pedunculate, terminal on the branches or on the simple stems, the peduncle enlarged below the flower. Ray- and disk-flowers yellow. Involucre hemispheric, the phyllaries in a single

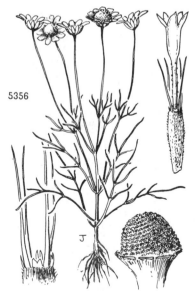

5356. Baeria fremontii

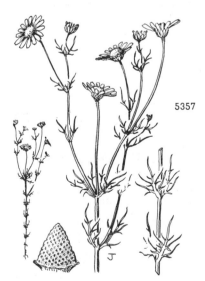

5357. Baeria californica

series and connate into a ciliate-toothed cup. Receptacle high-conical, muricate. Pistillate ray-flowers fertile; hermaphrodite disk-flowers regular, the tube slender, the throat widely campanulate, the lobes spreading. Anthers obtuse at base. Style-branches somewhat flattened, truncate. Achenes narrowly obovoid in outline, somewhat compressed and inconspicuously 2–4-angled; pappus none or of firm, awned or lacerate paleae. [A Greek courtesan, pupil of Plato.]

A genus of 3 species, two natives of the Pacific States and one of Chile. Type species, *Lasthenia obtusifolia* Cass.

Ligules of the ray-flowers inconspicuous, shorter than the involucre; pappus of erose or laciniate or awn-pointed
  scales.   1. *L. glaberrima*.
Ligules of the ray-flowers conspicuous, evidently surpassing the involucre; pappus none.   2. *L. glabrata*.

1. Lasthenia glabérrima DC. Smooth Lasthenia. Fig. 5358.

Lasthenia glaberrima DC. Prod. **5**: 664. 1836.
Lasthenia minima Suksd. Allg. Bot. Zeit. **12**: 7. 1906.

Plants mostly simple, 10–35 cm. long, weakly ascending and with few branches, rooting from the lower nodes, essentially glabrous throughout, sometimes puberulent about the involucres. Leaves 4–8 cm. long, linear, obtuse; heads on peduncles 2–15 cm. long; involucres 5–7 mm. high, the teeth acute, ciliate; ligules of the ray-flowers shorter than the involucres; disk-flowers about 1 mm. long, glandular, shorter than the achenes; achenes dull, somewhat compressed, the faces bluntly angled; pappus of 5–10 firm, erose-laciniate to awn-tipped paleae.

Vernal pools and muddy places, Upper Sonoran Zone; southwestern Washington mostly west of the Cascade Mountains southward through western Oregon and the Coast Ranges and Sacramento Valley, California, to Santa Clara and Alameda Counties. Type locality: California. Collected by Douglas. March–May.

2. Lasthenia glabràta Lindl. Yellow-rayed Lasthenia. Fig. 5359.

Lasthenia glabrata Lindl. Bot. Reg. **21**: *pl. 1780.* 1835.
Lasthenia californica DC. ex Lindl. loc. cit. under *pl. 1780.*

Erect plants 1.5–6 dm. high, simple-stemmed or less commonly branched from the base, corymbosely branched above, glabrous or more or less puberulent with curved hairs on the upper stems especially just below the heads. Leaves 3–10 cm. long, somewhat fleshy, more or less connate at base; heads many-flowered, borne on peduncles 1–10 cm. long, these enlarged at the summit; involucre 5–8 mm. high, broadly hemispheric, the acute tips of the phyllaries ciliate; ligules of the ray-flowers 5–10 mm. long; disk-flowers about 3 mm. long, the tube glandular; achenes narrowly oblong-ovoid, noticeably compressed, appearing smooth to the naked eye, occasionally a few achenes in the head bearing a few yellow, gland-like points, the thin cellular coat shining, epappose.

In heavy soil, borders of salt marshes and vernal pools in valleys and plains, Upper Sonoran Zone; Humboldt County to San Luis Obispo County, California, in the Coast Ranges and in the Sacramento and San Joaquin Valleys as far south as Kern County. Type locality: California. March–May.

Lasthenia glabrata var. **coùlteri** (Greene) A. Gray, Syn. Fl. N. Amer. **1**²: 324. 1884. (*Lasthenia coulteri* Greene, Bull. Calif. Acad. **1**: 192. 1885.) Habit of *L. glabrata* var. *glabrata*, differing in having the usually less compressed achenes more or less covered with yellow, gland-like points. Saline marshes and alkaline plains, southern Salinas Valley and southern Sacramento Valley to southern California where it is the typical form. Type locality: southern California. March–May.

51. CROCKÉRIA Greene ex A. Gray, Syn. Fl. N. Amer. 1²: 445. 1884.

Glabrous or sparsely pubescent herbs, mostly simple-stemmed below, branched above. Leaves opposite, linear, sessile, shortly connate at base, cymosely branched, the many-flowered heads borne on rather long peduncles. Involucre hemispheric, the phyllaries thin, united more than half their length, the tips triangular and sparsely pilose. Receptacle conic. Ray-flowers pistillate, fertile. Disk-flowers perfect, many, tube slender, longer than the expanded throat and lobes. Anthers appendaged above, rounded at base. Style-branches short. Achenes flat, striate, obovate, with callus at the base and ciliate with a marginal fringe of short, stout, glandular hairs. Pappus none. [Name in honor of Charles Crocker, a patron of California botany.]

A monotypic genus.

1. Crockeria chrysántha Greene. Crockeria. Fig. 5360.

Crockeria chrysantha Greene ex A. Gray, Syn. Fl. N. Amer. **1**²: 445. 1884.

Erect annuals, simple or with several stems arising from the base in vigorous plants, 8–28 cm. tall. Leaves linear, slightly fleshy, 1–5 cm. long, shorter than or equaling the internodes; heads radiate, many-flowered, golden yellow, the peduncles 3–6 cm. long, slightly enlarged and pubescent below the heads; involucre about 5 mm. high and 10 mm. or less broad, the triangular lobes 10–12, pilose on the inner surface and margin, about 2 mm. long; rays 6–10, narrowly oblong, 6–7 mm. long, with 2–3 shallow teeth at the apex; disk-flowers glandular-puberulent, about 2.5 mm. long; achenes about 2 mm. long, dull black and more or less granular, striate, with a dense marginal fringe of stout, clavate, glandular hairs and a white callus at base, epappose.

Fields, in alkaline soil, Lower Sonoran Zone; San Joaquin Valley, California, Stanislaus County south to Kings and Tulare Counties. Type locality: near Lake Tulare, Tulare County. Collected by Greene. March–April.

52. BAHÍA Lag. Gen. & Sp. Pl. 30. 1816.

Annual or perennial, pubescent but not at all woolly, often glandular herbs, sometimes suffrutescent at base but usually herbaceous. Leaves alternate or opposite, entire to much dissected. Heads terminating the branches, yellow, many-flowered, radiate or rarely discoid. Involucre hemispheric. Phyllaries in 1 or 2 series, subequal, herbaceous, flat, spreading in age and sometimes reflexed. Receptacle flat. Ray-flowers when present pistillate, fertile. Disk-flowers perfect, fertile, the tube glandular, the throat ample, the lobes about equaling the throat. Anthers obtuse at base, acute above. Style-branches truncate or obtuse. Achenes narrowly obpyramidal, 4-angled. Pappus-paleae with a callous base and thickened midrib as long as or shorter than the paleae, or paleae completely lacking. [Named in honor of Juan Francisco Bahi, a Spanish botanist.]

A genus of about 15 species, natives of southwestern United States, Mexico, and western South America. Type species, *Bahia ambrosioides* Lag.

Pappus wanting; stout biennials or perennials.	1. *B. dissecta.*
Pappus present; slender annuals.	2. *B. neomexicana.*

1. Bahia disséacta (A. Gray) Britt. Dissected Bahia. Fig. 5361.

Amauria ? dissecta A. Gray, Mem. Amer. Acad. II. **4**: 104. 1849.
Villanova chrysanthemoides A. Gray, Smiths. Contr. **5**[6]: 96. 1853.
Bahia chrysanthemoides A. Gray, Proc. Amer. Acad. **19**: 28. 1883.
Bahia dissecta Britt. Trans. N.Y. Acad. **8**: 68. 1889.
Amauriopsis dissecta Rydb. N. Amer. Fl. **34**: 37. 1914.

Biennial or short-lived perennial herbs 3–8 dm. high; stems striate, reddish, puberulent and glandular, simple to the loosely branching inflorescence. Leaves petioled, the basal leaves with

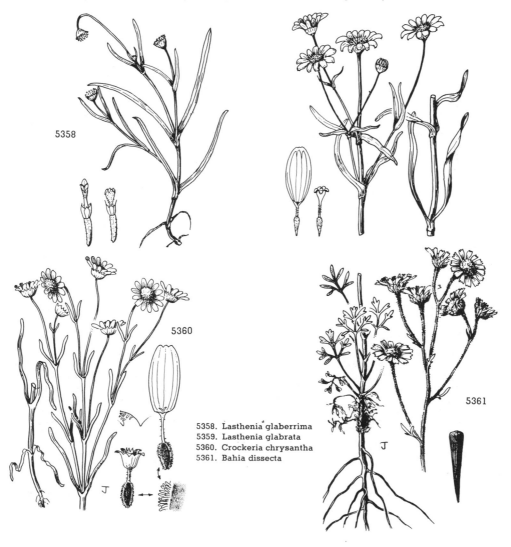

5358. Lasthenia glaberrima
5359. Lasthenia glabrata
5360. Crockeria chrysantha
5361. Bahia dissecta

longer petioles than the upper stem-leaves, ternately parted into oblong or linear, blunt lobes, 2–7 cm. long, puberulent; peduncles 2–4 cm. long, enlarged below the head, densely beset with gland-tipped hairs; involucre 5–6 mm. high, about twice as broad, glandular-hairy; phyllaries 16–20, narrowly obovate, abruptly acuminate above; rays narrowly cuneate, 3-cleft, 6–8 mm. long; achenes black, striate and 4-angled, glabrous or minutely glandular-puberulent especially above, epappose.

Gravelly places along streams and roadsides, Arid Transition Zone; San Bernardino and Santa Rosa Mountains, California, and mountains of northern Lower California east to southern Nevada, Wyoming, Colorado, New Mexico, and Chihuahua. Type locality: a few miles east of Mora River, New Mexico. Aug.–Oct.

2. Bahia neomexicàna A. Gray. New Mexico Bahia. Fig. 5362.

Schkurhia neomexicana A. Gray, Mem. Amer. Acad. II. **4:** 96. 1849.
Amblyopappus neo-mexicanus A. Gray in Torr. Pacif. R. Rep. **4:** 106. 1857.
Bahia neomexicana A. Gray, Proc. Amer. Acad. **19:** 27. 1883.
Achyropappus neo-mexicanus A. Gray ex Rydb. Fl. Colo. 377. 1906.
Cephalobembix neomexicana Rydb. N. Amer. Fl. **34:** 46. 1914.

Slender annuals 1–2 dm. high, single-stemmed below or branching from the base, the stems often reddish, sparsely pubescent with ascending or appressed hairs or glabrate, glandular-puberulent below the heads. Leaves opposite, often alternate and entire above, 2–3 cm. long, impressed-punctate and sparsely hispidulous, usually tripartite into linear-filiform divisions; heads 3-flowered or more, borne on the ends of the branchlets on peduncles 0.5–3 cm. long; involucre obconic; phyllaries 5–8, obovate, unequal in width, about 6 mm. long, herbaceous and often purplish-tinged on the margin, glandular-punctate or glandular-puberulent and also sparsely hirsute; ray-flowers none; disk-flowers pale yellow or whitish; achenes glandular, very narrowly obpyramidal, about 3–6 mm. long, sparsely strigose, densely white-hirsute at base with ascending hairs; pappus-paleae 7–8, equal, about 1.5 mm. long, obovate, obtuse at apex, hyaline but callous-thickened at the base, this often bearing a dark spot.

Sandy washes and slopes, Arid Transition Zone; known in California only from Clark Mountain, eastern San Bernardino County (*Roos 4955*); Arizona and northern Lower California to Colorado, New Mexico, and Chihuahua. Type locality: Santa Fe, New Mexico. Aug.–Oct.

53. MONOLÒPIA DC. Prod. **6:** 74. 1837.

White-lanate, erect, spring annuals simple or much branched at or above the base. Lower leaves opposite and narrowed to a petioliform base, the upper alternate, sessile, and reduced upward. Heads solitary, terminating the branches. Involucres hemispheric or campanulate. Phyllaries in 1 series, foliaceous, free or connate in a lobed cup. Receptacle conic, naked. Ray-flowers pistillate, equaling the phyllaries in number, fertile, yellow, sub-bilabiate, the posterior lobe minute, the tube glandular-hispidulous. Disk-flowers yellow, few to many, hermaphrodite, fertile, 5-lobed, the lobes hairy within, the tube glandular-hispidulous. Stamens 5, the anthers rounded below, bearing apical ovate appendages above. Style-branches of ray-flowers slender, obtuse, of the disk-flowers stoutish, obtuse or subtruncate. Achenes strigulose to glabrate, obpyramidal in outline, the ray-achenes triquetrous but little obcompressed, the disk-achenes more or less obcompressed and carinate or quadratish, epappose. [From the Greek meaning single and husk, alluding to the bracts of the involucre.]

A genus of 4 species, all natives of California. Type species. *Monolopia major* DC.

Phyllaries distinct to base.
　　Limb of ray-flowers rounded, apically emarginate or denticulate; disk-achenes essentially as wide as thick.
　　　　Peduncles strongly divergent; plants of the Coast Ranges, Contra Costa and San Mateo Counties to San
　　　　　　Luis Obispo County.　　　　　　　　　　　　　　　　　　　　　　1. *M. gracilens.*
　　　　Peduncles strict; plants of southern San Joaquin Valley and adjacent slopes.　　2. *M. stricta.*
　　Limb of the ray-flowers truncate, apically dentate; disk-achenes obcompressed.　　3. *M. lanceolata.*
Phyllaries united one-half their length into a lobed cup.　　　　　　　　　　　4. *M. major.*

1. Monolopia grácilens A. Gray. Woodland Monolopia. Fig. 5363.

Monolopia gracilens A. Gray, Proc. Amer. Acad. **19:** 20. 1883.
Monolopia major var. *gracilens* J. F. Macbride, Contr. Gray Herb. No. 56: 49. 1918.

Stems 1–4 dm. high, simple below or sometimes branching from the base, the branches of midstem divaricately spreading, white-lanate and somewhat deciduous in age. Basal and lower leaves crowded, narrowly or broadly oblanceolate, dentate, 3–7 cm. long, tardily withering; the upper broadly lanceolate, dentate to nearly entire, acute or acuminate, 2–10 cm. long, mostly reduced and becoming bract-like at the inflorescence; heads subcorymbose on divergently ascending peduncles, these 2.5–12 cm. long; phyllaries 7–11, free, black-lanate, narrowly or broadly ovate, 5–6 mm. long; ray-flowers 7–11, yellow, with 7–11 greenish veins, entire or shallowly emarginate at apex; disk-flowers yellow; achenes black or dark brown, essentially glabrous, the ray-achenes slightly convex and somewhat dorsally carinate, a little shorter than the disk-achenes, these 2 mm. long.

Open woods or grassy or rocky slopes, Upper Sonoran and Transition Zones; central California from Contra Costa County south through the Outer Coast Ranges to San Luis Obispo County. Type locality: New Almaden, Santa Clara County. April–June.

2. **Monolopia strícta** Crum. Crum's Monolopia. Fig. 5364.

Monolopia stricta Crum, Madroño **5**: 258. 1940.

Stems 1–6 dm. high, simple or much branched from the base and above, all the branches strictly ascending, white-lanate to floccose and becoming glabrate. Basal and lower leaves oblanceolate, entire, 3.5–8 cm. long and soon withering, upper stem-leaves lanceolate to oblong-lanceolate, apex obtuse or acutish, 3.5–6 cm. long, usually entire, equaling or shorter than the internodes; heads on the lower as well as upper branches, the peduncles about 3–5 cm. long; phyllaries about 8, free, ovate or ovate-lanceolate, usually black-lanate at the apex, 5–7 mm. long; ray-flowers about 8, 8–10-nerved, oblong, rounded at apex and entire or minutely few-toothed, 4–15 mm. long; disk-flowers yellow; achenes densely grayish-strigulose, 2.5 to nearly 3 mm. long, the ray-achenes obcompressed, the disk-achenes not obcompressed.

Plains or open slopes in the foothills, Lower Sonoran Zone; Inner Coast Ranges, San Benito County, California, to Kern County and eastward in the San Joaquin Valley to Tulare County. Type locality: Lost Hills, Kern County. March–April.

3. **Monolopia lanceolàta** Nutt. Common Monolopia. Fig. 5365.

Monolopia lanceolata Nutt. Proc. Acad. Phila. **4**: 21. 1848.
Monolopia major var. *lanceolata* A. Gray, Bot. Calif. **1**: 384. 1876.

Stems 1–4.5 dm. high, simple or branched from the base and above, the branches diffusely spreading, densely white-lanate and becoming somewhat glabrate. Basal and lower cauline leaves

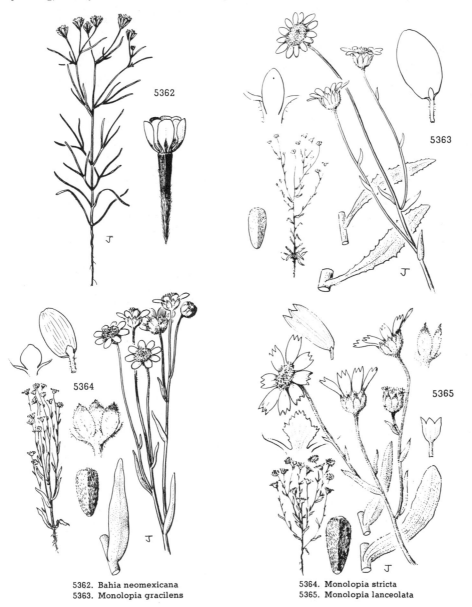

5362. Bahia neomexicana
5363. Monolopia gracilens
5364. Monolopia stricta
5365. Monolopia lanceolata

broadly lanceolate, obtuse at apex, 1.5–10 cm. long, soon withering, the upper stem-leaves linear to broadly lanceolate, obtuse or acute, usually entire, 3–11 cm. long, mostly much shorter than the internodes; heads terminating the upper branches, the divergent peduncles 1–13 cm. long; phyllaries about 8, free or 2 or 3 rarely united, lanceolate to ovate- or rhombic-lanceolate, 5–11 mm. long, white-lanate below and black-lanate at apex; ray-flowers about 8, yellow, 8–10-nerved, narrowly or broadly oblong, truncate at apex and 2–3-dentate, 9–21 mm. long, the disk-flowers yellow; achenes mostly gray-strigose, 0.5–4 mm. long, the ray-achenes obcompressed, convex and somewhat flattened and often subcarinate dorsally, the disk-achenes obcompressed and carinate dorsally and ventrally.

Open slopes and valleys, Upper and Lower Sonoran Zones; Inner Coast Ranges, San Joaquin County to San Luis Obispo County and the mountains of southern California in Riverside and Los Angeles Counties and the eastern edges of the San Joaquin Valley from Fresno County to the Tehachapi region; also the extreme western edge of the Mojave Desert. Type locality: Los Angeles. March–May. Closely related to *M. major*.

4. **Monolopia màjor** DC. Cupped Monolopia. Fig. 5366.

Monolopia major DC. Prod. **6**: 74. 1837.

Stems about 1–5 dm. high, simple or branched from the base and above, the branches diffusely spreading, densely white-lanate and becoming somewhat glabrate. Basal and lower cauline leaves oblanceolate, obscurely dentate to subentire, obtuse, acute or acuminate, from about 1–10 cm. long, the upper leaves about 1–10 cm. long, the uppermost reduced in size, the margins usually dentate, sometimes subentire; heads terminating the branches, the divergent peduncles 1–12 cm. long; phyllaries usually 8, united for about one-half their length, the involucre 8–13 mm. long, the lobes ovate or deltoid, 3–6 mm. long, white-lanate, black-lanate at apex; ray-flowers usually 8, yellow, 9–13-nerved, oblanceolate to cuneate-oblong, truncate at apex and 2–3-dentate, usually deeply so, 8–20 mm. long, the disk-flowers yellow; achenes black or blackish brown, somewhat strigose, becoming glabrate, 2.5–4 mm. long, the ray-achenes obcompressed, the disk-achenes obcompressed and more or less carinate dorsally and ventrally.

Open slopes and valleys, Upper and Lower Sonoran Zones; mostly in the Inner Coast Ranges from Tehama County, California, to Monterey County and the adjacent western borders of the Great Valley. Type locality: California. Collected by Douglas. March–May.

54. **ERIOPHÝLLUM*** Lag. Gen. & Sp. Pl. 28. 1816.

Shrubby or herbaceous perennials or annuals, the herbage tomentose or floccose. Leaves essentially alternate, usually lobed, toothed, or divided, more often entire in the annual species. Inflorescence various, the heads solitary on the branchs in leafy clusters, cymose or loosely or closely corymbose; disk-flowers yellow and ray-flowers yellow except in two annual species. Phyllaries in 1 series, firm, mostly carinate and somewhat concave on the inner surface, permanently erect, the tip only often reflexed, distinct or slightly united at the base (more in one species). Receptacle convex or conic to nearly flat (modified in *E. mohavense*), naked (with a few hyaline scales in one species and subulate processes attached basally to the phyllaries in another). Ray-flowers usually few, rarely wanting, pistillate, fertile, the ligules yellow, oval to oblong, toothed or lobed at the apex. Disk-flowers yellow, perfect, narrowly pubescent or glabrous. Anthers obtuse at base, bearing appendages above. Style-branches flattened, the apex obtuse or deltoid, or cuneate in *E. wallacei* and *E. lanosum*. Achenes slender, 4-nerved or -angled or 5-nerved or -angled in some annual species, sometimes somewhat compressed. Pappus of scarious, nerveless, usually erose or fimbriate paleae sometimes slender and awn-like, rarely epappose. [From the Greek words meaning wool and leaf.]

A genus of about 13 often variable species, all native of western North America. Type species, *Eriophyllum trollifolium* Lag.
For complete synonymy of the perennial species, see Constance, Univ. Calif. Pub. Bot. **18**: 69–136. 1937.

Perennials, somewhat shrubby or suffruticose at base, 1–10(15) dm. high.
  Heads 1.5–3 cm. broad, solitary or several on long peduncles.
    Plants shrubby only at base; rays 8–20 mm. long (less in some varieties); of widespread distribution.
                      1. *E. lanatum*.
    Plants shrubby throughout; rays 6–8 mm. long; central Inner Coast Ranges in California only.
                      2. *E. jepsonii*.
  Heads 1.5 cm. or less broad, several to many in sessile clusters or in short, pedunculate, branching clusters.
    Ultimate branches bearing inflorescences slender; phyllaries 4–6, overlapping; plants of various habitats.
                      3. *E. confertiflorum*.
    Ultimate branches bearing inflorescences stout; phyllaries 8–12, scarcely overlapping; strictly maritime plants.
      Leaves glabrous above, tomentose beneath, entire or pinnately parted into few lobes; rays 3–5 mm. long.
                      4. *E. staechadifolium*.
      Leaves equally tomentose on both surfaces, bipinnatifid into many lobes; rays 2 mm. long.
                      5. *E. nevinii*.
Small herbaceous annuals 2–20 cm. high (sometimes higher in *E. congdonii*).
  Pappus-paleae of unequal length, lanceolate or awn-tipped paleae alternating with short obtuse paleae.
    Rays yellow; plants of the Sierra Nevada foothills.   7. *E. congdonii*.
    Rays white or pink-tinged; plants of the desert regions.   9. *E. lanosum*.

---

* Treatment based in part on extensive notes made by Sherwin Carlquist.

Pappus-paleae all alike and of equal length or sometimes lacking.
    Heads many-flowered with never less than 9 or 10; rays present (except in *E. pringlei*).
        Receptacle more or less dome-shaped or conical; peduncles at least 1 cm. long (sometimes shorter in
            E. wallacei), generally very much longer.
            Rays scarcely exceeding the disk; plants of the higher Sierra Nevada.
                                                      6. *E. nubigenum*.
            Rays evidently exceeding the disk; plants of desert ranges, and desert slopes of the Sierra Ne-
              vada and the Tehachapi Range.
                Stems openly branched, single or several from the base; pappus-paleae, when present, hya-
                  line.   8. *E. ambiguum*.
                Stems many from the base but little branched; pappus-paleae, when present, opaque.
                                          10. *E. wallacei*.
        Receptacle flat or nearly so; heads in leafy clusters sessile or the peduncles usually not more than
            5 mm. long.
            Ray-flowers present.   11. *E. multicaule*.
            Ray-flowers absent.   12. *E. pringlei*.
    Heads few-flowered with never more than 4; rays absent.   13. *E. mohavense*.

1. **Eriophyllum lanàtum** (Pursh) Forbes. Common Woolly-sunflower or Eriophyllum. Fig. 5367.

Actinella lanata Pursh, Fl. Amer. Sept. 560. 1814.
Eriophyllum caespitosum Dougl. ex Lindl. Bot. Reg. **14**: *pl. 1167*. 1828.
Eriophyllum lanatum Forbes, Hort. Woburn. 183. 1833.
Bahia leucophylla DC. Prod. **5**: 657. 1836.

    Loosely tomentose perennial, erect or decumbent from a woody base, the stout stems (1)2–6 dm. high, the tomentum tending to be deciduous in age. Leaves 2–6 cm. long, permanently white-woolly below, glabrate or glabrous above, variable in size and outline, the lower spatulate or oblanceolate, entire or lobed, the stem-leaves oblong-lanceolate to obovate and pinnatifid, the uppermost quite narrow and usually entire; heads solitary or loosely corymbose on peduncles 5–10 cm. long; involucre hemispheric, 8–10 mm. high, 10–12 mm. broad; phyllaries 8–12, loosely floccose or glabrate, linear to ovate-lanceolate, strongly carinate, distinct to the base but with overlapping margins; ray-flowers 8–12, yellow, 10–20 mm. long; disk-flowers 3–4 mm. long with a glandular hairy tube; achenes 3–4 mm. long, oblong in outline, narrowed toward the base, angled, glabrous, the pappus-paleae 4–10, usually obtuse and erose, sometimes linear-lanceolate and acute, not exceeding 2 mm. in length.

    In thickets and open dry places, Transition Zone; common west of the Cascade Mountains from British Columbia to southern Oregon and adjacent Del Norte County, California, and on the eastern side in the drainage of the Columbia River in southeastern Washington and northeastern Oregon eastward to western Montana. Type locality: banks of the Kooskooskie. May–July.

    Eriophyllum lanatum var. **achillaeoìdes** (DC.) Jepson, Man. Fl. Pl. Calif. 1118. 1925. (*Bahia achillaeoides* DC. Prod. **5**: 657. 1836; *Eriophyllum ternatum* Greene, Pittonia **3**: 185. 1897; *E. idoneum* Jepson, Fl. W. Mid. Calif. 524. 1901; *E. greenei* Elmer, Bot. Gaz. **41**: 313. 1906; *E. cusickii* Eastw. ex Rydb. N. Amer. Fl. **34**: 93. 1915, as a synonym; *E. achillaeoides* var. *aphanactis* J. T. Howell, Leaflets West. Bot. **3**: 126. 1942.) Stems 3–7 dm. high, not strongly woody at base; leaves pinnately to tripinnately divided, the upper leaves sometimes entire or nearly so; heads loosely corymbose or solitary on slender peduncles, 3–10 cm. long; involucres 6–8 mm. high; rays 6–9 mm. long (lacking in var. *aphanactis*); achenes short, 2.5–3 mm. long, cuneate-oblong, hispidulous or glabrous; pappus present. Occasional in Oregon from Lane and Harney Counties to northern California and occurring more commonly in the California Coast Ranges as far south as the Santa Cruz Mountains, and in eastern California, though occurring less commonly from Modoc County to Mariposa County. Type locality: California.

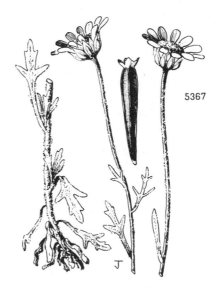

5366

5367

5366. Monolopia major   5367. Eriophyllum lanatum

Eriophyllum lanatum var. grandiflòrum (A. Gray) Jepson, Fl. W. Mid. Calif. 524. 1901. (*Bahia lanata* Benth. Pl. Hartw. 317. 1849, not *B. lanata* DC. 1836; *Egletes californicus* Kell. Proc. Calif. Acad. 1: 56. 1855; *B. lanata* var. *grandiflora* A. Gray, Bot. Calif. 1: 381. 1876; *Eriophyllum caespitosum* var. *grandiflorum* A. Gray, Proc. Amer. Acad. 19: 26. 1883; *E. speciosum* Greene, Erythea 1: 149. 1893.) Stems 3–10 dm. high from a short rootstock, stout, leafy below, scarcely so above; leaves woolly on both surfaces, entire and linear or lanceolate or laciniately toothed and sometimes pinnatifid; heads large, solitary at the ends of the stems (rarely more), the peduncles 1–3 dm. long; involucres 8–10 mm. high; rays 10–20 mm. long; achenes 4 mm. long, linear-clavate, more or less pubescent with soft hairs; pappus present. In the foothills of northern California in Del Norte, Siskiyou, and Shasta Counties south to Humboldt and Colusa Counties and in the foothills on the eastern side of the Sacramento Valley as far south as Mariposa County. Type locality: Sacramento Valley, California. Collected by Hartweg.

Eriophyllum lanatum var. arachnoìdeum (Fisch. & Avé.-Lall.) Jepson, Man. Fl. Pl. Calif. 1119. 1925. (? *Eriophyllum trollifolium* Lag. Gen. & Sp. Pl. 28. 1816, identity uncertain; *Bahia arachnoidea* Fisch. & Avé-Lall. Ind. Sem. Hort. Petrop. 9: 63. 1842; *B. latifolia* Benth. Bot. Sulph. 30. 1844; *B. lanata* var. *brachypoda* A. Gray, Bot. Calif. 1: 381. 1876; *Eriophyllum caespitosum* var. *latifolium* A. Gray, Proc. Amer. Acad. 19: 26. 1883, in part.) Stems 3–6 dm. high, branching from a woody base and usually decumbent, usually densely leafy to the corymbose inflorescence; herbage loosely floccose except for the glabrous upper surface of the leaves; leaves with 3–5 incised teeth or lobes; heads on peduncles 3–10 cm. long; involucres 8–11 mm. high; rays 8–10 mm. long; achenes 3–4 mm. long, turbinate, glabrous; pappus reduced to a crown of erose teeth. Principally in the redwood belt, Del Norte County, California, south to Monterey County. Type locality: California, presumably from the area around Fort Ross, Sonoma County.

Eriophyllum lanatum var. cròceum (Greene) Jepson, Man. Fl. Pl. Calif. 1118. 1925. (*Eriophyllum caespitosum* var. *latifolium* A. Gray, Proc. Amer. Acad. 19: 26. 1883, in part; *E. croceum* Greene, Erythea 3: 124. 1895.) Stems 1.5–6 dm. high, mostly simple, leafy throughout to the inflorescence; leaves silky-lanate beneath, green, lightly floccose to glabrate above, coarsely serrate or lobed; heads solitary or few at ends of the branches, the peduncles 3–8 cm. long; involucres 5–8 mm. high; rays 8–10 mm. long; achenes as in the preceding form. In the Arid Transition Zone on the western face of the Sierra Nevada, California, from Butte County to Tulare County. Type locality: "Amador and Calaveras County hills . . ." Much resembling *E. lanatum* var. *arachnoideum* and differing principally in the vesture of the leaves and to a less degree in the leaf-margins.

Eriophyllum lanatum var. hállii Constance, Proc. Nat. Acad. Sci. 20: 411. 1934. Plants 3–4 dm. high with rather leafy stems arising from the leafy base; leaves loosely floccose on both surfaces, pinnately incised or pinnatifid; heads solitary or few at the ends of the branches on slender peduncles 5–12 cm. long; involucres 8–12 mm. high; rays 10–13 mm. long; tube of disk-corollas glabrous, differing in this character from all the other related taxa; achenes 4–5 mm. long, narrowly oblong, glabrous or somewhat pubescent; pappus present. Known only from the vicinity of Fort Tejon, Kern County, California, the type locality.

Eriophyllum lanatum var. cuneàtum (Kell.) Jepson, Man. Fl. Pl. Calif. 1118. 1925. (*Bahia cuneata* Kell. Proc. Calif. Acad. 5: 49. 1873; *B. integrifolia* A. Gray, Bot. Calif. 1: 381 (in part). 1876, not *B. integrifolia* DC. 1836; *Eriophyllum chrysanthum* Rydb. N. Amer. Fl. 34: 89. 1915; *E. bolanderi* Rydb. op. cit. 91; *E. cineraria* Rydb. op. cit. 93.) Plants 2–4 dm. high, the stems decumbent or spreading from the woody base, tomentose or floccose throughout; leaves coarsely toothed or lobed, less so above; head solitary on the branches, the peduncles 5–10 cm. long; involucres 8–10 mm. high; rays 8–10 mm. long; achenes 3–4 mm. long, linear-clavate, glabrous or hairy; pappus evident but glabrous. In the Sierra Nevada, California, from southeastern Lassen County and Butte County to Placer County and in adjacent Washoe County, Nevada. Type locality: Cisco, Placer County, California.

Eriophyllum lanatum var. integrifòlium (Hook.) Smiley, Univ. Calif. Pub. Bot. 9: 378. 1921. (*Trichophyllum integrifolium* Hook. Fl. Bor. Amer. 1: 316. 1833; *T. multiflorum* Nutt. Journ. Acad. Phila. 7: 35. 1834; *Bahia gracilis* Hook & Arn. Bot. Beechey 353. 1840; *Eriophyllum caespitosum* var. *integrifolium* A. Gray, Proc. Amer. Acad. 19: 26. 1883, in part; *E. caespitosum* var. *leucophyllum* A. Gray, loc. cit., not *Bahia leucophylla* DC. 1836; *E. watsonii* A. Gray, loc. cit.; *E. lutescens* Rydb. N. Amer. Fl. 34: 87. 1915; *E. monoense* Rydb. loc. cit.; *E. trichocarpum* Rydb. op. cit. 89; *E. nevadense* Gandoger, Bull. Soc. Bot. Fr. 65: 40. 1918.) Stems 1–2 dm. high, many, erect or decumbent from a woody base or short caudex, the herbage persistently tomentose, canescent or floccose; lower leaves entire or 3–5-toothed or -lobed at the apex, the stem-leaves incised or pinnatifid above into 3 divisions; heads solitary or few on peduncles 3–10 cm. long; involucres 6–8 mm. high; rays 6–10 mm. long; achenes mostly hairy, clavate, the pappus present but extremely variable in this complex taxon. In the Pacific Northwest east of the Cascade Mountains in Oregon and Washington to Montana and Wyoming; in California it occurs in the northeastern counties southward in the Sierra Nevada, along the crest and on the eastern face, as far south as Tulare and Inyo Counties and also on the adjacent higher ranges in Nevada. Type locality: sources of the Columbia River.

Eriophyllum lanatum var. obovàtum (Greene) H. M. Hall, Univ. Calif. Pub. Bot. 3: 186. 1907. (*Eriophyllum caespitosum* var. *integrifolium* A. Gray, Proc. Amer. Acad. 19: 26. 1883, in part; *E. obovatum* Greene, Erythea 3: 123. 1895; *E. brachylepis* Rydb. N. Amer. Fl. 34: 88. 1915.) Stems few to several, 2–4 dm. high, leafy, erect or decumbent at the woody base, herbage persistently tomentose throughout; leaves entire or coarsely serrate toward the apex, narrowly or broadly oblanceolate, the lower leaves wider; heads mostly solitary on peduncles 3–10 cm. long; involucre 7–12 mm. high, campanulate; rays 6–10 mm. long; achenes 2–3 mm. long, cuneate-oblong, glabrous; pappus present. In the southern Sierra Nevada and Greenhorn Range in Tulare County and adjacent Kern County and in the San Bernardino Mountains of southern California. Type locality: San Bernardino Mountains.

Eriophyllum lanatum var. lanceolàtum (Howell) Jepson, Man. Fl. Pl. Calif. 1118. 1925. (*Eriophyllum lanceolatum* Howell, Fl. N.W. Amer. 355. 1900; *E. rixfordii* Eastw. Proc. Calif. Acad. IV. 20: 158. 1931.) Stems several, 2–4 dm. high, persistently woolly, stout; leaves entire or coarsely serrate, woolly on both surfaces; heads solitary to several, the peduncles stout and somewhat swollen below the heads; involucres 8–12 mm. high, more or less hemispheric; rays 7–10 mm. long; achenes hairy; pappus present. Not uncommon, Siskiyou region of Curry, Josephine and Jackson Counties, Oregon, south to Trinity and Humboldt Counties, California. Type locality: Jackson County, Oregon. One of the more distinct and less variable taxa in the *lanatum* complex except for its resemblance to var. *obovatum* of the southern Sierra Nevada and mountains of southern California.

2. Eriophyllum jepsònii Greene. Jepson's Eriophyllum. Fig. 5368.

Eriophyllum jepsonii Greene, Pittonia 2: 165. 1891.

Plants woody to well above the base, 5–8 dm. high, the stems closely white-tomentulose, the internodes long. Leaves alternate, 3–6 cm. long, floccose and glabrate above, pinnatifid into 5–7 linear obtuse lobes; heads few, loosely corymbiform at the ends of the branches, the peduncles 5–10 cm. long; involucres 5–7 mm. high, campanulate, as broad as long; phyllaries 6–8 and overlapping, ovate, acute, carinate below, loosely floccose, becoming glabrate; ray-flowers 6–8 with elliptic or oblong, yellow ligules 6–10 mm. long; disk-flowers 3–5 mm. long, the glandular-puberulent or hispid tube about the length of the funnelform throat; achenes narrowly clavate, 4-angled, more or less appressed-hairy; pappus-paleae unequal, the narrowly lanceolate paleae alternating with the oblong erose ones.

In open wooded slopes or edges of the chaparral, Upper Sonoran Zone; Mount Diablo, Contra Costa County, south through the Mount Hamilton Range to San Benito County, California. Type locality: between Arroyo Mocho and Arroyo Valle, Alameda County. April–May.

Though markedly different in its typical form, this taxon shows close affinities with *E. confertiflorum* var. *confertiflorum* and especially with *E. confertiflorum* var. *laxiflorum*. In the southern part of the Mount Hamilton

Range, Santa Clara County, and in San Benito County intermediates are to be found in which the inflorescence is more densely flowered and the peduncles are shorter.

Eriophyllum tanacetiflòrum Greene, Pittonia **2**: 21. 1889. (*Eriophyllum confertiflorum* var. *tanacetifolium* Jepson, Man. Fl. Pl. Calif. 1116. 1925.) Somewhat shrubby, 3–6 dm. high, with erect, mostly simple, leafy stems, tomentum densely pannose and persistent. Leaves like those of *E. jepsonii* but mostly shorter; inflorescence of close corymbose clusters with 3–10 heads, nearly sessile or with peduncles to 2.5 cm. long; involucre densely tomentose, 5–7 mm. high, 5–6 mm. broad, the phyllaries overlapping, ovate, broadly acute to obtuse, carinate at the base; rays 4–6 or lacking, 4–5 mm. long; disk-flowers 3.5–4 mm. long, glandular-hairy, tube short, throat narrowly campanulate; achenes 3.5 mm. long, strigose to nearly glabrous; pappus-paleae of unequal length. Foothills of the Sierra Nevada, California, in Calaveras and Mariposa Counties. Type locality: between Sheep Ranch and Murphy's, Calaveras County.

Eriophyllum latilòbum Rydb. N. Amer. Fl. **34**: 94. 1914. Plants perennial, stems leafy, 3–4 dm. high; leaves rhombic to obovate in outline, deeply 3-lobed, the divisions often toothed or lobed, glabrous above, tomentose beneath; heads 10 or more in loose clusters on peduncles 1–2.5 cm. long; involucres 4–5 mm. high, 7–10 mm. broad; phyllaries 6–8, ovate and somewhat overlapping, rays 6–8 mm. long, entire, oval or oblong; tubes of disk- and ray-flowers glandular; achenes 3 mm. long, glabrous or very sparsely hispidulous, the pappus-paleae lanceolate, laciniate. A very restricted endemic found in San Mateo County in the hills west of San Mateo, where it grows on slopes covered with mixed California live oak, buckeye, and shrubs.

3. Eriophyllum confertiflòrum (DC.) A. Gray. Yellow Yarrow. Fig. 5369.

Bahia confertiflora DC. Prod. **5**: 657. 1836.
Eriophyllum confertiflorum A. Gray, Prod. Amer. Acad. **19**: 25. 1883.
Eriophyllum confertiflorum var. *discoideum* Greene, Man. Bay Region 207. 1894.
Eriophyllum cheiranthoides Rydb. N. Amer. Fl. **34**: 95. 1915.
Eriophyllum biternatum Rydb. op. cit. 96.
Eriophyllum tridactylum Rydb. loc. cit.
Eriophyllum crucigerum Rydb. loc. cit.

Tomentose plants, the tomentum more or less deciduous, the stems woody below and branching, the upper stems erect and slender, 2–6 dm. high. Leaves 1–4 cm. long, the upper surface becoming glabrate, once- or twice-pinnatifid into slender linear lobes (sometimes broader earlier in the season), the upper leaves sometimes merely cleft; heads sessile or nearly so, in clusters at the ends of the branches; involucres campanulate, 3–5 mm. high and equally broad; phyllaries distinct, broadly elliptic, carinate, strongly overlapping; ray-flowers 2–4(5) mm. long, few, sometimes lacking, the rays oval or oblong; disk-flowers 2–3 mm. long, glandular-hirsute or puberulent, the tube shorter than the narrowly campanulate throat; achenes linear-clavate, 4-angled, hispidulous or sparsely glandular; pappus-paleae about 1 mm. long, few to several, erose.

In open chaparral and rocky places, mostly Upper Sonoran Zone; Outer and Inner Coast Ranges from Mendocino and Tehama Counties south to southern California, and in the foothills of the Sierra Nevada from Calaveras County to Tulare County and the mountain ranges of southern California. Type locality: California. Collected by Douglas. April–Aug.

As indicated by Constance (Univ. Calif. Pub. Bot. **18**: 107–109. 1937), this is a widespread and polymorphic complex in which the individuals are seemingly susceptible to ecological modification and the growth forms show much seasonal change. Various species and varieties have been based on such distinct characters as discoid heads and reduction of the number of pinnae of the leaves. These characters occur so sporadically throughout the range of *E. confertiflorum* var. *confertiflorum* that the named taxa which were described from the characters are here considered to be synonyms of the typical form. Two of the variants which show a definite geographical distribution associated with the characters that distinguish them are retained as taxonomic entities.

Eriophyllum confertiflorum var. **laxiflòrum** A. Gray. Proc. Amer. Acad. **19**: 25. 1883. (*Bahia tenuifolia* DC. Prod. **5**: 657. 1836; *Eriophyllum tenuifolium* Rydb. N. Amer. Fl. **34**: 96. 1915.) Habit and height of the name-bearing taxon; upper stems very slender; leaves with narrow divisions; inflorescence open, few-flowered, the peduncles 1–3 cm. long; heads small; rays often longer than in *E. confertiflorum* var. *confertiflorum*. Dry slopes, in California from Santa Clara County to the Tehachapi Mountains in the Inner Coast Ranges, the eastern face of the Santa Lucia Mountains, and the higher foothills of the Sierra Nevada in Mariposa County; also appearing sporadically in the mountains of southern California. Here the inflorescences usually have more heads and the peduncles tend to be of equal length. Type locality: California, presumably on drier slopes of the Santa Lucia Mountains. Collected by Douglas and also by Coulter.

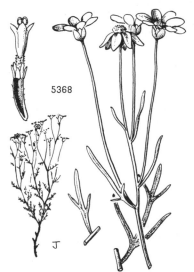

5368. Eriophyllum jepsonii

5369. Eriophyllum confertiflorum

Eriophyllum confertiflorum var. làtum H. M. Hall, Univ. Calif. Pub. Bot. **3**: 186. 1907. Of much the same habit as *E. confertiflorum* var. *confertiflorum* but differing in the leaves, which are 7–14 mm. wide, broader, spatulate or obovate, with blunt and short lobes; upper leaves sometimes entire. Local about San Bernardino, Riverside, and Pasadena, southern California. Type locality: plains near Riverside, Riverside County, California.

4. Eriophyllum staechadifòlium Lag. Seaside Woolly-sunflower or Lizard Tail. Fig. 5370.

Eriophyllum staechadifolium Lag. Gen. & Sp. Pl. 28. 1816.
Bahia staechadifolium var. *californica* DC. Prod. **5**: 656. 1836.
Eriophyllum staechadifolium var. *depressum* Greene, Bull. Calif. Acad. **2**: 404. 1887.

Shrubby perennial, much branched, the leafy branches 3–10 dm. long, spreading and decumbent, forming usually rounded shrubs or mats, tomentum tardily dehiscent; leaves linear or linear-oblanceolate, entire or with few linear, horizontal or divergent lobes, glabrous above, densely tomentose beneath, the margins revolute; heads short-peduncled in rather close clusters in loosely branched inflorescence; involucres campanulate, 4–5 mm. high; phyllaries 8–11, distinct to base, carinate, obtuse or acute, scarcely if at all overlapping; ray-flowers sometimes lacking, the ligules elliptic, entire or toothed at apex, 3–5 mm. long; disk-flowers glandular, about 4 mm. long; achenes linear-oblong in outline, angled, glandular and somewhat hispidulous; pappus-paleae 8–12, unequal, mostly obtuse, erose.

Adjacent to the sea, mostly on bluffs or beaches, Humid Transition Zone; Santa Cruz and Monterey Counties, California. Type locality: Monterey, California. April–Sept.

Growing with and intergrading with the following form.

Eriophyllum staechadifolium var. **artemisiaefòlium** (Less.) J. F. Macbride, Contr. Gray Herb. No. 59: 39. 1919. (*Bahia artemisiaefolia* Less. Linnaea **5**: 160. 1830, **6**: 253. 1831; *Bahia artemisiaefolium* var. *douglasii* DC. Prod. **5**: 657. 1836; *E. artemisiaefolium* O. Kuntze, Rev. Gen. Pl. **1**: 336. 1891.) Much more common than *E. staechadifolium* var. *staechadifolium* and probably not much more than a form as it differs only in the leaves which are always lobed, often deeply so; though the rachis is broader than the linear leaves of the preceding, the lobes are often incised or toothed. Intergrades are common where the distribution of the two coincide. Coos County, Oregon, southward along the coast to Santa Barbara County, California, and the Channel Islands. Type locality: San Francisco, California. Collected by Chamisso.

5. Eriophyllum nevínii A. Gray. Nevin's Eriophyllum. Fig. 5371.

Eriophyllum nevinii A. Gray, Syn. Fl. N. Amer. ed. 2. 1². : 452. 1886.

Shrubby perennial about 1 m. high; stems stout, decumbent, densely leafy, with a dense tomentum, becoming glabrate, bearing the essentially naked corymbiform, much-branched, many-flowered inflorescence. Leaves 8–20 cm. long, broadly ovate in outline, bipinnatifid into narrow obtuse lobes, tomentose on both surfaces, the petiole 2–4.5 cm. long; heads very short-peduncled to sessile; involucres campanulate, 6–7 mm. long; phyllaries 6–8, oblong, obtuse; ligules of ray-flowers 2 mm. long; disk-flowers 3–4 mm. long, throat longer than the glandular-pubescent tube; achenes 3–4 mm. long, narrowly obpyramidal, 4-angled and hispidulous on the angles; pappus-paleae 3–6, entire or erose.

Among rocks and on ocean bluffs, Upper Sonoran Zone; Santa Barbara, Santa Catalina, and San Clemente Islands, California. Type locality: San Clemente Island, Orange County. April–Sept.

6. Eriophyllum nubígenum Greene. Yosemite Eriophyllum. Fig. 5372.

Eriophyllum nubigenum Greene in Gray, Proc. Amer. Acad. **19**: 25. 1883.
Actinolepis nubigena Greene, Fl. Fran. 442. 1897.

Leafy annual 5–15 cm. high with extremely dense, persistent, loose tomentum throughout; stems simple or branching from the base, the branches erect. Leaves 1–2 cm. long, mostly entire, spatulate or oblanceolate; heads solitary at ends of branches, about 10-flowered, the peduncles about 1 cm. long; involucres narrow, 5–6 mm. high, 3–4 mm. broad; phyllaries free, 4–6, oblong, acute, carinate; ray-flowers few, the ligules elliptic, about 1 mm. long; disk-flowers 2 mm. long, the glandular-hairy tube shorter than the campanulate throat; achenes linear, 4-angled, appressed-hirsute; pappus-paleae erose.

Wooded slopes, Canadian Zone; in the mountains surrounding Yosemite Valley, Mariposa County, California. Type locality: Cloud's Rest, Yosemite National Park. June.

7. Eriophyllum congdònii Brandg. Congdon's Eriophyllum. Fig. 5373.

Eriophyllum congdonii Brandg. Bot. Gaz. **27**: 449. 1899.
Eriophyllum nubigenum var. *congdonii* Constance, Univ. Calif. Pub. Bot. **18**: 115. 1937.

Loosely floccose annual 1–3 dm. high, branched above the base with ascending branches, the internodes usually long. Leaves 1–3 cm. long, spatulate, rather narrowly so on the upper stems, the lower often with a pair of lobes or blunt teeth; heads solitary at the ends of the branches, the peduncles swollen at base of involucre, many-flowered; involucres 5–8 mm. high, 5–6 mm. broad; phyllaries 8–10, free to the base, oblong, acute at apex, firm in age and concave and somewhat carinate; ray-flowers yellow, the ligules 3–5 mm. long, mostly entire at the apex; disk-flowers with funnelform throat longer than the glabrous or glandular tube; achenes 1.5–2.5 mm. long, linear-clavate, 4-angled, clothed with hirsute, appressed or ascending hairs; pappus-paleae 6–10, of unequal shape, strongly erose, the lanceolate acute paleae alternated with somewhat shorter obtuse ones.

Hillsides, Upper Sonoran Zone; known only from Mariposa County, California. Type locality: near Henessey's, Mariposa County. May–June.

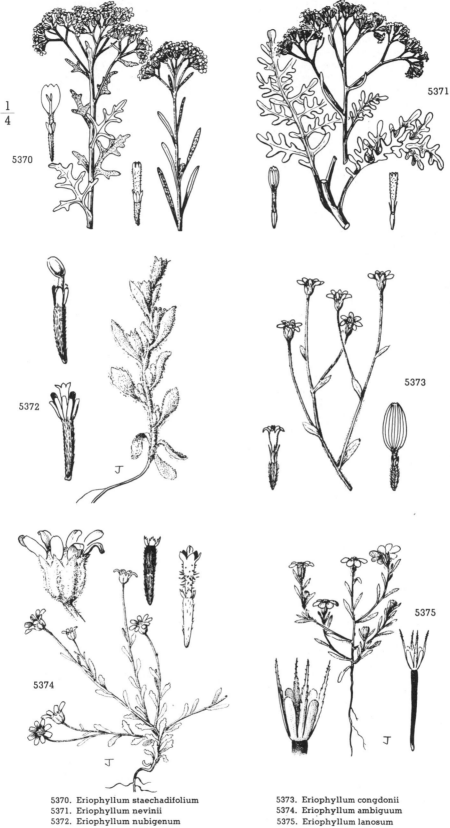

5370. Eriophyllum staechadifolium
5371. Eriophyllum nevinii
5372. Eriophyllum nubigenum

5373. Eriophyllum congdonii
5374. Eriophyllum ambiguum
5375. Eriophyllum lanosum

8. **Eriophyllum ambíguum** A. Gray. Woolly Daisy or Eriophyllum. Fig. 5374.

Bahia wallacei A. Gray, Proc. Bost. Soc. Nat. Hist. **7**: 146. 1859. Not *Eriophyllum wallacei* A. Gray, 1857.
Lasthenia ambigua A. Gray, Proc. Amer. Acad. **6**: 547. 1865.
Bahia parviflora A. Gray, Bot. Calif. **1**: 382. 1876.
Eriophyllum ambiguum A. Gray, Proc. Amer. Acad. **19**: 26. 1883.

Rather densely floccose-tomentose annuals 5–30 cm. high, branching from the base and above, the branches spreading, the lowest decumbent-ascending on more vigorous plants. Leaves alternate, sessile, 1–3 cm. long, oblanceolate or spatulate, entire or shallowly few-toothed at apex, commonly shorter than the internodes, sometimes surpassing or equaling them on dwarf plants; heads solitary on peduncles 1–5 cm. long; involucres 4–5.5 mm. high, rather narrowly campanulate, the phyllaries united into a cartilaginous cup with the herbaceous, sharply triangular lobes free and reflexed in age or in the shorter-rayed form oblong-ovate, scarious-margined, and free to the fused indurated base; receptacle conic, glabrous or nearly so; rays 5–10, yellow, oblong and usually shallowly notched at apex, 3–8 mm. long; disk-flowers 1.3-2 mm. long, the tube and the lower part of the narrowly funnelform throat with dense or sometimes scattered hirsute hairs and sometimes also a little glandular; achenes black, linear-clavate but with evident angles, more or less strigose, varying in the same head; pappus-paleae short, about 0.3 mm. long, truncate and erose or completely lacking.

Grassy, oak-covered slopes, Upper Sonoran Zone; Kern County, California, the southeastern part of the Greenhorn Range south through the Tehachapi Mountains to Fort Tejon, the type locality. Intergrading on the eastern border of its distribution with the more common *E. ambiguum* var. *paleaceum* (Brandg.) Ferris.

Eriophyllum ambiguum var. **paleàceum** (Brandg.) Ferris, Contr. Dudley Herb. **5**: 100. 1958. (*Eriophyllum paleaceum* Brandg. Bot. Gaz. **27**: 450. 1899; *E. parishii* H. M. Hall ex Constance, Madroño **2**: 114. 1934, as a synonym.) Differing from *E. ambiguum* var. *ambiguum* in the larger (5–7 mm. high) and broader involucres and broadly acute phyllaries, these essentially herbaceous and not indurated even in age, except for the carinate ridge and thickened base; also in having 1–6 linear hyaline scales on the summit of the conic receptacle and disk-flowers 2–3 mm. long, the short tube and narrowly funnelform throat beset with minute, gland-tipped hairs only; achenes as in the name-bearing variety but usually more strigose and the pappus-paleae commonly reduced to a vestigial crown, but occasionally longer. California, in the Walker Basin and desert slopes of the Mojave Desert eastward in the desert ranges to the Death Valley region and adjacent Nye County, Nevada, south to the Colorado Desert in Riverside County, California.

The form *E. parishii* H. M. Hall ex Constance listed in synonymy by Constance and discussed by Hall (Univ. Calif. Pub. Bot. **3**: 184. 1907.) differs from the above in having smaller heads and a consistently smooth receptacle, leaves usually more sharply pointed, and short pappus-paleae usually present. The disk-flowers, however, are longer and minutely glandular and not at all hairy. Found on the bordering slopes of the Coachella Valley.

9. **Eriophyllum lanòsum*** A. Gray. White Woolly Daisy or Eriophyllum. Fig. 5375.

Burrielia lanosa A. Gray, Pacif. R. Rep. **4**: 107. 1857.
Actinolepis lanosa A. Gray, Proc. Amer. Acad. **9**: 198. 1874.
Eriophyllum lanosum A. Gray, op. cit. **19**: 25. 1883.
Antheropeas lanosum Rydb. N. Amer. Fl. **34**: 98. 1915.

Loosely floccose annuals 3–15 dm. high, becoming glabrate, openly branching from the base, the stems often reddish, sometimes decumbent. Leaves 5–20 mm. long, linear to narrowly oblanceolate, entire or rarely lobed near the apex; heads terminal, solitary on peduncles 1–5 cm. long; involucres 5–6 mm. high, subcylindric; phyllaries 8–10, essentially distinct, carinate, linear-oblong, acute at apex and the thin tip reflexed; receptacle dome-shaped; ray-flowers 5–10, the ligules 3–5 mm. long, oval or oblong and toothed at the apex, often red-veined in anthesis, white or rose-tinged; disk-flowers 2–3 mm. long, the funnelform throat about equaling the glandular-puberulent tube in length; achenes 3–4 mm. or more long, linear and sparsely appressed-pubescent to glabrate; pappus-paleae hyaline, lanceolate and produced into slender awns, these alternating with oblong, obtuse, much shorter paleae.

Sandy soil, Sonoran Zones; eastern San Bernardino County to eastern Imperial County, California, south to central part of Lower California and eastward to southern Nevada, southern Utah, and Arizona. Type locality: "Gravelly hills near the Colorado of the West." Feb.–April.

10. **Eriophyllum wállacei** A. Gray. Wallace's Woolly Daisy. Fig. 5376.

Bahia wallacei A. Gray, Pacif. R. Rep. **4**: 105. 1857.
Actinolepis wallacei A. Gray, Proc. Amer. Acad. **9**: 198. 1874.
Eriophyllum wallacei A. Gray, op. cit. **19**: 25. 1883.
Antheropeas wallacei Rydb. N. Amer. Fl. **34**: 98. 1915.

Annual (1)2–10 cm. high, the dense tomentum commonly persistent, much branched from the base, often appearing tufted, the branches ascending. Leaves 8–20 mm. long, spatulate to obovate, entire or rarely 3-lobed; heads terminal, the peduncles 1–3 cm. long; involucres campanulate, 5–7 mm. high and nearly as broad; phyllaries 5–10, distinct or nearly so, ovate, carinate; receptacle dome-shaped; ligules of the ray-flowers oval, entire or 3-toothed at the apex, 3–4 mm. long and nearly as wide; disk-flowers with the funnelform throat a little longer than the glandular-puberulent tube; achenes black when ripe, linear, angled, glabrous or more or less strigosulose; pappus-paleae 6–10, opaque, white, shorter than the corolla-tube, oval or oblong, obtuse, the margin erose.

Sandy soil, Sonoran Zones; Inyo County, California, south through the Mojave and Colorado Deserts and adjacent mountain slopes to northern Lower California and east to southern Nevada and Utah and also northwestern Arizona. Type locality: Tujunga Wash, Los Angeles County, California. March-June.

---

* On the basis of his investigations, Carlquist (Madroño **13**: 226–239. 1956) concludes that *Eriophyllum lanosum* and *E. wallacei* belong to the genus *Antheropeas* Rydb. rather than *Eriophyllum*. This conclusion is based on the subulate anther-tips, the cuneate tips of the style-branches, and lower chromosome number.

Three weak variants have been named which vary from *Eriophyllum wallacei* var. *wallacei* by the characters listed below.

Eriophyllum wallacei var. **rubéllum** A. Gray, Proc. Amer. Acad. **19**: 25. 1883. (*Bahia rubella* A. Gray, Bot. Mex. Bound. 95. 1859.) Like the preceding in habit and distinguished only in having the ray-flowers pale purple, pinkish, or rarely white. Desert slopes above the Colorado Desert in Riverside and San Diego Counties, California. Type locality: near San Felipe, San Diego County.

Eriophyllum wallacei var. **calvéscens** Blake, Journ. Wash. Acad. **19**: 278. 1929. (*Eriophyllum aureum* Brandg. Bot. Gaz. **27**: 449. 1899.) Identical with var. *wallacei* except that the pappus-paleae are completely lacking or reduced to a mere ring. This form appears to be restricted, though the name-bearing variety is also found in this area, to the lower Owens Valley, Inyo County, California, southward along the western edge of the Mojave Desert to Victorville, San Bernardino County, which is the type locality.

Eriophyllum wallacei subsp. **austràle** (Rydb.) Wiggins, Contr. Dudley Herb. **3**: 304. 1944. (*Antheropeas australe* Rydb. N. Amer. Fl. **34**: 98. 1915.) As *E. wallacei* var. *wallacei* but having the paleae nearly as long as the corolla-tube and achenes with appressed hairs. The character of strigulose achenes, however, occurs sporadically throughout the range of the typical form and var. *rubellum*; the longer pappus-paleae may be rounded and scarcely to conspicuously erose. This form is more commonly found on the western slopes above the Colorado Desert and extends into adjacent Lower California. Type locality: "mountains of northern Lower California."

11. **Eriophyllum multicaùle** (DC.) A. Gray. Many-stemmed Eriophyllum. Fig. 5377.

Actinolepis multicaulis DC. Prod. **5**: 656. 1836.
Actinolepis multicaulis var. *papposa* A. Gray, Proc. Amer. Acad. **6**: 546. 1865.
Eriophyllum multicaule A. Gray, op. cit. **19**: 24. 1883.

Sparsely floccose, diffuse annual 2–15 cm. high, becoming glabrate below, much branched from the base, the branches decumbent or ascending. Leaves about 10 mm. long, mostly cuneate and shallowly 2–3-lobed at apex, pubescence sparse; heads in leafy-bracted clusters at tips of branches, nearly sessile; invólucres 3–4 mm. long, campanulate; phyllaries 5–7, elliptic, strongly concave and firm in age, partially enclosing the ray-achenes; receptacle nearly flat; ray-flowers 3–7, yellow, ligules obovate, shallowly toothed at apex, about 2 mm. long; disk-flowers less than 2 mm. long, the glandular-puberulent tube longer than the campanulate throat; achenes about 2 mm. long, black, sparsely strigose to glabrous; pappus of 10–15 narrow, unequal, laciniate, acuminate paleae or rarely obsolete, shorter than the corolla-tube.

In sandy soil, Upper Sonoran Zone: Monterey and San Benito Counties, California, south to San Diego County; also reported from Arizona by Kearney & Peebles. Type locality: California. Collected by Douglas. March–May.

12. **Eriophyllum prínglei** A. Gray. Pringle's Eriophyllum. Fig. 5378.

Eriophyllum pringlei A. Gray, Proc. Amer. Acad. **19**: 25. 1883.
Actinolepis pringlei Greene, Fl. Fran. 441. 1897.

Densely white-floccose-woolly annuals 2–5 cm. high, becoming somewhat glabrate in age on the lower stems, much branched from the base and more or less tufted. Leaves 3–8 mm. long, flabelliform or broadly spatulate, with 3 rounded lobes at the apex, revolute; heads sessile or very short-peduncled in leafy clusters at tips of branches, copiously woolly, about 12–25-flowered; involucres campanulate, 3–4 mm. high; phyllaries 6–8, oblong, concave; receptacle nearly flat; ray-flowers lacking; disk-flowers 2 mm. long, the glandular tube as long as the widely campanulate throat and acute lobes; achenes 5-nerved, about 2 mm. long, black, densely strigose; pappus-paleae 5–10, silvery-scarious, oblong-lanceolate or obtuse, with laciniate-fringed margin, a little shorter than the corolla-tube in length.

Gravelly mesas and slopes, Sonoran Zones; eastern San Luis Obispo County, California, Greenhorn Range, Kern County, and the eastern side of the Sierra Nevada and mountains of southern California from Mono and Inyo Counties to San Diego County, eastward to southern Nevada and western and southern Arizona. Type locality: Mojave Desert, California. March–May.

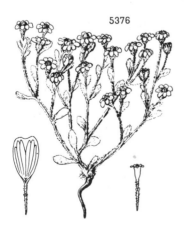

5376. Eriophyllum wallacei

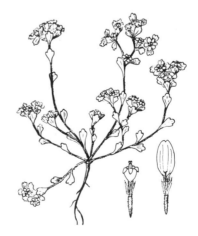

5377. Eriophyllum multicaule

13. Eriophyllum mohavénse (I. M. Johnston) Jepson. Barstow Eriophyllum. Fig. 5379.

Eremonanus mohavensis I. M. Johnston, Contr. Gray Herb. No. 68: 101. 1923.
Eriophyllum mohavense Jepson, Man. Fl. Pl. Calif. 1117. 1925.

Thinly and loosely floccose-woolly annuals 1.5–2.5 cm. high, tending to be glabrate on the stems, much branched and spreading from the base, forming rounded tufts. Leaves 3.5–10 mm. long, spatulate, attenuate to a long, slender, petiolar base, entire and cuspidate at apex or with 2 or usually 3 short, acute, cuspidate lobes, revolute in age; heads more or less clustered among the leaves on the ends of the branches, peduncles rather stout, 1.5–3.5 mm. long, 3–4-flowered; involucre cylindric, 3–4 mm. high; phyllaries 3 or 4, narrowly oblong to oblanceolate, concave; receptacle with slender subulate projections which are attached to the phyllaries and between which the flowers are attached to the receptacle; ray-flowers lacking; disk-flowers about 2 mm. long or more, the glandular tube as long as the throat and acute lobes; achene 5-nerved, about 2 mm. long, black, densely strigose; pappus-paleae 12–14, silvery-scarious, linear to obovate, about 1.5 mm. long.

A rarely collected plant growing on sandy washes and mesas, Upper Sonoran Zone; San Bernardino County, California, in the vicinity of Barstow and at Stoddard's Well (*Jepson 5902*). Type locality: Barstow. April–May.

55. PSEUDOBAHÍA Rydb. N. Amer. Fl. **34**: 83. 1915.

More or less floccose-tomentose spring annuals. Leaves alternate, entire, 3-lobed or pinnately or bipinnately parted. Heads yellow, terminal on the ends of the branches. Involucres broadly campanulate. Phyllaries in one series, herbaceous, slightly concave centrally at maturity, free except at the base or partly united and becoming indurated. Receptacle conical, naked. Rays present, about as many as the phyllaries, the ray-flowers pistillate, the tube of the ray-flowers very slender, glandular, and bearing at the summit a ring of villous hairs; disk-flowers many, hermaphrodite, fertile, tube slender, glandular, with villous hairs at the summit and extending to the lower part of the widely funnelform throat, the lobes triangular, glabrous. Anthers entire at base, the appendages at the apex ovate. Style-branches truncate. Achenes obovoid in outline, compressed and ridged on the flattened faces, pubescent with ascending hairs to nearly glabrous, epappose or sometimes with vestigial pappus-paleae. [Greek, meaning false, and *Bahia,* another helenioid genus.]

A genus of 3 species, all natives of California. Type species, *Monolopia bahiaefolia* Benth.

Leaves entire or 3-lobed.   1. *P. bahiaefolia.*
Leaves pinnately or bipinnately divided.
    Leaves bipinnatifid; stems coarse; lower part of involucre scarcely cartilaginous in age and lacking callous
        processes at the junction of the lobes.   2. *P. peirsonii.*
    Leaves pinnatifid; stems slender; lower part of involucre cartilaginous in age and usually bearing callous
        processes at the junction of the lobes.   3. *P. heermannii.*

1. Pseudobahia bahiaefòlia (Benth.) Rydb. Hartweg's Pseudobahia. Fig. 5380.

Monolopia bahiaefolia Benth. Pl. Hartw. 317. 1849.
Eriophyllum bahiaefolia Greene, Fl. Fran. 446. 1897.
Pseudobahia bahiaefolia Rydb. N. Amer. Fl. **34**: 83. 1915.

Floccose-tomentose annuals 6–15 cm. high, simple or branched below the middle with ascending branches. Leaves 8–25 cm. long, spatulate or oblanceolate in outline, narrowed to a petioliform base, entire or with 2–3 blunt lobes toward the apex, mostly a little longer than the internodes; heads solitary at the ends of the branches on peduncles 2–5 cm. long; involucre hemispheric, about 5 mm. high, nearly twice as broad (in pressed specimens); phyllaries about 8, narrowly ovate, united below the middle, herbaceous, the basal portion tending to become indurated; ray-flowers 6–7, tube slender, with a ring of villous hairs at the summit, the ligule 5–10 mm. long, ovate to oblong, obscurely toothed at the apex; disk-flowers about 2.5 mm. long, tube slender, throat ample, funnelform, lobes triangular, naked; achenes black, 1.5 mm. long, narrowly ovoid, obcompressed, the flattened surface ridged, rather densely covered with ascending hairs, mostly epappose but occasionally with minute pappus-paleae.

In depressions on low rolling hills, Lower Sonoran Zone; eastern side of the Sacramento–San Joaquin Valley, California, as far south as Madera County. Type locality: "In pascuis vallis superioris Sacramento." Collected by Hartweg. March–April.

2. Pseudobahia peirsònii Munz. Tulare Pseudobahia. Fig. 5381.

Pseudobahia peirsonii Munz, Aliso **2**: 84. 1949.

Loosely grayish-floccose annual 2–6 dm. high, branching from above the base, the rather stout stems often reddish and becoming glabrate. Leaves 2–5.5 cm. long, usually longer than the internodes, grayish-floccose, triangular-ovate in outline, bipinnatifid (pinnatifid in depauperate plants), the divisions 1–5 mm. broad, narrowed below to a flattened petiole one-third as long as to equaling the divided portion; heads solitary at the ends of the branches, the peduncles 2–8 cm. long; involucres hemispheric, the white tomentum persistent, 6–9 mm. high, much wider than high; phyllaries about 8, herbaceous, scarcely thickened below and united only at the base, concave in the center portion; ray-flowers about 8, the ligule 5–10 mm. long, broadly ovate, mostly entire at the apex; disk-flowers about 3 mm. long, tube slender, throat broadly funnelform;

achenes black, about 3 mm. long, narrowly obovoid and compressed, the flattened surfaces ridged, glabrous or with some appressed hairs, epappose or with vestigial pappus-paleae.

Grassy flats and rolling hills, Lower Sonoran Zone, southern Tulare County and Kern County, California. Type locality: Ducor, Tulare County. March–April.

3. Pseudobahia heermánnii (Durand) Rydb. Foothill Pseudobahia. Fig. 5382.

Monolopia heermannii Durand, Journ. Acad. Phila. II. **3**: 93. 1855.
Monolopia bahiaefolia var. *pinnatifida* A. Gray, Bot. Calif. **1**: 383. 1876.
Eriophyllum heermannii Greene, Fl. Fran. 445. 1897.
Pseudobahia heermannii Rydb. N. Amer. Fl. **34**: 83. 1915.

Loosely grayish-floccose annuals 1–3 dm. high, branching at and above the base, the stems slender, usually reddish and becoming glabrate or nearly so. Leaves 1–3 cm. long, pinnately lobed with linear divisions 0.5–1.5 mm. wide, narrowed below into a flattened petiole about one-third as long as the divided portion, the internodes shorter than the leaves; heads solitary on the ends of the branches, the peduncles 2–5 cm. long, often curved at maturity; involucres 5–6 mm. high, broader than long, the tomentum persisting; phyllaries 8 or more, ovate-acute, herbaceous above, usually fused about one-half their length, the lower part becoming cartilaginous in age and usually developing callous processes between the lobes; rays about 8, mostly oblong, usually obscurely toothed at apex, 6–8 mm. long; disk-flowers about 2.5 mm. long; achenes narrowly obovoid, black, more or less appressed-hairy, strongly compressed but ridged on the flattened surfaces, epappose or with vestigial pappus-paleae.

On open wooded slopes, Upper Sonoran Zone; slopes of the Sierra Nevada, California, from Butte County to Kern County; also occurring rarely in the Santa Lucia Mountains, Monterey County. Type locality: "Calaveras." Collected by Heermann. Aberrant forms occur in the southern part of the range. March–May.

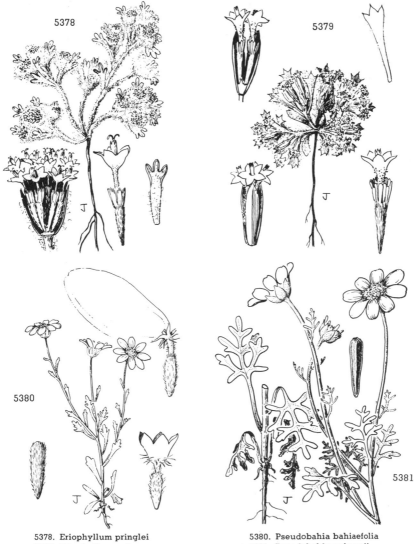

5378. Eriophyllum pringlei
5379. Eriophyllum mohavense
5380. Pseudobahia bahiaefolia
5381. Pseudobahia peirsonii

56. SYNTRICHOPÁPPUS A. Gray in Torr. Pacif. R. Rep. **4**: 106. 1857.

Low, usually diffuse annuals. Leaves alternate, sometimes opposite below, entire or 3-lobed at apex, narrowed to a petioliform base. Heads many, terminating the branchlets, short-peduncled. Phyllaries of the involucre few, partly enclosing the ray-achenes, in 1 series, erect or ascending in anthesis, spreading after achenes are shed. Receptacle flat, naked. Ray-flowers pistillate, fertile, as many as the phyllaries, the ligules yellow or white and pinkish-tinged, oval and toothed at apex. Disk-flowers perfect, fertile, yellow, tube much shorter than the trumpet-shaped throat and short lobes. Anthers obtuse at base, appendaged at apex. Style-branches flattened, acute. Achenes narrowly obpyramidal or clavate, 5-angled. Pappus none or the paleae dissected into many barbellate bristles united at base into a ring. [From the Greek meaning united, hair, and pappus.]

A genus of 2 species, natives of the deserts in California, southern Utah, Nevada, and adjacent Arizona. Type species, *Syntrichopappus fremontii* A. Gray.

Ligules of the ray-flowers yellow; pappus of numerous bristles united at the base.   1. *S. fremontii.*
Ligules of the ray-flowers white or pinkish; epappose.   2. *S. lemmonii.*

1. **Syntrichopappus fremóntii** A. Gray. Yellow Syntrichopappus or Fremont's Xerasid. Fig. 5383.

Syntrichopappus fremontii A. Gray in Torr. Pacif. R. Rep. **4**: 106. *pl. 15.* 1857.

Loosely floccose annuals 3–10 cm. high, diffusely much branched from the base, except in depauperate specimens, the branches ascending. Leaves 5–20 mm. long, linear-spatulate to spatulate, 3-toothed at apex or less often entire, narrowed to a petioliform base; heads many, terminating the branchlets, the peduncles short; involucre 5–6 mm. high, narrowly campanulate to nearly cylindric; phyllaries 5, oblong, acute and thin at apex, scarious-margined, the midsection strongly concave; ray-flowers 5, the ligules golden yellow, 3–5 mm. long, 5–7-veined; disk-corollas 20 or more, golden yellow, about 2.5-3 mm. long; achenes 3 mm. long, angled, obpyramidal, hirsute with ascending hairs, the paleae finely dissected into 35–40 white barbellate bristles united at the base and deciduous in a ring.

Sandy desert plains and slopes, Sonoran Zones; Death Valley region, Inyo County, south in the Mojave Desert to Los Angeles, Kern, and San Bernardino Counties, California, east to southern Nevada, Utah, and northwestern Arizona. Type locality: "somewhere between the Rocky Mountains and the Sierra Nevada." Collected by Fremont. April–May.

2. **Syntrichopappus lemmònii** A. Gray. Pink-rayed Syntrichopappus or Lemmon's Xerasid. Fig. 5384.

Actinolepis lemmonii A. Gray, Proc. Amer. Acad. **16**: 101. 1880.
Syntrichopappus lemmonii A. Gray, op. cit. **19**: 20. 1883.
Microbahia lemmonii Cockerell, Muhlenbergia **3**: 9. 1907.

Thinly floccose to glabrate annuals 2–8 cm. high with reddish slender stems, mostly branching above the base with ascending to erect branches. Leaves 3–8 mm. long, linear or linear-spatulate, obtuse; heads several to many, the peduncles 8–10 mm. long; involucre 4–5 mm. high, narrowly campanulate; phyllaries 6–8, narrowly oblong, scarious-margined, the midsection concave, the tip acute and thin, usually spreading in age; ligules of ray-flowers 6–8, white or pinkish above with darker veins conspicuous below, 2–3 mm. long, 3-veined; disk-flowers pale yellow; achenes clavate, about 2 mm. long, rather thinly strigose to nearly glabrous, epappose.

Sandy slopes, mostly lower part of Upper Sonoran Zone; western edge of the Mojave Desert in Kern, Los Angeles, and San Bernardino Counties, California. Type locality: Mojave Desert. April–May.

57. RIGIOPÁPPUS A. Gray, Proc. Amer. Acad. **6**: 548. 1865.

Slender, more or less hirsutulous annuals, simple or branched above, the lateral branches usually surpassing the main axis. Leaves alternate, linear, the lower usually deciduous. Heads many-flowered, terminating the branches. Involucre broadly turbinate, the phyllaries narrowly lanceolate, firm, subequal, in 2 series, subulate, narrowly hyaline-margined, partially clasping the outer achenes. Receptacle flat. Ray-flowers pistillate, the ligules short. Disk-flowers perfect, the tube shorter than the tubular throat, the teeth very short. Anthers obtuse at base, subentire. Style-branches short, with subulate hairy tips. Achenes linear, abruptly narrowed and truncate at apex, somewhat compressed, finely transverse rugose and strigose with broad-tipped hairs. Pappus of 3–5 firm, strigulose, awn-like paleae. [From the Greek meaning stiffened and pappus.]

A monotypic genus of western United States.

1. **Rigiopappus leptocládus** A. Gray. Rigiopappus. Fig. 5385.

Rigiopappus leptocladus A. Gray, Proc. Amer. Acad. **6**: 548. 1865.
Rigiopappus leptocladus var. *longiaristatus* A. Gray, Syn. Fl. N. Amer. **1²**: 339. 1884.
Rigiopappus longiaristatus Rydb. N. Amer. Fl. **34**: 64. 1914.

Gray-green annual 1–3 dm. high. Leaves narrowly linear, mostly erect, 1–3 cm. long; heads terminating the branches; involucres 4–7 mm. high; phyllaries hirsutulous; ray-flowers 5–15, pale yellow, often tinged with purple, 1.5–2 mm. long, as short as or shorter than the pappus-paleae; disk-flowers 5–35, shorter than the pappus-paleae, the teeth of the disk-corollas minute, 2–3-toothed; achenes brown, about 4 mm. high, the pappus-paleae as long as to half as long as the achenes.

Common on grassland or with sagebrush, Upper Sonoran and Transition Zones; central Washington and western Idaho south to Utah and Nevada and in the Pacific States to the mountains of Los Angeles County, California. Type locality: The Dalles of the Columbia River. Collected by Lyall. April–June.

58. TRICHOPTÍLIUM A. Gray in Torr. Bot. Mex. Bound. 97. 1859.

Low floccose plants much branched from the base, annuals or persisting more than one season. Leaves alternate or the lowest subopposite, broadly oblanceolate or spatulate, the margins incised-dentate, narrowed below to a long petiolar base. Heads on long, slender, naked peduncles much surpassing the leaves. Phyllaries in 2 nearly equal series. Receptacle subconvex, naked. Flowers perfect, fertile, the ray-flowers wanting. Corolla-tube short, the throat longer, expanding to the spreading lobes. Anthers oblong-lanceolate, minutely sagittate at base. Style-branches linear, subtruncate at apex. Achenes obpyramidal, the surface whitened with minute papillae, clothed with ascending hirsute hairs. Pappus-paleae broad, minutely scaberulous and dissected above into many bristles of unequal length which are as long as or longer than the length of the scale. [From the Greek words meaning hair and feather, in reference to the pappus-paleae.]

A monotypic genus of southwestern United States.

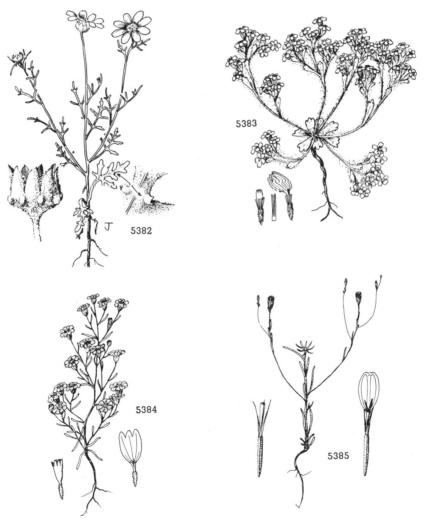

5382. Pseudobahia heermannii
5383. Syntrichopappus fremontii

5384. Syntrichopappus lemmonii
5385. Rigiopappus leptocladus

1. **Trichoptilium incìsum** A. Gray. Yellow-head. Fig. 5386.

Psathyrotes incisa A. Gray, Mem. Amer. Acad. II. **5**: 322. 1854.
Trichoptilium incisum A. Gray in Torr. Bot. Mex. Bound. 97. 1859.

Fragrant herbs branched from the base, more or less densely floccose-villous throughout except the peduncles, the stems 0.5–2.0 dm. high. Leaves clustered toward the base, 1–5 cm. long, 3–9 mm. wide; involucres 6–7 mm. high, hemispheric; phyllaries lanceolate to oblanceolate, glandular as well as floccose-villous; peduncles 8–10 cm. long, slender, reddish, somewhat glandular; flowers 35–80, the corollas 4 mm. long, puberulent and somewhat glandular; achenes 3 mm. long, sharply truncate at apex, 5-angled, the pappus-paleae as long as or longer than the achene, hyaline, stramineous or slightly brownish.

Gravelly soil, mesas and canyons, Lower Sonoran Zone; southern Nevada, eastern San Bernardino County, California, southward through the Colorado Desert to central Lower California and east to western Arizona. Type locality: "On the Californian desert near the Rio Colorado." Feb.–May; Oct.–Nov.

59. **CHAENÁCTIS** DC. Prod. **5**: 659. 1836.

Annual, biennial, or perennial herbs sometimes suffrutescent as base. Leaves basal, persisting in some perennial forms, and alternate above, subentire to more or less pinnately dissected. Heads large, terminal at the ends of branches and branchlets or subscapose. Involucres turbinate, campanulate, or hemispheric. Phyllaries herbaceous, in 1 or 2 series, subequal, a shorter outer series sometimes present, spreading or reflexed after the ripe achenes have fallen. Receptacle flat, alveolate, naked or rarely with few scales or hairs in one species. Flowers white, yellow, or pinkish, perfect, regular; tube very short; throat narrowly funnelform to almost tubular, the short spreading limb 5-cleft and usually densely glandular; marginal corollas larger mostly in the annual species, sometimes zygomorphic, the limb unequally cleft. Anthers truncate at base, appendaged above. Style-branches linear, narrowly to broadly acute at apex. Achenes dark, sometimes marked with paler suberized areas, narrowly clavate or obpyramidal, or terete and somewhat compressed, in most species clothed with appressed hairs. Pappus-paleae in 1 or 2 series, hyaline, erose-margined, mostly conspicuous, sometimes vestigial, the pappus-paleae of the marginal flowers markedly shorter than those of the disk. [From the Greek words meaning to gape and ray, in reference to the marginal flowers.]

A genus of about 25 species, all natives of western North America. Type species, *Chaenactis glabriuscula* DC. The distribution of many of the annual species of *Chaenactis* overlaps and there is evidence that the taxa will hybridize experimentally and also not infrequently in nature (see Stockwell, Contr. Dudley Herb. **3**: 89–168. 1940).

Perennials or stout biennials.
  Leaves not flattened, the divisions contorted or crisped and often revolute-margined.
    Leaves linear in outline, the pinnae very short, crisped and of equal length; mountains of southern California.   1. *C. santolinoides.*
    Leaves not linear in outline, the pinnae not of equal length; plants tall and erect or depressed alpines.   8. *C. douglasii* and varieties.
  Leaves plane, the divisions flat, little or not at all revolute on the margins (except *C. ramosa*).
    Plants mat-forming; flowering stems short, leafless; plants of the Sierra Nevada, California.
      Phyllaries glandular-puberulent.   5. *C. nevadensis.*
      Phyllaries tomentose, lacking glands.   6. *C. alpigena.*
    Plants suffrutescent at base or at least with a woody caudex; flowering stems more or less leafy.
      Phyllaries densely or sparingly glandular with coarse hairs.
        Heads large, 16–20 mm. high; plants of the Siskiyou region, California.   3. *C. suffrutescens.*
        Heads smaller, 11–14 mm. high; plants of the Wenatchee Mountains, Washington.   7. *C. ramosa.*
      Phyllaries without coarse glandular hairs.
        Heads 2–3 or solitary, borne well above the basal leaves on stems with much-reduced leaves; plants of southern California mountains.   2. *C. parishii.*
        Heads mostly solitary, not much surpassing the basal leaves, the stem-leaves not reduced in size; plants of north central Washington.   4. *C. thompsonii.*
Annuals or winter annuals.
  Phyllaries acute or obtuse at the apex, or if at all attenuate not with slender, terete, colored tips; receptacle naked.
    Pappus absent or rudimentary.
      Slender, yellow-flowered plants; achenes papillate; John Day Valley, Oregon.   10. *C. nevii.*
      Coarse, white-flowered plants; achenes somewhat strigose; cismontane southern California.   11. *C. artemisiaefolia.*
    Pappus present in 1 or 2 series.
      Stamens always included; corollas 10–14 mm. long.   12. *C. macrantha.*
      Stamens exserted; corollas 4–8(9) mm. long.
        Pappus-paleae usually 8 in 2 series.
          Inner and outer series of paleae not markedly different in length; plants of southeastern Oregon.   9. *C. cusickii.*
          Inner and outer series of paleae markedly different in length, the outer much reduced but equaling the inner in number.
            Flower-heads white; leaves entire or with 1 or 2 pairs of remote pinnae.   13. *C. xantiana.*

Flower-heads shades of yellow; leaves usually bipinnate, the pinnae more than 2 pairs
and more or less crowded (see also *C. glabriuscula heterocarpha*).
14. *C. tanacetifolia.*

Pappus-paleae 4(5) in 1 series.
Flower-heads yellow.
Stems rather stout; the reduced uppermost leaves usually entire; involucres broadly
turbinate, with phyllaries thin in age. 15. *C. glabriuscula.*
Stems slender and wiry; the reduced uppermost leaves definitely parted; involucres
hemispheric, with phyllaries thickened and firm in age; plants of San Diego
County. 16. *C. tenuifolia.*
Flower-heads white.
Leaves if pinnate with few long linear divisions; plants essentially glabrous.
17. *C. fremontii.*
Leaves pinnate or bipinnate, the lobes short and blunt; plants with a more or less
persistent tomentum.
Paleae of disk-achenes one-half or more the length of the corollas.
18. *C. stevioides.*
Paleae of disk-achenes about one-third the length of the corollas.
18. *C. steviodes brachypappa.*
Phyllaries narrowly linear, attenuate into slender, terete, colored tips; receptacles bearing very few to
several setiform bracts. 19. *C. carphoclinia.*

1. **Chaenactis santolinoìdes** Greene. Santolina Chaenactis. Fig. 5387.

Chaenactis santolinoides Greene, Bull. Torrey Club **9**: 17. 1882.
Chaenactis santolinoides var. *indurata* Stockwell, Contr. Dudley Herb. **3**: 107. 1940.

Perennial with a stout woody root and woody, often much-branched caudex, this often buried
in loose soil; stems very short, densely and loosely tomentose, with densely clustered leaves, the
naked glabrous inflorescences arising from the leaf-axils, 1–2.5 dm. high, scapiform, with a single
head, more rarely branched. Leaves 3–8 cm. long, densely woolly-tomentose, petiolate, pinnate,
the pinnae 4–5 mm. long, lobed and much contorted; involucres broadly turbinate, 9–12 mm.
high; phyllaries unequal, linear, obtuse or broadly acute, granular-glandular and often a little
ciliate on the margin; corollas white, often with pink lobes, about 7 mm. long, sparsely glandular-
hirsute without, more densely so on the lobes; achenes densely hirsute, mostly equaling the corollas; pappus-paleae 12–16, unequal in length and shorter than the corollas, linear-oblanceolate.

On sandy or rocky slopes in pinyon and yellow pine forest, Upper Transition and lower Canadian Zones;
Greenhorn and Tehachapi Mountains, Kern County, California, and the mountains of western San Bernardino
County and of Los Angeles County westward to the Mount Pinos region. Type locality: San Bernardino Moun-
tains. May–July.

2. **Chaenactis parìshii** A. Gray. Parish's Chaenactis. Fig. 5388.

Chaenactis parishii A. Gray, Proc. Amer. Acad. **20**: 299. 1885.

Perennial with pale herbage, from a woody root, with several stems 1.5–3.5 dm. high, woody
at base, arising from a usually branched crown, more densely leafy toward the base (internodes
shorter than the leaves), more sparsely leafy on midstem (with longer internodes), leaves becoming bract-like below the inflorescence. Leaves 1–3.5 cm. long, closely canescent, tending to
become glabrate, the petiole about equaling the blade, pinnately parted to the midrib, the pinnae

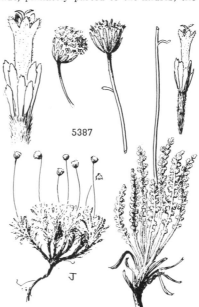

5386. Trichoptilium incisum   5387. Chaenactis santolinoides

linear (rarely with a few small lobes), obtuse, 5–9 mm. long; heads turbinate, solitary or 2–3 at ends of branches, the peduncles 2–8 cm. long; phyllaries few, unequal, 4–12 mm. long, linear, acute, somewhat canescent and granular-glandular; corollas white or pink, 7–8 mm. long, sparsely puberulent without, densely so on the lobes; achenes densely hirsute, 7–8 mm. long; pappus-paleae 14–18, unequal in length, the longest slightly shorter than the corollas, linear-lanceolate.

Dry rocky slopes and in chaparral, Upper Sonoran Zone; San Jacinto and Santa Rosa Mountains, Riverside County, California, southward through the mountains of San Diego County to adjacent Lower California. Type locality: "southern border of California." May–July.

3. Chaenactis suffrutéscens A. Gray. Shasta Chaenactis. Fig. 5389.

Chaenactis suffrutescens A. Gray, Proc. Amer. Acad. **16**: 100. 1880.
Chaenactis suffrutescens var. *incana* Stockwell, Contr. Dudley Herb. **3**: 108. 1940.

Perennial from a woody root with several to many stems, erect or decumbent and woody below, 2.5–4.5 dm. high, erectly branched above, the leafy branches terminated by large solitary heads, the lower stems densely and closely lanate. Leaves 5–10 cm. long, the petiole shorter than the blade, pinnately parted to the midrib, the linear divisions occasionally lobed, less lanate than the stems, sometimes becoming glabrate; heads large, many-flowered, on densely glandular, fistulous peduncles 10–20 cm. long, usually bearing a few linear bracts; phyllaries unequal, 6–18 mm. long, linear, obtuse or acute, densely glandular; corollas white, about 8 mm. long, puberulent without especially on the lobes; achenes densely hirsute, about the length of the corolla; pappus-paleae 10, oblanceolate, equal in length and a little shorter than the corolla.

Dry plains and slopes, Arid Transition Zone; mountains of Siskiyou and Trinity Counties, California. Type locality: upper Shasta Valley. Collected by Lemmon. May-June.

4. Chaenactis thompsònii Cronquist. Thompson's Chaenactis. Fig. 5390.

Chaenactis thompsonii Cronquist, Vasc. Pl. Pacif. Northw. **5**: 123. *fig.* 1955.

Perennial with a taproot and branching caudex, not at all matted, the numerous leafy, reddish stems laxly erect, 1–3 dm. high. Leaves all cauline, with a petioliform base, rather thick, grayish with a thin tomentum, 2–5 cm. long, evenly pinnatifid to the broad rachis, the blunt linear lobes narrower than the rachis; heads mostly solitary on the stems, the peduncles 2–4 cm. long, 12–14-flowered, the flowers all alike; phyllaries few, thin, 10–13 mm. long, linear, acute, thinly tomentose and not at all glandular; flowers about 9 mm. long, white with tomentose pinkish lobes about 1 mm. long; achenes 8 mm. long, hirsute, the pappus-paleae 10–16, about 5 mm. long, rounded at the apex.

Serpentine slopes, Boreal Zone; Wenatchee Mountains, Chelan and Kittitas Counties, Washington. Type locality: Three Brothers Peak, Chelan County. June–Aug.

5. Chaenactis nevadénsis (Kell.) A. Gray. Northern Sierra Chaenactis. Fig. 5391.

Hymenopappus nevadensis Kell. Proc. Calif. Acad. **5**: 46. 1873.
Chaenactis nevadensis A. Gray, Bot. Calif. **1**: 391. 1876.

Cespitose perennial with gray herbage, arising from a woody taproot and stout multicipital caudex, terminating in aerial reddish stems 8–10 cm. high. Leaves more or less clustered below, permanently and loosely tomentose, not glandular, about 2.5–4.5 cm. or more long, 12–26 mm. wide; petiole longer than the blade, this ovate or flabelliform in outline, flat, irregularly pinnatifid or sometimes bipinnatifid into rounded lobes; heads surpassing the leaves, on glandular naked peduncles 3–5 cm. long, arising from the upper leaf-axils; involucres reddish, turbinate; phyllaries 10–14 mm. long, linear, obtuse, densely glandular-puberulent; corollas about 7–8 mm. long, white with pinkish pubescent tips; achenes about equaling the corollas, hirsute; pappus-paleae 10–16, a little shorter than the corolla, linear and obtuse to oblanceolate.

Rocky slopes, Arctic-Alpine Zone; Lassen Peak, Shasta County, California, and the Sierra Nevada in Nevada and Alpine Counties. Type locality: "Above Summit, Sierra Nevada Mts." Aug.

6. Chaenactis alpígena C. W. Sharsmith. Southern Sierra Chaenactis. Fig. 5392.

Chaenactis alpigena C. W. Sharsmith, Contr. Dudley Herb. **4**: 319. 1955.

Cespitose or mat-forming perennial from a multiciptal caudex, the stems 2–7 cm. high. Leaves more or less clustered below, 1–2.5 cm. long, 3–8 mm. wide; petiole longer than the blade, this flabelliform, spatulate, ovate, or elliptic in outline, pinnately lobed, with a dense, grayish or yellowish tomentum, glandless; heads surpassing the leaves, on glandless, more or less tomentose peduncles 2–4 cm. long; phyllaries 8–14 mm. long, tomentose, glandless; corollas 5.5–8 mm. long, usually white; achenes about equaling the corollas, hirsute; pappus-paleae 8–20, about 5–8 mm. long, linear and obtuse or oblanceolate.

Rocky slopes, Arctic-Alpine Zone; in the Sierra Nevada of California in Eldorado County (and adjacent Washoe County, Nevada) south to Tulare County. Type locality: east face of Mount Conness, Mono County. July–Aug.

7. Chaenactis ramòsa Stockwell. Branching Chaenactis. Fig. 5393.

Chaenactis ramosa Stockwell, Contr. Dudley Herb. **3**: 117. *pl. 29, figs. 4–6.* 1940.

Short-lived perennial with several woody, red, spreading or more or less prostrate, rather slender stems, arising from a taproot, and with numerous laxly ascending, leafy branches 8–18 cm. high. Leaves all cauline, 1–6 cm. long, loosely and sparsely lanate, pinnatifid or subpinnatifid,

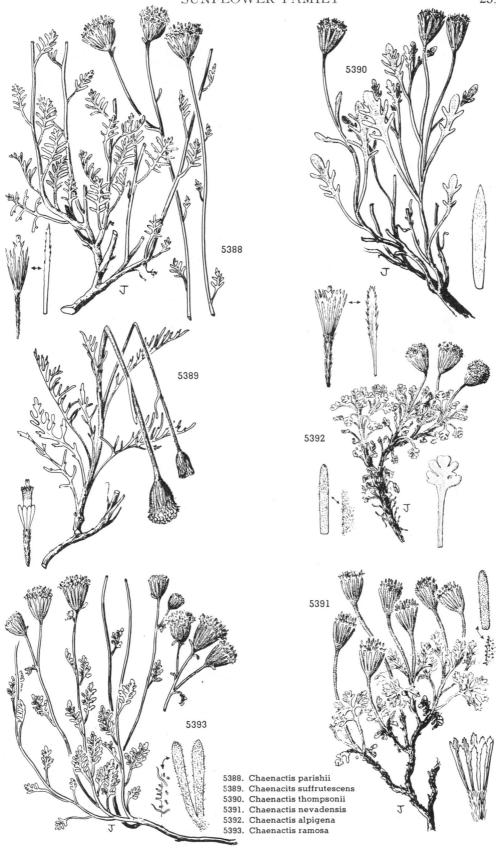

5388. Chaenactis parishii
5389. Chaenacits suffrutescens
5390. Chaenactis thompsonii
5391. Chaenactis nevadensis
5392. Chaenactis alpigena
5393. Chaenactis ramosa

the divisions somewhat contorted, not flat, the rachis about 1 mm. broad; heads 2 or 3, whitish, 12–14 mm. high, terminating the leafy stems, the peduncles short; phyllaries 10–12 mm. long, obtuse or spatulate, glandular-puberulent and sparsely lanate; corollas 6–7 mm. long, somewhat puberulent with occasional glands; achenes 6–7 mm. long, strigose; pappus-paleae 10–16, unequal, 2–5 mm. long, oblong or linear and rounded at the apex.

Rocky, often serpentine slopes, Canadian and Hudsonian Zones; Wenatchee Mountains in Chelan and Kittitas Counties, Washington. Type locality: mouth of Beverly Creek, Kittitas County. June–July.

8. Chaenactis douglásii (Hook.) Hook. & Arn. Hoary Chaenactis. Fig. 5394.

Hymenopappus douglasii Hook. Fl. Bor. Amer. **1**: 316. 1834.
Chaenactis douglasii Hook. & Arn. Bot. Beechey 354. 1840.
Macrocarphus douglasii Nutt. Trans. Amer. Phil. Soc. II. **7**: 376. 1841.
Chaenactis suksdorfii Stockwell, Contr. Dudley Herb. **3**: 108. *pl. 23, figs. 4–7.* 1940.

Stout, biennial or perennial herbs 1.5–6 dm. high, stems densely or sparsely tomentose, also somewhat glandular below the inflorescence, single or few from the base, unbranched below the corymbiform inflorescence. Basal leaves 5–12 cm. long, 2–3-pinnate, stem-leaves reduced and less dissected, all thick in texture, more or less tomentose, glandular-pitted, the ultimate lobes obtuse, contorted and somewhat revolute, the leaf thus not flat; heads large, 50–70-flowered, white or pinkish; phyllaries 11–17 mm. long, the outermost shorter, narrowly oblanceolate or oblong, granular-glandular, more or less woolly with soft, slender, jointed hairs and also bearing some coarse straight hairs on the back and margin and also on the peduncle; corolla-lobes hairy on the outer surface, the tube glandular; achenes 6–8 mm. long, densely hirsute; pappus-paleae unequal, 10–16, linear or oblong, about three-fourths the length of the corolla.

Arid, rocky or sandy, sagebrush areas, Arid Transition Zone; Yakima County, Washington, south to Jefferson County, Oregon. Type locality: near Celilo Falls on the Columbia River. June–July. Douglas' Pincushion.

A taxon which is part of an extremely variable and widespread complex which seemingly, from the available collecting data, is quite sensitive to ecological conditions; consequently ecological as well as regional forms are to be found. In eastern Washington and eastern Oregon plants occur that are intermediate in their characteristics between *C. douglasii* var. *douglasii* and *C. douglasii* var. *achilleaefolia*. These forms that otherwise might be considered to be *C. douglasii* var. *achilleaefolia* have more flowers per head and some coarse straight hairs on the phyllaries.

All of the taxa of the *douglasii* complex have to a greater or less degree a distinguishing leaf-character. The leaves do not present a plane surface but rather a contorted one, due partly to the orientation of the pinnae and a tendency to become revolute. This is specially noticeable in the reduced forms from higher elevations.

Chaenactis douglasii var. **achilleaefòlia** (Hook. & Arn.) A. Nels in Coult. & Nels. New Man. Bot. Rocky Mts. 577. 1909. (*Chaenactis achilleaefolia* Hook. & Arn. Bot. Beechey 354. 1840; *Macrocarphus achilleaefolia* Nutt. Trans. Amer. Phil. Soc. II. **7**: 376. 1841; *Chaenactis imbricata* Greene, Leaflets Bot. Obs. **2**: 222. 1912; *C. bracteata* Greene, op. cit. 224; *C. cheilanthoides* Greene, op. cit. 225.) Annual or biennial up to 6 dm. high, the stems all erect, single or a few from the base, typically rather densely floccose-canescent; inflorescence as in *C. douglasii* var. *douglasii*; heads about 40–50-flowered, the corollas white or pinkish; phyllaries 8–12 mm. long, moderately glandular-puberulent as well as loosely canescent; corollas, achenes, and pappus-paleae as in *C. douglasii* var. *douglasii*, but the pappus-paleae, especially of plants of the southern extension of the range, usually one-half or less the length of the corollas. Plains and dry slopes, British Columbia to Montana and south to northern Arizona and southward east of the Cascade Mountains in Washington and Oregon to northeastern California; also the Sierra Nevada principally on the eastern face and the desert ranges as far south as Inyo County.

As here identified var. *achilleaefolia* is the most common segregate of the *douglasii* complex and many variants are included under the name. Also intergrades with named segregates from higher elevations as well as those of lower dry canyons and plains are frequently found. Type locality: "Dry plains of the Snake Country." Collected by Tolmie.

Chaenactis douglasii var. **glandulòsa** Cronquist, Vasc. Pl. Pacif. Northw. **5**: 122. 1955. Annual or biennial (?) about 3 dm. or more high, differing from the two preceding taxa in greener foliage, stipitate-glandular involucres and peduncles, and consistently pink flowers; number of flowers in each head as in *C. douglasii* var. *achilleaefolia*. Dry canyons in southeastern Washington, northeastern Oregon, and adjacent Idaho in the drainage of the Snake River. Type locality: Blue Mountains southwest of Anatone, Asotin County, Washington. Dusty Maidens.

Chaenactis douglasii var. **montàna** M. E. Jones, Proc. Calif. Acad. II. **5**: 700. 1895. (*Chaenactis panamintensis* Stockwell, Contr. Dudley Herb. **3**: 113. 1940; *C. douglasii* var. *nana* Stockwell, op. cit. 117; *C. douglasii* var. *ramosior* Cronquist, Madroño **7**: 81. 1943.) Low, perennial, leafy-stemmed plants rather rarely more than 1 dm. high, sparsely to densely floccose-tomentose, stems often reddish, erect, usually freely branching, single or several arising from the root-crown or from a subterranean branched caudex, the taproot rather thick (attaining great thickness in plants of talus slopes); basal and lower leaves many, tending to wither in age, upper leaves reduced in size; inflorescence corymbosely branched as in the preceding taxa, the peduncles glandular; phyllaries unequal, 8–10 mm. long, granular-glandular only or intermixed with longer glandular hairs and often also woolly-tomentose especially toward the base; pappus-paleae of the inner as well as marginal flowers typically one-half or less the length of the corolla. Idaho and Montana south to Utah, and Washington south to Oregon, Nevada, and eastern California in the Sierra Nevada and the desert ranges. Type locality: "higher mountains from 7000′ to 9000′ alt., and ranges from the Rocky Mountains of Colorado to the Sierras."

Apparently only an extension at higher elevations of var. *achilleaefolia* which has developed the characters of short stature and perennial growth. *Chaenactis pedicularia* Greene (Pittonia **4**: 98. 1899), *C. angustifolia* Greene (Leaflets Bot. Obs. **2**: 223. 1912), *C. humilis* Rydb. (N. Amer. Fl. **34**: 72. 1910), and *C. cineria* Stockwell (Contr. Dudley Herb. **3**: 109. 1940) are probably synonyms of this subalpine phase.

Chaenactis panamintensis Stockwell is an apparent ecotype adapted to the arid alpine conditions of peaks of desert ranges. It simulates *C. douglasii* var. *alpina* in the woody, much-branched caudex and low stature but is here considered to be *C. douglasii* var. *montana* because of the leafy branching stems with evident internodes, moderate head size, and pappus-paleae which are one-half or less the length of the corolla. Telescope Peak, Panamint Mountains, California, and, according to Stockwell, southern Nevada. A specimen from Olancha Peak, Tulare County, California (*Munz 15287*), appears to be this form.

Chaenactis douglasii var. **rubricaùlis** (Rydb.) Ferris, Contr. Dudley Herb. **5**: 100. 1958. (*Chaenactis rubricaulis* Rydb. N. Amer. Fl. **34**: 72. 1914.) Perennial 1.5–3 dm. high, stems red, solitary or few (1–6), nearly glabrous with traces of a lanate tomentum especially about the bases of the leaves, glandular on the peduncles, the slender root often horizontal, the stems assurgent. Leaves mostly basal, about 3–5 cm. long, the stem-leaves few, somewhat reduced in size and much shorter than the internodes, bi- or tripinnatifid, lightly lanate, becoming glabrate; stems bearing few heads; involucres narrow, 17–23-flowered; phyllaries unequal, 7–12 mm. long, rather narrowly linear, obtuse at apex, usually densely glandular-puberulent; flowers tinged with pink, reddish-tipped; pappus of 10 nearly equal oblanceolate paleae, somewhat erose, more than half the length of the corolla. In the upper Arid Transition and Canadian Zones in the Sierra Nevada from Nevada County to Calaveras

County, California, and in adjacent Nevada where it is the predominant form, and occurring less commonly as far south as Tulare County, California. Type locality: Deer Park, Placer County, California.

In its characteristic form, one of the more distinct segregates of the *C. douglasii* complex. In the Siskiyou Mountains of northern California, the higher North Coast Ranges of California, and the northern Sierra Nevada occasional specimens have been collected that can truly be assigned to var. *rubricaulis*. However, other plants from those regions growing in similar habitats that have been called *rubricaulis* seem, for the most part, intermediate between var. *rubricaulis* and a Siskiyou form with large, many-flowered heads and a strong tendency to wooliness, in that they combine the typical growth habit of the former with the large heads of the latter. The same situation occurs in the southern Sierra Nevada where typical specimens of var. *rubricaulis* are to be found, as well as intergrading forms which appear to be intermediate with the southern extension of var. *achilleaefolia* which grows on the eastern face of the Sierra Nevada and the White, Inyo, and Panamint Mountains.

Chaenactis douglasii var. alpina A. Gray, Syn. Fl. N. Amer. 1^2: 341. 1884. (*Chaenactis alpina* M. E. Jones, Proc. Calif. Acad. II. **5**: 699. 1895; *C. leucopsis* Greene, Leaflets Bot. Obs. **2**: 221. 1912; *C. pumila* Greene, loc. cit.; *C. rubella* Greene, op. cit. 222; *C. minuscula* Greene, op. cit. 223.) Dwarf, more or less cespitose, alpine perennial 4–15 cm. high, loosely floccose-tomentose and usually becoming glabrate, arising from a thick, usually much-branched, subterranean caudex and woody taproot; leaves densely clustered on the short stems, 2–6 cm. long, rather thinly floccose-tomentose, mostly bipinnatifid, the divisions condensed and much contorted, occasionally more open; peduncles axillary, scapose, glandular-pubescent above, occasionally forked and bearing more than one head, mostly of equal length, the heads borne 1.5–5 cm. above the leaf-clusters; heads 40–50-flowered, the flowers white or roseate; phyllaries unequal, 8–12 mm. long, narrowly oblanceolate or linear-lanceolate, obtuse or acute at apex, the pubescence various, densely or sparsely granular-glandular only or mixed with spreading pointed hairs, also often sparsely tomentose toward the base; pappus-paleae 10, often roseate, of unequal length, oblanceolate, the margins somewhat erose, three-fourths to one-half the length of the corolla. Rocky or talus slopes, Arctic-Alpine Zone: Idaho and western Montana south to central Utah and western Colorado; Wallowa Mountains, Oregon, Mount Shasta, California, and occasional in the Sierra Nevada as far south as Sonora Pass, Tuolumne County, California. Type locality: not stated by A. Gray but a specimen from Alta, Wasatch Mountains, Utah (*M. E. Jones 1232*), has been named the lectotype by Stockwell (Contr. Dudley Herb. **3**: 113. 1940).

Perhaps the most distinct of the segregates of *C. douglasii* var. *douglasii* but occasional intergrades can be demonstrated throughout its range with var. *montana*.

9. **Chaenactis cusíckii** A. Gray. Cusick's Chaenactis. Fig. 5395.

Chaenactis cusickii A. Gray, Syn. Fl. N. Amer. ed. 2. 1^2: 452. 1886.

Annual 3.5–10 cm. high, branching from the base, the stems reddish, glabrous, with some cobwebby hairs about the heads and leaf-bases. Leaves 1–4 cm. long, 1–4 mm. wide, rather succulent, linear, entire, rarely with 1–3 minute blunt lobes; heads white-flowered, usually closely subtended by 1–2 leaves surpassing the head in length; phyllaries unequal, 5–7 mm. long, linear, obtuse at apex, rather succulent, the midvein scarcely evident; corolla white, 5–5.5 mm. long, puberulent without, the marginal flowers narrowly funnelform, the lobes longer than those of the disk; achenes 5–5.5 mm. long with ascending hairs; pappus-paleae 8–10, laciniate-margined, the longest about 2 mm., the shortest about 0.5 mm. long.

Clayey or sandy slopes, Upper Sonoran Zone; southwest Idaho and eastern Oregon in Baker and Malheur Counties. Type locality: sandy hills of the Malheur, Baker County. April–June.

10. **Chaenactis névii** A. Gray. John Day Chaenactis. Fig. 5396.

Chaenactis nevii A. Gray, Proc. Amer. Acad. **19**: 30. 1883.

Glandular-puberulent annual often red-stemmed, 10–25 cm. high, simple or branching from the base and at the midstem with ascending branches, glandular-puberulent, densely so below the heads and slightly arachnoid-villous. Lower leaves at flowering time 4.5 cm. long or less including

5394. Chaenactis douglasii

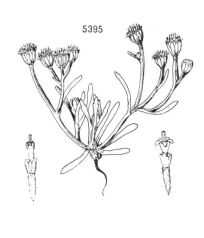

5395. Chaenactis cusickii

the petiolar base, pinnately divided nearly to the midrib with 3–5 widely spreading, linear, obtuse lobes, these 5–10 mm. long; upper leaves reduced above in length and often linear on unbranched plants, all leaves sparsely glandular-puberulent to glabrate; heads bright yellow, terminating the branches, the outer flowers regular but larger than the inner, the peduncles 1–6 cm. long; involucres 6–7 mm. high; phyllaries linear-lanceolate, glandular-puberulent with occasional cobwebby hairs, thin, the midrib becoming somewhat prominent in age; achenes black, about 5 mm. long, thickly beset with shining, coarse, blunt, spreading hairs; pappus of 10 vestigial paleae.

Barren dry slopes in heavy clay soil, Arid Transition Zone; Wheeler and Grant Counties, Oregon. Type locality: Idaho; probably an error in the collecting data, as the species seems to be limited to the John Day Basin, Oregon. April–May. Nevius' Chaenactis.

11. **Chaenactis artemisiaefòlia** (Harv. & Gray ex A. Gray) A. Gray. Artemisia-leaved Chaenactis. Fig. 5397.

Acarphaea artemisiaefolia Harv. & Gray ex A. Gray, Mem. Amer. Acad. II. 4: 98. 1849.
Chaenactis artemisiaefolia A. Gray, Proc. Amer. Acad. 10: 74. 1874.

Coarse, single-stemmed annual 2.5–10(20) dm. high, often reddish, leafy below, the leaves reduced or absent in the open cymose-paniculate inflorescence; stems usually mealy-pubescent below, glandular-hirsute above. Leaves petioled, 3–15(20) cm. long, usually mealy-pubescent, the flat blade ovate in outline, about two-thirds as wide, bi- or tripinnate, finely divided into small, oblong or linear divisions, the petioles one-half or more the length of the blade; heads whitish- or pinkish-flowered, hemispheric, on curved, ascending, glandular-hirsute peduncles 1.5–6 cm. long; phyllaries 5–9 mm. long, linear-lanceolate, acute, glandular-hirsute; corollas about 5 mm. long, the well-defined tube glandular-pubescent, the funnelform throat glabrous or nearly so, marginal corollas like the disk-corollas; achenes 5–7 mm. long, linear-clavate and somewhat flattened, the outer somewhat curved, black and microscopically striate, glabrous or nearly so; pappus-paleae lacking, if present rudimentary and soon deciduous.

Rather common on mountain slopes, often on burns and in disturbed areas. Upper Sonoran Zone; coastal ranges from Santa Barbara County to San Diego County and northern Lower California. Type locality: California. Collected by Coulter. April–July.

12. **Chaenactis macrántha** D. C. Eaton. Large-flowered Chaenactis. Fig. 5398.

Chaenactis macrantha D. C. Eaton, Bot. King Expl. 171. *pl. 18. figs. 1–5.* 1871.

Annual, 7–25 cm. high, stems rather stout, ascending, several from the base and branching, the herbage thinly floccose-tomentulose, often becoming more or less glabrate. Leaves 1.5–4 cm. long, reduced above, pinnatifid to bipinnatifid, the blade ovate in outline, the petioliform base longer than or at least equaling the blade; heads on peduncles 1.5–8 cm. long, white-flowered, the flowers spreading at night, closely united by day; involucre broadly turbinate, the phyllaries of unequal length, 9–14 mm. long, thin and flat, the midvein whitish, linear-lanceolate, the acute apex often spreading; corollas 10–12 mm. long, the lobes acute, about 2 mm. long, slender, cylindric, densely tomentulose without, anthers always included, the marginal flowers not differing from those of the disk; achenes about 6 mm. long, strigose; pappus-paleae in 2 series, the four inner linear-oblong, 5.5–6 mm. long, the four blunt outer ones about 1 mm. long.

On open deserts or in washes, Sonoran and Arid Transition Zones; southwestern Idaho and southeastern Oregon to Nevada, southern Utah, and northwestern Arizona and the Mojave Desert in San Bernardino County, California. Type locality: "Pahute Mountains, Nye County, Nevada." Collected by Watson. May–June. Mojave Pincushion.

13. **Chaenactis xantiàna** A. Gray. Xantus' Chaenactis. Fig. 5399.

Chaenactis xantiana A. Gray, Proc. Amer. Acad. 6: 545. 1865.
Chaenactis xantiana var. *integrifolia* A. Gray, loc. cit.

Annual, 1.5–3.5 dm. high, stems rather stout, reddish, the leaves branching at or a little above the base, ascending, the herbage glabrous or sparsely and loosely lanate about the heads and leaf-bases. Leaves 2–6 cm. long, rather equally distributed on the stem and not markedly reduced above, linear, entire or with 1–2 pairs of linear spreading pinnae, more or less succulent; heads medium to large, the flowers sordid-white, the peduncles 1–5 cm. long, enlarged below the head, fistulose; involucre broadly turbinate, the phyllaries 10–16 mm. long, linear-lanceolate, flat and rather lax, becoming firmer in age except for the spreading, broadly acute or obtuse, tomentose tip; corollas 6–8 mm. long, nearly tubular, lobes of marginal corollas but little longer than those of the disk; achenes black, 7–8.5 mm. long, strigose; inner pappus-paleae 4, lanceolate, equaling the corollas, outer pappus-paleae usually 4, about 1.5 mm. long, broad and rounded at the apex.

Sandy flats and open desert slopes, Upper Sonoran Zone; Harney and Malheur Counties, Oregon, south through western Nevada and adjacent California to northwestern Arizona and in the San Joaquin Valley, California, from eastern San Luis Obispo and western Merced Counties south to the inner northern ranges of southern California and the western Mojave Desert. Type locality: Fort Tejon, Kern County. April–June. Desert Chaenactis; Xantus' Pincushion.

14. **Chaenactis tanacetifòlia** A. Gray. Inner Coast Range Chaenactis. Fig. 5400.

Chaenactis tanacetifolia A. Gray, Proc. Amer. Acad. 6: 545. 1865.

Annual, 6–15 cm. high, stems few to several from and near the base, ascending, often curved, little or not at all branched above, lightly floccose, more densely so at the nodes and below the heads. Leaves basal or near the base, tending to remain at flowering, few above, 1–5 cm. long, about 5–10 mm. wide, fleshy, floccose to glabrate, the pinnae crowded and often contorted, redivided with short, often minute lobes; heads yellow-flowered, terminating the stems and branches, borne on

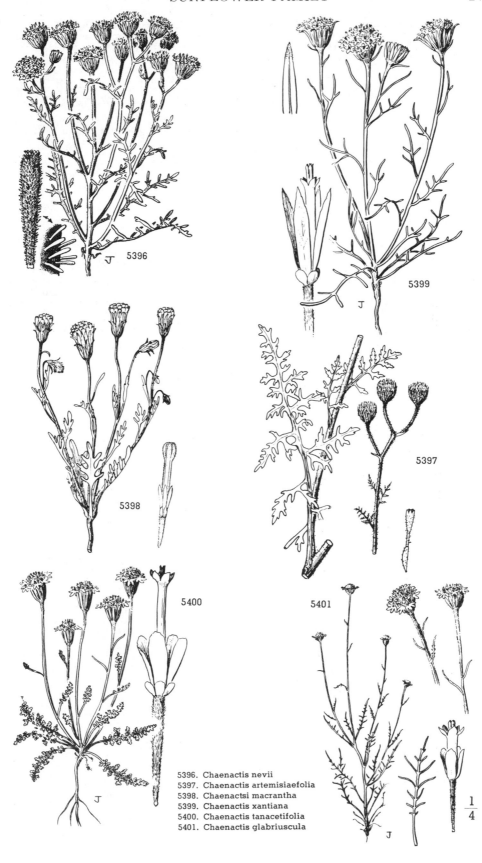

5396. Chaenactis nevii
5397. Chaenactis artemisiaefolia
5398. Chaenactsi macrantha
5399. Chaenactis xantiana
5400. Chaenactis tanacetifolia
5401. Chaenactis glabriuscula

peduncles 2–7 cm. long, the involucre broadly turbinate; phyllaries herbaceous, the midvein more prominent in age, 6–10 mm. long, up to 2 mm. wide, linear-lanceolate to linear-oblanceolate, broadly acute, more or less floccose, usually becoming glabrate; achenes about 4 mm. long, sparsely hirsute; inner pappus one-half to three-fourths the length of the corollas, oblong, often broadly so, obtuse, the outer series 4, about 0.5 mm. long, ovate or rounded.

Open rocky slopes and ridges, apparently preferring serpentine, Upper Sonoran Zone; northern Inner Coast Ranges in Sonoma, Lake, Napa, and Yolo Counties, the Hamilton Range and inner slope of the Sierra Azul Range in Santa Clara County, south through the Inner Coast Ranges of San Benito and Monterey Counties, California. Type locality: near Clear Lake, Lake County. April–June. Serpentine Chaenactis. Intergrading with *C. tanacetifolia* var. *gracilenta* in its northern range.

Chaenactis tanacetifolia var. **gracilénta** (Greene) Stockwell, Contr. Dudley Herb. **3**: 124. 1940. (*Chaenactis gracilenta* Greene, Fl. Fran. 447. 1897; *C. glabriuscula* var. *filifolia* Jepson, Man. Fl. Pl. Calif. 1124. 1925, as to northern plants only; *C. glabriuscula* var. *gracilenta* Keck, Aliso **4**: 101. 1958.) A minor variant of *C. tanacetifolia* var. *tanacetifolia* 15–25 cm. high, differing by its tendency toward erect habit, basal leaves soon withering, and leaves in which the pinnate divisions are more remote; pappus as in the preceding but the outer series often incomplete and much shorter. Open places, Yolo, Lake, and Napa Counties, California. Type locality: dry ridges of Napa County. Intermediate between *C. tanacetifolia* var. *tanacetifolia* and *C. glabriuscula* var. *heterocarpha*.

15. Chaenactis glabriúscula DC. Common Yellow Chaenactis. Fig. 5401.

Chaenactis glabriuscula DC. Prod. **5**: 659. 1836.

Annual, 1.5–4 dm. high, more or less leafy, typically single-stemmed and branching above with ascending branches, sometimes also branched from the base, the herbage thinly floccose, often becoming glabrate. Basal leaves soon withering, the stem-leaves 3–5 cm. long, longer than the internodes, pinnately divided to midrib into flat, remote, linear lobes 2–10 mm. long, the uppermost leaves more reduced and often entire; involucres broadly turbinate or subcampanulate, sparingly floccose except at the base, obscurely if at all granular-glandular; phyllaries 6.5–10 mm. long, up to 2 mm. wide, mostly linear-lanceolate, broadly acute to obtuse at the tip, herbaceous, rather thin in age; marginal flowers conspicuously enlarged; achenes about 5 mm. long, appressed-hispid, dark or some with pale suberized areas; pappus-paleae equal or unequal, one-half to three-fourths the length of the corolla, usually broadly lanceolate and sometimes obtuse at apex.

Open or wooded hill slopes and valleys, Upper Sonoran Zone; Monterey and San Benito Counties, California, south through the Coast Ranges to Los Angeles and western San Bernardino Counties. Type locality: California. Collected by Douglas. March–June.

Notes made by H. M. Hall of the type specimen state that great variability occurs in the length of the pappus-paleae. Part of a highly variable complex. This and the following taxa, though generally recognizable, intergrade freely on areas of contact. Those of the mountains of southern California, where *C. glabriuscula* var. *glabriuscula*, var. *denudata*, and var. *curta* occur, are particularly confusing.

Key to Varieties

Leaves few, not concentrated at the base; heads borne on branches of the typically single-stemmed plants.
  Outer series of pappus-paleae 1–4; plants of northern borders of the Sacramento Valley.
                                                           C. glabriuscula heterocarpha.
  Outer series of pappus-paleae lacking (sometimes vestigial); plants of central California southward.
    Heads large; phyllaries 8–11 mm. long.
      Paleae one-half or over the length of the corollas, often obtuse.   *C. glabriuscula glabriuscula.*
      Paleae nearly equaling the corollas, always acute.   *C. glabriuscula megacephala.*
    Heads small; phyllaries 5–7 mm. long.   *C. glabriuscula curta.*
Leaves many, basal or nearly basal; heads borne on few to several nearly naked stems.
  Heads (including the corollas) about 15 mm. high; plants about 2 dm. high. *C. glabriuscula denudata.*
  Heads (including the corollas) 10–11 mm. high; plants 1–1.5 dm. high. *C. glabriuscula lanosa.*

Chaenactis glabriuscula var. **heterocárpha** (Torr. & Gray ex A. Gray) H. M. Hall, Univ. Calif. Pub. Bot. **3**: 190. 1907. (*Chaenactis heterocarpha* Torr. & Gray ex A. Gray, Mem. Amer. Acad. II. **4**: 98. 1849.) Habit like that of *C. glabriuscula* var. *glabriuscula* but usually more woolly-floccose; some achenes of the heads often with pale corky areas; inner pappus-paleae broadly lanceolate, nearly as long as the disk-corollas, also having an outer series of 1–4 minute paleae. Northern borders of the Sacramento Valley, California, in Glenn and Tehama Counties and from Butte County south to Mariposa County. Type locality: Sacramento Valley. Collected by Fremont.

Chaenactis glabriuscula var. **megacéphala** A. Gray in Torr. Pacif. R. Rep. **4**: 104. 1857, not A. Gray, Proc. Bost. Nat. Hist. Soc. **7**: 146. 1861. (*Chaenactis heterocarpha* var. *megacephala* Stockwell, Contr. Dudley Herb. **3**: 123. 1940.) Distinguished from *C. glabriuscula* var. *glabriuscula* only by large heads with usually larger marginal flowers and broad phyllaries, and more robust habit, by floccose hairs of the involucre which are sometimes flecked with black, and by lanceolate pappus-paleae nearly equaling the disk-flowers (vestiges of outer pappus-paleae rarely found); some achenes with suberized areas as in *C. glabriuscula* var. *glabriuscula*. Commonly found in the lower foothills of the Sierra Nevada, California, from Amador County south to the Tehachapi Mountains, Kern County; also the western slopes of the San Joaquin Valley, where it merges completely with *C. glabriuscula* var. *glabriuscula*. Type locality: Knights Ferry, Stanislaus County. Weakly separable from the name-bearing taxon by the long paleae and larger size of plants.

Chaenactis glabriuscula var. **lanòsa** (DC.) H. M. Hall, Univ. Calif. Bot. **3**: 192. 1907. (*Chaenactis lanosa* DC. Prod. **5**: 659. 1836.) Plants 0.8–1.5(2) dm. high, densely woolly-lanate throughout, with several to many stems branched from the base; leaves basal or nearly basal, linear with few linear divisions or entire; heads smaller than *C. glabriuscula* var. *glabriuscula*, on long scapose stems. Inner Coast Ranges, California, from Monterey and San Benito Counties south to San Luis Obispo and northeastern Santa Barbara Counties. Type locality: California. Collected by Douglas.

Chaenactis glabriuscula var. **denudàta** (Nutt.) Munz, Man. S. Calif. 567. 1935. (*Chaenactis denudata* Nutt. Proc. Acad. Phila. II. **4**: 21. 1848.) Much like *C. glabriuscula* var. *lanosa* in habit but the plants taller, leaves more divided, heads larger, and stems more evidently branched. Southern coastal San Luis Obispo County, California, south to Los Angeles County and eastward to San Gorgonio Pass, Riverside County. Type locality: Los Angeles. Collected by Nuttall.

Chaenactis glabriuscula var. **cúrta** (A. Gray) Jepson, Man. Fl. Pl. Calif. 1124. 1925. (*Chaenactis heterocarpha* var. *curta* A. Gray, Syn. Fl. N. Amer. ed. 2. 1²: 452. 1886; *C. aurea* Greene ex Rydb. N. Amer. Fl. **34**: 68. 1914; *C. glabriuscula* var. *aurea* Stockwell, Contr. Dudley Herb. **3**: 131. 1940.) Plants 1–2.5 dm. high, simple or openly branched; leaves shorter and heads smaller than the name-bearing taxon; pappus-paleae oval to broadly oblong, all much shorter than the corolla or one of the four much longer. Greenhorn Mountains, Kern County, California, to Ventura County and south to northern San Diego County. Type locality: upper part of the Santa Clara Valley, Ventura County. In its southern range intergrading with *C. tenuifolia*.

16. Chaenactis tenuifòlia Nutt. San Diego Chaenactis. Fig. 5402.

Chaenactis tenuifolia Nutt. Trans. Amer. Phil. Soc. II. **7**: 375. 1841.
Chaenactis filifolia A. Gray, Mem. Amer. Acad. II. **4**: 98. 1849.
Chaenactis glabriuscula var. *tenuifolia* H. M. Hall, Univ. Calif. Pub. Bot. **3**: 191. 1907.
Chaenactis glabriuscula var. *filifolia* Jepson, Man. Fl. Pl. Calif. 1124. 1925, as to name only.

Annuals, (1.5)2–6 dm. high, stems slender and wiry, reddish at least at the base, mostly branching above the base with ascending branches, the herbage sparsely floccose to glabrate. Leaves 3–6(8), about 0.5–2 cm. wide, pinnate or often bipinnate, divided to the midrib in linear divisions of unequal length, the uppermost leaves reduced, pinnately parted and mostly much shorter than the internodes; heads mostly many, rather small, on slender peduncles; involucres 5–8 mm. high, hemispheric, many-flowered, the marginal flowers not greatly enlarged; phyllaries narrowly linear-lanceolate and acute, about 1 mm. wide, not conspicuously herbaceous and becoming rather firm in age, noticeably glandular-puberulent, sometimes densely so, slightly if at all woolly; achenes 3.5 mm. or more long, slender; pappus-paleae narrowly oblong or sometimes acute, one-half to two-thirds as long as the corollas.

Open places in the chaparral and sandy canyons, Sonoran Zones; Orange and San Diego Counties east to western Imperial County; also Riverside County, where it intergrades with *C. glabriuscula* var. *curta*. Type locality: San Diego, California. May–July.

Chaenactis tenuifolia var. orcuttiàna Greene, W. Amer. Sci. **3**: 157. 1887. (*Chaenactis orcuttiana* Parish, Erythea **6**: 92. 1898; *C. glabriuscula* var. *orcuttiana* H. M. Hall, Univ. Calif. Pub. Bot. **3**: 192. 1907.) Annuals, 1–3 dm. high, stems and branches more divaricate and stouter than *C. tenuifolia* var. *tenuifolia*; leaves more or less succulent, the ultimate lobes rather broad and obtuse; involucres hemispheric, 8–10 mm. high; phyllaries and peduncles more conspicuously glandular than the preceding taxon. An ecotype of dunes, beaches, and bluffs of the San Diego County coastline and that of adjacent Lower California; merging with *C. tenuifolia* var. *tenuifolia* away from the ocean.

Some plants from coastal Los Angeles County (Ballona; Redondo) have been collected which have the coarse glandular pubescence of this taxon but the habit and leaves of *C. tenuifolia* var. *tenuifolia*.

17. Chaenactis fremóntii A. Gray. Fremont Pincushion or Chaenactis. Fig. 5403.

Chaenactis fremontii A. Gray, Proc. Amer. Acad. **19**: 30. 1883.

Annuals, 10–30 cm. high, stems often reddish, usually rather stout, 1 to several from the base, branched above, the branches ascending, glabrous though sometimes sparsely and loosely lanate below the heads, the peduncles glandular-puberulent and often hispidulous. Leaves 1.5–5(10) cm. long, reduced above, succulent, glabrous, linear or with 1–2(5) divergent linear divisions; heads rather large, white- or pinkish-flowered, the lanceolate pappus visible in bud, the peduncles 2–8 cm. long; involucres broadly turbinate; phyllaries 8–10 mm. long, glabrous, broadly linear, acute at apex, rather firm, the midrib prominent; marginal corollas usually greatly enlarged with a dilated lobed limb; disk-corollas 5–6 mm. long; achenes slender, about 7 mm. long, black, strigose; pappus-paleae 4, occasionally 6, as long as the corollas, linear-lanceolate, those of the marginal corollas usually with 1 long and 3 short obtuse scales.

Open desert, mostly Lower Sonoran Zone; southwestern Nevada and the adjacent Death Valley region, California, southward through the Mojave and Colorado Deserts to Imperial and eastern San Diego Counties, and eastward to Arizona; also in the San Joaquin Valley in western Kern County. Type locality: California. Collected by Fremont on his second expedition. March-May.

18. Chaenactis stevioìdes Hook. & Arn. Broad-flowered Chaenactis. Fig. 5404.

Chaenactis stevioides Hook. & Arn. Bot. Beechey 353. 1839.
Chaenactis floribunda Greene, Pittonia **3**: 168. 1897.

Annual, 10–15(25) cm. high, single-stemmed or several-stemmed from the base, branching above the middle, the heads many, terminating the branches, the herbage finely grayish floccose-

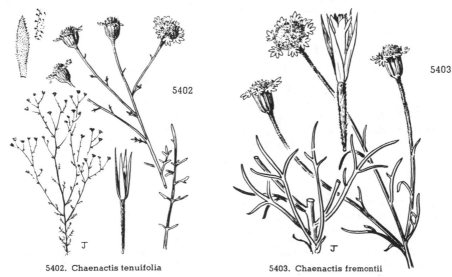

5402

5403

5402. Chaenactis tenuifolia   5403. Chaenactis fremontii

tomentulose, becoming more or less glabrate, the peduncles and involucres glandular-hispidulous as well as tomentose. Leaves 1–3.5 cm. long, the upper leaves rather few, equaling or shorter than the internodes, pinnately divided to the midrib with divergent lobes of unequal length, 5–15 mm. long, these often pinnatifid with short lobes; involucres subhemispheric, borne on peduncles 2–5 cm. long; phyllaries 5–7 mm. long, linear, obtuse to broadly acute at apex, firm, the midvein evident, becoming more so in age; marginal flowers moderately enlarged, nearly regular, creamy white; disk-flowers 4.5–6 mm. long, nearly tubular, creamy white, sometimes tinged with rose; achenes about 6 mm. long, strigose; pappus-paleae 4, oblong-lanceolate, somewhat unequal, the longest shorter than the disk-flowers.

Open, rocky or sandy deserts, Arid Transition and Sonoran Zones; southeastern Oregon, southern Idaho, Wyoming, Utah, and western Colorado south to New Mexico, Arizona, and Sonora; desert areas of eastern California to Lower California and also occurring in the southern San Joaquin Valley, California. Type locality: Idaho. Collected by Tolmie. March–June. Esteve Pincushion.

Chaenactis stevioides var. **brachypappa** (A. Gray) H. M. Hall, Univ. Calif. Pub. Bot. **3**: 194. 1907. (*Chaenactis brachypappa* A. Gray, Proc. Amer. Acad. **8**: 390. 1872. Heads often larger than the preceding; pappus of the disk-flowers as short as those of the marginal ones. Occurring with *C. stevioides* var. *stevioides* from the type locality in the Pahranagat Mountains, central Nevada, through the desert regions of California and also occasional in the southern San Joaquin Valley.

19. **Chaenactis carphoclìnia** A. Gray. Pebble Pincushion. Fig. 5405.

Chaenactis carphoclinia A. Gray, Bot. Mex. Bound. 94. 1859.
Chaenactis paleolifera A. Nels. Bot. Gaz. **47**: 434. 1909.
Chaenactis peirsonii Jepson, Madroño **1**: 259. 1929.

Slender, wiry-stemmed annuals, 1.5–3 dm. high, leafy, much-branched above the base, irregularly dichotomous, the heads many, terminating the spreading branches, glandular-puberulent throughout and often with glandular hirsute hairs on peduncles and involucres. Leaves 1–6 cm. long, petiolate, the upper leaves reduced, becoming bracteate, pinnate to bipinnatifid, the lobes remote and narrowly linear to filiform; heads white- or pinkish-flowered on slender peduncles 2–6 cm. long; involucres subcampanulate, the phyllaries 7–10 mm. long, firm, linear, caudate-attenuate with a pale to reddish, setaceous tip, this more noticeable on the inner phyllaries; receptacle with few to several persistent, rigid, filiform bracts as long as the phyllaries; corollas 4–5.5 mm. long, narrowly funnelform, the lobes of the marginal flowers but little enlarged; achenes about 4 mm. long, with ascending hairs; pappus-paleae of the disk-flowers broadly lanceolate, about one-half the length of the corollas, those of the marginal flowers truncate, less than one-half the length of the corollas.

Sandy and gravelly, desert flats, Lower Sonoran Zone; southern Nevada and the Death Valley region, California, south through the Mojave and Colorado Deserts to Lower California and western Arizona to Sonora. Type locality: Fort Yuma, Arizona. March–May.

Chaenactis carphoclinia var. **attenuàta** (A. Gray) M. E. Jones, Proc. Calif. Acad. II. **5**: 699. 1895. (*Chaenactis attenuata* A. Gray, Proc. Amer. Acad. **10**: 73. 1874.) Differs from *C. carphoclinia* var. *carphoclinia* by having all the pappus-paleae short and truncate and heads often, but not always, fewer-flowered and more narrow; receptacular bracts sometimes completely lacking. Growing with the species throughout its range. Type locality: Ehrenberg, Yuma County, Arizona.

Chaenactis latifòlia Stockwell, Contr. Dudley Herb. **3**: 128. *pl. 36, figs. 4–6.* 1940. Somewhat leafy-stemmed, succulent annual, 15–35 cm. high; leaves 3–7 cm. long, the winged petiole equaling the blade, the blade with a broad rachis and irregularly, pinnately divided with some lobes bipinnatifid; heads about 1 cm. high, the flowers cream-colored; phyllaries oblong-acute, hispid, glandular; corollas expanded near the base; achenes about 5 mm. long, the pappus-paleae 4, lanceolate, three-fourths the length of the corollas. Eastern San Diego County, California, and northern Lower California. Type locality: Jacumba, San Diego County.

Chaenactis mexicàna Stockwell, Contr. Dudley Herb. **3**: 129. *pl. 37, figs. 7–9.* 1940. Wiry-stemmed, much-branched, hispidulous annual, 10–25 cm. high; leaves about 1 cm. long, thick, with few short obtuse pinnae; heads about 1 cm. high, the flowers white; phyllaries linear to oblong, acute, glandular-puberulent; corollas expanded near the base; achenes 3–4 mm. long, the pappus-paleae 4, lanceolate, acute or obtuse at the apex. San Diego County, California, and northern Lower California. Type locality: northern Lower California. Collected by Orcutt.

The descriptions of the two preceding taxa have been adapted from the original descriptions, as no material has been found that completely conforms to the original diagnoses. Perhaps they are hybrids derived from some of the several species occurring in this area.

60. **OROCHAENÁCTIS** Coville, Contr. U.S. Nat. Herb. **4**: 134. *pl. 10.* 1893.

Slender annual with long internodes, cymosely branching above. Leaves alternate or opposite below, linear. Heads discoid, solitary or clustered in the leaf-axils, sessile or short-pedunculate. Involucre of few distinct phyllaries in 1 series, noticeably much shorter than the corollas, these equaling the phyllaries in number. Receptacle flat. Flowers perfect, fertile. Corolla-tube much shorter than the narrow, little-expanded throat, the lobes short, revolute in anthesis. Anthers sagittate. Style-branches linear, obtuse. Achenes clavate, striate, granular-glandular; the pappus-paleae thin, hyaline, narrowly spatulate, lacerate-dissected on the margins and apex. [From the Greek word meaining mountain, and the genus *Chaenactis*.]

A monotypic genus.

1. **Orochaenactis thysanocárpha** (A. Gray) Coville. Orochaenactis. Fig. 5406.

Chaenactis thysanocarpha A. Gray, Proc. Amer. Acad. **19**: 30. 1883.
Bahia palmeri S. Wats. op. cit. **24**: 83. 1889.
Orochaenactis thysanocarpha Coville, Contr. U.S. Nat. Herb. **4**: 134. 1893.

Slender, reddish-stemmed annual 5–35 cm. high, simple below and branching above or branching from the base, more or less granular-glandular throughout and bearing sparse, loose, cobwebby hairs about the nodes, leaves, and heads. Leaves 1.5–3 cm. long; phyllaries 4 or 5, herbaceous, 4–5 mm. long, oblanceolate, and rounded at the apex, densely glandular; corollas 3.5–4 mm. long, glandular without; achenes about 3.5 mm. long, the surface dull with darker spots; pappus-paleae 6–12, shining, about as long as the achenes.

Sandy flats and slopes, upper Canadian and Hudsonian Zones; southern Sierra Nevada in Inyo, Tulare, and Kern Counties, California. Type locality: southern Sierra Nevada. Collected by Rothrock. July–Aug.

61. **PALAFÓXIA** Lag. Gen. & Sp. Nov. 26. 1816.

Erect, branching, more or less pubescent, annual or perennial herbs, in some species arising from a woody base. Leaves alternate, thick, entire. Heads corymbose or paniculate, discoid or sometimes radiate, the flowers in shades of pink. Involucre turbinate to nearly cylindric, the phyllaries mostly in 1 series, herbaceous, sometimes with scarious tips. Receptacle small, flat, and naked. Ray-flowers when present deeply lobed. Disk-flowers with short tube, the 5 lobes longer or shorter than the throat. Anthers obtuse at base, entire or emarginate. Style-branches linear, hispidulous throughout. Achenes linear to narrowly obpyramidal, 4-angled, pubescent. Pappus-paleae 4–12, subequal, with strong midribs, those of the marginal flowers often shorter. [Name in honor of José Palafox, a Spanish general.]

A genus of about 10 or 12 species, natives of southern United States and adjacent Mexico. Type species, *Ageratum lineare* Cav.

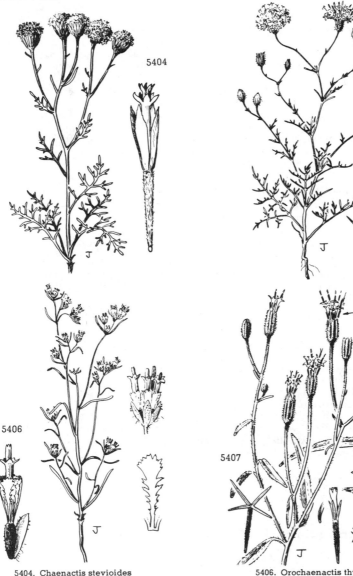

5404. Chaenactis stevioides
5405. Chaenactis carphoclinia
5406. Orochaenactis thysanocarpha
5407. Palafoxia linearis

1. Palafoxia lineàris (Cav.) Lag. Spanish Needle. Fig. 5407.

Ageratum lineare Cav. Ic. **3** : 3. *pl. 205.* 1795.
Stevia linearis Willd. Sp. Pl. **3** : 1774. 1804.
Stevia lavandulaefolia Schlecht. in Willd. Enum. Pl. Suppl. 57. 1813, as a synonym.
Palafoxia linearis Lag. Gen. & Sp. Nov. 26. 1816.
Paleolaris carnea Cass. Bull. Soc. Philom. **1818** : 47. 1818.

Hispid or scabrous annuals, glandular above, the herbage dark, 2–7 dm. high, branched above the base. Leaves sparsely canescent, linear to linear-lanceolate, narrowed to a short petiole, strongly 1-nerved, 2–6 cm. long; heads few, loosely corymbose on glandular peduncles, discoid, flowers pinkish ; involucre 15–18 mm. high, 10–20-flowered ; phyllaries 7–13, herbaceous, linear, acute, hispid and glandular, somewhat keeled at base; corollas perfect, 7–10 mm. long, the lobes shorter than the throat, the styles long-exserted ; mature achene dark, 4-angled, rather densely strigose, the body of the achene about as long as the phyllaries; pappus-paleae subequal, usually 4, spreading at maturity, a little more than one-half the length of the achene, lanceolate, with a strong midrib bordered by a narrow, fragile, hyaline margin, the outer flowers sometimes with the paleae unequal.

Sandy washes and flats, Lower Sonoran Zone; Inyo County, California, south to the Colorado Desert where it is common and adjacent Lower California, east to Arizona and northern Sonora. Type locality: Mexico. Feb.–Aug.

Palafoxia lineaɪis var. **gigantèa** M. E. Jones, Contr. West. Bot. No. 18: 79. 1933. (*Palafoxia linearis arenicola* A. Nels. Amer. Journ. Bot. **23** : 265. 1936.) Plants becoming shrubby, 1–2 m. tall, the herbage green and scabrous; leaves lanceolate, narrowed to a short petiole, strongly 3-nerved; phyllaries 15–20 mm. long, eglandular; achenes about the same length or longer, the pappus-paleae about 8, four about one-half the length of the achene, the other four much shorter. Type locality: Algodones sand hills west of Yuma in Imperial County, California.

62. AMBLYOPÁPPUS Hook. & Arn. Hook. Journ. Bot. **3** : 321. 1841.

Erect, branched, annual herb, glabrous and glandular-granuliferous, sweet-scented. Leaves alternate, linear, the lower pinnate with few linear lobes. Inflorescence many-flowered, cymose-paniculate, the discoid heads short-pedunculate. Heads 5–25-flowered. Receptacle small, conical. Phyllaries in 1 series from a broadened base, concave, herbaceous, obovate to obovate-oblong. Marginal pistillate flowers fertile, 2-toothed. Perfect flowers with short tube and ample 5-toothed limb. Anthers small, obtuse at base. Style-branches linear, little exserted, somewhat broadened distally, truncate, hairy. Achenes dark, 4-angled, narrowed below, the pappus of obtuse, scarious, often colored paleae. [From the Greek words meaning blunt and pappus.]

A monotypic genus.

1. Amblyopappus pùsillus Hook. & Arn. Amblyopappus. Fig. 5408.

Amblyopappus pusillus Hook. & Arn. Hook. Journ. Bot. **3** : 321. 1841.
Aromia tenuifolia Nutt. Trans. Amer. Phil. Soc. II. **7** : 396. 1841.
Infantea chilensis Remy in Gay, Fl. Chil. **4** : 259, *pl.* 48. 1849.

Plants yellowish, erect, 1–4 dm. high, stems leafy, simple or branched below, much branched above, the branches ascending. Leaves linear, entire, 1–3 cm. long, the lower somewhat longer and often pinnately 3–5-parted; heads many, yellowish ; involucre 3 mm. high, the phyllaries 4–6, thin ; marginal and disk-flowers minute, about twice the length of the pappus ; achenes 2 mm. long, roughened, shining, sparsely covered with coarse hairs, these mostly on the angles ; pappus-paleae 8–12, striate, shining, 0.5 mm. long.

On ocean bluffs and by salt marshes, Sonoran Zones; coast of San Luis Obispo County and on the Channel Islands, California, southward to Lower California; also Peru and Chile. Type locality: Coquimbo, Chile. April–June.

63. BLENNOSPÉRMA Less. Syn. Comp. 267. 1832.

Annual, nearly glabrous herbs branching from the base. Leaves alternate, pinnately parted, the lobes remote, or the lowest rarely entire. Heads terminating the branches, many-flowered, the base of the peduncle somewhat enlarged. Receptacle flat. Phyllaries in 1 series, thin and often colored at the tip, united at the base. Ray-flowers pistillate, fertile. Disk-flowers many, hermaphrodite, sterile, the tube slender, throat broadly campanulate. Anthers oval. Style of pistillate flowers with 2 flattened branches, those of the disk-flowers undivided. Achenes obovoid, obscurely 6–10-angled, the surface papillate, becoming mucilaginous when wet ; epappose. [From the Greek words meaning mucus and seed.]

A genus of 3 species, natives of California and Chile. Type species, *Blennosperma chilensis* Less.

Lower leaves 8–10-lobed, the lobes short; branches of the stigma yellow.   1. *B. nanum.*
Lower leaves entire or 3-lobed, the lobes long; branches of the stigma red.   2. *B. bakeri.*

1. Blennosperma nànum (Hook.) Blake. Common Blennosperma. Fig. 5409.

Chrysanthemum ? nanum Hook. Fl. Bor. Amer. **1** : 320. 1833.
Coniothele californica DC. Prod. **5** : 531. 1836.
Blennosperma californicum Torr. & Gray, Fl. N. Amer. **2** : 272. 1842.
Blennosperma nanum Blake, Proc. Biol. Soc. Wash. **39** : 144. 1926.

Herb 8–20 cm. high, the rather succulent stems diffusely branched from the base and above, rarely simple, glabrous except for scattered hairs on the peduncles and base of the involucre and tips of the phyllaries. Lower leaves 3–5.5 cm. long, pinnately parted into 8–10 short lobes, the upper much shorter, with 3–6 short lobes; involucres 5–6 mm. long, the free phyllaries ovate with reddish brown margin, curved over the receptacle after anthesis, usually spreading after ripe achenes are shed; ray-flowers rounded at apex, purplish brown on outer face, yellow within; disk-flowers and stigmas yellow; achenes 2.5–3 mm. long.

Wet places on hill slopes and in cultivated fields and vernal pools, Upper Sonoran Zone; Mendocino and Lake Counties, California, to San Luis Obispo County, and Butte and Glenn Counties south to Tulare County; less common in southern California, where it occurs in Los Angeles and San Diego Counties. Type locality: "North-West coast of America," probably San Francisco or Monterey. Collected by Menzies. Feb.–April.

Blennosperma nanum var. **robústum** J. T. Howell, Leaflets West. Bot. **5:** 108. 1948. Leaves as in *B. nanum* var. *nanum* but stouter; involucre 7 mm. long, the phyllaries ovate-lanceolate; achenes 3–4.5 mm. long, some achenes without the papillate coat. Sandy soil, Point Reyes peninsula, Marin County, California. Type locality: McClure Beach, Marin County.

2. Blennosperma bàkeri Heiser. Baker's Blennosperma. Fig. 5410.

Blennosperma bakeri Heiser, Madroño **9:** 103. 1947.

Rather succulent herbs up to 30 cm. high, branching from the base and above, glabrous except for scattered hairs on the peduncles and bases of the involucres. Lower leaves 10–15 cm. long, linear, entire or 2–3-lobed with linear lobes 1–3 cm. long, the upper leaves shorter, 3–5-lobed, the lobes about 1 mm. wide; involucre 8–9 mm. long, the free phyllaries 6–8 mm. long, reddish brown at the tip, scarcely spreading after achenes are shed; stigmas of ray-flowers red; achenes 3–4 mm. long, 4–6-angled.

In vernal pools, Upper Sonoran Zone; known only from the type locality. Type locality: Sonoma, Sonoma County, California. March–April.

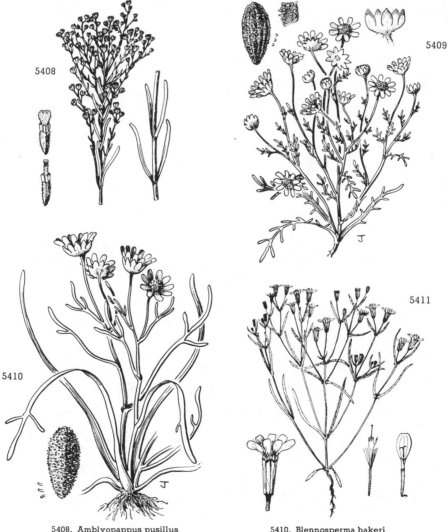

5408. Amblyopappus pusillus
5409. Blennosperma nanum

5410. Blennosperma bakeri
5411. Pectis papposa

64. PÉCTIS L. Syst. Nat. ed. 10. 1221. 1759.

Low, branching, aromatic, annual or perennial herbs. Leaves opposite, glandular-dotted, entire, usually with a few marginal setae. Inflorescence cymose, the heads radiate. Involucre various (in ours turbinate), the phyllaries 3–12 in 1 series and carinate below, often glandular-dotted. Receptacle naked. Ray-flowers perfect, usually equaling the bracts in number, the ligules yellow or tinged with red or purple. Disk-flowers perfect, few, with short tube and funnelform throat, the limb 5-lobed and spreading. Anthers entire, obtuse at base. Style-branches short, obtuse, hispidulous. Achenes terete or somewhat angled. Pappus of scales, awns, or bristles, sometimes lacking. [From the Greek word meaning to comb, referring to the setose margins of the leaves.]

A genus of about 70 species, natives of Mexico and South America and southwestern United States. Type species, *Pectis ciliaris* L.

1. Pectis pappòsa A. Gray. Chinch-weed. Fig. 5411.

Pectis papposa Harv. & Gray ex A. Gray, Mem. Amer. Acad. II. 4: 62. 1849.

Dichotomously much-branched annual 1–2.5 dm. high with yellowish green herbage. Leaves narrowly linear with 2–5 marginal setae near the base, 1–6 cm. long; heads yellow-flowered in leafy cymes, the peduncles 1–3 cm. long; phyllaries gland-dotted, 7–9, linear; involucre keeled and gibbous at base, obtuse and scarious at apex; ray-flowers 7–9, the ligules about 4–6 mm. long; disk-flowers 10–15; achenes black, linear-clavate, sparsely strigulose, 4–5 mm. long; pappus of disk-flowers of 12–20 short plumose bristles, these sometimes reduced to a crown, that of the ray-flowers a short crown of united scales sometimes produced into an awn.

Sandy and gravelly soil of the desert, Lower Sonoran Zone; Death Valley region, Inyo County, California, and eastern Mojave Desert south through the Colorado Desert to Lower California, east to Utah and New Mexico and south to adjacent Mexico. Type locality: California. Collected by Coulter. June–Oct.

65. NICOLLÉTIA A. Gray ex Torr. in Frem. Second Rep. 315. 1845.

Perennial glabrous herbs with slender rootstocks, and stems corymbosely few-branched above. Leaves alternate, pinnately parted. Heads large, terminating the branches. Involucres turbinate, the phyllaries in 1 series with a single gland at the tip, subtended by 1–5 calyculate bractlets. Receptacle convex. Ray-flowers few, pistillate, fertile, pinkish or purplish, entire or 3-toothed. Disk-flowers many, hermaphrodite, fertile, yellow, sometimes pinkish in age, with a short tube tapering to the elongated throat, the 5 lobes short, erect. Anthers truncate at base, entire. Style-branches of disk-flowers subulate, with a tuft of hairs at the apex. Achenes narrowly clavate, sparsely hirsute. Pappus in 2 series, the inner of 5 lanceolate, awn-tipped paleae, the outer of numerous capillary bristles. [Name in honor of Nicollet, an early American explorer.]

A genus of 3 species, natives of southwestern United States and Mexico. Type species, *Nicolletia occidentalis* A. Gray.

1. Nicolletia occidentàlis A. Gray. Nicolletia. Fig. 5412.

Nicolletia occidentalis A. Gray ex Torr. in Frem. Second Rep. 316. 1845.

Stout, glaucous, ill-smelling perennials with erect leafy stems 2–6 dm. high, these several, arising from a deep-seated root-crown. Leaves thick, 3–5 cm. long, pectinately pinnatifid, with awn-tipped lobes; involucres turbinate or campanulate; phyllaries 8–12, 11–13 mm. long, broadly linear, abruptly acute, hyaline-margined, bearing a large gland at the tip with rarely additional ones below, the calyculate series fewer, about one-third as long as the principal phyllaries; rays pinkish, 4–5 mm. long; achenes strigose-hispidulous, 8 mm. long, the awn-tipped pappus-paleae equaling the achenes, the bristles a little shorter.

Sandy soil, Lower Sonoran Zone; western edges of the Mojave and Colorado Deserts, California, from San Bernardino County to San Diego County. Type locality: banks of the Mojave River, San Bernardino County. April–May. Hole-in-the-sand Plant.

Tagetes minùta L. Sp. Pl. 887. 1753. Annual, 3–8 dm. high, with conspicuous glands; leaflets lanceolate-serrate; heads in a congested cyme; involucres cylindrical; rays yellow, minute. This species was collected at Riverside, Riverside County, California, in 1921 and apparently has not become established as a weed.

66. DYSSÓDIA Cav. Descr. 202. 1802.

Annual or perennial, strong-scented herbs often with a woody caudex or shrubs, the herbage with conspicuous, translucent, oil glands. Leaves opposite or alternate, entire or dissected. Inflorescence various, ours with peduncled heads terminating the branches. Involucre turbinate or hemispheric, usually with calyculate bractlets, the phyllaries in 2 equal series, united at the base or nearly to the apex. Receptacle naked or fimbrillate. Ray-flowers when present pistillate, fertile. Disk-flowers hermaphrodite, fertile, the throat trumpet-shaped, scarcely differentiated from the tube, lobes 5. Anthers obtuse at base. Style-branches slender, truncate or sometimes appendaged. Achenes narrowly obconic; pappus-paleae many, tipped with 1–3 bristles or dissected into many bristles. [From the Greek word meaning an evil smell.]

A genus of about 40 species, natives of southwestern United States and Mexico. Type species, *Tagetes papposa* Vent. (= *Dyssodia glandulosa* Cav. Descr. 202. 1802.)

Involucres 12–16 mm. high; stems sparsely leafy, the leaves alternate, much shorter to almost equaling the internodes.

    Phyllaries abruptly acute; leaves 3–5-parted into narrow lobes.   1. *D. porophylloides.*
    Phyllaries long-attenuate; leaves simple, spinulose-dentate.   2. *D. cooperi.*
Involucres 5–6 mm. high; stems densely leafy, the leaves opposite and mostly longer than the internodes.
                                                           3. *D. thurberi.*

1. **Dyssodia porophylloìdes** A. Gray. San Felipe Dyssodia. Fig. 5413.

Dyssodia porophylloides A. Gray, Mem. Amer. Acad. II. **5**: 322. 1854.
Lebetina porophylloides A. Nels. Bot. Gaz. **47**: 435. 1909.
Clomenocoma porophylloides Rydb. N. Amer. Fl. **34**: 166. 1915.

Ill-smelling low shrub or shrubby perennial, glabrous throughout, 3–6 dm. high, with numerous striate branching stems arising from the woody base. Leaves mostly alternate, much shorter than the internodes, 1–2 cm. long, thick, glandless, the lower narrowed to a petioliform base, parted into 3–5 cuneate to lanceolate, entire or incised divisions, the upper often entire or merely incised; peduncles 3–8 cm. long; involucres turbinate, 12–15 mm. long, subtended by attenuate calyculate bractlets bearing glands; phyllaries 14–20, not united, abruptly acute, terminal gland present and usually several lateral ones; rays orange-yellow, about 5 mm. long, sometimes lacking; disk-flowers orange or deep yellow, about 8 mm. long; achenes striate, 5 mm. long; pappus-paleae 10–12, longer than the achene, dissected into many bristles.

Gravelly or rocky slopes and alluvial fans, Lower Sonoran Zone; along the Colorado River in San Bernardino County, California, and the Colorado Desert in Riverside and San Diego Counties, and northern Lower California, east to southern Arizona and Sonora. Type locality: San Felipe, San Diego County. March–June.

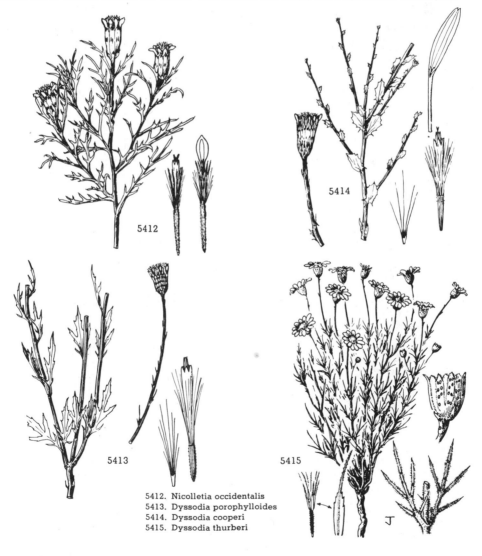

5412. Nicolletia occidentalis
5413. Dyssodia porophylloides
5414. Dyssodia cooperi
5415. Dyssodia thurberi

2. Dyssodia coòperi A. Gray. Cooper's Dyssodia. Fig. 5414.

Dysodia cooperi A. Gray, Proc. Amer. Acad. **9**: 201. 1874.
Lebetina cooperi A. Nels. Bot. Gaz. **47**: 435. 1909.
Clomenocoma cooperi Rydb. N. Amer. Fl. **34**: 166 1915.
Clomenocoma laciniata Rydb. loc. cit.

Ill-smelling shrubby perennial 3–4.5 dm. high with numerous striate, minutely scabridous or scabridulous stems bearing leafy ascending branches arising from a woody base. Leaves alternate, about equaling or shorter than the internodes, 1–2 cm. long, broadly ovate to lanceolate, the margin spinulose-dentate, rarely with small lobes; peduncles 3–8 cm. long; involucres turbinate, 14–18 mm. high, subtended by calyculate bractlets of varying lengths, these bearing glands; phyllaries 20–30, not united, narrowly lanceolate, attenuate, with or without glands; rays 8–12, orange-yellow, about 10 mm. long; disk-corollas about as long as the rays; achenes striate, 6–7 mm. high, glabrate at maturity; pappus-paleae 10–15, mostly longer than the achene, dissected into many bristles.

Rocky slopes and gravelly alluvial fans, Lower Sonoran Zone; southern Nevada and the adjacent Death Valley region, California, and Mohave County, Arizona, to the southern borders of the Mojave Desert, San Bernardino County, California. Type locality: on the eastern side of the Providence Mountains, California. Collected by Cooper. April–July.

3. Dyssodia thúrberi (A. Gray) Robinson. Thurber's Dyssodia. Fig. 5415.

Hymenantherum tenuifolium var. A. Gray, Smiths. Contr. **5**[6]: 93. 1853.
Hymenantherum thurberi A. Gray, Proc. Amer. Acad. **19**: 41. 1883.
Dysodia cupulata A. Nels. Bot. Gaz. **47**: 435. 1909.
Dysodia thurberi A. Nels. op. cit. 436, a provisional name.
Dyssodia thurberi Robinson, Proc. Amer. Acad. **49**: 508. 1913.

Pubescent perennial 1–2 dm. high with many slender, striate, densely leafy, branching stems arising directly from the perennial root or from a woody caudex. Leaves opposite, sessile, rigid, sparsely beset with glands, 9–16 mm. long, pinnately parted nearly to the midrib into 3–7 acicular, spreading divisions, the leaves thus appearing fascicled; heads many, terminating the branchlets, the slender peduncles 4–7 cm. long; involucres turbinate-campanulate, 4–5 mm. high, the calyculate bractlets few, short, the 2 series of phyllaries united to near the apex, coriaceous, thinner above, the outer linear series ciliate along the free margin, both series beset with small oval glands; rays yellow, oblong to oval, about 3 mm. long; disk-flowers about 2.5 mm. long; achenes 2–3 mm. long, sparsely hispidulous; pappus-paleae about 10, about equaling the achenes in length, the paleae all awn- or bristle-tipped.

Gravelly or rocky slopes, often growing on limestone, Sonoran Zones; southern Nevada and eastern San Bernardino County, California, eastward through Arizona to western Texas; also adjacent Mexico. Type locality: near El Paso, Texas. May–Sept.

Dyssodia pappòsa (Vent.) Hitchc. Trans. Acad. Sci. St. Louis **5**: 503. 1891. (*Tagetes papposa* Vent. Descr. Pl. Jard. Cels *pl. 36.* 1801; *Dyssodia glandulosa* Cav. Descr. 202. 1802; *Boebera chrysanthemoides* Willd. Sp. Pl. **3**: 2125. 1804; *B. papposa* Rydb. ex Britt. Man. 1012. 1901.) Ill-scented, much-branched, leafy annual, the leaves mostly opposite, pinnatifid or bipinnatifid with linear divisions; involucres subtended by linear bractlets, these and the phyllaries bearing a few conspicuous, linear or elliptic glands; achenes pubescent, the pappus-paleae dissected into bristles.

Roadsides, waste places, and cultivated areas; collected by Roos (*5056*) at Loma Linda, San Bernardino County, California, but not yet well established; widespread as a weed from Illinois west to Montana and Arizona. Type locality: Illinois.

67. POROPHÝLLUM [Vaill.] Adans. Fam. Pl. 2: 122. 1763.

Annual or perennial herbs or often low shrubs. Leaves simple, alternate or opposite with marginal oil glands, these sometimes present on the surface of the leaves. Heads discoid, solitary on the branches. Involucres cylindric or campanulate, the phyllaries in 1 series, 5–9, oblong, equal, bearing oil glands. Flowers perfect, fertile, purplish or yellow. Throat of the corolla funnelform, longer or shorter than the tube, the lobes reflexed, often irregularly cleft. Anthers rounded at base. Style-branches conspicuous, slender, hirsutulous, the apices subulate. Achenes slender, striate. Pappus of many scabrous or hirsutulous bristles. [From the Greek meaning pore and leaf, referring to the translucent oil glands.]

A genus of about 30 species, natives of southwestern United States, Mexico, and South America. Type species, *Cacalia porophyllum* L.

1. Porophyllum grácile Benth. Odora. Fig. 5416.

Porophyllum gracile Benth. Bot. Sulph. 29. 1844.
Porophyllum junciforme Greene, Leaflets Bot. Obs. **2**: 154. 1911.
Porophyllum vaseyi Greene, loc. cit.
Porophyllum caesium Greene, op. cit. 155.

Bushy perennial, woody at base, 2 dm. high, with many slender, erect, rush-like branches; herbage dark green, often purplish, glaucous, with a strong disagreeable odor from the scattered oil glands. Leaves few, linear-filiform, entire, 1–5 cm. long; involucres narrowly campanulate, 10–15 mm. long; phyllaries 5, often tinged with purple, oil glands few, linear-oblong, obtuse, the hyaline margin often pinkish, rounded on the back and somewhat gibbous at the base; corollas 7–8 mm. long, purplish white with purple lines, throat funnelform, much exceeding the tube, puberulent; achenes dark, 8–9 mm. long, hispidulous, the pappus about 6 mm. long, of coarse, hispidulous, often rose-tinged bristles.

Stony desert slopes, Lower Sonoran Zone; eastern San Bernardino County (Clark Mountain), California, south through the Colorado Desert and the mountains of San Diego and Orange Counties to Lower California, eastward to southern Nevada, Texas, and Sonora. Type locality: Magdalena Bay, Lower California. March–June.

Tribe 3. **ASTEREAE**[*]

Heads all with perfect disk-flowers, commonly radiate, not dioecious.
  Ray-flowers yellow, only exceptionally wanting; disk-flowers yellow.
    Pappus of paleae, scales, or flattened awns.
      Leaves usually serrate and more than 5 mm. wide; achenes glabrous; pappus of 2–8 firm deciduous awns.   68. *Grindelia.*
      Leaves entire, mostly much less than 5 mm. wide; achenes hairy; pappus of 10 or more persistent members.
        Leaves filiform to narrowly oblanceolate; pappus of 10–12 sordid, oblong, free scales; phyllaries not prominently scarious-margined.   69. *Gutierrezia.*
        Leaves broader in outline; pappus of 15–40 bright, white, narrow paleae; phyllaries broad, rounded, with prominent, scarious, erose margin.
          Heads glomerate; disk-flowers 4–7; plant spinescent.   70. *Amphipappus.*
          Heads solitary but sometimes approximate; disk-flowers mostly 20 or more; plant not spinescent.   71. *Acamptopappus.*
    Pappus of capillary bristles, sometimes with some short outer scales as well.
      Ray-flowers present.
        Pappus of disk-flowers double, the inner series of capillary bristles, the outer much shorter.
          Ray-flowers with copious pappus and pubescent achenes.   72. *Chrysopsis.*
          Ray-flowers epappose or the pappus reduced to a deciduous crown and with essentially glabrous achenes.   73. *Heterotheca.*
        Pappus of all flowers simple, generally unequal but not divided into two lengths.
          Disk-flowers fertile.
            Plants taprooted, never rhizomatous.
              Heads nodding in bud; very slender, usually diffuse annuals with slender or filiform stems; phyllaries thin with prominent scarious margin.   74. *Chaetopappa.*
              Heads erect or at least not nodding in bud; coarser, mostly perennial herbs or shrubs; heads usually few and relatively large, if small and panicled then the plants shrubby.   75. *Haplopappus.*
            Plants with numerous fibrous roots arising from a rhizome or caudex, without a taproot; heads usually small and very numerous; always herbaceous.   77. *Solidago.*
          Disk-achenes sterile.
            Plant annual; phyllaries not in vertical ranks, with recurved tip bearing a prominent gland.   76. *Benitoa.*
            Plant a tufted perennial herb; phyllaries in vertical ranks, the erect tip not bearing a prominent gland.   78. *Petradoria.*
      Ray-flowers none.
        Plants shrubby.
          Phyllaries in more or less distinct vertical ranks.   79. *Chrysothamnus.*
          Phyllaries not in vertical ranks.   75. *Haplopappus.*
        Plant a diffusely branched annual herb.   92. *Lessingia germanorum.*
  Ray-flowers white, pink, purple, or blue, sometimes very inconspicuous, occasionally lacking; disk-flowers yellow, white, pink, or reddish purple.
    Pappus none; receptacle conic; plants scapose.   80. *Bellis.*
    Pappus present; receptacle not conic; habit various.
      Pistillate flowers few to numerous, in 1 or more series, usually with conspicuous rays; disk-flowers few to generally numerous.
        Pappus of paleae or scales, sometimes with additional bristles.
          Disk-achenes winged; inflorescence of numerous heads in leafy panicles.   81. *Boltonia.*
          Disk-achenes not winged; flowers mostly solitary.
            Pappus of unequal bristles alternating with short paleae or of a scarious cup and 1 bristle; diminutive desert annuals.   82. *Monoptilon.*
            Pappus of numerous rigid narrow awns, or awns and scales, or scales only; ours mostly montane cushion plants.   83. *Townsendia.*
        Pappus of mostly capillary bristles.
          Ray-flowers present, sometimes very inconspicuous.
            Pappus of 5 firm bristles; phyllaries with broad scarious margins; small annual with white or purplish rays.   74. *Chaetopappa bellidiflora.*
            Pappus usually of numerous bristles, in *Erigeron* very rarely as few as 5 and then plants perennial.
              Achenes beaked; phyllaries deciduous; erect annual with filiform branches.   84. *Tracyina.*
               Achenes not beaked; phyllaries persistent.
                Pappus in ray-flowers obsolete or none; leaves mostly laciniate-pinnatifid; slender desert annual.   85. *Psilactis.*
                Pappus present in both ray- and disk-flowers.
                  Phyllaries usually obviously graduated and imbricated in 3 or more series.
                    Style-branches of disk-flowers not tipped with a prominent tuft of yellow hairs; ray-flowers fertile; pappus not reddish.
                      Perennials, usually rhizomatous or with fibrous roots, if annuals then leaves essentially entire.

---

[*] Key by David Daniels Keck.

Phyllaries conspicuously chartaceous and white at base, the tip green and usually lax; rays few, short and white. 86. *Sericocarpus.*

Phyllaries if indurated not conspicuous and white at base; rays pink or purple, mostly several to many, if short or vestigial not on plants from a woody caudex or stout rhizome.

Appendages of style-branches lanceolate to subulate, acute; phyllaries lacking an even scarious margin. 87. *Aster.*

Appendages of style-branches ovate or oblong, obtuse; low, tufted, desert perennial with bristle-tipped, subulate leaves. 88. *Leucelene.*

Annuals, biennials, or short-lived perennials with a taproot; leaves spinulose-tipped, entire or usually spinulose-dentate to pinnatifid or dissected. 89. *Machaeranthera.*

Style-branches of disk-flowers densely tufted with rigid yellow hairs; ray-flowers neutral; pappus reddish. 90. *Corethrogyne.*

Phyllaries only slightly or not at all graduated, in 1 or 2 series; style-branches of disk-flowers with lanceolate and acute, or usually with triangular and obtuse, appendages. 91. *Erigeron.*

Ray-flowers absent, but the marginal disk-corollas sometimes enlarged to resemble rays.

Marginal corollas enlarged, becoming palmately 5-lobed; pappus usually brownish to reddish. 92. *Lessingia.*

Marginal corollas not enlarged, regular; pappus not brownish or reddish. 87. *Aster.*

Pistillate flowers very numerous, several-seriate, tubular-filiform, rayless or the inconspicuous ray barely equaling the pappus. 93. *Conyza.*

Heads unisexual, discoid; dioecious; pappus of male flowers of clavellate bristles. 94. *Baccharis.*

68. GRINDÈLIA* Willd. Ges. Naturf. Fr. Berl. Mag. 1: 260. 1807.

Annual, biennial, or usually perennial herbs, mostly with a taproot, rarely suffrutescent at base, more or less resinous particularly on the involucre. Leaves alternate, punctate, usually serrate and sessile, often clasping. Heads medium to large, yellow, usually radiate, usually solitary at branch-tips. Involucre multiseriate, imbricate, the phyllaries thickish, with pale appressed base and narrow, often squarrose or revolute, herbaceous tip. Receptacle flattish, foveolate. Ray-florets 10–45, uniseriate, fertile. Disk-florets usually fertile. Style-branches with slender hispidulous appendages. Achenes compressed to subquadrangular, few-angled, glabrous; pappus of 2–8 stiff, often curved, deciduous, corneous or paleaceous awns. [Named for Professor David Hieronymous Grindel, 1776–1836, botanist of Dorpat and Riga.]

A genus of about 50 species, of western North and South America. Type species, *Grindelia inuloides* Willd.

Tips of phyllaries erect or spreading, some gradually curved but not sharply reflexed.
Fruticose; salt marshes of San Francisco Bay. 1. *G. humilis.*
Herbaceous, more or less woody at caudex.
Branches prostrate or decumbent; leaves obovate-cuneate; coastal. 2. *G. stricta.*
Branches ascending-erect; leaves not obovate-cunate; northern interior. 3. *G. integrifolia.*
Involucre pubescent; Coast Ranges, central California. 4. *G. hirsutula.*
Involucre glabrous; central California.
Stems 8–18 dm. high, erect, strictly branching above; interior valleys. 5. *G. procera.*
Stems 3–8 dm. high, ascending, openly branched; San Francisco. 6. *G. maritima.*
Tips of phyllaries (at least of some middle and outer ones) sharply reflexed or looped.
Coastal succulent plants.
Herbaceous to caudex; more or less decumbent or prostrate.
Leaves obovate-cuneate or rounded; mostly north of San Francisco Bay. 2. *G. stricta* and subspecies.
Leaves acutely or obtusely pointed at tip; mostly south of San Francisco Bay. 7. *G. latifolia.*
Woody below; nearly erect. 2. *G. stricta blakei.*
Interior nonsucculent plants; stems more or less erect.
Leaves callous-serrulate; rare introduction. 8. *G. squarrosa.*
Leaves sharply toothed or entire, not callous-serrulate.
Involucres mostly 1 cm. in diameter; ligules 5–8 mm. long; San Diego County, California. 9. *G. hallii.*
Involucres 12–25 mm. in diameter; ligules (when present) mostly 8–15 mm. long.
Tips of phyllaries spreading or reflexed (usually some sharply reflexed) but rarely looped; interior valleys of California. 10. *G. camporum.*
Tips of phyllaries looped back to form a tight ring.
Heads relatively small, the involucre 7–10 mm. high, 9–15 mm. thick; northern California northward. 11. *G. nana.*
Heads larger, the involucre 8–15 mm. high, 12–25 mm. thick; southern California. 12. *G. robusta.*

---

* Text contributed by David Daniels Keck.

1. **Grindelia hùmilis** Hook. & Arn. Marsh Grindelia. Fig. 5417.

Grindelia humilis Hook & Arn. Bot. Beechey 147. 1833.
Grindelia robusta var. *angustifolia* A. Gray, Bot. Calif. 1: 304. 1876.
Grindelia cuneifolia of most authors, not Nutt.

Frutescent, to 1.5 m. high, stout, the woody stems giving rise to herbaceous shoots of the season, the branches subcorymbose, glabrous to villous, few-headed. Leaves coriaceous, scarcely resinous, remotely serrulate, cuneate-oblanceolate to lance-oblong, 2–8 cm. long, 5–15 mm. wide, narrowed to a sessile base or amplexicaul; heads 3–5 cm. across; phyllaries of the involucre lanceolate, erect, with short, flat, appressed or slightly reflexed tip, or tip more subulate and strongly reflexed; rays 16–34, 12–18 mm. long.

Salt marshes, Humid Transition Zone; San Francisco, San Pablo, and Suisun Bays, California. Type locality: California. Principally Aug.–Oct. but found in flower throughout the year.

Grindelia paludòsa Greene, Man. Bay Reg. 172. 1894. (*Grindelia cuneifolia* var. *paludosa* Jepson, Fl. W. Mid. Calif. 556. 1901.) Stems herbaceous to the crown, stout, freely branching, glabrous, 5–15 dm. high. Leaves subcoriaceous, not resinous, remotely serrate, serrulate, or entire, lance-ovate, oblong, or oblanceolate, strongly clasping, the lower mostly up to 8 cm. long and 1.5 cm. wide, those near the heads much reduced; heads 3–4 cm. across, very resinous, the prominent green tips of the phyllaries erect or spreading, the outer ones strongly recurved; rays 20–35, 10–12 mm. long. Salt marshes, Upper Sonoran Zone; Suisun Bay, Solano County, California. In all probability this is a stabilized hybrid population of the parentage *G. humilis* and *G. camporum.* Type locality: Suisun marsh. July–Sept.

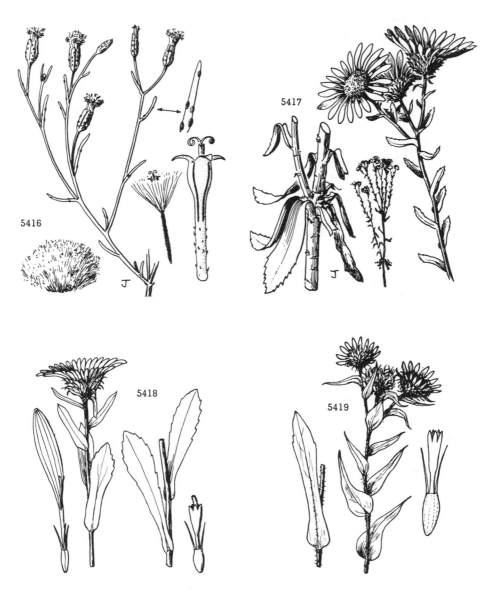

5416. Porophyllum gracile
5417. Grindelia humilis

5418. Grindelia stricta
5419. Grindelia integrifolia

2. Grindelia strícta DC.   Pacific Grindelia.   Fig. 5418.

Grindelia stricta DC. Prod. **7**: 278. 1838.
Grindelia oregana A. Gray, Syn. Fl. N. Amer. **1²**: 118. 1884.
Grindelia hendersonii Greene, Pittonia **2**: 18. 1889.
Grindelia lanata Greene, op. cit. 290. 1892.
Grindelia macrophylla Greene, op. cit. **3**: 297. 1898.
Grindelia oregana subsp. *wilkesiana* Piper in Piper & Beattie, Fl. Northw. Coast 363. 1915.
Grindelia andersonii Piper, Proc. Biol. Soc. Wash. **31**: 77. 1918.
Grindelia stricta var. *andersonii* Steyermark, Ann. Mo. Bot. Gard. **21**: 559. 1934.
Grindelia stricta var. *aestuarina* Steyermark, op. cit. 560.
Grindelia stricta var. *macrophylla* Steyermark, op. cit. 561.
Grindelia stricta var. *hendersonii* Steyermark, op. cit. 562.
Grindelia aggregata Steyermark, op. cit. 566.
Grindelia integrifolia var. *macrophylla* Cronquist, Vasc. Pl. Pacif. Northw. **5**: 207. 1955.

Herbaceous perennial to 0.5 dm. high; stems simple or corymbosely branching above, decumbent or ascending, 3–8 dm. long, glabrous to moderately or densely villous and sometimes glandular. Leaves thickish, the basal oblanceolate, acute to rounded, entire to serrulate toward apex, 10–25 cm. long, 1–4 cm. wide, narrowed to a margined petiole, the cauline broadly oblong to oblanceolate, obtuse or acute, entire or slightly serrulate, 3–12 cm. long, amplexicaul; heads 4–5 cm. across, often leafy-bracted; involucre 8–14 mm. high, 12–20 mm. wide, the linear-lanceolate phyllaries erect (occasionally the outer ones spreading), the slender, tapering, herbaceous tip erect or gradually recurving; rays 10–35, 12–20 mm. long.

Coastal dunes and salt marshes, Humid Transition Zone; southern Alaska and Vancouver Island to Lane County, Oregon. Type locality: Mulgrave (Yakutat Bay), Alaska, but quite possibly from farther south. June–Sept.

Grindelia stricta subsp. **venulòsa** (Jepson) Keck, Aliso **4**: 102. 1958. (*Grindelia venulosa* Jepson, Man. Fl. Pl. Calif. 1021. 1925; *G. arenicola* Steyermark, Ann. Mo. Bot. Gard. **21**: 224. 1934; *G. stricta* var. *procumbens* Steyermark, op. cit. 559; *G. arenicola* var. *pachyphylla* Steyermark, op. cit. 596.) Procumbent to decumbent, the stems usually whitish or yellowish; leaves often shorter, broader, and fleshier, rounded at tip; phyllaries with prominently reflexed or looped tips. Coastal salt marshes and seaside bluffs, Coos County, Oregon, to San Francisco Bay; rare in Monterey County, California. Type locality: Big Flat, Humboldt County.

Grindelia stricta subsp. **blàkei** (Steyermark) Keck, Aliso **4**: 102. 1958. (*Grindelia blakei* Steyermark, Ann. Mo. Bot. Gar. **21**: 567. 1934.) Suffrutescent, the stout red-brown stems of the season ascending from a woody base, up to 1 m. high; leaves subcoriaceous, usually entire; heads to 5.5 cm. across, the tapering, herbaceous, outer phyllaries with prominently reflexed tips. Restricted to the salt marshes bordering Humboldt Bay, California. Type locality: Eureka. July–Sept. Despite its woody axis and essentially erect branches, this plant is doubtless more closely related to forms of *G. stricta* from the Puget Sound area than to *G. humilis* of the San Francisco Bay area, which it also mimics.

3. Grindelia integrifòlia DC.   Puget Sound Grindelia.   Fig. 5419.

Grindelia integrifolia DC. Prod. **5**: 315. 1836.
Grindelia virgata Nutt. Trans. Amer. Phil. Soc. II. **7**: 314. 1840.
Grindelia integrifolia var. *virgata* Torr. & Gray, Fl. N. Amer. **2**: 248. 1842.
Grindelia collina Henry, Fl. S. Brit. Columbia 291. 1915.
Grindelia integrifolia var. *aestivalis* Henry, loc. cit.
Grindelia integrifolia var. *autumnalis* Henry, loc. cit.
Grindelia stricta var. *collina* Steyermark, Ann. Mo. Bot. Gard. **21**: 564. 1934.

Perennial with several erect or ascending stems from the ligneous caudex, 2–8 dm. high, glabrate or somewhat villous. Leaves membranous, the basal tufted, oblanceolate, rounded, moderately serrate or more commonly entire, up to 35 cm. long and 4 cm. wide, the cauline lanceolate or oblanceolate, usually acute, sessile or even auriculate-clasping; heads solitary to usually several in a corymb, 2.5–4 cm. across; involucre somewhat glutinous, the slender green tips of the phyllaries becoming loose or spreading or gradually recurving; rays 10–35, mostly 8–15 mm. or 20 mm. long.

Salt marshes and wet meadows west of the Cascade Mountains, Humid Transition Zone; British Columbia south through the Puget Trough, to the head of the Willamette Valley, Oregon. Type locality: Williamette Valley. June–Sept.

4. Grindelia hirsútula Hook. & Arn.   Hirsute Grindelia.   Fig. 5420.

Grindelia hirsutula Hook. & Arn. Bot. Beechey 147. 1833.
Grindelia hirsutula var. *brevisquama* Steyermark, Ann. Mo. Bot. Gard. **21**: 572. 1934.
Grindelia hirsutula var. *calva* Steyermark, op. cit. 575.
Grindelia hirsutula var. *subintegra* Steyermark, loc. cit.
Grindelia pacifica M. E. Jones, Bull. Torrey Club **9**: 31. 1882.

Herbaceous perennial 3–8 dm. high; stems erect, slender, simple or commonly corymbosely branching above, the branches monocephalous, more or less villous with crisped hairs, especially on peduncles. Leaves chartaceous, gray-green, the basal oblanceolate or spatulate, obtuse, remotely serrate or lobed, sometimes merely shallowly crenate, tapering to a narrowly margined petiole, 10–22 cm. long, 1–2.5 cm. wide, the cauline much reduced in size upward, from oblanceolate and sessile below to narrowly oblong, entire, bract-like and clasping above; heads 2.5–4 or 5.5 cm. across; involucre more or less pubescent, the multiseriate phyllaries erect or nearly so, not caudate; rays 20–35, 1–2 cm. long.

Arid slopes, Humid Transition Zone; Coast Ranges of central California from Napa County to Monterey County; also Ventura County. Type locality: California. April–July.

Grindelia hirsutula subsp. **rubricaùlis** (DC.) Keck, Aliso **4**: 102. 1958. (*Grindelia rubricaulis* DC. Prod. **5**: 316. 1836; *G. patens* Greene, Pittonia **2**: 290. 1892; *G. robusta* var. *patens* Jepson, Fl. W. Mid. Calif. 554. 1901.) Stems usually becoming densely villous above; peduncles prominently bracteate, the uppermost linear bracts crowded, retrorse, usually lanate, equaling or surpassing the disk; outer phyllaries often spreading with recurved tips. Morphologically well marked but sympatric with the typical subspecies; also in the Sierran foothills of Fresno County. Type locality: Berkeley.

5. Grindelia prócera Greene. Tall Grindelia. Fig. 5421.

Grindelia procera Greene, Man. Bay Reg. 172. 1894.
Grindelia camporum var. *parviflora* Steyermark, Ann. Mo. Bot. Gard. 21: 534. 1934.

Herbaceous perennial 8–18 dm. high; stems erect, pallid, unbranched below, rather strictly much-branched above, bearing numerous heads. Leaves chartaceous, dark green, principally cauline, oblong-obovate to lance-oblong, obtuse, clasping, shallowly toothed, 3–7 cm. long, up to 2 cm. wide; heads 2–3.5 cm. across; involucre slightly resinous, mostly under 8 mm. high, the short green phyllaries erect or slightly squarrose at tip; rays 21–45, 8–10 mm. long.

Deep soils, Upper Sonoran Zone; bottom lands of the San Joaquin River, Sacramento County to Kern County, California; also Marin County. Type locality: "the lower San Joaquin." July–Oct.

6. Grindelia marítima (Greene) Steyermark. San Francisco Grindelia. Fig. 5422.

Grindelia rubricaulis var. *maritima* Greene, Pittonia 2: 289. 1892.
Grindelia maritima Steyermark, Ann. Mo. Bot. Gard. 21: 576. 1934.

Stems erect or ascending from a ligneous caudex, 3–8 dm. high, slender, glabrous or rarely villous, loosely branching. Leaves thickish, mostly serrulate, the basal tufted, narrowly oblanceo-

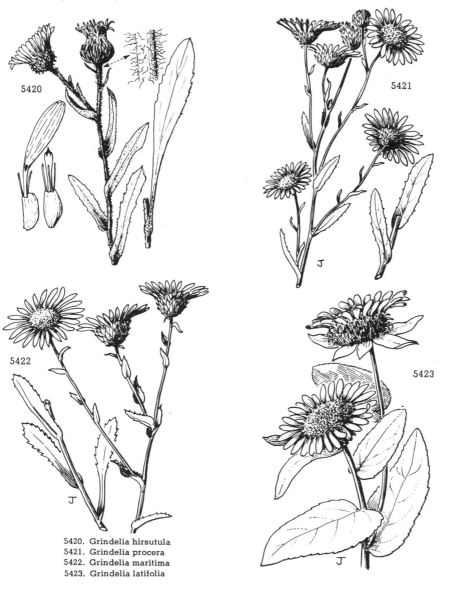

5420. Grindelia hirsutula
5421. Grindelia procera
5422. Grindelia maritima
5423. Grindelia latifolia

late, up to 18 cm. long, tapering to slender petioles, the cauline oblong to lanceolate, mostly 3–7 cm. long, 8–15 mm. wide, clasping; heads terminal on the slender branches, 2.5–4 cm. across; involucre more or less glandular, the phyllaries with mostly erect, green, relatively short tips; rays 30–40, 10–13 mm. long.

Ocean bluffs and open hillsides, Humid Transition Zone; San Francisco, California. Type locality: Point Lobos, San Francisco. Aug.–Sept.

7. **Grindelia latifòlia** Kell. Coastal Gum-plant. Fig. 5423.

Grindelia latifolia Kell. Proc. Calif. Acad. **5**: 36. 1873.
Grindelia robusta var. *latifolia* Jepson, Man. Fl. Pl. Calif. 1020. 1925.
Grindelia rubricaulis var. *latifolia* Steyermark, Ann. Mo. Bot. Gard. **21**: 227. 1934.

Very succulent and leafy, herbaceous perennial 4–6 dm. high, the stout, decumbent or ascending branches monocephalous or topped by a close cluster of 2 or 3 large heads, the heads often partially enveloped by subtending leaves. Leaves principally cauline, thick, irregularly serrate to regularly and sharply dentate, scabrociliate, lance-ovate to broadly oblong, amplexicaul to subcordate, rounded at apex, 3–8 cm. long, 1.5–4 cm. wide; heads 3–5 cm. across, milky-resinous, the outermost phyllaries foliaceous, their green tips and usually those of many inner ones squarrose; rays 30–45, 10–15 mm. long.

Salt marshes and dunes, Upper Sonoran Zone; Watsonville coast, Santa Cruz County, and Surf to Point Conception and Santa Barbara Islands, California. Type locality: Santa Rosa Island. May–Sept.

Grindelia latifolia subsp. **platyphýlla** (Greene) Keck, Aliso **4**: 102. 1958. (*Grindelia robusta* var. *platyphylla* Greene, Pittonia **2**: 289. 1892; *G. rubricaulis* var. *platyphylla* Steyermark, Ann. Mo. Bot. Gard. **21**: 227. 1934; *G. rubricaulis* var. *permixta* Steyermark, op. cit. 582.) Neither so succulent nor so leafy, the many simple or branched, decumbent or ascending, usually glabrous stems radiating from the crown, forming plants up to 6 dm. high and 1 m. in diameter, the milky-resinous heads sometimes subtended by several small leafy bracts but never enveloped by these. Beaches and low ground near the coast, the common form of the species from Marin County to Santa Barbara County, California. Type locality: Monterey. June–Sept.

8. **Grindelia squarròsa** (Pursh) Dunal. Resin-weed. Fig. 5424.

Donia squarrosa Pursh, Fl. Amer. Sept. 559. 1814.
Grindelia squarrosa Dunal, Mem. Mus. Paris **5**: 50. 1819.

Erect biennial or short-lived perennial 2–10 dm. high, openly branched above and bearing many heads. Leaves regularly callous-serrulate, sometimes sharply toothed or even entire, mostly oblong, 2–5 cm. long, the upper clasping; heads 2–3 cm. across, strongly resinous, the green tips of the phyllaries strongly rolled back; rays 25–40, 7–15 mm. long.

Dry fields and waste places, Upper Sonoran and Arid Transition Zones; sparingly introduced in eastern Washington, Oregon, and southern California; chiefly east of the Continental Divide to the Great Plains. Type locality: banks of the Missouri. July–Sept.

9. **Grindelia hállii** Steyermark. San Diego Grindelia. Fig. 5425.

Grindelia hallii Steyermark, Ann. Mo. Bot. Gard. **21**: 229. 1934.

Several glabrous herbaceous stems from the woody crown, 3–6 dm. high, corymbosely branching above. Leaves mostly sharply and regularly serrate, subcoriaceous, prominently resinous-punctate, the basal in often persistent rosettes, oblanceolate, 5–7 cm. long, the cauline oblong, much smaller; heads numerous, terminating the branches, 2–3 cm. across, strongly resinous, the green tips of the outer phyllaries strongly recurved or hooked; rays 13–21, 5–8 mm. long.

Dry flats and grassy mesas, Arid Transition Zone; Cuyamaca Mountains, San Diego County, California. Type locality: Cuyamaca Lake. July–Oct.

10. **Grindelia campòrum** Greene. Great Valley Grindelia. Fig. 5426.

Grindelia robusta var. *rigida* A. Gray, Bot. Calif. **1**: 304. 1876.
Grindelia camporum Greene, Man. Bay Reg. 171. 1894.
Grindelia robusta var. *davyi* Jepson, Fl. W. Mid. Calif. 554. 1901.
Grindelia rubricaulis var. *interioris* Jepson, Man. Fl. Pl. Calif. 1021. 1925.
Grindelia camporum var. *davyi* Steyermark, Ann. Mo. Bot. Gard. **21**: 536. 1934.
Grindelia camporum var. *interioris* Steyermark, op. cit. 538.

Several erect herbaceous stems from the subligneous caudex, 5–12 dm. high, simple or openly branched, glabrous or nearly so. Leaves glabrous or scabrous, rarely more hairy, subcoriaceous, very resinous, saliently dentate, narrowly oblong to broadly oblanceolate, the cauline 2–8 cm. long, 7–15 mm. wide; heads terminating the branches, 2.5–4 cm. across, strongly and translucently resinous, the green tips of the multiseriate elongated phyllaries strongly recurved or hooked; rays 18–35, 8–15 mm. long.

Dry banks, rocky fields and plains, low alkaline ground, Upper Sonoran Zone; the common *Grindelia* of the Great Valley of California and adjacent foothills south to northern Los Angeles County and west to San Francisco Bay. Type locality: east of Mount Diablo Range. May–Oct.

11. **Grindelia nàna** Nutt. Idaho Resin-weed. Fig. 5427.

Grindelia nana Nutt. Trans. Amer. Phil. Soc. II. **7**: 314. 1840.
Grindelia nana var. *integrifolia* Nutt. loc. cit.
Grindelia nana var. *altissima* Steyermark, Ann. Mo. Bot. Gard. **21**: 544. 1934.
Grindelia nana var. *turbinella* Steyermark, op. cit. 545.

Glabrous perennial with erect, usually branching stems, 2–8 dm. high. Leaves spinulose-toothed to entire, the lower oblanceolate, up to 15 cm. long and 3 cm. wide, tapering to slender petioles, the upper much smaller and clasping; heads 2–3 cm. across; involucre rarely very resinous, the green tips of the phyllaries strongly rolled back; rays mostly 12–25, 5–15 mm. long.

Open ground, dry hillsides, roadsides, and subalkaline flats, Upper Sonoran and Arid Transition Zones; northern California, eastern Oregon and Washington (where common), east to Montana. Type locality: probably northeastern Oregon. June–Oct.

Grindelia nana subsp. **columbiàna** Piper, Contr. U.S. Nat. Herb. **11**: 556. 1906. (*Grindelia discoidea* Nutt. Trans. Amer. Phil. Soc. II. **7**: 315. 1840, not Hook. & Arn. 1836; *G. nana* var. *discoidea* A. Gray, Syn. Fl. N. Amer. 1²: 119. 1884.) Heads discoid. Common in dry or moist ground in central Washington, where it replaces the typical subspecies, west to Portland; northern Idaho. Type locality: "on the banks of the Oregon." June–Aug.

12. **Grindelia robústa** Nutt. Big Grindelia. Fig. 5428.

Grindelia robusta Nutt. Trans. Amer. Phil. Soc. II. **7**: 314. 1840.
Grindelia cuneifolia Nutt. op. cit. 315.
Grindelia rubricaulis var. *elata* Steyermark, Ann. Mo. Bot. Gard. **21**: 227. 1934.
Grindelia rubricaulis var. *robusta* Steyermark, loc. cit.
Grindelia camporum var. *australis* Steyermark, op. cit. 228.

Stems few, erect from a subligneous crown, stout, usually corymbosely branching above, glabrous, 5–12 dm. high. Leaves sharply toothed to finely and remotely serrate or often entire, the basal oblanceolate, up to 18 cm. long (including the margined petiole) and 3 cm. wide, the cauline

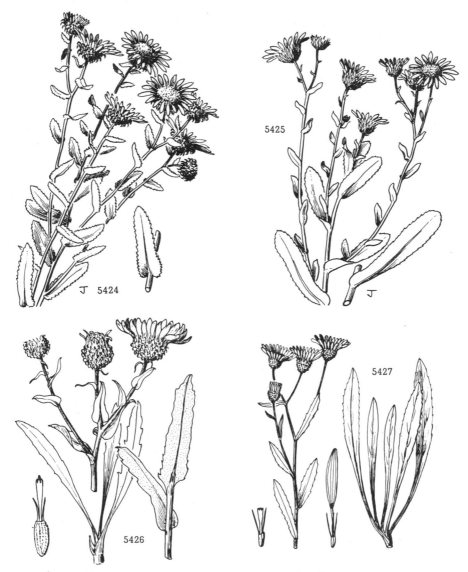

5424. Grindelia squarrosa
5425. Grindelia hallii

5426. Grindelia camporum
5427. Grindelia nana

much reduced, ovate-lanceolate to linear-oblong, broadly clasping; heads 3–5 cm. across, often strongly and translucently resinous, the long green tips of the phyllaries rolled back in a loop; rays 25–45, 8–15 mm. long.

Dry fields and banks, also borders of salt marshes and shores, Upper Sonoran Zone; not far inland, Santa Barbara County, California, to northern Lower California. Type locality: San Pedro. March–Sept.

Grindelia robusta var. **bracteòsa** (J. T. Howell) Keck, Aliso **4**: 102. 1958. (*Grindelia bracteosa* J. T. Howell, Madroño **2**: 22. 1931; *G. rubricaulis* var. *bracteosa* Steyermark, Ann. Mo. Bot. Gard. **21**: 227. 1934.) Heads discoid. Dry hills and canyons, Upper Sonoran Zone; away from the immediate coast, intermittent, Puente Hills, Los Angeles County, to Pine Hills, San Diego County, California. Type locality: Santa Ana Canyon. May–July.

69. GUTIERRÈZIA* Lag. Gen. & Sp. Pl. 30. 1816.

Perennial herb or subshrub, glutinous, glabrous to hirtellous. Leaves alternate, entire, filiform to narrowly oblanceolate, usually punctate-glandular. Heads very small, radiate, yellow, numerous, scattered or crowded in cymes or panicles. Involucre cylindric to turbinate-subglobose, the imbricated phyllaries coriaceous, appressed, whitish. Receptacle foveate, sometimes hairy. Ray-florets pistillate, fertile. Disk-florets perfect, sometimes sterile. Achenes obovoid or oblong, pubescent; pappus of 10–12 oblong, unequal, free scales, shorter on the ray-achenes. [Named for the Spanish nobleman, Pedro Gutierrez.]

A genus of about 25 species, mostly of western North and South America. Type species, *Gutierrezia linearifolia* Lag.

Heads flaring, more than 4-flowered.
    Involucre campanulate, 3–5 mm. thick; ray-florets 5–10; disk-florets 6–16; cismontane.
                                                1. *G. californica.*
    Involucre narrowly turbinate, 2–3 mm. thick; ray- and disk-florets each 3–6; transmontane.
                                                2. *G. sarothrae.*
Heads cylindric, about 1 mm. thick, 2–4-flowered.
                                                3. *G. microcephala.*

1. Gutierrezia califórnica (DC.) Torr. & Gray. San Joaquin Matchweed. Fig. 5429.

Brachyris californica DC. Prod. **5**: 313. 1836.
Gutierrezia californica Torr. & Gray, Fl. N. Amer. **2**: 193. 1842.
Xanthocephalum californicum Greene, Man. Bay Reg. 171. 1894.
Gutierrezia bracteata Abrams, Bull. Torrey Club **34**: 265. 1907.
Gutierrezia californica var. *bracteata* H. M. Hall, Univ. Calif. Pub. Bot. **3**: 36. 1907.

Subshrub 3–6 dm. high, nearly glabrous to densely hirtellous, the branches stiff and erect or sometimes divergent and spreading, very twiggy above. Leaves spreading to deflexed, up to 5 cm. long, mostly 1 mm. wide; heads usually solitary at ends of branchlets, rarely crowded; involucre broadly turbinate-obovoid or campanulate, 5–7 mm. high, 3–5 mm. thick, the usually broad blunt phyllaries with definite green tips; ray-florets 5–10; disk-florets 6–16.

Dry hills and plains, Sonoran Zones; Marin County, California, the Inner South Coast Range from San Francisco Bay to upper San Joaquin Valley, and cismontane southern California east to Arizona and Chihuahua. Type locality: California. May–Oct. Intergrading with *G. sarothrae* in southern California.

2. Gutierrezia saròthrae (Pursh) Britt. & Rusby. Common Matchweed or Snakeweed. Fig. 5430.

Solidago' sarothrae Pursh, Fl. Amer. Sept. **2**: 540. 1814.
Brachyris euthamiae Nutt. Gen. **2**: 163. 1818.
Gutierrezia euthamiae Torr. & Gray, Fl. N. Amer. **2**: 193. 1842.
Gutierrezia sarothrae Britt. & Rusby, Trans. N.Y. Acad. **7**: 10. 1887.
Gutierrezia divergens Greene, Pittonia **4**: 56. 1899, in large part.
Gutierrezia laricina Greene, Rep. Nov. Sp. **7**: 195. 1909.
Gutierrezia ionensis Lunell, Amer. Midl. Nat. **2**: 194. 1912.
Xanthocephalum sarothrae Shinners, Field & Lab. **18**: 29. 1950.

Subshrub 1.5–6 dm. high, mostly hirtellous, the numerous slender stems cymosely paniculate above. Leaves punctate, 2–5 cm. long, 1–2 mm. wide; inflorescence flat-topped, the numerous resinous heads scattered or usually in small glomerules; involucre narrowly turbinate, 4–5 mm. high, 2–3 mm. thick, the phyllaries with often obscurely thickened, herbaceous tips; ray- and disk-florets 3–8 each, the ligule about 3 mm. long; achenes subsericeous-pilose; pappus of ray about 0.7 mm. long, of disk about 1.5 mm. long.

Dry plains and mountainsides, Sonoran Zones; southeastern Washington and eastern Oregon south along the California border to southwestern Mojave Desert and San Diego County; more abundant eastward to the Great Plains, from Saskatchewan to northern Mexico. Type locality: plains of the Missouri River. May–Oct.

3. Gutierrezia microcéphala (DC.) A. Gray. Small-headed Matchweed. Fig. 5431.

Brachyris microcephala DC. Prod. **5**: 313. 1836. Not Hook. 1837.
Gutierrezia microcephala A. Gray, Mem. Amer. Acad. II. **4**: 74. 1849.
Gutierrezia euthamiae var. *microcephala* A. Gray, Syn. Fl. N. Amer. **1²**: 115. 1884.

---

* Text contributed by David Daniels Keck.

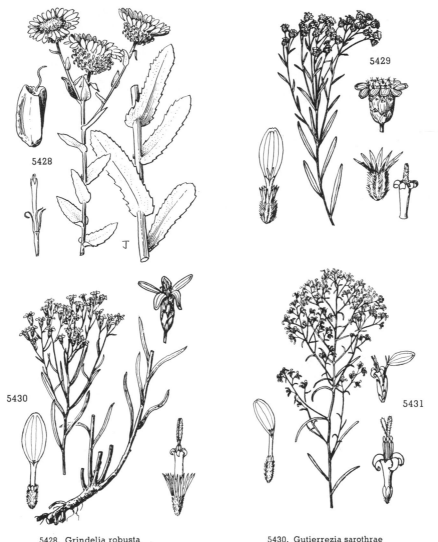

5428. Grindelia robusta
5429. Gutierrezia californica

5430. Gutierrezia sarothrae
5431. Gutierrezia microcephala

Xanthocephalum lucidum Greene, Pittonia **2**: 282. 1892.
Gutierrezia lucida Greene, Fl. Fran. 361. 1897.
Gutierrezia glomerella Greene, Pittonia **4**: 54. 1899.
Gutierrezia sarothrae var. *microcephala* L. Benson, Amer. Journ. Bot. **30**: 631. 1943.
Xanthocephalum microcephalum Shinners, Field & Lab. **18**: 29. 1950.

Many-stemmed, 3–6 dm. high, strongly resinous, essentially glabrous, the slender stems striate-angled, much-branched above. Leaves 2–5 cm. long, 0.5–2 mm. wide, spreading-deflexed; inflorescence cymose-paniculate, the tiny heads about 3 mm. high, 1 mm. wide, in small terminal glomerules of 2–5 or a few solitary; phyllaries yellowish white with yellow tips, the hyaline margin prominent, the inner row of only 2 members, as long as the disk-florets; ray- and disk-florets 1–2 each, the ligule up to 2.5 mm. long; ray-achenes fertile, appressed-pilose, their pappus of about 8 oblong paleae 0.8 mm. long; disk-pappus of about 12 paleae 1 mm. long.

Open desert, Sonoran Zones; Mojave Desert from White Mountains, Inyo County, to Palmdale and Little San Bernardino Mountains, California; east to Colorado, Texas, and northern Mexico. Type locality: Saltillo, Mexico. Aug.–Oct.

70. AMPHIPÁPPUS* Torr. & Gray, Bost. Journ. Nat. Hist. **5**: 107. 1845.

Low shrub with divaricate spinescent branchlets. Leaves alternate, small, entire, oval to elliptic or obovate, short-petioled. Heads radiate, few-flowered, small, glomerate at tips

---

* Text contributed by David Daniels Keck.

of branchlets. Involucre obovoid, about 3-seriate, strongly graduate, pale, the 7–12 broad rounded phyllaries thin, with scarious erose margins. Receptacle fimbrillate. Ray-florets 1–2, pale yellow, small, pistillate, fertile. Disk-florets 4–7, perfect, sterile. Anthers narrowly lance-tipped. Style-branches thick, short-subulate. Ray-achenes broadly oblanceolate, compressed, pilose, their pappus of about 15–20 short, basally united, unequal, white paleae. Disk-achenes undeveloped, glabrous or sparingly pilose, their pappus of about 25 tortuous, hispidulous, irregularly basally united, white paleae of unequal width, equaling the corolla. [Name Greek; *amphi,* both (kinds of), and *pappos,* pappus.]

A monotypic genus.

1. **Amphipappus fremóntii** Torr. & Gray. Chaff-bush. Fig. 5432.

Amphipappus fremontii Torr. & Gray, Bost. Journ. Nat. Hist. **5**: 108. 1845.
Amphiachyris fremontii A. Gray, Proc. Amer. Acad. **8**: 633. 1873.

Much-branched, white-barked shrub 3–6 dm. high with yellow-green cast, glabrous throughout, slightly glutinous around heads. Leaves 5–12 mm. long, 2–4 mm. wide, obtuse or acute, 1-nerved; heads 4–5 mm. high, the phyllaries closely appressed.

Open desert and alkaline flats, Lower Sonoran Zone; Death Valley region, California, from Inyo and Argus Ranges to Funeral Mountains and Pahrump Valley, Inyo and San Bernardino Counties; east to Nevada. Type locality: ". . . on the Mohave River." April–May.

Amphipappus fremontii subsp. spinòsus (A. Nels.) Keck, Aliso **4**: 102. 1958. (*Amphiachyris fremontii* var. *spinosa* A. Nels. Bot. Gaz. **47**: 431. 1909; *Amphipappus spinosa* A. Nels. Amer. Journ. Bot. **21**: 579, 1934; *A. fremontii* var. *spinosus* C. L. Porter, op. cit. **30**: 483. 1943.) Herbage densely scabrohispidulous. Similar sites, eastern Mojave Desert, California, and farther south in San Bernardino County; east to Utah and Arizona. Type locality: Moapa, Nevada.

71. **ACAMPTOPÁPPUS*** A. Gray, Proc. Amer. Acad. **8**: 634. 1873.

Low, much-branched, desert shrubs with white bark. Leaves alternate, small, usually spatulate or oblanceolate, entire, 1-nerved. Heads yellow, radiate or discoid, subglobose, solitary or cymosely arranged at tips of branches, the florets all fertile. Involucre broad, about 4-seriate, strongly graduate, the phyllaries broad, rounded, whitish with greenish tip, firm with prominent, thin, scarious, erose margin. Receptacle convex, alveolate-fimbrillate. Style-branches linear, the narrowly lanceolate appendages equaling the stigmatic portion. Achenes subturbinate, short, densely villous. Pappus persistent, of about 30–40 silvery flattened paleae and bristles of different widths, the broader ones usually somewhat dilated apically. [Name Greek, meaning stiff pappus.]

A genus of 2 known species. Type species, *Aplopappus sphaerocephalus* Harv. & Gray.

Heads discoid, small, gathered in small cymes.   1. *A. sphaerocephalus.*
Heads radiate, large, solitary at tips of branches.   2. *A. shockleyi.*

1. **Acamptopappus sphaerocéphalus** (Harv. & Gray) A. Gray. Goldenhead. Fig. 5433.

Aplopappus sphaerocephalus Harv. & Gray in A. Gray, Mem. Amer. Acad. II. **4**: 76. 1849.
Acamptopappus sphaerocephalus A. Gray, Proc. Amer. Acad. **8**: 634. 1873.

Round-topped shrub 2–9 dm. high, corymbosely branched, often densely twiggy, glabrous throughout or sparsely hispidulous on leaf-margins. Leaves linear to spatulate, 5–20 mm. long, 1.5–5 mm. wide, obtuse to acute, mucronulate, firm, sessile, gray-green; heads subglobose, 7–10 mm. high, mostly solitary but approximate at tips of cymosely arranged branchlets; involucre 5–6 mm. high.

Open desert, Lower Sonoran Zone; eastern Mojave Desert and western border of Colorado Desert, Inyo County to San Diego County, California, east to Nevada, Utah, and Arizona. Type locality: "California." April–June.

Acamptopappus sphaerocephalus var. hirtéllus Blake, Journ. Wash. Acad. **19**: 270. 1929. Stems and leaves densely hirtellous. Western borders of Mojave Dessert to Lone Pine, Inyo County, California; western Nevada. Type locality: Lone Pine. May–June.

2. **Acamptopappus shóckleyi** A. Gray. Shockley Goldenhead. Fig. 5434.

Acamptopappus shockleyi A. Gray, Proc. Amer. Acad. **17**: 208. 1882.

Rounded shrub 1.5–5 dm. high, spinescent-branched, the herbage finely hirtellous or hispidulous. Leaves spatulate, oblanceolate, or elliptic, 5–15 mm. long, usually mucronulate; heads globose, radiate, 1.5–3 cm. wide, solitary at ends of nearly naked peduncles; involucre 8–11 mm. high; rays 8–13, the oblong ligule about 1 cm. long.

Desert plains and washes, Lower Sonoran Zone; White Mountains, Inyo County, to Clark Mountain, San Bernardino County, California; southern Nevada. Type locality: near Candelaria, Nevada. April–June.

---

* Text contributed by David Daniels Keck.

72. CHRYSÓPSIS* (Nutt.) Ell. Bot. S.C. & Ga. 2: 333. 1824.

Low, perennial (rarely annual), pubescent herbs, sometimes suffrutescent. Leaves alternate, usually entire. Heads radiate or discoid, yellow, medium-sized, terminating the stems and branches. Involucre campanulate to hemispheric; phyllaries numerous, narrow and imbricated. Receptacle low-convex, foveolate. Ray-flowers pistillate, fertile; ligules narrow. Disk-flowers perfect, fertile, slender. Style-branches flattened, the appendages hairy, elongated, much longer than the stigmatic portion. Achenes oblong or obovate, more or less flattened, often twisted. Pappus usually double, brownish, the inner of numerous capillary bristles, the outer (when present) of short linear scales (or bristles). [Name Greek, meaning golden aspect, from the color of the heads.]

A genus of about 20 species, all native in temperate North America. Type species, *Chrysopsis mariana* (L.) Ell.

Heads radiate; outer pappus linear-squamellate.   1. *C. villosa.*
Heads discoid; outer pappus none or indistinct.
    Corolla filiform; pappus exceeding the involucres by most of its length; mostly below 3,000 feet.
                                                             2. *C. oregona.*
    Corolla funnelform; pappus exceeding the involucres by less than half its length; above 4,500 feet.
                                                             3. *C. breweri.*

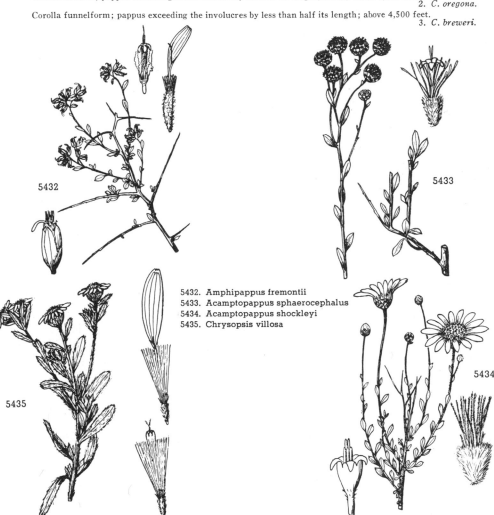

5432. Amphipappus fremontii
5433. Acamptopappus sphaerocephalus
5434. Acamptopappus shockleyi
5435. Chrysopsis villosa

1. Chrysopsis villòsa (Pursh) Nutt. ex DC. Hairy Golden-aster. Fig. 5435.

Amellus villosus Pursh, Fl. Amer. Sept. 564. 1814.
Diplopappus villosus Hook. Fl. Bor. Amer. 2: 22. 1834.
Chrysopsis villosa Nutt. ex DC. Prod. 5: 327. 1836.
Diplogon villosum Kuntze, Rev. Gen. Pl. 1: 334. 1891.

Gray-green perennial herb with several erect stems from an often more or less woody base

_____

* Text contributed by David Daniels Keck.

surmounting a taproot, 1–5 dm. high, the herbage canescently strigose and often somewhat glandular. Leaves oblong-spatulate, entire, 1–5 cm. long, the aestival leaves smaller, firmer, and more sessile than the early-deciduous vernal ones; heads paniculate or cymose in a short inflorescence; involucre 7–10 mm. high; phyllaries linear, acuminate, hirsutulous; rays 10–16, becoming revolute from the tip; achenes oblong-ovate, villous; outer pappus evident.

Open, well-drained slopes, often in sand, Upper Sonoran and Transition Zones; both sides of the Cascade Mountains from southern British Columbia to southern Oregon, reported from the northern Sierra Nevada in California east to the Great Plains. Type locality: "On the banks of the Missouri." June–Sept.

Chrysopsis villosa var. sessiliflòra (Nutt.) A. Gray, Syn. Fl. N. Amer. 1²: 123. 1884. (*Chrysopsis sessiliflora* Nutt. Trans. Amer. Phil. Soc. II. 7: 317. 1840; *C. californica* Elmer, Bot. Gaz. 39: 48. 1905; *Heterotheca sessiliflora* Shinners, Field & Lab. 19: 71. 1951.) Herbage mostly grayish, villous-canescent and glandular; leaves oblong or spatulate, crowded, 1–2 cm. long; heads mostly large and solitary, foliose-bracteate; outer pappus evident, squamellate. Brushy sand-flats, washes, etc., Upper Sonoran Zone; usually near the coast from Santa Barbara, California, to northern Lower California; reported north to Mendocino County, California. Type locality: "St. Barbara, Upper California." July–Sept.

Chrysopsis villosa var. bolánderi (A. Gray) A. Gray ex Jepson, Man. Fl. Pl. Calif. 1036. 1925. (*Chrysopsis bolanderi* A. Gray, Proc. Amer. Acad. 6: 543. 1865; *C. sessiliflora* var. *bolanderi* A. Gray, Bot. Calif. 1: 309. 1876; *C. arenaria* Elmer, Bot. Gaz. 41: 321. 1906.) Stems decumbent or erect, 1–3 dm. high, several from the branching woody root-crown, the leaves of lower half often soon shaded out by other vegetation; herbage densely hirsute and more or less densely beset with sessile glands; leaves oblanceolate or spatulate, rounded, mucronate, the lower petiolate, 2–4 cm. long, 5–10 mm. wide; heads large, mostly solitary, foliose-bracteate; involucre 10–15 mm. high; phyllaries linear-lanceolate, acuminate, villous and glandular; outer pappus of very slender scales. Grassy slopes or dunes on or near the coast, Humid Transition Zone; Mendocino County to San Francisco Bay, California. Type locality: "Hills of Oakland, near San Francisco." June–Nov.

Chrysopsis villosa var. fastigiàta (Greene) H. M. Hall, Univ. Calif. Pub. Bot. 3: 42. 1907. (*Chrysopsis fastigiata* Greene, Pittonia 3: 296. 1898.) Stems 3–10 dm. high, subsimple to multibranched, usually closely beset with small, ascending, linear-oblong to elliptic, obtuse or acute, mucronate leaves, these silky-tomentose especially dorsally, 1–2 or 3 cm. long, the margins often crisped; heads small, 1–3 at tips of stiff erect branches, forming a terminal cymose panicle; outer pappus absent, or present as short bristles, not as squamellae. Dry washes, rocky canyons and plains, Upper Sonoran Zone; California in the Sierran foothills of Tulare and Kern Counties and from Ventura County to Orange County. Type locality: vicinity of San Bernardino, San Bernardino County. July–Nov.

Chrysopsis villosa var. echioìdes (Benth.) A. Gray, Syn. Fl. N. Amer. 1²: 123. 1884. (*Chrysopsis echioides* Benth. Bot. Sulph. 25. 1844; *C. sessiflora* var. *echioides* A. Gray, Bot. Calif. 1: 309. 1876; *Heterotheca echioides* Shinners, Field & Lab. 19: 71. 1951.) Stems erect, 3–8 dm. high, virgate, simple or branching throughout; leaves crowded, spreading, densely hirsute-canescent, firm, the upper 1–2 cm. long, sessile; heads numerous, paniculate to cymose; phyllaries numerous, very slender, hispid-hirsute; outer pappus of short narrow squamellae. Dry sandy soils of open fields, Upper Sonoran Zone; Coast Ranges of California mostly away from the immediate coast from Sonoma and Solano Counties to San Diego County, and also adjacent Lower California. Type locality: Bodegas [Bodega Bay], Sonoma County. July–Nov.

Chrysopsis villosa var. híspida (Hook.) A. Gray ex D. C. Eaton, Bot. King Expl. 164. 1871. (*Diplopappus hispidus* Hook. Fl. Bor. Amer. 2: 22. 1834; *Chrysopsis hispida* DC. Prod. 7: 279. 1838; *C. columbiana* Greene, Erythea 2: 95. 1894; *C. hirsuta* Greene, Pittonia 3: 296. 1898.) Stems slender, virgate, 2–4 dm. high; herbage sparsely or densely spreading-hirsute, somewhat glandular; leaves linear-oblanceolate or spatulate, 1–4 cm. long, 2–10 mm. wide, often hirsute-ciliate; heads solitary to numerous, cymose; involucre 6–8 mm. high; phyllaries minutely glandular and sparsely hispid; outer pappus of very short, narrow squamellae. Dry rocky ground, Upper Sonoran and Arid Transition Zones; Saskatchewan to British Columbia east of the Cascade Mountains south to the Sierra Nevada and Colorado Desert of California (rare), east to Arizona and the Rocky Mountains. Type locality: Carlton House Fort, Saskatchewan. April–Nov.

Chrysopsis villosa var. camphoràta (Eastw.) Jepson, Man. Fl. Pl. Calif. 1036. 1925. (*Chrysopsis camphorata* Eastw. Zoe 5: 81. 1900.) Stems slender, from a slender branching rootstock, 3–8 dm. high; herbage moderately to rather densely hirsute and conspicuously glandular-pubescent, camphor-scented; leaves lance-oblong, not very crowded; heads usually many, cymose-paniculate, rather small, usually subtended by small foliaceous bracts; phyllaries sparsely hispid and ciliate or merely ciliolate from the fine dissection of the narrow hyaline margin, prominently glandular. Open sandy slopes and wooded banks, Humid Transition Zone; Santa Clara County to San Benito and Monterey Counties, California. Type locality: near Glenwood, Santa Cruz Mountains, Santa Cruz County. July–Oct.

2. Chrysopsis oregòna (Nutt.) A. Gray.   Oregon Golden-aster.   Fig. 5436.

Ammodia oregona Nutt. Trans. Amer. Phil. Soc. II. 7: 321. 1840.
Chrysopsis oregona A. Gray, Proc. Amer. Acad. 6: 543. 1865.
Diplogon oregona Kuntze, Rev. Gen. Pl. 1: 334. 1891.
Heterotheca oregona Shinners, Field & Lab. 19: 71. 1951.

Stems clustered from a woody base, much branched, erect, 3–6 dm. high, the herbage glabrate or sparsely hirsute, rather viscid, green. Leaves lanceolate to elliptic-ovate, sessile, ascending, 1-nerved, 2–4 cm. long, 4–10(15) mm. wide; heads corymbosely paniculate, discoid; involucre 9–12 mm. high; phyllaries 4-seriate, regularly imbricate, glandular-atomiferous but scarcely hairy, with white hyaline margin, acuminate; corolla nearly filiform; outer pappus setulose and obscure.

Sandy or gravelly stream banks, Humid Transition Zone; Washington chiefly west of the Cascade Mountains south to Marin County, California. Type locality: "On the sand and gravel bars of the Oregon [Columbia] and its tributary streams." June–Sept.

Chrysopsis oregona var. rùdis (Greene) Jepson, Fl. W. Mid. Calif. 558. 1901. (*Chrysopsis rudis* Greene, Man. Bay Reg. 174. 1894.) Stems 3–8 dm. high, often virgate, brittle; herbage hispid-hirsute and more or less viscid; leaves linear-lanceolate to broadly oblong or oblanceolate, mostly sessile, erect or spreading, 2–5 cm. long; involucre 8–11 mm. high, glandular and somewhat hispid. Dry gravelly places, Humid Transition Zone; California in the northern Sierra Nevada (Tehama County) and North Coast Ranges from Siskiyou and Humboldt Counties to San Francisco Bay. Type locality: ". . . along stream banks in Napa Co." July–Oct.

Chrysopsis oregona var. compácta Keck, Aliso 4: 102. 1958. Low rounded bushes 1.5–4 dm. high, with many stems from the woody base; herbage gray-green, densely hispid- or hirsute-canescent and glandular-atomiferous, the hairs from prominent pustulate bases; leaves typically small (1–2 cm.), densely crowded, spreading, lance-oblong; heads crowded toward the ends of the branches; involucre 8–10 mm. high, glandular but usually sparingly hairy. Dry sandy places, Upper Sonoran Zone; Central Valley of California from Tehama County to Fresno County. Type locality: Newville, Tehama County. June–Oct.

Chrysopsis oregona var. scabérrima A. Gray, Syn. Fl. N. Amer. 1²: 124. 1884. Tall, rather openly branching, the heads commonly corymbosely paniculate on slender peduncles; leaves in upper part of plant becoming bract-like; herbage viscidulous and and scabrous from the pustulate bases of the fractured hairs. Sandy roadsides and dry streamways, Upper Sonoran Zone; California in the South Coast Ranges and Sierran foothills (Tulare County). Type locality: Tulare County. June–Oct.

3. **Chrysopsis bréweri** A. Gray. Brewer's Golden-aster. Fig. 5437.

Chrysopsis breweri A. Gray, Proc. Amer. Acad. **6**: 542. 1865.
Chrysopsis wrightii A. Gray, Syn. Fl. N. Amer. **1²**: 445. 1884.
Heterotheca breweri Shinners, Field & Lab. **19**: 71. 1951.

Few to 20 or so erect stems from a woody caudex, 2–8 dm. high, simple or more commonly racemosely branching throughout, moderately hirtellous and glandular-puberulent. Leaves lanceolate or narrowly oblong to lance-ovate, mucronate, sessile, thin, 1–5 cm. long, 5–20 mm. wide, 3-nerved from base; heads solitary at the ends of the slender leafy branches, broadly campanulate, discoid; involucre 7–11 mm. high, 2–3-seriate; phyllaries lance-linear to lance-oblong, acuminate; corollas slender-funnelform; outer pappus of short fine bristles or none.

Open rocky slopes and coniferous forests, Canadian and Hudsonian Zones; Sierra Nevada from Shasta County to Tulare County and on San Gorgonio Peak, California; adjacent Nevada. Type locality: "Near Sonora Pass and Ebbett's Pass, in the Sierra Nevada." July–Sept.

Chrysopsis breweri var. **multibracteàta** Jepson, Man. Fl. Pl. Calif. 1037. 1925. (*Chrysopsis gracilis* Eastw. Bot. Gaz. **41**: 291. 1906.) Herbage sparingly arachnoid-villous; phyllaries somewhat thicker, narrowly ovate to lanceolate, acute or acuminate, sometimes purple-margined, 4–5-seriate. Dry rocky slopes, Canadian Zone; Siskiyou, Trinity, and Tehama Counties, California. Type locality: Sisson (Mt. Shasta City), Siskiyou County. July–Aug.

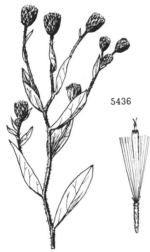

5436. Chrysopsis oregona

5437. Chrysopsis breweri

73. **HETEROTHÈCA*** Cass. Bull. Soc. Philom. 137. 1817.

Coarse erect herbs with yellow-flowered heads disposed in terminal corymbose panicles. Leaves alternate. Involucre hemispheric; phyllaries narrow, closely imbricate in several series, appressed. Ray-florets 1-seriate, fertile; disk-florets many, fertile. Ray-achenes triangular-compressed; pappus none or caducous. Disk-achenes cuneiform; pappus double, the copious inner bristles capillary, long, the outer setose, short. [Name Greek, meaning different and case (or ovary), from the unlike achenes of ray and disk.]

A genus of 3 or more species in the southern United States and Mexico. Type species, *Heterotheca lamarckii* Cass. (= *H. subaxillaris* (Lam.) Britt. & Rusby).

Upper leaves narrowed to a sessile base; heads relatively large; involucre 7–10 mm. high, glandular-pubescent
but not also canescent.   1. *H. grandiflora.*
Upper leaves subcordate-clasping at base; heads smaller; involucre 6–8 mm. high, glandular-puberulent and also
canescent.   2. *H. subaxillaris.*

1. **Heterotheca grandiflòra** Nutt. Telegraph Weed. Fig. 5438.

Heterotheca grandiflora Nutt. Trans. Amer. Phil. Soc. II. **7**: 315. 1840.
Heterotheca floribunda Benth. Bot. Sulph. 24. 1844.

Annual or biennial, the stout stem simple below, 5–20 dm. high, hirsute, the ample inflorescence glandular-pubescent and heavy-scented. Leaves thickish, villous, ovate to oblong or oblanceolate, 2–6 cm. long, obtuse, serrate, the lower petiolate, commonly with a pair of stipule-like lobes at base; heads medium large; involucre 7–9 mm. high; ray-florets 25–35, the corolla 6–8 mm. long, 1 mm. wide, revolute from the tip, the tube hairy; disk-florets 50–65, very slender, the stubby style-branches scarcely exserted; pappus brick-red, the outer series inconspicuous.

Sandy open places, Upper Sonoran Zone; coastal valleys of southern California northward as a ruderal to San Francisco Bay and the San Joaquin and Sacramento Valleys; also southern Arizona. Type locality: Santa Barbara. March–Jan.

* Text contributed by David Daniels Keck.

2. **Heterotheca subaxillàris** (Lam.) Britt. & Rusby. Camphor Weed. Fig. 5439.

Inula subaxillaris Lam. Encycl. Meth. **3**: 259. 1789.
Inula scabra Pursh, Fl. Amer. Sept. **2**: 531. 1814
Chrysopsis scabra Nutt. Gen. **2**: 151. 1818.
Heterotheca lamarckii Cass. Dict. Sci. Nat. **21**: 131. 1821.
Heterotheca scabra DC. Prod. **5**: 317. 1836.
Heterotheca subaxillaris Britt. & Rusby, Trans. N.Y. Acad. **7**: 10. 1887.

Annual or biennial, the moderately stout stem simple below or openly branching, 5–20 dm. high, hispid-hirsute, glandular above. Leaves rather coarsely hirsute, glandular, ovate to lance-oblong, serrate-dentate or subentire, the lower petiolate, the upper subcordate-clasping; heads relatively small, numerous; involucre 6-8 mm. high, glandular and somewhat canescent; ray-florets about 20–28; disk-florets 40–60; pappus rufous, the outer serie. sually conspicuous.

Sandy roadsides and ditches, Sonoran Zones; easternmost Mojave Desert in California; introduced at San Gabriel, California; common eastward to Florida and Delaware. Type locality: Carolina. Aug.–Oct.

74. **CHAETOPÁPPA*** DC. Prod. **5**: 301. 1836.

Ours low, very slender annuals with simple or diffusely branched stems and alternate, entire, chiefly linear leaves. Heads small, few- to many-flowered, terminating very slender peduncles, radiate, disciform, or discoid, all flowers potentially fertile, yellow, white, or reddish. Involucre turbinate to hemispheric, the phyllaries 2–5-seriate, graduate or equal, thin, green-centered, prominently scarious-margined, usually setulose-tipped, persistent. Receptacle convex, naked. Pistillate flowers 1–3-seriate, ligulate or tubular; hermaphrodite flowers very slender, 3–5-toothed. Achenes linear-fusiform, often compressed, pubescent. Pappus of 3 to many fragile slender bristles, sometimes dilated and more or less joined at very base, or wanting. [Name Greek, meaning bristle-pappus.]

A genus of about 15 species of the southwestern United States and Mexico. Type species, *Chaetanthera asteroides* Nutt.

Ray-flowers present, conspicuous; involucre broadly hemispheric.
  Ligules golden yellow; southern California.
    Involucre pubescent.   1. *C. lyonii.*
    Involucre glabrous.
      Phyllaries in strongly graduated series, tapering to the short-caudate tip.   2. *C. aurea.*
      Phyllaries in subequal series, obtuse, mucronulate.   3. *C. fragilis.*
  Ligules white or purplish; San Francisco Bay region.   4. *C. bellidiflora.*
Ray-flowers reduced to a filiform tube or absent; involucre turbinate.
  Stem simple or with few erect branches; disk-corollas dilated at throat, contracted at orifice.   5. *C. exilis.*
  Stem diffusely branched; disk-corollas narrowly linear.   6. *C. alsinoides.*

1. **Chaetopappa lyònii** (A. Gray) Keck. Lyon's Chaetopappa. Fig. 5440.

Pentachaeta lyonii A. Gray, Syn. Fl. N. Amer. **1²**: 445. 1884.
Chaetopappa lyonii Keck, Aliso **4**: 102. 1958.

Stem 1–5 dm. high, simple or branched, hirsute chiefly on the leaf-margins. Leaves narrowly linear or spatulate-linear, 2–5 cm. long; involucre about 5 mm. high, the phyllaries hirsute, subequal, lance-linear, acuminate, narrowly scarious-margined; ray-flowers about 30–50; disk-flowers about 80–100; achenes of ray and disk similar, dark brown, moderately strigose with short hairs; pappus of 10–12 very fragile, filiform bristles, flared at very base and forming a rudimentary corona.

In clayey soil of grassland areas, Upper Sonoran Zone; coastal.part of Los Angeles County, California, and Santa Catalina Island, Los Angeles County. Type locality: San Pedro, near Palos Verdes Mountain. March–April.

2. **Chaetopappa aùrea** (Nutt.) Keck. Golden Chaetopappa. Fig. 5441.

Pentachaeta aurea Nutt. Trans. Amer. Phil. Soc. II. **7**: 336. 1840.
Chaetopappa aurea Keck, Aliso **4**: 102. 1958.

Usually diffusely branched, 8–30 cm. high, entirely glabrous, or the leaf-margins usually ciliate. Leaves mostly narrowly linear, the lower 1–3.5 cm. long, up to 2 mm. wide, the upper much shorter and up to 1 mm. wide; heads solitary at tips of branches, 1–2.5 cm. wide; involucre broad, 4–7 mm. high, the phyllaries about 4–5-ranked, lance-ovate (outer) to oblong, from setaceous-acuminate to obtuse and apiculate, the greenish central portion scarcely wider than the shining scarious margin; ray-flowers 0–70, usually 10–40, the ligule 5–12 mm. long; disk-flowers usually numerous; achenes mahogany-brown, about 1 mm. long, sparsely and finely short-strigose; pappus of 5(–7) scabrous filiform bristles slightly thickened toward apex, flared at very base and often united, about equaling the corolla.

Dry open ground and grassy slopes up to 5,000 feet, mostly Upper Sonoran Zone; cismontane southern California from Los Angeles County to San Diego, San Diego County. Type locality: San Diego. April–July.

---

* Text contributed by David Daniels Keck.

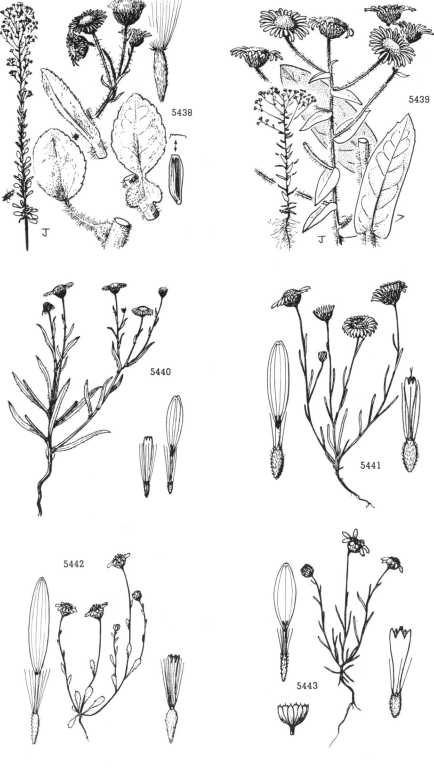

5438. Heterotheca grandiflora
5439. Heterotheca subaxillaris
5440. Chaetopappa lyonii

5441. Chaetopappa aurea
5442. Chaetopappa fragilis
5443. Chaetopappa bellidiflora

3. Chaetopappa frágilis (Brandg.) Keck. Fragile Chaetopappa. Fig. 5442.

Pentachaeta fragilis Brandg. in H. M. Hall, Univ. Calif. Pub. Bot. **6**: 170. 1915.
Chaetopappa fragilis Keck, Aliso **4**: 102. 1958.

Very slender, wiry, diffusely branched, glabrous except for the hirsutulous leaf-margins, the branches 4–10 cm. long. Leaves spatulate to linear-oblanceolate, the basal 8–15 mm. long, 2–3 mm. wide, obtuse, the cauline 2–6 mm. long, 1–1.8 mm. wide; involucre about 4 mm. high, the phyllaries subequal, oblong, subtruncate, rather broadly scarious-margined, lacerate-ciliate at apex; ray-flowers about 10, the ligule 5 mm. long; achenes sparsely pilose; pappus of about 20 very fragile, filiform bristles scarcely enlarged toward apex, not at all dilated at base, slightly shorter than the disk-corolla.

Dry, grassy, foothill slopes, Upper Sonoran Zone; southern Sierra Nevada and Greenhorn Mountains, Kern County and Inner South Coast Range, San Luis Obispo County, California. Type locality: Havilah, Kern County. May–June.

4. Chaetopappa bellidiflòra (Greene) Keck. White-rayed Chaetopappa. Fig. 5443.

Pentachaeta bellidiflora Greene, Bull. Calif. Acad. **1**: 86. 1885.
Chaetopappa bellidiflora Keck, Aliso **4**: 102. 1958.

Stem 6–15 cm. high, simple or with few erect branches, glabrous or nearly so. Leaves narrowly linear, 8–35 mm. long, about 1 mm. wide; heads 1–1.7 cm. wide; involucre 3–4 mm. high, the phyllaries subequal, often purplish, oblong, short-acuminate or apiculate from the truncate, lacerate-ciliate apex, broadly scarious-margined, glabrous dorsally; ray-flowers 5–16, white or purplish-tinged, the ligule 5 mm. long; disk-flowers numerous, yellow, the ample throat not contracted at orifice; achenes densely tawny-hirsute, rarely glabrous; pappus of 5 relatively firm, scabrous bristles, not dilated at base, shorter than the disk-corolla, occasionally lacking.

Open, dry, rocky slopes, Humid Transition Zone; Marin, San Mateo, and Santa Cruz Counties, California; also reported from Monterey County. Type locality: Corte Madera, Marin County. March–May.

5. Chaetopappa éxilis (A. Gray) Keck. Meager Chaetopappa. Fig. 5444.

Aphantochaeta exilis A. Gray, Pacif. R. Rep. **4**: 100. *Pl. 11. Fig. A.* 1857.
Pentachaeta exilis A. Gray, Proc. Amer. Acad. **8**: 633. 1873.
Pentachaeta exilis var. *aphantochaeta* A. Gray, Bot. Calif. **1**: 305. 1876.
Pentachaeta aphantochaeta Greene, Bot. Gaz. **8**: 256. 1883.
Pentachaeta exilis var. *grayi* Jepson, Man. Fl. Pl. Calif. 1039. 1925.
Chaetopappa exilis Keck, Aliso **4**: 103. 1958.

Stem 3–18 cm. high, simple or with few erect branches, often purplish, sparsely pubescent. Leaves filiform or nearly so, up to 25 mm. long, ciliate-pubescent toward base; peduncle white-villous beneath the head; involucre 3–5 mm. high, glabrous, often purplish, the phyllaries subequal, few, broad, oblong, weakly short-bristle-tipped at the obtuse or truncate, more or less laciniate apex, moderately scarious-margined; pistillate flowers 0–5, reduced to a short filiform tube; disk-flowers 4–8, purplish, distinctly widened at throat, contracted at orifice; achenes brown, moderately villous or rarely glabrous; pappus of 3–5 slender scabrid bristles, not dilated at base, or these reduced to triangular scales, or often entirely wanting.

Grassy slopes of valleys and foothills, Sonoran and Transition Zones; from Placer and Mendocino Counties to Mariposa and Monterey Counties, California. Type locality: Napa Valley, Napa County. April–May.

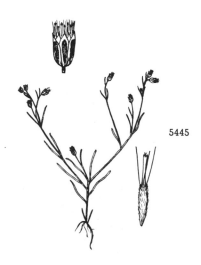

5444. Chaetopappa exilis   5445. Chaetopappa alsinoides

6. **Chaetopappa alsinoìdes** (Greene) Kéck. Tiny Chaetopappa. Fig. 5445.

Pentachaeta exilis var. *discoidea* A. Gray, Bot. Calif. 1: 305. 1876, in part.
Pentachaeta alsinoides Greene, Bull. Torrey Club 9: 109. 1882.
Chaetopappa alsinoides Keck, Aliso 4: 103. 1958.

Diffusely branched, 3–12 cm. high and wide, somewhat villous. Leaves narrowly linear or filiform, 1 mm. wide or less; heads tiny, not strictly solitary; involucre 2.6–3 mm. high, glabrous or somewhat hirsute, the phyllaries subequal, few (6–7), oblong or oval-oblong, obtuse or shortly setaceous-acuminate, green, narrowly scarious-margined, lacerate toward apex; pistillate flowers about 4–6, capillary, tubular or with minute, involute, erect ligule, not exceeding the 3–5 very similar, slightly thicker, reddish-tinged disk-flowers, these commonly with imperfect anthers; achenes brownish, lightly to moderately villous; pappus usually of 3 capillary bristles somewhat exceeding the florets, scarcely at all dilated at base.

Grassy foothill slopes, Upper Sonoran Zone; mostly near the coast, Napa and Marin Counties to Santa Cruz, Santa Barbara, and Tulare Counties, California. Very inconspicuous and probably much overlooked. Type locality: Berkeley, Alameda County. April–May.

75. **HAPLOPÁPPUS*** Cass. Dict. Sci. Nat. **56**: 168. 1828.

Herbs or shrubs, very varied in habit, often glandular. Leaves alternate, entire to bipinnatifid, often thickish, sometimes glandular-punctate. Heads radiate or discoid, large or small, solitary to numerous and cymose or paniculate, yellow, rarely creamy white. Involucre cylindric or turbinate to hemispheric, the phyllaries numerous, subequal to strongly graduate, usually narrow and indurate or chartaceous, at least below. Receptacle usually alveolate. Ray-florets pistillate, rarely sterile; disk-florets fertile, their style-branches ovate to subulate. Achenes terete or angled, linear-fusiform to turbinate, glabrous to silky-pilose. Pappus of numerous capillary, subequal or graduate bristles, usually persistent. [Name Greek, meaning simple and pappus.]

A genus of perhaps 150 species, all American, chiefly of the western United States, Mexico, and Chile. Type species, *Haplopappus glutinosus* Cass.

Achenes turbinate, 2–3 mm. long; leaves dentate to bipinnatifid, the teeth spinulose-tipped or bristle-tipped. (Section *Blepharodon*.)
   Annual; herbage, including involucre, hirsute with strigose hairs, the phyllaries also minutely glandular-puberulent.   1. *H. gracilis.*
   Perennials; involucres not strigose.
      Tufted with several slender erect stems from a suffrutescent base; leaves very narrow; heads radiate.
         Phyllaries prominently glandular-puberulent and scabrous; leaves 2–5 cm. long.
             2. *H. gooddingii.*
         Phyllaries beset with granular glands; leaves 1–2 cm. long.   3. *H. junceus.*
      Rigidly branched shrub; leaves mostly oval, merely dentate; heads discoid.   4. *H. brickellioides.*
Achenes nearly prismatic, subcylindric, or fusiform, 3 mm. or more long; leaves various but if toothed these not as above.
   Perennial herbs with shoots of the season arising from prominent basal rosettes of leaves surmounting a deep fusiform taproot. (Section *Pyrrocoma*.)
      Heads large, the involucre mostly 1.5–3 cm. high, the phyllaries mostly 3–8 mm. wide, the rays inconspicuous.
         Basal leaves 1–4 cm. wide; involucre 14–20 mm. high; widespread.   5. *H. carthamoides.*
         Basal leaves 5–15 cm. wide; involucre about 25 mm. high; plants very robust; upper end of Snake River Canyon.   6. *H. radiatus.*
      Heads smaller, the involucre mostly 5–15 mm. high, the phyllaries less than 3 mm. wide, the rays evident.
         Heads solitary or cymose, or if racemose not numerous in a long narrow inflorescence.
            Plants stipitate-glandular as well as hirsute or villous.   7. *H. hirtus.*
            Plants not stipitate-glandular.
               Heads usually solitary, terminating long peduncles.
                  Phyllaries typically herbaceous, thin, obscurely graduate; achenes pubescent.   8. *H. uniflorus.*
                  Phyllaries green only toward tip, firm, evidently graduate; achenes glabrous.   9. *H. apargioides.*
            Heads corymbose or cymose-paniculate, rarely solitary; phyllaries firm toward base, green-tipped; achenes sericeous.   10. *H. lanceolatus.*
         Heads usually numerous, spicate or racemose.
            Plants not strongly if at all glandular; phyllaries coriaceous and distinctly graduate, obtuse to acute.   11. *H. racemosus.*
            Plants vernicose from sessile glands; phyllaries coriaceous-herbaceous, obscurely graduate, acuminate to attenuate.   12. *H. lucidus.*
   Shrubs or subshrubs.
      Plants cespitose or tufted, usually mat-forming, with much-branched caudex; heads mostly solitary at ends of branches (usually several in *H. whitneyi*).
         Stems relatively leafy; herbage glandular-puberulent; plants of high mountains. (Section *Tonestus*.)
            Leaves entire; outer phyllaries lanceolate; achenes glabrous or sparsely pilose.   13. *H. lyallii.*
            Leaves toothed.
               Achenes glabrous; leaves sharply serrate; involucre campanulate-oblong; phyllaries linear-lanceolate; stems to 50 cm. high.   14. *H. whitneyi.*
               Achenes densely pubescent; leaves saliently dentate in outer half; involucre hemispheric; stems to 15 cm. high.

---

* Text contributed by David Daniels Keck.

Involucre 7.5–10 mm. high, 12–15 mm. wide (pressed), shorter than the disk-florets, the outer phyllaries spatulate-oblong to oblong-obovate, rounded or obtuse; disk-corolla 6–7 mm. long.   15. *H. eximius.*

Involucre 14–18 mm. high, 20–30 mm. wide (pressed), equaling the disk-florets, the outer phyllaries lanceolate or lance-oblong; obtuse or acute; disk-corolla 9–10 mm. long.   16. *H. peirsonii.*

Stems sparsely leafy or subscapose; herbage scarcely if at all glandular; leaves entire; mid-altitude plants. (Section *Stenotus.*)

Leaves soft, floccose-tomentose.   17. *H. lanuginosus.*

Leaves rigid, glabrous to scabrous-puberulent.

Leaves linear, 7–18 mm. long, 1 mm. or less wide.   18. *H. stenophyllus.*

Leaves linear-oblanceolate to spatulate, 10–60 mm. long, 1.5–7 mm. wide.   19. *H. acaulis.*

Plants not at all cespitose or mat-forming, but sometimes forming low rounded bushes.

Appendages of style-branches at least twice as long as the stigmatic portion.

Plants low, intricately branched shrubs mostly under 3 dm. high. (Section *Macronema.*)

Twigs closely white-tomentose; heads discoid; leaves densely glandular.   20. *H. macronema.*

Twigs glabrous or glandular, if rarely loosely tomentose then heads radiate.

Heads relatively large, 20–45-flowered; involucre broadly campanulate, about 2-seriate, the phyllaries not obviously ciliate or scarious-margined; leaves undulate-margined; herbage densely stipitate-glandular.   21. *H. suffruticosus.*

Heads smaller and narrower; involucre more than 2-seriate; leaves not crisped, very rarely stipitate-glandular.

Phyllaries villous-ciliate; leaves without axillary fascicles.

Involucre narrowly campanulate, 3–6-seriate, the outer phyllaries not squarrose-tipped; heads radiate, 6–25-flowered.

Leaves oblanceolate, 1.5–3.5 cm. long, 3–7 mm. wide; heads solitary or few, cymose.   22. *H. greenei.*

Leaves mostly linear, 2–6 cm. long, 0.5–3 mm. wide; heads several or numerous in a raceme or thyrsoid panicle.   23. *H. bloomeri.*

Involucre cylindric, 5–8-seriate, the outer phyllaries squarrose-tipped; heads discoid, mostly 5-flowered, solitary or few, openly cymose.   24. *H. ophitidis.*

Phyllaries glabrous (obscurely ciliate in *H. gilmanii*); leaves often with axillary fascicles.

Leaves mostly spatulate, spaced; phyllaries resinous, the outer squarrose; flowers white (or pale yellow); disk-florets 15–18; Death Valley region.   25. *H. gilmanii.*

Leaves mostly linear or filiform, crowded; phyllaries not resinous.

Outer phyllaries often squarrose; flowers white (or pale yellow); disk-florets 10–15, their narrow lobes 1–2 mm. long; Washington and Oregon.   26. *H. resinosus.*

Outer phyllaries erect; flowers yellow; disk-florets 5–10, their lobes 0.5 mm. long; Great Basin.   27. *H. nanus.*

Plant an herbaceous perennial from a woody base, mostly 3–6 dm. high. (Section *Hesperodoria.*)   28. *H. hallii.*

Appendages of style-branches only equaling or shorter than the stigmatic portion.

Involucre hemispheric, 10–18 mm. wide; herbage glandular-punctate; leaves entire. (Section *Stenotopsis.*)   29. *H. linearifolius.*

Involucre turbinate or subcylindric.

Disk-corolla abruptly dilated from narrow tube to much wider throat; heads discoid. (Section *Isocoma.*)

Phyllaries with green but thin tips; stems brownish; coastal.   30. *H. venetus.*

Phyllaries with thickened subepidermal resin-pocket near tip, not green; stems whitish; deserts.   31. *H. acradenius.*

Disk-corolla only slightly ampliate upward.

Heads large, the 8–15 mm. high involucres tightly imbricate, mostly 6–8-seriate; herbage without distinct resin-pits. (Section *Hazardia.*)

Herbage tomentose; insular.   32. *H. canus.*

Herbage not tomentose; mainland.   33. *H. squarrosus.*

Heads small, the 3–8 mm. high involucres loosely imbricate, 2–6-seriate; herbage with distinct resin-pits. (Section *Ericameria.*)

Ray-florets present; leaves filiform; heads more or less paniculate.

Outer phyllaries more or less caudate-tipped.

Leaves 10–35 mm. long, with shorter fascicles in the axils; achenes pilose.   34. *H. pinifolius.*

Leaves 4–12 mm. long, scarcely exceeding the axillary fascicles, ericoid; achenes glabrous.   35. *H. ericoides.*

Outer phyllaries acute to obtuse.

Heads solitary or few; involucre campanulate, 7–8 mm. high.   36. *H. eastwoodiae.*

Heads many; involucre turbinate, 5–6.5 mm. high.   37. *H. palmeri.*

Ray-florets reduced or wanting (see also *H. laricifolius*).

Leaves filiform to linear, less than 3 mm. wide.

Heads solitary or racemose-paniculate, discoid; San Diego County south.   38. *H. propinquus.*

Heads cymose.

Leaves 0.5–2 cm. long; broad rounded shrubs seldom more than 1 m. high; deserts.

Herbage glabrous; leaves subterete, without persistent fascicles in the old axils; ray-florets 3–11; fall-flowering.   39. *H. laricifolius.*

Herbage hairy; leaves flat, with persistent fascicles in the old
axils; ray-florets 0-2; spring-flowering.
40. *H. cooperi.*

Leaves 3-6 cm. long; arborescent shrubs mostly 1-3 m. high; cismontane. 41. *H. arborescens.*

Leaves oblanceolate to obovate, 3-10 mm. wide.

Heads 9-12-flowered, discoid; erect shrub 2-5 m. high; leaves 2-6 cm.
long, mostly tapering to base and apex. 42. *H. parishii.*

Heads 16-30-flowered, sometimes radiate; spreading subshrub mostly less
than 1 m. high; leaves cuneate. 43. *H. cuneatus.*

1. Haplopappus grácilis (Nutt.) A. Gray. Annual Bristleweed. Fig. 5446.

Dieteria gracilis Nutt. Journ. Acad. Phila. II. 1: 177. 1848.
Haplopappus gracilis A. Gray, Mem. Amer. Acad. II. 4: 76. 1849.
Aster dieteria Kuntze, Rev. Gen. Pl. 1: 315. 1891.
Eriocarpum gracile Greene, Erythea 2: 109. 1894.
Sideranthus gracilis A. Nels. Bot. Gaz. 37: 266. 1904.
Machaeranthera gracilis Shinners, Field & Lab. 18: 41. 1950.

Annual herb 6-35 cm. high, usually divaricately branching throughout, hirsute, usually strigose.
Leaves numerous, strigose, the lower oblanceolate, pinnatifid or bipinnatifid, 1.5-3 cm. long, 3-7
mm. wide, the upper linear, appressed, much reduced, serrate-dentate to serrulate, each tooth or
lobe and apex tipped with a prominent white bristle; heads cymose or solitary; involucre hemispheric, 6-7 mm. high, 8-12 mm. wide; phyllaries linear-lanceolate, well imbricated, green with
hyaline margin, cinereous or strigose and minutely glandular-puberulent, with appressed bristle-
tip; ray-florets 16-28, the ligules 7-12 mm. long; pappus of numerous tawny unequal bristles
slightly dilated below.

Sandy or rocky flats and slopes, Lower Sonoran Zone; eastern Mojave Desert from Clark Mountain to Providence Mountains, San Bernardino County, California, east to Colorado, Texas, and Mexico. Type locality: near
Santa Fe, New Mexico. April-June.

2. Haplopappus gooddíngii (A. Nels.) Munz & Jtn. Goodding's Bristleweed. Fig. 5447.

Sideranthus gooddingii A. Nels. Bot. Gaz. 37: 266. 1904.
Haplopappus gooddingii Munz & Jtn. Bull. Torrey Club 49: 44. 1922.
Haplopappus spinulosus gooddingii Blake, Contr. U.S. Nat. Herb. 25: 543. 1925.
Haplopappus spinulosus subsp. *gooddingii* H. M. Hall, Carnegie Inst. Wash. Publ. No. 389: 75. 1928.

Taprooted perennial with several stiffly erect or ascending, slender stems 2-6 dm. high, glabrate
to harshly glandular-puberulent, sometimes also canescent. Leaves scattered, lanate ventrally,
scabrid dorsally, pinnatifid, the rachis and remote lobes linear and bristle-tipped, 2-5 cm. long, the
upper becoming entire and much reduced; heads terminating long branches; involucre depressed-
hemispheric, 6-9 mm. high, 10-18 mm. wide; phyllaries linear-lanceolate, well imbricated, greenish,
prominently glandular-puberulent and scabrous, with short apical bristle; ray-florets 30-45, the
ligules 6-16 mm. long; pappus of numerous tawny unequal bristles.

Rocky mesas, canyon sides and cliffs, Lower Sonoran Zone; eastern Mojave Desert, California, to southern
Nevada and northeastern Arizona. Type locality: "The Pockets," southern Nevada. Feb.-May.

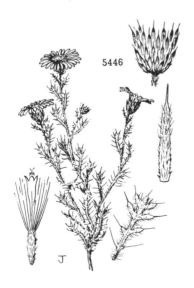

5446. Haplopappus gracilis

5447. Haplopappus gooddingii

3. **Haplopappus júnceus** Greene. Rush-like Bristleweed. Fig. 5448.

Haplopappus junceus Greene, Bull. Calif. Acad. **1**: 190. 1885.
Eriocarpum junceum Greene, Erythea **2**: 108. 1894.
Sideranthus junceus Davids. & Moxley, Fl. S. Calif. 377. 1923.
Machaeranthera juncea Shinners, Field & Lab. **18**: 40. 1950.

Stems tufted, 4–10 dm. high, suffrutescent at base, slender, branching, sparingly strigose, slightly glandular near the heads. Leaves chiefly linear, pinnatifid or serrate with bristle-tipped teeth, 1–2 cm. long, the upper reduced, entire, bristle-tipped; heads solitary on long, slender, scaly-bracted branches or in open cymes; involucre hemispheric, 5–8 mm. high, the closely imbricated phyllaries linear, covered with granular glands, bristle-tipped; ray-florets 15–25, the ligules 5–6 mm. long; pappus of numerous tawny unequal bristles.

Dry brushy hillsides, Lower Sonoran Zone; southern San Diego County, California; northwestern Mexico. Type locality: San Diego County. June–Oct.

4. **Haplopappus brickellioìdes** Blake. Brickellia-like Goldenweed. Fig. 5449.

Haplopappus brickellioides Blake, Proc. Biol. Soc. Wash. **35**: 173. 1922.

Rigidly branched shrub 25 cm. or more high, the older stems white-barked and more or less pilose, the branches yellowish, densely pilosulose, some hairs thickened and tipped with yellow glands. Leaves oval, elliptic, or obovate-cuneate, 1–3.5 cm. long, 5–25 mm. wide, acute, spinescent-tipped, dentate with 1–4 pairs of spinescent teeth, firm, triplinerved, the midnerve prominent, pilose, and yellow-glandular; heads discoid, about 12-flowered, rather small, sessile or subsessile in ones to threes toward tips of leafy branchlets; involucre 6–7 mm. high, the phyllaries 4–5-seriate, lanceolate, 1-ribbed, hispidulous and glandular, the tip greenish; achenes oblong, hispidulous; pappus sparse.

Rocky canyons, Lower Sonoran Zone; Death Valley region, California, and adjacent Nevada. Type locality: Ash Meadows (or Sheep Mountain), Nevada. April–Sept.

5. **Haplopappus carthamoìdes** (Hook.) A. Gray. Columbia Pyrrocoma. Fig. 5450.

Pyrrocoma carthamoides Hook. Fl. Bor. Amer. **1**: 307. *pl. 107*. 1833.
Haplopappus carthamoides A. Gray, Proc. Acad. Phila. **1863**: 65. 1864.
Aster carthamoides Kuntze, Rev. Gen. Pl. **1**: 317. 1891.
Pyrrocoma erythropappa Rydb. Bull. Torrey Club **27**: 624. 1900.
Pyrrocoma rigida Rydb. loc. cit. Not Philippi, 1856.
Hoorebekia carthamoides Piper, Contr. U.S. Nat. Herb. **11**: 559. 1906.
Haplopappus carthamoides subsp. *rigidus* H. M. Hall, Carnegie Inst. Wash. Publ. No. 389: 103. 1928.
Haplopappus carthamoides subsp. *erythropappus* H. M. Hall, op. cit. 105.

Stems few, stout, erect or ascending from a thick taproot, 1–6 dm. high, glabrate to canescent-villous, few-leaved. Basal leaves tufted, 1–2(–4) dm. long, 1–4 cm. wide, oblanceolate to oval, narrowly petiolate, entire or spinulose-serrate; cauline leaves oblanceolate, elliptic, or elliptic-ovate, usually sessile; heads solitary or sometimes several in a corymbiform or racemiform inflorescence, large; involucre hemispheric, 15–20 mm. high, often subtended by a few leafy bracts, the phyllaries coriaceous-herbaceous, lance-oblong, obtuse to acuminate, cuspidate, the margin obviously scarious and usually ciliate, entire to erose or spinescent-toothed; ray-florets 0–30, inconspicuous, seldom exceeding the pappus; achenes glabrous; pappus stiff, sordid.

Prairies and open hillsides, Upper Sonoran and Arid Transition Zones; Columbia River Gorge, Washington-Oregon, north and east to northern Washington and western Montana. Type locality: "North West Coast of America." June–Aug.

Haplopappus carthamoides subsp. **cusícki** (A. Gray) H. M. Hall, Carnegie Inst. Wash. Publ. No. 389; 104. 1928. (*Haplopappus carthamoides* var. *cusickii* A. Gray, Syn. Fl. N. Amer. **1²**: 126. 1884; *Pyrrocoma cusickii* Greene, Erythea **2**: 59. 1894; *Hoorebekia carthamoides* subsp. *cusickii* Piper, Contr. U.S. Nat. Herb. **11**: 560. 1906.) Generally lower, with stems more decumbent; heads mostly narrower, mostly turbinate-campanulate; phyllaries narrower, tapering from near base to the longer, more acute, herbaceous tip, looser and rarely imbricate, the margin obscurely if at all scarious. Scablands and barren rocky soils, Upper Sonoran Zone; Blue Mountains, Washington, and Wallowa foothills south through eastern Oregon to Lassen County, California; east to northern Nevada and central Idaho. Type locality: Union County, Oregon. June–Aug.

6. **Haplopappus radiàtus** (Nutt.) Cronquist. Snake Pyrrocoma. Fig. 5451.

Pyrrocoma radiata Nutt. Trans. Amer. Phil. Soc. II. **7**: 333. 1840.
Haplopappus carthamoides var. *maximus* A. Gray, Syn. Fl. N. Amer. **1²**: 126. 1884.
Haplopappus radiatus Cronquist, Vasc. Pl. Pacif. Northw. **5**: 223. 1955.

Stems few to several, stout, erect or ascending, 4–9 dm. high, essentially glabrous throughout. Basal leaves tufted, long-petiolate, the entire elliptic blades 10–20 cm. long, 5–15 cm. wide; cauline leaves numerous, often toothed, the lower oblanceolate, becoming sessile, the upper ovate, amplexicaul; heads solitary or several in an open corymb, hemispheric, very large; involucre about 25 mm. high, the phyllaries ovate-oblong, very firm, imbricate, the narrow thin margin denticulate; ray-florets 25–35, the ligules 6–12 mm. long; achenes glabrous; pappus 12–14 mm. long, sordid or reddish.

Dry hillsides, Upper Sonoran Zone; in and near the south end of the Snake River Canyon, Oregon and Idaho. Type locality: "Plains of Oregon, near Walla-Walla," collected by Nuttall; but probably from the Snake River near Huntington, Oregon. June–Aug.

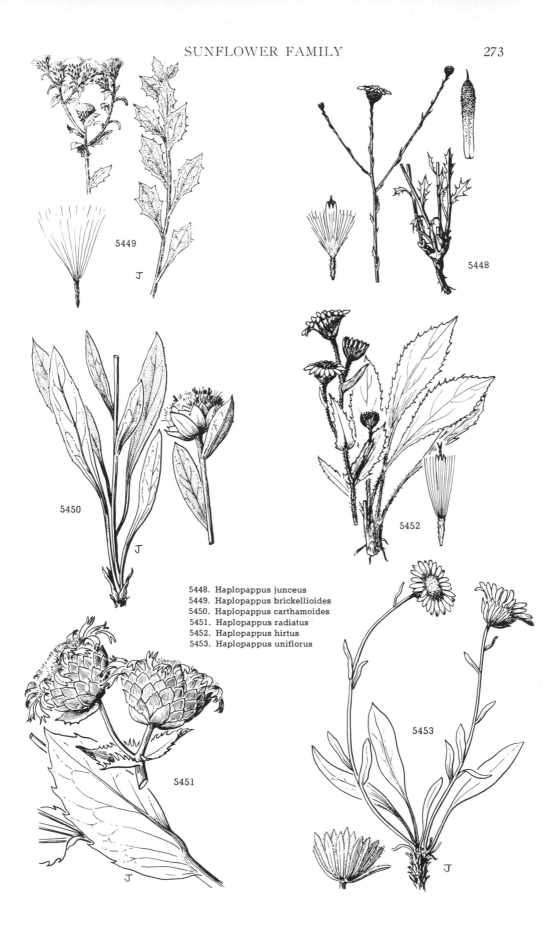

5449

5448

5450

5452

5448. Haplopappus junceus
5449. Haplopappus brickellioides
5450. Haplopappus carthamoides
5451. Haplopappus radiatus
5452. Haplopappus hirtus
5453. Haplopappus uniflorus

5451

5453

7. Haplopappus hírtus A. Gray. Sticky Pyrrocoma. Fig. 5452.

Haplopappus hirtus A. Gray, Syn. Fl. N. Amer. 1²: 127. 1884.
Aster grayanus Kuntze, Rev. Gen. Pl. 1: 316. 1891.
Pyrrocoma hirta Greene, Erythea 2: 69. 1894.
Hoorebekia hirta Piper, Contr. U.S. Nat. Herb. 11: 560. 1906.

Stems several, erect or ascending, often decumbent at base, 1.5–3 dm. high, from a woody, often slightly branched caudex surmounting a taproot, more or less equably leafy throughout, sparingly to rather densely villous-tomentose with jointed hairs, stipitate-glandular at least above. Basal leaves tufted, elliptic-lanceolate, the blade 3–8 cm. long, 8–25 mm. wide, short-petiolate, sharply pectinate-serrate or doubly serrate, rarely subentire, the cauline smaller, sessile; heads few or several, loosely racemose or subcymose; involucre broadly campanulate, 9–12 mm. high; phyllaries about 3-seriate, subequal, linear-lanceolate, herbaceous, with loose tip, stipitate-glandular and villous; ray-florets 13–25, the ligules 7–10 mm. long; achenes silky; pappus sordid.

Dry rocky places, meadows or open woods, Arid Transition Zone; northeastern Oregon to northeastern California east to northern Nevada. Type locality: Baker County, Oregon. July–Aug.

Haplopappus hirtus subsp. **sonchifòlius** (Greene) H. M. Hall, Carnegie Inst. Wash. Publ. No. 389: 125. 1928. (*Pyrrocoma sonchifolia* Greene, Leaflets Bot. Obs. 2: 18. 1909; *P. foliosa* Greene, loc. cit., not Gray, 1844; *Haplopappus hirtus* var. *sonchifolius* M. E. Peck, Man. Pl. Oregon 713. 1941.) Relatively robust; stems 1.5–4.5 dm. high; leaves ample, the basal elliptic-lanceolate to oval, the blade 7–15 cm. long, up to 45 mm. wide, on short or long petioles; involucre 12–14 mm. high, the phyllaries attenuate-tipped. Meadowy or dry places, Arid Transition Zone; central and southeastern Washington, northeastern Oregon, and adjacent Idaho. Type locality: Yakima region. June–July.

Haplopappus hirtus subsp. **lanulòsus** (Greene) H. M. Hall, Carnegie Inst. Wash. Publ. No. 389: 125. 1928. (*Pyrrocoma lanulosa* Greene, Leaflets Bot. Obs. 2: 16. 1909; *P. turbinella* Greene, op. cit. 17; *Haplopappus hirtus* var. *lanulosus* M. E. Peck, Man. Pl. Oregon 713. 1941.) Habit as in subsp. *hirtus* but the cauline leaves more obviously reduced in size, the herbage tending to be more obviously woolly and less glandular, the involucre 8–11 mm. high, the phyllaries imbricate, merely green-tipped, the tip not elongated nor spreading. Moist or dry soil of sagebrush flats or yellow pine woods, Arid Transition Zone; Grant County, Oregon, to Lassen County, California. Type locality: Bear Flat, eastern Oregon. June–Aug.

8. Haplopappus uniflòrus Hook.) Torr. & Gray. Single-headed Pyrrocoma. Fig. 5453.

Donia uniflora Hook. Fl. Bor. Amer. 2: 25. 1834.
Homopappus uniflorus Nutt. Trans. Amer. Phil. Soc. II. 7: 333. 1840.
Haplopappus uniflorus Torr. & Gray, Fl. N. Amer. 2: 241. 1842.
Haplopappus howellii A. Gray, Syn. Fl. N. Amer. ed. 2. 1²: 446. 1886.
Pyrrocoma uniflora Greene, Erythea 2: 60. 1894.
Pyrrocoma howellii Greene, op. cit. 70.
Hoorebekia uniflora M. E. Jones, Bull. Univ. Mont. Biol. Ser. No. 15: 49. 1910.
Haplopappus uniflorus subsp. *howellii* H. M. Hall, Carnegie Inst. Wash. Publ. No. 389: 150. 1928.
Haplopappus uniflorus var. *howellii* M. E. Peck, Man. Pl. Oregon 713. 1941.

Stems 1–3 dm. high, from a fibrous-coated, fusiform crown, decumbent to ascending-erect, usually anthocyanous, glabrate to tomentulose. Leaves mostly basal, narrowly to broadly lanceolate, tapering to base and apex, the blade 5–12 cm. long, 6–15 mm. wide, entire to laciniate-dentate, the cauline much reduced and sessile; heads solitary, terminating rather long peduncles; involucre hemispheric, 8–10 mm. high, the phyllaries 2–3-seriate, obscurely graduate, typically herbaceous at least medianly from apex to base, not thickened, the margin scarious, glabrous to tomentose; ray-florets 18–32, the ligules 6–9 mm. long; achenes sericeous.

Mountain meadows and marshes, often alkaline, Upper Sonoran and Arid Transition Zones; southeastern Oregon, Modoc and Mono Counties, California, east to northern Colorado, north to Saskatchewan. Type locality: Saskatchewan. June–Aug.

Haplopappus uniflorus subsp. **lineàris** Keck, Aliso 4: 103. 1958. Basal leaves grass-like, 8–12 cm. long, 1.5–3 or –4 mm. wide, entire; otherwise similar. Marshy land, Upper Sonoran and Arid Transition Zones; Harney County, Oregon, to Owyhee County, Idaho. Type locality: Mud Flats, Owyhee County. May–June. This is the taxon that has commonly gone under the name of *howellii*, but the type of that name is clearly referable to typical *uniflorus*.

Haplopappus uniflorus subsp. **gossýpinus** (Greene) H. M. Hall, Carnegie Inst. Wash. Publ. No. 389: 150. 1928. (*Pyrrocoma gossypina* Greene, Pittonia 3: 23. 1896; *Haplopappus gossypinus* H. M. Hall, Univ. Calif. Pub. Bot. 3: 49. 1907.) Stems 1–3 dm. high, decumbent, floccose-tomentose like the leaves; basal leaves lanceolate, the blade 2–8 cm. long, 7–20 mm. wide, serrate to entire; heads racemose or usually solitary; involucre 10–13 mm. high, the thin, linear-oblong, acuminate phyllaries rather loose, of unequal lengths but scarcely imbricate, somewhat arachnoid-pilose. Dry meadows, Arid Transition Zone; Bear Valley, San Bernardino Mountains, southern California. Type locality: Bear Valley. July–Sept.

9. Haplopappus apargioìdes A. Gray. Alpine Pyrrocoma. Fig. 5454.

Haplopappus apargioides A. Gray, Proc. Amer. Acad. 7: 354. 1868.
Aster apargioides Kuntze, Rev. Gen. Pl. 1: 317. 1891.
Pyrrocoma apargioides Greene, Erythea 2: 70. 1894.
Pyrrocoma demissa Greene, Leaflets Bot. Obs. 2: 10. 1909.

Stems several, decumbent to ascending-erect, 0.5–1.5(–2.5) dm. high, from a thick taproot that is sometimes branched at the crown, glabrate or somewhat villous, few-leaved or scapiform. Basal leaves tufted, linear-lanceolate to oblanceolate, acuminate, petiolate, 3–10 cm. long, 3–10 mm. wide, laciniate with spinescent teeth to entire, typically ciliate toward base with scabrous or rather coarse hairs, otherwise glabrous, coriaceous; heads usually 1, long-peduncled; involucre subhemispheric, 8–12 mm. high, the phyllaries loosely imbricated in few ranks narrowly (sometimes broadly) oblong, usually pungently acute, sometimes obtusish, firm, green toward tip with pale margin becoming hyaline below, glabrous; ray-florets 13–34, the ligules 6–9 mm. long; achenes flattened, glabrous, striate, 3–7 mm. long; pappus sordid.

Rocky slopes and meadows, Hudsonian and Arctic-Alpine Zones; eastern Sierra Nevada and White Mountains, California, and adjacent Nevada. Type locality: Soda Springs on the Tuolumne River, California. July–Sept.

10. Haplopappus lanceolàtus (Hook.) Torr. & Gray. Intermountain Pyrrocoma. Fig. 5455.

Donia lanceolata Hook. Fl. Bor. Amer. **2**: 25. 1834.
Haplopappus lanceolatus Torr. & Gray, Fl. N. Amer. **2**: 241. 1842.
Aster lanceolatus Kuntze, Rev. Gen. Pl. **1**: 313. 1891.
Pyrrocoma lanceolata Greene, Erythea **2**: 69. 1894.
Pyrrocoma cuspidata Greene, Leaflets Bot. Obs. **2**: 17. 1909.
Hoorebekia lanceolata M. E. Jones, Bull. Univ. Mont. Biol. Ser. No. 15: 49. 1910.

Stems few to several from a taproot and simple or forked, fibrous-coated crown, decumbent to ascending-erect, 1–5 dm. high, glabrous or slightly villous-tomentulose. Leaves spiny-toothed or entire, rather thin, glabrous to lanulose, the basal tufted, oblanceolate, acuminate at each end, the blade 5–15 cm. long, 5–15 cm. wide, much exceeding the slender petiole, the cauline much reduced; heads several or many in an open corymb or racemose panicle, rarely solitary; involucre subhemispheric, 7–10 mm. high, the phyllaries 3–4-seriate, distinctly to obscurely graduate, typically herbaceous toward tip, firm toward base, linear-lanceolate to linear-oblong, glabrate to loosely villous; ray-florets 13–34, the ligules 5–10 mm. long; achenes sericeous; pappus sordid.

Meadows and alkaline flats, Upper Sonoran and Arid Transition Zones; eastern Oregon to Plumas County, California, reappearing in Mono County, east to Saskatchewan, Nebraska, and Colorado. Type locality: Saskatchewan. June–Aug.

Haplopappus lanceolatus subsp. **tenuicaùlis** (D. C. Eaton) H. M. Hall, Carnegie Inst. Wash. Publ. No. 389: 118. 1928. (*Haplopappus tenuicaulis* D. C. Eaton, Bot. King Expl. 160. 1871; *H. lanceolatus* var. *tenuicaulis* A. Gray, Syn. Fl. N. Amer. **1**²: 129. 1884; *Pyrrocoma tenuicaulis* Greene, Erythea **2**: 69. 1894; *P. solidaginea* Greene, Proc. Acad. Phila. **1895**: 549. 1896; *Hoorebekia curvata* Piper, Proc. Biol. Soc. Wash. **31**: 77. 1918; *Haplopappus lanceolatus* subsp. *solidagineus* H. M. Hall, Carnegie Inst. Wash. Publ. No. 389: 118. 1928.) Heads smaller, the involucre 5–7 mm. high; stems typically slender, flexuous, decumbent at base. Intergrading with the species. The common form in eastern Oregon south to Sierra County, California, east to southwestern Idaho, northern Nevada. Type locality: Ruby Valley, Nevada.

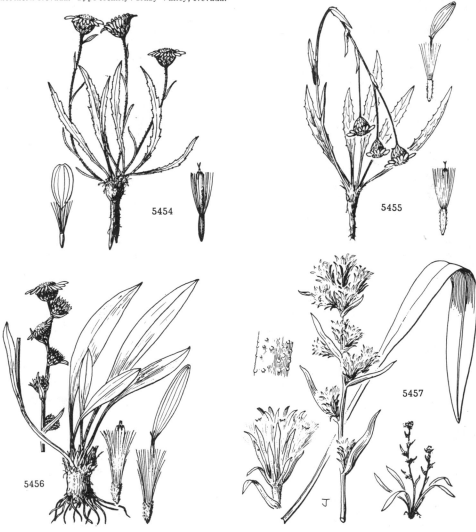

5454. Haplopappus apargioides
5455. Haplopappus lanceolatus

5456. Haplopappus racemosus
5457. Haplopappus lucidus

11. Haplopappus racemòsus (Nutt.) Torr. Racemose Pyrrocoma. Fig. 5456.

Homopappus racemosus Nutt. Trans. Amer. Phil. Soc. II. **7**: 332. 1840.
Pyrrocoma racemosa Torr. & Gray, Fl. N. Amer. **2**: 244. 1842.
Haplopappus racemosus Torr. in Sitgr. Rep. 162. 1854.
Haplopappus lanceolatus var. *strictus* A. Gray, Proc. Amer. Acad. **8**: 389. 1872.
Aster pyrrocoma Kuntze, Rev. Gen. Pl. **1**: 317. 1891.
Pyrrocoma elata Greene, Man. Bay Reg. 173. 1894.
Pyrrocoma longifolia Greene, Pittonia **3**: 183. 1897.
Hoorebekia racemosa Piper, Contr. U.S. Nat. Herb. **11**: 560. 1906.
Pyrrocoma balsamitae Greene, Leaflets Bot. Obs. **2**: 15. 1909.
Haplopappus longifolius Jepson, Man. Fl. Pl. Calif. 1027. 1925.
Hapolpappus racemosus subsp. *longifolius* H. M. Hall, Carnegie Inst. Wash. Publ. No. 389: 130. 1928.

Stems few to several, erect from a stout taproot and short, sometimes branched caudex, 3–10 dm. high, the herbage essentially glabrous. Leaves alternate, the basal tufted, oblanceolate to elliptic, 10–30 cm. long (including slender petiole), 10–25 mm. wide, entire to shallowly serrate, rarely spinulose-toothed, the cauline reduced, sessile or clasping; heads several to many, racemose or spicate or rarely solitary, not glomerate; involucre narrowly to broadly hemispheric, 9–12 mm. high; phyllaries 4–5-seriate, firm, herbaceous throughout or green-tipped, with narrow, whitish or hyaline margin, sometimes ciliate, otherwise glabrous, sharply acute to obtusish; ray-florets 13–30, the ligules 5–12 mm. long; achenes densely villous; pappus sordid, about 7 mm. long.

Coastal valleys, in neutral or saline soils, Upper Sonoran and Humid Transition Zones; Willamette Valley, Oregon, to San Benito County, California, east to Shasta Valley, Siskiyou County. Type locality: "Plains of the Wahlamet." June–Oct.

Haplopappus racemosus subsp. **congéstus** (Greene) H. M. Hall, Carnegie Inst. Wash. Publ. No. 389: 128 1928. (*Pyrrocoma congesta* Greene, Pittonia **3**: 23. 1896; *Haplopappus racemosus* var. *congestus* M. E. Peck, Man. Pl. Oregon 713. 1941.) Stems slender, curving upward from base, 2–7.5 dm. high, glabrous or sparingly tomentose. Leaves green, relatively thin, the lateral veins visible, entire or sparsely denticulate; heads sessile or rarely short-peduncled in a typically glomerate-spicate inflorescence; involucre broadly campanulate to hemispheric, 6–7.5 mm. high; phyllaries firm but not thick, oblong, abruptly narrowed to a short acute tip, bright yellow-green, glandular-ciliolate at apex, sparsely tomentulose throughout or at least on margin below. Dry or moist slopes in serpentine soils, Humid Transition Zone; western end of the Siskiyou Mountains, Josephine County, Oregon, to Del Norte County, California. Type locality: near Waldo, Oregon. Aug.–Sept.

Haplopappus racemosus subsp. **glomeràtus** (Nutt.) H. M. Hall, Carnegie Inst. Wash. Publ. No. 389: 132. 1928. (*Homopappus glomeratus* Nutt. Trans. Amer. Phil. Soc. II. **7**: 331. 1840; *H. argutus* Nutt. loc. cit.; *H. paniculatus* Nutt. loc. cit.; *Pyrrocoma arguta* Torr. & Gray, Fl. N. Amer. **2**: 244. 1842; *P. glomerata* Torr. & Gray, loc. cit.; *P. paniculata* Torr. & Gray, loc. cit.; *Haplopappus paniculatus* A. Gray, Proc. Amer. Acad. **7**: 354. 1868; *H. paniculatus* var. *virgatus* A. Gray, loc. cit.; *H. paniculatus* var. *stenocephalus* A. Gray, Bot. Calif. **1**: 312. 1876; *H. racemosus* var. *glomerellus* A. Gray, Syn. Fl. N. Amer. **1²**: 127. 1884; *H. racemosus* var. *stenocephalus* A. Gray, loc. cit.; *H. racemosus* var. *virgatus* A. Gray, loc. cit.; *Pyrrocoma eriopoda* Greene, Proc. Acad. Phila. **1895**: 549. 1896; *P. ciliolata* Greene, Pittonia **3**: 184. 1897; *P. brachycephala* A. Nels. Bot. Gaz. **37**: 265. 1904; *P. microdonta* Greene, Leaflets Bot. Obs. **2**: 11. 1909; *P. prionophylla* Greene, op. cit. 12; *P. halophila* Greene, op. cit. 16; *P. duriuscula* Greene, loc. cit.; *P. paniculata virgata* Davids. & Moxley, Fl. S. Calif. 377. 1923; *Haplopappus eriopodus* Blake, Contr. U.S. Nat. Herb. **25**: 544. 1925; *H. racemosus* subsp. *duriusculus* H. M. Hall, Carnegie Inst. Wash. Publ. No. 389: 129. 1928; *H. racemosus* subsp. *prionophyllus* H. M. Hall, op. cit. 131; *H. racemosus* subsp. *brachycephalus* H. M. Hall, op. cit. 134; *H. racemosus* subsp. *halophilus* H. M. Hall. op. cit. 136; *H. racemosus* var. *glomeratus, halophilus* [as "*halophiloides*"], *duriusculus,* and *brachycephalus* M. E. Peck, Man. Pl. Oregon 712–713. 1941.) Habitally similar to subsp. *congestus,* with stems 2–7 dm. high but herbage gray-green or glaucous, the foliage stiff and thickish, without obvious lateral veins, entire to sharply denticulate-serrate, even the basal leaves usually less than 15 mm. wide; heads sessile or short-pedunculate, sometimes more or less glomerate, in narrow racemes or spikes; involucre 6–8 mm. high; phyllaries firm, thickish, lance-oblong to oblong, acute to obtuse, pale below the dark green tip, glabrous or ciliate, rarely more hairy. Alkaline plains and meadows, Upper Sonoran Zone; northern Oregon to Inyo County, California, east of the Cascade-Sierran axis except in the Bakersfield region in California, east to southern Idaho and Utah. Type locality: probably in the Grande Ronde Valley, Oregon. July–Oct.

Haplopappus racemosus subsp. **sessiliflòrus** (Greene) H. M. Hall, Carnegie Inst. Wash. Publ. No. 389: 136. 1928. (*Pyrrocoma sessiliflora* Greene, Leaflets Bot. Obs. **2**: 11. 1909.) Stems slender, nearly prostrate to ascending-erect or erect, 2–5 dm. long; herbage blue-glaucous; leaves thickish, mostly entire, ciliolate, the basal tufts often grass-like, with leaves 5–10 cm. long, or oblanceolate and even up to 20 cm. long and 28 mm. wide, the cauline abruptly reduced, at least the upper amplexicaul, more or less recurving, the internodes very short; heads sessile or nearly so, spicate, often glomerate in twos or threes; involucre 5–7 mm. high; phyllaries thickish, broad, pale, and with hyaline margin below the prominent, green, often squarrose tip, ciliate. Alkaline flats or meadows, Sonoran Zones; southern Mono County to south of Death Valley, Inyo County, California, east to central Nevada. Type locality: Twin Springs, Nevada. July–Oct.

Haplopappus racemosus subsp. **pinetòrum** Keck. Madroño **5**: 166. 1940. (*Haplopappus racemosus* var. *pinetorum* J. T. Howell, Leaflets West. Bot. **6**: 86. 1950; *H. racemosus* var. *praticola* J. T. Howell, loc. cit.) Stems erect or decumbent at base, 2–6 dm. high, sparingly arachnoid-villous; leaves entire or spinulose-serrate, the basal narrowly lanceolate, up to 20 cm. long and 16 mm. wide, the cauline much reduced, amplexicaul, all more or less densely arachnoid-villous; heads spicate-racemose or sometimes paniculate; involucre campanulate to hemispheric, 10–15 mm. high, multiseriate, rather densely villous but not glandular. Rocky forested slopes and meadows. Transition and Canadian Zones; Siskiyou and Trinity Counties, California. Type locality: Scott Mountains, Siskiyou County. July–Aug.

Haplopappus racemosus subsp. **liatrifórmis** (Greene) Keck, Aliso **4**: 103. 1958. (*Pyrrocoma liatriformis* Greene, Leaflets Bot. Obs. **2**: 17. 1909; *P. suksdorfii* Greene, op. cit. 18; *P. scaberula* Greene, op. cit. 19; *Haplopappus integrifolius* subsp. *liatriformis* and *H. integrifolius* subsp. *scaberulus* H. M. Hall, Carnegie Inst. Wash. Publ. No. 389: 111. 1928.) Stems erect, 3–7 dm. high, villous with jointed hairs; leaves entire, undulate or sparingly toothed, villous especially on midrib beneath, the basal narrowly oblanceolate, with blades 6–15 cm. long, 1–3 cm. wide; heads in a raceme or narrow panicle, sometimes solitary; involucre campanulate to subhemispheric, 10–15 mm. high; phyllaries firm, herbaceous throughout or green-tipped, sharply acute, villous or at least ciliate and beset with sessile glands. Grasslands, in dry, often rocky soils, Upper Sonoran Zone; Palouse region or south eastern Washington and adjacent Idaho. Type locality: Pullman, Washington. July–Aug.

12. Haplopappus lùcidus (Keck) Keck. Sticky Pyrrocoma. Fig. 5457.

Haplopappus racemosus subsp. *lucidus* Keck, Madroño **5**: 167. 1940.
Haplopappus lucidus Keck, Aliso **4**: 103. 1958.

Stems few, slender or stout, erect from a thick taproot, 2–10 dm. high, the herbage very sticky throughout from numerous sessile glands, not pubescent. Leaves deep green, usually entire, sometimes sharply serrate-dentate, often scabrid-ciliolate, the basal (long- or short-petiolate) with narrowly to broadly lanceolate blades 6–18 cm. long, 5–30 mm. wide, the cauline much reduced but prominent, amplexicaul; inflorescence spiciform-paniculate, the heads solitary or clustered; involucre campanulate, 10–15 mm. high, the phyllaries 2–3-seriate, obscurely imbricate, coriaceous-herbaceous, vernicose, linear-lanceolate, acuminate to attenuate, the tip at length spreading; achenes sericeous.

Alkaline flats and forest openings, Arid Transition Zone; Plumas, Sierra, and Yuba Counties, California. Type locality: west of Portola, Plumas County. July–Sept.

13. Haplopappus lyállii A. Gray. Lyall's Tonestus. Fig. 5458.

Haplopappus lyallii A. Gray, Proc. Acad. Phila. **1863**: 64. 1864.
Aster jamesii Kuntze, Rev. Gen. Pl. **1**: 316. 1891.
Stenotus lyallii Howell, Fl. N.W. Amer. 300. 1900.
Pyrrocoma lyallii Rydb. Mem. N.Y. Bot. Gard. **1**: 382. 1900.
Tonestus lyallii A. Nels. Bot. Gaz. **37**: 262. 1904.
Hoorebekia lyallii Piper, Contr. U.S. Nat. Herb. **11**: 560. 1906.

Perennial herb with few to several erect leafy stems 5–15 cm. high from a subterranean branched caudex or rhizomes, densely stipitate-glandular like the leaves and involucre. Basal leaves tufted, oblanceolate to obovate, obtuse, mucronate, 1–5 cm. long, 4–13 mm. wide, entire, the cauline gradually reduced above; heads solitary; involucre subhemispheric, 8–11 mm. high, often with small leafy bracts at base; phyllaries 2–3-seriate, scarcely graduate, linear to oblong, herbaceous, often purplish; ray-florets 13–35, short; achenes glabrous or sparsely pilose.

Rocky ridges or talus slopes of high mountains, Boreal Zone; Blue and Wallowa Mountains, Oregon, north through the Cascade Mountains of Washington to British Columbia and east to Alberta and Colorado. Type locality: east side of the Cascade Mountains, near the Canadian boundary. July–Sept.

14. Haplopappus whítneyi A. Gray. Whitney Haplopappus. Fig. 5459.

Haplopappus whitneyi A. Gray, Proc. Amer. Acad. **7**: 353. 1868.
Aster whitneyi Kuntze, Rev. Gen. Pl. **1**: 318. 1891.
Hazardia whitneyi Greene, Pittonia **3**: 43. 1896.

Perennial herb, the several simple ascending stems from a woody root-crown, 2–5 dm. high, moderately pilose with septate hairs and stipitate-glandular. Leaves broadly oblong or slightly spatulate, the lower narrowed at base, the upper subamplexicaul, 2.5–5 cm. long, 7–16 mm. wide, mucronate-serrate, firm, glandular-scabrid, sometimes also pilose; heads solitary, spicate, race-mose, or cymose-clustered, leafy-bracted; involucre campanulate, 11–13 mm. high; phyllaries 4–6-seriate, loosely graduate, linear-lanceolate, acuminate, chartaceous, granular-glanduliferous, the herbaceous portion progressively less inward, the hyaline, often roseate margin increasingly prominent; ray-florets 5–18, yellow; disk-florets 15–30; achenes glabrous, 8–14-ribbed; pappus copious, brownish.

Rocky, openly forested slopes, Canadian and Hudsonian Zones; Sierra Nevada from Plumas County to Tulare County, California. Type locality: Mono Trail and Sonora Pass, altitude 9,000 feet. July–Sept.

Haplopappus whitneyi subsp. discoìdeus (J. T. Howell) Keck, Aliso **4**: 103. 1958. (*Haplopappus whitneyi* var. *discoideus* J. T. Howell, Leaflets West. Bot. **6**: 84. 1950.) Heads discoid; otherwise very similar to subsp. *whitneyi*. Rocky wooded slopes, Canadian and Hudsonian Zones; northern Lake County, California, through the Inner North Coast Ranges to Mount Eddy and the Siskiyou Mountains, southwestern Oregon. Type locality: Shackleford Creek Trail south of Sky High Valley, Marble Mountains, California. July–Sept.

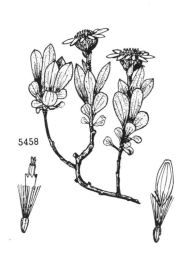

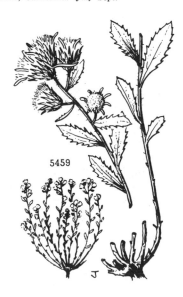

5458. Haplopappus lyallii   5459. Haplopappus whitneyi

15. **Haplopappus exímius** H. M. Hall. Tahoe Tonestus. Fig. 5460.

Haplopappus eximius H. M. Hall, Univ. Calif. Pub. Bot. **6**: 170. 1915.
Tonestus eximius Nels. & Macbr. Bot. Gaz. **65**: 70. 1918.

Perennial herb with few to several erect leafy stems 3–15 cm. high from a subterranean, branched, slender caudex or deep rhizomes, forming a loose mat up to 8 dm. across, the herbage glandular-puberulent. Leaves cuneate or spatulate, saliently dentate above middle, obtuse, mucronate, firm, 2–5 cm. long, 7–15 mm. wide; heads solitary; involucre hemispheric, 7.5–10 mm. high, 12–15 mm. wide (pressed); phyllaries shorter than the disk, scarcely graduate, unlike, the outer about 2-seriate, obovate, oblong, or oblanceolate, obtuse, herbaceous and glandular, the inner 2-seriate, subequal, lanceolate, attenuate, ciliolate, scarious, reddish above; ray-florets 15–20, the ligules 8 mm. long; achenes densely pubescent; pappus sordid.

Rocky summits near treeline, Hudsonian Zone; high Sierra Nevada from southern Washoe County, Nevada, to Eldorado County, California. Type locality: peak 1 kilometer south-southwest of Angora Peak, Eldorado County. July–Aug.

16. **Haplopappus peirsònii** (Keck) J. T. Howell. Inyo Tonestus. Fig. 5461.

Haplopappus eximius subsp. *peirsonii* Keck, Madroño **5**: 169. 1940.
Haplopappus peirsonii J. T. Howell, Leaflets West. Bot. **6**: 86. 1950.

Habit of *H. eximius* but more robust, the caudex branching but rarely producing elongate rhizomes, the stems to 20 cm. high. Leaves 3–8 cm. long, 10–25 mm. wide, less deeply toothed; heads larger, the involucre 14–18 mm. high, 20–30 mm. wide (pressed); phyllaries equaling the disk, the outer lanceolate or lance-oblong, mostly acute.

Rocky summits near treeline, Hudsonian and Arctic-Alpine Zones; principally east of the Sierran crest in Inyo County, California, and also in Fresno County. Type locality: Upper Rock Creek Lake Basin, Inyo County. July–Aug.

17. **Haplopappus lanuginòsus** A. Gray. Woolly Stenotus. Fig. 5462.

Haplopappus lanuginosus A. Gray in Torr. Bot. Wilkes Exp. 347. 1874.
Aster pickeringi Kuntze, Rev. Gen. Pl. **1**: 316. 1891.
Stenotus lanuginosus Greene, Erythea **2**: 72. 1894.
Hoorebekia lanuginosa Piper, Contr. U.S. Nat. Herb. **11**: 560. 1906.

Stems numerous from a closely branched, woody, rather fibrous-rooted caudex, 7–20 cm. high, monocephalous, subscapose, the herbage floccose-tomentose. Basal leaves densely tufted, narrowly oblanceolate to almost linear, 2–10 cm. long, 2–7 mm. wide; involucre hemispheric, 7–12 mm. high, the phyllaries 2–3-seriate, subequal or graduate, linear-lanceolate to lance-oblong, green with scarious margin, tomentulose; ray-florets 10–20, showy, the ligules 8–12 mm. long; achenes short-villous; pappus whitish, copious.

Exposed ridges and flats, in shallow, often rocky soils, Arid Transition Zone; central Washington to southern Oregon east of the Cascade Mountains east to Idaho. Type locality: upper part of North Fork of Columbia River, Washington. May–July.

18. **Haplopappus stenophýllus** A. Gray. Linear-leaf Stenotus. Fig. 5463.

Haplopappus stenophyllus A. Gray in Torr. Bot. Wilkes Exp. 347. 1874.
Aster stenophyllus Kuntze, Rev. Gen. Pl. **1**: 318. 1891.
Stenotus stenophyllus Greene, Erythea **2**: 72. 1894.
Hoorebekia stenophylla Piper, Contr. U.S. Nat. Herb. **11**: 561. 1906.

Densely cespitose from a much-dissected, rather slender, mat-like, woody caudex surmounting a taproot, the mats up to 45 cm. across, the numerous stems densely leafy in lower half, scapiform above and monocephalous, 3–8 cm. high, like the leaves densely hispidulous or hirtellous-scabrous and sometimes glandular. Leaves crowded, linear-spatulate to linear-filiform, 7–18 mm. long; involucre hemispheric, 5–9 mm. high, the linear to oblanceolate, acute or acuminate phyllaries 2-seriate, subequal, herbaceous, densely glandular-scabriusculous; ray-florets 8–12, the ligules 7–11 mm. long; achenes appressed-villous; pappus whitish, copious.

Rocky scablands and sagebrush slopes, Arid Transition Zone; central Washington east of the Cascade Mountains to northeastern California, occasional eastward to central Idaho and northern Nevada. Type locality: between Spipen (Naches) River and the North Fork of the Columbia, Washington. May–July.

19. **Haplopappus acaùlis** (Nutt.) A. Gray. Cushion Stenotus. Fig. 5464.

Chrysopsis acaulis Nutt. Journ. Acad. Phila. **7**: 33. *pl. 3, fig. 1.* 1834.
Chrysopsis caespitosa Nutt. loc. cit.
Stenotus acaulis Nutt. Trans. Amer. Phil. Soc. II. **7**: 334. 1840.
Haplopappus nevadensis Kell. Proc. Calif. Acad. **3**: 9. 1863.
Haplopappus acaulis A. Gray, Proc. Amer. Acad. **7**: 353. 1868.
Haplopappus acaulis var. *glabratus* D. C. Eaton, Bot. King Expl. 161. 1871.
Stenotus acaulis var. *kennedyi* Jepson, Man. Fl. Pl. Calif. 1028. 1925.
Haplopappus acaulis subsp. *glabratus* H. M. Hall, Carnegie Inst. Wash. Publ. No. 389: 166. 1928.

Stems scapiform, numerous, monocephalous, cespitose from a much-branched, woody caudex, 5–10 cm. high, densely clothed at base with the marcescent leaves, the whole mat may be up to several dm. across. Leaves linear-oblanceolate to spatulate, mostly erect, entire, obtuse to usually acuminate, cuspidate-tipped, veiny, pale green, densely hispidulous to glabrous except for the scabrid margin, 1–6 cm. long, 1.5–7 mm. wide; involucre hemispheric, 7–10 mm. high; phyllaries

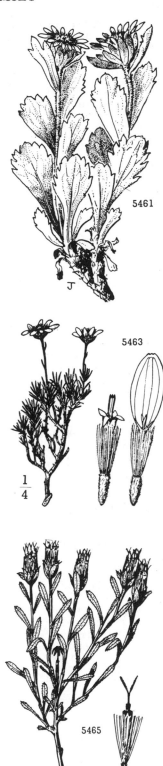

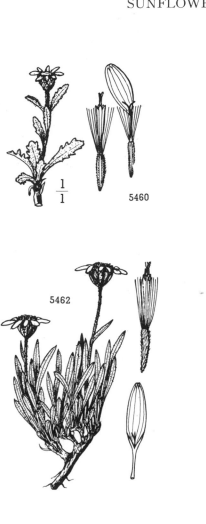

5460. Haplopappus eximius
5461. Haplopappus peirsonii
5462. Haplopappus lanuginosus

5463. Haplopappus stenophyllus
5464. Haplopappus acaulis
5465. Haplopappus macronema

2–3-seriate, broad, acute or acuminate, pallid; ray-florets 6–10, the ligules 6–10 mm. long; achenes densely sericeous to glabrous.

Dry ridges and plateaus, Arid Transition and Canadian Zones; central Oregon and eastern California south to Inyo County east to the Rocky Mountains and Saskatchewan. Type locality: "Little Goddin River," Idaho. May–Aug.

20. Haplopappus macronèma A. Gray. Discoid Macronema. Fig. 5465.

Macronema discoideum Nutt. Trans. Amer. Phil. Soc. II. **7**: 322. 1841.
Haplopappus macronema A. Gray, Proc. Amer. Acad. **6**: 542. 1865.
Aster macronema Kuntze, Rev. Gen. Pl. **1**: 318. 1891.
Bigelovia macronema M. E. Jones, Proc. Calif. Acad. II. **5**: 693. 1895.
Haplopappus discoideus Hall & Hall, Yosemite Fl. 246. 1912. Not DC. 1836.

Undershrub 1–4 dm. high with numerous erect stems of the season from a multibranched woody base, the twigs masked by a white tomentum. Leaves numerous, oblong or oblanceolate, sessile, entire or undulate-margined, 1–3 cm. long, 3–6 mm. wide, green, densely stipitate-glandular; heads discoid, yellow, usually solitary at branch-tips, rarely several and subracemose, turbinate or campanulate, 10–26-flowered; involucre 11–15 mm. high, glandular-puberulent, the phyllaries subequal, few-ranked, the outer broader and more herbaceous, the inner acuminate to attenuate, thin, dry; achenes appressed-pilose; pappus brownish.

Rocky, mostly open slopes at high elevations, Arctic-Alpine Zone; southeastern Oregon to Modoc County, California, and in the Sierra Nevada from Nevada County to Tulare County, California, east to Utah, Colorado, and Wyoming. Type locality: "Banks of Lewis' River, and other streams of the Oregon." July–Aug.

21. Haplopappus suffruticòsus (Nutt.) A. Gray. Big-head Macronema. Fig. 5466.

Macronema suffruticosum Nutt. Trans. Amer. Phil. Soc. II. **7**: 322. 1840.
Haplopappus suffruticosus A. Gray, Proc. Amer. Acad. **6**: 542. 1865.
Aster suffruticosus Kuntze, Rev. Gen. Pl. **1**: 318. 1891.
Macronema imbricatum Nels. & Macbr. Bot. Gaz. **62**: 150. 1916.
Haplopappus suffruticosus subsp. *tenuis* H. M. Hall, Carnegie Inst. Wash. Publ. No. 389: 190. 1928.
Haplopappus suffruticosus var. *tenuis* McMinn, Ill. Man. Calif. Shrubs 573. 1939.

Low compact subshrub up to 2 or even 4 dm. high with densely stipitate-glandular, fragrant herbage. Leaves very numerous on the brittle twigs, linear-oblanceolate to spatulate-oblong, 1–3 cm. long, 1.5–5 mm. wide, entire, usually crisped and with axillary fascicles; heads 1–4 at branch-tips, mostly solitary, broadly campanulate; involucre with several outer foliaceous oblong bracts often longer than the 2-seriate, chartaceous, lanceolate, acuminate, obscurely ciliolate, inner phyllaries, 10–14 mm. high, stipitate-glandular; ray-florets 3–6, showy; disk-florets 18–40; achenes villous, somewhat flattened; pappus stramineous.

Open rocky slopes and ridges, Hudsonian Zone; Wallowa Mountains, Oregon, east across northern Nevada and central Idaho to northwestern Wyoming and Montana and occurring in California in the high Sierra Nevada and the White Mountains. Type locality: banks of the Malade, a tributary of the Columbia, near the Blue Mountains. July–Sept.

22. Haplopappus greènei A. Gray. Greene's Macronema. Fig. 5467.

Haplopappus greenei A. Gray, Proc. Amer. Acad. **16**: 80. 1880.
Haplopappus mollis A. Gray, loc. cit.
Haplopappus greenei var. *mollis* A. Gray, Syn. Fl. N. Amer. **1²**: 135. 1884.
Aster greenei Kuntze, Rev. Gen. Pl. **1**: 318. 1891.
Macronema greenei Greene, Erythea **2**: 73. 1894.
Macronema molle Greene, loc. cit.
Hoorebekia greenei mollis Piper, Contr. U.S. Nat. Herb. **11**: 561. 1906.
Macronema pulvisculiferum Nels. & Macbr. Bot. Gaz. **62**: 150. 1916.
Macronema greenei var. *molle* Jepson, Man. Fl. Pl. Calif. 1030. 1925.
Haplopappus greenei subsp. *mollis* H. M. Hall, Carnegie Inst. Wash. Publ. No. 389: 195. 1928.

Undershrub 1–3 dm. high, glabrate to more or less densely tomentose, eglandular to more or less resinous from punctate or sessile (very rarely stalked) glands. Leaves very numerous, oblanceolate, 1.5–3.5 cm. long, 3–7 mm. wide, plane or rarely slightly crisped; heads clustered at ends of the twigs; involucre 8–12 mm. high, subtended by a few leafy bracts; phyllaries 3–4-seriate, subequal to somewhat imbricate, the outer with herbaceous ligulate tips three to four times as long as the body, the inner with proportionately shorter caudate tips or acuminate or even acute, the body prominently scarious-margined, villous-ciliate; ray-florets 1–7, showy; disk-florets 6–20.

Rocky flats and sparsely wooded slopes, Canadian and Hudsonian Zones; Cascade Mountains of Washington on the east slopes and on both slopes in Oregon, and mountains of northeastern Oregon to Idaho; in California in the Inner North Coast Ranges south to Mendocino County and also in Modoc County. Type locality: Scott Mountains, Siskiyou County, California. July–Sept.

23. Haplopappus bloómeri A. Gray. Bloomer's Macronema. Fig. 5468.

Haplopappus bloomeri A. Gray, Proc. Amer. Acad. **6**: 541. 1865.
Haplopappus bloomeri var. *angustatus* A. Gray, op. cit. **7**: 354. 1868.
Ericameria erecta Klatt, Abh. Natürf. Ges. Halle **15**: 326. 1882.
Haplopappus bloomeri var. *sonnei* Greene, Pittonia **2**: 17. 1889.
Aster bloomeri Kuntze, Rev. Gen. Pl. **1**: 317. 1891.

Chrysothamnus bloomeri Greene, Erythea **3**: 115. 1895.
Chrysothamnus bloomeri angustatus Greene, N. Amer. Fauna **16**: 166. 1899.
Macronema filiforme Nels. & Macbr. Bot. Gaz. **62**: 148. 1916.
Macronema glomeratum Nels. & Macbr. op. cit. 149.
Macronema scoparium Nels. & Macbr. loc. cit.
Macronema walpoleanum Nels. & Macbr. loc. cit.
Ericameria bloomeri J. F. Macbride, Contr. Gray Herb. No. 56: 36. 1918.
Haplopappus bloomeri subsp. *angustatus* H. M. Hall, Carnegie Inst. Wash. Publ. No. 389: 198. 1928.
Haplopappus bloomeri subsp. *sonnei* H. M. Hall, op. cit. 199.
Chrysothamnus bloomeri var. *pubescens* Henderson, Rhodora **32**: 27. 1930.

Low compact shrub broader than tall with woody trunk up to 3 cm. thick, 1.5–4(–8) dm. high, glabrate to sometimes more or less tomentose, eglandular to pruinose-glandular or occasionally glutinous from sessile glands. Leaves numerous, nearly filiform to narrowly oblanceolate, mostly 2–6 cm. long, 0.5–3 mm. wide, plane or rarely twisted; heads in small terminal clusters or commonly more numerous in subracemose spikes or panicles, narrowly campanulate; involucre 7–12 mm. high; phyllaries 3–6-seriate, clearly imbricate, linear-lanceolate to oblong, stramineous, the outer with caudate herbaceous tips, at least the inner prominently scarious-margined, villous-ciliate; ray-florets mostly 1–5, wanting from some heads, not very showy; disk-florets 4–13.

Sandy or rocky soils, openings in coniferous woods, Arid Transition and Canadian Zones; southern British Columbia east of the Cascade Divide south through central Washington and Oregon to the southern Sierra Nevada, California, and western Nevada. Type locality: Mount Davidson, Nevada. July–Oct.

24. **Haplopappus ophitìdis** (J. T. Howell) Keck. Serpentine Macronema. Fig. 5469.

Haplopappus bloomeri var. *ophitidis* J. T. Howell, Leaflets West. Bot. **6**: 85. 1950.
Haplopappus ophitidis Keck, Aliso **4**: 103. 1958.

Low, mat-like undershrub up to 20 cm. high, the stout woody trunk (up to 12 mm. thick) soon intricately and compactly branched, black-barked, the herbage resinous from sessile yellow glands.

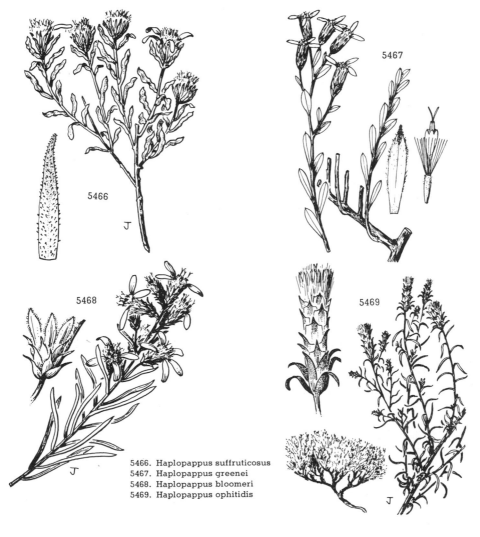

5466. Haplopappus suffruticosus
5467. Haplopappus greenei
5468. Haplopappus bloomeri
5469. Haplopappus ophitidis

Leaves numerous on the very slender stems, narrowly linear, apiculate, falcate-recurved, sulcate, very sparsely arachnoid-ciliate, 5–15 mm. long, 0.5–1 mm. wide; heads discoid, terminating leafy twigs, solitary or in small cymes, cylindric at anthesis; involucre 12–14 mm. high; phyllaries 5–8-seriate, strongly imbricate, the outermost linear-lanceolate, herbaceous, the inner broadly oblong with wide hyaline margin, truncate to the base of the lanceolate to deltoid, herbaceous, thickened, often squarrose tip, glutinous, somewhat ciliate; florets mostly 5(4–6), light yellow.

Serpentine soil, Arid Transition Zone; summit of Mount Tedoc, northwestern Tehama County, California, the type locality and only known station. July–Aug.

25. Haplopappus gilmánii Blake. Gilman's Macronema. Fig. 5470.

Haplopappus gilmanii Blake, Proc. Biol. Soc. Wash. **52**: 97. 1939.

Low, rounded, intricately branched, aromatic shrub 2–3.5 dm. high, the old stems pallid like the new, the very slender twigs rather rigid, sulcate and sticky. Leaves numerous, often with axillary fascicles, vernicose, spatulate, 6–12 mm. long, 2–3 mm. wide, often conduplicate and recurved at tip; heads solitary or cymose at branch-tips, narrowly campanulate; involucre 7–9 mm. high, resinous; phyllaries 4–6-seriate, clearly imbricate, the outer ovate-lanceolate, often with thickened, appendage-like, green, squarrose or reflexed tip, the inner oblong, chartaceous, with prominent, hyaline and ciliolate margin, appressed; ray-florets 4–6, white; disk-florets 15–18, white; achenes silky-sericeous; pappus ochroleucous.

High limestone ridges and walls, Arid Transition Zone; Panamint and Inyo Mountains, Inyo County, California. Type locality: summit of Telescope Peak, Panamint Range. Aug.–Sept.

26. Haplopappus resinòsus (Nutt.) A. Gray. Columbia Macronema. Fig. 5471.

Ericameria resinosa Nutt. Trans. Amer. Phil. Soc. II. **7**: 319. 1840.
Haplopappus resinosus A. Gray, Bot. Calif. **1**: 313. 1876.
Aster resinosus Kuntze, Rev. Gen. Pl. **1**: 317. 1891.
Chrysothamnus resinosus Howell, Fl. N.W. Amer. 303. 1900.
Haplopappus gummiferus Gandoger, Bull. Soc. Bot. Fr. **65**: 38. 1918.
Haplopappus hamatus Gandoger, loc. cit.

Intricately branched, rounded, aromatic shrub 2–3(–5) dm. high, the old wood very black, the new twigs very slender, resinous. Leaves numerous, often with axillary fasicles, glabrous, resinous, filiform to broadly linear, involute-thickened or conduplicate, recurving at tip, mostly less than 15 mm. long; heads solitary or loosely cymose, narrow; involucre 6–8 mm. high, glabrous; phyllaries 4–5-seriate, chartaceous throughout or with short green tip, the outer often squarrose; ray-florets 0–7, white or pale yellow (as also the disk); disk-florets 10–15, their narrow lobes 1–2 mm. long.

Rocky plains and banks, mostly on basalt, Upper Sonoran and Arid Transition Zones; central Washington from the eastern Cascade Mountains to northern and eastern Oregon and adjacent Idaho. Type locality: shelving rocks on the Blue Mountains of Oregon. July–Sept.

27. Haplopappus nànus (Nutt.) D. C. Eaton. Rubber Weed. Fig. 5472.

Ericameria nana Nutt. Trans. Amer. Phil. Soc. II. **7**: 319. 1840.
Haplopappus nanus D. C. Eaton, Bot. King Expl. 159. 1871.
Chrysoma nana Greene, Erythea **3**: 10. 1895.
Chrysothamnus nanus Howell, Fl. N.W. Amer. 302. 1900.

Habit and size of *H. resinosus*. Leaves very narrowly linear-spatulate to linear or involute and linear-filiform, to 20 mm. long, subulate-mucronate, obscurely if at all punctate but resinous, often with bud-like axillary fascicles; heads in small leafy cymes, yellow; involucre turbinate, 6–8.5 mm. high, glabrous; phyllaries 4–6-seriate, chartaceous and firm, only the outermost somewhat herbaceous, the tips obtuse to sharply acute or even aristate-caudate, not squarrose; ray-florets (0–)2–7; disk-florets 5–10, their lobes about 0.5 mm. long; pappus stramineous, fragile, scanty.

Dry rocky plains, cliffs, and crevices, Sonoran Zones; rare in Mono and Inyo Counties, California; common through the Great Basin to Utah and the Snake Plains of Idaho. Type locality: "Blue Mountains of Oregon," but more likely Idaho. July–Nov.

28. Haplopappus hállii A. Gray. Hesperodoria. Fig. 5473.

Haplopappus hallii A. Gray, Proc. Amer. Acad. **8**: 389. 1872.
Aster howellii Kuntze, Rev. Gen. Pl. **1**: 316. 1891.
Pyrrocoma hallii Howell, Fl. N.W. Amer. 299. 1900.
Hesperodoria hallii Greene, Leaflets Bot. Obs. **1**: 174. 1906.
Hoorebekia hallii Piper, Contr. U.S. Nat. Herb. **11**: 560. 1906.

Suffrutescent, 3–6 dm. high, the woody base branched, the virgate stems usually simple below the inflorescence, smooth or hirtellous. Leaves oblanceolate, sessile or the lowest petiolate, 2–5 cm. long, 3–8 mm. wide, sparsely scabrid, firm, entire, somewhat veiny; heads few in a close terminal cyme or raceme or cymose-panicled; involucre cylindroturbinate, 8–11 mm. high, glabrous but more or less glutinous; phyllaries 5–6-seriate, strongly graduate, lanceolate to linear-oblong, with short green tip; ray-florets 5–8; disk-florets 15–25; pappus stramineous.

Dry, grassy or openly wooded slopes, Transition Zone; overlooking the Columbia River Gorge, Washington and Oregon, south to Calapooia Mountains, Oregon. Type locality: bluffs at The Dalles, Oregon. Aug.–Oct.

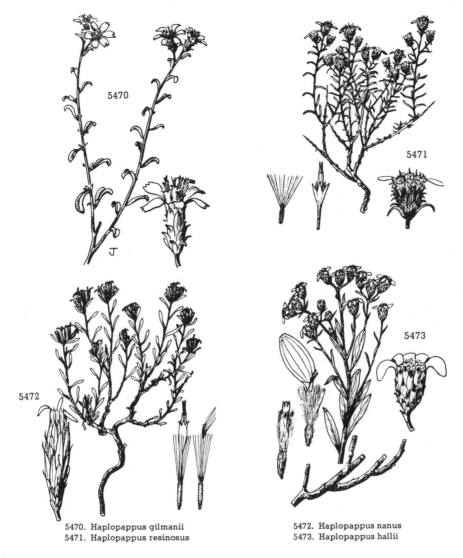

5470. Haplopappus gilmanii
5471. Haplopappus resinosus

5472. Haplopappus nanus
5473. Haplopappus hallii

29. **Haplopappus linearifòlius** DC. Stenotopsis. Fig. 5474.

Haplopappus linearifolius DC. Prod. **5**: 347. 1836.
Stenotus linearifolius Torr. & Gray, Fl. N. Amer. **2**: 238. 1842.
Aster linearifolius Kuntze, Rev. Gen. Pl. **1**: 318. 1891.
Haplopappus interior Coville, Proc. Biol. Soc. Wash. **7**: 65. 1892.
Stenotus interior Greene, Erythea **2**: 72. 1894.
Haplopappus linearifolius var. *interior* M. E. Jones, Proc. Calif. Acad. II. **5**: 697. 1895.
Stenotopsis linearifolia Rydb. Bull. Torrey Club **27**: 617. 1900.
Stenotopsis interior Rydb. loc. cit.
Stenotus linearifolius var. *interior* H. M. Hall, Univ. Calif. Pub. Bot **3**: 49. 1907.
Stenotopsis linearifolia var. *interior* J. F. Macbride, Contr. Gray Herb. No. 49: 59. 1917.
Haplopappus linearifolius subsp. *interior* H. M. Hall, Carnegie Inst. Wash. Publ. No. 389: 158. 1928.

Much-branched shrub 4–15 dm. high, essentially glabrous but usually puberulent at summit of peduncles, the twigs fastigiate, very leafy, resinous, glandular-punctate. Leaves nearly linear, narrowed toward base, 1–4 cm. long, 1–2.5 mm. wide, entire, flat or becoming subterete, sometimes fasciculate; heads numerous, solitary on nearly naked peduncles; involucre hemispheric, 8–14 mm. high, the phyllaries 2–3-seriate, scarcely graduate, lance-oblong to linear, acuminate, beset with granular glands, with greenish center and lacerate-ciliate, scarious margin; ray-florets 13–18, the ligules 8–15 mm. long; achenes silky-pilose; pappus white, deciduous.

Rocky or sandy soils of mountainsides and deserts. Arid Transition and Sonoran Zones; Marysville Buttes and Lake County, California, south through the Inner South Coast Range to interior cismontane southern California and Lower California east across the Mojave Desert to Inyo County, California, southern Utah, and western Arizona. Type locality: California. March–May.

The rather modest variation in head size and leaf length has led to persistent attempts to recognize a desert var. *interior*. The reduction in organ size in more arid areas, however, seems to be merely environmental modification rather than genetically governed variation.

30. Haplopappus venètus (H.B.K.) Blake subsp. vernonioìdes (Nutt.) H. M. Hall. Coastal Isocoma. Fig. 5475.

Pyrrocoma menziesii Hook. & Arn. Bot. Beechey 351. 1838.
Isocoma vernonioides Nutt. Trans. Amer. Phil. Soc. II. 7 : 320. 1840.
Haplopappus menziesii Torr. & Gray, Fl. N. Amer. 2: 242. 1842.
Bigelovia menziesii A. Gray, Proc. Amer. Acad. 8: 638. 1873.
Isocoma veneta var. *vernonioides* Jepson, Fl. W. Mid. Calif. 560. 1901.
Isocoma leucanthemifolia Greene, Leaflets Bot. Obs. 1: 171. 1906.
Isocoma microdonta Greene, loc. cit.
Isocoma villosa Greene, op. cit. 172.
Haplopappus venetus subsp. *vernonioides* H. M. Hall, Carnegie Inst. Wash. Publ. No. 389: 224. 1928.

Shrub 4–12 dm. high with erect, ascending or decumbent stems from a branched suffrutescent base, usually simple below the inflorescence, somewhat resinous, from nearly glabrous to pilose or tomentose, very leafy. Leaves linear to oblanceolate or spatulate-oblong, 1–4 cm. long, 2–8 mm. wide, spinulose-dentate throughout or only near apex, sometimes even lobed or essentially entire, usually with axillary fascicles; heads discoid, turbinate, 15–30-flowered, in rounded, terminal, usually compact cymes; involucre 5–7 mm. high, strongly graduate, the phyllaries oblong, obtuse or acute, firm, pale, with short, usually granulous-greenish, appressed tips; achenes silky; pappus brownish.

Coastal valleys, in sandy, often subsaline places, Upper Sonoran Zone; San Francisco to San Diego, California, and islands. Type locality: marshes near the sea, Santa Barbara. April–Dec.

Exceedingly variable in habit, pubescence, shape and cut of leaves, and involucre. Typical *H. venetus* grows in central Mexico; all other forms of the species occur in California and Lower California. The more noteworthy variants, all of which intergrade with subsp. *vernonioides,* are as follows:

Haplopappus venetus subsp. oxyphýllus (Greene) H. M. Hall, Carnegie Inst. Wash. Publ. No. 389: 225. 1928. (*Isocoma oxyphylla* Greene, Leaflets Bot. Obs. 1: 171. 1906; *Haplopappus venetus* var. *oxyphyllus* Munz, Man. S. Calif. 523. 1935.) Robust shrub 1–2 m. high, loosely villous to glabrate; leaves oblanceolate to narrowly spatulate, acute or acuminate, entire, 3–5 cm. long; cymes openly paniculate. Largely replacing the preceding in southern San Diego County and Lower California. Type locality: Jamul Valley back of San Diego.

Haplopappus venetus subsp. furfuràceus (Greene) H. M. Hall, Carnegie Inst. Wash. Publ. No. 389: 226. 1928. (*Bigelovia furfuracea* Greene, Bull. Calif. Acad. 1: 87. 1885; *Isocoma decumbens* Greene, Leaflets Bot. Obs. 1: 172. 1906; *I. veneta* var. *decumbens* Jepson, Man. Fl. Pl. Calif. 1029. 1925; *Haplopappus venetus* var. *decumbens* Munz, Man. S. Calif. 522. 1935; *H. venetus* var. *furfuraceus* Munz, op. cit. 523.) Stems slender, decumbent or curved, 3–5 dm. long; leaves crowded, narrow, small, few-toothed or entire, with prominent fascicles in the axils, mostly glabrous, sometimes woolly; heads loosely cymose, not in large glomerules. Dry sandy soils, southern San Diego County, California, Lower California, and the southern islands. Type locality: not known.

Haplopappus venetus var. sedoìdes (Greene) Munz, Man. S. Calif. 522. 1935. (*Bigelovia veneta* var. *sedoides* Greene, Bull. Calif. Acad. 2: 400. 1887; *Isocoma sedoides* Greene, Leaflets Bot. Obs. 1: 172. 1906; *I. latifolia* Greene, loc. cit.; *I. veneta* var. *sedoides* Jepson, Man. Fl. Pl. Calif. 1029. 1925.) Prostrate, almost glabrous, stout; leaves succulent, obovate, obtuse, toothed; heads large, glomerate. Santa Cruz and Santa Rosa Islands, California. Type locality: Santa Cruz Island.

Haplopappus venetus var. argùtus (Greene) Keck, Aliso 4: 103. 1958. (*Isocoma arguta* Greene, Man. Bay Reg. 175. 1894; *I. veneta* var. *arguta* Jepson, Fl. W. Mid. Calif. 500. 1901.) Low bush with erect stems 1.5–4 dm. high; leaves pinnately cleft into acute lobes or only saliently but pungently denate. North and south of Carquinez Straits, Solano and Contra Costa Counties, California. Type locality: subsaline plains east of the Vaca Mountains.

31. Haplopappus acradènius (Greene) Blake. Desert Isocoma. Fig. 5476.

Bigelovia acradenia Greene, Bull. Torrey Club 10: 126. 1883.
Aster acradenius Kuntze, Rev. Gen. Pl. 1: 317. 1891.
Isocoma acradenia Greene, Erythea 2: 111. 1894.
Isocoma veneta var. *acradenia* H. M. Hall, Univ. Calif. Pub. Bot. 3: 64. 1907.
Haplopappus acradenius Blake, Contr. U.S. Nat. Herb. 25: 546. 1925.

Similar to *H. venetus;* low shrub with numerous erect, more woody, brittle, white-barked, striate, glabrous stems 3–10 dm. high. Leaves linear-spatulate to oblong, 1–4 cm. long, 1–5 mm. wide, thick, entire, mostly mucronate and glabrous, minutely impressed-punctate, fewer in axillary fascicles; heads in smaller cymes, nearly sessile, 6–13-flowered; involucre 5–6.5 mm. high; phyllaries with conspicuous, thick, subepidermal resin-pocket near the rounded or blunt tip, the very narrow membranous margin fimbrillate.

Subalkaline or sandy flats, Lower Sonoran Zone; southwestern Mojave Desert (common in Antelope Valley) and southern approaches to Death Valley, California; rare in southern Nevada and western Arizona. Type locality: Mojave Desert. Aug.–Nov.

Haplopappus acradenius subsp. eremóphilus (Greene) H. M. Hall, Carnegie Inst. Wash. Publ. No. 389: 233. 1928. (*Isocoma eremophila* Greene, Leaflets Bot. Obs. 1: 171. 1906; *Haplopappus acradenius* var. *eremophilus* Munz, Man. S. Calif. 523. 1935.) Leaves denticulate to dentate or even saliently lobed, with some entire, 2–5 cm. long, to 7 mm. wide, often scabrid-hirtellous; heads 15–25-flowered; involucre 6–8 mm. high. Southernmost Mojave Desert and throughout the Colorado Desert, California, extending slightly into Arizona and Lower California. Type locality: southwestern part of the Colorado Desert.

Haplopappus acradenius subsp. bracteósus (Greene) H. M. Hall, Carnegie Inst. Wash. Publ. No. 389: 233. 1928. (*Isocoma bracteosa* Greene, Leaflets Bot. Obs. 1: 170. 1906; *Haplopappus acradenius* var. *bracteosus* McMinn, Ill. Man. Calif. Shrubs 574. 1939.) Leaves mostly denticulate, the lowermost 2–3 cm. long but the great majority only 6–10(–15) mm. long, horizontal or reflexed; heads 15–22-flowered; involucre 6–9 mm. high. Both sides of the San Joaquin Valley, California. Type locality: Tulare County.

32. Haplopappus cànus (A. Gray) Blake. Island Hazardia. Fig. 5477.

Diplostephium canum A. Gray, Proc. Amer. Acad. 11: 75. 1876.
Corethrogyne detonsa Greene, Bull. Torrey Club 10: 41. 1883.
Corethrogyne cana Greene, Bull. Calif. Acad. 1: 223. 1885.
Hazardia cana Greene, Pittonia 1: 29. 1887.
Hazardia detonsa Greene, loc. cit.

Hazardia serrata Greene, op. cit. 30.
Haplopappus canus Blake, Contr. U.S. Nat. Herb. **24**: 86. 1922.
Haplopappus traskae Eastw. Proc. Calif. Acad. IV. **20**: 156. 1931.

Openly branched shrub 6–25 dm. high, moderately to densely lanate-tomentose throughout, the foliage often glabrate above. Leaves obovate to oblanceolate, obtuse, petiolate or sessile, 4–12 cm. long, 1–5 cm. wide, entire to sharply serrate, thick; heads numerous, in large panicles or cymes, pedunculate or sessile; involucre broadly turbinate, 10–13 mm. high, the phyllaries multi-seriate, strongly graduate, appressed, oblong, acute, tomentose, or the inner tomentose only at tip or glabrous; ray-florets 6–14, inconspicuous, not exceeding disk, like the 20–54 disk-florets turning from yellow to purplish; achenes nerved, pilose; pappus tawny or brown.

Dry rocky slopes, Upper Sonoran Zone; Santa Rosa, Santa Cruz, and San Clemente Islands, California; Guadalupe Island, Lower California. Type locality: Guadalupe Island. June–Dec.

33. **Haplopappus squarròsus** Hook. & Arn. Common Hazardia. Fig. 5478.

Haplopappus squarrosus Hook. & Arn. Bot. Beechey 146. 1833.
Hazardia squarrosa Greene, Erythea **2**: 112. 1894.

Multistemmed shrub, woody at base, 3–10 dm. high, glabrate to densely hirtellous or, especially above, pilose. Leaves many, oblong to cuneate-obovate, obtuse, clasping at base, 1.5–4 cm. long, 1–2 cm. wide, sharply serrate throughout with mucronulate teeth, firm, resinous from punctate glands, glabrous or pubescent on midrib beneath; heads discoid, 15–30-flowered, spicate or race-mosely paniculate; involucre turbinate, 11–15 mm. high; phyllaries 8–10-seriate, strongly graduate,

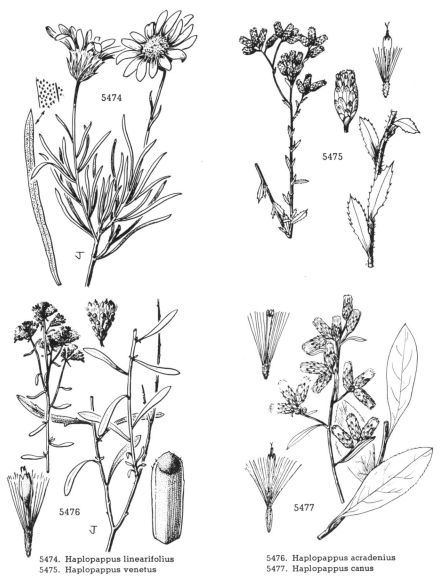

5474. Haplopappus linearifolius
5475. Haplopappus venetus

5476. Haplopappus acradenius
5477. Haplopappus canus

prominently glandular-scurfy at the green, obtuse to acute, usually squarrose tips; achenes glabrous or sparsely pilose; pappus yellow-tawny.

Coastal bluffs and montane canyons and ridges, often in chaparral, Upper Sonoran Zone; Monterey County to Santa Barbara County, California. Type locality: probably near Monterey. Aug.–Oct.

Haplopappus squarrosus subsp. grindelioìdes (DC.) Keck, Aliso 4: 103. 1958. (Pyrrocoma grindelioides DC. Prod. 5: 350. 1836; Aster grindelioides Kuntze, Rev. Gen. Pl. 1: 316. 1891.) More strongly hairy, the stems often tomentulose near the heads, even the upper surface of the leaves often somewhat hairy; heads smaller; involucre mostly 8–11 mm. high; phyllaries prominently cinereous on both faces of the green tip but glandular only marginally if at all; pappus red-brown. Cismontane southern California from Santa Barbara to San Diego and in Lower California. Type locality: "California." July–Oct.

Haplopappus squarrosus subsp. obtùsus (Greene) H. M. Hall, Carnegie Inst. Wash. Publ. No. 389; 253. 1928. (Hazardia obtusa Greene, Fl. Fran. 375. 1897; H. squarrosa var. obtusa Jepson, Man. Fl. Pl. Calif. 1030. 1925; Haplopappus squarrosus var. obtusus McMinn, Ill. Man. Calif. Shrubs 571. 1939.) Heads relatively large, 18–25-flowered; involucre broadly turbinate, 13–15 mm. high; phyllaries relatively broad, very blunt, mucronate, the resinous-granular, pallid, glabrous tips appressed. Canyons in the mountains west of Tejon Pass, Kern County, south through Ventura County to Nordhoff Peak, California. Type locality: San Emigdio Canyon. Sept.–Nov.

Haplopappus squarrosus subsp. stenolèpis H. M. Hall, Carnegie Inst. Wash. Publ. No. 389: 253. 1928. (Haplopappus squarrosus var. stenolepis McMinn, Ill. Man. Calif. Shrubs 571. 1939.) Usually woodier at base, very branched, forming dense shrubs to 1 m. high and 2 m. across, the twigs scabrid or hirsutulous; leaves relatively small; heads 4–8-flowered, densely spicate; involucre very narrowly turbinate, 13–17 mm. high; phyllaries 6–7-seriate, regularly but loosely imbricate, linear-acuminate, at length spreading but not squarrose, sparingly dotted with granular glands on the small herbaceous portion near tip, somewhat viscid but glabrous; pappus red-brown. Serpentine areas in Inner South Coast Range, Fresno and Monterey Counties, California to Cuyama River, Santa Barbara County. Type locality: Parkfield Grade. Sept.–Nov.

34. Haplopappus pinifòlius A. Gray. Pine-bush. Fig. 5479.

Haplopappus pinifolius A. Gray, Proc. Amer. Acad. 8: 636. 1873.
Aster pityphyllus Kuntze, Rev. Gen. Pl. 1: 316. 1891.
Chrysoma pinifolia Greene, Erythea 3: 12. 1895.
Ericameria pinifolia H. M. Hall, Univ. Calif. Pub. Bot. 3: 54. 1907.
Haplopappus illinitus Eastw. Proc. Calif. Acad. IV. 20: 155. 1931.

Stout shrub 6–25 dm. high, the main stems trunk-like, fastigiately branched, resinous, glandular-punctate, sometimes slightly pilose. Leaves linear-filiform, 1–3.5 cm. long, mucronate, subterete, with shorter leaves fascicled in the axils; heads solitary in the spring form, large, terminating leafy twigs, often surpassed by subtending leaves, in the autumnal form smaller, in short racemes or panicles or cymose-clustered at tips of branches; involucre of the latter turbinate, 6–8 mm. high, overlapped by leafy bracts; phyllaries loosely 3–5-seriate, lanceolate-acuminate to oblong, the outer often more or less caudate-tipped, the inner shorter-tipped or merely acute, the tips green, otherwise pale with scarious ciliate margin, the costa sometimes glandular-thickened above; ray-florets 5–10 (15–30 in vernal heads); disk-florets 12–18; achenes sparsely pilose; pappus buff or reddish.

Cismontane dry slopes and washes, Upper Sonoran Zone; cismontane southern California from northern Los Angeles County to southern San Diego County. Type locality: near Los Angeles. April–July; Sept.–Jan.

35. Haplopappus ericoìdes (Less.) Hook. & Arn. Mock Heather. Fig. 5480.

Diplopappus ericoides Less. Linnaea 6: 117. 1831.
Haplopappus ericoides Hook. & Arn. Bot. Beechey 146. 1833.
Ericameria microphylla Nutt. Trans. Amer. Phil. Soc. II. 7: 319. 1840.
Aster ericina Kuntze, Rev. Gen. Pl. 1: 313. 1891.
Chrysoma ericoides Greene, Erythea 3: 11. 1895.
Ericameria ericoides Jepson, Fl. W. Mid. Calif. 559. 1901.

Broad compact shrub with dense crown, 3–8(–15) dm. high, the twigs crowded, densely leafy, together with the leaves sparsely pilosulose and somewhat resinous. Leaves nearly filiform, divaricate, 4–12 mm. long, subterete, with dense fascicles of scarcely shorter leaves in the axils; heads cymose-paniculate, terminating leafy branches; involucre turbinate, 5–6 mm. high; phyllaries loosely 3–5-seriate, villous-ciliate, the outer ovate-lanceolate with short-caudate, greenish tip, the inner broadly oblong, acute, not herbaceous, the costa thickened above into a filiform gland; ray-florets 2–6; disk-florets 8–14; achenes glabrous.

Sand dunes on and near the coast, Upper Sonoran Zone; Point Reyes, Marin County, to Los Angeles County, California; reported from San Miguel Island. Type locality: California. Aug.–Nov.

Haplopappus ericoides subsp. blàkei C. B. Wolf, Occ. Papers Rancho Santa Ana Bot. Gard. 1: 87. 1938. Achenes moderately sericeous. Sand hills away from the immediate coast, up to 1,500 feet altitude, Upper Sonoran and Transition Zones; Santa Cruz Mountains, Santa Cruz and Monterey Counties, Santa Maria River Valley, San Luis Obispo and Santa Barbara Counties, and near Santa Barbara, California. Type locality: near Bonnie Doon School, Sant Cruz County. Sept.–Nov. The leaves are usually but not always longer in the southern material of this.

36. Haplopappus eastwoódiae H. M. Hall. Eastwood's Ericameria. Fig. 5481.

Chrysoma fasciculata Eastw. Bull. Torrey Club 32: 215. 1905.
Ericameria fasciculata J. F. Macbride, Contr. Gray Herb. No. 56: 36. 1918.
Haplopappus eastwoodae H. M. Hall, Carnegie Inst. Wash. Publ. No. 389: 258. 1928.

Stout, dense, fastigiately branched shrub 5–10 dm. high, glabrous, very resinous, glandular-punctate, densely leafy. Leaves linear, 8–20 mm. long, terete or somewhat flattened, mucronate, nearly all with axillary fascicles of smaller leaves; heads solitary or usually several in terminal cymes of close racemes; involucre campanulate, 7–8 mm. high; phyllaries about 5-seriate, strongly graduate, scarious and pale yellow, the outer ovate-lanceolate, sharply acute, the inner oblong, obtusish, none appendaged, villous-ciliate on the narrow hyaline margin, the costa somewhat glandular-thickened above; ray-florets 1–6; disk-florets 18–22; achenes densely silky-pilose.

Coastal sand dunes, Upper Sonoran Zone; Monterey and Carmel Bays, California. Type locality: near Monterey. July–Oct.

37. Haplopappus pálmeri A. Gray. Palmer's Ericameria. Fig. 5482.

Haplopappus pàlmeri A. Gray, Proc. Amer. Acad. 11: 74. 1876.
Aster nevinii Kuntze, Rev. Gen. Pl. 1: 316. 1891.
Chrysoma palmeri Greene, Erythea 3: 12. 1895.
Ericameria palmeri H. M. Hall, Univ. Calif. Pub. Bot. 3: 53. 1907.

Stout shrub 1–4 m. high with numerous ascending branches, obscurely puberulous, resinous, glandular-punctate, very leafy. Leaves filiform, 1.5–4 cm. long, subterete, fasciculate; heads many, in thyrsoid panicles; involucre turbinate, 5–6.5 mm. high; phyllaries 30–40, loosely 4–5-seriate, linear or nearly so, blunt, glabrous or the outer glandular-atomiferous, ciliolate at tip, the hyaline margin narrow, the costa thickened for most of its length into a linear-oblong gland; ray-florets 4–10; disk-florets 8–20; achenes moderately sericeous.

Dry mesas, Sonoran Zones; southwestern San Diego County, California, to northwestern Lower California. Type locality: Tecate Mountain, Lower California. Sept.–Nov.

Haplopappus palmeri subsp. **pachylepis** H. M. Hall, Carnegie Inst. Wash. Publ. No. 389: 267. 1928. (*Haplopappus palmeri* var. *pachylepis* Munz, Man. S. Calif. 522. 1935.) Smaller, 5–15 dm. high; leaves mostly 8–12 mm. long; involucre cylindroturbinate, 6–7 mm. high; phyllaries 16–25, somewhat thicker, 4-seriate, broadly lanceolate to oblong, the bullate costal gland in apical half only; ray-florets 1–4; disk-florets 5–10; achenes densely sericeous. Rather common on clayey plains, Upper Sonoran Zone; cismontane California from southeastern Ventura County to the desert borders, Riverside County; Santa Catalina Island. Type locality: summit of Box Springs Grade near Riverside. Aug.–Dec.

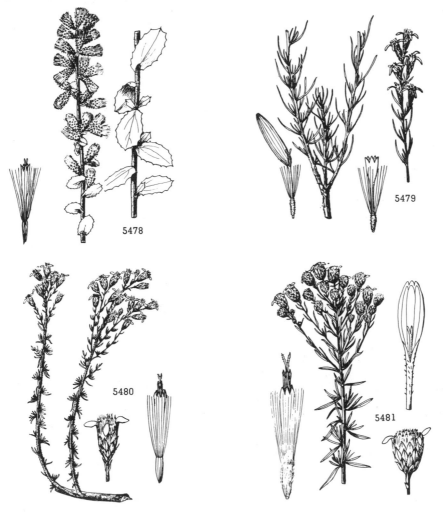

5478
5479
5480
5481

5478. Haplopappus squarrosus
5479. Haplopappus pinifolius

5480. Haplopappus ericoides
5481. Haplopappus eastwoodiae

38. Haplopappus propínquus Blake. Boundary Ericameria. Fig. 5483.

Bigelovia brachylepis A. Gray, Bot. Calif. 1: 614. 1876.
Aster brachylepis Kuntze, Rev. Gen. Pl. 1: 317. 1891.
Chrysoma brachylepis Greene, Erythea 3: 12. 1895.
Ericameria brachylepis H. M. Hall, Univ. Calif. Pub. Bot. 3: 56. 1907.
Haplopappus brachylepis H. M. Hall, op. cit. 7: 273. 1919. Not Phil. 1894.
Haplopappus propinquus Blake, Contr. U.S. Nat. Herb. 23: 1490. 1926.

Shrub 1–2 m. high, rigidly branched, glabrous, more or less resinous, glandular-punctate. Leaves crowded, linear-filiform, 1–2 cm. long, 0.5–1 mm. wide, flattish or subterete, often mucronate, sharply ascending, with axillary fascicles; heads discoid, 9–14-flowered, yellow, terminal or racemose; involucre turbinate, 4.5–5.5 mm. high; phyllaries 3–4-seriate, strongly graduate, ovate to linear-oblong, the outer grading into the scaly bracts of the peduncles, the costa prominently glandular-thickened throughout its length; achenes densely villous.

Desert slopes, Upper Sonoran Zone; southern San Diego County, California, and northern Lower California. Type locality: "Larkens' Station, 80 miles east by north of San Diego." Sept.–Dec.

39. Haplopappus laricifòlius A. Gray. Turpentine-bush. Fig. 5484.

Haplopappus laricifolius A. Gray, Smiths. Contr. 5⁶: 80. 1853.
Aster laricifolius Kuntze, Rev. Gen. Pl. 1: 318. 1891.
Chrysoma laricifolia Greene, Erythea 3: 11. 1895.
Ericameria laricifolia Shinners, Field & Lab. 18: 27. 1950.

Compact, fastigiately branched, broadly rounded shrub 3–10 dm. high, resinous, prominently impressed-punctate, glabrous. Leaves linear, 1–2 cm. long, 1–2 mm. wide, usually subterete, mucronate, sometimes with much smaller ones in axillary fascicles; heads in small leafy cymes; involucre broadly turbinate, 3–5 mm. high; phyllaries loosely 3–4-seriate, lance-acuminate, soft and ciliolate at the tip, rather firm, the costa thickened for most of its length into an olive-brown gland; ray-florets 3–11; disk-florets 10–16, much exceeding the involucre; achenes densely pilose.

Rocky desert slopes and mesas, Sonoran Zones; California, eastern Mojave Desert from Clark Mountain to Hackberry Mountain; east to western Texas and adjacent Mexico. Type locality: Guadalupe Pass, New Mexico. Sept.–Oct.

40. Haplopappus coóperi (A. Gray) H. M. Hall. Goldenbush. Fig. 5485.

Bigelovia cooperi A. Gray, Proc. Amer. Acad. 8: 640. 1873.
Haplopappus monactis A. Gray, op. cit. 19: 1. 1883.
Aster cooperi Kuntze, Rev. Gen. Pl. 1: 317. 1891.
Aster monactis Kuntze, op. cit. 318.
Ericameria monactis McClatchie, Erythea 2: 124. 1894.
Chrysoma cooperi Greene, op. cit. 3: 12. 1895.
Acamptopappus microcephalus M. E. Jones, Contr. West. Bot. No. 7: 30. 1898.
Chrysothamnus corymbosus Elmer, Bot. Gaz. 39: 50. 1905.
Tumionella monactis Greene, Leaflets Bot. Obs. 1: 173. 1906.
Ericameria cooperi H. M. Hall, Univ. Calif. Pub. Bot. 3: 56. 1907.
Haplopappus cooperi H. M. Hall, Carnegie Inst. Wash. Publ. No. 389: 275. 1928.

Low, flat-topped, fastigiately branched shrub 2.5–6(–15) dm. high, very woody, the bark becoming shreddy, the herbage glutinous, loosely pilosulose. Leaves linear-spatulate, 6–15 mm. long, up to 1.5 mm. wide, with glandular-punctate, diminutive fascicles in the lower axils that persist after the fall of the principal leaves; heads prominently peduncled in small, rounded, terminal cymes; involucre narrowly campanulate, 4–5 mm. high; phyllaries 2–3-seriate, few (9–15), the outer ovate, acute, the inner broadly oblong, obtuse, more or less puberulent, the glandular thickening of the costa obscure; ray-florets 0–2; disk-florets 4–7(–11), much exceeding the involucre; achenes silky-pilose.

Common in rocky desert basins and mesas, Lower Sonoran Zone; California across the Mojave Desert from southern Mono County to Lancaster Valley and Little San Bernardino Mountains; rare in interior cismontane southern California; adjacent Nevada. Type locality: eastern slope of Providence Mountains, southeastern California. March–June.

41. Haplopappus arboréscens (A. Gray) H. M. Hall. Golden Fleece. Fig. 5486.

Linosyris arborescens A. Gray in Torr. Bot. Mex. Bound. 79. 1859.
Bigelovia arborescens A. Gray, Proc. Amer. Acad. 8: 640. 1873.
Aster chrysothamnus Kuntze, Rev. Gen. Pl. 1: 315. 1891.
Ericameria arborescens Greene, Man. Bay Reg. 175. 1894.
Chrysoma arborescens Greene, Erythea 3: 10. 1895.
Haplopappus arborescens H. M. Hall, Univ. Calif. Pub. Bot. 7: 273. 1919.

Stout erect shrub 0.6–3 m. high, fastigiately branched, glabrous, resinous, prominently glandular-punctate. Leaves narrowly linear to filiform, 3–6 cm. long, up to 2 mm. wide, thick, crowded but rarely with axillary fascicles; heads discoid, 18–23-flowered, many, in rounded terminal cymes or cymose panicles; peduncles (3–8 mm. long) bearing scales like the phyllaries; involucre turbinate, 4–5 mm. high; phyllaries 4-seriate, graduate, lanceolate to linear, acuminate to acute, thin and chaffy except for the glandular-thickened costa; achenes turgid, obscurely 5-angled, less than 2 mm. long, softly sericeous; pappus very fragile.

Dry foothill slopes, in chaparral, mostly below 5,000 feet but up to 9,300 feet in Fresno County, Upper Sonoran and Transition Zones; west side of the Sierra Nevada from Nevada County to Tulare County; Outer Coast Ranges from Oregon line to Ventura County, California. Type locality: "California." Aug.–Nov.

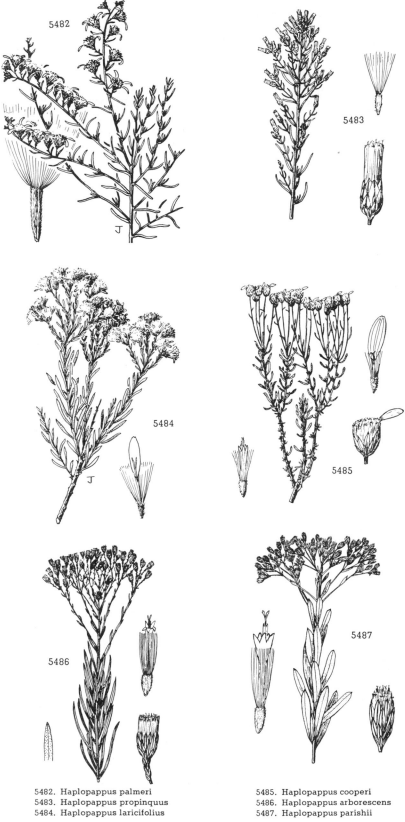

5482. Haplopappus palmeri
5483. Haplopappus propinquus
5484. Haplopappus laricifolius

5485. Haplopappus cooperi
5486. Haplopappus arborescens
5487. Haplopappus parishii

42. **Haplopappus paríshii** (Greene) Blake. Parish's Ericameria. Fig. 5487.

Bigelovia parishii Greene, Bull. Torrey Club **9**: 62. 1882.
Aster parishii Kuntze, Rev. Gen. Pl. **1**: 318. 1891.
Chrysoma parishii Greene, Erythea **3**: 10. 1895.
Ericameria parishii H. M. Hall, Univ. Calif. Pub. Bot. **3**: 55. 1907.
Haplopappus parishii Blake, Contr. U.S. Nat. Herb. **23**: 1491. 1926.

Erect shrub 2–5 m. high, glabrous, resinous, densely glandular-punctate, the stems trunk-like at base, very leafy above. Leaves linear-oblanceolate to lance-elliptic, tapering to base, acute or obtuse, 2–6 cm. long, 3–10 mm. wide, flat, coriaceous; heads discoid, 9–12-flowered, in compact rounded cymes, the short peduncles bracteate with small scales; involucre turbinate, about 5 mm. high; phyllaries 4-seriate, lanceolate to lance-oblong, acutish to acuminate, whitish, firm, carinate by the glandular-thickened costa; achenes appressed-pilosulose; pappus copious but fragile.

Outwash fans and dry hillsides in chaparral, not common, Upper Sonoran Zone; south slopes of San Gabriel and San Bernardino Mountains, California, south to Lower California. Type locality: Waterman Canyon, San Bernardino Mountains. July–Oct.

43. **Haplopappus cuneàtus** A. Gray. Cuneate-leaved Ericameria. Fig. 5488.

Haplopappus cuneatus A. Gray, Proc. Amer. Acad. **8**: 635. 1873.
Bigelovia spathulata A. Gray, op. cit. **11**: 74. 1876.
Aster cuneatus Kuntze, Rev. Gen. Pl. **1**: 317. 1891.
Ericameria cuneata McClatchie, Erythea **2**: 124. 1894.
Chrysoma cuneata Greene, op. cit. **3**: 11. 1895.
Chrysoma cuneata spathulata Greene, loc. cit.
Chrysoma merriami Eastw. Bull. Torrey Club **32**: 215. 1905.
Ericameria cuneata (var.) *spathulata* H. M. Hall, Univ. Calif. Pub. Bot. **3**: 52. 1907.
Haplopappus cuneatus spathulatus Blake, Contr. U.S. Nat. Herb. **23**: 1489. 1926.

Low spreading shrub 1–5(–12) dm. high, much branched, glabrous, balsamic-resinous, glandular-punctate. Leaves crowded, deep green, cuneate to suborbicular-obovate, entire, often undulate, apiculate at the obtuse or broadly rounded or often retuse apex, 5–20 mm. long, 3–10 mm. wide, coriaceous; heads compactly cymose; involucre turbinate, 5–7 mm. high; phyllaries 4–6-seriate, regularly imbricate, linear-oblong to lance-ovate, the costa glandular-thickened above, the outer passing into minute, ovate, thick scales of the peduncle; ray-florets 1–5 (in northern Sierra Nevada) or usually wanting; disk-florets 16–28; achenes densely appressed-pilose.

Cliffs and rocky slopes, Upper Sonoran and Arid Transition Zones; both slopes of the Sierra Nevada from Plumas County to Tulare County and ranges bordering the Mojave and Colorado Deserts from southern Mono County to Ventura and San Diego Counties, and eastern San Luis Obispo County to the San Gabriel Mountains, California; to Nevada, Arizona, and Lower California. Type locality: Bear Valley (Sierra Nevada), California. Sept.–Oct.

76. **BENITÒA*** Keck, Leaflets West. Bot. **8**: 26. 1956.

Annual herb from a taproot. Stem erect, cymose-paniculately branching above, the heads terminating the branchlets. Herbage heavy-scented from a harsh glandular pubescence. Leaves essentially entire, prominently reticulate-veiny. Flowers yellow, tinged with red. Involucre cylindroturbinate, 5–6-seriate, the phyllaries 35–50, corneous, linear-attenuate, glandular-atomiferous, the spreading or recurved tips of the outer ones bearing a prominent gland. Ray-florets fertile, 5–14, the ligule drying tightly circinate, the achene 3-angled; disk-florets sterile, 9–25, the corolla constricted at base of throat, its style-branches included, appressed, scarcely differentiated into appendages and stigmatic portion; pappus on all florets similar, of 2–8 brownish, very slender, readily deciduous bristles about equaling the achenes. [Named for San Benito County, California, where it occurs.]

A monotypic genus of California.

1. **Benitoa occidentàlis** (H. M. Hall) Keck. Benitoa. Fig. 5489.

Haplopappus occidentalis H. M. Hall, Carnegie Inst. Wash. Publ. No. 389: 214. 1928.
Benitoa occidentalis Keck, Leaflets West. Bot. **8**: 26. 1956.

Plant 3–10 dm. high, viscid throughout with harsh, septate, gland-tipped hairs interspersed with a denser glandular pubescence, the leaves often additionally somewhat hispid-hirsute, often more or less anthocyanous when mature. Leaves linear-lanceolate, obtuse or acute, narrowed to a clasping base, 5–9 cm. long, 6–15 mm. wide, becoming bract-like and apiculate in upper half of plant; involucre 8–10 mm. high, 3–5 mm. wide; ray-achenes olive, brown-maculate, finely sericeous, 3.5 mm. long.

Hot, dry, exposed, serpentine hillsides, Upper Sonoran Zone; Diablo Range in San Benito, Monterey, and Fresno Counties, California. Type locality: east side of Parkfield Grade, Fresno County. June–Nov.

* Text contributed by David Daniels Keck.

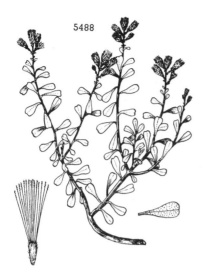

5488

5489

5488. Haplopappus cuneatus

5489. Benitoa occidentalis

77. **SOLIDÀGO*** L. Sp. Pl. 878. 1753.

Perennial herbs with leafy, usually simple stems arising from rhizomes or a caudex. Leaves alternate, entire or toothed. Heads radiate, yellow (in ours), small, campanulate to subcylindric, panicled, racemose, or cymose. Involucre few-seriate, graduate or subequal, the phyllaries usually with obscurely herbaceous tips. Receptacle usually alveolate. Ray-florets small, fertile. Disk-florets perfect, fertile, their anthers subentire at base, their style-branches with mostly lanceolate appendages. Achenes short, pubescent (in ours), usually few-nerved. Pappus setose, copious, whitish. [Name Latin, meaning to make whole, from its reputed medicinal value.]

A genus of about 100 species, chiefly North American, a few also in South America and Eurasia; the genus is best developed in the eastern United States. Type species, *Solidago virgaurea* L.

Plants with well-developed, creeping, herbaceous rhizomes; stems rather equably leafy, the lowest leaves not prominently different from the upper cauline and at length deciduous (except in *S. missouriensis*).

    Leaves glandular-punctate, lance-linear; inflorescence ample, copiously bracteate, the heads in terminal cymose clusters; ray-florets 15–25.   1. *S. occidentalis.*

    Leaves not punctate, broader; inflorescence more compact, not interrupted, the heads not glomerate in cymes; ray-florets 8–13.

        Stems puberulent at least above the middle.

            Leaves densely puberulent on both faces, the middle and upper usually elliptic and entire.   2. *S. californica.*

            Leaves puberulous chiefly on nerves beneath or chiefly glabrous, the middle and upper lanceolate, usually sharply toothed.

                Involucre evidently imbricate; inflorescence a compact thyrse to a broad panicle, 5–20 cm. long.   3. *S. canadensis.*

                Involucre obscurely imbricate; inflorescence smaller, short and compact; rare with us.   4. *S. lepida.*

        Stems glabrous below the inflorescence.

            Stems densely and nearly uniformly leafy throughout, glaucous, 5–20 dm. high; upper leaves lanceolate, sharply serrate.   5. *S. gigantea.*

            Stems with upper leaves reduced, not glaucous, 2–5 dm. high; upper leaves mostly linear or subulate, entire.   6. *S. missouriensis.*

Plants with rather short woody rhizomes or a branched caudex; stems with lower and basal petiolate leaves much larger than the upper reduced sessile ones; inflorescence without recurving branches.

    Plants of low or mid-elevations, usually more than 4 dm. high.

        Inflorescence not glutinous; involucre 3.5–5 mm. high; leaves entire (rarely remotely serrulate in *S. spectabilis*).

            Panicle subracemose, few-headed; stems slender; central California.   7. *S. guiradonis.*

             Panicle usually oblong and very dense; stems rather stout.

                  Phyllaries acute or acuminate; rays mostly 8; southern California.   8. *S. confinis.*

                  Phyllaries obtusish; rays mostly 13; Great Basin.   9. *S. spectabilis.*

        Inflorescence glutinous; involucre 5–6 mm. high; leaves crenate-serrate; coastal.   10. *S. spathulata.*

    Plants of high altitudes, mostly less than 3 dm. high.

        Leaves glabrous; phyllaries obtuse.   11. *S. decumbens.*

        Leaves ciliate at least toward petiole; phyllaries acute to acuminate.   12. *S. multiradiata.*

* Text contributed by David Daniels Keck.

1. Solidago occidentàlis (Nutt.) Torr. & Gray. Western Goldenrod. Fig. 5490.

Euthamia occidentalis Nutt. Trans. Amer. Phil. Soc. II. **7**: 326. 1840.
Solidago occidentalis Torr. & Gray, Fl. N. Amer. **2**: 226. 1842.
Euthamia linearifolia Gandoger, Bull. Soc. Bot. Fr. **65**: 41. 1918.
Euthamia californica Gandoger, loc. cit.

Stems stout, from creeping rhizomes, much branched, 6–20 dm. high, glabrous throughout, often glutinous above. Leaves lance-linear, sessile, entire, 3–5-nerved, 4–10 cm. long, 3–9 mm. wide, the margin often scabrid, glandular-punctate; inflorescence ample, leafy-bracteate, interrupted-elongate or rounded, the heads in small cymose clusters; involucre 4 mm. high, the phyllaries firm, lance-oblong to lance-linear, acute; ray-florets 15–25, 1.5–2.5 mm. long; disk-florets 7–14; achenes pilose.

Moist ground at low altitudes, Upper Sonoran and Transition Zones; British Columbia to southern California east to Alberta, Nebraska, and Texas. Type locality: "Banks of the Oregon and Wahlamet, and Lewis' River, in the Rocky Mountains." July–Nov.

Solidago graminifòlia var. **màjor** (Michx.) Fernald, Rhodora **46**: 330. 1944. (*Solidago lanceolata* var. *major* Michx. Fl. Bor. Amer. **2**: 116. 1803.) Leaves lance-linear, seven to eleven times as long as wide, entire, the margins scabrous-hirtellous; inflorescence compact, flat-topped, the heads mostly sessile in clusters of 2–5 at tips of the branchlets; involucre 4–5 mm. high, the phyllaries mostly broader, blunter, and often more obviously green-tipped than in *S. occidentalis*. Rare in our region (Aberdeen, Washington); British Columbia to Newfoundland, Virginia, and New Mexico. Type locality: Lake St. John, Quebec. Typical *S. graminifolia* occurs in the northeastern United States. July–Oct.

2. Solidago califòrnica Nutt. California Goldenrod. Fig. 5491.

Solidago californica Nutt. Trans. Amer. Phil. Soc. II. **7**: 328. 1840.
Solidago californica var. *nevadensis* A. Gray, Bot. Calif. **1**: 319. 1876.
Aster californicus Kuntze, Rev. Gen. Pl. **1**: 314. 1891.
Solidago californica var. *aperta* Henderson, Rhodora **32**: 28. 1930.

Stems from creeping rhizomes, 2–12 dm. high, like the leaves densely cinerous-puberulent. Basal and lower cauline leaves spatulate to obovate or oval, obtuse to acute, attenuate to base, firm, crenate or serrate, 5–12 cm. long, 1–3.5 cm. wide, upper cauline usually much reduced, elliptic, entire, sessile; inflorescence a narrow dense thyrse or sometimes broadened into a pyramidal panicle with spreading branches; involucre 3–4.5 mm. high; phyllaries lance-linear to narrowly oblong, sharply acute to obtuse, puberulent or glabrous; ray-florets 8–13; disk-florets 4–11; achenes hispidulous.

Common on dry or moist fields, clearings, and forest openings, Upper Sonoran and Transition Zones; southwestern Oregon through the Coast Ranges and the western flank of the Sierra Nevada to San Diego County, California; also Inyo County. Type locality: Santa Barbara. July–Oct.

3. Solidago canadénsis L. subsp. elongàta (Nutt.) Keck. Meadow Goldenrod. Fig. 5492.

Solidago elongata Nutt. Trans. Amer. Phil. Soc. II. **7**: 327. 1840.
Aster elongatus Kuntze, Rev. Gen. Pl. **1**: 318. 1891.
Solidago caurina Piper, Bull. Torrey Club **28**: 40. 1901.
Solidago lepida var. *elongata* Fernald, Rhodora **17**: 9. 1915.
Solidago lepida var. *caurina* M. E. Peck, Man. Pl. Oregon 717. 1941.
Solidago canadensis subsp. *elongata* Keck, Aliso **4**: 103. 1958.

Stems from creeping rhizomes, 3–10(15) dm. high, puberulent or pilosulose up toward (and always including) the inflorescence or throughout, densely leafy. Leaves nearly uniform, lanceolate or oblong-lanceolate, 5–12 cm. long, 1–2 cm. wide, tapering to base and apex, triplinerved, sharply serrate to entire, scabrid-margined, from essentially glabrous to scabrid-puberulent on both faces; panicle 5–20 cm. long, dense, usually rhombic and broad or oblong, the lower branches not obviously recurved nor the heads secund; involucre 3.5–5 mm. high; phyllaries thin, linear-lanceolate; ray-florets mostly about 13, little exceeding the disk; achenes hispidulous.

Meadows and moist openings in woods, from sea-level to 7,500 feet. Transition Zone; southern British Columbia to central coastal California and in the Sierra Nevada to Tulare County, east to the Rocky Mountains; also Lower California. Type locality: "Wappatoo Island and the plains of the Oregon." July–Oct. A variable species with rather poorly marked regional subspecies and covering most of the United States and Canada east to Newfoundland. Typical *canadensis* is northeastern.

Solidago canadensis subsp. **salebròsa** (Piper) Keck, Aliso **4**: 104. 1958. (*Solidago serotina* var. *salebrosa* Piper in Piper & Beattie, Fl. Palouse Reg. 185. 1901; *S. canadensis* var. *salebrosa* M. E. Jones, Bull. Univ. Mont. Biol. Ser. **15**: 49. 1910; *S. salebrosa* Rydb. Fl. Rocky Mts. 870. 1917; *S. gigantea* var. *salebrosa* Friesn. Butler Univ. Bot. Stud. **4**: 196. 1940.) Similar to subsp. *elongata*, differing in having large broad panicles, with obviously recurved-spreading lower branches bearing secund heads. Much less common in the Pacific States than the preceding; eastern Oregon and Washington to the Rocky Mountains. Type locality: Pullman, Whitman County, Washington.

4. Solidago lépida DC. Alaskan Goldenrod. Fig. 5493.

Solidago lepida DC. Prod. **5**: 339. 1836.
Solidago lepida var. *subserrata* DC. loc. cit.
Aster lepidus Kuntze, Rev. Gen. Pl. **1**: 318. 1891.
Solidago canadensis var. *subserrata* Cronquist, Vasc. Pl. Pacif. Northw. **5**: 305. 1955.

Stems mostly 4–8 dm. high. Leaves elliptic-lanceolate, usually sharply serrate-dentate and scarcely reduced up to the inflorescence; panicle short and compact, not at all secund; involucre not very imbricate, the outer phyllaries more than half as long as the inner. Otherwise similar to *S. canadensis* subsp. *elongata*.

Moist ground, Humid Transition Zone; Saddle Mountain, Clatsop County, Oregon; Vancouver Island north to southern Alaska. Type locality: Nootka or Multgrave (Yakutat Bay). July–Sept.

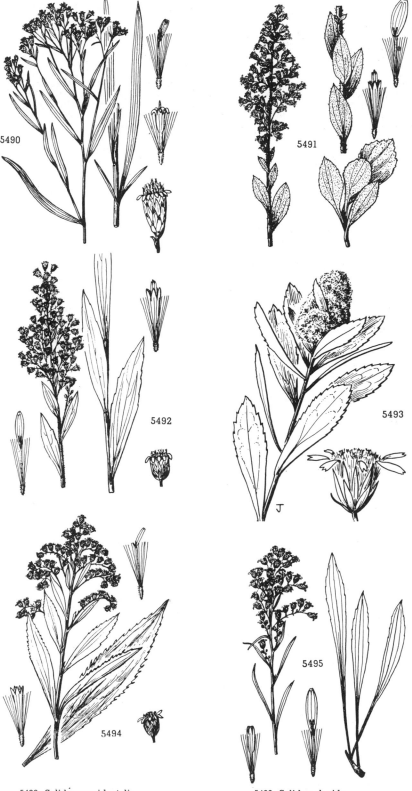

5490. Solidago occidentalis
5491. Solidago californica
5492. Solidago canadensis

5493. Solidago lepida
5494. Solidago gigantea
5495. Solidago missouriensis

5. **Solidago gigantèa** Ait. Smooth Goldenrod. Fig. 5494.

Solidago gigantea Ait. Hort. Kew. **3**: 211. 1789.
Solidago serotina Ait. loc. cit. Not Retz. 1781.
Solidago serotina var. *gigantea* A. Graý, Proc. Amer. Acad. **17**: 196. 1882.
Aster latissimifolius var. *serotinus* Kuntze, Rev. Gen. Pl. **1**: 314. 1891.
Solidago gigantea var. *leiophylla* Fernald, Rhodora **41**: 457. 1939.

Similar to *S. canadensis*; stems stout, 6–20 dm. high, glabrous and often glaucous below the inflorescence, the latter pilosulose. Leaves lanceolate to elliptic-oblong, 8–15 cm. long, 13–35 mm. wide, acuminate, sharply serrate above the entire cuneate base, scabrid-margined, essentially glabrous on both faces; panicle large, pyramidal, the heads secund on the recurved branches; involucre 3.5–4.5 mm. high, the phyllaries rather firm and often obtuse; ray-florets 9–16, distinctly surpassing the disk; achenes hispidulous.

Thickets and meadows at low elevations, Arid Transition Zone; southern British Columbia to Oregon (mostly east of the Cascade Mountains) east to Newfoundland, Georgia, and Texas. Type locality: North America (the type a garden plant). July–Sept.

6. **Solidago missouriénsis** Nutt. Missouri Goldenrod. Fig. 5495.

Solidago missouriensis Nutt. Journ. Acad. Phila. **7**: 32. 1834.
Solidago tolmieana A. Gray, Syn. Fl. N. Amer. **1²**: 151. 1884.
Aster missouriensis Kuntze, Rev. Gen. Pl. **1**: 318. 1891.
Aster tolmieanus Kuntze, loc. cit.
Solidago missouriensis var. *tolmieana* Cronquist, Vasc. Pl. Pacif. Northw. **5**: 307. 1955.

Stems from creeping rhizomes or clustered from a simple or branched caudex, 2–5(–9) dm. high, essentially glabrous below the sparsely puberulent inflorescence. Basal leaves crowded, oblanceolate, 5–20 cm. long, 5–20 mm. wide, entire or toothed above, tapering into margined petioles, scabrid-ciliolate, often lost early, the cauline strongly reduced above, the upper mostly linear or subulate, entire; panicle usually oblong or rhombic, with erectish branches, sometimes pyramidal with spreading branches, the heads then secund; involucre 3–5 mm. high, the phyllaries linear-oblong to broadly lanceolate, obtuse to acutish, rather firm, the costal gland prominent; ray-florets 8(–13), usually distinctly exceeding the disk; achenes hispidulous.

Rather dry open places, Upper Sonoran and Transition Zones; British Columbia to southern Oregon east to Ontario, Tennessee, Texas, and Arizona. A common and variable species from which a number of varieties have been segregated. Type locality: "on the upper branches of the Missouri and in Arkansas." July–Sept.

7. **Solidago guiradònis** A. Gray. Guirado's Goldenrod. Fig. 5496.

Solidago guiradonis A. Gray, Proc. Amer. Acad **6**: 543. 1865.

Stems slender, erect, from a woody rhizome, 8–10 dm. high, glabrous throughout. Basal leaves lanceolate, tapering to the petiole, 12–15 cm. long, 5–10 mm. wide, the cauline elongate but reduced, linear, above becoming linear-subulate bracts, all entire; panicle erect, slender, sub-racemose, few-headed, not secund, 10–20 cm. long; involucre about 4–5 mm. high, the phyllaries lance-linear, the midvein broad; ray-florets 8–10, little exceeding the disk; disk-florets 10–12; achenes puberulent.

Moist slopes, Upper Sonoran Zone; California, rare, in the Coast Ranges from Santa Clara County southward and in the southernmost Sierra Nevada. Type locality: base of San Carlos Peak, San Benito County. Sept.–Oct.

8. **Solidago confìnis** A. Gray. Southern Goldenrod. Fig. 5497.

Solidago confinis A. Gray, Proc. Amer. Acad. **17**: 191. 1882.
Solidago confinis f. *luxurians* H. M. Hall, Univ. Calif. Pub. Bot. **3**: 46. 1907.
Solidago confinis var. *luxurians* Jepson, Man. Fl. Calif. 1035. 1925.

Stems usually stout, terminating rather short woody rhizomes or sometimes clustered on a short caudex, 3–14 dm. high, glabrous throughout. Leaves thick, pale green, entire, glabrous or only the margin scabrid, the basal spatulate or oblanceolate to obovate, tapering to the petiole, 15 cm. or less long, the cauline gradually reduced, lance-elliptic to linear, sessile; panicle usually oblong and very dense, up to 25 cm. long, the branches erect, rarely spreading; involucre 3.5–4.5 mm. high, the phyllaries only slightly imbricate, linear-lanceolate, acute or acuminate, rather firm; ray-florets 6–10, scarcely surpassing the 11–21 disk-florets; achenes sparsely hispidulous to canescent.

Dry or moist banks, Upper Sonoran and Transition Zones; Ventura County to San Diego County, California, to the desert's edge, mostly in the mountains. Type locality: "southern borders of California." July–Oct.

9. **Solidago spectábilis** (D. C. Eaton) A. Gray. Basin Goldenrod. Fig. 5498.

Solidago guiradonis var. *spectabilis* D. C. Eaton, Bot. King Expl. 154. 1871.
Solidago spectabilis A. Gray, Proc. Amer. Acad. **17**: 193. 1882.

Stems rather stout, terminating a rather short woody rhizome or caudex, 4–13 dm. high, glabrous throughout or becoming somewhat hispidulous within the inflorescence. Leaves entire or rarely remotely serrate, the basal oblanceolate, acute or obtuse, tapering to a long, winged, clasping petiole, including the petiole 9–28 cm. long, 1.3–4 cm. wide, the upper cauline linear-lanceolate and often much reduced toward the inflorescence, scabrid-ciliolate, otherwise glabrous; panicle usually oblong and very dense, mostly less than 10 cm. long; involucre about 4 mm. high;

phyllaries linear-oblong, obtusish; ray-florets 11–15, little exceeding the 15–22 disk-florets; achenes puberulent.

Alkaline meadows or bogs, Sonoran and Arid Transition Zones; southeastern Oregon south to Death Valley, California, east to Utah. Type locality: mountains of western Nevada. July–Sept.

10. Solidago spathulàta DC. Dune Goldenrod. Fig. 5499.

Solidago spathulata DC. Prod. 5: 339. 1836.
Homopappus spathulatus Nutt. Trans. Amer. Phil. Soc. II. 7: 332. 1840.
Solidago spiciformis Torr. & Gray, Fl. N. Amer. 2: 202. 1842.
Aster candollei Kuntze, Rev. Gen. Pl. 1: 315. 1891.

Stems stout, from a caudex or woody rhizome, 2–6 dm. high, usually glabrous throughout, glutinous especially above. Basal leaves broadly obovate to spatulate-oblanceolate, mostly blunt or rounded, crenate-serrate, tapering to the petiole, the cauline similar but reduced, the uppermost acute and subsessile; heads in a simple or compound, sometimes racemiform thyrse 6–25 cm. long; involucre 5–6 mm. high, the phyllaries firm, oblong, very blunt; ray-florets 7–9, scarcely exceeding the 10–16 disk-florets; achenes densely pubescent.

Sandy coastal hills and dunes, Humid Transition Zone; Coos County, Oregon, to Monterey, California. Type locality: Monterey, California, erroneously given as "Mexico." Collected by Haenke. May–Oct.

Solidago spathulata subsp. glutinòsa (Nutt.) Keck, Aliso 4: 104. 1958. (*Solidago confertiflora* DC. Prod. 5: 339. 1836, not Nutt. 1834; *S. glutinosa* Nutt. Trans. Amer. Phil. Soc. II 7: 328. 1840; *S. vespertina* Piper in Piper & Beattie, Fl. Northw. Coast 365. 1915.) Stems 8 dm. high or less, glabrous, becoming glutinous and sparsely hispidulous within the inflorescence; basal leaves narrowly spatulate to narrowly obovate; heads in a narrow thyrse 40 cm. long or less; phyllaries linear or oblong-linear, obtuse or the innermost acutish; ray-florets 6–17; disk-florets 5–35. Coastal, British Columbia to northwestern Oregon. Type locality: "On the plains of the Oregon and Wahlamet Rivers," Oregon. June–Oct.

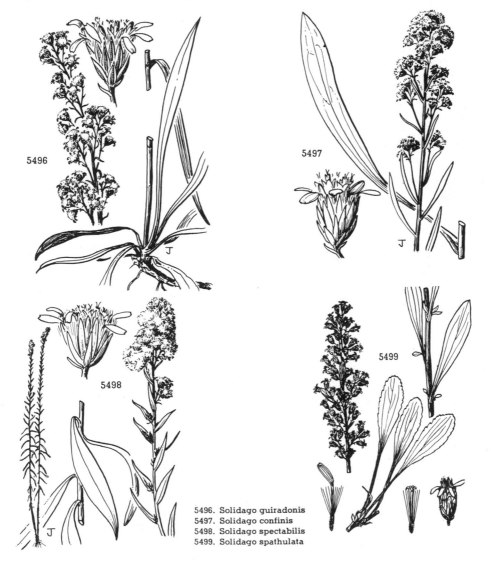

5496. Solidago guiradonis
5497. Solidago confinis
5498. Solidago spectabilis
5499. Solidago spathulata

11. Solidago decúmbens Greene. Dwarf Goldenrod. Fig. 5500.

Solidago humilis var. *nana* A. Gray, Syn. Fl. N. Amer. 1²: 148. 1884.
Solidago decumbens Greene, Pittonia 3: 161. 1897.
Solidago bellidifolia Greene, op. cit. 4: 100. 1899.
Solidago hesperia Howell, Fl. N.W. Amer. 303. 1900.
Solidago purshii var. *nana* Farwell, Amer. Midl. Nat. 12: 72. 1930.
Solidago glutinosa var. *nana* Cronquist, Rhodora 49: 76. 1947.
Solidago spathulata var. *nana* Cronquist, Vasc. Pl. Pacif. Northw. 5: 311. 1955.

Stems several from a ligneous branching caudex, 5–25 cm. high, glabrous or sometimes puberulent, more or less viscid and often hispidulous within the inflorescence. Basal and lower leaves obovate or spatulate, 2–7 cm. long (including petiole), obtuse or rounded, crenate-serrate above, thick, essentially glabrous; upper leaves few, reduced; heads comparatively few, corymbose-glomerate or in a short raceme or nearly simple thyrse; involucre 4–5.5 mm. high, the phyllaries linear-oblong, obtuse; ray-florets 8–12, not much exceeding the 11–19 disk-florets; achenes hispidulous.

Alpine ridges and rocky slopes, Boreal Zone; Cascade Range from Washington to northern Oregon; Rocky Mountains from Saskatchewan to New Mexico and Utah. Type locality: "Rocky Mountains of Colorado." July–Sept.

12. Solidago multiradiàta Ait. Alpine Goldenrod. Fig. 5501.

Solidago multiradiata Ait. Hort. Kew. 3: 218. 1789.
Solidago corymbosa Nutt. Trans. Amer. Phil. Soc. II. 7: 328. 1840. Not Poir. 1817, nor Ell. 1823.
Solidago virgaurea var. *multiradiata* Torr. & Gray, Fl. N. Amer. 2: 207. 1842.
Solidago multiradiata var. *scopulorum* A. Gray, Proc. Amer. Acad. 17: 191. 1882.
Aster multiradiatus Kuntze, Rev. Gen. Pl. 1: 318. 1891.
Solidago ciliosa Greene, Pittonia 3: 22. 1896.
Solidago humilis f. *glacialis* Gandoger, Bull. Soc. Bot. Fr. 50: 215. 1903.
Solidago scopulorum A. Nels. Bot. Gaz. 37: 264. 1904.
Solidago algida Piper in Piper & Beattie, Fl. Northw. Coast 365. 1915.
Solidago cusickii Piper, Proc. Biol. Soc. Wash. 29: 100. 1916.

Stems erect from a rather short woody rhizome or branching caudex, 0.5–4 dm. high, pilose or pilosulose at least above. Basal leaves oblanceolate to elliptic, obtuse or sometimes acute, serrate or crenate-serrate above to subentire, mostly 2–10 cm. long, 5–18 mm. wide, scabrid-margined, tapering to a ciliate petiole, otherwise glabrous; cauline leaves spatulate to lanceolate, usually acute, sessile, mostly entire, ciliate at least toward base; heads few to rather numerous, in a loose or usually dense terminal corymb or with 1 or 2 loose axillary clusters; involucre 4–6.5 mm. high, the phyllaries linear to lance-linear, acute to acuminate, not much imbricate, thin, more or less ciliolate; ray-florets commonly 13, distinctly exceeding the 13–34 disk-florets; achenes hispidulous.

Sunny, rocky or grassy places in the high mountains, Boreal Zone; in the Pacific States from Washington south through Oregon to the southern Sierra Nevada, California; also Alaska and adjacent Siberia, east to Labrador, south through the Rocky Mountains to New Mexico and Arizona. Type locality: Labrador. June–Sept.

78. PETRADÒRIA* Greene, Erythea 3: 13. 1895.

Low, tufted, perennial herb with a short pauci- to multibranched caudex surmounting a stout scaly taproot. Leaves narrow, rigid, sharp-pointed, entire. Heads yellow, small, numerous, in terminal flat-topped corymbs. Involucre cylindric, 4-seriate, strongly graduate, the phyllaries corneous, stramineous, in 4 or 5 vertical rows. Ray-florets 1–3, pistillate, the achene 10-striate, glabrous; disk-florets 3–5, staminate, the achene undeveloped. Pappus of ray and disk of numerous, very slender, stramineous bristles. [Name from the Greek, meaning rock goldenrod.]

A monotypic genus of the southwestern United States. It appears to combine the characters of *Chrysothamnus* and *Haplopappus*.

1. Petradoria pùmila (Nutt.) Greene. Rock Goldenrod. Fig. 5502.

Chrysoma pumila Nutt. Trans. Amer. Phil. Soc. II. 7: 325. 1840.
Solidago pumila Torr. & Gray, Fl. N. Amer. 2: 210. 1842. Not Crantz, 1766.
Aster pumilus Kuntze, Rev. Gen. Pl. 1: 318. 1891.
Petradoria pumila Greene, Erythea 3: 13. 1895.
Solidago petradoria Blake, Contr. U.S. Nat. Herb. 25: 540. 1925.

Rather rigid, flat-topped plant with numerous simple, erect, leafy stems from the multicipital caudex, 1.5–2.5 dm. high, glabrous, resiniferous, light green. Leaves densely crowded in basal rosettes, somewhat reduced on stems, obscurely punctate, prominently 3–5-nerved, the basal nearly linear to oblanceolate on very slender petioles, 5–8 cm. long, 3–7 mm. wide, the cauline sessile and becoming bract-like toward the inflorescence; involucre 5.5–8 mm. high, about 2.5 mm. wide; phyllaries lance-oblong to lanceolate, obtusish to acuminate, apiculate, more or less carinate, with or without a somewhat thickish green tip; ligules 2 mm. long; ray-achenes somewhat flattened, strongly differentiated from the undeveloped disk-achenes.

Dry stony hillsides, Upper Sonoran Zone; occasional on mountains of eastern Mojave Desert, San Bernardino County, California; common eastward across Arizona and Nevada to Texas and Wyoming. Type locality: "In open situations, on shelving rocks toward the western declivity of the Rocky Mountains." July–Oct.

* Text contributed by David Daniels Keck.

5500. Solidago decumbens   5501. Solidago multiradiata

79. CHRYSOTHÁMNUS* Nutt. Trans. Amer. Phil. Soc. II. **7**: 323. 1840.

Shrubs or subshrubs, usually much branched with erect stems. Leaves alternate, entire, narrow, not fasciculate, sometimes glandular-punctate. Heads numerous, strictly discoid, narrow, mostly 5-flowered and yellow, sometimes white, arranged in cymes, racemes, or panicles. Involucre cylindraceous; phyllaries strongly imbricate, usually arranged in 5 vertical ranks, chartaceous or coriaceous, mostly carinate and sometimes with a costal gland, the tip sometimes greenish. Style-branches flattened, the appendages usually longer than the stigmatic portion. Achenes slender, terete or angled or flattened, glabrous to densely sericeous. Pappus copious, of soft capillary bristles. [Name Greek, meaning gold and shrub.]

A genus of 13 well-defined species, except for one Mexican species, all confined to the western United States and adjacent borders. Type species, *Chrysothamnus pumilus* Nutt. (= *C. viscidiflorus* subsp. *pumilus* (Nutt.) Hall & Clem.]

Leaves resinous-punctate, terete.
  Flowers yellow.
    Phyllaries thin, not glandular-thickened apically.   1. *C. paniculatus.*
    Phyllaries tipped with a conspicuous, thick, green, glandular spot.   2. *C. teretifolius.*
  Flowers white.   3. *C. albidus.*
Leaves not resinous-punctate.
  Twigs not tomentose.
    Plants definitely shrubby; leaves not prominently ribbed; phyllaries not cuspidate at a truncate or retuse apex.
      Phyllaries in obscure vertical ranks.
        Flowers white; phyllaries with slender tapering tip.   3. *C. albidus.*
        Flowers yellow; phyllaries obtuse to acute.   4. *C. viscidiflorus.*
      Phyllaries in 5 sharply defined vertical ranks.
        Stems glabrous; leaves terete, 0.5 mm. wide; involucre 5–6 mm. high; achenes sericeous.
                                            5. *C. axillaris.*
        Stems scabrid-puberulent; leaves oblanceolate, 1.5–4 mm. wide; involucre 9–12 mm. high; achenes glabrate.   6. *C. depressus.*
    Plants largely herbaceous above a branched woody caudex; leaves longitudinally ribbed; phyllaries broad, cuspidate at the truncate or retuse apex.   7. *C. gramineus.*
  Twigs densely pannose-tomentose.
    Inflorescence mostly racemose or spicate; phyllaries very attenuate.   8. *C. parryi.*
    Inflorescence mostly cymose; phyllaries obtuse to moderately attenuate.   9. *C. nauseosus.*

1. **Chrysothamnus paniculàtus** (A. Gray) H. M. Hall. Punctate Rabbit-brush. Fig. 5503.

Bigelovia paniculata A. Gray, Proc. Amer. Acad. **8**: 644. 1873.
Aster asae Kuntze, Rev. Gen. Pl. **1**: 315. 1891.
Chrysothamnus paniculatus H. M. Hall, Univ. Calif. Pub. Bot. **3**: 58. 1907.
Ericameria paniculata Rydb. Fl. Rocky Mts. 853. 1917.

Loosely branched, broadly rounded shrub 6–20 dm. high, the herbage glabrous, resinous, strongly glandular-punctate, somewhat glaucescent. Leaves terete, mucronate-tipped, 1–3 cm.

---

* Text contributed by David Daniels Keck.

long, 0.5 mm. wide; heads paniculate, very numerous, 5–8-flowered; involucre subcylindric, 5.5–6.5 mm. high; phyllaries 4–5-seriate, graduate, the vertical ranks not sharply defined, oblong, obtuse, whitish, indurate, the costa narrow, obscure, scarcely or not at all glandular-thickened above; achenes appressed-villous or sericeous; pappus brownish white.

Stony open deserts, not common, Lower Sonoran Zone; Mojave and Colorado Deserts, California, east to southwestern Utah and Arizona. Type locality: "California." June–Dec.

2. Chrysothamnus teretifòlius (Dur. & Hilg.) H. M. Hall. Needle-leaved Rabbit-brush. Fig. 5504.

Linosyris teretifolia Dur. & Hilg. Journ. Acad. Phila. II. **3**: 41. 1855; also Pacif. R. Rep. **3**: 9. *pl. 7*. 1857.
Bigelovia teretifolia A. Gray, Proc. Amer. Acad. **8**: 644. 1873.
Aster durandii Kuntze, Rev. Gen. Pl. **1**: 316. 1891.
Chrysoma teretifolia Greene, Erythea **3**: 12. 1895.
Chrysothamnus teretifolius H. M. Hall, Univ. Calif. Pub. Bot. **3**: 57. 1907.
Ericameria teretifolia Jepson, Man. Fl. Pl. Calif. 1024. 1925.

Fastigiately branched, globose or spreading shrub 2–15 dm. high, the herbage glabrous, densely glandular-punctate, balsamic-resinous, dark green. Leaves terete, obtuse, not mucronate, 1–2.5 cm. long, 0.5–1 mm. wide; heads in short terminal spikes, 4–6-flowered; involucre subcylindric, 6–8 mm. high; phyllaries 4–5-seriate, strongly graduate in rather definite vertical ranks, oblong, obtuse, indurate, stramineous, obscurely carinate, tipped with a conspicuous, green, glandular spot; achenes appressed-villous; pappus stramineous.

Canyon walls and rocky slopes, Lower Sonoran Zone; Santa Rosa Mountains, Riverside County, to Tehachapi Mountains and across the Mojave Desert to southern Mono County and the Kingston Mountains, California; east to Utah and Arizona. Type locality: "all over the mountains around Tejon Valley, California." Sept.–Nov.

3. Chrysothamnus álbidus (M. E. Jones ex A. Gray) Greene. White-flowered Rabbit-brush. Fig. 5505.

Bigelovia albida M. E. Jones ex A. Gray, Proc. Amer. Acad. **17**: 209. 1882.
Chrysothamnus albidus Greene, Erythea **3**: 107. 1895.

Shrub 3–10 dm. high, fastigiately branched, white-barked, glabrous, resinous-viscid, aromatic. Leaves filiform, 1.5–3 cm. long, 0.5–1 mm. wide, terete, mucronate, impressed-punctate, crowded, with axillary fascicles; heads in small compact cymes at branch-tips, 4–6-flowered; involucre 7–9 mm. high; phyllaries about 4-seriate, graduate, in obscure vertical ranks, the outer lanceolate to ovate, herbaceous-thickened in outer half, abruptly narrowed to a subulate-attenuate, curved tip, the inner oblong, acuminate-tipped, the narrow hyaline margin somewhat erose; corollas white, 7–8 mm. long, the lobes about 2 mm. long; achenes pilose, glandular above; pappus copious, exceeding the corolla.

Dry, alkaline, sandy or silty soils, Sonoran Zones; Owens Valley and Death Valley region, California (rare), east to the Salt Lake Desert, Utah. Type locality: Wells, Nevada. Aug.–Oct.

4. Chrysothamnus viscidiflòrus (Hook.) Nutt. Sticky-leaved Rabbit-brush. Fig. 5506.

Crinitaria viscidiflora Hook. Fl. Bor. Amer. **2**: 24. 1834.
Bigelovia viscidiflora DC. Prod. **7**: 279. 1838.
Chrysothamnus viscidiflorus Nutt. Trans. Amer. Phil. Soc. II. **7**: 324. 1840.
Linosyris viscidiflora Torr. & Gray, Fl. N. Amer. **2**: 234. 1842.
Bigelovia douglasii A. Gray, Proc. Amer. Acad. **8**: 645. 1873.
Bigelovia douglasii var. *tortifolia* A. Gray, op. cit. 646.
Aster viscidiflorus Kuntze, Rev. Gen. Pl. **1**: 313. 1891.
Chrysothamnus viscidiflorus var. *tortifolius* Greene, Erythea **3**: 96. 1895.
Chrysothamnus tortifolius Greene, Fl. Fran. 368. 1897.
Chrysothamnus douglasii Clem. & Clem. Rocky Mt. Fls. 266. 1914.

Rounded, fastigiately branched, white-barked shrub usually less than 1 m. high, the twigs brittle. Leaves linear or linear-lanceolate, flat or twisted, erect, spreading, or reflexed, 2–5 cm. long, 2–5 mm. wide, 1–3-nerved, glabrous, sometimes scabrous-ciliolate, viscid, sometimes with punctate glands ventrally; heads about 5-flowered, in terminal broad cymes; involucre 5–7 mm. high; phyllaries linear-oblong to lanceolate, obtuse (but commonly mucronate) to acute, not keeled, strongly graduate but in obscure vertical ranks, chartaceous; achenes densely villous; pappus brownish white. A highly polymorphic species.

Dry plains and hillsides, at low or moderate elevations, Upper Sonoran and Arid Transition Zones; east of the Cascade-Sierran axis from southern British Columbia to southern California, east to New Mexico and Montana; this typical subspecies most frequent in the southern Rocky Mountains. Type locality: "barren plains of the Columbia, from the Great Falls to the Mountains." Collected by Douglas. July–Sept. Composed of several freely intergrading subspecies of overlapping distribution.

Key to Subspecies

Leaves glabrous (sometimes viscid or with punctate glands ventrally) at least on the faces, the margin sometimes scabrociliate.
    Shrubs mostly 4–12 dm. high (shorter in some wide-leaved plants); leaves mostly 2 mm. or more wide.
        Leaves linear to linear-lanceolate, tapering to base and apex, 2–5 mm. wide, 1–3-nerved, acute or acuminate; widespread.   subsp. *viscidiflorus*.
        Leaves broadly elliptic to elliptic-oblong, 6–12 mm. wide, 3–5-nerved, obtuse but mucronate; shrubs often only 2–4 dm. high; nearly limited to northeastern Nevada.   subsp. *latifolius*.

Shrubs mostly 1–3.5 dm. high; leaves linear or linear-filiform, 0.5–2 mm. wide.   subsp. *pumilus.*
Leaves more or less densely puberulent.
Leaves mostly 2.5–6 mm. wide, 3–5-nerved.   subsp. *lanceolatus.*
Leaves 1–2 mm. wide, 1-nerved.   subsp. *puberulus.*

Chrysothamnus viscidiflorus subsp. **latifòlius** (D. C. Eaton) Hall & Clem. Carnegie Inst. Wash. Publ. No. 326: 184. 1923. (*Linosyris viscidiflora* var. *latifolia* D. C. Eaton, Bot. King Expl. 157. 1871; *Bigelovia douglasii* var. *latifolia* A. Gray, Proc. Amer. Acad. **8**: 646. 1873; *Chrysothamnus viscidiflorus* var. *latifolius* Greene, Erythea **3**: 96. 1895; *C. latifolius* Rydb. Bull. Torrey Club **33**: 152. 1906.) Mountains of northeastern Nevada and adjacent Idaho. Reported from northeastern California by Hall and Clements but doubtfully a member of our flora.

Chrysothamnus viscidiflorus subsp. **pùmilus** (Nutt.) Hall & Clem. Carnegie Inst. Wash. Publ. No. 326: 182. 1923. (*Chrysothamnus pumilus* Nutt. Trans. Amer. Phil. Soc. II. **7**: 323. 1840; *Linosyris pumila* A. Gray, Smiths. Contr. 5⁶: 80. 1853; *Bigelovia douglasii* var. *stenophylla* A. Gray, Proc. Amer. .Acad. **8**: 646. 1873; *B. douglasii* var. *pumila* A. Gray, Syn. Fl. N. Amer. 1²: 140. 1884; *Chrysothamnus stenophyllus* Greene, Erythea **3**: 94. 1895; *C. viscidiflorus* var. *stenophyllus* H. M. Hall, Univ. Calif. Pub. Bot. **3**: 59. 1907; *C. viscidiflorus* subsp. *stenophyllus* Hall & Clem. Carnegie Inst. Wash. Publ. No. 326: 183. 1923; *C. viscidiflorus* var. *pumilus* Jepson, Man.,Fl. Pl. Calif. 1031. 1925.) Dry, sagebrush-covered plains and ridges, often in alkaline soils, from southern Washington to Inyo County and the San Bernardino Mountains, California, eastward to eastern Montana and Utah; rare in the southern Rocky Mountains and southern Great Basin. Type locality: "on the borders of Lewis' River and the Rocky Mountain plains."

Chrysothamnus viscidiflorus subsp. **lanceolàtus** (Nutt.) Hall & Clem. Carnegie Inst. Wash. Publ. No. 326: 181. 1923. (*Chrysothamnus lanceolatus* Nutt. Trans. Amer. Phil. Soc. II. **7**: 324. 1840; *Linosyris lanceolata* Torr. & Gray, Fl. N. Amer. **2**: 233. 1842; *Bigelovia lanceolata* A. Gray, Proc. Amer. .Acad. **8**: 639. 1873; *B. douglasii* var. *lanceolata* A. Gray, Syn. Fl. N. Amer. 1²: 140. 1884; *Chrysothamnus viscidiflorus* var. *lanceolatus* Greene, Erythea **3**: 95. 1895.) Dry foothills and sagebrush-covered slopes, sometimes alkaline, from eastern Washington (frequent) and eastern Oregon (rare) eastward to Montana and northern Colorado and Utah. Type locality: "In the Rocky Mountains, toward the sources of the Platte."

Chrysothamnus viscidiflorus subsp. **pubérulus** (D. C. Eaton) Hall & Clem. Carnegie Inst. Wash. Publ. No. 326: 182. 1923. (*Chrysothamnus pumilus euthamoides* Nutt. Trans. Amer. Phil. Soc. II. **7**: 323. 1840, apparently this; *Linosyris viscidiflora* var. *puberula* D. C. Eaton, Bot. King Expl. 158. 1871; *Bigelovia douglasii* var. *puberula* A. Gray, Proc. Amer. Acad. **8**: 646. 1873; *Chrysothamnus puberulus* Greene, Erythea **3**: 93. 1895; *C. humilis* Greene, Pittonia **3**: 24. 1896; *C. viscidiflorus* subsp. *humilis* Hall & Clem. Carnegie Inst. Wash. Publ. No. 326: 182. 1923; *C. viscidiflorus* var. *puberulus* Jepson, Man. Fl. Pl. Calif. 1031. 1925; *C. viscidiflorus* var. *humilis* Jepson, loc. cit.) Sagebrush-covered plains and slopes up to 10,500 feet, southern Washington through eastern Oregon and California to the San Bernardino Mountains; abundant through Nevada and southern Idaho to western Utah and Yellowstone Park. Type locality: near the Truckee and on the Hot Spring Mountains in western Nevada.

5502. Petradoria pumila
5503. Chrysothamnus paniculatus
5504. Chrysothamnus teretifolius
5505. Chrysothamnus albidus

5. **Chrysothamnus axillàris** Keck. Inyo Rabbit-brush. Fig. 5507.

Chrysothamnus axillaris Keck, Aliso 4: 104. 1958.

Low rounded shrub 6 dm. high, intricately branched, the branches slender, white-barked, glabrous, with small axillary buds in the axils of the oldest leaves. Leaves tightly involute, terete, green, spreading, filiform, 0.5 mm. wide, acicular-tipped; heads in rounded cymose panicles, 3–5-flowered; involucre turbinate, 5–6 mm. high; phyllaries 4-seriate, strongly graduate in 4 or 5 sharply defined vertical ranks, spreading at maturity, broadly linear, sharply acute or apiculate, the costa glandular-thickened above, the small herbaceous tip ciliolate; corolla about 5 mm. long, the lobes 1–1.2 mm. long; style-appendages longer than the stigmatic portion; achenes sericeous; pappus tawny, not very long or copious.

Desert slopes in granitic sand, Upper Sonoran Zone; eastern Inyo County, California, and adjacent Nevada. Type locality: head of Deep Springs Valley, Inyo County. Sept.–Oct.

Related to *C. albidus* (M. E. Jones) Greene, from which it differs in having too few leaves and these not at all impressed-punctate, and in the smaller involucres and florets. It is probably more distantly related to *C. greenei* (A. Gray) Greene subsp. *filifolius* (Rydb.) Hall & Clem., a plant known only from eastern Nevada and eastward, which has plane rather than tightly involute leaves.

6. **Chrysothamnus depréssus** Nutt. Long-flowered Rabbit-brush. Fig. 5508.

Chrysothamnus depressus Nutt. Journ. Acad. Phila. II. 1: 171. 1847.
Linosyris depressa Torr. in Sitgr. Rep. 161. 1854.
Bigelovia depressa A. Gray, Proc. Amer. Acad. 8: 643. 1873.

Depressed subshrub with many erect herbaceous stems from a much-branched, spreading, woody crown, 1–3 dm. high, cinereous with a dense scabrid puberulence. Leaves oblanceolate or spatulate, the lowermost rounded or obtuse, the upper becoming sharply apiculate, 7–20 mm. long, 1.5–4 mm. wide, firm; heads in compact terminal cymes, 5-flowered; involucre 9–12 mm. high; phyllaries usually 5-seriate, in 5 sharply defined vertical ranks, lance-acuminate, drawn to a soft mucro, strongly keeled, the outer herbaceous and minutely puberulent, the inner broader, scarious, with hyaline margins; achenes glabrate; pappus brownish white.

Dry canyons, Sonoran Zones; occasional on mountains of eastern Mojave Desert, San Bernardino County, California, east to southern Colorado and New Mexico. Type locality: Rocky Mountains, as the type specimen is labelled; not "in the Sierra of Upper California," as stated by Nuttall. Aug.–Oct.

7. **Chrysothamnus gramíneus** H. M. Hall. Charleston Rabbit-brush. Fig. 5509.

Chrysothamnus gramineus H. M. Hall, Muhlenbergia 2: 342. 1916.

Many-stemmed from a branched woody caudex, 2.5–6 dm. high, light green, essentially glabrous throughout, the striate-angled stems simple or erect-branched above. Leaves equally distributed, the larger ones narrowly linear-lanceolate, 3–7 cm. long, 3–8 mm. wide, acuminate, sessile, 3–5-ribbed, coriaceous; heads mostly subsessile in small terminal clusters, 4–5-flowered, pale yellow; involucre cylindric, 10–13 mm. high; phyllaries 4–6-seriate, strongly graduate, obscurely vertical-ranked, stramineous, oblong, 1-nerved, abruptly cuspidate at the truncate or retuse, ciliolate apex; corolla about 10 mm. long; achenes linear, glabrous, about 6 mm. long; pappus copious, stramineous, 8–9 mm. long.

Rocky wooded slopes, 7,500–9,500 feet. Arid Transition Zone; Inyo and Panamint Mountains, Inyo County, California; Charleston Mountains, Nevada. Type locality: head of Lee Canyon, Charleston Mountains. July–Aug.

8. **Chrysothamnus párryi** (A. Gray) Greene. Parry's Rabbit-brush. Fig. 5510.

Linosyris parryi A. Gray, Proc. Acad. Phila. 1863: 66. 1863.
Bigelovia parryi A. Gray, Proc. Amer. Acad. 8: 642. 1873.
Chrysothamnus parryi Greene, Erythea 3: 113. 1895.

Shrub up to 5 dm. high, the numerous pliable branches erect or spreading, densely clothed with a white, gray, or greenish yellow tomentum, very leafy. Leaves narrowly linear to elliptic, 1–8 cm. long, 0.5–8 mm. wide, 1–3-nerved; heads in short leafy racemes or racemiform panicles, yellow; involucre 9–14 mm. high; phyllaries 4–6-seriate, in more or less obscure vertical ranks, acuminate or attenuate, the outer often with herbaceous tip, ciliate and often somewhat tomentose; corolla 8–11 mm. long; achenes densely appressed-pilose; pappus brownish white, about equaling the corolla.

Mountainsides and flats, Arid Transition Zone; Nebraska to New Mexico and California. July–Sept.

Variable in habit, pubescence, foliage, inflorescence, and characters of head. The typical form, ranging from Wyoming and Colorado to eastern Nevada, has linear, green, 3-nerved leaves 1.5–3 mm. wide, 10–20-flowered heads, and 10–15 phyllaries in obscure vertical ranks. The following subspecies occur in our area.

Chrysothamnus parryi subsp. **bolánderi** (A. Gray) Hall & Clem. Carnegie Inst. Wash. Publ. No. 326: 199. 1923. (*Linosyris bolanderi* A. Gray, Proc. Amer. Acad. 7: 354. 1868; *Bigelovia bolanderi* A. Gray, op. cit. 8: 641. 1873; *Aster bolanderi* Kuntze, Rev. Gen. Pl. 1: 317. 1891; *Chrysothamnus bolanderi* Greene, Erythea 3: 114. 1895; *Macronema bolanderi* Greene, Leaflets Bot. Obs. 1: 81. 1904; *Chrysothamnus parryi* var. *bolanderi* Jepson, Man. Fl. Pl. Calif. 1033. 1925.) Leaves 3–4 cm. long, 4–5 mm. wide, green, viscid-glandular; heads 7–11-flowered, in short, sometimes branched racemes; involucre 9–10 mm. high, the phyllaries about 11–15, in obscure ranks. Known only from the type locality, Mono Pass, Sierra Nevada, California, at 9,000–10,000 feet altitude, Arctic-Alpine Zone.

Chrysothamnus parryi subsp. **làtior** Hall & Clem. Carnegie Inst. Wash. Publ. No. 326: 199. 1923. (*Chrysothamnus parryi* var. *latior* Jepson, Man. Fl. Pl. Calif. 1033. 1925, where improperly attributed to Hall.) Leaves elliptic or oblanceolate, 2–4 cm. long, 4–8 mm. wide, 3-nerved, green, scabroglandular; heads 5–7-flowered, in racemiform panicles; involucre 12–15 mm. high, the phyllaries 12–15, in rather distant vertical ranks, nearly smooth, with attenuate, rather pungnet tips; corolla 11–12 mm. long. Coniferous slopes, Arid Transition Zone; mountains of Siskiyou and Modoc Counties, California, at 3,000–4,500 feet altitude. Type locality: Wagon Creek, at the foot of Mount Eddy.

Chrysothamnus parryi subsp. **ímulus** Hall & Clem. Carnegie Inst. Wash. Publ. No. 326: 200. 1923. (*Chrysothamnus parryi* var. *imulus* Jepson, Man. Fl. Pl. Calif. 1033. 1925.) Spreading at base, the shoots about 1 dm. high; leaves narrowly spatulate, 1–1.5 cm. long, 2–3 mm. wide, gray-tomentose; heads few, 11–15-flowered; involucre 11–12 mm. high; phyllaries about 16, obscurely ranked, the outer tomentose. San Bernardino Mountains, California. Type locality: Bear Valley.

Chrysothamnus parryi subsp. **ásper** (Greene) Hall & Clem. Carnegie Inst. Wash. Publ. No. 326: 200. 1923. (*Chrysothamnus asper* Greene, Leaflets Bot. Obs. 1: 80. 1904; *C. parryi* var. *asper* Munz, Man. S. Calif. 524. 1935.) Stems 1.5–4 dm. high, white-tomentose; leaves linear, 2–5 cm. long, 1–3 mm. wide, green, finely stipitate-glandular and roughish; heads several, 5–10-flowered; involucre 11–15 mm. high; phyllaries 9–13, usually glandular-puberulent as well as arachnoid-ciliate below. Mountainsides above the desert, Mono County to Ventura County and San Bernardino Mountains, California, east to south-central Nevada. Type locality: Hockett Trail, Little Cottonwood Creek, Inyo County, California.

Chrysothamnus parryi subsp. **vulcánicus** (Greene) Hall & Clem. Carnegie Inst. Wash. Publ. No. 326: 200. 1923. (*Chrysothamnus vulcanicus* Greene, Leaflets Bot. Obs. 1: 80. 1904; *C. parryi* var. *vulcanicus* Jepson, Man. Fl. Pl. Calif. 1033. 1925.) Very similar to subsp. *asper*; leaves viscid but not obviously stipitate-glandular. Southern Sierra Nevada, California. Type locality: Volcano Creek, above Volcano Falls, Tulare County.

Chrysothamnus parryi subsp. **monocéphalus** (Nels. & Kenn.) Hall & Clem. Carnegie Inst. Wash. Publ. No. 326: 200. 1923. (*Chrysothamnus monocephalus* Nels. & Kenn. Proc. Biol. Soc. Wash. 19: 39. 1906; *C. nevadensis* var. *monocephalus* Smiley, Univ. Calif. Pub. Bot. 9: 357. 1921; *C. parri* var. *monocephalus* Jepson, Man. Fl. Pl. Calif. 1033. 1925.) Usually low, rigidly much branched; leaves 1–3 cm. long, 0.8–1.5 mm. wide, tomentulose and viscidulous; heads 1–4, crowded, 5–6-flowered; involucre 9–11 mm. high; phyllaries 8–12, thinly tomentose and arachnoid-ciliate, sometimes purplish. High mountains, eastern California and western Nevada. Type locality: Mount Rose, Washoe County, Nevada.

Chrysothamnus parryi subsp. **nevadénsis** (A. Gray) Hall & Clem. Carnegie Inst. Wash. Publ. No. 326: 201. 1923. (*Linosyris howardii* var. *nevadensis* A. Gray, Proc. Amer. Acad. 6: 541. 1865; *Bigelovia howardii* var. *nevadensis* A. Gray, op. cit. 8: 641. 1873; *B. nevadensis* A. Gray, Syn. Fl. N. Amer. 1²: 136. 1884; *Chrysothamnus nevadensis* Greene, Erythea 3: 114. 1895; *C. parryi* var. *nevadensis* Jepson, Man. Fl. Pl. Calif. 1032. 1925, where improperly attributed to Hall.) Up to 6 dm. high; branches white-, gray-, or greenish-tomentose; leaves linear to linear-oblanceolate, 1.5–4 cm. long, 0.5–3 mm. wide, gray-tomentulose to green and glandular; heads few to numerous, 4–6-flowered; involucre 12–15 mm. high; phyllaries 13–18, thinly tomentose and arachnoid-ciliate, the attenuate herbaceous tips tending to recurve. Dry mountainsides, 4,600–9,000 feet altitude, eastern California from Modoc County to Alpine County; east to eastern Nevada and northern Arizona. Type locality: Mount Davidson, Nevada.

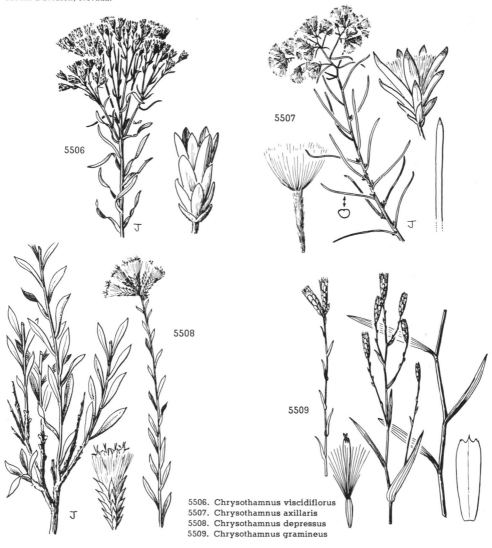

5506. Chrysothamnus viscidiflorus
5507. Chrysothamnus axillaris
5508. Chrysothamnus depressus
5509. Chrysothamnus gramineus

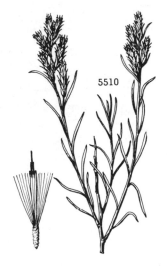

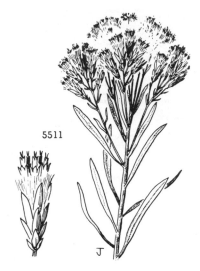

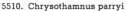

5510. Chrysothamnus parryi   5511. Chrysothamnus nauseosus

9. **Chrysothamnus nauseòsus** (Pall.) Britt. Common Rabbit-brush. Fig. 5511.

Chrysocoma nauseosa Pall. in Pursh, Fl. Amer. Sept. **2** : 517. 1814.
Chondrophora nauseosa Britt. Mem. Torrey Club **5** : 317. 1894.
Chrysothamnus nauseosus Britt. in Britt. & Brown, Ill. Fl. **3** : 326. 1898.

Shrub 3–20 dm. high, commonly with several fibrous-barked, main stems from the base, these much branched, the often ill-smelling, erect, usually densely leafy twigs clothed with a persistent, felt-like, gray, white, or greenish tomentum. Leaves variable, linear-filiform to narrowly linear-oblanceolate, 2–7 cm. long, 0.5–4 mm. wide, tomentose to subglabrous; heads in terminal, rounded, cymose clusters; involucre 6–13 mm. high; phyllaries usually 3–4-seriate, strongly graduate, in rather definite ranks, mostly lanceolate or linear-lanceolate, not green-tipped, usually with resin-ous-thickened costa; florets usually 5, yellow, the corolla 7–12 mm. long; pappus copious, dull white.

Plains and mountains, mostly in dry open places, often in alkaline soil, Sonoran and Transition Zones (oc-casionally higher); southern British Columbia to California, east to Saskatchewan, Texas, and northern Mexico. The typical form occurs principally east of the Continental Divide and on the Great Plains. The following ap-parently natural subspecies are distinguishable in our area.

Key to Subspecies

Achenes densely pubescent.
    Phyllaries, at least the outer, more or less pubescent or tomentose (sometimes only ciliate).
        Corolla-lobes lanceolate, 1.3–2.5 mm. long; style-appendage longer than the stigmatic portion.
            Involucre 7–10(–13) mm. high; phyllaries rarely glandular, the margin not prominently hyaline or
                fimbriate (but often more or less ciliate); corolla 8–11 mm. high.   subsp. *albicaulis.*
            Involucre 10–13 mm. high; phyllaries more or less glandular-atomiferous, the margin prominently
                hyaline and slashed-fimbriate (or merely ciliate); corolla 9–12 mm. long.
                                        subsp. *bernardinus.*
        Corolla-lobes ovate, 0.5–1 mm. long; style-appendage shorter than the stigmatic portion; involucre 6–7 mm.
        high.
            Herbage mostly grayish or whitish, not obviously glandular.   subsp. *hololeucus.*
            Herbage mostly yellow-green, the inflorescence including the involucres glandular-pubescent.
                                          subsp. *viscosus.*
Phyllaries not hairy but sometimes glandular or viscid; corolla-tube glabrous.
    Plants mostly 1–3 dm. tall; montane in northeastern Oregon.   subsp. *nanus.*
    Plants mostly 4–20 dm. tall.
        Phyllaries without a recurved mucronate tip; corolla 7 mm. or more long.
            Involucre 6.5–9 mm. high, not sharply angled, the phyllaries slightly if at all keeled.
                                        subsp. *consimilis.*
            Involucre 9–10 mm. high, sharply angled, the strongly keeled phyllaries in very distinct vertical
            rows.   subsp. *mohavensis.*
        Phyllaries with a very slender, recurved, mucronate tip; corolla less than 7 mm. long.
                                    subsp. *ceruminosus.*
                                    subsp. *leiospermus.*

     Chrysothamnus nauseosus subsp. **albicaùlis** (Nutt.) Hall & Clem. Carnegie Inst. Wash. Publ. No. 326: 212. 1923. (*Chrysothamnus speciosus* Nutt. Trans. Amer. Phil. Soc. II. **7**: 323. 1840; *C. speciosus* var. *albi-caulis* Nutt. op. cit. 324; *Linosyris albicaulis* Torr. & Gray, Fl. N. Amer. **2**: 234. 1842; *Bigelovia graveolens* var. *albicaulis* A. Gray, Proc. Amer. Acad. **8**: 645. 1873; *Chrysothamnus californicus* Greene, Erythea **3**: 111. 1895; *C. californicus* var. *occidentalis* Greene, op. cit. 112; *C. occidentalis* Greene, Fl. Fran. 369. 1897; *C. nau-seosus* var. *albicaulis* Rydb. Mem. N.Y. Bot. Gard. **1**: 385. 1900; *C. orthophyllus* Greene, Pittonia **5**: 62. 1902; *C. nauseosus* var. *occidentalis* H. M. Hall, Univ. Calif. Pub. Bot. **3**: 60. 1907; *C. nauseosus* var. *speciosus* H. M. Hall, op. cit. **7**: 169. 1919; *C. nauseosus* var. *californicus* H. M. Hall, op. cit. 174; *C. nauseosus* subsp. *speciosus* Hall & Clem. Carnegie Inst. Wash. Publ. No. 326: 211. 1923; *C. nauseosus* subsp. *occidentalis* Hall

& Clem. op. cit. 213; *C. nauseosus* var. *macrophyllus* J. T. Howell, Leaflets West. Bot. **2**: 58. 1937.) Shrub 5–20 dm. high; foliage gray or white with a rather copious tomentum; leaves 1(–4) mm. wide; involucre 7–13 mm. high, the phyllaries mostly acute or acuminate, at least the outer usually more or less tomentose but sometimes even these merely obscurely ciliate; corolla 8–11 mm. long, the tube loosely arachnoid-villous to puberulent, the lobes mostly 1.3–2 mm. long; appendages of the style-branches longer than the stigmatic portion. Common, often in moderately alkaline places, southern British Columbia to Fresno and Lake Counties, California, crossing the Cascade Mountains only in southern Oregon southward, east to northwestern Colorado and western Montana. Type locality: in the Rocky Mountain plains, near "Lewis' River." Collected by Nuttall. The most variable subspecies but not readily divisible into natural groups.

Chrysothamnus nauseosus subsp. **bernardinus** (H. M. Hall) Hall & Clem. Carnegie Inst. Wash. Publ. No. 326. 214. 1923. (*Chrysothamnus nauseosus* var. *bernardinus* H. M. Hall, Univ. Calif. Pub. Bot. **7**: 171. 1919.) Branches gray- or white-pannose; leaves usually green, 1–2 mm. wide; involucre 10–13 mm. high, the phyllaries sharply acuminate, with rather prominent hyaline, fimbriate, or merely somewhat ciliate margins, the outer often puberulent or rarely tomentose, together with the peduncles more or less glandular-atomiferous; corolla 9–10 mm. long, the lobes 1.5–2.5 mm. long, the tube puberulent. Mountains of southern California, from Los Angeles County to the San Jacinto Mountains. Type locality: Bluff Lake, San Bernardino Mountains.

Chrysothamnus nauseosus subsp. **hololeùcus** (A. Gray) Hall & Clem. Carnegie Inst. Wash. Publ. No. 326: 211. 1923. (*Bigelovia graveolens* var. *hololeuca* A. Gray, Proc. Amer. Acad. **8**: 645. 1873; *Chrysothamnus speciosus* var. *gnaphalodes* Greene, Erythea **3**: 110. 1895; *C. gnaphalodes* Greene, Pittonia **4**: 42. 1899; *C. nauseosus* var. *hololeucus* H. M. Hall, Univ. Calif. Pub. Bot. **7**: 166. 1919; *C. nauseosus* subsp. *gnaphalodes* H. M. Hall, op. cit. 167; *C. nauseosus* subsp. *gnaphalodes* Hall & Clem. Carnegie Inst. Wash. Publ. No. 326: 211. 1923.) Shrub 5–20 dm. high, the twigs white, gray, or yellowish green; herbage fragrant; leaves 0.5–1.5 mm. wide, gray- or white-tomentose; involucre 6–7 mm. high, the phyllaries rather obtuse, keeled; corolla 6.5–8 mm. long, the lobes 0.5–1 mm. long; appendages of the style-branches shorter than the stigmatic portion. Sandy nonalkaline slopes, Upper Sonoran Zone; eastern California in Mono and Inyo Counties and western ends of Mojave and Colorado Deserts; throughout Nevada. Type locality: "Owen's Valley, interior of California, Dr. Horn."

Chrysothamnus nauseosus subsp. **viscòsus** Keck, Aliso **4**: 104. 1958. Dense round shrub 5–20 dm. high; foliage yellowish (or grayish) green; leaves spreading or recurved, thickish, 1–2 mm. wide, permanently pannose, together with the mostly divaricate flowering branchlets and peduncles and involucres yellow-glandular; involucre 5–8 mm. high, the phyllaries narrow but obtuse, more or less glandular-tomentose, the costal gland prominent; corolla 7–8 mm. long, the tube sparsely puberulent, the lobes 0.4–0.6 mm. long; appendages of the style-branches shorter than the stigmatic portion. Sandy washes and flats, 4,000–8,000 feet altitude, Upper Sonoran Zone; southern White Mountains to southern Sierra Nevada (Walker Pass), California; adjacent Nevada. Type locality: 7.7 miles east of Laws, White Mountains.

Chrysothamnus nauseosus subsp. **nànus** (Cronquist) Keck, Aliso **4**: 104. 1958. (*Chrysothamnus nauseosus* var. *nanus* Cronquist, Vasc. Pl. Pacif. Northw. **5**: 129. 1955.) Plants 1–3 dm. high; stems white; leaves gray or greenish, 0.5–2 mm. wide; involucre 8–9 mm. high, the phyllaries mostly obtusish, essentially glabrous or the outer ones more or less tomentose; corolla 8–10 mm. long, the lobes 1.5–2 mm. long, the tube moderately pubescent. Rocky places, Blue and Wallowa Mountains, northeastern Oregon and adjacent Washington, east to central Idaho. Type locality: 10 miles west of Anatone, Asotin County, Washington.

Chrysothamnus nauseosus subsp. **consímilis** (Greene) Hall & Clem. Carnegie Inst. Wash. Publ. No. 326: 215. 1923. (*Chrysothamnus oreophilus* A. Nels. Bot. Gaz. **28**: 375. 1899; *C. consimilis* Greene, Pittonia **5**: 60. 1902; *C. tortuosus* Greene, op. cit. 63; *C. angustus* Greene, op. cit. 64; *C. oreophilus* var. *artus* A. Nels. Bot. Gaz. **54**: 413. 1912; *C. nauseosus* var. *oreophilus* H. M. Hall, Univ. Calif. Pub. Bot. **7**: 175. 1919; *C. nauseosus* var. *consimilis* H. M. Hall, op. cit. 176; *C. nauseosus* var. *viridulus* H. M. Hall, op. cit. 177; *C. nauseosus* subsp. *viridulus* Hall & Clem. Carnegie Inst. Wash. Publ. No. 326: 215. 1923; *C. nauseosus* var. *artus* Cronquist, Vasc. Pl. Pacif. Northw. **5**: 129. 1955.) Shrub 5–30 dm. high; tomentum gray, greenish yellow, or whitish; leaves mostly linear-filiform, less than 1 mm. wide, green or gray-tomentulose; inflorescence tending to be narrow and elongate rather than flat-topped; involucre 6.5–8.5 mm. high, glabrous, the phyllaries acute or obtuse, not sharply keeled; corolla 7–9.5 mm. long, the lobes 1–2.5 mm. long. Alkaline valleys and plains, Upper Sonoran Zone; northeastern Oregon south to Inyo County, eastern California, recurring in the mountains of southern California, east to Utah and Idaho. Type locality: Deeth, Elko County, Nevada.

Chrysothamnus nauseosus subsp. **mohavénsis** (Greene) Hall & Clem. Carnegie Inst. Wash. Publ. No. 326: 216. 1923. (*Bigelovia mohavensis* Greene in A. Gray. Syn. Fl. N. Amer. **1²**: 138. 1884; *C. mohavensis* Greene, Erythea **3**: 113. 1895; *C. nauseosus* var. *mohavensis* H. M. Hall, Univ. Calif. Pub. Bot. **7**: 179. 1919.) Shrub 6–20 dm. high, often fastigiately branched, the branches often leafless and rush-like, closely gray- or greenish yellow-tomentose; leaves narrowly linear, 1 mm. or less wide, tomentulose or nearly glabrous; inflorescence a rounded or somewhat elongate thyrse; involucre very narrow, 8–10.5 mm. long, glabrous, sharply 5-angled; phyllaries keeled, mostly obtuse, in very distinct vertical ranks, the costa usually dilated at apex into an oblong gland; corolla 8–10 mm. long, the lobes 1.5–2.5 mm. long. Well-drained, scarcely alkaline soils, Upper Sonoran Zone; California from Owens Valley to the western end of the Mojave Desert and the surrounding mountains, rare in the South Coast Ranges up to Mount Hamilton; east to Nevada. Type locality: Mojave Desert, California.

Chrysothamnus nauseosus subsp. **ceruminòsus** (Dur. & Hilg.) Hall & Clem. Carnegie Inst. Wash. Publ. No. 326: 216. 1923. (*Linosyris ceruminosus* Dur. & Hilg. Journ. Acad. Phila. II, **3**: 40. 1855; *Bigelovia ceruminosa* A. Gray, Proc. Amer. Acad. **8**: 643. 1873; *Chrysothamnus ceruminosus* Greene. Erythea **3**: 94. 1895; *C. nauseosus* var. *ceruminosus* H. M. Hall, Univ. Calif. Pub. Bot. **7**: 175. 1919.) Shrub 5–12 dm. high; tomentum yellowish green; leaves linear-filiform, less than 1 mm. wide, soon lost; involucre 7–9 mm. high, glabrous; phyllaries with abrupt, filiform recurved tip about 1 mm. long; corolla 6.5 mm. long, the lobes 1.5–2 mm. long. Southern Mojave Desert, Little San Bernardino Mountains·to Tejon Pass, California. Type locality: somewhere near Tejon Pass.

Chrysothamnus nauseosus subsp. **leiospérmus** (A. Gray) Hall & Clem. Carnegie Inst. Wash. Publ. No. 326: 217. 1923. (*Bigelovia leiosperma* A. Gray, Syn. Fl. N. Amer. **1²**: 139. 1884; *Chrysothamnus leiospermus* Greene, Erythea **3**: 113. 1895; *C. nauseosus* var. *leiospermus* H. M. Hall, Univ. Calif. Pub. Bot. **7**: 173. 1919.) Shrub 3–12 dm. high, with many short branches, usually very leafy, sometimes nearly leafless, white- (usually) or yellowish green-tomentose; leaves short and very narrow, essentially glabrous; heads in small terminal cymes; involucre 6–9 mm. high, glabrous; phyllaries obtuse, not obviously keeled; corolla 5–8 mm. long, the lobes 0.5 mm. long or less; achene glabrous or essentially so. Very arid slopes, Sonoran Zones; mountains of eastern Mojave Desert, California, to Nevada, southern Utah, and northern Arizona. Type locality: St. George, Utah.

80. **BÉLLIS** L. Sp. Pl. 886. 1753.

Herbs, sometimes scapose. Leaves alternate, sometimes all basal, toothed or entire. Heads small or medium-sized, solitary at tips of stems or branches, radiate, many-flowered; the rays white, pink, or violet; the disk yellow. Involucre broad; the phyllaries 2-seriate, equal or subequal, thin-herbaceous, flat. Receptacle convex or (in ours) conical, naked. Rays pistillate, spreading, narrow, obscurely 2-toothed or subentire; disk-corollas with short tube and subcampanulate throat, 5-toothed. Anthers entire at base. Style-branches with ovate obtusish appendages. Achenes obovate, strongly compressed, with thickened margin. Pappus none. [From the Latin *bellus*, meaning pretty.]

A genus of about 14 species, mostly Palearctic but with 1 native species in the southwestern United States and about 6 in Mexico. Type species, *Bellis perennis* L.

1. Bellis perénnis L. English Daisy. Fig. 5512.

Bellis perennis L. Sp. Pl. 886. 1753.

Scapose tufted perennial 6–15 cm. high. Leaves essentially basal, spatulate or obovate, 2–7 cm. long (including the narrowly margined petiole), obtuse to retuse, cuneate at base, shallowly few-toothed, thin, feather-veined, sparsely pilose on both sides; scapes several, 1-headed, naked, thickened below the head, spreading-pilose, densely appressed-pilose toward apex; head 1.3–2.5 cm. wide; involucre 2-seriate, equal, about 4 mm. high, the phyllaries ovate or oval, obtuse, appressed, blackish green, sparsely pilose, ciliolate toward apex; rays about 50, white or pinkish, about 8 mm. long; achenes obovate, 1.5 mm. long, whitish, with thickened margin, finely hispidulous on the sides.

An escape from gardens, naturalized in British Columbia and well established west of the Cascade Mountains in Oregon and Washington southward in the coastal counties of California to Santa Barbara County; also in eastern United States. A native of Europe. March–July.

81. BOLTÒNIA L'Her. Sert. Angl. 27. 1788.

Glabrous and glaucescent perennials, leafy-stemmed, resembling *Aster*. Leaves alternate, lanceolate to linear, entire, firm, sessile, often vertical by a basal twist. Heads medium or small, radiate, usually numerous and cymose-panicled, pedunculate, the rays white to purplish or violet, the disk yellow. Involucre broad, about 3-seriate, slightly graduate, appressed, the phyllaries linear to oblong, pale and subchartaceous, sometimes subherbaceous toward apex, with glandular-thickened costa. Receptacle rounded, alveolate. Ray-flowers pistillate, the tube short, the rays linear-elliptic; disk-corollas with cylindric-funnelform throat, 5-toothed. Anthers entire at base, with triangular terminal appendages. Style-branches short, oblong, with very short, merely papillose, obtusish appendages. Ray-achenes trigonous, 3-winged; disk-achenes broadly obovate, very strongly compressed, broadly or narrowly 2-winged. Pappus of 2 (or in the ray 3) subulate awns and a circle of short, slender, squamellate bristles. [Dedicated to James Bolton, an English botanist of the 18th century.]

A genus of 4 or 5 species, all natives of the United States. Type species, *Boltonia glastifolia* (Hill) L'Her. (=*B. asteroides* (L.) L'Her.)

1. Boltonia asteroìdes (L.) L'Her. Boltonia. Fig. 5513.

Matricaria asteroides L. Mant. 116. 1767.
Boltonia asteroides L'Her. Sert. Angl. 27. 1788.
Boltonia latisquama var. *occidentalis* A. Gray, Syn. Fl. N. Amer. 1²: 166. 1884.
Boltonia occidentalis Howell, Fl. N.W. Amer. 305. 1900.

Herbaceous perennial with fibrous roots, 0.6–2 m. high, glabrous and glaucescent, the stem striate-angled, leafy. Leaves lanceolate or elliptic-lanceolate, 4.5–13.5 cm. long, 7–18 mm. wide, obtuse or acute, apiculate, narrowed to the sessile base, entire, tuberculate-roughened on margin, thick, pale green, with strong midrib and few usually obscure basal veins; heads about 1.8 cm. wide; ray-flowers about 30, the rays white, pink, or purplish, about 8 mm. long; achenes broadly obovate, about 2.5 mm. long, sparsely hispidulous, broadly 2-winged, the wings ciliate; awns about 1.5 mm. long.

River bottoms, probably introduced, occurring in northeastern Oregon in Union County, in northern Idaho, and at the mouth of the Columbia River; eastward to the Atlantic seaboard. Type locality: Virginia. Aug.–Sept.

82. MONÓPTILON Torr. & Gray ex A. Gray, Bost. Journ. Nat. Hist. 5: 106. *pl. 13, figs. 1–6.* 1845.

Low desert annuals. Leaves alternate, spatulate to linear, entire. Heads small, radiate, solitary at tips of branches and branchlets, the rays white to purple, the disk yellow. Involucre broad, nearly 1-seriate, equal, appressed, the phyllaries linear, acuminate, herbaceous, somewhat indurate and 1-ribbed toward base. Receptacle broad, flat, naked. Ray-flowers rather numerous, pistillate, 1-seriate, with elliptic rays; disk-corollas with short tube and cylindric throat, 5-toothed. Achenes more or less compressed, obovate or obovoid, hispidulous. Pappus similar in ray and disk, of a short-toothed cup and a single apically plumose bristle, or of 1–12 nonplumose bristles alternating with shorter laciniate paleae. Anthers entire at base, with ovate-lanceolate terminal appendages. Style-appendages deltoid, obtuse, merely papillose, much shorter than the stigmatic area. [Name Greek, one feather, from the pappus of the original species.]

A genus of 2 known species, natives of the deserts of southwestern United States and northern Mexico. Type species, *Monoptilon bellidiforme* Torr. & Gray.

Pappus of a dentate cup and a single, apically plumose bristle; disk-corollas densely pilose below.
1. *M. bellidiforme*.

Pappus of 1–12 nonplumose bristles, alternating with lacerate paleae; disk-corollas subglabrous to sparsely pilose below.
2. *M. bellioides*.

1. **Monoptilon bellidifórme** Torr. & Gray. Desert Star. Fig. 5514.

Monoptilon bellidiforme Torr. & Gray ex. A. Gray, Bost. Journ. Nat. Hist. **5**: 106. *pl. 13, figs. 1–6.* 1845.

Slender branching annual, spreading-hirsute with white hairs throughout, 1–5 cm. high, the branches usually ascending. Leaves spatulate to linear-oblanceolate, 4–10 mm. long, 0.5–2.5 mm. wide, obtuse to acute, narrowed into a petioliform base, entire; heads about 1 cm. wide, solitary at tips of branches and branchlets; involucre about 3.5 mm. high, leafy-bracted at base, the proper phyllaries about 18, nearly 1-seriate, narrowly linear, acuminate, herbaceous, with 1 strong whitish rib toward base; ray-flowers about 18, the rays white, rosy-tinged, or purple, oval or obovate, about 5 mm. long, the tube densely pilose; tube and base of throat of disk-corollas rather densely pilose; achenes obovate, hispidulous, about 2 mm. long; pappus a short, few-toothed cup, split on one side, and a single bristle (very rarely 2) longer than achene and short-plumose toward tip, or the bristle rarely wanting.

Sandy or stony desert places, less common than the following species, Lower Sonoran Zone; Mojave Desert and northern border of Colorado Desert, California, east to northwestern Arizona, Nevada, and southern Utah. Type locality: unknown. March–May.

2. **Monoptilon bellioìdes** (A. Gray) Hall. Mohave Desert Star. Fig. 5515.

Eremiastrum bellioides A. Gray, Mem. Amer. Acad. II. **5**: 321. 1854.
Eremiastrum orcuttii S. Wats. Proc. Amer. Acad. **25**: 132. 1890.
Eremiastrum bellioides orcuttii Coville, Contr. U.S. Nat. Herb. **4**: 125. 1893.
Monoptilon bellioides H. M. Hall, Univ. Calif. Pub. Bot. **3**: 75. 1907.

Very similar to *M. bellidiforme* but often larger, the branches up to 15 cm. long. Leaves up to 2 cm. long; heads 1.3–1.8 cm. wide; involucre 4–6 mm. high; ray-flowers white or apparently violet, up to 7 mm. long, the tube rather sparsely pilose; base of disk-corollas obscurely puberulous to sparsely long-pilose; achenes somewhat thickened, obovoid, hispidulous; pappus of 1–12 merely hispidulous bristles and about as many lacerate to setose-dissected squamellae about half as long.

Common in sandy or stony desert places, Lower Sonoran Zone; Inyo County to Imperial County, California, east to Arizona, Nevada, Utah, and northwestern Mexico. Type locality: "On the Californian desert, not far west of the Colorado." Jan.–Sept.

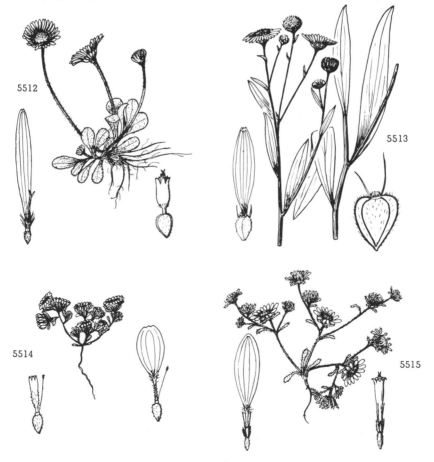

5512

5513

5514

5515

5512. **Bellis perennis**
5513. **Boltonia asteroides**

5514. **Monoptilon bellidiforme**
5515. **Monoptilon bellioides**

83. TOWNSÉNDIA Hook. Fl. Bor. Amer. 2: 16. *pl. 119*. 1834.

Depressed or low, many-stemmed herbs, sometimes scapose, usually cinereous-strigose, resembling *Aster*. Leaves alternate, the basal usually crowded, linear to spatulate, entire. Heads medium to large, solitary at tips of stems and branches, radiate, the rays white, pink, purple, or violet, the disk yellow. Involucre broad, few- to several-seriate, somewhat graduate, appressed, the phyllaries mostly lanceolate, with green center and white or pink-tinged, subscarious margin. Receptacle flat, broad, naked. Ray-flowers numerous, pistillate, with elliptic to linear-elliptic, 3-toothed rays; disk-corollas with short tube and cylindric throat, 5-toothed. Achenes obovate or oblong (those of the rays sometimes trigonous), strongly compressed, thick-margined, usually pubescent with often forked or glochidiate hairs. Pappus of rather numerous 1-seriate, narrow, barbellate awns or squamellae. Anthers subentire at base, with lance-ovate terminal appendages. Style-branches with ovate or triangular-ovate, acutish, hispidulous appendages, much shorter than the stigmatic area. [Named in honor of David Townsend of Philadelphia, associate of William Darlington.]

A genus of 21 species, one confined to Mexico and the others all of the western United States and Canada, two of them entering Mexico. Type species, *Aster ? exscapus* Richards.

Pappus of mature achenes persistent; plants not with long, loose, woolly-villous hairs.
    Plants subscapose or acaulescent, the heads pedunculate or sessile in the tufted basal leaves.
        Phyllaries linear to narrowly lanceolate, 5–7-seriate.   2. *T. leptotes.*
        Phyllaries broadly lanceolate to ovate or elliptic (more narrowly lanceolate in *T. scapigera*), 2–4(5)-seriate.
            Rays mostly blue to white; achenes often glabrous; alpine plants, Wallowa Mountains, Oregon.
                                                 1. *T. montana.*
            Rays pink or lavender; achenes pubescent; plants mostly of pinyon-juniper forest, eastern California.
                                                 5. *T. scapigera.*
    Plants with erect or suberect, leafy stems, the heads terminating the branches.
        Phyllaries acuminate, mostly 5-seriate.   3. *T. parryi.*
        Phyllaries acute, mostly 3–4-seriate.   4. *T. florifer.*
Pappus of mature achenes deciduous; plants with long, loose, woolly-villous hairs.   6. *T. condensata.*

1. Townsendia montàna M. E. Jones. Mountain Townsendia. Fig. 5516.

Townsendia montana M. E. Jones, Zoe 4: 262. 1893.
Townsendia alpigena Piper, Bull. Torrey Club 27: 394. 1900.
Townsendia dejecta A. Nels. Bot. Gaz. 37: 267. 1904.

Rosulate-pulvinate perennial with a taproot and branched often partly subterranean caudex; scapes to about 6.5 cm. high (rarely sessile), cinereous-pubescent with appressed or ascending hairs. Leaves 1–2.8 cm. long (including petiole), 2–6 mm. wide, spatulate, rounded to acute or apiculate, thick, rather sparsely strigose to nearly or quite glabrous on one or both sides; involucre 6–12 mm. high, 8–15 mm. wide, 3–4-seriate, graduate, the phyllaries oblong, obtuse or the innermost sometimes acutish, ciliate and along center substrigose, the margins purplish; ray-flowers about 13–18, the rays 5–7 mm. long, blue or white; ray-achenes about 4 mm. long, from nearly glabrous to rather sparsely pilose with forked or glochidiate hairs, the pappus-awns about 5 mm. long; disk-achenes nearly or quite glabrous except for a few hairs at base, their pappus 5 mm. long, equaling the corollas.

Rocky or gravelly places, Canadian and Transition Zones; subalpine ridges of the Wallowa Mountains, Oregon, and Idaho east to southwestern Montana, western Wyoming, and south to Utah. Type locality: Alta, Utah, above the Flagstaff mine, altitude about 2,900 meters. May–July.

2. Townsendia leptòtes (A. Gray) Osterhout. Common Townsendia. Fig. 5517.

Townsendia sericea var. *leptotes* A. Gray, Proc. Amer. Acad. 16: 85. 1880.
Townsendia leptotes Osterhout, Muhlenbergia 4: 69. 1908 (as *lepotes*).

Rosulate-pulvinate perennial with a taproot and usually much-branched, subterranean caudex, with sessile or pedunculate heads, the scapes to about 3 cm. high. Leaves up to 6 cm. long including the petiole and to 3.5 mm. wide, linear to oblanceolate or spatulate, usually involute, glabrous to densely strigose; involucre 5–15 cm. high, 8–23 cm. wide, 4–7-seriate, the phyllaries lanceolate to linear, mostly acute, glabrous or lightly strigose, the scarious margin ciliate; ray-flowers about 15–40, the rays 8–14 mm. long, whitish, pink, or blue; achenes about 4 mm. or more long, oblanceolate, the ray and disk similar, more or less hirsute with forked hairs, the pappus exceeding the corollas.

Mountain slopes and crests, Boreal Zone; in the Pacific States in the White Mountains of Mono County, California, and from western Montana to Idaho south to northwestern New Mexico and central Nevada. Type locality: Middle Park, Grand County, Colorado. June–Aug.

3. Townsendia párryi D. C. Eaton. Parry's Townsendia. Fig. 5518.

Townsendia parryi D. C. Eaton in Parry, Amer. Nat. 8: 212. 1874.

Biennial or short-lived perennial; stems simple, 1 to many, 2–20 cm. high, stout, rather sparsely leafy, cinereous-hirsute with subappressed to loosely ascending hairs, usually simple and 1-headed. Basal leaves tufted, spatulate or oblanceolate, 1–6 cm. long including petiole, 2–5 mm. wide, acute to obtuse, narrowed into long petioles, glabrous above, substrigose beneath; stem-leaves similar; involucre 1–1.4 cm. high, up to 4 cm. wide, about 5-seriate, graduate, the phyllaries lance-oblong or the outermost lanceolate, long-acuminate, lacerate-ciliate and along center substrigose; rays about 40 or more, blue to violet, 12–25 mm. long; ray-achenes about 4 mm. long, densely hispidulous with entire or bidentate hairs, their pappus of barbellate awns about 5 mm. long; disk-achenes similar, their pappus about 5 mm. long, about equaling the corolla.

Subalpine ridges, Canadian Zone; Wallowa Mountains, Oregon, in the Pacific States; British Columbia and Alberta south to Idaho and Wyoming. Type locality: Wind River Range, Wyoming, altitude 2,745 meters. May–Aug.

4. Townsendia flòrifer (Hook.) A. Gray. Showy Townsendia. Fig. 5519.

Erigeron ? florifer Hook. Fl. Bor. Amer. **2**: 20. 1834.
Townsendia florifer A. Gray, Proc. Amer. Acad. **16**: 84. 1880.
Townsendia scapigera var. *ambigua* A. Gray, loc. cit.
Townsendia watsoni A. Gray, loc. cit.

Many-stemmed biennial or winter annual, 5–20 cm. high, with spreading or ascending stems, cinereous throughout with appressed to loosely ascending hairs; stems leafy, usually branching. Basal leaves tufted, spatulate or rarely oblanceolate, 1.5–4 cm. long including petiole, 2–7 mm. wide, rounded to acute, narrowed into the long petiole; stem-leaves similar but much narrower; heads pedunculate; involucre 7–10 mm. high, 15–35 mm. wide, about 3-seriate, graduate, the phyllaries lanceolate or the outermost linear-lanceolate, acute, the margins often purplish-tinged; ray-flowers about 18–30, white or pink, the rays 6–12 mm. long; ray-achenes 4 mm. long, densely hirsute-pilose with bidentate hairs, their pappus of barbellate awns about 3–6 mm. long; disk-achenes similar, their pappus about 6 mm. long, exceeding the corolla.

Gravelly or sandy hillsides and deserts, Arid Transition Zone; east of the Cascade Mountains in Washington and Oregon, and from Idaho south to Nevada and Utah. Type locality: "near Priest's Rapids of the Columbia," Washington. May–July.

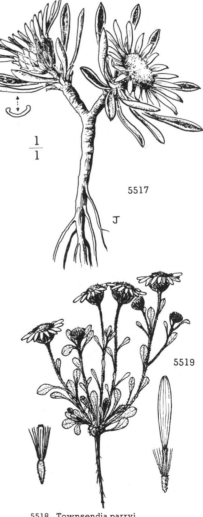

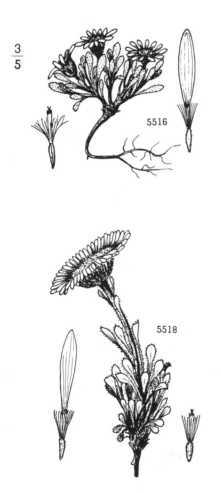

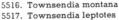

5516. Townsendia montana
5517. Townsendia leptotes

5518. Townsendia parryi
5519. Townsendia florifer

5. Townsendia scapígera D. C. Eaton. Ground-daisy. Fig. 5520.

Townsendia scapigera D. C. Eaton, Bot. King Expl. 145. *pl. 17. figs. 1–7.* 1871.
Townsendia scapigera var. *caulescens* D. C. Eaton, loc. cit.

Subscapose, taprooted perennial or biennial, 8 cm. high or less, with numerous simple stems from a much-branched, sometimes subterranean caudex, cinereous throughout (except on invo-lucres) with appressed or subappressed hairs. Basal leaves 1–4.5 cm. long including petiole, 2–6 mm. wide, spatulate, usually obtuse, tapering into the long petiole, strigose; stem-leaves wanting or few and subbasal; heads solitary, the peduncles up to 10 cm. high; involucre 7–14 mm. high, 12–32 mm. wide, about 3-seriate, graduate, the phyllaries oblong, acute to subacuminate, ciliate and along center substrigose, often purplish-tinged; ray-flowers about 15–35, the rays white, pinkish, or dull red, 6.5–16 mm. long; ray-achenes 5 mm. long, hirsute-pilose with forked hairs, the pappus 6 mm. long, of barbellate awns; disk-achenes similar, the pappus usually longer, equaling or sur-passing the corollas.

Rocky ridges in the pinyon-juniper forest, Arid Transition Zone; Modoc, Mono, and Inyo Counties, California, to Nevada. Type locality: "Dry rocky ridges in the Trinity and Pah-Ute Mountains, Nevada." May–June.

6. Townsendia condensàta D. C. Eaton. Cushion Townsendia. Fig. 5521.

Townsendia condensata D. C. Eaton in Parry, Amer. Nat. **8**: 213. 1874.
Townsendia anomala Heiser, Madroño **9**: 240. 1948.

Pulvinate-cespitose, short-lived perennial from a taproot with a branching caudex, densely woolly-villous throughout, 3–5 cm. high. Leaves crowded, spatulate, tapering to a petiole longer than the blade, 1–3.5 cm. long and about 2–4 mm. wide; heads sessile or subsessile, the involucres 8–18 mm. high, 10–40 mm. wide; phyllaries pinkish, lanceolate, acuminate, the inner weakly at-tenuate, narrowly scarious-margined and laciniate-ciliate; ray-flowers 12–100, the rays 1 cm. or more long, white, pink, or lavender; achenes pubescent, the pappus of ray- and disk-flowers similar, of slender setae as long as the disk-flowers.

Mountain slopes, Boreal Zone; western Montana, Wyoming, and Idaho, and in the Pacific States from the Sweetwater and White Mountains, Mono County, California. Type locality: "Black Hills of the Platte." Col-lected by Nuttall. June–Aug.

Townsendia mensàna var. **jònesii** Beaman, Contr. Gray Herb. No. 183: 88. 1957. A depressed acaulescent perennial with broadly lanceolate to narrowly obovate leaves which exceed the heads has been collected in the Charleston Mountains, Nevada. This may be expected to occur in adjacent high desert ranges in California.

84. TRACYÌNA* Blake, Madroño 4:74. 1937.

Slender annuals with narrow alternate leaves. Heads many-flowered, terminating slen-der, ascending or erect peduncles, heterogamous, the tiny rays erect. Involucre turbinate, the phyllaries 3–4-seriate, plane, linear, acuminate or attenuate, appressed, narrowly scari-ous-margined. Receptacle small, naked. Ray-flowers 1-seriate, the corollas filiform, slightly reddish-tinged; disk-flowers even more inconspicuous. Achenes linear-fusiform, subterete, 5-nerved, tapering above to a short sterile beak, dilated at very tip. Pappus of persistent but very fragile capillary bristles. [Named in honor of Joseph P. Tracy, Cali-fornia botanist.]

A monotypic California genus.

1. Tracyina rostràta Blake. Tracyina. Fig. 5522.

Tracyina rostrata Blake, Madroño **4**: 75. *fig. 1.* 1937.

Habit of *Rigiopappus leptocladus*, the plant 15–30 cm. high; stem solitary, simple or occasion-ally with few erect branches from base, usually with 2–4 filiform ascending branches subterminally, glabrous. Leaves appressed, narrowly lance-linear, up to 2 cm. long, ciliate-margined, callous-apiculate; peduncle loosely pilose beneath the head; involucre 6–7 mm. high, the phyllaries glabrous, deciduous; ray-flowers about 12–20, the corollas 4–5 mm. long; disk-flowers about 15–25, the corollas very slender, shorter than the rays; achenes brown, 5–5.5 mm. long, hirsutulous; pappus of about 36–38 graduated bristles, the majority longer than and nearly concealing the disk-corollas, not quite as long as the rays.

Dry grassy slopes at low elevations, Humid Transition Zone; locally frequent in southern Humboldt County, California. Type locality: "Alder Point, on Eel River." May–June.

85. PSILÁCTIS A. Gray, Mem. Amer. Acad. II. 4:71. 1849.

Desert annuals, more or less glandular and pubescent. Leaves alternate, entire to pin-natifid, the lower petioled, the upper sessile. Heads small, solitary at tips of stems and branches, radiate, the rays white, violet, or purplish, the disk yellow. Involucre hemi-spheric or campanulate, about 3-seriate, graduate, the phyllaries narrow, with pale dry base and acute, subequal, sometimes spreading, herbaceous tip. Receptacle flat, naked. Rays pistillate, often infertile, narrow, 3-denticulate; disk fertile, the corollas with cylindric-

_____
* Text contributed by David Daniels Keck.

funnelform throat, 5-toothed. Achenes nearly linear, somewhat compressed. Pappus in the ray-achenes none or an obscure ring, in the disk-achenes of unequal or subequal, capillary bristles. Anthers entire at base with narrowly triangular, terminal appendages. Style-branches with triangular, acute, hispidulous appendages, much shorter than the stigmatic area. [Name Greek, meaning naked ray.]

A genus of 5 species, occurring from Texas to California and Mexico. Besides the following, another species occurs from Texas to Arizona. Type species, *Psilactis asteroides* A. Gray.

1. **Psilactis coùlteri** A. Gray. Silver Lake Daisy. Fig. 5523.

Psilactis coulteri A. Gray, Mem. Amer. Acad. II, 4: 72. 1849.

Annual with ascending basal branches 5–30 cm. high, the stems stipitate-glandular and loosely spreading-pilose. Lower leaves petioled, 2–3 cm. long (including petiole), oblong in outline, pin-natifid or bipinnatifid, stipitate-glandular; stem-leaves narrow-oblong, about 1 cm. long, sessile and somewhat clasping, thick, pale green, laciniate-toothed or pinnatifid, the teeth spinescent-tipped; leaves of the branches smaller, sometimes entire; heads about 2 cm. wide; involucre about 4 mm. high, about 3-seriate, graduate, glandular-granular, the phyllaries oblong, pale below, with sub-equal, acuminate, herbaceous tips; rays about 30, lavender, about 6 mm. long; achenes pubescent, those of the ray epappose, those of the disk with a pappus of about 40 unequal bristles up to 2 mm. long.

In desert sand or gravel, Lower Sonoran Zone; Mojave Desert, San Bernardino County, California, southern Arizona, and Sonora. Type locality: Mexico. Collected by Coulter. March–April.

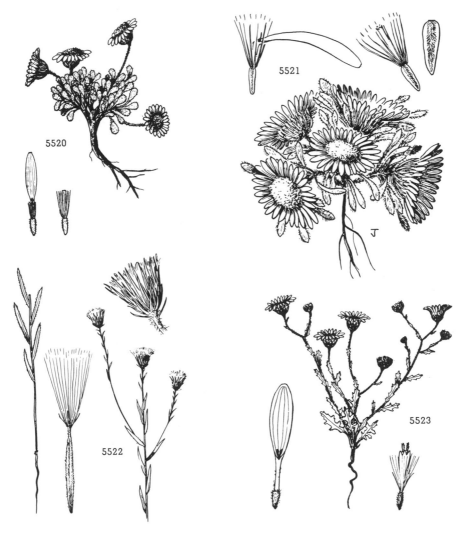

5520. Townsendia scapigera
5521. Townsendia condensata
5522. Tracyina rostrata
5523. Psilactis coulteri

86. SERICOCÁRPUS Nees, Gen. & Sp. Aster. 148. 1832.

Perennial herbs, leafy-stemmed, resembling *Aster*. Leaves linear to elliptic or obovate, entire or toothed, usually sessile, firm. Heads small, fasciculate-cymose in terminal panicles, radiate, white, the anthers often purple. Involucre narrow, few- to several-seriate, gradu-ate, the phyllaries mostly oblong, with indurate white base and (at least in the outer) abrupt herbaceous tip. Receptacle narrow, flattish, alveolate. Ray-flowers few, pistillate, with elliptic ligule; disk-flowers rather numerous, their corollas with funnelform throat about equal to tube, deeply 5-toothed. Anthers subsagittate at base, with lance-subulate ter-minal appendages, ours purple. Style-branches with elongate-triangular, papillose, basally hispidulous appendages much longer than the stigmatic area. Achenes turbinate to nearly linear, plump or those of the disk somewhat compressed, silky-pubescent, about 8-nerved. Pappus rather copious, about 3-seriate, somewhat graduate, the bristles capillary, hispidu-lous. [Name Greek, meaning silky fruit.]

A genus of 4 species, all of the United States and Canada. Type species, *Aster solidagineus* Michx. (=*S. linifolius* (L.) B.S.P.).

Rays shorter than the pappus, typically 2; inflorescence of heads in a solitary terminal cluster.
1. *S. rigidus.*

Rays exceeding the pappus, typically 5; inflorescence of clusters of heads paniculately disposed.
2. *S. oregonensis.*

1. **Sericocarpus rígidus** Lindl. Rigid White-topped Aster. Fig. 5524.

Sericocarpus rigidus Lindl. in Hook. Fl. Bor. Amer. 2 : 14. 1834.
Sericocarpus rigidus var. *laevicaulis* Nutt. Trans. Amer. Phil. Soc. II. 7 : 302. 1840.
Aster curtus Cronquist, Vasc. Pl. Pacif. Northw. 5 : 80. 1955.

Perennial herb from slender creeping rhizome, the leafy stems usually simple below and glabrous or sparsely hispidulous, 1–3 dm. high. Lowermost leaves usually soon deciduous, the larger above the base, 2.5–3.5 cm. long, 5–9 mm. wide, oblanceolate, tapering to the base, obtuse or acutish at apex, hispidulous-ciliate on the margin, tending to be 3-nerved; inflorescence usually a simple fasciculate cyme, more rarely branched, the involucres 7–9 mm. high, narrow; phyllaries imbricate in a series of 3–4, oblong or the inner linear, obtuse or acute, erect or the green tip somewhat spread-ing; rays white, 1–3, usually 2, shorter than the pappus, 1–3 mm. long; disk-flowers few, pale yellow with purple anthers; achenes canescent, about 2.5 mm. long, pappus white, about 7 mm. long.

Prairies, Humid and Arid Transition Zones; western Washington and adjacent Oregon; also occurring in the Klamath region of southern Oregon. Type locality: Columbia River. Collected by Scouler. July–Aug.

2. **Sericocarpus oregonénsis** Nutt. Oregon White-topped Aster. Fig. 5525.

Sericocarpus oregonensis Nutt. Trans. Amer. Phil. Soc. II. 7 : 302. 1840.
Aster oregonensis Cronquist, Vasc. Pl. Pacif. Northw. 5 : 91. 1955.

Perennial from a woody caudex sometimes elongated, stems 4–12 dm. high, glabrous or nearly

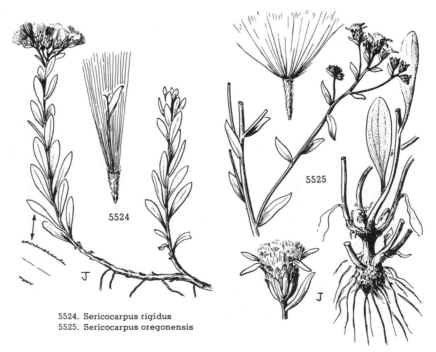

5524. Sericocarpus rigidus
5525. Sericocarpus oregonensis

so. Lowest leaves narrowed to a petiole, early deciduous, the stem-leaves 4–8 cm. long, oblanceolate, usually sharply acute above, narrowed below, midvein prominent, scabrous above, somewhat glandular and sometimes scabrous beneath; inflorescence usually many-flowered, paniculate, the heads in close, short-peduncled clusters on the branches; involucres 7–8 mm. high, more or less turbinate; midrib of the phyllaries slightly keeled, the green tips herbaceous, usually loose and spreading; rays 4–7, usually 5, 4–7 mm. long, surpassing the pappus; achenes about 2.5 mm. long, densely appressed-hairy.

In woodlands, Humid and Arid Transition Zones; western Washington through western Oregon to northwestern California as far south as Humboldt County. Type locality: near Fort Vancouver, Washington. Collected by Nuttall. July–Sept.

Sericocarpus oregonensis subsp. **califórnicus** (Durand) Ferris, Contr. Dudley Herb. **5**: 100. 1958. (*Sericocarpus californicus* Durand, Journ. Acad. Phila. II **3**: 90. 1855; *S. rigidus* var. *californicus* Blake, Proc. Amer. Acad. **51**: 515. 1916.) Habit as in the preceding subspecies but stem more or less densely hirsute or pilose with spreading, many-celled hairs; leaves often large, up to 9 cm. long and 1.5 cm. wide, and somewhat wavy-margined. Sierra Nevada of California from Plumas County to Tulare County. Type locality: Nevada City, California.

87. **ASTER** L. Sp. Pl. 872. 1753.

Ours perennial herbs, more rarely annuals, biennials, or plants basally suffrutescent, the stems arising from a creeping rhizome, or a more or less horizontal woody caudex with numerous fiberous roots, or from a taproot or taproot-like caudex. Leaves alternate, entire or toothed. Inflorescence corymbose, racemose, or paniculate, the heads rarely solitary. Involucres more or less definitely graduated, the phyllaries usually with herbaceous tips. Ray-flowers shades of purple, more rarely white, sometimes lacking; disk-flowers yellow, sometimes white or purplish. Achenes hairy or glabrous, the pappus of subequal, capillary, persistent bristles, sometimes with a few outer bristles. [Name Greek, meaning star, because of the star-like heads of flowers.]

A genus of 250 or more species, of world-wide distribution in temperate regions. Type species, *Aster amellus* L. (lectotype).

In most cases complete synonymy is not given; only those names applied to Pacific States entities are listed.

Plants perennial.
    Stems monocephalous, subscapose, arising from a short caudex terminating a deep conical root; alpine or subalpine plants with basal tufts of grass-like leaves.   IV. Oreostemma.
    Stems usually with several to many heads, not arising from a deep conical root.
        Heads solitary at tips of stems and branches; low-tufted plants with narrow, 1-nerved, veinless, cuspidate leaves; pappus double, the outer bristles shorter than the inner.   III. Ianthe.
        Heads several or numerous, if rarely solitary, not with leaves as above; pappus mostly in a single row.
            Plant woody toward base, glaucous and glabrous, paniculately much branched, with discoid heads; or else herbaceous, glabrous, striate-stemmed, often spiny, and paniculately much branched, the branches essentially naked.   V. Leucosyris.
            Plants otherwise.
                Phyllaries oblong to ovate, rarely lance-linear, chartaceous with narrow scarious margin and often herbaceous tip, more or less keeled by the strong midnerve.   II. Eucephalus.
                Phyllaries usually linear (varying to oblong), usually indurate below and herbaceous-tipped, or the outer herbaceous throughout, without strong midnerve.
                    Plants with leaves various, often toothed or petioled and if at all glandular, lacking sessile, grass-like leaves.   I. Euaster.
                    Plants (in ours) with sessile, grass-like leaves at base and above, glabrous except for the glandular inflorescence.   VI. Orthomeris.
Plants annual.
    Phyllaries distinctly graduate; rays in one series, evident and surpassing the pappus.
                VII. Oxytripolium.
    Phyllaries subequal or slightly graduate; rays if present in more than one series, very inconspicuous and equaling or shorter than the mature pappus.   VIII. Conyzopsis.

I. Euaster.

Involucre (at least on the margins of phyllaries) and usually the branches of inflorescence glandular.
    Leaves lanceolate to obovate; heads usually few, comparatively large.
        Leaves thick and firm, lance-ovate to obovate; phyllaries with indurate base.
            Leaves sharply serrate, mostly lance-ovate, scarcely clasping; phyllaries strongly graduate, the indurate base much longer than the herbaceous tip.   1. *A. conspicuus.*
            Leaves entire, at least the lower obovate, with conspicuously clasping base; phyllaries but little graduate, the indurate base shorter than the herbaceous tip. 2. *A. integrifolius.*
        Leaves thin, chiefly lanceolate; phyllaries herbaceous throughout.   3. *A. modestus.*
    Leaves linear or linear-lanceolate, 6 mm. wide or less; heads small, numerous.
                6. *A. campestris.*
Involucre and branches of inflorescence not glandular.
    Phyllaries usually with purple tips or margins, oblong to lanceolate, usually strongly graduate; leaves obovate to elliptic-oblong, firm, usually rough beneath or on both sides.
        Leaves sharply serrate; heads several or numerous, in a nearly naked cyme or cymose panicle.
                4. *A. radulinus.*
        Leaves entire to obtusely serrate; heads 1 to few, in axils of leafy bracts.
                5. *A. sibericus meritus.*
    Phyllaries not purple-tipped nor margined (except in *paludicola* and some forms of *foliaceus*), variable in shape; leaves not rough beneath (except *A. greatai*).
        Heads usually numerous, in an open, nearly naked panicle; larger leaves 2–4 cm. wide.
            Stem and leaves glaucescent or very pale green; leaves thick; achenes essentially glabrous.
                7. *A. laevis geyeri.*

Stem and leaves not glaucescent; leaves thin; achenes pubescent.
         8. *A. greatai.*
Heads, when numerous, in leafy-bracted panicles or nearly naked cymes or cymose panicles.
   Heads mostly not small, not racemosely arranged along the branches, the rays usually blue to purple; phyllaries not mucronulate, the outer appressed or spreading.
     Larger leaves obovate to oblong, usually more than 2 cm. wide, with conspicuously clasping bases.
       Heads crowded toward tips of stem and branches, usually short-pedicelled or subsessile; involucre 6–8 mm. high, rarely leafy-bracted, the phyllaries densely pilosulose.
         9. *A. jessicae.*
       Heads solitary or few at tips of stem and branches, distinctly peduncled.
         12. *A. foliaceus.*
     Larger leaves lanceolate or linear, 0.5–2 cm. wide, slightly if at all clasping.
       Plants of cold bogs; heads solitary at tips of widely divergent branches or branchlets.
         11. *A. paludicola.*
       Plants not of cold bogs; heads not solitary at tips of widely divergent branches or branchlets.
         Inflorescence a leafy panicle; heads many (sometimes few in *A. subspicatus*).
           Branches of inflorescence and peduncles not conspicuously bracteate with small leaves; involucre usually only slightly graduate (more so in *A. hesperius*).
             Leaves mostly linear or linear-lanceolate, entire, the lowest rarely serrate.
               Coarse plants 1–1.5 m. high; pubescence of stems and branchlets in decurrent lines from the leaf bases; involucre rarely if at all with subequal subtending bracts.   10. *A. hesperius.*
               Plants usually less than 1 m. high; pubescence of stem uniform, or if in lines at least uniform under the heads or nearly glabrate.
                 14. *A. eatonii.*
             Leaves chiefly oblanceolate or lanceolate, even the middle ones mostly serrate or serrulate.   13. *A. subspicatus.*
           Branches of the inflorescence and peduncles conspicuously bracteate with small leaves; involucre strongly graduate.
             Inflorescence an open, sometimes widely divergent panicle; involucre 7–10 mm. high; plants of northern and central California.
               18. *A. chilensis.*
             Inflorescence a close panicle or raceme; involucre 4–6 mm. high.
               Herbage including phyllaries pubescent with spreading hairs; plants of southern California.   19. *A. bernardinus.*
               Herbage essentially glabrous (sometimes scabrous on the leaf-margins); plants of western Washington and Oregon.
                 15. *A. hallii.*
         Inflorescence essentially a naked cyme or cymose panicle; heads usually few (often many in *A. adscendens*).
           Involucre definitely graduate, in 3, sometimes 4, series.
             Stem-leaves mostly linear, not reduced in size on the upper stems; branches of the inflorescence always ascending and often many-flowered.
               16. *A. adscendens.*
             Upper stem-leaves few and reduced in size; the cymose panicle open with spreading branches.   17. *A. occidentalis intermedius.*
           Involucre scarcely graduate, the phyllaries in 2 series.
             17. *A. occidentalis.*
   Heads small and very numerous, racemosely arranged on the usually spreading branches; rays white (rarely purple); phyllaries mucronulate, at least the outer spreading.
     Plants with well-developed creeping rhizomes; disk 6–7 mm. high, the rays 22–30.
         20. *A. falcatus crassulus.*
     Plants with stems clustered from a short rhizome or caudex; disk 4–5 mm. high, the rays 12–20.
         21. *A. pansus.*

II. Eucephalus.

Leaves conspicuously pale beneath with a more or less dense tomentum.
   Rays 0–4; inner phyllaries ovate, obtuse, canescent.   22. *A. brickellioides.*
   Rays 6–10; inner phyllaries lanceolate, glandular.   25. *A. ledophyllus.*
Leaves not conspicuously pale beneath, the undersurface glabrous, scabrous-puberulent, or sparsely pilose.
   Heads discoid or occasionally with 2–3 rays.
     Inflorescence corymbiform with elongated branchlets; peduncles glabrous or somewhat glandular beneath the involucre.   23. *A. siskiyouensis.*
     Inflorescence an elongated panicle with short branchlets; peduncles glandular and somewhat hirsute.
       24. *A. vialis.*
   Heads radiate with 6–18 rays.
     Involucre narrow, the phyllaries closely imbricate and appressed in age; leaves scabrous-puberulent on both surfaces and margin.   26. *A. perelegans.*
     Involucre broad, hemispheric, or campanulate, the phyllaries more or less loosely imbricate and spreading in age; leaves not scabrous-puberulent on both surfaces and margin.
       Inflorescence corymbiform, with several to many heads; plants robust, 6–15 dm. tall.
         Leaves glaucescent, about eight or more times as long as wide, the margins obscurely serrulate.
           27. *A. glaucescens.*
         Leaves not glaucescent, three to six times as long as wide, the margins entire.
           28. *A. engelmannii.*
       Heads solitary, rarely 3 or 4; plants 1.5–3.5(5) dm. tall.
         Plants from a woody caudex; phyllaries lanceolate-acuminate; plants of the Olympic Mountains.
           29. *A. paucicapitatus.*
         Plants from a creeping woody rhizome; phyllaries oblong-ovate to ovate; plants of the vicinity of Mount Jefferson, Oregon.   30. *A. gormanii.*

III. Ianthe.

Leaves linear, usually 2–3 cm. long, not definitely callous-margined; plant green.
         31. *A. stenomeres.*

Leaves chiefly elliptic or linear-oblong, usually under 1.5 cm. long, with whitish callous margin; plant subcinereous.
         32. *A. scopulorum.*

IV. Oreostemma.

Stems and involucres more or less woolly-pubescent to glabrous, not glandular.
> Stems and involucres more or less woolly-pubescent; plants 2–30 cm. tall (sometimes taller in subsp. *andersonii*).   33. *A. alpigenus.*
> Stems and involucres completely glabrous; plants 30–70 cm. tall.   34. *A. elatus.*
> Stems and involucres more or less glandular-scaberulous.   35. *A. peirsonii.*

V. Leucosyris.

Plant essentially herbaceous, not glaucous, often with spines in or above the leaf-axils; heads radiate, white; achenes glabrous.   36. *A. spinosus.*
Plant shrubby, glaucous, not spiny; heads discoid, yellowish; achenes pubescent.   37. *A. intricatus.*

VI. Orthomeris.

Pappus simple; leaves grass-like; a single species.   38. *A. pauciflorus.*

VII. Oxytripolium.

Annual, glabrous; leaves linear or lance-linear, usually entire; heads numerous, panicled; rays pink, little-exserted; a single species.   39. *A. exilis.*

VIII. Conyzopsis.

Phyllaries mostly oblanceolate or spatulate, obtuse; rays about 2 mm. long.   40. *A. frondosus.*
Phyllaries linear, acute; rays wanting or vestigial.   41. *A. brachyactis.*

1. Aster conspícuus Lindl. Showy Aster. Fig. 5526.

Aster conspicuus Lindl. in Hook. Fl. Bor. Amer. 2: 7. 1834.
Aster macdougali Coult. & Fisher, Bot. Gaz. 18: 301. 1893.

Rootstock elongate, woody, stoloniferous; stems usually solitary, stout, simple below the inflorescence, usually 30–60 cm. high, leafy, densely glandular, especially in the inflorescence, sparsely pilose or hirsute. Leaves oblong-ovate to oval or obovate-oval, the larger 8–17 cm. long, 2.5–8 cm. wide, acute, sessile and usually clasping, sharply toothed except toward base, usually firm, veiny, scabrous on both sides; heads few or rather numerous, 2.5–4 cm. wide, in a usually rounded cyme or cymose panicle; involucre broadly campanulate, strongly graduate, about 6-seriate, 7–10 mm. high, the phyllaries linear-oblong to lance-oblong or oblong, densely glandular, strongly ciliate, with pale, indurate, 1-ribbed base and usually shorter, often spreading, acute to acuminate, herbaceous tip, the outermost sometimes entirely herbaceous; rays about 20–30, violet, 1–1.5 cm. long; achenes appressed-pubescent.

Forests and open woods, Arid Transition and Canadian Zones; British Columbia south through eastern Washington to northeastern Oregon and east to Saskatchewan, Wyoming, and South Dakota. Type locality: "Carlton House on the Saskatchawan River to the Rocky Mountains." July–Sept.

2. Aster integrifòlius Nutt. Entire-leaved Aster. Fig. 5527.

Aster integrifolius Nutt. Trans. Amer. Phil. Soc. II. 7: 291. 1840.
Aster amplexifolius Rydb. Mem. N.Y. Bot. Gard. 1: 391. 1900.

Stems several, stout, ascending from a short, woody, fibrous-rooted rootstock, 2–7 dm. high, often purplish, from densely crisped-pilose to nearly glabrous below, densely stipitate-glandular and villous above. Lowest leaves obovate to elliptic-lanceolate, the blade acute or acuminate at each end, narrowed into a petiole, 5–18 cm. long, entire, sometimes crisped, veiny, firm, from essentially glabrous to short-pilose; lower stem-leaves obovate or oblanceolate, sessile and strongly clasping, the middle and upper usually much smaller, elliptic to lance-linear; heads few to several,

5526. Aster conspicuus

5527. Aster intergrifolius

2–4.5 cm. wide, mostly solitary at tips of peduncles, in an oblong false raceme or narrow panicle; involucre broadly campanulate, about 3–seriate, little graduate, the phyllaries mostly linear-oblong, with short indurate base and acute or acuminate, erect or spreading, herbaceous tip or the outer wholly herbaceous, densely glandular, often deep purple-tinged; rays about 10–18, purple or violet, about .1 cm. long; achenes short-pilose.

Dry, disintegrated, granite slopes and edges of meadows, Canadian and Hudsonian Zones; Blue Mountains of southeastern Washington, mountains of eastern Oregon, south to Tulare County in the Sierra Nevada, California; eastward to Montana and Colorado. Type locality: in the vicinity of Thornberg's Pass, Blaine County, Idaho. July–Sept.

3. **Aster modéstus** Lindl. Great Northern Aster. Fig. 5528.

Aster unalaschkensis β? major Hook. Fl. Bor. Amer. **2**: 7. 1834.
Aster modestus Lindl. in Hook. op. cit. 8.
Aster sayianus Nutt. Trans. Amer. Phil. Soc. II. **7**: 294. 1840.
Aster mutatus Torr. & Gray, Fl. N. Amer. **2**: 142. 1841.
Aster majus Porter, Mem. Torrey Club **5**: 325. 1894.

Stems usually solitary from a slender stoloniferous rootstock, slender, very leafy, 0.3–1 m. high, simple below the inflorescence, densely stipitate-glandular above or throughout, and usually sparsely or densely villous, sometimes glabrous below. Leaves nearly uniform, linear-lanceolate to oblong-lanceolate, 6–15 cm. long, 1–3 cm. wide, acuminate, sessile by a more or less clasping base, entire to sharply and remotely serrate, thin, rough or smooth above, glabrous or pubescent beneath; heads 1 to many, cymosely arranged, 2–3 cm. wide; involucre hemispheric, 2–3-seriate, subequal, about 7 mm. high, the phyllaries linear or linear-lanceolate, acuminate, spreading-tipped, densely stipitate-glandular, often purplish, the outer herbaceous throughout, the inner pale-margined below; rays about 20–30, purple or violet, about 1 cm. long; achenes strongly about 5-ribbed, sparsely pubescent.

Moist woods and stream banks, Arid Transition and Canadian Zones; Alaska to southwestern Oregon east to Montana, Minnesota, and western Ontario. Type locality: "Mountain woods, at the mouth of the Smoking River, lat. 56°." July–Sept.

4. **Aster radulìnus** A. Gray. Rough-leaved Aster. Fig. 5529.

Aster radulìnus A. Gray, Proc. Amer. Acad. **8**: 388. 1872.
Aster torreyi Porter, Bull. Torrey Club **17**: 37. *pl. 100.* 1890.
Aster eliasii A. Nels. Univ. Wyo. Publ. Sci. **1**: 68. 1924.

Stems solitary or several from a slender stoloniferous rootstock, 15–60 cm. high, usually simple below the inflorescence, ascending or erect, from rather densely spreading-hirsute to incurved-puberulous or strigillose, sometimes glabrous below, not glandular; leaves reduced to entire bracts above, the lowest obovate to oval, obtuse to acute, narrowed to short, ciliate, margined petioles, the lower and middle ones obovate to oval or lance-oblong, 4–12.5 cm. long, 1–7 cm. wide, acute, sessile, not clasping, sharply serrate or dentate except on the cuneate base, firm, scabrous-pubescent beneath, scabrous or smooth above; heads 1.5–2.5 cm. wide, several to numerous in a flat or rounded cyme or cymose panicle, the peduncles densely hirsutulous or hirsute-pilose; involucre 5–7-seriate, strongly graduate, 6–9 mm. high, the phyllaries oblong to (inner) linear, usually acute, hirsutulous on back, ciliate, with pale indurate base and shorter, appressed or very rarely spreading, herbaceous tip, at least the inner often purple-tipped or -margined; rays 10–15, pale violet to white, 5–11 mm. long, the white-rayed forms lacking purple color in the phyllaries; achenes sparsely hirsutulous; pappus brownish.

Dry wooded slopes in foothills and mountains, mostly Transition Zone, Vancouver Island and central Washington through Oregon (west of the Cascade Mountains and Ochoco Mountains on the eastern side) to San Luis Obispo County and to Mariposa County in the Sierra Nevada, California; also from Santa Cruz Island, Santa Barbara County. Type locality: Oregon. July–Oct.

The form described as *Aster eliasii* resembles to some degree robust forms of *A. sibiricus* var. *meritus* in habit but the ranges and habitats of the two do not overlap.

5. **Aster sibíricus** var. **méritus** (A. Nels.) Raup. Arctic Aster. Fig. 5530.

Aster sibiricus A. Gray, Syn. Fl. N. Amer. **1²**: 176. 1884. Not *A. sibiricus* L. 1753.
Aster meritus A. Nels. Bot. Gaz. **37**: 268. 1904.
Aster bakerensis St. John, in St. John & Hardin, Mazama **11**: 93. 1929.
Aster richardsonii var. *meritus* Raup, Contr. Arnold Arb. **6**: 204. 1934.
Aster sibiricus var. *meritus* Raup. Sargentia **6**: 240. 1947.

Stems 1 or several, often tufted, ascending or spreading to suberect, from slender stoloniferous rootstocks, 3–30 cm. high, simple or little branched, usually purple, pilosulose with usually loosely ascending hairs. Leaves gradually or abruptly reduced above, obovate to elliptic or the upper lanceolate, 1.5–7 cm. long, 0.5–2 cm. wide, acute, sessile and scarcely clasping or the lower narrowed to a ciliate, margined, petioliform base, entire, crenate, or depressed-serrate, firm, from nearly glabrous to cinerous-pilosulose beneath, smooth and essentially glabrous above; heads 2–3.5 cm. wide, solitary or few and cymose on monocephalous axillary peduncles, or rarely rather numerous in a leafy cymose panicle; involucre hemisphere, 5–9 mm. high, little graduate, the phyllaries oblong or lance-oblong, acute to acuminate, pilosulose on back (rarely glabrous), ciliate, with whitish indurate base and appressed or rarely loose herbaceous tip (the outermost sometimes herbaceous throughout), at least the inner with conspicuous purple tips or margins; rays about 15–25, purple to violet, 7–12 mm. long; achenes sparsely hirsutulous, strongly about 8-ribbed; pappus deeper brown than in *A. radulinus*, usually purplish-tinged.

Open rocky slopes, reaching 3,000 meters elevation, Hudsonian and Arctic-Alpine Zones; South Dakota and Wyoming to Alberta and British Columbia; known in our range from the northern Cascade Mountains in Washington, and the Wallowa Mountains, eastern Oregon. Type locality: Yellowstone National Park. July–Aug.

6. **Aster campéstris** Nutt. Western Meadow Aster. Fig. 5531.

Aster campestris Nutt. Trans. Amer. Phil. Soc. II. 7: 293. 1840.

Stems usually several, ascending from slender stoloniferous rootstocks, 1–5 dm. high, slender, simple or much branched, leafy, densely stipitate-glandular at least above and usually more or less strigillose, often glabrate below. Stem-leaves nearly uniform, linear to linear-spatulate, 2–5 cm. long, 2.6–8 mm. wide, obtuse to acuminate, sessile and often slightly clasping, entire, firm, pale green, rough-margined, glabrous to hispidulous, 1-nerved and often with a pair of weak lateral veins; branch-leaves much reduced, linear; heads 1 to many, 1.5–2 cm. wide, in a usually narrow cyme or panicle; involucre about 3-seriate, subequal or slightly graduate, 5–6 mm. high, the phyllaries linear to lance-linear, acute or acuminate, appressed or slightly spreading, densely glandular, with whitish base and shorter or longer herbaceous tip, the outer often herbaceous throughout; rays about 20–30, violet or purple, 5–8 mm. long; achenes silky.

Meadows and dry open slopes, often in alkaline soil, Arid Transition Zone; Montana and Wyoming to British Columbia and eastern Washington and Oregon. Type locality: "along the plains of Lewis' River." July–Oct.

Aster campestris var. bloòmeri A. Gray, Syn. Fl. N. Amer. 1²: 178. 1884. (*A. bloomeri* A. Gray, Proc. Amer. Acad. 6: 539. 1865; *A. campestris suksdorfii* Piper, Contr. U.S. Nat. Herb. 11: 572. 1906; *A. argillicolus* Peck, Proc. Biol. Soc. Wash. 47: 188. 1934.) Usually 20 cm. high or less; stem glandular at least above; involucres glandular as in the species; leaves spreading-hispidulous to spreading-hirsute. This rather poorly defined variety replaces *A. campestris* var. *campestris* from Klickitat County, Washington, southward (east of the Cascade Mountains) to the eastern side of the Sierra Nevada and adjacent Nevada in the region of Lake Tahoe. Intergrading forms are to be found particularly in the northern part of its range. Type locality: "high slopes of Mount Davidson, near Virginia City, Nevada."

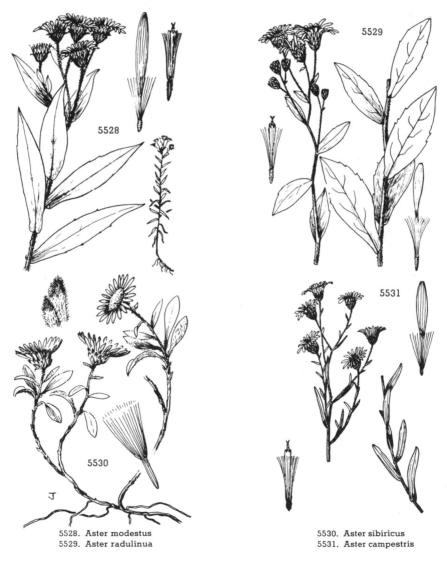

5528. Aster modestus
5529. Aster radulinua

5530. Aster sibiricus
5531. Aster campestris

7. Aster làevis var. gèyeri A. Gray. Geyer's Aster. Fig. 5532.

Aster laevis var. *geyeri* A. Gray, Syn. Fl. N. Amer. 1²: 183. 1884.
Aster brevibracteatus Rydb. Mem. N.Y. Bot. Gard. 1: 392. 1900.
Aster geyeri Howell, Fl. N.W. Amer. 1: 308. 1900.
Aster pickettiana Suksd. Werdenda 1: 42. 1927.

Stems erect from a stoloniferous rootstock, 0.3–1 m. high, simple or branched, glabrous and glaucescent throughout or the peduncles and branchlets very rarely (also the stem) pubescent in lines. Basal leaves usually obovate, obtuse, usually deciduous; lower stem-leaves oblanceolate to obovate, 8–18 cm. long, 1–3.5 cm. wide, acuminate, narrowed to a margined, clasping, petioliform base, entire or serrate, thick, pale green, rough-margined; middle stem-leaves linear-lanceolate to ovate or lanceolate, sessile and cordate-clasping, the upper reduced, mostly lanceolate, entire, those of the branchlets greatly reduced, usually subulate and somewhat clasping; heads few to many, 1.3–2.5 cm. wide, in an often ample corymbiform panicle; involucre about 5-seriate, strongly graduate, 5–7 mm. high, the phyllaries linear, erect, with white, indurate, 1-nerved base and shorter, rhombic or lance-rhombic, herbaceous tip, ciliolate; rays about 20–28, blue or violet, about 7–10 mm. long; achenes 4–5-nerved, essentially glabrous, the pappus usually reddish.

Usually in moist soil, Arid Transition Zone; South Dakota to Colorado, eastern Washington, and Alberta. Type locality: "Valleys of the Northern Rocky Mountains to Idaho, south to Wyoming, &c." Collected by Geyer. June–Sept.
Distinguished from the more eastern *A. laevis* L. only by the narrower and longer herbaceous tips of the phyllaries.

8. Aster greàtai Parish. Greata's Aster. Fig. 5533.

Aster greatai Parish, Bull. S. Calif. Acad. 1: 15. *fig. 2.* 1902.

Stems 5–10 dm. high from a creeping rootstock, leafy, sparsely coarse-pubescent, often glabrous and reddish below. Leaves longer than the internodes, the lower elliptic, narrowed to a petiolar base, deciduous at flowering time, the stem-leaves sessile, clasping to subauriculate, usually subtending a small leafy bract, oblong-obovate to elliptic, rough above, hispidulous-pilosulose beneath with margins entire or somewhat serrate above; usually bearing numerous heads 2–2.5 cm. wide in an open bracteate panicle; involucre 6–7 mm. high, strongly graduate or subequal; the phyllaries chiefly linear, with usually narrowly rhombic, acute to acuminate, sometimes loose, green tip, ciliolate, sometimes hispidulous on back; rays 25–40, light purple, 5–10 mm. long; achenes pubescent.

Damp shaded canyons, Upper Sonoran and Transition Zones; the San Gabriel Mountains, Los Angeles and San Bernardino Counties, southern California. Type locality: Eaton Canyon, San Gabriel Mountains, California. June–Oct.

9. Aster jéssicae Piper. Pullman Aster. Fig. 5534.

Aster jessicae Piper, Erythea 6: 30. 1898.
Aster latahensis Henderson, Contr. U.S. Nat. Herb. 5: 201. 1899.
Aster mollis Rydb. Bull. Torrey Club 28: 22. 1901.

Stems erect from a stoloniferous rootstock, 0.6–1.5 m. high, densely cinereous-pilose with spreading to loosely ascending hairs, the numerous erect branches floriferous toward apex. Leaves obovate to elliptic or the upper oblong, 5–15 cm. long, 1.5–3 cm. wide, acuminate or acute, the lower narrowed to a winged clasping base, the others sessile and clasping by slightly or not auriculate bases, entire or crenulate, often wavy-margined, firm, roughish-pubescent, or the hairs beneath soft; heads 1.5–2.5 cm. wide, crowded toward tips of branches and stem, usually short-peduncled or subsessile; involucre sometimes with 1 or 2 equal or longer, leafy bracts at base, about 5-seriate, 6–10 mm. high, subequal or distinctly graduate, the phyllaries linear to linear-oblong or somewhat oblanceolate, densely pilosulose, with whitish 1-ribbed base and usually shorter, erect, herbaceous tip, or the outer wholly herbaceous; rays about 20–28, violet or purple, 8–10 mm. long; achenes pubescent.

River banks, Arid Transition Zone; southeastern Washington and adjacent Idaho. Type locality: Pullman, Washington. Aug.–Sept.

10. Aster hespèrius A. Gray. Marsh Aster. Fig. 5535.

Aster coerulescens of authors. Not DC. 1836.
Aster hesperius A. Gray, Syn. Fl. N. Amer. 1²: 192. 1884.
Aster ensatus Greene, Pittonia 4: 223. 1900.
Aster foliaceus var. *hesperius* Jepson, Man. Fl. Pl. Calif. 1047. 1925.

Stem 1–2 m. high with numerous ascending branches, pubescent in lines above, very leafy. Leaves lance-linear, rather thick with midvein prominent below, 6–13 cm. long, 3–15 mm. wide, attenuate, clasping, entire or the lower leaves serrate, rough-margined; heads about 2 cm. wide, usually very numerous, in rather narrow or spreading panicles; involucre distinctly graduate, 6–7 mm. high, sometimes subtended by a few hispid-ciliate, chiefly herbaceous, linear bracts; the phyllaries linear, acuminate, ciliolate, erect or loose, the hyaline margin of the inner phyllaries extending to the tip or nearly so; rays about 25, white or purple, about 8 mm. long.

Marshy meadows and along streams and irrigation ditches, Upper Sonoran and Transition Zones; Alberta and North Dakota south to Texas and west to the east face of the Sierra Nevada, California, from Mono County south and throughout southern California. Type locality: not definitely stated. Aug.–Oct.

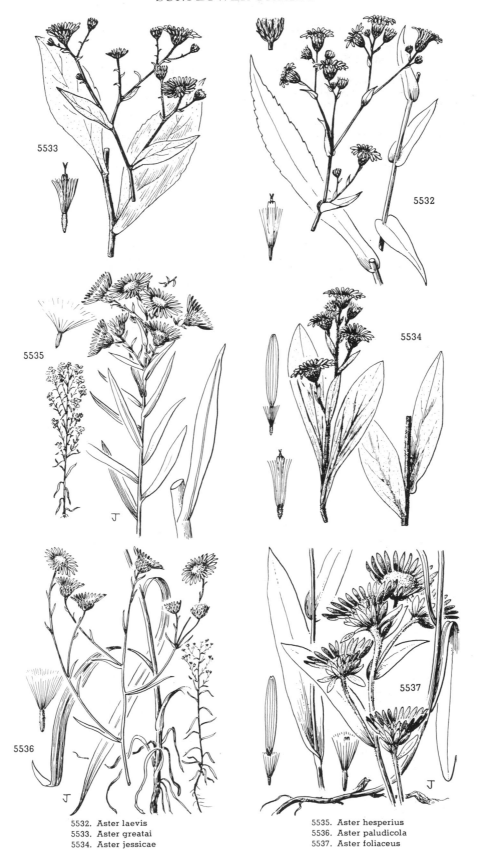

5533

5532

5535

5534

J

5536

J

5537

J

5532. Aster laevis
5533. Aster greatai
5534. Aster jessicae

5535. Aster hesperius
5536. Aster paludicola
5537. Aster foliaceus

11. **Aster paludícola** Piper. Western Bog Aster. Fig. 5536.

Aster paludicola Piper, Contr. U.S. Nat. Herb. **16**: 210. 1913.

Stem very slender, 1.5–8 dm. high or less, from slender rootstocks, usually pubescent in lines, with few or many branches above. Leaves narrowly linear, 3–15 cm. long, 2–6 mm. wide, sessile, subclasping, entire, rough-margined, more abundant at base of stems, reduced above; heads few to many, about 1–1.5 cm. wide, cymose, solitary at tips of branches and divaricate branchlets which bear appressed, lanceolate, leafy bracts; involucre about 5–7 mm. high, definitely graduate, the phyllaries erect, linear, acute or acuminate or the outer sometimes linear-oblong and obtusish, the base whitish, the thin green tips lanceolate or rhombic-lanceolate and usually purple-margined and -tipped; rays about 22–38, white, rose-colored, violet, or purple, about 8 mm. long; achenes glabrous or pubescent.

Bogs, Canadian Zone; Siskiyou Mountains of southern Oregon and northern California. Type locality: Eight Dollar Mountain, Josephine County, Oregon. July–Sept.

12. **Aster foliàceus** Lindl. ex DC. Leafy Aster. Fig. 5537.

Aster foliaceus Lindl. ex. DC. Prod. **5**: 228. 1836.

Plants 20–50 dm. tall from a creeping rootstock, the stems (in ours) reddish, glabrous below. Lower leaves oblanceolate, 12–20 cm. long, 16–24 mm. wide, narrowed to a petiolar base, margins entire, ciliate-appressed, the stem-leaves sessile and conspicuously clasping; inflorescence mostly monocephalous, sometimes subcymose bearing 4–6 heads, these up to 3.5 cm. wide, usually pubescent below the involucre; phyllaries green, 9–12 mm. high, the outer and inner subequal, additional foliaceous large bracts sometimes present, glabrous on the back, ciliate-margined, the outer enlarged and foliaceous, oblong, obtuse, or broadly acute at the apex; rays 10–17 mm. long, purple; achenes glabrous or sparsely pubescent.

Damp places, Boreal Zone; Alaska through British Columbia and Alberta south to the Olympic and Cascade Mountains, Washington, east to Montana. Type locality: Unalaska. July–Sept.

Aster foliaceus in the large sense breaks up into many recognizable related entities although at points of contact these forms intergrade freely. In addition to the extreme variability within the group itself, hybrids occur between *A. foliaceus* and *A. subspicatus* and also with *A. occidentalis*. A key to the varieties of *A. foliaceus* is given below based upon that of Arthur Cronquist (Amer. Midl. Nat. **29**: 429–468. 1943).

Outer foliaceous phyllaries conspicuous, broadly lanceolate to ovate, obtuse, or broadly acute.
    Stems usually monocephalous (sometimes bearing as many as 6 heads); Olympic and Cascade Mountains of Washington northward.   var. *foliaceus.*
    Stems with several heads in a cymose inflorescence.
        Stem-leaves strongly auriculate-clasping; plants of stream banks and meadows with the lower stem-leaves usually persistent.
            Phyllaries very narrow and acute or acuminate, the foliaceous ones if present linear or lanceolate or long-pointed.   var. *lyallii.*
            Phyllaries broader and often more blunt, foliaceous ones if present mostly lanceolate or ovate to acute or even obtusish.   var. *cusickii.*
        Stem-leaves less conspicuously auriculate-clasping; plants of drier habitats with the lower stem-leaves mostly deciduous.   var. *canbyi.*
Outer foliaceous phyllaries not conspicuous, linear or narrowly lanceolate, markedly acute.
    Plants 1.5–2.5 dm. tall, often monocephalous (sometimes bearing 4–6 heads); in alpine habitats.
                            var. *apricus.*
    Plants usually over 5 dm. tall; heads few to several in a cymose panicle; not in alpine habitats.
                            var. *parryi.*

Aster foliaceus var. **lyallii** (A. Gray) Cronquist, Amer. Midl. Nat. **29**: 443. 1943. (*Aster cusickii* var. *lyallii* A. Gray, Syn. Fl. N. Amer. **1**²: 195. 1884; *A. hendersonii* Fernald, Bull. Torrey Club, **22**: 273. 1895; *A. eriocaulis* Rydb. op. cit. **37**: 143. 1910; *A. kootenayi* Nels. & Macbr. Bot. Gaz. **56**: 477. 1913.) Much resembling var. *cusickii* but differing principally by the characters indicated in the key. Subalpine mountains of eastern Idaho and Montana and, less commonly, adjacent British Columbia, Washington, and northeastern Oregon. Type locality: between Kootenai and Pend Oreille Lake, Idaho. Collected by Lyall.

Aster foliaceus var. **cusickii** (A. Gray) Cronquist, Amer. Midl. Nat. **29**: 443. 1943. (*Aster cusickii* A. Gray, Proc. Amer. Acad. **16**: 99. 1880.) Plants 6–10 dm. high; leaves rather thin, the lower leaves not markedly exceeding the broadly clasping stem-leaves in length; outer phyllaries green and foliaceous. Southern British Columbia mostly east of the Cascade Mountains, and southeastern Washington and northeastern Oregon east to Montana. Type locality: small subalpine streams in the mountains of Union County, Oregon.

Aster foliaceus var. **cánbyi** A. Gray, Syn. Fl. N. Amer. **1**²: 193. 1884. (*Aster foliaceus* var. *burkei* A. Gray, loc. cit.; *A. majusculus* Greene, Pittonia **4**: 215. 1900; *A. tweedyi* Rydb. Bull. Torrey Club **31**: 655. 1904; *A. phyllodes* Rydb. op. cit. **37**: 145. 1910.) About 5 dm. high; leaves rather thick, the lower not much enlarged, and usually deciduous; the outer phyllaries much enlarged and usually obtuse. Usually growing in drier habitats than the last, eastern Washington and Oregon east to Montana and south to New Mexico. Type locality: White River, Colorado. Collected by Vasey.

Aster foliaceus var. **apricus** A. Gray, loc. cit. (*Aster apricus* Rydb. Mem. N.Y. Bot. Gard. **1**: 396. 1900; *A. incertus* A. Nels. Bot. Gaz. **37**: 269. 1904.) Plants 2.5 dm. or less high, cespitose, pubescent above, rootstocks branched; heads tending to be solitary; inflorescence sometimes subcymose. Alpine situations, southern British Columbia to northern California east to Montana and Colorado. Type locality: Union Pass, Colorado.

Aster foliaceus var. **párryi** (D. C. Eaton) A. Gray, Syn. Fl. N. Amer. **1**²: 193. 1884. (*Aster adscendens* var. *parryi* D. C. Eaton, Bot. King Expl. 139. 1871; *A. amplissimus* Greene, Proc. Acad. Phila. **1895**: 550. 1896; *A. frondeus* Greene, op. cit. 551; *A. glastifolius* Greene, Pittonia **4**: 218. 1900; *A. ciliomarginatus* Rydb. Mem. N.Y. Bot. Gard. **1**: 392. 1900; *A. diabolicus* Piper, Bull. Torrey Club **29**: 645. 1902; *A. vaccinus* Piper, op. cit. 646.) Plants about 5 dm. high with large, usually persistent, lower leaves and few stem-leaves; inflorescence corymbose, of few to several heads. In damp places, northern Washington and Montana south to northern California and the central Sierra Nevada and east to Colorado and New Mexico. A common form. Type locality: Rocky Mountains of Colorado. Collected by Parry.

13. **Aster subspicàtus** Nees. Douglas' Aster. Fig. 5538.

Aster subspicatus Nees, Gen. & Sp. Aster. 74. 1832.

Aster douglasii Lindl. in Hook. Fl. Bor. Amer. **2**: 11. 1834.

Tripolium oregonum Nutt. Trans. Amer. Phil. Soc. II. **7**: 296. 1840. Not *A. oreganus* of authors.

Aster elmeri Piper, Bull. Torrey Club **29**: 645. 1902. Not *A. elmeri* Greene, 1891.
Aster wattii Piper, loc. cit.
Aster umbraticus Sheldon, op. cit. **30**: 310. 1903.
Aster okanoganus Piper, Proc. Biol. Soc. Wash. **29**: 101. 1916.
Aster carterianus J. K. Henry, Ottawa Nat. **31**: 57. 1917.
Aster grayi Suksd. Werwenda **1**: 41. 1927.

Slender, about 1 m. high or less, with erect or ascending branches; stem pubescent in lines. Leaves rough-margined, practically glabrous, the lowest oblanceolate, narrowed to a winged petiolar base, serrate, the middle ones lanceolate, 7–13 cm. long, 1–2 cm. wide, slightly or not clasping, usually serrate above the middle; inflorescence cymose-panicled, the heads many to few, 2–3 cm. wide; involucre 5–6 mm. high, graduate to subequal, loosely imbricate, the phyllaries linear, acute or the outer often spatulate, with green, obtuse or acute, often loose, herbaceous tips; rays 20–30, violet, about 1 cm. long.

In moist places along the seashore and stream banks, mostly in the Humid Transition Zone; coastal Alaska and British Columbia south, mostly west of the Cascade Mountains, in Washington and Oregon and along the coast in California from Del Norte County to Monterey County; also occurring less abundantly east of the Cascade Mountains in Washington and east to Montana. Type locality: "crescit in insulis Multgräve," Alaska. June–Oct.

A variable species and apparently intergrading with other species such as *A. foliosus* and *A. chilensis* at points of contact. *A. grayi* Suksd., from Skamania and Klickitat Counties, Washington, and adjacent Oregon, is characterized by soft pubescence; in the plants adjacent to the seashore the outer phyllaries are conspicuously spatulate with loose herbaceous tips.

14. **Aster eatònii** (A. Gray) Howell. Eaton's Aster. Fig. 5539.

Aster foliaceus var. *eatoni* A. Gray, Syn. Fl. N. Amer. 1²: 194. 1884.
Aster eatonii Howell, Fl. N.W. Amer. 310. 1900.
Aster oreganus of authors, not Nutt.

Stem 1 m. high or less, pubescent usually in lines below the inflorescence, with numerous, erect or ascending branches. Leaves chiefly lance-linear or linear, rather thin, with the midvein prominent beneath, 5–13 cm. long, 0.5–2 cm. wide, acuminate, sessile with a somewhat truncate base, usually entire, rough-margined, often rough above; heads 1.5–3 cm. wide, very numerous in an oblong terminal panicle, subspicate or racemose on the branches and usually crowded toward their tips; involucre subequal or loosely graduate, 5–9 mm. high, occasionally subtended by herbaceous bracts, the phyllaries thin, oblanceolate or spatulate to (inner) linear, acute, often loose, ciliolate, glabrous on back, the herbaceous tips of the outer ones longer than the pale base and spreading; rays about 20–35, lavender or violet, about 7 mm. long.

Moist ground along streams, mostly Arid Transition Zone; southern British Columbia and Alberta southward to California mostly east of the Cascade Mountains and east to Idaho, Montana, and Wyoming south to Nevada and northern New Mexico; in California in the mountains of the Siskiyou region to Trinity County, and Modoc County south to Inyo County. Type locality: Virginia Mountains, Nevada. Collected by Watson. July–Sept.

15. **Aster hállii** A. Gray. Hall's Aster. Fig. 5540.

Aster hallii A. Gray, Syn. Fl. N. Amer. 1²: 191. 1884.
Aster mucronatus Sheldon, Bull. Torrey Club **30**: 309. 1903.
Aster chilensis subsp. *hallii* Cronquist, Amer. Midl. Nat. **29**: 462. 1943.

Slender, strict, 1 m. high or less, with short erect branches; stem sparsely pubescent all around, at least above. Leaves linear, 5–7 cm. long, 4–8 mm. wide, sessile, entire, rough-margined, usually scabrous above, glabrous beneath; heads small, 2 cm. wide, subracemose to cymose toward

5538. Aster subspicatus

5539. Aster eatonii

tips of the more or less leafy-bracted, ascending branches, forming a narrow elongate panicle; involucre more or less graduate, about 4–6 mm. high, the phyllaries narrow-oblong or the outer usually spatulate, erect, obtuse to acute, ciliolate, glabrous on back, with whitish base and oval to subrhombic or oblong green tip, the outermost sometimes herbaceous and equaling the inner; rays about 25, white to violet, 5–7 mm. long.

Alluvial soil or meadows, Transition Zones; Okanogan County, Washington, south to Wasco County, Oregon, and west of the Cascade Mountains from the lower Fraser River Valley, British Columbia, south to the Willamette Valley, Oregon. Type locality: Oregon. July–Aug.

16. **Aster adscéndens** Lindl. Long-leaved Aster. Fig. 5541.

Aster adscendens Lindl. in Hook. Fl. Bor. Amer. 2: 8. 1834.

Stems slender, 0.3–0.6 m. high, pubescent all around or only in lines above, erectish-branched above. Lower leaves narrowly oblanceolate, tapering to the petioliform base, the others lanceolate to linear, all usually entire, often half-clasping, rough-margined, glabrous to more or less pubescent, thickish; heads about 2.5 cm. wide, few to many, in a nearly naked, close or open cyme or cymose panicle; involucre 4–7 mm. hlgh, usually strongly graduate, erect, the phyllaries linear or linear-oblong, the outermost usually spatulate, obtuse to acute or the inner acuminate, ciliolate, glabrous or pubescent on back; rays 22–35, violet or purple, about 8 mm. long.

Moist or dry soil in various habitats, Arid Transition and Canadian Zones; British Columbia and southeastern Washington southward chiefly east of the Cascade Mountains and the Sierra Nevada to the San Bernardino and San Gabriel Mountains of southern California; Saskatchewan south to Utah and Colorado. On the west face of the mountains from Jackson County, Oregon, to Tulare County and southern California, it is represented by a densely leafy form having a rather close cyme and glabrous or pubescent herbage. Type locality: "Banks of the Saskatchewan." July–Sept.
A widespread complex group breaking up into ill-defined forms. All have strongly graduated involucres and a cymose or cymose-paniculate, mostly naked inflorescence with ascending branches but are highly variable as to pubescence and leaf-texture and habit.

17. **Aster occidentàlis** (Nutt.) Torr. & Gray. Western Mountain Aster. Fig. 5542.

Aster spathulatus Lindl. ex DC. Prod. 5: 231. 1836. Not *A. spathulatus* Lag. 1832.
Tripolium occidentale Nutt. Trans. Amer. Phil. Soc. II. 7: 296. 1840.
Aster occidentalis Torr. & Gray, Fl. N. Amer. 2: 164. 1841.
Aster adscendens δ *fremontii* Torr. & Gray, op. cit. 503. 1843.
Aster durbrowi Eastw. Proc. Calif. Acad. III. 2: 292. 1902.
Aster misellus Piper, Proc. Biol. Soc. Wash. 33: 105. 1920.

Stems usually 0.5 m. high or less, slender, arising from creeping rhizomes, glabrous below, usually pubescent in lines or all around above, at least toward tips of the rather few erectish branches. Lower leaves usually persisting, oblanceolate or linear-oblanceolate, tapering to a petioliform ciliate base, entire or serrulate, rough-margined, thickish; middle and upper leaves linear-lanceolate, 3–10 mm. wide, entire, scarcely clasping; heads 1 to few, about 2.5 cm. wide, in a nearly naked cyme or cymose panicle; involucre about 6 mm. high, not at all to slightly graduate, the phyllaries chiefly linear, acute or somewhat obtusish, appressed, ciliolate; rays about 30, violet or purple, 6–9 mm. long.

Mountain meadows and river thickets, Boreal Zone; British Columbia south to Tulare County in the Sierra Nevada, California, east to Colorado and Idaho. Type locality: "margins of muddy ponds in the Rocky Mountains." July–Sept.
Aster occidentalis var. **yosemitànus** (A. Gray) Cronquist, Amer. Midl. Nat. 29: 467. 1943. (*Aster adscendens* var. *yosemitanus* A. Gray, Syn. Fl. N. Amer. 1²: 191. 1884; *A. copelandi* Greene, Leaflets Bot. Obs. 1: 200. 1906.) Stems slender, leafy to the inflorescence; leaves alike, thin, 2–3 mm. long, linear, acute and gradually reduced upwards; heads solitary or several, 1.5–2 cm. broad, the phyllaries linear-lanceolate, loosely arranged. Meadows, southern Oregon to the Siskiyou region, California, and south in the Sierra Nevada to Tulare County. Type locality: Vernal Falls, Yosemite Valley. A variable and ill-defined form. Specimens are to be found which are evidently intergrades with *A. adscendens* and also with *A. occidentalis*.
Aster occidentalis var. **paríshii** (A. Gray) Ferris, Madroño 15: 128. 1959. (*Aster fremonti* var. *parishii* A. Gray, Syn. Fl. N. Amer. 1²: 192. 1884.) Plants about 3 dm. or more tall, the stems leafy, the leaves slightly clasping, linear-lanceolate and reduced upward, the basal leaves oblanceolate; heads in a short cymose panicle; involucres 5–8 mm. high, phyllaries mostly linear, loosely imbricate, acute. Mountain meadows of the southern California ranges south to the San Pedro Martir in Lower California. Type locality: Bear Valley, San Bernardino Mountains. Very close to *A. occidentalis* var. *occidentalis* of the Sierra Nevada. Evidently intergrading with the next form.
Aster occidentalis var. **delectábilis** (H. M. Hall) Ferris, Madroño 15: 128. 1959. (*Aster delectabilis* H. M. Hall, Univ. Calif. Pub. Bot. 3: 82. 1907.) Plants 1.5–4 dm. high; stems leafy, the leaves narrowly lanceolate, 12–20 cm. long, the leaf-bases definitely clasping; heads solitary or few, the involucres large, 8–10 mm. high, the phyllaries linear, acute, loosely imbricate. Mountain meadows, southern Sierra Nevada in Fresno and Tulare Counties, California, where it intergrades with other forms of *A. occidentalis* var. *occidentalis*, south to the San Jacinto and San Bernardino Mountains, and also the San Pedro Martir in Lower California. Type locality: Mill Creek, San Bernardino Mountains, California.
Aster occidentalis var. **intermèdius** A. Gray, Syn. Fl. N. Amer. 1²: 192. 1884. (*Aster delectus* Piper, Contr. U.S. Nat. Herb. 16: 210. 1913.) Plants 3–5 dm. high, herbage glabrate to densely cinereous-puberulent; leaves rather thick, the lower leaves 10–15 cm. long and often deciduous, the upper numerous but much reduced; inflorescence open with spreading branches, the heads mostly 5–6 mm. high, mostly few to many; phyllaries definitely graduate, the outer often obtuse. Meadows and drier places, eastern Washington in Okanogan and Klickitat Counties south along the eastern face of the Cascade Mountains, Oregon, to eastern Siskiyou County and Plumas County, California. Occasional specimens from the Sierra Nevada as far south as Tulare County appear to be this form. Type locality: Falcon Valley, Klickitat County, Washington. Collected by Suksdorf. Further studies will probably prove that this variable anomalous form should be removed from *A. occidentalis* from which it differs most markedly in its strongly graduate phyllaries.

18. **Aster chilénsis** Nees. Common California Aster. Fig. 5543.

Aster chilensis Nees, Gen. & Sp. Aster. 123. 1832.
Aster menziesii Lindl. in Hook. Fl. Bor. Amer. 2: 12. 1834.

Aster chamissonis A. Gray, Bot. Wilkes Exp. 341. 1874.
Aster militaris Greene, Proc. Acad. Phila. **1895**: 550. 1896.

Erect, 0.5–1 m. high, paniculately branched above; stem pubescent all around varying to nearly glabrous. Lowest leaves often obovate, narrowed to a somewhat clasping base, 12 cm. long, 3.5 cm. wide, the middle ones mostly lanceolate, 4–9 cm. long, 0.5–2 cm. wide, entire or serrulate, rough-margined, often rough above, essentially glabrous beneath; inflorescence racemose-paniculate, sometimes a simple raceme; heads usually numerous, 2–2.5 cm. wide, borne on leafy-bracteate branches; involucre 5–7 mm. high, 4–5-seriate, the phyllaries strongly and closely graduate (or the outer rarely herbaceous and equaling the inner in intermediate forms), linear-oblong to oblong, sometimes slightly broadened above, with whitish, 1-nerved base and shorter, somewhat rhombic, obtuse to acute, mucronulate, narrowly pale-margined, appressed, green tip, ciliolate, glabrous on back; rays about 20–30, violet, bluish, or white, 6–12 mm. long.

Usually in moist soil, Humid Transition and Upper Sonoran Zones; along the coast from northern Oregon south to Santa Barbara County and Santa Rosa Island, California. Type locality: California, erroneously published as Chile. June–Oct.

Aster chilensis subsp. *chilensis* as identified by the earlier botanists from the collections of Haenke, Chamisso, Menzies, and Hinds is the coastal form described above in which the larger stem-leaves are oblong-spatulate to lanceolate with the margins entire or crenate-dentate, sessile to somewhat clasping, and the pubescence of the stems and to a lesser degree the leaves varying from rather coarsely canescent to almost glabrous. It occurs on the Oregon coast southward to Santa Barbara, California, evidently more abundant from Humboldt County south to Monterey Bay along the coast and the Outer Coast Ranges than at the extremes of the range. The occurrence of *Aster subspicatus,* where its more southerly coastal distribution coincides with that of *chilensis,* may account for the presence of the troublesome forms which are evidently hybrids. Other segregates of this variable taxon are recognizable locally but intergrading forms are frequently found from the Oregon border to Tulare County, California.

Aster chilensis var. **léntus** (Greene) Jepson, Man. Fl. Pl. Calif. 1047. 1925. (*Aster lentus* Greene, Man. Bay Reg. 180. 1894.) Plants 1 m. or more in height, glabrous or nearly so; leaves linear-lanceolate; inflorescence ample, widely branching, the branches conspicuously leafy-bracted, the flowers large. Suisun marshes, Solano

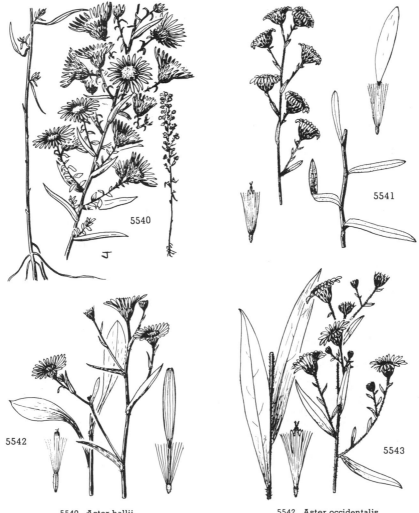

5540. Aster hallii
5541. Aster adscendens

5542. Aster occidentalis
5543. Aster chilensis

County, California, the type locality; also, specimens approaching this form have been collected in salt marshes along the coast in Humboldt and Mendocino Counties.

Aster chilensis var. **sonoménsis** (Greene) Jepson, loc. cit. (*Aster sonomensis* Greene, Man. Bay Reg. 180. 1894.) Plants 3–3.5 dm. high, glabrous and rather glaucous; basal leaves oblong, the stem-leaves reduced, lanceolate; inflorescence a cymose, leafy-bracted panicle, the heads few and solitary at the ends of the branches. Marshes at northern end of San Francisco Bay, California, Sonoma and Napa Counties, and in similar situations at the southern end in Santa Clara County. Type locality: "open plains of the Sonoma Valley."

Aster chilensis var. **mèdius** Jepson, loc. cit. Plants 1 m. or less high, glabrous or nearly so, very leafy, the leaves small, lanceolate or oblong-lanceolate, mostly spreading; inflorescence with widely divaricate branches, the leafy bracts somewhat smaller than the stem-leaves, the heads few. Along watercourses, lower foothills of Amador and Tuolumne Counties, and in Sacramento and San Joaquin Counties, California. Type locality: lower Sacramento River Valley.

Aster chilensis var. **invenùstus** (Greene) Jepson, loc. cit. (*Aster invenustus* Greene, Man. Bay Reg. 179. 1894.) Plants stout, up to 6 dm. high, herbage scabrous to short-hirsute; leaves lanceolate, the lower stem-leaves lanceolate-spatulate, 4.5–7 cm. long; inflorescence many-flowered in an ample cymose panicle. Described by Greene from material collected by him at Calistoga, Napa County, California. Leafy-stemmed specimens that approximate this form but which vary in pubescence from nearly glabrous to short-hirsute and have few to many heads in the inflorescence have been collected at various localities in the Inner Coast Ranges from Siskiyou County south to Sonoma County and the Sierra Nevada foothills as far south as Tulare County.

19. **Aster bernardìnus** H. M. Hall. San Bernardino Aster. Fig. 5544.

Aster bernardinus H. M. Hall, Univ. Calif. Pub. Bot. **3** : 79. 1907.
Aster deserticola J. F. Macbride, Contr. Gray Herb. No. 56: 36. 1918.

Erect perennial 3–10 dm. high, densely cinereous throughout, rarely somewhat glabrate, stems 1 to several from a woody root, densely leafy, the internodes short, simple below, the leaves early-deciduous, branched above. Leaves 3–5 cm. long, 3–6 mm. wide, sessile, linear to linear-lanceolate; heads in simple, short-branched, leafy- bracted racemes or an elongated narrow panicle with the uppermost heads congested on the branchlets; involucres 5–6 mm. high; phyllaries strongly graduate, pubescent to nearly glabrate, green-tipped, with the margins white-chartaceous and ciliate, the outermost phyllaries obtuse or rounded, the inner acute; rays 30–35, purple, 6–10 mm. long; achenes canescent, the pappus sordid.

Meadows and drainage ditches, Upper Sonoran Zone; Los Angeles and San Bernardino Counties south to San Diego County, California, reaching the western edge of the Mojave Desert in Los Angeles and San Bernardino Counties. Type locality: vicinity of San Bernardino, San Bernardino County. July–Nov.

Aster defoliàtus Parish, Bot. Baz. **38** : 461. 1904. Slender, rigid, short-pubescent perennial about 1 m. high; leafless below at flowering, except for the branchlets of the inflorescence, these clothed with linear acute leaflets about 1 cm. long and longer than the internodes; branchlets widely divaricate, 6–15 cm. long, the heads solitary, sometimes 2, at the tips of the branchlets; phyllaries strongly graduate, broadly scarious-margined below; pappus copious. Known only from the type (*Parish 5336*) collected in damp meadows, about 900 feet altitude, San Bernardino Valley, San Bernardino County, California, collected October 17, 1903.

20. **Aster falcàtus** var. **cràssulus** (Rydb.) Cronquist. Little Gray Aster. Fig. 5545.

Aster adsurgens Greene, Pittonia **4** : 216. 1900.
Aster crassulus Rydb. Bull. Torrey Club **28** : 504. 1901.
Aster commutatus crassulus Blake, Contr. U.S. Nat. Herb. **25** : 560. 1925.
Aster falcatus var. *crassulus* Cronquist, Bull. Torrey Club **74** : 144. 1947.

Stems usually 0.5 m. high or less with well-developed, creeping rhizomes, often much branched above, very leafy, becoming leafless below, spreading-hirsute or -hirsutulous or somewhat appressed. Leaves linear, 3–6 cm. long, 3–6 mm. wide, firm, entire, cuspidate, sessile, subclasping, usually hirsute or hirsutulous, those of the usually erect branches much reduced, numerous; heads 1–1.5 cm. wide, normally in racemes on stem or branches, solitary at tips of branchlets; involucre 5–7 mm. high, graduate, the phyllaries linear-oblong to linear or the outer usually spatulate, with a strongly indurate, whitish base and a rhombic (or in the inner lanceolate), thick-herbaceous, mucronulate, obtuse to acute, often loose tip, hispidulous and usually pubescent dorsally; rays 22–30, white or rarely violet-tinged, 5–8 mm. long.

Dry open places, Arid Transition Zone; eastern Washington and Oregon east to Saskatchewan and Colorado south to Utah, Arizona, and New Mexico. Type locality: La Veta, Colorado. Aug.–Oct.

21. **Aster pánsus** (Blake) Cronquist. Heath-like Aster. Fig. 5546.

Aster multiflorus var. *pansus* Blake, Rhodora **30** : 227. 1928.
Aster pansus Cronquist, Leaflets West. Bot. **6** : 45. 1950.

Perennial, about 0.6 m. high, fibrous-rooted from a short rhizome or woody caudex; stems clustered, strigose or in ours mostly with spreading-hirsute or hispidulous hairs. Leaves deciduous below, linear, 2–7 cm. long, 1–6 mm. wide, obtusish, callous-apiculate, sessile, rigid, rough-margined, strigillose, those of the branches crowded, abruptly reduced; heads small, usually numerous and racemosely arranged on the arching branches on short, minutely spreading-bracted branchlets, the disk 4–6 mm. high; involucre about 4 mm. high, graduate, the outermost phyllaries usually squarrose and spatulate, the others mostly linear, with indurate whitish base and abrupt, more or less rhombic, thick-herbaceous, often squarrose tip, stiff-ciliate and often pubescent on back, callous-mucronulate; rays 12–20, less commonly more, white, about 3 mm. long; achenes pubescent.

Usually in dry open places, Arid Transition Zone; rather common in British Columbia south to eastern Washington and northeastern Oregon and extending eastward to Colorado and Nebraska. Type locality: Ellensburg, Washington. Aug.–Oct.

Aster columbiànus Piper, Contr. U.S. Nat. Herb. **16** : 210. 1913. (*Aster multiflorus* var. *columbianus* Blake, Rhodora **30** : 227. 1928.) It has been suggested by Cronquist that this rarely occurring, violet-rayed form is of hybrid origin, having been derived from *A. compestris* and *A. pansus*. Type locality: Waitsburg, Walla Walla County, Washington.

22. Aster brickellioides Greene. Brickellbush Aster or Rayless Leafy Aster.
Fig. 5547.

Sericocarpus tomentellus Greene, Pittonia 1: 283. 1889. Not *Aster tomentellus* Hook. & Arn. 1833.
Aster brickellioides Greene, op. cit. 2: 16. 1889.
Eucephalus tomentellus Greene, op. cit. 3: 55. 1896.
Eucephalus bicolor Eastw. Proc. Calif. Acad. IV. 20: 157. 1931.

Stems 6–9 dm. high, arachnoid-tomentose or tomentulose, glabrescent, leafy, sometimes much branched. Leaves mostly spreading or deflexed, oval or elliptic-oblong to linear-oblong, 4–5.5 cm. long, 1–2 cm. wide, acute, the broader leaves obtuse, apiculate, rounded at the sessile base, entire, sometimes narrowly revolute-margined, subcoriaceous, glabrous or glabrescent above, canescent- or cinereous-tomentose and sometimes glabrescent in age, reticulate-veined, those of the branches greatly reduced, mostly linear-subulate or linear-lanceolate; heads numerous, cymose-clustered at tips of branches in a usually elongate-oblong panicle; involucre turbinate, about 6-seriate, strongly graduate, 7–9 mm. high, the phyllaries linear or ovate to linear-oblong or oblong, acute or the inner obtusish, with indurate-chartaceous, scarious-margined base, the short herbaceous tips of the outer spreading, the tips of the inner erect, not herbaceous, pilose-ciliate at least above, tomentose dorsally, sometimes glabrescent, the inner sometimes purplish; rays 0–5, light violet, about 6–8 mm. long; achenes compressed, sparsely pilose to essentially glabrous; pappus tawny, the inner pappus-bristles somewhat flattened apically, the outermost short, setulose.

Dry ridges and rocky slopes, Arid Transition Zone; Siskiyou Mountains, southwestern Oregon and adjacent California. Type locality: near Waldo, Josephine County, Oregon. July–Oct.

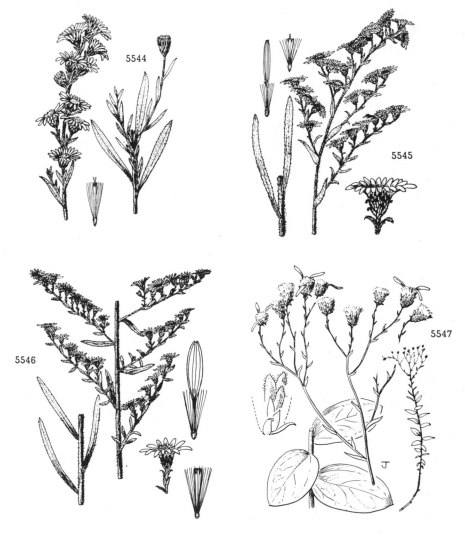

5544. Aster bernardinus
5545. Aster falcatus

5546. Aster pansus
5547. Aster brickellioides

23. **Aster siskiyouénsis** Nels. & Macbr. Siskiyou Rayless Aster. Fig. 5548.

Aster brickellioides var. *glabratus* Greene, Pittonia **2**: 17. 1889.
Eucephalus glabratus Greene, op. cit. **3**: 56. 1896.
Aster siskiyouensis Nels. & Macbr. Bot. Gaz. **56**: 477. 1913.
Eucephalus glandulosus Eastw. Proc. Calif. Acad. IV. **20**: 157. 1931.
Aster glabratus Blake ex M. E. Peck, Man. Pl. Oregon 726. 1941. Not Kuntze, 1891.

Stems 3–6 dm. high, simple, reddish below, arising from a woody, creeping, branched rhizome, usually glabrous or somewhat glandular especially beneath the involucres. Lower leaves reduced and scale-like, the others nearly uniform, somewhat smaller toward the inflorescence, 3–6 cm. long, ovate-lanceolate to oblong, margins entire, glabrous and sometimes roughish-puberulent beneath, occasionally thinly pilosulose in intergrading forms; heads few to several in a leafy-bracted, corymbiform panicle, the pedicels slender and spreading, the involucre campanulate, 4–5-seriate, and strongly graduate, 8–9 mm. high; outermost phyllaries loose, lanceolate, and green, the inner broader, indurated, pale with the tip greenish, the margin scarious, ciliate at least above, rarely tinged with purple, glabrous on the back or sometimes minutely glandular; rays absent, rarely 1 or 2; achenes appressed-pilose; pappus tawny.

Dry forested slopes, Arid Transition and lower Canadian Zones; Siskiyou Mountain area, Josephine and Jackson Counties, Oregon, and adjacent Siskiyou County, California. Type locality: "toward the summits of the Siskiyou Mountains, Oregon." Aug.–Sept.

The marked variation of this species suggests that hybridization may occur between *A. siskiyouensis* and the other species of the *Eucephalus* section that are found in the Siskiyou mountain area.

24. **Aster viàlis** (Bradshaw) Blake. Wayside Aster. Fig. 5549.

Eucephalus vialis Bradshaw, Torreya **20**: 122. 1921.
Aster vialis Blake, Rhodora **30**: 228. 1928.
Sericocarpus sipei Henderson, Madroño **2**: 105. 1933.

Stems about 1.2 m high, simple below the inflorescence, leafy, densely stipitate-glandular especially above and sparsely pilose. Lowest leaves minute, the others gradually reduced above, elliptic or oblong-elliptic, 3.5–6 cm. long, 0.8–2.3 cm. wide, acute or subacuminate, apiculate, rounded at the sessile base, entire or the larger rarely with a few sharp teeth, rather firm, green, and obscurely glandular above, beneath dull, stipitate-glandular, and sparsely pilose, 1-nerved, and with a pair of basal veins; heads discoid, 1–1.2 cm. high, several to numerous, cymose or in a narrow, oblong, leafy panicle, the peduncles short; involucre turbinate to campanulate, 5–6-seriate, strongly graduate, 7–8 mm. high; the phyllaries linear or linear-lanceolate to (in the inner) linear-oblong, acute or acuminate, erect or the outer somewhat loose, stipitate-glandular, obscurely if at all lacerate-ciliolate, with chartaceous base and much shorter, herbaceous tips, or some of the outermost essentially herbaceous, 1-nerved, the inmost sometimes slightly purplish-tinged; achenes compressed, 2–5-nerved, pilose; pappus graduate, the inmost bristles slightly dilated apically, the outermost short, setulose.

Rocky hillsides, Willamette Valley, Lane County to Douglas County, Oregon. Type locality: Eugene, Lane County. July–Aug.

25. **Aster ledophýllus** A. Gray. Cascade Aster. Fig. 5550.

Aster engelmanni var. *ledophyllus* A. Gray, Proc. Amer. Acad. **8**: 388. 1872.
Aster ledophyllus A. Gray, op. cit. **16**: 98. 1880.
Eucephalus ledophyllus Greene, Pittonia **3**: 55. 1896.
Eucephalus covillei Greene, op. cit. 162. 1897.
Aster covillei Blake ex M. E. Peck, Man. Pl. Oregon 725. 1941.

Stems several from a stout woody caudex, normally simple below the inflorescence, 20–90 cm. high, tomentose or sometimes only loosely pilose, leafy. Lowest leaves scale-like, the others nearly uniform, lance-elliptic or oblong to oblong-lanceolate, 1.5–5.5 cm. long, 4–20 mm. wide, acute or obtuse, callous-apiculate, rounded at the sessile nonclasping base, entire, green and essentially glabrous above, griseous- or cinereous-tomentose beneath, 1-nerved and with a pair of weaker basal veins; heads few or numerous, rarely solitary, 2.5–4 cm. wide, usually solitary at tips of corymbosely or thyrsoidly arranged, naked or bracteate, glandular and often tomentulose peduncles or branches; involucres broadly campanulate, 4–5-seriate, strongly graduate, 8–10 mm. high, the phyllaries finely glandular, obscurely ciliate, the outermost subulate to lanceolate, with short indurate base and much longer, loose, attenuate, herbaceous tip, the others lance-ovate to oblong, mostly acute, chartaceous, narrowly scarious-margined, with short or (in the inner) obsolete, herbaceous tip, usually purple-tipped or -margined; rays about 6–20, violet, purple, or pink shading to lavender, 1–1.5 cm. long; achenes 5-ribbed, thinly pilose; pappus graduate, the innermost bristles slightly dilated apically, the outermost very short, setulose.

Meadows and open woods, Canadian and Hudsonian Zones; Cascade Mountains of Washington south to southern Oregon and in a usually glabrate form to adjacent northern California as far south as the mountains of Trinity and Humboldt Counties. Type locality: "high up in the Cascade Mountains" [Mount Hood], Oregon. Collected by Hall. July–Sept.

The form described as *Eucephalus covillei* Greene from Crater Lake, Oregon, has granular-glandulose peduncles which completely lack the loose short-pilose pubescence. Some of the California specimens also exhibit this character, more evidence of its extreme variability in the Siskiyou region. The rays are usually fewer than in the northern forms. In its northern range *A. ledophyllus* intergrades with *A. Engelmannii* and specimens are to be found that are difficult to assign precisely to either species.

26. **Aster perélegans** Nels. & Macbr. Elegant Aster. Fig. 5551.

Eucephalus elegans Nutt. Trans. Amer. Phil. Soc. II. 7: 298. 1840.
Aster elegans Torr. & Gray, Fl. N. Amer. 2: 159. 1841. Not Willd. 1803.
Aster perelegans Nels. & Macbr. Bot. Gaz. 56: 477. 1913.

Stems several, erect from a woody caudex, roots fibrous, up to 65 cm. high, usually corymbosely branched in the inflorescence, slender, striate, finely incurved-puberulous, leafy. Basal leaves reduced, scale-like, the others nearly uniform, elliptic to lance-linear, 2.5–6 cm. long, 4–11 mm. wide, acute, rarely obtuse, rounded at the sessile base, entire, hispidulous-ciliolate, finely roughish-hispidulous on both sides, 1-nerved and with a pair of weaker veins arising at base; heads 2–2.8 cm. wide, 1 to many, solitary at tips of branches or branchlets, forming a short rounded cyme or a cymose panicle; involucre narrowly campanulate, 5–7-seriate, strongly graduate, 8–10 mm. high, the phyllaries appressed, ovate to (in the inner) linear-oblong, acute, whitish, chartaceous-indurate, narrowly scarious-margined, deep purple above, with strong green costa, densely villous-ciliate, puberulous or subglabrous on back, the inner ones somewhat deciduous; rays 6–11, usually about 8, violet or purple, 6–13 mm. long; achenes short-pilose, 5–nerved; pappus graduate, the innermost bristles slightly dilated apically, the outermost much shorter, setulose.

Dry mountain slopes, Arid Transition and Canadian Zones; mountains of northeastern Oregon eastward to southern Montana and Utah; also Nebraska. Type locality: "Oregon plains and the Blue Mountains of the west." July–Sept.

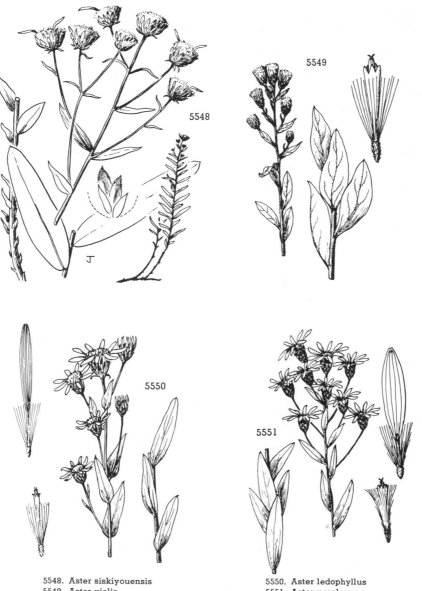

5549

5548

5550

5551

5548. Aster siskiyouensis
5549. Aster vialis

5550. Aster ledophyllus
5551. Aster perelegans

27. Aster glaucéscens (A. Gray) Blake. Klickitat Aster. Fig. 5552.

Aster engelmannii var. *glaucescens* A. Gray, Syn. Fl. N. Amer. 1²: 200. 1884.
Eucephalus serrulatus Greene, Pittonia 3: 55. 1896.
Eucephalus glaucescens Greene, op. cit. 56.
Eucephalus glaucophyllus Piper, Contr. U.S. Nat. Herb. 11: 570. 1906.
Aster glaucophyllus Frye & Rigg, Northwest Fl. 385. 1912.
Aster serrulatus Frye & Rigg, loc. cit.
Aster glaucescens Blake, Rhodora 30: 278. 1928.

Stems several from a stout caudex, up to 80 cm. high, erect and corymbosely branched above, slender, striate, glabrous, somewhat glaucous, leafy throughout. Lowest leaves reduced, scale-like, the others nearly uniform, lanceolate or linear-lanceolate, 3.5–9.5 cm. long, 4–15 mm. wide, acuminate to acute, mucronulate, narrowed at the sessile base, margins obscurely serrulate, sometimes entire, glabrous, glaucous, 1-nerved, and with a pair of weaker veins, somewhat venose; heads 6–14, solitary at tips of leafy branches, 2.5–3.5 cm. wide; involucre broadly campanulate, 4–5-seriate, strongly graduate, 6–8 mm. high, the outermost phyllaries subulate to lanceolate, with indurate base and longer, loose, herbaceous tip, the others ovate to oblong-lanceolate, attenuate, with whitish, indurate, narrowly scarious-margined base and shorter or obsolete, acuminate, herbaceous tip, carinately 1-ribbed, somewhat glandular-puberulous or subglabrous, obscurely lacerate-ciliate; rays usually about 13, purple, 1–1.2 cm. long; achenes appressed-pilose; pappus brownish, almost double, the inner bristles dilated apically, the outermost much shorter, setulose.

Open slopes, Arid Transition Zone; region around Mount Adams, Yakima and Klickitat Counties, Washington. Type locality: "Mount Paddo" [Mount Adams], Washington. July–Oct.

28. Aster engelmánnii (D. C. Eaton) A. Gray. Engelmann's Aster. Fig. 5553.

Aster elegans var. *engelmannii* D. C. Eaton, Bot. King Expl. 144. 1871.
Aster engelmannii A. Gray, Syn. Fl. N. Amer. 1²: 199. 1884.
Eucephalus engelmannii Greene, Pittonia 3: 54. 1896.

Stems several from a woody root, up to 1.5 m. high, corymbosely branched in the inflorescence, sparsely pilose or nearly glabrous below, puberulous above, leafy. Lowest leaves scale-like, the others nearly uniform, elliptic to oblong, sometimes oval or ovate, 4–11 cm. long, 1.5–3.5 cm. wide, acute or acuminate, rounded or narrowed at the sessile nonclasping base, entire or nearly so, glabrous above except usually on costa, sparsely pilose or pilosulose on veins beneath or also on surface, thin, 1-nerved, and with a pair of weaker, basal or subbasal veins, loosely venose; heads few to many, 2–4.5 cm. wide, solitary or several at tips of the often elongate branches, forming a rather short, rounded cyme or cymose panicle, the bracts narrowly lanceolate or subulate; involucre broadly campanulate, 5–6-seriate, strongly graduate, 7–10 mm. high, outermost phyllaries linear to subulate or lanceolate, with indurate base and much longer, loose, attenuate, herbaceous tip, the others lanceolate or lance-ovate to oblong or lance-oblong, acuminate or the inner acute, chartaceous, narrowly scarious-margined (the outer and middle with lanceolate herbaceous tip), usually purplish, pilose-ciliate especially toward apex, glabrous or pubescent on back; rays 9–13, white, becoming pinkish in age, sometimes purple, 1–1.7 cm. long; achenes appressed-pilose; pappus graduate, the innermost bristles scarcely enlarged apically, the outermost setulose.

Open forests, Canadian and Hudsonian Zones; Alberta to Colorado and Nevada, and British Columbia southward in the Cascade Mountains of Washington; also rarely occurring at high elevations in the Siskiyou Mountains, California (*Kildale 6577*). Most of the specimens from the Siskiyou and Trinity area usually assigned here appear to be variants of *Aster ledophyllus* rather than *A. engelmannii*. The leaves are smaller and clothe the stem more closely, the rays if present at all are purple, and the outer phyllaries are less attenuate and more appressed. Type locality: not definitely stated. June–Sept.

29. Aster paucicapitàtus Robinson. Olympic Aster. Fig. 5554.

Aster engelmannii var. *paucicapitatus* Robinson, Proc. Amer. Acad. 26: 176. 1891.
Aster paucicapitatus Robinson, op. cit. 29: 329. 1894.
Eucephalus paucicapitatus Greene, Pittonia 3: 56. 1896.

Stems several, erect or ascending from a woody root, 25–40 cm. high, simple, 1–4-headed, stipitate-glandular and thinly spreading-pilose, leafy. Lowest leaves reduced, scale-like, the others uniform, elliptic or elliptic-oblong, 1.5–3.4 cm. long, 4–10 mm. wide, acute or apiculate, narrowed at the sessile or subsessile base, entire, pubescent like the stem and ciliolate, 1-nerved and with a pair of weaker lateral veins, mostly erect; heads 2.5–4 cm. wide on peduncles 1–7 cm. long; involucres hemispheric-campanulate, solitary or few, about 3-seriate, subequal or scarcely graduate, 7–10 mm. high, the phyllaries linear-lanceolate, acuminate, obscurely glandular and sparsely pilosulose, ciliate above, with pale, shining-chartaceous base and shorter or longer, erect or rather loose, herbaceous tip, the outer often wholly herbaceous and narrowly purple-margined; rays 12–18, white turning pink, 8–14 mm. long; achenes flattened, 4-ribbed, appressed-pilose; pappus-bristles rather soft, graduate, brownish white.

Ridges and dry slopes, Boreal Zone; known only from the type locality, Olympic Mountains, Washington. Aug.–Sept.

30. Aster gormánii (Piper) Blake. Gorman's Aster. Fig. 5555.

Eucephalus gormanii Piper, Proc. Biol. Soc. Wash. 29: 101. 1916.
Aster gormanii Blake, Rhodora 30: 278. 1928.

Stems several from a perennial creeping rootstock, ascending, 11–15 cm. high, leafy throughout, simple or subsimple, 1-headed, sparsely stipitate-glandular. Lowest leaves scale-like, the others nearly uniform, elliptic or lance-elliptic, 1.8–3 cm. long, 4–10 mm. wide, obtuse or acute, apiculate, sessile, nonclasping, entire, firm, hispidulous-ciliolate, stipitate-glandular on both sides, 1-nerved and with a pair of basal veins; peduncle 1.5–2.5 cm. long; head 2–2.5 cm. wide; involucre hemispheric, 3–4-seriate, slightly graduate, 6–8 mm. high; phyllaries lanceolate (outermost) to ovate or oblong-ovate, acute or subacuminate or the innermost obtusish, with indurate-chartaceous base and shorter or longer, subherbaceous, loose tip, with narrow, lacerate-scarious margin, pilose-ciliate toward tip, essentially glabrous dorsally, sometimes purplish-tinged above; rays 8–13, white or sometimes pink, about 1 cm. long; achenes 3-nerved, pilose; pappus tawny, somewhat graduate, the inner bristles enlarged apically, the outermost short, setulose.

Dry rocky slopes, Boreal Zone; known only from the vicinity of Mount Jefferson, Lane and Jefferson Counties, Oregon. Type locality: Hanging Valley, Mount Jefferson, Cascade Mountains. July.

31. **Aster stenomères** A. Gray. Rocky Mountain Aster. Fig. 5556.

Aster stenomeres A. Gray, Proc. Amer. Acad. **17**: 209. 1882.
Ionactis stenomeres Greene, Pittonia **3**: 246. 1897.

Resembling *A. scopulorum*; stems numerous, tufted from a woody root, 12–30 cm. high, simple, rigid, monocephalous, densely leafy, loosely pilose or thinly subtomentose. Basal leaves reduced, spatulate, the others uniform, linear, 1.5–3 cm. long, 1.5–3.5 mm. wide, callous-pointed, sessile, entire, rigid, pale green, 1-nerved, scabrous-hispidulous especially along margin; peduncle 3–8 cm. long, pubescent like the stem; head 3–5 cm. wide; involucre hemispheric, 8–13 mm. high, 3–4-seriate, somewhat graduate, the phyllaries linear or lance-linear, acuminate, erect, pilose and hispidulous, with greenish or sometimes purplish midline and subscarious margin; rays about 12, violet, 1–2 cm. long; achenes stipitate-glandular and sparsely pilose; outer pappus setulose, 0.5–1 mm. long.

Dry mountain slopes, Canadian and Hudsonian Zones; southeastern British Columbia and northeastern Washington eastward through central and northern Idaho to adjacent Montana. Type locality: "Rocky Mountains of Montana and Idaho." June–Sept.

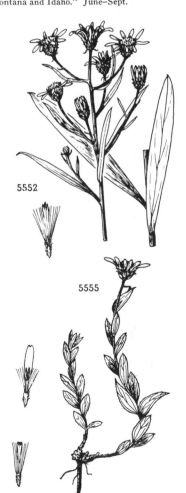
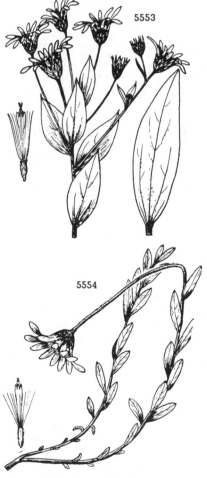

5552. Aster glaucescens
5553. Aster engelmannii
5554. Aster paucicapitatus
5555. Aster gormanii

32. **Aster scopulòrum** A. Gray. Lava Aster. Fig. 5557.

Chrysopsis alpina Nutt. Journ. Acad. Phila. **7**: 34. *pl. 3, fig. 2.* 1834. Not *A. alpinus* L. 1753.
Diplopappus alpinus Nutt. Trans. Amer. Phil. Soc. II. **7**: 304. 1840.
Aster scopulorum A. Gray, Proc. Amer. Acad. **16**: 98. 1880.
Ionactis alpina Greene, Pittonia **3**: 245. 1897.

Stems numerous, tufted, from a compactly branched, woody caudex, erect or ascending, simple, monocephalous, 5–12 cm. high, densely leafy, from sparsely and loosely pilose to subtomentose. Lowest leaves smaller, spatulate, the others elliptic to oblong or linear, 4–14 mm. long, 1–3 mm. wide, callous-cuspidulate, sessile, entire, rigid, 1-nerved, pale green, densly hispidulous-scabrous, with narrow whitish margin; peduncles 1–4 cm. long; heads 1.5–2.5 cm. wide; involucre campanulate-hemispheric, 7–10 mm. high, about 4-seriate, strongly graduate, the phyllaries erect, lanceolate to linear, acuminate or acute, carinately 1-ribbed, hirtellous and pilose, with greenish center and narrow subscarious margin; rays about 12–18,-violet or purple, 7–12 mm. long; achenes silky-pilose; outer pappus setulose, about 1 mm. long.

Dry mountainous regions, often in sagebrush, Arid Transition and Canadian Zones; eastern Oregon east to Wyoming and Montana and south to the Warner and White Moutains, California, and eastward to central Nevada. Type locality: dry prairies, not far from the Flathead River, Montana. May–July.

33. **Aster alpígenus** (Torr. & Gray) A. Gray. Alpine Aster. Fig. 5558.

Aplopappus ? alpigenus Torr. & Gray, Fl. N. Amer. **2**: 241. 1842.
Aster alpigenus A. Gray, Proc. Amer. Acad. **8**: 389. 1872.
Oreastrum alpigenum Greene, Pittonia **3**: 147. 1896.
Oreostemma alpigenum Greene, op. cit. **4**: 224. 1900.

Stems decumbent at base, several from a short, sometimes few-branched caudex arising from a long conical root, ascending, scapiform, 5–17 cm. high, glabrous below, pilose above, tomentose at apex of monocephalous stem. Basal leaves tufted, grass-like, linear-spatulate or oblanceolate, 3.5–15 cm. long, 5–15 mm. wide, broadly rounded or obtuse, rarely acutish, bluntly callous-tipped and with narrow callous margin, tapering to a petioliform base, entire, coriaceous, glabrous, rather obscurely 3–5-nerved; stem-leaves few, chiefly below the middle, linear or linear-spatulate, 0.6–4.5 cm. long, the upper ciliate; head 2.2–3.5 cm. wide; involucre hemispheric, 2–3-seriate, subequal or slightly graduate, 8–10 mm. high, the phyllaries linear or oblong, acute or obtusish, tomentulose, 1-nerved, often purplish, erect or rather loose, the outer subherbaceous throughout or with short pale base, the inner subscarious-margined, the margins often purple above; rays about 20–28, deep violet or purple, sparsely short-pilose toward the base on outer side, 6–10 mm. long; achenes nearly linear, compressed, about 5-nerved, pilose toward apex, glabrous below; pappus simple, the bristles strongly hispidulous.

Mountain ridges and meadows, Boreal Zone; the Cascade Mountains of Washington and Oregon and the higher mountains in northeastern Oregon. Type locality: Mount Rainier, Washington. June–Sept.

Aster alpigenus subsp. **haydénii** (Porter) Cronquist, Leaflets West. Bot. **5**: 77. 1948. (*Aster haydeni* Porter, Hayden Geol. Rep. Montana 1871: 485. 1872; *A. pulchellus* D. C. Eaton, Bot. King Expl. 143. 1871, not *A. pulchellus* Willd.; *Oreastrum haydeni* Rydb. Mem. N.Y. Bot. Gard. **1**: 398. 1900; *Oreostemma haydeni* Greene, Pittonia **4**: 224. 1900.) Decumbent perennial with habit of species but more slender stems, 3–15 cm. high; leaves linear to linear-elliptic, acute at apex, about 10 cm. long and to 5 mm. wide; involucre narrower than the name-bearing subspecies; rays 10–30; achenes hairy at apex or completely glabrous. Western Wyoming and Montana west to the Blue, the Wallowa, and the Steen Mountains, Oregon, south to Elko County, Nevada. Type locality: Upper Falls of the Yellowstone River, Wyoming.

Aster alpigenus subsp. **andersònii** (A. Gray) Onno, Bibl. Bot. 26[106]: 15. 1932. (*Erigeron andersonii* A. Gray, Proc. Amer. Acad. **6**: 540. 1865; *Aster andersonii* A. Gray, op. cit. **7**: 352. 1868; *Oreastrum andersonii* Greene, Pittonia **3**: 147. 1896; *Oreostemma andersonii* Greene, op. cit. **4**: 224. 1900.) Habit and general character of *A. alpigenus* but stems 8–40 cm. high, erect or ascending, more or less tomentose or pilose toward apex; basal leaves linear to linear-elliptic, 6–35 cm. long, 1.5–10 mm. wide, tapering at each end, definitely 3–5-nerved; stem-leaves greatly reduced; heads solitary, 2–3.8 cm. wide; involucre as in *A. alpigenus*, tomentulose to nearly glabrous, the inner phyllaries usually acuminate or subacuminate; rays purple, sparsely short-pilose without toward the base of the ray, the scarious margins usually purple above; achenes pilose throughout. Mountain meadows and bogs, Boreal Zone; in the Siskiyou Mountains, Josephine and Jackson Counties, Oregon, south to Humboldt County, California, and through the Sierra Nevada to Tulare County, California, and adjacent Nevada; also on Mount San Jacinto, southern California. Type locality: near Carson City, Nevada. June–Sept.

Specimens are to be found within the range of typical material in the Siskiyou region in California and also in Butte County that have achenes pilose only at the apex.

34. **Aster elàtus** (Greene) Cronquist. Plumas Alpine Aster. Fig. 5559.

Oreastrum elatum Greene, Pittonia **3**: 147. 1896. Not *A. elatus* Bert. ex Steudel. (Hyponym)
Aster elatus Cronquist, Leaflets West. Bot. **5**: 80. 1948.

Fibrous-rooted, glabrous perennial with stems 30–70 cm. long arising from the woody caudex. Basal and lower leaves linear-elliptic, 8–25 cm. long, 0.5–1 cm. wide at broadest part; the upper stem-leaves few and much reduced; flower-heads solitary on the stems; involucre imbricate, low-hemispheric, glabrous, 11–14 mm. high; phyllaries firm, broadly linear, the apex triangular-subulate, green apically and pale-chartaceous below; ray-flowers about 25, violet or lavender, 7–12 mm. long; achenes several-nerved, glabrous or sparsely hairy at the apex; pappus whitish, of about 40 bristles.

Meadows, Canadian Zone; Plumas County and probably Lassen County, California. Type locality: Mount Dyer, Plumas-Lassen County Line. July–Aug.

35. **Aster peirsònii** C. W. Sharsmith. Peirson's Aster. Fig. 5560.

Aster peirsonii C. W. Sharsmith, Leaflets West. Bot. **5**: 50. 1947.

Plants cespitose from a stout, branching, erect root, the stems 1 to several, arising from the woody caudex covered with persistent bases of dead leaves. Leaves many, mostly basal, firm, linear, sharply acute at apex, 1.5–5 cm. long, usually conduplicate, glabrous to sparsely glandular-scaberulous; flower-heads solitary usually on stems 1.5–7 cm. high; involucres purplish, turbinate to hemispheric, 7–11 mm. high, the phyllaries imbricate, lanceolate to linear-lanceolate, acute or acuminate, usually densely glandular-scaberulous; ray-flowers 8–18, sky blue to violet, 14–18 mm. long and 1.5–2.5 mm. wide; achenes 4–4.5 mm. long, terete or somewhat compressed, usually 10-nerved, glabrous below and sparsely pubescent or puberulent to almost glabrous above; pappus 8–9 mm. long, of 25–40 slender, white or sordid bristles.

Meadows and granitic gravels, Arctic-Alpine Zone; in the Sierra Nevada from southern Fresno County to Inyo and Tulare Counties, California. Type locality: northwest base of University Peak in the region of Kearsarge Pass, Fresno County. July–Aug.

36. **Aster spinòsus** Benth. Mexican Devil-weed. Fig. 5561.

Aster spinosus Benth. Pl. Hartw. 20. 1839.
Leucosyris spinosa Greene, Pittonia 2: 244. 1897.

Stems herbaceous from a woody root, 0.6–3 m. high, reedy, much branched, light green, glabrous, striate, often bearing subterete or flattened, axillary or supra-axillary spines 1.5 cm. long or less, essentially leafless above, the branches erectish, often broom-like. Lower leaves linear or linear-spatulate, usually 1–3 cm. long, 1–3 mm. wide, acute, sessile, entire, rough-margined, 1-nerved, those of the branches mostly reduced to minute scales; heads white, radiate, 0.8–1.5 cm. wide, solitary at tips of short or long branchlets, these racemosely or paniculately disposed; involucre campanulate-hemispheric, 4–5-seriate, strongly graduate, 3.5–6 mm. high, appressed, the phyllaries lanceolate or lance-ovate, mostly acuminate, with greenish center and subscarious, sometimes ciliolate margins; rays about 22, white turning brown, 3–4 mm. long; achenes glabrous, 4-nerved; style-appendages triangular, much shorter than the stigmatic region.

Stream banks, irrigation ditches, and moist places, Lower Sonoran Zone; Needles, San Bernardino County, California, and adjacent Nevada southward to Imperial and San Diego Counties, California, and Lower California, and eastward to Texas and Louisiana; also Mexico and Central America. Type locality: Aguascalientes, Mexico. June–Dec.

5556

5557

5558

5559

J

5556. Aster stenomeres
5557. Aster scopulorum

5558. Aster alpigenus
5559. Aster elatus

37. **Aster intricàtus** (A. Gray) Blake. Shrubby Alkali Aster. Fig. 5562.

Linosyris ? carnosa A. Gray, Smiths. Contr. 5[6]: 80. 1853.
Aster carnosus A. Gray ex Hemsl. Biol. Centr. Amer. Bot. 2: 120. 1881. Not Gilib. 1781.
Bigelovia intricata A. Gray, Proc. Amer. Acad. 17: 208. 1882.
Leucosyris carnosa Greene, Fl. Fran. 384. 1897.
Aster intricatus Blake, Journ. Wash. Acad. 27: 378. 1937.

Shrubby, 6–8 dm. high, rigidly and intricately branched, glaucescent and essentially glabrous throughout. Stem-leaves linear, 1–2 cm. long, 1–2 mm. wide, entire, mucronulate, fleshy, those of the branches reduced, mostly appressed, squamiform, 1–4 mm. long; heads solitary at tips of branches, discoid, 5–8 mm. thick, 8–10 mm. high; involucre turbinate-campanulate, 6–7 mm. high, about 5-seriate, strongly graduate, appressed, the phyllaries linear or lance-linear, acute or acuminate, the outer cuspidulate, glabrous or obscurely ciliolate, chartaceous, whitish, with greenish midline; corollas usually yellowish; achenes terete, many-ribbed, appressed-pilose; style-appendages lance-subulate, longer than the stigmatic region.

Occasional in alkaline meadows, Upper and Lower Sonoran Zones; Mono County, California, southward through the Mojave Desert to Los Angeles and San Bernardino Counties and in the San Joaquin Valley from Madera County to Kern County; also southern Nevada and western and southern Arizona. Type locality: Lancaster, Los Angeles County. Collected by Parry. June–Sept.

38. **Aster pauciflòrus** Nutt. Marsh Alkali Aster. Fig. 5563.

Aster pauciflorus Nutt. Gen. 2: 154. 1818.
Aster caricifolius H.B.K. Nov. Gen. & Sp. 4: 92. *pl. 333.* 1820.
Tripolium subulatum Nees, Gen. & Sp. Aster. 156. 1832.
Aster thermalis M. E. Jones, Proc. Calif. Acad. II. 5: 694. 1895.
Aster hydrophilus Greene ex Woot. & Standl. Contr. U.S. Nat. Herb. 16: 187. 1913.

Erect perennials from creeping rootstocks; stems simple or branched from the base, 2–9 dm. tall, completely glabrous except for the glandular inflorescence. Leaves somewhat fleshy, midvein not prominent, linear or lanceolate-linear, entire, sessile, acuminate at apex, 6–12 cm. long, 3–6 mm. wide, longer at base, reduced above and bract-like on the branches of the corymbiform inflorescence; involucres 6–12 mm. wide, 6–8 mm. long, the phyllaries linear-lanceolate, herbaceous except for a narrow hyaline margin, rather loose, of two or three lengths but scarcely graduate; rays pale purple or whitish, 6–10 mm. long; achenes appressed-pubescent.

In alkaline soil about springs and streams, and not common, Upper and Lower Sonoran Zones; Saskatchewan south to Texas and Mexico and west to Nevada and the Death Valley region, Inyo County, California. Type locality: "On the margins of saline springs, near Fort Mandan, on the Missouri." Aug.–Oct.

39. **Aster éxilis** Ell. Slim Aster. Fig. 5564.

Aster exilis Ell. Bot. S.C. & Ga. 2: 344. 1823.
Aster exilis var. *australis* A. Gray, Syn. Fl. N. Amer. 1[2]: 203. 1884.

Slender glabrous annual, paniculately much branched, 0.3–1.2 m. high. Stem-leaves oblanceolate to narrowly linear, 2.5–12 cm. long, 2–13 mm. wide, attenuate to acute or the lowest obtusish, sessile (the lower narrowed into a petioliform base), entire or the larger serrate or serrulate, fleshy, 1-nerved, roughish-margined, those of the branches mostly reduced to subulate bracts; heads usually very numerous, in flower 4–8 mm. wide; involucre turbinate, about 4-seriate, strongly graduate, 4–6 mm. high, the phyllaries lance-linear or subulate to linear, attenuate or acuminate, erect or somewhat loose, glabrous, with green, shining-chartaceous center and narrow subscarious margin, sometimes purple-tipped; rays about 15–40, pink, purple, or bluish, about 3 mm. long; achenes about 4-nerved, finely hirsutulous; pappus brownish white, soft, rather scanty.

In wet, often alkaline places, Sonoran Zones; Sacramento and San Joaquin Valleys to coastal southern California south to Mexico and South America and east to the Atlantic coast. Type locality: western Georgia. July–Oct.

40. **Aster frondòsus** (Nutt.) Torr. & Gray. Short-rayed Alkali Aster. Fig. 5565.

Tripolium frondosum Nutt. Trans. Amer. Phil. Soc. II. 7: 296. 1840.
Aster frondosus Torr. & Gray, Fl. N. Amer. 2: 165. 1841.
Brachyactis frondosa A. Gray, Proc. Amer. Acad. 8: 647. 1873.
Aster humistratus Gandoger, Bull. Soc. Bot. Fr. 65: 39. 1918.

Annual, erect or decumbent, 2.5–60 cm. high, simple or much branched, light green, the stem sparsely hispidulous or subglabrous. Leaves spatulate to linear, 1.5–6 cm. long, 1.5–9 mm. wide, obtuse or acute, mucronulate, the larger narrowed to a margined petioliform base, entire or depressed-serrulate, fleshy, 1-nerved, hispidulous-ciliolate, otherwise glabrous; heads few to very numerous, 6–10 mm. high, usually racemosely arranged along the branches or virgate-panicled; involucre hemispheric-campanulate, about 3-seriate, equal or somewhat graduate, 5–7 mm. high, the phyllaries appressed, linear-spatulate to obovate, obtuse to barely acute, slightly ciliolate, 1-nerved, the outer herbaceous throughout, the inner with subscarious margin below; rays pinkpurple, very numerous, in several ranks, narrow, exceeding the involucre (in anthesis) by 2 mm. or less; achenes appressed-pubescent; pappus simple, copious, soft, whitish.

Moist saline flats and borders of water, Upper Sonoran and Transition Zones; eastern Washington and Oregon, rarely occurring west of the Cascade Mountains, southward to northern California along the Sierra Nevada to the mountains of southern California and northern Lower California; eastward to Idaho and Wyoming and south to northern Arizona. Type locality: "near Lewis' River of the Shoshonee." June–Aug.

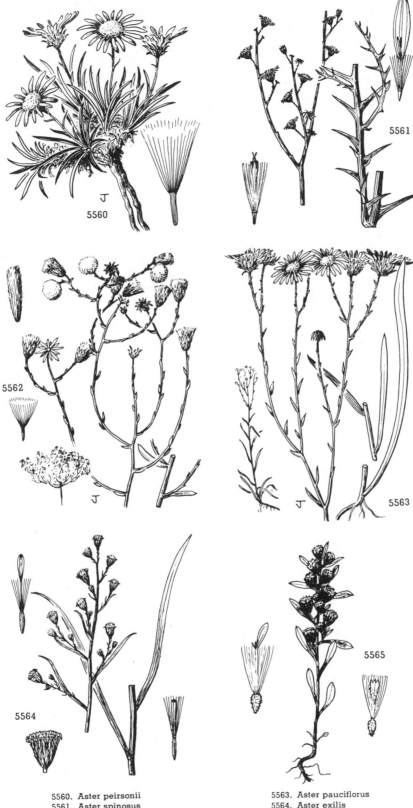

5560. Aster peirsonii
5561. Aster spinosus
5562. Aster intricatus

5563. Aster pauciflorus
5564. Aster exilis
5565. Aster frondosus

41. **Aster brachyáctis** Blake. Rayless Alkali Aster. Fig. 5566.

Tripolium angustum Lindl. ex Hook. Fl. Bor. Amer. **2**: 15. 1834.
Aster angustus Torr. & Gray, Fl. N. Amer. **2**: 162. 1841. Not Nees, 1818.
Brachyactis angustus Britt. in Britt. & Brown, Ill. Fl. **3**: 383. *fig. 3808.* 1898.
Aster brachyactis Blake, Contr. U.S. Nat. Herb. **25**: 564. 1925.

Erect annual, simple or much branched above the base, stems glabrous or occasionally hispidulous, 1–7 dm. high. Leaves linear, entire, acute at the apex, 1-nerved, sparsely ciliate on the margins, 3–8 cm. long; inflorescence virgate- to open-paniculate; peduncles short, the heads numerous, 8–11 mm. high; phyllaries 2–3-seriate, loosely arranged, linear to lanceolate, acute, the outer
green and shorter than the usually scarious-margined inner phyllaries, glabrous or remotely ciliate
on the margins, 5–7 mm. high; rays vestigial or absent; pappus copious, surpassing the phyllaries.

Moist, usually saline soil on lake and pond margins, Transition Zone; British Columbia and Alberta south
to northern Washington and to Nevada and Wyoming east to Minnesota and Missouri and adventive in other
eastern localities; also Siberia. Type locality: "Banks of the Saskatchawan." Collected by Drummond. Aug.–
Sept.

88. **LEUCELÈNE** Greene, Pittonia **3**: 147. 1896.

Tufted herbaceous perennials with a slender, creeping, subterranean rootstock, and
many leafy stems arising from partially subterranean, slender caudices. Leaves alternate,
linear or subulate. Heads radiate, solitary at the ends of the branches. Involucres turbinate, definitely graduate; phyllaries green, with a narrow scarious margin. Ray-flowers
pistillate, the rays white, sometimes tinged with rose. Disk-flowers perfect, fertile, yellow.
Achenes subcylindric, somewhat compressed when mature, the pappus longer than the
achenes. [Origin of name not known.]

A monotypic genus of southwestern United States and Mexico.

1. **Leucelene ericoìdes** (Torr.) Greene. Rose-heath or White Aster. Fig. 5567.

Inula ? ericoides Torr. Ann. Lyc. N.Y. **2**: 212. Not *Aster ericoides* L. 1753.
Diplopappus ericoides var. *hirtella* A. Gray, Mem. Amer. Acad. II. **4**: 69. 1849.
Aster ericaefolius Rothrock, Bot. Gaz. **2**: 70. 1877. Not Forsk. 1775.
Aster ericaefolius var. *tenuis* A. Gray, Syn. Fl. N. Amer. **1²**: 198. 1884.
Leucelene ericoides Greene, Pittonia **3**: 148. 1896.
Leucelene arenosa Heller, Cat. N. Amer. Pl. **8**. 1898.
Aster bellus Blake, Proc. Biol. Soc. Wash. **35**: 174. 1922.
Aster leucelene Blake, Contr. U.S. Nat. Herb. **25**: 562. 1925.
Aster hirtifolius Blake, loc. cit.

Stems numerous from branching woody caudices, 4–15 cm. high, slender, simple or erect-
branched, leafy, cinereous-strigose and more or less glandular. Lower leaves linear-spatulate, 4–
10 mm. long, 1–2 mm. wide, obtuse, cuspidate, narrowed to a petioliform base, 1-nerved, entire,
subcoriaceous, gradually passing to the linear or subulate upper ones, all more or less hispid or
strigose and glandular, at least the lower conspicuously hispid-ciliate; heads about 12 mm. wide,
solitary at tips of stems and branches; involucre campanulate or turbinate, 4-seriate or more,
graduate, 5–7 mm. high, the phyllaries appressed, lanceolate to lance-oblong, acute or acuminate,
with greenish or subchartaceous center and rather broad scarious margin, cuspidate, strigose or
hispid and more or less glandular, ciliate above, often purplish; rays 12–24, white, often turning
reddish or purplish, about 4 mm. long; disk-corollas pale yellow; achenes up to 3 mm. long,
pubescent, the hairs at summit the longer, about 5-nerved; pappus 4.5–5.5 mm. long, of subequal,
white, scabrous bristles.

Dry hills and gravelly soil, Upper Sonoran and Transition Zones; eastern face of the Sierra Nevada in Mono
County and in the New York, Providence, and Clark Mountains of San Bernardino County, California, eastward
through Nevada and Arizona to Kansas, Texas, and Mexico. Type locality: "On the Canadian?" Collected by
James. April–Sept.

Variable as to pubescence; specimens within our range exhibit the strigose nonglandular pubescence of *Aster
bellus,* while others from the same locality resemble *Aster hirtifolius,* which has hispid-ciliate as well as strigose
pubescence intermixed with some glands.

89. **MACHAERÁNTHERA** Nees, Gen. & Sp. Aster. 224. 1832.

Annual, biennial, or perennial herbs or shrubs with a well-defined taproot and in some
species a branching woody caudex. Leaves alternate, spinulose-tipped, spinulose-dentate,
pinnatifid or pinnately parted, more rarely entire. Heads few to many in a corymbose or
paniculate inflorescence or solitary, often large. Involucres turbinate to hemispheric, graduate; phyllaries in several series, green above and sometimes herbaceous, chartaceous or
coriaceous below. Ray-flowers pistillate, fertile, the rays shades of purple or white, rarely
lacking. Disk-flowers perfect. Achenes turbinate to linear, oblong, several-nerved, glabrous to woolly-villous; pappus of unequal barbellate bristles, often brownish. [From
the Greek, meaning dagger and flower.]

A genus of 26 species, natives of western United States and Mexico. Type species, *Aster tanacetifolia* H.B.K.

Annual, biennial, or perennial herbs; heads few to many on a branching inflorescence.
  Phyllaries narrowly linear, the subulate, loose, green, herbaceous tips longer than the indurated base.
                                                                    1. *M. tephrodes.*
    Phyllaries not narrowly linear, the erect or squarrose green tips shorter than the indurated base.
      Phyllaries not strongly imbricate, 3–4(5)-seriate; plants mostly low perennials with heads 6–8 mm. high.
                                                                    3. *M. shastensis.*
      Phyllaries strongly imbricate, 6–10-seriate; plants mostly tall biennials or perennials with heads 8–15 mm.
        high.
          Phyllaries not squarrose, not evidently glandular; lower leaves sparingly if at all spinose-toothed;
            plants of San Diego County.   2. *M. lagunensis.*
          Phyllaries squarrose, usually glandular; lower leaves usually conspicuously spinose-toothed; wide-
            spread species complex.   4. *M. canescens.*
  Plants shrubby; heads solitary at tips of branches.
    Stem glandular and hispid, often tomentose; leaves chiefly lanceolate or lance-linear, pubescent like the stem
      or rarely glabrous except for the ciliate margin; phyllaris densely glandular-hirtellous or pilose dorsally.
                                                                    5. *M. tortifolia.*
    Stem glabrous or slightly glandular toward apex; leaves chiefly oblong, glabrous or sometimes sparsely villous-
      ciliate; phyllaries essentially glabrous dorsally, usually stipitate-glandular on margin.
        Outer phyllaries linear-attenuate, equaling or exceeding the inner; peduncles glandular above.
                                                                    6. *M. cognata.*
        Outer phyllaries lanceolate, acuminate, much shorter than the inner, the involucre regularly graduate;
          peduncles glabrous.   7. *M. orcuttii.*

1. **Machaeranthera tephròdes** (A. Gray) Greene. Ash-colored Aster. Fig. 5568.

Aster canescens tephrodes A. Gray, Proc. Amer. Acad. **16**: 99. 1880.
Machaeranthera tephrodes Greene, Pittonia **4**: 24. 1899.
Aster tephrodes Blake, Contr. U.S. Nat. Herb. **25**: 563. 1925.

Erect biennial, single-stemmed, often paniculately branched, 2–8 dm. high; stem cinereous-puberulent or -pilosulose with mostly incurved hairs. Larger leaves oblanceolate or lanceolate, 3–10 cm. long, 4–10 mm. wide, acute, the lower tapering to a petioliform base, the others sessile and often slightly clasping, shallowly toothed with spinescent-mucronate teeth, pubescent chiefly along margin, rather firm, more or less distinctly triplinerved, the upper gradually reduced to small entire bracts; heads 2.5–4 cm. wide, solitary at tips of cymosely or paniculately arranged branches; involucre hemispheric, 8–10 mm. high, 6–8-seriate, graduate, the phyllaries narrowly linear-lanceolate, cinereous-pilosulose, with whitish-chartaceous base and shorter than the subulate-attenuate, mucronate, spreading or reflexed, herbaceous tip; rays about 23–40, violet or purple, 1–1.2 cm. long; achenes striate, finely pubescent; pappus stiffish, scarcely graduate.

Arid valleys and river bottomlands, Sonoran Zones; southern Nevada and eastern San Bernardino County, California, south and west to Riverside, San Diego, and Imperial Counties, and adjacent Lower California, eastward to Arizona, Sonora, and New Mexico. Type locality: "between the Del Norte and the waters of the Gila." April–Dec.

2. **Machaeranthera lagunénsis** Keck. Laguna Aster. Fig. 5569.

Machaeranthera lagunensis Keck, Brittonia **9**: 238. 1957.

Erect perennial with few to several leafy, cinereous-puberulent stems 3–7 dm. high from an erect branched caudex. Lower leaves 4–8 cm. long, about 8 mm. wide, rather firm, spatulate and narrowed to a petiole, sparsely puberulent to glabrous, entire or with occasional spinulose-denticulate teeth; upper leaves reduced, sessile, acute at apex, sparsely spinulose-denticulate, rather thinly

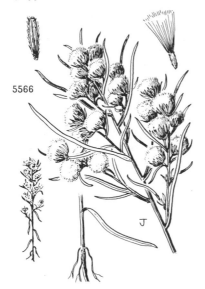

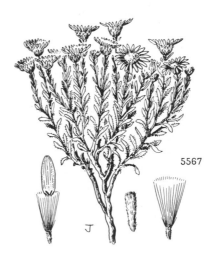

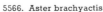

5566. Aster brachyactis

5567. Leucelene ericoides

cinereous-puberulent; heads large in a racemose or paniculate inflorescence; involucres 12–15 mm. high, campanulate or hemispheric; phyllaries up to 2 mm. wide, imbricate, appressed, cinereous-puberulent, 6–10-seriate, the short green tip triangular; rays 13–30, purple, 1–1.7 cm. long; achenes compressed, sparingly hairy, the pappus tawny.

Dry slopes in yellow pine forest, Upper Sonoran Zone; Laguna Mountains, San Diego County, California. Type locality: in recreation area at 5,200 feet, Laguna Mountains. July–Aug.

3. **Machaeranthera shastènsis** A. Gray. Shasta Aster. Fig. 5570.

Machaeranthera shastensis A. Gray, Proc. Amer. Acad. **6**: 539. 1865.
Aster shastensis A. Gray, Bot. Calif. **1**: 332. 1876.

Perennial with a woody root and slender-branched caudex, 5–20 cm. high, usually several-stemmed; stems ascending or decumbent, usually branching, often purple, densely cinereous-puberulous with incurved hairs and sometimes glandular above. Basal leaves spatulate to oblanceolate, 1–6.5 cm. long, 3–7 mm. wide, obtuse or acutish, cuspidate, tapering into a petioliform base, entire or few-toothed, cinereous-puberulous, subcoriaceous, 1-nerved; stem-leaves oblanceolate or oblong to linear, 0.7–5 cm. long, usually entire, those of the branches reduced to bracts; heads cymose or panicled, in reduced forms solitary, 1–2 cm. wide; involucre campanulate to turbinate, 4–5-seriate, 6–8 mm. high, the phyllaries linear or lance-linear to oblong, acuminate or acute, with indurate, usually purplish base and shorter, usually spreading, herbaceous tip (obsolete in the inner ones), cinereous-puberulous and rarely at all glandular; rays 8–20, violet, 5–7 mm. long, neutral or sometimes pistillate; disk-corollas often purple, short-pilose on the teeth; achenes pilose; pappus sordid, that of the ray-achenes often greatly reduced, that of the disk normal.

Dry slopes, Canadian and Hudsonian Zones; Crater Lake National Park, Oregon, to Mount Shasta, Siskiyou County, California. Type locality: Mount Shasta. July–Sept.

Machaeranthera shastensis var. **eradiàta** (Howell) Cronq. & Keck, Brittonia **9**: 238. 1957. (*Machaeranthera eradiata* Howell, Fl. N.W. Amer. 314. 1900; *M. inops* Nels. & Macbr. Bot. Gaz. **62**: 148. 1916; *M. inops* var. *atrata* Nels. & Macbr. loc. cit.) Plants perennial, 15–25 cm. high; stems few, stoutish, ascending, arising from a woody root, canescent–puberulent; lower leaves oblong-spatulate, up to 10 mm. broad, attenuate below to a margined petiole, 2–6 cm. long, upper sessile, somewhat reduced, oblanceolate, entire or somewhat toothed; heads hemispheric, the phyllaries in series of 3–4, densely glandular, the canescent pubescence sparse or lacking, the tips darkened and somewhat squarrose; achenes sparsely hairy. Rocky ridges, southern Oregon in the Siskiyou Mountains to Lake County, California. Type locality: Scott Mountain, Siskiyou County, California.

Machaeranthera shastensis var. **glossophýlla** (Piper) Cronq. & Keck, Brittonia **9**: 238. 1957. (*Aster glossophyllus* Piper, Bull. Torrey Club **29**: 646. 1902; *A. shastensis* var. *glossophyllus* Cronquist, Vasc. Pl. Pacif. Northw. **5**: 94. 1955.) Plants perennial from a woody root, the woody caudex little or not at all branched; stems few to several, ascending to erect, short-canescent throughout; leaves oblong-spatulate, the upper linear-oblanceolate, cuspidate at apex, 2–5 cm. long, 3–8 mm. wide, entire or occasionally toothed, the lower leaves mostly deciduous at flowering time; heads hemispheric, the rays few and often lacking; phyllaries in series of 3–4, canescent-puberulent and sometimes sparsely glandular, the inner ones purplish, the outer green-tipped and scarcely squarrose. Dry sandy soil, in sagebrush and adjacent forest; eastern base of the Cascade Mountains and adjacent ranges from Crook, Deschutes, and Jefferson Counties, Oregon, south in eastern California to the Lake Tahoe region, and eastward into Nevada. Rather variable and intergrading with the various local forms of *M. canescens.* Type locality: Black Butte, Jefferson County, Oregon.

Machaeranthera shastensis var. **latifòlia** (Cronquist) Cronq. & Keck, Brittonia **9**: 238. 1957. (*Aster shastensis* var. *latifolius* Cronquist, Vasc. Pl. Pacif. Northw. **5**: 94. 1955.) Perennial with a woody root and slender-branched caudex from which arise several rather lax, few-flowered stems 5–15 cm. long, canescent-puberulent; leaves entire or crenate-toothed; basal leaves many, obovate to spatulate, narrowed below to a margined petiole, 1.5–2 cm. long, 7–10 mm. broad; upper leaves few, more narrowly spatulate than the lower, mostly sessile; heads hemispheric, the pedicel glandular and canescent-puberulent; phyllaries mostly in series of 3, rather broad, glandular-margined, the rather thin, deltoid tip squarrose; achenes silky-haired as in *M. shastensis.* Loose rocky slopes at high altitudes in the Wallowa Mountains, Oregon, and in the Cascade Mountains in Deschutes County, Oregon. Type locality: middle fork of the Imnaha River, Wallowa Mountains.

Machaeranthera shastensis var. **montàna** (Greene) Cronq. & Keck, Brittonia **9**: 238. 1957. (*Machaeranthera montana* Greene, Pittonia **3**: 60. 1896, emended op. cit. **4**: 24. 1899.) Perennial with a woody root and stout, short, branched caudex, cespitose, many-stemmed, 5–12 cm. high, more or less glandular and cinereous; leaves spatulate in outline, canescent-puberulent, 2.5–3.5 cm. long, 3–6 mm. wide, usually deeply incised-dentate, the lower many, the upper fewer and reduced and narrower; heads radiate, broadly hemispheric, 1–3(4) on each branch; phyllaries usually 3-seriate, sometimes in series of 4, firm, rather broad, usually densely glandular and more sparsely canescent-puberulent, the green tips usually squarrose. Dry forests and ridges at high elevations. Hudsonian Zone, in the Sierra Nevada from Sonora Pass, Tuolumne County, and Sweetwater Mountains, Mono County, south to Inyo County, California, and on the higher ranges eastward in Nevada. The plants from Sonora Pass usually have narrower basal leaves and the phyllaries vary from glandular to nearly glandless. At lower elevations taller forms are found which apparently intergrade with the *M. leucanthemifolia* of related *M. canescens* complex. Type locality: "eastern slope of the California Sierra."

4. **Machaeranthera canéscens** (Pursh) A. Gray. Hoary Aster. Fig. 5571.

Aster canescens Pursh, Fl. Amer. Sept. 547. 1814.
Diplopappus incanus Lindl. Bot. Reg. **20**: *pl. 1693.* 1834.
Dieteria viscosa Nutt. Trans. Amer. Phil. Soc. II. **7**: 301. 1840.
Machaeranthera canescens A. Gray, Smiths. Contr. **3**⁵: 89. 1852.
Machaeranthera canescens var. *glabra* A. Gray, loc. cit.
Aster inornatus Greene, Erythea **3**: 119. 1895.
Aster leucanthemifolius Greene, loc. cit.
Machaeranthera spinulosa Greene, Pittonia **4**: 24. 1899.
Machaeranthera attenuata Howell, Fl. N.W. Amer. 314. 1900.
Machaeranthera pinosa Elmer, Bot. Gaz. **39**: 49. 1905.
Machaeranthera scoparia Greene, Leaflets Bot. Obs. **2**: 227. 1912.
Machaeranthera hiemalis A. Nels. Amer. Journ. Bot. **21**: 580. 1934.

Biennial (sometimes perennial), 15–60 cm. high, usually several-stemmed and paniculately branched; stems cinereous-puberulous, often stipitate-glandular especially above. Lower leaves usually oblanceolate, 1.5–8 cm. long, 3–13 mm. wide, acute or obtuse, narrowed into a petioliform base, sharply spinescent-toothed to nearly entire, subcoriaceous, usually densely puberulous; stem-leaves smaller, oblanceolate or lanceolate to linear, sessile, toothed or entire, those of the branches reduced to often minute bracts; heads usually numerous, cymosely, paniculately or sometimes spicately arranged; involucre turbinate or campanulate, 6–8-seriate, strongly graduate, 6–10 mm. high, the phyllaries linear to lance-linear or linear-oblong, acute to acuminate, with indurate base and a shorter herbaceous tip (especially the lower always squarrose), cuspidate, cinereous-puberulous to densely glandular especially above; rays 11–25, rarely wanting, violet, 4–10 mm. long; achenes somewhat compressed, obscurely striate, 2–4-nerved, finely pilose; pappus rather soft, somewhat graduate.

Usually in sterile places, Upper Sonoran and Transition Zones; southern British Columbia southward through eastern Oregon and California to the mountains of southern California; eastward to Saskatchewan and Colorado; sometimes occurring as a weed. Type locality: "on the banks of the Missouri." July–Oct.

A variable species as to involucral characters and growth habit. Certain segregates seem to be locally distinguishable but confusing intergrades are found in the same region. It is therefore here considered to be a species complex until intensive work is done on the group to clarify the status of the segregates. *Machaeranthera pinosa* (a probable synonym of *Diplopappus incanus* Lindl.) has a perennial root and a corymbose inflorescence of few heads; *M. leucanthemifolia* has a biennial root from which arise several divaricate stems and an inflorescence of many rather small heads; *M. attenuata* is a robust plant with large heads and appears to be close to true *canescens* as defined by A. Gray in the *Synoptical flora* (1²: 206); *M. inornatus* is a tall rayless form virgately branched, from the plains east of Mount Shasta, California.

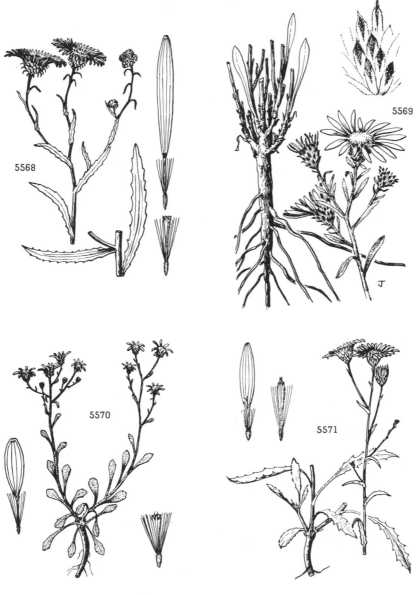

5568. Machaeranthera tephrodes
5569. Machaeranthera lagunensis

5570. Machaeranthera shastensis
5571. Machaeranthera canescens

5. **Machaeranthera tortifòlia** (Torr. & Gray) Cronq. & Keck. Desert or Mojave Aster. Fig. 5572.

Aplopappus tortifolius Torr. & Gray ex A. Gray. Bost. Journ. Nat. Hist. **5**: 109. 1845.
Aster tortifolius A. Gray, Proc. Amer. Acad. **7**: 353. 1868. Not Michx. 1803.
Aster mohavensis Coville, Contr. U.S. Nat. Herb. **4**: 126. 1893. Not Kuntze, 1891.
Aster tortifolius var. *funereus* M. E. Jones, Proc. Calif. Acad. II. **5**: 695. 1895.
Xylorrhiza tortifolia Greene, Pittonia **3**: 48. 1896.
Xylorrhiza lanceolata Rydb. Bull. Torrey Club **37**: 146. 1910. Not *Aster lanceolatus* Willd. 1803.
Aster abatus Blake, Contr. U.S. Nat. Herb. **25**: 562. 1925.
Machaeranthera tortifolia Cronq. & Keck, Brittonia **9**: 239. 1957.

Suffruticose, about 0.6 m. high, erect-branched, white-barked, loosely cinereous-pilose-tomentose throughout the green, stipitate-glandular, and sparsely pilose, the branches leafy. Leaves linear to lanceolate, oblanceolate, or oblong, 2–7 cm. long, 3–18 mm. wide, acuminate or acute, sessile and often slightly clasping, spinose–toothed and -tipped or rarely entire, coriaceous, often twisted, 1-nerved and veiny; heads solitary at tips of branches, 3–5 cm. wide, on naked or sparsely bracted peduncles usually 10–18 cm. long; involucres very broad, 4–5-seriate, graduate, 9–15 mm. high, pubescent like the stem; phyllaries linear-lanceolate or lance-subulate, attenuate, with indurate, whitish, carinately 1-ribbed base and longer or (in the inner) shorter, loose, herbaceous tip, the inner with only a greenish midline, the outermost often entirely herbaceous; rays about 40–60, lavender or violet, 1.5–2.5 cm. long; achenes elliptic-oblong, compressed, about 4-nerved, silky-pilose; pappus-bristles stiff, unequal, the inner somewhat flattened.

Usually among rocks in desert hilly or mountainous places, Sonoran Zones; Death Valley region to northern Colorado Desert, California, east to southern Utah and western Arizona south to Yuma County. Type locality: mountains of California. April–Oct.

6. **Machaeranthera cognàta** (H. M. Hall) Cronq. & Keck. Mecca Aster. Fig. 5573.

Aster cognatus H. M. Hall, Univ. Calif. Pub. Bot. **6**: 173. 1915.
Aster standleyi Davidson, Bull. S. Calif. Acad. **22**: 5. *pl.* 1923.
Xylorrhiza standleyi Davidson in Davids. & Moxley, Fl. S. Calif. 387. 1923.
Machaeranthera cognata Cronq. & Keck, Brittonia **9**: 239. 1957.

Similar to *M. orcuttii,* the branches somewhat arching. Leaves sparsely glandular, sessile on the lower part of the stem, clasping, those above and on the branchlets usually cuneate at the base, elliptic, acute and spine-tipped at apex, the margins with 3–6 coarse, spreading, spinose teeth on each side; peduncles 2.5 cm. long or less, glandular above; involucre about 1.5–2 cm. high, the phyllaries narrowly linear or lance-linear, attenuate, the outer sometimes elongate and equaling or surpassing the inner, sometimes much shorter, the involucre then regularly graduate, all sparsely glandular on back and especially on margin, the inner lacerate-ciliolate on the somewhat broadened scarious-margined base; rays blue or violet, 1.5–2 cm. long; achenes and pappus as in *M. orcuttii.*

In canyons, Lower Sonoran Zone; Coachella Valley, Colorado Desert, in the vicinity of Indio and Mecca, Riverside County, California. Type locality: near Indio, north side of Coachella Valley, Riverside County. May–June.

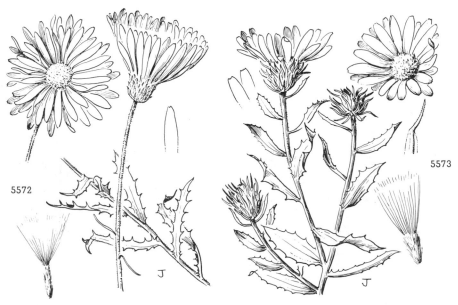

5572. Machaeranthera tortifolia   5573. Machaeranthera cognata

7. Machaeranthera orcúttii (Vasey & Rose) Cronq. & Keck. Orcutt's Aster. Fig. 5574.

Aster orcuttii Vasey & Rose, Bot. Gaz. **16**: 113. *pl. 11.* 1891.
Xylorrhiza orcuttii Greene, Pittonia **3**: 48. 1896.
Machaeranthera orcuttii Cronq. & Keck, Brittonia **9**: 239. 1957.

Suffruticose, 0.5–0.8 m. high, erect-branched, very leafy up to the heads, white-barked, glabrous. Leaves oblong to narrowly elliptic, 2.2–4.5 cm. long, 0.8–1.5 cm. wide, acute, sessile and slightly clasping, the margin with 7–10 rather slender spinose teeth on each side and at apex, coriaceous, light green, glabrous or sometimes loosely ciliate, 1-nerved and with a pair of lateral veins, somewhat veiny; heads solitary at tips of branches on peduncles 4 cm. long or less, about 5 cm. wide; involucre very broad, 1.7–2 cm. high, strongly graduate, about 6-seriate, the phyllaries lanceolate to oblong-lanceolate, acuminate with indurate whitish base and narrower, shorter or longer, slightly loose, herbaceous tip, glabrous except for the usually sparsely glandular-ciliolate margin; rays about 24–30, purple or lavender, about 2 cm. long; achenes densely silky-pilose; pappus-bristles stiff, unequal, the longer somewhat flattened.

Rocky places, Lower Sonoran Zone; western side of the Colorado Desert on the slopes of the Santa Rosa Mountains southward to northern Lower California. Type locality: Cariso Creek Wash, Imperial County, California. April.

90. CORETHRÓGYNE DC. Prod. **5**: 215. 1836.

Herbaceous perennials, sometimes suffrutescent at base, white-tomentose, often glabrate and glandular above, the stems and involucres often purplish-tinged. Leaves alternate, linear to obovate, toothed or entire, petioled or sessile. Heads heterogamous, radiate, solitary, cymose or panicled, the rays neutral, purple, violet, or pink, the disk yellow, sometimes becoming purplish. Involucre campanulate or hemispheric to turbinate, graduate, several- to many-seriate, the phyllaries chiefly cartilaginous, sometimes weakly so, or even scarious, linear, with herbaceous, often spreading tips, deflexed after shedding of achenes. Receptacle flattish, alveolate. Rays elliptic-linear, 2–3-toothed; disk-corollas tubular, 5-toothed, pubescent on teeth and base of throat. Anthers rounded or subcaudate at base, with linear-subulate terminal appendages. Style-branches with triangular to subulate, densely hispid appendages, half as long as the stigmatic region or less. Achenes subturbinate to linear, 5–7-ribbed, pilose, those of the ray abortive. Pappus of disk-achenes of numerous stiff, unequal, rufous or brownish, setiform awns and bristles, the outer shorter; of ray-achenes of few unequal awns and bristles. [From the Greek, meaning broom and female, referring to the densely hispid style-tips.]

An unstable genus in which many species and varieties have been described. These are all natives of California and, except for two forms which extend into northern Lower California and one which reaches southwestern Oregon, restricted to that state. The descriptions of the taxa listed below are based on authentic material. Relatively few individual collections in any given range completely conform to the type and the original description. Intermediate forms abound, particularly along the coast from southern Santa Cruz County, California, to northern Lower California and also the southern slope of the San Bernardino Mountains and the adjacent valleys. Type species, *Corethrogyne californica* DC.

Inflorescence of solitary heads or sometimes 2 or 3 on a stem.
  Stems prostrate except for the ascending pedunculate branches; leaves broadly spatulate or oval, rounded or truncate at the apex.   1. *C. obovata.*
  Stems erect except for the decumbent and unbranched, or decumbent and leafy-branched; leaves oblanceolate to linear-oblanceolate, the apex acute, sometimes mucronate.
    Stems erect from the decumbent base; flowering branches sparsely bracteate.
      Phyllaries 1–1.5 mm. wide, the tips erect.   2. *C. californica.*
      Phyllaries 2–2.5 mm. wide, the tips lax and deflexed.   2. *C. californica lyonii.*
    Stems much branched and often decumbent; flowering branches densely bracteate.   3. *C. leucophylla.*
Inflorescence of few to many heads, openly branched or congested.
  Lower leaves always linear or linear-lanceolate (1–8 mm. wide), entire or nearly so; heads large; plants of coastal San Diego County.
    Involucre coarsely and densely glandular, the phyllaries loosely imbricate and spreading.   5. *C. incana.*
    Involucre persistently white-woolly, the phyllaries closely imbricate, appressed.   4. *C. linifolia.*
  Lower leaves oblanceolate, oblong, or spatulate (about 8–20 mm. wide), mostly toothed above; heads small or medium except in *C. sessilis.*
    Involucre 8–12 mm. high; phyllaries white-floccose and weakly cartilaginous; plants of San Bernardino Mountains.   6. *C. sessilis.*
    Involucre 5–8(9) mm. high; phyllaries glandular and strongly cartilaginous in most forms.   7. *C. filaginifolia.*

1. Corethrogyne obovàta Benth. Prostrate Beach-aster. Fig. 5575.

Corethrogyne obovata Benth. Bot. Sulph. 22. 1844.
Corethrogyne spathulata A. Gray, Proc. Amer. Acad. **7**: 351. 1868.
Corethrogyne californica var. *obovata* Kuntze, Rev. Gen. Pl. **1**: 330. 1891.
Corethrogyne californica α *spathulata* Kuntze, loc. cit.

Perennial from a woody root with numerous more or less tomentose, leafy stems 1.5–5 dm. long and few ascending flowering branches. Leaves (1.5) 2–3.5 cm. long, 8–22 mm. wide, the lowermost withering and deciduous, broadly spatulate or obovate, often truncate at the apex, the petiolar base and lower part entire, the upper portion dentate, serrate or serrulate, thinly or densely white-tomentose, becoming partially glabrate in age; heads large, solitary on the branches, morely rarely in twos or threes, the glandular peduncles with few small, linear bracts; involucre 7–9 mm. high, hemispheric; phyllaries 4–5-seriate, loosely imbricate and spreading, linear, glandular, mostly not tomentose, the outer greenish- or purplish-tinged, the inner weakly cartilaginous at least below; rays 8–12 mm. long, lilac to violet; achenes about 4 mm. long, the pappus-bristles many, brownish, exceeding the achenes.

Grassy benches and bluffs above the ocean and sometimes on beaches, Humid Transition Zone; Curry County, Oregon, south to Marin County, California. Type locality: Bodega Bay, Sonoma County, California. June–Aug.

This taxon shows relatively little variation; the few collections having an erect habit have been made by Tracy in Humboldt County, California.

2. Corethrogyne califórnica DC. California Corethrogyne. Fig. 5576.

Corethrogyne californica DC. Prod. 5. 215. 1836.
Corethrogyne caespitosa Greene, Fl. Fran. 378. 1897.

Plants 9–20 cm. high, stems few from a woody root, closely white-tomentose, simple or branched near the base, very leafy below, decumbent at base, erect above and terminated by a single head. Leaves 2–3.5 cm. long, rather thin, much exceeding the internodes, linear-oblanceolate to narrowly oblanceolate, margin entire or few-toothed toward the acute apex, white-tomentose; heads large, borne on rather slender, sparsely bracteate, glandular peduncles; involucres hemispheric, 7–9(10) mm. high; phyllaries mostly 3-seriate, loosely imbricate, linear-lanceolate, not squarrose, glandular (sometimes floccose in intermediate forms), the apices greenish without a well-defined spot or purplish, the inner weakly cartilaginous to nearly scarious except for the tips; rays lavender; achenes about 3.5 mm. long, the reddish pappus longer than the achenes.

Open hill slopes, Humid Transition Zone; San Francisco County south to Monterey County, California. Type locality: "California," presumably Monterey. Collected by Douglas. April–June.

Rarely collected in its typical form. The type locality of the taxa *C. leucophylla, C. filaginifolia* var. *filaginifolia, C. filaginifolia* var. *rigida, C. tomentella,* and *C. viscidula* also is Monterey or its environs. The presence of numerous intermediate forms suggests interbreeding among the many taxa native to this locality.

Corethrogyne californica var. *lyònii* Blake, Journ. Wash. Acad. 33: 267. 1943. Perennial with several more or less erect, densely white-tomentose, leafy stems arising from a stout woody root; leaves obovate in outline, apiculate at apex, 3–4.5 cm. long, 1 or more cm. wide, crenate-serrate toward the acute apex, densely white-tomentose; heads large, hemispheric, the phyllaries glandular, differing from *C. californica* var. *californica* by the broadly linear, loose, herbaceous, outer phyllaries and the middle and inner ones with broadly linear, herbaceous tips; achenes as in *C. californica* var. *californica.* Open rocky slopes, San Carlos Range, Merced and San Benito Counties, California. Type locality: Cathedral Peak, 3,000 feet altitude.

3. Corethrogyne leucophýlla (Lindl.) Jepson. Branching Beach-aster. Fig. 5577.

Diplopappus leucophyllus Lindl. in DC. Prod. 5: 278. 1836.
Corethrogyne californica δ *leucophylla* Kuntze, Rev. Gen. Pl. 1: 330. 1891.
Corethrogyne leucophylla Jepson, Fl. W. Mid. Calif. 564. 1901 (attributed to Menzies).

Perennial 7–30 cm. high with leafy, densely white-tomentose stems, ascending to erect, sometimes decumbent at base, irregularly branching, more or less suffrutescent below and clothed with withering leaves. Leaves 1–4 cm. long, 6–10 mm. wide, longer than the internodes, broadly spatulate, narrowed to a petiolar base, thick, veiny beneath, the margin sharply dentate or serrate, permanently tomentose, the larger leaves more thinly so; heads several, solitary or 2 or 3 on the branches, the linear-bracteate peduncles 4 cm. or less long, sometimes nearly sessile on the stems; involucre (in pressed specimens) 11–15 mm. wide, 7–9 mm. high, essentially hemispheric; phyllaries loosely 4-seriate, thinly cartilaginous to scarious below, the outer with a deciduous tomentum, more or less marked above with green or purple-tinged, often to 1.5 mm. wide, the inner more narrow, glabrous or granular-glandular; rays 7–8 mm. long, lavender or violet; achenes about 4 mm. long, silky-villous, the pappus longer than the achenes, brownish.

Ocean cliffs and bluffs and also sand dunes, Humid Transition Zone; southern Santa Cruz County, California, to southern Monterey County and probably farther south. Type locality: Monterey, California. Aug.–Dec. (April; June).

The type specimen of *C. leucophylla* was one of the few plants in flower available for collecting in December, the time when the ships of Vancouver's voyage were anchored at Monterey (see transcript of Menzies' Journal). The seasonal changes of this coastal form are quite marked.

4. Corethrogyne linifòlia (H. M. Hall) Ferris. Del Mar Sand-aster. Fig. 5578.

Corethrogyne filaginifolia var. *linifolia* H. M. Hall, Univ. Calif. Pub. Bot. 3: 71. 1907.
Corethrogyne linifolia Ferris, Contr. Dudley Herb. 5: 100. 1958.

Plants 4–5.5 dm. high, stems erect, rather slender above and divaricately branched, densely leafy below, more sparsely so above, floccose throughout. Leaves 2–5 cm. long, mostly 1–3 mm. wide, sessile, much reduced above and becoming bracteate on the flowering branches, linear, entire, loosely tomentose to floccose; heads several, large, terminating the slender branches; involucre 8–10 mm. high, 9–11 mm. broad (in pressed specimens), hemispheric to broadly turbinate; phyllaries 6–8-seriate, closely imbricate, obtuse-angled at apex, rather thin-cartilaginous, the exposed tip permanently white-tomentose, concealing the sometimes granular-glandular surface; rays 20–25, violet, 6–8 mm. long; achenes about 4 mm. long, sparsely silky-villous, shorter than the brownish pappus.

Dry slopes above the ocean, Upper Sonoran Zone; apparently restricted to the coast of northern San Diego County, California. Type locality: near Del Mar. Collected by K. Brandegee. June–Sept.

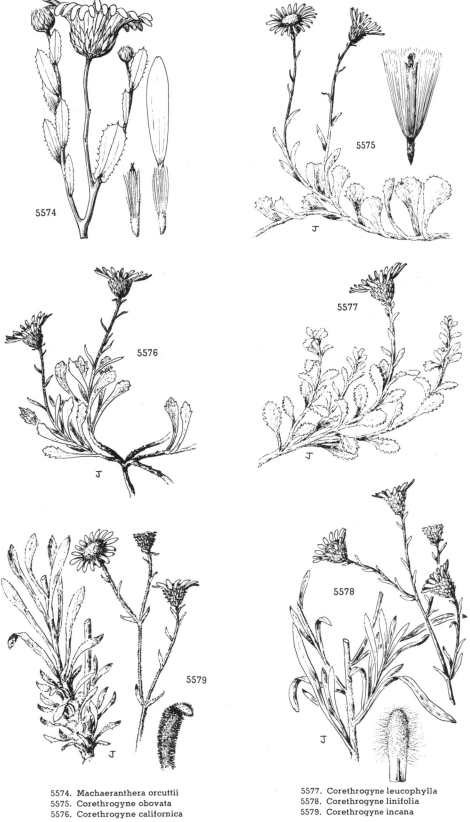

5574. Machaeranthera orcuttii
5575. Corethrogyne obovata
5576. Corethrogyne californica

5577. Corethrogyne leucophylla
5578. Corethrogyne linifolia
5579. Corethrogyne incana

5. **Corethrogyne incàna** Nutt. San Diego Sand-aster. Fig. 5579.

Corethrogyne incana Nutt. Trans. Amer. Phil. Soc. II. **7**: 290. 1840.
Corethrogyne californica β *incana* Kuntze, Rev. Gen. Pl. **1**: 330. 1891.
Corethrogyne filaginifolia var. *pacifica* H. M. Hall, Univ. Calif. Pub. Bot. **3**: 73. 1907.
Corethrogyne filaginifolia var. *incana* Canby, Bull. S. Calif. Acad. **26**: 14. 1927.

Plants 3–7.5 dm. high, fastigiately branched from about the middle especially in taller forms, the stems 1 to several, erect, more or less stout, leafy with short internodes and more densely so at the base, floccose throughout and also stipitate-glandular on the flowering branches. Leaves 2–6 cm. long, 2–8 mm. wide, the uppermost reduced, sessile, linear-lanceolate or lanceolate, acute, entire or nearly so, loosely tomentose; heads many, terminating the branches, large to medium; involucre 9–12 mm. high, hemispheric, the phyllaries 4–7-seriate, loosely imbricate, linear-lanceolate, less than 1 mm. wide, rather thin-cartilaginous and greenish toward the tip, conspicuously stipitate-glandular; rays 20–26, lavender or violet, 8–12 mm. long; achenes about 4 mm. long, silky-villous, the pappus reddish, longer than the achenes.

Bluffs and sandy slopes above the ocean, Upper Sonoran Zone; San Diego County, California. Type locality: San Diego. Collected by Nuttall. May–July.
Not common; intergrading freely with *C. filaginifolia* var. *virgata* and *C. linifolia*.

6. **Corethrogyne séssilis** Greene. San Bernardino Corethrogyne. Fig. 5580.

Corethrogyne sessilis Greene, Leaflets Bot. Obs. **2**: 25. 1910.
Corethrogyne filaginifolia var. *sessilis* Canby, Bull. S. Calif. Acad. **26**: 15. 1927.

Stems 2–6(7) dm. high, few from a woody base, leafy, densely white-tomentose. Leaves 1.5–4 cm. long, 1–2 cm. wide, obovate to oblong, sessile with broad clasping base, those at the base of stems narrowed below and spatulate, white-tomentose, entire or shallowly toothed at apex; heads many-flowered, sessile and either solitary or in clusters of 2 or 3 in the upper leaf-axils or pedunculate on the spreading branches of the inflorescence; involucre 9–13 mm. high, 9–10 mm. wide (in pressed specimens), campanulate, the phyllaries 5–6-seriate, the larger over 1 mm. wide, typically glandless, closely imbricated, appressed or somewhat spreading, not squarrose, rather weakly cartilaginous, the exposed portion permanently tomentose except for the broadly angled, green tip; rays 10–12 mm. long, violet or lavender; ripe achenes not seen.

Dry pine forests and open hillsides, Upper Sonoran and Transition Zones; San Bernardino Mountains, San Bernardino County, California, the type locality. Collected by Parish. Aug.–Sept.
Intergrading with *C. filaginifolia* var. *bernardina* or with *C. filaginifolia* var. *glomerata* of the adjacent lower elevations south of the San Bernardino Mountains, as indicated by the occasional occurrence of plants having glandular, somewhat squarrose phyllaries and more narrow acute leaves.

7. **Corethrogyne filaginifòlia** (Hook. & Arn.) Nutt. Common Corethrogyne. Fig. 5581.

Aster ? filaginifolius Hook. & Arn. Bot. Beechey 146. 1833.
Aster ? tomentellus Hook. & Arn. loc. cit.
Aplopappus? haenkei DC. Prod. **5**: 349. 1836.
Corethrogyne filaginifolia Nutt. Trans. Amer. Phil. Soc. II. **7**: 290. 1840.
Corethrogyne tomentella Torr. & Gray, Fl. N. Amer. **2**: 99. 1841.
Corethrogyne filaginifolia var. *tomentella* A. Gray, Bot. Calif. **1**: 321. 1876.
Corethrogyne californica ζ *tomentellus* Kuntze, Rev. Gen. Pl. **1**: 330. 1891.

Stems 3.5–8 dm. high, erect or ascending, few to several from the perennial root, slender, often reddish, sparsely leafy, the upper part with longer internodes, loosely white-tomentose, more or less glabrescent in age. Leaves of the lower stem 4–7 cm. long, usually early-deciduous, the lowest

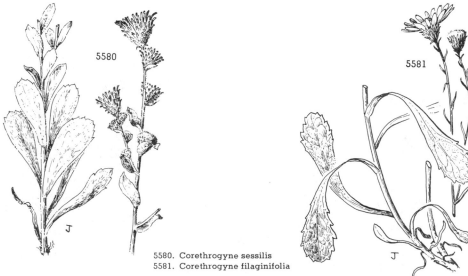

5580. Corethrogyne sessilis
5581. Corethrogyne filaginifolia

narrowed to a petiolar base, the upper tending to be sessile, oblanceolate to linear-oblanceolate, more or less toothed toward the apex and sometimes entire, the upper progressively smaller, all thinly white-tomentose; inflorescence openly branched, the heads few to several on the nearly naked, paniculate branches; involucre 5.1–7 mm. high, broadly turbinate or turbinate-campanulate; phyllaries 3–4(5)-seriate, 1 mm. or less wide, erect or spreading, not at all squarrose, weakly cartilaginous, the tip completely glandless, more or less greenish-spotted and sometimes lightly floccose, glandless or granular-glandular (especially in inland forms); rays 5–8 mm. long, lavender or violet; achenes hairy; pappus ample, reddish.

Pine woods, among oaks or on open hillsides, Upper Sonoran and Transition Zones; Hamilton Range, Santa Clara County, California, southward through the Inner Coast Ranges to Santa Barbara County and along the coast from Santa Cruz County through Monterey County. Type locality: California. Collected on Beechey's voyage. June–Sept. Cudweed Aster.

Intergrading freely with other taxa of this group at points of contact. The type of the form described as *Aster ? tomentellus* is an imperfect specimen densely leafy up to the few small heads and having linear to linear-oblong, appressed, white-tomentose leaves. Among the many diverse forms of *Corethrogyne* growing in the Monterey Bay region, specimens are occasionally collected that conform to the type of *tomentellus* but they are here consigned to *C. filaginifolia* var. *filaginifolia* because of their slender habit and the involucres which are glandless and of the shape and texture of those of var. *filaginifolia*.

Key to Principal Varieties

Involucres and inflorescence scarcely or not at all glandular, often somewhat floccose.
  Stems slender; phyllaries loosely imbricate; 3–4-seriate; coast and Inner and Outer Coast Ranges of central California.   *C. filaginifolia filaginifolia.*
  Stems woody below; phyllaries closely imbricate, 5-seriate; southern California coast, Ventura County northward.   *C. filaginifolia latifolia.*
Involucres, inflorescence, and sometimes the upper stem densely glandular, not or scarcely floccose (involucre and peduncle just below the head glandular only in *C. filaginifolia bernardina*).
  Involucre turbinate (broadly so in *C. filaginifolia bernardina*); phyllaries squarrose.
    Only the involucre and upper peduncle glandular; plants of San Bernardino Valley.   *C. filaginifolia bernardina.*
    Involucre and entire inflorescence glandular; plants not so restricted.
      Stem slender; inflorescence open-paniculate; plants of southern seacoast.   *C. filaginifolia virgata.*
      Stem stout; inflorescence glomerate or a short-branched raceme or panicle; plants of inland mountains.   *C. filaginifolia glomerata.*
  Involucre essentially hemispheric; phyllaries scarcely or not at all squarrose.
    Erect plants from a woody root, the inflorescence several- to many-flowered; central California coast.   *C. filaginifolia rigida.*
    Decumbent suffruticose plants, the inflorescence few-flowered; Channel Islands.   *C. filaginifolia robusta.*

Corethrogyne filaginifolia var. *hamiltonensis* Keck (Aliso **4**: 104. 1958), a plant with the habit and inflorescence of *C. filaginifolia* var. *filaginifolia*, has involucres obconic, 7.5–11 mm. high, 6–7-seriate, the phyllaries appressed, somewhat tomentose at anthesis and obscurely if at all glandular. It is found on the western side of the Mount Hamilton Range in Santa Clara and San Benito Counties and apparently merges with the eastward extension of *C. filaginifolia* var. *filaginifolia*, and was described from material collected on Mount Hamilton (*Hall 9865*).

Corethrogyne filaginifolia var. **rígida** A. Gray, Syn. Fl. N. Amer. 1²: 170. 1884, as to type. (*?*) *Corethrogyne viscidula* Greene, Fl. Fran. 378. 1897; *? C. viscidula* var. *greenei* Jepson, Fl. W. Mid. Calif. 564. 1901; *C. rigida* Heller, Muhlenbergia **2**: 256. 1906; *C. filaginifolia* var. *viscidula* Keck, Aliso **4**: 105. 1958.) Plants 2.5–5 dm. high with stems stouter than *C. filaginifolia* var. *filaginifolia*, erect or decumbent at the extreme base, usually leafy to the inflorescence with rather short internodes, floccose below, becoming glabrate and usually densely glandular above including the inflorescence; leaves more or less white-tomentose and sometimes glandular, narrowly or broadly oblanceolate, acute at apex and usually toothed on the upper third; heads few to several; involucre 5–6(7) mm. high, hemispheric to broadly turbinate, densely glandular (except in intermediate forms) and often stipitate-glandular, the phyllaries rather loosely 3–4(5)-seriate, not squarrose. Sand hills and bluffs along the ocean and open hill slopes adjacent to the coast; Alameda County south to Monterey County, California. Type locality: sand hills near Monterey. Collected by Hartweg.

Photographs of the type collection (*Hartweg 1772*) show a rather stout, erect, leafy-stemmed plant, floccose below and glandular above, with an open, many-flowered inflorescence. *Corethrogyne filaginifolia* var. *rigida* differs from *C. filaginifolia* var. *glomerata*, with which it sometimes has been identified, in having turbinate-campanulate to hemispheric involucres, phyllaries which are seldom if ever squarrose, and a more open inflorescence with longer pedunculate branches. *Corethrogyne viscidula* var. *greenei* Jepson and some specimens annotated by Greene as *C. viscidula* differ considerably from *C. filaginifolia* var. *rigida*. Further investigation may prove that these plants which have large heads with distinctly hemispheric involucres and rather broad phyllaries represent a distinct taxon. The large heads and the habitat in which they grow suggest relationship with the rarely collected *C. californica*.

Corethrogyne filaginifolia var. **robústa** Greene, Pittonia **1**: 89. 1887. Stems stout, ascending, densely leafy; lower leaves and lower part of stem tomentose, the branches and inflorescence glandular and lacking tomentum; leaves large, ovate to oblong, more or less toothed above, sessile, the upper much reduced, glandular, and of the same shape; inflorescence loosely paniculate or corymbiform; heads rather few, about as broad as high in pressed specimens; phyllaries glandular with a broad green tip, not squarrose. San Miguel, Santa Rosa, and Santa Cruz Islands, Santa Barbara County, and apparently occasional, though not in the typical form, along the coast of San Luis Obispo and Santa Barbara Counties. Type locality: San Miguel Island.

Having some characteristics in common with *C. filaginifolia* var. *rigida* but differing from that taxon in growth form.

Corethrogyne floccosa Greene (Leaflets Bot. Obs. **2**: 25. 1910), described from material collected by Alice Eastwood in September at Elwood near Santa Barbara, is shrubby in habit, with slender stems and filiform, glandular, pedunculiform branches. The leaves are oblanceolate or oblong and somewhat toothed. It is possibly a hybrid between the above and *C. filaginifolia* var. *virgata*.

Corethrogyne flagellaris Greene (op. cit. 26), described from material collected by Braunton in May at Redondo, Los Angeles County, and also collected by Abrams and Braunton at Playa del Rey, has white-tomentose, linear or linear-oblanceolate leaves, the habit of *C. filaginifolia* var. *virgata* especially in older plants, and rather large hemispheric involucres which resemble those of *C. filaginifolia* var. *robusta*.

Corethrogyne filaginifolia var. **latifòlia** H. M. Hall, Univ. Calif. Pub. Bot. **3**: 70. 1907. Stout woody stems, the leaves and stems and involucres tomentose and not at all glandular; leaves 1–4 cm. long, 5–10 mm. wide, sessile, mostly oblong, sometimes narrowed at the base, somewhat toothed toward the rounded apex; heads rather few, the branches of the inflorescence bracteate; involucres turbinate, the phyllaries but little if at all squarrose, floccose at the tips and not glandular. Occasional along the coast of Santa Barbara and Ventura Counties, California, and apparently intergrading freely with the glandular forms in the same area. Type locality: Oxnard, Ventura County.

Corethrogyne filaginifolia var. **virgàta** (Benth.) A. Gray, Bot. Calif. 1: 321. 1876. (*Corethrogyne virgata* Benth. Bot. Sulph. 23. 1844; *C. californica* η *virgata* Kuntze, Rev. Gen. Pl. 1: 330. 1891; *C. lavandulacea*

Greene, Leaflets Bot. Obs. 2: 27. 1910.) Tall, slender-stemmed plants 6–10 dm. high, widely paniculate-branched above; heads many, borne on short, divaricate, pedunculate branchlets; lower stems loosely floccose, rather leafy below, the tomentum nearly lacking on the upper stem and inflorescence, these short-stipitate-glandular; lower leaves as in *C. filaginifolia* var. *filaginifolia,* the upper much reduced, bracteate; involucre small, turbinate, the phyllaries closely 5–7-seriate, squarrose, cartilaginous, marked at the tip with a well-defined, green or brownish green spot. Occurring at or near the coast from Santa Barabara County, California, to coastal northern Lower California. Type locality: San Pedro, Los Angeles County. Collected by Hinds.

Intermediate forms between this taxon and other forms of the *C. filaginifolia* complex are not infrequently found in cismontane southern California.

Corethrogyne filaginifolia var. *peirsonii* Canby (Bull. S. Calif. Acad. **26:** 14. 1927), a local form seemingly related to the above and to *C. filaginifolia* var. *glomerata,* is found growing in the vicinity of Newhall, Los Angeles County. The plants so designated are stout, have a many-flowered inflorescence, and have involucres which are 6–8-seriate.

Corethrogyne scabra Greene (Leaflets Bot. Obs. **2:** 25. 1910), another local species collected by Hasse in 1890 at Los Angeles, is slender-stemmed, leafy to the few-flowered inflorescence, and has green and scabrous herbage.

Corethrogyne filaginifolia var. **glomeràta** H. M. Hall, Univ. Calif. Pub. Bot. **3:** 72. 1907. (*Corethrogyne brevicula* Greene, Leaflets Bot. Obs. **2:** 26. 1910; *C. racemosa* Greene, loc. cit.; *filaginifolia* var. *pinetorum* I. M. Johnston, Bull. S. Calif. Acad. **18:** 21. 1919; *C. filaginifolia* var. *brevicula* Canby, op. cit. **26;** 12. 1927; *C. filaginifolia* var. *rigida* A. Gray, in part; *C. filaginifolia* var. *tomentella* A. Gray, of authors, not Torr. & Gray.) Stems 1.5–3.5 dm. high, stout or more slender, erect or divaricately ascending, persistently leafy at and near the suffrutescent base; leaves 0.5–3.5 cm. long, persistently white-tomentose; heads sessile or nearly so in the upper leaf-axils or few to several in narrow, short-branched panicles or racemes; involucres 7–9 mm. high, turbinate; most phyllaries less than 1 mm. wide, well imbricated in 4–6 ranks, more or less squarrose, strongly cartilaginous below, the uppermost distinctly marked with a glandular green tip. Dry forests and meadows in the Upper Sonoran and Transition Zones; Inner Coast Ranges in San Benito and Monterey Counties and in the Sierra Nevada from Mariposa County southward to the mountains of southern California and adjacent Lower California. Type locality: Oak Glen, Yucaipe Ranch, near Redlands, San Bernardino County. Intermediate forms between this taxon and *C. filiginifolia* var. *virgata* are frequently found in southern California.

In the inflorescence of the type of *C. filaginifolia* var. *glomerata,* the heads are axillary on the upper stem, either singly or in clusters and are nearly sessile. *Corethrogyne racemosa* also has this type of inflorescence. *Corethrogyne filaginifolia* var. *glomerata* was defined by Hall to include plants with a less leafy inflorescence on which the heads are borne on short branches. The more open inflorescence is by far the more common. The type of *C. filaginifolia* var. *brevicula,* though short in stature, does not differ essentially from the taller plants to the north. *Corethrogyne filaginifolia* var. *pinetorum,* perhaps an ecological form of higher elevations, has few heads and a more naked inflorescence and occurs from the southern Sierra Nevada to the Lower California border.

Corethrogyne filaginifolia var. **bernardìna** (Abrams) H. M. Hall, Univ. Calif. Pub. Bot. **3:** 71. 1907. (*Corethrogyne virgata* var. *bernardina* Abrams, Fl. Los Ang. 401. 1904.) Erect plants 6–8 dm. high, with heads rather few on divaricate branches; stems densely white-tomentose throughout except for the glandular involucres and the upper part of the peduncles; involucres turbinate-campanulate, 5–7 mm. high, the phyllaries squarrose. Common on dry plains especially in the San Bernardino Valley. Type locality: Mentone, San Bernardino County.

The densely white-tomentose stems and somewhat larger heads suggest relationship with *C. sessilis.*

91. ERÍGERON* L. Sp. Pl. 863. 1753.

Annual, biennial, or perennial herbs with alternate (or sometimes all basal) leaves. Heads solitary to numerous, hemispheric to turbinate, usually radiate, the few to usually more or less numerous pistillate flowers usually bearing evident and often narrow rays, these chiefly of varying shades of pink, blue, or purple, to white, or in a few species yellow; in a few species the pistillate flowers rayless and the heads thus disciform, in a few others the pistillate flowers wanting and the heads thus discoid. Involucral bracts narrow, varying from herbaceous and equal to scarcely herbaceous and evidently imbricate, the loss of herbaceousness either uniform throughout their length or more prominent toward the tip. Receptacle flat or a little convex, naked. Disk-flowers more or less numerous, generally yellow; some species with rayless pistillate flowers between the disk-flowers and the ray-flowers. Anthers entire or nearly so at the base. Style-branches flattened, with introrsely marginal stigmatic lines and short (up to 0.5 mm.), externally minutely hairy, lanceolate and acute to more often broadly triangular and obtuse appendages, or the appendage rarely (*E. annuus*) obsolete. Achenes 2- to many-nerved; pappus of capillary and often fragile bristles, with or without a short outer series of minute bristles or scales. [Name Greek, meaning early old man, or old man in the spring, presumably referring to the early flowering and fruiting of many of the species.]

A genus of nearly 200 species, of North and South America, Europe, and Asia, nearly all of temperate or boreal regions, or mountainous areas in tropical America. Type species, *E. uniflorus* L.

Pistillate corollas very numerous, filiform, with very narrow, short, erect rays, these sometimes not exceeding the disk (inner pistillate corollas sometimes rayless).   V. TRIMORPHAEA.

Pistillate corollas few to numerous (or in some species absent), the tube generally cylindrical; rays well developed and spreading or sometimes reduced or absent but not short, narrow, and erect.

     Pappus of the ray- and disk-flowers unlike, that of the disk-flowers composed of bristles and short outer setae, that of the ray-flowers lacking the bristles; weedy, mostly annual plants.   IV. PHALACROLOMA.

     Pappus of the ray- and disk-flowers alike, of bristles, sometimes also with outer setae or scales; plants mostly perennial, a few species biennial or casually annual, seldom weedy except for *E. philadelphicus* and *E. divergens.*

         Internodes very numerous and short; leaves linear or narrowly oblong, essentially uniform from the base to near the top of the plant, the basal ones, if present, not markedly larger than the cauline ones; phyllaries markedly imbricate.   III. PYCNOPHYLLUM.

         Internodes not excessively numerous nor usually very short; leaves variously shaped, sometimes linear, but then the basal ones obviously larger than the cauline ones; phyllaries equal or imbricate.

            Achenes 4–8(10)-nerved; desert and foothill plants with the phyllaries evidently imbricate and the leaves silvery-strigose.   II. WYOMINGIA.

---

* Text contributed by Arthur John Cronquist.

Achenes in most species 2-nerved; a few species that sometimes or regularly have more numerous nerves occur in montane, woodland, or maritime habitats, have the phyllaries equal or nearly so, and are not at all silvery-strigose. I. EUERIGERON.*

I. EUERIGERON

Plants maritime and submaritime; stems mostly curved-ascending; heads large, hemispheric, the disk 14–35 mm. wide, the disk-corollas 4.7–7 mm. long.

Heads radiate, the rays about 9–15 mm. long; cauline leaves ample in well-developed specimens; Oregon and California. 8. *E. glaucus.*

Heads discoid, the flowers all tubular and perfect; cauline leaves narrow, mostly lance-linear to narrowly oblong, or the lower somewhat oblanceolate; Humboldt and Mendocino Counties, California. 9. *E. supplex.*

Plants not maritime or submaritime (save perhaps occasionally in *E. sanctarum*?); heads various but mostly smaller except in some tall erect species.

Cauline leaves ample, usually lanceolate or broader, entire or toothed, never trilobed; phyllaries equal or subequal; achenes 2–7-nerved.

Plants, at least when well developed, tall and erect (up to nearly 1 m. tall), somewhat *Aster*-like; achenes 2–7-nerved.

Rays relatively broad, mostly 2–4 mm. wide (1–2 mm. in *E. coulteri*, which has the hairs of the involucre with black cross-walls near the base); pappus simple or nearly so.

Hairs of the involucre without black cross-walls; rays mostly 2–4 mm. wide, white or colored; leaves hairy or glabrous.

Achenes 4–7-nerved; leaves glabrous or occasionally hairy; involucre mostly merely glandular, sometimes hirsute or with the bracts more or less ciliate-margined but not hirsute below and glandular above; disk-corollas mostly 4–6 mm. long.

Basal leaf-blades usually tapering to the petiole; cauline leaves various but rarely both thin and strongly clasping; rays colored or sometimes white; widespread, at moderate to high elevations in the mountains. 1. *E. peregrinus.*

Basal leaf-blades abruptly contracted to the petiole; cauline leaves thin, conspicuously clasping; rays white; south side of Columbia River Gorge in Oregon. 2. *E. howellii.*

Achenes 2–4-nerved; leaves hirsute; involucre hirsute below, glandular above; disk-corollas mostly 3–4 mm. long; Olympic, Oregon Cascade, and Klamath regions. 3. *E. aliceae.*

Hairs of the involucre with black cross-walls near the base; rays 1–2 mm. wide, white; leaves hairy; achenes 2-nerved; mountains of northeastern Oregon to California and eastward. 4. *E. coulteri.*

Rays narrower, about 1 mm. wide or less; hairs of the involucre, if present, without black cross-walls; pappus, except in *E. philadelphicus,* mostly double.

Rays not excessively numerous, commonly about 75–150, about 1 mm. wide; disk-corollas mostly 4–5 mm. long; achenes 2–4-nerved; true perennials.

Leaves glabrous or nearly so except for the ciliate margins; stem glabrous below the inflorescence or merely bearing a few scattered hairs; involucre with few or no long hairs; widespread. 5. *E. speciosus.*

Leaves, stem, and involucre more or less long-hairy; Washington and eastward. 6. *E. subtrinervis.*

Rays very numerous, commonly 150–350 or more, narrow, about 0.2–0.6 mm. wide; disk-corollas mostly 2.5–3.2 mm. long; achenes 2-nerved; biennial or short-lived perennial; widespread, somewhat weedy. 7. *E. philadelphicus.*

Plants low and often spreading or ascending, 0.5–3 dm. tall, scarcely *Aster*-like; achenes 2-nerved.

Basal leaves, like the cauline ones, entire or nearly so; pappus-bristles straight.

Leaves and stem glabrous or (especially above) somewhat glandular, sometimes with a few long hairs; involucre glutinous to glandular, sometimes with a few long hairs; Oregon to northern California.

Leaves rather equally distributed, the middle cauline about as large as the lower, the upper ones gradually reduced; Del Norte County, California, and adjacent Oregon. 11. *E. delicatus.*

Leaves inequably distributed, the basal ones obviously larger than the few and progressively reduced cauline ones.

Rays blue or purple; disk-corollas usually more or less puberulent at least below the middle; involucre finely glandular and not at all hairy; stem ordinarily wholly glabrous except directly under the head, where glandular; Siskiyou Mountains of northwestern California and adjacent Oregon. 12. *E. cervinus.*

Rays usually white; disk-corollas essentially glabrous; involucre usually with a few hairs; stem ordinarily with a few hairs or glands or both; Cascade and Klamath regions of Oregon but more northern or eastern than *E. cervinus.* 13. *E. cascadensis.*

Leaves, involucre, and usually the stem with obvious long hairs as well as stipitate-glandular; rays blue or purple; Cascade Mountains of Washington. 14. *E. leibergii.*

Basal leaves coarsely toothed or incised; pappus-bristles characteristically twisted and curled for at least the upper half; Columbia River Gorge in Oregon and Washington. 15. *E. oreganus.*

Cauline leaves mostly not very well developed, commonly linear or oblanceolate, sometimes linear-oblong or narrowly lance-oblong (sometimes larger in *E. caespitosus,* which has an evidently imbricate involucre, and in *E. basalticus,* which has the leaves evidently trilobed), those of *E. decumbens, E. corymbosus,* and a few other species, while narrow, sometimes not much smaller than the basal ones; achenes normally 2-nerved.

Leaves, or many of them, trilobed or two to four times ternate.

Stem evidently leafy, the cauline leaves more or less deeply trilobed; Yakima County, Washington. 16. *E. basalticus.*

Stem scapose or subscapose; the cauline leaves if present few and much reduced, mostly linear and entire.

---

* The perhaps artificial section *Olygotrichium,* consisting of short-lived, mostly biennial species, is here for convenience submerged in *Euerigeron.*

Caudex divided into several or many long, slender, rhizome-like branches; leaves mostly merely trilobed, the lobes short, broad, rounded, and obtuse; Sierra Nevada, Wallowa Mountains, and eastward (absent from the Cascade Mountains).
17. *E. vagus.*

Caudex stout, occasionally branching, but the branches mostly relatively stout and usually short, not slender and diffuse (caudex approaching that of *E. vagus* in some plants from the Cascade Mountains of Washington); leaves trifid or more often two to four times ternate, mostly with relatively slender lobes; widespread.
18. *E. compositus.*

Leaves all entire or nearly so, not at all trilobed.

Involucre woolly-villous with multicellular hairs; alpine and subalpine plants with solitary heads, always radiate; Oregon and Washington.

Rays yellow; Cascade Mountains of Washington and northward.
19. *E. aureus.*

Rays blue or pink to white, not yellow.

Rays about 50–125, pink or blue; northeastern Oregon and eastward.
20. *E. simplex.*

Rays about 25–50, white; northwestern Washington.
21. *E. flettii.*

Involucre variously hairy or glandular to glabrous but not woolly-villous; heads radiate to disciform or discoid; habit and distribution various.

Pubescence of the stem widely spreading, sometimes scanty (or largely appressed in three species; of these, *E. multiceps* has the pubescence distinctly spreading under the heads, *E. piperianus* and *E. engelmannii* have conspicuous, coarse, spreading hairs on the margins or petioles of at least the lower leaves).

Alpine and subalpine plants with solitary or rarely 2 heads, always radiate; California.

Plants with several more or less well-developed cauline leaves.

Stem viscid or glandular as well as hirsute; phyllaries about equal; taproot poorly or scarcely developed.
22. *E. petiolaris.*

Stem not viscid or glandular; phyllaries more or less imbricate; taproot well developed.
23. *E. clokeyi.*

Plants scapose or very nearly so.

Basal leaves oblanceolate or linear-oblanceolate, up to 2.5 mm. wide, mostly tapering to the petiole; rays blue or purple; disk-corollas 3.7–5.3 mm. long; Sierra Nevada.
24. *E. pygmaeus.*

Basal leaf-blades subrotund to broadly oblanceolate, rather abruptly contracted to and much shorter than the petiole, sometimes as much as 8 mm. wide; rays pink or white; disk-corollas 2.7–3.4 mm. long; south and east of the Sierra Nevada.
25. *E. uncialis.*

Plants chiefly of the valleys, foothills, and moderate elevations in the mountains, the rayless species *E. aphanactis* ascending to rather high elevations in southern California; heads 1 to many, widespread.

Basal leaves not triple-nerved.

Achenes densely long-hairy, the hairs completely covering the surface of the achenes and sometimes obscuring the outer pappus; heads radiate, with pink or purple to deep blue or violet rays, or rarely disciform, with the pistillate flowers rayless; central Oregon to southern British Columbia and western Idaho, east of the Cascade Mountains.
27. *E. poliospermus.*

Achenes sparsely or moderately hairy, the surface exposed between the hairs, the outer pappus if any not obscured by the hairs; rays various.

Rays blue or purple to pink or white, or wanting, but not yellow.

Heads evidently radiate.

Pappus simple or nearly so; disk-corollas 5–6 mm. long; Santa Barbara and San Luis Obispo Counties, California.
10. *E. sanctarum.*

Pappus distinctly double; disk-corollas mostly 3–5 mm. long; more northern or eastern.

Perennials; disk-corollas except in *E. multiceps* 3–5 mm. long; rays and pappus various.

Pubescence of the stem spreading throughout.

Rays mostly 1.5–2.3 mm. wide, white, turning pink or pinkish; disk-corollas relatively bulky, with short broad tube and heavy limb, the lower part of the limb only slightly indurated; heads solitary; near the Snake River in eastern Oregon, extreme southeastern Washington, and adjacent Idaho.
28. *E. disparipilus.*

Rays mostly 0.7–1.5 mm. wide, colored or occasionally white; disk-corollas relatively slender, with slender, comparatively long tube and narrow limb, the lower part of which is strongly indurated; heads 1 or usually more; widespread.
29. *E. pumilus.*

Pubescence of the stem appressed or closely ascending except sometimes directly under the heads.

Inner pappus-bristles about 12–20; disk-corollas about 3.3–4.5 mm. long; rays about 1.1–2.2 mm. wide; pubescence of the stem appressed even under the heads; at least the lower leaves with some coarse spreading hairs on the margins or petioles; northeastern Oregon and eastward.
30. *E. engelmannii.*

Inner pappus-bristles 5–8; disk-corollas about 2–3 mm. long; rays less than 1 mm. wide; pubescence of the stem spreading under the heads; pubescence of the leaves wholly appressed; southern Sierra Nevada.
31. *E. multiceps.*

Biennial; disk-corollas 2–3 mm. long; rays about 75–150, 0.5–1.2 mm. wide; inner pappus-bristles 5–12; widespread.
32. *E. divergens.*

Heads disciform, the pistillate flowers present but essentially rayless.
    Biennials; southern California.
        Inner pappus about 5–12 fragile bristles; disk-corollas 2–3 mm.
           long.   32. *E. divergens.*
        Inner pappus about 15–20 firm bristles; disk-corollas 4–5 mm.
           long.   33. *E. calvus.*
    Perennials; widespread.
        Outer pappus of evident scales; stem often leafy, subnaked
           chiefly in forms from southern California; southeastern
           Oregon to southern California and eastward.
                                 34. *E. aphanactis.*
        Outer pappus setose and obscure, or wanting; stem subnaked;
           northern Nevada and California to extreme southeastern
           Washington.   35. *E. chrysopsidis.*
Rays yellow.
    Pubescence of the stem spreading throughout; leaves narrowly oblance-
        olate, straight or slightly arcuate, scarcely flexuous; stem sub-
        naked, the leaves all or nearly all in a basal cluster; extreme
        southeastern Washington to northern California and Nevada.
                               35. *E. chrysopsidis.*
    Pubescence of the stem mostly appressed, at least above; leaves linear,
        flexuous; stem more or less leafy, at least below the middle;
        Washington east of the Cascade Mountains.
                             36. *E. piperianus.*
Basal leaves more or less strongly triple-nerved.
    Involucre more or less spreading-hirsute as well as glandular; rays about 10–25,
        or none; northern California.   42. *E. lassenianus.*
    Involucre canescent with fine white hairs (sometimes sparsely so), sometimes also
        glandular, but not spreading-hirsute; rays about 30–100.
        Basal leaves acute; phyllaries only slightly or obscurely thickened on the
           back; stem ordinarily purplish at the base; east of the Cascade Moun-
           tains from Oregon to southern British Columbia and eastward.
                               43. *E. corymbosus.*
        Basal leaves usually rounded or obtuse at the tip; phyllaries evidently thick-
           ened on the back; stem rarely purplish at the base; extreme eastern
           Washington and eastward.   44. *E. caespitosus.*
Pubescence of the stem and leaves more or less closely appressed or wanting, never definitely
    spreading.
    Heads discoid, the flowers all tubular and perfect; widespread.
                               37. *E. bloomeri.*
    Heads evidently radiate.
    Base of the stem conspicuously enlarged, shining, and somewhat indurated, straw-
        colored or purplish.
    Rays yellow; pappus-bristles 10–20; widespread.   38. *E. linearis.*
    Rays blue, purple, or pink; pappus-bristles mostly 20–50.
    Leaves linear, 1 mm. wide or less; involucre 3.5–5 mm. high; disk 6–11 mm.
        wide; pappus-bristles 20–30; northeastern Oregon to Siskiyou, Modoc,
        and Lassen Counties, California.   39. *E. elegantulus.*
    Leaves oblanceolate, the basal ones 1.5–5 mm. wide; involucre 5.5–9 mm.
        high; disk 13–18 mm. wide; pappus-bristles mostly 30–40, sometimes
        50; Lassen County to Fresno County, California.
                              40. *E. barbellulatus.*
    Base of the stem not conspicuously enlarged, shining, and indurated (alpine specimens
      of *E. peregrinus* would be sought here, except for the 4–7-nerved achenes).
    Basal leaves narrow, linear or rather narrowly oblanceolate, the blade if distin-
        guishable tapering very gradually to the petiole.
        Basal leaves with their bases neither enlarged nor of different texture from
           the blades; leaves linear or linear-filiform; stem more densely hairy
           toward the base than above; southern British Columbia to Sierra
           County, California, and eastward.   41. *E. filifolius.*
        Basal leaves with their bases commonly somewhat enlarged, membranous or
           indurated, of different texture from the blades; leaves linear or more
           often oblanceolate; stem not more densely hairy toward the base than
           above.
        Plants neither pulvinate-cespitose nor scapose; basal leaves mostly triple-
           nerved except in *E. flexuosus.*
        Disk-corollas mostly 5–7 mm. long; style-appendages mostly 0.3–
           0.4 mm. long; pappus coarse and copious, the inner of 25–40
           bristles; involucre 7.5–10 mm. high; eastern California to
           Nevada.   45. *E. nevadincola.*
        Disk-corollas mostly 2.5–5 mm. long; style-appendages 0.1–0.25
           mm. long; pappus relatively scanty and fragile, the inner of
           10–25 bristles; involucre 3.5–8 mm. high.
           Cauline leaves mostly abruptly reduced and smaller than the
              basal ones; rays mostly white; east of the Cascade and
              Sierra Nevada summits except in northern California.
                              46. *E. eatonii.*
           Cauline leaves only gradually reduced; rays mostly colored;
              chiefly west of the Cascade summits.
              Basal leaves more or less evidently triple-nerved; heads
                mostly few or solitary and more or less naked-pe-
                dunculate; rays mostly 20–50; involucre hairy but
                scarcely glandular, and in California forms mostly
                6–8 mm. high; western Oregon and northern Cali-
                fornia.   47. *E. decumbens.*
             Basal leaves only obscurely if at all triple-nerved; heads
                several or rather numerous on leafy branches; rays
                mostly 12–20; involucre glandular as well as hairy,
                4–5 mm. high; Trinity Mountains, California.
                            48. *E. flexuosus.*

Plants pulvinate-cespitose, scapose; leaves not triple-nerved; Inyo Mountains, California, and eastward.   49. *E. compactus.*
Basal leaves relatively broad, broadly oblanceolate or usually broader, the blade well defined, usually more or less abruptly contracted to the petiole.
Leaves glabrous; involucral bracts subequal, finely glandular, not hairy; Siskiyou Mountains of northwestern California and adjacent Oregon.
12. *E. cervinus.*
Leaves strigose, sometimes sparsely so; involucral bracts distinctly imbricate, finely spreading-hairy as well as glandular; Sierra Nevada and southern Cascade Mountains eastward.   26. *E. tener.*

II. WYOMINGIA

Achenes mostly 4-nerved, rarely 6-nerved; basal leaves often withered by flowering time, not forming a conspicuous persistent tuft; San Bernardino County, California, and eastward.
Outer pappus of evident narrow scales; stem silvery white with pubescence; San Bernardino Mountains region.   50. *E. parishii.*
Outer pappus of inconspicuous setae; stem merely gray-green except near the base; Providence Mountains and eastward.   51. *E. utahensis.*
Achenes mostly 6–8(10)-nerved; basal leaves tufted and persistent, the cauline ones reduced; Inyo County, California, and eastward.   52. *E. argentatus.*

III. PYCNOPHYLLUM

Heads evidently radiate.
Root-crown mostly subterranean, giving rise to slender rhizomatous stems rarely more than 1.5 mm. thick which become aerial stems; plants relatively slender, usually with solitary or few heads; stem generally branched into equal parts near the ground-level; pubescence of the stem except in one reduced and slender variety spreading or retrorse.
Stem finely glandular, as well as hirsute with relatively long spreading hairs; Tulare County, California.   53. *E. aequifolius.*
Stem not glandular, though the hairs may be a little viscid; pubescence of the stem relatively short or wanting; Shasta County to Riverside County, California, and eastward.   54. *E. breweri.*
Root-crown superficial or nearly so; base of stem, if rhizomatous, either very short or relatively stout or both; plants variously slender or stout, with solitary to generally several or many heads; stem except in var. *confinis* generally not branching near the ground-level into equal parts; pubescence of the stem varying from absent to strigose to spreading-hirtellous or even spreading-hirsute; southern Oregon to Lower California.   55. *E. foliosus.*
Heads discoid, the flowers all tubular and perfect.
Stem and leaves conspicuously spreading-villous or spreading-villosulous, occasionally rather sparsely so.
Larger leaves generally 2–4 cm. long; outer pappus obscure or wanting; Klamath region of northern California and adjacent Oregon and in the Coast Ranges of California.   56. *E. petrophilus.*
Larger leaves generally 1–2 cm. long; outer pappus evident; Sierra Nevada near Donner Pass.   57. *E. miser.*
Middle and upper part of stem not spreading-villous, varying from glabrous to glandular and sometimes ascending-hirtellous or appressed-hairy; leaves not spreading-villous, though sometimes otherwise short-hairy; southeastern Washington to southern California.   58. *E. inornatus.*

IV. PHALACROLOMA

Foliage ample; plants mostly 6–15 dm. tall; pubescence of the stem long and spreading.
59. *E. annuus.*
Foliage sparse; plants mostly 3–7 dm. tall; pubescence various.   60. *E. strigosus.*

V. TRIMORPHAEA

Rayless pistillate flowers wanting; inflorescence racemiform, the peduncles erect or nearly so or the head solitary.
61. *E. lonchophyllus.*
Rayless pistillate flowers present between the rays and disk-flowers; inflorescence corymbiform, the peduncles arcuate or obliquely ascending or the head solitary.   62. *E. acris.*

1. **Erigeron peregrìnus** (Pursh) Greene. Wandering Daisy. Fig. 5582.

Aster peregrinus Pursh, Fl. Amer. Sept. **2**: 556. 1814.
Erigeron peregrinus Greene, Pittonia **3**: 166. 1897.

Fibrous-rooted perennial from a short rhizome or short stout caudex, up to 7 dm. tall, amply leafy or in small forms subscapose. Basal and lower cauline leaves well developed, with linear-oblanceolate to broadly oblanceolate or spatulate blade tapering to the petiole, the others smaller, mostly sessile and often slightly clasping; heads solitary or few, the disk 10–25 mm. wide; involucre 7–11 mm. high, the phyllaries linear, attenuate, loose, mostly rather herbaceous and about equal, about 1 mm. wide; rays 30–80, white to purple, 8–25 mm. long, 2–4 mm. wide; disk-corollas mostly 4–6 mm. long; style-appendages acute, 0.2–0.4 mm. long; pappus of 20–30 bristles, occasionally with a few short and inconspicuous, outer setae; achenes asymmetrically 4–7-nerved, most commonly 5-nerved. $n = 9$.

Moist meadows, streamsides, or bogs at moderate to high elevations in the mountains, Canadian and Hudsonian Zones; mountains of California (Sierra Nevada), Utah, and northern New Mexico north to about 56° in the Canadian Rockies and extending along the Canadian and Alaskan coast through the Aleutian Islands to the Commander Islands. Type locality: Unalaska. (June)July–Aug.
The nomenclaturally typical phase does not occur in our range. The foregoing description is drawn to cover all forms of the species.

KEY TO SUBSPECIES AND VARIETIES

Phyllaries villous on the back or sometimes merely ciliate on the margins and glutinous on the back, not at all glandular; rays commonly rather pale or even white; herbage often soft-pubescent, the stem usually sparsely villous, the peduncular hairs rather loose; leaves often toothed; coastal and Cascade Mountains from central Washington northward, not extending to the interior ranges.   subsp. *peregrinus.*
Phyllaries hairy on the back, often also more or less long-ciliate on the margins; rays usually colored, sometimes white; foliage ample to rather scanty.

Upper cauline leaves reduced and distant; northern Washington to Unalaska.   var. *dawsonii.*

Upper cauline leaves either ample or closely set; chiefly from the Alaska panhandle northward, occasional in British Columbia.   var. *peregrinus.*

Phyllaries merely glutinous on the back, ciliate on the margins; rays ordinarily white; foliage scanty; bogs along the southern fringe of the Olympic Mountains in Washington.   var. *thompsonii.*

Phyllaries densely glandular on the back, rarely with a few long hairs; rays commonly rich rose-purple or darker; herbage (except in var. *hirsutus*) usually glabrous except for the closely villous peduncles; leaves ordinarily entire; interior mountains of British Columbia and Alberta southward to Washington, California, Utah, and New Mexico.   subsp. *callianthemus.*

Leaves hirsute on both surfaces; peduncles spreading-hirsute; central and southern Sierra Nevada of California extending into adjacent Nevada.   var. *hirsutus.*

Leaves glabrous on both surfaces, sometimes with a few hairs along the midvein, or rarely short-villous; peduncles with appressed or ascending, often curled or twisted hairs.

Reduced alpine plants, less than 2 dm. tall, with relatively ample, apically obtuse, basal leaves and very much smaller cauline leaves, often subscapose; range of the subspecies except for the Sierra Nevada.   var. *scaposus.*

Larger, mostly subalpine or merely montane plants, up to 7 dm. tall (smaller in alpine phases of var. *angustifolius* but then with narrow, acute, basal leaves).

Leaves narrow, the basal oblanceolate or narrower, the cauline linear or lanceolate; Sierra Nevada of California and adjacent Nevada extending northward less commonly to southern British Columbia.   var. *angustifolius.*

Leaves more ample, the basal oblanceolate or broader, the cauline mostly ovate and not greatly reduced; range of the subspecies except for the Sierra Nevada.   var. *callianthemus.*

Erigeron peregrinus subsp. **peregrinus** var. **dawsònii** Greene, Pittonia **3:** 166. 1897. Type locality: Queen Charlotte Islands, British Columbia.

Erigeron peregrinus subsp. **peregrinus** var. **thompsònii** (Blake) Cronquist, Brittonia **6:** 144. 1947. (*Erigeron thompsoni* Blake ex Thompson, Rhodora **34:** 238. 1932.) Type locality: Lake Quinault, Grays Harbor County, Washington.

Erigeron peregrinus subsp. **calliánthemus** (Greene) Cronquist, Rhodora **45:** 264. 1943, var. **callianthemus.** (*Erigeron membranaceus* Greene, Pittonia **3:** 294. 1898; *E. callianthemus* Greene, Leaflets Bot. Obs. **2:** 197. 1912; *E. peregrinus* subsp. *callianthemus* var. *callianthemus* f. *membranaceus* Cronquist, Brittonia **6:** 145. 1947, a form with thin, broad, strongly clasping leaves; *E. peregrinus* subsp. *callianthemus* var. *callianthemus* f. *subvillosus* Cronquist, op. cit. 146, a form with shortly villous leaves; *E. peregrinus* subsp. *callianthemus* var. *callianthemus* f. *dentatus* Cronquist, loc. cit., a form with toothed leaves.) Type locality: Centennial, Albany County, Wyoming.

Erigeron peregrinus subsp. **callianthemus** var. **angustifòlius** (A. Gray) Cronquist, Brittonia **6:** 147. 1947. (*Aster salsuginosus* var. *angustifolius* A. Gray, Bot. Calif. **1:** 325. 1876; *Erigeron salsuginosus* var. *angustifolius* A. Gray, Proc. Amer. Acad. **16:** 93. 1880; *E. angustifolius* Rydb. Bull. Torrey Club **24:** 295. 1897; *E. hesperocallis* Greene, Leaflets Bot. Obs. **2:** 200. 1912; *E. loratus* Greene, op. cit. 202; *E. regalis* Greene, op. cit. 205.) Type locality: Sierra County, California.

Erigeron peregrinus subsp. **callianthemus** var. **hirsùtus** Cronquist, Brittonia **6:** 147. 1947. Type locality: Lake Tenaya, Yosemite National Park, California.

Erigeron peregrinus subsp. **callianthemus** var. **scapòsus** (Torr. & Gray) Cronquist, Brittonia **6:** 146. 1947. (*Aster gracilis* Nutt. Trans. Amer. Phil. Soc. II. **7:** 291. 1840; *A. salsuginosus* var. *scaposus* Torr. & Gray, Fl. N. Amer. **2:** 503. 1843; *Erigeron salsuginosus* var. *glacialis* A. Gray, Syn. Fl. N. Amer. **1²:** 209. 1884; *E. glacialis* A. Nels. Bot. Gaz. **37:** 207. 1904; *E. suksdorfii* Greene, Leaflets Bot. Obs. **2:** 203. 1912; *E. ciliolatus* Greene, loc. cit.) Type locality: Wind River Range, Sublette County, Wyoming.

2. **Erigeron howéllii** A. Gray. Howell's Daisy. Fig. 5583.

Erigeron salsuginosus var. *howellii* A. Gray, Proc. Amer. Acad. **16:** 93. 1880.
Erigeron howellii A. Gray, Syn. Fl. N. Amer. **1²:** 209. 1884.

Fibrous-rooted perennial from a short rhizome, 2–5 dm. tall, scantily short-villous under the heads. Leaves entire or irregularly few-toothed, thin, glabrous, the lowermost ones with elliptic or suborbicular blade 2–8 cm. long and 1.5–5 cm. wide, abruptly contracted to the 2–12 cm. petiole;

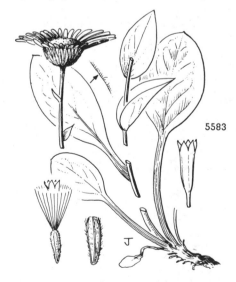

5582. Erigeron peregrinus   5583. Erigeron howellii

middle cauline leaves ample, ovate to cordate, strongly clasping; upper leaves similar but smaller; heads solitary, the disk 12–20 mm. wide; phyllaries loose, equal, acuminate or attenuate, glandular, somewhat herbaceous; rays 30–50, white, 13–25 mm. long, 2–4 mm. wide; disk-corollas 4–5 mm. long, more flaring than in *E. peregrinus*; style-appendages acute, 0.3–0.4 mm. long; achenes mostly asymmetrically 5-nerved; pappus of 20–30 bristles.

Moist, often rocky places, Humid Transition Zone; south side of the Columbia Gorge, Oregon. Type locality: Cascade Mountains, Oregon. April–June.

3. Erigeron alíceae Howell. Eastwood's Daisy. Fig. 5584.

Erigeron aliceae Howell, Fl. N.W. Amer. 317. 1900.
Erigeron amplifolius Howell, loc. cit.
Erigeron nemophilus Greene, Leaflets Bot. Obs. **2**: 210. 1912.

Fibrous-rooted perennial from a rather short rhizome or woody caudex, 3–8 dm. tall, amply leafy. Leaves hirsute on both sides, entire or coarsely toothed, the lowermost ones up to 20 cm. long (including the petiole) and 3.5 cm. wide, the middle ones sessile, narrowly lanceolate to oblong or ovate; heads 1 to several, the disk mostly 12–20 mm. wide; phyllaries loose, attenuate, subequal, conspicuously white-hirsute on the lower one-fourth to three-fourths, glandular thence to the tip; rays 45–80, mostly 10–15 mm. long and 2–3 mm. wide, white to pink-purple; disk-corollas mostly 3–4 mm. long; style-appendages acutish, about 0.25 mm. long; achenes 2-nerved or sometimes 4-nerved; pappus of 20–30 bristles.

Moist or fairly dry soil in shady or open places in the mountains, Canadian and Hudsonian Zones; Mount Hood southward in the Cascade Mountains, and in the Siskiyou Mountains of Oregon and adjacent California; also in the Olympic Mountains, Washington. Type locality: top of Siskiyou Mountains near Waldo, Oregon. June–Aug.

4. Erigeron coùlteri Porter. Coulter's Daisy. Fig. 5585.

Erigeron coulteri Porter in Port. & Coult. Fl. Colo. 61. 1874.
Erigeron frondeus Greene, Fl. Fran. 387. 1897.
Erigeron leucanthemoides Greene, Leaflets Bot. Obs. **2**: 211. 1912.

Fibrous-rooted perennial from a slender rhizome or branching caudex; stems 1–6 dm. tall, usually amply leafy, spreading-hirsute at least above. Leaves hirsute, at least the lower ones generally toothed, the middle cauline ones mostly broadly lanceolate to oblong or ovate, tending to be clasping at the base, up to about 9 cm. long and 3 cm. wide, larger or smaller than the persistent or deciduous, mostly more petiolate, lower ones; heads 1–4, the involucre 7–10 mm. high, its thin green phyllaries equal, attenuate, villous-hirsute, the hairs with black cross-walls toward the base; rays 50–100, white, 9–24 mm. long, 1.2–1.7 mm. wide; disk-corollas 3–4.4 mm. long; style-appendages more or less acute, 0.2–0.35 mm. long; inner pappus of 20–25 bristles, the outer obscure or wanting; achenes 2-nerved.

Meadows and stream banks, mostly at rather high elevations in the mountains, Boreal Zone; northern Idaho and adjacent Montana to northeastern Oregon southward irregularly to the southern Cascade Mountains and Sierra Nevada in California, the Wasatch Mountains of Utah, and the mountains of Colorado and northern New Mexico. Type locality: Weston's Pass, Colorado. July–Aug.

5. Erigeron speciòsus (Lindl.) DC. Showy Daisy. Fig. 5586.

Stenactis speciosa Lindl. Bot. Reg. **19**: *pl. 1577*. 1833.
Erigeron speciosus DC. Prod. **5**: 284. 1836.

More or less fibrous-rooted perennial from a woody caudex, the stems clustered, 1.5–8 dm. tall, amply leafy, generally glabrous below the inflorescence. Leaves entire, glabrous or nearly so except for the commonly ciliate margins, often triple-nerved, the lower oblanceolate or spatulate, petiolate, mostly deciduous, the others becoming sessile but fairly ample, the uppermost ones mostly lanceolate; heads 1–13, the involucre 6–9 mm. high, glandular and commonly with a very few long hairs; rays 65–150, blue or rarely white, 9–18 mm. long, about 1 mm. wide; disk-corollas 4–5 mm. long; style-appendages acute, 0.15–0.2 mm. long; pappus of 20–30 bristles and some short outer setae; achenes 2-nerved or occasionally 4-nerved.

Open woods and openings in wooded areas, mostly in the foothills and at moderate elevations in the mountains, Transition Zones; southern British Columbia, Washington, northwestern Oregon, northern Idaho, and northwestern Montana. Type locality: "California"; probably actually in northern Oregon or southern Washington June–Aug.

A few of our specimens resemble the otherwise more eastern and southern var. *macranthus* (Nutt.) Cronquist, of the Rocky Mountains and eastern Great Basin region, characterized by its broader (mostly ovate) upper leaves that tend to be less strongly ciliate; likewise, occasional specimens throughout the range of var. *macranthus* resemble typical *E. speciosus*.

6. Erigeron subtrinérvis Rydb. var. conspícuus (Rydb.) Cronquist. Three-veined Daisy. Fig. 5587.

Erigeron conspicuus Rydb. Mem. N.Y. Bot. Gard. **1**: 400. 1900.
Erigeron villosulus Greene, Leaflets Bot. Obs. **2**: 215. 1912.
Erigeron subtrinervis subsp. *conspicuus* Cronquist, Bull. Torrey Club **70**: 271. 1943.
Erigeron subtrinervis var. *conspicuus* Cronquist, Vasc. Pl. Pacif. Northw. **5**: 193. 1955.

Closely related to *E. speciosus* and not always sharply distinct, but more pubescent, the stem, leaves, and especially the involucre evidently pubescent with long spreading hairs, the pubescence of the leaves sometimes confined chiefly to the margins and larger veins.

Open woodlands, usually in drier places than *E. speciosus*, Canadian and Arid Transition Zones; southern

British Columbia and Alberta to Washington, central Idaho, western Montana, and northwestern Wyoming. Type locality: Jack Creek, Montana. (June)July–Aug.

Typical *E. subtrinervis*, characterized by its shorter and denser pubescence, occurs from Wyoming, Utah, and New Mexico to western Nebraska and South Dakota.

7. Erigeron philadélphicus L. Philadelphia Daisy. Fig. 5588.

Erigeron philadelphicus L. Sp. Pl. 863. 1753.
Erigeron purpureus Ait. Hort. Kew. **3**: 186. 1789.
Erigeron purpureus var. *attenuatus* Nutt. Trans. Amer. Phil. Soc. II. 7: 307. 1840.
Tessenia philadelphica Lunell, Amer. Midl. Nat. **5**: 59. 1917.
Erigeron philadelphicus f. *purpureus* Farwell, op. cit. 11: 70. 1928.

Biennial or short-lived perennial or rarely apparently annual, mostly fibrous-rooted, commonly 2–7 dm. tall; herbage long-spreading-hairy to occasionally subglabrous. Basal leaves mostly oblanceolate, toothed or lobed to sometimes entire, seldom over 15 cm. long and 3 cm. wide; cauline leaves becoming sessile, mostly ample and more or less clasping, less so in smaller specimens; heads 1 to many, involucre 4–6 mm. high, the phyllaries subequal, light greenish or brownish, the brown midvein more or less hirsute with flattened hairs, or nearly glabrous, the broad hyaline margins occasionally purplish; disk 6–15 mm. wide; rays 150–400, deep pink to white, 0.2–0.6 mm. wide, 5–10 mm. long; disk-corollas 2.5–3.2 mm. long; style-appendages obtuse or acutish, about 0.1 mm. long; pappus of 20–30 fragile bristles; achenes 2-nerved.

In a wide variety of habitats, most commonly in moist, often also disturbed soil; throughout the United States and most of Canada. Type locality: Canada. Mostly May–July or sometimes until fall. Highly variable in size and luxuriance, according to the habitat.

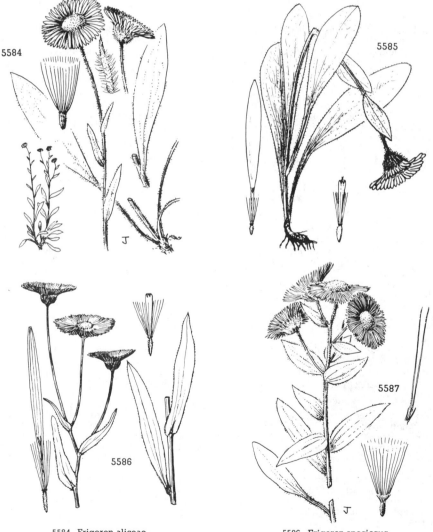

5584. Erigeron aliceae
5585. Erigeron coulteri
5586. Erigeron speciosus
5587. Erigeron subtrinervis

8. **Erigeron glaùcus** Ker-Gawl. Seaside Daisy. Fig. 5589.

Erigeron glaucus Ker-Gawl, Bot. Reg. **1**: *pl. 10.* 1815.
Aster bonariensis Spreng. Syst. **3**: 528. 1826.
Aster californicus Less. Linnaea **6**: 121. 1831.
Stenactis glauca Nees, Gen. & Sp. Aster. 275. 1832.
Woodvillea calendulacea DC. Prod. **5**: 318. 1836.
Erigeron hispidus Nutt. Trans. Amer. Phil. Soc. II. **7**: 310. 1840.
Erigeron maritimus Nutt. loc. cit.
?Erigeron squarrosus Lindl. Bot. Reg. **27**: 44. 1841.

Somewhat succulent, maritime perennial from a stout rhizome or branched caudex, highly variable according to exposure, 5–50 cm. high, the stem curved at the base, more or less spreading-villous or somewhat glandular to subglabrous. Basal leaves up to 15 cm. long and 5 cm. wide, with obovate to broadly spatulate blade, usually toothed above the middle; cauline leaves smaller but usually ample; heads 1–15 on leafy, often long branches; disk 1.5–3.5 cm. wide; phyllaries numerous, equal, acuminate or attenuate, scarcely herbaceous, sparsely to densely long-villous and often viscid; rays mostly about 100, blue or white, 9–15 mm. long, 1–2 mm. wide; disk-corollas about 5 mm. long, narrow, with short lobes about 0.5 mm. long; style-appendages obtuse, about 0.2 mm. long; pappus usually of 20–30 bristles, sometimes with some inconspicuous, short, outer setae; achenes mostly 4(2–6)-nerved.

Sea bluffs and sandy beaches along the coast, under the influence of salt water, Canadian and Humid Transition Zones; Clatsop County, Oregon, southward to San Luis Obispo County and the Santa Barbara Islands, California. Type locality: originally thought to be South America; doubtless actually California. April–Aug.

9. **Erigeron súpplex** A. Gray. Supple Daisy. Fig. 5590.

Erigeron supplex A. Gray, Proc. Amer. Acad. **7**: 353. 1868.

Submaritime perennial with a short branched caudex which sometimes surmounts a taproot, 15–40 cm. tall, the stems decumbent at the base, sparsely to moderately pubescent with long spreading hairs and sometimes obscurely glandular. Leaves entire, ciliate-margined, otherwise subglabrous to moderately villous-hirsute, relatively numerous and crowded, scarcely tufted at the base, the lower ones oblanceolate, petiolate, up to 6 cm. long and 7 mm. wide, those above similar but progressively reduced, becoming sessile and lance-linear or narrowly oblong; head solitary on a subnaked peduncle 2–10 cm. long, the disk 14–20 mm. wide, very broad in shape; involucre 7–11 mm. high, sparsely to densely villous-hirsute with long, spreading, white hairs, also viscid or finely glandular, the phyllaries subequal, green, with paler thin margins and thin, attenuate or acuminate tip; pistillate flowers wanting; disk-corollas bright orange-yellow, 4.7–7.0 mm. long, the lobes about 1 mm.; style-appendages obtuse, 0.2–0.25 mm. long; pappus of 17–30 firm bristles, with a few short and inconspicuous outer setae; achenes 2-nerved.

Near the seacoast, Humid Transition Zone; Mendocino and Humboldt Counties, California. Type locality: near Mendocino, Mendocino County. Mostly May–July.

10. **Erigeron sanctàrum** S. Wats. Saints' Daisy. Fig. 5591.

Erigeron sanctarum S. Wats. Proc. Amer. Acad. **24**: 83. 1889.

Perennial, apparently from a slender rhizome which may perhaps connect to a subterranean root-crown, 5–35 cm. tall, the stem ascending or erect from a mostly curved base, sparsely pubescent, especially above, with spreading or retrorse hairs. Leaves entire, shortly hispid or hirsute on at least the upper surface and generally also ciliate-margined, the basal ones 2–5 cm. long, 3–10 mm. wide, petiolate, oblanceolate, or with well-defined elliptic blade, the cauline ones evidently reduced upward, becoming linear and sessile; heads solitary (2–3) on a naked or subnaked peduncle 2–10 cm. long; disk 12–17 mm. broad; involucre 6–9 mm. high, evidently spreading-hirsute, the phyllaries subequal, narrow, attenuate, green with evident brown midvein; rays 45–90, blue, 7–13 mm. long, 1.3–1.9 mm. wide; disk-corollas 5–6.5 mm. long; style-appendages obtuse or acutish, 0.1–0.25 mm. long; pappus of 18–25 bristles, sometimes with some short and inconspicuous outer setae; achenes 2-nerved.

Hills near the seacoast, Upper Sonoran Zone; Santa Barbara and San Luis Obispo Counties, California. Type locality: Santa Inez Mountains near Santa Barbara. March–June.

11. **Erigeron delicàtus** Cronquist. Del Norte Daisy. Fig. 5592.

Erigeron delicatus Cronquist, Brittonia **6**: 216. 1947.

Fibrous-rooted perennial from a short rhizome or slender caudex, glabrous below, finely stipitate-glandular above, the stems 2–3 dm. tall, slender, arising singly or few together. Leaves entire, rather numerous and equably distributed, the middle cauline about as large as the lower, the upper gradually reduced, the lowermost ones with oblanceolate, obtuse, or rounded blade 1–3 cm. long and 4–8 mm. wide tapering to the 1–4 cm. petiole, the others progressively less petiolate or sessile; heads solitary or 2 on short, stipitate-glandular peduncles; disk 9–13 mm. wide; involucre 5–6 mm. high, the phyllaries glandular or glabrous, subequal, greenish-stramineous, the tips acute or acuminate, sometimes purplish; rays about 40, blue, 7–10 mm. long, 1.5–2 mm. wide; disk-corollas about 3.5 mm. long, broad-based, slightly flaring above the middle; style-appendages acutish, 0.2 mm. long; pappus of about 15 slender bristles and some inconspicuous, short, slender setae; achenes 2-nerved.

At lower elevations, Humid Transition Zone; Del Norte County, California, and adjacent Oregon. Type locality: Smith River, mouth of Mill Creek, Del Norte County. June–July.

5588. Erigeron philadelphicus
5589. Erigeron glaucus
5590. Erigeron supplex

5591. Erigeron sanctarum
5592. Erigeron delicatus
5593. Erigeron cervinus

12. Erigeron cervìnus Greene. Siskiyou Daisy. Fig. 5593.

Erigeron cervinus Greene, Pittonia **3**: 163. 1897.

Fibrous-rooted perennial from a branching caudex, 1–3 dm. tall, the herbage glabrous except immediately under the heads, where usually finely glandular. Leaves entire, the basal ones broadly oblanceolate to elliptic or obovate, up to 12 cm. long (petiole included) and 15 mm. wide, usually mucronate at the rounded apex; cauline leaves several, scattered, moderately reduced; heads solitary or sometimes 2–4, the disk 10–15 mm. wide; involucre finely or obscurely glandular, 3–7 mm. high; phyllaries subequal, herbaceous, purple-tipped, the outer acute, the inner acuminate; rays 20–45, blue or sometimes pink-purple, 7–10 mm. long, 1.3–2.7 mm. wide; disk-corollas 3.5–5 mm. long, the limb somewhat puberulent below the middle; style-appendages obtuse or acutish, 0.2–0.25 mm. long; pappus of about 12–15 bristles, with a few short outer setae; achenes 2-nerved.

Meadows and open places, Canadian Zone; Siskiyou Mountains region of northwestern California and adjacent Oregon. Type locality: Deer Creek Mountains, Josephine County, Oregon. June–early Aug.

13. Erigeron cascadénsis Heller. Cascade Daisy. Fig. 5594.

Erigeron spatulifolius Howell, Fl. N.W. Amer. 317. 1900. Not *E. spathulifolius* Rydb. 1899.
Erigeron cascadensis Heller, Muhlenbergia **1**: 6. 1900.
Erigeron pachyrhizus Greene, Leaflets Bot. Obs. **2**: 216. 1912.

Perennial from a stout branched caudex which may surmount a short taproot, 5–15 cm. tall, the herbage glabrous or with a few scattered glands and/or spreading hairs. Basal leaves oblanceolate to spatulate or sometimes obovate, 2.5–9 cm. long, 5–17 mm. wide, rounded or broadly obtuse, entire or obscurely toothed; cauline leaves few, narrowly lanceolate to ovate, sessile, 1–3.5 cm. long, 2–9 mm. wide; heads solitary (2–3), the disk 8–14 mm. wide; involucre 5–8 mm. high, its bracts subequal, glandular-viscid or glutinous and very sparsely to moderately villous-hirsute, green with purplish acuminate or attenuate tip; rays 30–50, mostly white, 6–10 mm. long, 1.2–2.0 mm. wide; disk-corollas essentially glabrous, 3.7–4.4 mm. long; style-appendages acute, 0.2–0.3 mm. long; pappus double, the inner of 15–20 firm bristles, the outer of short narrow scales; achenes 2-nerved.

Often in rocky places, moderate to high elevations in the mountains, Canadian and Hudsonian Zones; Cascade and Calapooia Mountains, Oregon, from about the 43rd to the 45th parallel. Type locality: Pansy Camp, Cascade Mountains. June–Aug.

14. Erigeron leibérgii Piper. Leiberg's Daisy. Fig. 5595.

Erigeron leibergii Piper, Bull. Torrey Club **28**: 41. 1901.
Erigeron chelanensis St. John, Research Stud. St. Coll. Wash. **1**: 107. 1929.

Perennial from a stout branched caudex which may surmount a taproot, 7–25 cm. tall; herbage and involucre sparsely to moderately villous-hirsute with flattened, flexuous hairs and more or less stipitate-glandular. Basal leaves broadly oblanceolate to obovate or elliptic, sometimes triplenerved, entire or denticulate, up to 12 cm. long and 2 cm. wide; cauline leaves several, fairly well developed, ovate to oblong or lanceolate; heads solitary or up to 5, the disk 7–14 mm. wide; involucre 5–8 mm. high, the phyllaries subequal, thin, loose, green, the attenuate or acuminate tip often purplish; rays 20–45, mostly blue or pink, rarely white, 5–12 mm. long, 1.3–2.0 mm. wide; disk-corollas 3.0–4.3 mm. long; style-appendages acute or acutish, mostly 0.2–0.25 mm. long; pappus of 12–16 bristles, sometimes with a few short, slender, outer setae; achenes 2-nerved.

Cliffs and rocky places at moderate to high elevations, Boreal Zone; Cascade and Wenatchee Mountains of Okanogan, Chelan, and Kittitas Counties, Washington. Type locality: Mount Stuart, Chelan County. June–Aug

15. Erigeron oregànus A. Gray. Gorge Daisy. Fig. 5596.

Erigeron oreganus A. Gray, Proc. Amer. Acad. **19**: 2. 1882.

Perennial with a stout, mostly simple caudex and stout taproot; herbage glandular and loosely viscid-villous; stems lax, 5–15 cm. long. Basal leaves tufted, spatulate to obovate, coarsely toothed or incised, up to 9 cm. long and 2.5 cm. wide; cauline leaves well developed, broadly lanceolate to elliptic or ovate, up to 4 cm. long and 1 cm. wide; heads 1 to several in a leafy inflorescence, the disk 9–13 mm. wide; involucre 5–7 mm. high, glandular and viscid-villous, the phyllaries loose, equal, thin, green, attenuate; rays mostly 30–60, bluish to more often pink or white, 5–8 mm. long; disk-corollas 3.4–4.7 mm. long; pappus simple, of 15–20 bristles which are characteristically curled and twisted for at least the upper half.

Moist shady cliffs and ledges, Humid Transition Zone; Columbia River Gorge, most frequently collected on the Oregon side. Type locality: Columbia River, Oregon. June–Sept.

16. Erigeron basálticus Hoover. Basalt Daisy. Fig. 5597.

Erigeron basalticus Hoover, Leaflets West. Bot. **4**: 40. 1944.

Stems several from a perennial taproot, sprawling or pendent, branched, leafy especially near the tip; herbage spreading-hirsute and finely glandular. Leaves cuneate to obovate, up to about 4 cm. long and 1.5 cm. wide, more or less deeply and often irregularly trilobed, the lobes broad, often slightly lobed again; heads terminating the branches, the disk 8–12 mm. wide; involucre 5–6 mm. high, densely glandular and sometimes sparsely long-hairy, the phyllaries subequal or slightly imbricate, green, the slender acuminate tips commonly purplish; rays 25–30, pink or pink-purple, 5–7 mm. long and 1.5 mm. wide; disk-corollas 3–4 mm. long; style-appendages obtuse, 0.1–0.15 mm. long; pappus of 10–15 bristles, with some inconspicuous, short, outer setae; achenes 2-nerved.

Cliff-crevices in rocky canyons at low elevations, Arid Transition Zone; Yakima County, Washington. Type locality: Selah Creek and vicinity, Yakima County. May–Oct.

17. **Erigeron vàgus** Payson. Loose Daisy. Fig. 5598.

Erigeron vagus Payson, Univ. Wyo. Pub. Sci. 1: 179. 1926.

Perennial from a diffuse, slenderly branching caudex which sometimes connects to an eventual taproot; herbage and involucre more or less evidently spreading-hirsute and glandular. Leaves crowded on the short aerial stems, mostly 3-lobed, with short, broad, rounded lobes; head solitary, borne on short scapiform peduncle up to 5 cm. long; disk 8–16 mm. wide; involucre 5–7 mm. high, the phyllaries subequal, purple at least toward the tips; rays 25–35, white or pink, 4–7 mm. long, 1–2 mm. wide; disk-corollas 3.0–3.8 mm. long; style-appendages acute, 0.2–0.25 mm. long; pappus of about 20 bristles; achenes 2-nerved.

Shifting talus slopes at high altitudes in the mountains, Arctic-Alpine Zone; southwestern Colorado, Utah, Nevada, California (Sierra Nevada), and northeastern Oregon (Wallowa Mountains). Type locality: Mt. Tomaski, LaSal Mountains, Utah. July–Aug.

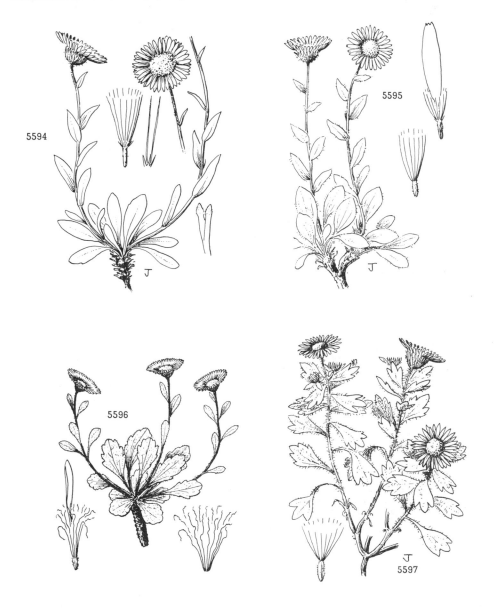

5594. Erigeron cascadensis
5595. Erigeron leibergii

5596. Erigeron oreganus
5597. Erigeron basalticus

18. **Erigeron compósitus** Pursh. Cut-leaved Daisy. Fig. 5599.

Erigeron compositus Pursh, Fl. Amer. Sept. **2**: 535. 1814.
Cineraria lewisii Richards. in Frankl. 1st Journ. Bot. App. 748. 1823.
Erigeron compositus var. *grandiflorus* Hook. Fl. Bor. Amer. **2**: 17. 1834.
Erigeron compositus var. *submontanus* M. E. Peck, Torreya **28**: 56. 1928.

Perennial from a taproot and branching caudex, 1–2.5 dm. tall, the herbage densely glandular and spreading-hirsute to subglabrous. Basal leaves ternately dissected, the better-developed ones mostly three to four times ternate, often irregularly so, with relatively long and linear divisions; cauline leaves few, mostly linear and entire, occasionally some of them trifid; heads solitary, the involucre mostly 7–10 mm. high, with thin subequal phyllaries which are purplish at least at the tips; pistillate flowers mostly 30–60, with white, pink, or blue ligules up to 12 mm. long and 2 mm. wide or the ligules reduced or wanting; disk-corollas 3–5 mm. long; style-appendages acute or acutish, about 0.2 mm. long; pappus of 12–20 rather coarse bristles; achenes 2-nerved.

Sandy river banks at low elevations, Arid Transition Zone; Washington, Oregon, and adjacent Idaho. Type locality: "banks of the Kooskoosky"; probably near the present site of Lewiston, Idaho. April–May.

Erigeron compositus var. **glabrátus** Macoun, Cat. Can. Pl. **2**: 231. 1884. (*Erigeron multifidus* Rydb. Mem. N.Y. Bot. Gard. **1**: 402. 1900; *E. multifidus* var. *nudus* Rydb. loc. cit.; *E. multifidus* var. *incertus* A. Nels. Bot. Gaz. **30**: 198. 1900; *E. compositus* var. *incertus* A. Nels. in Coult. & Nels. New Man. Bot. Rocky Mts. 529. 1909; *E. compositus* var. *nudus* A. Nels. loc. cit.; *E. compositus* var. *multifidus* Macbride & Payson, Contr. Gray. Herb. No. 49: 75. 1917; *E. compositus* var. *petraeus* Macbride & Payson, op. cit. 76.) Best-developed leaves mostly regularly two to three times ternate, the segments shorter and relatively broader than in typical *E. compositus;* plants averaging smaller than in typical *E. compositus;* herbage evidently glandular and spreading-hirsute to subglabrous; heads evidently radiate to disciform. Rocky or sandy places in the mountains, often at high elevations; Alaska to Greenland southward in the cordillera to California (Sierra Nevada), northern Arizona, Colorado, and South Dakota; also in Quebec. Type locality: South Kootanie Pass, 49th parallel. June–Aug.

Erigeron compositus var. **discoídeus** A. Gray, Amer. Journ. Sci. II, **33**: 237. 1862. (*Erigeron trifidus* Hook. Fl. Bor. Amer. **2**: 17. 1834; *E. pedatus* Nutt. Trans. Amer. Phil. Soc. II. **7**: 308. 1840; *E. compositus* var. *trifidus* A. Gray, Proc. Amer. Acad. **16**: 90. 1880; *E. multifidus* var. *discoideus* Rydb. Mem. N.Y. Bot. Gard. **1**: 402. 1900; *E. trifidus* var. *discoideus* A. Nels. in Coult. & Nels. New Man. Bot. Rocky Mts. 529. 1909; *E. trifidus* var. *prasinus* Macbride & Payson, Contr. Gray Herb. No. 49: 78. 1917; *E. trifidus* var. *deficiens* Macbride & Payson, loc. cit.; *E. compositus* var. *trifidus* f. *discoideus* Vict. & Rousseau, Contr. Inst. Bot. Univ. Montreal **36**: 61. 1940; *E. compositus* var. *trifidus* f. *deficiens* Vict. & Rousseau, op. cit. 63; *E. compositus* var. *discoideus* f. *trifidus* Fernald, Rhodora **50**: 238. 1948.) Leaves mostly only once ternate; plant commonly smaller than the other varieties and sometimes becoming strongly pulvinate. Total distribution nearly as in var. *glabratus*, but in our range rare and confined to the highest elevations. Type locality: alpine ridges lying east of Middle Park, Colorado. June–Aug.

19. **Erigeron aùreus** Greene. Alpine Yellow Daisy. Fig. 5600.

Aplopappus brandegei A. Gray, Syn. Fl. N. Amer. **1²**: 132. 1884. Not *Erigeron brandegei* A. Gray, 1884.
Aster brandegei Kuntze, Rev. Gen. Pl. **1**: 317. 1891.
Erigeron aureus Greene, Pittonia **2**: 169. 1891.
Stenotus brandegei Howell, Fl. N.W. Amer. 300. 1900.

Fibrous-rooted to weakly taprooted perennial from a branching caudex, 2–15 cm. tall, the herbage finely pubescent with short, appressed or loose hairs. Basal leaves petiolate, with elliptic to obovate or subrotund, mostly broadly rounded or obtuse blade up to 13 mm. wide; cauline leaves few and reduced; head solitary, the disk 7–16 mm. wide; involucre 5–8 mm. high, its bracts loose, equal, herbaceous, sometimes anthocyanic, sparsely to densely woolly-villous, the hairs sometimes with purple cross-walls; rays mostly 25–70, yellow, 6–9 mm. long, 1.4–2.5 mm. wide; disk-corollas 3.6–4.9 mm. long; style-appendages acute, 0.3–0.5 mm. long; pappus of 10–20 bristles and some short outer setae or narrow scales; achenes 2-nerved. $n = 9$.

Rocky places at high elevations in the mountains, Arctic-Alpine Zone; southern Alberta, southern British Columbia, and the Cascade Mountains of Washington. Type locality: "Mountains of Washington Terr., in the Yakima district." July–Aug.

20. **Erigeron símplex** Greene. Alpine Daisy. Fig. 5601.

Erigeron simplex Greene, Fl. Fran. 387. 1897.

Fibrous-rooted perennial with few or solitary, more or less viscid-villous, mostly erect stems 2–20 cm. tall arising from a simple or moderately branched caudex. Basal leaves oblanceolate or spatulate, up to 8 cm. long and 13 mm. wide; cauline leaves few and reduced; head solitary, the disk 8–20 mm. wide; involucre 5–8 mm. high, moderately to densely white-woolly-villous with flattened multicellular hairs, the phyllaries thin, equal, green or often purplish, often some of the outer loose or reflexed; rays 50–125, blue, pink, or rarely white, 7–11 mm. long, 1–2.5 mm. wide; disk-corollas 3–3.6 mm. long; style-appendages acute, about 0.2 mm. long; pappus of 10–15 capillary bristles and some short, outer, setose scales; achenes 2-nerved.

Open slopes and dry meadows at high altitudes in the mountains, Arctic-Alpine Zone; northeastern Oregon (Wallowa Mountains) to Montana, northeastern Nevada, northern Arizona, and New Mexico. Type locality: Colorado. July–Aug.

21. **Erigeron fléttii** G. N. Jones. Olympic Daisy. Fig. 5602.

Erigeron flettii G. N. Jones, Univ. Wash. Pub. Biol. **5**: 244. 1936.

More or less fibrous-rooted perennial from a stout branched caudex, 5–15 cm. tall, the stems sparsely spreading-hairy. Leaves largely basal, spatulate or oblanceolate, up to 5 cm. long and 12 mm. wide, glabrous or somewhat hirsute, marginally ciliate; cauline leaves few and reduced; head solitary, the disk mostly 10–15 mm. wide; involucre 6–8 mm. high, sparsely to moderately long-villous and sometimes also viscid, the phyllaries about equal, thin, green, purplish above; rays

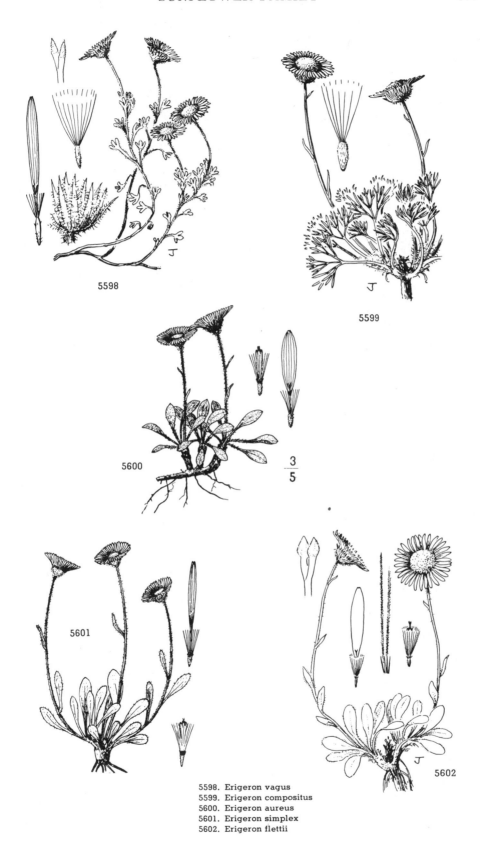

5598. Erigeron vagus
5599. Erigeron compositus
5600. Erigeron aureus
5601. Erigeron simplex
5602. Erigeron flettii

mostly 25–50, white, 7–10 mm. long, 1.5–2.5 mm. wide; disk-corollas 3.5–4.5 mm. long; style-appendages acute, about 0.25 mm. long; pappus of 15–20 capillary bristles and numerous short outer setae; achenes 2-nerved.

Cliffs and other rocky places, at high altitudes, Arctic-Alpine Zone; Olympic Mountains of Washington. Type locality: Olympic Mountains. July–Aug.

22. **Erigeron petiolàris** Greene. Sierra Daisy. Fig. 5603.

Erigeron petiolaris Greene, Leaflets Bot. Obs. **2**: 205. 1912.
Erigeron algidus Jepson, Man. Fl. Pl. Calif. 1052. 1925.

Fibrous-rooted to weakly taprooted perennial from a branching caudex, 2–30 cm. tall, the stem loosely spreading-hirsute and also viscid or glandular. Leaves shortly spreading-hairy or occasionally glabrate, the basal ones oblanceolate to elliptic or obovate, up to about 7 cm. long and 12 mm. wide, obtuse or rounded at the apex, the blade commonly rather abruptly contracted to the petiole; cauline leaves few, reduced, linear; head solitary, the disk 8–18 mm. wide; involucre 5–8 mm. high, glandular and more or less hirsute or villous-hirsute with spreading hairs, the phyllaries about equal, green with purple margins and tips, or purplish throughout; rays 25–125, blue, rose-purple, or pink, 5–13 mm. long, 1.2–2.6 mm. wide; disk-corollas 3.0–4.8 mm. long; style-appendages acutish, 0.15–0.3 mm. long; pappus of 10–15(20) coarse but rather fragile bristles, with or without a few obscure outer setae; achenes 2-nerved.

Rocky places and meadows near and above timberline, Boreal Zone; Sierra Nevada from Mount Rose, Nevada, to Mount Whitney, California. Type locality: Alta Meadows, Tulare County, California. July–Aug.

23. **Erigeron clòkeyi** Cronquist. Clokey's Daisy. Fig. 5604.

Erigeron clokeyi Cronquist, Brittonia **6**: 214. 1947.

Perennial with a stout taproot and usually branched caudex; stems lax, spreading or suberect, 3–20 cm. long, moderately or densely pubescent with short spreading hairs, not glandular. Leaves uniformly pubescent with short spreading hairs, sometimes a few coarser hairs along the petioles; basal leaves mostly oblanceolate, up to about 8 cm. long and 6(10) mm. wide, the blade sometimes slightly triple-nerved at the base; cauline leaves several or rather numerous, reduced, mostly linear or lance-linear; head solitary or occasionally 2, hemispherical to turbinate, the disk 8–11 mm. wide; involucre 4–7 mm. high, the phyllaries slightly thickened and somewhat imbricate, glandular, and especially on the midrib more or less spreading-hirsute; rays 20–50, blue or pink, 6–10 mm. long, 1–2.2 mm. wide; disk-corollas 2.9–4.9 mm. long; style-appendages blunt, 0.1–0.15 mm. long; pappus of 13–25 bristles and some slender and inconspicuous outer setae; achenes 2-nerved.

Dry rocky places at moderate to more often high elevations in the mountains, often in dry meadows or flats, Boreal Zone; summits and east slope of the Sierra Nevada from Mono County to Inyo County, California, eastward to Nye and Clark Counties, Nevada, and Beaver County, Utah. Type locality: Lee Canyon, Charleston Mountains, Clark County, Nevada. July–Aug.

24. **Erigeron pygmaèus** (A. Gray) Greene. Dwarf Daisy. Fig. 5605.

Erigeron nevadensis var. *?pygmaeus* A. Gray, Proc. Amer. Acad. **8**: 649. 1873.
Erigeron pygmaeus Greene, Fl. Fran. 390. 1897.

Pulvinate-cespitose perennial with a stout taproot and a stout branched caudex; stem up to 6 cm. high, more or less pubescent with short spreading hairs and usually also viscid or finely glandular. Leaves copiously hirtellous or coarsely strigose with short, appressed or spreading hairs, the basal ones oblanceolate or linear-oblanceolate, up to 2.5 cm. long and 2.5 mm. wide; cauline leaves absent or occasionally a few reduced linear ones present near the base; head solitary, turbinate, the disk 6–13 mm. wide; involucre 4–7 mm. high, glandular and more or less spreading-hirsute; phyllaries slender, subequal or somewhat imbricate, deep blackish purple or with green body and purple tip; rays 15–35, blue or purple, 4–7 mm. long, 1.1–1.8 mm. wide; disk-corollas 3.7–5.3 mm. long; style-appendages acute, 0.3–0.5 mm. long; pappus of 15–25 firm bristles, with a few intermingled short setae; achenes 2-nerved.

Rocky places at high elevations in the mountains, Arctic-Alpine Zone; Sierra Nevada from Mount Rose, Nevada, to Mount Whitney, California. Type locality: Mono Pass, Mono County, California. July–Aug.

25. **Erigeron unciàlis** Blake. Cliff Daisy. Fig. 5606.

Erigeron uncialis Blake, Proc. Biol. Soc. Wash. **47**: 173. 1934.

Perennial with a taproot and branching caudex; stem 1–5 cm. high, naked, hirsute or villous-hirsute especially under the head, with spreading hairs. Leaves all basal, hirsute or hirsute-strigose with loose or appressed hairs, sometimes glabrate on one or both surfaces, up to about 4 cm. long and 8 mm. wide, the blade subrotund to broadly oblanceolate, rounded to acute at the apex, rather abruptly contracted to and usually much shorter than the petiole; head solitary, the disk 6–11 mm. wide; involucre 4.5–5 mm. long, finely and sometimes obscurely glandular, and more or less villous-hirsute especially toward the base; phyllaries acuminate or attenuate, slightly imbricate, thin, green or with green midrib, the margins often purplish; rays 15–40, white or light rose, 4–6 mm. long, 0.7–1.4 mm. wide; disk-corollas 2.7–3.4 mm. long; style-appendages acutish, 0.15–0.2 mm. long; pappus of 13–22 firm bristles, with some inconspicuous short setae; achenes 2-nerved.

Cliff-crevices, lower Boreal Zone; mountains of San Bernardino and Inyo Counties, California. Type locality: Clark Mountain, eastern San Bernardino County. June–July.

The var. *conjugans* Blake, with the hairs of the petioles and lower part of the stem mostly appressed or ascending, occurs in Clark and Nye Counties, Nevada.

26. Erigeron téner A. Gray. Slender Daisy. Fig. 5607.

Erigeron cespitosus var. *tenerus* A. Gray, Bot. Calif. 1: 328. 1876.
Erigeron tener A. Gray, Proc. Amer. Acad. **16**: 91. 1880.
Erigeron copelandii Eastw. Bot. Gaz. **41**: 291. 1906.

Perennial with a taproot and branched caudex, slender, lax, 3–15 cm. tall, the herbage moderately or sparsely strigose. Basal leaves petiolate, with oblanceolate to obovate or rhombic, mostly acute or acutish blade 7–25 mm. long and 2–7 mm. wide; cauline leaves few and linear; heads 1–3, the involucre 3.5–5 mm. high, finely glandular and usually shortly spreading-hairy; phyllaries imbricate, firm, somewhat thickened especially the outer, mostly greenish brown with darker midrib, the inner with thin, often purplish margins; rays 15–40, blue or sometimes purple, 4–8 mm. long, 1–1.7 mm. wide; disk-corollas 2.7–4.1 mm. long; inner pappus of 15–30 bristles, the outer obscure or wanting; achenes 2-nerved.

Rock crevices and dry rocky soil, Boreal Zone; mountains of California (Cascade Mountains and Sierra Nevada) and eastern Oregon to western Wyoming and central Utah; rare in our range. Type locality: Silver Mountain, near Ebbett Pass, Alpine County, California. June–Aug.

27. Erigeron poliospérmus A. Gray. Hairy-seeded Daisy. Fig. 5608.

Erigeron poliospermus A. Gray, Syn. Fl. N. Amer. 1²: 210. 1884.
Erigeron poliospermus f. *disciformis* Cronquist, Brittonia **6**: 194. 1947. A form with the pistillate flowers eligulate.

Perennial with a taproot and short branched caudex, the several stems seldom over 15 cm. tall; herbage and involucre evidently spreading-hairy and glandular, more hairy than glandular, or the

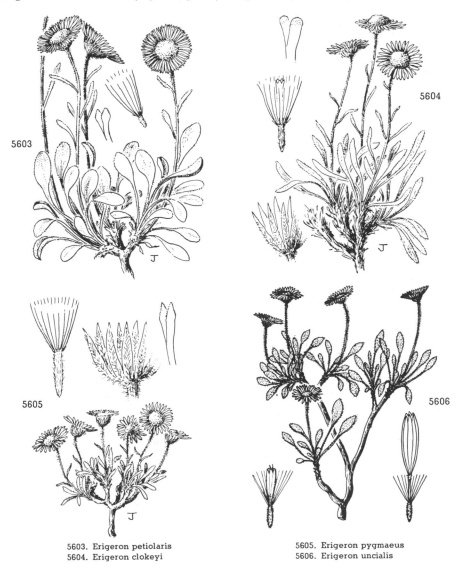

5603. Erigeron petiolaris
5604. Erigeron clokeyi

5605. Erigeron pygmaeus
5606. Erigeron uncialis

involucre more glandular than hairy. Basal leaves linear-oblanceolate to spatulate, up to 8 cm. long and 12 mm. wide, the cauline ones generally more or less reduced; heads hemispheric, mostly solitary, the disk 9–20 mm. wide; involucre 5–9 mm. high, the phyllaries equal or nearly so, green, sometimes with light brown midrib and narrow scarious margins; rays 15–45, pink or purple to deep violet, 5–14 mm. long, 1.3–3.6 mm. wide; disk-corollas (3.5)4–5.5(6.5) mm. long; inner pappus of 20–30 bristles; achenes 2-nerved, very densely long-silky, the hairs sometimes obscuring the setose outer pappus.

Dry places in the plains and foothills, often among sagebrush, Arid Transition Zone; Washington from the Cascade Mountains eastward, extending into adjacent Idaho and southernmost British Columbia, and southward to northern Harney County, Oregon. Type locality: Umatilla, Umatilla County, Oregon. April–June.

Erigeron poliospermus var. **cèreus** Cronquist, Brittonia 6: 194. 1947. Herbage and involucre densely glandular and only sparsely or moderately hairy; plants often branched and several-headed and then with fairly well-developed cauline leaves. Kittitas County and adjacent parts of Grant and Chelan Counties, Washington. Type locality: Rock Island, Kittitas County.

28. **Erigeron disparipílus** Cronquist. Snake River Daisy. Fig. 5609.

Erigeron disparipilus Cronquist, Brittonia 6: 194. 1947.

Perennial with a taproot and short branched caudex; stems 3–12 cm. high, with spreading, very unequal hairs. Leaves nearly all in a basal cluster, finely hirsute, linear or linear-oblanceolate, up to 4 cm. long and 2 mm. wide; head solitary, the involucre 5–7 mm. high, spreading-hirsute and often glandular; phyllaries subequal or slightly imbricate, acuminate, green, with or without a darker midrib and sometimes with narrow scarious borders; rays 30–60(100), white, or rarely pinkish in age, 5–10 mm. long, 1.5–2.3 mm. wide; disk-corollas 3.5–4.3 mm. long, coarser and less indurate than in *E. pumilus;* pappus of 15–25 capillary bristles and some inconspicuous outer setae; achenes 2-nerved, rather densely short-hairy.

Open rocky places in the foothills and at moderate elevations in the mountains, Arid Transition Zone; extreme southeastern Washington and presumably adjacent Idaho southward in the vicinity of the Snake River to Owyhee County. Type locality: 14 miles west of Clarkston, Asotin County, Washington. May–July.

29. **Erigeron pùmilus** Nutt. subsp. **intermèdius** Cronquist. Hairy Daisy. Fig. 5610.

Erigeron hispidissimus Piper, Contr. U.S. Nat. Herb. 11: 565. 1906.
Erigeron pumilus subsp. *intermedius* Cronquist, Brittonia 6: 180. 1947.

Perennial (sometimes short-lived) with a taproot and simple or branched caudex; stems clustered, mostly 2–5 dm. tall, relatively robust, commonly more than 1.5 mm. thick at the base and bearing 5 to many heads; herbage copiously spreading-hirsute and sometimes a little glandular. Leaves oblanceolate, up to 8 cm. long and 8 mm. wide, cauline as well as basal; heads mostly hemispheric, the disk mostly 7–15 mm. wide; involucre 4–7 mm. high, spreading-hairy and very finely glandular, the phyllaries subequal, narrow, acuminate or attenuate, green, with brown midrib and sometimes very narrow scarious margins; rays mostly 50–100, blue, pink, or white, 6–15 mm. long, mostly 0.7–1.5 mm. wide; disk-corollas 3.5–5 mm. long, slender, the limb pale and indurated below, the indurated portion often finely puberulent; pappus of 13–20 slightly sordid bristles and some short, outer, coarse bristles or very narrow scales; achenes 2-nerved, sparsely or moderately short-hairy.

Open places in the foothills, valleys, and plains, often among sagebrush, Arid Transition Zone; east of the Cascade Mountains from Washington to northwestern Montana, southward to northern Oregon and less commonly to northeastern California and adjacent Nevada. Type locality: near Lewiston, Idaho. May–July.

Erigeron pumilus subsp. **intermedius** var. **gracílior** Cronquist, Brittonia 6: 180. 1947. Plants smaller and more slender than in typical subsp. *intermedius,* 0.5–3 dm. tall, the larger stems seldom over 1.5 mm. thick at the base and rarely with as many as 5 heads. Central and southern Idaho and adjacent Wyoming to central and northeastern Oregon, northeastern California, and adjacent Nevada. Type locality: United States Sheep Farm, Clark County, Idaho.

Erigeron pumilus subsp. **concinnoìdes** Cronquist, Brittonia 6: 181. 1947. (*Distasis concinna* Hook. & Arn. Bot. Beechey 350. 1838; *Erigeron concinnus* Torr. & Gray, Fl. N. Amer. 2: 174. 1841; *E. concinnus* var. *eremicus* Jepson, Man. Fl. Pl. Calif. 1057. 1925.) Mostly similar to var. *gracilior* in habit or a little more robust; rays nearly always colored; indurated portion of the disk-corollas consistently puberulent; inner pappus of 7–12(15) coarse bristles. the outer of evident, relatively broad scales. Dry open (but scarcely desert) places in the foothills and at moderate elevations in the mountains; southern California in Inyo and San Bernardino Counties; also southern Idaho and western Wyoming south to northern New Mexico and southern Nevada. Type locality: Kyle Canyon, Charleston Mountains, Clark County, Nevada.

30. **Erigeron engelmánnii** A. Nels. var. **davísii** Cronquist. Engelmann's Daisy. Fig. 5611.

Erigeron engelmanni subsp. *davisii* Cronquist, Leaflets West. Bot. 3: 167. 1942.
Erigeron engelmanni var. *davisii* Cronquist, Vasc. Pl. Pacif. Northw. 5: 177. 1955.

Erect perennial with a taproot and shortly branched caudex, 10–30 cm. tall, appressed-hairy. Basal leaves linear-oblanceolate, up to 10 cm. long and 4 mm. wide, with some coarse, spreading, marginal hairs toward the base; cauline leaves linear and reduced; heads solitary or few, on long scapiform peduncles; involucre (5)6–8 mm. high, spreading-hairy and usually finely glandular, the phyllaries subequal, narrow, green, usually with brown midrib and pale margins and tip; disk-corollas 3.3–4.5 mm. long; rays 35–55, mostly white, 8–14 mm. long, 1.1–2.2 mm. wide; pappus of 12–20 capillary bristles and some inconspicuous outer setae or narrow scales; achenes 2-nerved, finely hairy.

Woods, meadows, and open hillsides in the foothills and at moderate elevations in the mountains, Arid Transition Zone; Lewis and western Idaho Counties, Idaho, and Wallowa County, Oregon. Type locality: Whitebird summit, Idaho County, Idaho. May–June. Typical *E. engelmannii* is smaller in all parts and occurs in Wyoming, Colorado, Utah, and southeastern Idaho.

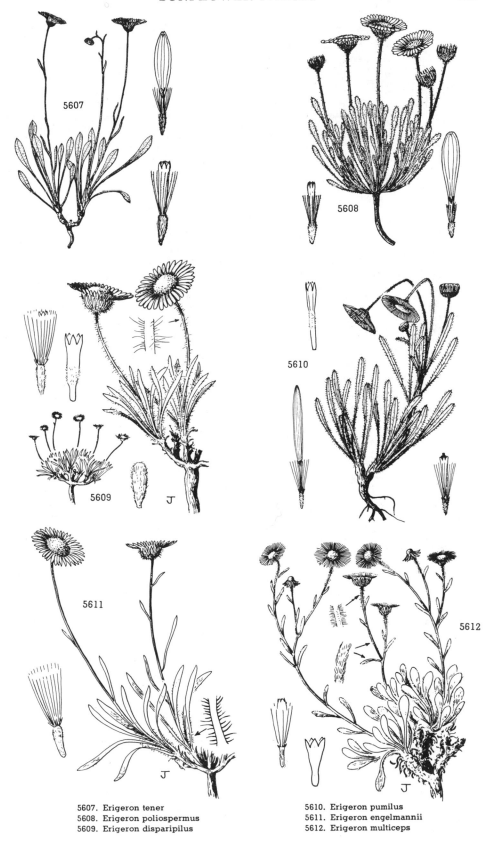

5607. Erigeron tener
5608. Erigeron poliospermus
5609. Erigeron disparipilus

5610. Erigeron pumilus
5611. Erigeron engelmannii
5612. Erigeron multiceps

31. Erigeron múlticeps Greene. Kern River Daisy. Fig. 5612.

Erigeron multiceps Greene, Pittonia 2: 167. 1891.

Perennial from a stout taproot and short branched caudex; stems slender, branched, up to 2 dm. tall, hirsute with appressed hairs except under the heads where the hairs are spreading. Leaves appressed-hairy, the basal oblanceolate to obovate, up to 4 cm. long, the blade shorter than the petiole; cauline leaves linear or the lower oblanceolate; heads several, the involucre 4–5 mm. high, finely glandular and moderately or sparsely hirsute with short spreading hairs; phyllaries greenish with slightly darker midrib and sometimes paler margins, the outer a little shorter and evidently narrower than the relatively broad, thin, inner ones; rays about 75, lilac or white, 5–7 mm. long, 0.6 mm. wide; disk-corollas about 2.5 mm. long, the limb white and strongly indurated near the base, flaring above; style-appendages less than 0.1 mm. long; pappus double, the inner of 5–8 slender bristles, the outer of evident and sometimes more or less united short scales; achenes 2-nerved.

Gravelly spots near river banks, mostly Arid Transition Zone; north fork of the Kern River, Tulare County, California. Type locality: near north fork of the Kern River, near Kernville. June.

32. Erigeron divérgens Torr. & Gray. Diffuse Daisy. Fig. 5613.

Erigeron divaricatus Nutt. Trans. Amer. Phil. Soc. II. 7: 311. 1840. Not Michx. 1803.
Erigeron divergens Torr. & Gray. Fl. N. Amer. 2: 175. 1841.
Erigeron incomptus A. Gray, Syn. Fl. N. Amer. 1²: 218. 1884.
Erigeron californicus Jepson, Bull. Torrey Club 18: 324. 1891.
Erigeron cinereus var. *aridus* M. E. Jones, Proc. Calif. Acad. II. 5: 695. 1895.
Erigeron tephrodes Greene, Leaflets Bot. Obs. 1: 222. 1906.
Erigeron divergens f. *incomptus* Cronquist, Brittonia 6: 261. 1947. A form with reduced rays.

Taprooted biennial or short-lived perennial, 1–7 dm. tall, freely branched, the herbage covered with spreading hairs mostly well under 1 mm. long. Basal leaves petiolate, with oblanceolate or spatulate blade up to 2.5 cm. long, usually deciduous; cauline leaves numerous, smaller, often linear; heads mostly numerous, the involucre 4–5 mm. high, glandular and spreading-hairy; phyllaries light yellow-greenish with broad brown midvein, the outer a little shorter and evidently narrower than the relatively broad, thin-margined, inner ones; rays 75–150, blue, pink, or white, 5–10 mm. long, 0.5–1.2 mm. wide, or rarely much reduced; disk-corollas 2–3 mm. long, the limb white and strongly indurated toward the base; style-appendages less than 0.1 mm. long; pappus double, the inner of 5–12 very fragile bristles, the outer of conspicuous, setose, short scales; achenes 2-nerved or sometimes obscurely 4-nerved, the nerves usually inconspicuous.

Dry and waste places in the valleys, foothills, and lower mountains, especially in sandy soils, Upper Sonoran and Arid Transition Zones; southern British Columbia and Montana to Lower California, Texas, and Oklahoma. Type locality: "Rocky Mountains and plains of Oregon." April–July, sometimes continuing until fall.

33. Erigeron cálvus Coville. Inyo Daisy. Fig. 5614.

Erigeron calvus Coville, Proc. Biol. Soc. Wash. 7: 69. 1892.

Taprooted biennial about 1 dm. tall, branching widely from the base, the herbage hirsute with long spreading hairs and the stems finely glandular especially under the heads. Basal leaves numerous, with oblong to obovate blade 1–1.5 cm. long tapering to a petiole twice as long; cauline leaves reduced, oblanceolate or spatulate; heads terminating the branches; disk about 13 mm. wide; involucre about 5 mm. high, coarsely spreading-hirsute and finely glandular; phyllaries moderately imbricate, acuminate, greenish, sometimes with chaffy margins; pistillate flowers present but the ligules wanting or inconspicuous and much shorter than the style; disk-corollas 4–5 mm. long; style-appendages 0.15–0.25 mm. long; pappus double, the inner of 15–20 firm sordid bristles, the outer of several relatively long, firm setae; achenes 2-nerved.

Upper Sonoran Zone; Inyo Mountains, California. Type locality: foot of Inyo Mountains about 4 miles north of Keeler, Inyo County. May.

34. Erigeron aphanáctis (A. Gray) Greene. Basin Rayless Daisy. Fig. 5615.

Erigeron concinnus var. *aphanactis* A. Gray, Proc. Amer. Acad. 6: 540. 1865.
Erigeron aphanactis Greene, Fl. Fran. 389. 1897.

Perennial with a taproot and branching caudex, mostly 1–3 dm. tall, the herbage densely spreading-hairy. Basal leaves linear-oblanceolate to spatulate, long-petiolate, up to 8 cm. long and 6 mm. wide, persistent or deciduous, the cauline similar but nearly sessile, or linear and reduced; heads several or solitary, yellow, sometimes turning brownish in age; involucre 4–6 mm. high, glandular and spreading-hirsute, the phyllaries subequal or slightly imbricate, slender, acuminate, green or greenish brown, with thickened midrib; pistillate flowers present, tubular and eligulate or sometimes bearing inconspicuous ligules shorter than the disk; disk-corollas 2.8–4.9 mm. long, the limb commonly with a firm, whitish, often puberulent portion below the middle; style-appendages 0.1–0.25 mm. long; pappus double, the inner of 7–20 bristles, the outer of evident, sometimes narrow scales; achenes 2-nerved.

Mostly in hot dry places in the foothills and deserts, Arid Transition Zone; southwestern Idaho and southeastern Oregon to Colorado, Arizona, and southern California, largely or wholly replaced by the following variety in the San Bernardino Mountains. Type locality: Carson City, Ormsby County, Nevada. May–July.

Erigeron aphanactis var. congéstus (Greene) Cronquist, Brittonia 6: 177. 1947. (*Erigeron congestus* Greene, Leaflets Bot. Obs. 2: 218. 1912.) Plants about 1 dm. tall or less, essentially scapose; corolla-lobes and often the upper part of the limb turning reddish or purplish at least in age. Dry places at moderate to high elevations in the San Bernardino Mountains of California. Type locality: Gold Hill, Bear Valley, San Bernardino Mountains, San Bernardino County. June–Aug.

35. Erigeron chrysópsidis A. Gray. Golden Daisy. Fig. 5616.

Chrysopsis hirtella DC. Prod. **5**: 237. 1836. Not *Erigeron hirtellus* DC. 1836.
Erigeron ochroleucus var. *hirtellus* A. Gray, Proc. Amer. Acad. **16**: 90. 1880.
Erigeron chrysopsidis A. Gray, Syn. Fl. N. Amer. **1²**: 210. 1884.
Erigeron curvifolius Piper, Bull. Torrey Club **27**: 396. 1900, as to type only.

Perennial from a taproot and branching caudex, 6–16 cm. tall; herbage spreading-hairy or the pubescence of the leaves occasionally partly appressed. Leaves all or nearly all in a basal cluster (low-cauline as well as strictly basal), linear-oblanceolate, straight or slightly arcuate, mostly (2)3–9 cm. long and 1–3 mm. wide; head solitary, hemispheric, the disk mostly 11–17 mm. wide; involucre 5–7.5 mm. high, spreading-hairy and sometimes slightly glandular, the phyllaries subequal, narrow, green or greenish at least along the midrib, often with more chartaceous or hyaline margins, gradually acuminate; pistillate flowers 20–50, the rays bright yellow, 5–10 mm. long and 1–2.5 mm. wide; disk-corollas mostly 4–5 mm. long; pappus of 15–25 slender bristles, commonly with a few short, slender, inconspicuous, outer setae; achenes 2-nerved.

Dry open places, often sagebrush, extending up to a little over 4,000 feet altitude in the ponderosa pine forest, Arid Transition Zone; east of the Cascade Mountains in northern Oregon from Grant (and Crook?) Counties northward, barely extending into southeastern Washington. Type locality: Columbia River. May–June (July).

Erigeron chrysopsidis var. **brevifòlius** Piper, Bull. Torrey Club **27**: 395. 1900. Smaller in all respects than typical *E. chrysopsidis*; plants 3–6(9) cm. tall; basal leaves 1–3(4) cm. long; involucre 4–5 mm. high; disk 6–11 mm. wide; disk-corollas 3–4 mm. long; pubescence of leaves often more or less appressed. Wallowa Mountains of Oregon, at higher altitudes than typical *E. chrysopsidis*. Type locality: Wallowa Mountains.

Erigeron chrysopsidis subsp. **austíniae** (Greene) Cronquist, Brittonia **6**: 196. 1947. (*Erigeron austiniae* Greene, Erythea **3**: 100. 1895.) Pistillate flowers present but the rays wanting or small and inconspicuous; otherwise resembling typical *E. chrysopsidis* or smaller and more compact. Lake, Harney, and Malheur Counties, Oregon, southward to the vicinity of Mount Lassen, California, and eastward to Twin Falls County, Idaho, and Elko County, Nevada. Type locality: Davis Creek, Modoc County, California.

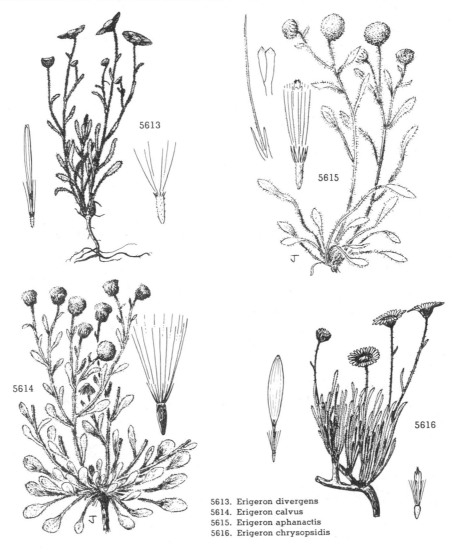

5613. Erigeron divergens
5614. Erigeron calvus
5615. Erigeron aphanactis
5616. Erigeron chrysopsidis

36. **Erigeron piperiànus** Cronquist. Piper's Daisy. Fig. 5617.

Erigeron curvifolius Piper, Bull. Torrey Club **27**: 396. 1900, pro max. part., excl. type.
Erigeron piperianus Cronquist, Brittonia **6**: 197. 1947.

Slender perennial from a taproot and short branched caudex; stem 3–10 cm. high, seldom much exceeding the leaves, hirsute with mostly appressed or ascending hairs. Leaves numerous, cauline and basal, linear or nearly so, lax and usually curved, up to 4 cm. long and 1.5 mm. wide, hispid-ciliate on the margins, especially below, and appressed-hairy on the surface, the lower with conspicuously enlarged, whitish-indurated base; heads solitary or few, small, the disk 5–10 mm. wide; involucre 3–5 mm. high, spreading-hirsute with long white hairs, the phyllaries subequal or imbricate, green or greenish, with darker thickened midrib, acuminate at the tip; rays 25–40, yellow, 4–9 mm. long, 1.0–1.8 mm. wide; disk-corollas mostly 2.8–4.2 mm. long; pappus of 15–25 bristles and often some short outer setae; achenes 2-nerved.

Dry open places, often among sagebrush, Arid Transition Zone; Columbia River plains of Washington from Douglas County to Benton County. Type locality: north of Soap Lake in the Grand Coulee, Grant County, Washington. May–June.

37. **Erigeron bloòmeri** A. Gray. Bloomer's Daisy. Fig. 5618.

Erigeron bloomeri A. Gray, Proc. Amer. Acad. **6**: 540. 1865.
Erigeron filifolius bloomeri A. Nels. Bot. Gaz. **54**: 413. 1912.

Perennial from a taproot and much-branched caudex, 5–15 cm. tall, the herbage finely white-strigose. Leaves all or nearly all in a basal cluster, linear, 2–7 cm. long, 0.7–2 mm. wide; head solitary, turbinate to subhemispheric, the disk 7–20 mm. wide; involucre 5–10 mm. high, strigose to villous especially toward the base, the phyllaries broad, acute, firm, imbricate, usually green; pistillate flowers wanting; disk-corollas 4.5–7.0 mm. long; style-appendages acute, 0.3–0.5 mm. long; pappus-bristles 25–40, generally unequal, sometimes with a few inconspicuous outer setae; achenes 2-nerved, glabrous below, short-hairy above.

Dry, often rocky places in the mountains and foothills, Arid Transition and Canadian Zones; Kittitas County, Washington, to Siskiyou (Marble Mountains) and Nevada Counties, California, eastward to Lemhi and Cassia Counties, Idaho, and Elko and Lander Counties, Nevada. Type locality: near Virginia City, Storey County, Nevada. June–Aug.

Erigeron bloomeri var. **nudàtus** (A. Gray) Cronquist, Brittonia **6**: 199. 1947. (*Erigeron nudatus* A. Gray, Proc. Amer. Acad. **20**: 297. 1885.) Stem and leaves glabrous or finely and sparsely strigose; involucre glabrous or very nearly so; style-appendages acutish, 0.2–0.3 mm. long. Serpentine areas in the Klamath region of southwestern Oregon and adjacent California. Type locality: Waldo, Josephine County, Oregon. Specimens of *E. bloomeri* from serpentine in Grant County, Oregon, approach but do not match var. *nudatus*.

Erigeron bloomeri var. **pùbens** Keck, Aliso **4**: 105. 1958. Differs from *E. bloomeri* and *E. bloomeri* var. *nudatus* in being soft-pilose throughout with spreading hairs. In the Marble Mountains, Siskiyou County, California. Type locality: King's Castle, Marble Mountains.

38. **Erigeron lineàris** (Hook.) Piper. Desert Yellow Daisy. Fig. 5619.

Diplopappus lineàris Hook. Fl. Bor. Amer. **2**: 21. 1834.
Erigeron peucephyllus A. Gray, Proc. Amer. Acad. **16**: 89. 1880.
Erigeron luteus A. Nels. Bull. Torrey Club **27**: 33. 1900.
Erigeron linearis Piper, Contr. U.S. Nat. Herb. **11**: 567. 1906.
Erigeron yakimensis A. Nels. Amer. Journ. Bot. **23**: 268. 1936.

Perennial from a stout taproot and branched caudex, 5–30 cm. tall; herbage finely gray-strigose; bases of stems and of basal leaves conspicuously indurated and somewhat enlarged, stramineous to sometimes purplish. Leaves linear or nearly so, 1.5–9 cm. long, 0.5–3 mm. wide, basal and cauline or nearly all basal; heads solitary or few, the disk mostly 8–13 mm. wide; involucre 4–7 mm. high, strigose-villous and sometimes finely glandular, the phyllaries subequal or imbricate, green or greenish-stramineous, often with darker midrib, acute, firm, thickened on the back, the inner sometimes purple-tipped; rays 15–45, yellow, 4–11 mm. long, 1.3–2.5 mm. wide; disk-corollas 3.5–5.3 mm. long; pappus of 10–20 bristles and some short, often narrow, outer scales; achenes 2-nerved, short-hairy.

Dry, often rocky soil from the plains and foothills to moderate elevations in the mountains, often with sagebrush, Arid Transition Zone; southern British Columbia through eastern Washington and Oregon to Yosemite National Park, California, eastward to Park County, Montana, Yellowstone National Park, Wyoming, and Elko County, Nevada. Type locality: "near the 'Priest's Rapid' of the Columbia, and also on Lewis and Clarke's River." May–July(Aug.).

39. **Erigeron elegántulus** Greene. Volcanic Daisy. Fig. 5620.

Erigeron elegantulus Greene, Erythea **3**: 65. 1895.
Erigeron linearis var. *elegantulus* J. T. Howell, Leaflets West. Bot. **1**: 205. 1936.

Perennial from a taproot and much-branched caudex, 3–15 cm. tall; herbage sparsely or moderately gray-strigose; bases of stems and basal leaves conspicuously smooth, shining, somewhat enlarged and indurated, usually stramineous. Leaves narrowly linear, mostly 1.5–6 cm. long and 0.5–1 mm. wide, mostly or all in a basal cluster; head solitary, small, the disk 7–11 mm. wide; involucre 3.5–5 mm. high, strigose and glutinous; phyllaries imbricate, the short outer ones narrow and rather firm, the inner ones broader, with scarious and usually erose margins, acute or abruptly acuminate; rays 15–25, blue or pink to almost white, 6–9 mm. long, 1.3–1.8 mm. wide; disk-corollas 3.4–5.0 mm. long; pappus of 20–30 slender bristles, occasionally with a few slender, outer, short setae; achenes 2-nerved, glabrous, or strigose above or throughout.

Open, often rocky places in the plains and foothills, especially in basaltic or volcanic rocks, Arid Transition and Canadian Zones; Baker, Union, and Klamath Counties, Oregon, to Modoc, Siskiyou, and Lassen Counties, California. Type locality: Dixie Valley, Lassen County, California. June–July.

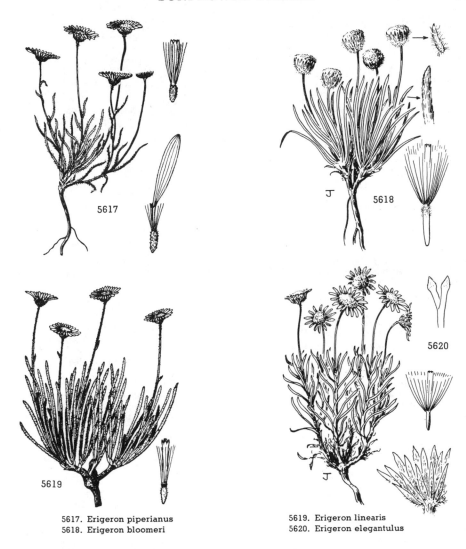

5617
5618
5620
5619

5617. **Erigeron piperianus**
5618. **Erigeron bloomeri**

5619. **Erigeron linearis**
5620. **Erigeron elegantulus**

40. **Erigeron barbellulàtus** Greene. Shining Daisy. Fig. 5621.

Erigeron barbellulatus Greene, Erythea **3**: 65. 1895.

Perennial, 5–15 cm. tall, from a much-branched caudex, the taproot often poorly or scarcely developed; herbage finely strigose; bases of stems and of basal leaves conspicuously smooth, shining, somewhat enlarged and indurated, usually purplish. Leaves oblanceolate, mostly 1.5–4 cm. long and 1.5–5 mm. wide, mostly or all in a basal cluster (basal and near-basal); head solitary, the disk 13–18 mm. wide; involucre 5.5–9 mm. high, hirsute-villous or loosely strigose, often sparsely so, and somewhat glutinous; phyllaries subequal or slightly imbricate, long-acuminate, usually a little thickened on the back; rays 15–35, blue or purple to white, 7–12 mm. long, 1.5–3 mm. wide; disk-corollas 4.4–7.0 mm. long; pappus of 30–40(50) slender bristles, occasionally with a few slender, outer, short setae; achenes 2-nerved, sparsely strigose, or glabrous below.

Rocky places at moderate to fairly high elevations in the mountains, Canadian and Hudsonian Zones; Lassen County to Fresno County, California. Type locality: Silver Lake, Lassen County. May–Aug.

41. **Erigeron filifòlius** Nutt. Thread-leaved Daisy. Fig. 5622.

Diplopappus filifolius Hook. Fl. Bor. Amer. **2**: 21. 1834.
Chrysopsis canescens DC. Prod. **5**: 328. 1836.
Erigeron filifolius Nutt. Trans. Amer. Phil. Soc. II. **7**: 308. 1840.

Perennial from a taproot and branching woody caudex, 1–5 dm. tall; herbage finely strigose, the stem densely so at the base. Leaves linear or linear-filiform, 1–8 cm. long, 0.3–2.5(3.5) mm. wide, well distributed along the stem but the basal ones commonly longer or more numerous or both; heads usually several, the disk 5–15 mm. wide; involucre 4–6 mm. high, closely villous-strigose or finely glandular or both; phyllaries firm, somewhat thickened on the back, subequal or imbricate, usually greenish, often with darker midrib; rays 15–50(75), blue to pink or white, 3–11 mm. long, 1–2 mm. wide; disk-corollas 2.5–4.4 mm. long; style-appendages 0.1–0.15 mm. long; pappus of 20–30 slender white bristles, sometimes with a few very slender and inconspicuous, short, outer setae; achenes 2-nerved, hairy or subglabrous.

Dry, often sandy places in the valleys and foothills, commonly with sagebrush, Arid Transition Zone; southern British Columbia and extreme northwestern Montana through eastern Washington and Oregon to Sierra County, California, and adjacent Nevada eastward through the Snake River plains to Rexburg, Idaho, and thence southward to Logan, Utah. Type locality: "In the Rocky Mountain range, in Oregon." May–July.

Erigeron filifolius var. **robústior** M. E. Peck, Proc. Biol. Soc. Wash. **50**: 123. 1937. Relatively stout; head usually solitary, the disk 12–18 mm. wide; involucre 5–7 mm. high; rays 50–125, 6–13 mm. long; disk-corollas 3.7–5.5 mm. long; style-appendages 0.15–0.3 mm. long. Kittitas County, Washington, southward to Wasco and Gilliam Counties, Oregon. Type locality: Columbia River near Rowena, Wasco County, Oregon.

42. **Erigeron lasseniànus** Greene. Lassen Daisy. Fig. 5623.

Erigeron lassenianus Greene, Fl. Fran. 389. 1897.
Erigeron lassenianus var. *deficiens* Cronquist, Brittonia **6**: 171. 1947. A discoid form.

Slender perennial with an evident slender taproot and a simple or subsimple crown or short caudex; herbage finely spreading-hirsute; stems weak, sprawling or decumbent at the often purplish base 5–20 cm. long. Basal leaves linear-oblanceolate, more or less triple-nerved, acute, tapering gradually to the slender petiole, up to 11 cm. long and 5 mm. wide; cauline leaves several, reduced, linear, occasionally triple-nerved; heads usually several (up to 20), occasionally solitary, the disk 5–11 mm. wide; involucre 5–6 mm. high, glandular and spreading-hirsute; phyllaries scarcely to evidently imbricate, green or greenish, often thin-margined, acuminate; rays 10–25, blue or pink to sometimes white, 5–6 mm. long, 1.5–2.0 mm. wide, or occasionally wanting; disk-corollas 3–4 mm. long; pappus of 15–25 fragile bristles and a few short outer setae; achenes 2-nerved.

Dry places at moderate elevations, Arid Transition Zone; Butte and Plumas Counties to Eldorado County, California. Type locality: Mount Dyer, Plumas County. June–Aug.

43. **Erigeron corymbòsus** Nutt. Foothill Daisy. Fig. 5624.

Erigeron corymbosus Nutt. Trans. Amer. Phil. Soc. II. **7**: 308. 1840.

Perennial with a taproot and not much-branched caudex, 1–5 dm. tall, suberect, generally purplish at the base; herbage densely and rather shortly spreading-hairy. Basal leaves triple-nerved, elongate, acute, tapering gradually below, up to 25 cm. long and 1 cm. wide; cauline leaves gradually or strongly reduced; heads 1–16, the disk 7–13 mm. wide; involucre 5–7 mm. high, canescently villous-hirsute with short subappressed hairs; phyllaries somewhat imbricate, only slightly or obscurely thickened on the back, green or tan, with darker midrib and purplish, usually close tips; rays 35–65, deep blue or occasionally pink, 7–13 mm. long, 1.2–2 mm. wide; disk-corollas (3.0)4.0–5.3 mm. long; style-appendages acute, mostly 0.2–0.25 mm. long; pappus of 20–30 firm bristles and some evident, short, outer setae or narrow scales; achenes 2-nerved.

Open, usually dry places, often among sagebrush, Arid Transition Zone; southern British Columbia to central and southeastern Oregon east of the Cascade Mountains eastward to western Montana and western Wyoming. Type locality: "Rocky Mountains, towards the Oregon." June–July.

44. **Erigeron caespitòsus** Nutt. Gray Daisy. Fig. 5625.

Diplopappus canescens Hook. Fl. Bor. Amer. **2**: 21. 1834.
Diplopappus grandiflorus Hook. loc. cit.
Erigeron caespitosus Nutt. Trans. Amer. Phil. Soc. II. **7**: 307. 1840.
Erigeron canescens Torr. & Gray, Fl. N. Amer. **2**: 179. 1841. Not Hook. & Arn. 1836.
Erigeron caespitosus var. *grandiflorus* Torr. & Gray, loc. cit. Not *E. grandiflorus* Hook. 1834.
Erigeron subcanescens Rydb. Bull. Torrey Club **24**: 294. 1897.

Perennial with a stout taproot and stout, usually branched caudex, sometimes also creeping below ground; stems curved at the base, 5–30 cm. high; herbage densely pubescent with short spreading hairs. Basal leaves triple-nerved (obscurely so in depauperate specimens), oblanceolate or spatulate, usually rounded or obtuse at the tip, up to 12 cm. long and 15 mm. wide; cauline leaves ovate-oblong to linear, sometimes nearly as large as the basal ones; heads 1–10, the disk 9–18 mm. wide; involucre 4–7 mm. high, glandular and generally canescent with short white hairs; phyllaries more or less imbricate, appressed, firm, evidently thickened on the back, scarcely herbaceous; rays 30–100, blue, pink, or white, 5–15 mm. long, 1–2.5 mm. wide; disk-corollas 3.2–4.4 mm. long; style-appendages 0.1–0.15(0.2) mm. long; pappus of 15–25 firm bristles and some evident, outer, setose scales; achenes 2-nerved.

Dry, open, often rocky places from the foothills to high elevations in the mountains, Arid Transition and lower Boreal Zones; interior Alaska and Yukon southward through British Columbia and extreme eastern Washington to Idaho, Utah, Arizona, and New Mexico eastward to Saskatchewan, North Dakota, and Nebraska. Type locality: "summits of dry hills in the Rocky Mountain range, on the Colorado of the West." June–Aug.

45. **Erigeron nevadíncola** Blake. Nevada Daisy. Fig. 5626.

Erigeron nevadensis A. Gray, Proc. Amer. Acad. **8**: 649. 1873. Not Wedd. 1857.
Erigeron nevadincola Blake, Proc. Biol. Soc. Wash. **35**: 78. 1922.

Coarse perennial with a strong taproot and simple or slightly branched caudex; stems usually decumbent at base, 10–30 cm. high; herbage strigose or short-hirsute with appressed or ascending hairs. Basal leaves triple-nerved, acute or sometimes obtuse at the apex, tapering gradually to the petiole, up to 20 cm. long and 1 cm. wide; cauline leaves few and abruptly reduced; head solitary, usually on a long subnaked peduncle, the disk 10–20 mm. wide; involucre 7.5–10 mm. high,

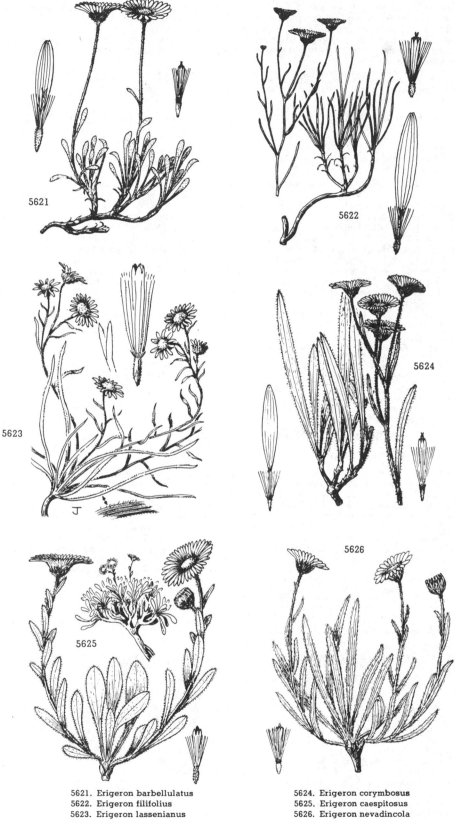

5621. Erigeron barbellulatus
5622. Erigeron filifolius
5623. Erigeron lassenianus

5624. Erigeron corymbosus
5625. Erigeron caespitosus
5626. Erigeron nevadincola

villous-hirsute, scarcely or not at all glandular; phyllaries subequal, broad, dark green, usually somewhat indurate and brownish at the base, the tips scarious, merely acute; rays 20–40, white or blue, 5–11 mm. long, 1.5–3 mm. wide; disk-corollas mostly 5–7 mm. long; style-appendages acute, 0.3–0.4 mm. long; pappus of 25–40 rather coarse, firm bristles and a few slender and inconspicuous, short, outer setae; achenes 2-nerved, evidently hairy above, usually glabrous or glabrate below.

Dry open places in the foothills and at moderate elevations in the mountains, mostly Arid Transition Zone; Plumas County to Mono County, California (east of the Sierra crest), eastward to Elko and Nye Counties, Nevada. Type locality: near Virginia City, Nevada. May–Aug.

46. Erigeron eatònii A. Gray var. villòsus Cronquist. Eaton's Daisy. Fig. 5627.

Erigeron robertianus Greene, Pittonia 3: 293. 1898.
Erigeron pacificus Howell, Fl. N.W. Amer. 319. 1900.
Erigeron eatoni subsp. *villosus* Cronquist, Brittonia 6: 172. 1947.
Erigeron eatoni var. *villosus* Cronquist, Vasc. Pl. Pacif. Northw. 5: 175. 1955.

Perennial with a taproot and crown or short, slightly branched caudex, 5–30 cm. tall, the stems usually decumbent at the mostly reddish purple base; herbage strigose to villous-hirsute with appressed or closely ascending hairs. Basal leaves well developed and conspicuous, narrow, acute, tapering gradually to the petiole, up to 15 cm. long and 10 mm. wide, triple-nerved (occasionally linear and 1-nerved in depauperate individuals); cauline leaves several, more or less strongly reduced; heads solitary or as many as 7, more or less naked-pedunculate, the disk 8–15 mm. wide; involucre 5–7(8) mm. high, evidently white-villous-hirsute especially toward the base; phyllaries subequal, green, often purple-tipped and sometimes with a narrow dark midvein; rays 20–50, mostly white, occasionally light blue, 5–10 mm. long, 1–3 mm. wide; disk-corollas mostly 3.5–5 mm. long; style-appendages obtuse to acutish, 0.1–0.2 mm. long; pappus of 15–25 fragile bristles with a few very fine and inconspicuous, short, outer setae; achenes mostly 2-nerved, shortly villous-hirsute.

Open places in the mountains and foothills, mostly Canadian Zone; east (and occasionally just west) of the Cascade Mountains summit in Oregon, extending northward to the Wenatchee Mountains of Washington and eastward through the Blue Mountains region (including extreme southeastern Washington) to the mountains of central Idaho (as far east as Custer County). Type locality: Wallowa Mountains, Oregon. May–Aug.

Typical *E. eatonii*, with distinctly imbricate, evidently glandular and not very hairy involucre, is more eastern, occurring from southwestern Montana to Colorado and Arizona.

Erigeron eatonii A. Gray var. **plantagineus** (Greene) Cronquist, Vasc. Pl. Pacif. Northw. 5: 175. 1955. (?*Erigeron sonnei* Greene, Pittonia 1: 218. 1888; *E. plantagineus* Greene, op. cit. 3: 292. 1898; *E. eatoni* subsp. *plantagineus* Cronquist, Brittonia 6: 173. 1947.) Pubescence of the leaves, stems, and achenes averaging shorter, coarser, more appressed, and less copious than in var. *villosus;* phyllaries only sparsely hirsute, sometimes a little glandular, subequal or slightly imbricate. Southwestern Jackson County to Malheur County, Oregon, southward in the Cascade-Sierra region to Mono County, California. Type locality: Modoc County, California.

47. Erigeron decúmbens Nutt. Willamette Daisy. Fig. 5628.

Erigeron decumbens Nutt. Trans. Amer. Phil. Soc. II. 7: 309. 1840.

Perennial with a taproot and crown or short, slightly branched caudex, mostly 1.5–5 dm. tall, the stems decumbent at the mostly reddish purple base; herbage strigose. Basal leaves and some or most of the only gradually reduced cauline ones triple-nerved, the basal ones up to 25 cm. long (including the long petiole) and 1 cm. wide; heads 1–5(20), rather shortly naked-pedunculate, the disk 8–15 mm. wide; involucre 3.5–6 mm. high, more or less hirsute and usually obscurely viscid; phyllaries slightly or scarcely imbricate, light or dark green to brownish; rays 20–50, blue or lilac, 6–12 mm. long, 1–2 mm. wide; disk-corollas 2.5–4.0(4.5) mm. long; style-appendages obtuse to acutish, 0.1–0.2 mm. long; pappus of 12–16 fragile bristles, occasionally with a few very fine and inconspicuous, short, outer setae; achenes 2-nerved, finely hairy.

Open places, Humid Transition Zone; Willamette River drainage of western Oregon. Type locality: "Rocky Mountains, towards the Oregon." June–July.

Erigeron decumbens Nutt. var. **robústior** Cronquist, Vasc. Pl. Pacif. Northw. 5: 175. 1955. (*Erigeron decumbens* subsp. *robustior* Cronquist, Brittonia 6: 174. 1947.) Stouter and more robust than typical *E. decumbens;* involucre 6–8 mm. high; phyllaries often broader than in typical *E. decumbens,* up to 1 mm. or more wide; rays up to 20 mm. long and 3 mm. wide; disk-corollas 3.5–5.0 mm. long. Chiefly in Humboldt and Trinity Counties, California, extending to Plumas County, California, and also southern Klamath County, Oregon. Type locality: valley of South Yager Creek, Humboldt County, California.

48. Erigeron flexuòsus Cronquist. Flexuous Daisy. Fig. 5629.

Erigeron flexuosus Cronquist, Brittonia 6: 174. 1947.

Perennial with a taproot and short, slenderly branched caudex, mostly 1–3 dm. tall, the stems slender and often flexuous, decumbent at the base; herbage strigose, the hairs under the heads tending to be spreading. Leaves linear or linear-oblanceolate, 1-nerved or the lowermost sometimes obscurely 3-nerved, the lower ones evidently petioled, up to 10 cm. long and 3 mm. wide, but not forming a conspicuous basal tuft; cauline leaves less petiolate and gradually reduced upward; heads several or rather numerous on leafy branches, the disk 5–10 mm. wide; involucre 4–5 mm. high, finely glandular and more or less spreading-hairy; phyllaries equal or nearly so; rays 12–20, pink, 5–7 mm. long, 1.4–2.0 mm. wide; disk-corollas 2.7–3.5 mm. long; style-appendages acutish, 0.1–0.2 mm. long; pappus of 15–25 fragile bristles, commonly with a few slender and inconspicuous, short, outer setae; achenes 2-nerved, finely hairy.

In rocky places in forest, Canadian Zone; Trinity Mountains, California. Type locality: Trinity Alps Resort, Trinity County. June–July.

49. **Erigeron compáctus** Blake. Cushion Daisy. Fig. 5630.

Erigeron pulvinatus Rydb. Fl. Rocky Mts. 911. 1917. Not Wedd. 1857.
Erigeron compactus Blake, Proc. Biol. Soc. Wash. **35**: 78. 1922.

Pulvinate-cespitose perennial with a short, much branched caudex and a stout taproot. Leaves all basal, finely strigose, linear, 4–20 mm. long, 0.6–1.1 mm. wide; involucre 5–6 mm. high, finely strigose and obscurely glutinous; phyllaries slightly imbricate, stramineous or light greenish with brown midrib, the outer narrow and slightly thickened, the inner broader and thinner, sometimes purple-tipped, all firm and acute; rays 15–30, white or pinkish, 6–9 mm. long, about 2 mm. wide; disk-corollas 4.5–6.0 mm. long; style-appendages acute, 0.2–0.3 mm. long; pappus of 30–40 slender, firm, slightly sordid bristles and some evident outer setae; achenes 2-nerved, conspicuously long-ciliate on the margins, otherwise glabrous.

Dry places in the valleys and foothills, Arid Transition Zone; Inyo Mountains, California, to northeastern Nevada and northwestern Utah. Type locality: Deep Creek, Tooele County, Utah. May–June.

50. **Erigeron parishii** A. Gray. Parish's Daisy. Fig. 5631.

Erigeron parishii A. Gray, Syn. Fl. N. Amer. 1²: 212. 1884.

Perennial with a stout taproot and branching caudex; stems several, 10–35 cm. high, stout, silvery-white with a dense, villous-strigose, appressed pubescence, more densely so toward the base, where the pubescence may be looser and villous. Leaves pubescent like the stem but less densely so and merely grayish, all linear or linear-oblanceolate, the basal ones up to 6 cm. long and 4 mm. wide, often deciduous before anthesis, the cauline ones a little smaller, plentiful but not

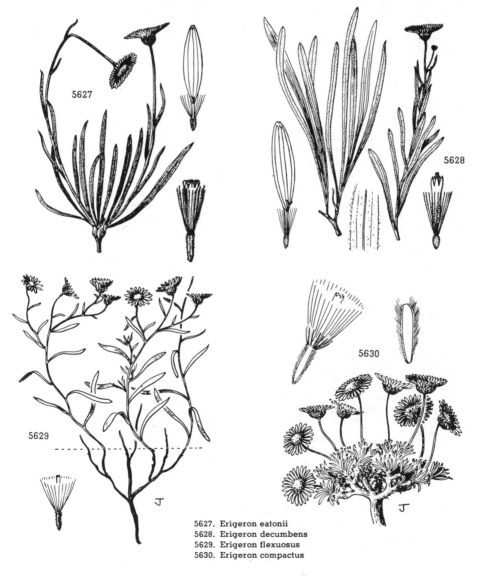

5627. Erigeron eatonii
5628. Erigeron decumbens
5629. Erigeron flexuosus
5630. Erigeron compactus

368 COMPOSITAE

crowded; heads solitary or up to 10, the disk 10–15 mm. wide; involucre 5–7 mm. high, glandular and usually also with a few short white hairs near the base; phyllaries evidently imbricate, olive-greenish, the inner sometimes with brown midrib and stramineous margins; rays 30–55, pink or white, 6–13 mm. long, 1.5–2.6 mm. wide; disk-corollas 3.5–5.0 mm. long; style-appendages obtuse, 0.05–0.15 mm. long; pappus 18–26 coarse and firm, white or sordid bristles and some conspicuous, slender, outer scales; achenes 4-nerved, hairy.

Sandy or rocky places in the foothills of the mountains, Upper Sonoran Zone; mountains of San Bernardino County, California, and to be expected in adjacent Riverside County. Type locality: border of Mojave Desert at Cushenberry Spring, north side of San Bernardino Mountains, San Bernardino County. May–June.

51. Erigeron utahénsis A. Gray. Utah Daisy. Fig. 5632.

Erigeron stenophyllus var. *tetrapleuris* A. Gray, Proc. Amer. Acad. **8**: 650. 1873.
Erigeron utahensis A. Gray, op. cit. **16**: 89. 1880.
Erigeron tetrapleuris Heller, Bull. Torrey Club **25**: 628. 1898.

Perennial with a stout woody taproot and branching caudex; stems several, 10–50 cm. high, more or less strigose and gray-green, generally more densely strigose and whitish at the base. Leaves strigose and gray-green, the basal and lowermost cauline ones linear-oblanceolate, up to 10 cm. long and 6 mm. wide, commonly withered or deciduous by flowering time; cauline leaves linear, up to about 7 cm. long and 3 mm. wide, usually longer than the internodes, often spreading; heads 1–10, the disk 8–15 mm. wide; involucre 4–6 mm. high, more or less strigose, especially below, and often finely glandular, especially above; phyllaries strongly imbricate, the outer greenish, the inner more stramineous and usually with brown midrib, the margins and tips occasionally purplish; rays 10–40, blue, pink, or white, 9–18 mm. long, 1.4–2.7 mm. wide; disk-corollas 3.5–4.6 mm. long; style-appendages obtuse, 0.1–0.25 mm. long; pappus double, the outer of inconspicuous, though well-developed setae; achenes 4-nerved or rarely 6-nerved, hairy.

Dry places, especially on sandstone, mostly Upper Sonoran Zone; Providence Mountains, San Bernardino County, California, to southern Utah, southwestern Colorado, and northern Arizona. Type locality: Kanab, Kane County, Utah. May–June.

52. Erigeron argentàtus A. Gray. Silvery Daisy. Fig. 5633.

Erigeron argentatus A. Gray, Proc. Amer. Acad. **8**: 649. 1873.
Wyomingia argentata A. Nels. in Coult. & Nels. New Man. Bot. Rocky Mts. 531. 1909.

Perennial with a taproot and short branching caudex; stems several 6–40 cm. high, finely strigose and more or less silvery or grayish. Leaves silvery-strigose, sometimes becoming greener at maturity, the basal ones tufted, oblanceolate or linear-oblanceolate, up to 7 cm. long and 6 mm. wide, the blade ill-defined and mostly shorter than the petiole; cauline leaves reduced, scattered, linear or linear-oblanceolate; head solitary, the disk 10–18 mm. wide; involucre 5.5–9 mm. high; phyllaries strongly imbricate, the outer silvery-strigose, the inner sparsely strigose and often finely glandular, thinner than the outer and usually with stramineous margins, all sharply acute or acuminate; rays 20–50, blue to sometimes lavender, pink, or perhaps white, 9–15 mm. long, 1.6–2.8 mm. wide; disk-corollas 3.8–5.6 mm. long; style-appendages obtuse, 0.1–0.15 mm. long; pappus double, the inner of 25–40 fine white bristles, the outer of well-developed but inconspicuous short bristles; achenes rather long-hairy, with 6, 7, or, most commonly, 8 prominent nerves, reputedly sometimes 10-nerved.

Foothills or at moderate elevations in the mountains, mostly Arid Transition Zone; Inyo County, California, to Nevada and western Utah. Type locality: Pah-Ute Mountains, northern Nevada. Mostly June.

53. Erigeron aequifòlius H. M. Hall. Hall's Daisy. Fig. 5634.

Erigeron aequifolius H. M. Hall, Univ. Calif. Pub. Bot. **6**: 174. 1915.

Perennial with a deep-seated root-crown which gives rise to very slender, branched, rhizomatous stems, these becoming aerial stems on reaching the surface; aerial stems slender, erect or ascending, 10–20 cm. high; herbage sparsely or moderately spreading-hirsute with relatively long hairs, also finely glandular. Leaves oblanceolate or narrowly elliptic-oblong, exceeding the internodes, all about alike, without a basal tuft and only gradually reduced upward, 5–20 mm. long, 1.5–3 mm. wide; heads 1–3, the disk 8–13 mm. wide; involucre 4–5 mm. high, finely glandular and sometimes also sparsely hirsute; phyllaries strongly imbricate, with brown midvein, the outer otherwise greenish, the inner more chartaceous and with narrow scarious margins; rays 20–40, blue, about 6 mm. long and 1.3 mm. wide; disk-corollas 3.6–4.3 mm. long; style-appendages obtuse, 0.1 mm. long; pappus double, the inner of 20–35 slender white bristles, the outer of a few well-developed setae; achenes 2-nerved, sparsely hairy.

Rocky places and dry ridges at 6,000–7,000 feet altitude, Canadian Zone; Kern River region of the Sierra Nevada of eastern Tulare County, California. Type locality: Trout Meadows, Sierra Nevada, Tulare County. July.

54. Erigeron brèweri A. Gray. Brewer's Daisy. Fig. 5635.

Erigeron breweri A. Gray, Proc. Amer. Acad. **6**: 541. 1865.

Perennial with a taproot and generally subterranean root-crown, this giving rise to slender rhizomatous branches rarely more than 1.5 mm. thick, which become aerial stems on reaching the surface; aerial stems slender, trailing to more often suberect, 1–3(5) dm. tall, branched at or near the ground level into nearly equal parts, sometimes also branched above but then with an evident

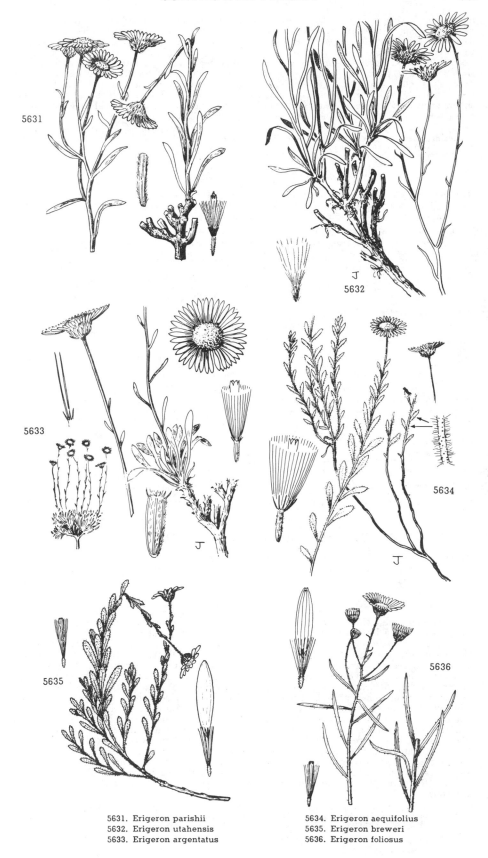

5631. Erigeron parishii
5632. Erigeron utahensis
5633. Erigeron argentatus
5634. Erigeron aequifolius
5635. Erigeron breweri
5636. Erigeron foliosus

main axis; herbage rather densely retrorse-hirtellous or partly spreading-hirsute to (in var. *elmeri*) glabrous or strigose. Leaves numerous, narrowly linear to oblong or oblong-oblanceolate, 4–40 mm. long and 1–8 mm. wide, acute to rounded at the tip, nearly uniform in size, surpassing the internodes, without a basal tuft, only very gradually reduced upward; heads several or occasionally solitary, the disk 6–15 mm. wide; involucre 3–8 mm. high, rather densely glandular and sometimes also hispidulous; phyllaries evidently imbricate, usually with brown midrib, otherwise greenish or chartaceous; rays 10–45, blue or pink to white, 4–10 mm. long, 1.0–2.0 mm. wide; disk-corollas 3.5–6.1 mm. long; style-appendages obtuse, 0.05–0.15 mm. long; pappus double, the inner of 20–50 white or usually brownish, coarse bristles of very unequal diameter, the outer of a few inconspicuous setae; achenes 2-nerved, sparsely hairy.

Dry places in the foothills and mountains, often among rocks, Arid Transition and Canadian Zones; Sierra Nevada and the mountains of southern California, northward occasionally as far as the vicinity of Mount Shasta eastward to central Nevada. Type locality: Yosemite Valley, California. June–Aug.
The foregoing description is drawn to cover all the varieties of the species.

Stem and leaves subglabrous or sparsely appressed-pubescent; Sierra Nevada.   var. *elmeri*.
Stem and leaves more or less densely spreading-hirtellous, or the leaves with longer hairs and somewhat hirsute.
     Leaves mostly less than 12 mm. long; local in the mountains of southern California.
         var. *jacinteus*.
     Leaves mostly more than 12 mm. long.
         Involucre generally hispidulous as well as glandular; phyllaries relatively narrow, long-acuminate, without evident *Aster*-like, green areas near the tips; many of the axillary buds of the middle part of the stem usually developed into floriferous branches or short leafy shoots; leaves generally linear; mostly in Nevada and southeastern California.   var. *porphyreticus*.
         Involucre generally merely glandular; phyllaries or at least the outer ones relatively broad, abruptly acute or acuminate, usually with somewhat *Aster*-like, though ill-defined, greenish areas near the tips; axillary buds of the middle part of the stem usually remaining undeveloped and inconspicuous; leaves linear or generally broader; mostly in the Sierra Nevada and the mountains of southern California.   var. *breweri*.

Erigeron breweri A. Gray var. **porphyreticus** (M. E. Jones) Cronquist, Brittonia 6: 283. 1947. (*Erigeron petrocallis* Greene, Erythea 3: 21. 1895; *E. porphyreticus* M. E. Jones, Contr. West. Bot. No. 8: 33. 1898; *E. foliosus* var. *porphyreticus* Compton, Bull. S. Calif. Acad. 33: 53. 1934.) Characters as given in the key; heads larger than in the other varieties, the involucre 5–8 mm. high and the disk 10–15 mm. wide (as contrasted to 4–6 mm. and 7–15 mm. in typical *E. breweri*); rays generally blue, averaging more numerous than in typical *E. breweri*. Foothills and moderate elevations in the mountains; western Nevada and eastern and southern California. Type locality: Hawthorne, Big Indian Canyon, Mineral County, Nevada.

Erigeron breweri A. Gray var. **jacinteus** (H. M. Hall) Cronquist, Brittonia 6: 284. 1947. (*Erigeron jacinteus* H. M. Hall, Univ. Calif. Pub. Bot. 1: 127. 1902.) Prostrate or ascending plant; stems about 1 dm. long or less; pubescence averaging longer and finer than in the other varieties; leaves not over about 12 mm. long, crowded; heads generally solitary, not large for the species; phyllaries often purplish especially at the margins and tips; rays often light-colored. Talus slopes and rocky ridges at high elevations in the San Jacinto, San Antonio, San Gabriel, and San Bernardino Mountains, California. Type locality: near Tauquitz, San Jacinto Mountains, Riverside County, California.

Erigeron breweri A. Gray var. **élmeri** (Greene) Jepson, Man. Fl. Pl. Calif. 1056. 1925. (*Aster elmeri* Greene, Pittonia 2: 170. 1891; *Erigeron elmeri* Greene, Fl. Fran. 393. 1897.) Very slender, prostrate or ascending plant not over about 15 cm. high; herbage subglabrous or sparsely appressed-pubescent; leaves relatively distant, not over 15 mm. long, not much longer than the internodes; heads solitary or few, small for the species; involucre glandular; phyllaries with or without a well-developed, subapical, green spot; rays mostly pink or lavender, sometimes darker. Cliffs and rock crevices at high altitudes in the mountains; from Yosemite National Park to Sequoia National Park, California. Type locality: Grand Canyon of the Tuolumne River, California.

55. Erigeron foliòsus Nutt. Leafy Daisy. Fig. 5636.

Erigeron foliosus Nutt. Trans. Amer. Phil. Soc. II. 7: 309. 1840.
Erigeron douglasii Torr. & Gray, Fl. N. Amer. 2: 177. 1841.
Erigeron mariposanus Congdon, Erythea 7: 185. 1900.
Erigeron striatus Greene, Bull. S. Calif. Acad. 1: 39. 1902.
Erigeron foliosus f. *grinnellii* Cronquist, Brittonia 6: 279. 1947. A broad-leaved form.

Perennial with a stout taproot and superficial or nearly superficial root-crown; stems erect or nearly so, often a little woody at base, 1–10 dm. tall, usually unbranched at base except in var. *confinis* and sometimes var. *hartwegii*, the base occasionally somewhat rhizomatous but if so then either very short, or relatively stout and more than 1.5 mm. thick, or both; herbage varying from glabrous to more commonly rather coarsely appressed-hairy or even spreading-hirtellous or spreading-hirsute. Leaves numerous, more or less crowded, generally several times longer than the internodes, all nearly alike, without a conspicuous basal tuft, only very gradually reduced upward, varying from linear-filiform and less than 1 mm. wide to narrowly oblong and 5–6 mm. wide, rarely as much as 10 mm. wide, and from 1–8 cm. long, the base only slightly narrowed and not at all cartilaginous; heads generally several or numerous in a corymbiform, leafy or subnaked inflorescence, often solitary in var. *confinis;* disk 10–18 mm. wide; involucre 4–7 mm. high, subglabrous to densely spreading-hairy or glandular or both; phyllaries evidently imbricate, narrow, linear-subulate, acuminate, greenish or the inner more stramineous, with brown midrib; rays 15–65, mostly blue, 5–15 mm. long, 1.0–2.7 mm. wide; disk-corollas 4.0–6.5 mm. long; style-appendages subtruncate to acute, 0.05–0.2 mm. long; pappus double, the inner of 20–33 sordid barbellate bristles, the outer of a few inconspicuous setae; achenes 2-nerved, sparsely hairy or sometimes glabrous.

Dry, often rocky places or in waste land, mostly Transition Zones; southwestern Oregon through California to Lower California. Type locality: near Santa Barbara, Santa Barbara County, California. April–July.
The foregoing description is drawn to cover all the varieties of the species.

Plant 25 cm. high or usually less; leaves very crowded, the internodes of the middle part of the stem averaging
    3 mm. long or usually less; chiefly in the Klamath region.   var. *confinis.*
Plant usually over 25 cm. high; leaves less crowded, the internodes often averaging more than 3 mm. long.
    Stem glabrous to more or less strigose or hirsute-pubescent with appressed hairs, or the hairs near the base of
        the stem spreading; leaves glabrous to variously hairy.
        Leaves linear to narrowly oblong, the larger ones mostly over 1.5 mm. wide or if occasionally less than
            1.5 mm. wide then flat and without subconical hairs.
            Leaves and often also the stem more or less hairy; subconical hairs often present; chiefly west of
                the Sierra Nevada and south of San Francisco.   var. *foliosus.*
            Leaves and stem subglabrous or only sparsely strigose; subconical hairs absent; chiefly north of San
                Francisco and Yosemite National Park northward to Oregon.   var. *hartwegii.*
        Leaves linear-filiform or narrowly linear, 1.5 mm. wide or less, and either folded or with some of the
            marginal hairs enlarged toward the base and subconical; Fresno and Monterey Counties south-
            ward.   var. *stenophyllus.*
    Stem and leaves densely and finely puberulent, or hirtellous with distinctly spreading hairs.
        Achenes glabrous; stem puberulent with fine curly hairs; sand dunes near the seacoast in Santa Barbara
            and San Luis Obispo Counties.   var. *blochmaniae.*
        Achenes more or less hairy; stem hirtellous with short, generally straight hairs; deserts and dry moun-
            tains of the interior of southern California, extending into the Sierra Nevada.
                                    var. *covillei.*

    Erigeron foliosus var. **hartwègii** (Greene) Jepson, Man. Fl. Pl. Calif. 1056. 1925. (*Erigeron hartwegi*
Greene, Erythea **3**: 21. 1895; *E. blasdalei* Greene, op. cit. 124; *E. mendocinus* Greene, Leaflets Bot. Obs. **2**: 9.
1909.) Plants 2.5–5(8) dm. tall; herbage subglabrous or sparsely strigose; leaves 1–4 mm. wide, flat, without
any subconical hairs; rays averaging larger and more conspicuous than in some other varieties; achenes hairy.
Southern Oregon and occasionally as far north as Linn County, south to Yosemite National Park and San Francisco, California. Type locality: California.
    Erigeron foliosus var. **stenophýllus** (Nutt.) A. Gray, Bot. Calif. 1: 330. 1876. (*Erigeron stenophyllus*
Nutt. Journ. Acad. Phila. II. **1**: 176. 1847; *E. foliosus* var. *tenuissimus* A. Gray, Syn. Fl. N. Amer. **1²**: 215.
1884; *E. tenuissimus* Greene, Pittonia **3**: 25 1896; *E. nuttallii* Heller, Bull. Torrey Club **25**: 628. 1898; *E.
setchellii* Jepson, Fl. W. Mid. Calif. 568. 1901; *E. fragilis* Greene, Bull. S. Calif. Acad. **1**: 39. 1902.) Plants
generally tall and stout, seldom less than 4 dm. high; leaves as noted in the key, the surfaces commonly subglabrous, occasionally stiffly short-hairy; stem generally subglabrous; rays averaging shorter than in some of the
other varieties; achenes hairy. Fresno and Monterey Counties to San Diego County, California. Type locality:
near Monterey, California.
    Erigeron foliosus var. **blochmániae** (Greene) H. M. Hall, Univ. Calif. Pub. Bot. **3**: 91. 1907. (*Erigeron
blochmaniae* Greene, Pittonia **3**: 125. 1896.) Tall coarse plants; herbage finely and rather densely puberulent with
short, loose, crinkled hairs; leaves 1–3.5 mm. wide; heads averaging larger than in the other varieties; involucre
more or less densely puberulent or hirtellous; inflorescence usually congested; achenes glabrous or nearly so. Sand
dunes in Santa Barbara and San Luis Obispo Counties, California. Type locality: northern Santa Barbara County.
    Erigeron foliosus var. **covíllei** (Greene) Compton, Bull. S. Calif. Acad. **33**: 51. 1934. (*Erigeron covillei*
Greene, Erythea **3**: 20. 1895.) Plants commonly tall and stout, becoming smaller in intergrades with *E. breweri*; stem and leaves finely and usually densely hirtellous, the hairs of the leaves loose or spreading, those of the
stem spreading or retrorse; involucre generally hirtellous or in more northern plants merely glandular; achenes
sparsely short-hairy. In the foothills and at moderate elevations in the mountain ranges of southern California,
extending northward in the Sierra Nevada as far as the vicinity of Lake Tahoe, in forms usually approaching
E. breweri.
    Erigeron foliosus var. **confínis** (Howell) Jepson, Man. Fl. Pl. Calif. 1056. 1925. (*Erigeron confinis* How-
ell, Erythea **3**: 35. 1895.) Low plants with numerous slender, erect or suberect stems and very crowded, linear-
filiform to linear-oblanceolate leaves 0.5–2.5 mm. wide and less than 4 cm. long; herbage varying from subglabrous
(in forms approaching var. *hartwegii*) to usually more or less strigose-hirsute, to sometimes coarsely spreading-
hirsute; heads solitary or few; rays often longer and more showy than in some of the other varieties; achenes
sparsely hairy. Siskiyou region of northwestern California and southwestern Oregon, and also extending northward
as far as Mount Jefferson, Linn County, Oregon. Type locality: Siskiyou Mountains near Waldo, Josephine County,
Oregon.

56. **Erigeron petróphilus** Greene. Rock Daisy. Fig. 5637.

Erigeron petrophilus Greene, Pittonia **1**: 218. 1888.

    Perennial with a stout taproot and root-crown; stems 10–35 cm. high, moderately to densely
spreading-villous or -villosulous, often also more or less glandular, the hairs averaging shorter and
often coarser than those of *E. miser*. Leaves numerous, longer than the internodes, all nearly
alike, without a basal tuft, only gradually reduced upward, mostly broadly linear or narrowly
oblanceolate, the larger ones mostly 2–4 cm. long and 2–5 mm. wide, more or less hairy like the
stem and sometimes also strongly glandular, apt to be less hairy when more glandular, sometimes
conspicuously long-ciliate on the margins; heads several or sometimes solitary, the disk 9–15 mm.
wide; involucre 5–9 mm. high, densely glandular; phyllaries strongly imbricate, firm, acuminate,
greenish, sometimes with a brown midrib, often purplish above and sometimes with a fairly well-
developed, *Aster*-like, green area near the tip; pistillate flowers wanting; disk-corollas 4.5–6.5
mm. long, yellow to more often pink or reddish; style-appendages obtuse or acutish, 0.15–0.2 mm.
long; pappus of (20)25–40 firm, white or sometimes sordid bristles, generally with a few short
and inconspicuous, outer setae; achenes 2-nerved, sparsely hairy.
    Rocky places, usually at moderate elevations in the mountains, Upper Sonoran and Transition Zones; north-
ern California and southwestern Oregon and the California Coast Ranges as far south as Monterey County. Type
locality: "California Coast Range, from near Berkeley"; perhaps from Wildcat Creek, Contra Costa County.
July–Sept.

57. **Erigeron mìser** A. Gray. Starved Daisy. Fig. 5638.

Erigeron miser A. Gray, Proc. Amer. Acad. **13**: 272. 1878.

    Perennial with a stout taproot and root-crown or short caudex; stems numerous, ascending,
mostly 5–15(25) cm. long; herbage more or less densely spreading-villous with long white hairs,
sometimes also finely and obscurely glandular. Leaves numerous, oblanceolate or broadly linear,
all nearly alike, without a basal tuft, only slightly reduced upward, the lower ones sometimes with
more evident distinction into blade and petiole than the others, the larger ones mostly 1–2 cm. long
and 2–3 mm. wide; heads solitary or few, the disk 7–14 mm. wide; involucre 4–6 mm. high, glandu-

lar; phyllaries evidently imbricate, firm, acuminate, green or greenish especially the outer, occasionally purplish above; pistillate flowers wanting; disk-corollas 3.7–5.0 mm. long; style-appendages obtuse, 0.1 mm. long; pappus double, the inner of 12–28 white or brownish bristles, the outer of evident short setae; achenes 2-nerved, sparsely hairy.

Rocky places, Canadian Zone; vicinity of Donner summit, Sierra Nevada, California. Type locality: Donner Lake, Nevada County. July–Sept.

58. **Erigeron inornàtus** A. Gray. California Rayless Daisy. Fig. 5639.

Erigeron douglasii var. *? eradiatus* A. Gray, Pacif. R. Rep. **12**: 52. 1860.
Erigeron foliosus var. *inornatus* A. Gray, Bot. Calif. **1**: 330. 1876.
Erigeron inornatus A. Gray, Proc. Amer. Acad. **16**: 88. 1880.
Erigeron eradiatus Piper, Contr. U.S. Nat. Herb. **11**: 568. 1906.

Perennial with a generally stout taproot and root-crown; short stem-base often rhizomatous in appearance; stems 1–9 dm. tall, generally spreading-hirsute at the base, soon becoming glabrous upward, or sometimes shortly appressed-hairy to near the summit, or hairy below and becoming glandular upward, or even glandular throughout but not, except near the base, approaching a spreading-villous condition. Leaves numerous, longer than the internodes, only very gradually reduced upward, without a basal tuft, firm, linear-filiform to narrowly oblong or sometimes oblong-oblanceolate, commonly with a short, slightly cartilaginous area near the base, up to 5 cm. long and 6 mm. wide, those of the inflorescence often reduced and distant, generally all ciliate on the margins, otherwise glabrous to short-hairy or glandular but not villous; heads several or many, rarely solitary, the disk 8–17 mm. wide; involucre 3–9 mm. high, often much shorter than the disk, glabrous to densely glandular, rarely also a little hairy; phyllaries strongly imbricate, narrow, sharply acute to long-acuminate, with brown midrib, otherwise greenish or more commonly yellow-greenish, often purplish above the middle; pistillate flowers absent; disk-corollas 4.2–8.0 mm. long; style-appendages obtuse or acutish, 0.1–0.2 mm. long; pappus double, the inner of 25–60 coarse brownish bristles, the outer of more or less evident short setae; achenes 2-nerved, sparsely hairy.

Dry, often rocky places from the lowlands to moderate elevations in the mountains, Upper Sonoran and Transition Zones; southeastern Washington through eastern Oregon to California as far south as San Mateo and Tulare Counties and in adjacent Nevada. Type locality: Eel River, Mendocino County, California. June–Aug. The foregoing description is drawn to include all the varieties of the species.

KEY TO VARIETIES

Involucre not glandular or only very finely, sparsely, and inconspicuously so; nearly the range of the species.
            var. *inornatus*.
Involucre distinctly glandular.
  Broader leaves mostly over 2 mm. wide; stem commonly either spreading-hairy at the base or glandular or both.
    Plants about 20 cm. high or less; leaves relatively short and broad, the longer ones seldom over 3 cm. long, the broader ones commonly not over eight times as long as broad; stem and leaves generally appressed-hairy (stem spreading-hairy at the base), not at all or only very obscurely glandular; Siskiyou County and vicinity.   var. *viscidulus*.
    Plants mostly over 20 cm. high; leaves longer, the longer ones generally well over 3 cm. long, often but not always more than eight times as long as broad; stem and leaves generally more or less glandular; commonest in the California Coast Ranges, extending also to Siskiyou County, and probably also in the Sierra Nevada.   var. *biolettii*.
  Leaves narrower, the broader ones not more than 2 mm. wide; stem subglabrous, neither glandular nor spreading-hairy at the base.
    Plants comparatively robust, 2–9 dm. tall; longer leaves mostly over 2.5 cm. long; California Coast Ranges.   var. *angustatus*.
    Plants low and slender, mostly 5–15(20) cm. tall; leaves short, the longer ones not over 2.5 cm. long; northern Sierra Nevada.   var. *reductus*.

Erigeron inornatus var. **biolèttii** (Greene) Jepson, Fl. W. Mid. Calif. 569. 1901. (*Erigeron biolettii* Greene, Man. Bay Reg. 181. 1894.) Plants mostly 3–9 dm. tall; leaves generally well over 3 cm. long and 2 mm. wide, sometimes less than eight times as long as wide but generally more; herbage usually more or less evidently glandular; involucre strongly glandular, mostly 6–8 mm. high; phyllaries broader than in typical *E. inornatus*, generally purplish above the middle. Coast Ranges of California from near San Francisco to Humboldt County; occasionally also in Siskiyou County and probably in the Sierra Nevada. Type locality: Mount Hood, Sonoma County.

Erigeron inornatus var. **viscidulus** A. Gray, Syn. Fl. N. Amer. 1²: 215. 1884. (*Erigeron viscidulus* Greene, Pittonia **1**: 174. 1888; *E. decumbens* Eastw. Bot. Gaz. **41**: 290. 1906, not Nutt. 1840.) Plants about 20 cm. tall or less with relatively short broad leaves seldom over 3 cm. long, the broader ones not more than eight times as long as broad; pubescence at the base of the stem inclined to be shorter than in typical *E. inornatus*; herbage scarcely or obscurely glandular, usually more pubescent than in typical *E. inornatus*, sometimes subglabrous; involucre 5–7 mm. high, distinctly glandular; phyllaries a little broader than in typical *E. inornatus*, often purplish near the tip. Mountains of northern California, particularly in the vicinity of Mount Eddy, Siskiyou County. Type locality: mountains about the headwaters of the Sacramento River, Siskiyou County.

Erigeron inornatus var. **angustàtus** A. Gray, Syn. Fl. N. Amer. 1²: 215. 1884. (*Erigeron angustatus* Greene, Bull. S. Calif. Acad. **1**: 88. 1885.) Characters as given in the key; heads relatively large, the involucre mostly 5–9 mm. high; phyllaries often purplish above the middle, averaging narrower than in var. *viscidulus* and var. *biolettii* but broader than in typical *E. inornatus;* leaves generally subglabrous. Coast Ranges of California from Trinity County to San Mateo County. Type locality: Red Mountain, Mendocino County.

Erigeron inornatus var. **redúctus** Cronquist, Brittonia **6**: 288. 1947. Characters as given in the key; heads smaller than in var. *angustatus*, the involucre mostly 4–7 mm. high; phyllaries greener than in the other varieties, only occasionally purple-tipped. Sierra Nevada from Placer County to Plumas County, California. Type locality: Yuba River near Cisco, Placer County.

59. **Erigeron ánnuus** (L.) Pers. Annual Daisy. Fig. 5640.

Aster annuus L. Sp. Pl. 875. 1753.
Erigeron annuus Pers. Syn. Pl. **2**: 431. 1807.

Annual or occasionally biennial, 6–15 dm. tall, the stem sparsely to copiously long-spreading-hairy below the inflorescence. Leaves ample, generally toothed, the basal ones when present with elliptic to broadly ovate or suborbicular blade up to 10 cm. long and 7 cm. wide (or sometimes even larger), mostly rather abruptly contracted to a petiole of about the same length; cauline leaves numerous, broadly lanceolate or usually broader, soon becoming short-petiolate or subsessile but otherwise only gradually reduced upward, more or less hirsute, especially the lower; inflorescence large and usually leafy, with more or less numerous heads; involucre 3–5 mm. high, finely glandular and sparsely long-hairy; phyllaries subequal, light greenish or light brownish, acuminate or attenuate; rays 80–125, white or occasionally bluish, 5–10 mm. long, 0.5–1.0 mm. wide; disk-corollas 2.0–2.8 mm. long; style-branches truncate or nearly so, scarcely appendiculate; pappus double, the outer of slender short scales, the inner of 10–15 very fragile bristles, these wanting from the ray-flowers; achenes 2-nerved, hairy.

A weed in moist ground or waste places, widespread in northern United States and adjacent Canada; known from Washington, Oregon, and northern California but more common to the east of our range. Type locality: Canada. June–Sept.

60. Erigeron strigòsus Muhl. Branching Daisy. Fig. 5641.

Doronicum ramosum Walt. Fl. Carol. 205. 1788.
Erigeron strigosus Muhl. ex Willd. Sp. Pl. **3**: 1956. 1803.
Diplemium strigosum Raf. Fl. Tell. **2**: 50. 1838.
Stenactis strigosa DC. Prod. **5**: 299. 1836.
Erigeron strigosus var. *gracilis* Nutt. Trans. Amer. Phil. Soc. II. **7**: 311. 1840.
Erigeron ramosus B. S. P. Prel. Cat. N.Y. 27. 1888. Not Raf. 1817.
Tessenia ramosa Lunell, Amer. Midl. Nat. **5**: 59. 1917.
Stenactis ramosa Domin. Preslia **13–15**: 226. 1935.

Annual or occasionally biennial, mostly 3–7 dm. tall, the stem evidently to very sparsely strigose, or shortly spreading-hirsute near the base; foliage appearing scanty as contrasted to

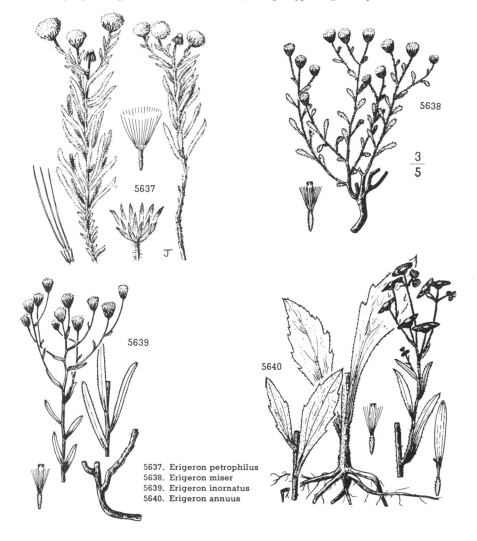

5637. Erigeron petrophilus
5638. Erigeron miser
5639. Erigeron inornatus
5640. Erigeron annuus

374   COMPOSITAE

E. annuus. Basal leaves mostly oblanceolate or elliptic, tapering to the petiole, entire or toothed, the blade and petiole up to 15 cm. long and 25 mm. wide, often deciduous; cauline leaves linear to lanceolate, becoming sessile, entire, or the lower and sometimes even the middle ones slightly toothed, strigose or shortly spreading-hairy to subglabrous; heads several or numerous in an often leafy-bracteate inflorescence, the disk 5–12 mm. wide; involucre 2.5–4 mm. high, finely and obscurely glandular and rather stiffly short-hairy, the hairs less than 1 mm. long; phyllaries mostly subequal, acute or acuminate, with brown midrib and light greenish or stramineous margins; rays 50–100, mostly white, 3–6 mm. long, 0.4–1.0 mm. wide; disk-corollas 1.5–2.6 mm. long; style-appendages blunt, only about 0.05 mm. long; pappus double, the outer of slender short scales, the inner of 10–15 very fragile bristles, these wanting from the ray-flowers; achenes 2-nerved, hairy.

A weed over most of the United States and southern Canada, often in drier places than *E. annuus,* sometimes occurring in relatively undisturbed habitats such as open woodland; in our range known from Washington to central California. Type locality: Pennsylvania. June–Sept.

Erigeron strigosus var. **septentrionàlis** (Fern. & Wieg.) Fernald, Rhodora **44**: 340. 1942. (*Erigeron ramosus* var. *septentrionalis* Fern. & Wieg. Rhodora **15**: 60. 1913.) Foliage a little more ample than in typical *E. strigosus* but much less so than in *E. annuus;* involucre up to 5 mm. high, its hairs long and flattened, shiny, over 1 mm. long; hairs of the stem long and spreading except sometimes on the upper part of the stem and in the inflorescence, or the stem sometimes nearly glabrous. Chiefly in southeastern Canada and adjacent United States, where apparently constituting a self-perpetuating natural population that intergrades more with *E. strigosus* than with *E. annuus;* similar specimens from elsewhere in the range of *E. strigosus* (including our area) may represent recent hybrids. Type locality: Harry's River, Newfoundland.

Both *E. annuus* and *E. strigosus* are commonly apomictic, but there is evidently some sexual reproduction as well.

61. **Erigeron lonchophýllus** Hook. Short-rayed Daisy. Fig. 5642.

Erigeron lonchophyllus Hook. Fl. Bor. Amer. **2**: 18. 1834.
Erigeron glabratus var. *minor* Hook. loc. cit.
Erigeron armerifolius sensu Amer. authors, perhaps not Turcz. ex DC. Prod. **5**: 291. 1836.
Erigeron racemosus Nutt. Trans. Amer. Phil. Soc. II. **7**: 312. 1840.
Erigeron minor Rydb. Bull. Torrey Club **24**: 295. 1897.
Erigeron acris racemosus Clem. & Clem. Rocky Mt. Fl. 269. 1914.
Tessenia racemosa Lunell, Amer. Midl. Nat. **5**: 60. 1917.

Weak-rooted biennial or short-lived perennial, erect, (2)10–60 cm. tall, the herbage spreading-hirsute or the leaves glabrate. Basal leaves mostly oblanceolate, up to 15 cm. long and 12 mm. wide, the cauline ones mostly linear and often elongate; heads solitary to more often several or rather many, borne on nearly erect peduncles, usually some of the lower ones at least equaled or surpassed by their subtending leaves; involucre 4–9 mm. high, sparsely or moderately hirsute, not glandular; phyllaries thin, light green, commonly purplish near the sharply acute or acuminate but scarcely attenuate tip, usually evidently imbricate; rays numerous but inconspicuous, 2–3 mm. long, 0.25–0.5 mm. wide, white or sometimes pinkish, erect, only slightly surpassing the disk; disk-corollas 3.5–5.0 mm. long; pappus of 20–30 slender, generally white bristles, evidently surpassing the disk-corollas, sometimes with a few slender and inconspicuous, short, outer setae; achenes 2-nerved, sparsely hirsute.

Meadows and other moist places, chiefly at moderate to rather high elevations in the mountains, Transition and Boreal Zones; Alaska to Quebec southward in the western cordillera to northern New Mexico, southern Utah, and southern California; in our range occurring from the Cascade Mountains of Washington and Oregon, mostly not extending west of the summits, southward to California through the Sierra Nevada and adjacent easterly ranges to the San Bernardino Mountains. Type locality: Saskatchewan. July–Aug.

62. **Erigeron àcris** L. var. **asteroìdes** (Andrz. ex Bess.) DC. Northern Daisy. Fig. 5643.

Erigeron droebachensis O. Muell. Fl. Dan. *pl. 874.* 1782.
Erigeron asteroides Andrz. ex Bess. Enum. Pl. Volh. 33. 1822.
Erigeron elongatus Ledeb. Fl. Alt. **4**: 91. 1833.
Erigeron acris var. *asteroides* DC. Prod. **5**: 290. 1836.
Erigeron kamtschaticus DC. loc. cit.
Erigeron acris var. *droebachensis* Blytt, Norges Fl. **1**: 562. 1861.
Erigeron angulosus var. *kamtschaticus* Hara, Rhodora **41**: 389. 1939.

Biennial or perennial, usually with a short, simple or slightly branched caudex, 3–8 dm. tall; herbage subglabrous to spreading-hirsute, becoming evidently glandular in the inflorescence. Basal leaves mostly oblanceolate, up to 15 cm. long and 12 mm. wide, seldom much tufted; cauline leaves mostly narrower and becoming sessile but still fairly well developed; heads several or rather numerous in a more or less corymbiform inflorescence, the peduncles arcuate or obliquely ascending and generally well surpassing their subtending leaves; involucre 5–12 mm. high, evidently glandular and often also hirsute; phyllaries subequal or slightly imbricate, green or more or less purplish, the inner long-attenuate, often almost caudate, the outer merely acuminate or acuminate-attenuate; pistillate flowers numerous, in several series, of 2 kinds, the outer with long filiform tube and narrow, pink to purplish or white, erect rays 2.5–3.5 mm. long and 0.2–0.4 mm. wide, the inner nearly or quite rayless with the slender tubular corolla from nearly as long as the style to less than half as long; disk-corollas 4.2–6.2 mm. long; pappus of 25–35 slender, white to reddish bristles, sometimes with a few inconspicuous, short, outer setae; achenes 2-nerved, sparsely hairy.

Rocky places in the mountains, Boreal Zone; circumpolar, extending southward to northern Oregon, northern Utah, Colorado, Minnesota, and Maine. Type locality: eastern Europe. June–Aug.

Erigeron acris var. **débilis** A. Gray, Syn. Fl. N. Amer. **1²**: 220. 1884. (*Erigeron nivalis* Nutt. Trans. Amer. Phil. Soc. II. **7**: 311. 1840; *E. jucundus* Greene, Pittonia **3**: 165. 1897; *E. debilis* Rydb. Mem. N.Y. Bot. Gard. **1**: 408. 1900.) Plants 0.2–3 dm tall, the stem often curved at the base; leaves relatively wider than

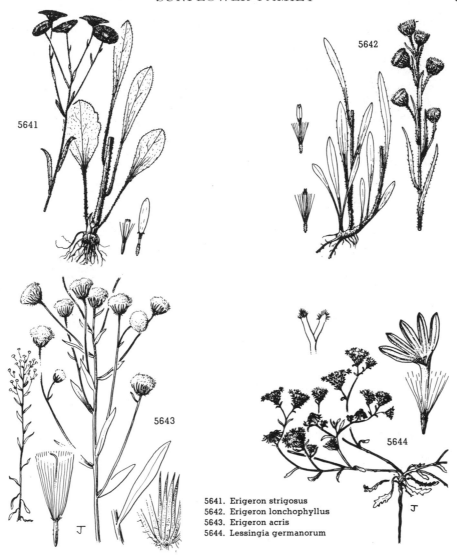

5641. Erigeron strigosus
5642. Erigeron lonchophyllus
5643. Erigeron acris
5644. Lessingia germanorum

in var. *asteroides*, the basal ones generally clustered, and the cauline ones often much reduced; heads few or solitary; rays generally pinkish, a little larger than in var. *asteroides* (up to 4.5 mm. long and 0.55 mm. wide) and more evidently surpassing the pappus; rayless pistillate flowers often few. Alaska to northern California (Mount Shasta) and southern Colorado. Type locality: Woodruff's Falls west of Upper Marias Pass, Rocky Mountains of Montana.

92. LESSÍNGIA* Cham. Linnaea 4: 203. 1829.

Annuals, blooming in summer or late spring, the stems simple or branching, arising from a taproot, the herbage white-tomentose to glabrous, mostly glandular with conspicuous, stipitate, tack-shaped or sessile glands and often with punctate glands. Leaves alternate, entire, toothed, or pinnatifid, those of the branches sessile and bractiform above. Heads solitary or clustered at the tips of stems and branches, in some species spicate in the axils of leaves or sessile among the basal leaves, homogamous, but the outer flowers often enlarged and palmately 5-cleft and reflexed. Involúcres cylindric, turbinate, or campanulate, the phyllaries graduate in several series, usually much appressed. Corollas lavender, pink, yellow, or rarely white. Anthers rounded at base, bearing a terminal appendage above. Style-branches with subtruncate or triangular, hairy appendages, these often bearing a terminal cusp. Achenes turbinate to obconic, silky-pilose; pappus of many bristles or awns, sometimes connate in phalanges, or reduced to slender paleae, tan, brownish, or reddish. [Name in honor of the Lessings, a German family of scientists and authors.]

* Based largely on the earlier published work of John Thomas Howell and later discussions with him.

A genus of 11 or 12 species, except for one species which extends into Arizona, limited to California. Type species, *Lessingia germanorum* Cham.

Plants depressed or dwarfs; tips of inner phyllaries white and cartilaginous.   5. *L. nana.*
Plants not depressed or dwarfs; tips of inner phyllaries herbaceous or scarious.
    All corollas yellow (heads with some corollas purplish or white in *L. tenuis*); involucres campanulate or, if turbinate, broad in relation to length.
        Style-appendages subulate, bearing a long cusp, nearly or quite as long as the stigmatic portion; plants of Mojave Desert (occurring rarely in Mount Pinos and Cuyama Valley regions).
            4. *L. lemmonii.*
        Style-appendages less than half the length of the stigmatic portion, the cusp if present short (longer in some forms of *L. glandulifera*); plants not of desert regions (except *L. glandulifera* var. *tomentosa*).
            Glands few, confined to the phyllaries; San Francisco County.   1. *L. germanorum.*
            Glandular, the glands if few not confined to the phyllaries; widely distributed.
                Phyllaries loosely imbricate, about 4-seriate; slender spring-flowering annuals.
                    2. *L. tenuis.*
                Phyllaries closely imbricate, about 6-seriate; often stout summer-flowering annuals.
                    3. *L. glandulifera.*
    Corollas pink, lavendar, or sometimes white, never yellow; involucres turbinate to cylindro-turbinate, long in relation to width (broad in *L. ramulosa* var. *ramulosa*).
        Inflorescence spicate, the heads sessile in the leaf-axils, rarely glomerate on axillary branches.
            6. *L. virgata.*
        Inflorescence open, the heads solitary on the branchlets or more rarely glomerate.
            Basal leaves relatively few and withering at or before anthesis; stems relatively slender at base; punctate glands present.
                Phyllaries woolly-tomentose (see also *L. micradenia arachnoidea*); punctate glands only present, larger stipitate glands lacking.   9. *L. leptoclada.*
                Phyllaries not woolly-tomentose (except *L. micradenia arachnoidea*); punctate and larger stalked glands present, either sparse or abundant.
                    Punctate glands abundant, evident; widespread species.   7. *L. nemaclada.*
                    Punctate glands very few and inconspicuous; serpentine areas in Marin, San Mateo, and Santa Clara Counties.   8. *L. micradenia.*
            Basal leaves many and tending to persist through anthesis; stems thickened at base; punctate glands lacking.
                Upper cauline leaves with tack-shaped glands; outer corollas only slightly enlarged and not palmately spreading.
                    Heads 6–10 mm. broad; pappus-bristles many, nearly free.   10. *L. ramulosa.*
                    Heads 3–5 mm. broad; pappus of few bristle-tipped awns or completely paleaceous.
                      10. *L. ramulosa adenophora.*
                Upper cauline leaves lacking tack-shaped glands; outer corollas much enlarged and palmately spreading.   11. *L. hololeuca.*

1. Lessingia germanòrum Cham. San Francisco Lessingia. Fig. 5644.

Lessingia germanorum Cham. Linnaea **4**: 203. 1829.

Annual, 1–3 dm. high, rather slender, diffusely branched from the base, procumbent, rarely erect, the herbage and young stems loosely grayish-tomentose, glandless, glabrescent, and darkened in age. Lower leaves 1–3 cm. long, oblanceolate, acute or obtuse, tapering to a petioliform base; stem-leaves shorter, sessile, pinnatifid or pinnatisect, the uppermost subentire, bracteate; heads 25–38-flowered, solitary on the slender divergent branchlets; involucres 5–7 mm. high, campanulate, the phyllaries about 6-seriate, loosely imbricate with recurved tips, glandular above, inodorous; corollas deep lemon-yellow with a brownish or purplish band at the throat, the outer corollas palmately cleft; appendages of style-branches short-deltoid, with or without a short cusp; pappus-bristles about 20–30, equal, essentially free to the base.

Coastal dunes, Humid Transition Zone; San Francisco County, the type locality. Aug.–Nov.

2. Lessingia ténuis (A. Gray) Coville. Spring Lessingia. Fig. 5645.

Lessingia ramulosa var. *tenuis* A. Gray, Bot. Calif. **1**: 307. 1876.
Lessingia tenuis Coville, Contr. U.S. Nat. Herb. **4**: 124. 1893.
Lessingia parvula Greene, Fl. Fran. 376. 1897.
Lessingia heterochroma H. M. Hall, Univ. Calif. Pub. Bot. **3**: 67. 1904.
Lessingia tenuis var. *jaredii* Jepson, Man. Fl. Pl. Calif. 1041. 1925.
Lessingia germanorum var. *tenuis* J. T. Howell, Univ. Calif. Pub. Bot. **16**: 16. 1929.

Annual, 3–15 cm. high, diffusely branched from the base, the stems and the very slender often reddish divergent branches glandular, loosely tomentose below when young, older plants glabrous. Basal leaves 1–3.5 cm. long, spatulate, narrowed to a petiole, not persisiting, entire or shallowly to deeply and irregularly pinnately parted, tomentose on both faces; stem-leaves 0.3–1.5 cm. long, sessile, oblanceolate to obovate, entire (some of the plants of the Mount Hamilton Range with irregularly toothed stem-leaves), mostly persistently tomentose on both sides but sometimes glabrate, also glandular; heads solitary on slender branchlets, these naked or with few reduced stem-leaves, 13–22-flowered; involucre 4–5 mm. high, campanulate, loosely 4-seriate, the phyllaries spreading, herbaceous-tipped, the inner purplish and stipitate-glandular and the outermost sparsely tomentose; corollas all yellow with a purplish ring at the throat or the outermost palmately parted corollas yellow, rose-colored, or suffused with purple; appendages of the style-branches short-deltoid, without a cusp; pappus-bristles 15–25, mostly free to the base, reduced to 5–8 paleaceous awns in forms described as var. *jaredii*.

Dry slopes, open or in chaparral, Upper Sonoran Zone; Inner Coast Ranges, Santa Clara and Stanislaus Counties, south to the Mount Pinos region, Ventura County. Type locality: Piru Creek, Ventura County. April–July.

3. Lessingia glandulífera A. Gray. Valley Lessingia. Fig. 5646.

Lessingia glandulifera A. Gray, Proc. Amer. Acad. **17**: 207. 1882.
Lessingia germanorum var. *vallicola* J. T. Howell, Univ. Calif. Pub. Bot. **16**: 19. 1929.
Lessingia germanorum var. *tenuipes* J. T. Howell, op. cit. 20.
Lessingia germanorum var. *glandulifera* J. T. Howell, op. cit. 22.

Annual, (1)2–7.5 dm. high, more or less slender to rather stout, branched below the middle with stiffly spreading stems, these much branched above, the herbage in younger plants loosely tomentose below often up to the branching inflorescence but becoming glabrate in age, very glandular and strongly scented. Lowest leaves 2–6 cm. long, obovate or oblanceolate, pinnate to pinnately lobed (rarely bipinnatifid), narrowed to a petioliform base; stem-leaves smaller, sessile, oblong to ovate, entire or toothed, the uppermost bract-like and rather densely beset on the more slender flowering branches, usually with tack-shaped marginal glands, the entire surface sometimes glandular; heads 17–33-flowered; involucre 5–8 mm. high, broadly turbinate, the phyllaries closely imbricate, about 6-seriate, glandular also with tack-shaped glands; corollas yellow with a purplish band at the throat, the outer palmately cleft; appendages of style-branches deltoid to narrowly triangular, with or without a short cusp; pappus-bristles 30 or more, free to the base.

Grasslands, Sonoran Zones; San Joaquin County in the Central Valley and the adjacent Coast Ranges to cismontane southern California and Lower California; also the foothill area of the Sierra Nevada from Placer County southward. Type locality: not definitely stated. May–Nov.

Lessingia glandulifera var. **pectinàta** (Greene) Jepson, Man. Fl. Pl. Calif. 1041. 1925. (*Lessingia pectinata* Greene, Proc. Acad. Phila. **1895**: 548. 1896; *L. germanorum* var. *pectinata* J. T. Howell, Univ. Calif. Pub. Bot. **16**: 18. 1929.) Plants very diffuse and branched from the base, glabrous or soon glabrate, more or less glandular throughout; cauline leaves pectinate-pinnatifid with cuspidate segment; phyllaries with conspicuous glands, the tips squarrose. Dunes near the coast, Monterey County south to northern Santa Barbara County. Type locality: Monterey.

Lessingia glandulifera var. **tomentòsa** (Greene) Ferris, Contr. Dudley Herb. **5**: 101. 1958. (*Lessingia tomentosa* Greene, Leaflets Bot. Obs. **2**: 32. 1910; *L. germanorum* var. *tomentosa* J. T. Howell, Univ. Calif. Pub. Bot. **16**: 24. 1929.) Plants depressed, densely tomentose throughout except the involucres, these 5–6 mm. high, with large marginal glands. Southwestern part of the Colorado Desert, San Diego County. Collected by Orcutt.

4. Lessingia lemmònii A. Gray. Lemmon's Lessingia. Fig. 5647.

Lessingia lemmonii A. Gray, Proc. Amer. Acad. **21**: 412. 1886.
Lessingia germanorum var. *lemmonii* J. T. Howell, Univ. Calif. Pub. Bot. **16**: 25. 1929.

Annual, 6–30 cm. high, branching from the base when young with few divaricate branches, becoming much branched above in age, the branches grayish-tomentose when young, soon green and finely glandular-puberulent. Lower leaves 3–3.5 cm. long, spatulate or oblong-ovate; stem-leaves about 1 cm. long, the uppermost smaller and narrowly ovate to linear, the margins entire with some large sessile glands, all thinly tomentose; heads solitary on the branches, 12–22-flowered; involucres 5–6 mm. high, turbinate to campanulate, the phyllaries graduate, not closely imbricate, obtuse or acute, glandular-puberulent and bearing large marginal glands; corollas yellow, without a purplish band at the throat, the outer palmately divided; appendages of the style-branches 0.7–1 mm. long, slender-subulate; pappus-bristles mostly free, abundant.

Dry sandy soil, Sonoran Zones; northwestern Arizona west through San Bernardino County, California, and also in Inyo County, to the Mount Pinos region. Ventura and Los Angeles Counties; occurring sparingly in eastern Santa Barbara County. Type locality: Ashfork, Yavapai County, Arizona. June–Sept.

Lessingia lemmonii var. **ramulosíssima** (A. Nels.) Ferris, Contr. Dudley Herb. **5**: 101. 1958. (*Lessingia ramulosissima* A. Nels. Univ. Wyo. Publ. Sci. **1**: 138. 1926; *L. germanorum* var. *ramulosissima* J. T. Howell, Univ. Calif. Pub. Bot. **16**: 24. 1929.) Like *L. lemmonii* var. *lemmonii* in stature and branching but older plants more intricately branched; herbage including phyllaries more glandular; differs in having narrower

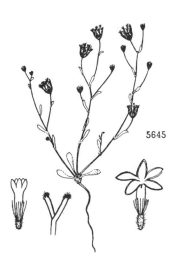

5645

5645. Lessingia tenuis

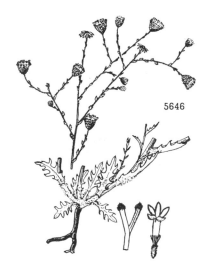

5646

5646. Lessingia glandulifera

involucres and more bracteate peduncular branches. Owens Valley, Inyo County, California, and in the north-western Mojave Desert. Type locality: Hinckley, San Bernardino County.

Lessingia lemmonii var. **peirsònii** (J. T. Howell) Ferris, Contr. Dudley Herb. **5**: 101. 1958. (*Lessingia germanorum* var. *peirsonii* J. T. Howell, Univ. Calif. Pub. Bot. **16**: 26. 1929.) Like *L. lemmonii* var. *lemmonii* in growth habit, differing in having a dense matted tomentum which is only partly deciduous in age and by persistently woolly involucres. Western Mojave Desert and the adjacent mountains of Kern and northern Los Angeles Counties. Type locality: Kings Canyon, Liebre Mountains, Los Angeles County.

5. **Lessingia nàna** A. Gray. Dwarf Lessingia. Fig. 5648.

Lessingia nana A. Gray in Benth. Pl. Hartw. 315. 1849.
Lessingia nana var. *caulescens* A. Gray, Syn. Fl. N. Amer. **1**²: 163. 1884.
Lessingia parryi Greene, Bull. Calif. Acad. **1**: 191. 1885.

Dwarf plants, stemless or nearly so, 1–2 cm. high or with a few erect or decumbent, sometimes branching stems 2.5–10 cm. high, densely white-lanate throughout with long tangled hairs, the leaves becoming glabrate in age. Basal leaves rosulate, up to 3.5 cm. long, petiolate, linear-oblanceolate, acute and spinescent-tipped, entire or spinescent-toothed, prominently veined, the stem-leaves similar, glandular-punctate, about 1 cm. long; heads clustered at the base or at ends of branches, 10–15-flowered; involucres 7–10 mm. high, the outer phyllaries densely woolly, lanceolate, mucronate, green-tipped, the inner longer, scarious-chartaceous, with a stiffly erect, cartilaginous, awned tip; corollas rose-colored or violet, all essentially regular; style-branches with or without a short cusp; achenes 2–3 mm. long; pappus-bristles showy, rose-colored or rufous, about twice as long as the achenes.

Open places on plains and slopes, Sonoran Zones; Sacramento Valley and adjacent eastern slopes from Tehama County south through the San Joaquin Valley to the Tehachapi Mountains in Kern County; also natividad, Monterey County (*Abbott*). Type locality: near Sacramento. Collected by Pickering. June–Oct.

6. **Lessingia virgàta** A. Gray. Virgate Lessingia. Fig. 5649.

Lessingia virgata A. Gray in Benth. Pl. Hartw. 315. 1849.
Lessingia subspicata Greene, Leaflets Bot. Obs. **2**: 29. 1910.

Erect, often rather stout annual, 4–6 dm. high, branching at and also above the base with long virgate branches bearing sessile heads in the axils of the leaves (axillary short branchlets sometimes present with heads in small glomerules), the herbage densely woolly-tomentose, becoming glabrate. Lowest leaves oblanceolate, acute, soon withering, the cauline leaves 7–10 mm. long, the uppermost scarcely reduced, usually longer than the internodes, appressed, oblong-ovate to ovate; heads 3–6-flowered; involucres 5–7 mm. high, cylindro-turbinate, the phyllaries 3–4-seriate, closely imbricate, glandular, loosely woolly below, and purple-tipped; corollas lavender, the outer palmately cleft; appendages of the style-branches with a short or a slender subulate cusp; pappus-bristles 27–32, about equaling the achene, tending to be united at the base.

Dry plains and foothills, principally in the Lower Sonoran Zone; Sacramento Valley and bordering foot-hills from Tehama and Butte Counties south in the northern San Joaquin Valley to San Joaquin and Stanislaus Counties. Type locality: near Sacramento. June–Oct.

7. **Lessingia nemáclada** Greene. Slender-stemmed Lessingia. Fig. 5650.

Lessingia leptoclada var. *microcephala* A. Gray, Proc. Amer. Acad. **7**: 351. 1868.
Lessingia nemaclada Greene, Bull. Calif. Acad. **1**: 191. 1885.
Lessingia ramulosa var. *microcephala* Jepson, Man. Fl. Pl. Calif. 1041. 1925.

Erect, very slender annual, 1–3 dm. high, either simple below and branched above or branched from the base, the ultimate ascending or spreading and filiform, glabrous. Lower leaves oblong-oblanceolate, narrowed to a petioliform base, becoming sessile above and conduplicate, loosely woolly; uppermost leaves bract-like, 1–2 mm. long, appressed, punctate- and stipitate-glandular, woolly within, glabrous without; heads solitary at the ends of the branchlets; involucres 5–6 mm. high, cylindro-turbinate, the phyllaries about 5-seriate, linear-oblong, acute, glabrous except for sessile and stipitate glands; corollas lavender, 1 or 2 palmately cleft; the cusped appendages of the style-branches about 0.5–0.7 mm. long; pappus of 5 simple or bristly, paleaceous awns or about 25 partially united bristles.

Stony ground, Upper Sonoran Zone; occurring rather locally from Eldorado County to Mariposa County and also sparsely in Glenn and Lake Counties in the Inner Coast Range. Type locality: Sweetwater Creek, Eldorado County. July–Oct.

Lessingia nemaclada var. **mendocína** (Greene) J. T. Howell, Univ. Calif. Pub. Bot. **16**: 28. 1929. (*Lessingia mendocina* Greene, Leaflets Bot. Obs. **2**: 28. 1910; *L. cymulosa* Greene, op. oit. 30; *L. fastigiata* Greene, op. cit. 31; *L. paleacea* Greene, loc. cit.) Though highly variable in form, stouter and more leafy than *L. nemaclada* var. *nemaclada* and the upper branches, though slender, not filiform; heads 4–11-flowered; stipitate glands on phyllaries more abundant; cusped appendages of style-branches a little longer; pappus-bristles 14–23, free above, more or less united in 5 groups, paleaceous at the base. Abundant in the northern Inner Coast Ranges extending to Lake County; also in the Mount Hamilton Range (*H. Sharsmith*) and the foothills of the Sierra from Butte County to Fresno County.

Lessingia nemaclada var. **albiflòra** (Eastw.) J. T. Howell, Univ. Calif. Pub. Bot. **16**: 29. 1929. (*Lessingia albiflora* Eastw. Bull. Torrey Club **32**: 217. 1905; *L. glandulifera* var. *albiflora* Jepson, Man. Fl. Pl. Calif. 1041. 1925.) A slender plant with slender divaricate branches and heads 6–8-flowered with white corollas or white with a purple throat, the lobes suffused with pale lavender. Inner Coast Ranges, southern Monterey County to the Tehachapi Mountains and the southern Sierra foothills in Kern County.

8. **Lessingia micradènia** Greene. Tamalpais Lessingia. Fig. 5651.

Lessingia micradenia Greene, Leaflets Bot. Obs. **2**: 28. 1910.
Lessingia ramulosa var. *micradenia* J. T. Howell, Univ. Calif. Pub. Bot. **16**: 39. 1929.

Slender annual, 3–4 dm. high, stems simple below, paniculately branched from about midstem with divaricately spreading branches, essentially glabrous but bearing tack-shaped glands below the heads. Basal leaves 3–4 cm. long, narrowed to a petioliform base, irregularly toothed, soon deciduous; upper leaves progressively shorter, narrowly oblanceolate to linear-lanceolate, mostly entire, bearing an apical cusp, loosely tomentose, the rather few bract-like leaves of the branchlets more densely so on the upper surface; heads 5–10-flowered; involucre 5–6 mm. long, narrowly turbinate, the phyllaries closely imbricate, acute, beset with stipitate or tack-shaped glands, the outer more densely glandular; corollas lavender, the outer corollas scarcely palmately spreading; appendages of the style-branches evidently cusped; pappus united into 5 paleaceous awns.

Thin gravelly soil, mostly on serpentine, Upper Sonoran Zone; hills of Marin County, especially on Mount Tamalpais, the type locality. Sept.

This taxon and the following taxa are very limited in distribution. They appear to be akin to the *L. nemaclada* complex.

Lessingia micradenia var. **glabràta** (Keck) Ferris, Contr. Dudley Herb. **5**: 101. 1958. (*Lessingia ramulosa* var. *glabrata* Keck, Aliso **4**: 105. 1958.) Like *L. micradenia* var. *micradenia* in habit and leaf-shape, the uppermost bract-like leaves persistently tomentose on upper surface; heads 3–5-flowered; involucre about 4.5–6 mm. long, cylindro-turbinate; phyllaries with very few small stipitate glands, the more abundant, larger glands sessile or short-stalked, one large terminal gland usually present; pappus-bristles more or less united below. On serpentine, inner slopes of the Coast Range, Santa Clara County. Type locality: east of Los Gatos on road to Almaden.

5648

5647

5649

5650

5647. Lessingia lemmonii
5648. Lessingia nana

5649. Lessingia virgata
5650. Lessingia nemaclada

Lessingia micradenia var. arachnoìdea (Greene) Ferris, Contr. Dudley Herb. **5**: 101. 1958. (*Lessingia arachnoidea* Greene, Leaflets Bot. Obs. **2**: 29. 1910; *L. hololeuca* var. *arachnoidea* J. T. Howell, Univ. Calif. Pub. Bot. **16**: 42. 1929; *L. leptoclada* var. *arachnoidea* Blake, Journ. Wash. Acad. **19**: 271. 1929.) Habit and leaf-shape of the two preceding taxa; plants glabrate, loosely woolly below the heads; uppermost bract-like leaves persistently tomentose on the upper surface; heads 7–18-flowered; involucre 5–8 mm. high, about 5-seriate; phyllaries closely imbricate, acute, loosely woolly, very sparsely beset with sessile or short-stalked, large glands; outer corollas palmately cleft; pappus-bristles short, free or united in sets. On serpentine, vicinity of Crystal Springs Lakes, San Mateo County, the type locality. Perhaps specifically distinct on the basis of the abundance of wool on the involucres and the somewhat larger, more abundantly flowered heads.

9. Lessingia leptóclada A. Gray. Sierra Lessingia. Fig. 5652.

Lessingia leptoclada A. Gray, Proc. Amer. Acad. **7**: 351. 1868.
Lessingia leptoclada var. *tenuis* A. Gray, loc. cit.

Erect slender annual, 3–8 dm. high except depauperate forms, commonly simple at the base and branching about the midstem with very slender divergent branches, sometimes basally branched, woolly when young, becoming glabrate. Basal leaves 0.8–5 cm. long, oblanceolate or spatulate, mostly narrowed to a petioliform base, entire or irregularly and few-toothed, woolly; stem-leaves 2.5–4 cm. long, sessile, becoming fewer and much reduced below the heads, glandular-punctate; heads solitary at the ends of the branches, sometimes in clusters of 2 or more, rarely somewhat spike-like, (6)12–22-flowered; involucres 5–10 cm. high, rather broadly turbinate, closely imbricate; phyllaries 5–7(8)-seriate, the loose tomentum mostly persisting, punctate glands sparse; corollas lavender to bluish purple, the outer palmate; appendages of the style-branches triangular, with or without a cusp; pappus-bristles 18–40, typically distinct to the base, more rarely somewhat basally united in groups.

Openings in forest, mostly Arid Transition Zone; Sierra Nevada, Eldorado County to Kern County. Type locality: Yosemite Valley, Mariposa County. July–Oct.

10. Lessingia ramulòsa A. Gray. Sonoma Lessingia. Fig. 5653.

Lessingia ramulosa A. Gray in Benth. Pl. Hartw. 314. 1849.
Lessingia bicolor Greene, Leaflets Bot. Obs. **2**: 28. 1910.

Annuals, 2.5–4.5 dm. high from a thickened tomentose base, the main stem erect and divaricately branched, also branching from the base, finely and rather densely stipitate-glandular. Basal and lowest leaves 3–7 cm. long, mostly oblong-oblanceolate, some narrowed to a petioliform base, margin irregularly dentate to nearly entire, loosely tomentose, persisting; upper and bract-like leaves progressively smaller, ovate, sessil, and often clasping, densely and persistently woolly on upper surface, the lower glabrate, more or less densely beset with tack-shaped glands; heads solitary at the ends of the branches, about 13–flowered; involucres 5–7 mm. high, 6–10 mm. wide, broadly turbinate to campanulate; phyllaries about 4–5-seriate, rather loosely imbricate, thin, the inner purplish-tipped, stipitate-glandular, with some larger tack-shaped glands; corollas rose-colored or pale pink, the outermost scarcely enlarged and not palmately cleft; appendages of the style-branches very short and not cusped; pappus- bristles 25–37, mostly free at the base.

Clay soil on slopes or in valleys, Upper Sonoran Zone; North Coast Ranges, Mendocino County to Sonoma County. Type locality; near Sonoma, Sonoma County. Aug.-Nov.

Lessingia ramulosa var. adenóphora (Greene) A. Gray. Syn. Fl. N. Amer. ed. 2. 1². : 446. 1886. (*Lessingia adenophora* Greene, Bull. Calif. Acad. **1**: 190. 1885.) Plants smaller, stems less thickened at the tomentose base but basal and lower leaves mostly persistent; stem-leaves ovate, woolly, margin densely beset with tack-shaped glands; heads typically with fewer flowers, the involucres 3–5 mm. broad; pappus paleaceous or united into 5 bristle-tipped awns. In the Inner Coast Range in Lake, Napa, and Colusa Counties. Type locality: near Epperson's, Lake County.

11. Lessingia hololeùca Greene. Woolly-headed Lessingia. Fig. 5654.

Lessingia hololeuca Greene, Fl. Fran. 377. 1897.
Lessingia bakeri Greene, Leaflets Bot. Obs. **2**: 27. 1910.
Lessingia imbricata Greene, op. cit. 29.
Lessingia leptoclada var. *hololeuca* Jepson, Man. Fl. Pl. Calif. 1040. 1925.

Erect or ascending annual, 0.5–3 dm. high. high from a thickened tomentose base, the stems simple below or branching from the base, rather stout and densely tomentose, the branches more slender, stiffly ascending and tardily glabrescent. Lowest leaves 1.5–11.5 cm. long, obovate to oblanceolate, acute, sessile or narrowed to a petioliform base, persisting at anthesis, entire or irregularly toothed; cauline leaves reduced, obovate to oblongish, sessile, becoming bract-like below the heads, densely tomentose and becoming glabrous in age; heads mostly solitary on the ends of the branches, sometimes sessile in the leaf-axils, the flowers 13–18, pinkish or lavender; involucres 10–12 mm. long, turbinate; phyllaries 5–6-seriate, spreading, lanceolate, sharply acute, glandless, the outer woolly, the inner glabrous or nearly so, tinged apically with purple; marginal flowers palmately cleft and enlarged; style-branches with subulate appendages; pappus-bristles 40–55, up to 8 mm. long, not united at the base.

Grasslands on hills and in valleys, Upper Sonoran Zone; Yolo, Napa, and Sonoma Counties south to San Mateo County. Type locality: San Rafael, Marin County. July–Nov.

93. CONỲZA L. Sp. Pl. 861. 1753; emend. Less. Syn. Compos. 203. 1832.
Nomen conservandum.

Annual or perennial herbs. Leaves alternate, pinnatifid or bipinnatifid to subentire. Heads small, usually numerous, disciform, whitish or minutely radiate with white or purple rays. Involucre (in ours) about 2-seriate, subequal, of narrow subherbaceous phyllaries. Receptacle flat, naked. Pistillate flowers numerous, several-seriate, their corollas tubular-filiform, without distinct ligule, much shorter than or equaling the style and pappus; disk-flowers few, their corollas (in ours) with a slender tube and slender throat abruptly enlarged above, 5-toothed. Anthers entire at base, with lanceolate terminal appendages. Style-branches hispidulous above with short, ovate, hispidulous appendages. Achenes small, oblong, compressed, the pappus (in ours) of rather few and fragile, unequal, capillary bristles. [Name Greek, applied to some kind of fleabane.]

A genus of about 50 species, most of which are natives of tropical or subtropical regions in both hemispheres. Type species, *Conyza chilensis* Spreng.

All leaves cuneate-obovate, sharply and coarsely toothed.   1. *C. coulteri.*
Leaves linear or linear-lanceolate, entire or the lower sometimes laciniate.
    Involucres densely hirsutulous, 5–8 mm. high; pistillate flowers without ligules.   2. *C. bonariensis.*
    Involucres nearly glabrous, 3–4 mm. high; pistillate flowers with minute ligules.   3. *C. canadensis.*

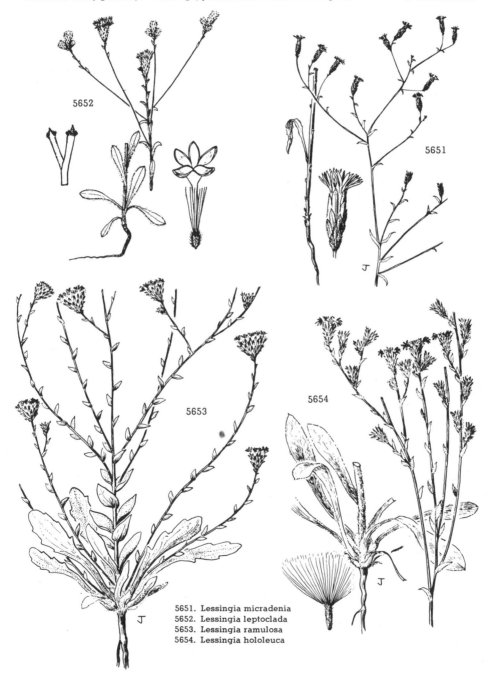

5651. Lessingia micradenia
5652. Lessingia leptoclada
5653. Lessingia ramulosa
5654. Lessingia hololeuca

1. Conyza coùlteri A. Gray. Coulter's Conyza. Fig. 5655.

Conyza coulteri A. Gray, Proc. Amer. Acad. **7**: 355. 1868.
Erigeron discoidea Kell. Proc. Calif. Acad. **5**: 55. 1873.
Conyzella coulteri Greene, Fl. Fran. 386. 1897.
Eschenbachia coulteri Rydb. Bull. Torrey Club **33**: 154. 1906.

Annual, usually 1-stemmed, 0.2–1 m. high, usually simple below the much-branched inflorescence, stipitate-glandular and spreading-pilose with many-celled hairs throughout (including involucres), very leafy. Basal leaves cuneate-obovate, up to 10 cm. long, sharply toothed, acute, tapering to the margined petiole; stem-leaves narrowly oblong, 1–6.5 cm. long, 2–20 mm. wide, sessile and somewhat clasping, sharply serrate to coarsely and doubly serrate, with a strong midrib; heads small, 4–5 mm. high in fruit, very numerous in a virgate or broad and pyramidal panicle; involucre 2–3 mm. high, about 2-seriate, subequal, the phyllaries linear, acuminate, with green center and whitish margin, the outer more herbaceous, the inner subscarious; pistillate flowers very numerous, their corollas whitish, about half as long as the style and pappus; perfect flowers 5–10; pistillate achenes elliptic-oblong, 0.7 mm. long, hispidulous, their pappus fragile, rather sparse, white, 3 mm. long; disk-achenes puberulous with glanduliform hairs.

Valley flats and plains, often in alkaline soil, frequently occurring as a weed, Upper and Lower Sonoran Zones; central California in Santa Clara, Santa Cruz, and Amador Counties south, including the Channel Islands, to San Diego County and Lower California, east to Colorado and Texas and south to Mexico. Type locality: not definitely stated. Collected by Coulter. April–Oct.

2. Conyza bonariénsis (L.) Cronquist. South American Conyza. Fig. 5656.

Erigeron bonariense L. Sp. Pl. 863. 1753.
Erigeron crispus Pourr. Mém. Acad. Toul. **3**: 318. 1788.
Erigeron undulatus Moench, Meth. 598. 1794.
Erigeron linifolium Willd. Sp. Pl. **3**: 1955. 1804.
Conyza ambigua DC. Fl. Franc. **6**: 468. 1815.
Conyza crispa Rupr. Bull. Acad. St. Pétersb. **14**: 235. 1856.
Conyzella linifolia Greene, Fl. Fran. 386. 1897.
Conyza bonariensis Cronquist, Bull. Torrey Club **70**: 632. 1943.

Annual, usually 5 dm. high or less, subsimple or with erect leafy branches as long as or longer than the main axis; stem and leaves subcinereously strigose or incurved-pubescent and also hirsute or hispid. Leaves linear to lance-linear, or the lower oblanceolate, mostly 8 cm. long or less, 8 mm. wide or less, the lower often laciniate-toothed, the upper entire; heads several or rather numerous, 5–6 mm. high, racemose or in a narrow panicle; involucre similar to that of *C. canadensis*, 4 mm. high, the phyllaries densely hirsutulous along middle; pistillate corollas very numerous, whitish, filiform, slightly dilated and acutely 3-toothed at the often purplish apex, at maturity much surpassing the style, equalling the pappus; perfect flowers about 8–25; achenes hirsutulous, 2-nerved; pappus simple, brownish white, rufescent in old specimens.

Waste places near towns; adventive at Portland, Oregon, and occurring more commonly from central California and the Bay Region to southern California, east to Florida and New Jersey, south to South America; introduced from South America. Type locality: Buenos Aires. June–Aug.

Conyza floribúnda H.B.K. Nov. Gen. & Sp. **4**: 73. 1820. (*Conyza albida* Willd. ex Spreng. Syst. **3**: 514. 1826; *Erigeron floribundus* Schultz-Bip. Bull. Soc. Bot. Fr. **12**: 81. 1865.) Lateral branches of the inflorescence not surpassing the main axis; pubescence resembling *C. bonariensis*; heads somewhat smaller than that species. Adventive at Portland, Oreland. Type locality: near Quito, Ecuador.

Conyza bilbaoàna Remy in Gay, Fl. Chil. **4**: 76. 1849. Differs from *C. canadensis* in having obliquely tubular outer flowers in the heads and in being perennial or at least persisting more than one season. Adventive in San Francisco County. Type locality: South America.

3. Conyza canadénsis (L.) Cronquist. Horseweed. Fig. 5657.

Erigeron canadensis L. Sp. Pl. 863. 1753.
Conyzella canadensis Rupr. Mém. Acad. St. Pétersb. VII. **14**: 51. 1869.
Leptilon canadense Britt. & Brown, Ill. Fl. **3**: 391. 1898.
Conyza canadensis Cronquist, Bull. Torrey Club **70**: 632. 1943.
Marsea canadensis Badillo, Bol. Soc. Venez. Cienc. Nat. **10**: 256. 1946.

Annual, strict, 2 m. high or less, simple below, with many erectish branches above, leafy into the inflorescence; stem and leaves (especially on margin) hispid or hirsute. Leaves numerous, linear to linear-lanceolate or the lower oblanceolate, 2–10 cm. long, usually 2–8 mm. wide, entire or the lower serrate, sessile or the lower subpetiolate, thin, feather-veined; heads small, 3–4.5 mm. high, abundant in a usually cylindric panicle; involucre graduate, 3–4 mm. high, the phyllaries linear or narrowly lance-linear, acute, glabrous or sparsely hispidulous, with conspicuous, central glandular area and narrow subscarious margin; rays numerous (not more than 20), white, short and inconspicuous, exceeding the disk by 1 mm. or less; achenes hirsutulous; pappus simple, dirty-whitish.

Cultivated and waste places, nearly throughout temperate North America south to South America; naturalized in Europe. Type locality: "Canada, Virginia." June–Sept. Hogweed.

Conyza canadensis var. **glabràta** (A. Gray) Cronquist, Bull. Torrey Club **74**: 150. 1947. (*Erigeron canadensis* var. *glabratus* A. Gray, Bost. Journ. Nat. Hist. **6**: 220. 1850.) Differs from *C. canadensis* var. *canadensis* in having the stems glabrous or nearly so. More common in our area than the typical form but having much the same distribution. Type locality: Texas.

94. BÁCCHARIS L. Sp. Pl. 860. 1753.

Dioecious shrubs, rarely only suffrutescent. Leaves alternate, entire or toothed, usually thick and firm. Heads discoid, small or medium, usually numerous and many-flowered.

Involucre few- to several-seriate, of usually thickened, dryish phyllaries. Receptacle flat, naked, very rarely conical or paleaceous. Pistillate corollas tubular-filiform, shorter than the style; hermaphrodite (functionally staminate) flowers with tubular, deeply 5-toothed corollas. Achenes of the pistillate flowers narrow, 4–10-ribbed, of the staminate flowers infertile. Pappus of the pistillate flowers of usually copious capillary bristles, often elongated in fruit; of the staminate achenes of stiff, scabrous, often twisted and apically dilated bristles. Anthers entire at base. Styles of the staminate flowers often 2-parted and hispidulous. [Named for Bacchus, without evident application.]

An American genus of about 300 species, the great majority of the species in South America. Type species, *Baccharis halimifolia* L.

Stems and branches distinctly puberulous or pubescent.
Stem finely puberulous; leaves entire, linear or squamiform. 1. *B. brachyphylla*.
Stem sordidly pilose; leaves sharply and closely toothed, linear-oblong or elliptic-oblong. 2. *B. plummerae*.
Stems and branches glabrous but often resinous-granular or glutinous.
Plants fastigiately much branched, with very numerous erect, broom-like branches, usually leafless at flowering time.
Pappus of the pistillate flowers short, about 3 mm. long, not surpassing the style; receptacle somewhat chaffy; larger leaves obovate. 3. *B. sergiloides*.
Pappus of the pistillate flowers in fruit 6–11 mm. long, much surpassing the style; receptacle naked; larger leaves linear. 4. *B. sarothroides*.
Plants not fastigiately much branched (except sometimes in *B. emoryi*), not broomlike, persistently leafy.
Leaves cuneate to oblong, 1-nerved or with a pair of weak lateral veins; heads partly or wholly in sessile or pedunculate axillary glomerules.
Inflorescence sparsely leafy, its leaves chiefly oblong-oblanceolate or cuneate-oblanceolate. 5. *B. emoryi*.
Inflorescence densely leafy, its leaves oval, obovate, or broadly cuneate. 6. *B. pilularis*.
Leaves linear-lanceolate to ovate, usually rather strongly triplinerved; heads in small or large panicles at tips of stems or branches.
Stems essentially herbaceous; leaves ovate to lanceolate; phyllaries thin; fertile achenes hirtellous. 7. *B. douglasii*.
Stems woody; leaves chiefly linear or linear-lanceolate; phyllaries thick and firm; fertile achenes glabrous.
Heads panicled at apex of stem. 8. *B. glutinosa*.
Heads mostly in panicles at tips of short, lateral, flowering branches. 9. *B. viminea*.

1. **Baccharis brachyphýlla** A. Gray. Short-leaved Baccharis. Fig. 5658.

Baccharis brachyphylla A. Gray, Smiths. Contr. 5⁶: 83. 1853.

Suffrutescent, intricately branched shrub less than 1 m. high, spreading-puberulous or hirtellous throughout, not glutinous, diffusely branched, slender; branches sulcate-striate. Larger leaves linear, 1.8 cm. long or less, acute, entire, sessile, those of the branches reduced and scale-like; heads numerous, loosely panicled; pistillate heads 1 cm. high in fruit, the involucre about 6 mm. high, about 3-seriate, graduate, the phyllaries lance-oblong or lanceolate, acute or acuminate, with narrow green center and whitish or pink-tinged, subscarious margin; staminate heads 4–5 mm. high, their involucres similar but shorter; fertile achenes about 2.5 mm. long, 4–5-nerved, sparsely puberulous, their pappus brownish-tinged, about 7 mm. long, rather copious, much exceeding the style.

Rocky ground, Lower Sonoran Zone; San Diego and southern Riverside Counties, California, east to Arizona and south to Sonora. Type locality: "Between Conde's Camp and the Chiricahui Mountains, in stony soil." Collected by Wright. Aug.–Nov.

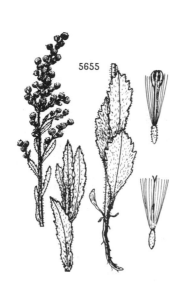

5655. Conyza coulteri
5656. Conyza bonariensis

2. Baccharis plúmmerae A. Gray. Plummer Baccharis. Fig. 5659.

Baccharis plummerae A. Gray, Proc. Amer. Acad. 15: 48. 1879.

Herbaceous from a woody base, 1 m. high or less, usually little branched below the inflorescence; stem striate, sordidly crisped-pilose. Leaves linear to elliptic-oblong or narrow-oblong, 1.5–4.5 cm. long, 3–12 mm. wide, acute to obtuse, narrowed at base, sessile, sharply and closely serrate, 3-nerved, glabrous above except for costa, densely crisped-pilose beneath; heads loosely or closely panicled; pistillate heads 1–1.2 cm. high in fruit, their involucres 6–8 mm. high, graduate, the phyllaries linear or linear-lanceolate, acute or acuminate, with narrow or broad, crisped-pilose, green center and ciliate subscarious margin; staminate heads 5–6 mm. high, their involucre similar but shorter; fertile achenes 3 mm. long, sordidly subglandular-puberulous, 4–5-nerved, their pappus becoming rufous, rather copious, 7 mm. long.

Mountain ravines, Upper Sonoran Zone; coast of Santa Barbara and Los Angeles Counties, California, and Santa Cruz Island. Type locality: Glen Loch ravine, near Santa Barbara, California. Aug.–Oct.

3. Baccharis sergiloìdes A. Gray. Squaw Waterweed or Desert Baccharis. Fig. 5660.

Baccharis sergiloides A. Gray in Torr. Bot. Mex. Bound. 83. 1859.

Erect shrub, 1–2 m. high, the stem and branches green and strongly striate-angled, the branchlets numerous, fastigiate, broom-like, essentially leafless or sometimes (especially the sterile ones) leafy. Larger leaves obovate, 1–3 cm. long, 3–8(12) mm. wide, entire, rarely with 1 or 2 teeth on the largest leaves, obtuse, apiculate, cuneate to the sessile base, thick, with strong costa and a pair of weaker nerves arising from base, the leaves of the branchlets mostly squamiform; heads numerous, densely panicled, sessile or pedunculate; pistillate heads 4–5 mm. high in fruit, their involucres 3 mm. high, about 5-seriate, strongly graduate, the phyllaries ovate or lance-ovate, obtuse to acutish, with greenish center and whitish margin; staminate heads 3.5–4 mm. high, with similar involucre; receptacle in both sexes sometimes conic, bearing few to many linear paleaceous bracts; pistillate corollas with 5–6 slender teeth; fertile achenes about 1.5 mm. long, glabrous, 10-ribbed, their pappus sparse, 2.5–3 mm. long, sordid, shorter than the style; pappus-bristles of staminate flowers dilated and subbarbellate at apex.

Dry creek beds and canyons, Lower Sonoran Zone; the Death Valley region, Inyo County, and the Mojave and Colorado Deserts, California, east to southwestern Utah and Arizona, and south to Sonora. Type locality: not definitely stated, but from the Colorado Desert region in Arizona or California. April–June.

4. Baccharis sarothroìdes A. Gray. Broom Baccharis or Hierba del Pasmo. Fig. 5661.

Baccharis sarothroides A. Gray, Proc. Amer. Acad. 17: 211. 1882.
Baccharis arizonica Eastw. Proc. Calif. Acad. IV. 20: 155. 1931.

Erect glabrous shrub 2–4 m. high with very dense, fastigiate, broom-like branches, nearly leafless; stem brown-barked, ridged, the branchlets green, strongly striate-angled. Leaves linear, 2 cm. long or less, 1–5 mm. wide, 1-nerved, those of the branchlets squamiform; heads numerous, mostly solitary at tips of short or long, naked, 4-angled branchlets, cymosely or racemose-paniculately arranged; pistillate heads 1–1.7 cm. high in fruit, their involucres oblong, (3)6–8 mm. high, strongly graduate, about 6-seriate, the phyllaries ovate (the inner linear, somewhat deciduous), obtuse, chiefly whitish and indurate, with greenish subapical spot; staminate heads 4–6 mm. high, their involucres similar, 2.5–4.5 mm. high; pistillate corollas obscurely denticulate; fertile achenes glabrous, 10-striate, 2–2.5 mm. long, their pappus copious, silky, rufidulous, 6–11 mm. long, greatly exceeding the styles; pappus-bristles of the staminate flowers dilated but not bearded at tip.

Washes and desert places, often in alkaline soil, Sonoran Zones; San Diego County, California, and south to Lower California, east to southwestern New Mexico and south to Sinaloa. Type locality: San Diego County, California. Feb.–Sept. Desert Broom.

5. Baccharis émoryi A. Gray. Emory Baccharis. Fig. 5662.

Baccharis emoryi A. Gray in Torr. Bot. Mex. Bound. 83. 1859.

Glabrous, somewhat resinous-granular, evergreen shrub 1–4 m. high, the branches erect, sometimes subfastigiate; stem brown-barked in age, multistriatulate, the branches green, striate, somewhat angled, like the stem usually persistently leafy. Larger leaves cuneate to lance-oblong, the upper mostly linear, 1.5–8 cm. long (including petiole, this 5 mm. long or less), 3–20 mm. wide, acute or obtuse, tapering to base, entire or usually sharply or sinuately few-toothed, pale green, thick, 1-nerved and with a more or less distinct pair of lateral veins; heads usually numerous, in glomerules of 1–5 at tips of axillary branchlets or spicately or racemosely clustered at tips of branches, the whole forming usually broad and pyramidal panicles; pistillate heads 1.2–1.5 cm. high, their involucres 5–8 mm. high, about 6-seriate, strongly graduate, the phyllaries ovate and obtuse (outer) to lance-linear and subacuminate (innermost), indurate and whitish, with or without greenish subapical spot or line; staminate heads 6 mm. high, the phyllaries nearly all obtuse, with more conspicuous greenish spot, the inner scarcely elongate; fertile achenes glabrous, about 10-nerved, about 1.5 mm. long, their pappus slightly rufescent, silky, 1 cm. long; pappus-bristles of staminate flowers with lacerate-bearded, somewhat dilated tips.

Along watercourses, Sonoran Zones; Kern County (Bakersfield), California, south to northern Lower California eastward to the Colorado Desert and southern Nevada, southern Utah, and Arizona to western Texas. Type locality: along the Gila River, Arizona. Collected by Emory. Aug.–Dec.

5657. Conyza canadensis
5658. Baccharis brachyphylla
5659. Baccharis plummerae

5660. Baccharis sergiloides
5661. Baccharis sarothroides
5662. Baccharis emoryi

6. **Baccharis pilulàris** DC. Dwarf Chaparral-broom. Fig. 5663.

Baccharis pilularis DC. Prod. **5**: 407. 1836.

Matted and widely spreading shrub 12–30 cm. high, the stems prostrate or nearly so, glabrous, resinous-granular; stems brown-barked, striate, the branchlets striate-angled, very leafy throughout. Leaves obovate-cuneate or angulate-oval, 1–4 (rarely 8) cm. long (including petiole, this 1–8 mm.' long or wanting), 0.4–1.5 (rarely 3) cm. wide, obtuse or sometimes acute, cuneate at base, repand-serrate above the entire base or the upper subentire, thick, 1-nerved and with an obscure or evident pair of lateral veins, somewhat feather-veined above; heads numerous, in sessile or pedunculate axillary glomerules and also spicate or clustered at tips of branches; pistillate heads in fruit 8–10 mm. high, their involucres 4–5 mm. high, graduate, the phyllaries ovate-oval and obtuse (outer) to linear (innermost), indurate, whitish, with subapical greenish spot and narrow, ciliolate, scarious margin; staminate heads 4–5 mm. high, their involucres 3–4 mm. high; fertile achenes glabrous, 10-ribbed, about 1.3 mm. long, their pappus rather sparse, rufidulous, 6–10 mm. long; pappus-bristles of staminate flowers with dilated lacerate tips.

Hill slopes, dunes and bluffs along the ocean, Humid Transition Zone; Sonoma County to Point Sur, Monterey County, California. Type locality: California. Collected by Douglas. Feb.–Aug.

Baccharis pilularis var. **consanguínea** (DC.) Kuntze, Rev. Gen. Pl. **1**: 319. 1891. (*Baccharis consanguinea* DC. Prod. **5**: 408. 1836; *B. pilularis* subsp. *consanguinea* C. B. Wolf, Occ. Papers Rancho Santa Ana Bot. Gard. No. 1: 21. 1935.) Much-branched evergreen shrub 2–4 m. high, differing from *B. pilularis* var. *pilularis* only in habit and of much more common occurrence than that form. Hills and canyons of the Outer Coast Ranges from Tillamook County, Oregon, south to northern San Diego County, California, and the islands off the coast of southern California; also occurring locally in the foothills of the central Sierra Nevada from Nevada County to Tuolumne County. Type locality: California. Collected by Douglas. Chaparral Broom or Coyote-brush.

7. **Baccharis douglásii** DC. Salt Marsh or Douglas' Baccharis. Fig. 5664.

Baccharis douglasii DC. Prod. **5**: 400. 1836.

Herbaceous essentially to base, 2 m. high or less, subsimple below the inflorescence or erect-branched, glabrous, resinous-glutinous, the stem and branches green, striate. Leaves ovate to lanceolate or the small upper ones linear, the larger 4–12 cm. long, 0.7–3 cm. wide, acute or acuminate, cuneate at base into the petiole (1 cm. long or less), entire or serrulate, rather thin, punctate, triplinerved; heads in small, close, cymose clusters at tips of stem and branches, usually aggregated into a panicle, the branches nearly naked above; pistillate heads 4–7 mm. high in fruit, their involucres 4–5 mm. high, about 4-seriate, graduate, the phyllaries lanceolate to linear-lanceolate, acute to acuminate, thin, with narrow greenish center and ciliate, whitish, scarious margins, sometimes purplish above; staminate heads about 5 mm. high, their phyllaries mostly oblong or lance-oblong; receptacle broadly conical; fertile achenes about 0.8 mm. long, hirtellous, about 5-nerved, their pappus rather sparse, 3–5 mm. long, dull whitish or at length rufidulous, the bristles hispidulous, about equaling the style; pappus-bristles of staminate flowers hispidulous, slightly widened apically.

Watercourses and thickets, often in salt marshes, Humid Transition and Upper Sonoran Zones; Humboldt County to San Diego County in the Coast Ranges and occurring less commonly in the Great Valley and the Sierra Nevada foothills in Amador County southward to western San Bernardino County, California; also in southern Curry County, Oregon. Type locality: California. Collected by Douglas. July–Dec.

8. **Baccharis glutinòsa** Pers. Water-wally or Seep-willow. Fig. 5665.

Baccharis glutinosa Pers. Syn. Pl. **2**: 425. 1807.
Baccharis coerulescens DC. Prod. **5**: 402. 1836.

Woody at least below, 1–3 m. high, forming thickets, the stems simple below the inflorescence or branched, very leafy, glabrous, glutinous, striate. Leaves lance-linear to narrow lanceolate or linear, 5–15 cm. long, 0.5–2.3 cm. wide, acuminate, tapering into the short petiole (8 mm. long or less), firm, sharply serrate to entire, strongly triplinerved, bright green, punctate; heads numerous in small or sometimes rather large cymes or cymose panicles at apex of stem and principal branches; pistillate heads 6–8 mm. high in fruit, their involucres 3–4.5 mm. high, about 4-seriate, graduate, the phyllaries ovate (outer) to lance-linear, obtuse to acute or subacuminate (innermost), stramineous, indurate, with narrow, cross-ciliolate, scarious margins, the midnerve sometimes obscurely greenish above; staminate heads 5–6 mm. high, their involucres similar; receptacle flat; fertile achenes glabrous, 4–5-nerved, about 0.8 mm. long, the pappus whitish, 4–5 mm. long, slightly exceeding the style; pappus-bristles of staminate flowers hispidulous, slightly dilated at apex.

Moist places and along watercourses, Sonoran Zones; Fresno County to Kern County in the San Joaquin Valley, California, south to Lower California; eastward in the Colorado and Mojave Deserts to Colorado and Texas and south to Mexico; also South America. Type locality: South America. May–Oct. Sticky Baccharis.

9. **Baccharis viminèa** DC. Mule Fat. Fig. 5666.

Baccharis viminea DC. Prod. **5**: 400. 1836.

Similar to *B. glutinosa* but more woody, up to 4 m. high, with numerous, short, lateral, flowering branches. Leaves more usually entire, rarely denticulate, less glutinous; heads in small cymes or cymose panicles at tips of the numerous lateral branches as well as at apex of stem; pistillate involucre 4–5 mm. high, about 5-seriate, graduate, the phyllaries ovate to (innermost) linear, obtuse to acute or subacuminate, usually purplish above, the outer often with distinctly green center; staminate heads similar; achenes and pappus similar.

Borders of streams and canyons, Sonoran Zones; Tehama and Butte Counties and in the Sacramento Valley and adjacent Coast Ranges and Sierra Nevada foothills south to San Diego County, California, the Channel Islands, and northern Lower California; occurring less commonly in western Arizona and southwestern Utah. Type locality: California. Collected by Douglas. Jan.–June.

Tribe 4. **ANTHEMIDEAE**

All flowers of the heads with corollas, all discoid or the marginal ones rayed.

Receptacle chaffy at least toward the center, with cartilaginous or membranous scales.

Rays present (absent in cultivated variety of *Anthemis nobilis*); annual or perennial herbs.

Heads solitary at the tips of the branches; rays elongate.   95. *Anthemis.*

Heads closely corymbiform; rays short and broad.   96. *Achillea.*

Rays absent; subshrubs.   97. *Santolina.*

Receptacle naked (with slender hairs in *Tanacetum potentillioides*; scaly in *Artemisia palmeri*).

Pappus (in ours) paleaceous and conspicuous.

Flowering heads yellow; pappus-paleae not awn-tipped.   98. *Hymenopappus.*

Flowering heads white or purplish; pappus-paleae awn-tipped.   99. *Hymenothrix.*

Pappus absent or a short crown.

Heads solitary or in corymbiform inflorescences.

Receptacle flat or somewhat convex.

Rays conspicuous, white or yellow (pistillate flowers lacking and thus rayless in *Chrysanthemum balsamita*).   100. *Chrysanthemum.*

Rays absent or very small in proportion to the head.   101. *Tanacetum.*

Receptacle conspicuously hemispheric or conical.   102. *Matricaria.*

Heads in spiciform, racemose, or paniculate inflorescences.   105. *Artemisia.*

Marginal flowers only of the heads without corollas.

Heads peduncled; marginal flowers pedicellate on the receptacle; style deciduous, only the inconspicuous base remaining.   104. *Cotula.*

Heads sessile; marginal flowers sessile; style persisting as a conspicuous stout spine.   103. *Soliva.*

5663. Baccharis pilularis
5664. Baccharis douglasii

5665. Baccharis glutinosa
5666. Baccharis viminea

95. ÁNTHEMIS L. Sp. Pl. 893. 1753.

Annual or perennial, aromatic herbs, usually with leafy stems. Leaves alternate, laciniately incised to pinnately or bipinnately parted. Heads solitary at the ends of the branches, radiate (in ours), the rays white or yellow. Involucre saucer-shaped, the short phyllaries many, subequal or imbricated, dry and scarious. Receptacle convex, hemispheric or conical, receptacular bracts present at least at the summit, hyaline, chaffy or cartilaginous, slender or broader, sometimes enclosing the achene. Ray-flowers pistillate, fertile or sterile. Disk-flowers perfect, tube cylindric or compressed, throat narrowly funnelform to campanulate, lobes short. Style-branches truncate or obtuse, sometimes penicillate. Anthers entire at base, ovate, acute, or obtuse at apex. Achenes ellipsoid, terete or obcompressed, ribbed, nearly smooth or tuberculate. Pappus wanting or a short crown. [Ancient name of the chamomile.]

A genus of about 60 species, natives of Europe, Asia, and Africa. Type species, *Anthemis nobilis* L.

Annuals; rays white.
    Receptacular bracts subulate or awn-tipped, firm.
        Bracts of the receptacle present only toward the middle; ribs of the achenes tuberculate.   1. *A. cotula.*
        Receptacle bracteate throughout; ribs of the achenes smooth.   2. *A. arvensis.*
    Receptacular bracts oblong, hyaline.   3. *A. fuscata.*
Perennials; rays yellow.   4. *A. tinctoria.*

1. **Anthemis cótula** L. Mayweed. Fig. 5667.

Anthemis cotula L. Sp. Pl. 894. 1753.
Anthemis foetida Lam. Fl. Franc. 2: 164. 1778.
Maruta vulgaris Bluff & Fingerh. Comp. Fl. Germ. 2: 392. 1825.
Maruta cotula DC. Prod. 6: 13. 1837.
Matricaria pubescens Schultz-Bip. Bonplandia 8: 369. 1860

Ill-scented, leafy, erect annual 1–6 dm. high, usually much branched. Leaves 2–6 cm. long, sparsely hairy, bi- or tripinnatifid with linear divisions; heads rather numerous, short-cylindric at maturity, solitary at the ends of the branches, the peduncles short, 3–6 cm. long; phyllaries loosely short-villous, with a yellowish or greenish, thickened midsection and hyaline margins; receptacle conic, the receptacular bracts present only above the middle, slender, subulate, firm, about 2.5 mm. long; ray-flowers sterile, the rays 6–11 mm. long, oblong; disk-flowers tubular-funnelform; achenes 1–1.5 mm. long, obscurely angled, with about 10 tuberculate ribs; pappus none.

An introduced European weed very common in disturbed areas and waste places in the Pacific States; also common throughout the United States. May–Sept. Dog Fennel. Stinking Chamomile.

2. **Anthemis arvénsis** L. Field Chamomile. Fig. 5668.

Anthemis arvensis L. Sp. Pl. 894. 1753.

Leafy, more or less branching, villous-puberulent annuals 1–6 dm. high, often decumbent at the base, not ill-scented. Leaves 3–5 cm. long, bipinnatifid, equaling or shorter than the internodes; heads solitary at the ends of the branches, the peduncles 4–11 cm. long, phyllaries villous-tomentose, narrowly hyaline-margined; receptacle low-conic, chaffy throughout, the receptacular bracts persistent, equaling or shorter than the disk-flowers, narrowly oblanceolate and awn-tipped; ray-flowers pistillate, 15–20, the rays white, 7–13 mm. long; disk-flowers tubular-funnelform; achenes 1.5–2 mm. long, quadrangular and broader above than below, 8–10-ribbed, the ribs smooth; pappus a minute crown or none.

A native of Europe not uncommon in cultivated land and waste places west of the Cascade Mountains in Washington and Oregon and occurring in Whitman County, Washington, and Garberville, Humboldt County, California; also established in eastern United States. May–July.

3. **Anthemis fuscàta** Brot. Dusky Chamomile. Fig. 5669.

Anthemis fuscata Brot. Phyt. Lusit. 1: 15. *pl. 28.* 1816.
Maruta fuscata DC. Prod. 6: 14. 1837.

Erect annual 1.5–2.5 dm. high, branching from the base, glabrous or sparsely puberulent. Leaves 1.5–4 cm. long, pinnately parted to the midrib, the narrow divisions pinnatifid into few linear acute lobes; heads rather few, solitary at the ends of the branches; phyllaries about 3-seriate, ovate-oblong, the thickened greenish midportion triangular in outline, the wide outer part hyaline, dark brown; receptacle hemispheric, the receptacular bracts 3 mm. or more long, about equaling the disk-flowers, oblong, hyaline, tinged with brown; rays 12–15 mm. long, narrowly oblong, white with a yellowish base; disk-flowers tubular; achenes about 1 mm. long, ellipsoid, finely striate; pappus none.

A native of the Mediterranean region well established near Asti, Sonoma County, California. March–April.

Anthemis nòbilis L. Sp. Pl. 894. 1753. Aromatic, mat-forming, villous perennial; leaves bipinnatifid with finely dissected divisions; heads borne on naked peduncles; phyllaries with broad scarious margins; rays white; receptacle conic, the receptacular bracts hyaline, concave, oblong, obtuse; achenes 1–1.5 mm. long, turbinate, obtusely 3-angled; pappus none. A European plant much grown in gardens. It has been collected at Fort Bragg, Mendocino County, California. A discoid form of *A. nobilis* is used extensively as a ground cover and perhaps may be found as a garden escape.

4. Anthemis tinctòria L. Yellow Chamomile. Fig. 5670.

Anthemis tinctoria L. Sp. Pl. 896. 1753.
Cota tinctoria J. Gay ex Guss. Fl. Sic. Syn. **2**: 867. 1844.

Erect perennial 3–9 dm. high, simple-stemmed or sparingly branched, whitish throughout with a villous pubescence. Leaves 2–5 cm. long, about equalling the internodes, evenly pinnately divided nearly to the midrib, the segments incised; heads large, the peduncles 6–12 cm. long; phyllaries lanceolate, the inner obtuse, scarious-margined; receptacle low, hemispheric, the receptacular bracts firm, rather narrow, abruptly awned; ray-flowers pistillate, the rays 20–30, about 10 mm. long, elliptic, yellow; tube of disk-corollas somewhat compressed, the throat funnelform; achenes 2 mm. long, quadrangular, smooth to striate; pappus a short crown.

A native of Europe cultivated in gardens and occasionally escaping; Prattville, Plumas County, and Berkeley, Alameda County, California, and British Columbia; also sparingly naturalized in eastern United States. July–Aug. Yellow Marguerite.

Anthemis míxta L. Sp. Pl. 894. 1753. (*Ormenis mixta* Dumort. Fl. Belg. 69. 1827; *O. bicolor* Cass. Dict. Sci. Nat. **36**: 356. 1825.) Fragrant annual; stems 1–4 dm. long, leafy, divaricately spreading; leaves coarsely pinnatifid with entire or toothed divisions; ray-flowers yellow, perfect; disk-flowers with cylindric tube with a spur-like basal appendage and a campanulate throat; achenes about 1 mm. long, ellipsoid, 3-ribbed on the inner side. A native of Italy that has been collected on ballast at Portland, Oregon, and at Nanaimo, Vancouver, Island, British Columbia.

Anthemis altíssima L. Sp. Pl. 893. 1753. (*Cota altissima* J. Gay ex Guss. Fl. Sic. Syn. **2**: 867. 1844.) A European annual, much resembling *A. tinctoria* but differing in the bipinnate leaves and white ray-flowers, has been reported from Oregon.

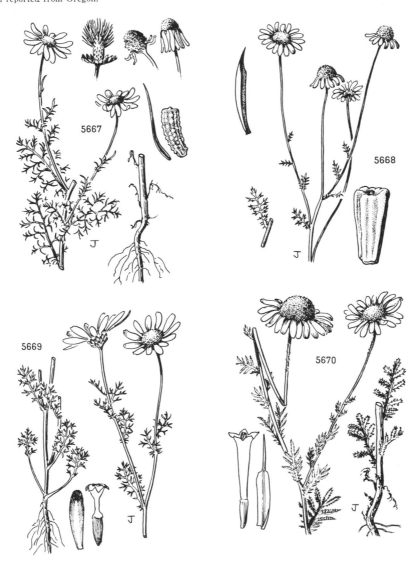

5667. Anthemis cotula
5668. Anthemis arvensis

5669. Anthemis fuscata
5670. Anthemis tinctoria

96. ACHILLÈA* L. Sp. Pl. 896. 1753.

Perennial herbs with serrate to tripinnatifid, alternate leaves. Heads several to many, aggregated into terminal corymbiform panicles, generally radiate. Involucre campanulate to ovoid; bracts imbricate with broad, scarious, fimbriate margins, in 3–4 unequal series. Receptacle conic to convex, with chaffy scales subtending and nearly equaling the disk-flowers. Ray-florets pistillate and fertile; disk-florets perfect. Achenes oblong to ovate, strongly obcompressed, callous-margined, glabrous; pappus none. [Name in honor of Achilles.]

A genus of perhaps 75 species native to the northern hemisphere, mainly in the Old World. Type species, *Achillea santolina* L.

For detailed information on the biological variation and cytology of the *Achillea millefolium* complex, see the following references:

Clausen, J., David D. Keck, and William Hiesey. 1940. Experimental studies on the nature of species. I. Effect of varied environments on western North American plants. Carnegie Inst. Wash. Publ. No. 520: 296–324.

Clausen, J., David D. Keck, and William Hiesey. 1948. Experimental studies on the nature of species. III. Environmental responses of climatic races of *Achillea*. Carnegie Inst. Wash. Publ. No. 581: 1–129.

Lawrence, W. E. 1947. Chromosome numbers in *Achillea* in relation to geographic distribution. Amer. Journ. Bot. **34**: 538–45.

Turesson, Göte. 1939. North American types of *Achillea millefolium* L. Bot. Notiser **1939**: 813–16.

1. Achillea millefòlium L. Common Yarrow or Milfoil. Fig. 5671.

Achillea millefolium L. Sp. Pl. 899. 1753.
Alitubus millefolium Dulac, Fl. Hautes-Pyr. 500. 1867.
Achillea borealis Bong. Mem. Acad. St. Pétersb. VI. **2**: 149. 1832.
Achillea lanulosa Nutt. Journ. Acad. Phila. **7**: 36. 1834.

Plants aromatic, rhizomatous to cespitose. Leaves tripinnately dissected, the basal petioled, the cauline auriculate and sessile; heads numerous, radiate, aggregated into flat to round-topped, corymbiform panicles; rays few, 3–8, broadly ovate to orbicular, generally white, occasionally pink; disk-flowers 15–40, of the same color as the rays.

A highly polymorphic circumboreal species found in habitats from marshes to semideserts, and from coastal bluffs to arctic and alpine conditions. In North America the species is represented by two chromosome forms: the tetraploid, $n=18$ ($=A. lanulosa$ Nutt.), and the hexaploid, $n=27$ ($A. borealis$ Bong.). Because these chromosome types show parallel morphological variation, are indistinguishable on any criteria other than cytology, and intergrade into European forms of *A. millefolium* L., they are here considered to be units of this single large polymorphic species. In the Pacific States, the species is represented by 10 recognizable, though continuously intergrading, ecogeographic forms here designated as varieties. (The foregoing description covers all the forms of the species listed below.)

Key to Varieties

Stems stout, 10–20 dm. high.
  Plants cespitose from a deep taproot; herbage gray, villous. var. *gigantea.*
  Plants rhizomatous, with fibrous roots forming a buttress at the base of the stem; herbage dark green, puberulent to sparsely villous. var. *puberula.*
Stems generally less than 10 dm. high.
  Leaf-segments thick and fleshy, the terminal ones ovate to ovate-lanceolate, densely congested.
    Herbage dark green, sparsely villous. var. *litoralis.*
    Herbage gray, densely white-villous. var. *arenicola.*
  Leaf-segments thin, not fleshy, oblanceolate to acicular.
    Herbage gray, villous to woolly.
      Margins of phyllaries light brown to black; roots fibrous from slender rhizomes.
         var. *alpicola.*
      Margins of phyllaries light brown to straw-colored; nearly cespitose, with short stout rhizomes and a deep central taproot. var. *lanulosa.*
    Herbage usually green, moderately villous.
      Margins of involucral bracts dark brown to black; stems tawny-villous; leaf-segments acicular.
         var. *borealis.*
      Margins of involucral bracts pale greenish white to straw-colored.
        Leaf-segments oblanceolate, all segments tending to be oriented in the same plane.
           var. *millefolium.*
        Leaf-segments linear, spreading and ascending.
          Involucre generally 5–6 mm. high. var. *pacifica.*
          Involucre generally 6–9 mm. high. var. *californica.*

Achillea millefolium var. **lanulòsa** (Nutt.) Piper in Piper & Beattie, Fl. Palouse Reg. 196. 1901. (*Achillea tomentosa* Pursh, Fl. Amer. Sept. 563. 1814, not *A. tomentosa* L. 1753; *A. lanulosa* Nutt. Journ. Acad. Phila. **7**: 36. 1834; *A. millefolium lanulosa* Piper, Mazama **2**: 97. 1901; *A. millefolium* [subsp.] *lanulosa* Piper, Contr. U.S. Nat. Herb. **11**: 584. 1906; *A. lanulosa* subsp. *typica* Keck, Carnegie Inst. Wash. Publ. No. 520: 299. 1940.) Stems 3–5 dm. high, arising from short stout rhizomes (nearly cespitose); herbage gray, woolly; leaves 5–15 cm. long, linear-lanceolate, finely dissected into linear segments, the ultimate ones tapering into prominent spine-tips; heads numerous, in dense, compound, flat-topped corymbs, the phyllary-margins light brown. Upper Sonoran and Arid Transition Zones; on the eastern side of the Cascade Mountains and Sierra Nevada eastward to the Rocky Mountains and from southern British Columbia to northern Mexico. Type locality: "banks of the Kooskoosky" [Clearwater River, Idaho]. May–Aug. $n=18$.

Achillea millefolium var. **alpícola** (Rydb.) Garrett, Spring Fl. Wasatch Reg. 101. 1911. (*Achillea lanulosa alpicola* Rydb. Mem. N.Y. Bot. Gard. **1**: 426. 1900; *A. subalpina* Greene, Leaflets Bot. Obs. **1**: 145. 1905; *A. alpicola* Rydb. Bull. Torrey Club **33**: 157. 1906; *A. fusca* Rydb. N. Amer. Fl. **34**: 221. 1916; *A. borealis* Bong. in Jepson, Man. Fl. Pl. Calif. 1137. 1925 (of Jepson, not Bong. 1832); *A. millefolium* var. *fusca* G. N. Jones, Univ. Wash. Publ. Biol. **5**: 250. 1936; *A. lanulosa* subsp. *alpicola* Keck, Carnegie Inst. Wash. Publ. No. 520: 300. 1940.) Stems simple and erect, 0.5–3 dm. high, arising from long slender rhizomes; roots shallow, fibrous, distributed over the whole rhizome system; herbage gray, white-villous; leaves 5–10 cm.

---

* Text contributed by Malcolm Anthony Nobs.

long, linear-lanceolate, finely dissected into linear-lanceolate segments, prominently spine-tipped; heads few, 5–7 mm. high, in dense, flat-topped corymbs, the margins of phyllaries dark brown to black. Canadian Zone to Arctic-Alpine Zone; Alaska southward to Inyo and Tulare Counties, California, eastward to the Rocky Mountains. Type locality: Teton Forest Reserve, Wyoming. July–Oct. $n=18$.

Achillea millefolium var. **pacífica** (Rydb.) G. N. Jones, Univ. Wash. Publ. Biol. **5**: 250. 1936. (*Achillea pacifica* Rydb. N. Amer. Fl. **34**: 222. 1916.) Stems stout, 3–10 dm. high from long slender rhizomes; roots fibrous from the entire rhizome system; herbage green, moderately villous; leaves oblanceolate, 15–35 cm. long, dissected into fine, linear, spine-tipped segments; heads many, 5–6 mm. long, in loose, broad, round-topped corymbs, the phyllary-margins hyaline to light straw color. Interior area, Arid and Humid Transition Zones; Alaska south through British Columbia, Washington, and Oregon along the western slopes of the Cascade Mountains and in the Coast Ranges of Humboldt County, California, extending eastward along the western slopes of the Sierra Nevada south to San Diego County. Type locality: Grenada, Siskiyou County, California. May–Sept. $n=18$. A variable taxon containing many forms, some indistinguishable from form of var. *californica*.

Achillea millefolium var. **pubérula** (Rydb.) Nobs in Ferris, Contr. Dudley Herb. **5**: 102. 1958. (*Achillea puberula* Rydb. N. Amer. Fl. **34**: 223. 1916.) Stems stout, 10–18 dm. high, branched, arising from long slender rhizomes; roots fibrous, concentrated into a dense buttress at the base of the stems; herbage dark green, sparsely villous to puberulent; leaves linar-lanceolate, 10–20 cm. long, finely dissected into linear, spine-tipped segments; upper surface nearly glabrous; heads 5–7 mm. high in dense, compound, flat-topped corymbs, the phyllary-margins greenish white. Salt and brackish marshes, Humid Transition Zone; San Francisco Bay region in Marin, Solano, and Contra Costa Counties, California. Type locality: Suisun Marsh, Solano County. July–Aug. $n=18$.

Achillea millefolium var. **litoràlis** Ehrendorfer ex Nobs in Ferris, Contr. Dudley Herb. **5**: 101. 1958. Stems stout, 1–4 dm. high, ascending to decumbent, arising from well-developed rhizomes; roots essentially fibrous from the rhizomes but retaining a strong central taproot; herbage dark green, sparsely villous throughout; leaves abundant, congested, 10–15 cm. long, oblanceolate, primary pinnae borne at right angles to the rachis, ultimate segments highly congested, ovate, very fleshy, tapering into prominent callous spine-tips; heads many, 4–6 mm. high, densely aggregated into broad, round-topped corymbs, the phyllary-margins greenish white. Exposed coastal bluffs, Curry and Coos Counties, Oregon. Type locality: on bluffs south of Port Orford, Curry County. May–Sept. $n=18$. A maritime form of the tetraploid closely allied to var. *pacifica*, resembling var. *arenicola* in-growth habit.

Achillea millefolium var. **califórnica** (Pollard) Jepson, Man. Fl. Pl. Calif. 1137. 1925. (*Achillea californica* Pollard, Bull. Torrey Club **26**: 369. 1899; *A. millefolium* f. *californica* H. M. Hall, Univ. Calif. Pub. Bot. **3**: 211. 1907; *A. borealis* subsp. *californica* Keck, Carnegie Inst. Wash. Publ. No. 520: 299. 1940; *A. borealis* var. *californica* J. T. Howell, Leaflets West Bot. **5**: 107. 1948.) Stems stout, 5–12 dm. high, arising from thick, well-developed rhizomes; root-system a well-developed taproot, secondary fibrous roots from the rhizomes; herbage green, occasionally grayish, moderately villous; leaves 10–30 cm. long, linear-lanceolate, finely dissected into slender, linear, spine-tipped segments; heads numrous, up to 9 mm. high, in many-branched, irregular, loose corymbs, the phyllary-margins light brown. Coastal regions, Upper Sonoran and Humid Transition Zones; Grays Harbor County, Washington, south to Lower California and eastward to the foothills of the Sierra Nevada in central California (Eldorado County). Type locality: sea coast at Santa Ysabel, San Diego County. March–July. $n=27$.

Achillea millefolium var. **arenícola** (Heller) Nobs in Ferris, Contr. Dudley Herb. **5**: 102. 1958. (*Achillea arenicola* Heller, Muhlenbergia **1**: 61. 1904: *A. millefolium* var. *maritima* Jepson, Man. Fl. Pl. Calif. 1137. 1925; *A. borealis* subsp. *arenicola* Keck, Carnegie Inst. Wash. Publ. No. 520: 299. 1940.) Stems stout, 1–6 dm. high, ascending to decumbent, arising from well-developed rhizomes; root-system mainly fibrous from the rhizomes but retaining a strong central taproot; herbage gray, densely white-villous throughout; leaves oblanceolate, 10–20 cm. long, the primary pinnae at right angles to the rachis, secondary and tertiary pinnae densely crowded, ovate to oblanceolate, fleshy, the tips mucronate; heads 5–6 mm. high, densely packed in round-topped corymbs, the phyllary-margins light brown. Coastal dunes and sea bluffs; Del Norte County south to Monterey County, California. Type locality: sand hills at the upper edge of Bodega Bay, Sonoma County. March–Nov. $n=27$.

Achillea millefolium var. **gigantèa** (Pollard) Nobs in Ferris, Cont. Dudley Herb. **5**: 102. 1958. (*Achillea gigantea* Pollard, Bull. Torrey Club **26**: 370. 1899.) Stems robust, 10–20 dm. high, generally branched, arising from a cespitose base; roots stout, a deep vertical taproot system; herbage gray, villous throughout; leaves 15–40 cm. long, linear-lanceolate, finely dissected into linear-lanceolate segments, with white mucronate tips; heads 5–6 mm. high, numerous, in congested, many-branched, round-topped corymbs, the phyllary-margins pale greenish white. On borders of marshes and watercourses, Lower Sonoran Zone; Fresno, Kings, Tulare, and Kern Counties, California. Type locality: near Tulare Lake, Tulare County. May–July. $n=27$.

Achillea millefolium L. var. *millefolium*. Stems delicate. 2–6 dm. high, arising from slender rhizomes; roots fibrous, arising from the entire rhizome system; herbage sparingly villous; leaves linear-lanceolate, 5–12 cm. long, the ultimate leaf-segments short, broadly oblanceolate, all tending to be oriented in the same plane, the tips mucronate; heads numerous. 4–5 mm. long, in loose, flat-topped corymbs, the phyllary-margins pale. Native of northern Europe, reputed to be adventive in waste lands and fields; west of the Cascade Mountains in Washington and Oregon, possibly also in California. Type locality: "Europe." $n=27$.

Achillea millefolium var. **boreàlis** (Bong.) Farwell, Amer. Midl. Nat. **11**: 268. 1929. (*Achillea borealis* Bong. Mem. Acad. St. Pétersb. VI. **2**: 149. 1832; *A. borealis* subsp. *typica* Keck, Carnegie Inst. Wash. Publ. No. 520: 299. 1940.) Stems 1–4 dm. high, arising from slender rhizomes; roots shallow, fibrous, over the entire rhizome system; herbage dark green, the stems tawny-villous; leaves linear-lanceolate, 5–20 cm. long, finely dissected into acicular, spine-tipped segments; heads 5–6 mm. high, in dense, round-topped corymbs, the phyllary-margins dark brown to nearly black. A maritime variety occurring from the Aleutian Islands and Alaska south to coastal British Columbia; may later be found possibly in western Washington. Type locality: island of Sitka, Alaska. $n=27$. This variety is commonly confused in the literature with the tetraploid, high mountain var. *alpicola* (Rydb.) Garrett from which it may be distinguished by its darker green leaves.

97. **SANTOLÌNA** L. Sp. Pl. 842. 1753.

Aromatic, glabrous or tomentose shrubs or perennial herbs. Leaves alternate, pinnately toothed, lobed, or finely parted. Heads solitary at the ends of the branches, eradiate, many-flowered, the flowers yellow, rarely white, only disk-flowers present. Involucre campanulate, the phyllaries scarious on the margins or throughout, imbricated in 2–4 series. Receptacle convex, chaffy. Disk-flowers tubular with a gradually expanding throat, 5-lobed. Anthers entire at base. Style-branches flattened, truncate, penicillate. Achenes 3–5-angled, glabrous. Pappus lacking. [Derivation of name obscure.]

A genus of about 8 species, natives of the Mediterranean region. Type species, *Santolina chamaecyparissus* L.

1. **Santolina chamaecyparíssus** L. Lavender Cotton. Fig. 5672.

Santolina chamaecyparissus L. Sp. Pl. 842. 1753.
Santolina incana Lam. Fl. Franc. **2**: 43. 1778.

Small, much-branched, compact shrubs 2–5 dm. high, silvery short-tomentose throughout, the stems striate. Leaves persistent, 1–2 cm. long, 2–3 mm. wide, pectinately pinnatifid into minute or short rounded segments; heads yellow, globular, 8–17 mm. in diameter, on peduncles 3–9 cm. long; phyllaries 3-seriate, 2.5–3 mm. long, lanceolate, strongly ribbed; receptacle convex, the receptacular bracts cartilaginous with a conspicuous midrib, loosely villous-tomentose on the obtuse tips; disk-flowers about 2 mm. long, funnelform, the lobes lanceolate; achenes 2.5 mm. long or a little longer, 5-angled, glabrous; pappus none.

A Mediterranean plant much grown in gardens and occasionally escaping and becoming locally established; Sonoma and Monterey Counties, California. June–Aug.

98. HYMENOPÁPPUS L'Her. Diss. Hymenop. 1. 1788.

Perennial, biennial, or sometimes annual herbs, sparingly to densely tomentose, often glabrate. Leaves basal and alternate, simple to bipinnately compound. Inflorescence corymbiform. Heads medium to large, discoid or in some species radiate. Phyllaries herbaceous, in 2 or 3 subequal series, conspicuously scarious-margined, often tinged with yellow or red, spreading or reflexed in age. Receptacle smooth, flat. Disk-flowers perfect, yellow or white, the tube slender, throat abruptly expanded, lobes reflexed. Anthers minutely sagittate. Style-branches with obtuse flattened lobes, the tips papillate. Achenes obpyramidal, pubescent, striate and 4-angled, somewhat compressed and contorted in the marginal flowers. Pappus-paleae several to many, often vestigial, hyaline or membranous, median nerve obscure or lacking, never produced into an awn. [From the Greek words meaning membrane and pappus.]

A genus of 10 species, all natives of temperate North America. Type species, *Hymenopappus scabiosaeus* L'Her.

1. Hymenopappus filifòlius* Hook. Columbia Cutleaf. Fig. 5673.

Hymenopappus filifolius Hook. Fl. Bor. Amer. 1: 317. 1833.
Hymenopappus columbianus Rydb. N. Amer. Fl. 34: 52. 1914.

Perennial from a taproot, 3–10 dm. high, the thinly or densely tomentose stems erect with ascending branches, several arising from the stout, branched, densely leafy, tufted caudex. Basal leaves many, 10–20 cm. long including the long petiole, dissected to the midrib into linear-filiform divisions about 10–50 mm. long and rather crowded, sometimes bipinnatifid, thinly tomentose but the broadened leaf-bases and axils conspicuous with a persistent tomentum; upper leaves very few, similar but reduced in size, usually sessile; inflorescence corymbiform, the heads 20–45-flowered, several to many, on peduncles 1–15 cm. long; involucres 6–10 mm. high, broader than long at anthesis; phyllaries obovate, spreading or reflexed in age, thin with a broad, yellowish, hyaline margin and tip; flowers yellow, 2.5–4.5 mm. long, tube densely glandular-pubescent, as long as or surpassing the abruptly expanded, more or less cylindric throat, three to five times longer than the recurved lobes; corolla appearing urceolate in age, anthers 2–3 mm. long; achenes 4.5 mm. or more long, with densely silky, hirsute pubescence which almost obscures the minute pappus-paleae.

Dry washes and sagebrush slopes, Arid Transition Zone; Columbia River drainage in central Washington and central Oregon. Type locality: near Walla Walla, Walla Walla County, Washinton. Collected by Douglas. May–June. Low Tufted Hymenopappus.

Hymenopappus filifolius var. eriopòdus (A. Nels.) B. L. Turner, Rhodora 58: 225. 1956. (*Hymenopappus eriopoda* A. Nels. Bot. Gaz. 37: 274. 1904.) Plants from a stout woody caudex, sparsely leafy as in *H. filifolius* var. *filifolius*, 4–8 dm. high. Leaves 10–20 cm. long, the linear divisions of the blade not greatly crowded, green and glabrous or nearly so, the leaf-axils densely woolly-tufted; heads few, long-peduncled; phyllaries 7–10 mm. long; flowers 4–5 mm. long, cream-colored; anthers 3–4 mm. long. In calcareous soil, dry rocky slopes; Providence and Clark Mountains, San Bernardino County, California, eastward through the mountains of Nevada to southern Utah. Type locality: Diamond Valley, Utah.

Hymenopappus filifolius var. lùgens (Greene) Jepson, Man. Fl. Pl. Calif. 1128. 1925. (*Hymenopappus lugens* Greene, Pittonia 4: 43. 1899.) Scapose plants from a stout woody caudex, the nearly naked stems floccose, becoming glabrate, 2.5–6 dm. high; leaves floccose-tomentose, becoming glabrate, 6–9 cm. long, linear divisions of blade crowded, the leaf-axils densely woolly-tufted; heads few, usually long-peduncled; corollas 4–6 mm. long, yellow; anthers 3–4 mm. long; phyllaries 6–10 mm. long, the outer much shorter and usually reddish. Dry rocky slopes, San Bernardino Mountains, San Bernardino County, California, south through the mountains of San Diego County to northern Lower California; also in southwestern Utah, Nevada in the Virgin Mountains, and in Arizona. Type locality: San Bernardino Mountains.

Hymenopappus filifolius var. megacéphalus B. L. Turner, Rhodora 58: 227. 1956. Habit as in the preceding taxa; leaves 8–20 cm. long, the divisions of the leaf-blades rather woolly, not crowded, linear but not at all filiform; phyllaries 8–14 mm. long; corollas 4–7 mm. long; anthers 3–4 mm. long. In sandy or gravelly desert soil, western Colorado and southern Utah and adjacent eastern Arizona to Clark County, Nevada, and the Providence Mountains, San Bernardino County, California. Type locality: Las Vegas, Nevada.

Hymenopappus filifolius var. nànus (Rydb.) B. L. Turner, Rhodora 58: 240. 1956. (*Hymenopappus nanus* Rydb. N. Amer. Fl. 34: 53. 1914.) Usually scapose plants from a stout, branched, woody caudex, the naked, white-floccose stems 1–3 dm. high, bearing long-peduncled heads; leaves densely white-tomentose throughout, 2–6 cm. or more long, linear divisions of blade crowded, leaf-axils densely woolly-tufted; phyllaries 6–9 mm. long; corollas 3–4 mm. long, light yellow; anthers 2–3 mm. long. Occurring in calcareous soil in the pinyon-juniper belt; Inyo Mountains, Inyo County, California, and in Mohave County, Arizona, eastward in the desert ranges of central Nevada to Utah. Type locality: Cave Creek Post Office, Elko County, Nevada.

99. HYMENÓTHRIX A. Gray, Mem. Amer. Acad. II. 4: 102. 1849.

Slender, erect, annual or perennial herbs. Leaves alternate, petioled, biternately or triternately compound. Inflorescence cymose-panicled with numerous heads. Heads yellow,

---

* Treatment based upon the recent monograph of Billie Lee Turner.

purple, or white, medium, discoid or radiate. Phyllaries in 2 unequal series, thin, obtuse, with a broad, white or pinkish margin, spreading in age and often deciduous. Receptacle small, flat. Ray-flowers when present pistillate. Disk-flowers perfect, fertile, zygomorphic, tube slender, glandular, throat campanulate, much shorter than to nearly equaling the narrow spreading lobes, these of unequal length. Anthers rounded at base, with ovate appendages above. Style-branches flat, linear, with short conic tips. Achenes obpyramidal, 4–5-angled, more or less pubescent. Pappus-paleae several, hyaline, with a prominent midrib prolonged into an awn. [From the Greek words meaning membrane and bristle.]

A genus of 5 or 6 species, natives of southwestern United States and Mexico. Type species, *Hymenothrix wislizenii* A. Gray.

1. **Hymenothrix wrìghtii** A. Gray. Wright's Hymenothrix. Fig. 5674.

Hymenothrix ? wrightii A. Gray, Smiths. Contr. 5[6]: 97. 1853.
Hymenopappus wrightii H. M. Hall, Univ. Calif. Pub. Bot. **3**: 179. 1907.
Trichymenia wrightii Rydb. N. Amer. Fl. **34**: 56. 1914.
Hymenopappus wrightii var. *viscidulus* Jepson, Man. Fl. Pl. Calif. 1128. 1925.

Perennial with several erect reddish stems from a woody root, 2–6 dm. high, conspicuously leafy to midstem, the internodes short, more or less spreading-hirsute below and sometimes glandular-viscid, nearly glabrous above except for the sparsely or densely glandular branches of the corymbose-paniculate inflorescence. Leaves 3–4.5 cm. long, bi- or triternately divided into

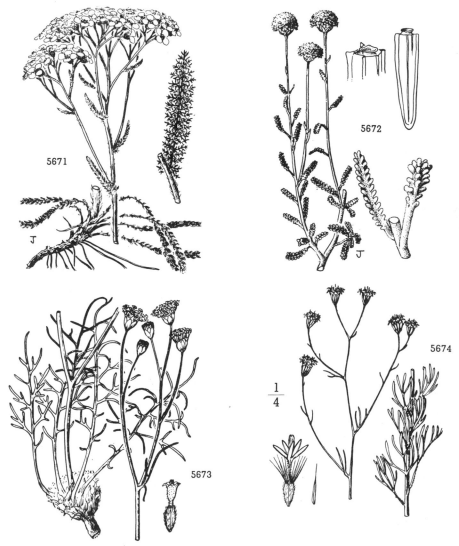

5671. Achillea millefolium
5672. Santolina chamaecyparissus
5673. Hymenopappus filifolius
5674. Hymenothrix wrìghtii

linear divisions, the blades about equaling the petiole, this tending to persist on the lower stem after blade has shattered, often densely hirsute below to nearly glabrous above, becoming reduced and bract-like on the branching inflorescence; heads discoid, broader than long, flowers white or purplish; involucres turbinate, 7–8 mm. high; phyllaries few, of unequal length, fragile, obovate or oblong, with a wide, scarious, pink or white margin; flowers 4.5–5 mm. long, unequally 5-cleft nearly to the glandular tube into linear-oblong spreading lobes, the anthers thus completely exserted; achenes about 5 mm. long, hirsute, the pappus-paleae 12–20, about 4 mm. long, lanceolate, with the strong midrib prolonged into a scabrous awn.

Slopes in the oak and pine belt, Upper Sonoran and Arid Transition Zones; southern New Mexico and central and southern California and the mountain ranges of San Diego County, California, south to adjacent Sonora and Lower California. Type locality: between Babocomori and Santa Cruz, Sonora. Aug.–Oct.

100. CHRYSÁNTHEMUM L. Sp. Pl. 887. 1753.

Annual or perennial, often aromatic herbs, sometimes woody at base. Leaves alternate, entire, crenate, dentate, or pinnately or bipinnately incised. Heads solitary on the branches or in corymbiform clusters, radiate, the rays rarely lacking. Involucre hemispheric or saucer-shaped; phyllaries many, in 2–5 series, dry, scarious or hyaline at least on the margins or tip. Receptacle flat or convex, naked. Ray-flowers if present pistillate and fertile, color various, in ours white or yellow. Disk-flowers perfect and fertile, numerous, yellow, tube flattish, expanding into the narrowly funnelform throat, 5-lobed. Anthers ovate to ovate-obtuse at apex, entire at base. Style-branches flattish, truncate, penicillate. Achenes subterete or angled in outline, striate, ribbed or somewhat wing-angled. Pappus a short crown or none. [From the Greek, meaning golden flower.]

A genus of about 100 species, mostly natives of Europe and Asia. Type species, *Chrysanthemum coronarium* L.

Ray-flowers present and conspicuous, white or yellow; herbage variously aromatic but not mint-scented.
   Perennials; rays white.
      Rays 14–21 mm. long; upper leaves incised.   1. *C. leucanthemum.*
      Rays 4–8 mm. long; upper leaves pinnately or bipinnately parted.   4. *C. parthenium.*
   Annuals; rays yellow.
      Leaves pinnatifid, merely incised or toothed above; achenes of disk-flowers not at all winged.
                                                   2. *C. segetum.*
      Leaves all more or less bipinnatifid; achenes of disk-flowers winged at the apex.   3. *C. coronarium.*
  Rays lacking or very short-rayed; herbage mint-scented.   5. *C. balsamita.*

1. Chrysanthemum leucánthemum L. Ox-eye Daisy. Fig. 5675.

Chrysanthemum laucanthemum L. Sp. Pl. 888. 1753.
Leucanthemum vulgare Lam. Fl. Franc. **2**: 137. 1778.
Chrysanthemum leucanthemum var. *pinnatifidum* Lecoq & Lamotte, Cat. Pl. France 227. 1847.
Leucanthemum leucanthemum Rydb. N. Amer. Fl. **34**: 235. 1916.

Perennial, the stems erect, 2–8 dm. high from a creeping rhizome, simple or sparingly branched, glabrous throughout or sparsely hairy. Basal leaves 4–10 cm. long including the petiole, spatulate or obovate, the blades shorter than the petioles, often incised-pinnatifid toward the base, crenate or toothed toward the apex; cauline leaves shorter or a little longer than the internodes, reduced above and becoming bracteate, sessile, clasping, incised; heads borne on long bracteate peduncles, 3–5 cm. broad including the rays; phyllaries 2- or 3-seriate, coriaceous with a scarious, reddish-tinged margin, this narrow below and widening above into a rounded erose apex; receptacle flattish; rays many, white, linear and rounded at apex, 14–21 mm. long; disk-flowers yellow, cylindric; achenes about 2 mm. long, irregularly ellipsoid, dark with 8–10 thickened white ribs; pappus none.

A native of Eurasia occurring commonly in the Pacific States in fields and roadsides throughout Washington and Oregon and in northern California as far south as San Mateo County and Eldorado County; also adventive in Ventura and western San Bernardino Counties. June–Sept. Whiteweed.

Chrysanthemum anethifólium Brouss. ex Willd. Enum. Hort. Berol. 904. 1809. A native of the Canary Islands, sometimes grown in cultivation, has been found growing spontaneously in Santa Barbara County, California, by C. F. Smith. It resembles the marguerite or Paris daisy (*C. frutescens* L.) differing from that taxon by the glaucous, more finely cut and shorter, petioled leaves.
The Shasta daisy, a horticultural form of *C. maximum* Ramond (Bull. Soc. Philom. **2**: 140. 1800), has been reported as an escape west of the Cascade Mountains in Washington and Oregon; also Bolinas Lagoon, Marin County, California, *J. T. Howell.*

2. Chrysanthemum ségetum L. Corn Chrysanthemum. Fig. 5676.

Chrysanthemum segetum L. Sp. Pl. 889. 1753.
Chrysanthemum laciniatum Gilib. Fl. Lithuan. 218. 1781–82.
Matricaria segetum Schrank, Baier. Fl. **2**: 406. 1789.
Pyrethrum segetum Moench, Meth. 597. 1794.

Erect, mostly glabrous annual from a fibrous root, 3–5 dm. high, branched from the base or single-stemmed, branching above. Lower leaves and those of the midstem 3.5–6 cm. long, sessile, obovate to broadly spatulate in outline, pinnatifid mostly on the upper half of the leaf, the divisions not reaching the midrib; upper leaves 1.5–3.5 cm. long, sessile, clasping, incised or irregularly toothed or incised only at the tip; heads solitary at the ends of the branches, 12–15 mm. broad not including the rays, the peduncles 2–10 cm. long; phyllaries usually 3-seriate, 6–10 mm. long, broadly elliptic to ovate, the central area firm, yellowish, rounded, the broad hyaline margin widest at the apex. brownish; rays 12–15 mm. long, elliptic, toothed at the apex; disk-flowers about 3.5 mm. long, tunnelform, lobes acute; ray-achenes about 2 mm. long, somewhat compressed, 2-winged, the body of the achene ribbed; disk-achenes pale brown or straw-colored, subcyclindric, conspicuously and evenly 10-ribbed; pappus none.

An attractive introduced European weed, roadsides and fields near the coast; Del Norte County to Santa Barbara County, California, more common in its northern range. Type locality: Skåne, Sweden. May–July. Cornmarigold.

3. Chrysanthemum coronàrium L. Garland Chrysanthemum. Fig. 5677.

Chrysanthemum coronarium L. Sp. Pl. 890. 1753.
Matricaria coronaria Desr. in Lam. Encycl. **3**: 737. 1791.
Pyrethrum breviradiatum Ledeb. Mém. Acad. St. Pétersb. **5**: 577. 1815.

Erect annuals 3–5 dm. high from a fibrous root, glabrous or slightly pubescent mostly on the leaves, single-stemmed or branched from the base, the stems branching above with ascending branches. Leaves sessile, 2.5–6 cm. long, the upper but little reduced in size, obovate in outline, all pinnately parted almost to the midrib, the divisions pinnately or bipinnately incised with acute lobes ending in short cusps; heads solitary at the ends of the branches, about 2 cm. broad not including the rays, the peduncles 3.5–11 cm. long; phyllaries about 8–10 mm. long, mostly ovate, acute, or obtuse, the center firm and acute in outline, the hyaline margin widest at the apex; rays yellow, 10–15 mm. long, elliptic, more or less toothed at the apex; disk-flowers 3.5–4 mm. long, tube slender, with funnelform throat and acute lobes; ray-achenes about 2.5 mm. long, more or less triquetrous, the angles bearing wings broad and truncate above, attenuate below; disk-achenes more or less 4-angled but laterally compressed, one angle with a wing about 1 mm. broad and truncate at the apex, attenuate at the base, the opposite angle somewhat winged, the other two ridged, the space between the angles usually glandular-tuberculate and ribbed; pappus none.

A native of the Mediterranean region, roadsides and fields; Marin County, California, south to San Diego County; also in Alameda County. May–July.

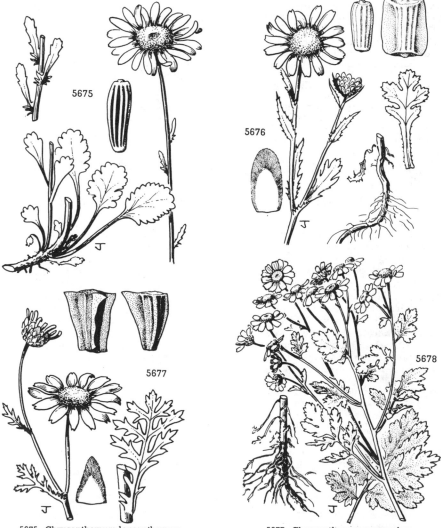

5675. Chrysanthemum leucanthemum
5676. Chrysanthemum segetum

5677. Chrysanthemum coronarium
5678. Chrysanthemum parthenium

4. Chrysanthemum parthènium (L.) Bernh. Feverfew. Fig. 5678.

Matricaria parthenium L. Sp. Pl. 890. 1753.
Matricaria odorata Lam. Fl. Franc. **2**: 135. 1778.
Chrysanthemum parthenium Bernh. Syst. Verz. Erf. 145. 1800.
Pyrethrum parthenium J. E. Smith, Fl. Brit. 900. 1800–4.
Matricaria vulgaris S. F. Gray, Nat. Arr. Brit. Pl. **2**: 454. 1821.

Much-branched, leafy, aromatic perennial 2.5–10 dm. high, sparsely puberulent to glabrous, simple-stemmed or branched from the base from a taproot or stout caudex, branching paniculately above into a corymbose inflorescence. Leaves petioled, broadly ovate or ovate-oblong in outline, 4–10 cm. long, the petiole one-third or less the length of the blade, the leaves somewhat reduced above, bipinnately or pinnately parted, the segments acute or rounded; heads many, about 14–18 mm. wide including the white rays; phyllaries 2- or 3-seriate, coriaceous and keeled, lanceolate, sparsely pubescent, with a narrow hyaline margin sometimes enlarged to a lobe at the apex; rays about 10–20, many more in double forms, 4–8 mm. long, oblong; disk-flowers 2 mm. or more long, narrowly funnelform, somewhat glandular below and thickened at the base; achenes subterete, 8–10-nerved; pappus lacking or a minute crown.

A native of Europe grown in gardens and often escaping; widely distributed but not common throughout the Pacific States; also in central and eastern United States. Forms with double flowers as well as the usual single-flowered forms are often met with as adventives. May–Aug.

5. Chrysanthemum balsámita (L.) Baillon. Costmary. Fig. 5679.

Tanacetum balsamita L. Sp. Pl. 845. 1753.
Balsamita major Desf. Act. Soc. Hist. Nat. Paris `1`: 3. 1792.
Pyrethrum balsamita Willd. Sp. Pl. **3**: 2153. 1804.
Pyrethrum balsamita var. *tanacetoides* Boiss. Fl. Orien. **3**: 346. 1875.
Chrysanthemum balsamita Baillon, Hist. Pl. **8**: 311. 1882.
Chrysanthemum balsamita var. *tanacetoides* Boiss. ex W. Miller in Bailey, Cyclop. Hort. 313. 1900.
Balsamita balsamita Rydb. N. Amer. Fl. **34**: 238. 1916.

Coarse perennial 6–12 dm. high with 1 to several erect leafy stems arising from a stout caudex, the herbage mint-scented, silvery-strigose throughout, becoming more or less glabrate especially below. Basal leaves many, oblong or elliptic with a rounded apex, petioled, the blade 15–30 cm. long, about 3–7 cm. broad, crenate-margined, the petioles a little shorter than the blade; stem-leaves sessile or with a very short petiole, longer than or about equaling the internodes, reduced above, 4–11 cm. long, crenate-margined, simple or sometimes 2-lobed at the base; heads many, 4–7 mm. wide, in corymbiform inflorescences on the upper branches, yellow and in ours rayless, these white when present; phyllaries about 3-seriate, coriaceous, lanceolate, subacute, tipped with a conspicuous, erose, hyaline lobe; disk-flowers glabrous, about 2 mm. long, tubular-funnelform, expanding to the 5 short lobes; achenes about 2 mm. long with about 10 ribs; pappus a small crown.

A native of western Asia often grown in gardens and occasionally becoming established locally in the Pacific States; Whitman County, Washington, and in Jackson and Josephine Counties, Oregon. Aug.–Sept. Mint Geranium.

101. TANACÈTUM [Tourn.] L. Sp. Pl. 843. 1753.

Aromatic, stout, erect, annual or perennial herbs from a rootstock, or low, cespitose, woody-based herbs, or sometimes subshrubs. Leaves alternate, entire to 1–3-pinnatifid. Inflorescence corymbiform or capitate or sometimes solitary at the ends of the branches. Heads heterogamous, many-flowered. Involucres hemispheric or campanulate. Phyllaries imbricate to subequal in 2–3 series, more or less scarious. Receptacle flat, convex or low-conic, naked or hairy. Flowers with throat and tube scarcely differentiated, the marginal flowers pistillate, fertile, with or without a short ray. Disk-flowers regular, perfect, more or less cylindric, the limb with 5 short lobes. Anthers entire at base, the appendages broadly acute to obtuse. Styles of the ray-flowers or marginal ones exserted, the branches short, obtuse or truncate and minutely penicillate at the tip, those of the disk-flowers included or shortly exserted. Achenes commonly glandular, 4–5-angled or 5-veined, truncate at apex. Pappus none or coroniform. [Origin of the name obscure.]

A genus of about 50 species, natives of the northern hemisphere. Type species, *Tanacetum vulgare* L.

Coarse-stemmed herbs arising from a creeping rhizome.
    Heads numerous, 20–200, 5–10 mm. broad; herbage glabrous or nearly so; introduced weed.
                                                           1. *T. vulgare.*
    Heads few, 6–20, about 12–18 mm. broad; herbage more or less villous-tomentose; native dune plants.
        Marginal flowers without a ray; herbage densely villous-tomentose.   2. *T. camphoratum.*
        Marginal flowers rayed; herbage thinly villous-tomentose to glabrate.   3. *T. douglasii.*
Stems slender; herbs from a thickened caudex, or subshrubs.
    Leaves simple or 3–5-cleft; receptacle naked.   4. *T. canum.*
    Leaves bi- or tripinnatifid; receptacle covered with pilose hairs.   5. *T. potentilloides.*

1. Tanacetum vulgàre L. Tansy. Fig. 5680.

Tanacetum vulgare L. Sp. Pl. 844. 1753.
Tanacetum officinarum Crantz, Inst. **1**: 273. 1766.
Tanacetum elatum Salisb. Prod. Stirp. 190. 1796.
Tanacetum crispum Steudel, Nom. 825. 1821.

Coarse, aromatic, glabrous or sparsely hairy perennials 4–15 dm. high from a stout rhizome, the striate stems densely leafy to the compound, many-flowered, corymbiform inflorescence. Leaves punctate, 1–2.5 dm. long, much surpassing the short internodes and scarcely diminished in size on the upper stem, sessile or short-petiolate, pinnate to the winged rachis, the pinnae laciniately incised or toothed; heads many, yellow; involucre broadly hemispheric, about 6–10 mm. wide, about 4 mm. high; phyllaries imbricate in about 3 series, carinate, firm except for the scarious margin and tip, the outermost lanceolate, the inner oblong, obtuse; marginal flowers glandular, with a 3-lobed limb; disk-flowers shortly 5-parted into obtuse lobes; achenes sparsely glandular-dotted, 5-angled, about 1 mm. long; pappus a very minute, toothed crown.

A common weed in Washington and western Oregon and occurring less commonly in California along the coast as far south as Humboldt County and in Plumas County; also widespread throughout the United States. Naturalized from Europe. July–Sept.

2. Tanacetum camphoràtum Less. Dune Tansy. Fig. 5681.

Tanacetum camphoratum Less. Linnaea 6: 521. 1831.
Omalanthus camphoratus Less. Syn. Comp. 260. 1832.
Omalotes camphorata DC. Prod. 6: 84. 1837.
Tanacetum elegans Decne. Fl. Serres 12: 19. 1857.

Stout aromatic perennial 3–5 dm. high from a branched creeping rhizome with striate, erect or ascending, leafy stems, loosely villous-tomentose throughout, densely so on new growth and

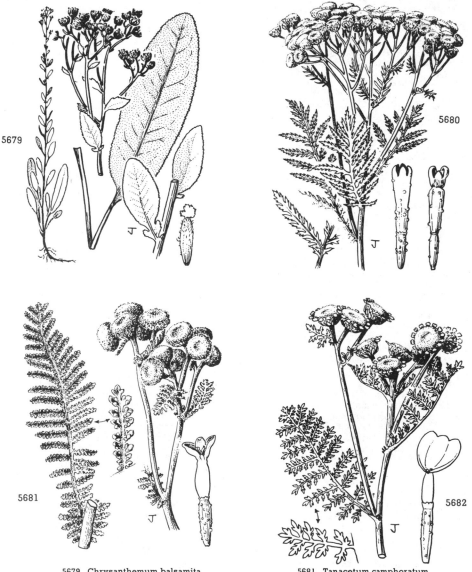

5679. Chrysanthemum balsamita
5680. Tanacetum vulgare
5681. Tanacetum camphoratum
5682. Tanacetum douglasii

somewhat deciduous in age on stems and upper surfaces of leaves. Leaves thick, 7–25 cm. long, 2–6 cm. wide, with an expanded clasping base, mostly bipinnatifid, the divisions of the pinnae parted to near the midrib, 1–3 mm. long, overlapping or nearly so when expanded but usually strongly revolute, the pinnae thus appearing pectinate, the margin entire or with minute rounded lobes; inflorescence a dense corymbiform cluster on the simple or branched stems, the peduncles of the mature heads very short or up to 1 cm. long; heads about 6–10, yellow-flowered, the mature heads 10–15 mm. broad; phyllaries 4–6 mm. long, lanceolate to oblong, firm except for the scarious margin and tip; marginal pistillate flowers few, inconspicuous, about 2 mm. long, saccate-tubular, obliquely split, the short terminal portion with 3 triangular lobes; disk-flowers tubular, 2 mm. long, with 5 triangular lobes; achenes about 3 mm. long, truncate, 5-angled, glandular-dotted; pappus a somewhat lobed crown.

Dunes, Humid Transition Zone; Marin and San Francisco Counties, California. Type locality: San Francisco. Collected by Chamisso. June–Sept.
Closely related to the following species.

3. Tanacetum douglásii DC. Northern Dune Tansy. Fig. 5682.

Tanacetum douglasii DC. Prod. 6: 128. 1837.

Stout aromatic perennial 2–6 dm. high from a branched creeping rhizome, with stout, erect or ascending, leafy, sometimes reddish stems, thinly and sparsely villous-tomentose to glabrate. Leaves rather thick, 8–20 cm. long, 2–5 cm. wide, with an expanded clasping base, mostly tripinnatifid, more rarely bipinnatifid, the divisions of the pinnae not strongly revolute, 3–5 mm. long, divided to the midrib, the segments not at all overlapping, usually deeply redivided into acute or obtuse, mucronulate lobes; inflorescence corymbiform on the simple or branched stems, the peduncles of the mature heads 1–2 cm. long; heads 8–18, yellow-flowered, the mature heads 10–15 mm. broad; phyllaries 4–6 mm. long, lanceolate to oblong, firm except for the scarious margin and tip; marginal pistillate flowers few, 5 mm. long, with a short but distinct, emarginate ray about 3 mm. long, the throat tubular-saccate; disk-flowers about 4 mm. long, 5-toothed; achenes 3 or 4 mm. long, truncate, 5-angled, glandular-dotted; pappus a somewhat lobed crown.

Dunes along the coast, Humid Transition Zone; British Columbia south to Humboldt County, California. Type locality: western North America. Collected by Douglas. June–Sept.
Some of the collections from southern Oregon and northern California have less divided leaves more suggestive of *T. camphoratum* than *T. douglasii*, but are placed here because of the presence of a definite ligule on the pistillate flowers and the lack of copious villous tomentum.

4. Tanacetum cànum D. C. Eaton. Gray Tansy. Fig. 5683.

Tanacetum canum D. C. Eaton, Bot. King Expl. 179. *pl. 29, figs. 8–14.* 1871.
Sphaeromeria cana Heller, Muhlenbergia 1: 7. 1900.

Subshrub from a woody root, 1.5–3 dm. high, markedly branched and woody below with many erect leafy stems. Leaves 5–12 mm. long, sessile, oblanceolate, entire or cleft above into 3–5 divisions; heads rayless, short-peduncled, in terminal clusters of 2–8, sometimes solitary, the flowers pale yellow; involucre hemispheric, 3 mm. high, somewhat wider than high; phyllaries ovate, canescent, scarious-margined; receptacle low-conic, glandular; disk-flowers about 2 mm. long, subcylindric, glandular-dotted without, the short lobes villous; achenes stramineous, truncate, 1.5 mm. long, 10-ribbed; pappus none.

Cliffs and talus slopes, Hudsonian and Arctic-Alpine Zones; ranges of Harney and Malheur Counties, Oregon, southward through the ranges of western Nevada; in California occurring in the Sierra Nevada from Mono County to Inyo and Tulare Counties and in the Inyo and Panamint Mountains. Type locality: eastern Humboldt Mountains, Nevada. July–Aug.

5. Tanacetum potentilloìdes A. Gray. Cinquefoil Tansy. Fig. 5684.

Artemisia potentilloides A. Gray, Proc. Acad. 6: 551. 1865.
Tanacetum potentilloides A. Gray, op. cit. 9: 204. 1874.
Sphaeromeria potentilloides Heller, Muhlenbergia 1: 7. 1900.
Vesicarpa potentilloides Rydb. N. Amer. Fl. 34: 242. 1916.

Silvery-canescent herbs, the decumbent or spreading, sparsely leafy stems 1–3 dm. long, arising from a stout, sometimes branched caudex. Basal leaves clustered, 5–9 mm. long, petiolate, the blades bi- or tripinnately dissected into narrow lobes, the stem-leaves sessile, few, resembling the basal but less divided, mostly shorter than the internodes; heads rayless, yellow-flowered, several to numerous, short-peduncled, clustered at the ends of the stems and branchlets; involucres 6–8 mm. broad; phyllaries 3–4 mm. long, ovate to broadly oblong, hairy except for the broad scarious margin; flowers about 2 mm. long, tubular, the lobes acute, glandular-dotted; receptacle convex, covered with pilose hairs; achenes obovoid, thin-textured, swelling when wet.

In moist alkaline meadows, Arid Transition Zone; Wheeler County, Oregon, southward into eastern California to Mono County and adjacent Nevada. Type locality: Carson City, Nevada. June–Aug.

102. MATRICÀRIA [Tourn.] L. Sp. Pl. 890. 1753.

Annual or perennial, erect or decumbent herbs, often aromatic. Leaves alternate, bi- or tripinnatifid into linear divisions. Inflorescence corymbiform or the heads solitary on the ends of the branches. Heads many-flowered, sometimes radiate, the rays white. Involucre saucer-shaped, the phyllaries in 2 or 3 series, not much imbricated, dry with scarious margins. Receptacle naked, hemispheric or long-conical. Marginal corollas pistillate, with or without white rays. Disk-corollas yellow or green, perfect and fertile, tubular, with a gradually expanding, somewhat inflated throat, the lobes 4 or 5, ovate or lanceolate. Anthers entire or nearly so at base. Style-branches flattened, truncate, minutely pencillate. Achenes glabrous, asymmetrical, 3–5-ribbed or -nerved on one side, usually smooth on the other. Pappus coroniform. [From the Latin word meaning womb, so named for its supposed medical virtues.]

A genus of about 40 species, natives of the northern hemisphere and also South Africa. Type species, *Matricaria chamomilla* L.

Heads bearing white ray-flowers.
 Receptacle conic; mature achenes smooth on the outer face, thus lacking glands at its apex.
                              1. *M. chamomilla.*
 Receptacle dome-shaped; mature achenes rugose on the outer face, this bearing 2 rounded glands at its apex.
                              2. *M. maritima.*
Heads discoid.
 Pappus-crown minute, entire, the achenes with 2 glandular lines which extend the length of the achene but not into the minute pappus-crown; mature heads usually 6–9 mm. high.   3. *M. matricarioides.*
 Pappus-crown evident, with 2 short lobes or teeth, these bearing oblanceolate or elliptic glands which scarcely extend onto the body of the achene; mature heads usually 10–11 mm. high.   4. *M. occidentalis.*

1. **Matricaria chamomílla** L. Sweet False-chamomile. Fig. 5685.

Matricaria chamomilla L. Sp. Pl. 891. 1753.
Matricaria suaveolens L. Fl. Suec. ed. 2. 297. 1755.
Leucanthemum chamaemelum Lam. Fl. Franc. **2**: 139. 1778.
Matricaria patens Gilib. Fl. Lithuan. 220. 1781–82.
Chamomilla vulgaris S. F. Gray, Nat. Arr. Brit. Pl. **2**: 454. 1821.
Matricaria obliqua Dulac, Fl. Hautes-Pyr. 505. 1867.
Chamomilla chamomilla Rydb. N. Amer. Fl. **34**: 231. 1916.

Sweet-scented, glabrous, rather slender annuals usually much branched above, the stems 2–8 dm. high, striate. Leaves 2–6 cm. long, bipinnatifid into linear-filiform divisions about as long as or a little longer than the internodes; heads several on the upper branches; involucres about 3 mm. high, 6–10 mm. broad; phyllaries subequal, mostly oblong, broadly scarious, obtuse or rounded at apex; rays 10–20, reflexed in age, 4–10 mm. long; disk-flowers many, yellow, about 2 mm. long; receptacle about 4 mm. high, narrowly dome-shaped; achenes asymmetrical, smooth except for 5 narrow, wing-like ribs; pappus none or sometimes a short crown.

A native of Europe, occasional around gardens and waste places in the Pacific States and occurring more commonly in eastern United States; also introduced in Arizona. May–June.

2. **Matricaria marítima** L. False-chamomile. Fig. 5686.

Matricaria maritima L. Sp. Pl. 891. 1753.
Matricaria inodora L. Fl. Suec. ed. 2. 297. 1755.

Perennial or biennial, leafy, coarse, striate-stemmed herbs 1.5–6 dm. high, glabrous or nearly so, branching above, not aromatic. Leaves 2–7 cm. long, much longer than the internodes, bipinnatifid, the leaf-segments 1.5–5 mm. long, mostly linear-filiform; heads radiate, pedunculate; involucre 4–5 mm. high, about 11–14 mm. broad; phyllaries subequal, mostly oblong, obtuse at apex, with a broad scarious margin and central, dark greenish or brownish area; rays 12–25, narrowly oblong, 6–13 mm. long; disk-flowers about 1 mm. long, yellow, with 5 short acute lobes; receptacle 5–8 mm. high, dome-shaped; achenes about 2 mm. long, asymmetrical, with 3 light-colored, rounded, calloused ribs on the inner face, the intervals brown and rugose, the outer flattened, brown and rugose, and bearing 2 glands at the apex; pappus a minute denticulate crown.

A native of Europe; introduced along the Atlantic seaboard and locally established along the coast in Washington and along the lower Columbia River in northern Oregon. July–Aug.

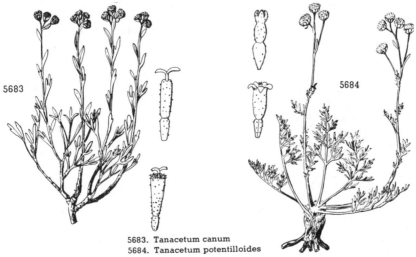

5683. Tanacetum canum
5684. Tanacetum potentilloides

3. **Matricaria matricarioìdes** (Less.) Porter. Pineapple Weed. Fig. 5687.

Santolina suaveolens Pursh, Fl. Amer. Sept. 520. 1814. Not *Matricaria suaveolens* L.
Artemisia matricarioides Less. Linnaea **6**: 210. 1831.
Matricaria discoidea DC. Prod. **6**: 50. 1837.
Matricaria matricarioides Porter, Mem. Torrey Club **5**: 341. 1894.

Pineapple-scented, nearly glabrous annuals 1–3 dm. high, usually much branched from the base, the leafy striate stems divaricate or ascending. Leaves 1–5 cm. long, bipinnatifid with short, linear, acute divisions, longer than the internodes or equalling them; heads rayless, several to many on rather stout peduncles, 0.5–1.5 cm. long; involucres about 3 mm. high; phyllaries subequal, broadly oblong to oval, with scarious margins and a central, brownish or greenish area; disk-flowers 1.25–1.5 mm. long, yellowish green, 4-toothed; receptacle 4–6 mm. high, narrowly dome-shaped to conical; achenes 1 mm. long, asymmetrical, usually indistinctly 1-nerved on the inner face and having 2 marginal nerves, the outer face smooth and usually rounded, the margins bearing linear, brown, glandular lines extending nearly or completely the length of the body of the achene and not extending into the minute pappus-crown.

A common widespread weed of waste places and cultivated areas throughout the Pacific States; probably a native of western North America and introduced elsewhere in the United States and Europe. Type locality: Unalaska. Collected by Chamisso. Rayless Chamomile.

4. **Matricaria occidentàlis** Greene. Valley Pineapple Weed. Fig. 5688.

Matricaria occidentalis Greene, Bull. Calif. Acad. **2**: 150. 1886.
Chamomilla occidentalis Rydb. N. Amer. Fl. **34**: 232. 1916.

Rather stout (rarely slender), glabrous, not strongly scented (?) annuals 1.5–5.5 dm. high, single-stemmed or branching from the base, the stems normally erect, branched above the middle with erect or ascending branches. Leaves 1–6 cm. long, bipinnatifid with short, linear-acute divisions, somewhat longer than the internodes or equaling them; heads rayless, few to several on stout peduncles 1–3 cm. long; involucre 4(3) mm. high; phyllaries subequal, broadly oblong to oval, with scarious margins and a central, greenish or brownish area; disk-flowers about 1.25 mm. long, yellowish green, 4-toothed; receptacle about 8 mm. high except in depauperate specimens, dome-shaped, sometimes narrowly so; achenes about 1 mm. long, asymmetrical, with usually 2 distinct, wing-like, whitish nerves on the inner face, 2 marginal which are topped by the elliptic or oblanceolate glands extending to the crown, and 1 rather indistinct nerve on the slightly rounded outer face; pappus-crown bearing 2 short but distinct lobes, each bearing brown glands that extend slightly on the body of the achene.

In vernal pools and fields in the Sacramento and San Joaquin Valleys and in valleys of the Inner Coast Ranges but becoming a weed in other localities; Lake County, Oregon, and Eureka, Humboldt County, California, and in southern California. Type locality: not definitely stated but collected by Greene in Contra Costa and Solano Counties, California. April–May.

Much resembling *M. matricarioides* but differing in the distinctive achene characters and growth habit.

103. **SOLÌVA** Ruiz & Pav. Fl. Peruv. Prod. 113. *pl. 24.* 1794.

Small, diffusely branched annuals. Leaves alternate, petioled, bi- or tripinnately parted. Heads few- to many-flowered, sessile in the leaf-axils. Phyllaries thin, free, rotate, subequal, in 2 series. Receptacle convex or low-conic, naked. Marginal flowers of the heads in 2 or 3 series, pistillate, fertile, without corolla. Disk-flowers few, hermaphrodite, sterile, the corolla 4-lobed. Anthers with obtuse lobes. Style stout, exserted, scarcely bilobed. Achenes compressed, variously sculptured, beaked by the persistent 4-lobed style and having 2 either broad or narrow, marginal wings produced above (in ours) into cuspidate tips that are free from the persistent, spine-like style. [Name in honor of Salvador Soliva, a Spanish physician.]

A genus of 9 species, natives of temperate South America. Type species, *Soliva sessilis* Ruiz & Pav.

Wings of the achenes entire.   1. *S. sessilis.*
Wings of the achenes conspicuously lobed near the base.   2. *S. pterosperma.*

1. **Soliva séssilis** Ruiz & Pav. Common Soliva. Fig. 5689.

Soliva sessilis Ruiz & Pav. Fl. Peruv. Prod. 113. *pl. 24.* 1794.
Gymnostyles chilensis Spreng. Syst. **3**: 500. 1826.
Soliva daucifolia Nutt. Trans. Amer. Phil. Soc. II. **7**: 403. 1841.
Soliva microloma Phil. Anal. Univ. Chile **27**: 331. 1865.

Fibrous-rooted annual, stems few to several from the base, 3–20 cm. long, decumbent, dichotomously branching, the branches divaricate, thinly villous throughout. Leaves 1–4 cm. long, bipinnately dissected into narrow oblanceolate lobes, the blade orbicular to oval in outline, the petioles winged at the base, longer than the blade especially on the basal leaves to usually about equalling the blade on the stem-leaves; heads sessile in the basal rosette and at the nodes, few-flowered; phyllaries 2–2.5 mm. long, broadly ovate and abruptly acute, thin, green above and hyaline below; pistillate flowers fertile, marginal in 2 or 3 series; central flowers 2–4, perfect, sterile, subcylindric; body of the achenes 2–2.5 mm. or more long, obovate in outline, planoconvex, hispidulous on both surfaces, sometimes becoming glabrate, the wings plane or incurved, very narrow to nearly 1 mm. wide at its greatest width, glabrous and somewhat cartilaginous, sometimes

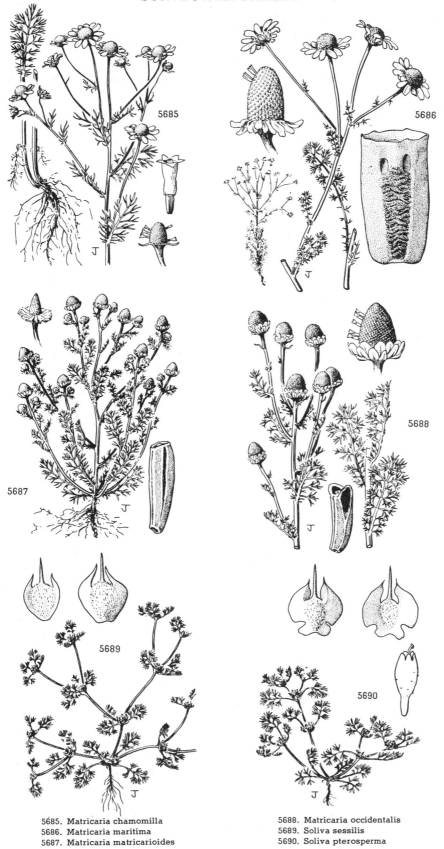

5685. Matricaria chamomilla
5686. Matricaria maritima
5687. Matricaria matricarioides

5688. Matricaria occidentalis
5689. Soliva sessilis
5690. Soliva pterosperma

segment402COMPOSITAE

narrowed at the base but never notched, extending above the body of the achene with incurved cuspidate lobes much shorter than the indurated style.

A native of Chile, well established as a weed along the coast and adjacent ranges; Lincoln County, Oregon, southward in California to Santa Barbara County; also adventive in lawns in southern California. Type locality: Chile. April–June.

Soliva daucifolia was originally described by Nuttall from specimens collected at Santa Barbara in which the wings of the achenes are quite narrow. The width of the wing is quite variable and specimens with wide and with narrow wings are found in the same colony.

2. Soliva pterospérma (Juss.) Less. South American Soliva. Fig. 5690.

Gymnostyles pterosperma Juss. Ann. Mus. Paris **4**: 262. *pl. 61. fig. 3.* 1804.
Ranunculus alatus Poir. Encycl. **6**: 127. 1804.
Soliva pterosperma Less. Syn. Comp. 268. 1832.
Soliva barclayana DC. Prod. **6**: 143. 1837.
Soliva sessilis of authors, not Ruiz & Pav.

Habit, height, leaves, and inflorescence as in *S. sessilis*. Body of the achene about 3 mm. long or less not including the indurated style, obovate in outline, planoconvex, hispidulous on both faces, the wings plane or incurved, each about 1 mm. wide at its greatest width, glabrous and somewhat cartilaginous, becoming thinner on the extreme edge, deeply notched near the base and extended above the body of the achene into cuspidate-tipped lobes that are shorter than the indurated style.

Occasional along roadsides and waste places; a South American herb introduced in the Sierra Nevada foothills of California from Eldorado County to Tuolumne County; also adventive in Australia and New Zealand. Type locality: Uruguay. April.

104. CÓTULA [Tourn.] L. Sp. Pl. 891. 1753.

Low, diffuse or creeping, strong-scented, annual or perennial herbs. Leaves alternate, toothed, lobed, or dissected. Heads pedunculate, small or medium, disciform, solitary, terminal on the branches or axillary, hemispheric or globose. Involucres in 1–3 series, the phyllaries somewhat unequal, membranous or subherbaceous, mostly with a scarious margin. Receptacle flat or convex, smooth. Marginal flowers pistillate, fertile, in 1 or more series, often stipitate, the corolla vestigial or lacking. Disk-flowers perfect, fertile or sometimes sterile, the corollas regular, tubular, the 4-parted limb very short. Anthers obtuse at base. Style included, branches short, obtuse or truncate. Achenes with pedicels persisting on the receptacle, compressed, margined or winged. Pappus a short crown or lacking. [From the Greek words meaning a small cup.]

A genus of about 50 species of widespread distribution in the tropical and subtropical regions of the world. Type species, *Cotula coronopifolia* L.

Leaves pinnately cleft or entire, with a sheathing base, glabrous.   1. *C. coronopifolia.*
Leaves pinnately or bipinnately dissected, without a sheathing base, more or less villous.   2. *C. australis.*

1. Cotula coronopifòlia L. Brass Buttons. Fig. 5691.

Cotula coronopifolia L. Sp. Pl. 892. 1753.
Lancisia coronopifolia Rydb. N. Amer. Fl. **34**: 286. 1916.

Rather succulent, glabrous perennial branching from the base, decumbent and spreading, rooting from the nodes, 5–30 cm. long. Leaves sessile with a sheathing, somewhat scarious base, linear, lanceolate, or oblong, entire, toothed, or with narrow lobes, 1–6 cm. long; heads bright yellow, solitary from the upper leaf-axils, on peduncles 1–5 cm. long; involucres flat-hemispheric, 5–11 mm. broad; phyllaries in 3 subequal series, oblong-lanceolate or elliptic; marginal pistillate flowers in a single series, corolla lacking; disk-flowers tubular; achenes or marginal flowers shorter than the pedicel, conspicuously winged and emarginate above and below, the achenes of the disk-flowers short-pedicellate, not winged; pappus none.

Common in tidal flats along the coast or extending inland in wet places; British Columbia south to northern Lower California; also on the Atlantic Coast and widespread elsewhere especially in the southern hemisphere. A native of South Africa. March–Dec.

2. Cotula austràlis (Sieber) Hook f. Australian Cotula. Fig. 5692.

Anacyclus australis Sieber ex Spreng. Syst. **3**: 497. 1826.
Cotula australis Hook. f. Fl. Nov. Zeland. **1**: 128. 1853.
Lancisia australis Rydb. N. Amer. Fl. **34**: 286. 1916.

Low slender annual 3–20 cm. high, diffusely branching from the base, sparsely villous throughout. Leaves with a petiolar base, pinnately or bipinnately dissected into linear or lanceolate, acute divisions 2–4 cm. long; heads terminal on the branches, pale yellow, on peduncles 2–5 cm. long; involucres flat-hemispheric, 5–6 mm. broad; phyllaries in 2 series, rounded above, brownish-tipped and scarious-margined; marginal pistillate flowers in 2 or 3 series, pedicelled, corolla lacking; disk-flowers tubular with a very slightly flaring throat, nearly sessile; achenes of the marginal flowers almost equaling the pedicel, winged and broadly oval in outline, scarcely or not at all emarginate; achenes of the disk-flowers nearly sessile, oblong and merely margined.

In disturbed places and about habitations, widespread and becoming common; western Oregon and California; also reported from Utah and Maine. Introduced from Australia. Jan.–March.

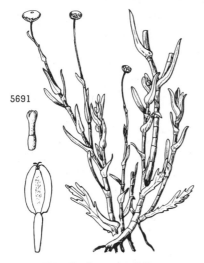

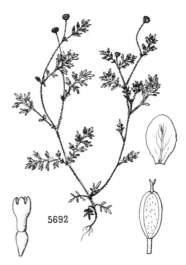

5691. Cotula coronopifolia

5692. Cotula australis

105. ARTEMÍSIA* L. Sp. Pl. 845. 1753.

Annual, biennial, and perennial herbs or shrubs, usually aromatic, the roots fibrous, or with taproot; perennial herbs often with rootstock or caudex. Leaves entire to finely divided. Inflorescence paniculiform, sometimes reduced and raceme-like or spiciform. Phyllaries dry, imbricate, at least the inner ones scarious or with scarious margins. Heads small, discoid or disciform. Ray-florets pistillate and fertile or wanting; corolla tubular and usually tapered upward with 2, 3, or 4 teeth; style 2-cleft, more or less exserted, the somewhat flattened branches usually recurved and rounded, or truncate and entire, or erose at the tip. Disk-florets perfect, fertile or the pistil sometimes sterile by an abortive ovary; corolla campanulate or funnelform, regular and 5-toothed; style usually 2-cleft into somewhat flattened, more or less recurved branches, these truncate and erose or penicillate at the tip. Achenes ellipsoid or ovoid to nearly prismatic, 2–5-angled or -ribbed, or with numerous faint striae, usually glabrous or resinous-granuliferous, rarely hairy. Receptacle flat to hemispheric, naked or with many long hairs, with paleae in only 1 species. [Name Greek, from ancient name of mugwort, in memory of Artemisia, Queen of Caria and wife of Mausolus.]

A circumboreal genus of 150–280 species, depending upon interpretation. Type species, *Artemisia vulgaris* L.

Pistillate marginal florets present (may be absent in some heads of *A. bigelovii*).
    Disk-florets perfect and fertile.
        Receptacle glabrous (Section *Abrotanum*).
            Plants herbaceous, sometimes a little woody at the base.
                Plants perennial from rhizome or caudex, sometimes taprooted.
                    Leaves concentrated basally, fewer and reduced upward.
                        Pubescence of leaves loosely villous or wanting; plants mostly 2–5 dm. high.
                                1. *A. norvegica saxatilis*.
                        Pubescence of leaves sericeous; plants mostly under 2 (rarely to 3) dm. high.
                                2. *A. trifurcata*.
                  Leaves all chiefly cauline.
                    Leaves finely divided, at least some of them bipinnatifid, some of the segments again
                      toothed.   8. *A. michauxiana*.
                  Leaves entire to bipinnatifid with entire lobes, these usually much broader than above.
                    Involucre higher than broad or height and breadth about equal.
                      Leaves 1 cm. or less in width (rarely to 1.5 cm.) exclusive of lobes when
                      present; plants rarely over 1 m. tall.
                        Plants suffrutescent at base, tending to be taprooted, without well-devel-
                      oped rhizomes.   7. *A. lindleyana*.
                        Plants herbaceous to base, never taprooted, freely spreading by creeping
                      rhizomes.   3. *A. ludoviciana*.
                    Leaves wider, 1–5 cm. exclusive of lobes when present; plants often over
                  1 m. tall.
                    Involucre campanulate, more or less tomentose.
                          4. *A. douglasiana*.
                    Involucre cylindric or narrow-ovoid, glabrous and shining.
                        5. *A. suksdorfii*.
             Involucre broader than high.   6. *A. tilesii unalaschcensis*.

---

* Text contributed by George Henry Ward.

Plants annual or biennial, taprooted; leaves essentially glabrous.   9. *A. biennis.*
Plants shrubby.
    Tall shrub, 5–25 dm. high; leaves ternately to bipinnately divided into narrow linear lobes.
                                                10. *A. californica.*
    Low spreading shrub, 2–4 dm. high; leaves narrowly cuneate, tridentate.
                                                11. *A. bigelovii.*
Receptacle covered with long hairs (Section *Absinthium*).
    Divisions of leaves oblong or oblanceolate, ultimate segments usually 1.5–4 mm. wide; plants usually 4–12 dm. tall.   12. *A. absinthium.*
    Divisions of leaves linear, ultimate segments about 1 mm. wide; plants usually 1–4 dm. tall.
                                              13. *A. frigida.*
Disk-florets perfect but pistil infertile, the ovary abortive (Section *Dracunculus*).
Plants herbaceous; achenes essentially glabrous.
    Leaves mostly entire, occasionally to 3-parted.   14. *A. dracunculus.*
    Leaves finely divided into narrow, linear or linear-oblanceolate lobes.
        Pubescence when present appressed and silky, some forms sparsely pubescent or glabrate.
            Heads relatively small, the involucre usually 2–3 mm. tall; inflorescence paniculiform, more or less condensed; plants to 10 dm. in height.   15. *A. campestris pacifica.*
            Heads larger, the involucre usually 3–4 mm. tall; inflorescence reduced, often spiciform or nearly so; plants 1–4 dm. in height.   15. *A. campestris borealis.*
        Pubescence loosely silky-villous, very dense over all parts of the plant; found only along coast of California, and in southern Oregon.   16. *A. pycnocephala.*
Plants shrubby and spinescent; achenes long-hairy.   17. *A. spinescens.*
Pistillate marginal florets absent, all florets perfect and fertile (Section *Seriphidium*).
Receptacle naked; plants woody shrubs.
    Leaves mostly cuneate or flabelliform with 3–7 teeth at the apex, or sometimes linear, entire and truncate or rounded (a few leaves of *A. rothrockii* may be entire and acute).
        Plants seldom root-sprouting; heads small to large, 3–12 florets per head; without entire leaves or nonflowering shoots, or if such leaves present then these truncate or rounded; leaves deeply divided only on some montane forms of *A. arbuscula.*
            Leaves mostly three or more times as long as wide, 1.5–5 cm. long, entire or 3–7-toothed at apex, linear to cuneate or with divergent apical lobes; habit variable.
                Achenes glandular-granuliferous, rarely sparsely short-villous; leaves cuneate, 3–5-dentate at tip or sometimes linear and entire with a rounded or truncate apex.
                                    18. *A. tridentata tridentata.*
                Achenes arachnoid short-villous, rarely less pubescent; leaves linear, with an obscurely notched, truncate or rounded apex, or with divergent lobes at apex.
                                      18. *A. tridentata parishii.*
            Leaves mostly as long or three times as long as wide, 0.5–1.5 cm. long (spring leaves sometimes longer), broadly cuneate or flabelliform, 3–5-toothed or -lobed at apex; low spreading shrubs.   19. *A. arbuscula.*
        Plants often root-sprouting; heads large, 8–20 florets per head; often with many leaves entire and acute, leaves also often shallowly to deeply parted; found at high elevations in the Sierra Nevada and San Bernardino Mountains.   22. *A. rothrockii.*
Leaves mostly linear or oblanceolate, entire with an acute or acuminate apex, or leaves deeply divided into 3 or more linear or linear-oblanceolate lobes.
    Leaves narrowly oblanceolate, mostly entire or with a few irregular teeth or lobes (some leaves may be cleft up to one-third their length in some plants).   21. *A. cana.*
    Leaves mostly divided into 3 or more narrow lobes, entire linear leaves also may be present.
        At least the upper heads exceeding their subtending bracts; plants evergreen.
                                    20. *A. tripartita.*
        Heads all exceeded by their subtending bracts; leaves deciduous.   23. *A. rigida.*
Receptacle with chaff, all or most of the florets subtended by bracts; plants with shrubby base and long herbaceous wands.   24. *A. palmeri.*

1. **Artemisia norvégica** subsp. **saxátilis** (Bess.) Hall & Clem.   Mountain Sagewort. Fig. 5693.

Artemisia arctica Less. Linnaea **6**: 213. 1831.
Artemisia chamissoniana var. *saxatilis* Bess. in Hook. Fl. Bor. Amer. **1**: 324. 1833.
Artemisia norvegica var. *pacifica* A. Gray, Syn. Fl. N. Amer. 1²: 371. 1884.
Artemisia saxicola Rydb. Bull. Torrey Club **32**: 128. 1905.
Artemisia norvegica subsp. *saxatilis* Hall & Clem. Carnegie Inst. Wash. Pub. No. 326: 58. 1923.
Artemisia norvegica var. *saxatilis* Jepson, Man. Fl. Pl. Calif. 1141. 1925.

Tufted perennial from a branching caudex, 1.5–6 dm. tall; plant loosely villous to essentially glabrous. Basal leaves persistent, petiolate, the broad blade 2–10 cm. long, pinnately dissected, the ultimate segments narrow and acute; cauline leaves reduced, the upper ones sessile; inflorescence spiciform to narrowly paniculiform, the heads relatively large; involucre hemispheric, glabrous to moderately woolly-villous, 4–6 mm. high, the phyllaries dark-margined; ray-florets 6–12; disk-florets 30–80, fertile, the corolla long-hairy at least near the base; achenes nearly cylindric, glabrous or granuliferous; receptacle glabrous.

Talus slopes, rock outcrops, and open woods, Boreal Zone; Cascade and Olympic Mountains of Washington southward in the mountains to the southern Sierra Nevada, in the Rocky Mountains south to Colorado; northward through Canada and Alaska to eastern Siberia. Type locality: Rocky Mountains. July–Sept.

2. **Artemisia trifurcàta** Steph. ex Spreng.   Three-forked Sagewort. Fig. 5694.

Artemisia trifurcata Steph. ex Spreng. Syst. **3**: 488. 1826.
Artemisia heterophylla Bess. Nuov. Mem. Soc. Nat. Mosc. **3**: 74. 1834.
Artemisia tacomensis Rydb. N. Amer. Fl. **34**: 362. 1916.
Artemisia norvegica subsp. *heterophylla* Hall & Clem. Carnegie Inst. Wash. Publ. No. 326: 59. 1923.
Artemisia trifurcata subsp. *tacomensis* Hultén, Fl. Alaska & Yukon 1577. 1950.

Tufted perennial, from a branching caudex, 0.5–3 dm. tall; leaves canescent-sericeous, the stem usually less so. Basal leaves persistent, petiolate, the blade 0.5–3 cm. long, trifid or palmately divided into 3 or more narrow and rather blunt segments, these often again divided; cauline leaves fewer and reduced, sometimes entire; inflorescence spiciform or racemiform, the heads few; involucre hemispheric, densely woolly-villous, 3–4.5 mm. high, the phyllaries more or less dark-margined; ray-florets 5–10; disk-florets 10–25 (or more), fertile, the corolla sparsely to densely villous-hirsute at the summit; achenes nearly cylindric, sparsely long-hairy; receptacle glabrous.

Open rocky ledges and talus slopes at high elevations, Boreal Zone; Olympic and Cascade Mountains of Washington to Alaska, Manchuria, and Kurile Islands. Type locality: Soongaria, Manchuria. July–Sept.

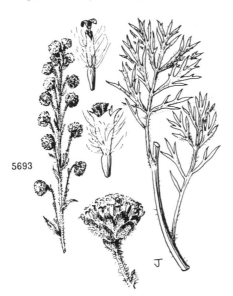

5693. Artemisia norvegica   5694. Artemisia trifurcata

3. Artemisia ludoviciàna Nutt. Western Mugwort. Fig. 5695.

Artemisia ludoviciana Nutt. Gen. **2**: 143. 1818.
Artemisia gnaphalodes Nutt. loc. cit.
Artemisia ludoviciana var. *gnaphalodes* Torr. & Gray, Fl. N. Amer. **2**: 420. 1843.
Artemisia vulgaris var. *gnaphalodes* Kuntze, Rev. Gen. Pl. **1**: 39. 1891.
Artemisia vulgaris var. *ludoviciana* Kuntze, loc. cit.
Artemisia vulgaris subsp. *ludoviciana* Hall & Clem. Carnegie Inst. Wash. Publ. No. 326: 76. 1923.
Artemisia vulgaris subsp. *gnaphalodes* Hall & Clem. op. cit. 77.
Artemisia ludoviciana subsp. *typica* Keck, op. cit. No. 520: 330. 1940.

Perennial herb, 3–10 dm. tall; stems herbaceous to base, slender to moderately stout. Leaves numerous, 3–10 cm. long, densely white-tomentose on both sides or loosely floccose to green and glabrate above, linear to lanceolate, oblanceolate or elliptic, entire or few-toothed or -lobed especially near the apex; inflorescence paniculiform, usually rather elongate and compact, the heads often in glomerules; involucre ovoid to campanulate, 2–3 mm. wide, usually densely tomentose, occasionally glabrate; ray-florets 5–12; disk-florets 6–21, fertile; achenes ellipsoid, glabrous, sometimes resinous-granuliferous; receptacle glabrous. $n = 18$.

Lowlands and mountains up to middle elevations, Arid Transition and Upper Sonoran Zones; Washington south to California and north and east to Alberta, Indiana, and Arkansas; rare west of the Cascade Mountains and Sierra Nevada. Type locality: on the banks of the Mississippi River, near St. Louis. July–Sept.

Artemisia ludoviciana subsp. **álbula** (Woot.) Keck, Proc. Calif. Acad. IV. **25**: 446. 1946. (*Artemisia microcephala* Woot. Bull. Torrey Club **25**: 455. 1898, not Hillebr. 1888; *A. albula* Woot. Contr. U.S. Nat. Herb. **16**: 193. 1913.) Leaves mostly 1–2 cm. long, lanceolate, usually white-tomentose on both sides, the margins often narrowly revolute; involucre about 3 mm. high, densely tomentose; ray-florets 8–11; disk-florets 8–13. Mountains of southern California east to New Mexico and southern Colorado south to northern Lower California, Sonora, and Chihuahua. Type locality: Organ Mountains, Dona Ana County, New Mexico. May–Oct.

Artemisia ludoviciana subsp. **cándicans** (Rydb.) Keck, Proc. Calif. Acad. IV. **25**: 447. 1946. (*Artemisia ludoviciana* var. *latiloba* Nutt. Trans. Amer. Phil. Soc. II. **7**: 400. 1841; *A. candicans* Rydb. Bull Torrey Club **24**: 296. 1897; *A. latiloba* Rydb. Mem. N.Y. Bot. Gard. **1**: 429. 1900; *A. platyphylla* Rydb. N. America Fl. **34**: 275. 1916; *A. vulgaris* subsp. *candicans* Hall & Clem. Carnegie Inst. Wash. Publ. No. 326: 73. 1923; *A. vulgaris* var. *candicans* M. E. Peck, Man. Pl. Oregon 766. 1941.) Principal leaves 5–10 cm. long, more or less parted or divided, often deeply pinnatifid with some of the lobes again toothed or divided, white-tomentose on both sides, the tomentum sometimes thinner above, rarely green above; involucre 3.5–4 mm. high, the phyllaries tomentose; ray-florets 7–12; disk-florets 17–42. Lowlands and mountains up to middle elevations in eastern Washington and Oregon, and northern California east to Montana, Wyoming, and northern Utah. Type locality: Little Belt Mountains, Lewis and Clark National Forest, Meagher County, Montana. June–Sept.

Artemisia ludoviciana subsp. **incómpta** (Nutt.) Keck, Carnegie Inst. Wash. Publ. No. 520: 327. 1940. (*Artemisia incompta* Nutt. Trans. Amer. Phil. Soc. II. **7**: 400. 1841; *A. discolor* var. *incompta* A. Gray, Syn. Fl. N. Amer. 1²: 373. 1884; *A. atomifera* Piper, Contr. U.S. Nat. Herb. **11**: 588. 1906; *A. ludoviciana* var. *atomifera* M. E. Jones, Bull. Univ. Mont. Biol. Ser. **15**: 48. 1910; *A. vulgaris* subsp. *michauxiana* var. *incompta* St. John, Research Stud. St. Coll. Wash. **1**: 106. 1929.) Stems 3–9 dm. tall; principal leaves 2–8 cm.

long, more or less parted into linear or lanceolate, forward-projecting lobes, some of these again toothed or lobed, commonly glabrate above and white-tomentose beneath but sometimes tomentose throughout and rarely essentially glabrous throughout; involucre 3–3.5 mm. high, the phyllaries sericeous-tomentose to glabrate and shiny; ray-florets 6–10; disk-florets 15–30 or rarely more. Mountains of southern and eastern California to eastern Oregon east to Montana and central Utah. Usually at somewhat higher elevations than *A. ludoviciana* subsp. *candicans* when growing in the same area. $2n=18$. Type locality: "Thornberg's Pass," probably Custer County, Idaho. July–Sept.

Artemisia vulgàris L. Sp. Pl. 848. 1753. Aromatic perennial herb; stems 5–15 dm. tall, erect, simple or branched above, usually in small clumps from a shallow rhizome; leaves smooth, green, and glabrous or nearly so above, densely white-tomentose beneath, the principal ones mostly 5–10 cm. long, 3–7 cm. wide, obovate or broadly oblanceolate in outline, cleft into unequal lobes of which the terminal is the larger and which are again cleft or only toothed; inflorescence paniculiform, leafy and open or occasionally reduced and compact; involucre campanulate, mostly 3.5–4.5 mm. high, more or less tomentose; 6–12 pistillate ray-florets, 8–18 perfect and fertile disk-florets; receptacle glabrous. $n=8, 9$. Introduced from Europe, established in eastern United States and adjacent Canada, but in our area known only from near ballast dumps at our largest ports. Type locality: Europe.

4. Artemisia douglasiàna Bess. Douglas' Mugwórt. Fig. 5696.

Artemisia douglasiana Bess. in Hook. Fl. Bor. Amer. 1: 323. 1833.
Artemisia vulgaris var. *californica* Bess. Linnaea 15: 91. 1841.
Artemisia heterophylla Nutt. Trans. Amer. Phil. Soc. II. 7: 400. 1841, in part. Not Bess. 1834.
Artemisia ludoviciana var. *douglasiana* D. C. Eaton, Bot. King Expl. 183. 1871.
Artemisia kennedyi A. Nels. Proc. Biol. Soc. Wash. 18: 175. 1905.
Artemisia vulgaris subsp. *heterophylla* Hall & Clem. Carnegie Inst. Wash. Publ. No. 326: 76. 1923.
Artemisia vulgaris var. *heterophylla* Jepson, Man. Fl. Pl. Calif. 1142. 1925.
Artemisia vulgaris subsp. *douglasiana* St. John, Fl. S.E. Wash. 422. 1937.

Perennial herb, 5–15 dm. tall (sometimes taller); stems herbaceous to base, simple or with erect to divergent branches. Leaves lanceolate to elliptic and entire, or oblanceolate to obovate and coarsely few-lobed or -toothed toward the apex, 7–15 cm. long, 1–5 cm. wide exclusive of salient lobes, gradually reduced in inflorescence, all sparsely tomentulose to glabrous and green above, gray-tomentose beneath; inflorescence paniculiform, leafy and open or dense and elongated, the heads erect or nodding; involucre campanulate, 3–4 mm. high, 2–3 mm. wide, the phyllaries more or less tomentose; ray-florets 6–10; disk-florets 10–25, fertile; achenes ellipsoid, glabrous; receptacle glabrous. $n = 27$.

Transition Zones; Washington south to northern Lower California; rare east of the Cascadean-Sierran axis except near Lake Tahoe and rarely to 6,000 feet elevation. Type locality: "North-West America." June–Oct.

5. Artemisia suksdórfii Piper. Suksdorf or Coastal Mugwort. Fig. 5697.

Artemisia heterophylla Nutt. Trans. Amer. Phil. Soc. II. 7: 400. 1841, in part. Not Bess. 1834.
Artemisia vulgaris var. *litoralis* Suksd. Deutsch. Bot. Monatss. 18: 98. 1900.
Artemisia suksdorfii Piper, Bull. Torrey Club 28: 42. 1901.
Artemisia vulgaris subsp. *litoralis* Hall & Clem. Carnegie Inst. Wash. Publ. No. 326: 76. 1923.

Perennial herb 6–15 dm. tall; stems stout, herbaceous or somewhat suffrutescent at base. Leaves broadly lanceolate to elliptic or some broadly oblanceolate, entire or with a few coarse teeth or lobes toward apex, 8–15 cm. long, 1.5–5 cm. wide exclusive of lobes, gradually reduced within inflorescence, deep green and glabrous or glabrate above, white-tomentose beneath; inflorescence paniculiform, very dense and terete or narrowly ovoid, with many very narrow heads; involucre terete or nearly ovoid, 3–4 mm. high, the phyllaries lightly tomentulose or glabrate, yellow-green and shining; ray-florets 3–7; disk-florets 2–8, fertile; achenes ellipsoid, glabrous; receptacle glabrous. $n = 9$.

Coastal Canadian and Humid Transition Zones; Vancouver Island southward along the coast and somewhat inland along some of the rivers to Sonoma County, California, not ascending to over 300 feet elevation. Type locality: stony sea-beaches near Fairhaven, Whatcom County, Washington. June–Aug.

6. Artemisia tilèsii subsp. unalaschcénsis (Bess.) Hultén. Aleutian Mugwort. Fig. 5698.

Artemisia vulgaris var. *vulgatissima* Bess. in Hook. Fl. Bor. Amer. 1: 322. 1833.
? *Artemisia vulgaris* var. *americana* Bess. Linnaea 15: 105. 1841.
Artemisia tilesii var. *unalaschcensis* Bess. op. cit. 106.
Artemisia tilesii var. *elatior* Torr. & Gray, Fl. N. Amer. 2: 422. 1843.
? *Artemisia obtusa* Rydb. N. Amer. Fl. 34: 274. 1916.
Artemisia tilesii var. *elatior* f. *pubescens* G. N. Jones, Univ. Wash. Publ. Biol. 5: 254. 1936.
Artemisia tilesii subsp. *unalaschcensis* Hultén, Fl. Aleut. Isl. 327. 1937.
? *Artemisia ludoviciana* var. *americana* Fernald, Rhodora 47: 248. 1945, in part.
Artemisia tilesii subsp. *elatior* Hultén, Fl. Alaska & Yukon 1573. 1950.

Perennial herb, 3–15 dm. tall; stems herbaceous to base, floccose when young but soon glabrate. Leaves variable, 5–15 cm. long, linear-lanceolate to oblanceolate, ovate or subrotund in outline, the narrowest sometimes entire, others few-toothed or -lobed to deeply pinnately or subpalmately divided, some of the segments again toothed or parted, tomentose beneath, green or rarely tomentulose above, those of the inflorescence entire or nearly so; inflorescence paniculiform, more or less narrowed and spike-like, well exceeding the leaves, the heads erect or nodding; involucre campanulate or turbinate, 3–5 mm. high, 4–8 mm. wide, the phyllaries purplish or sometimes green, glabrate or tomentulose; ray-florets 8–15; disk-florets 20–40, fertile; achenes ellipsoid, glabrous; receptacle glabrous. $n = 27$.

Open, rocky or gravelly sites, Boreal and Transition Zones; northern Oregon east to Montana and north to Hudson's Bay, southern Alaska, the Aleutian Islands, and Japan. Type locality: island of Unalaska. July–Sept.

Typical *A. tilesii* is more northerly and has a shorter, more compact inflorescence which is usually over-topped by the leaves. In parts of our area *A. tilesii* subsp. *unalaschcensis* seems to intergrade to some extent with *A. ludoviciana* subsp. *candicans* and with *A. douglasiana*.

7. Artemisia lindleyàna Bess. Columbia River Mugwort. Fig. 5699.

Artemisia lindleyana Bess. in Hook. Fl. Bor. Amer. **1** : 322. 1833.
Artemisia lindleyana var. *brevifolia* Bess. in Hook. loc. cit.
Artemisia lindleyana var. *subdentata* Bess. in Hook. loc. cit.
Artemisia lindleyana var. *coronopus* Bess. in Hook. loc. cit.
Artemisia pumila Nutt. Trans. Amer. Phil. Soc. II. **7** : 399. 1841. Not Link, 1822.
Artemisia arachnoidea Sheldon, Bull. Torrey Club **30** : 310. 1903.
Artemisia leibergii Rydb. N. Amer. Fl. **34** : 267. 1916.
Artemisia vulgaris subsp. *lindleyana* Hall & Clem. Carnegie Inst. Wash. Publ. No. 326: 79. 1923.
Artemisia vulgaris var. *lindleyana* Jepson, Man. Fl. Pl. Calif. 1142. 1925.

Stems several, 2–7 dm. tall, suffrutescent at the base. Leaves linear-oblanceolate, rarely wider, 2–5 cm. long, mostly under 1 cm. wide, entire or with a few teeth or narrow lobes, white-tomentose beneath, usually green and glabrate above ; inflorescence paniculiform, usually short, narrow, and spike-like ; involucre campanulate, 3 mm. high, 2–3 mm. wide, the phyllaries usually lightly

5695. Artemisia ludoviciana
5696. Artemisia douglasiana

5697. Artemisia suksdorfii
5698. Artemisia tilesii

tomentulose; ray-florets 5–9; disk-florets 10–30, fertile; achenes ellipsoid, glabrous; receptacle glabrous.

Shores of streams below the high-water level, Transition Zones; mostly along the Columbia River and its tributaries in Washington and Oregon sporadically to western Montana and southern British Columbia. Type locality: "North-West coast of America." July–Sept.

Artemisia prescottiàna Bess in Hook. Fl. Bor. Amer. 1: 324. 1833. Differs from *A. lindleyana* mainly in its more fragile appearance and in its narrower and finely cut leaves, these 2–5 cm. long, pinnatifid with 3–7 filiform lobes, less than 1 mm. wide, green, the margins closely revolute. This is the form found in the eastern part of the Columbia Gorge. Plants with characters intermediate between this and *A. lindleyana* are also common in this area, and it is probably at best only a variety of that species. Type locality: "North-West America."

8. Artemisia michauxiàna Bess. Michaux's Mugwort. Fig. 5700.

Artemisia michauxiana Bess. in Hook. Fl. Bor. Amer. 1: 324. 1833.
Artemisia discolor Dougl. ex Bess. Bull. Soc. Nat. Mosc. 9: 46. 1836.
Artemisia vulgaris subsp. *discolor* Hall & Clem. Carnegie Inst. Wash. Publ. No. 326: 74. 1923.
Artemisia vulgaris var. *discolor* Jepson, Man. Fl. Pl. Calif. 1141. 1925.
Artemisia vulgaris subsp. *michauxiana* St. John, Research Stud. St. Coll. Wash. 1: 106. 1929.
Artemisia vulgaris subsp. *michauxiana* var. *typica* St. John, loc. cit.
Artemisia discolor var. *glandulifera* Henderson, Rhodora 32: 27. 1930.
Artemisia vulgaris var. *glandulifera* M. E. Peck, Man. Pl. Oregon 766. 1941.

Perennial herb, 2–4 dm. tall, sometimes taller; stems several from a woody caudex. Leaves tomentose beneath at least when young, generally glabrate and green above, rather crowded, 2–5 cm. long, bipinnatifid with secondary lobes again toothed, the lobes linear, widely spreading and acute, reduced upward, the uppermost sometimes entire; inflorescence narrowly paniculiform or spiciform-thyrsoid, the heads nodding at least at first; involucre hemispheric, 3.5–4 mm. high, 2–4 mm. wide, the phyllaries glabrous or sparingly tomentose; ray-florets 9–12; disk-florets 15–35, fertile; achenes ellipsoid, glabrous; receptacle glabrous.

Rocky sites in the mountains, mostly Boreal Zone; Cascade Mountains of Washington north to British Columbia, Alberta, and northern Utah, also sporadically in southeastern Oregon, California, and Nevada. Type locality: "near Kettle Falls and sources of the Columbia." May–Aug.

This species, at least in the southern part of its range, seems to intergrade somewhat with *A. ludoviciana* subsp. *incompta*.

9. Artemisia biénnis Willd. Biennial Sagewort. Fig. 5701.

Artemisia biennis Willd. Phytog. 11. 1794.

Taprooted annual or biennial herb, 3–30 dm. tall; stems simple below, erect, striate, glabrous. Leaves widely petiolate, 5–15 cm. long, glabrous, pinnatifid or bipinnatifid, the segments lanceolate, often sharply toothed; inflorescence compact, leafy, spiciform or of spiciform branches, the heads numerous, erect, scarcely peduncled; involucre hemispheric, glabrous, 2–3 mm. high, slightly broader, the outer phyllaries narrow and green, the inner ones with only a green midrib and broad scarious margin; ray-florets 6–25; disk-florets 15–40, fertile; achenes ellipsoid, glabrous; receptacle glabrous.

Sandy waste places and stream banks, Upper Sonoran and Transition Zones; introduced occasionally throughout our area and eastward as a weed; native in British Columbia and the northern Rocky Mountains. Type locality: erroneously given as "New Zealand." Aug.–Oct.

Artemisia ánnua L. Sp. Pl. 847. 1753. Taprooted annual, 3–30 dm. tall; stem simple below, erect or flexuous, striate, glabrous. Leaves petiolate, 2–10 cm. long, glabrous, pinnatifid to tripinnatifid, the segments lanceolate or linear; inflorescence broad and open, paniculiform, leafy, the heads peduncled, often nodding; involucre glabrous, 1–2 mm. high, slightly broader, outer phyllaries narrow and green, the inner ones with only a green midrib and a broad scarious margin; ray-florets 5–10; disk-florets 5–20, fertile; achenes narrowly turbinate, glabrous; receptacle glabrous. $n = 9$. Fields and waste places; native of Eurasia, naturalized in central and eastern United States, rare and probably not established in our area. Type locality: Siberia. Annual Sagewort.

10. Artemisia califórnica Less. Coast Sagebrush or Old Man. Fig. 5702.

Artemisia californica Less. Linnaea 6: 523. 1831.
Artemisia fisheriana Bess. Nuov. Mem. Soc. Nat. Mosc. 3: 21. 1834.
Artemisia fisheriana vegetior Bess. op. cit. 88.
Artemisia foliosa Nutt. Trans. Amer. Phil. Soc. II. 7: 397. 1841.
Artemisia abrotanoides Nutt. op. cit. 399.
Chrossostephium foliosum Rydb. N. Amer. Fl. 34: 243. 1916.
Chrossostephium californicum Rydb. loc. cit.

Rounded shrub, 5–20 dm. high, rarely higher; stems freely branched, the older with brown fibrous bark, the twigs erect, stout, striate, and covered with a gray to nearly white, dense canescence. Leaves sessile, ternately, pinnately, or sometimes bipinnately divided into long narrow segments 0.5–1 mm. wide, minutely but densely cinereous or canescent, with strongly revolute margins, usually some borne in fascicles in the axils of others, fascicled and upper leaves reduced in size; inflorescence paniculiform, leafy-bracted, the heads nodding; involucre hemispheric, 2.5–4 mm. high, 3–5 mm. wide, the outer phyllaries short, thick, and herbaceous, the inner ones with a thick back and broad scarious margin, canescent; ray-florets 6–10; disk-florets 15–30, fertile; achenes oblong-turbinate, with minute squamellate crown, resinous-granuliferous; receptacle essentially glabrous.

Exposed slopes, Upper Sonoran and Arid Transition Zones; Coast Ranges from north of San Francisco Bay south to northern Lower California and on adjacent islands, extending eastward in southern California to edge of Colorado Desert. Type locality: California. Aug.–Oct.

Artemisia californica var. insulàris (Rydb.) Munz, Man. S. Calif. 575. 1935. (*Crossostephium insulare* Rydb. N. Amer. Fl. 34: 244. 1916.) Differs from *A. californica* var. *californica* in that the leaves are little or not at all revolute, the lobes therefore 1–3 mm. in width, lobes also generally fewer; ray-florets to 15, disk-florets to 40. San Clemente Island, Orange County, and San Nicolas Island, Ventura County, California. Type locality: "along Pot's Trail, San Clemente Island." Collected by Blanche Trask.

11. Artemisia bigelòvii A. Gray. Bigelow's Sagebrush. Fig. 5703.

Artemisia bigelovii A. Gray in Torr. Pacif. R. Rep. **4**: 110. 1857.
Artemisia petrophila Woot. & Standl. Contr. U.S. Nat. Herb. **16**: 193. 1913.

Low evergreen shrub, 2–4 dm. tall; stems many, spreading below and sometimes rooting when in contact with the soil, the bark grayish brown, sheathing or becoming shredded on the larger branches, twigs densely canescent, striate. Leaves of vegetative shoots sessile by a narrow base or short-petioled, narrowly cuneate, 1–2 cm. long, 2–5 mm. wide, sharply 3-toothed at the truncate apex or sometimes entire; leaves of the inflorescence mostly entire, oblanceolate, acute, or slightly obtuse; flowering shoots numerous, erect or the tips drooping, narrowly paniculiform, dense, the heads several on each short recurved branch; involucre turbinate, 2–4 mm. long, 1.5–2.5 mm. broad, the inner phyllaries twice as long as the outer, all densely tomentose, the margins scarious; ray-florets 0–2, usually 1, both lacking and present in heads of the same plant; disk-florets 1–3, usually 2, fertile; achenes ellipsoid, glabrous; receptacle glabrous. $n = 9$.

Rocky soils, Upper Sonoran Zone; occurs in California at the east end of the Mojave Desert, to Colorado and Texas. Type locality: "along the upper Canadian" in western Texas. Aug.–Oct.

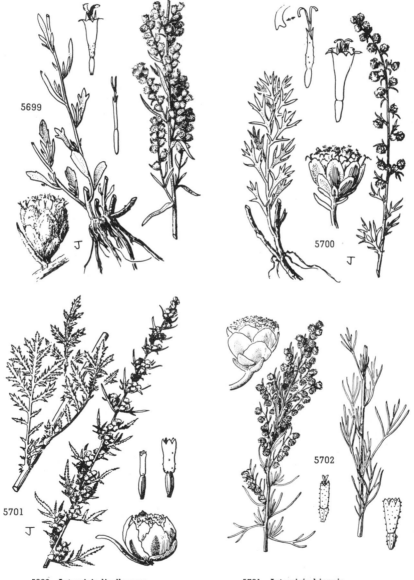

5699. Artemisia lindleyana
5700. Artemisia michauxiana
5701. Artemisia biennis
5702. Artemisia californica

12. **Artemisia absínthium** L. Absinthe or Wormwood. Fig. 5704.

Artemisia absinthium L. Sp. Pl. 848. 1753.

Perennial herb sometimes slightly woody at base, 4–12 dm. tall; aromatic stems clustered, simple below, striate, sericeous or glabrate. Lower leaves long-petiolate, the blade 3–8 cm. long, rounded-ovate in outline, bi- to tripinnatifid, with oblong or oblanceolate, obtuse lobes, these often toothed, pubescence silky-sericeous or sometimes subglabrate; upper leaves progressively shorter-petiolate and less divided, the lobes often lanceolate and acute; inflorescence paniculiform, profuse, leafy; involucre hemispheric, 2–3 mm. high, finely and densely sericeous, the outer phyllaries linear, the inner broadly elliptic with broad scarious margins; ray-florets 9–20; disk-florets 30–50, fertile; achenes nearly cylindric, narrowed at the base, glabrous; receptacle thickly beset with long hairs. $n = 8, 9$.

Fields and waste places, Arid Transition Zone; introduced in eastern Oregon and Washington eastward across the continent; native to Europe. July–Sept.

13. **Artemisia frigida** Willd. Prairie Sagewort. Fig. 5705.

Artemisia frigida Willd. Sp. Pl. **3**: 1838. 1804.
Absinthium frigidum Bess. Bull. Soc. Nat. Mosc. **1**: 251. 1829.

Perennial, mat-forming herb often woody at base and even somewhat shrubby, 1–4 dm. high, fragrant. Leaves small and numerous, crowded at the base and well distributed along the stem, sericeous-tomentose, twice or thrice ternately divided into linear or linear-filiform lobes 1 mm. or less wide, usually with a pair of trifid, stipule-like divisions at the base, petiolate or the upper ones subsessile, the blade 5–12 mm. long; inflorescence paniculiform or much reduced and racemiform on depauperate plants, the heads sessile or short-peduncled, nodding; involucre hemispheric, 2–3 mm. high, the phyllaries loosely tomentose; ray-florets 10–17; disk-florets 25–50, fertile; achenes subcylindric, glabrous; receptacle thickly beset with long hairs.

Dry open places in plains and foothills, Arid Transition and Boreal Zones; northeastern Washington to Arizona, Wisconsin, Alaska, and Siberia; characteristic of the high plains. Type locality: Davuria, eastern Siberia. July–Sept.

14. **Artemisia dracúnculus** L. Dragon Sagewort or Tarragon. Fig. 5706.

Artemisia dracunculus L. Sp. Pl. 849. 1753.
Artemisia glauca Pall. ex Willd. Sp. Pl. **3**: 1831. 1804.
Artemisia dracunculoides Pursh, Fl. Amer. Sept. 742. 1814.
Artemisia dracunculus var. *glauca* Bess. in Hook. Fl. Bor. Amer. **1**: 326. 1833.
Artemisia aromatica A. Nels. Bull. Torrey Club **27**: 273. 1900.
Artemisia dracunculus subsp. *typica* Hall & Clem. Carnegie Inst. Wash. Publ. No. 326: 115. 1923.
Artemisia dracunculus subsp. *glauca* Hall & Clem. op. cit. 116.
Artemisia dracunculoides var. *glauca* Munz, Man. S. Calif. 575. 1935.

Perennial herb, 5–15 dm. tall, strongly odorous to nearly inodorous; stems several but not crowded, erect, simple to the inflorescence. Leaves linear to somewhat lanceolate, mostly 3–8 mm. long, 2–10 mm. wide, glabrous or occasionally villous-puberulent and glabrate, entire or a few of them cleft; inflorescence paniculiform, leafy-bracted and with ascending branches, the heads generally nodding; involucre hemispheric, 2–3 mm. high, 2–4 mm. wide; ray-florets 6–30; disk-florets 10–30, the pistil sterile; achenes glabrous, those of the disk-florets abortive; receptacle glabrous. $n = 9$.

Open places, plains to moderate elevations in the mountains, Arid Transition Zone; eastern Washington south to Lower California, to Texas, Illinois, Yukon, and Eurasia. Type locality: Siberia. July–Oct.

15. **Artemisia campéstris** subsp. **pacífica** (Nutt.) Hall & Clem. Pacific Sagewort. Fig. 5707.

Artemisia desertorum var. *douglasiana* Bess. in Hook. Fl. Bor. Amer. **1**: 325. 1833.
Artemisia desertorum var. *scouleriana* Bess. in Hook. loc. cit.
Artemisia pacifica Nutt. Trans. Amer. Phil. Soc. II. **7**: 410. 1841.
Artemisia scouleriana Rydb. Bull. Torrey Club **33**: 157. 1906.
Artemisia camporum Rydb. N. Amer. Fl. **34**: 254. 1916.
Artemisia campestris var. *pacifica* M. E. Peck, Man. Pl. Oregon 768. 1941.
Artemisia campestris var. *scouleriana* Cronquist, Leaflets West. Bot. **7**: 20. 1953.

Perennial herb, 3–10 dm. high, with somewhat spreading base; stems several to numerous, erect, very leafy below, somewhat leafy to the inflorescence. Leaves bipinnately divided into narrow linear divisions about 1 mm. wide, canescent or silky, rarely glabrous, 4–12 cm. long including the petiole, reduced upward; inflorescence paniculiform, usually with numerous closely ascending branches, the heads horizontal or erect, nodding only when young; involucre hemispheric, 2–3 mm. high, about 2.5 mm. wide, the phyllaries greenish and glabrous or sparingly tomentulose; ray-florets 8–20; disk-florets 10–25, the pistil sterile; achenes glabrous, those of the disk-florets abortive; receptacle glabrous.

Open places, usually in sandy soil, Canadian and Transition Zones; Oregon and Washington to Yukon, western Nebraska, and New Mexico. Type locality: "shores of the Pacific at the outlet of the Oregon [Columbia River]." July–Sept.

Artemisia campestris subsp. boreàlis (Pall.) Hall & Clem. Carnegie Inst. Wash. Publ. No. 326: 122. 1923. (*Artemisia borealis* Pall. Reise **3**: 755. 1776; *A. spithamea* Pursh, Fl. Amer. Sept. 522. 1814; *A. borealis* var. *purshii* Bess. in Hook. Fl. Bor. Amer. **1**: 326. 1833; *A. borealis* var. *wormskioldii* Bess. in Hook. op. cit. 327; *A. borealis* var. *besseri* Torr. & Gray, Fl. N. Amer. **2**: 417. 1843; *A. borealis* var. *spithamea*

Torr. & Gray, loc. cit.; *A. borealis* subsp. *wormskioldii* Piper, Contr. U.S. Nat. Herb. **11** : 587. 1906; *A. ripicola* Rydb. N. Amer. Fl. **34** : 256. 1916; *A. campestris* subsp. *spithamea* Hall & Clem. Carnegie Inst. Wash. Publ. No. 326: 123. 1923; *A. campestris* var. *spithamea* M. E. Peck, Man. Pl. Oregon 768. 1941; *A. borealis* f. *wormskioldii* Rousseau, Le Nat. Canadien **71** : 189. 1944; *A. campestris* var. *wormskioldii* Cronquist, Leaflets West. Bot. **6** : 43. 1950.) Perennial herb, 1–4 dm. high, often with a spreading base; stems crowded, erect or ascending, densely leafy below, less so up into the inflorescence. Leaves once or twice ternately divided into mostly linear or linear-filiform lobes or sometimes pinnately divided, glabrous to sericeous or villous, the upper surface sometimes glabrate, 2–10 cm. long, 0.7–4 cm. wide, reduced upward, the uppermost often entire; inflorescence spiciform, recemiform or narrowly paniculiform, the heads erect to nodding; involucre hemispheric, 3–4 mm. tall, 3.5–5 mm. broad, the phyllaries brownish or yellowish green with a brownish median line, glabrous to densely villous; ray-florets 10–25; disk-florets 15–30; achenes glabrous, those of the disk-florets abortive; receptacle glabrous.

Open places in sandy soil, along the Columbia River between Oregon and Washington and in the Cascade Mountains of Washington to northern Montana and northward, and in Siberia. Type locality: Siberia.

Artemisia campestris subsp. **caudàta** (Michx.) Hall & Clem. Carnegie Inst. Wash. Publ. No. 326: 122. 1923. (*Artemisia caudata* Michx. Fl. Bor. Amer. **2** : 129. 1803; *A. forwoodii* S. Wats. Proc. Amer. Acad. **25** : 133. 1890.) Root biennial, stem usually single; leaves in our plants mostly canescent when young, glabrous; otherwise much like subsp. *pacifica*. A few plants of this are found in the Puget Sound and Columbia Gorge areas; common from Rocky Mountains eastward. Type locality: sandy banks of the Missouri River.

The plants here included in *A. campestris* L. belong to a number of different races. The entire group is poorly understood and in need of clarification. *Artemisia campestris* subsp. *campestris* is European and has become established in North America only in the eastern part. Our western plants are mostly members of two taxa here referred to as *A. campestris* subsp. *borealis* and *A. campestris* subsp. *pacifica*, which have been variously treated as species, subspecies, and varieties, and even split into a number of species. The limits of these taxa are not at all clear and have been variously drawn. Under the circumstances it seems best to follow closely (but with some modification) the treatment of Hall and Clements. Although inadequate, this is the latest treatment of a monographic nature. The form with pubescent involucre is separated by some as a separate taxon (*A. spithamea* Pursh, *A. borealis* var. *purshii* Bess., *A. campestris* subsp. *spithamea* (Pursh) Hall & Clem.), but this seems to be a character with little if any geographical significance and is not well correlated with other characters. The Columbia River form also has been separated—probably with more justification, at least on ecological grounds—as *A. ripicola* Rydb., *A. borealis* var. *wormskioldii* Bess., or *A. campestris* var. *wormskioldii* (Bess.) Cronquist.

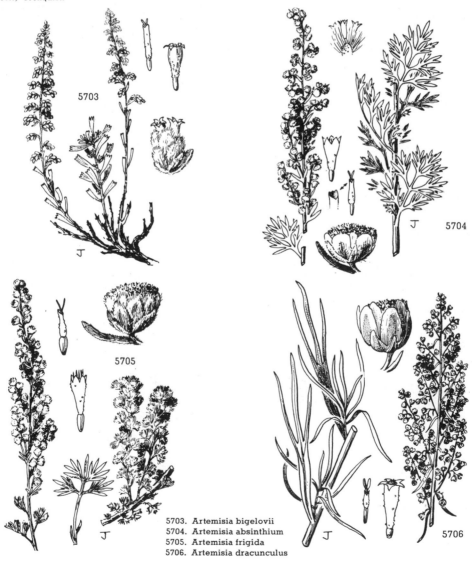

5703. Artemisia bigelovii
5704. Artemisia absinthium
5705. Artemisia frigida
5706. Artemisia dracunculus

16. **Artemisia pycnocéphala** (Less.) DC. Beach Sagewort. Fig. 5708.

Oligosporus pycnocephalus Less. Linnaea **6**: 524. 1831.
Artemisia pycnocephala DC. Prod. **6**: 99. 1837.
Artemisia pachystachya DC. op. cit. 114.
Artemisia pycnostachya Nutt. Trans. Amer. Phil. Soc. II. **7**: 401. 1841.
Artemisia campestris subsp. *pycnocephala* Hall & Clem. Carnegie Inst. Wash. Publ. No. 326: 123. 1923.
Artemisia campestris var. *pycnocephala* M. E. Peck, Man. Pl. Oregon 768. 1941.

Perennial herb, 2–6 dm. high; stems several or numerous on a stout woody caudex, erect, or ascending near the base, very leafy up to the inflorescence. Leaves twice or thrice pinnately divided into narrow, linear or linear-spatulate divisions, densely and permanently silky-villous, 2–7 cm. long, reduced upward; inflorescence paniculiform but very dense and spike-like or with spike-like branches, leafy, the heads sessile and erect; involucre hemispheric, 3.5–4.5 mm. high, about as broad, the phyllaries densely villous; ray-florets 8–15; disk-florets 12–25, the pistil sterile; achenes subcylindric, glabrous, those of the disk-florets abortive; receptacle glabrous. $n = 9$.

Sandy ocean beaches, Humid Transition Zone; from central Oregon south to Point Sur, California. Type locality: California. Aug.–Sept.

17. **Artemisia spinéscens** D. C. Eaton. Bud Sage or Spring Sagebrush. Fig. 5709.

Picrothamnus desertorum Nutt. Trans. Amer. Phil. Soc. II. **7**: 417. 1841. Not *Artemisia desertorum* Spreng. 1826.
Artemisia spinescens D. C. Eaton, Bot. King Expl. 180. 1871.

Rounded shrub, 0.5–5 dm. tall, pungently aromatic, stout and much branched from the base, somewhat spiny by the persistence of the axes of previous years' inflorescences. Leaves up to about 2 cm. long including the petiole, pedately 3- to 5-lobed, the lobes again cleft into linear-spatulate lobes, densely villous, reduced and less divided upward; inflorescence racemiform, 1–5 cm. long, leafy-bracted, sometimes reduced to a single head; involucre broadly turbinate, 2–3.5 mm. high, 3–4.5 mm. wide, the scarcely unequal phyllaries thick and herbaceous with narrow scarious margins, densely villous; ray-florets 2–8, the corolla long-hairy; disk-florets 5–15, the corolla long-hairy, the pistil sterile, style undivided with an expanded penicillate summit; achenes ellipsoid, densely arachnoid-hairy, those of the disk-florets abortive and essentially absent; receptacle glabrous.

Arid plains and hills, Sonoran Zones; southeastern Oregon and eastern California north of the San Bernardino Mountains to southwestern Montana and New Mexico. Type locality: "Rocky Mountain plains, in arid deserts, towards the north sources of the Platte." April–June.

18. **Artemisia tridentàta** Nutt. Sagebrush. Fig. 5710.

Artemisia tridentata Nutt. Trans. Amer. Phil. Soc. II. **7**: 398. 1841.
Artemisia tridentata var. *angustifolia* A. Gray, Proc. Amer. Acad. **19**: 49. 1883.
Artemisia angusta Rydb. N. Amer. Fl. **34**: 283. 1916.
Artemisia vaseyana Rydb. loc. cit.
Artemisia tridentata subsp. *typica* Hall & Clem. Carnegie Inst. Wash. Publ. No. 326: 136. 1923.

Strong-scented evergreen shrub, rounded or somewhat spreading and flattened on top, rarely root-sprouting, 1–50 (usually 3–30) dm. high, with a single short trunk or with few ascending branches from the base; older stems covered with shredding, gray or light brown bark, younger stems covered with a dense tomentum, striated. Leaves of vegetative shoots sessile or short-petioled, silvery-canescent and slightly viscid, typically cuneate with 3 blunt teeth at the truncate apex, sometimes 4–9-toothed or shallowly lobed, or linear and entire with a blunt or rounded apex, 1–4 cm. long or rarely longer, 0.2–1.3 cm. wide, mostly three to six times as long as wide; leaves of the flowering shoots the same or more often entire and linear to oblanceolate, rounded to acute; inflorescence well exceeding to barely exceeding the leafy shoots, narrowly to broadly paniculiform, the branches erect or sometimes drooping; involucre 3–3.8 mm. high, 2–2.8 mm. wide, the outer phyllaries short, orbicular-ovate, canescent, the inner elliptic, somewhat narrowly so, obtuse, canescent or sometimes nearly glabrous; ray-florets absent; disk-florets usually 3–6 (or as many as 12 in some high-elevation forms); achenes cylindric-turbinate, resinous-granuliferous, rarely sparsely short-villous; receptacle glabrous. $n = 9, 18$.

On dry plains and hills and extending into the mountains to timberline mostly on rocky open ground, Upper Sonoran Zone to Hudsonian Zone; Washington, Oregon, and California mostly east of the Cascadian-Sierran axis into the Coast Ranges of southern California to northern Lower California and to the western edge of the Great Plains. Type locality: "Plains of the Oregon [Columbia], and Lewis' [Snake] River." July–Oct.

A highly polymorphic species with great variation and numerous ecological races, some of which may be recognized as varieties, if sufficient correlation of ecologic and morphologic characters can be discovered. To date this has not been done.

Artemisia tridentata subsp. **paríshii** (A. Gray) Hall & Clem. Carnegie Inst. Wash. Publ. No. 326: 137. 1923. (*Artemisia parishii* A. Gray, Proc. Amer. Acad. **17**: 220. 1882; *A. tridentata* var. *parishii* Jepson, Man. Fl. Pl. Calif. 1140. 1925.) Rounded evergreen shrub, usually 10–20 dm. tall; leaves mostly linear or narrowly spatulate, entire or shallowly notched or toothed at the apex or frequently with 3 short divergent lobes, rarely more deeply divided or pinnately 5-lobed; achenes glandular and sparingly short-villous to arachnoid-hairy; otherwise much the same as *A. tridentata* subsp. *tridentata*. $n = 18$. Western Mojave Desert of California, extending toward the Pacific Coast through the Santa Clara Valley. Type locality: Newhall, Los Angeles County.

19. **Artemisia arbúscula** Nutt. Dwarf Sagebrush. Fig. 5711.

Artemisia arbuscula Nutt. Trans. Amer. Phil. Soc. II. **7**: 398. 1841.
Artemisia tridentata subsp. *arbuscula* Hall & Clem. Carnegie Inst. Wash. Publ. No. 326: 138. 1923.
Artemisia tridentata var. *arbuscula* McMinn, Ill. Man. Calif. Shrubs 608. 1939.

Low, spreading, evergreen shrub, 1–4 dm. high, not root-sprouting, odor strong to mild, bark light brown to nearly black, shredding on older branches; twigs densely canescent, often becoming nearly glabrous in late summer, light greenish yellow or reddish brown, striated. Leaves of vegetative shoots broadly cuneate or flabelliform, 0.5–1.5 cm. long, 0.3–1 cm. wide, length one and one-half to three times (usually about twice) the width, usually 3–5-toothed but sometimes deeply divided into 3–5 narrow lobes or rarely a few leaves entire, canescent; leaves of flowering shoots linear-oblanceolate and entire to cuneate with a truncate, 3-toothed apex, or sometimes more deeply divided with 3 narrow lobes; inflorescence spiciform to narrowly paniculiform, when branched the branches few and erect; involucre narrowly to broadly campanulate, 1.5–3 mm. broad, 3–4.2 mm. high, the phyllaries canescent or the innermost sometimes nearly glabrous; ray-florets absent; disk-florets 6–11, or sometimes fewer in dry locations, fertile; achenes cylindric-turbinate, resinous–granuliferous; receptacle glabrous. $n = 9, 18$.

Usually on rocky or gravelly soil, Upper Sonoran, Arid Transition, and Canadian Zones; arid plains and hills and rocky openings in the mountains in eastern California from Mono County northward, eastern Oregon, and isolated stations in the Wenatchee Mountains of Washington and the Siskiyou Mountains and Coast Ranges of California eastward to southwestern Montana and northwestern Colorado. Type locality: "arid plains of Lewis' [Snake] River." Aug.–Oct.

Artemisia arbuscula subsp. **nòva** (A. Nels.) G. Ward, Contr. Dudley Herb. **4**: 183. 1953. (*Artemisia nova* A. Nels. Bull. Torrey Club **27**: 274. 1900; *A. tridentata* subsp. *nova* Hall & Clem. Carnegie Inst. Wash. Publ. No. 326: 137. 1923; *A. tridentata* var. *nova* McMinn, Ill. Man. Calif. Shrubs 608. 1939; *A. arbuscula* var. *nova* Cronquist, Vasc. Pl. Pacif. Northw. **5**: 58. 1955.) Differs from *A. arbuscula* subsp. *arbuscula* principally in having small heads with mostly 3–5 florets and inner phyllaries glabrous or rarely with a few hairs, shiny; there is also a tendency, particularly to the east of our range, toward leaves of a dark greenish color and very glandular; leaves also average somewhat smaller than in the typical subspecies. $n = 9, 18$. East of the Sierra Nevada from Mono County, California, southward in the San Bernardino and Mojave Desert ranges to northern New Mexico and southern Montana. Type locality: Medicine Bow, Carbon County, Wyoming.

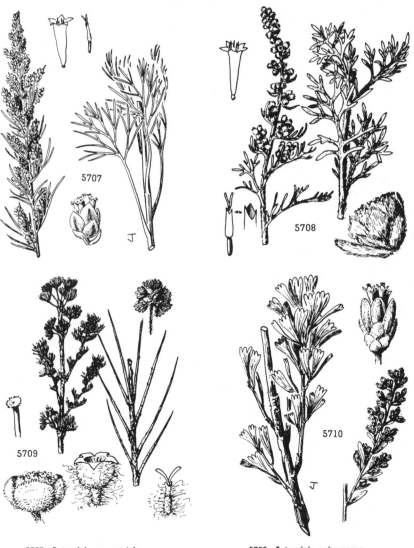

5707. Artemisia campestris
5708. Artemisia pycnocephala
5709. Artemisia spinescens
5710. Artemisia tridentata

20. **Artemisia tripartìta** Rydb. Cutleaf Sagebrush. Fig. 5712.

Artemisia trifida Nutt. Trans. Amer. Phil. Soc. II. 7 : 398. 1841. Not Turcz. 1832.
Artemisia tripartita Rydb. Mem. N.Y. Bot. Gard. 1 : 432. 1900.
Artemisia tridentata subsp. *trifida* Hall & Clem. Carnegie Inst. Wash. Publ. No. 326: 137. 1923.
Artemisia tridentata var. *trifida* McMinn, Ill. Man. Calif. Shrubs 608. 1939.

A low, rounded, evergreen shrub, 2–8 dm. high with a short central trunk or with numerous branches ascending from the base, frequently root-sprouting, mildly aromatic, bark light brown or grayish, shredding on older branches; twigs canescent. Leaves of vegetative shoots 0.5–3 cm. long or rarely longer, canescent, deeply divided into 3 linear or narrowly linear-lanceolate lobes which may sometimes themselves be 3-cleft or some of the leaves often linear and entire; leaves of flowering shoots deeply 3-cleft into linear lobes or linear and entire; inflorescence narrowly to broadly paniculiform; involucre campanulate, 3–4 mm. high, 2–3 mm. wide, the phyllaries canescent, the outer ones broadly ovate, sometimes with a narrow herbaceous tip, the inner ones oblong; ray-florets absent; disk-florets 4–7 (rarely more), fertile; achenes turbinate or nearly cylindrical, resinous-granuliferous; receptacle glabrous. $n = 9, 18$.

Dry hills and plains, Upper Sonoran and Arid Transition Zones; eastern Washington and northeastern Oregon through southern Idaho to Wyoming, northern Utah, and western Montana. Type locality: "Plains of the Rocky Mountains." July–Sept.

21. **Artemisia càna** Pursh. Hoary Sagebrush. Fig. 5713.

Artemisia cana Pursh, Fl. Amer. Sept. 521. 1814.
Artemisia columbiensis Nutt. Gen. 2 : 142. 1818.

Low, rounded, evergreen (at least in our area) shrub, 4–9 (or rarely to 15) dm. high, often root-sprouting, freely branching, somewhat fastigiate, mildly aromatic; twigs striate, canescent; older branches covered with a brown fibrous bark. Leaves of vegetative shoots 2–6 cm. long, 0.1–0.7 cm. wide, linear or linear-oblanceolate, entire and acute or acuminate or sometimes with 1–2 irregular teeth or lobes, silvery-canescent, sometimes becoming viscidulous with age; leaves of the inflorescence similar, reduced above; inflorescence paniculiform or sometimes reduced, in the extreme spike-like; involucre campanulate, 2.3–4.5 mm. wide, 3.5–5 mm. high, the outer phyllaries broadly ovate with an acute or acuminate tip and densely canescent or only sparsely tomentose, the inner phyllaries elliptic to linear-obovate with a broad scarious margin, canescent to nearly glabrous; ray-florets absent; disk-florets 8–20, rarely fewer, fertile; achenes cylindric-turbinate, granuliferous; receptacle glabrous. $n = 9, 18$.

Mostly on low ground, Upper Sonoran, Transition, and Canadian Zones; southeastern Oregon and eastern California south to Mono County east to Saskatchewan, Nebraska, and Colorado; mostly on moister sites than *A. tridentata* and also more tolerant of alkali. Type locality: bluffs of the Missouri River. Collected by Lewis. Aug.–Sept.

Artemisia cana subsp. **bolánderi** (A. Gray) G. Ward, Contr. Dudley Herb. **4**: 192. 1953. (*Artemisia bolanderi* A. Gray, Proc. Amer. Acad. **19**: 50. 1883; *A. tridentata* subsp. *bolanderi* Hall & Clem. Carnegie Inst. Wash. Publ. No. 326: 139. 1923; *A. tridentata* var. *bolanderi* McMinn, Ill. Man. Calif. Shrubs 609. 1939.) Shrub 2–6 dm. high, much branched, often root-sprouting; leaves mostly 1–3 cm. long, 0.1–0.25 cm. wide, linear or linear-oblanceolate and entire or sometimes dilated at the tip and divided into 3 narrow lobes, silvery-canescent; involucre densely canescent; disk-florets 8–16. This subspecies differs from *A. cana* subsp. *cana* mainly in its smaller size and usually somewhat more divided leaves. $n = 9$. Limited to the area around Mono Lake, Mono County, California. Type locality: Mono Pass. Mono-Tuolumne County line, Sierra Nevada.

22. **Artemisia rothróckii** A. Gray. Timberline Sagebrush. Fig. 5714.

Artemisia rothrockii A. Gray, Bot. Calif. 1 : 618. 1876.
Artemisia tridentata subsp. *rothrockii* Hall & Clem. Carnegie Inst. Wash. Publ. No. 326: 139. 1923.
Artemisia tridentata var. *rothrockii* McMinn, Ill. Man. Calif. Shrubs 608. 1939.

Low, spreading, evergreen shrub, 2–6 dm. high, often root-sprouting, mildly aromatic, bark light gray or straw-colored to dark grayish brown, fibrous on older stems; twigs striated, covered with a dense tomentum at least when young. Leaves of vegetative shoots 0.5–5 cm. long, 0.2–1.5 cm. wide, mostly broadly cuneate or flabelliform, 3-toothed or -lobed, or sometimes lanceolate or oblanceolate and entire, densely canescent, sometimes becoming glabrate or viscidulous with age; leaves of flowering shoots similar or all lanceolate and entire; inflorescence spiciform to narrowly paniculiform; involucre campanulate, 3–5 mm. wide, 4–5.5 high, the outer phyllaries ovate with an acute or acuminate tip and canescent or only sparsely tomentose, the inner phyllaries elliptic to narrowly obovate and usually sparsely tomentose to glabrous; ray-florets absent; disk-florets 8–20 (mostly 10–16), fertile; achenes cylindric-turbinate, granuliferous; receptacle glabrous. $n = 18, 27$, about 36.

Rocky sites or sometimes in moist pockets at lower elevations, Boreal Zone; mostly at high elevations in the Sierra Nevada and the White and San Bernardino Mountains of California. Type locality: Monache Meadows, Tulare County, Sierra Nevada. Aug.–Sept.

23. **Artemisia rígida** (Nutt.) A. Gray. Scabland Sagebrush. Fig. 5715.

Artemisia trifida var. *rigida* Nutt. Trans. Amer. Phil. Soc. II. 7 : 398. 1841.
Artemisia rigida A. Gray, Proc. Amer. Acad. **19** : 49. 1883.

Low, spreading, deciduous shrub, 1–4 dm. tall, probably never root-sprouting, pungently aromatic; stems thick and rigid, the older covered with dark gray, shredding bark, the twigs striated and covered with a dense tomentum at least when young. Leaves of vegetative shoots silvery-canescent, 1.5–4 cm. long, 0.1–1 cm. wide, mostly spatulate in outline, divided into 3–5 linear lobes or some linear and entire; leaves of flowering shoots similar, only slightly reduced above, all longer than the heads; inflorescence spiciform or rarely with 1 or 2 spiciform branches from near the base; involucre campanulate, 4–5 mm. high, 2.5–3.5 mm. wide, canescent, the outer phyllaries broadly orbicular and acute, the inner elliptic to spatulate and obtuse; ray-florets absent; disk-florets 5–16, fertile; achenes somewhat prismatic, glabrous; receptacle glabrous. $n = 9, 18$.

Steep rocky hillsides and rocky scablands, Upper Sonoran Zone; central and southeastern Washington, northeastern Oregon, and western Idaho. Type locality: "plains of Lewis' [Snake] River, in the Rocky Mountains." Sept.–Oct.

24. Artemisia pálmeri A. Gray. San Diego Sagewort. Fig. 5716.

Artemisia palmeri A. Gray, Proc. Amer. Acad. **11**: 76. 1876.
Artemisiastrum palmeri Rydb. N. Amer. Fl. **34**: 285. 1916.

Shrubby at the base with long, herbaceous, wand-like stems 10–30 dm. high, in clusters up to 10 dm. across, aromatic, the lower woody stems covered with a grayish yellow bark, herbaceous stems green or yellowish below, reddish above, heavily striated, glabrous or minutely puberulent. Leaves 5–12 cm. long, pinnately parted into 3–9 linear or linear-lanceolate lobes, some entire, linear or linear-lanceolate, margins closely revolute, glabrous or minutely puberulent above, densely tomentose beneath; inflorescence broadly paniculiform, the heads sessile or on peduncles to 5 mm. long; involucre hemispheric or campanulate, 2.5–4 mm. high, 2–3.5 mm. wide, sparingly pubescent or glabrous, the inner phyllaries little longer than the outer; ray-florets absent; disk-florets 12–30, nearly prismatic, granuliferous; receptacle chaffy, each floret subtended by an elliptic obtuse bract, or center florets without bracts. $n = 9$.

Ravines and moist banks, Upper Sonoran Zone; southwestern San Diego County, California, and northwestern Lower California. Type locality: Jamul Valley, San Diego County. July–Aug.

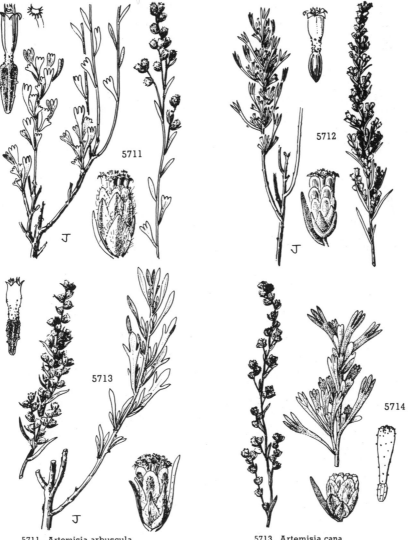

5711. Artemisia arbuscula
5712. Artemisia tripartita
5713. Artemisia cana
5714. Artemisia rothrockii

Tribe 5. **SENECIONEAE.***

Flowers (at least those of the disk) yellow or orange, sometimes tinged with purplish (some spp. of *Luina* merely dull yellowish); plants with all or nearly all of the disk-flowers fertile and bearing divided styles; involucral bracts† 4 or more; pappus well developed at least in the disk-flowers.
    Cauline leaves opposite (or sometimes some of the reduced uppermost ones offset); perennial herbs with the style of *Senecio*.   108. *Arnica*.
    Cauline leaves all or nearly all alternate (or the leaves all basal); styles and habit diverse.
        Pappus evidently plumose; style-branches with elongate, slender, hairy appendages; perennial herbs, discoid or less often radiate.   109. *Raillardella*.
        Pappus not at all plumose; style-branches with deltoid or shorter appendages, or exappendiculate.
            Style-branches penicillate, truncate or nearly so, and essentially exappendiculate.
                Involucre of 1 or 2 series of equal, erect-connivent bracts, sometimes with some smaller, outer, calyculate ones; plants of diverse habit, but not depressed and dichotomously branched.
                    Heads radiate or less often discoid, the pistillate flowers if present ligulate (though sometimes very shortly so); herbs or shrubs.   110. *Senecio*.
                    Heads disciform, the 2 or more rows of marginal flowers pistillate, with tubular-filiform, eligulate corolla; herbs.   115. *Erechtites*.
                Involucre of 2 distinct, subequal series of bracts, the outer rather loose; ours depressed, subdichotomously branched, rosette-forming plants; heads discoid.   113. *Psathyrotes*.
            Style-branches not penicillate (or barely so in some spp. of *Tetradymia*), with or without well-developed appendages; heads discoid, except in *Crocidium*.
                Heads radiate; style-branches with a flat, deltoid, externally short-hairy appendage; diminutive annual.   111. *Crocidium*.
                Heads discoid; style branches nearly or quite exappendiculate, mostly merely papillate and with the stigmatic lines extending nearly or quite to the blunt or acute tip; perennial herbs or shrubs.
                    Shrubs with narrow leaves (less than 5 mm. wide), these sometimes reduced to scales, or modified into spines.
                        Involucral bracts conspicuously imbricate in several series (unique among our genera of the *Senecioneae* in this respect).   118. *Lepidospartum*.
                        Involucral bracts uniseriate or nearly so.
                          Involucral bracts only 4–6.   117. *Tetradymia*.
                          Involucral bracts more numerous, commonly about 13.   116. *Peucephyllum*.
                  Herbs (sometimes slightly woody) with broader leaves, at least the better-developed ones commonly 1 cm. or more wide.   114. *Luina*.
Flowers white, sometimes tinged with pink or purple, not yellow or orange; plants with sterile disk-flowers and undivided or nearly undivided styles, except in *Dimeresia*.
    Involucral bracts and flowers more than 3; pappus not plumose; erect perennial herbs.
        Pappus of numerous capillary bristles; heads unlike, the plants subdioecious.   112. *Petasites*.
        Pappus none; heads all alike, the marginal flowers pistillate and eligulate.   106. *Adenocaulon*.
    Involucral bracts and flowers only 2 or sometimes 3; pappus plumose; depressed annuals.
             107. *Dimeresia*.

106. **ADENOCAÚLON** Hook. Bot. Misc. **1**: 19. 1830.

Heads disciform. Flowers whitish, tubular, the outer about 3–7, pistillate, the inner as many or a few more, staminate, with undivided style. Involucre small, of less than 10 nearly equal, green bracts. Receptacle naked. Anthers rather strongly sagittate but scarcely caudate. Pappus none. Annual or perennial herbs with large alternate leaves and ample subnaked inflorescence, its branches and achenes more or less stipitate-glandular. [Name from the Greek *adcn,* gland, and *kaulon,* stem.]

    A genus of 4 species, one in western North America, one in eastern Asia, one in Guatemala, and one in Chile. Type species, *Adenocaulon bicolor* Hook.

1. **Adenocaulon bìcolor** Hook. Trail-plant. Fig. 5717.

Adenocaulon bicolor Hook. Bot. Misc. 1: 19. *pl. 15.* 1830.
Adenocaulon integrifolium Nutt. Trans. Amer. Phil. Soc. II. **7**: 289. 1840.
Adenocaulon bicolor β integrifolium Torr. & Gray, Fl. N. Amer. **2**: 94. 1841.

Fibrous-rooted, slender perennial up to nearly 1 m. tall. Leaves mostly near the base, long-petiolate, large and thin, deltoid-ovate to cordate or subreniform, mostly 3–15 cm. wide, essentially glabrous above, closely white-woolly beneath, entire to more often coarsely toothed or shallowly lobed; involucral bracts only about 2 mm. long or less, reflexed in fruit and eventually deciduous; achenes clavate, becoming 6–8 mm. long, coarsely stipitate-glandular above.

    Moist shady woods, chiefly at moderate and low elevations, sometimes ascending to above 6,000 feet in the more southern portions of its range, mostly Transition Zone; southern British Columbia to northern Idaho and northwestern Montana, the Blue and Ochoco Mountains of northeastern and central Oregon, and southward in and west of the Cascade Mountains to northern California, thence southward in the Sierra Nevada to Tulare County, and in the Coast Ranges to Santa Cruz County; also in northern Michigan. Type locality: Straits of Juan de Fuca, and about Fort Vancouver on the Columbia River. June–Sept.

107. **DIMERÈSIA** A. Gray, Syn. Fl. N. Amer. ed. 2. **1**²: 448. 1886.

Heads discoid, 2–3-flowered, the flowers tubular and perfect. Involucre of 2 or sometimes 3 herbaceous bracts, united at the base, each broadly rounded on the back and loosely

---

    * Text contributed by Arthur John Cronquist.
    † In the Senecioneae, the term "involucral bract" is used instead of "phyllaries."

embracing a flower. Anthers sagittate but scarcely caudate. Style-branches flattened, slightly broader upward, papillate-puberulent over the outer surface, the stigmatic lines ventromarginal, extending completely around the rounded apex. Achenes glabrous, striate; pappus of about 20 coarse, sparsely long-plumose bristles, united at the base and deciduous in a ring. Compact little annuals with entire leaves. [Name from the Greek *dimeres*, of two parts or members.]

A monotypic genus of northwestern United States.

1. Dimeresia howéllii A. Gray. Doublet. Fig. 5718.

Dimeresia howellii A. Gray, Syn. Fl. N. Amer. ed. 2. 1²: 449. 1886.
Ereminula howellii Greene, Pittonia 2: 248. 1892.

Nearly acaulescent, cushion-like, dwarf annuals somewhat arachnoid at the base, more glandular upwards. Leaves spatulate to elliptic or ovate, often punctate, up to 1 cm. wide and 3 cm. long including the petiolar base, aggregated around the 1 or several compact clusters of heads; involucre about 4–6 mm. high; flowers white to pinkish or purplish.

Dry, gravelly or rocky soil in the foothills and at moderate elevations in the mountains, sometimes on serpentine, Arid Transition Zone; northeastern Baker County, Oregon, south to northwestern Nevada and Modoc and Siskiyou Counties, California. Type locality: Steen Mountains, Oregon. May–Aug.

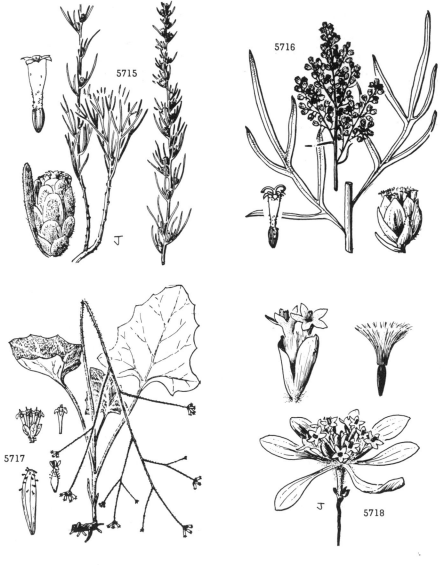

5715. Artemisia rigida
5716. Artemisia palmeri
5717. Adenocaulon bicolor
5718. Dimeresia howellii

108. **ÁRNICA** L. Sp. Pl. 884. 1753.

Heads radiate or discoid, the rays when present pistillate and fertile, yellow or orange, relatively few and broad. Involucral bracts herbaceous, more or less evidently biseriate but subequal and connivent. Receptacle convex or nearly flat, naked. Disk-flowers perfect and fertile, yellow or orange. Anthers entire to minutely sagittate. Style-branches more or less flattened, truncate, penicillate, with introrsely marginal stigmatic lines extending to the tip or sometimes very shortly appendiculate. Achenes subterete, 5–10-nerved; pappus of numerous white to tawny, barbellate to subplumose, capillary bristles. Fibrous-rooted perennial herbs from a rhizome or caudex. Leaves simple, opposite, or the reduced upper-most ones occasionally alternate. Heads rather large, turbinate to hemispheric, solitary to rather numerous. [Name of uncertain derivation, perhaps a corruption of *Ptarmica*.]

A genus of about 30 species, of circumboreal distribution but most highly developed in western North America. Type species, *Arnica montana* L.

Three species—*A. alpina, A. chamissonis*, and *A. diversifolia*—have been reported to show apomixis, and the last-named of these, at least, may well prove to be largely or wholly apomictic. Some of the problems in specific delimitation suggest the existence of a polyploid-apomictic complex, but this remains to be demonstrated.

Heads characteristically radiate (occasional rayless forms of some of these species occur, chiefly in company with the normal radiate plants).
  Cauline leaves relatively numerous and well developed, mostly 5–12 pairs.
    Involucral bracts obtuse or merely acutish, bearing a tuft of long hairs at or just within the tip; rhizomes elongate and nearly naked, the stems solitary.   1. *A. chamissonis.*
    Involucral bracts more or less sharply acute, the tip not markedly more hairy than the body.
      Leaves entire or nearly so; plants densely tufted (often in very large clones), the rhizome commonly shortened into a branching caudex.   2. *A. longifolia.*
      Leaves more or less toothed; plants seldom much tufted, the rhizomes mostly more elongate.   3. *A. amplexicaulis.*
  Cauline leaves few, mostly 2–4(5) pairs, not including those, if any, of the basal cluster.
    Rays short, mostly 7–15 mm. long; lower petioles and lower part of the stem densely and conspicuously pubescent with long, loosely spreading hairs; young heads nodding; mostly in California.   6. *A. parryi.*
    Rays longer, mostly 1.5–3 cm. long or occasionally a little shorter in some plants that do not have any conspicuous long hairs; heads erect; widespread.
      Pappus evidently subplumose, more or less tawny; rhizomes freely rooting, often shortened into a mere caudex.
        Heads broad, mostly subhemispheric; cauline leaves variable in shape; lower leaves generally the best developed.   4. *A. mollis.*
        Heads narrow, more or less turbinate; cauline leaves relatively broad, mostly ovate or deltoid to broadly elliptic or broadly lance-elliptic, the middle ones commonly the best developed.   5. *A. diversifolia.*
      Pappus merely barbellate, or becoming weakly subplumose in some species with elongate, nearly naked rhizomes, generally white or nearly so.
        Leaf-blades relatively narrow, mostly (2.5)3–10 times as long as wide; basal leaves often densely tufted but not always so; rhizomes short and freely rooting except sometimes in *A. rydbergii.*
          Heads relatively large, hemispheric or nearly so, with mostly 10–23 (about 13 or about 21) rays; lower cauline leaves generally petiolate; foothills and moderate elevations in the mountains.
            Old leaf-bases with dense tufts of long brown wool in the axils; disk-corollas with some spreading glandless hairs as well as commonly stipitate-glandular.   7. *A. fulgens.*
            Old leaf-bases without axillary tufts or the hairs few and white; disk-corollas stipitate-glandular, generally not otherwise hairy.   8. *A. sororia.*
          Heads smaller, turbinate-campanulate, with mostly 7–10 (about 8) rays; lower cauline leaves tending to be sessile; alpine and subalpine.   9. *A. rydbergii.*
        Leaf-blades relatively broad, the larger ones mostly 1–2.5(3) times as long as wide; basal leaves sometimes persistent but not densely tufted; rhizomes (except in forms of *A. latifolia*) elongate and subnaked, sometimes subapically branched and more scaly.
          Achenes generally glabrous below or glabrous throughout; basal leaves (those on separate short shoots) seldom cordate when present, the cauline ones even less frequently so; leaves generally more or less toothed; involucre with few or no long hairs.
            Pappus merely barbellate; heads 1 to several; plants widespread, not occurring on serpentine or at least not so restricted, commonly flowering from June to August.   10. *A. latifolia.*
            Pappus weakly subplumose; heads strictly solitary; plants occurring on serpentine in and near the Klamath region, flowering mostly in April and May (June).   11. *A. cernua.*
          Achenes mostly short-hairy (or glandular) nearly or quite to the base; leaves and involucre various but mostly not presenting the foregoing combination of characters.
            Involucral bracts obtuse or merely acutish, bearing a subapical tuft of hairs within; heads (3)5–7; leaves thinly tomentulose, ovate to ovate-elliptic, not cordate; rare species of the Sierra Nevada and Klamath areas.   12. *A. tomentella.*
            Involucral bracts more or less sharply acute, mostly without any very prominent subapical tuft of hairs; heads commonly 1–3, or sometimes more in plants with cordate leaves; leaves either glandular (rather than tomentulose) or cordate; common species.
              Involucre with few or no long hairs; leaves entire or nearly so, the lower broadly rounded to subcordate at the base; lateral setae of the pappus-bristles more prominent than in either *A. latifolia* or *A. cordifolia*; common in the Sierra Nevada, extending northward, less commonly, to northern Washington.   13. *A. nevadensis.*
              Involucre sparsely to usually copiously provided with long white hairs, especially toward the base; leaves toothed to sometimes entire, the basal and generally also the lower cauline ones commonly more or less strongly cor-

date (or scarcely so in reduced alpine forms); common in Oregon and Washington, extending southward, less commonly, to the southern end of the Sierra Nevada. 14. *A. cordifolia.*

Heads characteristically discoid (marginal corollas occasionally ampliate, rarely truly radiate).

At least the lowermost leaves more or less petiolate, the petiole sometimes (*A. spathulata*) broadly winged.

Young heads nodding; involucral bracts narrow, gradually and slenderly acute; Oregon and Washington. 6. *A. parryi.*

Heads erect; involucral bracts broader, merely acutish or abruptly acute.

Lowermost leaves with mostly ovate to subcordate or cordiate blade more or less abruptly contracted to the narrowly or scarcely winged petiole, rarely spatulate and wing-petiolate; achenes usually but not always hairy as well as stipitate-glandular; southern Washington to southern California, including the Klamath region. 15. *A. discoidea.*

Lowermost leaves more or less spatulate, usually with broadly winged, often poorly marked petiole; achenes stipitate-glandular; Klamath region only. 16. *A. spathulata.*

Leaves all broadly sessile, numerous, rather small; local species of northern California and southwestern Oregon.

Leaves toothed, reticulate-veiny; achenes hairy, not glandular. 17. *A. venosa.*

Leaves entire, merely 3–5-nerved; achenes stipitate-glandular. 18. *A. viscosa.*

1. Arnica chamissònis subsp. chamissonis var. intèrior Maguire. Meadow Arnica. Fig. 5719.

Arnica chamissonis subsp. *chamissonis* var. *interior* Maguire, Madroño 6: 154. 1942.

Perennial from long, nearly naked rhizomes; stems solitary, 2–10 dm. tall; herbage more or less villous-puberulent to villous-hirsute, becoming glandular or viscid above. Cauline leaves mostly 5–10 pairs, not much reduced upward, lanceólate to oblanceolate, sessile, or the lowermost usually connate-petiolate, evidently toothed or subentire, 5–20 cm. long, 1–4.5 cm. wide; heads generally several, campanulate or hemispheric, the involucre mostly 8–12 mm. high, its bracts obtuse or merely acutish, bearing an apical or internally subapical tuft of long white hairs; hairs at the base of the involucre with very prominent cross-walls; rays commonly about 13, usually pale, about 1.5 cm. long; achenes shortly hairy and glandular to subglabrous; pappus tawny, subplumose.

In meadows and other wet places, Boreal and Transition Zones; Alaska and Mackenzie to southern British Columbia and Alberta extending southward, mostly in forms transitional to subsp. *foliosa*, to central Washington and west-central Montana. Type locality: Palliser, British Columbia. June–Aug.

Typical *A. chamissonis*, a maritime phase of coastal Alaska and the Aleutian Islands, has larger and mostly fewer (typically 1–3) heads, with the lower leaves hardly if at all petiolate.

Arnica chamissonis subsp. **foliòsa** (Nutt.) Maguire, Rhodora **41**: 508. 1939. (*Arnica foliosa* Nutt. Trans. Amer. Phil. Soc. II. 7: 407. 1841.) Leaves mostly entire or merely denticulate; hairs at the base of the involucre less prominently septate; pappus stramineous or whitish, merely barbellate, though often strongly so; otherwise as in var. *interior* of the typical subspecies. Southern Mackenzie to Ontario south to southern California and northern New Mexico. Type locality: "On the alluvial flats of the Colorado of the West, particularly near Bear River, of the lake Timpanagos [Salt Lake]." This is a polymorphic subspecies, composed of several varieties. Those occurring in our range may be keyed as follows:

Involucral bracts very blunt and relatively broad; herbage often but not always rather densely tomentose; San Bernardino Mountains. var. *bernardina.*

Involucral bracts a little narrower and more pointed; more northern plants.

Herbage conspicuously silvery-tomentose; an ecotype of very wet places (typically growing in shallow water). var. *incana.*

Herbage less densely hairy, scarcely silvery; moderately dry or usually merely moist habitats but sometimes in habitats as extreme as those characteristic of var. *incana*. var. *foliosa.*

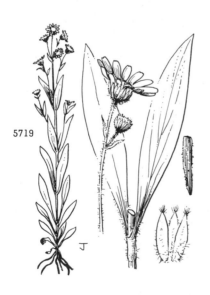

5719

5720

5719. Arnica chamissonis   5720. Arnica longifolia

Arnica chamissonis subsp. foliosa var. bernardìna (Greene) Maguire, Amer. Midl. Nat. 37: 140. 1947. (*Arnica bernardina* Greene, Pittonia 4: 170. 1900; *A. foliosa* var. *bernardina* Jepson, Man. Fl. Pl. Calif. 1157. 1925.) San Bernardino Mountains, California. Type locality: Bear Valley.

Arnica chamissonis subsp. foliosa var. incàna (A. Gray) Hultén, Fl. Alaska & Yukon 1591. 1950. (*Arnica foliosa* var. *incana* A. Gray in C. C. Parry, Amer. Nat. 8: 214. 1874; *A. incana* Greene, Pittonia 4: 169. 1900; *A. cana* Greene, Ottawa Nat. 15: 282. 1902.) Common from the southern end of the Sierra Nevada to the Cascade Mountains of southern Washington, extending eastward occasionally into central Utah and southwestern Idaho and apparently sometimes northward to southern Yukon. Type locality: Sierra Valley, California.

Arnica chamissonis subsp. foliosa var. foliosa. Range of the subspecies except the mountains of southern California.

An additional form of *A. chamissonis* subsp. *foliosa*, which approaches *A. parryi* var. *sonnei* in its somewhat narrower and more pointed involucral bracts and in its more or less reduced cauline leaves, has been described on the basis of a few collections from the Sierra Nevada as var. *jepsoniana* Maguire, Amer. Midl. Nat. 37: 140. 1947. The real biological nature of these specimens remains to be determined.

2. Arnica longifòlia subsp. myriadènia (Piper) Maguire.
Seep-spring Arnica. Fig. 5720.

Arnica myriadenia Piper, Proc. Biol. Soc. Wash. 33: 106. 1920.
Arnica longifolia subsp. *myriadenia* Maguire, Brittonia 4: 470. 1943.

Plants densely tufted, often in very large clones, with many smaller, sterile, leafy stems in addition to the floriferous ones, these commonly 3–6 dm. tall, leafy, the rhizome commonly shortened into a branching caudex. Leaves mostly 5–7 pairs, one or more of the lowermost pairs connate-sheathing, with reduced blade, the others sessile or shortly connate-petiolate, not much reduced upward, narrowly lanceolate or lance-elliptic, gradually tapering to the acute or acuminate tip, entire or sometimes slightly toothed, mostly 5–12 cm. long and 1–2 cm. wide; no well-developed basal leaves produced, either on the stem or on separate short shoots; herbage more or less scabrid-puberulent, at least upward, and usually a little viscid; heads several to rather numerous, campanulate, sometimes rather narrowly so; involucre mostly 7–10 mm. high, the bracts sharply acute or acuminate, glandular-puberulent and generally with a few intermingled, longer, conspicuously septate hairs, especially below; rays mostly 8–13, 1–2 cm. long; achenes subglabrous or glandular and hairy; pappus more or less tawny, barely subplumose.

In well-drained soil (or rocks) about seeps and springs, and along moist cliffs and river banks, at moderate to high elevations in the mountains, Boreal Zone; Cascade Mountains of central Washington south to the southern end of the Sierra Nevada, east to northwestern Nevada, eastern Oregon, and rarely northwestern Montana and adjacent Alberta. Type locality: Mount Rainer National Park, Washington. July–Sept.

Barely or scarcely distinguishable from the nomenclaturally typical phase of the species, which occurs in the Rocky Mountains and Great Basin, and which tends to differ in its merely glandular-puberulent involucre and paler, merely barbellate pappus.

3. Arnica amplexicaùlis Nutt. Streambank Arnica. Fig. 5721.

Arnica amplexicaulis Nutt. Trans. Amer. Phil. Soc. II. 7: 408. 1841.
Arnica macounii Greene, Pittonia 4: 160. 1900.
Arnica amplexifolia Rydb. Mem. N.Y. Bot. Gard. 1: 434. 1900.
Arnica aspera Greene, Ottawa Nat. 15: 281. 1902.
Arnica ciliaris Rydb. N. Amer. Fl. 34: 351. 1927.
Arnica macounii var. *aspera* G. N. Jones, Univ. Wash. Publ. Biol. 5: 256. 1936.

Perennial from rather coarse, freely rooting rhizomes, mostly 3–8 dm. tall, more or less glandular and hairy, especially upward, or subglabrous, leafy. Leaves all cauline, mostly 5–12 pairs, the lowermost often reduced, the others well developed and generally not much reduced upward, narrowly lance-elliptic to lance-ovate or ovate, sessile or the lowermost short-petiolate, more or less toothed, commonly 4–12 cm. long and 1–3.5(4) cm. wide; heads generally several, campanulate; involucre mostly 9–15 mm. high, its bracts sharply acute or acuminate; rays 8–14, pale yellow, 1–2 cm. long; achenes sparsely hirsuite and sometimes glandular; pappus tawny, subplumose.

Stream banks and moist woods, Canadian and Transition Zones; Alaska to western Montana, northeastern Nevada, and California to the southern Sierra Nevada. Type in locality: "On the rocks of the Wahlamet, at the Falls." July–Aug.

Arnica amplexicaulis var. pìperi St. John & Warren in St. John, Proc. Biol. Soc. Wash. 44: 36. 1931. (*A. hirticaulis* Rydb. N. Amer. Fl. 34: 349. 1927.) Leaves very broad, up to 6 or reputedly 8 cm. wide; involucre more coarsely hairy and less glandular; otherwise much as in typical *A. amplexicaulis* but less variable. Columbia River Gorge in Oregon and Washington. Type locality: Cape Horn, Washington.

4. Arnica móllis Hook. Cordilleran Arnica. Fig. 5722.

Arnica mollis Hook. Fl. Bor. Amer. 1: 331. 1834.
Arnica merriamii Greene, Pittonia 4: 36. 1899.
Arnica rivularis Greene, op. cit. 163. 1900.
Arnica scaberrima Greene, op. cit. 165.
Arnica mollis var. *scaberrima* Smiley, Univ. Calif. Pub. Bot. 9: 386. 1921.

Perennial from freely rooting rhizomes, which may be shortened into a loosely branched caudex, 2–6 dm. tall, variously puberulent to long-hairy and glandular. Cauline leaves mostly 3 or 4(5) pairs, the lower commonly the largest, all sessile, or the lower often short-petiolate, variously ovate, elliptic, or lanceolate to oblanceolate or obovate, irregularly denticulate (or dentate) to entire; well-developed, petiolate, basal leaves often produced; heads few or solitary, hemispheric-campanulate or broader, the disk sometimes as much as 3.5 cm. wide; involucre 10–16 mm. high, the bracts more or less acuminate, long-hairy at the base, more glandular above; rays mostly 12–18, 1.5–2.5 cm. long; pappus tawny, subplumose.

Moist places at moderate to high elevations in the mountains, Canadian and Hudsonian Zones; Alberta and British Columbia to southern Colorado, southern Utah, and the southern end of the Sierra Nevada in California. In all the higher mountains of Washington and Oregon but absent from the California Coast Ranges, and from the Oregon Coast Ranges south of Clatsop County. Type locality: "Alpine rivulets of the Rocky Mountains [Canada]." June–Sept.

5. **Arnica diversifòlia** Greene. Lawless Arnica. Fig. 5723.

Arnica latifolia var. *viscidula* A. Gray, Syn. Fl. N. Amer. 1²: 381. 1884.
Arnica diversifolia Greene, Pittonia 4: 171. 1900.

Perennial from freely rooting rhizomes; stems solitary or in loose tufts, 1.5–4 dm. tall, glandular-puberulent to subglabrous. Cauline leaves mostly 3–4 pairs, sessile or at least the lower generally shortly wing-petiolate, ovate or deltoid to elliptic or lance-elliptic, irregularly toothed, the middle ones commonly the largest, with blades 4–8 cm. long and 2–6 cm. wide; well-developed, petiolate, basal leaves sometimes produced; heads generally several, narrow, more or less turbinate, the disk-flowers much fewer than in characteristic *A. mollis*; involucre 10–14 mm. high, its bracts acute or acuminate, shortly stipitate-glandular throughout and sometimes sparsely long-hairy as well; rays commonly about 8 or about 13, 1.5–2 cm. long; pappus stramineous to tawny, subplumose.

Rocky places at moderate to high elevations in the mountains, Canadian and Hudsonian Zones; Alaska to Montana, Utah, and the northern Sierra Nevada in California; not common. Type locality. Eagle Creek [Wallowa] Mountains, Oregon. July–Sept.

Arnica diversifolia is merely a convenient name for a complex series of apparent hybrids and hybrid progeny, involving *A. mollis* or *A. amplexicaulis*, on the one hand, and *A. cordifolia* or *A. latifolia* on the other. The pollen is irregular and often very scanty, and apomixis has been reported.

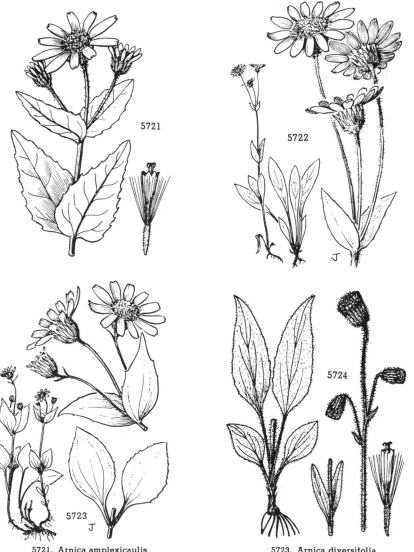

5721. Arnica amplexicaulis
5722. Arnica mollis
5723. Arnica diversifolia
5724. Arnica parryi

6. **Arnica párryi** A. Gray. Nodding Arnica. Fig. 5724.

Arnica augustifolia ? var. *eradiata* A. Gray, Proc. Acad. Phila. **1**: 68. 1863.
Arnica parryi A. Gray in C. C. Parry, Amer. Nat. **8**: 213. 1874.
Arnica-eradiata Heller, Cat. N. Amer. Pl. 7. 1898.

Perennial from freely rooting to nearly naked rhizomes; stems mostly solitary, 2–6 dm. tall, often somewhat woolly-villous toward the base, becoming glandular at least above. Cauline leaves mostly 2–4 pairs, strongly reduced upward, the lowermost petiolate, with lanceolate or lance-ovate blade 5–20 cm. long and 1.5–6 cm. wide; well-developed basal leaves similar to the lower cauline ones often produced on separate short shoots; heads generally several, nodding in bud, later more or less erect, campanulate, sometimes narrowly so, ordinarily discoid, the marginal corollas sometimes ampliate, or rarely shortly radiate; involucre mostly 10–14 mm. high, its rather narrow bracts sharply acute or acuminate; achenes glabrous to glandular or hairy; pappus tawny, strongly barbellate to weakly subplumose.

Open woods, drier meadows, and moist slopes, in the foothills and at moderate elevations in the mountains, mostly Canadian Zone; southern Alberta and British Columbia to Colorado, Utah, and southern Oregon, rarely extending into northern California; common in the Cascade and Olympic Mountains of Washington but seldom collected in Oregon. Type locality: Clear Creek, Colorado. July–Aug.

Arnica parryi var. **sònnei** (Greene) Cronquist in Ferris, Contr. Dudley Herb. **5**: 102. 1958. (*A. sonnei* Greene, Pittonia **3**: 104. 1896; *A. foliosa* var. *sonnei* Jepson, Man. Fl. Pl. Calif. 1157. 1925; *A. parryi* subsp. *sonnei* Maguire, Brittonia **4**: 482. 1943.) Lower part of the stem and lower petioles more consistently and more densely and conspicuously woolly-villous than in the typical subspecies; heads ordinarily radiate, the rays short, mostly 7–15 mm. long. Sierra Nevada of California and adjacent Nevada, extending northward, rarely, to the Cascade Mountains of southern Oregon. Type locality: near Truckee River, California.

7. **Arnica fùlgens** Pursh. Hillside Arnica. Fig. 5725.

Arnica fulgens Pursh, Fl. Amer. Sept. 527. 1814.
Arnica montana β *fulgens* Nutt. Gen. **2**: 164. 1818.
Arnica pedunculata Rydb. Bull. Torrey Club **24**: 297. 1897.

Perennial from short, freely rooting, densely scaly rhizomes, the basal leaves and persistent old leaf-bases with conspicuous axillary tufts of long brown wool; stem stout, 2–6 dm. tall, stipitate-glandular and often also hairy, more densely so upward. Leaves 3–5-nerved, entire or nearly so, sparsely to densely hairy and glandular, the basal ones petiolate, with narrowly to broadly oblanceolate or narrowly elliptic blade mostly 3–12 cm. long and 1–4 cm. wide, those above the basal cluster mostly 2–4 pairs, becoming sessile, progressively reduced, and distant; heads solitary or sometimes 3, broadly hemispheric; involucre mostly 10–15 mm. high, glandular and hairy, the bracts mostly narrowly elliptic or lance-elliptic and tapering from the middle to the obtuse or acutish tip, more or less ciliate upward, the tip hairy within; rays commonly 10–23, 1.5–2.5 cm. long and 4–10 mm. wide; disk-corollas with some spreading white hairs as well as generally glandular; achenes densely hairy; pappus whitish or stramineous, barbellate.

Open places, chiefly in the foothills and at moderate elevations in the mountains, Arid Transition and Canadian Zones; southern British Columbia to Saskatchewan south to Colorado, northern Nevada, and northeastern California. Type locality: "On the banks of the Missouri." May–July.

8. **Arnica soròria** Greene. Twin Arnica. Fig. 5726.

Arnica sororia Greene, Ottawa Nat. **23**: 213. 1910.

Similar to *A. fulgens* but averaging more slender, the rhizomes less scaly, more slender, and often more elongate. Leaves averaging narrower and fewer, the wool nearly or quite absent from the axils of the basal leaves, white when present; involucral bracts narrower and less hairy, generally broadest at or near the base, tapering thence to the more acute tip; disk-corollas stipitate-glandular, not otherwise hairy; pappus white or nearly so.

Open, often rather dry places in the foothills and at moderate elevations in the mountains, Arid Transition Zone; southern British Columbia and Alberta to Wyoming, northern Utah, and northeast California (southward nearly to Donner Pass in the Sierra Nevada). Type locality: between the Kettle and Columbia Rivers, Cascade, British Columbia. May–July.

9. **Arnica rydbérgii** Greene. Subalpine Arnica. Fig. 5727.

Arnica rydbergii Greene, Pittonia **4**: 36. 1899.
Arnica aurantiaca Greene, Torreya **1**: 42. 1901.
Arnica sulcata Rydb. N. Amer. Fl. **34**: 344. 1927.

Stems more or less clustered from scaly, often short and branched rhizomes, 1–3 dm. tall; herbage variously glandular and short-hairy or subglabrous. Petiolate basal leaves with oblanceolate or spatulate, 3–5-nerved blade up to 7 cm. long and 1.5 cm. wide frequently produced on separate short shoots; cauline leaves mostly 3–4 pairs, sessile, or the lower shortly and broadly wing-petiolate, oblanceolate or spatulate to lanceolate or sometimes a little broader, entire or nearly so, 3–5-nerved, mostly aggregated on the lower part of the stem, 3–10 cm. long and 5–25 mm. wide, or the upper smaller and distant; heads solitary or few, turbinate-campanulate; involucre mostly 9–13 mm. high, its bracts glandular and sparsely long-hairy to subglabrous on the back, ciliate or scaberulous on the margins, moderately acute; rays mostly about 8, 1–2 cm. long, the tip minutely toothed or entire; achenes densely short-villous throughout, the upper hairs the longest; pappus white, barbellate.

Dry meadows and open slopes, mostly at high elevations in the mountains, Hudsonian Zone; British Columbia and Alberta to Colorado, northern Utah, eastern Oregon, and (rarely) northern California (Scott Mountains). Type locality: Little Belt Pass, Montana. July–Aug.

10. **Arnica latifòlia** Bong. Mountain Arnica. Fig. 5728.

Arnica latifolia Bong. Mém. Acad. St. Pétersb. VI. **2**: 147. 1832.
Arnica menziesii Hook. Fl. Bor. Amer. **1**: 331. 1834.
Arnica platyphylla A. Nels. Bot. Gaz. **31**: 407. 1901.
Arnica eriopoda Gandoger, Bull. Soc. Bot. Fr. **65**: 38. 1918.
Arnica aphanactis Piper, Proc. Biol. Soc. Wash. **35**: 105. 1920.
Arnica glabrata Rydb. N. Amer. Fl. **34**: 335. 1927.
Arnica membranacea Rydb. op. cit. 338.

Perennial from elongate, mostly nearly naked rhizomes which may be branched and more scaly at the apex; stems mostly solitary or few together, 1–6 dm. tall. Broad (occasionally cordate), long-petiolate, basal leaves sometimes produced on separate short shoots; cauline leaves mostly 2–4 pairs, variously glandular and hairy to subglabrous, sessile or petiolate, lance-elliptic or generally broader, rarely cordate, more or less toothed, commonly 3–14 cm. long and 1.5–8 cm. wide, the middle ones commonly as large as or larger than those below, the stem thus generally appearing more leafy than in *A. cordifolia*; heads 1–3 or sometimes more, turbinate to sometimes

5725. Arnica fulgens
5726. Arnica sororia

5727. Arnica rydbergii
5728. Arnica latifolia

subhemispheric; involucre mostly 10–18 mm. high, more or less glandular and sometimes with a few long hairs; achenes generally glabrous, at least toward the base, often glandular and/or shortly hairy above; pappus white, barbellate.

Moist woods, meadows, and moist open places in the mountains, seldom at very high elevations, mostly Canadian and Hudsonian Zones; Alaska and Yukon to Colorado, Utah, and northern California (southward occasionally to Eldorado County). Type locality: island of Sitka. June–Aug.

Arnica latifolia var. **grácilis** (Rydb.) Cronquist, Vasc. Pl. Pacif. Northw. **5**: 51. 1955. (*Arnica gracilis* Rydb. Bull. Torrey Club **24**: 297. 1897; *A. betonicaefolia* Greene, Pittonia **4**: 163. 1900; *A. betonicaefolia* var. *gracilis* M. E. Jones, Bull. Univ. Mont. Biol. Ser. **15**: 48. 1910.) Small tufted plants (1–3 dm. tall) with several slender stems arising from the rhizome, which is commonly shortened into a loosely branched, scaly caudex. Leaves seldom over 2.5 cm. wide; heads mostly 3–9 (occasionally solitary), narrow, small, the involucre 7–13 mm. high. Rocky places at moderate or usually high elevations in the mountains; Alberta and British Columbia to Washington, northeastern Oregon, central Idaho, northern Utah, and Colorado. Type locality: Spanish Peaks, Madison County, Montana.

11. **Arnica cérnua** Howell. Serpentine Arnica. Fig. 5729.

Arnica cernua Howell, Fl. N.W. Amer. 373. 1900.
Arnica chandleri Rydb. N. Amer. Fl. **34**: 339. 1927.

Perennial from elongate, nearly naked rhizomes; stems solitary or few together, 1–3 dm. tall. Broad (more or less ovate-cordate), long petiolate, basal leaves sometimes produced on separate short shoots; cauline leaves mostly 3–4 pairs, glabrous or scabrous-puberulent, mostly petiolate and lance-ovate to obovate or even subcordate, the lower generally the larger, up to about 5 cm. long and 3 cm. wide; heads normally solitary, often long-pedunculate, turbinate-campanulate; involucre 10–18 mm. high, puberulent and often also glandular, the bracts ciliolate-margined; achenes evidently hirsute above, glabrous below; pappus white, with the lateral setae of the bristles better developed than in *A. latifolia* and *A. cordifolia,* less developed than in *A. mollis* and *A. amplexicaulis.*

Serpentine slopes, Arid Transition Zone; Josephine County, Oregon, and in Siskiyou, Humboldt, and Del Norte Counties, California. Type locality: near Waldo, Josephine County, Oregon. April–May (June).

12. **Arnica tomentélla** Greene. Recondite Arnica. Fig. 5730.

Arnica tomentella Greene, Pittonia **4**: 155. 1900.

Perennial from well-developed, long, nearly naked rhizomes; stems solitary, 2–5 dm. tall, glandular and hairy. Leaves apparently all cauline, mostly 3–4 pairs, thinly tomentulose-puberulent, the better developed ones petiolate, with ovate to ovate-elliptic, toothed or sometimes entire blade about 2.5–7 cm. long; heads (3)5–7, campanulate-hemispheric; involucre mostly 10–13 mm. high, its bracts acutish or obtuse, tomentulose-puberulent and ciliate, and bearing a tuft of longer hairs internally near the tip; rays 12–20 mm. long; achenes short-hairy throughout, and very sparsely stipitate-glandular; pappus white to stramineous, strongly barbellate or shortly subplumose.

Open slopes or open woods below timberline in the mountains, Canadian Zone; Josephine County, Oregon, and adjacent Siskiyou County, California, reappearing in the Sierra Nevada along the Truckee River and in Tulare County, California. Type locality: Middle Tule River, California.

Plants referred to *A. tomentella* seem to combine the features of *A. cordifolia* and *A. chamissonis* subsp. *foliosa.*

13. **Arnica nevadénsis** A. Gray. Sierra Arnica. Fig. 5731.

Arnica nevadensis A. Gray, Proc. Amer. Acad. **19**: 55. 1883.

Perennial from long, nearly naked rhizomes, these often apically branched and more scaly; stems solitary, or few loosely clustered together, 1–3 dm. tall; herbage more or less strongly glandular or glandular-puberulent and sometimes with scattered longer hairs. Cauline leaves mostly 2–3 pairs, the lower obviously the larger, these petiolate, the blade broadly ovate to sometimes rhombic or broadly elliptic, with broadly rounded or truncate to occasionally subcordate base, mostly 3–8 cm. long and 1.5–4 cm. wide, entire or occasionally denticulate; long-petiolate basal leaves, similar to those of the lower part of the stem, often produced on separate short shoots; heads 1–3, broadly turbinate to campanulate; involucre mostly 10–18 mm. high, the bracts more or less sharply acute, densely short- stipitate-glandular, and sometimes with some longer hairs below but these less conspicuous than in *A. cordifolia*; rays about 12–20 or 25 mm. long; achenes mostly uniformly glandular or short-hairy or both; pappus white to stramineous, strongly barbellate or shortly subplumose, the lateral setae of the bristles better developed than in *A. cordifolia* and *A. latifolia,* less developed than in *A. mollis* and *A. amplexicaulis.*

Open rocky slopes at high altitudes in the mountains, Canadian and Hudsonian Zones; common in the Sierra Nevada of California and adjacent Nevada, northward less commonly to the Cascade and Olympic Mountains of Washington. Type locality: Lassen Peak, California. July–Aug.

In the aforestated area *A. nevadensis* appears to be well characterized and readily separable from its near relative *A. cordifolia.* In the Rocky Mountains and Great Basin, however, reduced alpine extremes of *A. cordifolia* var. *pumila* (Rydb.) Maguire are barely if at all to be distinguished from Sierran *A. nevadensis,* from which they differ chiefly in distribution and in the slightly shorter barbels of the pappus-bristles. These plants, which represent the closest approach of *A. cordifolia* to *A. nevadensis,* have been treated as *A. cordifolia* var. *humilis* (Rydb.) Maguire.

14. **Arnica cordifòlia** Hook. Heart-leaved Arnica. Fig. 5732.

Arnica cordifolia Hook. Fl. Bor. Amer. **1**: 331. 1834.
Arnica macrophylla Nutt. Trans. Amer. Phil. Soc. II. **7**: 408. 1841.
Arnica austiniae Rydb. N. Amer. Fl. **34**: 340. 1927.
Arnica cordifolia var. *macrophylla* Maguire, Amer. Midl. Nat. **37**: 137. 1947.

Perennial from long, nearly naked rhizomes, these often apically branched and more scaly; stems solitary, or few loosely clustered together, mostly 2–6 dm. tall, glandular-puberulent to more often loosely white-hairy. Long-petiolate, cordate, basal leaves commonly produced on separate short shoots; cauline leaves mostly 2–4 pairs, the lower obviously the larger, these petiolate, with more or less cordate blade (sometimes very deeply so), commonly 4–12 cm. long and 3–9 cm. wide, entire or more often toothed; heads 1–3 or sometimes up to 7, broadly turbinate to campanulate; involucre mostly 13–20 mm. high, sparsely to copiously provided with long, spreading, white hairs, especially toward the base and generally glandular as well; rays 1.5–3 cm. long, mostly 10–15, or rarely wanting in individual plants; achenes mostly uniformly short-hairy or glandular (or both); pappus white or whitish, barbellate.

Mostly in woodlands, from the foothills to moderate or sometimes rather high elevations in the mountains (where it passes into the var. *pumila*), Arid Transition Zone to Canadian Zone; Alaska and Yukon to South Dakota, northern New Mexico, central Arizona, and California to the southern end of the Sierra Nevada; rare west of the Cascade summits in Oregon and Washington; also in northern Michigan. Type locality "Alpine woods of the Rocky Mountains, on the east side, [Alberta]." April–June (Sept.).

Arnica cordifolia var. **pùmila** (Rydb.) Maguire, Madroño 6: 154. 1942. (*A. pumila* Rydb. Mem. N.Y. Bot. Gard. 1: 433. 1900; *A. humilis* Rydb. N. Amer. Fl. 34: 341. 1927; *A. cordifolia* var. *humilis* Maguire, Amer. Midl. Nat. 37: 138. 1947.) A dwarf ecotype mostly less than 2 dm. tall with narrower, more often entire, only slightly or scarcely cordate leaves commonly 2–5 cm. long, with more glandular achenes, and often with less notably long-hairy involucre. Alpine and subalpine stations nearly or quite throughout the range of the species but rare in our area where it is mostly replaced by *A. nevadensis* (quod vide). Type locality: Gray's Peak, Colorado.

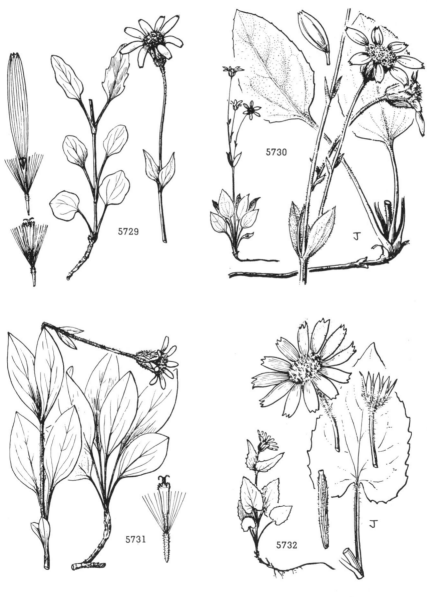

5729. Arnica cernua
5730. Arnica tomentella

5731. Arnica nevadensis
5732. Arnica cordifolia

15. **Arnica discoìdea** Benth. Rayless Arnica. Fig. 5733.

Arnica discoidea Benth. Pl. Hartw. 319. 1849.

Perennial from long, nearly naked rhizomes, the stems mostly solitary, 3–6 dm. tall, glandular-puberulent and often also more or less long-hairy: Leaves sparsely to rather copiously long-hairy on both sides, and often also glandular, rather coarsely toothed or occasionally subentire, the basal ones long-petiolate, with narrowly ovate to deltoid or even subcordate blade up to about 8 cm. long and 3.5 cm. wide, rarely broadly wing-petiolate and with ill-defined blade almost as in *A. spathulata*; cauline leaves several pairs, evidently and progressively reduced, mostly sessile or with broad winged petiole and rather ill-defined blade; stem commonly branched above or even from near the base and producing several or rather many (up to about 30) heads, the reduced upper leaves often alternate; heads turbinate-campanulate to subhemispheric, discoid (the marginal corollas rarely ampliate); involucre 9–13 mm. high, glandular and spreading-villous, the bracts obtuse or somewhat acute, carrying their width well above the middle or nearly to the tip; achenes glandular and hairy throughout, or glabrate below; pappus rather strongly barbellate, white or stramineous.

Open woodlands, Upper Sonoran and Humid Transition Zones; in the Outer Coast Ranges of California from Mendocino County to San Luis Obispo County (the Franciscan area of Jepson). Type locality: woods near Monterey, California. June–July.

Arnica discoidea var. **eradiàta** (A. Gray) Cronquist, Vasc. Pl. Pacif. Northw. **5**: 49. 1955. (*A. parviflora* A. Gray, Proc. Amer. Acad. **7**: 363. 1868; *A. cordifolia* var. *eradiata* A. Gray, Syn. Fl. N. Amer. **1²**: 381. 1884; *A. grayi* Heller, Muhlenbergia **1**: 5. 1900; *A. falconaria* Greene, Ottawa Nat. **23**: 215. 1910; *A. cusickii* Rydb. N. Amer. Fl. **34**: 343. 1927, in part.) Similar to typical *A. discoidea* but the middle and lower cauline leaves mostly similar to the basal ones (though smaller) in being more or less sharply divisible into ovate or deltoid to subcordate blade and narrowly or scarcely winged petiole; achenes sometimes merely glandular. Klamath region southward, chiefly in the Inner Coast Ranges of California, to Alameda County and northward rarely through the Cascade Mountains to the east end of the Columbia River Gorge in Oregon and Washington, where again common. Type locality: Hood River, Oregon. May–Aug.

Arnica discoidea var. **alàta** (Rydb.) Cronquist in Ferris, Contr. Dudley Herb. **5**: 102. 1958. (*A. alata* Rydb. N. Amer. Fl. **34**: 342. 1927; *A. parviflora* subsp. *alata* Maguire, Brittonia **4**: 455. 1943.) Similar to var. *eradiata* but with the basal leaves generally distinctly cordate and the heads averaging a little larger, the involucre up to 15 mm. high; sometimes much resembling *A. cordifolia* (which produces rare rayless individuals) but tending to be taller, leafier, and more branched, with more numerous heads which have somewhat blunter involucral bracts, as well as differing sharply from ordinary forms of *A. cordifolia* in being rayless. Moderate and lower elevations in the Sierra Nevada, northward less commonly to the Klamath region, also common in the Coast Ranges just south of San Francisco Bay and occasionally found northward in the Inner Coast Ranges of California. Type locality: Yosemite, California. June–Aug.

16. **Arnica spathulàta** Greene. Klamath Arnica. Fig. 5734.

Arnica spathulata Greene, Pittonia **3**: 103. 1896.
Arnica cusickii Rydb. N. Amer. Fl. **34**: 343, in part. 1927.

Perennial from branched, nearly naked rhizomes, the stems commonly solitary, mostly 1.5–4.5 dm. tall; herbage glandular and spreading-hairy. Leaves coarsely toothed or occasionally subentire, the basal ones more or less spatulate, with a broad, winged, petiolar base and poorly defined blade, the blade and petiole together up to about 10 cm. long and 3.5 cm. wide; cauline leaves mostly 3–5 pairs, sessile and broad-based or the lowermost similar to the basal ones; stems commonly branched above and producing several (up to about 20) heads, these turbinate-campanulate, discoid; involucre 12–17 mm. high, glandular and hairy-spreading like the herbage, more coarsely and conspicuously so than in *A. discoidea*, the bracts merely acutish, carrying their width well beyond the middle; achenes stipitate-glandular throughout, generally not otherwise hairy; pappus barbellate, white or nearly so.

Open woods, mostly fairly well up in the mountains, mostly Arid Transition Zone; in the Klamath region of southwestern Oregon and adjacent California. Type locality: Glendale, Oregon. April–July.

Arnica spathulata subsp. **eastwoòdiae** (Rydb.) Maguire, Brittonia **4**: 458. 1943. (*Arnica eastwoodiae* Rydb. N. Amer. Fl. **34**: 343. 1927.) More slender and less evidently hairy than the typical form of the species, and more inclined to branch (sometimes even from near the base), and commonly with the stem or leaves or both more or less suffused with red (anthocyanin); petioles sometimes less winged. Foothill and seacoast region in Curry and Josephine Counties, Oregon, and Del Norte and Siskiyou Counties, California; occurring commonly at lower elevations and farther west than the typical form of the species. Type locality: Gasquet, California.

This lowland ecotype appears to be taxonomically comparable to groups treated elsewhere in the genus as varieties, but I do not wish to propose a new combination at this time.

17. **Arnica venòsa** H. M. Hall. Veiny Arnica. Fig. 5735.

Arnica venosa H. M. Hall, Univ. Calif. Pub. Bot. **6**: 174. 1915.

Habitally similar to *A. viscosa*, usually less branched, and less conspicuously pubescent and less glandular. Leaves larger (up to 7 cm. long and 3 cm. wide) and firmer, evidently reticulate-veiny (at least beneath) as well as 3–5-nerved, sharply and irregularly toothed; heads a little larger, terminating the branches; disk corollas densely long-hairy below, more sparsely so above, not glandular; achenes shortly hispid, not glandular.

Hot dry slopes in the foothills, Arid Transition Zone; Shasta County (Salt Creek and Iron Mountain), California. Type locality: Salt Creek, Shasta County. May–June.

18. **Arnica viscòsa** A. Gray. Shasta Arnica. Fig. 5736.

Arnica viscosa A. Gray, Proc. Amer. Acad. **13**: 374. 1878.
Raillardella paniculata Greene, Erythea **3**: 48. 1895.
Chrysopsis shastensis Jepson, Man. Fl. Pl. Calif. 1037. 1925.

Perennial with numerous freely branched stems arising from a branching caudex, 2–5 dm. tall; herbage copiously pubescent with hairs of varying length, many or most of which are gland-

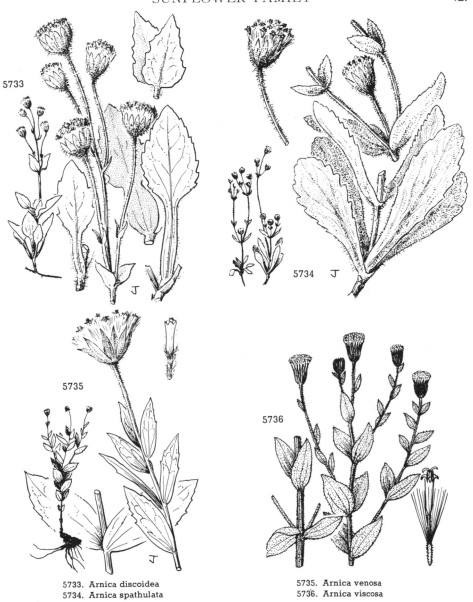

5733. Arnica discoidea
5734. Arnica spathulata

5735. Arnica venosa
5736. Arnica viscosa

tipped, or the shorter hairs merely viscid. Leaves numerous and equably distributed along the stems, the lowermost ones reduced and bract-like, the others nearly all alike, sessile, elliptic or oblong-ovate to oblong-obovate, small, 2–5.5 cm. long and 1–2 cm. wide, entire, evidently or rather weakly 3–5-nerved, not obviously veiny; no separate basal leaves produced; heads several or rather numerous, short-pedunculate, discoid; involucre 8–11 mm. high, glandular-hairy like the herbage; corollas sparsely long-stipitate-glandular below, glabrous above; achenes uniformly long-stipitate-glandular; pappus white or creamy, strongly barbellate.

Rocky places near timberline, Hudsonian Zone; Mount Shasta, California, and Mount Mazama (Crater Lake), Oregon. Type locality: Mount Shasta. Aug.

109. RAILLARDÉLLA Benth. & Hook. Gen. Pl. 442. 1873.

Heads discoid or sometimes with more or less well-developed, fertile, yellow rays. Involucral bracts about equal, subherbaceous, uniseriate or nearly so. Receptacle naked, flat or nearly so. Anthers truncate or nearly so at the base. Style-branches flattened, externally hispidulous, with introrsely marginal stigmatic lines and an elongate, slender, externally and internally hispidulous appendage. Pappus of more or less numerous, evidently plumose bristles that are somewhat flattened toward the base, or of narrow and elongate, bristle-like plumose scales. Achenes linear, somewhat compressed, several-nerved. Perennials with simple, entire or sparingly toothed, alternate or basal leaves (or the reduced lowermost leaves sometimes opposite), and solitary or few, yellow or orange-colored heads. [Named for its similarity to *Raillardia,* of the Hawaiian Islands.]

A genus of 5 species, native to the mountains of California (and adjacent Nevada) and Oregon. Type species, *Raillardia argentea* A. Gray.

Stems evidently leafy and generally branched; no well-developed basal leaves present.
Leaves more or less obtuse, the lowermost ones opposite, reduced and somewhat scale-like, more or less connate-sheathing at the base; heads ordinarily with a few short rays; Inner North Coast Ranges of California.   1. *R. scabrida.*
Leaves acute, apparently all alternate, none sheathing; heads strictly discoid; southern Sierra Nevada.
2. *R. muirii.*
Stems simple and monocephalous, scapose or subscapose; leaves chiefly or more often entirely basal.
Leaves green, either glabrous or glandular, not at all tomentose.
Heads with about 20 or more (probably commonly about 21) involucral bracts and 6–10 well-developed rays; leaves strictly glabrous; vicinity of Mount Eddy, California.   3. *R. pringlei.*
Heads mostly with about 8 or about 13 involucral bracts, discoid or sometimes with 1–3(5) irregularly developed rays; leaves sparsely to densely stipitate-glandular and sometimes inconspicuously hairy as well; Cascade Mountains of central Oregon south through the Sierra Nevada and on San Gorgonio Peak in southern California.   4. *R. scaposa.*
Leaves gray, silky-tomentose; Cascade Mountains of central Oregon south through the Sierra Nevada.
5. *R. argentea.*

1. **Raillardella scábrida** Eastw. Leafy Raillardella. Fig. 5737.

Raillardella scabrida Eastw. Bull. Torrey Club **32**: 216. 1905.
Raillardiopsis scabrida Rydb. N. Amer. Fl. **34**: 320. 1927.

Stems several from a stout woody root and branching caudex, often curved at the base, 1–5 dm. tall, commonly branched and bearing several heads; herbage hirsute and stipitate-glandular, more hirsute below, more glandular above. Leaves rather numerous and well distributed along the stems, principally alternate, small, linear or nearly so, more or less obtuse, mostly 1–3 cm. long and 1.5–4 mm. wide, the lowermost ones reduced and more or less scale-like, opposite, connate-sheathing; heads naked-pedunculate at the ends of the branches, discoid or more commonly at least some of them with 2 or 3 short broad rays up to about 6 mm. long; involucre 9–12 mm. high, coarsely long-stipitate-glandular, the broad bracts mostly about 8 or 10; rays epappose, their achenes glabrous or sparsely hairy; disk-flowers with a papus of about 15 or more elongate, plumose awn-scales, shorter than the pubescent linear achenes.

Open, stony places, Canadian Zone; in the Inner Coast Ranges of Lake, Mendocino, and Tehama Counties, California, at elevations up to 7,500 feet. Type locality: Snow Mountains, Lake County. July–Aug.

2. **Raillardella muírii** A. Gray. Muir's Raillardella. Fig. 5738.

Raillardella muirii A. Gray, Bot. Calif. **1**: 618. 1876.
Raillardiopsis muiru Rydb. N. Amer. Fl. **34**: 320. 1927.

Habitally similar to *R. scabrida.* Leaves acute, apparently all alternate, none sheathing; herbage less hairy and less glandular; heads discoid, the involucre with about 13 bracts; pappus- members only about 10, longer than the achenes.

Rare and local in open places at moderate elevations, Canadian and Hudsonian Zones; in the southern Sierra Nevada, California. Type locality: in the Sierra Nevada, supposedly near Yosemite, perhaps actually farther south. July–Aug.

3. **Raillardella prínglei** Greene. Showy Raillardella. Fig. 5739.

Raillardella pringlei Greene, Bull. Torrey Club **9**: 17. 1882.
Raillardella scaposa var. *pringlei* Jepson, Man. Fl. Pl. Calif. 1146. 1925.

Scapose perennial with a stout branching rhizome or caudex. Leaves oblanceolate or linear-elliptic, 3–10 cm. long, 6–8 mm. wide, wholly glabrous, firm, some of them remotely and shallowly serrate-dentate; scape 2–4 dm. tall, stipitate-glandular; heads solitary, broader and with more numerous flowers than in *R. scaposa*; involucre about 1 cm. high, shorter than the disk, the bracts relatively numerous, reputedly 20–30, coarsely stipitate-glandular and more or less long-ciliate as well; rays well developed, about 6–10 cm. long, deeply trifid, salmon or brick-colored; pappus similar in ray and disk, of about 15 stout, flattened, plumose bristles; achenes sparsely hairy above.

Stream banks and wet meadows, Hudsonian Zone; in the vicinity of Mount Eddy, northern California. Type locality: mountains west of Mount Shasta. Aug.-Sept.

4. **Raillardella scapòsa** A. Gray. Green-leaved Raillardella. Fig. 5740.

Raillardia scaposa A. Gray, Proc. Amer. Acad. **6**: 551. 1865.
Raillardella scaposa A. Gray, Bot. Calif. **1**: 417. 1876.
Raillardella scaposa var. *eiseni* A. Gray, Syn. Fl. N. Amer. **1²**: 380. 1884.
Raillardella nevadensis Nels. & Kenn. Proc. Biol. Soc. Wash. **19**: 38. 1906.
Raillardella scaposa nevadensis Blake, Contr. U.S. Nat. Herb. **25**: 605. 1925.

Fibrous-rooted, scapose or subscapose perennial from a branching caudex. Leaves narrowly oblanceolate, mostly 3–15 cm. long (including the petiolar base) and 3–8 mm. wide, entire, green, sparsely to densely stipitate-glandular (as also the scape) and sometimes inconspicuously hairy as well; scape 2–40 cm. tall; heads solitary, discoid or not infrequently with 1–3(5) irregularly devel-

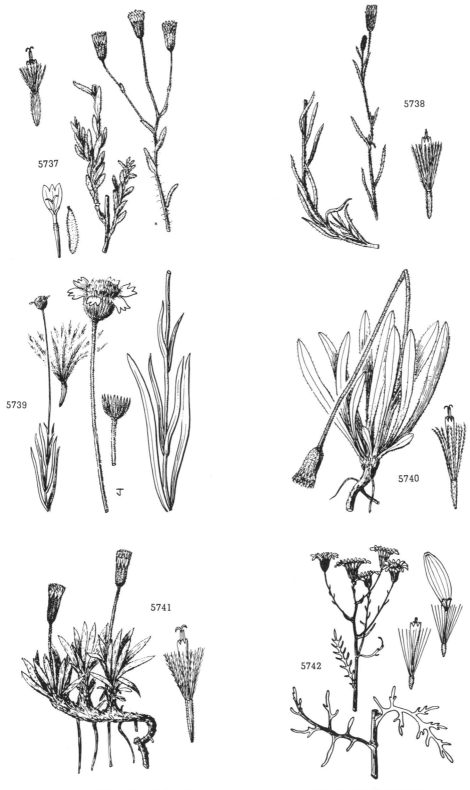

5737. Raillardella scabrida
5738. Raillardella muirii
5739. Raillardella pringlei

5740. Raillardella scaposa
5741. Raillardella argentea
5742. Senecio lyonii

oped short rays; involucre 12–16 mm. high, rather narrow, shorter than the disk, the bracts mostly about 8 or about 13, stipitate-glandular and often with some intermingled longer glandless hairs; pappus of about 15 or more stout, flattened, plumose bristles, that of the rays, when these are present, similar to that of the disk; achenes pubescent.

Open places at high altitudes in the mountains, rarely descending as low as 6,500 feet, Hudsonian Zone; Cascade Mountains of central Oregon (the Three Sisters) south through the Sierra Nevada. Type locality: on a peak of the Sierra Nevada northeast of Soda Springs. July–Aug.

5. **Raillardella argéntea** A. Gray. Silky Raillardella. Fig. 5741.

Raillardia argentea A. Gray, Proc. Amer. Acad. **6**: 550. 1865.
Raillardella argentea A. Gray, Bot. Calif. **1**: 417. 1876.
Raillardella minima Rydb. N. Amer. Fl. **34**: 319. 1927.

Habitally similar to *R. scaposa* but more strictly scapose and differing sharply in its conspicuously silky-tomentose, not at all glandular leaves; plants sometimes bearing well-developed creeping rhizomes; scape shorter, mostly 1–12 cm. tall; heads strictly discoid; involucre averaging a little smaller, mostly 9–15 mm. high, occasionally composed of only 5 bracts.

Open rocky places at high altitudes in the mountains, rarely as low as 7,000 feet, Hudsonian and Arctic-Alpine Zones; Cascade Mountains of central Oregon (the Three Sisters) south through the Sierra Nevada and on San Gorgonio Peak in the San Bernardino Mountains. Type locality: Sonora Pass, California. July–Sept.

110. **SENÈCIO** L. Sp. Pl. 866. 1753.

Heads radiate or sometimes discoid, the rays pistillate and fertile, yellow to orange or occasionally reddish, purple, or white or wanting. Involucral bracts herbaceous or subherbaceous, essentially equal, uniseriate or by lateral overlapping subbiseriate, seldom fewer than about 8, often with some smaller bracteoles at the base. Receptacle flat or convex, naked. Disk-flowers perfect and fertile, yellow to orange or reddish. Anthers entire to minutely sagittate. Style-branches more or less flattened, truncate, penicillate, with introrsely marginal stigmatic lines extending to the tip or rarely short-appendiculate. Achenes subterete, 5–10-nerved; pappus of numerous, usually white, entire or barbellulate, capillary bristles. Annual, biennial, or perennial herbs or shrubs, vines, or in some extralimital species even arborescent, with alternate (or all basal), entire to variously toothed or divided leaves and solitary to numerous, mostly small to medium-sized, cylindric to campanulate or hemispheric heads. [Name from the Latin *senex*, an old man, probably referring to the white pappus or hoary pubescence of some species.]

One of the largest genera of plants, containing probably well over 1,000 species, of very wide geographic distribution, perhaps 100 in North America north of Mexico. Type species, *Senecio vulgaris* L.

Many of the species are closely related, often, indeed, with only about the degree of distinctness more commonly associated with varieties. This is especially true of the species here treated under the groups *Aurei*, *Lobati*, and *Tomentosi*, which collectively form a huge complex with few or no clear-cut specific lines. The species-groups indicated in the key are in large part natural and correspond in some degree to the sections of Greenman and the species-groups of Rydberg; for purposes of this treatment they are, however, considered to be groups of convenience, without taxonomic status.

Senecio élegans L. Sp. Pl. 869. 1873, a low species with lyrate or pinnatifid leaves and purple rays, is adventive in sandy places about San Francisco. It is native to South Africa.

Stems erect or ascending, never twining or climbing; leaves various but scarcely ivy-like; heads various.
   Plants perennial (or in *S. jacobaea* sometimes merely biennial).
      Leaves well distributed along the stems, only slightly or gradually reduced upwards; no tuft of basal leaves present.
         Leaves two to three times pinnatifid, the ultimate segments evidently broader than linear; short-lived, not at all woody, introduced weed.   VIII. JACOBAEAE.
         Leaves entire to toothed or pinnatifid or in the woody insular species, *S. lyonii*, often bipinnatifid; native species.
            Leaves either narrowly linear or entire or evidently pinnatifid (or in *S. lyonii* often bipinnatifid) with linear, mostly elongate segments; plants taprooted, with numerous stems from the base, tending to become woody below.   I. SUFFRUTICOSI.
            Leaves evidently broader than linear, merely toothed or entire (or irregularly pinnatilobate in *S. clarkianus*); herbs, fibrous-rooted except for *S. fremontii*.   II. TRIANGULARES.
      Leaves basally disposed, the basal or lower cauline ones well developed, often tufted, the others evidently and progressively reduced (or the stem scapose).
         Plants with the numerous fibrous roots arising from a very short, erect, short-lived crown without (except sometimes in *S. crassulus*) any more elongate caudex or rhizome; leaves entire to dentate, or in forms of *S. integerrimus* occasionally irregularly sublobed; herbage crisp-hairy to essentially glabrous, not floccose or tomentose.   III. COLUMBIANI.
         Plants taprooted, or more often fibrous-rooted from an ascending or horizontal, simple or often branched caudex or from a creeping rhizome; herbage variously floccose or tomentose to glabrous, but not crisp-hairy.
            Leaves all sharply dentate or callous-denticulate to occasionally (in forms of *S. sphaerocephalus*) all entire, never crenate, lobed, or pinnatifid; bracteoles tending to be fairly well developed, as are also at least the lower cauline leaves; plants thinly tomentose or arachnoid-villous, at least when young.   IV. LUGENTES.
            Leaves otherwise (some or all of them crenate, serrate, lobed, or pinnatifid), or all entire in a few species which differ in other respects from the foregoing group; habit and vesture diverse.
               Basal leaves entire or toothed to sometimes pinnately irregularly few-lobed, ordinarily neither pinnatifid, lyrate, nor at all palmately lobed.
                  Herbage more or less tomentose at flowering time, sometimes only thinly or rather obscurely so; species of dry habitats at various altitudes.   V. TOMENTOSI.

Herbage glabrous from the first, or sometimes lightly floccose-tomentose when young and glabrous by flowering time except often for a little inconspicuous tomentum at the base and in the axils; species mostly of moist or montane habitats. VI. AUREI.

Basal leaves either pinnately dissected, deeply pinnatifid, lyrate-pinnatifid or tending to be shallowly palmately lobulate, in the latter case the lobes usually tending to be again toothed and the blade more or less cordate; heads several or many. (Forms of *S. ionophyllus* might be sought here.) VII. LOBATI.

Plants annual or winter-annual. IX. ANNUI.

Stems twining, climbing; leaves petiolate, cordate, ivy-like; heads small, discoid; introduced species of coastal California. X. MIKANIOIDEAE.

I. SUFFRUTICOSI.

Involucral bracts mostly about 21, rarely only 13; usually many of the leaves with some long lateral segments.

Heads relatively small, the involucre 5–7 mm. high; leaves more dissected; bracteoles short and inconspicuous, much less than half as long as the involucre; plants bearing dense tufts of persistent wool in the axils; islands off the coast of southern California. 1. *S. lyonii.*

Heads relatively large, the involucre about 7–11 mm. high; leaves less dissected; bracteoles relatively long and conspicuous, some of them generally at least half as long as the involucre; plants without tufts of axillary wool; continental and insular, from central California southward. 2. *S. douglasii.*

Involucral bracts mostly about 8 or about 13; bracteoles short and inconspicuous; leaves mostly entire, sometimes a few of them with lateral segments.

Plants about 6–12 dm. tall, rather strongly woody, not dying back to the base each year; leaves very narrow, commonly about 1(2) mm. wide or less; Santa Barbara and San Luis Obispo Counties, California. 3. *S. blochmaniae.*

Plants mostly 2.5–8 dm. tall, often somewhat woody below but apparently dying back to the base each year; leaves a little wider, the larger ones commonly 2 mm. wide or more; San Bernardino Mountains, White Mountains, and eastward. 4. *S. spartioides.*

II. TRIANGULARES.

Plants dwarf, often freely branched; leaves about 4 cm. long or less.

Leaves mostly obovate to spatulate or broadly oblanceolate, up to about 2 cm. wide; plants eventually taprooted, although the branches of the caudex may become elongate and rhizome-like; widespread in the mountains. 5. *S. fremontii*

Leaves oblanceolate to linear-oblong, 5 mm. wide or less; perennial from a system of slender rhizomes, without a taproot; mountains of Mono County, California. 6. *S. pattersonensis.*

Plants taller, mostly 2–15 dm. tall or more, the stems generally simple below the inflorescence; leaf-blades mostly over 4 cm. long.

Some or most of the leaves irregularly pinnatilobate or incised; leaves neither triangular nor regularly serrate; Sierra Nevada. 7. *S. clarkianus.*

Leaves coarsely dentate to serrate or occasionally subentire, never pinnatilobate or incised; widespread species.

Leaves, except sometimes for the reduced upper ones, petiolate, or tapering to a narrow petioliform base.

At least the lower leaves triangular with deltoid to subcordate base. 8. *S. triangularis.*

Leaves all tapering to the base, not at all triangular. 9. *S. serra.*

Some or all of the cauline leaves sessile with broad, more or less clasping base; northeast Oregon and eastward. 10. *S. crassulus.*

III. COLUMBIANI.

Herbage loosely crisp-villous or arachnoid-villous with evidently multicellular hairs, much of the pubescence sooner or later deciduous; widespread, mostly in well-drained soil.

Heads usually radiate (discoid in var. *vaseyi* of Washington and northern Oregon), and with 20–50 or more flowers; widespread. 11. *S. integerrimus.*

Heads discoid or sometimes with 1 or 2 short rays and with about 15–20 flowers; California. 12. *S. aronicoides.*

Herbage essentially glabrous from the first, sometimes with a very little arachnoid tomentum in the inflorescence.

Heads few, seldom as many as a dozen, apparently always radiate; plants of dry to moderately moist habitats; northeastern Oregon and eastward. 10. *S. crassulus.*

Heads more numerous, seldom less than a dozen in well-developed plants, often very numerous, radiate or more often discoid; plants mostly of wet meadows and other wet places.

Robust, more or less glaucous plants 4–20 dm. tall; leaves generally entire, occasionally irregularly toothed; plants widespread, tolerant of alkali. 14. *S. hydrophilus.*

Less robust, scarcely glaucous plants 3–10 dm. tall; leaves generally saliently toothed, rarely subentire; plants intolerant of alkali. 13. *S. foetidus.*

IV. LUGENTES.

Heads discoid; San Luis Obispo County to San Bernardino County, California. 15. *S. astephanus.*

Heads radiate; more northern or eastern species.

Heads about 3–30, small or medium-sized (involucre 3.5–12 mm. high), erect or in *S. elmeri* sometimes slightly nodding.

Involucral bracts not dark-tipped; rays elongate, mostly 1.5–2 cm. long; involucre 9–12 mm. high; plants 3–7 dm. tall; California. 37. *S. layneae.*

Involucral bracts usually with minute to prominent, black or brownish tips, or the plants otherwise different from *S. layneae* as characterized above.

Plants mostly single-stemmed from a short rhizome, thinly tomentulose when young, later often more or less glabrate.

Involucral bracts with minute, black or brownish tips; species not boreal, in our range not extending as far north as Washington.

Leaves slightly denticulate or entire; plants mostly 3–8 dm. tall; involucral bracts mostly about 21, rarely only 13; achenes hispidulous or sometimes glabrous; Oregon, northeastern Nevada, and northeastward. 16. *S. sphaerocephalus.*

Leaves mostly saliently dentate, rarely subentire; plants mostly 2–5 dm. tall; involucral bracts mostly about 13, rarely as many as 21; achenes glabrous; Sierra Nevada and White Mountains, California. 17. *S. scorzonella.*

Involucral bracts with very conspicuous black tips, commonly about 13, rarely up to 21; boreal species, extending south in our range to the Olympic Mountains of Washington. 18. *S. lugens.*

Plants with several stems from a well-developed branching caudex, subtomentosely arachnoid-villous at first, only thinly so (or even glabrate) at flowering time; leaves dentate or denticulate; Cascade and Wenatchee Mountains of central Washington north to southern British Columbia. 19. *S. elmeri.*

Heads solitary or 2, large (involucre 11–17 mm. high), nodding; Olympic Mountains of Washington. (The Californian, flame-flowered *S. greenei* might mistakenly be sought here except for the erect heads.) 20. *S. neowebsteri.*

V. TOMENTOSI.

Heads several or rather many.

Leaves nearly all tufted at the base, the cauline ones few and abruptly very much reduced, the stems appearing subscapose; basal leaves regularly toothed to occasionally subentire; mountains of southern California. 23. *S. bernardinus.*

Leaves not all tufted at the base, the cauline ones though evidently smaller than the basal relatively better developed, and the stem not appearing subscapose; basal leaves entire to sometimes irregularly subpinnately lobed, not regularly toothed; more northern plants.

Tomentum relatively thin or even obscure; robust plants, mostly 2–7 dm. tall, with relatively long and slender leaves; west of the Cascade Mountains in Oregon and Washington. 21. *S. macounii.*

Tomentum relatively dense though sometimes partly deciduous; smaller plants, mostly 1–3(4) dm. tall; leaves various; plants occurring in and more especially east of the Cascade Mountains and Sierra Nevada, extending farther west only in the Klamath region of southern Oregon. 22. *S. canus.*

Heads solitary or sometimes 2–6.

Heads large, the involucre 8–14 mm. high, the disk mostly 1.5–3 cm. wide.

Flowers yellow; mountains of southern California. 24. *S. ionophyllus.*

Flowers flame or red-orange; North Coast Ranges of California. 25. *S. greenei.*

Heads smaller, the involucre 5–9 mm. high, the disk 7–15 mm. wide.

Stems clustered on a branching caudex; plants about 1(1.5) dm. tall or less; Sierra Nevada and eastward. 26. *S. werneriaefolius.*

Stems solitary from a short, simple or subsimple caudex; heads notably long-pedunculate; plants about 1–3 dm. tall; Josephine County, Oregon. 27. *S. hesperius.*

VI. AUREI.

Heads solitary or occasionally 2 or 3, ordinarily radiate (but discoid forms of *S. resedifolius* are not uncommon to the north of our area). (*S. hesperius* and *S. werneriaefolius* of the *Tomentosi* might be sought here, but are evidently tomentose at the top of the peduncle and base of the involucre.)

Plants with 1 or more, more or less well-developed cauline leaves, scarcely scapose, seldom less than 1 dm. tall; rhizome short.

Rootstock very slender; widespread in wet alpine and subalpine meadows. 28. *S. subnudus.*

Rootstock short and thick; drier, more rocky and exposed alpine habitats; boreal species, extending south in our range to central Washington. 29. *S. resedifolius.*

Plants essentially scapose, less than 1 dm. tall, the peduncle merely bearing 1 or 2 minute bracts; rhizome elongate, branched; Wallowa Mountains of Oregon. 30. *S. porteri.*

Heads ordinarily several, sometimes only 1 or 2 in the mostly discoid species, *S. pauciflorus.*

Heads characteristically discoid (rare radiate forms occur, chiefly in company with the normal discoid plants).

Heads 1–6, rarely 12, orange or reddish; leaves relatively thick and firm, crenate; alpine and subalpine plants; boreal species, extending south into the high Sierra Nevada. 31. *S. pauciflorus.*

Heads more numerous, mostly 6–40, yellow; leaves relatively thin and lax, serrate to incised-pinnatifid; stream banks and moist woods; northern species, in our range rarely extending south of northern Washington. 32. *S. indecorus.*

Heads characteristically radiate (rare discoid forms of some of these species occur, chiefly in company with the normal radiate plants).

Basal leaves, or some of them, cordate or subcordate, sharply toothed; cauline leaves mostly laciniate-pinnatifid, at least toward their bases; leaves thin. 33. *S. pseudaureus.*

Basal leaves not cordate or subcordate, though sometimes subrotund.

Heads relatively large, the involucre 9–12 mm. high, the rays mostly 15–20 mm. long; Eldorado County, California. 37. *S. layneae.*

Heads smaller, the involucre 5–9 mm. high, the rays mostly 5–12 mm. long.

Plants strongly glaucous, relatively robust, up to about 7 dm. tall, mostly with rather numerous heads; leaves all entire, or in one variety some of the cauline leaves toothed; plants occurring on serpentine in California. 36. *S. clevelandii.*

Plants slightly or not at all glaucous, commonly but not always smaller and fewer-headed than *S. clevelandii*; some or all of the leaves generally toothed to pinnatifid; widespread plants mostly not inhabiting serpentine.

Leaves relatively thin and lax, the basal ones mostly elliptic or oblanceolate, crenate or serrate to subentire. 34. *S. pauperculus.*

Leaves relatively thick and firm, the basal ones mostly elliptic to subrotund, coarsely crenate to shallowly lobulate, wavy, or subentire. 35. *S. cymbalarioides.*

VII. LOBATI.

Plants fibrous-rooted from a simple or branched caudex or short rhizome; herbage glabrous or very nearly so.

Plants robust, leafy-stemmed, commonly 4–8 dm. tall; leaves all pinnatifid, the lower lyrate; heads relatively large, the involucres mostly 7–11 mm. high and glabrous; Alameda County, California, south to San Luis Obispo County and inland to the Tehachapi Mountains. 41. *S. breweri.*

Plants otherwise, differing in one or more (usually more) features from the above; more northern species.

Plants commonly only 0.5–2(4) dm. tall; basal leaves mostly lyrate-pinnatifid; stem appearing nearly naked; Olympic Mountains, and in the Cascade Mountains near Mount Rainier, Washington. 38. *S. flettii.*

Plants usually taller, 1–6 dm. tall; basal leaves more often shallowly palmately lobulate, sometimes somewhat lyrate.

Involucre glabrous; leaves relatively thin and lax; Cascade region from southern Washington to southern Oregon and also, less commonly, in the Willamette Valley and Oregon Coast Range well back from the ocean. 39. *S. harfordii.*

Involucre usually provided with few to numerous coarse, conspicuously multicellular hairs, rarely glabrous; leaves relatively thick and firm; along the coast from the estuary of the Columbia River to northern California. 40. *S. bolanderi.*

Plants with a more or less well-developed taproot, which may be surmounted by a branching caudex; herbage generally thinly tomentulose, varying to sometimes essentially glabrous.

Heads relatively large, the involucre mostly 8–12 mm. high, the dry disk-corollas mostly 7–10 mm. long, the achenes 3.5–6.5 mm. long; heads mostly 3–30; species of northern cismontane California.
42. *S. eurycephalus.*

Heads smaller, the involucre mostly 4–8(9) mm. high, the dry disk-corollas mostly 4.5–7.5 mm. long, the achenes 2–3.5(4) mm. long; heads sometimes more numerous; transmontane species, extending west into southern and northeastern California.
43. *S. multilobatus.*

VIII. Jacobaeae.

A single species in our range.
44. *S. jacobaea.*

IX. Annui.

Rays well developed and conspicuous.
45. *S. californicus.*

Rays very short and inconspicuous, or wanting.

Bracteoles well developed, evidently black-tipped; rays wanting; involucral bracts mostly about 21; introduced weed.
46. *S. vulgaris.*

Bracteoles few and inconspicuous, not black-tipped, or wanting; rays present but minute and inconspicuous, or in *S. mohavensis* sometimes wanting; involucral bracts commonly about 8 or about 13.

Involucral bracts mostly about 13.

Leaves coarsely dentate, cordate-clasping by a broad base; herbage glabrous; native desert species.
47. *S. mohavensis.*

Leaves more or less pinnatifid, not notably expanded at the base; herbage more or less evidently puberulent; introduced weed.
48. *S. sylvaticus.*

Involucral bracts mostly about 8; herbage glabrous or nearly so; native species of the South Coast Ranges of California and adjacent islands.
49. *S. aphanactis.*

X. Mikanioideae.

A single species in our range.
50. *S. mikanioides.*

1. **Senecio lyònii** A. Gray. Island Butterweed. Fig. 5742.

Senecio lyonii A. Gray, Syn. Fl. N. Amer. ed. 2. 1²: 454. 1886.

Taprooted perennial with numerous woody stems arising from the base, commonly 3–12 dm. tall and dying back only part way to the ground each year; stems and leaves thinly tomentulose when young, generally soon glabrate or subglabrate except for the conspicuous dense tufts of persistent wool in the axils. Leaves well distributed along the stems, 4–12 cm. long, pinnatifid or (especially the larger ones) bipinnatifid, with rather blunt, narrow, and often short, more or less linear segments; heads several in a corymbiform inflorescence, the disk 8–15 mm. wide; involucre 5–7 mm. high, the principal bracts mostly about 21; bracteoles inconspicuous, few, short; rays about 1 cm. long or less; achenes canescent.

Ocean bluffs, Upper Sonoran Zone; islands off the coast of southern California and adjacent Lower California. Type locality: San Clemente Island. March–June.

2. **Senecio douglásii** DC. Shrubby Butterweed. Fig. 5743.

Senecio douglasii DC. Prod. **6**: 429. 1837.
Senecio regiomontanus DC. loc. cit.
Senecio douglasii var. *tularensis* Munz, Aliso **4**: 99. 1958.

Taprooted perennial with numerous woody stems arising from the base, commonly 0.5–2 m. tall and dying back only part way to the ground each year; leaves, young twigs, and often also the involucres thinly tomentulose at first, often eventually (rarely early) more or less glabrate. Leaves well distributed along the stems, narrowly linear, 4–12 cm. long and 0.5–5 mm. wide, tapering to a slender point, generally many or most of them with a few long lateral segments; principal leaves often with fascicles of somewhat smaller leaves (or short leafy branches) in the axils; heads several or sometimes rather numerous in a corymbiform inflorescence, relatively large, the disk mostly 1–1.5 cm. wide; involucre 7–11 mm. high, the principal bracts usually about 21, rarely only 13; bracteoles well developed, unequal, generally some of them at least half as long as the involucre or longer; rays showy, about 1–1.5 cm. long; achenes canescent.

Dry washes and open slopes and plains, Sonoran Zones; cismontane California from the lower slopes of the Sierra Nevada in Amador County west to Napa and San Mateo Counties, south to the mountains of southern California and adjacent Lower California; passing into the var. *monoensis* on the desert or basin side of its range in southern California. Type locality: California. May–Oct.

Senecio douglasii var. **monoénsis** (Greene) Jepson, Man. Fl. Pl. Calif. 1149. 1925. (*Senecio monoensis* Greene, Leaflets Bot. Obs. **1**: 221. 1906.) Plants glabrous or only rarely slightly tomentulose, averaging a little smaller, more slender, and less shrubby than the typical form of the species, often dying back to the ground each year. Leaves averaging a little more dissected and bracteoles averaging a little shorter than in typical *S. douglasii.* Southern Great Basin from southwestern Utah to the White Mountains of Mono County, California, and southward into northern Sonora and Lower California, Mexico. Type locality: White Mountains. March–May, and sometimes again in the fall.

3. **Senecio blochmániae** Greene. Blochman's Butterweed. Fig. 5744.

Senecio blochmaniae Greene, Erythea **1**: 7. 1893.

Perennial with numerous woody stems arising from the base, 4–12 dm. tall, apparently not dying back to the ground each year; herbage glabrous or with some inconspicuous evanescent tomentum. Leaves numerous, well distributed along the stems, linear-filiform, entire, commonly 3–8 cm. long and about 1(2) mm. wide or less, the lower ones drying and persistent on the branches; heads several or rather numerous in a corymbiform inflorescence, the disk 7–13 mm. wide; involucre 7–10 mm. high, the principal bracts mostly about 13; bracteoles few and short, inconspicuous; rays about 1 cm. long or less; achenes densely canescent.

Sand dunes and flood plains near the coast, Upper Sonoran Zone; Santa Barbara and San Luis Obispo Counties, California. Type locality: Santa Maria River, San Louis Obispo County. June–Oct.

4. **Senecio spartiòides** Torr. & Gray. Narrow-leaved Butterweed. Fig. 5745.

Senecio spartioides Torr. & Gray, Fl. N. Amer. **2**: 438. 1843.
Senecio serra var. *sanctus* H. M. Hall, Univ. Calif. Pub. Bot. **3**: 230. 1907.

Taprooted perennial with numerous stems arising from the base, commonly 2–8 dm. tall, somewhat woody below but commonly dying back to the ground each year; herbage glabrous. Leaves numerous, well distributed along the stems, linear and entire (rarely a few with some lateral segments), the lowermost reduced, the others mostly 3–10 cm. long and 1.5–5 mm. wide; heads numerous in a paniculate-corymbiform inflorescence, the disk mostly 5–10 mm. wide; involucre 5–11 mm. high, the principal bracts about 13 or sometimes only 8; bracteoles inconspicuous, few, and short; rays 7–15 mm. long; achenes canescent or sometimes glabrous.

Dry open slopes, Arid Transition and Canadian Zones; southern California (San Bernardino Mountains) and northern Lower California in the Sierra San Pedro Martir, north and east to New Mexico, western Nebraska, and Wyoming. Type locality: Sweetwater River, Wyoming. July–Aug.

Senecio spartioides var. **granulàris** Maguire & Holmgren ex Cronquist in Ferris, Contr. Dudley Herb. **5**: 102. 1958. Herbage villous-puberulent with short, crisped, multicellular hairs and obscurely viscidulous; achenes apparently subglabrous; otherwise similar to low, small-headed plants of the typical variety (plants 3 dm. tall; involucre 5–7 mm. high). Upper altitudes in California in Mono County south to the White Mountains and the adjacent mountains of Mineral County, Nevada. Type locality: head of Crooked Creek, White Mountains, Mono County, California.

5. **Senecio fremóntii** Torr. & Gray. Dwarf Mountain Butterweed. Fig. 5746.

Senecio fremontii Torr. & Gray, Fl. N. Amer. **2**: 445. 1843.
Senecio ductoris Piper, Contr. U.S. Nat. Herb. **11**: 601. 1906.

Glabrous perennial from a taproot and branching caudex (caudical branches sometimes elongate and rhizome-like), freely branched from the decumbent base, commonly about 1–1.5 dm. tall or the stems sometimes longer and sprawling with ascending branches. Leaves thickish, somewhat succulent in life, more or less toothed, generally sharply so, commonly 1–4 cm. long and up to 2 cm. wide, well distributed along the stems, the lowermost ones reduced, the others mostly obovate to spatulate or broadly oblanceolate and tapering to a narrower, shortly petiolate base; heads terminating the branches, usually shortly naked-pedunculate; involucre about 7–12 mm. high, the principal bracts mostly about 13; bracteoles few, usually short and broad, occasionally narrower and more elongate; rays 6–10 mm. long; achenes glabrous or occasionally strigose-hispidulous.

Talus slopes and other rocky places at high elevations in the mountains, Hudsonian and Arctic-Alpine Zones; southern Alberta and British Columbia to the high mountains of northern and eastern Oregon, northeastern California in the Warner Mountains, central Idaho, northeastern Utah in the Uintah Mountains, and northern and western Wyoming. Type locality: Wind River Mountains, Wyoming. July–Sept.

Replaced in the southern Rocky Mountains and on the eastern rim of the Great Basin by the well-marked var. *blitoides* (Greene) Cronquist, Vasc. Pl. Pacif. Northw. **5**: 290. 1955 (*S. blitoides* Greene, Pittonia **4**: 123. 1900).

Senecio fremontii var. **occidentàlis** A. Gray, Bot. Calif. **1**: 618. 1876. (*Senecio occidentalis* Greene, Pittonia **4**: 122. 1900.) Similar to the typical form of the species and perhaps scarcely to be distinguished but generally more slender and more flexuous and often somewhat taller. Leaves tending to be more bluntly and less conspicuously toothed. Sierra Nevada of California and adjacent Nevada and in the San Bernardino Mountains. Some plants from Mount Rose much resemble var. *fremontii*. Type locality: Mount Whitney and vicinity. July–Aug.

6. **Senecio pattersonénsis** Hoover. Mono Butterweed. Fig. 5747.

Senecio revolutus Hoover, Leaflets West. Bot. **3**: 256. 1943. Not Kirk, 1899.
Senecio pattersonensis Hoover, op. cit. **5**: 60. 1947.

Resembling a small form of *S. fremontii* but narrower-leaved and lacking a taproot; plants perennial from a branching system of slender rhizomes, about 1 dm. tall or less. Leaves oblanceolate to linear-oblong, up to about 3.5 cm. long and 5 mm. wide.

Open rocky places at very high elevations in the mountains, Hudsonian and Arctic-Alpine Zones; Sweetwater Mountains and Sierra Nevada, in Mono County, California. Type locality: Mount Patterson, Mono County. July–Aug.

7. **Senecio clarkiànus** A. Gray. Clark's Butterweed. Fig. 5748.

Senecio clarkianus A. Gray, Proc. Amer. Acad. **7**: 362. 1868.

Fibrous-rooted, apparently single-stemmed perennial about 6–12 dm. tall, the stem and sometimes also the leaves obscurely to evidently crisp-pubescent with multicellular hairs as in *S. integerrimus*. Leaves numerous, well distributed along the stem, up to 25 cm. long and 6 cm. wide, gradually reduced upwards, the lower tapering to a petiolar base, the middle and upper sessile; some or all of the leaves pinnately incised or lobed, mostly irregularly so, with pointed segments, the others sharply and remotely dentate; heads several in a corymbiform inflorescence, the terminal peduncle shortened and surpassed by the others; involucre about 7–11 mm. high, the bracts about 21 or sometimes only 13, not black-tipped; bracteoles well developed, loose, and elongate; rays about 8 or 13, 1 cm. long or more; achenes glabrous.

Meadows and other moist places, Canadian Zone; in the Sierra Nevada from Mariposa County to Tulare County, California. Type locality: "Clark's meadow, below the Mariposa Big-tree Grove," California. Aug–Sept.

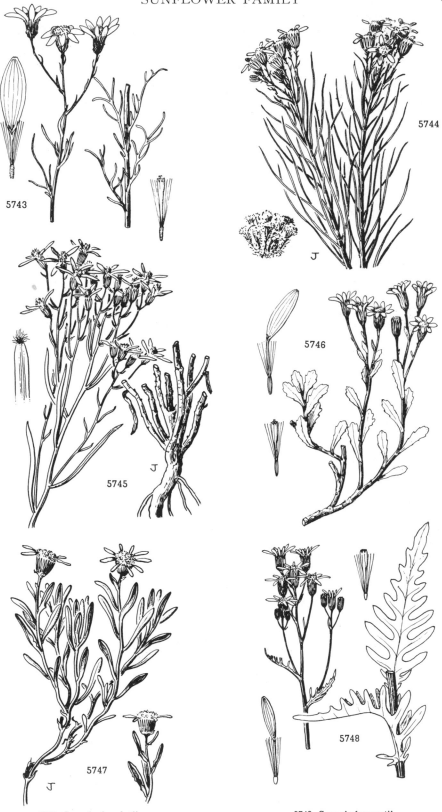

5743. Senecio douglasii
5744. Senecio blochmaniae
5745. Senecio spartioides

5746. Senecio fremontii
5747. Senecio pattersonensis
5748. Senecio clarkianus

8. **Senecio triangulàris** Hook. Arrowhead Butterweed. Fig. 5749.

Senecio triangularis Hook. Fl. Bor. Amer. **1**: 332. 1834
Senecio longidentatus DC. Prod. **6**: 428. 1837.
Senecio subvestitus Howell, Erythea **3**: 35. 1895.
Senecio triangularis var. *hanseni* Greene, op. cit. 124.
Senecio trigonophyllus Greene, Pittonia **3**: 106. 1896.
Senecio saliens Rydb. Bull. Torrey Club **24**: 298. 1897.
Senecio triangularis var. *subvestitus* Greenm. Monog. Gatt. Senecio **1**: 25. 1901.
Senecio triangularis var. *trichophyllus* St. John & Hardin, Mazama **11**: 95. 1929.
Senecio triangularis var. *trigonophyllus* M. E. Peck, Man. Pl. Oregon 780. 1941.

Several-stemmed, often rather coarse and lush, fibrous-rooted perennial, mostly 3–15 dm. tall, glabrous or obscurely (rarely evidently) villous-puberulent. Leaves numerous, neither tufted at the base nor ordinarily very strongly reduced upwards, the lower broadly or narrowly triangular to triangular-hastate or triangular-cordate, rather long-petiolate, the upper with shorter petioles or becoming sessile, often relatively narrower and less evidently or scarcely triangular; leaf-blades commonly 4–20 cm. long and 2–10 cm. wide, generally strongly toothed; heads few or rather numerous in a short, flat-topped inflorescence; involucre 7–10 mm. high, its principal bracts about 13 or sometimes only 8, often black-tipped; bracteoles few, narrow and generally somewhat elongate; rays mostly about 8, sometimes only 5, 7–13 mm. long; achenes glabrous.

Stream banks and other moist places in the mountains, Canadian and Hudsonian Zones; Alaska and southern Yukon to Saskatchewan, northern New Mexico, and California; in all the higher mountains of our range, northward sometimes also at lower elevations. Type locality: "Moist Prairies among the Rocky Mountains." [Canada] June–Sept.

Senecio triangularis var. **angustifòlius** G. N. Jones, Univ. Wash. Pub. Biol. **5**: 257. 1936. A rather small, slender form with relatively narrow leaves, the upper strongly reduced and becoming linear, often only the lowermost ones triangular or subtriangular. Sphagnum bogs at low elevations near the coast from Vancouver Island to southern Oregon. Type locality: Raft River, Grays Harbor County, Washington.

Senecio gibbónsii Greene, Pittonia **2**: 20. 1899. Morphologically similar to normal, fairly robust *S. triangularis* and seemingly not to be separated from it but differing markedly from that species in its habitat and worthy of further investigation. Salt marshes at the mouth of the Columbia River in both Oregon and Washington.

9. **Senecio sérra** Hook. Tall Butterweed. Fig. 5750.

Senecio serra Hook. Fl. Bor. Amer. **1**: 333. 1834.
Senecio andinus Nutt. Trans. Amer. Phil. Soc. II. **7**: 409. 1841.
Senecio lanceolatus Torr. & Gray. Fl. N. Amer. **2**: 440. 1843.
Senecio serra var. *integriusculus* A. Gray, Syn. Fl. N. Amer. **1²**: 287. 1884.
Senecio serra [var.] *andinus* Rydb. Mem. N.Y. Bot. Gard. **1**: 439. 1900.
Senecio serra [subsp.] *lanceolatus* Piper, Contr. U.S. Nat. Herb. **11**: 601. 1906.
Senecio millikeni Eastw. Bot. Gaz. **41**: 293. 1906.
Senecio serra var. *altior* Jepson, Man. Fl. Pl. Calif. 1150. 1925.

Stout, fibrous-rooted perennial 5–15 or 20 dm. tall, glabrous or somewhat puberulent especially toward the base; stems clustered. Leaves numerous, not tufted at the base, not much reduced upwards, the lower mostly oblanceolate, short-petiolate and deciduous, rarely larger and more persistent, the others lanceolate or lance-elliptic, tapering or rather abruptly contracted to the short petiole or petioliform base, commonly 7–15 cm. long and 1–4 cm. wide, sharply toothed or occasionally subentire; heads numerous on slender peduncles, narrow, almost cylindric, the disk commonly only 3–7 mm. wide; involucre 6–8 mm. high, its bracts commonly about 8 or about 13, often black-tipped; bracteoles few, narrow and generally rather elongate; rays few, commonly about 5 or 8, 5–8 mm. long; achenes glabrous.

Meadows and other open places in the foothills and at moderate elevations in the mountains, in favorable habitats sometimes extending out onto the plains, Arid Transition and Canadian Zones; western Montana to the eastern base of the Cascade Mountains in Washington, south to northern and western Wyoming, Utah, and northern California, and extending south at middle altitudes in the Sierra Nevada to Tulare County. Type locality: "banks of the Wallawallah, Flathead, and Spokan Rivers." June–Aug.

In the southern Rocky Mountains the typical form of the species is replaced by the var. *admirabilis* (Greene) A. Nels. with larger and mostly fewer heads.

10. **Senecio cràssulus** A. Gray. Mountain Meadow Butterweed. Fig. 5751.

Senecio crassulus A. Gray, Proc. Amer. Acad. **19**: 54. 1883.
Senecio crassulus var. *cusickii* (Piper) Greenm. ex M. E. Peck, Man. Pl. Oregon 780. 1941, without Latin diagnosis.

Glabrous, fibrous-rooted perennial 2–7 dm. tall from a short, erect or ascending caudex, this often branched and bearing several stems. Leaves thickish, entire to sharply dentate, the basal and lower cauline ones petiolate, with an elliptic to broadly oblanceolate blade commonly 2.5–12 cm. long and 1–5 cm. wide; cauline leaves becoming sessile and more or less clasping, the middle ones sometimes larger than those below, sometimes more or less strongly reduced; heads several (rarely solitary) in a corymbiform, usually open inflorescence, the terminal head little if at all overtopped by the others; disk about 5–15 mm. wide; involucre about 5–9 mm. high, sometimes much shorter than the disk, the broad bracts commonly about 13, varying to 8 or 21, thickened on the back, rather abruptly contracted to the villosulose, usually blackish tip; bracteoles few, narrow and often somewhat elongate; rays commonly about 8, varying to 5 or 13, 6–13 mm. long; achenes glabrous.

In dry or rather moist soil in open woods, meadows, and other open places from the foothills to rather high elevations in the mountains, Canadian and Hudsonian Zones; southwestern Montana to northeastern Oregon south to South Dakota, New Mexico, and Utah. Type locality: mountains of Colorado. June–Aug.

11. Senecio integérrimus var. exaltàtus (Nutt.) Cronquist. Single-stemmed Butterweed. Fig 5752.

Senecio exaltatus Nutt. Trans. Amer. Phil. Soc. II. **7**: 410. 1841.
Senecio lugens var. *exaltatus* D. C. Eaton, Bot. King. Expl. 188. 1871.
Senecio columbianus Greene, Pittonia **3**: 170. 1897.
Senecio condensatus Greene, op. cit. 298. 1898.
Senecio atriapiculatus Rydb. Mem. N.Y. Bot. Gard. **1**: 442. 1900.
Senecio arachnoideus Rydb. loc. cit.
Senecio integerrimus var. *exaltatus* Cronquist, Leaflets West. Bot. **6**: 48. 1950.

Stout, fibrous-rooted perennial from a very short, erect crown; stems solitary, 2–7 dm. tall; herbage hirsute to arachnoid-villous with crisp loose hairs when young, generally more or less glabrate in age. Leaves entire to irregularly dentate, the basal ones petiolate, the blade and petiole commonly 6–25 cm. long and 1–6 cm. wide; cauline leaves progressively reduced upwards, becoming sessile; heads several or rather numerous in an often congested inflorescence, the peduncle of the terminal head thickened and much shorter than the others; involucre about 5–10 mm. high, the principal bracts commonly about 21, sometimes only 13, shortly but evidently black-tipped; bracteoles few, mostly narrow and rather elongate; rays 6–15 mm. long, yellow; achenes glabrous or hispidulous.

Moderately dry to rather moist open places and open woods, from the valleys to near timberline, ordinarily in well-drained soils, Arid Transition and Canadian Zones; southern Alberta and British Columbia to Colorado, Nevada, and California; in our range found chiefly east of the Cascade and Sierra Nevada summits but occasional in the Willamette Valley and in the Klamath region of southwestern Oregon and northwestern California, extending nearly or quite to the coast. Type locality: "plains of Oregon, near the outlet of the Wahlamet." May–July.

Typical *S. integerrimus* is a plant of the Great Plains. In our range, several additional varieties are found, as follows:

5749. Senecio triangularis
5750. Senecio serra
5751. Senecio crassulus
5752. Senecio integerrimus

Senecio integerrimus var. ochroleùcus (A. Gray) Cronquist, Leaflets West. Bot. 6: 48. 1950. (*Senecio cordatus* Nutt. Trans. Amer. Phil. Soc. II. 7: 411. 1841; *Senecio lugens* var. *ochroleucus* A. Gray, Syn. Fl. N. Amer. 1²: 388. 1884; *S. exaltatus* [subsp.] *ochraceus* Piper, Contr. U.S. Nat. Herb. 11: 600. 1906.) Basal leaves tending to have deltoid or subcordate blade though sometimes as narrow as in typical var. *exaltatus* (which also runs to occasional forms with deltoid or subcordate blade); rays white or creamy; otherwise as in var. *exaltatus*. Commonly in slightly more moist habitats than var. *exaltatus*, often in woodlands; western Montana, across northern Idaho to the Cascade Mountains of Washington, south to northern Oregon and reputedly to northern California. Type locality: Columbia River, Klickitat County, Washington.

Senecio integerrimus var. màjor (A. Gray) Cronquist, Aliso 4: 100. 1958. (*Senecio eurycephalus* var. *major* A. Gray in Torr. Pacif. R. Rep. 4: 111. 1857; *S. mendocinensis* A. Gray, Proc. Amer. Acad. 7: 362. 1868; *S. whippleanus* A. Gray, Syn. Fl. N. Amer. 1²: 384. 1884; *S. caulanthifolius* Davy, Erythea 3: 117. 1895; *S. sonnei* Greene, Fl. Fran. 467. 1897; *S. major* Heller, Muhlenbergia 1: 118. 1905; *S. mesadenia* Greene, Leaflets Bot. Obs. 2: 227. 1912; *S. fodinarum* Greene, op. cit. 228; *S. lugens* var. *megacephalus* Jepson, Man. Fl. Pl. Calif. 1152. 1925.) Leaves subentire to toothed or even slightly lobed; heads mostly few and relatively large, the involucre 7–12 mm. high, its bracts few or sometimes somewhat purplish tips; rays up to 2 cm. long; otherwise as in var. *exaltatus*. On the west slope of the Sierra Nevada and in the North Coast Ranges of California; also in the White Mountains in Nevada. Type locality: "near Murphy's [Calaveras County], California."

Senecio integerrimus var. vàseyi (Greenm.) Cronquist, Vasc. Pl. Pacif. Northw. 5: 293. 1955. (*Senecio vaseyi* Greenm. in Piper, Contr. U.S. Nat. Herb. 11: 600. 1906.) Heads not very numerous; principal involucral bracts commonly about 21, evidently black-tipped; rays wanting; otherwise as in var. *exaltatus*. Cascade and Wenatchee region of central Washington; also in the Wallowa Mountains of northeastern Oregon. Type locality: Washington.

12. Senecio aronicoídes DC. California Butterweed. Fig. 5753.

Senecio aronicoides DC. Prod. 6: 426. 1837.
Senecio exaltatus var. *uniflosculus* A. Gray, Pacif. R. Rep. 4: 111. 1857.
Senecio rawsonianus Greene, Pittonia 2: 166. 1891.
Senecio leptolepis Greene, Fl. Fran. 468. 1897.

Stout, fibrous-rooted perennial from a very short, erect crown; stems solitary, 3–9 dm. tall; herbage arachnoid-villous with crisp loose hairs at least when young. Leaves highly variable, subentire to more often toothed or even shallowly incised-lobulate, the basal and lower cauline ones 8–30 cm. long (petiole included) and 2–12 cm. wide, the blade often deltoid or subcordate, often much narrower; cauline leaves progressively reduced upward, becoming sessile; heads few to more often rather numerous in a loose or compact inflorescence, the terminal peduncle shorter than the others; involucre 4–8 mm. high, with mostly about 8 or about 13 bracts, these with pale to purplish or blackish tips; flowers in each head relatively few, commonly 15–20, the rays none or occasionally 1 or 2, short.

Rather dry places in open woods in the foothills and lower mountains, Upper Sonoran and Arid Transition Zones; around the rim of the Great Valley of California from the vicinity of Mount Shasta south to the San Francisco Bay region and thence along the western front of the Sierra Nevada to Fresno County. Type locality: California. April–July.

13. Senecio foètidus Howell. Sweet-marsh Butterweed. Fig. 5754.

Senecio foetidus Howell, Fl. N.W. Amer. 377. 1900.

Glabrous but scarcely glaucous, fibrous-rooted perennial from a very short erect crown, 3–10 dm. tall, the stems solitary or more commonly clustered. Leaves thickish and somewhat succulent, generally sharply dentate, the basal and lowermost cauline ones petiolate, with mostly elliptic or broadly oblanceolate blade 6–25 cm. long and 2–7 cm. wide; middle and upper leaves few, strongly and progressively reduced, becoming sessile; heads more or less numerous (rarely as few as 8) in a congested inflorescence; involucre 6–9 mm. high, the bracts commonly about 13, sometimes only 8, minutely black-tipped; bracteoles few, narrow and rather elongate; rays few, typically about 5, up to about 8 mm. long or more often wanting; achenes glabrous.

Wet meadows in the mountains and foothills, Arid Transition Zone; Washington east of the Cascade Mountains and adjacent southern British Columbia east to western Montana, south to northeastern California, northern Nevada, and central and southwestern Idaho. Type locality: Klickitat Valley, Washington. May–July.

Senecio foetidus var. hydrophiloìdes (Rydb.) T. M. Barkley ex Cronquist in Ferris. Contr. Dudley Herb. 5: 102. 1958. (*Senecio hydrophiloides* Rydb. Mem. N.Y. Bot. Gard. 1: 441. 1900.) Single-stemmed; leaves thickish or sometimes thin; heads in an open, corymbiform inflorescence, the terminal one short-pedunculate and overtopped by the others; rays mostly 5 or 8, 5–10 mm. long; otherwise like typical *S. foetidus*. Wet meadows and moist woods in the mountains and foothills, western Montana to northeastern Washington and northeastern Oregon. Type locality: Forest, Nez Perce County, Idaho.

14. Senecio hydróphilus Nutt. Alkali-marsh Butterweed. Fig. 5755.

Senecio hydrophilus Nutt. Trans. Amer. Phil. Soc. II. 7: 411. 1841.
Senecio hydrophilus var. *pacificus* Greene, Pittonia 1: 220. 1888.
Senecio pacificus Rydb. Fl. Rocky Mts. 998. 1917.

Glabrous, more or less glaucous, stout, hollow-stemmed, fibrous-rooted perennial from a short erect crown, mostly 4–20 dm. tall, the stems often (? regularly) clustered. Leaves thick and firm, somewhat succulent, commonly entire or nearly so, sometimes more evidently callous-toothed, the basal or lowermost cauline ones long-petiolate, with large, mostly narrowly elliptic blade commonly 10–20 cm. long and 2–5 cm. wide or even larger; middle and upper leaves few, strongly and progressively reduced, becoming sessile; heads usually rather numerous and crowded; involucre 5–8 mm. high, rather narrow, its bracts commonly about 8, sometimes 13, often black-tipped; bracteoles few, narrow, and short; rays few, about 4–8 mm. long or often wanting; achenes glabrous.

Swampy places in the valleys and foothills, tolerant of salt and alkali, Upper Sonoran and Transition Zones; southern British Columbia to the San Francisco Bay region, east to Montana, South Dakota, and Colorado. Type locality: "Ham's Fork of the Colorado of the West [Wyoming]." May–July.

15. Senecio astéphanus Greene. San Luis Obispo Butterweed. Fig. 5756.

Senecio astephanus Greene, Pittonia **1**: 174. 1888.
Senecio ilicetorum Davidson, Erythea **2**: 85. 1894.

Coarse, apparently single-stemmed perennial about 6 dm. tall or probably often taller, conspicuously but rather thinly tomentulose nearly throughout. Leaves sharply dentate, the lowermost ones the largest, up to 30 cm. long (including the petiole) and 7 cm. wide; middle and upper leaves strongly and progressively reduced, becoming sessile, not numerous; heads several in an open or very compact inflorescence, rather large for the genus, discoid, yellow; achenes glabrous.

In the mountains, Upper Sonoran Zone; San Luis Obispo County to San Bernardino County, California; insufficiently known. Type locality: mountains of San Luis Obispo County. May–June.

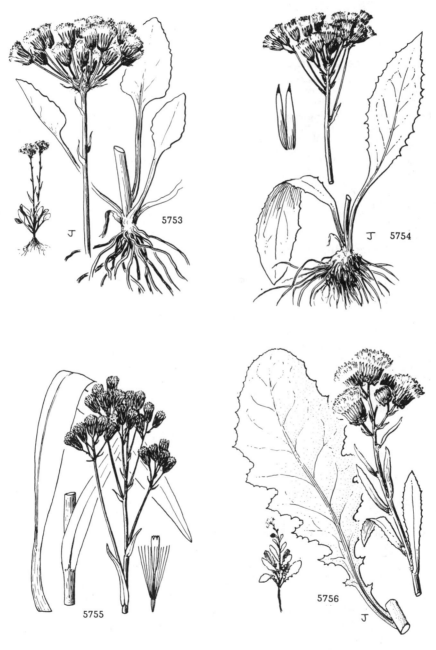

5753. Senecio aronicoides
5754. Senecio foetidus

5755. Senecio hydrophilus
5756. Senecio astephanus

16. Senecio sphaerocéphalus Greene. Mountain-marsh Butterweed. Fig. 5757.

Senecio sphaerocephalus Greene, Pittonia **3**: 106. 1896.

Fibrous-rooted perennial from a short, thick, horizontal or sometimes ascending rhizome 3–8 dm. tall, thinly tomentulose and often eventually (rarely early) glabrate; stems mostly arising singly from the rhizome. Leaves slightly denticulate or entire, the lowermost ones oblanceolate or elliptic, petiolate, well developed, the blade and petiole 5–25 cm. long and 1–4 cm. wide; middle and upper leaves few, strongly and progressively reduced, becoming sessile; heads about 3–25 in a usually rather compact cyme, the terminal peduncle often shortened and thickened; involucre 4–8 mm. high, its bracts commonly about 21, rarely only 13, bearing a minute, brownish or black tip; bracteoles few, narrow, short or rather elongate; disk 8–15 mm. wide; rays 6–10 mm. long; achenes usually hispidulous but in our forms glabrous.

Wet meadows in and near the mountains, Canadian Zone; Montana to northeastern Oregon (Wallowa Mountains) south to Colorado and northeastern Nevada. Type locality: Deeth, Humboldt River, Nevada. June–Aug. **Senecio oregànus** Howell, Fl. N.W. Amer. 377. 1900. Similar to *S. sphaerocephalus*, to which it perhaps should be reduced, but early-glabrate and with elongate, relatively narrow, lower leaves. Type locality: Lake Labish near Salem, Oregon, where it has been collected several times. June.

17. Senecio scorzonélla Greene. Sierra Butterweed. Fig. 5758.

Senecio scorzonella Greene, Pittonia **3**: 90. 1896.
Senecio covillei Greene, Fl. Fran. 469. 1897.
Senecio covillei var. *scorzonella* Jepson, Man. Fl. Pl. Calif. 1152. 1925.

Stems mostly arising singly from a short, thick, horizontal or ascending rhizome, 2–5 dm. tall, rarely taller; herbage thinly tomentulose when young, later often more or less glabrate. Basal leaves tufted, oblanceolate or broader, saliently dentate or rarely subentire, tapering to the petioliform base or short petiole, commonly 5–15 cm. long and 1.5–3 cm. wide; cauline leaves few, strongly and progressively reduced, becoming sessile; heads about 10–30 in a compact cyme, the terminal one borne on a short thick peduncle and overtopped by the others; involucre 3.5–6.5 mm. high, the minutely black-tipped bracts commonly about 13, sometimes 8 or 21; bracteoles few but often fairly well developed; rays scarcely 1 cm. long or sometimes wanting; achenes glabrous.

Meadows and moist slopes below timberline, Canadian and Hudsonian Zones; Sierra Nevada and adjacent southern Cascade Mountains from Mount Lassen to Tulare County and in the White Mountains of Inyo County, California. Type locality: ". . . northeastern California." July–Aug.

18. Senecio lùgens Richards. Black-tipped Butterweed. Fig. 5759.

Senecio lugens Richards. in Frankl. 1st Journ. Bot. App. 747. 1823.

Fibrous-rooted perennial with the stems mostly arising singly from a short, thick, ascending or horizontal rhizome, 1–5 dm. tall; herbage thinly tomentulose at first, generally subglabrate by flowering time. Lowermost leaves finely (or sometimes more coarsely) but not closely calloustoothed, well-developed, mostly 5–20 cm. long and 8–30 mm. wide, the oblanceolate to sometimes elliptic or nearly obovate blade tapering to the petiole or petioliform base; cauline leaves few, strongly and progressively reduced upwards, becoming sessile; heads mostly 3–21 in a compact cyme, the terminal peduncle equaling or often shorter and thicker than the others; involucre 5–8 mm. high, the bracts very conspicuously black-tipped, commonly about 13, rarely 21; bracteoles narrow but well developed, evidently black-tipped; disk 8–13 mm. wide; rays 7–15 mm. long; achenes glabrous.

Wet meadows, grassy alpine or arctic slopes, and rich northern woods, Boreal Zone; Alaska and Yukon south to the Olympic Mountains of Washington and possibly to northern Wyoming. Not known from the Washington Cascade Mountains. Type locality: Bloody Fall, on the Coppermine River, Yukon Territory. July–Aug.

19. Senecio élmeri Piper. Elmer's Butterweed. Fig. 5760.

Senecio elmeri Piper, Erythea **7**: 173. 1899.

Perennial from a well-developed branching caudex, 1–3 dm. tall, subtomentosely arachnoid-villous at first, only thinly so (or even glabrate) at flowering time. Basal or lower cauline leaves the largest, the basal sometimes tufted cn separate short shoots, oblanceolate, sometimes broadly so, up to about 20 cm. long (including the wing-petiole) and 3 cm. wide, sharply dentate or denticulate; cauline leaves well developed but usually evidently reduced upwards and becoming sessile; heads several in a corymbiform inflorescence, the peduncles often flexuous so that the heads may be a little nodding; involucre 7–12 mm. high, bracts mostly about 13, villosulose at the often dark or blackish tip; bracteoles narrow but well developed; disk 8–15 mm. wide; rays 8–16 mm. long; achenes glabrous.

Talus slopes and other rocky places, Hudsonian and Arctic-Alpine Zones; at alpine stations in the Cascade and the Wenatchee Mountains of central and northern Washington, extending into southern British Columbia. Type locality: "North Fork of Bridge Creek, Okanogan County, Wash." July–Aug.

20. Senecio neowébsteri Blake. Olympic Butterweed. Fig. 5761.

Senecio websteri Greenm. Bot. Gaz. **53**: 511. 1912. Not J. D. Hook. 1846.
Senecio neowebsteri Blake, Leaflets West. Bot. **8**: 143. 1957.

Perennial from a well-developed, often-branched caudex or short rhizome, 0.5–2 dm. tall, more or less arachnoid-villous at first and often also somewhat glandular, later more or less glabrate. Basal or lower cauline leaves the largest, the basal sometimes tufted on separate short shoots, petiolate, the blade broadly oblanceolate to subrotund, up to about 7 cm. long and 4 cm. wide, dentate or denticulate; cauline leaves few, more or less strongly reduced or the middle ones sometimes nearly as large as those below; heads solitary (rarely 2), mostly nodding, large, the involucre 11–17 mm. high, the disk mostly 1.5–2.5 cm. wide; involucral bracts commonly about 21, villosulous at the usually pale tip; bracteoles well developed and conspicuous, more than half as long as the involucre; rays yellow, 10–15 mm. long; achenes glabrous.

Talus slopes, Hudsonian and Arctic-Alpine Zones; Olympic Mountains of Washington. Type locality: Mount Angeles, Clallam County. Aug.–Sept.

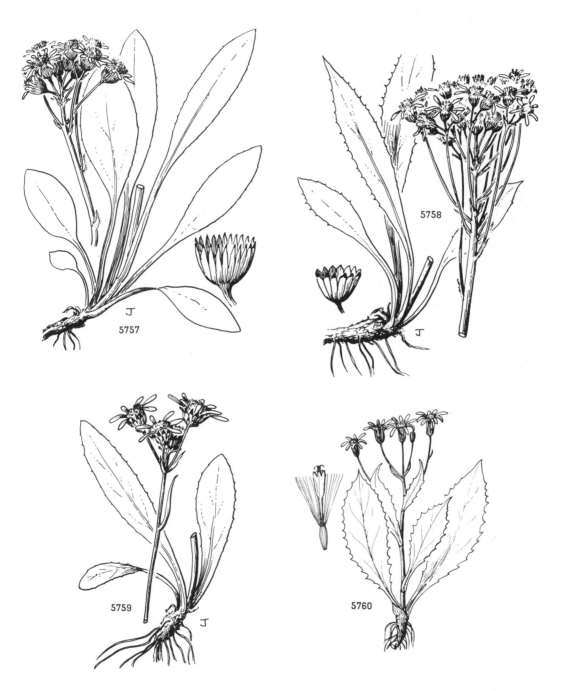

5757. Senecio sphaerocephalus
5758. Senecio scorzonella

5759. Senecio lugens
5760. Senecio elmeri

21. Senecio macoùnii Greene. Puget Butterweed. Fig. 5762.

Senecio fastigiatus Nutt. Trans. Amer. Phil. Soc. II. **7**: 410. 1841. Not Schweinf. 1824.
Senecio macounii Greene, Pittonia **3**: 169. 1897.
Senecio spatuliformis Heller, Bull. Torrey Club **26**: 552. 1899.
Senecio leucocrinus Greene, Leaflets Bot. Obs. **2**: 14. 1909.

Similar to *S. canus* but taller and more robust, commonly 2–7 dm. tall, more thinly tomentose (the tomentum sometimes obscure by flowering time), and with consistently narrow leaves, the basal ones commonly merely oblanceolate; heads averaging larger, the involucre up to 1 cm. high.

Open woods and dry open places, Upper Sonoran and Humid Transition Zones; west of the Cascade Mountains from Vancouver Island to southern Oregon, mostly in the Puget trough. Type locality: Mount Benson, Vancouver Island. June–July.

22. Senecio cànus Hook. Woolly Butterweed. Fig. 5763.

Senecio canus Hook. Fl. Bor. Amer. **1**: 333. 1834.
Senecio purshianus Nutt. Trans. Amer. Phil. Soc. II. **7**: 412. 1841.
Senecio howellii Greene, Bull. Torrey Club **8**: 98. 1881.
Senecio ligulifolius Greene, Leaflets Bot. Obs. **2**: 14. 1909.
Senecio canus [var.] *purshianus* A. Nels. in Coult. & Nels. New Man. Bot. Rocky Mts. 581. 1909.
Senecio howellii var. *lithophilus* Greenm. Bot. Gaz. **48**: 148. 1909.
Senecio kernensis Greenm. Ann. Mo. Bot. Gard. **1**: 286. 1914.
Senecio oreopolus Greenm. op. cit. 268.
Senecio oreopolus f. *aphanactis* Greenm. op. cit. 269.

Several-stemmed perennial from a more or less branching caudex, often with an evident short taproot, 1–3(4) dm. tall, more or less strongly white-tomentose, often less so in age, the upper surfaces of the leaves sometimes glabrate. Basal and lowermost cauline leaves more or less tufted, from narrowly oblanceolate to broadly elliptic or ovate, the blade mostly 1–4 or 5 cm. long, 4–25(45) mm. wide, entire to sometimes irregularly subpinnately lobed, borne on a short or elongate petiole; middle and upper leaves few, often coarsely toothed or subpinnatifid, strongly and progressively reduced, becoming bract-like, but the stem generally not appearing scapiform; heads several; involucre 4–8 mm. high, the bracts about 13 or about 21; bracteoles wanting or few and inconspicuous; disk 6–13 mm. wide; rays mostly 6–13 mm. long or rarely wanting; achenes glabrous.

Dry, open, often rocky places, from the plains and foothills to timberline or above, Arid Transition and Boreal Zones; southern British Columbia to Saskatchewan, south to California, Colorado, and Nebraska; in our range occurring east of the Cascade summits in Washington and northern Oregon, extending west to near the coast in the Klamath region of southern Oregon and southward along the Cascade Mountains and Sierra Nevada to Tulare County, California. Type locality: banks of the Saskatchewan River. May–Aug.

23. Senecio bernardìnus Greene. San Bernardino Butterweed. Fig. 5764.

Senecio bernardinus Greene, Pittonia **3**: 298. 1898.
Senecio ionophyllus var. *bernardinus* H. M. Hall, Univ. Calif. Pub. Bot. **3**: 232. 1907.

Perennial with 1 to several erect stems from a caudex, 1–3 dm. tall, rather thinly tomentulose or in part glabrate. Leaves tufted at the base, petiolate, the elliptic to obovate or subrotund blade 0.5–2.5 cm. long and 5–15 mm. wide, more or less toothed (often regularly so) or subentire; cauline leaves few and very much reduced, the stems somewhat scapiform; heads about 3–20; involucre 5–8 mm. high, the bracts about 21 or about 13; bracteoles inconspicuous or wanting; disk about 8–14 mm. wide; rays 6–10 mm. long; achenes glabrous or hispidulous.

Dry open places, Arid Transition and Canadian Zones; San Bernardino Mountains, California. Type locality: Bear Valley, San Bernardino Mountains. June–Aug.

24. Senecio ionóphyllus Greene. Tehachapi Butterweed. Fig. 5765.

Senecio ionophyllus Greene, Pittonia **2**: 20. 1899.
Senecio sparsilobatus Parish, Bot. Gaz. **38**: 462. 1904.
Senecio ionophyllus var. *sparsilobatus* H. M. Hall, Univ. Calif. Pub. Bot. **3**: 232. 1907.
Senecio bernardinus var. *sparsilobatus* Greenm. Ann. Mo. Bot. Gard. **5**: 46. 1918.
Senecio ionophyllus var. *intrepidus* Greenm. loc. cit.

Several-stemmed perennial from a branching caudex, 1.5–4 dm. tall, thinly tomentulose or sometimes nearly glabrous. Basal and lowermost cauline leaves with obovate to rotund or flabellate, coarsely blunt-toothed, sometimes more or less lyrate blade up to about 2.5 cm. long and 2 cm. wide, long–petiolate; cauline leaves strongly and progressively reduced, borne mostly below the middle of the stem, becoming sessile; heads 1–4, yellow, large, the disk 1.5–2.5 cm. wide; involucre about 8–10 mm. high, the principal bracts about 21 or sometimes only 13; bracteoles few but sometimes well developed or none; dry disk-corollas about 8–10 mm. long; rays about 8–14 mm. long; achenes 5–6.5 mm. long, glabrous.

Dry open slopes in the mountains, Arid Transition and Canadian Zones; mountains of southern California from Tehachapi southward. Type locality: near the summit of the mountains south of Tehachapi. June–July.

25. Senecio grèenei A. Gray. Flame Butterweed. Fig. 5766.

Senecio greenei A. Gray, Proc. Amer. Acad. **10**: 75. 1874.

Perennial from a branching, rhizome-like caudex, 1–3 dm. tall, thinly floccose-tomentose, eventually more or less glabrate. Basal and lower cauline leaves well developed, the basal often

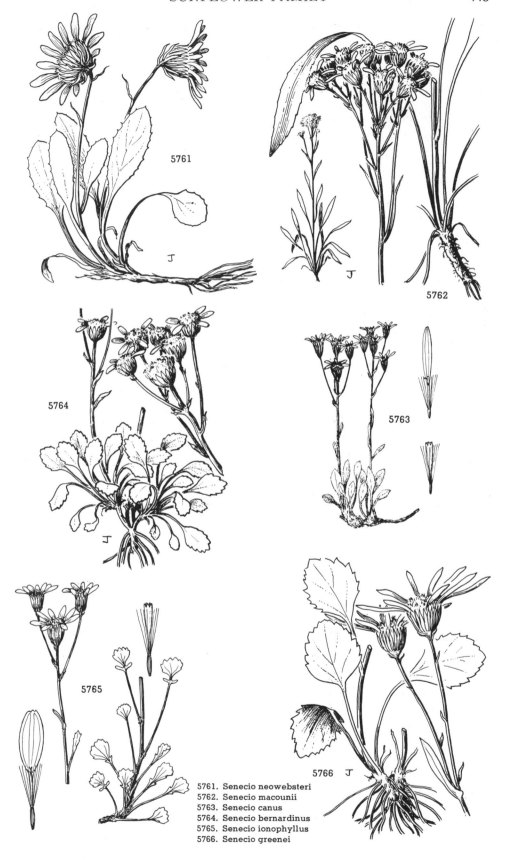

5761. Senecio neowebsteri
5762. Senecio macounii
5763. Senecio canus
5764. Senecio bernardinus
5765. Senecio ionophyllus
5766. Senecio greenei

tufted on separate short shoots, petiolate, the blade subrotund or broadly ovate, up to about 6 cm. long and broad, more or less dentate; middle and upper cauline leaves few and much reduced; heads 1–3, naked-pedunculate, erect, relatively very large, the involucre 10–14 mm. high, often much shorter than the broad (1.5–3 cm.) disk, the bracts about 13 or about 21; bracteoles few or none; rays 1–3 cm. long, flame or red-orange like the disk; achenes glabrous.

Brushy slopes, commonly on serpentine, Upper Sonoran and Transition Zones; Coast Ranges of northern California. Type locality: near the Geysers, Lake County, California. May–June.

26. **Senecio werneriaefòlius** A. Gray. Alpine Rock Butterweed. Fig. 5767.

Senecio aureus var. *werneriaefolius* A. Gray, Proc. Acad. Phila. **1863**: 68. 1863.
Senecio muirii Greenm. Ann. Mo. Bot. Gard. **5**: 56. 1918. Not *S. muirii* L. Bolus, 1915.
Senecio speculicola J. T. Howell, Leaflets West. Bot. **4**: 64. 1944.

Lax, several-stemmed perennial from a loosely branching caudex, up to 1.5 dm. tall, thinly tomentulose and often eventually glabrate. Leaves tufted at the base, oblanceolate to lance-elliptic or occasionally broader, tapering to the petiole or petiolar base, entire or nearly so, the whole about 1.5–7 cm. long and up to about 1 cm. wide; cauline leaves few, much reduced and bract-like, the stem scapiform; heads 1–6; involucre 6–9 mm. high, the bracts about 21 or about 13; bracteoles poorly developed or wanting; disk 1–1.5 cm. wide; rays about 1 cm. long or less; achenes glabrous.

Dry, open, often rocky places at high elevations in the mountains, Arctic-Alpine Zones; Sierra Nevada east to northern New Mexico, Colorado, western South Dakota, and western Montana. Type locality: ". . . on and near the Rocky Mountains, in Colorado Territory, lat. 39°–41°." June–Aug.

The California plants of *S. werneriaefolius* have been distinguished as *S. muirii* Greenm. or *S. speculicola* J. T. Howell but can easily be matched among specimens of true *S. werneriaefolius* from Colorado. Occasional plants from California approach or match the chiefly more northeastern, broader-leaved *S. saxosus* Klatt, another segregate from *S. werneriaefolius*.

27. **Senecio hespérius** Greene. Siskiyou Butterweed. Fig. 5768.

Senecio hesperius Greene, Pittonia **2**: 166. 1891.
Senecio auleticus Greene, Leaflets Bot. Obs. **2**: 15. 1909.

Fibrous-rooted, single-stemmed perennial from a short simple caudex, mostly 1–3 dm. tall; herbage tending to be slightly tomentulose, evidently so at the summit of the peduncles and base of the involucre. Basal leaves oblanceolate or elliptic to obovate-rotund, 1–6 cm. long (including the petiole) and 5–18 mm. wide; cauline leaves few, distant, and progressively reduced, sometimes somewhat pinnatifid; heads 1–3, all long-pedunculate, the disk 8–14 mm. wide; involucre 5–9 mm. high, the bracts about 13 or about 21, sometimes purplish toward the tip; bracteoles few, slender and elongate, or none; disk 1–1.5 cm. wide; rays about 9–14 mm. long, yellow; achenes glabrous.

Serpentine areas, Arid Transition Zone; Josephine County, Oregon. Type locality: near Kerby, Oregon. May–June.

28. **Senecio subnùdus** DC. Alpine Meadow Butterweed. Fig. 5769.

Senecio subnudus DC. Prod. **6**: 428. 1837.
Senecio aureus var. *subnudus* A. Gray, Syn. Fl. N. Amer. **1²**: 391. 1884.
Senecio pauciflorus var. *subnudus* Jepson, Man. Fl. Pl. Calif. 1154. 1925.

Glabrous, fibrous-rooted perennial from a short slender rhizome, 0.5–3 dm. tall. Leaves small, thin or slightly thickish, the basal ones commonly subrotund or broadly obovate, crenate-toothed, the blade up to about 2.5 cm. long and 2 cm. wide; cauline leaves few and reduced, sometimes more or less pinnatifid, the stem often subnaked; heads solitary or sometimes 2, the disk 8–15 mm. wide, yellow; involucre 5–8 mm. high, the bracts about 21 or rarely only 13, sometimes purplish toward the tip; bracteoles few, slender and rather elongate, or none; rays about 7–14 mm. long; achenes glabrous.

Alpine and subalpine wet meadows, Hudsonian and Arctic-Alpine Zones; Washington to Montana and southwestern Alberta, south to California (throughout the Sierra Nevada) and Wyoming. Type locality: Columbia River. July–Sept.

29. **Senecio resedifòlius** Less. Dwarf Arctic Butterweed. Fig. 5770.

Senecio resedifolius Less. Linnaea **6**: 243. 1831.
Senecio lyallii Klatt, Ann. Nat. Hofm. Wien **9**: 365. 1894. Not Hook. f. 1853.
Senecio ovinus Greene, Pittonia **4**: 110. 1900.
Senecio conterminus Greenm. Ann. Mo. Bot. Gard. **3**: 101. 1916.

Perennial from a branched caudex or short, rather thick rhizome, about 0.5–2 dm. tall, essentially glabrous or occasionally with a little persistent tomentum at the base and in the axils of the leaves. Leaves thickish, the basal ones mostly orbicular-ovate to reniform or even obovate, blunt-toothed or subentire, the blade up to about 2.5 cm. long and wide, generally abruptly contracted to the petiole; cauline leaves reduced upwards and becoming sessile, in the larger plants more or less pinnatifid; heads solitary or occasionally 2, the disk about 11–18 mm. wide, yellow to more often orange or even reddish; involucre about 7–8 mm. high, the bracts about 21 or rarely only 13, purplish at least distally; bracteoles averaging shorter and broader than in *S. subnudus*; rays variously developed, sometimes as much as 14 mm. long and 4 mm. wide or occasionally wanting; achenes glabrous.

Exposed rocky situations at alpine or occasionally subalpine stations in the high mountains, mostly upper Boreal Zone; Eurasia and Alaska to Newfoundland, south to central Washington, Montana, and northwestern Wyoming. Type locality: St. Lawrence Island in the Bering Sea. July–Sept.

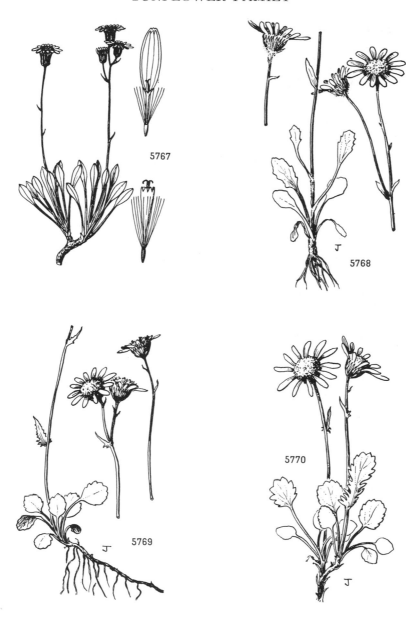

5767. Senecio werneriaefolius
5768. Senecio hesperius

5769. Senecio subnudus
5770. Senecio resedifolius

30. **Senecio pórteri** Greene. Porter's Butterweed. Fig. 5771.

Senecio renifolius Porter in Port. & Coult. Fl. Colo. 83. 1874. Not Schultz-Bip. 1845.
Senecio porteri Greene, Pittonia **3**: 186. 1897.

Fibrous-rooted perennial from a well-developed, slender, branched, creeping rhizome, glabrous throughout, essentially scapose, the 3–8 cm. peduncle naked or with 1 or 2 minute bracts.
Leaves petiolate, the thick, mostly reniform, crenate-lobulate blade up to 2.5 cm. wide; heads
solitary, fairly large, the somewhat purplish involucre about 1 cm. high.

Open rocky places (? talus slopes) in the mountains, Arctic-Alpine Zone; Wallowa Mountains of northeastern
Oregon (where known only from a single collection with the collector's note "very little seen") and in Colorado.
Type locality: White House Mountains, Colorado. Aug.
Our plant appears to be identical with that from Colorado.

31. **Senecio pauciflòrus** Pursh. Rayless Alpine Butterweed. Fig. 5772.

Senecio pauciflorus Pursh, Fl. Amer. Sept. 529. 1814.

Fibrous-rooted perennial from a simple or slightly branched caudex, 1.5–4 dm. tall, glabrous or lightly floccose-tomentose when young. Leaves thickish, somewhat succulent in life, the basal ones mostly elliptic-ovate to subrotund, abruptly contracted to the truncate or shallowly cordate base, crenate, petiolate; cauline leaves reduced and becoming sessile, bluntly toothed or more or less pinnatifid with blunt lobes; heads mostly 2–6, rarely 12, orange or reddish, discoid or rarely with short rays; involucre 6–8 mm. high, its bracts about 21 or about 13, generally suffused with reddish purple, at least above the middle; bracteoles inconspicuous or wanting; disk 1–1.5 cm. wide; achenes glabrous.

Alpine and subalpine meadows and moist cliffs, Boreal Zone; Alaska and Yukon to Labrador, south to Quebec, and to northern Wyoming, northern Idaho, and northern Washington, and throughout the Sierra Nevada of California. Type locality: Labrador. July–Aug.

The plants from California tend to be more robust than the northern, more typical forms of the species, with somewhat more numerous heads, and more often produce radiate forms. Many of the California specimens are wholly characteristic of *S. pauciflorus*, however, and no taxonomic segregation seems possible. Not known to me from Oregon.

32. **Senecio indécorus** Greene. Rayless Mountain Butterweed. Fig. 5773.

Senecio indecorus Greene, Fl. Fran. 470. (Aug.) 1897.
Senecio idahoensis Rydb. Bull. Torrey Club **27**: 183. 1900.
Senecio pauciflorus subsp. *fallax* Greenm. in Piper, Contr. U.S. Nat. Herb. **11**: 597. 1906.
Senecio pauciflorus f. *fallax* Fernald, Rhodora **30**: 225. 1928.

Fibrous-rooted perennial from a simple or slightly branched caudex, 3–8 dm. tall, glabrous or lightly floccose-tomentose when young. Leaves relatively thin, not succulent, the basal ones mostly elliptic or broadly ovate, tapering or subtruncate at the base, serrate or sometimes incised, petiolate; cauline leaves sharply incised-pinnatifid, the lobes irregularly again few-toothed, reduced and becoming sessile upwards; heads mostly 6–40, yellow, discoid or rarely with short rays; involucre mostly 7–10 mm. high, its bracts mostly about 21, sometimes only 13, often purple-tipped; bracteoles short and inconspicuous; disk 6–14 mm. wide; achenes glabrous.

Moist woodlands, stream banks, swales, and bogs, Canadian Zone; Alaska and Yukon to northern Washington and Wyoming, and apparently in northern California. Type locality: Pine Creek, Lassen County, California. July–Aug.

It is possible that the type of *S. indecorus* Greene is merely an aberrant form of *S. pseudaureus* or a hybrid between *S. pseudaureus* and *S. pauciflorus* and that the proper name for the species here described is *S. idahoensis* Rydb.

33. **Senecio pseudaùreus** Rydb. Streambank Butterweed. Fig. 5774.

Senecio pseudaureus Rydb. Bull. Torrey Club **24**: 298. (June) 1897.
Senecio pauciflorus var. *jucundulus* Jepson, Man. Fl. Pl. Calif. 1154. 1925.

Fibrous-rooted perennial from a short, horizontal or ascending rhizome or caudex, 3–7 dm. tall, lightly floccose-tomentose at first, soon essentially glabrous. Leaves relatively thin, not succulent, the basal ones long-petiolate, the blade serrate, tending to be cordate, subcordate, or truncate at the base; cauline leaves few and progressively reduced upwards, becoming sessile, more or less laciniate-pinnatifid at least toward their bases; heads several or many, the disk about 8–13 mm. wide; involucre 5–8 mm. high, the bracts about 13 or about 21; bracteoles inconspicuous or wanting; rays about 6–10 mm. long; achenes glabrous.

Stream banks, wet meadows, and moist woodlands in and near the mountains, mostly Arid Transition Zone; British Columbia to Saskatchewan south to California in the southern Sierra Nevada and in New Mexico. Type locality: Little Belt Mountains, Montana. June–Aug.

Only the var. *pseudaureus*, as described above, occurs in our range. The var. *flavulus* (Greene) Greenm., with more crenate lower leaves, replaces the var. *pseudaureus* in the southern Rocky Mountains and on the northern Great Plains.

34. **Senecio paupérculus** var. **thomsoniénsis** (Greenm.) Boiv. Canadian Butterweed. Fig. 5775.

Senecio multnomensis Greenm. Ottawa Nat. **25**: 115. 1911.
Senecio balsamitae var. *thomsoniensis* Greenm. op. cit. 116.
Senecio flavovirens var. *thomsoniensis* Greenm. Ann. Mo. Bot. Gard. **3**: 169. 1916.
Senecio pauperculus var. *thomsoniensis* Boiv. Nat. Can. **75**: 214. 1948.

Fibrous-rooted perennial with a rather short, simple or slightly branched caudex, occasionally with some very short, slender stolons, mostly 2–7 dm. tall, lightly floccose-tomentose when young, soon glabrate except frequently at the base and in the leaf-axils. Basal leaves petiolate, the blade oblanceolate to elliptic or occasionally suborbicular, crenate or serrate to subentire; cauline leaves more or less pinnatifid, the lower sometimes larger than the basal, the others conspicuously reduced and becoming sessile, all relatively thin and not at all succulent; heads several, the disk 7–12 mm. wide; involucre mostly 6–9 mm. high, the bracts about 13 or about 21; bracteoles inconspicuous or wanting; rays about 5–10 mm. long or very rarely wanting; achenes glabrous or sometimes hispidulous.

Stream banks, swamps, meadows, moist woods, and moist cliffs, mostly in the foothills and valleys, sometimes at moderate elevations in the mountains, Transition and Canadian Zones; southern Yukon to northern Oregon and western Montana. Type locality: North Thompson River, British Columbia. May–Oct.

The species as a whole ranges from Labrador to southern Yukon, south to Virginia, New Mexico, and northern Oregon.

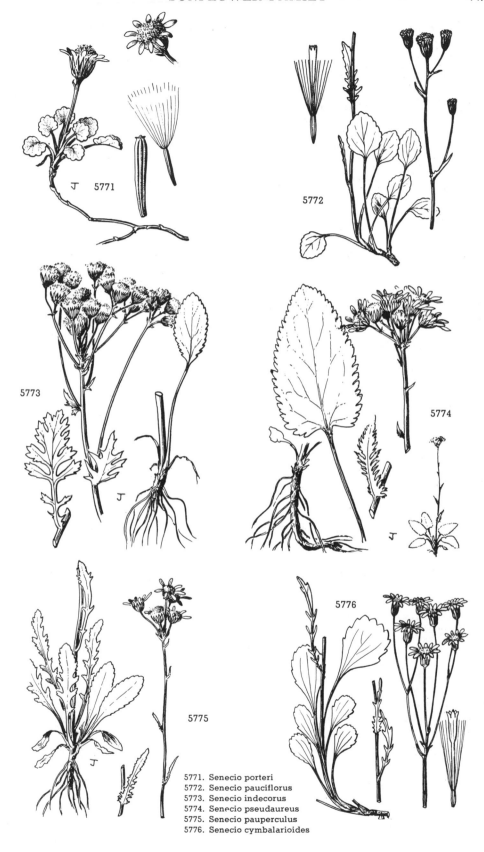

5771. Senecio porteri
5772. Senecio pauciflorus
5773. Senecio indecorus
5774. Senecio pseudaureus
5775. Senecio pauperculus
5776. Senecio cymbalarioides

35. Senecio cymbalarioides Nutt. Rocky Mountain Butterweed. Fig. 5776.

Senecio cymbalarioides Nutt. Trans. Amer. Phil. Soc. II. **7**: 412. 1841.
Senecio aureus var. borealis Torr. & Gray, Fl. N. Amer. **2**: 442. 1843.
Senecio laetiflorus Greene, Pittonia **3**: 88. 1896.
Senecio adamsi Howell, Fl. N.W. Amer. 379. 1900. Not S. adamsii Cheesm. 1896.
Senecio fraternus Piper, Contr. U.S. Nat. Herb. **11**: 598. 1906.
Senecio chapacensis Greene, Leaflets Bot. Obs. **2**: 14. 1909.
Senecio suksdorfii Greenm. Bot. Gaz. **53**: 511. 1912.
Senecio cymbalarioides var. borealis Greenm. Ann. Mo. Bot. Gard. **3**: 177. 1916.

Fibrous-rooted perennial from a caudex or short rhizome, 1–5 dm. tall, glabrous or lightly floccose-tomentose when young. Leaves thickish, somewhat succulent in life, the basal ones long-petiolate, with mostly elliptic to subrotund or rotund-obovate blade, wavy or coarsely crenate to shallowly lobulate or entire; cauline leaves few and reduced, becoming sessile, usually somewhat pinnatilobate, at least toward their bases; heads several or rather many, the disk 8–14 mm. wide; involucre 5–7 mm. high, the bracts about 13 or about 21, rarely only 8; bracteoles inconspicuous; rays 6–12 mm. long; achenes glabrous.

Woodlands and fairly moist to moderately dry open places in the mountains, Arid Transition and Boreal Zones; southern Yukon, British Columbia, and Alberta south to California and New Mexico; common in the higher mountains of Oregon and Washington, less frequent in the Sierra Nevada. Type locality: "In Oregon." June–Aug.
A form characteristic of alkaline meadows in southeastern Oregon and northern California has been distinguished as S. laetiflorus Greene but appears to be morphologically indistinguishable from many of the specimens occurring in more typical habitats.

36. Senecio clevelándii Greene. Serpentine Butterweed. Fig. 5777.

Senecio clevelandii Greene, Bull. Torrey Club **10**: 87. 1883.

Perennial with 1 or several stems from a short caudex, 3–7 dm. tall, glabrous and strongly glaucous throughout. Leaves thick and firm, scarcely veiny, all entire, the lowermost 1–2 dm. long and 5–30 mm. wide, with oblanceolate or elliptic blade equaling or more often shorter than the petiole; middle and upper leaves progressively reduced and becoming sessile; heads several or more often rather numerous, the disk about 1–1.5 cm. wide; involucre 6–7 mm. high, the bracts about 13 or about 21; bracteoles small but evident; rays small, about 5–7 mm. long; achenes glabrous.

Moist places, generally on serpentine, Upper Sonoran Zone; North Coast Ranges of California, especially in Lake County. Type locality: Indian Valley, Lake County. June–July.
Senecio clevelandii var. heteróphyllus Hoover, Leaflets West. Bot. **2**: 132. 1938. Robust plants resembling typical S. clevelandii but with some of the upper cauline leaves more or less pinnatilobate, especially toward the base. On serpentine in the foothills of the Sierra Nevada in Tuolumne County, California. Type locality: near Chinese Camp, Tuolumne County.

37. Senecio làyneae Greene. Layne's Butterweed. Fig. 5778.

Senecio layneae Greene, Bull. Torrey Club **10**: 87. 1883.
Senecio fastigiatus var. layneae A. Gray, Syn. Fl. N. Amer. 1²: 390. 1884.

Perennial with 1 or several stems from a well-developed caudex, 3–7 dm. tall, slightly tomentulose when young, soon essentially glabrous. Leaves rather thick and firm, scarcely veiny, dark green above, sometimes paler and glaucous beneath, entire or with irregularly scattered, sharp teeth or some of the cauline ones with a few narrow lobes; lowermost leaves well developed and more or less persistent, 8–20 cm. long and 5–12 mm. wide, the oblanceolate or narrowly elliptic blade tapering to a petiole often as long; cauline leaves progressively reduced and less petiolate upward, the upper often slightly expanded at the sessile or subsessile base; heads relatively large, 5–19 in an open inflorescence, the central one commonly overtopped by the lateral ones; disk 1–2 cm. wide; involucre 9–12 mm. high, the bracts commonly about 21 or sometimes only 13, relatively broad and often some of them partly connate; bracteoles few but well developed and often almost like the bracts, or none; rays few, commonly 5 or 8, showy, golden, mostly 1.5–2 cm. long; achenes glabrous, 4–5 mm. long.

Dry banks in the Pinus sabiniana-Quercus douglasii belt; foothills of the Sierra Nevada, Eldorado County, California. May.

38. Senecio fléttii Wiegand. Flett's Butterweed. Fig. 5779.

Senecio flettii Wiegand, Bull. Torrey Club **26**: 137. 1899.

Perennial from a well-developed, branching, rhizomatous caudex, 0.5–2(4) dm. tall, glabrous throughout, or with a little inconspicuous, persistent, tawny, floccose tomentum at the base and in the leaf-axils. Basal leaves well developed and persistent, up to 10 cm. long (including the well-developed petiole) and 3.5 cm. wide, mostly pinnatifid or lyrate-pinnatifid, the broad segments again toothed or lobulate, varying to occasionally simple, cordate, and merely palmately lobulate as in S. bolanderi and S. harfordii; cauline leaves few and much reduced, with mostly narrower segments than the basal leaves, the stem appearing nearly naked; heads several in a compact cymose cluster; involucre 5–9 mm. high, glabrous, the bracts about 13 or sometimes only 8; bracteoles few and well developed, or wanting; rays 5–10 mm. long; achenes glabrous.

Open rocky places, especially on talus slopes, principally Boreal Zone; from the foothills to the tops of the Olympic Mountains in Washington and also at rather high elevations in the Cascade Mountains in the vicinity of Mount Rainier, Washington. Type locality: headwaters of the Quilcene River, Olympic Mountains. June–Aug.

39. Senecio harfórdii Greenm. Cascade Butterweed. Fig. 5780.

Senecio harfordii Greenm. in Piper, Contr. U.S. Nat. Herb. 11: 597. 1906.

Very similar to *S. bolanderi* but with somewhat thinner, often more evidently lobed leaves and slightly smaller heads, the glabrous involucre mostly 4–6 mm. high.

Moist rocky woods and banks, Humid Transition Zone; Cascade region from southern Washington (Skamania County) to southern Oregon, especially in and about the Columbia Gorge; also, less commonly, in the Willamette Valley and Oregon Coast Range, well back from the ocean. Type locality: Cascade Mountains, Oregon, presumably in the Columbia Gorge. May–July.

40. Senecio bolánderi A. Gray. Seacoast Butterweed. Fig. 5781.

Senecio bolanderi A. Gray, Proc. Amer. Acad. 7: 362. 1868.

Perennial from a branching rhizome, mostly 1–6 dm. tall; herbage glabrous or with a little inconspicuous, persistent, tawny, floccose tomentum at the base and in the leaf-axils. Basal leaves well developed and persistent, rather thick and firm, petiolate, the blade commonly cordate, up to about 7 cm. long and wide, generally shallowly palmately lobulate, with the lobes again toothed or angled, or in smaller specimens sometimes merely undulate-toothed; some smaller lateral segments sometimes present, well removed from the terminal segment, and the blade thus lyrate; cauline leaves several, more or less reduced upwards, tending to be pinnatifid or lyrate-pinnatifid; heads several in a compact cymose cluster, the disk 6–13 mm. wide; involucre 5–8 mm. high, the bracts about 13 to about 21, usually with few to numerous loose, conspicuously multicellular hairs and often with a few well-developed, slender bracteoles; rays 6–12 mm. long; achenes glabrous.

Bluffs, woodlands, and beaches, Humid Transition Zone; along the coast from the estuary of the Columbia River to northern California (Mendocino County). Type locality: near Mendocino City, California. June–July.

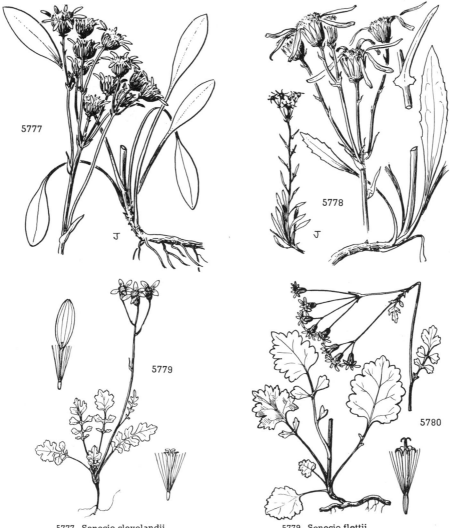

5777. Senecio clevelandii
5778. Senecio layneae

5779. Senecio flettii
5780. Senecio harfordii

41. **Senecio brèweri** Davy. Brewer's Butterweed. Fig. 5782.

Senecio breweri Davy, Erythea **3**: 116. 1895.
Senecio breweri var. *contractus* Greenm. Ann. Mo. Bot. Gard. **4**: 31. 1917.

Leafy-stemmed perennial with solitary stems from a short caudex, mostly 4–8 dm. tall; herbage essentially glabrous or with a little persistent tomentum in the axils. Leaves thin, all pinnatifid, the large lower ones generally lyrate, with enlarged, lobulate-toothed, terminal segment up to 10 cm. long and 8 cm. wide; cauline leaves more or less reduced upward, becoming sessile and more dissected, with smaller ultimate segments of which the terminal one is not enlarged; uppermost leaves often much reduced and the stem thus appearing naked above; heads several or rather many in a corymbiform inflorescence, relatively large; the glabrous involucre mostly 7–11 mm. high, the broad bracts about 21 or sometimes only 13; bracteoles few and small or wanting; disk 1–1.5 cm. wide; rays 8–15 mm. long; achenes glabrous.

Open or lightly wooded slopes in the valleys and foothills, Upper Sonoran Zone; California Coast Ranges from Contra Costa County to San Luis Obispo County and in the Tehachapi and Greenhorn Mountains, Kern County. Type locality: Alameda County, California. April–June.

42. **Senecio eurycéphalus** Torr. & Gray. Cut-leaved Butterweed. Fig. 5783.

Senecio eurycephalus Torr. & Gray ex A. Gray, Mem. Amer. Acad. II. **4**: 109. 1849.
Senecio austiniae Greene, Bull. Calif. Acad. **1**: 93. 1885.
Senecio eurycephalus var. *austiniae* Jepson, Man. Fl. Pl. Calif. 1154. 1925.
Senecio lewisrosei J. T. Howell, Leaflets West. Bot. **3**: 141. 1942.

Perennial, with several or many stems arising from a taproot and branched woody caudex, 2.5–6 dm. tall; herbage rather thinly tomentulose and often eventually more or less glabrate, or occasionally early-glabrate. Basal or lower cauline leaves well developed, pinnatifid or pinnately dissected to merely lyrate, with sharply toothed segments or rarely merely laciniate-toothed, 4–14 cm. long; middle and upper (or all) cauline leaves progressively reduced and becoming sessile, though frequently more dissected than the lower ones; heads about 3–30 in an open inflorescence, relatively large, the disk commonly 12–25 mm. wide; involucre 8–12 mm. high, the bracts mostly about 21, sometimes only 13; bracteoles small and inconspicuous or wanting; dry disk-corollas about 7–10 mm. long; rays 1–2 cm. long; achenes 3.5–6.5 mm. long, glabrous.

Dry open places, Upper Sonoran and Transition Zones; southern Cascade Mountains in Shasta, Butte, and Modoc Counties, California. south through the Inner Coast Ranges to San Francisco Bay. Type locality: California. April–June.

43. **Senecio multilobàtus** Torr. & Gray. Basin Butterweed. Fig. 5784.

Senecio multilobatus Torr. & Gray ex A. Gray, Mem. Amer. Acad. II. **4**: 109. 1849.
Senecio lynceus Greene, Erythea **3**: 22. 1895.
Senecio nelsonii [var.] *uintahensis* A. Nels. Bull. Torrey Club **26**: 484. 1899.
Senecio uintahensis Greenm. Monog. Gatt. Senecio **1**: 24. 1901.
Senecio stygius Greene, Leaflets Bot. Obs. **2**: 21. 1909.
Senecio prolixus Greenm. Ann. Mo. Bot. Gard. **1**: 264. 1914.
Senecio ionophyllus var. *stygius* Jepson, Man. Fl. Pl. Calif. 1154. 1925.

Perennial, sometimes rather short-lived, with 1 or several stems arising from a short taproot which is usually surmounted by a short branched caudex, 1–5 dm. tall; herbage thinly tomentulose and often eventually glabrous or nearly glabrous from the first. Basal leaves well developed, pinnatifid or more often lyrate-pinnatifid, commonly with toothed segments, up to about 11 cm. long and 2.5 cm. wide; cauline leaves more or less reduced but still generally fairly well developed, often more dissected than those below and less often lyrate; heads several or commonly rather numerous, small to medium-sized, the disk commonly 6–15 mm. wide; involucre 4.5–8(9) mm. high, the bracts about 13 or sometimes 21; bracteoles small and inconspicuous or wanting; dry disk-corollas about 4.5–7.5 mm. long; rays 5–10 mm. long or occasionally wanting; achenes 2–3.5(4) mm. long, glabrous or hispidulous.

Dry open places in the deserts and foothills, sometimes extending to high elevations in the drier southern mountains, Arid Transition and Sonoran Zones; southwestern Wyoming to western Texas and adjacent Mexico, west to southeastern Oregon, northeastern California, Nevada, and southern California south of the Sierra Nevada Type locality: "on the Uintah River, in the interior of California." April–July.

44. **Senecio jacobàea** L. Tansy Ragwort. Fig. 5785.

Senecio jacobaea L. Sp. Pl. 870. 1753.

Biennial or rather short-lived perennial with a poorly developed to evident taproot; stems solitary or several, erect, simple up to the inflorescence, 2–10 dm. tall; pubescence thinly floccose-tomentose but evanescent and generally nearly or quite wanting by flowering time, except frequently in the inflorescence. Leaves equably distributed, mostly 2–3 times pinnatifid, about 4–20 cm. long and 2–6 cm. wide, the lower petiolate and often deciduous, the upper becoming sessile; heads several or rather numerous in a short broad inflorescence, the disk about 7–10 mm. wide; involucre about 4 mm. high, its bracts about 13, over 1 mm. wide, generally dark-tipped; bracteoles narrow but sometimes rather well developed; rays commonly about 13, mostly 4–10 mm. long; achenes of the disk-flowers minutely pubescent, those of the rays glabrous.

A weed in pastures and other disturbed situations; native of Europe, now established in parts of the United States and Canada and becoming important as a weed west of the Cascade Mountains in Oregon and Washington and along the coast in northern California. Type locality: Europe. July–Sept.

Poisonous to livestock, the effect cumulative.

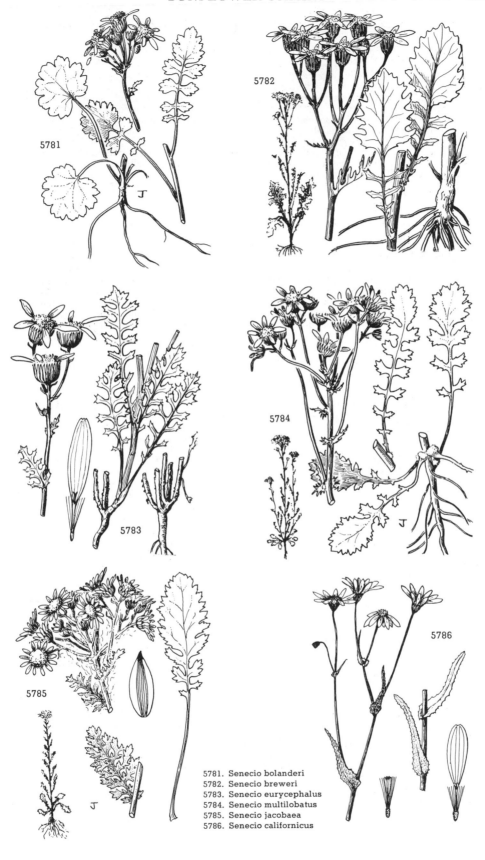

5781. Senecio bolanderi
5782. Senecio breweri
5783. Senecio eurycephalus
5784. Senecio multilobatus
5785. Senecio jacobaea
5786. Senecio californicus

45. Senecio califórnicus DC. California Butterweed. Fig. 5786.

Senecio californicus DC. Prod. **6**: 426. 1837.
Senecio californicus β *laxior* DC. loc. cit.
Senecio coronopus Nutt. Trans. Amer. Phil. Soc. II. **7**: 413. 1841.
Senecio ammophilus Greene, Bull. Calif. Acad. **1**: 193. 1885.
Senecio californicus var. *ammophilus* Greenm. Ann. Mo. Bot. Gard. **2**: 590. 1915.

Taprooted annual, glabrous or partly arachnoid-villous, 0.5–5 dm. tall, simple or branched. Leaves sessile (or the lower petiolate) and more or less clasping, subentire or more often coarsely toothed to pinnatifid or even subpinnatifid, sometimes broad-based as in *S. mohavensis*, sometimes narrower and more nearly resembling *S. aphanactis*, 1–7 cm. long and up to 3.5 cm. wide; heads several in an open inflorescence or solitary in small plants, the disk 5–13 mm. wide; involucre 4–7 mm. high, the principal bracts about 21 or sometimes only 13, usually minutely black-tipped; rays well developed and conspicuous, commonly about 1 cm. long; achenes strigillose-canescent.

Sandy and dry soils, Sonoran Zones; sand dunes along the coast of California from Monterey County south to northern Lower California, and in dry, open, often sandy places farther inland (east to Kern and Tulare Counties). Type locality: California. Feb.–May.

46. Senecio vulgàris L. Common Groundsel or Old Man in the Spring. Fig. 5787.

Senecio vulgaris L. Sp. Pl. 867. 1753.

Simple or strongly branched annual or winter annual with a more or less evident taproot; stem 1–4 dm. tall, leafy throughout; herbage sparsely crisp-hairy or subglabrous. Leaves coarsely and irregularly toothed to more often pinnatifid, about 2–10 cm. long and 5–45 mm. wide, the lower tapering to the petiole or petiolar base, the upper sessile and clasping; heads several or numerous, strictly discoid, the flowers all tubular and perfect; disk about 5–10 mm. wide; involucre about 5–8 mm. high, the principal bracts mostly about 21, often black-tipped; bracteoles short but well developed, evidently black-tipped; pappus very copious, equaling or generally surpassing the corollas; achenes strigillose-hirtellous, chiefly along the angles.

A weed in disturbed soil and waste places; native of the Old World but now widely distributed throughout most of the temperate zone; commoner west of the Cascade Mountains and Sierra Nevada than elsewhere in our area. Type locality: Europe. Flowering nearly the year around.

47. Senecio mohavénsis A. Gray. Mohave Groundsel. Fig. 5788.

Senecio mohavensis A. Gray, Syn. Fl. N. Amer. **1²**: 446. 1884.

Taprooted annual, 1.5–4 dm. tall, freely branched, the herbage wholly glabrous. Leaves sessile (or the lowermost subpetiolate) and cordate-clasping, ovate to broadly oblong, 2–6 cm. long, 1–4 cm. wide, coarsely and irregularly toothed; heads more or less numerous, slender-pedunculate, the disk 5–9 mm. wide; involucre 5–7 mm. high, the principal bracts commonly about 13; bracteoles few and small, not evidently black-tipped; pistillate flowers with very much reduced and inconspicuous rays or wanting; pappus copious; achenes strigillose-canescent.

Desert washes and flats, Sonoran Zones; Panamint Mountains of California south to western Arizona and northern Sonora. Type locality: Mojave region, California. April–May.

48. Senecio sylváticus L. Wood Groundsel. Fig. 5789.

Senecio sylvaticus L. Sp. Pl. 868. 1753.

Annual weed with a more or less evident taproot; stem about 1.5–8 dm. tall, generally simple up to the inflorescence, leafy throughout; herbage sparsely or moderately pubescent with crisp loose hairs, scarcely or not at all glandular. Leaves all more or less pinnatifid and irregularly toothed, commonly 2–12 cm. long and 4–40 mm. wide; heads several or numerous, the disk about 3–7 mm. wide; involucre about 5–7 mm. high, the rather narrow, 1–2-nerved bracts mostly about 13; bracteoles inconspicuous or wanting, not black-tipped; rays very much reduced and inconspicuous, less than 2 mm. long; pappus very copious, equaling or surpassing the slender disk-corollas; achenes strigillose-canescent.

A weed in disturbed soil and waste places; native of Europe, introduced in parts of the United States and Canada; in our region largely confined to the area west of the Cascade Mountains in Oregon and Washington and west of the Great Valley in California. Type locality: Europe. June–Sept.

49. Senecio aphanáctis Greene. California Groundsel. Fig. 5790.

Senecio aphanactis Greene, Pittonia **1**: 220. 1888.

Taprooted slender annual, simple or branched, 0.5–2.5 dm. tall, slightly arachnoid in the inflorescence, otherwise essentially glabrous. Leaves sessile, small, 1–4 cm. long and 1–12 mm. wide, mostly coarsely toothed or (especially the larger ones) pinnately lobed; heads several, narrow, tending to be a little constricted at or above the middle; involucre about 5 mm. high, the bracts about 8, relatively broad, 2–4-nerved; bracteoles wanting or inconspicuous, not black-tipped; rays very much reduced and inconspicuous, barely or scarcely surpassing the pappus; pappus copious, about equaling the slender disk-corollas; achenes densely strigillose-canescent.

Dry open places, Upper Sonoran Zone; in the California Coast Ranges and on adjacent islands from San Francisco Bay to northern Lower California. Type locality: Mare Island, San Francisco Bay. Feb.–March.

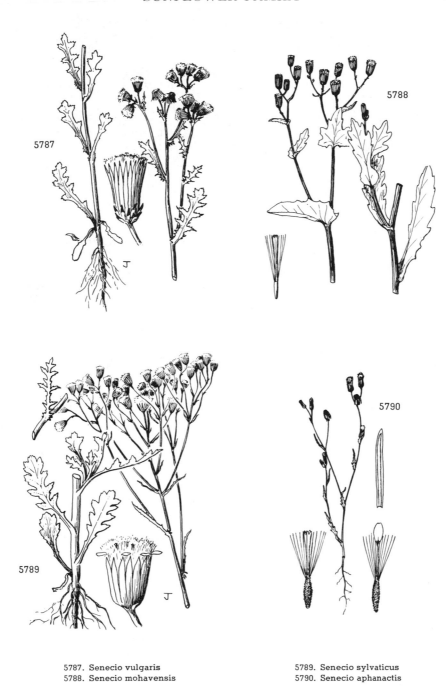

5787. **Senecio vulgaris**
5788. **Senecio mohavensis**

5789. **Senecio sylvaticus**
5790. **Senecio aphanactis**

50. **Senecio mikanioìdes** Otto. German Ivy. Fig. 5791.

Senecio mikanioides Otto, Allg. Gartenz. **10**: 168. 1842, nomen subnudum; ex Walp. op. cit. **13**: 42. 1845.

Glabrous perennial with twining stems, climbing to a height of 5 m. or more. Leaf-blades thin, somewhat ivy-like, cordate, sharply lobe-angled, up to 10 cm. long and about as wide; petioles about as long as the blades or longer, those of the larger leaves commonly bearing a pair of stipule-like appendages at the base; heads borne in small, condensed, pedunculate, corymbiform clusters, small and few-flowered, the disk only about 5 mm. wide; involucre 3–4 mm. high, much shorter than the disk, the principal bracts commonly about 8; rays wanting; achenes glabrous.

Along streams and gullies near the coast of California from Alameda and Marin Counties southward; native of South Africa. Type locality: horticultural specimens, mistakenly thought to have been of probable Mexican origin. Jan.–March.

111. **CROCÍDIUM** Hook. Fl. Bor. Amer. **1**: 335. 1834.

Heads radiate, the rays pistillate and often fertile, yellow. Involucre a single series of rather broad, herbaceous, equal bracts. Receptacle strongly conic, naked. Disk-flowers perfect and fertile, yellow. Anthers entire or nearly so at the base. Style-branches flattened, with marginal stigmatic lines and a well-developed, flat, deltoid, externally minutely papillate-hairy appendage. Achenes covered with thick, papillae-like hairs, becoming mucilaginous when wet; pappus of more or less numerous, very fragile, deciduous, white bristles or sometimes wanting from the ray-flowers. Delicate annuals with small, alternate and basal, entire or few-toothed leaves and rather small, long-pedunculate heads. [Name a diminutive derived from the Greek *croce*, loose thread or wool, referring to the persistent axillary tomentum.]

A single species, of doubtful affinities.

1. **Crocidium multicáule** Hook. Spring Gold. Fig. 5792.

Crocidium multicaule Hook. Fl. Bor. Amer. 1: 335. 1834.
Crocidium pugetense St. John, Torreya 28: 74. 1928.

Delicate, generally several-stemmed annual up to 1.5 or rarely 3 dm. tall, bearing loose tufts of axillary wool, otherwise glabrous or glabrate. Leaves slightly fleshy, the basal ones oblanceolate or broader, up to about 2.5 cm. long (including the petiolar base) and 1 cm. wide, often coarsely few-toothed; cauline leaves reduced and rather few, scarcely more than mere linear bracts; heads naked-pedunculate, solitary at the ends of the simple unbranched stems; rays 5-13, typically 8, 4-10 mm. long, fertile or sterile, individually subtended by the thin and membranous involucral bracts, these 3-7 mm. long; disk about 1 cm. wide or less.

Sand plains, cliff ledges, and other dry open places at low elevations, Upper Sonoran and Transition Zones; in the Puget trough from southern Vancouver Island to southern Oregon, extending eastward along and near the Columbia River to the base of the Blue Mountains in Walla Walla County, Washington, and Umatilla County, Oregon, northward near the eastern base of the Cascade Mountains to Kittitas County, Washington, and southward to northern Deschutes County; in the Klamath region of southwestern Oregon and adjacent California, thence southward in the California Coast Ranges to Santa Clara County and along the western base of the Sierra Nevada to Mariposa County. Type locality: "About Fort Vancouver, on the Columbia." March–May. Very abundant on the sand plains near the Columbia in eastern Oregon and Washington.

112. **PETASÍTES** [Tourn.] Mill. Gard. Dict. abr. ed. 4. 1754.

Heads radiate or discoid, subdioecious, the flowers in the female heads all or nearly all pistillate and fertile, with or without rays, those in the male heads chiefly or entirely hermaphrodite but sterile. Involucre a single series of equal, more or less herbaceous bracts, sometimes with a few reduced bracteoles at the base. Receptacle flat, naked. Pistillate flowers with filiform corollas, with or without a ligule. Hermaphrodite flowers tubular, with 5-cleft limb. Anthers entire or slightly sagittate at the base. Style puberulent, undivided or nearly so. Achenes linear, 5-10-ribbed; pappus of numerous capillary bristles, elongating in fruit, that of the sterile flowers more or less reduced. More or less white-tomentose or woolly perennial herbs with large basal leaves, merely bracteate stems (the bracts alternate), and several or numerous medium-sized, purple, white, or rarely yellowish heads. [Name from the Greek *petasos*, a broad-brimmed hat, referring to the large basal leaves.]

A genus of about 12 species, native to the cooler parts of the northern hemisphere. Type species, *Tussilago petasites* L. (= *Petasites hybridus* (L.) Gaertn., Mey. & Scherb.)

Leaves evidently lobed, varying to coarsely few-toothed with 5-15 teeth on each side; chiefly in the Cascade region and westward (and in the California Coast Ranges). 1. *P. frigidus.*
Leaves varying from merely a little wavy and callous-denticulate to more commonly conspicuously dentate with 20-45 teeth on each side; northeastern Washington northward and eastward. 2. *P. sagittatus.*

1. **Petasites frígidus** (L.) Fries. Sweet Coltsfoot. Fig. 5793.

Tussilago frigida L. Sp. Pl. 865. 1753.
Tussilago corymbosa R. Br. Chlor. Melv. 21. 1823.
Nardosmia frigida Hook. Fl. Bor. Amer. 1: 307. 1833.
Nardosmia corymbosa Hook. loc. cit.
Petasites frigidus Fries, Summa Veg. Scand. 182. 1845.
Nardosmia frigida var. *corymbosa* Herder, Bull. Soc. Impér. Nat. Mosc. 38: 372. 1865.
Petasites corymbosa Rydb. Bull. Torrey Club 37: 460. 1910.
Petasites warrenii St. John, Research Stud. St. Coll. Wash. 1: 109. 1929.
Petasites frigidus var. *corymbosus* Cronquist, Rhodora 48: 123. 1946.

Perennial from a creeping rhizone. Basal leaves expanding with or shortly after the flowers, long-petioled, glabrous or glabrate above, loosely white-tomentose beneath, sometimes eventually glabrate, of various sizes up to 1.5 or rarely 2 dm. wide, coarsely few-toothed or very shallowly lobed, pinnipalmately veined; stem erect, 1-5 dm. tall, with approximately or imbricate parellel-veined bracts mostly 2.5-6 cm. long, the lower sometimes with an abortive blade at the end, the upper commonly reduced and more distant; heads several or rather numerous in a corymbiform or racemiform inflorescence, campanulate; involucre about 5–9 mm. high; flowers whitish, the pistillate with short rays.

Typically in tundra and moist subalpine meadows and bogs, Boreal Zone; circumboreal, extending south into the high mountains of southern British Columbia and, in a form approaching var. *nivalis*, to the Wenatchee Mountains of Washington (where found at low elevations). Type locality: mountains of Lapland. April–June.

Petasites frigidus var. **nivàlis** (Greene) Cronquist, Leaflets West. Bot. **7**: 30. 1953. (*Petasites nivalis* Greene, Pittonia **2**: 18. 1889; *P. vitifolia* Greene, Leaflets Bot. Obs. **1**: 180. 1906; *P. hyperboreus* Rydb. N. Amer. Fl. **34**: 312. 1927.) Leaves pinnipalmately lobed and veined, seldom more than 2 dm. wide, seldom evidently wider than long, the sinuses seldom (though sometimes) extending more than half way to the base; short, slender, callous teeth few or wanting. Mostly in Canada and Alaska extending southward at high altitudes to the Olympic Mountains of Washington and the Cascade Mountains of northern Oregon, and to northern Minnesota. Type locality: Mount Rainier, Washington. July–Aug.

Petasites frigidus var. **palmàtus** (Ait.) Cronquist, Rhodora **48**: 124. 1946. (*Tussilago palmata* Ait. Hort. Kew. **3**: 188. 1789; *Nardosmia palmata* Hook. Fl. Bor. Amer. **1**: 308. 1833; *N. speciosa* Nutt. Trans. Amer. Phil. Soc. II. **7**: 288. 1841; *N. frigida* var. *palmata* Herder, Bull. Soc. Impér. Nat. Mosc. **38**: 372. 1865; *Petasites palmata* A. Gray, Bot. Calif. **1**: 407. 1876; *P. speciosa* Piper, Mazama **2**: 97. 1901.) Leaves palmately lobed and veined, often very large (up to 4 dm. wide), tending to be broader than long, the lobes commonly extending at least half way to the base, often much deeper; short, slender, callous teeth rather freely produced in addition to the larger teeth. Stream banks, boggy ground, and moist woods, from the lowlands to moderate elevations in the mountains; transcontinental in Canada but more southern than the other varieties, rarely reaching Yukon, and extending south to Massachusetts, Michigan, and Minnesota; in our area occurring in and west of the Cascade Mountains in Washington and Oregon and extending south in the California Coast Ranges to Monterey County. Type locality: Newfoundland. Feb.–May.

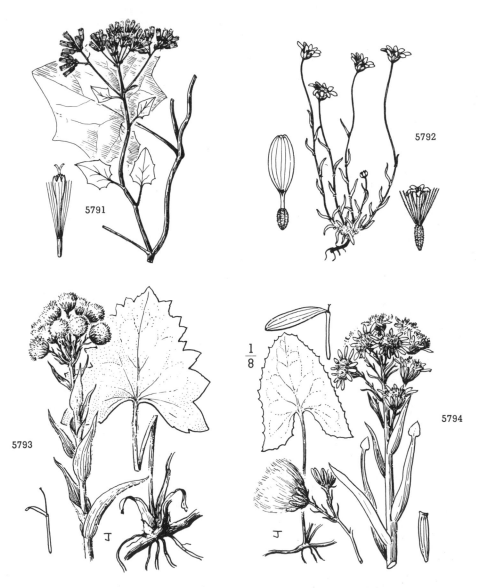

5791. Senecio mikanioides
5792. Crocidium multicaule
5793. Petasites frigidus
5794. Petasites sagittatus

2. Petasites sagittàtus (Banks) A. Gray. Arrowhead Coltsfoot. Fig. 5794.

Tussilago sagittata Banks ex Pursh, Fl. Amer. Sept. 531. 1814.
Nardosmia sagittata Hook. Fl. Bor. Amer. 1: 307. 1833.
Petasites sagittatus A. Gray, Bot. Calif. 1: 407. 1876.

Similar to *P. frigidus* but the basal leaves merely dentate, with 20–45 teeth on each side or even, especially in small forms, subentire, pinnipalmately veined, cordate or more commonly sagittate, sometimes as much as 30 cm. long and 25 cm. wide; bracts of the stem averaging longer and narrower, more often with abortive blades.

Wet places, Boreal Zone; Alaska to Labrador south to northeastern Washington in Okanogan County, northern Idaho, Montana, and Colorado. Type locality: Hudson Bay. April–June.

113. PSATHYRÒTES A. Gray, Smiths. Contr. 5⁶: 100. 1853.

Heads discoid, the flowers all tubular and perfect, yellow, often turning purple in age. Involucral bracts biseriate, the outer more herbaceous and often a little shorter than the inner. Receptacle flat, naked. Anthers minutely sagittate at the base. Style-branches flattened, truncate or nearly so, minutely penicillate. Achenes hairy, more or less turbinate, subterete or angled; pappus of relatively few and short, rather firm, capillary bristles. Strongly odorous annuals or perennials with alternate (or all basal), petiolate, small but relatively broad leaves. [Name a direct transliteration of the Greek word for brittleness, referring to the brittle stems and branches.]

A genus of 4 species, native to the deserts and drier mountains of southwestern United States and adjacent Mexico. Type species, *Bulbostylis annua* Nutt.

Psathyrotes, Peucephyllum and some other small southwestern genera have customarily been excluded from the *Heliantheae* (*Helenieae*) because of their capillary pappus, and referred to the *Senecioneae* instead. Their relationship to the main bulk of the *Senecioneae* is doubtful, however, and further investigation of the affinities of these genera is well warranted.

Outer involucral bracts relatively broad, more or less oblong-obovate, the expanded green tip commonly 1.5–3 mm. wide; plants shortly woolly as well as scurfy. 1. *P. ramosissima.*
Outer involucral bracts relatively narrow, nearly linear, commonly a little constricted at or above the middle, the slightly expanded green tip generally 0.4–1.3 mm. wide; plants scurfy-pubescent, not evidently woolly. 2. *P. annua.*

1. Psathyrotes ramosíssima (Torr.) A. Gray. Velvet Rosettes. Fig. 5795.

Tetradymia ramosissima Torr. in Emory, Notes Mil. Rec. 145. 1848.
Psathyrotes ramosissima A. Gray, Proc. Amer. Acad. 7: 363. 1868.

Winter annual, or in any case short-lived, depressed and subdichotomously much branched, the stems up to about 15(20) cm. long; herbage scurfy-pubescent and evidently short-woolly as well. Leaves broadly ovate to more often broader than long, up to about 2.5 cm. wide or with a few coarse teeth or angles; heads evidently pedunculate, borne singly in the forks or on pseudolateral branches; involucre 5–8 mm. high, the disk less than 1 cm. wide; outer involucral bracts equaling or a little shorter than the inner, relatively broad, more or less oblong-obovate, the expanded green tip commonly 1.5–3 mm. wide and often turned back; corollas long-hairy above, yellow, often turning purple in age; achenes densely long-hairy; pappus fulvous.

Dry, open, often sandy places in the deserts and desert mountains, Sonoran Zones; Colorado and Mojave Deserts of southern California south to Lower California, east to southwestern Utah, western Arizona, and northern Sonora; rarely extending as far north as Reno, Nevada. Type locality: "Hills bordering the Gila" [River, Arizona]. March–May, and sometimes again in the fall or winter. Turtleback.

2. Psathyrotes ánnua (Nutt.) A. Gray. Mealy Rosettes. Fig. 5796.

Bulbostylis annua Nutt. Journ. Acad. Phila. II. 1: 179. 1848.
Psathyrotes annua A. Gray, Smiths. Contr. 5⁶: 100. 1853.

Annual or winter annual, resembling *P. ramosissima*, differing chiefly in the characters given in the key.

Dry, open, often sandy or alkaline places, Upper Sonoran Zone; typically at somewhat higher elevations or farther north than *P. ramosissima*, apparently rare in California; southwestern Idaho to southern Utah, southern Nevada, northwestern Arizona, and the Mojave Desert in California, and reputedly to Sonora and Lower California. Type locality: "Rocky Mountains, near Santa Fé" but probably actually taken farther west. May–Aug.

114. LÙINA Benth. in Hook. Ic. Pl. III. 2: 35. 1876.

Heads discoid, the flowers all perfect and fertile, yellow or yellowish. Involucre a single series of rather firm, equal, scarcely herbaceous to subherbaceous bracts. Receptacle naked. Anthers tapering to an entire or minutely sagittate base. Style-branches flattened, externally merely papillate or papillate-puberulent, with broad, introrsely marginal, stigmatic lines and a thickened, very short and blunt, papillate appendage. Achenes prominently several-nerved; pappus of numerous capillary bristles. Perennials with simple, entire to deeply cleft, alternate leaves. [Name an anagram of *Inula*.]

The genus consists of the following 4 habitally very different but nonetheless allied species, all natives of the Pacific States and British Columbia. Type species, *Luina hypoleuca* Benth.

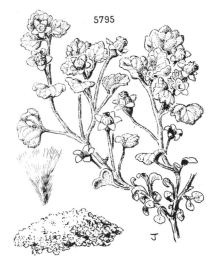

5795. Psathyrotes ramosissima   5796. Psathyrotes annua

Leaves palmately cleft; heads large, the disk 12–40 mm. wide, the flowers numerous (more than 30).
                                                                                    1. *L. nardosmia.*
Leaves entire or slightly toothed; heads smaller, the disk mostly about 1 cm. wide or less, the flowers less than 30
  in each head.
    Inflorescence short, corymbiform or subumbelliform; leaves white-tomentose on the lower surface; stem rather
      equably leafy; basal leaves wanting.
        Principal leaves petiolate, narrowly elliptic or lanceolate, 7–13 cm. long (petiole included), five to eleven
          times as long as wide; involucre 8–10 mm. high, composed of 10–17 (averaging 13) bracts; heads
          about 15–29- (averaging about 21-) flowered; Grant County, Oregon.   2. *L. serpentina.*
        Leaves all sessile, rather broadly elliptic or ovate, 2–6 cm. long, one and one-half to three and one-half
          times as long as wide; involucre 5–8 mm. high, composed of about 8–10 bracts; heads about 10–17-
          (averaging 13-) flowered; Cascade Mountains and westward.   3. *L. hypoleuca.*
    Inflorescence elongate, thyrsoid-racemiform; leaves glabrous, the basal and lowermost cauline ones large and
      persistent, the middle and upper ones progressively reduced; involucral bracts mostly about 5 or 6
      and flowers about 5 in each head.   4. *L. stricta.*

1. **Luina nardòsmia** (A. Gray) Cronquist. Cut-leaf Luina. Fig. 5797.

Cacalia nardosmia A. Gray, Proc. Amer. Acad. **7**: 361. 1868.
Adenostyles nardosmia A. Gray, op. cit. **8**: 631. 1873.
Cacaliopsis nardosmia A. Gray, op. cit. **19**: 50. 1883.
Luina nardosmia Cronquist, Vasc. Pl. Pacif. Northw. **5**: 257. 1955.

Perennial from a woody rhizome, 2.5–6 dm. tall, leafy chiefly toward the base, the middle and
cauline leaves strongly reduced. Principal leaves long-petiolate, with a broad, palmately cleft blade
up to nearly 2 dm. wide, the lobes seldom extending beyond the middle, commonly coarsely toothed
or again cleft, the teeth or segments mostly rather blunt; leaves thinly tomentose above when
young, generally soon glabrate, more persistently but still rather thinly gray-tomentulose beneath;
heads yellow, above 1–7 in a corymbiform inflorescence, relatively large, the involucre 12–20 mm.
high, the disk 17–40 mm. wide; involucral bracts commonly about 13 or up to 21, tomentulose or
subglabrous.

Open woodlands, sometimes on serpentine, Upper Sonoran and Transition Zones; Klamath area of southwestern Oregon and adjacent California and southward chiefly in the Inner Coast Ranges to Sonoma County.
Type locality: near the Geysers in Sonoma County, California. April–June.

Luina nardosmia var. **glabràta** (Piper) Cronquist, Vasc. Pl. Pacif. Northw. **5**: 257. 1955. (*Cacaliopsis
nardosmia* [var.] *glabrata* Piper, Bull. Torrey Club **29**: 222. 1902; *C. nardosmia* [subsp.] *glabrata* Piper, Contr.
U.S. Nat. Herb. **11**: 594. 1906; *C. glabrata* Rydb. N. Amer. Fl. **34**: 316. 1927.) Plants taller and more robust,
up to 1 m. high. Blades of the principal leaves up to 2.5 dm. wide, averaging more deeply cleft and with narrower, more acute segments, generally glabrous or nearly so above and often eventually glabrate beneath; heads
smaller and more numerous, mostly 5–18, the inflorescence sometimes becoming somewhat elongate; involucre
mostly 10–17 mm. high; disk 12–30 mm. wide. Meadows and open woods along the summit and the east side of
the Cascade Mountains in Washington and immediately south of the Columbia River in Oregon; perhaps irregularly southward in the Cascade Mountains of Oregon; less commonly in the Klamath region and the adjacent
southern Willamette Valley. Type locality: Klickitat County, Washington. May–July. A well-marked geographical race but not fully distinct.

2. **Luina serpentìna** Cronquist. Colonial Luina. Fig. 5798.

Luina serpentina Cronquist, Vasc. Pl. Pacif. Northw. **5**: 257. 1955.

Perennial from a stout, branching, woody base, which may be wholly prostrate and rooting
or may have branches arising as much as 2 dm. from the ground, forming clones several meters
across; stems of the season numerous, densely white-tomentose, 3–5 dm. tall, rather densely and
equably leafy. Leaves densely tomentose beneath, thinly so above, lanceolate or narrowly elliptic,
entire, acute, the middle and lower ones tapering to an often ill-defined petiolar base 1–3 cm. long,

the upper more or less sessile; leaves five to eleven times as long as wide (petiole included), those near or shortly below the middle the largest, 7–13 cm. long and 1–2 cm. wide, the upper and lower gradually reduced; basal leaves wanting; heads several in a short corymbiform inflorescence, rather bright yellow at first, later dull, the disk well surpassing the involucre, about 1 cm. wide; involucre white-tomentose, 8–10 mm. high, composed of 10–17 (averaging 13) bracts; flowers 15–29 (averaging about 21) in each head.

Steep serpentine slopes, Arid Transition Zone; Grant County, Oregon. Type locality: Fields Creek, 17 miles southeast of Dayville, Grant County. July.

3. Luina hypolèuca Benth. Little-leaf Luina. Fig. 5799.

Luina hypoleuca Benth. in Hook. Ic. Pl. III. **2**: 36. *pl. 1139*. 1876.
Luina hypoleuca var. *californica* A. Gray, Proc. Amer. Acad. **9**: 206. 1874.
Luina californica Rydb. N. Amer. Fl. **34**: 316. 1927.

Perennial from a stout, branched, woody caudex, which sometimes apparently surmounts a taproot; stems several or many, 1.5–4 dm. tall, white-tomentose throughout, equably leafy. Leaves all (except the reduced lowermost ones) reasonably similar in size and shape, sessile, more or less broadly elliptic or ovate, commonly 2–6 cm. long and 7–35 mm. wide, entire or nearly so, white-tomentose on the lower surface, green and thinly tomentulose or bright green and glabrous on the upper; basal leaves wanting; heads slender-pedunculate in a short, corymbiform or sub-umbelliform inflorescence, dull yellowish, smallish, the disk about 1 cm. wide or less, the thinly tomentose or glabrate involucre mostly 5–8 mm. high, composed of about 8–10 bracts, and commonly 10–17-flowered.

Cliff crevices, talus slopes, and similar rocky places, occasionally on serpentine, mostly Transition and Canadian Zones; commonly at 3,000 to 7,000 feet elevation but sometimes approaching sea level in the Klamath region and southward; southern British Columbia and Washington, from the Cascade Mountains westward, southward less commonly in the Coast Ranges to Lake and Mendocino Counties, California. Type locality: Lake Chilukweyuk, Cascade Mountains, Washington. June–Sept.

4. Luina strícta (Greene) Robinson. Tongue-leaf Luina. Fig. 5800.

Prenanthes stricta Greene, Pittonia **2**: 21. 1889.
Luina piperi Robinson, Bot. Gaz. **16**: 43. 1891.
Psacalium strictum Greene, Pittonia **2**: 228. 1892.
Rainiera stricta Greene, op. cit. **3**: 291. 1898.
Luina stricta Robinson, Proc. Amer. Acad. **49**: 514. 1913.

Fibrous-rooted perennial from a short stout rhizone or branched caudex, mostly 5–10 dm. tall, essentially glabrous except for the thinly tomentose-puberulent involucres and peduncles. Basal and lowermost cauline leaves large and persistent, broadly oblanceolate or broader, commonly 1.5–3.5 dm. long (including the petioliform base) and 2–7 cm. wide; middle and upper cauline leaves progressively reduced and becoming sessile; inflorescence 1–4 dm. long, thyrsoid-racemiform, its bracts seldom conspicuous; involucre narrow, its bracts mostly about 5–6, 7–9 mm. long, the midvein thickened; flowers few, typically about 5.

Meadows and moist open slopes at high altitudes in the mountains, Hudsonian Zone; common in the vicinity of Mount Rainier, Washington, southward at scattered stations in the Cascade Mountains to southern Lane County, Oregon. Type locality: Mount Rainier. July–Aug.

115. ERECHTÌTES Raf. Fl. Ludov. 65. 1817.

Heads disciform, dull yellow or whitish. Involucre a single series of narrow, equal, more or less herbaceous bracts, sometimes with a few minute bracteoles at the base. Receptacle flat, naked. Outer flowers pistillate, filiform-tubular, eligulate, in 2 to several series; inner flowers hermaphrodite but sometimes sterile. Corolla narrowly tubular, 4–5-toothed. Anthers entire or slightly sagittate at the base. Style-branches flattened, minutely penicillate about the subtruncate or very shortly appendiculate tip. Achenes 5-angled or 10–20-nerved; pappus of numerous capillary bristles. Erect annual or perennial herbs with alternate, entire to pinnately dissected leaves and cylindric to ovoid heads. [Name given by Dioscorides to a plant perhaps related to this.]

A genus of about 12 species, native to America and Australasia; only one species is native to North America north of Mexico and none is native to our range. Type species, *Senecio hieracifolius* L.

Heads relatively large, the involucre commonly 10–17 mm. high.   1. *E. hieracifolia.*
Heads much smaller, the involucre commonly 5–7 mm. high.
    Leaves more or less lobed or pinnatifid.   2. *E. arguta.*
    Leaves finely and sharply dentate, not at all lobed or pinnatifid.   3. *E. prenanthoides.*

1. Erechtites hieracifòlia (L.) Raf. ex DC. Eastern Fireweed. Fig. 5801.

Senecio hieracifolius L. Sp. Pl. 866. 1753.
Erechtites hieracifolia Raf. ex DC. Prod **6**: 294. 1837.

Annual weed up to 2.5 m. tall, glabrous or sometimes more or less spreading-hairy throughout. Leaves numerous, well distributed along the stem, of various sizes up to 20 cm. long and 8 cm. wide, sharply serrate and sometimes also irregularly lobed, the lower oblanceolate to obovate, often more or less petiolate, the middle and upper becoming sessile and often auriculate-clasping;

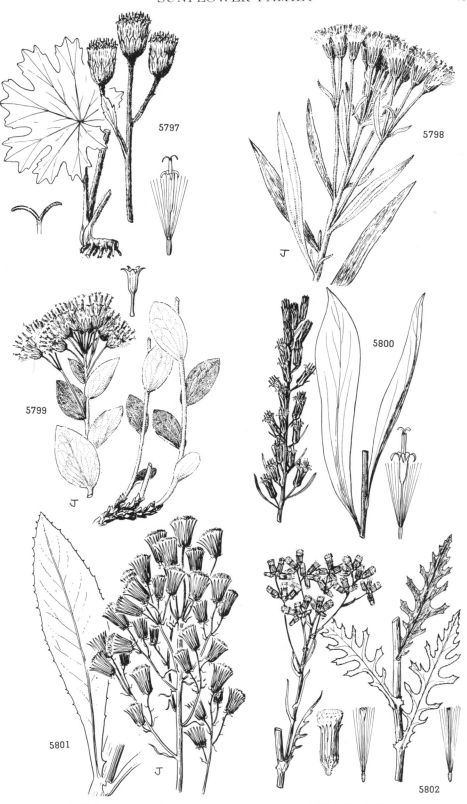

5797. Luina nardosmia
5798. Luina serpentina
5799. Luina hypoleuca

5800. Luina stricta
5801. Erechtites hieracifolia
5802. Erechtites arguta

heads whitish, several or numerous in a flat-topped or elongate inflorescence, or in depauperate plants often solitary, turbinate-cyclindric, with swollen base when fresh; involucre about 10–17 mm. high.

A weed in waste places, native from eastern United States to tropical America and introduced elsewhere in the world; in our range so far known only from a single collection from Seattle, Washington, but to be expected to spread into western Washington and Oregon. Type locality: North America. Aug.–Sept.

2. Erechtites argùta DC. Cut-leaved Coast Fireweed. Fig. 5802.

Senecio argutus A. Rich. Fl. N. Zeal. 258. 1832. Not *S. argutus* H.B.K.
Erechtites arguta DC. Prod. **6**: 296. 1837.

Annual, or in any case short-lived, weed up to 2 m. tall, thinly and somewhat deciduously villous-tomentulose. Leaves numerous, well distributed along the stem, up to 15 cm. long and 4 cm. wide, more or less deeply pinnately lobed or pinnatifid and often also irregularly toothed, the lower commonly more or less petiolate, the upper scarcely so and often auriculate; heads several or numerous in an often flat-topped inflorescence, small (the involucre only 5–7 mm. high) and narrow, nearly cylindric; flowers dull yellow.

A weed in waste places near the coast from Clatsop County, Oregon, as far south as San Mateo and northern San Benito Counties, California; native of Australia and New Zealand. Type locality: New Zealand. June–Aug.

3. Erechtites prenanthòides (A. Rich.) DC. Toothed Coast Fireweed. Fig. 5803.

Senecio prenanthoides A. Rich. Sert. Astrolab. 96. 1834.
Erechtites prenanthoides DC. Prod. **6**: 296. 1837.

Similar to *E. arguta*, averaging less hairy, differing most markedly in the sharply, regularly, and rather finely dentate, not at all lobed or pinnatifid, evidently auriculate leaves; inflorescence often larger and broader.

A weed in waste places along the coast from Lincoln County, Oregon, south at least to Santa Cruz County, California; native of Australia and New Zealand. Type locality: New Zealand. July–Sept.

116. PEUCEPHÝLLUM A. Gray, Bot. Mex. Bound. 74. 1859.

Heads discoid, the flowers all tubular and perfect, yellow, sometimes tinged with purple. Involucral bracts equal, uniseriate or subbiseriate, narrow, herbaceous. Receptacle flat, naked. Corolla-lobes short. Anthers sagittate. Style-branches flattened, externally glandular-papillate, with introrsely marginal stigmatic lines extending nearly to the bluntly rounded tip. Achenes densely hairy, obscurely striate; pappus of more or less numerous capillary bristles and intermingled, very slender and elongate, bristle-like scales. Shrubs with alternate, narrow, glandular-punctate leaves and numerous heads which are individually solitary at the ends of the numerous leafy branches. [Name from the Greek *peuce*, the fir, and *phyllon*, leaf, from some likeness in the foliage.]

A monotypic genus.

1. Peucephyllum schóttii A. Gray. Desert-fir or Pigmy-cedar. Fig. 5804.

Peucephyllum schottii A. Gray, Bot. Mex. Bound. 74. 1859.
Psathyrotes schottii A. Gray, Proc. Amer. Acad. **9**: 206. 1874.
Inyonia dysodioides M. E. Jones, Contr. West Bot. No. 8: 42. 1898.

Much-branched shrub up to 1.5 or 2 m. tall; young twigs finely glandular, especially upwards. Leaves numerous, subterete, 0.5–3 cm. long, 1 mm. wide or less, sometimes some of them with a few lateral teeth or short segments; heads subsessile or barely pedunculate at the ends of the branches, slightly if at all surpassing the upper leaves, the disk commonly about 1 cm. wide or less; involucral bracts commonly about 13, 7–9 mm. long, the upper parts often punctate like the leaves.

In desert canyons and foothills, Lower Sonoran Zone; Mojave and Colorado Deserts of southern California north to the Panamint Mountains in Inyo County, south to Lower California, and east to southern Nevada, western Arizona, and Sonora. Type locality: Colorado River in Sonora, Mexico. Feb.–May.

117. TETRADÝMIA DC. Prod. **6**: 440. 1837.

Heads discoid, cylindric, yellow, 4–9-flowered. Involucre of 4–6 erect equal bracts. Receptacle small, naked. Corolla-lobes longer than the throat. Anther strongly sagittate, almost caudate. Style-branches varying from as in *Luina* to nearly as in *Senecio*. Achenes terete, obscurely 5-nerved, glabrous to densely long-hairy; pappus of numerous white or whitish, capillary bristles. More or less canescent, branching, low shrubs with alternate, and often fascicled, narrow, entire, frequently spinose leaves. [Name from the Greek *tetra*, four, and *dymos*, together, referring to the tetramerous heads of several species.]

A genus native to western North America which consists of the following 6 species plus 2 additional ones which occur to the east of our range. Type species, *Tetradymia canescens* DC.

Achenes glabrous or pubescent with hairs that are much shorter than the conspicuous pappus; heads borne in terminal inflorescence.

Plants without spreading spines; flowers and involucral bracts normally 4.
Primary leaves foliaceous, sometimes minutely spinulose-tipped, not appressed; herbage densely, closely, and generally permanently white-tomentose. 1. *T. canescens.*
Primary leaves appressed, weakly spinescent, scarcely foliaceous; herbage thinly tomentulose or finally glabrate. 2. *T. glabrate.*
Plants with most or all of the primary leaves modified into rigid spreading spines; flowers and involucral bracts normally 5.
Tomentum of the twigs of the season borne in well-defined longitudinal strips that are separated by subglabrous intervals; achenes glabrous; spines pointing forward, commonly forming an angle of 30–60° with the stem. 3. *T. argyraea.*
Tomentum of the twigs of the season uniformly or nearly uniformly distributed, not separated into lines; achenes more or less canescent; spines spreading at a wide angle to the stem.
4. *T. stenolepis.*

Achenes provided with numerous very long hairs which nearly or quite equal and almost conceal the pappus-bristles; involucral bracts 5–6 and flowers 5–9, except in forms of *T. comosa.*
Heads borne in terminal inflorescences; leaves only weakly or scarcely spinose. 5. *T. comosa.*
Heads axillary to the primary leaves, distributed along the branches; primary leaves transformed into rigid, spreading or recurved spines. 6. *T. spinosa.*

1. **Tetradymia canéscens** DC. Spineless Horsebrush. Fig. 5805.

Tetradymia canescens DC. Prod. **6**: 440. 1837.
Tetradymia inermis Nutt. Trans. Amer. Phil. Soc. II. **7**: 415. 1841.
Tetradymia canescens var. *inermis* A. Gray, Bot. Calif. **1**: 408. 1876.

Unarmed, much-branched shrubs mostly 2–6 dm. tall. Leaves, involucres, and twigs conspicuously and closely white-tomentose; primary leaves linear or oblanceolate, 1–3 cm. long, 1–4 mm. wide, sometimes minutely spinulose-tipped and sometimes bearing axillary fascicles of shorter and proportionately broader leaves; heads in small cymose clusters terminating the numerous short branches; involucre 7–10 mm. high, of 4 or reputedly sometimes 5 bracts; flowers 4 in each head; achenes densely silky but the hairs not excessively long and not obscuring the well-developed and copious pappus.

Dry open places in the foothills and plains, southward extending to moderate elevations in the mountains, Arid Transition and Canadian Zones; Montana to New Mexico, westward to British Columbia, in Washington and Oregon east of the Cascade Mountains, and in California (eastern slopes of the Sierra Nevada and of the San Bernardino Mountains). Type locality: Columbia River. June–Sept.
Near the western borders of its range *T. canescens* commonly has glabrous rather than silky achenes, but there do not appear to be any correlated differences, and occasional hairy-fruited plants occur through much of the area in which the fruits are more characteristically glabrous.

2. **Tetradymia glabràta** A. Gray. Little-leaf Horsebrush. Fig. 5806.

Tetradymia glabrata A. Gray, Pacif. R. Rep. **2**: 122. 1854.
Tetradymia glabrata f. *calva* Payson, Univ. Wyo. Pub. Sci. **1**: 107. 1924.

Abundantly branched shrubs, mostly 3–10 dm. tall; twigs often rather persistently tomentose in lines, the herbage otherwise only thinly and deciduously tomentose or essentially glabrous. Primary leaves more or less appressed, linear-subulate, tending to be weakly spinescent, 6–10 mm. long, often deciduous; secondary leaves fascicled in the axils, blunt, linear, thickish or subterete, about 1 cm. long or less; heads in small cymose clusters terminating the branches; involucre 7–10 mm. high, of 4 or reputedly sometimes 5 thin-margined bracts; flowers 4 in each head; achenes densely hairy but the hairs not excessively long and not obscuring the well-developed and copious pappus.

Dry open places in the foothills and plains, Upper Sonoran and Transition Zones; central Oregon and central Idaho to Utah, southward along the east side of the Cascade Mountains and Sierra Nevada to the Mojave Desert and the Tahachapi Mountains. Type locality: Sierra Nevada. May–July.

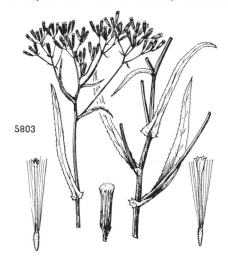

5803. Erechtites prenanthoides

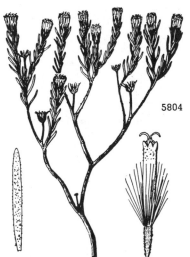

5804. Peucephyllum schottii

3. **Tetradymia argýraea** Munz & Roos. Striped Horsebrush. Fig. 5807.

Tetradymia argyraea Munz & Roos, Aliso 2: 237. 1950.

Much-branched shrubs, mostly 6–15 dm. tall; young twigs densely and persistently tomentose in lines that are separated by subglabrous intervals. Primary leaves thinly tomentulose and glabrate, modified into stiff, ascending-spreading spines 8–30 mm. long, commonly forming an angle of about 30–60° with the stem; secondary leaves fascicled in the axils, blunt, linear, thickish, about 1 cm. long or less, glabrous or tomentose; heads in small cymose clusters terminating the branches; involucre 6–8 mm. high, of 5 closely and densely tomentose bracts; flowers 5 in each head; achenes glabrous; pappus copious and well developed.

Dry rocky slopes at about 5,000 to 6,500 feet, Upper Sonoran and Arid Transition Zones; Kingston and Clark Mountains of San Bernardino County, California. Type locality: north side of the Kingston Mountains. July–Sept.

4. **Tetradymia stenolèpis** Greene. Mojave Horsebrush. Fig. 5808.

Tetradymia stenolepis Greene, Bull. Calif. Acad. 1: 92. 1885.

Much-branched shrubs, mostly 4–8 dm. tall; young twigs densely and persistently tomentose all around, rarely with more thinly pubescent lines. Primary leaves mostly modified into stout divaricate spines 2–3 cm. long that spread at a wide angle to the twigs, or the lowermost ones oblanceolate and sharply mucronate; secondary leaves fascicled in the axils, blunt, linear, about 1 cm. long or less or often quite wanting; heads in small cymose clusters terminating the branches; involucre 10–12 mm. high, of 5 densely and closely tomentose bracts; flowers 5 in each head; achenes more or less canescent but the hairs often partly deciduous; pappus copious and well developed.

Desert slopes, Sonoran Zones; Mojave Desert in San Bernardino and Kern Counties, extending to Los Angeles and Inyo Counties, California, and northern Lower California. Type locality: "A short distance southwest of the Southern Pacific railroad between Cameron and Mohave stations." Aug.–Sept.

5. **Tetradymia comòsa** A. Gray. Cotton-thorn. Fig. 5809.

Tetradymia comosa A. Gray, Proc. Amer. Acad. 12: 60. 1876.

Much-branched shrubs mostly 6–12 dm. tall with many erect virgate branches; herbage densely and generally permanently white-tomentose. Earlier primary leaves soft, linear, 2.5–6 cm. long, often 2 mm. wide, the later ones narrower, more rigid, sharply pointed and more or less spine-like; fascicles of smaller and softer secondary leaves sometimes borne in the axils; heads in small cymose clusters terminating the branches; involucre 6–10 mm. high, of 5 or 6 broad blunt bracts, tomentose like the herbage; flowers about 6–9 in each head; achenes very densely pubescent with very long, erect, white hairs which equal or exceed and conceal the scanty and fragile pappus.

Dry places at elevations below 5,000 feet, mostly Sonoran Zones; in interior cismontane southern California, less commonly on the Mojave Desert, and extending north, on the east side of the Sierra Nevada, as far as Reno, Nevada. Type locality: southeastern borders of California; and San Diego County, California. June–Oct.

The typical form of the species, as described above, is largely replaced in eastern Nevada by a form with mostly 4 involucral bracts and 4 flowers per head that has been described as *T. comosa* subsp. *tetrameres* (Blake, Proc. Biol. Soc. Wash. 35: 176. 1922). Although not yet definitely known from California, plants of this nature are to be expected near the eastern boundary of the state.

6. **Tetradymia spinòsa** Hook. & Arn. Catclaw Horsebrush. Fig. 5810.

Tetradymia spinosa Hook. & Arn. Bot. Beechey 360. 1838.

Much-branched shrubs mostly 5–12 dm. tall; twigs densely, persistently, and rather loosely white-tomentose. Primary leaves transformed into spreading, often more or less recurved spines about 5–13(16) mm. long, these bearing axillary fascicles of blunt, linear, subterete, green, and essentially glabrous leaves up to about 1 cm. long; heads distributed along the branches, borne on peduncles up to about 2.5 cm. long that are axillary to the primary leaves; involucre 8–11 mm. high, of 5 or 6 conspicuously white-tomentose bracts; flowers 5–9 in each head; achenes very densely pubescent with very long, erect, white hairs which nearly or quite equal and more or less conceal the rather scanty and fragile pappus.

Dry open places in the foothills and plains, mostly Arid Transition Zone; southwestern Montana and central Idaho to southeastern Oregon in Harney County and northern Malheur County, south to southwestern Colorado, southern Utah, Reno, Nevada, and Lassen County, California. Type locality: "Snake Country" [presumably in Idaho]. May–July.

Tetradymia spinosa var. **longispìna** M. E. Jones. Proc. Calif. Acad. II. 5: 698. 1895. (*Tetradymia axillaris* A. Nels. Bot. Gaz. 37: 277. 1904; *T. longispina* Rydb. Bull. Torrey Club 37: 471. 1910.) Differing from typical *T. spinosa* in its longer, more often straight spines, at least the longer ones generally 1.5 cm. long or commonly longer (up to 5 cm. long). Dry open places in the foothills and rarely occurring in southeastern Oregon in Lake and Malheur Counties and more common eastward and southward through central and western Nevada to southern Utah (and reputedly Arizona) and the eastern end of the San Bernardino Mountains, California. Type locality: Meadow Valley Wash, southern Nevada. April–July.

118. **LEPIDOSPÁRTUM** A. Gray, Proc. Amer. Acad. 19: 50. 1883.

Heads discoid, the flowers all tubular and perfect, yellow. Involucral bracts chartaceous or partly subherbaceous, evidently imbricate in several series. Receptacle small, flat, naked. Corolla-lobes equaling or longer than the throat. Anthers distinctly sagittate. Style-branches flattened, minutely papillate externally, scarcely hairy, without any well-defined appendage, the introrsely marginal stigmatic lines extending nearly or quite to

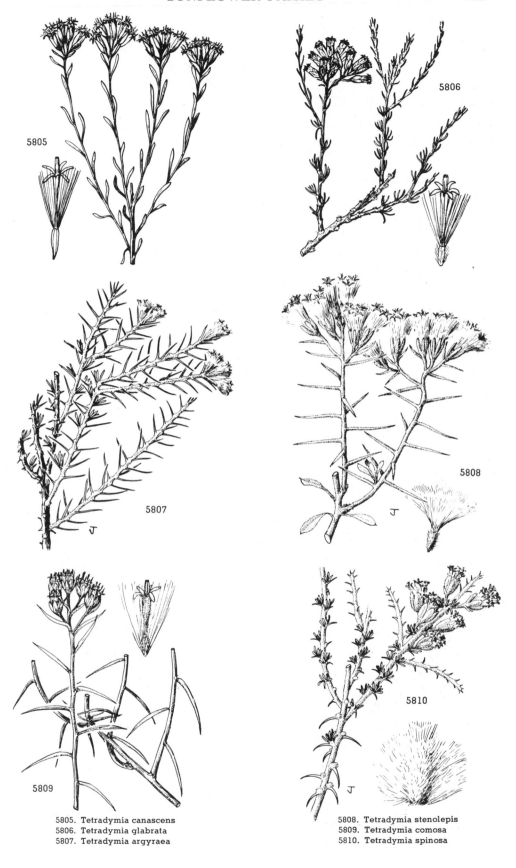

5805. **Tetradymia canascens**
5806. **Tetradymia glabrata**
5807. **Tetradymia argyraea**

5808. **Tetradymia stenolepis**
5809. **Tetradymia comosa**
5810. **Tetradymia spinosa**

the acute tip. Achenes thick. Shrubs with alternate narrow leaves which may be reduced to mere scales, and with numerous heads in corymbiform to racemose-paniculiform inflorescences. [Name from the Greek *lepis*, scale, and *sparton*, the broom plant.]

A genus of 2 species, native to Nevada, California, and northern Lower California. Type species, *Tetradymia squamata* A. Gray.

Lepidospartum is anomalous in the *Senecioneae* because of its evidently imbricate involucre, which is suggestive of the *Astereae*. The styles, however, would be equally anomalous in the *Astereae*; they more nearly resemble the styles of *Peucephyllum* and to a lesser extent those of *Luina* and some species of *Tetradymia*. Clarification of the relationships of the genus awaits further study.

Heads narrow, subcylindric, about 5–7-flowered; involucre mostly 8–10 mm. high; achenes densely long-hairy; leaves mostly linear.   1. *L. latisquamum.*
Heads broader, more campanulate, about 9–13- (reputedly to 18-) flowered; involucre mostly 5–8 mm. high; achenes glabrous or nearly so; leaves mostly reduced to mere short scales.   2. *L. squamatum.*

1. **Lepidospartum latisquàmum** S. Wats. Nevada Broom-shrub. Fig. 5811.

Lepidospartum latisquamum S. Wats. Proc. Amer. Acad. **25**: 133. 1890.
Lepidospartum striatum Coville, Proc. Biol. Soc. Wash. **7**: 73. 1892.

Shrub 6–25 dm. tall; twigs prominently striate, the narrow ridges glabrous or early glabrate, the intervals thinly but persistently tomentose. Leaves narrowly linear or linear-filiform, subterete, 1–4 cm. long, 1 mm. wide or less, glabrous or at first slightly tomentulose; heads borne in short compact clusters at the ends of the numerous branches, narrow, subcyclindric, about 5–7-flowered; involucre mostly 8–10 mm. high, the broad chartaceous bracts thinly tomentulose or subglabrate, tending to be striate, strongly and regularly imbricate in several series, the outermost ones mucronate-acuminate or tipped with a firm slender appendage up to 5 mm. long, the others broadly rounded at the tip; style-branches very gradually and finely acute; achenes densely long-hairy.

Dry rocky hillsides and sandy washes at elevations of 4,500 to 7,000 feet, commonly associated with pinyon-juniper woodland and with sagebrush, Upper Sonoran and Arid Transition Zones; southern Nevada, extending to Inyo County, California, and less commonly to the eastern end of the San Gabriel Mountains in San Bernardino County. Type locality: "Soda Spring Cañon, Esmeralda County, Nevada." May–Sept.

2. **Lepidospartum squamàtum** A. Gray. California Broom-shrub. Fig. 5812.

Linosyris squamata A. Gray, Proc. Amer. Acad. **8**: 290. 1870.
Linosyris squamata var. *breweri* A. Gray, loc. cit.
Linosyris squamata var. *palmeri* A. Gray, loc. cit.
Tetradymia squamata A. Gray, op. cit. **9**: 207. 1874.
Tetradymia squamata var. *breweri* A. Gray, Bot. Calif. **1**: 408. 1876.
Tetradymia squamata var. *palmeri* A. Gray, loc. cit.
Lepidospartum squamatum A. Gray, Proc. Amer. Acad. **19**: 50. 1883.
Baccharis sarothroides var. *pluricephala* Jepson, Man. Fl. Pl. Calif. 1059. 1925.
Lepidospartum squamatum var. *obtectum* Jepson, op. cit. 1159. 1925.
Lepidospartum squamatum var. *palmeri* L. C. Wheeler, Rhodora **40**: 322. 1938.

Shrub 6–20 dm. tall; vigorous young twigs tomentose, with oblanceolate, loosely tomentose, mostly soon-deciduous leaves 5–20 mm. long; principal twigs glabrous, slightly or scarcely striate with reduced, thickened, scale-like, glabrous leaves 1–4 mm. long, tomentose in the axils. Heads campanulate or turbinate-campanulate, about 9–13- (reputedly to 18–) flowered, racemosely distributed along the numerous branches, on scaly-bracted peduncles, the upper bracts of which tend to pass into the outer bracts of the involucre; involucre mostly 5–8 mm. high, glabrous or nearly so, the relatively narrow bracts well imbricate, scarious-margined, greenish toward the middle, often slightly striate, all more or less acute; style-branches rather abruptly acute; achenes glabrous or sometimes some of them with some readily deciduous pubescence.

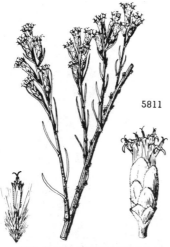

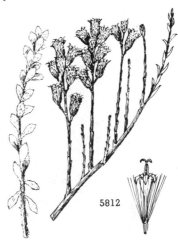

5811. Lepidospartum latisquamum

5812. Lepidospartum squamatum

Dry washes and other sandy or gravelly places from near sea level to nearly 5,000 feet elevation, Upper Sonoran Zone; cismontane California from Alameda and Fresno Counties to Lower California, extending eastward occasionally onto the Mojave and Colorado Deserts and reputedly to Nevada and Arizona. Type locality: "Low hills of the Sierra Santa Monica, Los Angeles Co., California." Aug.–Dec.

Tribe 6. CALENDULEAE

119. CALÉNDULA L. Sp. Pl. 921. 1753.

Annual or perennial herbs with leafy stems, the leaves simple and alternate. Involucres with disk- and ray-flowers; disk-flowers sterile. Phyllaries free, in 1 or 2 rows, scarious-margined. Peduncle enlarged at the base of the head. Receptacle flat, naked. Achenes of the ray-flowers incurved. Pappus none. [From the Latin meaning through the months.]

A genus of about 15 species, natives of the Mediterranean region and eastward to Iran. Type species, *Calendula officinalis* L.

1. Calendula arvénsis L. Field Marigold. Fig. 5813.

Calendula arvensis L. Sp. Pl. ed. 2. 1303. 1763.

Annuals, mostly simple-stemmed, sparingly branched above, the internodes long, herbage green, glandular-hirsutulous throughout. Leaves narrowly oblanceolate, the lower narrowed to a winged petiole, the upper sessile, 2–3.5 cm. long, the margins entire or somewhat denticulate; involucres solitary on the branchlets, the flowers yellow or orange; phyllaries in 2 rows, spreading in the ripened heads, about 7 mm. high, lanceolate, scarious-margined, the usually purple attenuate tip bearing multicellular, purplish, glandular hairs; achenes tan or brownish, smooth on the inner surface, with many transverse pectinate ridges on the outer face, incurved in half or almost complete circles.

Introduced weed in orchards and fields, locally abundant in Sonoma and Santa Barbara Counties, California; a native of central Europe and the Mediterranean region. Type locality: Europe. Feb.–April.

Calendula officinàlis L. Sp. Pl. 921.1753. The pot-marigold of the gardens, a coarse annual with thickish, oblong, sessile, usually clasping leaves and showy, pale yellow or orange flowers. Occasionally found occurring spontaneously in the vicinity of gardens in central California near the ocean.

Tribe 7. INULEAE

Heads with numerous yellow rays; involucres 4–5 cm. broad, at least the outer phyllaries completely foliaceous.
120. *Inula.*
Heads discoid; involucres 1.5 cm. broad or much narrower, the phyllaries not foliaceous.
Phyllaries dry but scarcely scarious, some chartaceous; stout glandular herbs without woolly herbage or silky-pubescent shrubs.
121. *Pluchea.*
Phyllaries completely or in part scarious, or inconspicuous or absent; marginal and inner bracts of the receptacle variously modified; low annual or perennial herbs or if up to 1 m. high more or less woolly.
Plants dioecious or subdioecious.
Basal leaves persisting in a rosette; heads nearly sessile or borne on stems usually less than 2 dm. high, the stem-leaves much reduced; plants strictly dioecious.
123. *Antennaria.*
Basal leaves soon withering; stems leafy, 2 dm. or more high; pistillate plants commonly with a few staminate flowers in each head.
124. *Anaphalis.*
Plants not dioecious; all flowers of the heads fertile or the outer ones pistillate and fertile, the inner hermaphrodite and often sterile.
Phyllaries numerous, imbricated; pistillate and hermaphrodite flowers all fertile, pappose; perennials or annuals.
122. *Gnaphalium.*
Phyllaries few, inconspicuous or absent, the outer receptacular bracts sometimes simulating an involucre; only pistillate flowers fertile, pappus absent (inner pistillate flowers in *Filago* sometimes with pappus); annuals.
Pappus lacking on all flowers of each head (vestigial capillary bristles present in some species of *Stylocline*).
Stem-leaves opposite; style lateral.
127. *Psilocarphus.*
Stem-leaves alternate; style terminal (except in *Micropus*).
Receptacular bracts subtending the pistillate flowers completely enclosing and deciduous with them.
Plants simple, rarely branched above; involucre shorter than the bracts of the receptacle.
128. *Micropus.*
Plants branched from the base, sometimes also above; involucre absent, the heads subtended by stem-leaves.
125. *Stylocline.*
Receptacular bracts subtending the pistillate flowers plane or concave, persistent.
129. *Evax.*
Pappus present on central sterile flowers of each head.
126. *Filago.*

120. ÍNULA L. Sp. Pl. 881. 1753.

Perennial, mostly tomentose or woolly herbs. Leaves basal and alternate above. Heads large, of tubular and yellow ray-flowers, hemispheric or campanulate. Phyllaries imbricate in several series, the outer foliaceous. Receptacle flat or convex, naked. Ray-flowers pistillate, the ligules 3-toothed. Disk-flowers hermaphrodite, fertile, 5-toothed. Anthers sagittate at the base with caudate auricles. Style-branches of the disk-flowers linear, obtuse. Achenes 4–5-ribbed, the pappus of capillary bristles. [The ancient Latin name.]

A genus of about 90 species, natives of Europe, Asia, and Africa. Type species, *Inula helenium* L.

1. Inula helénium L. Elecampane. Fig. 5814.

Inula helenium L. Sp. Pl. 881. 1753.

Coarse herb up to 2 m. high, more or less tufted from thick roots; stems simple, rarely branched. Basal leaves large, 2.5–5 dm. long including the petiole, ovate to oblong, the margin denticulate, rough-pubescent above, velvety-pubescent beneath; stem-leaves smaller, cordate-clasping, acute at the apex; heads 4–5 cm. broad with stout peduncles, terminal, few or solitary; phyllaries in several series, 1.5 to nearly 2 cm. long, the inner oblong, herbaceous at the tip, the outer completely foliaceous, broadly ovate, obtuse or sometimes acute at the apex; ray-flowers about 3 cm. long, many, narrow; achenes 5 mm. long, glabrous, the pappus-bristles longer than the achene, spreading at maturity.

Adventive from Eurasia, along roadsides; found in the Pacific States in western Oregon, principally in the Willamette Valley; well established in southeastern Canada, and south to North Carolina and Missouri. July–Aug. Horsehead. Yellow Starwort.

121. PLUCHEA Cass. Bull. Soc. Philom. 1817: 31. 1817.

Herbs or shrubs, usually pubescent, with alternate, toothed or entire leaves and cymes or cymose panicles of small, disciform, purplish heads. Involucre (in ours) campanulate, graduate, of dry acute phyllaries, at least the inner subscarious. Receptacle broad, flat to concave, naked. Outer flowers pistillate, very numerous, many-seriate, their corollas filiform, 3–4-toothed. Central flowers about 10–28, hermaphrodite, mostly sterile, their corollas tubular,.5-toothed. Achenes small, 4–5-nerved, their pappus of a single series of capillary bristles, sometimes clavellate-dilated at tip. Anthers caudate-sagittate at base, the auricles of adjacent anthers united. Style in the hermaphrodite flowers hispidulous above, undivided or bifid. [For Abbé N. A. Pluche, an amateur naturalist of the late eighteenth century.]

A genus of about 40 species, in warm regions throughout the world. Beside the following, several others occur in the eastern United States. Type species, *Conyza marilandica* Michx.

Green annual, glandular and pubescent; leaves mostly oblong-ovate, toothed; pappus-bristles not dilated at tip.
1. *P. camphorata.*
Silky-pubescent shrub; leaves mostly lance-linear, entire; pappus-bristles dilated at tip, especially in the hermaphrodite flowers.
2. *P. sericea.*

1. Pluchea camphoràta (L.) DC. Salt-marsh Fleabane. Fig. 5815.

Erigeron camphoratum L. Sp. Pl. ed. 2. 1212. 1763.
Pluchea camphorata DC. Prod. 5: 452. 1836.

Erect annual, usually 1 m. high or less, usually branched, leafy, pilose or pilosulose throughout with several-celled sordid hairs and gland-dotted, heavy-scented, the stem striate, usually stout. Leaves lance-oblong to ovate or oval, 4–15 cm. long, 1–5 cm. wide, acute or acuminate, cuneate at base, serrulate to coarsely serrate, feather-veined, dull green, thin, the lower tapering into short petioles, the upper sessile or subsessile; heads numerous, in rounded, terminal, cymose panicles, short-peduncled; involucre broadly campanulate, 4–6 mm. high, strongly graduate; outer phyllaries triangular-ovate, the inner oblong or linear, all acute or acuminate, the outer subherbaceous above, the inner scarious and purplish; pistillate flowers very numerous, their corollas filiform, with 3–4 short teeth; hermaphrodite flowers 10 or more, tubular, their teeth glandular above; achenes of pistillate flowers hispidulous and glandular, of disk-flowers glabrous or glandular; pappus-bristles not dilated above.

In salt marshes and wet places, Sonoran Zones; California west of the Sierra Nevada from the San Francisco Bay region mostly near the coast to Lower California (also on Catalina Island) east through Arizona, New Mexico, and Texas to the Atlantic coast north to Massachusetts and south into Mexico. Type locality: Virginia. Aug.–Oct. Camphorweed.

2. Pluchea serícea (Nutt.) Coville. Arrow-weed. Fig. 5816.

Polypappus sericeus Nutt. Journ. Acad. Phila. II. 1: 178. 1848.
Tessaria borealis Torr. & Gray ex A. Gray, Mem. Amer. Acad. II. 4: 75. 1849. (Nomen nudum.)
Pluchea sericea Coville, Contr. U.S. Nat. Herb. 4: 128. 1893.
Berthelotia sericea Rydb. Bull. Torrey Club 33: 154. 1906.

Şlender, willow-like shrub up to 5 m. high, essentially silvery-silky throughout, densely leafy. Leaves linear-lanceolate to lanceolate, 1–4.5 cm. long, 2–9 mm. wide, acute or acuminate at each end, entire, sessile, leathery, 1-nerved, the few lateral veins usually concealed by the tomentum; heads in small, terminal, cymose clusters; involucre about 7 mm. high, strongly graduate; outer and middle phyllaries triangular-ovate to oblong, acute, coriaceous, densely villous especially toward margin, the inner linear, thinner, essentially glabrous, lacerate-ciliate at tip, deciduous, some or all often purplish-tinged; pistillate flowers very numerous; hermaphrodite flowers 28 or fewer; achenes glabrous; pappus-bristles dilated at tip, especially in the hermaphrodite flowers.

Along watercourses in desert regions, Lower Sonoran Zone; from eastern Santa Barbara County in the Cuyama Valley and Inyo County, California, to Lower California and east to Utah and Texas; also on Catalina Island. Type locality: "Rocky Mountains of Upper California." Nearly throughout the year.

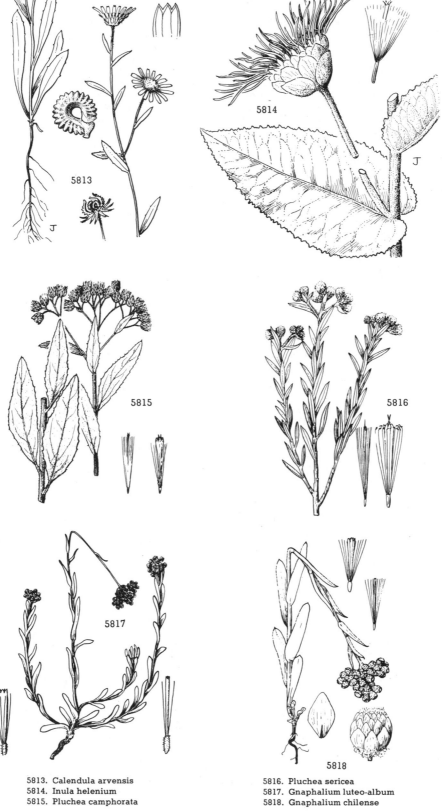

5813. Calendula arvensis
5814. Inula helenium
5815. Pluchea camphorata

5816. Pluchea sericea
5817. Gnaphalium luteo-album
5818. Gnaphalium chilense

122. GNAPHÀLIUM L. Sp. Pl. 850. 1753.

More or less woolly herbs often aromatic. Leaves alternate, entire, usually narrow, often decurrent. Heads small, disciform, sessile or subsessile, closely clustered at the tips of branches in usually cymose panicles, sometimes spicately arranged. Involucre uually campanulate, several-seriate, slightly or strongly graduate, appressed, the phyllaries with a thickened base and scarious or hyaline, white or colored, usually conspicuous tips, usually radiately spreading in age. Receptacle flat, naked. Pistillate flowers very numerous, several-seriate, their corollas filiform, about 3-toothed. Hermaphrodite flowers few, fertile, funnelform or subcyclindric, with a slightly dilated apex, 5-toothed. Corollas yellow, whitish, purplish, or reddish. Anthers with somewhat lacerate, sagittate-caudate bases. Style-branches in hermaphrodite flowers linear or with slightly dilated, truncate, minutely hispidulous apex. Achenes small, oblong. Pappus of numerous 1-seriate, soft, merely scabrous bristles sometimes united at base and deciduous in a ring. [Greek name for these or similar plants.]

A cosmopolitan genus of 120 or more species; beside the following, several other species occur in the United States and Canada, and many in Mexico. Type species, *Gnaphalium luteo-album* L.

Inflorescence paniculate or cymose-paniculate or corymbose, the branches spreading or much congested, the solitary or glomerate heads not leafy-bracted.
  Leaves persistently woolly above as well as below, grayish or white.
    Annuals (*G. chilense* sometimes persisting along sea coast); heads 150–230-flowered.
      Phyllary-tips almost hyaline, whitish faintly tinged with brown; heads 3–3.5 mm. high; corollas reddish at tip.   1. *G. luteo-album.*
      Phyllary-tips scarious, yellowish or straw-colored; heads 4–6 mm. high; corolla yellow.   2. *G. chilense.*
    Perennials, herbaceous or somewhat woody below; heads 35–90-flowered.
      Leaves above the base of the plant not at all decurrent, divergent, oblanceolate, 6 mm. or more wide; inflorescence congested or with divaricately spreading branches.   3. *G. microcephalum.*
      Leaves above the base of the plant shortly but noticeably decurrent, erect, linear-oblanceolate or linear, 2–4 mm. wide; inflorescence usually with ascending branches.
        Leaves of the young shoots oblanceolate, up to 5.5 cm. long; slender plants of the higher mountains.   4. *G. thermale.*
        Leaves of the young shoots linear, up to 10 cm. long; stout plants of the lower foothills.   5. *G. beneolens.*
  Leaves green and glandular above, or if at all floccose thinly so and usually becoming glabrate.
    Heads campanulate to globose; phyllaries white or tawny.
      Leaves of midstem linear, acuminate, 2–5 mm. wide; phyllaries not shining, opaque and white.   7. *G. leucocephalum.*
      Leaves of midstem lanceolate or oblanceolate, acute, 6–12 mm. wide; phyllaries shining, pearly white or tawny.
        Undersurface of leaves and also the stems densely and permanently white-woolly; leaves of the inflorescence conspicuously auriculate-clasping.   6. *G. bicolor.*
        Undersurface of the leaves and also the stems glabrate or grayish with a loose tomentum; leaves of the inflorescence not or inconspicuously auriculate.
          Outer phyllaries obtuse or rounded, pearly white except in extreme age; plants of the Coast Ranges from Oregon southward and the Sierra foothills.   8. *G. californicum.*
          Outer phyllaries more or less strongly acute, yellowish or tawny and becoming dingy; plants from the Cascade Mountains and Sierra Nevada eastward.   9. *G. macounii.*
    Heads turbinate or narrowly ovate; phyllaries pink or tinged with pink.   10. *G. ramosissimum.*
Inflorescence spiciform or capitate and leafy-bracted, or of leafy-bracted glomerules at the tips of the branches, the leaves surpassing the heads.
  Plants perennial, reproducing by creeping rootstocks and leafy runners.   13. *G. collinum.*
  Plants fibrous-rooted annuals or short-lived perennials with a taproot.
    Pappus-bristles falling separately; phyllaries but little imbricate.
      Plants floccose-tomentose; leaves spatulate to oblanceolate or oblong.   11. *G. palustre.*
      Plants appressed-tomentose; leaves linear-spatulate.   12. *G. uliginosum.*
    Pappus-bristles falling in a persistent or easily fragmented ring; phyllaries definitely imbricate.
      Heads in dense globose clusters.   14. *G. japonicum.*
      Heads in dense or interrupted spikes.
        Plants greenish, more or less loosely floccose; phyllaries yellowish or greenish.   15. *G. peregrinum.*
        Plants pannose-tomentose, markedly so on the stems and undersurface of the leaves; phyllaries brownish or purplish.   16. *G. purpurem.*

1. Gnaphalium luteo-álbum L. Weedy Cudweed. Fig. 5817.

Gnaphalium lutco-album L. Sp. Pl. 851. 1753.

Slender decumbent annual (rarely persisting more than one season) 1.5–4 dm. high, usually several-stemmed, normally corymbosely branched above, slightly or scarcely fragrant, densely and rather closely whitish-tomentose throughout or the upper surface of the leaves becoming greenish (but not glabrate), densely leafy below, sparsely so above. Lower leaves obovate, 1–3 cm. long, 2–8 mm. wide, obtuse, the others oblanceolate to linear-oblong or linear, 1–2 cm. long, 2–4 mm. wide, obtuse to acute, the auriculate-clasping base with short-decurrent auricles; heads small, 3–3.5 mm. high, campanulate or subglobose-campanulate, about 150–65-flowered, in small glomerules 1–2 cm. thick terminating the stem or the tips of branches of the inflorescence; phyllaries with thin-scarious, almost hyaline tips, these whitish and faintly brownish tinged, the base with a rhombic, greenish or brownish spot, ovate through obovate to linear-spatulate or linear, all obtuse or rounded, spreading in age; pistillate flowers 143–60, hermaphrodite 5–10; corollas reddish-tipped; achenes minutely papillose.

An introduced European weed of gardens, roadsides, and irrigation ditches, occurring locally around Portland, Oregon, and Klickitat County, Washington; more widespread from Marin County, California, southward to Orange County and also in the San Joaquin Valley in San Joaquin, Merced, and Madera Counties. April–Oct.

2. Gnaphalium chilénse Spreng. Cotton-batting Plant. Fig. 5818.

Gnaphalium chilense Spreng. Syst. **3**: 480. 1826.
Gnaphalium sprengelii Hook. & Arn. Bot. Beechey 150. 1833.
Gnaphalium gossypinum Nutt. Trans. Amer. Phil. Soc. II. **7**: 403. 1841.
Gnaphalium luteo-album var. *occidentale* Nutt. loc. cit.
Gnaphalium chilense var. *confertifolium* Greene, Fl. Fran. 400. 1897.
Gnaphalium sulphurescens Rydb. Mem. N.Y. Bot. Gard. **1**: 415. 1900.
Gnaphalium proximum Greene, Ottawa Nat. **15**: 279. 1902.

Annual or biennial with a strong taproot, 1–6 dm. high, often several-stemmed, mostly erect or sometimes decumbent in the coastal forms, rather loosely, floccosely, and persistently gray-woolly throughout, usually aromatic. Basal and lower stem-leaves spatulate to obovate or oblong, 2–9.5 cm. long, 3–10 mm. wide, usually obtuse; upper stem-leaves oblong, oblanceolate to linear, 1.5–6 cm. long, 1–8 mm. wide, obtuse to acute, often with broadened and clasping base, all more or less conspicuously short-decurrent; heads about 182–228-flowered, terminating the stem in loose, more or less capitate glomerules or clustered on short corymbose branches; involucres sub-globose or campanulate-subglobose, 4–6 mm. high; phyllaries pearly white when young to straw-colored or distinctly yellowish, somewhat brownish in age, thin-scarious and shining, oval or obovate through oblong to (innermost) linear, all broadly rounded or the innermost acutish; pistillate flowers 160–200, hermaphrodite 19–28; corollas yellow.

In various habitats, mostly moist, Transition and Sonoran Zones; Washington south, including the Channel Islands, to northern Lower California, east to Montana and southward to Arizona, New Mexico, and western Texas. Type locality: California, erroneously attributed to Chile. Collected by Chamisso. March–Oct.

A species variable in growth forms. Plants in exposed situations along the ocean, on cliffs or in sand dunes, persist more than one season. They have rather short decumbent stems, which are densely clothed with oblong leaves and arise from woody roots. The form described by Greene as *G. chilense* var. *confertifolium* found on the coast and inland has much the same habit. It is, however, a biennial and has linear-acute leaves.

3. Gnaphalium microcéphalum Nutt. White Everlasting. Fig. 5819.

Gnaphalium microcephalum Nutt. Trans. Amer. Phil. Soc. II. **7**: 404. 1841.
Gnaphalium albidum I. M. Johnston, Contr. Gray Herb. No. 70: 84. 1924.

Perennial with herbage densely white-tomentose throughout, odorless or nearly so, with several stems 5–10 dm. high arising from the woody base, these rather stout and usually branching. Leaves on the young shoots and the basal leaves oblanceolate, obtuse, or acute, narrowed at the base and sessile, 3–8 cm. long, 6–18 mm. wide; upper leaves reduced upward and becoming more narrow, longer than the internodes, divergent in fresh material, not at all or obscurely decurrent; inflorescence cymose-paniculate, 2–15 cm. long, often wider than long, the branches divaricately spreading or on some stems in scarcely expanded, oblong, head-like clusters; heads 35–40-flowered, the hermaphrodite flowers 4–6; involucres 4.5–5.5 mm. high, cylindric-campanulate, usually 3-seriate, the outer series woolly; phyllaries thin, hyaline and shining, the outer ovate to ovate-oblong, acute, occasionally obtuse, the inner oblong, sometimes apiculate; pappus-bristles distinct, falling separately.

Alluvial fans and open places in chaparral, Sonoran Zones; local in Napa County and the Mount Hamilton Range, Santa Clara County, California (*R. J. Smith*); more common from Monterey County south to San Bernardino and San Diego Counties, and on the Santa Cruz Islands; also the Sierra Nevada foothills from Madera County southward. Type locality: San Diego. Collected by Nuttall. June–Oct.

The photograph of Nuttall's type in the British Museum shows a single stem topped by a head-like inflorescence and bears the characteristic leaves. The specimen probably was collected in spring (see Madroño **2**: 146–46. 1934). This would account for the weather-beaten appearance (darkened foliage and rubbed tomentum) of this plant which normally flowers in summer and early fall.

Gnaphalium wrightii A. Gray, Proc. Amer. Acad. **17**: 214. 1882. Stems leafy, the leaves oblanceolate and not decurrent; heads about 35–40-flowered; involucres small, 2–3-seriate; phyllaries few, thin, hyaline and shining, the outer usually obtuse, the inner sharply acute. A specimen from Campo, San Diego County (*Wiggins 1032*) corresponds to the Arizona material of *G. wrightii*. Specimens from the San Bernardino Mountains appear to be intermediate between this species and the coarser *G. microcephalum*.

4. Gnaphalium thermàle E. Nels. Slender Cudweed. Fig. 5820.

Gnaphalium thermale E. Nels. Bot. Gaz. **30**: 121. 1900.
Gnaphalium williamsii Rydb. Bull. Torrey Club **37**: 324. 1910.
Gnaphalium johnstonii G. N. Jones, Univ. Wash. Pub. Bot. **7**: 176. 1939.

Perennial 2–4 dm. high, loosely woolly throughout and somewhat sweet-scented, stems several from the woody taproot, erect and simple below the inflorescence, decumbent at base, more or less densely leafy below, the internodes longer above. Leaves on the new shoots and basal leaves oblanceolate to spatulate, 2.5–5.5 cm. long; stem-leaves gradually reduced upward, sessile, oblanceolate to linear, acute at apex, 1–3 cm. long, noticeably decurrent in the bract-like leaves of the inflorescence; inflorescence usually short, cymose or cymose-paniculate, the branches ascending; heads about 40–50-flowered, the hermaphrodite flowers 2 or 3, glomerulate on the ultimate branches; involucres woolly only at the extreme base and on the short peduncle, 4–5 mm. high, usually 3-(sometimes 4-)seriate; phyllaries thin, hyaline, and shining, rarely with an opaque tip (southern forms), acute and often cuspidate at apex, the outer ovate, the inner narrowly lanceolate or linear; pappus-bristles distinct, falling separately.

In rocky or sandy soil, Transition and Canadian Zones; British Columbia southward, mostly in and east of the Cascade Mountains, to California where it occurs in the Siskiyou and Trinity Mountains of northern California and through the Sierra Nevada to the mountains of southern California, though differing from its typical form in its southern extension; east to Wyoming, Montana, and Colorado. Type locality: Yellowstone National Park, Wyoming. July–Sept.

5. **Gnaphalium beneòlens** Davidson. Fragrant Everlasting. Fig. 5821.

Gnaphalium beneolens Davidson, Bull. S. Calif. Acad. **17**: 17. *pl.* 1918.

Rather stout perennial 4.5–10 dm. high, persistently white-woolly throughout, sometimes greenish yellow, sweet-scented, the stems few from a usually woody base and usually branched. Leaves of the new shoots and the basal leaves linear, narrowly acute, 4–10 cm. long, 2–3 mm. wide; stem-leaves but little reduced upward on the stems, more so below and in the inflorescence, longer than the internodes, evidently and often conspicuously decurrent; inflorescence usually a rather narrow panicle (or cymose-paniculate) up to 3 dm. long with ascending branches, the heads glomerulate on the ultimate branches; heads 45–55-flowered, pale yellow, the hermaphrodite flowers few; involucres campanulate, 5–6 mm. high, 4–5-seriate; phyllaries obtuse and rounded to broadly acute, the inner more narrow and sometimes cuspidate, the exposed tips papery and opaque, the short hyaline area above the thickened base concealed; pappus-bristles distinct, falling separately.

Valleys and foothills, mostly Upper Sonoran Zone; Josephine County, Oregon, south through the Coast Ranges of California to cismontane southern California; also occurring in the foothills of the Sierra Nevada. Type locality: La Crescenta, Los Angeles County, California. July–Oct.

Some plants of the lower western slopes of the Sierra Nevada are intermediate between this species and *G. thermale*, having the opaque phyllary-tips and larger heads of the former and the growth habit and shorter inflorescence of the latter.

6. **Gnaphalium bìcolor** Bioletti. Bioletti's Cudweed. Fig. 5822.

Gnaphalium bicolor Bioletti, Erythea **1**: 16. 1893.

Ascending or erect, stout perennial up to 1 m. high, lignescent below, corymbosely branched above, sweet-scented, very leafy, the stem and lower leaf-surface densely white-woolly. Leaves thin, lanceolate to oblanceolate, 3–8 cm. long, 5–13 mm. wide, acute, sessile by a rather broad, clasping, not decurrent base, bright green and glandular or rarely gray-tomentose above; inflorescence comparatively dense and short, rounded or flattish; heads usually glomerate at tips of branches, about 60–85-flowered, about 5–6 mm. high; phyllaries white or faintly tinged with straw color in age, sometimes with a pinkish tinge, thin-scarious, shining, ovate through obovate to linear, obtuse or the inner often apiculate, glandular below; pistillate flowers 53–73, hermaphrodite 8–11; corollas yellowish.

Dry sandhills and foothills, Transition and Upper Sonoran Zones; Madera County to Tulare County on the western slope of the Sierra Nevada and from San Benito and southern Santa Cruz Counties to San Diego and San Bernardino Counties, California, the Santa Barbara Islands, and northern Lower California; also reported from Paradise Cove, Marin County. Type locality: San Diego, California. April–June (Oct.).

7. **Gnaphalium leucocéphalum** A. Gray. Sonora Everlasting. Fig. 5823.

Gnaphalium leucocephalum A. Gray, Smiths. Contr. **5**[6]: 99. 1853.

Slender erect perennial (or sometimes biennial?), often several-stemmed, 0.3–0.6 m. high, corymbosely branched at apex, sweet-scented, densely white-woolly on stem and lower leaf-surface. Leaves linear or narrowly linear-lanceolate, 3–10 cm. long, 2–6 mm. wide, acuminate, sessile by a scarcely broadened, shortly decurrent base, deep green and glandular above, the midrib evident beneath; inflorescence short and dense, flattish, up to 13 cm. wide; heads rather large, about 80–100-flowered, campanulate, 5–6 mm. high, glomerate at tips of branches; phyllaries white, rather thick and opaque, ovate through obovate to linear, obtuse or the inner apiculate; pistillate flowers 66–83, hermaphrodite 13–14; corollas yellowish.

Sandy creek beds, Sonoran Zones; Los Angeles County, California, to southern Arizona, western Texas, and Sonora; not recorded from New Mexico. Type locality; near Santa Cruz, Sonora. July–Sept.

8. **Gnaphalium califórnicum** DC. California Cudweed or Green Everlasting. Fig. 5824.

Gnaphalium californicum DC. Prod. **6**: 224. 1837.
Gnaphalium decurrens var. *californicum* A. Gray, Bot. Calif. **1**: 341. 1876.

Erect, rather stout biennial often flowering the first year, 3 dm. high, corymbosely branched above, glandular-pilose essentially throughout and often thinly gray-woolly, sweet-scented. Leaves thin, oblanceolate or the lower obovate to oblanceolate or linear-oblong, or the uppermost linear, gradually reduced above, 3.5–10 cm. long, 0.5–2.3 cm. wide, obtuse to acuminate, sessile and shortly but broadly decurrent, green and glandular on both sides, sometimes thinly gray-lanate; panicle corymbiform, usually short and flattish or rounded, sometimes 25 cm. wide; heads clustered at tips of branches, comparatively large, about 115–50-flowered, subglobose, about 6 mm. high; phyllaries pearly white or in age straw-colored, rarely pinkish, at first somewhat papery, rather thin-scarious and shining in age, broadly ovate to obovate and innermost linear, obtuse or the innermost apiculate, glandular below; pistillate flowers about 106–37, hermaphrodite 7–11; corollas yellowish; pappus-bristles distinct, falling separately.

Dry hills and canyons, Humid Transition and Upper Sonoran Zones; Lincoln County, Oregon, south in the Coast Ranges to San Diego County, California, and the Channel Islands and adjacent Lower California; also in the Sierran foothills from Tuolumne County to southern Fresno County. Type locality: California. May–July.

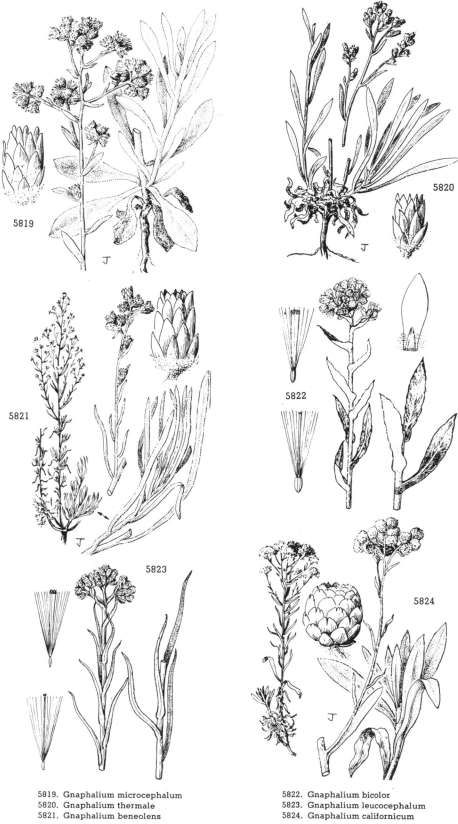

5819. Gnaphalium microcephalum
5820. Gnaphalium thermale
5821. Gnaphalium beneolens

5822. Gnaphalium bicolor
5823. Gnaphalium leucocephalum
5824. Gnaphalium californicum

Along the coast (Marin, Monterey, and Santa Barbara Counties) forms are occasionally found that appear to be of hybrid origin. In Marin County and in Monterey County these forms suggest relationship with *G. chilense*. The inflorescence is compact as in that species but the number of flowers per head and the size of the heads themselves suggest *G. californicum*, and the leaves, though loosely woolly above, correspond to *G. californicum*. On the other hand, aberrant forms are found in which the leaves have broad clasping bases resembling those of *G. bicolor*, which also commonly occurs in the area farther south.

9. Gnaphalium macoùnii Greene. Winged Cudweed. Fig. 5825.

Gnaphalium decurrens Ives, Amer. Journ. Sci. 1: 380. *pl. 1.* 1819. Not *G. decurrens* L. 1759.
Gnaphalium macounii Greene, Ottawa Nat. **15**: 278. 1902.
Gnaphalium ivesii Nels. & Macbr. Bot. Gaz. **61**: 46. 1916.

Annual or biennial, usually single-stemmed, 4–9 dm. high, corymbosely branched above, sweet-scented, the stem glandular-pilose and with thin, deciduous, grayish tomentum, more tomentose about the inflorescence. Leaves thin, lanceolate or oblanceolate or the uppermost linear, 3–10 cm. long, 3–13 mm. wide, usually acute or more rarely acuminate, sessile and shortly but usually broadly decurrent, deep green and glandular-pubescent above, beneath densely and persistently grayish-tomentose (very rarely glabrescent); basal leaves of the second year 3.5–4.5 cm. long, oblanceolate or obtuse; panicle comparatively short, up to 15 cm. wide; heads clustered at tips of branches, 5–6 mm. high, campanulate-subglobose, about 132–47-flowered; phyllaries much imbricated, dingy, straw-colored to pale brownish, thin-scarious and shining, woolly at base, ovate through oblong to linear, acutish or acute, or the outer obtusish; pistillate flowers 121–32, hermaphrodite 11–14; corollas yellowish; pappus-bristles distinct and falling separately.

Dry open ground in forested places, Transition Zone; British Columbia, and in Washington and Oregon mostly east of the Cascade Mountains, less common southward to California in Trinity and Plumas Counties and in the southern Sierra Nevada, eastward across the United States and Canada to the Atlantic seaboard. Type locality: Chilliwack Valley, British Columbia. July–Oct.

10. Gnaphalium ramosíssimum Nutt. Pink Everlasting. Fig. 5826.

Gnaphalium ramosissimum Nutt. Journ. Acad. Phila. II. **1**: 173. 1848.

Erect, rather slender biennial up to 1.5 m. high, loosely paniculate-branched above, herbage greenish, glandular-pubescent throughout, thinly grayish-lanate and becoming glabrate, very sweet-scented. Leaves linear to lanceolate or the lower lance-oblong, gradually reduced above, the larger 3–6.5 cm. long, 2–8 mm. wide, usually acuminate, sessile and conspicuously decurrent, often strongly revolute-margined and wavy, glandular-pubescent on both sides; inflorescence paniculate, usually elongate, subcyclindric or pryamidal, sometimes 45 cm. long; heads rather small, about 60-flowered, narrowly campanulate or turbinate, about 5 mm. high, clustered at tips of branches and branchlets; phyllaries pink or rarely white (at least in age), thin-scarious, somewhat shining, ovate or ovate-oblong to linear, at least the inner usually acute or acutish, glandular below; pistillate flowers 45–58, hermaphrodite about 6; corollas yellowish.

Dry wooded slopes at low altitudes near the coast, Transition and Upper Sonoran Zones; Humboldt County to Orange County, California; also on Santa Cruz Island. Type locality: Monterey, Monterey County. July–Sept.

11. Gnaphalium palústre Nutt. Lowland Cudweed. Fig. 5827.

Gnaphalium palustre Nutt. Trans. Amer. Phil. Soc. II. **7**: 403. 1841.
Gnaphalium palustre var. *nanum* Jepson, Man. Fl. Pl. Calif. 1067. 1925.

Slender, erect or diffuse annual 2–30 cm. high, simple or branched especially at the base, loosely and floccosely gray-woolly throughout or the leaves glabrescent, not sweet-scented. Leaves spatulate or oblanceolate to obovate or oblong, 0.5–4.5 cm. long, 3–8 mm. wide, usually obtuse and apiculate, narrowed to the sessile, not decurrent, often petioliform base; heads small, about 120–30-flowered, 3 mm. high, glomerate at tips of stem and branches and often in the axils, usually surpassed by the subtending leaves; involucre campanulate, woolly to the middle, the phyllaries linear, obtuse, greenish-centered below, with shorter or longer, exserted, scarious and shining, white tips, these often brownish-tinged below apex or essentially throughout in less densely floccose forms; pistillate flowers 110–20, hermaphrodite 6–8; corollas whitish; achenes smooth or papillose, the pappus-bristles distinct, falling separately.

Moist open ground and stream banks, Sonoran Zone to Canadian Zone; British Columbia and Alberta southward on the Pacific Slope to southern California and Lower California, and east to the Rocky Mountains and New Mexico. Type locality: "Rocky Mountains, Oregon, California and Chili." April–Oct.

12. Gnaphalium uliginòsum L. Marsh Cudweed. Fig. 5828.

Gnaphalium uliginosum L. Sp. Pl. 856. 1753.

Slender, erect or diffuse annual, usually much branched, 3–25 cm. high, closely whitish- or grayish-woolly throughout or the leaves glabrescent, not sweet-scented. Leaves narrowly oblanceolate or linear-spatulate to nearly linear, 0.8–5 cm. long, 1–4 mm. wide, obtuse or acute, apiculate, sessile by a narrowed, often petioliform base, not decurrent; heads in small clusters at tips of stem and branches, imbedded in wool at base, surpassed by the leaves, campanulate, about 2.5 mm. high, about 100–15-flowered; phyllaries ovate through lance-oblong to lanceolate or lance-linear, obtuse or the inner acutish, with pale greenish base and brownish or blackish brown tips, these rarely whitish at first; pistillate flowers 94–108, hermaphrodite 5–7; corollas whitish; achenes glabrous or minutely hispidulous, the pappus-bristles distinct and falling separately.

Growing in low, usually moist places and found on the Pacific Slope principally in the Puget Sound area and along the Columbia River in Klickitat County, Washington, and in the vicinity of Portland, Oregon. Well established in eastern United States as a weed. Type locality: Europe. June–Dec.

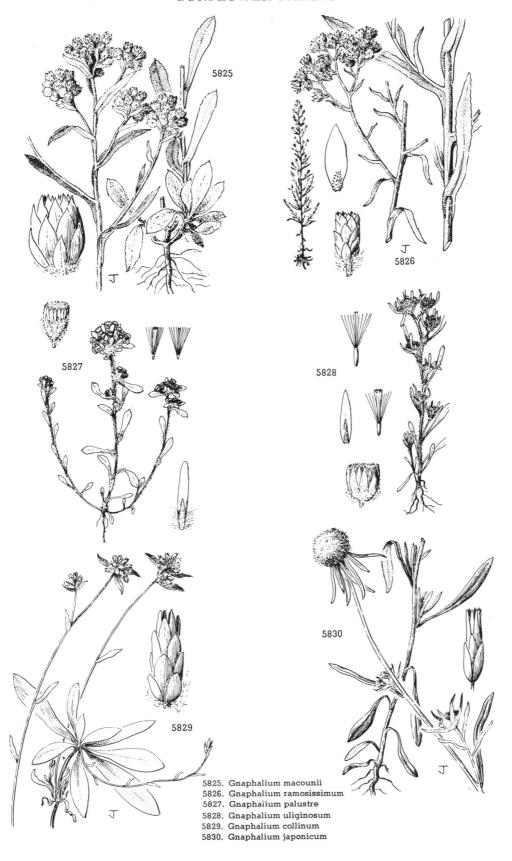

5825. Gnaphalium macounii
5826. Gnaphalium ramosissimum
5827. Gnaphaiium palustre
5828. Gnaphalium uliginosum
5829. Gnaphalium collinum
5830. Gnaphalium japonicum

13. Gnaphalium collìnum Labill. Creeping Cudweed. Fig. 5829.

Gnaphalium collinum Labill. Pl. Nov. Holl. **2**: 44. *pl. 189.* 1806.

Perennial, stems 6–30 cm. high, arising from a creeping rhizome, this producing leafy stolens rooting at the nodes. Basal leaves persistent, 2.5–10 cm. long, oblanceolate, narrowed to a long, slender, petioliform base, thick in texture, the margin revolute, glabrous above, pannose-tomentose below and rather silvery; upper leaves sessile, 1.5–2 cm. long, the internodes long; leaves subtending the terminal capitate clusters few, as the stem-leaves but shorter; involucre cylindric, 4 mm. high, woolly only at base, 20–30-flowered, the hermaphrodite flowers 2 or 3; phyllaries about 3-seriate, oblong, obtuse at apex and often erose, tawny and often tinged with rose, hyaline and shining; achenes papillose, the pappus cohering in an easily fragmented ring.

A native of Australasia, locally abundant in Humboldt County, California. Type locality: Tasmania. March–Nov.

14. Gnaphalium japónicum Thunb. Japanese Cudweed. Fig. 5830.

Gnaphalium japonicum Thunb. Fl. Japonica 311. 1784.

Erect annual, stems appressed-woolly, erect, 1–4 dm. high, simple or branched from the base, branchlets few, arising from the leafy axils. Leaves thick, pannose-tomentose, green above, the margins revolute; lower stem-leaves 3–5 cm. long, narrowly spatulate to oblanceolate, narrowed to a slender petioliform base, the upper leaves sessile, becoming linear and but little reduced in length; inflorescence terminal on the stems and branchlets, the heads in a dense capitate cluster subtended by leaves, conspicuously woolly below the involucres and on leaf-bases; involucres imbricate, cylindric, about 4 mm. high, about 20-flowered, the hermaphrodite flowers 1 or 2; phyllaries glabrous, hyaline and fragile, tawny or brownish or purplish-tinged, the inner phyllaries linear, obtuse; achenes papillose, the pappus-bristles cohering in an easily fragmented ring.

An introduced weed in grassy open places in wooded areas well established in Curry County, Oregon, south in the Coast Ranges of California to Trinity and Humboldt Counties; also reported from San Joaquin County. A native of Japan widely distributed in the Orient and Australasia. July–Oct.

15. Gnaphalium peregrìnum Fernald. Wandering Cudweed. Fig. 5831.

Gnaphalium spathulatum Lam. Encycl. **2**: 758. 1786. Not Burm. f. 1768.
Gnaphalium peregrinum Fernald, Rhodora **45**: 479. *pl. 495.* 1943.

Annual 2–4 dm. high, the leafy stems simple or branching from the base. Leaves spatulate, attenuate to a petioliform base, 3.5–4.5 cm. long, about 1 cm. wide, scarcely reduced above, thin, greenish, loosely woolly, more densely so on the undersurface; inflorescence an interrupted spike of close, nearly sessile glomerules; involucres 3–4 mm. high, buried in wool, about 80–120-flowered; phyllaries fragile, hyaline, faintly tinged with brownish yellow, the outer with greenish mid-section, narrowly oblong, acute at apex; pappus-bristles deciduous in a ring.

A widespread subtropical weed of both hemispheres; occurring in southeastern United States and found sporadically in the Pacific States in cismontane southern California; also occurring locally in Mendocino County. Type locality: Pineville, Louisiana. March–May.

16. Gnaphalium purpùreum L. Purple Cudweed. Fig. 5832.

Gnaphalium purpureum L. Sp. Pl. 854. 1753.
Gnaphalium ustulatum Nutt. Trans. Amer. Phil. Soc. II. **7**: 404. 1841.
Gnaphalium pannosum Gandoger, Bull. Soc. Bot. Fr. **65**: 42. 1918. Not Schultz-Bip. 1845 or A. Gray, 1883.

Annual or biennial 15–40 cm. high, erect or decumbent, 1- to several-stemmed, normally simple, densely and closely pannose-tomentose throughout with usually somewhat silvery wool or the upper leaf-surface usually dark green and glabrescent, not sweet-scented. Leaves obovate, oblanceolate, or spatulate, gradually reduced above, usually narrowed into a petioliform base, 1.5–10 cm. long, 2–16 mm. wide, thickish, obtuse, apiculate, inconspicuously or not at all decurrent; inflorescence a spiciform, more or less leafy-bracted, scarcely interrupted panicle 1–18 cm. long, the heads campanulate, 4–5 mm. high, about 90–130-flowered; phyllaries with greenish base lightly floccose below and dark or light brown or often purplish, thin-scarious and shining tips, ovate through oblong-ovate and oblong to spatulate or linear, acute or acuminate; pistillate flowers about 88–121, hermaphrodite 4–7; corollas purplish; achenes minutely papillose, pappus united at base, deciduous in a complete ring.

Widespread native weed growing in various habitats, Transition and Sonoran Zones; occurring in the Pacific States from British Columbia south to southern California and adjacent Lower California mostly west of the Cascade Mountains and the Sierra Nevada; common on the eastern seaboard and in central United States and southward in South America; also Australia and New Zealand, where it is said to be introduced. Type locality: "Carolina, Virginia, Pensylvania."

123. ANTENNÁRIA* Gaertn. Fruct. **2**: 410. *pl. 167.* 1791.

Mainly white-woolly, dioecious perennials. Leaves mostly entire, both the basal and cauline alternate, the former more conspicuous, the latter usually reduced upward. Heads mostly several to numerous, discoid; phyllaries scarious at least toward the tips, imbricate,

---

* Text contributed by Carl William Sharsmith.

often colored. Receptacle convex or flat, naked. Staminate flowers with anthers caudate; style mostly entire; pappus of relatively few bristles, these in a single row and mostly dilated upward or sometimes merely barbellate. Pistillate flowers with corollas tubular-filiform; style bifid; pappus rather copious, the bristles capillary, mostly united at the base and falling as a ring. [Name Latin, from the resemblance of the staminate pappus to the antennae of some insects.]

A complex polymorphic genus of about 35 or 100 species, depending upon the taxonomic interpretation of the numerous apomictic forms; circumpolar at the north but mainly developed, especially in the cordillera, in North America and extending to southern South America. Type species, *Gnaphalium dioicum* L.

Heads several to many.
 Basal leaves as densely pubescent on their upper as on their lower surface, and if at all glabrate then so only in extreme age (except rarely in *A. media*).
  Plants not stoloniferous, seldom (except in *A. argentea*) mat-forming; basal leaves erectish or erect, not forming a rosette.
   Involucres pubescent from the base upward to about the middle, glabrous only distally.
    Basal leaves 2–15 cm. long, narrowed into a petiole or petioliform base; involucres campanulate, loosely pubescent.
     Stems 20–55 cm. high; basal leaves 6–15 cm. long; phyllaries distally white.
              1. *A. anaphaloides.*
     Stems 4–15(–20) cm. high; basal leaves mostly 2–6 cm. long; phyllaries distally brown or blackish green.   7. *A. lanata.*
    Basal leaves 1–3 cm. long, sessile; involucres turbinate-cylindric, invested in a closely interwoven tomentum concealing all but the very tips of the white or pink phyllaries.
               5. *A. geyeri.*
   Involucres glabrous from the very base to the apex.
    Phyllaries greenish or straw-colored, shining; bristles of staminate pappus dilated upward and with subentire or serrulate tips.
     Plant long-rhizomatous, forming loose mats; heads in a corymbiform cyme.
               2. *A. argentea.*
     Plant cespitose or short-rhizomatous, rarely mat-forming.
      Heads in a usually close corymbiform cyme; leaves mostly 3-nerved, the basal and lower 3–10 cm. long, the cauline mostly little reduced upward.
             3. *A. luzuloides.*
      Heads in a loose or close panicle; leaves 1-nerved or the midvein obscure, the basal and lower 2–4.5 cm. long, the cauline mostly well and abruptly reduced upward.
             4. *A. microcephala.*
    Phyllaries mainly dark or blackish brown; bristles of staminate pappus barbellate and scarcely dilated upward.   6. *A. stenophylla.*
  Plants stoloniferous and mat-forming; basal leaves mostly spreading or depressed, forming a rosette.
   Distal scarious portion of phyllaries blackish green or brownish.
    Tips of phyllaries mostly blackish green, acute; tips of upper cauline leaves with flattened scarious appendage.   8. *A. media.*
    Tips of phyllaries mostly brownish, obtuse; tips of upper cauline leaves with subulate subscarious acumination.   9. *A. umbrinella.*
   Distal scarious portion of phyllaries white or roseate.
    Phyllaries greenish or at most merely brownish at the base.
     Heads 5–7(–9) mm. high, the phyllaries mostly roseate.   10. *A. rosea.*
     Heads 6–9 (in the pistillate 9–13) mm. high, the phyllaries mostly white.
              12. *A. parvifolia.*
    Phyllaries with conspicuous dark or blackish spot at the base. 11. *A. corymbosa.*
 Basal leaves distinctly less pubescent on their upper than on their lower surface, or promptly glabrate above.
  Cespitose, mostly low perennial 2–15(–27) cm. high; basal and stolon leaves 1–3 cm. long; uppermost cauline leaves glabrate and glandular-puberulent.   13. *A. marginata.*
  Herbaceous, mostly tall perennials up to 40 cm. high or higher; basal and stolon leaves 2.5–10 cm. long.
   Heads in a usually close corymbiform cyme; peduncles of heads persistently tomentose, not glandular.
              14. *A. howellii.*
   Heads mostly in open racemes or panicles; peduncles of heads and rachis of inflorescence glabrate and densely stipitate-glandular.   15. *A. racemosa.*
Heads solitary, terminal, or at most 2.
 Plants distinctly perennial, densely cespitose and mat-forming.
  Basal leaves obtuse or emarginate, quickly glabrous and green above, persistently tomentose beneath.
              16. *A. suffrutescens.*
  Basal leaves acute, uniformly and persistently tomentose on both sides. 17. *A. dimorpha.*
 Plants seemingly biennial, propagating by slender, elongate, naked stolons, which terminate in a small leafy tuft.
              18. *A. flagellaris.*

1. **Antennaria anaphaloìdes** Rydb. Tall Everlasting. Fig. 5833.

Antennaria anaphaloides Rydb. Mem. N.Y. Bot. Gard. 1: 409. 1900.

Perennial from a short-branched caudex or lignescent rhizomatous base, thinly grayish-tomentose (the tomentum somewhat floccose with age), with erect tufts of basal leaves and simple, erect, leafy stems 20–55 cm. high. Basal leaves lanceolate or oblanceolate to elliptic-oblong, 6–15 cm. long including the often elongate petiole, 6–20 mm. wide, obscurely to strongly 3–5-nerved; cauline leaves becoming narrower and strongly reduced upward, leaving the upper part of stem nearly naked, the often darkened callous tips of the upper ones bearing a filiform to linear or lanceolate, acute or obtuse, scarious appendage 1–4 mm. long, this brown throughout or white with brown base; heads several to many, tomentose toward base, glomerate or in a rounded cymose panicle broader than long; involucres campanulate, 5–7 mm. high, the phyllaries with pale to dark brown submedial spot and conspicuous, ivory-white, obovate or oval, obtuse or rounded, scarious tips, those of the staminate heads slightly broader than those of the pistillate; achenes glabrous; bristles of staminate pappus with narrowly to more broadly spatulate-dilated tips.

Prairies, hillsides, meadows, and open woods, Upper Sonoran and Arid Transition Zones; Alberta and British Columbia south through eastern Washington to southern Oregon (Steen Mountains) and northeastern Nevada, east to Colorado, Wyoming, and Montana. Type locality: Spanish Basin, Montana. May–Aug. The Washington records of *A. pulcherrima* (Hook.) Greene belong to this species.

2. **Antennaria argéntea** Benth. Silvery Everlasting. Fig. 5834.

Antennaria argentea Benth. Pl. Hartw. 319. 1849.
Antennaria luzuloides var. *argentea* A. Gray, Pacif. R. Rep. 4 (ed. 2, pt. 5)⁴: 110. 1857.

Perennial from a rather strongly rhizomatous base, closely or somewhat loosely (in age sometimes floccosely) and subsericeously gray-tomentose throughout except on involucre, with short, erectish or ascending, sterile, leafy branches forming an often loose basal mat, and erect simple stems 15–60 cm. high. Basal and lower leaves obovate to spatulate or oblanceolate, 2.5–6.5 cm. long including petiole, 0.5–1.7 cm. wide, acute or mostly obtuse, apiculate, 3-nerved; cauline leaves similar, the upper much reduced, mostly linear, with attenuate scarious tips, the upper part of stem nearly naked; heads numerous in a subcapitate or mostly short, close, rounded, corymbiform or paniculate cyme; involucres essentially glabrous to the very base; pistillate involucres subcylindric to narrowly campanulate, 5–6 mm. high, subequaling or exceeded by the pappus, the shining phyllaries with pale greenish base and hyaline or thinly scarious and whitish (very rarely rosy) obtuse tips, the inner often acutish; staminate involucres broader, the phyllaries with more conspicuous and somewhat dilated, rounded, white, scarious tips; pistillate achenes finely puberulous, the pappus-bristles united at very base and deciduous in a ring; bristles of staminate pappus with clavately dilated tips.

Dry open woods and hillsides, Arid Transition Zone; Oregon (Wheeler and Crook Counties) southward to Lake and Fresno Counties, California, and western Nevada. Type locality: "In montibus Sacramento." May–July.

3. **Antennaria luzuloìdes** Torr. & Gray. Silvery-brown Everlasting. Fig. 5835.

Antennaria luzuloides Torr. & Gray, Fl. N. Amer. 2: 430. 1843.
Antennaria oblanceolata Rydb. Mem. N.Y. Bot. Gard. 1: 409. 1900.
Antennaria luzuloides var. *oblanceolata* M. E. Peck, Madroño 6: 136. 1941.

Cespitose perennial from a branching, rather ligneous caudex, closely though often rather thinly subsericeous-tomentose (sometimes floccose with age) and whitish or mostly grayish or greenish, with erect basal tufts of leaves, and slender, simple, erect or ascending, leafy stems 10–50 cm. high. Basal and lower leaves varying from narrowly linear to linear-oblanceolate or oblanceolate, often falcate, 3–10 cm. long including the petioliform base, 1–5 (sometimes 10) mm. wide, 3-nerved except when narrowest, usually acuminate; cauline leaves often only gradually reduced upward, the upper ones with twisted, attenuate, scarious tips; heads several to usually numerous in a close, sometimes subcapitate or mostly corymbiform cyme or rounded cymose panicle; involucres essentially or quite glabrous except at very base; phyllaries pale greenish or pale brownish, with age at least the outermost often shrivelling into concentric wrinkles, obtuse or the inner occasionally acutish; staminate involucres shorter and broader, the phyllaries with more conspicuous, somewhat dilated, rounded, scarious but rather firm, ivory-white or very rarely pink tips; pistillate achenes finely puberulous, their pappus-bristles united only at extreme base, mostly deciduous separately; bristles of staminate pappus with narrowly or more broadly spatulate-dilated, serrulate or subentire tips.

Prairies and hillsides, Arid Transition Zone; British Columbia south through eastern Washington and eastern Oregon to Lassen (and "Plumas") County, California, east to Montana, Wyoming, and Colorado. Type locality: "Oregon or Rocky Mountains." May–July. At its southern limits in our range occasionally intergrading with *A. microcephala* A. Gray.

4. **Antennaria microcéphala** A. Gray. Small-headed Everlasting. Fig. 5836.

Antennaria microcephala A. Gray, Proc. Amer. Acad. 10: 74. 1874.
Antennaria argentea subsp. *aberrans* E. Nels. Bot. Gaz. 34: 124. 1902.
Antennaria pyramidata Greene, Leaflets Bot. Obs. 2: 145. 1911.

Cespitose perennial from a branching caudex, closely and rather densely subsericeous- and whitish-tomentose except on involucre with short, often tufted, erectish or ascending, leafy branches at base, and erectish stems 10–22 cm. high, unbranched below the often elongate inflorescence. Basal and lower leaves spatulate or oblanceolate, 2–4.5 cm. long including the petioliform base, 2–7 mm. wide, acute or obtuse, 1-nerved or the midvein obscured by the dense tomentum and apiculately callous-tipped; cauline leaves narrower, usually rather abruptly reduced upward, the uppermost occasionally with attenuate scarious tips; heads rather numerous in a usually loose panicle and mostly aggregated at the tips of its branches or occasionally more densely disposed in a close panicle; pistillate involucres turbinate-cylindric, 4–5.5 mm. high, glabrous except at very base, the shining phyllaries pale green or pale brownish, hyaline and with age at least the outermost shrivelling into concentric wrinkles, obtuse or acutish and mostly with inconspicuous whitish, or the inner sometimes with pinkish, tips; staminate involucres shorter and broader, the phyllaries broader and blunter; pistillate achenes finely puberulous, their pappus-bristles united only at extreme base, mostly deciduous separately; bristles of staminate pappus clavately dilated above.

Dry meadows and slopes, Arid Transition Zone; Klamath and Lake Counties, Oregon, and from Lassen to Nevada Counties, California, and adjacent Nevada, and in Trinity and Glenn Counties, California. Type locality: "Washoe Valley, Nevada, [and] Sierra County, California." June–July.

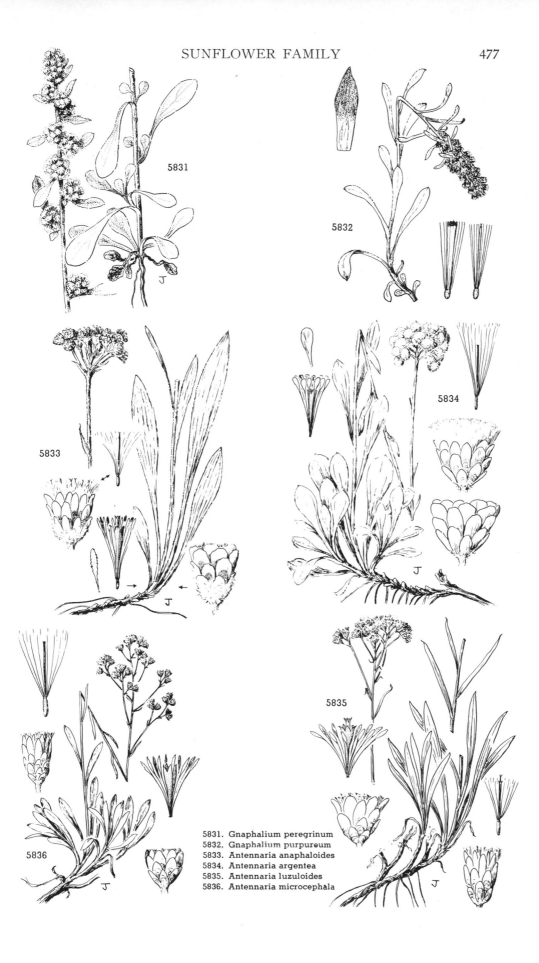

5831. Gnaphalium peregrinum
5832. Gnaphalium purpureum
5833. Antennaria anaphaloides
5834. Antennaria argentea
5835. Antennaria luzuloides
5836. Antennaria microcephala

5. **Antennaria geỳeri** A. Gray. Geyer's Everlasting. Fig. 5837.

Gnaphalium alienum Hook. Lond. Journ. Bot. **6**: 251. 1847. Not Hook. & Arn. 1841.
Antennaria geyeri A. Gray, Mem. Amer. Acad. II. **4**: 107. 1849.

Perennial and mat-forming from a lignescent branched caudex or rhizomatous base, densely and rather closely gray-tomentose, with numerous ascending or erect, often crowded, leafy stems suffrutescent below and 3–15 cm. high. Leaves narrowly oblanceolate to obovate-spatulate, 1–3 cm. long, 1.5–8 mm. wide, obtuse to acute, sessile, those of the stem numerous and rather little reduced upward; heads several in a close corymbose, sometimes bracteate cyme; involucres densely tomentose with a close interwoven tomentum except for the scarious tips of the phyllaries; phyllaries ovate to oblong or linear, obtuse to acute or appearing acute when dry, the tips of the outer ones pale brownish and of the inner elongated and pale brownish or whitish to often roseate or sometimes ivory-white; pistillate involucres subcylindric to turbinate-cylindric, 7–9 mm. high; staminate involucres shorter and broader, 4–8 mm. high, the phyllaries usually blunter; pistillate achenes loosely villous or merely glandular-puberulous when young, glandular-puberulous when mature; bristles of staminate pappus only slightly dilated upward and there barbellate or denticulate.

Dry hills and open woods, Arid Transition Zone; eastern Washington southward through Oregon to Lake, Colusa, and Eldorado Counties, California, and western Nevada. Type locality: "Arid sandy woods near Tshimakaine, Spokan country," Washington. June–Sept.

6. **Antennaria stenophỳlla** A. Gray. Narrow-leaved Everlasting. Fig. 5838.

Antennaria alpina var. *? stenophylla* A. Gray in Torr. Bot. Wilkes Exp. **17**: 366. 1874.
Antennaria stenophylla A. Gray, Proc. Amer. Acad. **17**: 213. 1882.
Antennaria leucophaea Piper, Bull. Torrey Club **29**: 221. 1902.

Densely cespitose perennial from the crown of a short-branched caudex, closely and subsericeously gray-tomentose throughout except on involucre, with tufts of erect basal leaves, and erect or rarely decumbent stems 3–15 cm. high. Leaves very narrowly linear-lanceolate or -oblanceolate to mostly linear, callous-apiculate at tip, often falcate, the basal 1.5–8 cm. long, 0.5–2.5 mm. wide, the cauline rather numerous and gradually reduced upward; heads 3–25, usually congested into a single, terminal, subcapitate glomerule about 7–15 mm. wide or occasionally in a wider corymbose cyme of glomerules; involucres glabrous except at extreme base, their phyllaries largely scarious throughout and pale or mainly dark or blackish brown with brownish to whitish or white, hyaline tips; pistillate involucres subcylindric to turbinate- or oblong-campanulate, 5–6.5 mm. high, the phyllaries ovate to oblong, obtuse to mostly acute; staminate involucres somewhat shorter and broader, their phyllaries obtuse; achenes papillate; bristles of staminate pappus barbellate and only slightly if at all dilated above.

Dry meadows, hill slopes and ridges, Arid Transition Zone; central and eastern Washington southward to northern Nevada, east to Idaho. Type locality: "Spipen [Naches] River," Washington. May.

7. **Antennaria lanàta** (Hook.) Greene. Woolly Everlasting. Fig. 5839.

Antennaria carpatica β *lanata* Hook. Fl. Bor. Amer. **1**: 329. 1834.
Antennaria lanata Greene, Pittonia **3**: 288. 1898.

Perennial from a thickish, short-branched caudex or lignescent rhizomatous base, densely gray-lanate-tomentose, with erect tufts of basal leaves and stoutish, erect, leafy stems 4–15(–20) cm. high. Basal leaves linear-oblanceolate to oblanceolate or obovate, 2–6(–10) cm. long including the petioliform base, 2–14 mm. wide, obscurely to distinctly 3-nerved; cauline leaves lanceolate to linear-lanceolate, becoming sessile and moderately reduced above, the upper ones attenuate into an often darkened tip bearing a linear to spatulate, acute to obtuse, brownish, scarious, often flaglike appendage 1–5 mm. long; heads 6–12, in a close umbelliform cyme or terminal glomerule, the involucres densely tomentose on the lower half; pistillate involucres campanulate, 5–8 mm. high, the phyllaries mostly greenish at base, the outer ones with their scarious portion dark brown or mostly blackish green either throughout or paler and brownish toward the acute or obtuse tip, the inner usually with whitish tips; staminate involucres shorter and broader, the inner phyllaries with more conspicuous, obtuse or rounded, whitish tips; achenes glabrous; bristles of staminate pappus with narrowly dilated, serrulate tips.

Open slopes, Boreal Zone; Alberta and British Columbia south to Washington and northeastern Oregon, east to Wyoming and Montana. Type locality: "Summits of the most elevated among the Rocky Mountains, lat. 52°." July–Aug.

8. **Antennaria mèdia** Greene. Alpine Everlasting. Fig. 5840.

Antennaria media Greene, Pittonia **3**: 286. 1898.
Antennaria media subsp. *ciliata* E. Nels. Proc. U.S. Nat. Mus. **23**: 700. 1901.
Antennaria tomentella E. Nels. op. cit. 701.
Antennaria pulchella Greene, Leaflets Bot. Obs. **2**: 149. 1911.
Antennaria scabra Greene, op. cit. 150.
Antennaria densa Greene, op. cit. 151.
Antennaria candida Greene, loc. cit.
Antennaria alpina var. *media* Jepson, Man. Fl. Pl. Calif. 1070. 1925.
Antennaria alpina var. *scabra* Jepson, loc. cit.
Antennaria gormanii St. John in St. John & Hardin, Mazama **11**: 92. 1929.

Cespitose low perennial, gray-tomentose, with rosulate basal leaves, decumbent leafy stolons up to 5 cm. long, forming mats or sometimes compact and pulvinate, with erect leafy stems

1–10(–14) cm. high. Basal and stolen leaves oblanceolate, spatulate-obovate or occasionally spatulate, (3–)6–25 mm. long, 1.5–4.5(–7) mm. wide, obtuse or acutish, usually apiculate, 1-nerved, densely or somewhat loosely gray-tomentose persistently on both sides, rarely becoming partly or wholly glabrate by the second or even in the first season, then green and glandular-puberulent; stems and cauline leaves similar in pubescence, the cauline leaves small, mostly linear, the upper ones, except in excessively dwarf plants, tipped with a dark brown, scarious, flattened, linear to spatulate, obtuse appendage 1–2 mm. long; heads several in a close corymbiform or sub-capitate cyme; pistillate involucres turbinate-campanulate, loosely woolly at base, 3–6 mm. high, the phyllaries with greenish or brownish base, often brown or blackish medial spot, and oblong or lanceolate to linear, obtuse or acuminate, usually scarious and cellular-reticulate, blackish- or brownish-green tips, or tips occasionally brown or white in the innermost or very rarely roseate or lurid blackish red; staminate involucres shorter, their phyllaries broader, blunter, with tips more frequently brown or white; achenes mostly papillose, occasionally glabrous; bristles of staminate pappus with clavately dilated tips.

Dry or sometimes moist, grassy or rocky mountain slopes, mostly Boreal Zone; Alberta and British Columbia south through Washington and Oregon and through the Sierra Nevada to the high summits of the San Bernardino Mountains, California, east to Colorado, Wyoming, and Montana. Type locality: "Mountains above Coldstream, Placer County, California," as designated by Elias Nelson (Proc. U.S. Nat. Mus. **23**: 700. 1901). (June–)July–Sept.

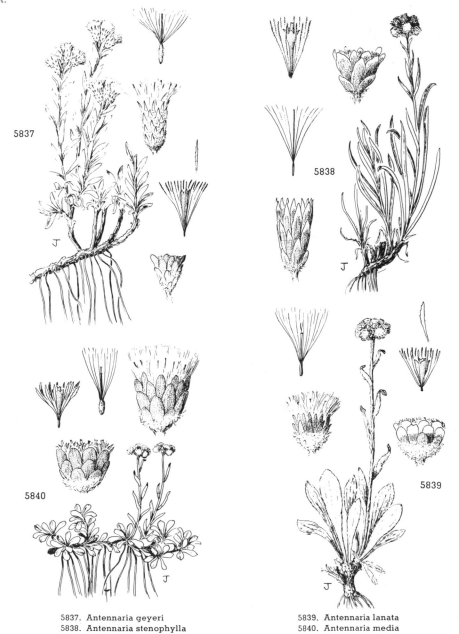

5837. Antennaria geyeri
5838. Antennaria stenophylla

5839. Antennaria lanata
5840. Antennaria media

9. **Antennaria umbrinélla** Rydb. Brown or Isabella Everlasting. Fig. 5841.

Antennaria umbrinella Rydb. Bull. Torrey Club **24**: 302. 1897.
Antennaria confinis Greene, Pittonia **4**: 40. 1899.
Antennaria flavescens Rydb. Mem. N.Y. Bot. Gard. **1**: 411. 1900.
Antennaria dioica var. *kernensis* Jepson, Man. Fl. Pl. Calif. 1071. 1925.

Loosely or densely cespitose perennial, gray-tomentose essentially throughout (the basal and stolon leaves sometimes yellowish), with rosulate basal leaves and erect leafy stems (2.5)6–15(–20) cm. high. Stolons 3.5–5 cm. long, decumbent to erectish, arising on relatively elongate branches of a rhizomatous caudex, or stolons more abbreviated and densely crowning a short-branched caudex; basal and stolon leaves oblanceolate to spatulate-obovate, 5–25 mm. long, 1.5–5 mm. wide, obtuse or acute, 1-nerved; cauline leaves narrowly lanceolate to oblanceolate, usually acute or acuminate, 1–3.5 mm. wide, tipped with a short, colored, callous point or the upper with a subulate (not flattened) subscarious tip; heads in a rounded capituliform glomerule or in a close rounded cyme; pistillate involucres turbinate-campanulate, 4–6 mm. high, woolly at base, the phyllaries with greenish or greenish brown base, brownish or darker medial spot, and linear-oblong to obovate, mostly obtuse (sometimes acute), brownish or brownish white, relatively firm tips or tips sometimes partly or wholly blackish- or smoky-green, scarious and coarsely cellular, or sometimes white or slightly roseate, firm and finely striate; staminate involucres shorter, the tips broader, blunter, firm, brownish or white; achenes mostly glabrous; bristles of staminate pappus clavately dilated above.

Dry prairies, meadows, hill slopes, and ridges, Upper Sonoran Zone to Boreal Zone; Alberta and British Columbia south through Washington and Oregon and through the Sierra Nevada to Tulare and Inyo Counties, California, east to Arizona, Colorado, Wyoming, and Montana. Type locality: Long Baldy, Little Belt Mountains, Montana. May–Sept. When occurring in proximity of *A. media* Greene or of *A. rosea* Greene, sometimes in character very closely approaching the one, sometimes the other.

10. **Antennaria ròsea** Greene. Rosy Everlasting. Fig. 5842.

Antennaria microphylla Rydb. Bull. Torrey Club **24**: 303. 1897. Not Gandoger. 1887.
Antennaria rosea Greene, Pittonia **3**: 281. 1898.
Antennaria imbricata E. Nels. Bot. Gaz. **27**: 211. 1899.
Antennaria angustifolia Rydb. Bull. Torrey Club **26**: 546. 1899.
Antennaria concinna E. Nels. Proc. U.S. Nat. Mus. **23**: 705. 1901.
Antennaria rosea var. *angustifolia* E. Nels. op. cit. 706.
Antennaria rosea var. *imbricata* E. Nels. op. cit. 707.
Antennaria hendersoni Piper, Bull. Torrey Club **29**: 221. 1902.

Cespitose perennial, closely or in age floccosely tomentose essentially throughout, with rosulate basal leaves, slender, leafy, decumbent to erectish, often lignescent stolons up to 6 cm. long, and erect leafy stems 7–35 cm. high. Basal and stolon leaves narrowly spatulate or oblanceolate, 1–3.5 cm. long, 2–6 mm. wide, acutish or obtuse, apiculate, 1-nerved, sometimes smaller and more cuneate-obovate and blunter or sometimes more obovate-spatulate; cauline leaves rather numerous, narrowly oblanceolate or spatulate to linear, callous-apiculate, mostly acute or acuminate, at least the upper usually with falcate, attenuate, colored tip; heads several in a small glomerule or usually close cyme; pistillate involucres campanulate, 4–7 mm. high, woolly at base, the phyllaries with greenish or sometimes brown base and firm, elliptic to oval or obovate, obtuse, erose, rosy or sometimes whitish tips, or the inner sometimes linear and acute, the outer sometimes with deep brown subterminal spot; staminate involucres shorter and broader, their phyllaries broader and with blunter, "broadly elliptic, pale pink".or white tips; achenes glabrous; bristles of staminate pappus clavately dilated above.

Mostly dry meadows, open woods, and hills, Transition Zone to Boreal Zone; "Alaska," "Yukon," and "Mackenzie"; Alberta and British Columbia south through Washington and Oregon, reaching Trinity and Tehama Counties, California, through the Coast Ranges and the San Bernardino and San Jacinto Mountains through the Sierra Nevada, east to Colorado, Wyoming, and Montana. Type locality: "Utah or Nevada." June–Aug.
Material which belongs to the form known as *A. microphylla* Rydb. is the source for the above description of staminate involucres and pappus except that quoted from Rydberg (Fl. Rocky Mts. 918. 1917) for *A. rosea* Greene. Typically *A. microphylla* differs from *A. rosea* in possessing spatulate, closely whitish or silvery-white, sericeous-tomentose, basal and stolon leaves mostly 1–2 cm. long and phyllaries white-tipped (with age often fading yellowish). Furthermore, it is abundantly sexual whereas in typical *A. rosea* staminate plants appear to be unknown. On these grounds *A. microphylla* Rydb., which occurs from eastern Washington east to North Dakota and south to Colorado (type locality Manhattan, Montana), probably should be maintained as distinct. Until its relationship to still other forms described as species is effectively clarified, a provisional substitute for the preoccupied name *A. microphylla* Rydb. is *A. solstitialis* Lunell ex A. Nels. Proc. Biol. Soc. Wash. **20**: 39. 1907 (*A. microphylla* var. *solstitialis* Lunell, Amer. Midl. Nat. **5**: 61. 1917).
On examination, collections of staminate plants labelled as "*A. rosea* Greene" seem invariably to belong either to *A. microphylla* Rydb. or to *A. umbrinella* Rydb. or sometimes even to *A. corymbosa* E. Nels., or to intergrade between one of these and *A. rosea*, those between *A. umbrinella* and *A. rosea* being the most abundant, at least from the Pacific States.

11. **Antennaria corymbòsa** E. Nels. Meadow Everlasting. Fig. 5843.

Antennaria corymbosa E. Nels. Bot. Gaz. **27**: 212. (March) 1899.
Antennaria nardina Greene, Pittonia **4**: 82 (December) 1899.
Antennaria hygrophila Greene, Leaflets Bot. Obs. **2**: 144. 1911.
Antennaria dioica var. *corymbosa* Jepson, Man. Fl. Pl. Calif. 1071. 1925.

Herbaceous, somewhat cespitose, loosely mat-forming perennial from a slender branching caudex, with rosulate basal leaves and slender, leafy, decumbent or ascending stolons up to 7 cm. long, and erect leafy stems 7–25 cm. high. Basal and stolon leaves ascending, thinnish, narrowly and distinctly oblanceolate or sometimes a few narrowly spatulate, 1.5–4 cm. long including the indistinct petiole, 2–5 mm. wide, acute and apiculate, rather thinly subsericeous-tomentose and mostly greenish, 1-nerved or weakly 3-nerved; cauline leaves gradually more narrowed and reduced upward, the upper acuminate and bearing on their often darkened tips a linear or filiform, acuminate, scarious appendage about 1.5 mm. long; stem thinly tomentose, often floccose; heads several, glomerate or in a corymbiform cyme about 2.5 cm. wide; pistillate involucres woolly from base upward to about the middle, 3–5 mm. high, the phyllaries with mostly greenish base, a conspicuous, dark brown or blackish medial spot, and firm, oblong or obovate, rounded, ivory-white tips, the outer sometimes with brownish-tinged tips, the innermost sometimes linear and acutish; staminate involucres similar, the phyllaries with broader, mostly obovate tips; achenes sparsely puberulous; bristles of staminate pappus clavately dilated above.

Moist or wet, mostly subalpine and alpine meadows and meadowy sites, Transition Zone to Boreal Zone; Colorado and Utah north to Montana and Idaho; Oregon (Crook County); and in the Sierra Nevada from ("Washoe") and Ormsby Counties, Nevada, south to Inyo and Fresno Counties, California. Type locality: Battle Lake, Sierra Madre Mountains, Carbon County, Wyoming. June–Aug.

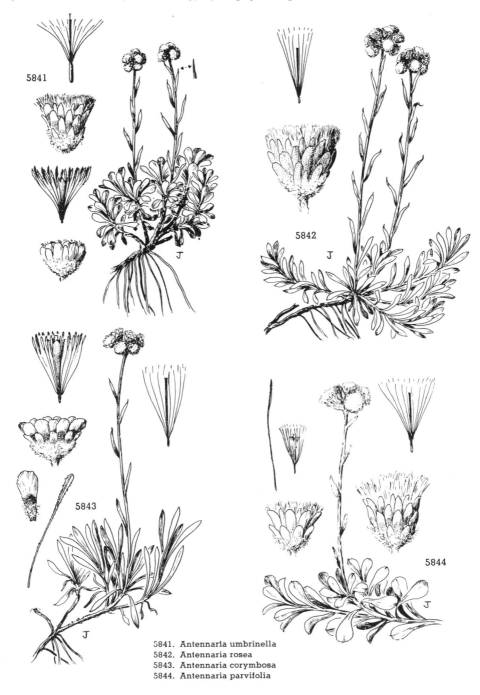

5841. Antennaria umbrinella
5842. Antennaria rosea
5843. Antennaria corymbosa
5844. Antennaria parvifolia

12. **Antennaria parvifòlia** Nutt. Nuttall's Everlasting. Fig. 5844.

Antennaria parvifolia Nutt. Trans. Amer. Phil. Soc. II. **7**: 406. 1841.
Antennaria aprica Greene, Pittonia **3**: 282. 1898.
Antennaria rhodantha Suksd. Allg. Bot. Zeit. **12**: 6. 1906.

Cespitose perennial, mostly and closely gray-tomentose, with rosulate basal leaves, decumbent or ascending leafy stolons up to 7 cm. long, and erect leafy stems 2–15 (rarely –18) cm. high, the staminate stems usually lower than the pistillate. Basal and stolon leaves spatulate-obovate or spatulate, 1–2.5(–3.5) cm. long including the petioliform base, 2.5–7(–13) mm. wide, obtuse or rounded, callous-apiculate, 1-nerved or the larger 3-nerved, persistently tomentose but often less densely so above or the older ones floccose or sometimes even glabrate; cauline leaves oblanceolate to linear, obtuse or acutish, with usually a callous apiculation, the uppermost sometimes with scarious appendages up to 1.5 mm. long; branches of inflorescence tomentose, nonglandular; heads 2–9, glomerate or closely cymose; pistillate involucres campanulate, 7–11 mm. high, woolly at base, the phyllaries with greenish or brown base, sometimes with a purplish brown submedial spot, and scarious or firm, mostly oval or obovate, obtuse or rounded (or the innermost occasionally linear, acute or acuminate), white or ivory-white tips, or the tips occasionally pinkish; staminate involucres shorter than but about as broad as the pistillate, their phyllaries similar but fewer-seriate and with scarious, firm or somewhat chartaceous, broad, rounded tips; achenes glandular-papillose or glabrous; bristles of staminate pappus merely and sparsely scaberulous upward, not barbellate, and with narrowly if at all dilated tips.

Dry plains, hill slopes, and open woods, Arid Transition Zone to Boreal Zone; Manitoba to Oklahoma west to New Mexico, Arizona, Utah, Washington (Skamania, Spokane, and Stevens Counties), and British Columbia. Type locality: "On the Black Hills and plains of the upper part of the Platte." May–July(–Aug.).

13. **Antennaria marginàta** Greene. White-margined Everlasting. Fig. 5845.

Antennaria marginata Greene, Pittonia **3**: 290. 1898.
Antennaria dioica var. *marginata* Jepson, Man. Fl. Pl. Calif. 1071. 1925.

Cespitose perennial, gray-tomentose throughout except the usually glabrate upper surface of leaves, with rosulate basal leaves, decumbent to ascending, usually lignescent, comparatively long-persistent, leafy stolons up to 11 cm. long, and erect, sparsely leafy stems 2–27 cm. high, these mostly distinctly glandular above and in the inflorescence, the staminate stems usually much lower than the pistillate. Basal and stolon leaves spatulate or spatulate-obovate, 1–3 cm. long, 3–7 mm. wide, obtuse, callous-apiculate, 1-nerved, above quickly glabrate and usually glandular-puberulent, the dense tomentum of the lower leaf-surface showing as a narrow white margin bordering the glabrate and bright green upper surface, or rarely persistently tomentose above; cauline leaves mostly linear and acute or acuminate, glabrate, green, and glandular-puberulent above, particularly the uppermost ones; heads 3–6, glomerate or in age closely cymose; pistillate involucres campanulate, woolly at base, 6–8 mm. high, the phyllaries with pale green base, often brown or purplish submedial spot, and oval or obovate to lanceolate, obtuse to acuminate, scarious, white or very rarely roseate tips; staminate involucre shorter than but about as broad as the pistillate, their phyllaries similar but fewer seriate and with firm or somewhat chartaceous, mostly obovate to rhombic, obtuse or mostly acute tips; achenes glandular; bristles of staminate pappus barbellate upward and with clavately dilated, crenate tips.

Slopes and rocky ridges, Arid Transition Zone; California in the San Bernardino Mountains east to New Mexico, north to Colorado and Utah. Type locality: New Mexico, probably about Santa Fe. May–Aug. Seemingly intergrading with and perhaps not more than a subspecies of *A. parvifolia* Nutt.

14. **Antennaria howéllii** Greene. Howell's Everlasting. Fig. 5846.

Antennaria howellii Greene, Pittonia **3**: 174. 1897.
Antennaria concolor Piper, Contr. U.S. Nat. Herb. **11**: 604. 1906.
Antennaria neglecta var. *howellii* Cronquist, Leaflets West. Bot. **6**: 43. 1950.

Herbaceous perennial, with rosulate basal leaves, prostrate or decumbent, short or elongate, leafy stolons up to 20 cm. long, and erect stems 15–40 cm. high. Basal and stolon leaves oblanceolate, obovate, cuneate-obovate, or spatulate, 2.5–5.5 cm. long including the petioliform base, (6–)8–20 mm. wide, acute or mostly obtuse, apiculate, quickly green and glabrous above or sometimes remaining thinly tomentose, persistently gray- or whitish-tomentose beneath, the larger 3-nerved; cauline leaves similar in pubescence, variable, broad, or narrow and reduced, lanceolate or linear, the upper tipped with a usually dark and nearly filiform appendage 1–2 mm. long; stem and peduncles in the inflorescence tomentose; heads glomerulate or usually in a close rounded cyme, or the lower on sometimes elongate peduncles; pistillate involucres turbinate-campanulate, 7–8 mm. high, loosely arachnoid-tomentose especially below, the phyllaries mostly lanceolate or linear, with pale greenish or brownish base and pale brownish or dull whitish, scarious tips, sometimes a purplish medial spot, the outer obtuse or acute, the inner acuminate or attenuate; achenes glabrous or papillate.

Open woods, mainly Transition Zone; Alberta and British Columbia south through Washington and central Oregon to Siskiyou and Trinity Counties, California, east to Idaho, Wyoming, and Montana. Type locality: Mt. St. Helens, Skamania County, Washington. May–Aug. Staminate plants appear to be unknown.

15. **Antennaria racemòsa** Hook. Racemose Everlasting. Fig. 5847.

Antennaria racemosa Hook. Fl. Bor. Amer. **1**: 330. 1834.
Antennaria pedicellata Greene, Pittonia **3**: 175. 1897.
Antennaria piperi Rydb. Bull. Torrey Club **28**: 21. 1901.

Herbaceous stoloniferous perennial with rosulate basal leaves, the stolons procumbent, leafy mainly toward the tip, the leafy stems erect, 10–60 cm. high. Basal and stolon leaves elliptic to elliptic-obovate or ovate or oval, 2.5–10 cm. long including the petiole, 1–5.5 cm. wide, usually obtuse and apiculate, thinnish, 1–3-nerved, glabrous or nearly so and green above, thinly or occasionally somewhat densely gray-tomentose persistently beneath; cauline leaves similar in pubes– cence, variable, broad or narrow, usually lanceolate, with acuminate callous tips; stem at first thinly arachnoid, later stipitate-glandular particularly and often densely so above and in the inflorescence; heads in open racemes or loose, paniculate or subcorymbiform cymes, the lower on peduncles often 2–5 cm. long, or heads occasionally more crowded into close cymes; pistillate involucres cylindric-turbinate or subcampanulate, essentially glabrous or thinly arachnoid, 6–8 mm. high, the phyllaries rather narrow, with pale greenish or pale brownish body and inconspicuous, obtuse to acuminate, whitish, hyaline tips; staminate involucres shorter and broader, the phyllaries with obtuse, broader and somewhat more conspicuous, pale brownish or whitish tips; achenes glabrous; bristles of staminate pappus slightly or sometimes scarcely at all dilated above.

Open woods and mountain slopes, Transition Zone to Canadian Zone; Alberta and British Columbia south through Washington and Oregon to Siskiyou and Trinity Counties, California; east to Idaho, Wyoming, and Montana. Type locality: "Alpine woods of the Rocky Mountains." May–Aug.

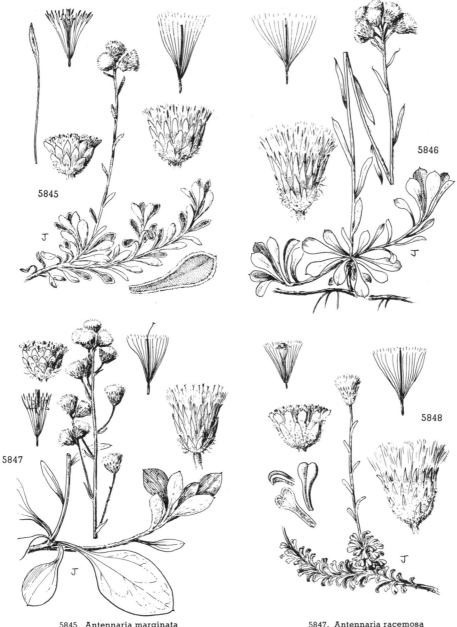

5845. Antennaria marginata
5846. Antennaria howellii

5847. Antennaria racemosa
5848. Antennaria suffrutescens

16. **Antennaria suffrutéscens** Greene. Evergreen or Siskiyou Everlasting. Fig. 5848.

Antennaria suffrutescens Greene, Pittonia **3**: 277. 1898.

Depressed, surculose-proliferous, densely mat-forming perennial from a branched lignescent caudex or rhizomatous base, the prostrate stems and decumbent or ascending stolons suffrutescent, densely leafy, the flowering stems herbaceous, erect, sparsely leafy, 1–2-headed, 3–11 cm. high. Basal and stolon leaves narrowed downward into their sessile base by the recurved margins of the blade, sulcate by the impressed and concealed midvein, spatulate or cuneate-spatulate, 4–12 mm. long, 2–4.5 mm. wide, obtuse or rounded but appearing emarginate or even bifurcate by reflexing of the tip, coriaceous, long-persistent, quickly green and glabrous above except for a minute glandular-puberulence which mostly disappears with age, densely and persistently white-tomentose beneath; stem slender, thinly tomentose, beneath puberulous with often purplish, sometimes glandular hairs; cauline leaves narrower, mostly linear, otherwise similar to the basal; pistillate head solitary, the involucre campanulate, 9–10 mm. high, thinly arachnoid-tomentose except above, the phyllaries with greenish body, brownish or mostly purplish supramedial spot, and obtuse to acuminate, scarious, whitish tips; staminate head solitary (or if heads 2 then glomerate), the involucre shorter and broader, the inner phyllaries with more conspicuous, dilated, obtuse, erose, white tips, the innermost linear; achenes sparsely puberulous or glabrous; bristles of staminate pappus with narrowly if at all dilated tips.

Dry ridges of the Coast Ranges, Arid Transition Zone; Josephine and Curry Counties, Oregon, south to Del Norte and Humboldt Counties, California. Type locality: near Waldo, Josephine County, Oregon. June–July.

17. **Antennaria dimórpha** (Nutt.) Torr. & Gray. Low Everlasting. Fig. 5849.

Gnaphalium dimorphum Nutt. Trans. Amer. Phil. Soc. II. **7**: 405. 1841.
Antennaria dimorpha Torr. & Gray, Fl. N. Amer. **2**: 431. 1843.
Antennaria latisquama Piper, Bull. Torrey Club **28**: 41. 1901. Not Greene, 1905.

Densely cespitose, low perennial from a compactly multicipital caudex, forming small mats, gray- or subsericeous-tomentose throughout except on involucre, with erect basal leaves and erect, few-leaved, 1-headed stems 1–4(–10) cm. high or heads subsessile in the basal leafy tuft. Basal leaves linear-oblanceolate to oblanceolate or spatulate-obovate, 8–25(–40) mm. long including the petioliform base, 1.5–5 mm. wide, acute or obtuse; cauline leaves similar, a few often reduced ones at base of involucre; pistillate involucre cylindric-turbinate, 10–15(–18) mm. high, thinly woolly at base, the outer phyllaries ovate or ovate-lanceolate, with brownish center and acute to acuminate, hyaline tips, the inner linear-lanceolate, with pale greenish brown center and attenuate hyaline tips; staminate heads broader, their involucres 5–9 mm. high, the phyllaries ovate to oblong, with dark brown or blackish body or medial spot and blackish green or brownish or whitish, obtuse to acute or subacuminate, scarious or hyaline tips; pistillate achenes with persistent crown of pappus and with body finely puberulous, the hairs bifurcate when moistened; bristles of staminate pappus obscurely if at all thickened upward, barbellate and entire or some of them forking above.

Dry, open hillsides and stony slopes, usually Arid Transition Zone; British Columbia and Alberta south through eastern Washington and eastern Oregon to California (Siskiyou and Glenn Counties; Modoc County south through the Sierra Nevada to Alpine County but recurring also in Tulare County; east of the Sierra Nevada in Mono and Inyo Counties; and in the Transverse Ranges of southern California in Ventura, Los Angeles, and San Bernardino Counties), east to Colorado, Nebraska, Wyoming, and Montana. Type locality: "On the Black Hills of the Platte." (March–)April–July.

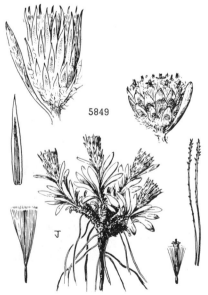

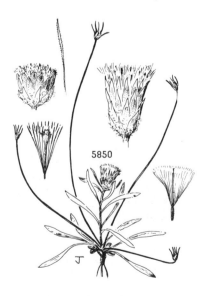

5849. Antennaria dimorpha

5850. Antennaria flagellaris

18. **Antennaria flagellàris** (A. Gray) A. Gray. Flagellate Everlasting. Fig. 5850.

Antennaria dimorpha var. *flagellaris* A. Gray in Torr. Bot. Wilkes Exp. **17**: 366. 1874.
Antennaria flagellaris A. Gray, Proc. Amer. Acad. **17**: 212. 1882.

Plants in small tufts or solitary, from a slender caudex, 1–3.5 cm. high, slender, few-leaved, 1-headed, the base of the plant emitting filiform, purplish, at first erectish and later spreading or prostrate stolons, these naked, up to about 15 cm. long and bearing a terminal tuft of small, subulate, ovate-based leaves enclosing a propagative bud by which the plant seemingly propagates chiefly as a biennial. Leaves rather closely and subsericeously gray-tomentose, the stolons glabrate; lowermost leaves oblong or obovate-oblong, about 6 mm. long, 3 mm. wide, obtuse, similar to those in the center of the stolon-buds; inner basal leaves and cauline leaves narrowly linear-spatulate, 1–3.5 cm. long including the petioliform base, 1–2 mm. wide, acutish, a few smaller leaves at base of involucre; pistillate involucre subcylindric to narrowly campanulate, 7–12(–13) mm. high, thinly woolly below, the phyllaries lanceolate-ovate to lanceolate, acute to acuminate, greenish medially, brown-bordered, and with brownish white hyaline tips; staminate involucre broader, 4–7 mm. high, the phyllaries mainly blackish brown, acuminate or the inner obtuse; pistillate achenes papillate with nonbifurcate hairs, the ring of pappus deciduous; bristles of staminate pappus barbellate upward, entire and slightly widened or some of them forking above.

Dry plains and open slopes, Upper Sonoran Zone to Canadian Zone; eastern Washington south to central and northeastern Oregon east to Idaho and northern Wyoming. Type locality: "Between Spipen [Naches] River and the north fork of the Columbia," Washington. (April–)May–July.

124. **ANAPHÁLIS** DC. Prod. **6**: 271. 1837.

Tomentose perennial herbs with running branching rootstocks; stems equably leafy, solitary or tufted. Leaves alternate, the lowest scale-like, the others lanceolate to linear, entire, tomentose beneath and often above, 1-nerved or triplinerved. Heads (in ours) numerous, in usually small, rounded, cymose panicles, mostly short-pedicelled, disciform or discoid, polygamodioecious or dioecious, many-flowered; in some plants composed only of hermaphrodite but sterile flowers, in others of many pistillate flowers with a few (in ours about 4) hermaphrodite fertile flowers. Involucre subglobuse, closely graduate, several-seriate, woolly at base, the phyllaries papery, the tips mostly ovate and obtuse, milk-white, sometimes with blackish brown bases, radiating in age. Receptacle flattish or slightly convex, scrobiculate. Corollas yellowish, those of hermaphrodite flowers slenderly cylindric-funnelform, 5-toothed, with glandular teeth; of pistillate flowers filiform, the limb irregularly 3–4-cleft, the teeth glandular. Achenes small, oblong-obovoid. Pappus in pistillate flowers of soft bristles free or united at base in small groups; in hermaphrodite flowers of bristles slightly thickened at tip, less so in hermaphrodite flowers in pistillate heads. Anthers caudate at base, the tails acuminate, those of adjacent anthers connate. Style of hermaphrodite flowers either undivided or 2-branched, the branches linear, truncate, hispidulous; in pistillate flowers the branches linear, obtuse, smooth, [Said to be an ancient Greek name of a plant allied to *Gnaphalium*.]

A genus of about 30 species, all Asiatic, a single one occurring in North America. Type species, *Anaphalis nubigena* DC.

1. **Anaphalis margaritàcea** (L.) A. Gray. Pearly Everlasting. Fig. 5851.

Gnaphalium margaritaceum L. Sp. Pl. 850. 1753.
Antennaria margaritaceum R. Br. Trans. Linn. Soc. **12**: 123. 1818.
Antennaria margaritacea var. *subalpina* A. Gray, Proc. Acad. Phila. **1863**: 67. 1863.
Anaphalis margaritacea A. Gray, Proc. Amer. Acad. **8**: 653. 1873.
Anaphalis margaritacea var. *subalpina* A. Gray, Syn. Fl. N. Amer. **1²**: 233. 1884.
Anaphalis margaritacea var. *occidentalis* Greene, Fl. Fran. 399. 1897.
Anaphalis margaritacea var. *revoluta* Suskd. Allg. Bot. Zeit. **12**: 7. (Jan.) 1906.
Anaphalis sierrae Heller, Muhlenbergia **1**: 147. (July) 1906.

Perennial, from slender running rootstocks, the stems usually solitary and simple, 20–80 cm. high, gray-tomentose. Leaves linear to lance-linear or the lowest oblanceolate, 5–15 cm. long, 2–20 mm. wide, acuminate to obtuse, sessile and more or less clasping, the larger triplinerved, early glabrate and green above and persistently tomentose below or soon deciduous on the basal leaves; heads numerous, in a rounded cymose panicle; involucre subglobose, 5–7 mm. high, woolly at base; phyllaries with mostly ovate and obtuse, papery, milk-white tips, the base sometimes blackish brown; achenes finely hispidulous.

In dry open places along streams or coastal ravines, Transition and Canadian Zones; occurring in the Pacific States south through Washington and Oregon to California, where it is found in the Coast Ranges as far south as Monterey County and in the Sierra Nevada to Tulare County and also in the San Bernardino Mountains, San Bernardino County; also from Alaska east to Newfoundland south to Pennsylvania and to Kansas and the Rocky Mountain and Great Basin areas as far south as New Mexico and Northern Arizona. Native in eastern Asia and introduced in Europe. Type locality: "Habitat in America *septentrionali, kamtschatca*." July–Oct.

Variable in foliage, represented with us mainly by the following ill-defined varieties: var. *revoluta* Suksd., with leaves narrowly linear, 2–5 mm. wide, glabrate above; var. *subalpina* A. Gray, low and with rather few broad leaves, these woolly above, found mostly in the mountains; and var. *margaritacea*, with the broad leaves (mostly 8–20 mm. wide) quickly glabrate and bright green above, found more commonly along the coast.

125. **STYLOCLÌNE** Nutt. Trans. Amer. Phil. Soc. II. **7**: 338. 1841.

Low, slender, erect or diffuse, usually much-branched, woolly annuals with small, narrow, entire, alternate leaves and small, clustered, leafy-bracted, ovoid to subglobose, disciform heads. Involucre (aside from the bracts subtending and enclosing the pistillate flowers) wanting except in two species (*S. gnaphalioides* and *S. amphibola*) where composed of about 5 suborbicular, flat, scarious, green-centered phyllaries. Receptacle oblong-cylindric to slender-cylindric. Pistillate flowers numerous or in one species only 5–9, closely, or in one species loosely, imbricated on the receptacle, each loosely enclosed in its subtending bract and deciduous with it, this with a woolly, firm, usually ovoid, boat-shaped body with inflexed margins, produced at apex or throughout its length into an ovate to suborbicular, plane, hyaline tip or border, this glabrous except for the greenish center; corollas filiform; pappus none. Hermaphrodite (functionally staminate) flowers few (about 3–5), at apex of receptacle, surrounded by about 5 oblong or linear, hyaline bracts, or in one species by about 5 large, firm, open-boat-shaped bracts tipped by a rigid, incurved-uncinate cusp and at maturity persistent and stellately spreading; corollas tubular, 4–5-toothed; ovaries abortive; pappus of about 2–5 deciduous bristles (thickened upward in one species) or wanting. Anthers sagittate-caudate at base. Style-branches in hermaphrodite flowers hispidulous, truncate or subtruncate. Achenes of pistillate flowers slenderly obovoid or elliptic, smooth, glabrous, few-nerved, slightly compressed or obcompressed, epappose, bearing the corolla at the symmetrical summit. [Name Greek, meaning column-bed, from the form of the receptacle.]

A genus of 7 species, six from western and southwestern United States and adjacent Mexico and one from Afghanistan. Type species, *Stylocline gnaphalioides* Nutt.

Receptacular bracts surrounding the staminate flowers inconspicuous, hyaline, oblong or linear, neither uncinate nor stellately spreading in age; staminate flowers with 2–5 deciduous pappus-bristles (lacking in *S. psilocarphoides*).
    Marginal bracts enclosing the pistillate flowers broadly hyaline-winged throughout their length.
                                                              1. *S. gnaphalioides.*
    Marginal bracts enclosing the pistillate flowers with a terminal hyaline appendage, the margin not hyaline-winged or with a narrow wing much shorter than the body of the bract.
        Hyaline appendage of the outer row of pistillate bracts horizontally inflexed; plants of the San Francisco Bay region.   2. *S. amphibola.*
        Hyaline appendage of all the pistillate bracts erect; plants of the Great Basin and southwestern deserts.
            Plants loosely branched from the base, open; leaves subtending the clustered heads attenuate at apex and usually conspicuously cusped.   3. *S. micropoides.*
            Plants much branched from the base, dense; leaves subtending the clustered heads merely acute and inconspicuously if at all cusped.   4. *S. psilocarphoides.*
Receptacular bracts surrounding the staminate flowers conspicuous, rigidly uncinate-cuspidate, stellately spreading in age; staminate flowers epappose.   5. *S. filiginea.*

1. **Stylocline gnaphalioìdes** Nutt. Everlasting Stylocline or Nest-straw. Fig. 5852.

Stylocline gnaphalioides Nutt. Trans. Amer. Phil. Soc. II. 7: 338. 1841.
Stylocline arizonica Coville, Proc. Biol. Soc. Wash. 7: 79. 1892.

Erect to diffuse, gray-wooly annual 15 cm. high or less, usually much-branched from base or throughout. Leaves spatulate or narrowly oblong to nearly linear, 3–8 mm. long, usually obtuse, apiculate, those subtending the heads larger and broader; heads clustered at tips of stem and branches, conspicuously scarious; proper involucre of about 5 orbicular scarious phyllaries with thinly woolly green center; pistillate flowers numerous, their bracts with narrow, greenish, compressed, dorsally woolly body bordered throughout by a broad, brownish white, scarious, glabrous, spreading wing, the whole suborbicular-ovate, obtuse; staminate flowers with a pappus of few ascending, somewhat thickened bristles; achenes slender, straight on inner side, convex on outer, somewhat compressed laterally.

Dry plains and open hillsides, Upper and Lower Sonoran Zones; Santa Cruz County and the San Joaquin Valley, California, south to northern Lower California east to Arizona. Type locality: Monterey, California. March–May.

2. **Stylocline amphíbola** (A. Gray) J. T. Howell. Mt. Diablo Cottonweed or Stylocline. Fig. 5853.

Micropus amphibolus A. Gray, Proc. Amer. Acad. 17: 214. 1882.
Gnaphalodes amphibola Greene, Man. Bay Reg. 183. 1894.
Stylocline amphibola J. T. Howell, Leaflets West. Bot. 5: 91. 1948.

Floccose-woolly plants 17 cm. high or less, the stems simple or shortly branched above. Pistillate flowers 9–11, imbricated on an oblong to nearly cylindric receptacle, their enclosing bracts thinner, nearly membranaceous, the upper part with a broad, ovate, hyaline margin throughout its length, in the outer flowers horizontally inflexed and at maturity about half as long as the body or more, in the inner flowers merely inflexed or erectish; hermaphrodite flowers about 3, subtended by thin linear bracts, and with a pappus of about 3 deciduous hispidulous bristles; achene 1.3 mm. long, bearing the corolla centrally or on the inner point of the truncate-rounded apex.

Low hills and rocks, Upper Sonoran Zone; Lake County to the northern San Francisco Bay region, California, in Marin, Contra Costa, and Alameda Counties, and south to Santa Cruz County. Type locality: based on a collection of Kellogg and Hartford (no. *416*) from an unknown locality and a collection made by Brewer at Walnut Creek, Contra Costa County, California. April–May.

3. Stylocline micropoìdes A. Gray. Woolly Stylocline or Desert Nest-straw. Fig. 5854.

Stylocline micropoides A. Gray, Smiths. Contr. **5**[6]: 84. 1853.

Erect to diffuse, gray-woolly annual, simple-stemmed or usually branched, about 4–13 cm. high. Leaves narrowly spatulate to nearly linear, 4–12 mm. long, 0.5–1.5 mm. wide, usually acute or acuminate, apiculate, those subtending the heads larger, often lanceolate; heads clustered at tips of stem and branches, leafy-bracted, appearing less scarious and more woolly than in *S. gnaphalioides*; proper involucre essentially wanting; pistillate flowers numerous, their bracts with a boat-shaped, densely long-woolly, thin, subcoriaceous body tapering into a subulate green tip, the latter bordered for most of its length by an ovate hyaline margin shorter than the body of the bract; staminate flowers with a pappus of about 3–5 deciduous bristles; achene ellipsoid, scarcely compressed, short-stipitate, pale, 1.5 mm. long.

In desert places, often around rocks, Lower Sonoran Zone; Inyo County, California, south through the Mojave and Colorado Deserts, California, east to southern Utah, Arizona, and New Mexico, and southern Sonora. Type locality: "Hills near Frontera, New Mexico." March–April. Some collections from the higher elevations in the Mojave Desert suggest *S. psilocarphoides* in growth form.

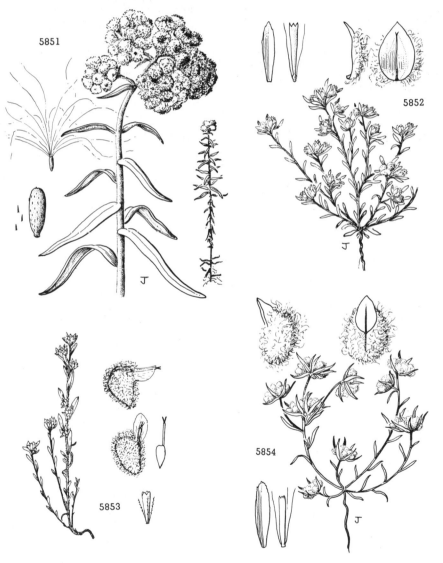

5851. Anaphalis margaritacea
5852. Stylocline gnaphalioides
5853. Stylocline amphibola
5854. Stylocline micropoides

4. **Stylocline psilocarphoìdes** Peck. Peck's Stylocline. Fig. 5855.

Stylocline psilocarphoides Peck, Leaflets West. Bot. **4**: 185. 1945.

Depressed, much-branched, woolly annual somewhat loosely branched in younger plants, 3–5 cm. high. Leaves rarely acute, linear or oblong, 4–8 mm. long, those subtending the heads larger; heads in small, dense, globose clusters; pistillate flowers 8–13, the enclosing bracts woolly, about 3 mm. long, with short, expanded, hyaline, acute to acuminate tips, narrowly hyaline-winged a short distance below the apex; staminate flowers 2–3, sometimes with additional neutral flowers, the subtending bracts hyaline, as long as or shorter than the corollas; achenes narrowly obovoid, scarcely compressed, rounded above, narrowed below, somewhat curved, dark, 1–1.2 mm. long, the pappus lacking.

Dry open places, Upper Sonoran Zone; Malheur County, Oregon, to Nevada and southwestern Utah. Type locality: 15 miles north of McDermitt, Malheur County, Oregon. May–June.

5. **Stylocline filagínea** A. Gray. Northern or Hooked Stylocline. Fig. 5856.

Ancistrocarphus filagineus A. Gray, Proc. Amer. Acad. **7**: 356. 1868.
Stylocline filaginea A. Gray, op. cit. **8**: 652. 1873.
Stylocline filaginea var. *depressa* Jepson, Man. Fl. Pl. Calif. 1063. 1925.

Dwarf, gray-woolly, slender annual, the tomentum not noticeably floccose, several-stemmed from base, erect or depressed, 1–11 cm. long. Leaves spatulate-oblanceolate to linear, 6–18 mm. long, 0.5–3 mm. wide, obtuse to acute, apiculate, entire, those surrounding the heads broader; heads terminating stem and branches, solitary or usually glomerate, leafy-bracted, 3–11 mm. thick; proper involucre none; pistillate flowers 5–9, each loosely included by its boat-shaped, obcompressed, firm, woolly phyllary, this with inflexed margins meeting in the center and with an abrupt, suborbicular, somewhat inflexed, hyaline tip about one-fourth its length, the whole deciduous together; staminate flowers about 3–4, subtended by a circle of about 5 shallowly boat-shaped, open, woolly bracts tipped by a rigid, incurved-uncinate cusp, at maturity subcoriaceous, about 4 mm. long, and stellately spreading at maturity; achenes obovoid, somewhat obcompressed, smooth, pale, 2 mm. long.

Dry, open, often stony hillsides, Upper Sonoran and Arid Transition Zones; Baker County, eastern Oregon, and adjacent Idaho south to northeastern Nevada and California; on the western face of the mountains from Jackson County, Oregon, to San Luis Obispo County, California, in the Coast Ranges and on the Sierran slope from Butte County to Kern County and in the Mount Pinos region and the San Bernardino Mountains. Type locality: Round Valley, Mendocino County, California. Collected by Bolander. April–June.

126. **FILÀGO** [Tourn.] L. Sp. Pl. 927. 1753.

Ours more or less floccose-woolly annuals with small, narrow, entire, alternate leaves and small clustered heads. Proper involucre of 5 or fewer elliptic, flat, green-centered, hyaline-margined phyllaries half as long as head or less, or wanting. Receptacle more or less obconic, bracteate below, its flattish (subulate in one species) naked center surrounded by about 5 plane or merely concave, nearly glabrous, mostly oblong-lanceolate bracts with hyaline margin and apex. Outer pistillate flowers about 3–13, each loosely (or in one species closely) enclosed in the boat-shaped, thin-subherbaceous (or in one species firm and sub-indurated) woolly body of the subtending bract, this with hyaline tip and inflexed margins usually not meeting in the center; these flowers with smooth or papillose, slenderly obovoid achenes; corollas terminal. Center of receptacle bearing about 4–20 flowers, the 2–5 central ones hermaphrodite with slender-tubular, 4–5-toothed corollas and pappose usually papillose achenes, the others pistillate, with filiform corolla and usually papillose and pappose achenes. Achenes slenderly obovoid, sometimes weakly about 4-nerved, subterete. Anthers sagittate-caudate at base, with conspicuous auricles. Style-branches in the hermaphrodite flowers with rather slender, hispidulous, rounded or subtruncate tips. Pappus when present of about 20 deciduous bristles somewhat united at extreme base. [Name Latin, meaning thread, referring to the characteristic wool.]

A genus of about 12 species, natives of both hemispheres. Type species, *Filago pyramidata* L.

Plant proliferately branching from the lowest capitate clusters of heads; outer receptacular bracts cuspidate; receptacle subulate. 1. *F. germanica.*
Plant not so branched, the heads terminal or axillary; outer receptacular bracts not cuspidate; receptacle merely convex or nearly flat.
  Outer pistillate flowers tightly enclosed in the somewhat indurated subtending receptacular bracts, these with nearly horizontal body and long, abruptly bent, erectish, hyaline-appendaged beak; erect plant with mostly linear-subulate leaves, those subtending the heads conspicuously surpassing them. 2. *F. gallica.*
  Outer pistillate flowers loosely enclosed in the not indurated subtending receptacular bracts, these not with abruptly bent beak; plants usually diffuse, if erect not with leaves conspicuously surpassing the heads.
    Diffuse plant with very slender naked internodes, the leaves practically wanting; flowers inside the inner circle of receptacular bracts about 4–8, all hermaphrodite. 3. *F. arizonica.*
    Plant erect or diffuse, the stems normally leafy at least below; flowers inside the inner circle of receptacular bracts about 12–20, only about 2–4 of them hermaphrodite.
      Plant diffuse or merely ascending; achenes all smooth; hyaline appendage of the outer receptacular bracts about as long as the body. 4. *F. depressa.*
      Plant normally erect; inner achenes papillose; hyaline appendage of the outer receptacular bracts about half as long as the body or less.
        Plant branched from the base or simple; branches when present few and short; outer receptacular bracts of the pistillate flowers 7–10, markedly boat-shaped. 5. *F. californica.*
        Plant fastigiately branched above the base; outer receptacular bracts of the pistillate flowers 3–5, merely concave with a hyaline tip. 6. *F. arvensis.*

1. **Filago germánica** L. Herba Impia. Fig. 5857.

Gnaphalium germanicum L. Sp. Pl. 857. 1753.
Filago germanica L. op. cit. ed. 2. 1311. 1763.
Gifola germanica Dumort. Fl. Berg. 68. 1827.

Erect, leafy-stemmed annual woolly throughout, 1–3.5 dm. high, simple or branched from the
base. Leaves sessile, ascending, longer than the internodes, about 1 cm. long, oblanceolate, cuspidate at the apex; inflorescence a dense capitate cluster of heads borne terminally, additional heads
borne on proliferating branches arising from the lowest cluster; involucres about 5 mm. high;
receptacle subulate; pistillate flowers filiform, subtended by strongly concave receptacular bracts,
these woolly without at the base, glabrous above, and attenuate into a somewhat divaricate cusp,
the achenes of the outermost flowers without pappus, the next series also pistillate but with pappus; hermaphrodite flowers tubular, the receptacular bracts surrounding them less concave,
scarious, without wool, and scarcely cuspidate-tipped, their achenes obscurely papillose, ellipsoid,
minute, about 0.5 mm. long.

Dry slopes; adventive in Douglas County, Oregon, and Mendocino County, California; commonly distributed
in eastern United States; Europe. Type locality: Europe. May–June. Cotton–rose.

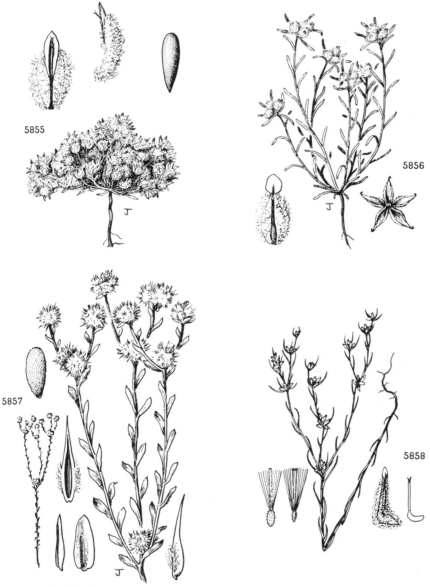

5855

5856

5857

5858

5855. Stylocline psilocarphoides
5856. Stylocline filaginea

5857. Filago germanica
5858. Filago gallica

2. **Filago gállica** L. Narrow-leaved Filago. Fig. 5858.

Filago gallica L. Sp. Pl. ed. 2. 1312. 1763.
Logfia subulata Cass. Dict. Sci. Nat. **27**: 116. 1826.
Logfia gallica Coss. & Germ. Ann. Sci. Nat. II. **20**: 291. *pl. 13.* 1843.

Slender, erect, subsericeous-woolly annual, simple or erect branched even from base, about 2 dm. high or less, leafy throughout. Leaves narrowly linear or linear-subulate, 0.5–2 cm. long, 0.3–1.5 mm. wide, acuminate to an obtuse callous tip, often revolute-margined, erect; heads in a small cluster at apex of stem, successively surpassed by 1 or more series of branches bearing similar terminal clusters, leafy-bracted, the leafy bracts rather broader than the stem-leaves, usually two or three times as long as the heads; heads pentagonal-ovoid, about 4 mm. high; 5 outer pistillate flowers completely enclosed in their bracts, these with horizontal indurated body about 1 mm. long and erect beak about 2.8 mm. long, the latter subherbaceous below, bordered for about its upper half by an ovate, obtuse, hyaline appendage, all portions of the bract except the appendage densely long-woolly; receptacle inside these barely convex, bearing about 10 pistillate and about 4 hermaphrodite flowers, surrounded by about 5 open, concave, lance-oblong, obtuse receptacular bracts, narrowly hyaline-margined and scarious-tipped; achenes of the 5 outer pistillate flowers glabrous, epappose; other achenes papillose and with a pappus of 20 bristles united at extreme base and deciduous together.

　　Adventive from Europe on dry slopes; Humboldt, Shasta, and Butte Counties, California, south to Santa Barbara County in the Coast Ranges and Mariposa County in the Sierra Nevada. April–Sept.

3. **Filago arizónica** A. Gray. Arizona Filago. Fig. 5859.

Filago arizonica A. Gray, Proc. Amer. Acad. **8**: 652. 1873.

Thinly woolly annual, soon much branched and diffuse, the branches about 20 cm. long or less with elongated naked internodes, repeatedly proliferous, the leaves almost wanting except for those bracting the heads. Leaves linear to lance-linear or narrowly spatulate, 4–13 mm. long, 1–2 mm. wide, callous-apiculate, clustered about the heads and usually about twice as long; heads in small clusters at tips of stem and branches; pistillate flowers about 13–15, each loosely inserted in an open boat-shaped receptacular bract with thin, densely woolly body (greenish above) and a much shorter, ovate, obtuse, hyaline tip; center of receptacle flattish, bearing 4–8 hermaphrodite flowers surrounded by about 4 thin, nearly or quite glabrous, flattish receptacular bracts; achenes of pistillate flowers smooth and epappose, of hermaphrodite flowers papillose and pappose.

　　Arid places, Sonoran Zones; the Channel Islands and Ventura County, California, south to northern Lower California, eastward in the Colorado and southern Mojave deserts to Arizona and adjacent Sonora. Type locality: Mesa Verde, Arizona. April–May.

4. **Filago depréssa** A. Gray. Dwarf Filago. Fig. 5860.

Filago depressa A. Gray, Proc. Amer. Acad. **19**: 3. 1883.

Depressed or ascending, whitish-woolly annual, much branched from base, 18 cm. long or less, leafy. Leaves spatulate or narrowly obovate to nearly linear, 3–8 cm. long, 1–1.5 mm. wide, usually obtuse, apiculate, those subtending the heads not evidently larger than the others and not or scarcely surpassing the heads; heads clustered at tips of stem and branches, occasionally axillary, 4 mm. high; outer pistillate flowers 5–6, rather loosely enclosed in their bracts, these ovoid, boat-shaped with thin, densely woolly body about equaled by the ovate, hyaline, terminal appendage, one or two of the inner flowers often pappose; center of receptacle surrounded by about 5 concave or flattish, lance-oblong receptacular bracts with hyaline margin and tip and thinly woolly on back above, about 12-flowered (2–3 of these hermaphrodite), the flowers all pappose; achenes all smooth, slenderly obovoid, about 4-nerved, 0.7 mm. long.

　　Desert places, Lower Sonoran Zone; Colorado and Mojave Deserts and Inyo County, California, east to southeastern Arizona. Type locality: Palm Springs, Riverside County, California. April–May.

5. **Filago califórnica** Nutt. California Filago. Fig. 5861.

Filago californica Nutt. Trans. Amer. Phil. Soc. II. **7**: 405. 1841.
Oglifa californica Rydb. Fl. Rocky Mts. 914. 1917.

Slender, erect, grayish-woolly annual 15–40 cm. high, simple or much branched from the base, leafy throughout. Leaves narrowly spatulate to oblanceolate or nearly linear, usually 0.8–2 cm. long, 1–3 mm. wide, usually obtuse, callous-apiculate; heads clustered at tips of stem and short upper branches, sometimes also axillary, not conspicuously surpassed by the involucrating leaves; outer pistillate flowers about 10, loosely enclosed in thin, boat-shaped, densely woolly receptacular bracts, these stellately spreading in age and having inflexed margins not meeting in the center, terminating in an ovate hyaline appendage about half as long as the body (continued to its base in some), the outermost (about 6) achenes smooth and epappose, a few inner (about 2–3) papillose and pappose; center of receptacle flattish, surrounded by about 5 concave, thinly pilose receptacular bracts, about 20-flowered (about 3 of these hermaphrodite), these achenes all papillose and with a pappus of deciduous, basally united bristles; central corollas often rosy and inner receptacular bracts sometimes rosy-margined.

　　Dry hillsides and open places, Transition and Sonoran Zones; Mendocino and Butte Counties to southern California and northern Lower California, more abundant toward the coast but also in the Mojave and Colorado Deserts, east to Utah and Arizona. Type locality: near Santa Barbara, California. March–May.

6. **Filago arvénsis** L. Field Filago. Fig. 5862.

Gnaphalium arvense L. Sp. Pl. 856. 1753.
Filago arvensis L. op. cit. Addenda

Erect, leafy-stemmed annual, floccose-woolly throughout, 30–50 cm. high, fasciculately branched well above the base, the branches ascending. Leaves about 1.5–3 cm. long, erect, alternate, linear, acute, longer than the internodes; heads 3–5 mm. high, in glomerules toward the ends of the branches; proper involucre of few phyllaries, these shorter than the outer receptacular bracts subtending the 3–5 pistillate flowers; outer bracts strongly concave, stellately spreading in age, scarious only at the extreme tip; achenes of the pistillate flowers epappose; receptacle flattish, the receptacular bracts surrounding the fertile and few sterile hermaphrodite flowers longer than the outer bracts, scarious and less woolly, glabrous above, the achenes papillate and bearing a pappus of capillary bristles.

Adventive from Europe, becoming common in overgrazed areas; southeastern British Columbia and adjacent Washington to Idaho and Montana. July–Aug.

127. **PSILOCÁRPHUS** Nutt. Trans. Amer. Phil. Soc. II. **7**: 340. 1841.

Low, floccose-woolly annuals, branched except in depauperate specimens, with narrow, opposite, entire stem-leaves and small, sessile, subglobose, disciform heads, solitary or clustered in the forks and at tips of stem and branches, bracted by whorled leaves. Proper involucre essentially none. Receptacle subglobose or truncately obpyriform or sometimes lobed, bluntly muricate. Pistillate flowers numerous (about 20–50), closely imbricate, each loosely enclosed by and deciduous with its subtending receptacular bract, this obliquely

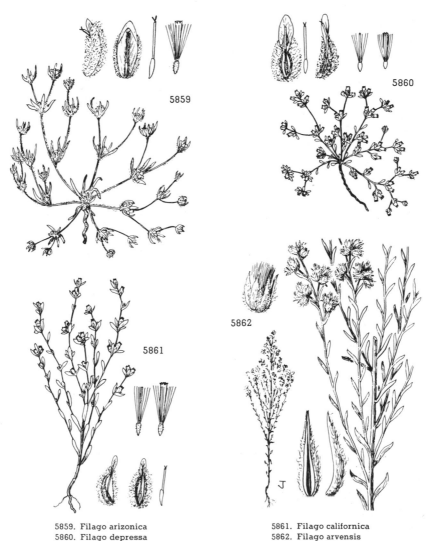

5859. Filago arizonica
5860. Filago depressa

5861. Filago californica
5862. Filago arvensis

obovoid in outline, somewhat compressed, saccate, reticulate, with sides meeting in the center, woolly, bearing below the rounded tip on inner side a small, ovate, horizontally introrse or sometimes deflexed, rarely erect, scarious appendage; corollas filiform; achenes subcylindric or obovoid, subterete or slightly compressed, smooth, glabrous, epappose. Hermaphrodite flowers few (about 4–9), borne in center of receptacle, not subtended by receptacular bracts; corollas with slender tube and funnelform or funnelform-campanulate throat, 4–5-toothed; ovaries abortive, epappose. Anthers sagittate-auriculate, with short acute auricles, those of adjacent anthers connate. Style-branches in hermaphrodite flowers slender, obtuse, hispidulous. [Name Greek, meaning naked chaff; wrongly explained by Nuttall as meaning slender chaff, in allusion to the membranous pales.]

A genus of about 5 species. In addition to the following, one occurs in Chile. Type species, *Psilocarphus globiferus* Nutt., as to the North American plants.

Bracts inclosing the achenes comparatively thickly and loosely woolly, usually about 3 mm. long; tomentum of whole plant comparatively long and loose; heads larger, 3–8 mm. thick, usually few.
  Plants usually prostrate; leaves mostly 5–15 mm. long; achenes oblanceolate.   1. *P. brevissimus.*
  Plants more or less erect; leaves mostly 12–25 mm. long; achenes narrowly oblong or elliptic-oblong.
                                                                                              2. *P. elatior.*
Bracts inclosing the achenes comparatively thinly or closely woolly, about 2 mm. long; tomentum of whole plant comparatively close; heads smaller, about 3–5 mm. thick, usually numerous.
  Plants closely subsericeous-woolly; leaves mostly 10–22 mm. long, linear or linear-oblanceolate; achenes ellipsoid-cylindric.   3. *P. oregonus.*
  Plants rather thinly grayish-woolly; leaves 4–10 mm. long, spatulate, oblong (ovate to elliptic-ovate in the variety); achenes obovoid-ellipsoid.   4. *P. tenellus.*

1. Psilocarphus brevíssimus Nutt. Round or Dwarf Woolly-heads. Fig. 5863.

Psilocarphus brevissimus Nutt. Trans. Amer. Phil. Soc. II. **7**: 340. 1841.
Psilocarphus globiferus Nutt. loc. cit. Not *Micropus globiferus* Bert. ex DC.
Bezanilla chilensis Remy in Gay, Fl. Chil. **4**: 110. 1849; Atlas, *pl. 46.* 1854.
Psilocarphus chilensis A. Gray, Syn. Fl. N. Amer. ed. 2. **1²**: 448. 1886.
Psilocarphus oreganus var. *brevissimus* Jepson, Fl. W. Mid. Calif. 549. 1901.

Dwarf, loosely whitish-woolly annual, simple and 1-headed or with prostrate branches up to 20 cm. long. Leaves spatulate to lanceolate, 5–15 mm. long, 1.5–3 mm. wide, obtuse, bluntly callous-apiculate; heads solitary or sometimes clustered in the forks and at tips of stem and branches, loosely long-woolly (more heavily so than in *P. elatior*), about 5–7 mm. thick, the involucrating leaves usually about equaling the heads, sometimes twice as long; pistillate flowers about 20–34, occasionally more, their enclosing bracts 2.5–3.2 mm. long, woolly, bearing on inner side near the top an ovate, horizontally inflexed or often erect, scarious appendage about 0.5 mm. long; hermaphrodite flowers about 6–11; achenes subcylindric, terete, 1.3–2 mm. long.

Dried beds of vernal pools and moist places, Arid Transition and Sonoran Zones; less common in eastern Washington, southern Idaho, and western Montana and common in eastern and southern Oregon southward through California to cismontane southern California and northern Lower California; also Chile and Argentina. Type locality: "Plains of the Oregon [Columbia] River, in inundated tracts." April–July.

Psilocarphus brevissimus var. multiflòrus Cronquist, Research Stud. St. Coll. Wash. **18**: 80. 1950. Plants tending to be erect, the leaves more or less linear-oblong; woolly pubescence of the heads rather sparse and close; pistillate flowers of the heads many, about 100. In the same habitat as the preceding, Solano County to Alameda and Santa Clara Counties, California. Type locality: dried vernal pools, Suisun, Solano County.

2. Psilocarphus elàtior A. Gray. Tall Woolly-heads. Fig. 5864.

Psilocarphus oreganus var. *elatior* A. Gray, Proc. Amer. Acad. **8**: 652. 1873.
Psilocarphus elatior A. Gray, Syn. Fl. N. Amer. ed. 2: **1²**: 448. 1886.

Gray-woolly annual 15 cm. high or less, erect or sometimes diffuse, with few branches, depauperate plants sometimes 1 cm. high. Leaves oblanceolate or the upper linear or linear-oblong, the larger 12–35 mm. long, 2–6 mm. wide, acute or obtuse, callous-apiculate, gradually or the upper scarcely narrowed to base; heads solitary or clustered in the forks and at tips of stem and branches, 4–8 mm. thick, usually about half as long as the involucrating leaves; pistillate flowers about 40–50 or more, the receptacular bracts rather loosely woolly especially above, 2.6–3 mm. long, bearing on the inner side below the tip an ovate, horizontal, introrse, scarious appendage 0.5–1 mm. long; staminate flowers 5–8, their corollas usually reddish-tipped; achenes subcylindric, slightly compressed, about 1.2 mm. long.

Fields and moist places, Transition Zone; Vancouver Island, British Columbia, south along the Pacific Slope to southern Oregon where it occurs abundantly; less common in eastern Washington and adjacent Idaho. Type locality: [Portland], Oregon. Collected by Hall. May–Aug.

3. Psilocarphus oregònus Nutt. Oregon Woolly-heads. Fig. 5865.

Psilocarphus oregonus Nutt. Trans. Amer. Phil. Soc. II. **7**: 341. 1841.

Slender, much-branched annual at first erectish, becoming diffuse and forming mats up to 25 cm. across, closely subsericeous-woolly. Leaves narrowly spatulate, mostly 8–22 mm. long, 1–3 mm. wide, gradually narrowed into a petioliform base, obtuse or acutish, callous-apiculate; heads mostly solitary in the forks and at tips of stem and branches, 3–5 mm. thick, much surpassed by the involucrating leaves; pistillate flowers much as in *P. tenellus*, about 20–80 in each head, their bracts somewhat larger, at maturity usually 2–2.4 mm. long, the scarious appendage rather smaller, often borne near the middle of the inner face of the bract, either horizontally introrse or deflexed; achenes more slender, ellipsoid-cylindric, terete, 0.8–1.2 mm. long.

Dried beds of vernal pools and dry open places, Arid Transition and Sonoran Zones; eastern Washington, adjacent Idaho, and eastern Oregon south in the valleys of the Inner Coast Ranges and the counties of the Sacramento and lower San Joaquin Valleys, California; also reported as occurring locally in Santa Barbara and San Diego Counties and in northern Lower California. Type locality: In inundated places "near the Oregon [Columbia River] and outlet of the Wahlamet [Willamette River]," though probably collected in eastern Oregon according to Cronquist (Research Stud. St. Coll. Wash. **18**: 85. 1950). April–July.

4. Psilocarphus tenéllus Nutt. Slender Woolly-heads. Fig. 5866.

Psilocarphus tenellus Nutt. Trans. Amer. Phil. Soc. II. 7 : 341. 1841.

Slender, much-branched, thinly floccose-woolly annual at first often erectish, soon diffusely spreading, forming mats 5–30 cm. wide, the filiform internodes usually longer than the leaves. Leaves spatulate, oblanceolate, or oblong or sometimes nearly linear, 4–5 mm. long, 1.5–3 mm. wide, obtuse, apiculate, gradually narrowed to base; heads mostly solitary in the forks and at tips of stem and branches, 3–4 mm. thick, surpassed by the involucrating leaves; pistillate flowers about 25–35, sometimes more, their enclosing bracts at maturity obliquely obovoid, (0.6) 1.5–2.3 mm. long, humpbacked, rather thinly woolly chiefly above, produced on inner side considerably below the rounded top into an ovate, horizontal, introrse, scarious appendage about 0.5 mm. long; hermaphrodite flowers 5–6, usually reddish above; achenes obovoid-ellipsoid, 0.6–1.2 mm. long.

Dried beds of vernal pools and dry open places, Transition and Sonoran Zones; occurring in isolated localities, Vancouver Island, British Columbia, Washington, eastern Oregon, and western Idaho but abundant in southwestern Oregon and the Coast Ranges of California (including Catalina Island) and northern Lower California; also abundant in the Sacramento and San Joaquin Valleys, California. Type locality: near Santa Barbara, California. March–June.

Psilocarphus tenellus var. ténuis (Eastw.) Cronquist, Research Stud. St. Coll. Wash. **18**; 88. 1950. (*Psilocarphus tenuis* Eastw. Bot. Gaz. **41**: 292. 1906.) Differs from *P. tenellus* var. *tenellus* in having the involucrating leaves and to a less extent those of the stem elliptic-ovate to ovate or broadly oblong; receptacular bracts averaging smaller than the name-bearing variety. Dry open spaces and dried vernal pools, Monterey and San Luis Obispo Counties, California, east to the San Joaquin Valley, Calaveras County to Kern County. Type locality: Monterey, Monterey County.

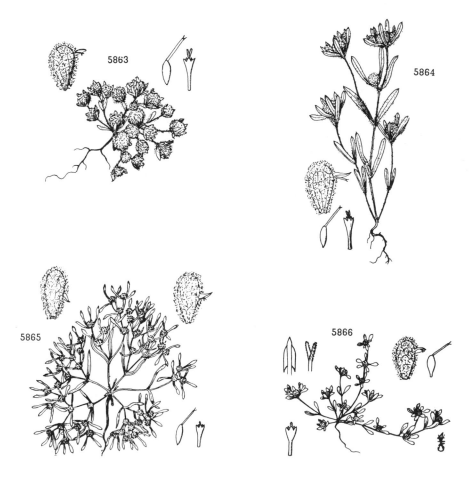

5863. Psilocarphus brevissimus
5864. Psilocarphus elatior

5865. Psilocarphus oregonus
5866. Psilocarphus tenellus

128. MÍCROPUS L. Sp. Pl. 927. 1753.

Low, floccose-woolly annuals with alternate, narrow, entire leaves and small, disciform, clustered heads. Involucre (in ours) of about 5 obovate or oval, flat phyllaries with greenish center and broad scarious margin. Receptacle small, flattish. Outer flowers pistillate, 1-seriate or imbricate, 5–6 (in our species), enclosed in conduplicate, subherbaceous, at length more or less indurate, laterally compressed, obliquely obovoid, woolly receptacular bracts, these straight on inner side, rounded and convex on outer, borne on short thick stipes, the inner margin terminated by a usually erect, scarious-tipped beak; corollas filiform, borne on inner side of ovary below the rounded tip. Hermaphrodite flowers (in ours) about 2–5, sterile, naked; their corollas slenderly tubular, 4–5-toothed. Achenes of pistillate flowers slenderly obovoid, compressed, smooth, glabrous, epappose, deciduous with the enclosing bracts; of disk-flowers abortive, epappose or with a few caduous bristles. Anthers rather shortly sagittate-caudate at base. Style in hermaphrodite flowers hispidulous above, usually bifid, with slender branches. [Name Greek, meaning small foot, in allusion perhaps to its likeness on a small scale to *Leontopodium*, lion's-foot.]

A genus of about 5 species of Europe, Asia, and North America. Type species, *Micropus supinus* L.

1. Micropus califórnicus Fisch. & Mey. Slender Cottonweed. Fig. 5867.

Micropus californicus Fisch. & Mey. Ind. Sem. Hort. Petrop. **1835**: 42. 1835.
Micropus angustifolius Nutt. Trans. Amer. Phil. Soc. II. 7: 339. 1841.
Micropus californicus var. *subvestitus* A. Gray, Bot. Calif. 1: 335. 1876.
Gnaphalodes californica Greene, Man. Bay Reg. 183. 1894.

Slender, erect, gray-woolly annual 5–35 cm. high, simple or erect-branched, leafy. Leaves nearly uniform, linear to linear-lanceolate or narrowly oblanceolate, 0.5–2.5 cm. long, 0.7–4 mm. wide, acute or acuminate, callous-pointed, shortly decurrent, entire; heads in small, dense, axillary and terminal clusters, at maturity about 3 mm. high and 5 mm. thick; involucre of about 8 green-centered, scarious, obovate obtuse or rounded, loosely villous phyllaries; receptacular bracts enclosing the 5–6 pistillate flowers deciduous at maturity, obliquely obovoid, densely long-woolly or sometimes with short appressed wool, greenish, becoming indurate, the body 2–3 mm. long, bearing at apex of straight inner side and thus below the rounded top an erectish beak about 1–1.5 mm. long, flattened and scarious for about its upper half; hermaphrodite flowers about 2–5, epappose, not subtended by receptacular bracts; achenes obovoid, greenish, about 1.8 mm. long.

Dry or damp ground, Upper Sonoran Zone; Marion County, Willamette Valley, southwestern Oregon, south (west of the Sierra Nevada) to northern Lower California. Type locality: probably Fort Ross, Sonoma County, California. April–June.

129. ÈVAX Gaertn. Fruct. 2: 393. *pl. 165.* 1791.

Small, erect, diffusely branched or acaulescent, more or less woolly annual (some European species perennial). Leaves entire, alternate along the stem or clustered beneath the heads. Heads in glomerules at the ends of the branches or solitary along the stem in the leaf-axils. Receptacle nearly flat to conical, in ours the apical portion bearing the hermaphrodite flowers produced into a slender column. Pistillate flowers fertile, several to many, in series on the outer part of the receptacle, each subtneded by a plane or slightly concave, scarious or chartaceous, usually persistent receptacular bract. Hermaphrodite flowers sterile (in ours), few, central, subtended by receptacular bracts, these plane or nearly so. Pistillate corolla filiform; hermaphrodite corolla tubular, the limb somewhat expanded and 4–5-dentate. Achenes smooth or minutely papillose, more or less obcompressed; pappus absent. [Name of an Arabian chief.]

A genus of 12 or 14 species, natives of both hemispheres. Type species, *Evax umbellata* Gaertn.

Flower-heads solitary (rarely 2) in leaf-axils or at ends of branches; upper stem-leaves 4–25 mm. long.
    Plants erect; heads markedly longer than broad, mostly in leaf-axils.   1. *E. sparsiflora.*
    Plants prostrate; heads nearly as broad as long, terminal on the branches (axillary heads rarely present).
                                                                       2. *E. acaulis.*
Flower-heads clustered (6–20) at apex of stem; upper stem-leaves 4.5–8.5 cm. long.   3. *E. caulescens.*

1. Evax sparsiflòra (A. Gray) Jepson. Erect Evax. Fig. 5868.

Evax caulescens var. *sparsiflora* A. Gray, Syn. Fl. N. Amer. 1²: 229. 1884.
Evax caulescens var. *brevifolia* A. Gray, loc. cit.
Hesperevax sparsiflora Greene, Fl. Fran. 402. 1897.
Hesperevax brevifolia Greene, loc. cit.
Evax sparsiflora Jepson, Fl. W. Mid. Calif. 549. 1901.

Slender, erect or ascending annual 3–10 cm. high, simple or usually branched at base, thinly floccose-woolly. Leaves spatulate, 5–25 mm. long, 1–6 mm. wide, the blade obovate or suborbicular, obtuse, callous-apiculate, narrowed into a petioliform indurated base sometimes nearly or quite twice as long as itself; heads slender, subcylindric, about 4–5 mm. high, much longer than broad, sessile, solitary in the leaf-axils, not whorled-involucrate, or sometimes 2 or 3 at tip of stem subtended by 3–4 leaves; receptacle with convex long-villous base bearing the pistillate flowers and their bracts, prolonged in the center into a slender cylindric column (about 1.5 mm. high in fruit), bearing at the slightly enlarged apex a circle of about 5 receptacular bracts subtending the hermaphrodite flowers, the bracts acute, woolly on the inner surface; pistillate flowers about 8–10, their bracts shallowly convex, somewhat infolded at base, the outer of the texture of the petioliform leaf-base and strongly costate, the inner chartaceous-scarious becoming subcoriaceous, short-pointed, long-villous on margin and back above, the hairs fragile; hermaphrodite flowers 3–4 with reddish tips and without ovary, surrounded by a circle of about 5 ovate, subobtuse, herbaceous-coriaceous receptacular bracts about 1.5 mm. long, connate at least toward base, tomentose on inner face, glabrous on outer; achenes obovoid, obcompressed, smooth, 1.7 mm. long.

Dry fields and hillsides, Upper Sonoran and Transition Zones; southern Oregon in Curry and Josephine Counties south to San Luis Obispo County, California; also in the foothills of Amador and Mariposa Counties, California. Type locality: nnot specifically given. Probably from the collections made by Brewer and Parry in the central coastal counties of California. April–June.

Some dwarfed specimens, taken to represent the variety *brevifolia* as originally defined by A. Gray, appear to be intermediate between *E. sparsiflora* and *E. acaulis.* The plants are erect but the heads tend to concentrate at the apex of the shortened stem, and due to the shortened internodes, the leaves subtending the heads are also conspicuously concentrated toward the apex of the stem. The leaves are usually narrower than the typical leaves of *E. sparsiflora.*

5867. Micropus californicus

5868. Evax sparsiflora

2. Evax acaùlis (Kell.) Greene. Dwarf Evax. Fig. 5869.

Stylocline acaule Kell. Proc. Calif. Acad. **7**: 112. 1877.
Evax acaulis Greene, Bot. Gaz. **8**: 257, 1883; as to name, not as to type.
Evax caulescens var. *minima* A. Gray, Syn. Fl. N. Amer. **1²**: 229. 1884.
Evax caulescens var. *brevifolia* A. Gray, loc. cit.
Hesperevax acaulis Greene, Fl. Fran. 402. 1897.

Dwarf woolly annual about 1 cm. high, acaulescent or branching from the base, the branches forking, forming mats up to 7 cm. wide in more vigorous plants. Leaves rather densely woolly, oblanceolate, 4–18 mm. long, the petiole shorter than or about equaling the blade, the involucrating leaves subtending the heads the longer, these with the base of the petiole expanded and becoming indurated and ribbed; heads about 2–2.5 mm. high, a little less than 2 mm. broad, solitary at the ends of the branches subtended by 4–5 involucrating leaves, or on longer stems 1–2 on the stem below the apex; pistillate flowers 6–7 in each head, the subtending receptacular bracts as in the preceding species and reflexed at flowering; hermaphrodite flowers reddish-tipped, borne on the slender prolongation of the receptacle encircled by 5 ovate, acute, herbaceous-coriaceous receptacular bracts, these woolly on the inner surface and spreading but not concealing the tips of the reflexed bracts subtending the pistillate flowers; achenes yellowish, obovoid, about 1 mm. high.

Dry fields and vernal pools, Sonoran Zones; Sacramento and San Joaquin Valleys and foothills of the Sierra Nevada, California, from Colusa County to Santa Clara County, and from Placer County to Fresno County. Type locality: Fresno, Fresno County. Collected by Eisen. April–May.

3. Evax caulèscens (Benth.) A. Gray. Involucrate Evax. Fig. 5870.

Psilocarphus caulescens Benth. Pl. Hartw. 319. 1849.
Evax caulescens A. Gray in Torr. Pacif. R. Rep. **4**: 101. *pl. 10.* 1857.
Hesperevax caulescens A. Gray, Proc. Amer. Acad. **7**: 356. 1868.
Evax caulescens var. *petiolata* A. Gray, Syn. Fl. N. Amer. **1²**: 229. 1884.
Evax involucrata Greene, Man. Bay Reg. 185. 1894.
Hesperevax humilis Greene, Fl. Fran. 401. 1897.

Caulescent or acaulescent, arachnoid-lanate annual 8 cm. high or less, simple or with ascending branches chiefly from the base, the stem when present leafy, erect or with decumbent base, the heads glomerate at enlarged apex of stem and branches and conspicuously rosulate-involucrate. Stem-leaves similar to the involucrating leaves but smaller; leaves subtending the heads numerous, spatulate, 4.5–8.5 cm. long, 5–17 mm. wide, the blade obovate to suborbicular, obtuse or acutish, callous-apiculate, tapering into a petioliform base usually two to four times as long, midvein scarcely costate, this dilated and indurate at base, brownish and smooth on the inner surface; heads subcylindric, long-villous at base, 8–20 (less in depauperate plants), crowded in terminal glomerules 7–12 mm. thick; details of structure essentially as in *E. sparsiflora* but prolongation of receptacle shorter and thicker; pistillate flowers few, with subtending obovate receptacular bracts, these concealed in the glomerules by the densely woolly, spreading receptacular bracts encircling the 4–7 hermaphrodite flowers, these with reddish tips; achenes smooth, irregularly obovoid, 2.5 mm. high or less.

Drying beds of vernal pools, Sonoran Zones; Butte County, California, south in the Sacramento Valley to San Joaquin County; also occurring in the adjacent Inner Coast Ranges from Napa County to Alameda County. Type locality: Sacramento Valley. Collected by Hartweg. April–May.

Evax multicaulis DC. Prod. 5: 459. 1836. This species of the southwest has been reported in various floras as growing in the Mojave Desert within the borders of California but no specimens have been found confirming its occurrence. An unauthenticated specimen collected by Lemmon is apparently the source of the presence of this name in the floras.

Tribe 8. EUPATORIEAE

Achenes 5-ribbed.
  Pappus, at least in part, of awns or scales.
    Involucre 2-seriate, of subequal and subherbaceous phyllaries.   130. *Trichocoronis.*
    Involucre imbricated in several series, of dry and longitudinally striped phyllaries.
      Achenes linear-fusiform, distinctly narrowed at apex; annuals; leaves linear.   131. *Malperia.*
      Achenes cylindric-prismatic or linear, not contracted toward the apex; suffrutescent or shrubby plants;
        leaves not linear.   132. *Hofmeisteria.*
  Pappus of slender bristles only.   133. *Eupatorium.*
Achenes 10-ribbed; involucral bracts striately nerved.   134. *Brickellia.*

130. **TRICHOCORÒNIS** A. Gray, Mem. Amer. Acad. II. **4**: 65. 1849.

Low herbs of wet places. Stems weak, usually ascending or sometimes floating. Leaves opposite or alternate, sessile, serrate or 3-lobed, thin. Heads small, discoid, cymose-panicled, or solitary. Involucre 2-seriate, subequal, of subherbaceous phyllaries. Receptacle rounded or highly convex, naked. Corollas with slender tube abruptly dilated into the throat, 5-toothed. Achenes prismatic, 5-angled. Pappus a minute setulose crown or of 5 short, weak awns with alternating lacerate squamellae. Anthers subentire at base. Style-branches linear, the linear, obtuse, hispidulous appendages equaling the stigmatic portion. [Name Greek, meaning hair-apex, referring to the pappus.]

A genus of 3 species, of the southwestern United States and Mexico. Type species, *Ageratum wrightii* Torr. & Gray.

1. **Trichocoronis wrìghtii** (Torr. & Gray) A. Gray. Trichocoronis. Fig. 5871.

Ageratum wrightii Torr. & Gray, Proc. Amer. Acad. **1**: 46. 1848.
Trichocoronis wrightii A. Gray, Mem. Amer. Acad. II. **4**: 65. 1849.
Biolettia riparia Greene, Pittonia **2**: 216. 1891.
Trichocoronis riparia Greene, Erythea **1**: 42. 1893.

Annual, simple or branched from base, the stems ascending to procumbent, 11–35 cm. long, villous with lax many-celled hairs. Leaves below the inflorescence opposite, oblong or rhombic-oblong, 1–2.5 cm. long, obtuse, sessile and clasping, bluntly serrate, villous on the nerves beneath; heads tiny, about 3 mm. high, cymose-panicled; involucre essentially glabrous; corollas purplish, the teeth white; achenes hispidulous above, 1.3 mm. long; pappus-bristles 5, weak, about 0.4 mm. long, alternating with minute lacerate squammelae.

Naturalized in wet ground along the Sacramento and San Joaquin Rivers, California, and at Cienega near Beaumont, Riverside County; also in Texas where it is native. Type locality: "Prope flumen Colorado Texas." Collected by Wright. April–Sept.

131. **MALPÈRIA** S. Wats. Proc. Amer. Acad. **24**: 54. 1889.

Slender annual, slightly pubescent, freely branched. Lowest leaves opposite, the others alternate, linear, entire, sessile. Heads discoid, rather small, numerous, loosely cymose-panicled. Involucre about 5-seriate, strongly graduate, the phyllaries chiefly lanceolate to linear, acute to acuminate, thin, scarious-margined, 2-ribbed. Receptacle flat, naked. Corollas tubular, slender, 5-toothed. Achenes linear-fusiform, 5-ribbed. Pappus of 2–5 (usually 3) slender bristles, longer than the achene, often chaffy-dilated at base, and 3–6 very short, oblong to obovate, erose, scarious squamellae. Anthers entire at base. Style-branches linear-clavate, elongate. [Anagram formed from the name of the collector, Edward Palmer.]

A monotypic genus of California and adjacent Mexico.

1. **Malperia ténuis** S. Wats. Malperia or Brown Turbans. Fig. 5872.

Malperia tenuis S. Wats. Proc. Amer. Acad. **24**: 54. 1889.
Hofmeisteria tenuis I. M. Johnston, Proc. Calif. Acad. IV. **12**: 1188. 1924.

Plants slender, about 25 cm. high; stem minutely and sparsely strigillose. Leaves linear or linear-lanceolate, 1.8–5 cm. long, 1–3 mm. wide, bluntly callous-tipped, entire, revolute-margined, 1-nerved, puberulous; heads nearly cylindric, 1 cm. high; phyllaries appressed, puberulous; corollas ochroleucous or whitish; achenes hispidulous on the ribs, 3.5 mm. long; pappus-bristles 6 mm. long, the squamellae 0.5 mm. long.

In desert washes, Lower Sonoran Zone; Split Mountain and Carriso Mountain in the Colorado Desert, Imperial County, California, to Lower California and Sonora. Type locality: near Los Angeles Bay, Lower California. April–Dec.

132. HOFMEISTÉRIA Walp. Rep. 6: 106. 1846–47.

Plants suffrutescent or shrubby, branching. Leaves alternate or opposite, usually fleshy, dentate to pinnatisect, the petioles usually much longer than the blades. Heads discoid, solitary to cymose-panicled, medium-sized, many-flowered, white to pink or lilac. Involucre several-seriate, strongly graduate, of narrow, acuminate, dryish, striate phyllaries, the outer sometimes with short herbaceous tips. Receptacle flat, naked. Corollas all tubular, slender, 5-toothed. Achenes linear or prismatic, 2–5-ribbed, truncate at apex. Pappus

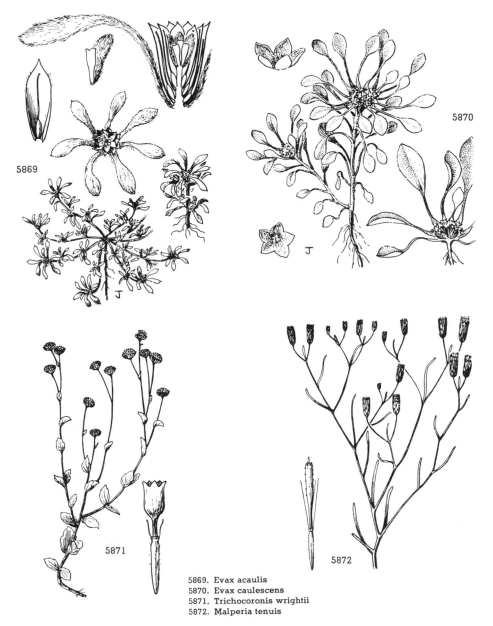

5869. Evax acaulis
5870. Evax caulescens
5871. Trichocoronis wrightii
5872. Malperia tenuis

longer than achenes, of 2–15 bristles, sometimes chaffy-dilated at base, alternating with scarious scales, these sometimes dissected into bristles. Anthers subentire at base. Style-branches linear-clavate, elongate. [Named for Wilhelm Friedrich Benedict Hofmeister of Leipzig, 1824–1877, botanical morphologist and physiologist.]

A genus of about 6 species, of the southwestern United States and northwestern Mexico. Type species, *Helogyne fasciculata* Benth.

1. Hofmeisteria pluriseta A. Gray. Arrow-leaf. Fig. 5873.

Hofmeisteria pluriseta A. Gray. in Torr. Pacif. R. Rep. **4**: 97, *pl. 9*, 1857.
Hofmeisteria viscosa A. Nels. Bot. Gaz. **37**: 263. 1904.

Slender shrub, up to 80 cm. high with tangled branches, glandular-pubescent, the bark white. Leaves opposite below, alternate above; petioles 2–6 cm. long; blades lanceolate to ovate, acute, entire or usually sharply toothed, 2–10 mm. long; heads few in pedunculate cymes or cymose panicles, rarely solitary, about 9 mm. high; involucre glandular-puberulous, the phyllaries dry, often purplish-tinged, striate, at least the outer with acute to acuminate, mucronate, spreading, herbaceous tips; corolla white; achenes densely hispidulous; pappus of about 12 slender bristles, sometimes paleaceous-dilated toward base, alternating with about as many shorter bristles or linear-attenuate scarious squamellae.

In desert places, Lower Sonoran Zone; Nevada and Utah to Arizona, southeastern California, and Lower California. Type locality: Bill Williams Fork, Arizona. Oct.–May.

133. EUPATÒRIUM L. Sp. Pl. 836. 1753.

Herbs or shrubs of varied habit. Leaves usually opposite at least below, commonly petioled and toothed. Heads discoid, many-flowered, never yellow. Involucre 2–many-seriate, equal or graduate, of usually dryish phyllaries. Receptacle flat or convex, naked. Corollas all tubular, 5-toothed. Achenes usually cylindric, 5-ribbed. Pappus of numerous, 1-seriate, rigid, capillary bristles. Stamens entire at base. Style-branches elongate, blunt, usually clavellate. [Named after Eupator Mithridates, King of Pontus, who is said to have used the plant in medicine.]

A very large genus of at least 500 species, chiefly of the Western Hemisphere. Besides the following, numerous species occur in the United States. Type species, *Eupatorium cannabinum* L.

Plants not conspicuously if at all glandular; heads 8–16-flowered, the achenes glandular or hispidulous.
Leaves chiefly alternate; flowers crimson to pink. 1. *E. occidentale.*
Leaves opposite; flowers white. 2. *E. herbaceum.*
Plants conspicuously glandular; heads 20–30-flowered, the achenes glabrous. 3. *E. adenophorum.*

1. Eupatorium occidentàle Hook. Western Eupatorium. Fig. 5874.

Eupatorium occidentale Hook. Fl. Bor. Amer. **1**: 305. 1833.
Kyrstenia occidentalis Greene, Leaflets Bot. Obs. **1**: 9. 1903.

Many-stemmed perennial from a woody root, 25–80 cm. high, the stems slender, finely incurved-puberulent. Leaves chiefly alternate; petioles 3–15 mm. long; blades ovate, 1.8–6 cm. long, usually acute, at base cuneate to truncate or subcordate, acutely serrate-dentate, gland-dotted and slightly puberulous beneath; heads 9–12-flowered, about 8 mm. high, crowded in small cymose panicles terminating stem and branches; involucre 2-seriate, equal, 3–4 mm. high, the phyllaries linear, acute or acuminate, puberulous; corollas crimson, pale purple or white; achenes 3 mm. long, sessile-glandular.

On stream banks and among rocks, Transition and Canadian Zones; Idaho to Utah, western Nevada and Washington south to the mountains of northern California and in the Sierra Nevada to Tulare County. Type locality: "On the low hills between the north and south branch of Lewis and Clarke's River," northern Idaho. July–Sept.

2. Eupatorium herbàceum (A. Gray) Greene. Desert Eupatorium. Fig. 5875.

Eupatorium ageratifolium var. *? herbaceum* A. Gray, Smiths. Contr. **5**⁶: 74. 1853.
Eupatorium occidentale var. *arizonicum* A. Gray, Syn. Fl. N. Amer. **1**²: 101. 1884.
Eupatorium herbaceum Greene, Pittonia **4**: 279. 1901.
Eupatorium arizonicum Greene, op. cit. 280.

Opposite-leaved perennial, 3–4.5 dm. high, from a woody root, herbaceous or somewhat suffrutescent at base, the stems minutely scaberulous, especially on the upper part. Leaves glabrous or somewhat granular beneath, 1–4.5 cm. long, the petiole 4–11 mm. long, the blade rather firm, veins prominent on older leaves, deltoid to ovate-deltoid, acute at apex, truncate or subcordate at base, the margin crenate except at the broad base; inflorescence of short-pedunculate dense cymes, terminal or terminating the upper branchlets; heads 12–16-flowered, white, about 5 mm. high, the peduncles bracteate, densely scaberulous; phyllaries nearly equal in length, lanceolate, more or less scaberulous, 3.5–4 mm. long; achenes hispidulous on the angles.

Canyon slopes and open forest on desert mountains, Upper Sonoran Zone; Colorado and Utah south to New Mexico and Arizona and the New York Mountains (*Munz*), and Clark Mountain (*Roos*), San Bernardino County, California. Type locality: Santa Cruz, Sonora. Collected by Wright. June–Oct.

3. Eupatorium adenóphorum Spreng. Sticky Eupatorium. Fig. 5876.

Eupatorium glandulosum H.B.K. Nov. Gen. & Sp. 4: 122. *pl. 346.* 1820. Not *E. glandulosum* Michx. 1803.
Eupatorium adenophorum Spreng. Syst. 3: 420. 1826.
Eupatorium pasadenense Parish, Zoe 5: 75. 1900.

Herbaceous opposite-leaved perennial with glandular puberulent stems 3–10 dm. high. Leaf-blades 3–7 cm. long, slightly longer than the petiole, ovate-deltoid, sharply acute at apex, prominently 3-veined at the base and truncate or somewhat cuneate at the base, the leaf-margins crenate-dentate from the widest portion of the leaf-blade to the apex; inflorescence of compact cymes on dichotomous peduncles longer than the leaves; heads 20–30-flowered, white, about 5 mm. long; the peduncles bracteate, glandular; phyllaries about 4 mm. long, lanceolate, glandular-puberulent; achenes glabrous.

Moist shrubby slopes, frequently escaping from cultivation but not abundant in the Coast Ranges from Marin and Alameda Counties to Los Angeles County, California; also introduced in the Hawaiian Islands. Type locality: Mexico. Feb.–April; Oct.

Eupatorium maculàtum var. **brùneri** (A. Gray) Breitung, Can. Field Nat. 61: 98. 1947. (*Eupatorium bruneri* A. Gray, Syn. Fl. N. Amer. 1²: 96. 1884.) Tall perennial up to 20 dm. high, differing from *E. maculatum* in the sharply serrate, more firm leaves densely covered beneath with short, spreading, curly hairs. Within our range, collected only in swamps at Sumas, Whatcom County, Washington, but occurring in British Columbia south to Utah and eastward. Type locality: Fort Collins, Colorado. Joe-pye Weed.

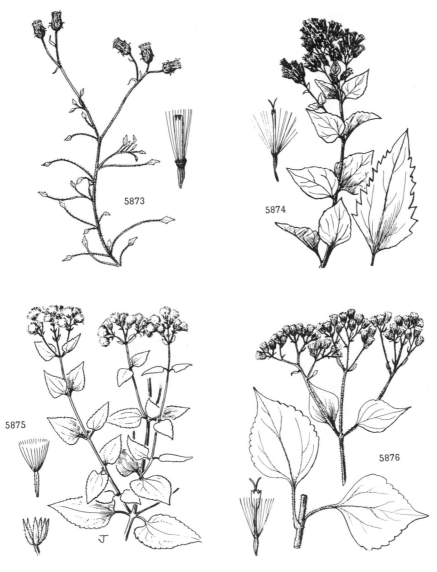

5873. Hofmeisteria pluriseta
5874. Eupatorium occidentale

5875. Eupatorium herbaceum
5876. Eupatorium adenophorum

134. **BRICKÉLLIA** Ell. Bot. S. C. & Ga. 2 : 290. 1823.

Herbs or shrubs. Leaves opposite or alternate, sessile or petioled. Heads discoid, 3-many-flowered, small to large, solitary to panicled, white, ochroleucous, or purplish. Involucre several-seriate, strongly graduate, the phyllaries dry, conspicuously striate, the outer herbaceous-tipped. Receptacle flattish, naked. Corollas all slender-tubular, 5-toothed. Achenes cylindric-prismatic, 10-ribbed. Pappus of numerous capillary bristles, usually white. Anthers rounded at base. Style-branches linear-clavate. [Named in honor of John Brickell, naturalist and physician, a contemporary of Muhlenberg and Elliott.]

A genus of 93 species, all natives of the Americas. Besides those here described. others occur in southwestern and southeastern United States. Type species, *Brickellia cordifolia* Ell.

Heads 3–7-flowered, very small.
  Leaves linear or narrowly linear-lanceolate, 1–10 mm. wide.   1. *B. longifolia.*
  Leaves lanceolate to ovate.
    Leaves entire or subentire; phyllaries nearly glabrous.   2. *B. multiflora.*
    Leaves serrate; phyllaries puberulous above.   3. *B. knappiana.*
Heads 8–many-flowered, usually medium or large.
  Leaves mostly spatulate and entire, usually 14 mm. long or less; intricately branched shrub.
           4. *B. frutescens.*
  Leaves usually ovate and toothed, larger, sometimes elliptic and entire or subentire.
    Leaves white-tomentose; heads large (2 cm. high or more), solitary, naked-peduncled.
           5. *B. incana.*
    Leaves green, or else heads much smaller and on leafy peduncles.
      Leaves firmly coriaceous, spinescent-toothed; heads solitary, naked-peduncled.
        Outer phyllaries broadly ovate-oblong; peduncles glabrate to finely glandular.
           6. *B. atractyloides.*
        Outer phyllaries ovate-lanceolate; peduncles with dense, short, spreading pubescence.
           7. *B. arguta.*
      Leaves neither firmly coriaceous nor spinescent-toothed; heads clustered or leafy-peduncled.
        Outer phyllaries with caudate-attenuate tips.   8. *B. grandiflora.*
        Outer phyllaries not with caudate-attenuate tips.
          Heads closely clustered in short-peduncled axillary and terminal cymules.
            Petioles 5–20 mm. long; leaf-blades 1–5 cm. long, not cinerous-puberulous.
                 9. *B. californica.*
            Petioles 1–3 mm. long; leaf-blades 3–14 mm. long, usually cinereous-puberulous.
                 10. *B. desertorum.*
          Heads otherwise.
            Stem and leaves gray-lanate.   11. *B. nevinii.*
            Stem and leaves not gray-lanate.
              Heads not leafy-bracted.
                Leaves ovate, usually distinctly toothed; heads numerous, solitary, or in clusters of 2–7 at tips of divergent bracteate branches.
                  Stem and pedicels pilose with wide-spreading gland-tipped hairs.
                     12. *B. microphylla.*
                  Stem and pedicels lanulose with mostly eglandular hairs.
                     13. *B. watsonii.*
                Leaves mostly elliptic to oval, entire; stems simple or corymbosely branched; heads 1 to many, cymosely arranged at tips of erect branches.
                   14. *B. oblongifolia.*
              Heads conspicuously leafy-bracted at base.   15. *B. greenei.*

1. **Brickellia longifòlia** S. Wats. Willow-leaved Brickellia. Fig. 5877.

Brickellia longifolia S. Wats. Amer. Nat. **7**: 301. 1873.
Coleosanthus longifolius Kuntze, Rev. Gen. Pl. **1**: 328. 1891.

Much branched shrub, the stem glabrous, viscid, white-barked. Leaves alternate, the blades linear or narrowly linear-lanceolate, 2–13 cm. long, 1–10 mm. wide, attenuate, entire or obscurely denticulate, gland-dotted, short-petioled; heads about 8 mm. high, 3–5-flowered, in small, short-peduncled, axillary and terminal, umbelliform cymules, forming a leafy thyrse; involucres 5 mm. high, the phyllaries 10–12, ovate to linear-oblong, obtuse, dry, glabrous, viscid; corollas whitish; achenes hispidulous.

Rare on canyon slopes and in washes, Sonoran Zones; southern Utah to northern Arizona and Nevada, and Inyo County, California. Type locality: southern Nevada. Aug.–Oct.

2. **Brickellia multiflòra** Kell. Gum-leaved Brickellia. Fig. 5878.

Brickellia multiflora Kell. Proc. Calif. Acad. **7**: 49. 1877.
Coleosanthus multiflorus Kuntze, Rev. Gen. Pl. **1**: 328. 1891.

Branching shrub 5–9 dm. high, more or less viscid, the stem white, glabrous. Leaves alternate, the blades ovate-lanceolate, the upper narrowly lanceolate, 3.5–8.7 cm. long, 1.2–2.8 cm. wide, acuminate, entire or obscurely denticulate, sparsely hispidulous, gland-dotted; phyllaries about 15–20; heads and inflorescence much as in *B. longifolia*, the leaves subtending the flowering branchlets shorter.

Rather common on canyon slopes and in gravelly or sandy washes, Sonoran Zones; southern Sierra Nevada, California, mostly on the eastern slopes eastward in the desert ranges to adjacent Nevada. Type locality: Kings Canyon, Sierra Nevada, Fresno County, California. Aug.–Oct.

3. Brickellia knáppiana Drew. Knapp's Brickellia. Fig. 5879.

Brickellia knappiana Drew, Pittonia 1: 260. 1888.
Coleosanthus knappianus Greene, Erythea 1: 54. 1893.

Branching shrub, 0.4–2.4 m. high, viscid, somewhat hispidulous, the stem white. Leaves alternate, the blades broadly lanceolate to ovate, 2.5–3.6 cm. long, 1–1.6 cm. wide, usually somewhat serrate, gland-dotted; heads 7 mm. high, 5–7-flowered, in a very leafy panicle; phyllaries puberulous especially toward apex; achenes hispidulous.

Rare, along streams, Sonoran Zones; Panamint Mountains, Inyo County, and San Bernardino County. Type locality: near the Mojave River, San Bernardino County, California.

4. Brickellia frutéscens A. Gray. Shrubby Brickellia or Brickellbush. Fig. 5880.

Brickellia frutescens A. Gray, Proc. Amer. Acad. 17: 207. 1882.
Coleosanthus frutescens Kuntze, Rev. Gen. Pl. 1: 328. 1891.

Aromatic, intricately branched, low shrub with spinescent branches and whitish bark, the stem and leaves griseous-tomentellous, sometimes scabrid. Primary leaves (rarely seen) up to 4 cm. long, 1 cm. wide, remotely cuspidate-denticulate, the others spatulate, oval, or elliptic, 3–14 mm. long, obtuse, entire, sessile or short-petioled; heads 1.3–1.5 cm. high, about 26-flowered, solitary at the tips of rather short, usually bracteate peduncles; involucre 9–12 mm. high, the phyllaries chiefly linear, obtuse or acutish, griseous-puberulous, the outer with short, sometimes squarrose, herbaceous tips; corollas purple; achenes hispidulous.

In desert places and among rocks, Lower Sonoran Zone; western edge of the Colorado Desert and adjacent mountains in California southward on the eastern and western slopes of the San Pedro Martir, Lower California, and Esmeralda County (*Monnet*), Nevada. Type locality: not definitely stated. April–Oct.

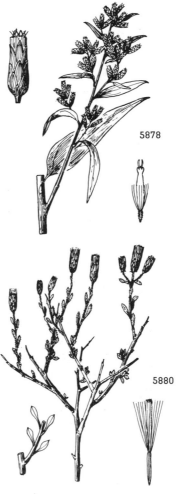

5877. Brickellia longifolia
5878. Brickellia multiflora

5879. Brickellia knappiana
5880. Brickellia frutescens

5. Brickellia incàna A. Gray. Woolly Brickellia or Brickellbush. Fig. 5881.

Brickellia incana A. Gray, Proc. Amer. Acad. **7**: 350. 1868.
Coleosanthus incanus Kuntze, Rev. Gen. Pl. 1: 328. 1891.

Much branched shrub, 3–6 dm. high, white-tomentose on leaves, involucre, and young branches, the stem glabrate, white-barked. Leaves alternate, ovate, 1–2 cm. long, acute to obtuse, sessile, serrulate to entire, glabrate at length; heads mostly solitary at tips of stem and branches, 2–2.8 cm. high, about 60-flowered, naked-peduncled; involucre 1.7–2.2 cm. high, the phyllaries oblong-ovate to linear, obtuse to acuminate, apiculate; corollas apparently whitish or ochroleucous; achenes densely subsericeous-hispidulous.

Sandy washes and canyon bottoms in the desert, Sonoran Zones; Inyo County, California, in the Death Valley region and south through the Mojave Desert to Riverside County; also southern Nevada and western Arizona. Type locality: Providence Mountains, San Bernardino County, California. May–Oct.

6. Brickellia atractylòides A. Gray. Spear-leaved Brickellia. Fig. 5882.

Brickellia atractyloides A. Gray, Proc. Amer. Acad. **8**: 290. 1870.
Coleosanthus atractyloides Kuntze, Rev. Gen. Pl. 1: 328. 1891.
Coleosanthus venulosus A. Nels. Bot. Gaz. **37**: 262. 1904.

Low, leafy, branching shrubs with white-barked, rather stiff branches and flexuous ascending branchlets. Leaves alternate, short-petioled or subsessile, the blades 1–2.5 cm. long, ovate and narrowly acute to acuminate at the spinescent apex, spinescent-toothed on the margin, coriaceous, bright green and glabrous, conspicuously reticulate; heads about 50-flowered, several, solitary on the ends of the branchlets, 13–16 mm. long; peduncles naked, glabrate or finely glandular below the heads; the outer phyllaries broadly ovate-oblong, resembling the leaves in texture, loose, entire; the inner phyllaries linear, acuminate, stramineous; corolla white or yellowish; achenes hispidulous.

Rocky hillsides and slopes, mostly Upper Sonoran Zone; southern Nevada and Utah to eastern Inyo and San Bernardino Counties, California, and Yuma County, Arizona. Type locality: "Utah, near the Rio Colorado." Collected by Palmer. March–Sept.

7. Brickellia argùta Robinson. California Spear-leaved Brickellia. Fig. 5883.

Brickellia arguta Robinson, Mem. Gray Herb. 1: 102. *fig. 79.* 1917.
Brickellia atractyloides var. *arguta* Jepson, Man. Fl. Pl. Calif. 1015. 1925.
Coleosanthus argutus Blake in Standley, Contr. U.S. Nat. Herb. **23**: 1483. 1926.

Low, branching shrub, closely resembling *B. atractyloides* in general appearance, densely spreading-hirtellous with mostly gland-tipped hairs on the leaves, involucre, and young branches, the stem glabrate, white-barked. Leaves alternate, short-petioled, the blades ovate, 1–2.3 cm. long, acute or acuminate, spinescent-toothed and -tipped, rigidly coriaceous, reticulate, puberulent on the veins beneath and the surface somewhat glandular; heads 1.2–1.5 cm. high, about 50-flowered, solitary at tips of branches, the naked peduncle densely glandular-puberulent with spreading hairs; involucre 1–1.3 cm. high, the outermost phyllaries usually rigid-herbaceous, more or less puberulent with spreading hairs especially at the base, lanceolate to ovate, rather loose, entire or rarely with 1–2 small teeth; the inner phyllaries linear or lance-linear, acuminate; corollas apparently whitish; achenes hispidulous.

On desert slopes and washes, Lower Sonoran Zone; western Mojave and Colorado Deserts, California, from the Death Valley region in Inyo County south to northern Lower California. Type locality: not stated. May–June.
Intergrading forms are to be found on the eastern limits of distribution which show strongly the close relationship of this species with *B. atractyloides.*

Brickellia arguta var. **odontólepis** Robinson (Mem. Gray Herb. 1: 103. 1917) differs from *B. arguta* var. *arguta* in the outer phyllaries which are broad and conspicuously toothed. Death Valley region, Inyo County, California (*Gilman 2337, 3721*), and southern Colorado Desert, Imperial County, California, to the San Felipe Desert in northeastern Lower California. Type locality: Colorado Desert. Collected by Orcutt.

8. Brickellia grandiflòra Nutt. Large-flowered Brickellia. Fig. 5884.

Eupatorium? grandiflorum Hook. Fl. Bor. Amer. **2**: 26. 1834.
Brickellia grandiflora Nutt. Trans. Amer. Phil. Soc. II. **7**: 287. 1840.
Brickellia grandiflora var. *petiolaris* A. Gray, Proc. Amer. Acad. **17**: 207. 1882.
Coleosanthus grandiflorus Kuntze, Rev. Gen. Pl. 1: 328. 1891.
Coleosanthus gracilipes Greene, Pittonia **4**: 237. 1901.

Herbaceous perennial 2–7 dm. high, from long, tuberous-thickened, fusiform roots, the stem simple or branched, incurved-puberulous. Leaves chiefly alternate; petioles 1–7 cm. long; blades triangular-ovate, 2–11 cm. long, acuminate, at base truncate to cordate, crenate-serrate or sharply serrate, thin, gland-dotted beneath, more or less scabrid-puberulous; heads 1–1.3 cm. high, 20–38-flowered, nodding or erect, in umbelliform clusters at tips of stem and branches; involucre 8–11 mm. high, the outer phyllaries with conspicuously caudate-attenuate tips, the inner linear, acute; corollas whitish; achenes hispidulous.

In canyons and on cliffs, Transition Zone; Washington and Montana to Nebraska, Missouri, and Arkansas south to New Mexico, Arizona, the Sierra Nevada of California and northern Lower California. Type locality: "On the low hills between the north and south branches of Lewis and Clarke's River," northern Idaho. July-Oct.

9. Brickellia califórnica A. Gray. California Brickellia or Brickellbush. Fig. 5885.

Bulbostylis californica Torr. & Gray, Fl. N. Amer. **2**: 79. 1841.
Brickellia californica A. Gray, Mem. Amer. Acad. II. **4**: 64. 1849.

Brickellia wrightii A. Gray, Smiths. Contr. 5[6]: 72. 1853.
Coleosanthus californicus Kuntze, Rev. Gen. Pl. 1: 328. 1891.
Coleosanthus albicaulis Rydb. Bull. Torrey Club 31: 646. 1905.
Brickellia californica var.? *jepsonii* Robinson, Mem. Gray Herb. 1: 71. 1917.

Branching shrub 5–10 dm. high, the stem tomentellous to scabrid-puberulous, whitish barked. Leaves chiefly alternate; petioles 5–20 mm. long; blades usually deltoid-ovate, 1–5 cm. long, obtuse or acute, truncate to subcordate at base, coarsely crenate or crenate-serrate, firm, crisped-puberulous, gland-dotted beneath, reticulate; heads 7–10 mm. high, 8–18-flowered, closely clustered in axillary and terminal cymules, forming leafy thyrses; involucre about 8 mm. high, the phyllaries dry, obtuse to acutish, ovate to linear; corollas ochroleucous, sometimes purplish; achenes hispidulous.

Dry canyons and streambeds, Sonoran Zones; Josephine County, Oregon, southward to Lower California and eastward through the desert ranges to Colorado and New Mexico, adjacent Texas, and Chihuahua. Type locality: California. July–Oct.

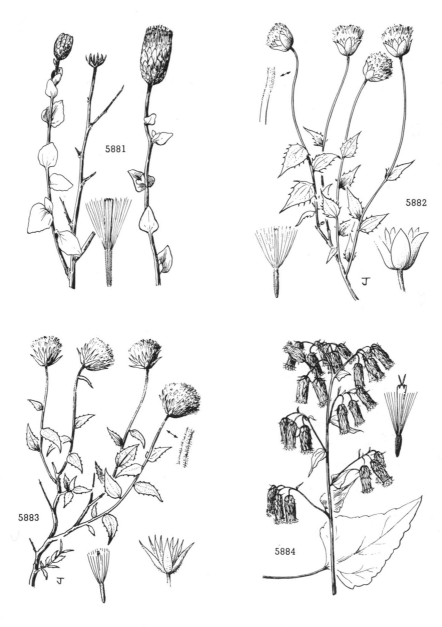

5881. Brickellia incana
5882. Brickellia atractyloides
5883. Brickellia arguta
5884. Brickellia grandiflora

10. **Brickellia desertòrum** Coville. Desert Brickellia or Brickellbush. Fig. 5886.

Brickellia desertorum Coville, Proc. Biol. Soc. Wash. **7**: 68. 1892.
Coleosanthus desertorum Coville, Contr. U.S. Nat. Herb. **4**: 119. 1893.
Brickellia californica var. *desertorum* Parish ex H. M. Hall, Univ. Calif. Pub. Bot. **3**: 33. 1907.

A much branched shrub 1 m. high, the stem and branches slender, white-barked, whitish-tomentellous, glabrate. Leaves opposite or alternate; petioles 1–3(6) mm. long; blades ovate, 3–14 mm. long, obtuse, rounded or subtruncate at base, crenate-serrate, cinereous-puberulous or greenish; heads 7–9 mm. high, 8–12-flowered, in small axillary and terminal glomerules; involucre about 6 mm. high, the phyllaries ovate to linear, obtuse to acutish, puberulous; achenes hispidulous.

Rocky slopes of desert ranges, Sonoran Zones; western Riverside County, California, eastward to southern Nevada and northwestern Arizona. Type locality: between Banning and Seven Palms on the Southern Pacific railroad, California. Nov.–March.

11. **Brickellia nevínii** A. Gray. Nevin's Brickellia. Fig. 5887.

Brickellia nevinii A. Gray, Proc. Amer. Acad. **20**: 297. 1885.
Coleosanthus nevinii Heller, Cat. N. Amer. Pl. 8. 1898.

Suffrutescent, low, the stems slender, branching, gray-lanate, glabrescent. Leaves alternate; petioles 2 mm. long or less; blades ovate, 5–16 mm. long, acute or obtuse, serrate, reticulate, grayish lanate, somewhat glabrescent; heads 1.2–1.5 cm. high, about 23-flowered, 1–3 at tips of usually minutely leafy branches; involucre 8–12 mm. high, more or less densely lanulose, the phyllaries ovate or lanceolate to linear, acute to acuminate, the outer with spreading tips; corollas whitish or purplish; achenes slightly hispidulous.

In dry places, Upper Sonoran Zone; coastal slope of the mountains of southern California from western Kern County to Ventura and Los Angeles Counties. Type locality: near Newhall, California. May–Oct.

12. **Brickellia microphýlla** A. Gray. Little-leaved Brickellia or Brickellbush. Fig. 5888.

Bulbostylis microphylla Nutt. Trans. Amer. Phil. Soc. II. **7**: 286. 1840.
Brickellia microphylla A. Gray, Smiths. Contr. **3**[5]: 85. 1852.
Brickellia cedrosensis Greene, Bull. Torrey Club **10**: 86. 1883.
Coleosanthus microphyllus Kuntze, Rev. Gen. Pl. **1**: 328. 1891.

Shrubby, 4–6 dm. high, much branched, glandular-pilose on stem and leaves, the bark whitish or grayish. Leaves alternate, greenish, petioles 1–5 mm. long, blades broadly ovate, 1–2 cm. long, acute or obtuse, coarsely few-toothed to nearly entire, veiny; heads 1–1.4 cm. high, about 22-flowered, in groups of 1–7 at tips of the divergent, minutely leafy branches; involucre about 1 cm. high, the phyllaries ovate to linear, glandular-viscid, acute or acuminate, the outer with squarrose herbaceous tips; corollas ochroleucous or purplish; achenes hispidulous above, 4–5 mm. long.

In rocky places, Upper Sonoran and Transition Zones; eastern Oregon in the Blue Mountains eastward to Idaho and southward to Utah and Nevada and in California on the eastern face of the Sierra Nevada from Sierra County to Tulare County and in the White and Inyo Mountains and San Gabriel Mountains, Los Angeles County; also on Cedros Island, Lower California. Type locality: Blue Mountains of Oregon. Aug.–Sept. Foliage said to be pineapple-scented.

13. **Brickellia watsònii** Robinson. Sweet Brickellia or Brickellbush. Fig. 5889.

Brickellia watsonii Robinson, Mem. Gray Herb. **1**: 42. *fig. 19.* 1917.
Coleosanthus watsonii Rydb. Fl. Rocky Mts. 843. 1917.

Similar to *B. microphylla*; stem and pedicels lanulose or crisped-puberulous with few or no spreading, gland-tipped hairs. Leaves not conspicuously veiny; heads 9–11 mm. high, about 18-flowered; involucre 7–8 mm. high; achenes hispidulous, about 3.5 mm. long.

In rocky places, Upper Sonoran and Transition Zones; Utah, southeastern Nevada, and the desert ranges of San Bernardino County and eastern slope of Santa Rosa Mountains, Riverside County, California. Type locality: not stated. Aug.–Sept.

Brickellia scàbra (A. Gray) A. Nels. in Coult. & Nels. New Man. Bot. Rocky Mts. 487. 1909, in synonymy as *scaber.* (*Brickellia microphylla* var. *scabra* A. Gray, Proc. Amer. Acad. **11**: 74. 1876.) Differing from the two preceding taxa in having the stems and peduncles with a short, glandular, spreading pubescence. May be expected in eastern Inyo County, California, as it has been collected in adjacent Esmeralda and Nye Counties, Nevada.

14. **Brickellia oblongifòlia** Nutt. Narrow-leaved Brickellia. Fig. 5890.

Brickellia oblongifolia Nutt. Trans. Amer. Phil. Soc. II. **7**: 288. 1840.
Coleosanthus oblongifolius Kuntze, Rev. Gen. Pl. **1**: 328. 1891.

Suffrutescent at base, 1–5 dm. high, the stems numerous, erect, whitish, hispidulous and glandular, usually simple below the inflorescence. Leaves alternate, elliptic to oval, 1–4 cm. long, acute or apiculate to obtuse, subsessile, entire or rarely few-toothed, pubescent like the stem; heads 1 to many, cymosely arranged at apex of stem and branches, 1.4–1.8 cm. high, 40–50-flowered; involucre about 1.5 cm. high, the phyllaries lance-oblong to linear, acute to acuminate, glandular-puberulous, striped green and white; corollas purplish or ochroleucous; achenes glandular, sometimes also with a few eglandular hairs.

In sandy or gravelly places, Upper Sonoran and Transition Zones; British Columbia to Oregon, Nevada, and Utah. Type locality: gravelly bars of the Columbia and tributary streams and along the Willamette. June–Aug.

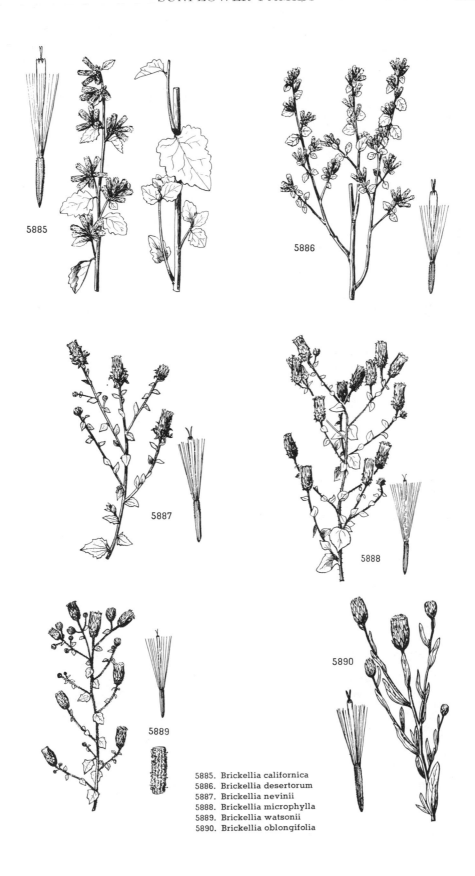

5885
5886
5887
5888
5889
5890

5885. Brickellia californica
5886. Brickellia desertorum
5887. Brickellia nevinii
5888. Brickellia microphylla
5889. Brickellia watsonii
5890. Brickellia oblongifolia

Brickellia oblongifolia var. linifòlia (D. C. Eaton) Robinson, Mem. Gray Herb. **1**: 104. 1917. (*Brickellia linifolia* D. C. Eaton, Bot. King Expl. **5**: 137. *pl. 15, figs. 1–6.* 1871; *B. mohavensis* A. Gray, Syn. Fl. N. Amer. **1**²: 104. 1884; *Coleosanthus linifolius* Kuntze, Rev. Gen. Pl. **1**: 328. 1891; *C. oblongifolius linifolius* (D. C. Eaton) Blake, Contr. U.S. Nat. Herb. **25**: 534. 1925.) Achenes densely hispidulous, the glands few or none. In more desert places, Upper Sonoran Zone; Colorado to New Mexico, northern Arizona, Nevada, and the Death Valley region of Inyo County, California, south to eastern San Bernardino County. Type locality: American Fork, Jordan Valley, Utah.

15. Brickellia grèenei A. Gray. Greene's Brickellia. Fig. 5891.

Brickellia greenei A. Gray, Proc. Amer. Acad. **12**: 58. 1876.
Coleosanthus greenei Kuntze, Rev. Gen. Pl. **1**: 328. 1891.

Stems numerous from a woody caudex, erect, 2.5–4.5 dm. high, simple or corymbosely branched above, spreading-pilose with short and long gland-tipped hairs throughout, the herbage with a strong sweetish odor. Leaves alternate; petioles 4 mm. long or less; blades ovate, 1–3 cm. long, obtuse or acute, serrate to entire, glandular-pilose; heads 1.5–2 cm. high, about 60-flowered, solitary at tips of stem and branches, involucrate-bracted by the uppermost leaves; proper phyllaries linear-lanceolate, acuminate, essentially glabrous; corollas ochroleucous; achenes hispidulous above.

Gravelly river beds and rocky mountain sides, Transition Zone; Josephine County, Oregon, south through the Siskiyou region to Tehama, Lake and eastern Humboldt Counties, California. Type locality: south branch of Scott River, California. July–Aug.

Tribe 9. CYNAREAE*

Heads 1-flowered, numerous, aggregated into dense, spherical, capitate clusters.   135. *Echinops.*
Heads few- to many-flowered, not forming spherical clusters.
    Attachment-scar of achenes ("hilum") basal; pappus present, deciduous (or the inner series of bristles more or less persistent in *Saussurea*); heads homogamous, the flowers alike and perfect.
        Leaves neither spinose nor pinnately lobed, the plants not thistle-like; pappus-bristles free at least in the outer series.
            Phyllaries linear or linear-lanceolate, the outer and middle tipped with an uncinate spine; receptacle densely setose.   136. *Arctium.*
            Phyllaries broadly ovate to lanceolate, acute, unarmed; receptacle naked or fimbrillate.   137. *Saussurea.*
        Leaves usually spinose and pinnately lobed to pinnatifid, the plants commonly thistle-like; pappus-bristles united at the base into a low ring.
            Receptacle deeply pitted, the pits membranous-bordered, not densely setose.   139. *Onopordum.*
            Receptacle densely setose.
                Pappus-bristles scabrous or setulose, not plumose.
                    Leaves mottled with white along the veins; phyllaries with a foliaceous, pinnately spinose appendage.   138. *Silybum.*
                    Leaves not white-mottled; phyllaries entire, gradually narrowed to a spinose apex.   143. *Carduus.*
                Pappus-bristles plumose.
                    Leaves very large; heads large, the phyllaries numerous, broad, imbricate; receptacle fleshy.   141. *Cynara.*
                    Leaves usually small or medium; heads small to large, the phyllaries usually narrower, frequently spreading; receptacle not fleshy.   142. *Cirsium.*
    Attachment-scar of achenes lateral; pappus, when present, consisting of free bristles or paleae, persistent or deciduous.
        Phyllaries with a foliaceous blade, the phyllaries and leaves usually spinose or spinulose-dentate; heads homogamous; achenes quadrangular, thick.   140. *Carthamus.*
        Phyllaries not foliaceous, either spinose or unarmed; heads generally heterogamous with the outermost flowers sterile, or sometimes the head homogamous; achenes terete or compressed.
            Achenes compressed, smooth or rugulose, not 10-toothed at apex; pappus present or lacking, not in series of 10; leaves not spinose or bristly.   144. *Centaurea.*
            Achenes terete, 20-ribbed, 10-toothed at apex; pappus in 2 series of 10 bristles each; leaves spinulose.   145. *Cnicus.*

135. ECHÌNOPS L. Sp. Pl. 814. 1753.

Annual, biennial, or perennial, arachnoid-tomentose herbs with more or less spiny leaves. Leaves pinnately divided with the larger divisions again once- or twice-lobed or -parted, the upper sessile by a broad clasping base, the lower with blades narrowly decurrent along a petiole-like base. Heads 1-flowered, numerous, aggregated to form a compact, spherical, capitate inflorescence that is terminal on peduncle-like stems, the heads deciduous at maturity. Phyllaries very numerous in many series attached to an elongate, stipe-like receptacle, the lowest phyllaries bristle-like or scale-like, chartaceous, passing gradually into broader, herbaceous-tipped phyllaries, the middle and uppermost phyllaries attenuate-spinose, pectinate-fimbriate with marginal bristles. Corolla deeply 5-parted, regular, the proper throat very short, the tube elongate, slender. Anthers with setose-hairy, basal appendages. Style hairy at the base of the branches. Achenes elongate, little-compressed, hairy, remaining attached to the receptacle and falling enclosed by the husk-like involucre. Pappus paleaceous, forming a low crown on the top of the achene, the paleae

---

* Text of *Cynareae* contributed by John Thomas Howell.

hairy or barbellate-scabrous, distinct or united to form a more or less erose membrane. [From the Greek, meaning urchin-like.]

An Old World genus of about 70 (or 125, according to Lemée) species, most common in the eastern Mediterranean region. Type species, *Echinops sphaeroccphalus* L.

1. Echinops ruthénicus M. Bieb. Ruthenian Globe Thistle. Fig. 5892.

Echinops ruthenicus M. Bieb. Fl. Taur. Caucas. 3 : 597. 1819.
Echinops ritro var. *ruthenicus* Halácsy, Consp. Fl. Graecae 2 : 91. 1902.

Perennial herb with erect stem 1–1.5 m. tall, simple below and few-branched above, the stem clothed with arachnoid tomentum that may be more or less deciduous especially on the lower part. Leaves arachnoid-tomentose and pale on lower side, green and sparsely visciduolous-hairy on the upper side, lower and middle cauline leaves up to 4 dm. long and 2 dm. wide, divided nearly or quite to the midrib into narrow or broad, elliptic-oblong, pinnately arranged divisions which are again more or less pinnately lobed or parted, the terminal and lateral segments attenuate into a weak or strong, spinose tip, the lower leaves with a narrowly bordered petiole-like base, the middle sessile with a more or less clasping base, the upper leaves much smaller, lanceolate to ovate-lanceolate, pinnately divided or serrate-lobed; inflorescence 3–4 cm. in diameter, the heads in flower 1.5–2 cm. long, the lowest bristle-like phyllaries white and shining, the middle and innermost phyllaries narrowly lanceolate with elongate attenuate tips, the outside lavender-tinged, scabrous-roughened on exposed parts, not glandular, the inside brownish-tinged; corolla-lobes lavender-blue, linear-oblong, 7 mm. long; filaments short, violet, anther-tube conspicuous, 6 mm. long; achene oblongish, a little constricted above the middle, 6 mm. long, thinly covered with appressed, ascending, bristle-like, straw-colored hairs; pappus-paleae united to form a shallowly lobed, membranous crown 1 mm. long, the lobes barbellulate.

Established in weedy places in Whitman County, Washington. Introduced from Europe. Aug.–Sept.

Echinops commutàtus Juratzka, Verh. Zool. Bot. Ges. Wien 8 : 17. 1858. A plant collected in 1954 in Cowlitz County, Washington, appears to be this central European globe thistle. It may be distinguished from *E. ruthenicus* M. Bieb. by the silvery sheen on the gray-green middle and innermost phyllaries, by the fewer sordid, bristle-like phyllaries and by the whitish corollas. Perhaps only a fugitive from cultivation.

Echinops sphaerocéphalus L. Sp. Pl. 814. 1753. Differing from *E. ruthenicus* and *E. commutatus* in the glandular pubescence on the exposed backs of the middle and innermost phyllaries, in the glandular hairs protruding from the arachnoid tomentum on the stems and lower leaf-surfaces, and in the denser glandular-scabridous pubescence on the upper leaf-surfaces. Locally established in Modoc County, California, east of Tulelake, Siskiyou County; native in Europe and Asia.

136. ÁRCTIUM L. Sp. Pl. 816. 1753.

Large biennial herbs with unarmed stems and leaves. Leaves broadly ovate, deeply cordate, long-petiolate. Heads homogamous, many-flowered, medium-sized, subcorymbosely or subracemosely arranged on short to long peduncles. Phyllaries very numerous in many series, narrowly lanceolate, tapering into a stiff, usually uncinate spine, the outer and middle phyllaries reflexed or spreading above the appressed base. Receptacle coarsely setose. Achenes compressed with a basal hilum, the sides tending to be spongy-rugulose or

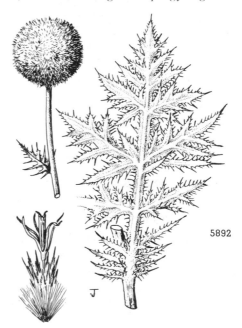

5891. Brickellia greenei   5892. Echinops ruthenicus

-ribbed. Pappus-bristles in several series, free, a little unequal, strongly scabrous-setulose, deciduous. [Ancient name derived from the Greek word for bear, perhaps in reference to the spiny involucre.]

A genus of about 6 species found chiefly in Eurasia; becoming naturalized elsewhere. Type species, *Arctium lappa* L.

Heads mostly 2.5 cm. broad or less, subracemosely arranged along the branches; inner phyllaries usually purplish-tinged, the margins serrulate with very fine, upwardly pointing, crustaceous teeth; pappus-bristles about 2 mm. long.   1. *A. minus.*
Heads mostly 3 cm. broad or more, subcorymbosely arranged at the ends of stems and branchlets; inner phyllaries usually green, the margins ciliolate with small reflexed or outwardly pointing hairs; pappus-bristles to 5 or 6 mm. long.   2. *A. lappa.*

1. **Arctium mìnus** (Hill) Bernh. Common Burdock. Fig. 5893.

Lappa minor Hill, Veg. Syst. 4: 28. 1762.
Arctium minus Bernh. Syst. Verz. Erf. 154. 1800.

Stems to 1 or 1.5 m. tall, the branches substrictly erect. Basal leaves ovate to round-ovate, deeply cordate, the earliest blades small, the second-year blades becoming as much as 6 dm. long and 4 dm. wide with stout hollow petioles 3 dm. long or more, green and glabrate above, pale and arachnoid-tomentose below, the margin undulate or shallowly undulate-lobed, acute or obtusish at the apex, the upper leaves much smaller, ovate, subcordate, truncate, or cuneate at base; heads subglobose, 1.5–3.5 cm. wide, sessile or pedunculate, racemose or subracemose along the branchlets, tending to be capitate-congested near the ends; phyllaries more or less arachnoid, slender, the outer and middle tapering from the lanceolate appressed base to the spreading, stiffish, uncinate spine, the inner chartaceous, attenuate, purplish-tinged, tipped by a straight or uncinate spine, the lower margins of the middle phyllaries and the upper margins of the inner phyllaries minutely serrulate, the teeth strongly ascending, crustaceous, sometimes a little glandular; corollas equaling or exceeding the phyllaries, generally pink or purplish; achenes oblong-cuneate, truncate at base and apex, 5–7 mm. long, the sides slightly spongy-wrinkled or -ribbed, pappus buff, the setulose bristles mostly 1.5–2.5 mm. long.

Fields, roadsides, and waste ground in towns, occasional but widespread, sometimes locally abundant; Washington to central California; British Columbia; east to the Atlantic. Naturalized from Europe. July–Oct.
Two sorts of plants are here treated as *A. minus*, one with heads about 2 cm. broad and the other with heads about 3 cm. broad. The latter plant may be the one called *A. nemorosum* in eastern United States, but with us the two appear to intergrade completely.

2. **Arctium láppa** L. Great Burdock. Fig. 5894.

Arctium lappa L. Sp. Pl. 816. 1753.
Lappa major Gaertn. Fruct. 2: 379. 1791.

Rank herb 1.5–3.5 m. tall. Leaves ovate, deeply cordate, the basal leaves of the flowering season very large, to 5 dm. long and 3 dm. wide, green and glabrate above, white-tomentose below except on the principal veins, margins repand-dentate or dentate, obtuse, petiole generally solid, to 2 dm. long or longer, the middle and upper cauline leaves smaller, cordate, rounded, or broadly cuneate at base; heads subglobose, 2.5–4.5 cm. in diameter, long- or short-pedunculate, corymbosely arranged at the ends of the main branches or on short lateral branches; phyllaries green or rarely purplish-tinged, glabrous, the outer and middle lanceolate, the base appressed, the upper part spreading or reflexed, tapering to the slender uncinate spine, the inner phyllaries stiff-chartaceous, ascending or squarrose, the spine straight or uncinate, the margin of the inner phyllaries and of the base of the outer and middle phyllaries narrowly scarious, minutely ciliolate with horizontally spreading or reflexed, glandular hairs; corollas equaling or shorter than the phyllaries, purplish red; achenes oblong-oblanceolate or somewhat cuneiform, 6–7 mm. long, brown or grayish brown, nearly smooth or spongy-wrinkled or -ribbed, pappus ivory-white or buff, the longest bristles 5–6 mm. long.

Occasional weed of pastures, roadsides, and waste ground; mostly west of the Cascade Mountains in Washington and in the Coast Ranges in California south to southern California; British Columbia; central and eastern United States and Canada. Native of Eurasia. June–Aug.

137. **SAUSSÙREA** DC. Ann. Mus. Paris 16: 156, 196. *pls. 10–13*. 1810.
Nomen conservandum.

Perennial unarmed herbs with alternate, entire, toothed, or pinnatifid leaves. Heads medium or large, solitary or corymbosely clustered, homogamous, the flowers all perfect. Involucre round-ovate, cylindric-campanulate, or turbinate, the phyllaries numerous in many series, appressed or squarrose, unarmed, acute. Receptacle plane or convex, naked or bearing chaffy scales or bristles. Corollas tubular, 5-lobed, bluish or purplish. Achenes oblong, more or less angled, hilum basal. Pappus-bristles in 1 or 2 series, plumose, united at the base and generally deciduous, an outer, shorter, dissimilar series present or lacking. [Named for Horace Benedict de Saussure and his son, Theodore, well-known naturalists of Geneva, Switzerland.]

A genus of about 130 species, mostly Arctic or high montane, occurring chiefly in Asia, with several species in Europe and North America and one in Australia. Type species, *Serratula alpina* Willd.

5893. Arctium minus   5894. Arctium lappa

1. Saussurea americàna D. C. Eaton. American Sawwort. Fig. 5895.

Saussurea americana D. C. Eaton, Bot. Gaz. 6: 283. 1881.
Saussurea alpina var. cordata Kurtz, Bot. Jahrb. 19: 354. 1894.

Plants with simple, erect, leafy stems from the crown of a short stout rootstock, sparsely puberulent or tomentulose and a little glandular, particularly on the lower side of the leaves; stems 3–10 (or 20) dm. tall, terete, mostly unwinged or sometimes the leaf-bases a little decurrent. Leaves thin-herbaceous, triangular-ovate to oblongish or lanceolate, saliently dentate, the lower and middle leaves cordate, truncate, or cuneate at base, attenuate and acute at apex, 5–15 cm. long, 2–7 cm. wide, with narrowly or widely bordered petioles, the upper leaves smaller, generally sessile, cuneate; heads subcapitately or openly corymbose, at the end of the stem or on short branches from the axils of the uppermost leaves, the heads medium-sized, mostly 10–20-flowered, cylindric or oblong-turbinate, about 2 cm. long; phyllaries numerous in many series, closely appressed, acute, sparsely arachnoid- or floccose-tomentose, coriaceous, the outer and middle broadly triangular-ovate or ovate, the inner lanceolate, somewhat membranous on the margin, exposed parts of the phyllaries frequently tinged with dark brown or purplish black; receptacle naked or fimbrillate; flowers lavender-blue or purple, rarely albino; achenes narrowly oblong, somewhat curved, a little turgid or compressed, several-angled, 4–6 mm. long, light brown, sparsely glandular-dotted or smooth; pappus in 2 unlike series, the outer bristle-like, setulose, deciduous, to 7 mm. long, the inner series more paleaceous and more or less connate at the base, about 1 cm. long, plumose, light brown, apparently persistent.

Moist meadowy or rocky slopes in the mountains, Canadian and Hudsonian Zones; Olympic and Cascade Mountains, Washington, south and east to the Blue and Siskiyou Mountains, Oregon, and to the Siskiyou Mountains, California; east to Idaho and Montana; southern Alaska and perhaps adjacent British Columbia. Type locality: "mountains of Union Co., Oregon." Collected by Cusick in 1877. July–Sept.

Saussurea nùda var dénsa (Hook.) Hultén, Fl. Alaska & Yukon, 1627. 1950. (Saussurea alpina β densa Hook. Fl. Bor. Amer. 1: 303. 1833; S. densa Rydb. Bull. Torrey Club 37: 541. 1910.) Dwarf plant mostly 2 dm. tall or less with narrow elliptic leaves that are scarcely reduced upward and frequently surpass the densely capitate-clustered heads. Mountain slopes, Hudsonian or Arctic-Alpine Zones; reported from Mount Benson, Vancouver Island; Rocky Mountains of British Columbia and Alberta. Type locality: "elevated parts of the Rocky Mountains." Collected by Drummond.

138. SÍLYBUM [Vaill.] Adans. Fam. Pl. 2: 116. 1763.

Annual or biennial herbs with leafy stems and alternate spiny leaves. Heads pedunculate, medium or large, many-flowered, homogamous. Phyllaries numerous in many series, the outer and middle bearing a stiff, spreading, spinose, foliaceous appendage. Receptacle fleshy, setose. Achenes compressed, nearly smooth, hilum basal. Pappus-bristles in several series, setulose, united at the base into a ring, deciduous. [From the Greek name of a thistle that was used for food.]

A genus of 2 species indigenous to the Mediterranean region. Type species, Carduus marianus L.

1. **Silybum mariànum** (L.) Gaertn. Milk Thistle. Fig. 5896.

Carduus marianus L. Sp. Pl. 823. 1753.
Silybum marianum Gaertn. Fruct. 2: 378. 1791.

Plants with erect leafy stems, sparsely arachnoid-pubescent and mealy-puberulent, glabrescent, 1.5 dm. tall in depauperate individuals or as much as 2–3 m. tall in robust specimens; stems simple or few-branched, terete, not winged. Leaves generally conspicuously white-mottled, becoming very large, up to 8 dm. long and 3 dm. wide, basal leaves subrosulate, these and the lower cauline leaves oblanceolate or elliptic, saliently spinose-dentate or pinnately lobed or divided, the lobes spine-tipped and spinose-dentate, petiolate, the middle and upper cauline leaves sessile with a broad, clasping, shortly decurrent base, shallowly or deeply lobed, spinose-dentate, the uppermost leaves much reduced with strongly spiny basal lobes and attenuate tip; head medium to large, 3–6 cm. long, oblong to globose, terminal on slender peduncles; phyllaries puberulent, the outer and middle with a foliose and divaricately spreading appendage above the appressed base, the appendage pinnately spiny at the base and along the sides, the end attenuate and rigidly spine-tipped, equaling or a little longer than the flowers, the inner phyllaries erect, coriaceous, bearing a reduced appendage or narrowed to an acute innocuous apex; flowers numerous, purplish, rarely albino; achene elliptic-obovate, 7 mm. long, minutely transversely rugulose, longitudinally lined and dotted with buff and dark brown; pappus-bristles ivory-white, a little unequal, 1–2 cm. long, the outer setiform, the inner broader and a little paleaceous, the top of the pappus-ring bearing fine erect hairs to 2 mm. long.

Occasional in waste ground and along roads west of the Cascade Mountains in Washington and Oregon; a common and widespread weed at lower elevations in California west of the Sierra Nevada; Vancouver Island; occasional east to the Atlantic; South America; Australia. Native of the Mediterranean region. April–Aug.

139. **ONOPÓRDUM** [Vaill.] L. Sp. Pl. 827. 1753.

Tall herbs with erect stems and alternate, simple, spiny leaves, the cauline leaves decurrent and the stems conspicuously spiny-winged. Heads homogamous, large, mostly solitary at the ends of the branches. Involucre broad, the phyllaries numerous in many series, entire, narrowed into a stiff spine. Receptacle fleshy, not setose, alveolate, the pits membranous-bordered. Achenes somewhat compressed, pubescent, transversely rugulose, hilum basal. Pappus setose, the bristles slender, scabrous, in several series, united at the base into a low ring, the whole deciduous. [Name from the Greek, meaning donkey flatulence, the plants believed to cause a flatulent state in donkeys.]

About 20 species, native in Europe, Asia, and northern Africa. Type species, *Onopordum acanthium* L.

1. **Onopordum acánthium** L. Scotch Thistle. Fig. 5897.

Onopordum acanthium L. Sp. Pl. 827. 1753.

Plants biennial, sometimes short and slender in depauperate specimens but frequently very robust and 2 m. tall or more, openly, fastigiately and virgately branched, greenish or canescent with a close arachnoid tomentum; stems broadly winged by the decurrent leaf-bases, the wings as much as 5 cm. broad and ribbed with stiff horizontal spines. Lowest leaves rosulate, oblanceolate, to 3 dm. long and 1 dm. wide, shallowly pinnately lobed, the lobes coarsely spinose-dentate, greenish above and gray or whitish below, narrowed at the base into a short petiole, the lower cauline leaves oblanceolate to elliptic or oblong-ovate, to 4 dm. long and 2 dm. wide or more, sessile, the upper leaves with much reduced blades, narrowly lanceolate or linear, frequently narrower than the stem-wings immediately beneath; heads large, subglobose, 3–5 cm. in diameter, solitary at the ends of the branches; phyllaries lanceolate, the base appressed, the upper part erect or spreading, attenuate and rigidly spine-tipped; flowers reddish purple, exceeding the involucre; achenes narrowly obovate, 5–6 mm. long, somewhat compressed, transversely rugulose, more or less pubescent especially below the middle, brownish gray, pale below, almost black at the top; pappus of numerous slender bristles, about 8 mm. long.

Fields and roadsides, becoming common in the Snake River Canyon; southeastern Washington to central and eastern Oregon, and Modoc, Lassen, and Lake Counties, California; western Idaho and Nanaimo, British Columbia; widespread in the United States. Native of Europe and western Asia. June–Sept.

140. **CARTHÁMUS** [Tourn.] L. Sp. Pl. 830. 1753.

Annual, biennial, or perennial herbs with unarmed, corymbosely branched stems and usually spinescent or spiny leaves. Heads medium to large, homogamous, many-flowered, solitary at the ends of leafy branches. Phyllaries numerous in many series, the outer and middle foliaceous with a spreading, spiny, coriaceous blade, the inner phyllaries entire and spine-tipped or bearing a denticulate or pinnately parted, spinose appendage. Receptacle coarsely setose. Achenes turgidly quadrangular, hilum obliquely lateral. Pappus none or the paleaceous bristles numerous in many series, unequal, free, persistent, setulose, occasionally rudimentary. [Name probably derived from the Arabic name for the safflower plant.]

A genus of about 20 species occurring chiefly in the Mediterranean region. Type species, *Carthamus tinctorius* L.

Leaves and outer phyllaries pinnately parted or lobed, rigidly spinose; stems and leaves more or less glandular-pubescent and arachnoid, especially near the inflorescence; corollas yellow, the upper part of the tube urceolate-inflated.

Heads broadly ovate, the phyllaries shorter or a little longer than the flowers.   1. *C. lanatus.*

Heads narrowly ovate or oblongish, the outer phyllaries attenuate and much longer than the flowers.
          2. *C. baeticus.*

Leaves and outer phyllaries spinulose-serrate or sometimes entire; stems and leaves glabrous; corollas orange-red, the tube slender throughout.   3. *C. tinctorius.*

1. **Carthamus lanàtus** L. Distaff Thistle. Fig. 5898.

Carthamus lanatus L. Sp. Pl. 830. 1753.

Plants annual or perhaps sometimes biennial with slender elongate taproots, leafy, stiffly erect, few-branched above the middle; stems 1–10 dm. tall, pubescent below with sparse jointed hairs,

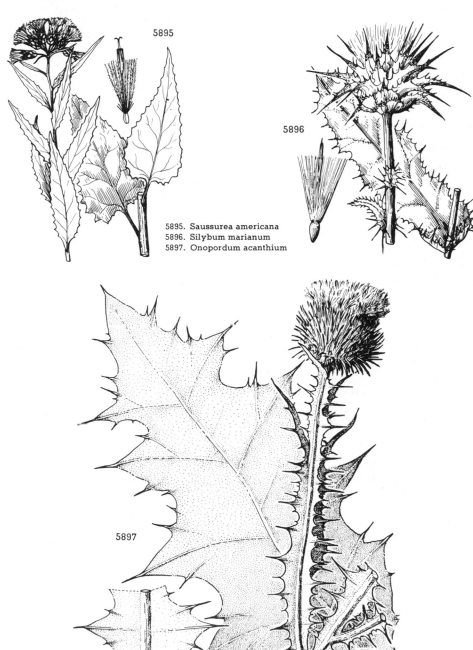

5895

5896

5895. Saussurea americana
5896. Silybum marianum
5897. Onopordum acanthium

5897

pubescent and arachnoid-pubescent above, as well as a little glandular, terete and wingless. Lowest leaves herbaceous, to 7 cm. long, oblanceolate, petiolate or sessile, pinnately parted with acute or spinose-tipped, triangular-lanceolate divisions, the margins bearing a few sharp or spinose serrations, the upper leaves lanceolate, 2–4 cm. long, sessile by a broad base, stiffly coriaceous, prominently veined, viscidulous-hairy, pinnately lobed or saliently dentate, the lobes and teeth rigidly spinose; heads solitary at the ends of the stems or branches, large, about 3 cm. long, ovate; involucres arachnoid-pubescent and viscidulous-hairy, the outer and middle phyllaries foliose, shorter or a little longer than the flowers, divaricately spreading or squarrose, pinnately lobed or dentate, rigidly spinose, the inner phyllaries erect, entire, coriaceous except for the somewhat scarious margin and appendage, the appendage with a terminal spine and lateral spinules or the sides merely denticulate; flowers many, yellow, the throat inflated; achenes 4–5 mm. long and about as broad, pale brown or buff, mottled or speckled with dark brown or black, the sides irregularly horizontally pitted, the lower rim of the pits and the truncate apex of the achenes dentate or denticulate, or the longitudinal angles of the achene produced upward into a cusp; pappus none on the outer achenes, present and persistent on the inner, purplish brown, the paleae in several series, unequal, the shorter outer ones emarginate, the inner longest ones acute, the innermost sometimes abbreviated.

Dry open hills and along roads; locally common in central California and around San Francisco Bay; reported from southern California. Native of the Old World. April–Oct.

2. **Carthamus baèticus** (Boiss. & Reut.) Nyman. Smooth Distaff Thistle. Fig. 5899.

Kentrophyllum baeticum Boiss. & Reut. Pugil. Pl. Nov. 65. 1852.
Carthamus baeticus Nyman, Conspec. Fl. Eur. 419. 1878.
Kentrophyllum lanatum β baeticum Battandier, Fl. Algérie 508. 1888–90.
Carthamus nitidus of California authors. Not Boiss.

Similar to *C. lanatus* but the stems whitish and more shiny. Leaves frequently longer and narrower, attenuate at the tip into a stiff slender spine; heads more slender, oblong-ovate, and fewer-flowered; outer and middle phyllaries divaricately spreading, the tips long-attenuate and spinose, much longer than the flowers, the inner phyllaries narrower, the appendage narrow and denticulate or the phyllaries entire and attenuate; the outer epappose achenes darker and more evenly and regularly rugulose, the inner pappus-bearing achenes with smoother, less denticulate sides.

Open or brushy hills and roadsides in central and southern California; foothills of the Sierra Nevada in Nevada and Tuolumne Counties, and local in San Joaquin County; occasional from San Luis Obispo County to San Diego County. Naturalized from the Mediterranean region. June–July.

It is not always easy to distinguish this species from *C. lanatus* and some European botanists have treated it as a variety. Cytological studies by Amram Ashri, however, have shown that $2n = 44$ in *C. lanatus* and $2n = 64$ in *C. baeticus*, clearly indicating that *C. baeticus* is specifically distinct.

3. **Carthamus tinctòrius** L. Safflower. Fig. 5900.

Carthamus tinctorius L. Sp. Pl. 830. 1753.

Plant annual with erect stems narrowly branching above the middle, leafy, glabrous; stems generally 4–6 dm. tall, smooth, terete, whitish or straw-colored. Leaves coriaceous, spinulose-serrate, the lowest narrowly ovate, attenuate into a short petiole, the middle and upper cauline leaves narrowly ovate to oblong or narrowly lanceolate, to 10 cm. long and 4 cm. wide, sessile by a cuneate or clasping base, the apex weakly spinose; heads solitary at the ends of branches, large, about 4 cm. long and sometimes twice as broad because of the spreading phyllaries; involucre puberulent and sparsely arachnoid, the outer and middle phyllaries foliaceous, weakly spinose at the tip and along the margins, rarely the margins smooth and entire, the base appressed, the foliaceous part spreading; inner phyllaries erect, ovate or lanceolate, stiffly coriaceous, narrowed above to a rigid spine, the margins entire; flowers many, saffron-red or reddish orange; achenes ivory-white, nearly smooth, subobovate, about 6 mm. long and 4 mm. wide, generally without pappus or with a few short rudimentary paleae, occasionally the pappus well developed with paleae numerous in many series, buff, unequal in length, mostly shorter than the achene.

Occasional fugitive from cultivation on roadsides and along the edge of fields, probably not persisting; Sacramento Valley to Los Angeles County, California. Of widespread occurrence in warmer regions of the Old and New Worlds wherever the plant is cultivated; originally described from Egypt. June–Oct.

141 **CÝNARA** [Vaill.] L. Sp. Pl. 827. 1753.

Robust perennial herbs with large basal leaves and stout erect stems. Heads large, pedunculate, many-flowered, homogamous. Involucre globose, the phyllaries imbricate in many series, appressed at base, subrigidly coriaceous, narrowed into a stout terminal spine (or rarely the phyllaries rounded and innocuous). Receptacle fleshy, long-setose. Flowers perfect. Achenes a little turgid, glabrous and smooth, hilum basal, apical areola not rimmed. Pappus-bristles numerous in several series, about equal, strongly plumose, slightly paleaceous, united at the base and deciduous. [Name derived from the Greek word for dog, early applied to a kind of artichoke.]

A genus of 11 species, natives of the Mediterranean region and Canary Islands. Type species, *Cynara cardunculus* L.

1. **Cynara cardúnculus** L. Cardoon or Thistle Artichoke. Fig. 5901.

Cynara cardunculus L. Sp. Pl. 827. 1753.

Plants perennial with broad loose rosettes of large basal leaves and erect thick stems, the stems and lower sides of the leaves gray or white with an arachnoid tomentum; stems striate, 0.5–2 m. tall. Basal leaves oblong or broadly lanceolate, to over 1 m. long and about half as broad, bipinnatisect, divided to the decurrently winged rachis, the segments irregularly parted or divided into narrow, lobed or toothed laciniae, the lobes and teeth bearing a short, stiff, yellow spine, glabrate and green above, tomentose below, middle and upper cauline leaves much smaller, sessile, a little decurrent, those in the inflorescence reduced to linear or oblong, spine-toothed bracts to 5 cm. long; heads large, 7–10 cm. long, solitary at the ends of short, cymose-clustered branchlets; involucre broadly ovate to globose, the phyllaries subglabrous, smooth, stiff-coriaceous, the lowest spreading or reflexed, the middle and upper appressed-imbricate, the middle narrowed into a short, stout, rigid spine, the uppermost bearing a rounded or subtruncate, denticulate, shortly spiny appendage; flowers bluish purple; achenes 6–7 mm. long, oblongish, light brown mottled with brown and black; pappus 2–3 cm. long, ivory-colored.

Occasionally escaping from cultivation; central to southern coastal California, sometimes becoming a rampant noxious weed, as on open hills at the north end of San Francisco Bay; Lower California; southern South America. Native of the Mediterranean region and Canary Islands. June–July.

Cynara scòlymus L. Sp. Pl. 728. 1753. (*Cynara cardunculus* subsp. *scolymus* Hegi, Fl. Mitt. Europa 6²: 924. 1929.) Differing from *C. cardunculus* in the spineless leaves and in the blunt or acute, spineless or nearly spineless phyllaries. Commonly cultivated in coastal California where it is an occasional escape. Probably a cultigen of *C. cardunculus*. Globe or Garden Artichoke.

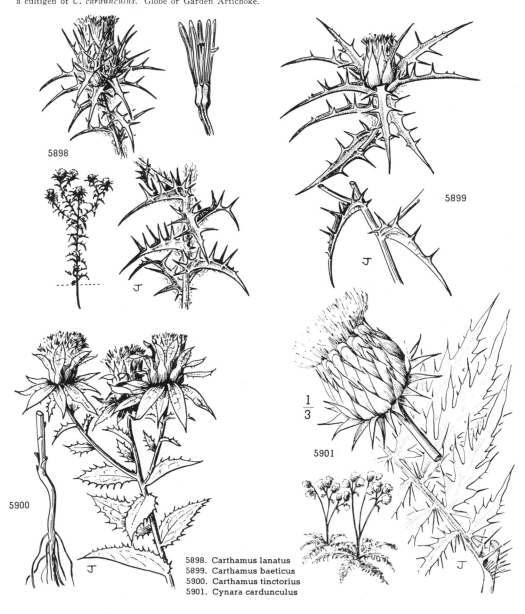

5898. Carthamus lanatus
5899. Carthamus baeticus
5900. Carthamus tinctorius
5901. Cynara cardunculus

142. CÍRSIUM [Tourn.] Adans. Fam. Pl. **2** : 116. 1763 ; emend. DC. Prod. **6** : 634. 1837.

Annual, biennial, or perennial, more or less arachnoid-tomentose herbs with slender or stout taproots, the stems of the perennial species arising from the caudex-like crown of the taproot or from buds produced on widely spreading, horizontal, rhizome-like roots. Leaves alternate, generally spiny-lobed or -divided, sessile by a petiole-like or amplexicaul base, the base sometimes more or less decurrent along the stem as a spiny wing. Heads large or medium-sized, many-flowered, solitary or loosely to densely clustered. Phyllaries in several to many series, appressed-imbricate or looser and more or less spreading, usually spine-tipped, sometimes expanded at the top into a chartaceous, lacerate-dentate appendage, occasionally developing a glandular line or spot on the back. Receptacle thickly bristly-hairy. Flowers perfect or rarely unisexual and the plants nearly or quite dioecious ; corollas whitish, yellowish, rose, or red and tubular with a slender tube and 5-cleft limb. Achenes narrowly obovate or oblongish, compressed, smooth, the hilum basal. Pappus of numerous plumose bristles (or sometimes a few merely barbellulate), these united at the base and deciduous in a ring. [Name from the Greek, referring to the use of thistles as a remedy for swollen veins.]

A genus of about 250 species, in the northern hemisphere. Type species, *Carduus heterophyllus* L.

For authorship of generic name, see Regnum Vegetabile **8** : 278. 1956.
One of the reasons for difficulties encountered in treating our native species of *Cirsium* is that natural lines of demarcation become blurred, if not all but obliterated, due to interspecific hybridization. Not only does hybridization occur between closely related species but it sometimes occurs between species which would seem to be only distantly related (cf. hybrids between *C. fontinale* and *C. quercetorum*, between *C. brevifolium* and *C. utahense*, etc.). During the course of our thistle studies in field and herbarium, hybrids (suspected or otherwise) have been noted as occurring in Washington, Oregon, and California between the following entities: *andrewsii* and *quercetorum, andrewsii and remotifolium* (? cf. *C. mendocinum* Petrak), *brevifolium and foliosum, brevifolium* and *undulatum, brevifolium* and *utahense, californicum* and *cymosum, californicum* var. *bernardinum* and *foliosum* (= *C. quercetorum* × *californicum* f. *parishii* Petrak), *callilepis* and *quercetorum* (cf. *C. amblylepis* Petrak), *callilepis* and *remotifolium, callilepis* and *cymosum. callilepis* var. *pseudocarlinoides* and *cymosum. callilepis* var. *pseudocarlinoides* and *hallii, canovirens* and *undulatum, canovirens* and *utahense* (cf. "*C. subniveum*" of Oregon), *ciliolatum* and *cymosum, cymosum* and *pastoris* (?), *cymosum* and *proteanum, cymosum* and *remotifolium, douglasii* and *quercetorum, douglasii* var. *canescens* and *cymosum, foliosum* and *undulatum, fontinale* and *quercetorum. occidentale* and *proteanum, occidentale* and *quercetorum, pastoris* and *proteanum, proteanum* and *remotifolium.*
In the following account of the genus *Cirsium* on the Pacific Coast, only synonymy pertinent to the region or immediately pertinent to the accepted name has been given. For further synonymy as well as for many other data, see Franz Petrak, *Die Nordamerikanischen Arten der Gattung Cirsium* (Beih. Bot. Centralbl. **35** : 223–567. 1917) ; also, *Cirsium* by Arthur Cronquist (Vasc. Pl. Pacif. Northw. **5** : 133–34. 1955).

Plants introduced from the Old World, either dioecious with unisexual flowers in small heads or monoecious with perfect flowers in medium-sized heads, if monoecious the upper surface of the leaves scabrous or hispid.
  Plants dioecious, perennial ; heads commonly numerous, small, 1.5–2.5 cm. long ; leaves not scabrous-hispid above.   1. *C. arvense.*
  Plants monoecious, biennial ; heads medium-sized, mostly 3–4 cm. long ; leaves scabrous-hispid on the upper surface.   2. *C. vulgare.*
Plants native in the New World, monoecious ; upper surface of leaves not scabrous-hispid.
  Phyllaries not pectinate on the margins (rarely the outer phyllaries spiny-margined in *C. callilepis. C. douglasii, C. hookerianum,* and *C. nidulum* ; uppermost spiny-margined leaves sometimes closely subtend the head and simulate phyllaries).
    Back of at least some of the phyllaries with a glandular spot or ridge.
      Pubescence of stems and leaves entirely arachnoid.
        Flowers bright purplish red or carmine, straight in the head ; taprooted perennial from the southern Sierra Nevada eastward.   14. *C. nidulum.*
        Flowers whitish, lavender, pink, or purplish, the outermost in the head spreading.
          Plants perennial with new rosettes arising from horizontal roots ; heads medium-sized or large, (2 or)3–5 cm. long ; plants of open hills and forest borders, chiefly of Oregon and Washington (and northward and eastward).
            Stems mostly 1–2 m. tall ; heads 3–4 cm. long, somewhat constricted above the subglobose base ; spines on phyllaries 1–2 mm. long ; south-central Oregon and adjacent California.   5. *C. ciliolatum.*
            Stems mostly 1 m. tall or less ; spines on phyllaries mostly 3–5 mm. long ; northern Oregon northward and eastward.
              Flowers pink or purplish, rarely white ; leaves arachnoid above and below ; heads 4–5 cm. long.   3. *C. undulatum.*
              Flowers whitish, sordid ; leaves mostly bicolored, green above ; heads (2 or)3–4.5 cm. long.   4. *C. brevifolium.*
          Plants monocarpic or if sometimes perennial the new rosettes arising from the caudex-like crown of the taproot ; cauline leaves sometimes markedly decurrent ; heads small or medium-sized, 2–3.5 cm. long ; plants of wet places, chiefly in California.
           Plants thinly arachnoid or glabrescent, the upper side of the leaves glabrous or nearly so ; plants of central California.   7. *C. hydrophilum.*
           Plants white-tomentose, scarcely glabrescent, the upper side of the leaves with persistent arachnoid tomentum ; plants of desert ranges and oases.   8. *C. mohavense.*
      Pubescence of stems and leaves crispy-puberulent as well as arachnoid (the crispy-puberulent hairs are multicellular, translucent, shining, and often iridescent ; sometimes they are quite scant and sometimes are covered by a thick coating of arachnoid tomentum).
        Phyllaries closely appressed except for the erect or spreading, spinose or chartaceous tip (i.e., all or almost all of the herbaceous part of the phyllary appressed) ; glandular spot on phyllaries conspicuous.
          Spine at tip of phyllaries widely spreading, mostly 3–5 mm. long ; taprooted, monocarpic or perennial plants of wet places from southern Oregon south to central California.   6. *C. douglasii.*

Spine at tip of phyllaries erect or somewhat spreading, mostly 2–3 mm. long; taprooted perennial of dry slopes east of the Cascade Mountains in central and northern Oregon.   9. *C. canovirens.*

Phyllaries not closely appressed, loosely imbricate and erect or more or less widely spreading above the appressed base (or phyllaries appressed in *C. californicum* var. *bernardinum*); glandular spot frequently inconspicuous or lacking on many phyllaries in a head.

Phyllaries tipped by a short spine mostly 1–3 mm. long.

Plants perennial, the new leaf-rosettes arising from rhizome-like roots or from the branched crown of the taproot.   10. *C. cymosum.*

Plants monocarpic, winter annuals or biennials with taproots.
                                                             11. *C. californicum.*

Phyllaries attenuate into an elongate slender spine mostly 3–10 mm. long; plants generally monocarpic, biennials, rarely perennials, with taproots.

Heads often broader than long, hemispheric; phyllaries with spines mostly 5–10 mm. long, the outer phyllaries often reflexed.   12. *C. neomexicanum.*

Heads subglobose; phyllaries with spines mostly 3–7 mm. long, the outer phyllaries spreading or ascending.   13. *C. utahense.*

Back of phyllaries without a glandular spot or ridge.

Corolla-throat gradually narrowed downward into the tube, the corolla without a distinct separation between the throat and the tube; flowers usually straight in the head, pinkish, purplish, or reddish (or white in albinos); pubescence arachnoid or inconspicuously villous-arachnoid.

Phyllaries thinly arachnoid, more or less glabrescent; perennials with stout taproots; southern Sierra Nevada eastward to desert mountains.   14. *C. nidulum.*

Phyllaries thickly and persistently tomentose or arachnoid, or in *C. proteanum* the tomentum sometimes thin and the phyllaries glabrescent; winter annuals or biennials, rarely perennials; southern Oregon to western Nevada and southern California.

Heads often as broad as long or broader; phyllaries generally straight, almost acicular above the appressed base, festooned with a fine, grayish white webbing; corolla 2.5–3.5 cm. long, scarcely exceeding the involucre; coastal slopes of central California to cismontane southern California.   15. *C. occidentale.*

Heads generally longer than broad; phyllaries generally curved downward, outward or inward, linear to ovatish but not acicular; corollas to 4 cm. long, generally conspicuously exceeding the involucre.

Plants thinly to thickly arachnoid-tomentose, rarely white-tomentose; phyllaries often subglabrate; widespread and common in California hills and mountains from Ventura and Kern Counties northward.   16. *C. proteanum.*

Plants snowy-white with dense, felt-like tomentum; phyllaries white-arachnoid-tomentose except for the glabrous yellowish spine; southwestern Oregon to central California and adjacent Nevada.   17. *C. pastoris.*

Corolla-throat more or less distinctly contracted at the base, the throat and tube generally distinct; flowers straight in the head or often the outermost widely spreading, whitish, lavender, pink, rose, or sometimes reddish; pubescence, at least on the stem, consisting of crispy-puberulent and arachnoid hairs (the crispy-puberulent hairs shining, multicellular, sometimes viscidulous, sometimes abundant, lacking in *C. mohavense*).

Phyllaries (at least the inner ones) ending in a more or less expanded chartaceous appendage with the margins often fimbriate, lacerate, spinulose-denticulate, or crisped (phyllaries sometimes a little expanded in intermediates between *C. callilepis* and *C. remotifolium*, and between *C. foliosum* and *C. hookerianum*).

Plants monocarpic, mostly biennial with taproots; widespread in western North America, mostly near the Cascade-Sierran crest and eastward.   26. *C. foliosum.*

Plants perennial with rhizome-like roots; California Coast Ranges and Oregon chiefly west of the Cascade Mountains.

Heads medium-sized or larger, 3.5–5 cm. long, mostly clustered on short stout stems; phyllaries closely appressed and graduate-imbricate; South Coast Ranges in California.   24. *C. quercetorum xerolepis.*

Heads medium-sized, 2–4 cm. long, solitary or clustered on elongate slender stems; phyllaries closely appressed or more commonly rather loosely ascending, rarely squarrose above the middle; central California north to Oregon chiefly west of the Cascade-Sierran crest.   29. *C. callilepis.*

Phyllaries without an expanded, scarious or chartaceous appendage.

Phyllaries densely and conspicuously pubescent.

Pubescence on phyllaries mostly crispy-puberulent; plants reaching northern Washington east of the Cascade Mountains from the north and the east.
                                                             27. *C. hookerianum.*

Pubescence on phyllaries mostly or entirely arachnoid.

Plants not succulent, often less than 1 m. tall; flowers whitish or pinkish; mostly east of the Cascade-Sierran crest.

Heads usually broader than long, hemispheric; outer phyllaries often reflexed.   12. *C. neomexicanum.*

Heads subglobose; outer phyllaries spreading or ascending.
                                                             13. *C. utahense.*

Plants succulent, mostly 1–2.5 m. tall; flowers rose-red or pinkish (rarely white in albinos); mostly west of the Cascade-Sierran crest.

Leaves arachnoid-tomentose on lower side; corolla-lobes 2–4 mm. long; style shortly exserted from anther-tube; achenes 3.5–4.5 mm. long.
                                                             19. *C. brevistylum.*

Leaves subglabrous on lower side; corolla-lobes (3.5) 4–10 mm. long; style exserted 3–5 mm. beyond anther-tube; achenes 4–6 mm. long.
                                                             20. *C. edule.*

Phyllaries subglabrous to thinly arachnoid or floccose-tomentose, not densely and conspicuously festooned with arachnoid tomentum.

Phyllaries well graduated and imbricate, closely appressed or subappressed. (See also *C. andersonii* with phyllaries spreading-ascending.)

Plants white-arachnoid-tomentose (crispy-puberulent hairs absent), biennial and monocarpic, or if perennial the new rosettes arising at the top of the taproot; desert oases in southeastern California.   8. *C. mohavense.*

Plants crispy-puberulent as well as arachnoid, often glabrescent.

Plants perennial with rhizome-like roots; Coast Ranges of central California.
                                                             24. *C. quercetorum.*

Plants monocarpic, annual or biennial with a taproot.
Plants nearly or quite glabrous, caulescent; heads solitary or clustered at the ends of branches; phyllaries loosely imbricate, mostly linear- or ovate-lanceolate; rare on coastal hills and flats in southern California. 25. *C. loncholepis.*
Plants more or less pubescent, caulescent or acaulescent; heads generally in clusters that are conspicuously surpassed by subtending leaves; phyllaries more or less appressed-imbricate, mostly lanceolate or ovate; widespread in western United States and Canada. 26. *C. foliosum.*
Phyllaries not closely appressed, loosely imbricate and erect or more or less spreading above the appressed base.
Pubescence crispy-puberulent and arachnoid, not viscidulous; heads not nodding.
Phyllaries mostly loosely ascending, sometimes a little spreading, sometimes more appressed. (See also *C. cymosum* with phyllaries more or less glandular.)
Flowers whitish, probably the outermost spreading in the head; phyllaries not strongly graduated; biennial or perennial with taproot (and perhaps also rhizome-like roots); Coast Ranges of California north to Washington. 31. *C. remotifolium.*
Flowers deep pink to purplish red, straight in the head; phyllaries graduated; perennial with rhizome-like roots; in the mountains from California to Idaho. 32. *C. andersonii.*
Phyllaries more or less widely spreading or reflexed above the appressed base (or subappressed to the tip in *C. californicum* var. *bernardinum*); flowers white, lavender, or pink, the outermost widely spreading in the head; plants monocarpic, mostly biennials with taproots (sometimes perennial in *C. utahense*).
Phyllaries thinly arachnoid-tomentose, subglabrescent, tipped by a short spine mostly 1–3 mm. long. 11. *C. californicum.*
Phyllaries usually more persistently and thickly arachnoid-tomentose.
Heads usually broader than long, hemispheric; phyllaries tipped by spines mostly 5–10 mm. long, the outer phyllaries often reflexed. 12. *C. neomexicanum.*
Heads subglobose; spines on phyllaries mostly 3–7 mm. long, the outer phyllaries spreading or ascending. 13. *C. utahense.*
Pubescence of stems and leaves viscidulous, as well as arachnoid-tomentose; heads or clusters of heads nodding in flower; outer phyllaries widely spreading or recurved above the appressed base; rare perennial plants along streams in areas of serpentine rock in the South Coast Ranges in California.
Outer phyllaries narrowly ovate, gradually narrowed to a strong spiny tip. 33. *C. campylon.*
Outer phyllaries broadly ovate, abruptly narrowed to a short-spinose tip. 34. *C. fontinale.*
Phyllaries (outer and sometimes middle) more or less pectinate-spinose on the margins.
Flowers whitish or yellowish; stems and leaves lanate with a close, suede-like tomentum; rare plant of southern California coastal dunes. 18. *C. rhothophilum.*
Flowers lavender, pink, or reddish, rarely whitish; stems and leaves not lanate.
Phyllaries (outer and middle) attenuate into long, stiff, yellow spines that exceed the inner phyllaries; flowers straight in the head; rare plant of Harney County, Oregon. 28. *C. peckii.*
Phyllaries spinose-acute or spiny, the spines generally less than 1 cm. long; outermost flowers in the head more or less outwardly spreading; mostly Coast Range plants of California, Oregon, and Washington.
Phyllaries conspicuously pectinate, the inner phyllaries chartaceous or scarious at the tip and glandular along the midvein below the tip; perennial with rhizome-like roots; southwestern Oregon and northwestern California. 30. *C. acanthodontum.*
Phyllaries rather inconspicuously pectinate, not glandular or chartaceous-expanded.
Phyllaries short-spinose, the spines less than 5 mm. long; heads often nodding after anthesis; taprooted perennials from Coos Bay, Oregon, to Washington. 21. *C. hallii.*
Phyllaries tipped with spines 5 mm. long or longer; heads not nodding; monocarpic plants of California.
Phyllaries widely spreading at about the middle; corollas purplish rose; plants of usually wet coastal slopes. 22. *C. andrewsii.*
Phyllaries appressed except at the spreading spinose tip; corollas lavender-rose; plants of stream-banks in the San Joaquin Valley. 23. *C. crassicaule.*

1. Cirsium arvénse (L.) Scop. Canada or Creeping Thistle. Fig. 5902.

Serratula arvensis L. Sp. Pl. 820. 1753.
Cirsium arvense Scop. Fl. Carn. ed. 2. 2: 126. 1772.

Plants mostly dioecious, perennial with erect shoots arising from widely creeping, horizontal roots; stems mostly simple below and corymbosely branching above, 0.3–1.5 or 2 m. tall, unwinged or sometimes more or less interruptedly winged with narrow decurrent leaf-bases. Basal leaves loosely clustered, oblong- or elliptic-oblanceolate, narrowed below to a short, petiole-like base, undulate or pinnately lobed, spinose-bristly on lobes and marginal teeth; cauline leaves elliptic-obovate or oblanceolate, narrowed at base or the base amplexicaul, to 2 dm. long, 3–7 cm. wide, shallowly to deeply pinnately lobed or divided, the margins spinose-bristly, the lobes and divisions spine-tipped; uppermost leaves reduced and sometimes bract-like; all leaves glabrous above and glabrous or more or less arachnoid-tomentose below; heads 1.5–2.5 cm. long, short-stalked, solitary or in small terminal clusters that form an irregular corymb; involucres 1–2 cm. long, subglobose to campanulate; phyllaries numerous in many series, the outer ovate, the middle ovate-lanceolate, both tipped by a short, stiff, spreading spine and ciliate-arachnoid on the margins, the inner linear-lanceolate with tip flattened and somewhat twisted; corollas purplish, rarely white, 12–15 mm. long, the tube 8–11 mm. long, the lobes about 4 mm. long; achenes light to dark brown, narrowly oblong or a little widened upward, 3–4 mm. long, about 1 mm. wide; pappus of staminate flowers about 8 mm. long, of pistillate flowers becoming 2 cm. long in fruit, pale buff.

Widespread and noxious weed of cultivated ground and pastures, common in western Washington and Oregon and northern California and occasional southward to southern California; widely distributed in northern United States and southern Canada; native of the Old World. June–Sept.

Plants are variable in pubescence and lobing of leaves, and European botanists have named many varieties and forms. In the western United States the most common plant is var. *horridum* Wimmer & Grabowski, the form with glabrous, very spiny, divided leaves; much less common are var. *argenteum* (Vest) Fiori, in which the lower side of the pinnately divided leaves is heavily white-tomentose, and var. *mite* Wimmer & Grabowski, in which the nearly glabrous leaves are undulate-margined or very shallowly lobed. According to Cronquist (Vasc. Pl. Pacif. Northw. 5: 135. 1955), the plant named by Linnaeus may have been the form with nearly entire leaves, but most recent European and American authors continue to accord it varietal recognition.

2. **Cirsium vulgàre** (Savi) Tenore. Common or Bull Thistle. Fig. 5903.

Carduus lanceolatus L. Sp. Pl. 821. 1753.
Carduus vulgaris Savi, Fl. Pis. 2: 241. 1798.
Cirsium vulgare Tenore, Syll. Fl. Neap. Appendix 5, 39. 1842.
Cirsium lanceolatum Scop. Fl. Carn. ed. 2. 2: 130. 1772. Not Hill, 1769.

Biennial herb with erect stem 0.3–2 m. tall, simple or branched at base, few-branched or many-branched above, the stems conspicuously spiny-winged by decurrent leaf-bases or on the upper-most branches the leaf-bases only shortly decurrent. Lower side of leaves thinly to heavily arachnoid-tomentose, upper side harshly setose-spinulose; seedling-leaves oblong or oblanceolate, undulate-margined or shallowly lobed; cauline leaves oblong-lanceolate or oblong-oblanceolate or ovatish, up to 30 cm. long and 10 cm. wide, shallowly to deeply pinnately lobed, the lobes ovate-deltoid and shallowly parted or deeply divided into elongate lanceolate divisions, the lobes and divisions produced into stiff yellowish spines, the lower frequently narrowed to margined petiole-like base, the upper with base broadly clasping; heads 3–5 cm. long and nearly as broad, terminating short branches, closely subtended by the uppermost, usually much-reduced leaves; involucre cylindric-campanulate to subglobose, 2–4 cm. long; phyllaries numerous in many series, lanceolate to linear-lanceolate, appressed below, ascending-spreading above, the outer and middle strongly spinose-tipped, glabrous and smooth dorsally and ventrally, conspicuously arachnoid-tomentose on the margins, the inner with somewhat flattened, twisted, spinulose tips and with midrib and margins scabrous-roughened; corollas rose-purple, rarely white, about 2.5–3.5 cm. long, the throat and lobes each about 5 mm. long; achenes oblongish, a little widened upward, 3–3.5 mm. long, light brown or buff with blackened lineal markings; pappus 2–3 cm. long.

Widely distributed and sometimes common in the Pacific states in waste places, pastures, and disturbed ground; widespread in North America; introduced from the Old World. June–Oct.

A variant with small aggregated heads and smaller flowers but with somewhat longer achenes (about 4 mm. long) has been collected in October and November in Santa Clara and Riverside Counties, California. This may be the plant listed by Hegi (Ill. Fl. Mitt.-Eur. 6²: 875. 1929) as *C. lanceolatum* var. *australe* Murr.

Cirsium scàbrum (Poir.) Bonnet & Barratte, Cat. Pl. Vasc. Tunisie 238. 1896. (*Carduus scaber* Poir. Voy. Barb. 2: 231. 1789; *C. giganteus* Desf. Fl. Atlantica 2: 245. 1800.) Robust plants with stems to 4 m. tall and about 5 cm. in diameter at the base; leaves broad, sinuate-pinnatifid, amplexicaul, the upper side scabrous, the lower side lanate-tomentose; heads numerous in twos and threes at the ends of branchlets, 3–3.5 cm. long, 2–3 cm. wide; phyllaries ovate to ovate-lanceolate, numerous, closely appressed and imbricate, somewhat spreading at the short-spiny tip, a little arachnoid-tomentose; flowers reddish. Spontaneous near Glenwood, Santa Cruz County, California, at about the turn of the century. Native of the Mediterranean region. Aug.

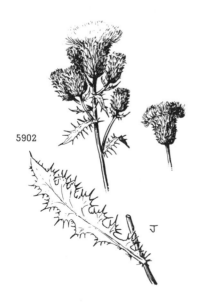

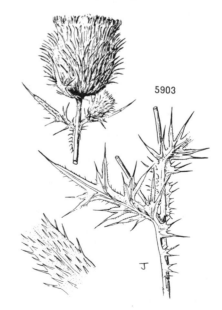

5902. Cirsium arvense   5903. Cirsium vulgare

3. **Cirsium undulàtum** (Nutt.) Spreng. Wavy-leaved Thistle. Fig. 5904.

Carduus undulatus Nutt. Gen. **2**: 130. 1818.
Cirsium undulatum Spreng. Syst. **3**: 374. 1826.
Cnicus undulatus var. *megacephalus* A. Gray, Proc. Amer. Acad. **10**: 42. 1874.

Perennial herb with a stout taproot and with fine, spreading, lateral roots from which new rosettes and shoots arise, the root-crown generally unbranched, the stems usually simple and solitary at the base, paniculately branching above or nearly simple; stems erect, 3–7.5 dm. tall, rather stout, somewhat angled to subterete, rather densely and evenly white-lanate. Earliest leaves entire, dentate or shallowly lobed, oblanceolate, spinescent or ciliate-spinulose, giving way to a large, many-leaved rosette that develops through two or perhaps three seasons and that may have disappeared before the aerial stem develops; rosette-leaves 3–7 dm. long, up to 1.2 dm. wide, oblongish or sometimes narrowly elongate, pinnately parted or divided into deltoid- or linear-lobed segments, the divisions and lobes tipped by short spines, the blade narrowed below into a longer or shorter, petiole-like base (up to 3 dm. long), the blades arachnoid above and below, nearly white beneath; cauline leaves smaller, more strongly spiny, becoming shorter-stalked or sessile with a semi-amplexicaul, shortly decurrent, spiny-lobed base; uppermost leaves reduced to a narrow, spine-bordered rachis; heads mostly 4–5 cm. long and usually somewhat broader, the flowers well exceeding the involucre and outwardly spreading; involucre campanulate or subglobose, usually distinctly constricted about the middle; lower and middle phyllaries in 5–7 series, lanceolate to oblong-lanceolate, well graduated and closely imbricate, appressed except at the divergently spreading tip, the tip narrowed into a short slender spine, the back of the phyllaries mostly glabrous with a conspicuous, elevated, glandular area below the tip, the margin usually more or less floccose-arachnoid, the inner phyllaries elongate, spreading or recurved above the middle, slightly or not at all glandular, attenuate into a chartaceous or spinescent tip, the chartaceous part minutely ciliolate or sometimes even a little lacerate; corollas generally pink, lavender-pink, or purplish, rarely white, the tube 1.5–2 cm. long, the throat 8–12 mm. long, the lobes 6–8 mm. long; achenes brown, narrowly oblong-oblanceolate, little compressed, 8 mm. long, 2.5 mm. wide; pappus white, 2–3 cm. long.

Rocky, sandy, or clayey soil of open rolling hills or bluffs and slopes of volcanic or granitic rocks, Arid Transition Zone; east of the Cascade Mountains, British Columbia south to northern Oregon; east to the central United States, and south to Arizona; introduced in southern California and perhaps elsewhere. Type locality: "On the calcareous islands of Lake Huron, and on the plains of Upper Louisiana." June–Sept.

Cirsium ochrocéntrum A. Gray, Mem. Amer. Acad. II. **4**: 110. 1849. Very spiny plants from thick, cord-like, long-creeping roots. Leaves bicolored, thinly arachnoid and green above, densely white-tomentose or lanate below, deeply pinnately lobed or parted, the segments armed with usually long, stiff, yellowish spines; heads few, large, 4–6 cm. long; phyllaries numerous, appressed, well imbricate, ovate to ovate-lanceolate, abruptly narrowed to short, stiff, spreading spines, glabrous except for a floccose tomentum on the margins, the lower and middle with a conspicuous median gland, the inner attenuate into a slender, more or less curved or twisted, spinose tip; corollas red or white; anther-tips subulate. Occurring as a weedy introduction in southern California; widespread in the western United States from Arizona east to Texas and Nebraska. Type locality: mountainsides around Santa Fe, Santa Fe County, New Mexico. May–Oct.

4. **Cirsium brevifòlium** Nutt. Palouse Thistle. Fig. 5905.

Cirsium brevifolium Nutt. Trans. Amer. Phil. Soc. II. **7**: 421. 1841.
Cirsium palousense Piper in Piper & Beattie, Fl. S.E. Wash. 260. 1914.

Plants perennial, spreading by slender horizontal roots from which new plants arise, the new plants developing stout elongate taproots, the stems solitary or few in a cluster; stems erect, 0.3–1 m. tall, slender or rather thick and hollow, somewhat angled, thinly or densely arachnoid-tomentose. Leaves mostly bicolored, white-tomentose below, green and thinly arachnoid or glabrescent above; earliest leaves narrowly oblanceolate to elliptic, subentire, sparsely dentate, or shallowly lobed; rosette-leaves rather sparse, narrowly to widely oblanceolate, to 5 dm. long and 1–2 dm. wide, pinnately divided, the divisions subentire and narrowly oblong or linear or deeply 2-parted, the sinuses between the divisions wide and extending into a rachis-like border along the midvein, the lobes and margins rather weakly spiny, the blade narrowed into an elongate, scarcely bordered, petiole-like base; cauline leaves smaller, becoming more spiny, the lower developing small, spiny, shortly de-current auricles at the base, the upper becoming sessile by a narrowly auriculate base; heads solitary at the ends of leafy branches and closely subtended by 1 or few bract-like leaves, subglobose, 3–4.5 cm. long and as wide or a little wider; involucre subglabrate, the margins of the numerous well-imbricate phyllaries sometimes more or less persistently floccose-tomentose, the outer and middle phyllaries appressed except at the divergently spreading, short-spinose tip, the tips of the inner phyllaries long-attenuate into a spreading, chartaceous, undulate or twisted tip, the backs of the outer and inner phyllaries scarcely or not at all glandular, the others with a conspicuous, elevated, oblongish, median gland; flowers usually only shortly exceeding the involucre, usually not widely spreading, ivory-white or sometimes slightly tinged with pink or lavender, the corolla-tube 9–13 mm. long, the throat 8–11 mm. long, the lobes 5–8 mm. long; achenes oblongish, 5–6 mm. long, 2 mm. wide, light brown except for the upper part which is buff; pappus sordid or nearly white, 2–2.5 cm. long.

Deep loam of open hills and meadows or loose rocky soil of cuts and canyonsides, Arid Transition Zone; east of the Cascade Mountains from central Washington south to central (perhaps southern?) Oregon east to Idaho. Type locality: "In the Rocky Mountain plains." June–Oct.

Where *C. brevifolium* grows with *C. undulatum* in Washington and with *C. utahense* in Oregon, suspected hybrids between those species and *C. brevifolium* have been observed.

5. **Cirsium ciliolàtum** (Henderson) J. T. Howell. Ashland Thistle. Fig. 5906.

Cirsium undulatum var. *ciliolatum* Henderson, Bull. Torrey Club **27**: 348. 1900.
Carduus ciliolatus Heller, Muhlenbergia **1**: 5. 1900.

Cirsium botrys Petrak, Beih. Bot. Centralbl. **35**: 483. 1917.
Cirsium howellii Petrak, op. cit. 486.
Cirsium ciliolatum J. T. Howell, Leaflets West. Bot. **9**: 9. 1959.

   Perennial herb reproducing by seed and by aerial stems arising from slender horizontal roots; stems erect, to 2 m. tall, fastigiately to divaricately branched, nearly terete, thinly to thickly arachnoid or white-lanate. Earliest leaves entire or shallowly undulate-dentate, finely spinulose-ciliate; leaves of basal rosette rather few, pinnately lobed with sinuses extending about halfway to the midrib, white-arachnoid below, gray-green-arachnoid above; lower cauline leaves oblong-elliptic or oblanceolate, to 2.5 dm. long, 1 dm. wide, pinnately parted into deltoid or ovatish, subentire or lobed divisions, the divisions spinulose-ciliate and tipped by short weak spines, the blade narrowed to a slender, petiole-like base or the leaves sessile by a broadened base that is shortly decurrent as spinulose wings; upper leaves shorter, less deeply parted and sessile by a broader-winged base; uppermost leaves on the peduncle-like branches or immediately under the heads reduced and bract-like; heads of medium size, 3–4 cm. long, 2–5 cm. broad; involucre campanulate or somewhat constricted near the top above a subglobose base, thinly arachnoid, glabrescent; phyllaries numerous, well graduated and closely imbricate in about 6 series, subappressed-ascending or with short, spreading, herbaceous tips, the tip attenuate into a short stiff spine, the backs of all phyllaries (except pos-

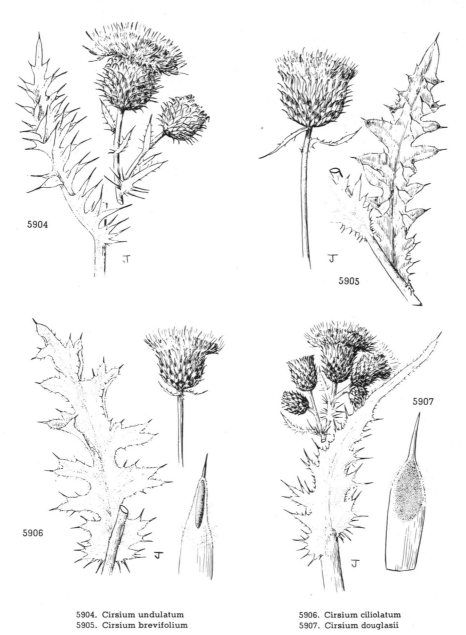

5904. Cirsium undulatum
5905. Cirsium brevifolium

5906. Cirsium ciliolatum
5907. Cirsium douglasii

sibly the outer and inner) with a conspicuous, linear or oblongish, glandular area, the marginal area on the backs of the phyllaries scabrous-papillate, the margins minutely ciliolate, the outer and middle phyllaries ovatish to lanceolate, the inner attenuate into a chartaceous-spinulose tip; corollas ochroleucous becoming sordid-brownish, the tube 9–11 mm. long, the throat 6–8 mm. long, the lobes 4–8 mm. long; achenes brown, narrowly obovate, compressed, 5–6 mm. long; pappus sordid-whitish, 1.5–2 cm. long.

Dry, grassy and rocky slopes, Arid Transition Zone; south-central Oregon and possibly in adjacent northern California. Type locality: dry hillsides near Ashland, Jackson County, Oregon. The type of *C. botrys* is labeled as having come from Montague, Siskiyou County, California, but it, too, may have been collected in southern Oregon. June–Aug.

In the mountains between Ashland and Klamath Falls, *C. ciliolatum* intergrades with *C. cymosum*. Plants similar in appearance to some of these intergrades occur occasionally in the range of *C. cymosum* in the North Coast Ranges in California.

6. Cirsium douglásii DC. Douglas or Swamp Thistle. Fig. 5907.

Cirsium douglasii DC. Prod. **6**: 643. 1837.
Carduus undulatus var. *douglasii* Greene, Proc. Acad. Phila. **1892**: 360. 1893.
Cirsium breweri var. *wrangelii* Petrak, Beih. Bot. Centralbl. **35**: 461. 1917.

Plants monocarpic, often developing a large basal rosette at the top of the stout taproot; stems erect, to about 1.5 m. tall, widely branched from near the base or above, mostly stout and hollow, crisp-puberulent beneath the thick arachnoid tomentum or the tomentum becoming thin and somewhat floccose. Rosette-leaves pinnately lobed or divided, oblanceolate, 2–10 dm. long, 3–25 cm. wide, the divisions deltoid or ovatish, or often oblongish to obovate and more or less deeply 2-parted, the lobes and divisions tipped by a short stout spine, the margin rather evenly and finely spinose-dentate between the larger spines, the blade narrowed below into a long, spinose-fringed, petiole-like base; cauline leaves smaller, generally more pinnatifid and more spiny, becoming sessile with a broadly auriculate or semi-amplexicaul base, the auriculate lobes very spiny, generally a little decurrent as short spiny wings; uppermost leaves much reduced, closely subtending the heads; heads small to medium, (2)2.5–3.5 cm. long and about as wide, rather numerous in loose to dense clusters at the ends of short branches; involucres short-campanulate or bowl-shaped, sparsely arachnoid or glabrous except for sparse floccose tomentum along the margins of the phyllaries; phyllaries numerous, closely appressed and imbricate except at the spreading spinose tip, the outer ovatish, the middle ovate-oblong, the inner linear-oblong or -lanceolate, the middle phyllaries with a conspicuous, dark, dorsal, glandular area below the slender terminal spine, the inner attenuate into a more or less twisted, chartaceous, attenuate-spinulose tip; flowers spreading in the head, dark purplish red, the corollas 1.8–2.1 cm. long, the tube 8–9 mm. long, the lobes and throat equal, each 5–6 mm. long; anther-tips sharply acute or subacuminate; achenes brown or almost blackish, plump, obovate-cuneate, 4–5 mm. long, 2–2.5 mm. wide; pappus light brown, 1.5–2 cm. long.

Wet or marshy places on coastal slopes and in canyons and valleys, Humid Transition Zone; Coast Ranges of central California from Mendocino County to Monterey County. Type locality: California. Collected by Douglas. May–Aug. On the Sonoma County coast, *C. douglasii* hybridizes with *C. quercetorum*.

Cirsium douglasii var. **canéscens** (Petrak) J. T. Howell, Leaflets West. Bot. **9**: 11. 1959. (*Cnicus breweri* A. Gray, Proc. Amer. Acad. **10**: 43. 1874; *Cirsium breweri* Jepson, Fl. W. Mid. Calif. 507. 1901; *C. breweri* var. *canescens* Petrak, Beih. Bot. Centralbl. **35**: 462. 1917; *C. breweri* var. *lanosissimum* Petrak, loc. cit. as to first-cited specimen.) Stems generally taller, to 2.5 m. tall; leaves usually less spiny and with the decurrent bases of the cauline leaves longer; heads generally numerous, small, 2–3 cm. long, cylindric to subglobose. Wet places, chiefly in the Arid Transition Zone; mountains of northern California southward in the Coast Ranges to the vicinity of Clear Lake, Lake County; also from Lake County, Oregon, southward to the Sierra Nevada in the vicinity of Lake Tahoe, California, and adjacent Nevada. Type locality: Sisson (Mount Shasta City), Siskiyou County, California; lectotype, *Eastwood 1199*. The vicinity of Sisson ("Strawberry Valley") is also the type locality of *Cnicus breweri*. June–Sept.

7. Cirsium hydrophílum (Greene) Jepson. Suisun Thistle. Fig. 5908.

Carduus hydrophilus Greene, Proc. Acad. Phila. **1892**: 358. 1893.
Cirsium hydrophilum Jepson, Fl. W. Mid. Calif. 507. 1901.
Cirsium vaseyi var. *hydrophilum* Jepson, Man. Fl. Pl. Calif. 1165. 1925.

Plant biennial, the stems stout and succulent, to 2 m. tall, thinly arachnoid-tomentose and early glabrate. Earliest leaves narrowly elliptic-oblong, subentire, undulate, or shallowly to rather deeply lobed with broad triangular lobes, the margin almost devoid of spines or with short spinose teeth; leaves of the basal rosette becoming quite large, up to 9 dm. long and 1.5 dm. wide, thinly arachnoid and glabrate above, thinly white-arachnoid-tomentose below, elongate-oblanceolate, deeply pinnately parted or divided, the divisions ovatish and again lobed or parted, the segments ending in a short stiff spine and the margins sometimes tending to be spinulose-ciliate, the blade gradually narrowed below to a spiny-margined, petiole-like base; cauline leaves much smaller, becoming broadly sessile and shortly decurrent, frequently developing an enlarged, auriculate, semi-amplexicaul base, the auricles rounded and radiately spinose-dentate or -lobed; uppermost leaves along the flowering branchlets and subtending the heads becoming reduced and sometimes bract-like, strongly spinose-pinnate; heads solitary or loosely clustered on short, peduncle-like branches, mostly 2.5–3 cm. long, cylindric-campanulate; involucre short-campanulate or urceolate-campanulate, thinly arachnoid, glabrescent; phyllaries appressed except for the spreading spinescent tip, conspicuously glandular along the midrib above the middle, the outer phyllaries broadly triangular-ovate, the middle narrowly ovate or lanceolate, the inner narrowly oblong-lanceolate to linear, the outer and middle tipped by short stout spines, the inner with thinner, twisted, spinescently acute tips, all evenly and minutely spinulose-ciliate; flowers lavender-rose; corollas somewhat spreading in the head, the tube 8–10 mm. long, the throat 5–6 mm. long, the lobes 5 mm. long; achenes plump, oblong-cuneate, about 5 mm. long, dark gray-brown or almost black, somewhat shining; pappus sordid-whitish, 1.5 cm. long.

Estuarine marshes that border the lower Sacramento River and that are tidally inundated by salt or brackish water, Upper Sonoran Zone; a local endemic known only in the Suisun Marshes, Solano County, California. July–Sept.

Cirsium hydrophilum var. **vàseyi** (A. Gray) J. T. Howell, Leaflets West. Bot. **9**: 11. 1959. (*Cnicus breweri* var. *vaseyi* A. Gray, Syn. Fl. N. Amer. 1²: 404. 1884; *Cirsium montigenum* Petrak, Beih. Bot. Centralbl. **35**: 454. 1917; *C. vaseyi* Jepson, Man. Fl. Pl. Calif. 1165. 1925.) Plants more widely branching from below the middle; heads somewhat larger, 3–3.5 cm. long; achenes oblongish or elliptic, a little cuneate, 4–5 mm. long. In wet ground in areas of serpentine rock, Mount Tamalpais and adjacent ridges, Marin County, California. May–July.

8. **Cirsium mohavénse** (Greene) Petrak. Mojave Thistle. Fig. 5909.

Carduus mohavensis Greene, Proc. Acad. Phila. **1892**: 361. 1893.
Cirsium mohavense Petrak, Bot. Tidsskr. **31**: 68. 1911.

Plants with stout taproots, generally biennial and monocarpic, or perhaps sometimes perennial and developing a compact crown at the top of the taproot; stems erect, slender or stout, 0.5–2.5 m. tall, simple below and branching above the middle or branching from near the base, thinly to thickly white-arachnoid-tomentose. Earliest leaves narrowly oblanceolate, undulate or undulate-lobed, subentire or spinulose-denticulate; rosette-leaves pinnately lobed to divided, oblanceolate, to 6 dm. long and 1.5 dm. wide, the segments widely spaced or approximate, deltoid or quadrate with 2 lobes, the lobes and divisions stoutly spine-tipped, the margins rather strongly spiny, narrowed below into a longer or shorter, spinose-margined, petiole-like base; cauline leaves becoming obviously sessile with the base usually prominently decurrent as a spiny wing; upper leaves very spiny, much reduced and sometimes becoming a fascicle of stout elongate spines, the base not always decurrent; all leaves white-arachnoid-tomentose, the tomentum thinner on the upper side, persistent; heads solitary at the ends of short, peduncle-like stems or the heads more or less congested at the ends of the branches, sometimes the heads developing racemosely along the branches below the terminal head or cluster, usually small, 2–3 or 3.5 cm. long, cylindric to subglobose; involucre thinly to densely floccose-tomentose; phyllaries usually glabrous and pale green on the back and floccose along the margins but sometimes more or less floccose over the back except for the glabrous yellowish spine, and closely appressed except for the spreading or ascending spine, the outer ovate, the middle ovate-lanceolate or lanceolate, the inner narrowly oblongish, tapering into a twisted or recurved, chartaceous, acute or spinescent tip, the middle phyllaries glandular-ridged (or in late heads the phyllaries only green-veined), the outer and inner phyllaries usually nonglandular; flowers spreading in the heads, pale lavender to pink, the corolla-tube 7–11 mm. long, the throat 5–7 mm. long, the lobes 5–7 mm. long; anther-tips attenuate-subulate; achenes plump, oblongish, stramineous to brown, 3.5–4.5 mm. long, 1.5–2 mm. wide; pappus nearly white, 1.5 cm. long, the bristles in the outermost flowers merely barbellulate.

Saline soil near springs at oases, along streams in canyons, and in low places in open valleys, Lower Sonoran Zone; deserts of southeastern California and perhaps eastward to Nevada and Utah. Type locality: Rabbit Springs, Mojave Desert, San Bernardino County, California. June–Oct.

9. **Cirsium canovìrens** (Rydb.) Petrak. Gray-green Thistle. Fig. 5910.

Carduus canovirens Rydb. Mem. N.Y. Bot. Gard. **1**: 450. 1900.
Cirsium canovirens Petrak, Bcih. Bot. Centralbl. **35**: 540. 1917.

Plants biennial and perhaps sometimes monocarpic, or becoming perennial with 1 or few stems arising from the scaly crown of a stout taproot, plants not reproducing from root-buds; stems

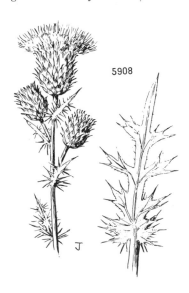

5908. Cirsium hydrophilum   5909. Cirsium mohavense

4–10 dm. tall, fastigiately few- to many-branched, thinly arachnoid and glabrescent below, gray-
or white-tomentose above, the lower stems also sparsely villous-pubescent with spreading, shining,
multicellular hairs; rosette-leaves up to 4 dm. long and to 1.2 dm. wide, oblanceolate in general
outline, pinnately parted or divided, the divisions deeply lobed, the lobes and margins short-spiny,
narrowed below into a spiny-lobed, petiole-like base, the upper surface greenish, thinly arachnoid,
the lower surface gray-green, arachnoid and villous; lower cauline leaves similar to the basal
leaves but smaller; middle and upper cauline leaves becoming linear-oblongish or narrowly lanceolate, shallowly lobed, sessile by a somewhat broadened, very spiny, shortly decurrent base, uppermost leaves reduced to spiny bracts on the peduncle-like stems or closely subtending the heads;
heads solitary at the ends of short or longer branches, small, 2.5–3.5 cm. long, oblongish to subglobose; involucre campanulate, thinly floccose-arachnoid, subglabrescent, light green; outer and
middle phyllaries appressed at base, loosely ascending or divergently spreading about the middle
or above, narrowly lanceolate, tipped by a short stout spine, usually all except the inner with a
conspicuous elongate gland, the inner phyllaries narrowed above into an attenuate, chartaceous,
spinescent, more or less twisted tip, margin of all phyllaries usually pectinate-ciliolate; flowers
sordid-whitish, corolla-tube 7 mm. long, the throat 5–7 mm. long, the lobes 5–7 mm. long; achenes
plump, oblanceolate-oblong, 6 mm. long, about 2 mm. wide; pappus slightly sordid, 1.5–2 cm. long.

  Open dry slopes and flats in clayey or rocky soil, Upper Sonoran and Arid Transition Zones; east of the
Cascade Mountains in central and northern Oregon east to Montana, Wyoming, and perhaps Utah. Type locality: "Jack Creek, Montana." June–Aug.

10. **Cirsium cymòsum** (Greene) J. T. Howell. Peregrine Thistle. Fig. 5911.

Carduus cymosus Greene, Fl. Fran. 480. 1897.
Cirsium triacanthum Petrak, Beih. Bot. Centralbl. **35**: 481. 1917.
Cirsium cymosum J. T. Howell, Amer. Midl. Nat. **30**: 37. 1943.

  Perennial with widely spreading, horizontal roots from which plants with stout taproots and
erect stems may develop; stems solitary or sometimes several from a branching crown, slender
or stout, 3–7.5 dm. tall, simple and few-headed·or cymosely branching above and many-headed,
angled or nearly terete, arachnoid-pubescent or glabrescent. Earliest leaves small, about 5 cm.
long and 1 cm. wide, oblanceolate or narrowly elliptic, entire or shallowly lobed, spinulose on the
margin; leaves of the basal rosette few to numerous, 1.5–6 dm. long, 0.6–1.5 dm. wide, oblanceolate or elliptic, shallowly to deeply parted into deeply lobed divisions, the divisions and lobes
oblongish, deltoid, obovate, rarely nearly linear, thinly to densely arachnoid-tomentose above and
below, less tomentose above or glabrescent, narrowed below the middle into a spiny-edged, petiole-
like base; lower cauline leaves similar but smaller and sessile by a spinose, slightly amplexicaul
base; uppermost leaves reduced, elongate-linear with prominent spines on the margins and on
divergent lobes or with broad, very spiny, clasping bases below a short deltoid blade; heads generally solitary on short branches or sometimes nearly sessile, small to large, campanulate to subglobose, 2.5–4.5 cm. high, 1.5–5 cm. in diameter, usually subtended by 1 or several bract-like
leaves; involucre variable in aspect and vestiture, thinly or densely arachnoid, the tomentum tending to be close and felt-like; phyllaries numerous, usually rather loosely appressed or somewhat
spreading-ascending, rarely more closely appressed and evenly graduated, the outer ovate-lanceolate, the middle and inner linear-lanceolate to linear, the outer and middle tipped by a stiffish spine,
the inner attenuate into a twisted chartaceous point or weak spine, the middle with an elongate
glutinous ridge along the midrib below the spine, the middle and inner spinulose-ciliolate on the
margins; corollas ivory-whitish or rarely pinkish, becoming tan and sordid in age, the tube 8–13
mm. long, the throat 7–10 mm. long, the lobes 5–7 mm. long; achene dark brown, oblong-oblanceolate, flattened, 7 mm. long; pappus sordid-whitish, 1.5–2 cm. long.

  In clayey or rocky soil derived from either igneous or sedimentary rocks on open, brushy, or wooded slopes,
widespread and variable but not very common, Upper Sonoran and Arid Transition Zones; mountains and sagebrush flats of south-central Oregon and northern California southward in California to the northern Sierra Nevada
and, in the Inner Coast Ranges, to the Diablo Range. Type locality: "Dry hills of Alameda and Contra Costa
counties," California; "Walnut Creek," Contra Costa County, on type in Greene Herbarium. April–July.

11. **Cirsium califórnicum** A. Gray. Bigelow Thistle. Fig. 5912.

Cirsium californicum A. Gray, Pacif. R. Rep. **4**: 112. 1856.
Cnicus lilacinus Greene, Bull. Calif. Acad. **2**: 404. 1887.

  Plant a winter annual or biennial with a strong taproot; stem erect, rather few-branched,
0.5–2 m. tall, thinly arachnoid and somewhat crisp-pubescent. Earliest leaves sinuately pinnately
lobed, the lobes broadly deltoid with apex and lateral teeth bearing short, slender, yellow spines;
basal leaves developing a large rosette, up to 4 dm. long and 1 dm. wide, narrowly elliptic-oblanceolate, prominently pinnately parted, the sinuses broad and rounded, the divisions broadly triangular
or subquadrate, 2–4-lobed, the lobes and teeth tipped by rather weak, slender, yellowish spines,
the blade thinly floccose-arachnoid or glabrescent above, thinly grayish- or white-arachnoid on
the lower side, narrowed below into a short, more or less definitely winged, petiole-like base;
lower cauline leaves similar to the basal leaves, the base not at all auriculate-expanded but shortly
decurrent as a spiny line, the middle and upper cauline leaves smaller and less deeply parted, the
decurrent bases forming narrow spiny wings; uppermost leaves bracteate; heads mostly solitary
on the ends of usually slender, elongate, peduncle-like branches, subglobose to hemispheric, usually
broader than long, 3–4.5 cm. long and to 6 cm. in diameter; involucre broadly bowl-shaped or
subglobose, thinly arachnoid-tomentose, subglabrescent; phyllaries narrowly lanceolate, appressed
at the base, spreading above, the outer spreading-reflexed, the middle horizontally spreading with
the tip strongly upwardly or introrsely curved, the inner sinuate-ascending or substrictly erect,
the bases chartaceous, the tips carinate-thickened and herbaceous, the outer and middle phyllaries

tipped by a short stiff spine, the inner attenuate into a flattened, chartaceous, spinescent tip, the middle phyllaries frequently glandular along the midrib, all phyllaries ciliolate; flowers white, pink, or lavender, the outermost conspicuously outwardly spreading, corolla-tube 8–13 mm. long, the throat 5–6 mm. long, the lobes 5–9 mm. long; achenes oblongish, slightly narrowed at base, 6 mm. long, 2–2.5 mm. wide, light to dark brown; pappus white, 1.5–2 cm. long.

Clayey, sandy, or rocky soil of grassy, brushy, or openly wooded slopes, Upper Sonoran and Transition Zones; California from the northern Sierra Nevada and South Coast Ranges south to coastal southern California and Santa Cruz Island. Type locality: "near Knight's Ferry, on the Stanislaus, California." March–July.

Cirsium californicum var. **bernardinum** (Greene) Petrak, Beih. Bot. Centralbl. **35**: 477. 1917. (*Carduus bernardinus* Greene, Proc. Acad. Phila. **1892**: 361. 1893.) Plants more densely white-tomentose; phyllaries ovatish, deltoid-narrowed into the shortly spinescent tip, rather closely appressed. Typical in southern California but similar forms occasional northward. Type locality: Little Bear Valley, San Bernardino Mountains, San Bernardino County.

Cirsium californicum subsp. *pseudoreglense* Petrak (Beih. Bot. Centralbl. **35**: 479. 1917), to be typified by *Braunton 416* from Elysian Park, Los Angeles County, seems to be a form of var. *bernardinum*.

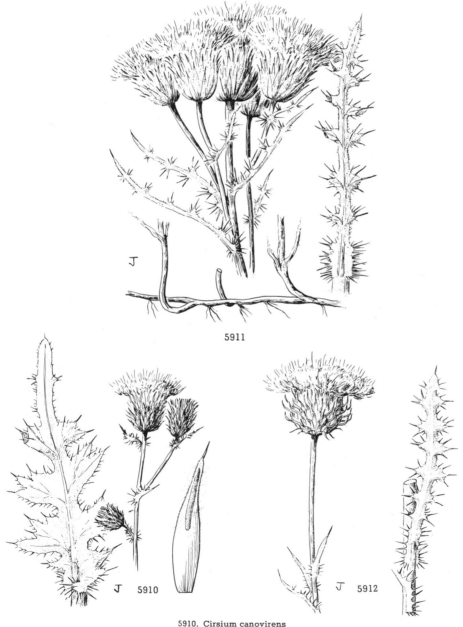

5911

5910. Cirsium canovirens
5911. Cirsium cymosum
5912. Cirsium californicum

12. Cirsium neomexicànum A. Gray. New Mexico Thistle. Fig. 5913.

Cirsium neomexicanum A. Gray, Smiths. Contr. 5[6]: 101. 1853.

Plants biennial or short-lived perennial from a stout taproot, generally monocarpic; stems 5–10 (or 20) dm. tall, stout with rather few elongate slender branches, white-arachnoid-tomentose and a little crisp-puberulent, the tomentum becoming thin and floccose on the lower part. Basal leaves forming a dense, many-leaved rosette that probably develops through more than one season, narrowly oblanceolate, the earliest shallowly sinuate-lobed or deltoid-lobed, becoming more deeply parted, up to 3 dm. long and 6 cm. wide, the divisions 2- or 3-lobed with the lobes terminating in a stoutish yellow spine, the blades narrowed to a short, more or less conspicuously spiny-bordered, petiole-like base, tomentose on both sides but greener and subglabrescent above; lower cauline leaves similar to the basal leaves but smaller and becoming sessile by a strongly decurrent base, the stem sometimes spiny-winged nearly to the next node; upper leaves becoming much smaller and bract-like, very spiny but scarcely decurrent; heads solitary, usually at the ends of long slender branches, usually broader than long and hemispheric, 3–6 cm. long, to 8 cm. wide; involucre campanulate to broadly bowl-shaped, more or less persistently tomentose or the tomentum sometimes tending to become floccose and deciduous; phyllaries linear-lanceolate, the base appressed, the elongate tips of the outer phyllaries reflexed, of the middle phyllaries divergently spreading or loosely ascending, straight or the upper ones sigmoid-curved, the base chartaceous, the tips herbaceous and often canaliculate-folded along the middle, sometimes glandular, tipped by a strong yellow spine, the inner phyllaries flat, attenuate into a slender chartaceous spinule; flowers white, pink, or lavender, the outermost generally outwardly spreading, the corolla-tube 1–2 cm. long, the throat 5–8 mm. long, the lobes 8–10 mm. long; achene purplish brown, elliptic-oblong or -obovate, 5 mm. long, 2–2.5 mm. wide; pappus 1.5–2.5 cm. long, white, sometimes with dark-colored processes along sides of setae near the base.

Dry, gravelly or rocky slopes and canyons, Lower and Upper Sonoran Zones; mountains in the eastern part of the Mojave Desert, California, east to Colorado and New Mexico. Type locality: "Side of the Organ Mountains," New Mexico. April–Sept.

13. Cirsium utahénse Petrak. Intermountain Thistle. Fig. 5914.

? Carduus nevadensis Greene, Pittonia **3**: 26. 1896. Not *Cirsium nevadense* Willk. 1859.
? Cirsium humboldtense Rydb. Fl. Rocky Mts. 1007, 1068. 1917 [1918].
Cirsium utahense Petrak, Beih. Bot. Centralbl. **35**: 470. 1917.
Cirsium wallowense M. E. Peck, Madroño **5**: 247. 1940.

Plants with a long, slender or stout taproot, the rosette probably developing through two to four years, monocarpic or sometimes perennial with new rosettes arising from the branched, caudex-like crown of the taproot; stems erect, 5–10 dm. tall, terete or slightly angled, thinly arachnoid-tomentose and crisp-puberulent. Earliest leaves elliptic to oblanceolate, almost immediately denticulate and with numerous small spines fringing the margin; rosette-leaves becoming numerous, narrowly oblanceolate, to 40 cm. long and 8 cm. wide, pinnately parted with the broad sinuses extending one-half to two-thirds the distance to midvein, the divisions varying from broadly deltoid to subquadrate, the marginal teeth and secondary lobes tipped with numerous slender spines, the lower side white-arachnoid-tomentose, the upper side green and thinly floccose, the blade narrowed below into a short, spiny-margined, petiole-like base; cauline leaves similar to the basal leaves but becoming shorter and more obviously sessile; upper leaves conspicuously decurrent by spiny-lobed wings, the uppermost bract-like and closely subtending the head; heads solitary at the ends of longer or shorter, leafless or leafy-bracteate branchlets, subglobose, 3.5–5 cm. long; involucre campanulate, conspicuously arachnoid-tomentose or the tomentum tending to

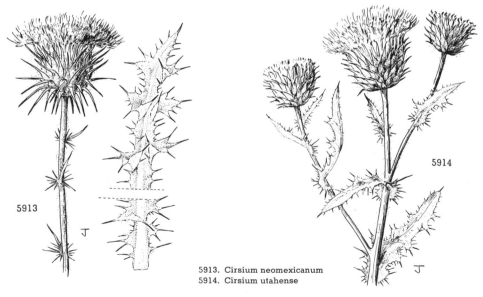

5913. Cirsium neomexicanum
5914. Cirsium utahense

become somewhat floccose; phyllaries appressed at the base, loosely ascending or widely spreading above, the broad bases chartaceous-bordered, attenuate above into the elongate spiny tip, the middle phyllaries with or without a narrow glandular line along the midrib, the inner phyllaries with flattened, purplish-tinged tips that are attenuate into a straw-colored spinule, the tips more or less twisted or curving with spinulose-ciliate margins or sometimes the margin a little chartaceous-expanded and somewhat lacerate; flowers whitish, lavender-tinged, or pinkish, the corolla-tube 13 mm. long, the throat 9–10 mm. long, the lobes 7 mm. long; achenes oblong-oblanceolate, brownish, 7 mm. long; pappus whitish, 2–3 cm. long.

Gravelly and rocky slopes, meadowy flats, and open coniferous woodland, Sonoran and Boreal Zones; mostly east of the Cascade-Sierran crest from southeastern Washington to middle California eastward to the Rocky Mountains and southward to Utah and perhaps to Arizona. Type locality: Silver Reef, Washington County, Utah. May–Aug.

Plants from the east side of the Sierra Nevada in California and Nevada are quite variable in habit, pubescence, and flowers. In all, however, the ovate to ovate-lanceolate phyllaries are subappressed to above the middle and are attenuate into a free tip ending in a slender, stiff, spreading or ascending spine. These plants may be the same as *C. humboldtense* Rydb. (*Carduus nevadensis* Greene; type locality, West Humboldt Mountains, Nevada), and they may prove to be specifically distinct from *C. utahense*.

In the plant of northeastern Oregon (*C. wallowense* M. E. Peck), the phyllaries are scarcely ever glandular, whereas in the typical form from southern Utah, the glandular line or ridge is present and conspicuous. In other parts of its range (as in eastern California), *C. utahense* may be variable in this character. In northeastern Oregon, where *C. utahense* intergrades with *C. brevifolium* and with *C. canovirens*, the intermediates seem to be the result of hybridization.

Carduus inamoenus Greene, Fl. Fran. 479. 1897. (*Carduus undulatus* var. *nevadensis* Greene, Proc. Acad. Phila. **1892**: 361. 1893.) Plant to about 1 m. tall; stem stout, thinly floccose-arachnoid; leaves green and glabrate on the upper side, white-arachnoid on the lower side; heads small, about 2.5 cm. long; phyllaries appressed except at the slender-spinose tip, closely imbricate in 6 or 7 rows, not glandular, the outer lanceolate to ovate-lanceolate, the inner linear-lanceolate; flowers whitish. This plant, definitely known only from Greene's collection made in 1883 near Truckee, Nevada County, California, may be a small-headed form of *Cirsium utahense* Petrak, as that species occurs along the east side of the Sierra Nevada; or the Truckee thistle may represent a hybrid derivative between the Utah thistle and *Cirsium douglasii* var. *canescens*.

Cirsium subniveum Rydb. Fl. Rocky Mts. 1006 (1004, 1068). 1917 [1918]. Plants from northeastern Oregon that are related to *C. utahense* but with smaller heads have been referred to this species (Cronquist, Vasc. Pl. Pacif. Northw. **5**: 141. 1955). Specimens of this sort that we have seen from Wallowa County, Oregon, were growing with plants of *C. utahense* (i.e., *C. wallowense* form) and with plants approaching *C. canovirens*, and it seems probable that the "*subniveum*" plants may have been derived from hybridization between those species. Type locality of *C. subniveum* is Jackson's Hole, Wyoming.

14. **Cirsium nídulum** (M. E. Jones) Petrak. Nidulous Thistle. Fig. 5915.

Cnicus nidulus M. E. Jones, Proc. Calif. Acad. II. **5**: 705. 1895.
Cirsium nidulum Petrak, Beih. Bot. Centralbl. **35**: 553. 1917.

Plants perennial with long woody taproot that may be once or several times branched at the top to produce a leafy crown and 1 or 2 erect leafy stems, the base of the stems closely invested with the dried scaly bases of rosette-leaves; stems 2.5–10 dm. tall, simple or few-branched above, thinly arachnoid-tomentose. Basal leaves few to numerous, narrowly oblanceolate, 7–20 cm. long and to 8 cm. wide, undulate-lobed or commonly more deeply parted or divided, the segments narrowly triangular to ovate-deltoid or sometimes 2-parted and subquadrate, the lobes and divisions tipped by shorter or longer, stiff, yellow spines, the blade narrowed below to a narrowly or broadly spiny-bordered, petiole-like base; lower cauline leaves similar but shorter, the middle and upper cauline leaves becoming sessile by an auriculate, slightly decurrent base, the uppermost reduced bract-like leaves closely subtending the heads; all leaves pale green with thin arachnoid tomentum above and below, or the tomentum on the upper side becoming floccose and deciduous; heads solitary or 2 to several in an open, cymose-paniculate cluster, cylindric or cylindric-campanulate, 4–5 cm. long; involucre thinly arachnoid at first, the phyllaries becoming more or less glabrescent except for the floccose tomentum that persists along the margins and sometimes on the inner faces; phyllaries with bases appressed, loosely ascending or somewhat spreading above the base, the outer sometimes intergrading with the uppermost bract-like leaves and with sides spinose-pectinate, generally the sides not spinose, the base deltoid-triangular, attenuate into an elongate spinose tip, the middle phyllaries lanceolate-attenuate, with a long, yellow, stiff spine, the inner phyllaries flattened, chartaceous, reddish, tipped by a short spinose point, the margins and backs of the middle and inner phyllaries often scabrous, the middle phyllaries occasionally with a glandular line on the back below the spine; flowers rose-red to dark red or dull scarlet, the corollas straight, (2.7)3–3.5 cm. long, the tube 7–11(13) mm. long, the throat (7)9–15 mm. long, the lobes 10–15 mm. long, the tip of the lobes often more or less glandular; achenes 6–7 mm. long, 2–2.5 mm. wide, strongly flattened, light to dark brown; pappus whitish, 2–3 cm. long.

Rocky slopes and flats from desert canyons and mesas to subalpine moraines and meadow borders, Sonoran Zones to Boreal Zone; southern Sierra Nevada and desert ranges of eastern California eastward to Utah and Arizona. Type locality: "in red alkaline sand, along the bottoms of the Pahria River, at Pahria [Paria]," Kane County, Utah. June–Oct.

15. **Cirsium occidentàle** (Nutt.) Jepson. Cobweb Thistle. Fig. 5916.

Carduus occidentalis Nutt. Trans. Amer. Phil. Soc. II. **7**: 418. 1841.
Cirsium coulteri Harv. & Gray in A. Gray, Mem. Amer. Acad. II. **4**: 110. 1849.
Cirsium occidentale Jepson, Fl. W. Mid. Calif. 509. 1901.
Cirsium occidentale var. *coulteri* Jepson, Man. Fl. Pl. Calif. 1167. 1925.

Biennials or winter annuals, developing a slender or stout taproot; stems erect, leafy, usually 0.5–1 m. tall but sometimes as short as 7 cm. or as tall as 1.5 m., thinly to thickly arachnoid- or lanate-tomentose, rarely glabrescent, inconspicuously crisp-puberulent. Earliest leaves oblanceolate-elliptic, sparsely spinose-dentate; rosette-leaves numerous, at first sinuately lobed, 1–1.5 dm. long, becoming 3 dm. long and 0.8 dm. wide, narrowly oblanceolate, rather deeply lobed, the lobes broadly deltoid and toothed, the tips of the lobes and the teeth bearing short spines, narrowed below into a spine-bordered, petiole-like base; basal leaves and also the cauline thinly arachnoid to thickly white-lanate on the lower side, arachnoid-floccose, or arachnoid on the upper side, the cauline oblongish or lanceolate, shallowly to deeply spiny-lobed, becoming broadly sessile with longer or shorter, decurrent, spiny lobes or wings below the point of attachment; uppermost leaves quite reduced but scarcely bracteate, conspicuously spiny; heads solitary at the ends of more or less leafy stems, medium-sized to large, 3–5 cm. long and to 6 or 7 cm. broad; involucres campanulate to broadly hemispheric, the numerous phyllaries interconnected by a webbing of fine, grayish white, filmy tomentum, the tomentum tending to become floccose in age; phyllaries acicular, linear or linear-lanceolate, the outer tending to be reflexed and downward-pointing from near the base, the middle phyllaries spreading horizontally or ascending from a short appressed base, the spreading part thickish, subterete, attenuate into the short, slender but stiffish spine, the inner phyllaries flattened toward the tip, sharply acute or weakly spine-tipped, spinulose-ciliolate and a little twisted, the phyllaries straight or rarely more or less curving outward and upward, not glandular; flowers purplish red or rarely white, scarcely exceeding the involucre; corollas 2.5–3.5 cm. long, the tube 13–18 mm. long, gradually widening into the throat, the lobes 7–10 mm. long, linear, appearing linear-filiform due to inrolling of margins; achenes oblong to obovate, cuneate, 5–6 mm. long, 2–2.6 mm. wide, brownish, somewhat shiny; pappus white or sordid, 2–3 cm. long.

Clayey, sandy, or gravelly soil of coastal dunes, grassy mesas, and brushy washes, Upper Sonoran and Humid Transition Zones; Sonoma County, California, southward to cismontane and insular southern California. Type locality: Santa Barbara, Santa Barbara County (which may also be the type locality of *C. coulteri*). April–July.

The cobweb thistle is remarkable and distinctive in its typical form but in places where its range approaches or overlaps that of *C. proteanum* a seeming genetic blending results that would account for numerous intermediate forms.

16. Cirsium proteànum J. T. Howell. Red or Venus Thistle. Fig. 5917.

Carduus venustus Greene, Proc. Acad. Phila. **1892**: 359. 1893. Not *Cirsium venustum* Porta, 1909.
Cirsium occidentale var. *venustum* Jepson, Man., Fl. Pl. Calif. 1167. 1925.
Cirsium proteanum J. T. Howell, Leaflets West. Bot. **9**: 14. 1959.

Plants from a slender or stout taproot, biennials or winter annuals, the stems and leaves arachnoid-tomentose and inconspicuously crisp-puberulent, the tomentum usually rather thin but sometimes white-lanate, tending to become floccose and deciduous, thinner on the upper side of the leaves than on the lower side; stems erect, 3–15 dm. tall, slender and monocephalous to stout, much-branched, and many-headed, the branches fastigiate-ascending. Early rosette-leaves rather broadly oblancoleate, spinescently dentate or sinuately lobed, the later rosette-leaves generally sparse, linear-oblong or -oblanceolate, to 4.5 dm. long and to 1 dm. wide, sinuately pinnately lobed with the lobes broadly triangular and the sinuses wide and shallow, or the blades deeply parted with the divisions spinescently lobed, narrowed below into a narrowly or broadly spiny-margined, petiole-like base; cauline leaves similar to the basal but becoming small and sessile by semi-amplexicaul, shortly decurrent, spiny bases; uppermost leaves much reduced and bract-like; heads generally few and rather large but sometimes smaller and more numerous, generally 4–5 cm. long but sometimes 2.5 cm. or 7 cm. long, solitary at the ends of peduncle-like stems, elongate-cylindric to hemispheric; involucre exceedingly variable, subcylindric or campanulate to broadly bowl-shaped; phyllaries varying from short to long and from ascending and subappressed to widely divaricate and reflexed, typically the outer ones short, ovate-lanceolate, more or less reflexed, the middle ones elongate-lanceolate, canaliculate and carinate-thickened, divergently spreading above the appressed base, the inner ones ascending and more or less introrse, the innermost erect, flattened, chartaceous, scabrous-ciliolate, all except the innermost shortly spine-tipped and floccose-arachnoid-tomentose, persistently pubescent or glabrescent, none glandular; flowers usually crimson-red but varying to purplish red and rose, rarely white, straight; corollas about 3 cm. long, the tube and throat 18–20 mm. long, the lobes 10–12 mm. long; achenes brown, oblong-obovate, 6-7 mm. long, 3 mm. wide, truncate at base; pappus white, about 2 cm. long.

Clayey, sandy, or rocky soil of grassland, brush, or open woodland, widespread and common in the drier hills and mountains, Upper Sonoran, Transition, and Canadian Zones; common in the Coast Ranges and occasional in the Sierra Nevada, California, Shasta County south to Kern and Ventura Counties; perhaps local in western middle Nevada. Type locality: "hills of the inner Coast Range of California, from Vacaville southward." April–Aug.

17. Cirsium pastóris J. T. Howell. Snowy Thistle. Fig. 5918.

Carduus candidissimus Greene, Proc. Acad. Phila. **1892**: 359. 1893.
Cirsium occidentale var. *candidissimum* J. F. Macbride, Contr. Gray Herb. No. 53: 22. 1918.
Cirsium candidissimum Davids. & Moxley, Fl. S. Calif. 438. 1923. Not Dammer, 1898.
Cirsium pastoris J. T. Howell, Amer. Midl. Nat. **30**: 38. 1943.

Plant biennial with a stout taproot, the stems and leaves white-lanate with a close, felt-like tomentum (also inconspicuously crisp-puberulent), or the upper side of the leaves sometimes only arachnoid and subglabrescent; stems erect, stout, 7–15 dm. tall, divaricately branching above the middle. Early rosette-leaves shallowly or sinuately pinnately lobed, the lobes rounded and spinescent-dentate, later rosette-leaves oblongish to oblanceolate, 1.5–3 dm. long and to 1 dm. wide, narrowed below to a petiole-like base, the blade becoming deeply lobed or parted with the divisions again saliently 3–5-lobed, the lobes ending in stiff, shining, yellowish spines; lower and middle cauline leaves similar to the basal but smaller and more obviously sessile with spiny wings extending downward below the point of attachment; upper leaves lanceolate, subentire above a spiny-lobed, decurrent base, the leaves on the flowering branches almost bracteate; heads rather few, solitary on longer or shorter, peduncle-like stems, oblongish to hemispheric, typically large and

4–6 cm. long, rarely (in forms intermediate to *C. proteanum*) more numerous, smaller, subglobose, 2–3 cm. long; involucre bowl-shaped to campanulate, densely white-tomentose except for the short glabrous spines, the tomentum adhering closely to the phyllaries or more or less arachnoid-festooned between them; phyllaries mostly linear to narrowly lanceolate or the lowest sometimes ovatish, very loosely spreading or deflexed above the appressed base, the outer deflexed or arcuate-recurved as much as 4.5 cm., the middle arcuate-divaricate to ascending, sometimes introrsely ascending, the inner attenuate into a flattened, acute or spinescent, barbellulate tip; flowers deep pink to bright red or crimson, straight, the corolla usually 3–4 cm. long, the tube and throat 1.5–2.5 cm. long, the lobes 1–1.5 cm. long, the ends of the lobes capitellate-thickened below the acute tip; achenes brown, oblong-cuneate, 7–8 mm. long and about 3 mm. wide; pappus white, 2–3 cm. long.

Dry rocky slopes and flats, Transition Zone; southwestern Oregon southward in the Coast Ranges to Mendocino County, California, and in the Sierra Nevada to the Lake Tahoe region in California and Nevada. Type locality: "extreme northern California." June–Oct.

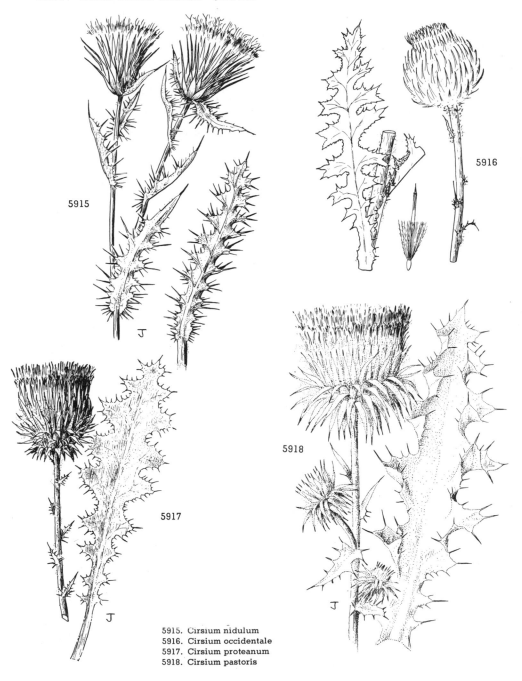

5915. Cirsium nidulum
5916. Cirsium occidentale
5917. Cirsium proteanum
5918. Cirsium pastoris

18. **Cirsium rhothophílum** Blake. Surf Thistle. Fig. 5919.

Carduus maritimus Elmer, Bot. Gaz. **39**: 45. 1905.
Cirsium maritimum Petrak. Beih. Bot. Centralbl. **35**: 288. 1917. Not Makino, 1910.
Cirsium rhothophilum Blake, Journ. Wash. Acad. **21**: 336. 1931.

Biennial or perennial, leafy herb with stout taproot; stems and leaves succulent-fleshy, everywhere covered by a white, persistent, close, suede-like tomentum, the stems erect, 0.1–1 m. long, simple or "much branched from the base, giving the plant a rounded bushy appearance." Earliest leaves elliptic, obtuse, entire or undulate, unarmed except for small spinescent mucro at apex or with short spiny tips to the undulations or shallow lobes, cuneate below and narrowed into a slender, petiole-like base; rosette-leaves oblanceolate to narrowly obovate, 1–1.3 dm. long; 4.5–6 cm. wide, cuneate and narrowed into a petiole-like base, the blade pinnately parted into rounded, ovatish, or broadly elliptic segments, the segments entire or undulate or shallowly lobed, tipped by a short stiff spine and bearing several similar spines on the margins; lower cauline leaves similar; upper cauline leaves shorter, oblongish, more acutely lobed, more strongly spiny, below narrowed to a winged, petiole-like base that develops larger or smaller, rounded, spiny, semi-amplexicaul auricles; uppermost leaves closely clustered below the heads; heads solitary or 2 to 4 on short branches in rather loose terminal clusters, globose-campanulate, 3–4 cm. long; involucre 2.5–3.5 cm. long, densely arachnoid-tomentose; phyllaries numerous, strongly ascending, loosely imbricate or more or less appressed, the sides of all except the inner bearing several stiff, spreading, straight or slightly curved spines and the apices ending in a stout yellowish spine, the outer and middle phyllaries linear-lanceolate or linear, the inner phyllaries thinner but ending in a stiffish spine, the sides chartaceous-margined and ciliate-spinulose, the spinules sometimes more or less confluent and the phyllary tending to be laterally crested; corollas whitish or lemon-yellow, 2–2.5 cm. long, the tube 11–13 mm. long, the throat 5–6 mm. long, the lobes 5–8 mm. long; achenes narrowly cuneate or oblongish, 5–7 mm. long, compressed or plumply turgid, with lateral angles evident, chestnut-brown or light brown; pappus buff, 1.5–2 cm. long.

A remarkable endemic known only from maritime dunes, Upper Sonoran Zone; San Luis Obispo and Santa Barbara Counties, California. Type locality: Surf, Santa Barbara County. April–July.

19. **Cirsium brevistỳlum** Cronquist. Indian Thistle. Fig. 5920.

Cirsium edule of authors in large part, not Nutt.
Cirsium brevistylum Cronquist, Leaflets West. Bot. **7**: 26. 1953.

Erect leafy biennial (or rarely perennial) herb with stout or slender, fusiform taproot; stems slender and nearly simple or becoming tall, robust, and much branched with fastigiately ascending branches, generally 1–2.5 m. tall, rarely only 0.25 m. or as much as 3.5 m., thinly to thickly arachnoid and crisp-villous. Rosette-leaves of first year oblanceolate or oblong-oblanceolate, 1–3 dm. long, narrowed to petiole-like base, subentire to pinnately lobed or divided, the younger bristly-ciliate, the older spinose-bristly, sparsely crisp-puberulent above, thinly to thickly tomentose below; lower cauline leaves similar to the rosette-leaves, above becoming broadly sessile, auriculate, or semi-amplexicaul, not decurrent, the uppermost leaves reduced but generally foliaceous and subtending the heads; heads subglobose, 2–4 cm. long, generally closely clustered at the ends of branches, rarely solitary and then either terminal or racemosely arranged along the stems; involucre campanulate, conspicuously arachnoid-tomentose; phyllaries numerous, loosely imbricate in many series, the outer and middle lanceolate to linear, attenuate and spinose-tipped, the inner elongate-attenuate, acute or subspinose, flattened and somewhat twisted-undulate, sometimes a little lacerate-denticulate; corollas purplish rose, rarely white, 2–2.5 cm. long, the throat 3.5–6 mm. long, the lobes 2–4 mm. long, these minutely cucullate-capitellate at the tip; achenes elliptic or oblongish, slightly cuneate, 4–4.5 mm. long, thickish, brown; pappus buff or light brown, about 2 cm. long, the bristles a little enlarged at the tip.

Moist places in woods and brush or on meadow borders and in forest clearings, Upper Sonoran, Transition, and Lower Canadian Zones; British Columbia to southern California in the Coast Ranges and Cascade Mountains; east to Montana and Idaho. Type locality: near Montesano, Grays Harbor County, Washington. April–Sept.

20. **Cirsium édule** Nutt. Edible Thistle. Fig. 5921.

Cirsium edule Nutt. Trans. Amer. Phil. Soc. II. **7**: 420. 1841.
Carduus macounii Greene, Ottawa Nat. **16**: 38. 1902.

Biennial or perennial monocarpic herb with fusiform taproot; stems erect, widely and openly branched above, frequently stout and hollow, up to 2 m. tall, more or less purplish-tinged, thinly arachnoid-pubescent or glabrescent. Early rosette-leaves rather numerous, oblanceolate, 0.3–1.5 dm. long, deeply pinnately lobed, thinly crisp-pubescent above and below; basal leaves of the flowering season becoming 4 dm. long and 1.2 dm. wide, pinnately divided, the divisions lobed or again divided into oblong or deltoid, spine-tipped lobes, the blade narrowed below to a spinose-margined, petiole-like base, weakly arachnoid- or crisp-pubescent above and below, or glabrescent; middle and upper cauline leaves shorter, much more stoutly spiny, usually less deeply lobed and sessile by a broadly rounded, semi-amplexicaul base; uppermost leaves reduced to strongly spiny bracts closely subtending the heads; heads solitary or densely congested in clusters of 2–5 at the ends of slender branches, globose or broadly campanulate, 3–4 cm. in diameter; involucre broadly campanulate, conspicuously arachnoid-tomentose, the tomentum more or less concealing the phyllaries; phyllaries numerous, purplish, loose with spreading tips, the outer with an appressed, ovate-lanceolate or lanceolate base, abruptly narrowed above into a linear-attenuate, spinose, divaricately spreading tip, the inner phyllaries flattened, weakly spinose, ciliate-spinulose; corolla bright rose-purple or lavender-rose, more or less radiately spreading, the tube about 1 cm. long, the throat about 8 mm. long, the lobes 4–10 mm. long; achenes cuneate-oblong, flattened, 4–6 mm. long, dark purplish brown; pappus buff, about 1.5 cm. long or a little shorter.

Moist deep soil of woods and meadows or high, sandy rocky ridges, Transition, Canadian, and Hudsonian Zones; southern British Columbia and Washington in the Cascade Mountains and Coast Ranges. Type locality: "plains of Oregon and the Blue Mountains; common . . . The young stems, stripped of their bark, are commonly eaten raw by the aborigines, and have a somewhat pleasant sweetish taste." June–Aug.

21. Cirsium hállii (A. Gray) M. E. Jones. Hall Thistle. Fig. 5922.

Cnicus hallii A. Gray, Proc. Amer. Acad. **19**: 56. 1883.
Cirsium hallii M. E. Jones, Bull. Univ. Mont. Biol. Ser. No. 15: 47. 1910.

Plants perennial, probably short-lived, first blooming the second or third year, thereafter developing at the summit of the taproot a several-branched crown that may produce in robust plants as many as 5 or 7 flowering shoots at one time, green and subglabrous, the stems and leaves crisply puberulent and sometimes thinly arachnoid, or the lower sides of the leaves grayish-arachnoid; stems erect, 0.2–2 m. tall, virgate and branching near the top, or bushy and branching from near the base. Earliest leaves broadly to narrowly oblanceolate, merely dentate-spinose or shallowly undulate-lobed, or pinnately parted; rosette-leaves rather few, to 5 dm. long and 1.2 dm. wide, oblanceolate, pinnately divided nearly to the midvein, the divisions crowded or distant, triangular or quadrangular, again lobed or parted with the lobes tipped by short weak spines, the

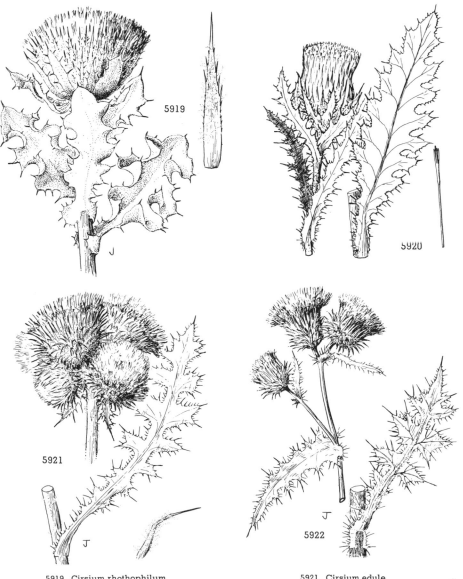

5919. Cirsium rhothophilum
5920. Cirsium brevistylum

5921. Cirsium edule
5922. Cirsium hallii

margin of the lobes and of the rachis-like border along the midrib spinulose-ciliate, the blade narrowed below to a spinulose-ciliate, petiole-like base; lower cauline leaves similar to the rosette-leaves but more spiny and with the base developing rounded, semi-amplexicaul, spiny, shortly decurrent auricles; upper leaves reduced, sessile by a broadly clasping base, shallowly to deeply lobed; uppermost leaves closely subtending the heads, very spiny, passing into the outermost pectinate-spinescent phyllaries; heads solitary or clustered at the ends of elongate branches or in more condensed plants the heads subcongested and leafy-bracteate, hemispheric to subglobose, 2.5–3.5 cm. long, in the more openly branched plants the heads nodding after anthesis; involucre thinly to densely arachnoid, shallowly campanulate to broadly hemispheric, the phyllaries spreading above the shortly appressed base, the free part of the outer and middle ones linear and either somewhat acicular or oblongish, widely spreading, ending in a short slender spine, several of the outer and sometimes 1 or more of the middle ones with 1 or several short spines along the margin, the inner phyllaries suberect, the tip more or less twisted, attenuate into an acute or subspinose point or tending to become a little chartaceous-expanded and lacerate-fimbriate along the margin; flowers lavender-pink to purplish or whitish, more or less spreading in the heads, the corolla-tube 8–10 mm. long, the throat abruptly expanding above the tube, 5–6 mm. long, the lobes 3–6 mm. long; achenes 5 mm. long, 2 mm. wide, narrowly obovate, brownish; pappus a little sordid, about 1.5 cm. long.

Deep moist soil of meadows, forest borders, and maritime slopes, Humid Transition and Canadian Zones; Oregon west of the Cascade Mountains, particularly in the Coast Ranges, from the Columbia River south to Coos Bay. Type locality: Salem, Marion County. April–Aug.
Some of the coastal plants are reminiscent of *C. andrewsii* and some of the montane plants resemble *C. edule*, but all are referred to *C. hallii* pending further study.

22. Cirsium andrèwsii (A. Gray) Jepson. Franciscan Thistle. Fig. 5923.

Cnicus andrewsii A. Gray, Proc. Amer. Acad. 10: 45. 1874.
Cnicus amplifolius Greene, Pittonia 1: 70. 1887.
Cirsium andrewsii Jepson, Fl. W. Mid. Calif. 506. 1901.

Plant monocarpic, biennial or perhaps the rosette sometimes developing through two years; stems erect, 0.6–2 m. tall, in robust plants up to 7 cm. in diameter and succulent, usually widely much-branched from the base, thinly arachnoid but early glabrate. Earliest leaves oblanceolate, spinulose-dentate, undulate-margined or soon becoming pinnately lobed; rosette-leaves large and numerous, to 7.5 dm. long and 2 dm. wide, oblanceolate, pinnately divided, the divisions broad and scarcely lobed or the divisions again divided into 2 or 3 deltoid to narrowly elliptic parts, the lobes and divisions terminating in a stiff spine, the margins sparsely and more finely spinose, the blade narrowed below to a longer or shorter, petiole-like base, thinly arachnoid and early glabrate above, white-arachnoid-tomentose below; cauline leaves becoming gradually smaller and the base rather broadly semi-amplexicaul with rounded, spinose-margined auricles; upper leaves short, lanceolate to ovate, fiercely spiny with divaricate, linear, spine-tipped lobes, the uppermost leaves closely subtending the head; heads solitary at the ends of branchlets or rarely loosely clustered, subglobose or even somewhat obpyramidal, 3–4 cm. long; involucre thinly to densely arachnoid-tomentose, globose-conical; phyllaries appressed at the broadened base, widely divaricate-spreading at about the middle into an attenuate spinose tip, the tips of the outer and middle phyllaries bearing 1–6 stiff lateral spines, the tips of the inner phyllaries flattened and less spinose or merely acute, minutely spinulose-denticulate on the margins or occasionally bearing on elongate lateral spine; corolla rosy purple, more or less spreading in the heads, the tube 8–9 mm. long, the throat 4–5 mm. long, the lobes 5–7 mm. long; achene cuneate-oblong, flattened, 4–5 mm. long, dark purple-brown; pappus buff, about 1.5 cm. long.

Coastal slopes and arroyos, frequently in wet or marshy ground along streams or around seepages, Humid Transition Zone; central California (Sonoma County south to San Mateo County). Type locality: "California . . . probably not far from San Francisco or Sacramento . . ." May–July.
In Marin County, California, *C. andrewsii* hybridizes with *C. quercetorum*. This hybrid has been called *C. hallii* × *quercetorum* f. *psilophyllum* Petrak (Beih. Bot. Centralbl. 35: 527. 1917). Type locality: Tennessee Valley, Marin County.

23. Cirsium crassicaùle (Greene) Jepson. Slough Thistle. Fig. 5924.

Carduus crassicaulis Greene, Proc. Acad. Phila. 1892: 357. 1893.
Cirsium crassicaule Jepson, Fl. W. Mid. Calif. 506. 1901.

Stout annual or biennial herb with a short fusiform root; stems strictly erect, 1 to nearly 3 m. tall, hollow and up to 1 dm. thick near the base, thinly arachnoid-tomentose and crisp-pubescent, more or less floccose, paniculately branching only near the top or in age developing short leafy flowering branches below. Basal leaves arachnoid and glabrescent above, arachnoid-canescent below, narrowly lanceolate, remotely sinuate-pinnatifid, the segments broadly ovate, spine-tipped and spinulose-ciliate; cauline leaves crisply arachnoid or glabrescent above, whitish-arachnoid-tomentose below, narrowly oblanceolate to linear-oblong or -elliptic, to 1.5 dm. long and 3 cm. wide, pinnately parted, the segments strongly spiny and spiny-ciliate, narrowed below to a spiny-margined base with small rounded auricles; upper leaves reduced and strongly spiny; heads subtended by a few spiny, bract-like leaves, solitary on a longer or shorter, peduncle-like stem or rarely closely congested, campanulate, 2.5 or 3–4 cm. long; involucres glabrous or perhaps rarely thinly arachnoid, the phyllaries not glandular, closely appressed and imbricate except for the ascending-spreading, spiny tips, the outer phyllaries ovate-lanceolate, attenuate into the long strong spine, frequently bearing 1 or more stiff, unequal, lateral spines, rarely the lateral spines again spinulose or chartaceous-expanded and more or less lacerate, the middle phyllaries elongate-lanceolate, attenuate into a slender spine, the spiny tip sometimes bearing a small lateral spine, the inner phyllaries with flattened, somewhat twisted, scabrous tips that end in a small spine or are merely spinosely acute, the margins of all phyllaries finely spinulose-ciliate; flowers lavender-rose, spreading a little in the head, the corolla 2.5 cm. long, the tube 10–12 mm. long, the throat 5–6 mm. long, the lobes 7–9 mm. long; achene narrowly oblongish-cuneate, 5–5.5 mm. long, blackish brown; pappus whitish buff, 1.5–2 cm. long.

Wet soil or shallow water along sloughs and canals, Upper and Lower Sonoran Zones; at two widely separated localities in the northern and southern San Joaquin Valley in San Joaquin and Kern Counties, California. Type locality: "low, grassy, and occasionally inundated river bottoms of the lower San Joaquin, near Lathrop, California." May–Sept.

24. Cirsium quercetòrum (A. Gray) Jepson. Brownie Thistle. Fig. 5925.

Cnicus quercetorum A. Gray, Proc. Amer. Acad. 10: 40. 1874.
Cirsium quercetorum Jepson, Fl. W. Mid. Calif. 507. 1901.
Cirsium quercetorum var. *mendocinum* Petrak, Beih. Bot. Centralbl. 35: 364. 1917.

Perennial with widely spreading, rhizome-like roots from which leafy flowering shoots arise, the shoots developing a stout taproot with a caudex-like crown; stems usually low, stout, and densely leafy or becoming taller, more slender, and more sparsely leafy and up to 2.5 (or 3) dm. tall, thinly arachnoid-tomentose and more or less crisp-puberulent, glabrescent, rarely subglabrous.

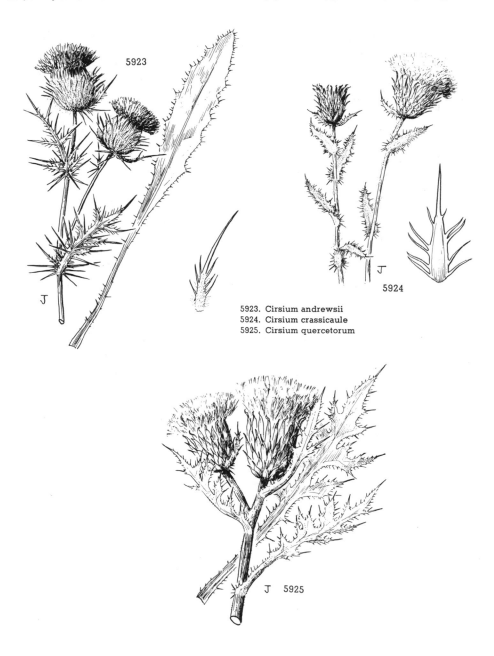

5923. Cirsium andrewsii
5924. Cirsium crassicaule
5925. Cirsium quercetorum

Earliest leaves small, elliptic, spinulose-denticulate or undulate-lobed; leaves of the basal rosette oblanceolate, to 3 dm. long and 1 dm. wide, arachnoid and puberulent beneath, thinly so above and early glabrescent, pinnately parted or divided, the divisions mostly broadly obovate and again 2- or 3-lobed or -parted, the lobes often narrow and elongate and strongly spine-tipped, margins entire or spiny, the blade cuneate-narrowed below to a petiole-like base; cauline leaves similar to the basal leaves, pinnatisect and spiny, the petiole-like base scarcely or not at all auriculate-lobed; 1 or few of the uppermost leaves usually subtending the heads (or in variants growing near *C. andrewsii*, the heads subtended by numerous very spiny leaves); heads on short, peduncle-like stems, solitary or generally in loose to dense clusters, campanulate, 3.5–5.5 cm. long; involucre thinly arachnoid, early glabrescent; phyllaries not glandular, numerous and closely imbricate in 5 to 10 or more series, the outer ovate-lanceolate to ovate, the middle and inner gradually longer and more oblongish, the outer and middle with a short, stiff, terminal, ascending or spreading spine and with the herbaceous center outlined or bordered by a bony, stramineous, spinulose-denticulate margin, the tips of the inner phyllaries thinner, more or less curved or wavy, sometimes spinulose-fimbriate below the acute or weakly spinescent apex; flowers usually sordid-whitish or buff, rarely lavender-rose; corollas 2.5–3 cm. long, the tube 1–1.5 cm. long, the lobes unequal with the sinuses between them 5–8 mm. deep; anther-appendages subacuminate-tipped; achenes narrowly obovate or oblongish, somewhat cuneate, more or less flattened, light to dark brown, 5–6.5 mm. long, 2–2.5 mm. wide; pappus 2–4 cm. long, buff, the tips scarcely enlarged.

On open, grassy or brushy slopes, rarely in open woodland, Upper Sonoran and Transition Zones; central California Coast Ranges. Type locality: "hills at Oakland and elsewhere near San Francisco, California." April–July.

Hybrid-like variants obviously related to *C. quercetorum* can be found where this species grows in the vicinity of *C. andrewsii, C. douglasii, C. fontinale, C. occidentale, C. callilepis,* and perhaps even other species.

Cirsium quercetorum var. **walkeriànum** (Petrak) Jepson, Man. Fl. Pl. Calif. 1164. 1925. (*Cirsium walkerianum* Petrak, Beih. Bot. Centralbl. **35**: 405. 1917.) Stems erect, 3–7 (or 9) dm. tall; heads more loosely clustered on longer peduncles. Chiefly in the region of San Francisco Bay, California. Type locality: Thousand Oaks district, Berkeley, Alameda County.

Cirsium quercetorum var. **xerolépis** Petrak, Beih. Bot. Centralbl. **35**: 363. 1917. Plants generally caulescent and 3–6 dm. tall; phyllaries numerous and graduate-imbricate as in the species but with margins more scarious and spinescently lacerate and with enlarged, spreading, scarious, lacerate, short-spiny tips. California Coast Range hills of San Mateo County and adjacent San Francisco County; perhaps also in the Santa Lucia Mountains, Monterey County. In San Mateo County, plants of this variety form a variable population and exhibit characters of *C. quercetorum, C. callilepis,* and possibly *C. fontinale.* Type locality: northwest of San Bruno, San Mateo County.

25. Cirsium loncholèpis Petrak. Gracious Thistle. Fig. 5926.

Cirsium loncholepis Petrak, Beih. Bot. Centralbl. **35**: 375. 1917.

Plants with a stout or slender taproot, annual or perhaps sometimes living more than one season, generally stout and succulent, subglabrous or sparsely crisp-puberulent and arachnoid; stems erect, hollow, subsimple, 0.3–1 m. tall, vascular-striate, commonly reddish-tinged. Leaves light green, slightly puberulent or glabrate above, thinly gray-arachnoid and more or less puberulent or subglabrate below; basal leaves unknown; lower cauline leaves oblanceolate, to 3.5 dm. long and 1 dm. wide, divided nearly to the midrib into several-lobed segments, the lobes tipped by short slender spines, narrowed below into an elongate, spiny-margined, petiole-like base, the base extending a short distance downward as a spiny wing; upper cauline leaves much more spiny, the divisions and lobes rather narrowly deltoid and extending into a prolonged stiff spine; uppermost leaves closely subtending the heads, becoming reduced to a narrow ligulate rachis bearing small clusters of stiff spines; heads campanulate or subcylindric, 3–4 cm. long, solitary or densely clustered in twos to fours at the top of the nearly simple stems; involucre glabrous, the phyllaries loosely appressed-ascending, sometimes a little scabrous and the margin frequently ciliolate, the outer phyllaries narrowly deltoid-ovate, tapering from the base to a short stiff spine, the middle phyllaries ovate-lanceolate, narrowed gradually into a short, stiff, somewhat shining spine, the body of the bract brownish or purplish lavender, dull, bordered by a narrow, shining, stramineous line near the margin, the inner phyllaries elongate, attenuate, shortly and weakly spinescent, chartaceous, a little twisted; flowers whitish tinged with purple; corollas 2.5–3 cm. long, the tube very slender, 13–17 mm. long, the throat 5–8 mm. long, the lobes 5–7 mm. long; achenes compressed or a little turgid, obovate-oblong, 3–4 mm. long, 1.5 mm. wide, purplish brown or buff mottled with brown; pappus light brown, 2–2.5 cm. long.

Wet soil of coastal hills and flats, often associated with willows or growing in the open with marsh plants, Upper Sonoran Zone; San Luis Obispo and Santa Barbara Counties, California. Type locality: near La Graciosa, Santa Barbara County. June–Aug. This rare thistle which was little known until it was studied recently in the field by Clifton F. Smith, is closely related to the *C. foliosum* complex.

26. Cirsium foliòsum (Hook.) DC. Leafy or Dwarf Thistle. Fig. 5927.

Carduus foliosus Hook. Fl. Bor. Amer. **1**: 303. 1833.

Cirsium foliosum DC. Prod. **6**: 654. 1837.

Cirsium drummondii Torr. & Gray, Fl. N. Amer. **2**: 459. 1843.

? *Carduus validus* Greene, Fl. Fran. 479. 1897.

Cnicus tioganus Congdon, Erythea **7**: 186. 1900.

Cirsium americanum K. Schum. in Just, Bot. Jahresb. **29¹**: 566. 1903.

Cirsium acaulescens K. Schum. loc. cit.

Carduus brownii Eastw. ex Heller, Muhlenbergia **2**: 160. 1906. (Nomen nudum.)

Cirsium drummondii subsp. *lanatum* Petrak, Beih. Bot. Centralbl. **35**: 354. 1917.

Cirsium drummondii subsp. *lanatum* var. *oregonense* Petrak, op. cit. 355.

Cirsium drummondii subsp. *latisquamum* Petrak, op. cit. 358.

Cirsium drummondii subsp. *vexans* Petrak, op. cit. 359.

Cirsium quercetorum var. *citrinum* Petrak, op. cit. 363.

Plants leafy, biennials or short-lived perennials with a thick taproot, acaulescent or with stems to 1 m. tall; stems often stout, succulent, strongly striate, and hollow; pubescence sparse to dense, made up almost entirely of crisp, elongate, shining, translucent, multicellular hairs, sometimes a little arachnoid near the top of the stem, or the pubescence more arachnoid throughout. Rosette-leaves rather numerous, narrowly oblanceolate, up to 5 dm. long and 5 cm. wide, entire, shallowly undulate-lobed or more saliently pinnately lobed, the margin spinulose-ciliate with short, slender, yellowish spines, the spines more elongate and sometimes quite stout at the tips of the lobes, narrowed below to a spiny-bordered, petiole-like base, the upper side light green, sparsely to densely crisp-puberulent, the lower side arachnoid-tomentose but usually crisp-puberulent along the midvein or the leaves glabrescent; lower and middle cauline leaves in caulescent plants similar to the basal leaves but not so large, more regularly pinnately lobed or parted, the margins more prominently spiny; uppermost leaves that subtend the heads becoming bracteate, elongate-ligulate, much exceeding the heads, sometimes conspicuously pectinate-spinose-ciliate and occasionally tinged with shades of yellow, lavender, or purple; heads either solitary at the top of the stem or arranged in a subspicate manner in the axils of the upper leaves, or more or less congested at the top of the stem, or in acaulescent plants congested in the rosette-leaves, the heads campanulate to subglobose, 3–5 cm. long; involucre bowl-shaped, closely subtended by the uppermost bract-like leaves, the phyllaries closely to rather loosely imbricate, the backs glabrous but often more or less scabrous, the margins a little crisp-puberulent, arachnoid-tomentose, scabrous-ciliolate, or smooth, the outer phyllaries narrowly to widely triangular-ovate, attenuate to the short, stiff, erect spine, the middle phyllaries similar but more elongate, sometimes long-spinose, the inner phyllaries with flattened chartaceous tips, attenuate into a sharply acute or subspinose point, or more or less expanded into a narrow or broad, lacerate-margined, subscarious appendage; flowers whitish to lavender-pink or rose, not curving outward, the corolla 2–3 cm. long, the tube 5–15 mm. long, the throat 5–10 mm. long, the lobes 4–10 mm. long; achenes light brown, 5–7 mm. long, 2–2.5 mm. wide, oblongish, a little widened upward; pappus pale brownish buff, 2–3 cm. long.

Dry or wet places in mountain meadows, sagebrush-covered valleys, open forests, and subalpine slopes, Transition, Canadian, and Hudsonian Zones; along the Cascade-Sierran ranges and to the east in Washington, Oregon, and California, rare or absent in the Coast Ranges; widespread in the mountains and high intermountain valleys of western North America from Yukon Territory to northern Lower California east to the Rocky Mountains. Type locality: "prairies of the Rocky Mountains." May–Sept.

The leafy thistles constitute a polymorphic aggregate in need of critical study before they can be satisfactorily aligned on either specific or subspecific levels.

27. **Cirsium hookeriànum** Nutt. Hooker Thistle. Fig. 5928.

Carduus discolor β Hook. Fl. Bor. Amer. 1: 302. 1833.
Cirsium hookerianum Nutt. Trans. Amer. Phil. Soc. II. 7: 418. 1841.

Probably a short-lived perennial with a taproot; stems erect, 0.6–1.5(2) m. tall, nearly simple or narrowly branched above, thinly arachnoid-tomentose and sparsely crisp-pubescent. Basal

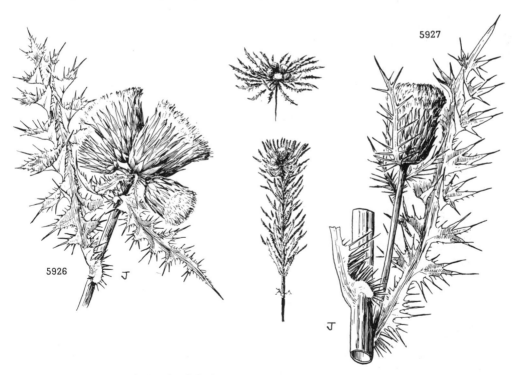

5926. Cirsium loncholepis
5927. Cirsium foliosum

leaves narrowly oblanceolate, to 3 dm. long or more, to 1.5 dm. wide, pinnately parted or divided, the divisions mostly 2-lobed with deltoid or ovatish segments, the segments shortly tipped with stiffish yellow spines, the margin sparsely spinulose-ciliate, tapering below to a spiny-bordered, petiole-like base; cauline leaves similar but the leaves becoming shorter and more obviously sessile upward, the base narrowly spiny-auriculate and more or less decurrent as narrow, spinose-pectinate fringes below the point of attachment; the upper cauline leaves much reduced, very spiny, linear or oblongish, the uppermost closely subtending the heads, the basal and cauline leaves thinly grayish-arachnoid-tomentose and a little puberulent along the veins, glabrescent above except for the sparse crisp puberulence; heads racemosely or subspicately arranged along the stem from the middle or above, becoming closely clustered at the top of the plant and also occasionally at the ends of abbreviated lateral branches, subglobose, 3–4 cm. long; involucre shortly campanulate, arachnoid and crisply puberulent-arachnoid; phyllaries rather loose, the outer and middle widely spreading and somewhat upward-curving above the appressed base, ovate-lanceolate, ending in a short, stiff, yellow spine, the inner linear, attenuate into a slender, twisted, chartaceous, spinose tip, the margins ciliolate and scabridous, the outer phyllaries sometimes laterally spinescent, intergrading in this character with the uppermost bract-like, spiny leaves; flowers whitish or yellowish, becoming sordid-tan, somewhat spreading in the heads; corollas 2–2.5 cm. long, the tube 9–12 mm. long, the throat 6–8 mm. long, the lobes 5–6 mm. long; achenes oblongish, a little narrowed downward, strongly flattened, 5–6 mm. long, 2 mm. wide; pappus buff, 1.5–2 cm. long.

Moist slopes and flats of mountains, thriving in fields and along roads, Canadian and Boreal Zones; northern Washington east of the Cascade Mountains north to British Columbia and Alberta east to the Rocky Mountains. Type locality: "Prairies of the Rocky Mountains." July–Aug. According to Cronquist (Vasc. Pl. Pacif. Northw. 5: 140. 1955), the specimens south of the Canadian border "vary in the direction of *C. foliosum.*"

28. **Cirsium péckii** Henderson. Steen Mountain Thistle. Fig. 5929.

Cirsium peckii Henderson, Madroño **5**: 97. 1939.

Plant yellowish green, simple below, branching above the middle, the stem stout, fistulous, vascular-striate, in dried specimens brownish or shining and golden, sparsely loose-pubescent with elongate, shining, translucent, multicellular hairs, glabrescent below. Basal leaves unknown; cauline leaves narrowly oblanceolate or elliptic-oblanceolate, becoming oblong-lanceolate above, to 2.5 dm. long and 5 cm. wide, parted or divided into deeply deltoid-lobed segments, the lobes tipped by slender or stoutish, yellow spines; lower leaves narrowed below to a narrowly bordered, spiny, petiole-like base; upper leaves sessile with small, very spiny, basal auricles, not decurrent; uppermost leaves reduced to an elongate, pectinately fringed rachis closely subtending the heads, sparsely pubescent on the upper and lower sides particularly along the midrib and veins or the leaves subglabrous; heads 4–5 cm. long, hemispheric, solitary at the ends of longer or shorter branches or nearly sessile in leaf-axils, sometimes in close terminal clusters; involucre campanulate to bowl-shaped, loosely crisp-puberulent and more or less arachnoid; phyllaries nearly equal, ascending and somewhat spreading, the outer pectinate-spinose, these and the middle lanceolate, attenuate upward into the long yellow spine, the spine straight, canaliculate on the inner face near its base but becoming subterete at the tip, the inner phyllaries shorter than the spines of the middle phyllaries, chartaceous and flattened but somewhat carinate on the outer side, subspinose; flowers straight, pale lavender, the corolla 2–3 cm. long, the tube 7–12 mm. long, the throat about 1 cm. long, the lobes 7–8 mm. long; achenes light brown, oblongish, a little widened upward, 6–7 mm. long, 2–2.5 mm. wide; pappus a little sordid, 2 cm. long.

Moist soil and dry stream banks in sagebrush country, Upper Sonoran Zone; eastern slopes of the Steen and Pueblo Mountains, Harney County, Oregon. Type locality: Alvord Ranch, east base of Steen Mountains, Harney County. June–July.

Cirsium praetériens J. F. Macbride, Contr. Gray Herb. No. 53: 19. 1918. Stems very leafy, arachnoid-tomentose with a few multicellular, shining, translucent hairs intermixed; cauline leaves oblong-elliptic, deeply parted into narrowly deltoid or linear, strongly spiny lobes, sparsely pubescent above with multicellular shining hairs, white-tomentose below; heads spicately congested along the stem in the axils of the uppermost cauline leaves and closely subtended by several reduced, bract-like leaves, large, 5.5–7 cm. high, subglobose; involucre rather densely grayish-arachnoid, campanulate-bowl-shaped; phyllaries loosely ascending, straight, the outer and middle phyllaries attenuate into a long stiff spine, the outer pectinate-spinulose on the margin, the inner flattened and somewhat curving and twisted, strongly scabridous below the slender spinulose tip; corolla "whitish," 3–3.5 cm. long, the tube 1.5 cm. long, the throat 8 mm. long, the lobes 1 cm. long; pappus about 3.5 cm. long, light brown or buff, copiously plumose from the base to the scarcely dilatate, barbellulate tip. Known only from two collections made at Palo Alto, Santa Clara County, California, by J. W. Congdon in July, in 1897 and in 1901.

29. **Cirsium callilèpis** (Greene) Jepson. Fringed-bract Thistle. Fig. 5930.

Carduus callilepis Greene, Proc. Acad. Phila. **1892**: 358. 1893.
Cirsium callilepis Jepson, Fl. W. Mid. Calif. 507. 1901.
Cirsium americanum var. *callilepis* Jepson, Man. Fl. Pl. Calif. 1164. 1925.

Perennial herbs, the aerial shoots arising from the simple or branched, woody crown of a slender or stout root, the new shoots developing from seeds or from buds on the widely spreading roots; stems erect, generally rather slender, few-branched and sparsely leafy, 4–12 dm. tall, terete or rather strongly striate-angled, crispy-puberulent with elongate, multicellular, shining hairs and also more or less arachnoid. Earliest leaves small, subentire, prickly-ciliate, narrowly elliptic, the later rosette-leaves few to many, glabrescent above, arachnoid-tomentose below, up to 4 dm. long and 1 dm. wide, elliptic or narrowly ovatish, pinnately divided almost to the midrib into lobed or deeply parted segments, the lobes and margins dentate-spiny or spiny-tipped, the spines short and weak, the blade narrowed to a more or less spiny-bordered, petiole-like base, this base and the midrib both above and below crispy-puberulent or arachnoid; lowest cauline leaves similar but developing a spiny-lobed, auriculate base, the middle cauline leaves becoming semi-amplexicaul with a more or less enlarged, rounded, spiny, shortly decurrent base; uppermost leaves much reduced and bract-like; heads subglobose or broadly campanulate, 2–4 cm. long, solitary or clustered on elongate, slender, somewhat spreading, peduncle-like branches; involucre subglabrous or thinly arachnoid and early glabrescent; phyllaries numerous in many series, closely appressed and well imbricate or (in var. *pseudocarlinoides*) frequently loosely ascending and less graduated, the outer and middle phyllaries narrowly obovate or oblanceolate-obovate with roundish or ovate, widely margined, green-herbaceous tips, the margin yellowish, scarious-cartilaginous, and deeply lacerate or spinescent-fimbriate, the lower margins minutely spinescent-serrulate, the inner phyllaries oblong-lanceolate with a thinnish, subscarious, elongate-deltoid, scabridous tip that is finely and irregularly serrulate-lacerate on the margins and is attenuate into a weak, twisted or curved, spine-like point; flowers sordid, buff or light brownish, corolla 2–2.5 cm. long, the lobes unequal, 4–7 mm. long, the tube about 1 cm. long; achenes narrowly oblanceolate or nearly elliptic, 7 mm. long, grayish brown; pappus light brown, 1.5–2 cm. long, the bristles scarcely enlarged at the barbellulate tips.

Grassy places on the edge of brush or woods, Humid Transition Zone; central California Coast Ranges just north of San Francisco Bay. Type locality: "Western California"; and "Mt. Tamalpais and northward" (Fl. Fran. 477. 1897.)

Cirsium remotifolium var. odontolèpis Petrak, Beih. Bot. Centralbl. **35**: 298. 1917. Tips of the phyllaries spreading or recurved. Known only from the type, which was collected on Mount Tamalpais, Marin County.

Cirsium callilepis var. oregonénse (Petrak) J. T. Howell, Leaflets West. Bot. **9**: 10. 1959. (*Cirsium remotifolium* subsp. *oregonense* Petrak, Beih. Bot. Centralbl. **35**: 300. 1917.) Phyllaries numerous, in many series, subappressed, conspicuously scarious or chartaceous, the outer shortly spinose at the emarginate or obcordate tip, the margins finely fimbriate-lacerate nearly to the base, the middle and inner phyllaries with more extended and expanded, scarious, lacerate tips, the inner tips light rose-color, ovate-deltoid, sharply acute or spinulose; flowers pink. Open forests of central Oregon, chiefly in the Cascade Mountains. Type locality: Camp Polk, Crook County. July–Aug.

Cirsium callilepis var. pseudocarlinoìdes (Petrak) J. T. Howell, loc. cit. (*Cirsium remotifolium* subsp. *pseudocarlinoides* Petrak, Beih. Bot. Centralbl. **35**: 297. 1917.) ·Phyllaries oblanceolate or oblong to oblong-linear, mostly loosely ascending, the outer phyllaries entire and gradually narrowed to the spiny tip or frequently the upper part chartaceous and sparsely spinose-dentate, the tip of the middle phyllaries scarious- or chartaceous-expanded, the margin deeply and irregularly lacerate-fimbriate, the inner phyllaris with scarious, curved or twisted tips that are serrulate-fimbriate and spinulose-attenuate. Grassland and woods of valleys and mountains, mostly in the Humid Transition Zone or Canadian Zone; from central California north to northwestern Oregon. Type locality: Mount Tamalpais, Marin County (cf. Wasmann Journ. Biol. 11: 248. 1953).

On Mount Tamalpais, plants have been found that are perhaps derived from a cross with *C. quercetorum*. At Willits, Mendocino County, a plant with the inner phyllaries glandular may represent a cross with *C. cymosum*.

Cirsium amblylèpis Petrak, Beih. Bot. Centralbl. **35**: 312. 1917. Similar to *C. callilepis* var. *pseudocarlinoides* but with phyllaries appressed and imbricate and with the phyllary-tips less broadly scarious-margined; a hybrid-suspect between *C. callilepis* var. *pseudocarlinoides* and *C. quercetorum*. Type locality: Mount Tamalpais, Marin County.

30. Cirsium acanthodóntum Blake. Klamath Thistle. Fig. 5931.

Cirsium acanthodontum Blake, Contr. Gray Herb. No. 53: 28. 1918.
Cirsium oreganum Piper, Proc. Biol. Soc. Wash. **32**: 43. 1919.
Cirsium remotifolium var. *rivulare* Jepson, Man. Fl. Pl. Calif. 1164. 1925.

Perennial herb, the aerial shoots arising singly or severally from the simple or somewhat branched woody crown of the stout or slender taproot, these crowns developing from seedlings or produced along elongate, horizontally spreading roots; stems slender or stout, 0.5–1.1 m. tall (or dwarfed and less than 1 dm. tall), simple or few-branched above or in more robust individuals openly fastigiately branched from just below the middle, rather strongly angled, arachnoid and

5928. Cirsium hookerianum

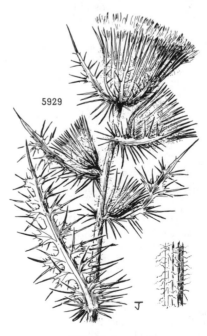

5929. Cirsium peckii

crisp-pubescent or glabrescent. Earliest leaves small, narrowly elliptic-oblanceolate, entire or sparsely spinulose-denticulate, the leaves of the basal rosette usually few, varying in one rosette from subentire to deeply pinnately divided, narrowly to broadly oblanceolate, 1.5–4 dm. long, the subentire leaves 1.5–3 cm. broad, the lobed or divided leaves 3–10 cm. broad, the lobes and divisions short and deltoid to elongate and again lobed or divided, the lobes terminating in slender weak spines, the leaf-margin more or less spinulose-ciliate, especially toward the narrowed, petiole-like base; lower cauline leaves similar to the basal leaves but the expanded, blade-like part extending downward toward or to the leaf-base and the leaf-base becoming auriculate; middle and upper cauline leaves entire or lobed, lanceolate to linear-lanceolate, auriculate-amplexicaul, sometimes broadly so, scarcely spiny and merely weakly spinulose along the margins, or rarely the cauline leaves quite spiny; uppermost leaves much reduced, linear and bract-like; all leaves thinly arachnoid but early glabrescent above and white-arachnoid-tomentose below; heads solitary or rarely clustered at the end of branches, campanulate, 2.5–3.5 cm. long; involucre more or less arachnoid-tomentose; phyllaries numerous, loosely appressed, the outer and middle linear-oblong, acutely spine-tipped, the sides pectinately spinulose, the spinules ascending or more or less arcuate-spreading, the inner phyllaries elongate, linear-oblong, the sides smooth or spinulose-denticulate, the tip more or less chartaceous-scarious and lacerate, triangular-acute or sometimes flabellate, the median line of the inner phyllaries glandular; corollas widely spreading in the head, rosy-purple or lavender-rose, the tube 8–9 mm. long, the throat 6–7 mm. long, the lobes 4–5 mm. long; achene elliptic-oblanceolate, somewhat flattened, 5 mm. long, light brownish or buff with purple-brown markings; pappus about 1.5 cm. long, sordid-whitish.

Moist deep soil of woods and meadow borders, or on rocky cuts and exposed maritime slopes, Humid Transition Zone; Coast Ranges from Curry County, Oregon, south to Humboldt County, California. Type locality: "dry rocky soil, 6.4 km. north of Agness, Curry Co.," Oregon. June–Sept.

31. **Cirsium remotifòlium** (Hook.) DC. Remote-leaved Thistle. Fig. 5932.

Carduus remotifolius Hook. Fl. Bor. Amer. 1: 302. 1833.
Cirsium remotifolium DC. Prod. 6: 655. 1837.
Cirsium stenolepidum Nutt. Trans. Amer. Phil. Soc. II. 7: 419. 1841.

Biennial or short-lived perennial herb with a rather stout taproot (or perhaps with new rosettes from rhizome-like roots); stems erect, 0.3–1.5 m. tall, remotely to rather closely leafy, terete but somewhat longitudinally ridged, purplish-tinged above, pubescent with loose, elongate, multicellular, shining, pustular-based hairs, also thinly arachnoid above, scarcely arachnoid near the base. Basal leaves to 4 dm. long and 1 dm. wide, narrowly elliptic, deeply pinnately lobed or parted, the segments broadly oblongish and generally 1-lobed on the outer side, the segments and lobes shortly and weakly spine-tipped, the margin sparsely spinulose-ciliate, thinly arachnoid-tomentose below with shining multicellular hairs along midrib and veins, glabrate above or with scattered, shining multicellular hairs, the blade narrowed gradually to a broad subacute tip, attenuate below into an elongate, spinescently bordered, petiole-like base; cauline leaves smaller and much more spiny, the lower with a petiole-like base with small spiny auricles, the upper with larger rounded auricles, the leaf-bases semi-amplexicaul, not decurrent, the upper surface glabrate or with elongate, shining, pustular-based hairs; heads subglobose, 3–4 cm. high, mostly solitary at the ends of short, naked or sparsely leafy-bracted, peduncle-like stems or rarely some of the heads subsessile in the axils of the upper leaves; involucre thinly arachnoid and becoming nearly glabrate; phyllaries numerous, loosely ascending or somewhat spreading, not glandular, neither strongly graduated nor imbricate, the outer and middle phyllaries linear-lanceolate, gradually narrowed to the short-spiny tip, the margins mostly entire or sometimes sparsely denticulate-ciliolate below the tip, the inner phyllaries attenuate into a thin, slender, somewhat twisted, spinescent tip, the margins spinulose-ciliate, or the tips a little expanded and lacerate-fringed; flowers whitish, sordid, the corollas 2 cm. long, the tube 8–9 mm. long, the lobes unequal, 4–6 mm. long; achenes dark or light brown speckled with dark purplish brown, scarcely shining, 4–5 mm. long, oblongish-cuneate; pappus 2 cm. long, sordid-whitish, the barbellulate tips not expanded.

Occasional or rare on open grassy slopes and valley lands or in open woods, Humid Transition Zone; Washington and Oregon chiefly west of the Cascade Mountains southward to the Coast Ranges of northern California. Type locality: "common in the valley of the Columbia." June–Aug. Almost everywhere this thistle appears to intergrade, probably through long-continued hybridization, with *C. callilepis.* Perhaps typical *C. remotifolium* is to be found only in western Washington and northwestern Oregon.

A hybrid-suspect between *C. remotifolium* and *C. cymosum* has been collected on Sexton Mountain, Josephine County, Oregon. It resembles *C. remotifolium* but it is more tomentose and many of the phyllaries are glandular along the midrib.

Cirsium mendocinum Petrak, Beih. Bot. Centralbl. **35**: 339. 1917. Similar to *C. remotifolium* but the leaves much more spiny, the involucres more copiously arachnoid, and the flowers pink or purplish due either to colored corolla or colored anther-tips. These characters are reminiscent of *C. andrewsii.* Brushy and wooded places on the coastal plain of Mendocino County, California, the type locality.

32. **Cirsium andersónii** (A. Gray) Petrak. Anderson's Thistle. Fig. 5933.

Cnicus andersonii A. Gray, Proc. Amer. Acad. **10**: 44. 1874.
Cirsium andersonii Petrak, Bot. Tidsskr. **31**: 68. 1911.

Perennial herb with widely creeping, slender, horizontal roots from which flowering plants with stout taproots, woody and sometimes branched caudices, and erect stems develop; stems usually 4–7 dm. tall, rarely depauperate and 1.5 dm. tall or as much as 10 dm. tall, sparsely crisp-puberulent and arachnoid-tomentose or glabrescent, frequently purplish. Leaves of the first season few, small, oblanceolate, obtuse or subacute, spinose-dentate or somewhat undulate and spiny on the margin, in succeeding year (or perhaps after several years?) developing a large rosette from which the flowering shoot grows; rosulate leaves up to 3.5 dm. long and 1 dm. wide, oblong-oblanceolate, pinnately divided with the divisions again lobed, rather sparsely and weakly spiny particularly on the acute tips of the lobes, the blades narrowed below to a petiole-like base; cauline leaves frequently rather sparse, oblanceolate or oblongish, to 2 dm. long, pinnately lobed or divided into 1–3-parted, ovatish lobes or divisions or in depauperate plants the cauline leaves subentire or shallowly lobed, the lobes and margins spiny and sometimes prominently so, the base of the leaves slender and petiole-like below, amplexicaul-clasping above, scarcely decurrent; uppermost leaves much reduced and bract-like; all leaves arachnoid-tomentose below, thinly arachnoid and glabrescent above; heads generally solitary or sometimes in small clusters, cylindric-campanulate, (3.5)4–5(6) cm. long, closely subtended by few or several reduced leaves with spiny margins or rarely the subtending leaves foliaceous; involucre thinly floccose-tomentose or glabrescent, 3–5 cm. long; phyllaries numerous, appressed or somewhat spreading, the outer narrowly ovate to lanceolate, the middle linear-lanceolate, both the outer and middle phyllaries tipped by a short stiff spine, the inner phyllaries elongate-linear, thinner, with attenuate, twisted or plane, muticous or spinescent tips, the upper part of the inner phyllaries finely scabrous-pubescent and purplish; corollas purplish rose, 3–4.5 cm. long, the lobes about 1 cm. long, the throat 1–1.6 cm. long; achenes strongly flattened, oblong-cuneate, 6-7 mm. long, brown; pappus sordid-white, 2.5–4 cm. long.

Open coniferous forests and dry meadows, Arid Transition Zone to Hudsonian Zone, 2,700–10,000 feet; mountains of northern California southward through the Sierra Nevada to Tulare County; southwestern Idaho and western Nevada. Type locality: "Sierra Nevada, California, and adjacent part of Nevada." June–Oct.

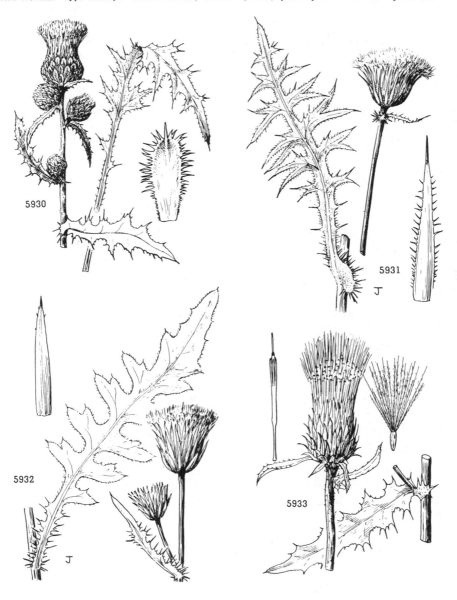

5930. Cirsium callilepis
5931. Cirsium acanthodontum

5932. Cirsium remotifolium
5933. Cirsium andersonii

33. **Cirsium campỳlon** H. K. Sharsmith. Hamilton Thistle. Fig. 5934.

Cirsium campylon H. K. Sharsmith, Madroño **5**: 85. 1939.

Stout, succulent, perennial herb with a thick woody root-crown; stems erect, thick, hollow, 0.3–2.1 m. tall, greenish or sometimes purplish-tinged, arachnoid-tomentose and also sparsely glandular-puberulent, the tomentum mostly deciduous. Earliest leaves small, thin, oblanceolate, undulate-lobed and spinulose-dentate, subglabrous; later basal leaves forming a small or large, loose rosette, oblanceolate or oblanceolate-elliptic, up to 7 dm. long and 2 dm. wide, pinnately parted or divided into broad lobed segments, the lobes ending in a short stiff spine and the margins sparsely spinulose, the upper surface with thin arachnoid tomentum and viscidulous puberulence, or the tomentum more or less lacking, the lower surface both tomentose and glandular-puberulent with the tomentum persisting and sometimes thick, the blades narrow below to an elongate, more or less spiny-bordered petiole; cauline leaves smaller, oblanceolate or more frequently oblongish or lanceolate, the petiole-like base spiny-lobed to the bottom or the cauline leaves broadly sessile with semi-amplexicaul base, the basal lobes more or less decurrent and strongly spiny; uppermost leaves becoming reduced and bract-like but not closely subtending the heads; heads solitary or usually paniculately clustered at the ends of branchlets, the heads or cluster nodding, the heads broader than long, 2.5–3.5 cm. long and 3–4.5 cm. broad; phyllaries greenish or more or less tinged with red, thinly puberulent but not arachnoid, the outer strongly recurved, narrowly ovate, gradually narrowed to a strong spiny tip, the margin undulate or plane, erose or ciliate, the middle phyllaries widely spreading, deltoid-triangular, and spine-tipped above the shortly appressed base, the inner phyllaries erect except for the short, abruptly spreading, sharply acute or spinescent tip; flowers sordid-white or somewhat pinkish-tinged, becoming sordid-brownish in age, the corolla about 2 cm. long, the tube 5–6 mm. long, the throat about 10 mm. long, the lobes more or less unequal, 4–5 mm. long; achenes oblong-cuneate, 4–5 mm. long, light to dark brown; pappus rather sparse, light brownish, 1.2–1.5 cm. long.

Restricted to the Mount Hamilton Range in the South Coast Ranges of California in Stanislaus and Santa Clara Counties, where the plant grows along streams and on springy slopes in areas of serpentine in the Upper Sonoran Zone. Type locality: Del Puerto Creek, at an altitude of 1,700 feet, Stanislaus County. April–Sept.

34. **Cirsium fontinàle** (Greene) Jepson. Fountain Thistle. Fig. 5935.

Cnicus fontinalis Greene, Bull. Calif. Acad. **2**: 151. 1886.
Carduus fontinalis Greene, Proc. Acad. Phila. **1892**: 363. 1893.
Cirsium fontinale Jepson, Fl. W. Mid. Calif. 505. 1901.

Stout, succulent, leafy, perennial herb; stem purplish-tinged, erect, thick, hollow, 0.6–1.2 m. tall, branching widely from the base and above, sparsely crisp-puberulent and somewhat viscidulous. Basal leaves forming a large rosette, up to 7 dm. long and 1.5 dm. wide, oblanceolate-elliptic, deeply pinnatifid, the segments broad and somewhat quadrate, more or less deltoid-lobed or coarsely deltoid-dentate, the lobes shortly but stiffly spine-tipped, the margin spinose-dentate, the upper surface dull green and viscidulous-puberulent, the lower surface paler and arachnoid-viscidulous, the leaf narrowed below to a slender, elongate, petiole-like base; cauline leaves mostly strongly spiny, becoming sessile by a broad, semi-amplexicaul base, the basal lobes auriculate, decurrent, and radiately spiny; heads as broad as long or broader, 2.5–3.5 cm. long, subtended by the uppermost leaves, solitary or more often closely clustered, the heads or clusters nodding; phyllaries loosely imbricate, strongly purplish-tinged, more or less puberulent but not arachnoid, the outer widely spreading or reflexed, broadly ovate, abruptly narrowed to a short spinose tip, the margin thinnish and somewhat erose, the middle phyllaries lanceolate or pandurate, appressed below, the upper half widely spreading, the margin erose and sometimes a little spinescent-toothed, the inner narrowly oblong, erect except at the flattened, spinescent, finely ciliate tips; flowers somewhat spreading in the heads, whitish or more or less pinkish-lavender-tinged, becoming sordid-brownish in age, the corolla 2–2.2 cm. long, the tube 1 cm. long, the throat and lobes each 5 or 6 mm. long; achenes oblong-cuneate, 4.5–5 mm. long, light brown; pappus buff, 1.5 cm. long, some of the bristles with slightly enlarged, flattened, barbellulate tips.

Along streams and about seepages in open grassy places on serpentine, Upper Sonoran Zone; near Crystal Springs Lake, San Mateo County, California, the only known locality. May–Oct. Suspected hybrids between *C. fontinale* and *C. quercetorum* have been named by Petrak (Beih. Bot. Centralbl. **35**: 522, 525. 1917).

Cirsium fontinale var. **obispoénse** J. T. Howell, Leaflets West. Bot. **2**: 71. 1938. Stems and leaves arachnoid-tomentose; corolla about 2 cm. long, the tube 7 mm. long, the throat 8 mm. long, the lobes 5 mm. long; achenes darker, purplish-brown-tinged. Wet places on serpentine near San Luis Obispo, California. Type locality: Chorro Creek, San Luis Obispo County.

143. **CÁRDUUS** [Vaill.] L. Sp. Pl. 820. 1753.

Annual, biennial, or perennial herbs resembling *Cirsium*. Stems frequently spiny-winged by decurrent leaf-bases. Heads homogamous, several- to many-flowered, medium to large, solitary or closely clustered. Phyllaries numerous in many series, the outer and middle spine-tipped. Receptacle setose. Achenes compressed with basal hilum. Pappus-bristles numerous in several series, united at the base, deciduous, setulose, not plumose. [Ancient Latin name for several thistle-like plants, particularly the teasel.]

A genus of about 120 species, widely distributed in Eurasia southward to tropical Africa; several species introduced as weeds in Australia and the New World. Type species, *Carduus nutans* L.

Plants biennial; heads nodding, solitary, broadly campanulate or hemispheric, 1.5–5 cm. long, 1.5–8 cm. wide, many-flowered; phyllaries (except the inner) strongly spreading or reflexed near the middle.
1. *C. nutans.*

Plants annual; heads erect, densely clustered, rarely solitary, cylindric, 1.5–2.5 cm. long, 1–1.5 cm. wide, rather few-flowered; phyllaries spreading slightly near the tip.
Phyllaries more or less scarious-margined, the tips glabrous and smooth except on the subciliate margin.
2. *C. tenuiflorus.*

Phyllaries not scarious-margined, the tips roughened on the margin and back by short upwardly appressed trichomes.
3. *C. pycnocephalus.*

1. **Carduus nùtans** L. Musk Thistle. Fig. 5936.

Carduus nutans L. Sp. Pl. 821. 1753.

Plants biennial, thinly tomentulose or glabrous; stems erect, to 1.5 (or 2) m. tall, simple or with a few elongate branches above the base, the lower leafy part spinose-winged, the peduncles with narrow interrupted wings or wingless. Cauline leaves lanceolate, oblong, or oblanceolate, to 2.5 dm. long and 1 dm. wide or sometimes larger, pinnately parted, the segments coarsely spinose-dentate or -lobed on the margins, rigidly spiny-tipped, the uppermost leaves much reduced and

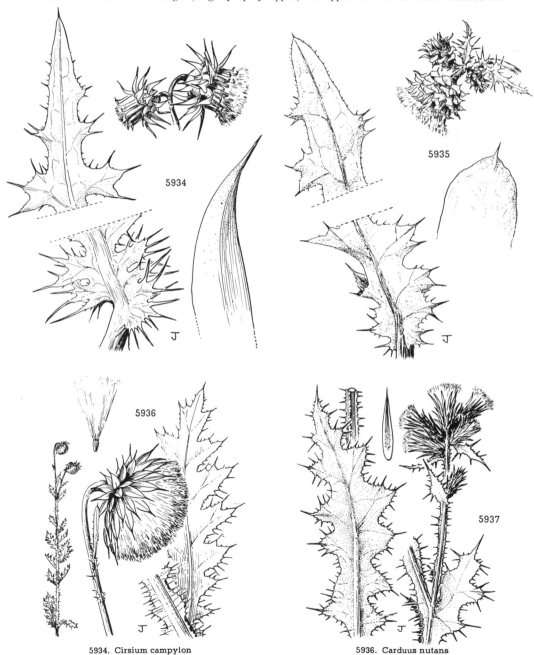

5934. Cirsium campylon
5935. Cirsium fontinale

5936. Carduus nutans
5937. Carduus tenuiflorus

bract-like, the basal leaves short-petiolate, the cauline sessile and decurrent in the spiny wings; heads solitary, nodding, large, mostly 3–5 cm. long and 4–8 cm. wide; involucre hemispheric, the phyllaries numerous in many series, thinly pubescent or glabrate, the outer and middle coriaceous, ovate-lanceolate with base appressed and with upper part reflexed or spreading, gradually narrowed to the stiff terminal spine, the innermost narrowly lanceolate to linear, chartaceous, attenuate, scarcely spinose, purplish; flowers purple; achenes obovate, 3–4 mm. long, transversely wrinkled, grayish; pappus-bristles numerous, whitish, 1.5–2.5 cm. long.

Rare and local weed along roads and in grassy places; Whatcom and Whitman Counties, Washington; on ballast, Linnton, Multnomah County, Oregon; Walnut, Los Angeles County, California; occasional northward to British Columbia (Alexis Creek) and eastward to the Atlantic; Argentina. Native of the Old World. May–Sept.

Carduus acanthoides L. Sp. Pl. 821. 1753. Plant biennial, the erect stems spinose-winged; leaves glabrescent, spiny-margined; heads ovate-campanulate or hemispheric, solitary or few-clustered, 1.5–2.5 cm. long and about as wide; phyllaries lanceolate or linear-lanceolate, the outer and middle with spreading or recurved, short-spinose tips; corolla reddish purple, nearly 2 cm. long; achene oblong-obovate, 3–3.5 mm. long, brown, obscurely longitudinally nerved and finely transversely rugulose; pappus whitish, 1–1.5 mm. long. Near Lake Waha, Nez Perce County, western Idaho; occasional east to the Atlantic; Argentina. Native of Europe.

2. **Carduus tenuiflòrus** Curtis. Slender-flowered Thistle. Fig. 5937.

Carduus tenuiflorus Curtis, Fl. Lond. Fasc. 6, *pl. 55.* 1790–98.
Carduus neglectus of California authors, in part. Not Tenore.

Plants annual, more or less tomentose, the upper side of the leaves glabrescent; stems erect, 3–20 dm. tall, narrowly to broadly spiny-winged, simple or fastigiately few-branched from near the base. Leaves oblanceolate to elliptic or obovate, to 1.5 dm. long and 7 cm. wide, the basal shortly petioled and sometimes only spinose-dentate, the cauline sessile and decurrent from the cuneate or amplexicaul base, pinnately parted, the segments spiny-lobed or -dentate, the segments and principal lobes strongly spine-tipped; heads cylindric, 1.5–2 cm. long, usually numerous and densely clustered at the ends of stems and branches, frequently more than 5 in a cluster; phyllaries numerous in several series, rather loosely imbricate, the outer and middle ovate to lanceolate, the base tomentulose, subappressed, scarious-margined, attenuate above into a squarrose spiny tip, which on the spine and back is glabrous and smooth except for the subciliate margin; flowers pinkish or mauve, the corolla-lobes generally one and one-half to two and one-half times as long as the corolla-throat; achenes oblong, 4–5 mm. long, brownish, usually finely 10–13-nerved; pappus sordid-whitish, 1–1.5 cm. long.

Grassy slopes, open woods, and roadsides; Coast Ranges from Coos County, Oregon, to central California; sparingly in the Sierra Nevada foothills and southward to Riverside County; Texas. Native of Europe. April–Aug.

3. **Carduus pycnocéphalus** L. Italian Thistle. Fig. 5938.

Carduus pycnocephalus L. Sp. Pl. ed. 2 1151. 1763.

Plants annual, gray-green to white-tomentose, the upper side of the leaves glabrescent, scabrid; stems erect, 1.5–10 (or 20) dm. tall, simple or few-branched, rather narrowly spiny-winged, the wing-spines divaricate-horizontal, often long and rigid. Leaves oblanceolate or lanceolate to oblongish and rarely ovate, to 15 cm. long and 8 cm. wide, pinnately parted into spiny-toothed or -lobed segments, the terminal spine of the segments and lobes the most prominent and rigid, the basal and lowest cauline leaves short-petioled, the middle cauline leaves narrowly cuneate, the upper amplexicaul; heads cylindric, 2–2.5 cm. long, generally 2–5-clustered at the ends of stems and branches, or sometimes solitary; phyllaries numerous in several series, rather loosely imbricate, the outer and middle ovate to lanceolate, rigid, rather thickly tomentose below the spinose tips which are scabrous-roughened by short ascending trichomes on margin, back, and raised midrib, the inner phyllaries thinner, less rigid and nearly innocuous; flowers reddish purple, the corolla-lobes about three times as long as the corolla-throat; achenes oblong, buff or pale brown, shining, 6 mm. long, about 20-nerved; pappus 1.5–2 cm. long.

On grassy flats and slopes and along roads; widespread and common in the San Francisco Bay area, California; local in Lane County, Oregon, and in Eldorado and San Diego Counties, California. Native of the Mediterranean region. April–June.

144. **CENTAURÈA** L. Sp. Pl. 909. 1753.

Perennial, biennial, and annual herbs with mostly erect stems and alternate, simple or pinnatifid leaves. Heads small, medium or large, generally pedunculate and paniculately or cymosely disposed, heterogamous with the outermost flowers sterile and the inner perfect or homogamous with the sterile flowers lacking. Involucre oblong, ovate, or globose; phyllaries appressed, graduated in many series, the margin and apex bordered or appendaged with prickles, spines, or a scarious or coriaceous membrane. Receptacle setose. Perfect flowers with a regular tubular corolla, the sterile marginal flowers with a modified corolla, the limb of the marginal flowers sometimes expanded and falsely radiate. Achenes oblong or obovate, compressed, glabrous or finely pubescent, generally smooth and shining, or occasionally rugulose and pitted, attachment-areola obliquely suprabasal. Pappus present or variously reduced or absent, setose or paleaceous, the bristles generally numerous and graduated, the innermost frequently shortened and connivent, or the innermost occasionally longest. [An ancient Greek name, signifying the centaurs' plant, of obscure application.]

A genus of about 400 species, found largely in the Mediterranean region but occurring also in North and South America and Australia. All of ours are introduced. Type species, *Centaurea centaurium* L.

Flowers purple, blue, pink, or white; stems more or less angled but not winged.

Pappus-bristles numerous, subplumose, deciduous, the innermost the longest and to 1 cm. long; flowers perfect; phyllaries scarious-margined, not spine-tipped or pectinately fringed.   1. *C. repens.*

Pappus-bristles, if present, persistent, frequently paleaceous, 5 mm. long or less; outermost flowers in head generally sterile, rarely all fertile.

Phyllaries with a long or short terminal spine.

Heads 2–3 cm. long; involucral spines stout and spreading, 1–3 cm. long; flowers pink to purple.

Pappus present, the paleaceous bristles 1–2 mm. long.   2. *C. iberica.*

Pappus none.   3. *C. calcitrapa.*

Heads 1–1.5 cm. long, narrowly oblong-ovate; involucral spines short and squarrose, 1–3 mm. long; flowers white or rarely pink.   4. *C. diffusa.*

Phyllaries not spine-tipped or, if the middle terminal division of the appendage is rigid and spinescent, then the spinule shorter than the lateral divisions.

Leaves pinnately to bipinnately parted or divided; flowers pink; plants biennial or perennial.
5. *C. maculosa.*

Leaves entire or with a few salient lobes or divisions; flowers blue or purple, rarely white or pink.

Plants annual; leaf-blades or -lobes linear to narrowly oblanceolate, 1 cm. wide or less; pappus well developed, to 4 mm. long.   6. *C. cyanus.*

Plants perennial; leaves oblanceolate or elliptic, few-lobed or entire, to 3 cm. wide; pappus generally poorly developed or none, if present 1 mm. long or less.

Appendages of middle phyllaries broad and rounded, scarious and lacerate, not pectinate, covering the base of the phyllaries; sterile marginal flowers enlarged; pappus none or represented by small scales.   7. *C. jacea.*

Appendages of middle phyllaries pectinate-fringed; pappus usually present.

Involucre as wide as long; appendage of middle phyllaries large, completely covering the base of the phyllaries, the laciniae setulose, mostly once to three times the width of the undivided body of the phyllary.

Appendages of phyllaries light to dark brown, the laciniae about as long as the undivided body of the phyllary; sterile marginal flowers with radiately enlarged corollas.   8. *C. pratensis.*

Appendages of phyllaries blackish, the laciniae two to three times the width of the body of the phyllary; sterile marginal flowers lacking.
9. *C. nigra.*

Involucre cylindric, longer than wide; appendage of middle phyllaries small, not entirely covering the base of the phyllaries; sterile marginal flowers enlarged; pappus reduced to minute scales or none.   10. *C. nigrescens.*

Flowers yellow, the outermost sterile; stems winged by the decurrent leaf-bases; phyllaries strongly spine-tipped.

Heads about 1.5 cm. long; spines on the middle phyllaries 0.5–1 cm. long, slender, purplish- or brownish-tinged, canaliculate on the upper side.   11. *C. melitensis.*

Heads about 2 cm. long; spines on the middle phyllaries 1–2 cm. long, stout, yellow, terete.
12. *C. solstitialis.*

1. **Centaurea rèpens** L. Russian Knapweed or Turkestan Thistle. Fig. 5939.

Centaurea repens L. Sp. Pl. ed. 2. 1293. 1763.
Centaurea picris Pall. ex Willd. Sp. Pl. **3**: 2302. 1804.
Acroptilon picris Boiss. Fl. Orien. **3**: 612. 1875.

Plants perennial with widely spreading, deep-seated rootstocks from which the aerial shoots arise; stems erect, openly branched, leafy, terete, wingless, 2–10 dm. tall. Leaves thinly tomentulose or glabrate, pale green or canescent, scabrous-margined, the basal leaves rosulate, oblong-oblanceolate, to 8 cm. long and 2.5 cm. wide, acute, deeply and irregularly pinnately lobed or divided, the divisions again lobed or only serrate, acute, petiole shorter than the blade, the lower cauline leaves somewhat smaller, pinnately lobed or divided or saliently serrate, the upper leaves narrowly oblong, 1–3 cm. long, 2–7 mm. wide, narrowed to a sessile base, abruptly acute and

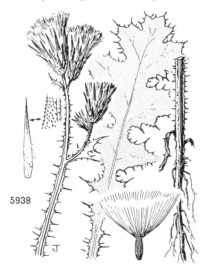

5938. Carduus pycnocephalus

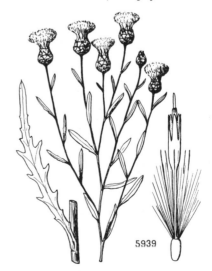

5939. Centaurea repens

mucronate, entire or serrate; heads solitary at the ends of the branches or sometimes a few in a loose cluster, ovate, 1.5–2 cm. high; involucres thinly tomentulose or glabrous, the outer and middle phyllaries with a rounded herbaceous part capped by a broad hyaline appendage, the appendage obtuse or acute, thinly pilose and ciliate, the inner phyllaries becoming narrower and longer, oblong or lanceolate, the appendage elongate and acute to acuminate, in the innermost phyllaries pilose-hairy and penicillate; flowers many and similar, lavender-blue to pink; achenes 3–3.5 mm. long, oblongish, somewhat compressed or obcompressed, ivory-white; pappus-bristles to 1 cm. long, numerous, early deciduous, the outer shorter and more slender, the inner longer and becoming a little paleaceous at base, all more or less plumose.

Cultivated fields, pastures, and roadsides, widespread and locally common at lower elevations and in warmer drier parts; Washington to southern California; British Columbia; east to the central United States; Argentina. Native in southern Russia and central Asia. May–Oct.

2. **Centaurea ibèrica** Trev. Iberian Star-thistle. Fig. 5940.

Centaurea iberica Trev. in Spreng. Syst. **3**: 406. 1826.

Plants biennial with a stout taproot; stems erect, 5–10 dm. high or more, much branched and bushy from below, terete or a little angled. Leaves puberulent or thinly tomentulose, the upper somewhat scaberulous, the basal leaves rosulate, oblanceolate, to 4.5 dm. long, once or twice deeply lobed or divided, the segments oblong to elliptic, obtuse or acute, mucronulate-serrulate or entire, narrowed below to a slender petiole, the cauline leaves variously pinnately lobed or divided or subentire, the uppermost leaves sessile, few-lobed or generally entire, narrowly oblong to elliptic; heads sessile or shortly pedunculate, ovate or roundish-ovate, 2–3 cm. long; involucres glabrous, the phyllaries nerveless, very stiff-coriaceous, pale green with narrow, white-scarious margins, the outer phyllaries small and nearly spineless, the middle phyllaries produced into a rigid, spreading, stramineous spine 1–2.5 cm. long, the spine channeled on the upper side and bearing 1–3 pairs of small spines near the base, the inner phyllaries spineless, tipped by a broad, denticulate or lacerate, scarious appendage; flowers pinkish or purplish, the outer not enlarged; achene oblong, 3–4 mm. long; pappus present, the bristles flattened and in several series, the outer short, the inner longer and narrower and about half the length of the achene.

An uncommon weed, occurring locally in Sonoma, Santa Barbara, and San Diego Counties, California. Native of southern Europe and western Asia. July–Sept.

3. **Centaurea calcítrapa** L. Purple Star-thistle. Fig. 5941.

Centaurea calcitrapa L. Sp. Pl. 917. 1753.

Plants biennial with a stout taproot; stems branching from the base, erect or spreading, 2–10 dm. tall, rounded or a little angled. Leaves puberulent or at length glabrate, basal leaves rosulate, oblanceolate, about 1 dm. long, 1–3 cm. wide, once- or twice-pinnatifid, the segments oblongish, acute, and serrulate, the blade narrowed gradually into a petiole somewhat shorter than the blade, the cauline leaves short-petiolate or sessile, not decurrent, deeply pinnately divided into oblong-linear segments, the uppermost leaves sessile, entire or serrate or serrate-lobed, the apex and teeth mucronulate-tipped; heads sessile or shortly pedunculate, ovate or oblong-ovate, 2–2.5 cm. long; involucres glabrous, the phyllaries nerveless, stiff, coriaceous, pale green with narrow, white, scarious margins, the outer phyllaries with a short spine, the middle phyllaries produced into a stout, spreading, stramineous spine 1.5–3 cm. long, the spine channeled on the upper side, bearing 2 or 3 pairs of short, stiff, lateral spines at the base, the inner phyllaries bearing a small spine or a spineless chartaceous tip; flowers purple, the outer scarcely enlarged; achenes subobovate, 2.5–3 mm. long; pappus none.

Fields, pastures and roadsides, common in middle western California; Humboldt County to Santa Clara County, California; Ellensburg, Washington; on ballast, Linnton, Oregon; Nanaimo, Vancouver Island; eastern United States. Native in the Old World from southern Sweden to North Africa and central Russia. May–Nov.
In Humboldt and Riverside Counties, California, a variant of *C. calcitrapa* is found in which the heads are somewhat smaller, the spines more slender and shorter, and the inner achenes (which are fertile) pappus-bearing. These plants may be *C. calcitrapoides* L.
Centaurea dilùta Ait. Hort. Kew. **3**: 261. 1789. Plants annual (or perennial?), glabrous or thinly pubescent above; stems to 2 m. tall, stout; lower cauline leaves lyrate-pinnatifid, to 15 cm. long and 10 cm. wide, sessile by a broad, clasping and shortly decurrent base, the upper leaves irregularly pinnately lobed or parted, the uppermost a little dentate or entire; heads rather large, 2.5–4 cm. long; involucre globose-conic, constricted at the top, 1.5–2 cm. long, the phyllaries broadly ovate, appressed, pale green, the hyaline margin lacerate-pectinate, the appendage brown, shortly palmate-spinose, the central spine about 1–2 mm. long, the laterals usually shorter, the innermost phyllaries with an expanded, brown, scarious, lacerate appendage; flowers pink or lavender-pink, the outer sterile corollas much enlarged; achenes buff, 3–3.5 mm. long; pappus of graduated white bristles to 5 mm. long, none on outer sterile achenes. Locally established in central and southern California; San Francisco, Santa Barbara, and in Los Angeles County near Watts and Whittier. Native of North Africa. May–June.
Centaurea muricàta L. Sp. Pl. 918. 1753. (*Amberboa muricata* DC. Prod. **6**: 559. 1837.) Plants annual; lower stems smooth, the branchlets and peduncles scabrous; lower leaves pinnately divided or parted, long-petioled, the upper sessile, the uppermost entire and much reduced; involucres ovate-conic or roundish, 1.5–2 cm. long, the phyllaries appressed, pilose and scabrous, ovate, gradually narrowed into a spreading or reflexed spine, the spines rather weak, yellowish, 2–5 mm. long, the base of the spine and margin of the phyllary blackish; flowers pink or purplish, the outer sterile and radiately much enlarged; achenes about 3.5 mm. long, grayish brown, finely pitted between the longitudinal ribs, the base of the achene and margin of the attachment-scar smooth, bony, ivory; pappus-bristles paleaceous, graduated in size, to 3 mm. long, pale brownish. An established garden escape in Santa Barbara, California, since 1904. Native in the western part of the Mediterranean region. April–June.
Centaurea moschàta L. Sp. Pl. 909. 1753. (*Amberboa moschata* DC. Prod. **6**: 560. 1837.) Plants annual with thin smooth foliage; heads large and showy on long naked peduncles; phyllaries imbricate in several series, hard-coriaceous, broadly rounded, unarmed, the inner with a more or less spreading, scarious appendage; flowers yellowish, purplish, or whitish, the outer sterile and much enlarged; achenes nearly smooth, brownish black, thinly villous-pubescent, 4 mm. long; pappus-bristles paleaceous, the innermost the longest and 3 mm. long. An old-fashioned garden favorite, sometimes becoming established in waste ground; native in southwestern Asia. May–July. Sweet Sultan.

4. Centaurea diffùsa Lam. Tumble Knapweed. Fig. 5942.

Centaurea diffusa Lam. Encycl. 1: 675. 1783.

Plants biennial with an elongate taproot, canescent or pale green, pubescent or thinly floccose-tomentulose becoming glabrate, scabrous; stems erect, (1 or) 5–8 dm. tall, angled but not winged, branched near or above the base, the branchlets numerous and intricate. Basal leaves rosulate, bipinnate to bipinnatifid, oblanceolate to oblong, to 20 cm. long and 5 cm. wide, short-petioled, the pinnules or ultimate segments narrowly oblong to elliptic, usually acute and cuneate, discrete or the uppermost more or less confluent, the cauline leaves sessile, the lower ones bipinnate or bipinnatifid, the upper much reduced and pinnately lobed or divided, the uppermost bract-like and either entire or with tiny lobes; heads solitary, more or less clustered at the ends of branches, 1.5 cm. long; involucre narrowly ovate or oblong, about 1 cm. long, tomentulose at first, becoming glabrate and granular, the phyllaries coriaceous, nerved, the outer and middle broadly to narrowly ovate, pale yellowish green with a buff or light brown margin, the upper part narrowed into a short, squarrose, stiff spine, the base of the spine and upper part of the phyllary spinose-pectinate on the sides, the inner phyllaries lanceolate, tipped by a scarious or coriaceous, lacerate appendage, spiny or spineless; flowers usually white, occasionally pink or lavender, the outermost sterile flowers inconspicuous with filiform corolla-lobes; achenes oblongish, 2.5 mm. long, dark or blackish brown, striate with several conspicuous or faint, pale or ivory lines; pappus none or on the inner achenes represented by white paleaceous scales less than 1 mm. long.

Fields, roadsides, and waste ground, locally common; Washington and northern Oregon east of the Cascade Mountains; Idaho and southern British Columbia; occasional in central and eastern United States. Native in southeastern Europe and western Asia. June–Sept.

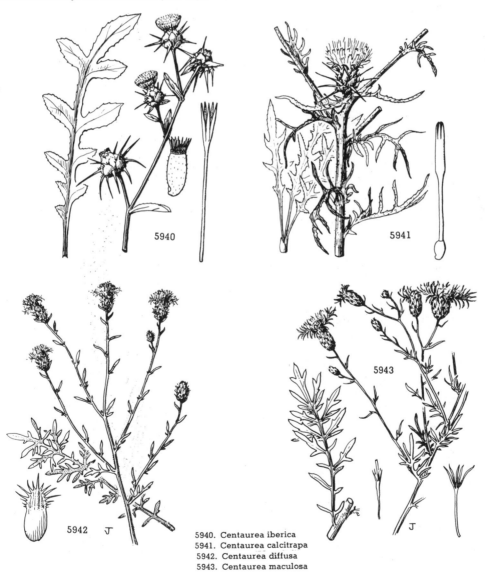

5940. Centaurea iberica
5941. Centaurea calcitrapa
5942. Centaurea diffusa
5943. Centaurea maculosa

Centaurea virgàta var. **squarròsa** Boiss. Fl. Orien. **3**: 651. 1875. (*Centaurea squarrosa* Willd. Sp. Pl. **3**: 2319. 1804.) Plants perennial, much branched from the crown of a stout taproot; stems about 5 dm. tall, branching widely above, scabrous; lower leaves pinnate or bipinnate with linear segments, long-petioled, the middle and upper leaves sessile, pinnate or simple, margins revolute; heads small, 1–1.3 cm. long; involucres ovate-oblong, about 7 mm. long, pale green and stramineous, a little violet-tinged, the middle phyllaries pectinately fringed with short spines along the sides below the terminal squarrose spine which is 1–2(or 3) mm. long, the innermost phyllaries spineless or nearly so, the tip narrowed and hyaline; flowers pinkish or lavender, the outer ones sterile and somewhat enlarged; achenes 2.5–3.5 mm. long; pappus-bristles slender, white, to 2.5 mm. long. Occurring locally in grain fields and along roads in Big Valley, Lassen County, California, and in adjacent Shasta County; Juab County, Utah. Native of middle eastern Europe to central Asia and Iran. July–Aug. In typical *C. virgata* Lam. the terminal spine of the involucral bracts is shorter or even lacking.

5. Centaurea maculòsa Lam. Spotted Knapweed. Fig. 5943.

Centaurea maculosa Lam. Encycl. **1**: 669. 1783.

Biennial or short-lived perennial with 1 to several stems arising from the top of a stout taproot, gray or pale green, arachnoid-tomentulose or glabrate, scabrous; stems erect, 2–8 (or 15) dm. tall, angled but not winged, branching near the middle or above, the branches not very numerous and frequently short. Basal leaves of seedlings rosulate, the lowest leaves of stems from the root-crown scattered, the lower leaves pinnately to bipinnately parted or divided, petioled or becoming sessile above, oblanceolate to broadly elliptic, to 10 cm. long and 3 cm. wide, the segments narrowly oblong to linear, acute and mucronulate, the margin more or less revolute, the upper side scabrous and glabrate, the lower side thinly tomentose, the uppermost leaves few-lobed or entire, linear, mostly less than 1 cm. long; heads 1.5–2 cm. long, solitary at the ends of short or elongate, peduncle-like branchlets which are cymosely arranged to form a close or open cluster; involucre ovate or oblong, 1–1.5 cm. long, thinly tomentose or glabrate, the phyllaries not spinose, coriaceous, scarious-margined, nerved, the outer and middle broadly to narrowly ovate, pale green and brownish below the black-spotted apex, the apex and upper lateral margins pectinately fringed or laciniate, the terminal division or lacinia spinescent but much shorter than the lateral divisions, the inner phyllaries oblong-lanceolate, lacerate or irregularly pectinate at the scarious tip; flowers pink, rarely white, the outermost sterile flowers rather numerous, with elongate filiform corolla-lobes; achenes narrowly obovate, 3 mm. long, brown or blackish with several faint pale longitudinal lines; pappus present, the slender paleaceous bristles varying in length, the longest 2–3 mm. long.

Fields, roadsides, and waste ground; central Oregon east of the Cascade Mountains north through western and central Washington to British Columbia; Idaho and Montana; widespread in central and eastern United States and Canada. Native in Europe and western Siberia. June–Oct.

Centaurea cinerària L. Sp. Pl. 912. 1753. Plants perennial, bushy, to 1 m. tall, the stems and leaves white-tomentose or pale green. All leaves except the youngest and the uppermost bipinnate or bipinnately divided, the segments narrowly oblongish and obtuse, the uppermost leaves pinnately divided or subentire; heads rather large, 2.5 cm. long, cymosely arranged; involucres globose-ovate, about 1.5 cm. long, the phyllaries pale green except for the brownish or blackish margin and appendage, appressed, the upper margins pectinate-ciliate, the terminal seta stiff and subspinose, shorter than the lateral setae and less than 1 mm. long, squarrose, the innermost phyllaries with a denticulate scarious tip; flowers purple, the sterile outer corollas enlarged; achenes 3 mm. long; pappus of numerous whitish bristles 4 mm. long. Occasionally spontaneous near gardens in central and southern California. Native of southern Europe and North Africa. May–July. Dusty Miller.

Centaurea salmántica L. Sp. Pl. 918. 1753. (*Microlonchus salmanticus* DC. Prod. **6**: 563. 1837.) Plants perennial. Lower leaves pinnate, the lateral leaflets few and rather remote, irregularly and runcinately pinnatifid or saliently and irregularly dentate, scabrid-pubescent, the upper leaves much smaller than the lower, pinnatifid or spinose-serrate; heads 2–2.5 cm. long; involucre globose-conic, strongly constricted at the top, 1–2 cm. long, the phyllaries appressed, triangular-ovate, subglabrous, pale yellow-green, blackish at the top, tipped by a short spinose mucro about 0.5 mm. long or less, the innermost phyllaries attenuate to an acute or acuminate apex; flowers purplish or white; achenes about 3.5 mm. long, light brown with vertically arranged, irregular or transverse, dark brown areolae, the base of the achene and margin of the attachment-scar smooth, bony, and yellowish; pappus-bristles graduated, to 3 mm. long except the innermost, single, paleaceous, acuminate bristle which is 3.5 mm. long. Local in Healdsburg, Sonoma County, California, where it has not been recently collected; dooryard weed, Jerome, Arizona. Native of the Mediterranean region. April–Sept. Escobilla.

6. Centaurea cỳanus L. Cornflower or Bachelor's Button. Fig. 5944.

Centaurea cyanus L. Sp. Pl. 911. 1753.
Centaurea cyanus var. *denudata* Suskd. Werdenda **1**: 43. 1927.

Plants annual, floccose-tomentose, usually thinly so, rarely almost glabrous, a little scabrid; stems erect, 1–7 dm. tall, simple or more commonly openly branched, the stems and branches slender, terete or somewhat angled, not winged. Basal and lower cauline leaves oblanceolate or linear, prominently pinnately few-lobed or lyrate-pinnatifid, short-petioled, to 10 cm. long and 1 cm. wide, the middle and upper cauline leaves narrowly oblanceolate or linear, sessile, dentate or entire, sharply acute or acuminate, the lower side arachnoid-tomentose, above more thinly so or glabrate, the margin revolute; heads solitary at the ends of slender, peduncle-like branches, 1.5–2 cm. long, about 3 cm. across; involucre ovate or oblong-ovate, thinly tomentulose, the phyllaries coriaceous, spineless, the outer and middle ovate, with thin-scarious margins, the margin pectinately parted into sharp triangular lobes and forming a short, serrate-parted appendage at the tip, the innermost phyllaries oblong, entire except at the more or less serrate-lacerate tip; flowers blue varying to white, pink, and purple, the outermost sterile flowers with much elongated, palmately expanded corollas, the ray-like limb about 1 cm. long above the funnelform throat; achenes 4 mm. long, oblong-elliptic, light brown or buff, thinly pubescent above, densely so at the very base; pappus to 4 mm. long, consisting of numerous, strongly graduated, brown bristles, narrowly oblong, paleaceous, and a little setose.

Grassy and brushy hills and valleys as well as in fields and along roads; widespread and naturalized in Washington, Oregon, and northern California, occasional southward to southern California as an escape from cultivation; British Columbia; east to the Atlantic coast. Native in the Mediterranean region; introduced in many parts of the world. May–Oct.

7. **Centaurea jàcea** L. Brown Knapweed. Fig. 5945.

Centaurea jacea L. Sp. Pl. 914. 1753.

Plants perennial, the stems solitary or clustered from a woody root-crown, pale green, thinly arachnoid-tomentose and pubescent, becoming glabrate, scabrid; stems erect, mostly 5–10 dm. tall, more or less angled but not winged, branching near the middle or above, the branches rather few and elongate. Basal and lower cauline leaves rather sparse, oblanceolate or narrowly elliptic, the blade to 15 cm. long and 3 cm. wide, entire or coarsely few-lobed, pubescent above and below, scabrous on the margins, long-petiolate, the petiole slender, nearly as long as the blade or longer, the middle cauline leaves lanceolate, oblong or oblanceolate, entire, serrulate or shallowly few-lobed, sessile, the uppermost leaves similar but smaller and generally entire; heads 2–2.5 cm. high, 3–4 cm. broad, solitary at the ends of the branches; involucres ovate or bowl-shaped, 1.5–2 cm. long and nearly as wide, light to dark brown, the phyllaries a little pubescent or the outermost phyllaries sparsely tomentulose, the outermost phyllaries and appendages of the middle phyllaries with broad scarious or chartaceous margins more or less lacerate, the appendages generally entirely covering the base of the middle phyllaries, the innermost phyllaries scarious at the tip, entire to deeply and irregularly dentate or lacerate; flowers rose to purple, rarely white, the outermost corollas enlarged and conspicuous; achenes oblongish, 3–3.5 mm. long, light brown with a few lighter longitudinal lines; pappus none or represented by a few very short scales less than 0.5 mm. long.

Occasional in waste ground and along roads; Washington and Oregon, chiefly west of the Cascade Mountains; east to the Atlantic. Naturalized from Europe. June–Aug.

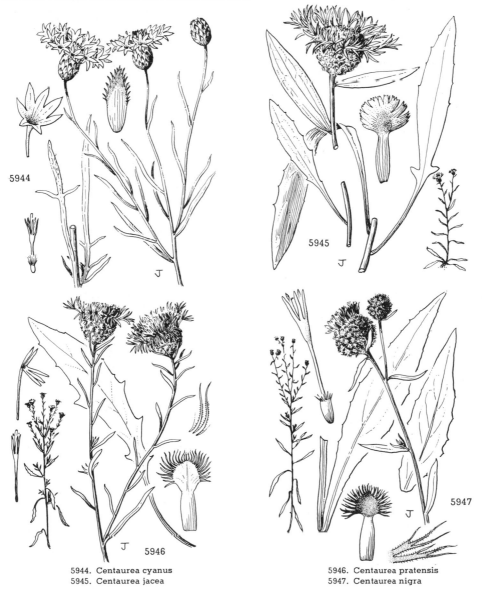

5944. Centaurea cyanus
5945. Centaurea jacea

5946. Centaurea pratensis
5947. Centaurea nigra

8. **Centaurea praténsis** Thuill. Protean Knapweed. Fig. 5946.

Centaurea pratensis Thuill. Fl. Paris ed. 2. 444. 1799.
Centaurea nigra var. *radiata* DC. Fl. Franc. **6**: 460. 1815.

Plants similar to *C. jacea*; involucre broadly elliptic or subglobose, about 1.5 cm. long and as wide oʼr wider, the appendages of the phyllaries light to dark brown, roundish with broad, pectinately fringed margins, the laciniae filiform or triangular-attenuate, more or less setulose, about as long as the width of the rachis of the appendage, the appendages completely covering the base of the middle phyllaries; marginal flowers with radiately enlarged corollas, sterile; pappus usually represented by a few or numerous unequal paleae, mostly 0.5 mm. long or less, rarely obsolete.

Fields and roadsides, widespread and common particularly at lower elevations west of the Cascade Mountains; Washington to northwestern California; British Columbia; eastern United States. Cultivated around Roseburg, Oregon, for winter forage.

This plant, introduced from Europe, is generally regarded as a hybrid of *C. jacea* and *C. nigra*, a fact which may account for its variability. Here it is accorded specific recognition because in our area it is by far the most common form of the *C. jacea-nigra* complex.

9. **Centaurea nìgra** L. Black Knapweed. Fig. 5947.

Centaurea nigra L. Sp. Pl. 911. 1753.
Centaurea jacea var. *nigra* Briq. in Hegi, Fl. Mitt.-Eur. **6**: 954. 1929.

Plants similar to *C. jacea*; involucre roundish-cupshaped, about 1.5 cm. long and as wide as long, the appendages of the phyllaries dark brown or nearly black, roundish with broad, pectinately fringed margins, the laciniae stiffish, attenuate, setulose, as much as three times as long as the width of the rachis of the appendage, the appendages completely covering the base of the middle phyllaries; sterile marginal flowers lacking; pappus usually present and well developed in several series, about 1 mm. long, rarely obsolete.

Occasional in weedy places and in fields; Washington and northwestern Oregon; adventive in central California; eastern United States. Introduced from Europe. June–Sept.

10. **Centaurea nigréscens** Willd. Short-fringed Knapweed. Fig. 5948.

Centaurea nigrescens Willd. Sp. Pl. **3**: 2288. 1804.
Centaurea dubia Suter, Fl. Helvet. **2**: 202. 1802. Not Gmelin.

Plants similar to *C. jacea*; cauline leaves lanceolate, attenuate at base, sessile; involucre broadly oblong or cylindric, longer than broad, about 1.5 cm. long, the appendages of the phyllaries smaller, not covering the greenish or straw-colored bases, triangular, pectinate-fringed, the segments about as long as the width of the undivided part of the phyllaries; sterile marginal flowers generally present and radiately enlarged; pappus reduced to small scales or none.

Widespread but rather rare, ruderal or pastoral; Washington and northern Oregon; Idaho; eastern United States. Introduced from Europe. June–Sept.

11. **Centaurea meliténsis** L. Napa Thistle or Tocalote. Fig. 5949.

Centaurea melitensis L. Sp. Pl. 917. 1753.

Plants annual, scabrid and thinly araneous-pubescent, mostly 2–6 dm. tall; stems erect, rather openly few-branched generally above the base, the branches stiffish, narrowly winged by decurrent leaf-bases, the wings entire or repand-dentate. Basal leaves rosulate, pinnately parted with the terminal lobe largest and broadly elliptic or roundish, or sometimes the basal leaves shallowly few-lobed to dentate and oblanceolate, the petiole much shorter than the blade, the cauline leaves sessile and decurrent, narrowly oblong or oblong-oblanceolate, up to 5 cm. long and 1 cm. wide, the margin entire or generally more or less dentate, the uppermost leaves small but not scale-like; heads about 1.5 cm. long, solitary or closely clustered at the ends of the branchlets and sometimes occurring subspicately lower down on short axillary branchlets; involucre tomentulose, the phyllaries obscurely nerved, coriaceous, the lowest and middle phyllaries tipped by pinnately branched spines 0.5–1 cm. long, the terminal segment much longer than the lateral particularly on the middle phyllaries, slender but subrigid, divergent, canaliculate on upper side, purplish- or brownish-tinged, the innermost phyllaries narrowed into a stiff or coriaceous, elongate-triangular, spine-tipped appendage; flowers yellow, numerous, the outermost with filiform lobes and inconspicuous; achenes 2.5 mm. long, light brown or buff, all with pappus; pappus-bristles to 2.5–3 mm. long, slender, white.

Grassy and brushy hills and valleys as well as in cultivated and waste ground; common and widespread from central Oregon to southern California west of the Cascade Mountains and Sierra Nevada, rare in southern British Columbia, Washington, northern Oregon, and Lower California; occasional east to the Atlantic coast; Hawaiian Islands; South America; South Africa; Australia. Native in the Mediterranean region. April–Sept.

12. **Centaurea solstitiàlis** L. Barnaby's Thistle or Yellow Star-thistle. Fig. 5950.

Centaurea solstitialis L. Sp. Pl. 917. 1753.

Plants annual, thinly arachnoid-tomentose and scabrid, generally 3–10 dm. tall; stems erect, openly branched from near the base, the branches stiff and somewhat virgate, winged by the decurrent leaf-bases, the wings narrow, nearly entire, and scabrid. Basal leaves rosulate, lyrately pinnatifid, the lateral lobes much smaller than the terminal, ovate-triangular one, the petiole shorter than the blade, the cauline leaves sessile and decurrent, linear or linear-lanceolate, as much as 10 cm. long and 5 mm. wide near the base but much reduced and even scale-like on the uppermost branchlets, entire, frequently undulate; heads solitary at the ends of the branchlets, ovate, about 2 cm. long; involucres thinly tomentulose, early glabrescent, the phyllaries nerveless, coriaceous, the lowest phyllaries tipped by short, weak, palmately disposed spines, the middle phyllaries bearing a stiff, terete, yellow, divergent, terminal spine 1–2 cm. long and 1 or 2 pairs of very short lateral spines at the base of the terminal spine, the innermost phyllaries expanded into a scarious, denticulate or lacerate, obovate appendage, spineless or nearly so; flowers yellow, numerous, the outermost with filiform lobes and inconspicuous; achenes 2.5–3 mm. long, oblong, either blackish brown or light brown and mottled, some of the outer without pappus or with poorly developed pappus, the inner with numerous, fine, white pappus-bristles to 3 mm. long.

Roadsides, fields, and waste ground about towns; very common and widespread from southern Oregon to central California at lower elevations, less common or rare in Lower California, southern California, northern Oregon, and southern and eastern Washington; occasional and local in the central and eastern United States. Native in southern Europe and the Mediterranean region. June–Nov.

Centaurea sulphùrea Willd. Enum. Hort. Ber. 930. 1809. (*Centaurea sicula* of California authors. Not L.) Stems conspicuously winged. Lower leaves coarsely pinnately parted or divided, or the lower and upper leaves entire or spinose-serrulate, the uppermost spinulose-tipped, those under the head tipped by a pinnately branched, spinose appendage; heads rather large, 3–4 cm. high; involucre nearly glabrous, ovate, 2–3 cm. high, strongly constricted at the top, the appendages on the phyllaries horizontally spreading or reflexed, brownish- or blackish-tinged, palmately spinose, the central spine of the middle phyllaries 2–2.5 cm. long, the innermost phyllaries narrowed to a bony acute or slightly spinose tip; flowers yellow; achenes 5 mm. high, strongly graduated pappus-paleae and -bristles to 1 cm. long. Locally established near Folsom, Sacramento County, California. Native of the western part of the Mediterranean region. May–June.

Centaurea erióphora L. Sp. Pl. 916. 1753. Stems conspicuously winged. Lower leaves pinnately parted or divided, the uppermost entire or dentate, abruptly acute or mucronate; heads rather large, about 3 cm. high; involucre conspicuously loosely arachnoid-tomentose, about 2 cm. long, broadly ovate or roundish, constricted at the top, the appendages on the phyllaries spreading, yellowish or brownish, pinnately spinose, the spine of the middle phyllaries 1–2 cm. long, the lateral spines short and stout, 2–4 on each side, the innermost phyllaries narrowed to a coriaceous acute tip; flowers yellow; achenes compressed, 4–5 mm. long; pappus of strongly graduated paleae and bristles, rose to purple and dark brown, 5–8 mm. long. Along roadside, Highland Park, Los Angeles, California; perhaps not persisting. Native of the western part of the Mediterranean region. May–June.

145. **CNÌCUS** [Tourn.] L. Sp. Pl. 826. 1753.

Annual herb with leafy stems and alternate, parted lobed or dentate, spinulose leaves. Heads medium-sized, sessile or terminating the branches, heterogamous with the outermost flowers sterile and the inner perfect. Involucre ovate; the phyllaries appressed, the middle and inner phyllaries bearing an elongate, pinnately branched spine. Receptacle densely long-setose. Perfect flowers with a 5-lobed, irregularly cleft corolla; the sterile flowers with a deeply 3-parted corolla. Achene strongly 20-ribbed, crowned by a bony 10-toothed ring; hilum lateral, large, deeply excavated. Pappus persistent, in 2 unlike series of 10 bristles each, the outer alternate with the inner and with the teeth on the achene-rim, subrigidly setose, subterete; the inner much shorter, subrigid, and a little paleaceous-flattened. [Name derived from the Greek and applied to the safflower in Latin.]

A monotypic genus, native of the Mediterranean region and adjacent parts of Asia.

1. **Cnicus benedíctus** L. Blessed Thistle. Fig. 5951.

Cnicus benedictus L. Sp. Pl. 826. 1753.

Plants with erect or spreading, leafy stems, sparsely arachnoid-tomentose and villous as well as punctate-glandular; primary stem sometimes lacking or very short with elongate sprawling

5948. Centaurea nigrescens

5949. Centaurea melitensis

branches arising from the basal or lower cauline leaf-axils, or the primary stem erect and un-branched for as much as 2 dm. and branching above, somewhat angled, not winged. Basal leaves subrosulate, these and the lower cauline leaves oblanceolate or oblongish, pinnately lobed to pinnately parted, narrowed to a longer or shorter petiole, the middle and upper cauline leaves sessile by a broad, decurrent base, oblong, up to 25 cm. long and 6 cm. wide but usually much smaller, sharply spinulose-dentate to pinnately parted, bright green and venulose-chartaceous, the uppermost leaves (or subbasal leaves if head is stemless) clustered beneath the head and exceeding it in length; heads narrowly to broadly ovate, 2–4 cm. long; phyllaries arachnoid-tomentose, appressed, in several series, the outermost triangular-ovate, abruptly tipped by an elongate slender spine, the middle oblong-ovate phyllaries bearing a spreading, pinnate, spinose appendage to about 2 cm. long, the lateral spines in about 5–7 pairs, the appendage in the innermost elongate phyllaries much reduced; flowers yellow, the outermost sterile ones inconspicuous; achenes oblong-obovate, about 8 mm. long, grayish brown, scarcely compressed, slightly constricted at the truncate apex below the dentate rim; pappus brownish-tinged, the outer bristles about 10 mm. long, denticulate and a little roughened, the inner bristles 2–4(5) mm. long, bearing a few pinnately disposed processes along the sides.

Fields, pastures, and roadsides, flourishing in rich cultivated soils, widely distributed but not very common; central Washington south to southern California; occasional eastward to the Atlantic. Introduced in many parts of the world from the Mediterranean region. April–Sept.

Tribe 10. **ARCTOTIDEAE**[*]

146. **GAZÀNIA** Gaertn. Fruct. 2: 451. 1791. Nomen conservandum.

Plants herbaceous, mostly perennial, low and spreading, with subrosulate radical leaves and erect scapes, or rarely caulescent with alternate scattered leaves. Heads medium to rather large, solitary on leafless peduncles or scapes, radiate, heterogamous, the ray-flowers sterile, the disk-flowers perfect. Involucre campanulate to ovate or globose, the phyllaries in 2 to several series, united at the base to form a cup or bowl, the tips free, linear, acute or spinescent. Receptacle plane or convex, shallowly honeycombed. Flowers showy, yellow or orange, the rays variously tinged or marked with blue, purple, or brown, the disk-flowers sometimes dark. Achenes obovate, turbinate, long-villous. Pappus-bristles in 2 series, narrowly acuminate, thin, scarious, inconspicuous and concealed by the long copious hairs of the achene. [Probably in honor of Theodore of Gaza who translated the botanical works of Aristotle and Theophrastus into Latin in the Fifteenth Century.]

A South African genus of about 24 species. Type species, *Gorteria rigens* L.

1. **Gazania longiscàpa** DC. Treasure Flower. Fig. 5952.

Gazania longiscapa DC. Prod. 6: 513. 1837.

Scapose perennial herb, the basal stems branching and forming loose leafy mats, the scapes equaling or exceeding the leaves. Leaves entire or pinnately divided, narrowly oblanceolate to elliptic, acute, attenuate at base into a long petiole, the blade and petiole 1–1.5 dm. long, the upper side smooth and glabrous, the lower side (except the midrib) densely lanate-tomentose, the revolute margin scabrous-ciliate; involucre about 2 cm. long, the cup glabrous, the phyllary-lobes attenuate-acute or spinescent, the outer (lower) lobes scabrous-ciliate; heads 3–6 cm. across, the flowers bright, closing in shadow and in late afternoon; pappus-bristles 3–4 mm. long, hidden by the hairs of the achene which are about twice as long.

Escaping from cultivation in southern California but perhaps not persisting. Introduced from South Africa. Feb.–May.

Arctòtis L. Sp. Pl. 922. 1753. Acaulescent or caulescent herbs or shrubs. Leaves alternate, unarmed; heads radiate, the ray-flowers pistillate and fertile, the disk-flowers perfect; phyllaries in several series, distinct; receptacle deeply pitted, fimbrillate; achenes winged dorsally, the wings appearing to form 2 pits, the base of the achene furnished with a long tuft of hairs; pappus in 2 series, the outer scales scarious, paleaceous, obtuse, the inner series sometimes very small. **A. stoechadifòlia** var. **grándis** (Thunb.) Less. Synop. Gen. Comp. 26. 1832. (*A. grandis* Thunb. Prod. Pl. Cap. 706. 1820.) Annual tomentose herb with erect stems and undulate, dentate or lobed leaves; phyllaries very unequal, the outer herbaceous, oblong-elliptic, the middle and inner elongate-oblong, scarious or scarious-margined; achenes plumply obovate, 2.5 mm. long, blackish except for the buff, incurved, dentate, lateral margins of the dorsal pits, the basal hairs light brown, copious, 4 mm. long; outer pappus-paleae hyaline, 3 mm. long, the inner series consisting of rudimentary bristles. Occasional escape from cultivation in southern California probably not persisting. Native of the Cape region, South Africa.

Bérkheya Ehrh. Beitr. 3: 137. 1788. South African herbs or shrubs with more or less spiny or spinose-dentate leaves; heads discoid or radiate; phyllaries in several series, free or more or less united by their bases into a cup, the free ends spine-tipped; receptacle honeycombed or deeply pitted, more or less fimbrillate; pappus-bristles in 1 or 2 series, paleaceous, obtuse or acuminate. **B. heterophýlla** O. Hoffm. Ann. Nat. Hofmus. Wien 24: 314. 1910. (*Stobaea heterophylla* Thunb. Prod. Pl. Cap. 622. 1820.) Thistle-like herb with pinnately divided leaves; heads discoid; phyllaries divaricate, spinose on margin and at apex; achenes enclosed within receptacle-pits; pappus-paleae roundish, 0.25–0.3 mm. long, obtuse or subtruncate, denticulate. On ballast near Portland, Oregon. Native of South Africa.

Venídium Less. Linnaea 6: 91. 1831. Acaulescent or caulescent annuals and perennials resembling the closely related *Arctotis*, differing in glabrous achenes that are usually epappose, rarely minutely pappose. **V. fastuòsum** (Jacq.) Stapf, Bot. Mag. 152: *pl. 9127.* 1928. (*Arctotis fastuosa* Jacq. Hort. Schoenbr. 2: 21. *pl. 166.* 1797; *V. wyleyi* Harvey, Fl. Cap. 3: 463. 1865.) Annual herb to 1 m. tall with thin cobwebby pubescence; lower leaves more or less rosulate and petioled, the cauline leaves becoming sessile with an auriculate base, all lanceolate to oblanceolate or oblongish and irregularly pinnately lobed; heads solitary at the ends of branches; phyllaries dimorphic, the outer linear with a slender spreading herbaceous tip, the inner obovate-spatulate with a rounded scarious appendage; rays orange with dark purplish base; achenes smooth or wrinkled; pappus present as a minute crown or lacking. Occasionally escaping from cultivation in southern California. Native of South Africa.

---

[*] Text of *Arctotideae* contributed by John Thomas Howell.

Tribe 11. **MUTISEAE**

Involucres several- to many-flowered; leaves alike.
    Shrubs; flowers yellow.   147. *Trixis.*
    Herbs; flowers pink or white.   148. *Perezia.*
Involucres 1-flowered; stem-leaves linear-lanceolate, floral leaves ovate.   149. *Hecastocleis.*

147. **TRÍXIS** P. Br. Nat. Hist. Jamaica 312. 1756. (Hyponym); Crantz, Inst. Herb. **1** : 329. 1766.

Shrubs or rarely herbs with alternate and entire or dentate leaves. Inflorescence cymose or paniculate. Involucres 2-seriate, 9–12-flowered, the outer phyllaries few and usually shorter than the inner. Flowers perfect, yellow. Corolla bilabiate, the outer lip erect, 2-cleft, the inner spreading, 3-toothed or subentire. Anthers with tail-like appendages. Style-branches flattened and truncate at apex. Achenes subcyclindric, the scabrous pappus-bristles copious, white to tawny, arising from the enlarged disk-like summit of the achene. [From the Greek meaning threefold, referring to the 3-cleft corollas.]

A genus of about 35 species, natives of southwestern United States, Mexico, and Central and South America. Type species, *Trixis suffrutescens* P. Br.

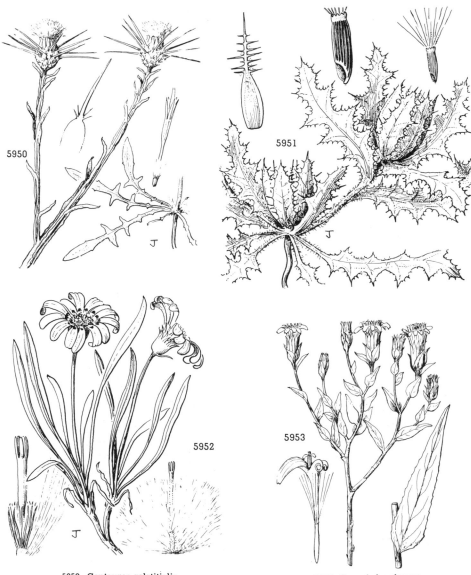

5950. Centaurea solstitialis
5951. Cnicus benedictus

5952. Gazania longiscapa
5953. Trixis californica

1. **Trixis califórnica** Kell. California Trixis. Fig. 5953.

Trixis californica Kell. Proc. Calif. Acad. **2**: 182. *fig. 53.* 1862.
Trixis suffruticosa S. Wats. Bot. Calif. **2**: 459. 1880.
Trixis angustifolia var. *latiuscula* A. Gray, Syn. Fl. N. Amer. 1²: 410. 1884.

Much branched shrub about 0.5–1 m. high, leafy above to the heads, lower stems gray, the upper branchlets striate, tan, glandular-puberulent. Leaves alternate, sessile or subsessile, lanceolate to ovate-lanceolate, 6–15(20) mm. broad, 2–8 cm. long, sparingly appressed-puberulent on both surfaces, densely glandular especially beneath, midvein prominent below, the margins entire or denticulate, inrolled on the older leaves; inflorescence corymbose, the heads 1.5–2 cm. high, 12–14-flowered; inner phyllaries thin except at the thickened and somewhat keeled base, linear-lanceolate, about 15 mm. long, glandular and sparingly puberulent to glabrate, ciliate on the upper margin, becoming straw-colored, the outer series shorter and broader, remaining green; corolla yellow, about 10 mm. long; achenes dark, 5–8 mm. long, cyclindric and narrowed above, densely glandular-hispidulous, the pappus copious, straw-colored.

Rocky slopes and canyons walls, Lower Sonoran Zone; eastern San Bernardino County and south through the Colorado Desert, California, to Lower California and east to western Texas and northern Mexico. Type locality: Cedros Island, Lower California. Feb.–June and Oct.–Nov.

148. **PERÈZIA** Lag. Amen. Nat. **1**: 31. 1811.

Perennial herbs or rarely annuals; stems (in ours) erect and leafy, arising from a more or less woody caudex more or less covered with felty, rust-colored wood. Leaves alternate, mostly sessile. Inflorescence single and terminal or of few to many heads in panicles, or corymbose or thyrsoid cymes. Involucres strongly graduated. Flowers perfect, mostly pink, lavender, or white. Corollas bilabiate, the outer lip 3-dentate, the inner with 2 free or strongly recurved segments. Anthers with tail-like appendages. Style-branches flattened and truncate at apex. Achenes linear-cylindric or fusiform, the pappus copious, of scabrous, white to tawny or brownish bristles. [Name in honor of Lorenzo Perez, a pharmacist of the sixteenth century.]

A genus of about 75 species, natives of southwestern United States, Mexico, and Central America and of western South America. Type species, *Perdicium magellanicum* L. f.

1. **Perezia microcéphala** (DC.) A. Gray. Sacapellote. Fig. 5954.

Acourtia microcephala DC. Prod. **7**: 66. 1838.
Perezia microcephala A. Gray, Smiths. Contr. **3**⁵: 127. 1852.
Perezia sericophylla Millsp. & Nutt. Field Mus. Bot. Ser. **5**: 297. 1923. (Incidental mention)

Stout erect perennials 6–14 dm. high, the striate stems densely leafy and glandular-scabrid. Leaves oblong-ovate to elliptic-oblong, 5–15 cm. long, sessile with a broad base, the larger cordate-clasping, the margin spinose-denticulate, chartaceous to coriaceous in texture and somewhat reticulate-veined, glandular-scabrid or puberulent on both surfaces; inflorescence rather broad, cymose-paniculate, leafy-bracted; heads about 10–12 mm. high, 10–20-flowered; phyllaries oblong-oblanceolate, mucronulate, glandular-puberulent, loosely imbricated in about 3 series; corollas lavender-pink to white; achenes glandular-puberulent, the pappus white.

Common on dry slopes, Upper Sonoran Zone; San Luis Obispo County and the Channel Islands south to San Diego County, California, and adjacent Lower California. Type locality: California. Collected by Douglas. June–July.

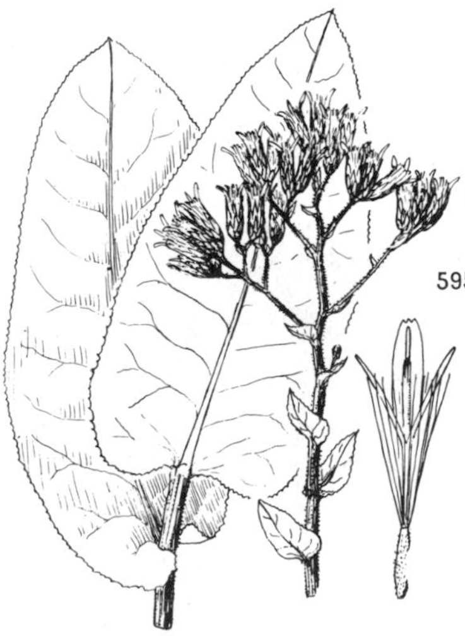

5954. Perezia microcephala

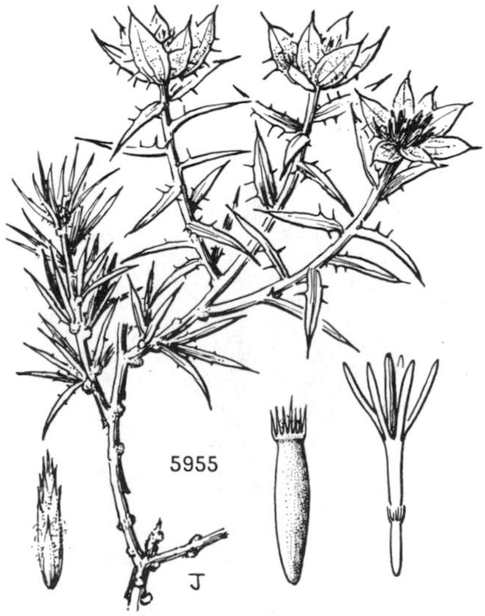

5955. Hecastocleis shockleyi

149. HECASTÓCLEIS A. Gray, Proc. Amer. Acad. 17: 220. 1882.

Shrubs with cauline leaves alternate, appearing singly on younger shoots, later developing axillary fascicles of smaller leaves in the axils. Involucres in terminal clusters, subtended by whorls of 3–4 oval or ovate, reticulate, stramineous, floral leaves, each involucre 1-flowered with narrowly lanceolate, spinose-tipped phyllaries. Corolla whitish, tubular, deeply divided into 5 linear spreading lobes firm in texture and widely spreading in anthesis. Stamens firm in texture with a glabrous appendage. Style purplish red, surpassing the anthers, the stigma emarginate. Achene glabrous in age, the pappus coroniform and laciniate-dentate, coriaceous. [Name Greek, referring to the separate enclosure of each flower in its involucre.]

A monotypic genus, native in the desert regions of southwestern Nevada and the Death Valley region of adjacent California.

1. Hecastocleis shóckleyi A. Gray. Prickle-leaf. Fig. 5955.

Hecastocleis shockleyi A. Gray, Proc. Amer. Acad. 17: 221. 1882.

Stout, divaricately branched shrub 5–7 dm. high, with rigid leafy branches, bark of the older stems dark gray or blackish, somewhat shreddy, that of the younger stems stramineous, shining, the ultimate branchlets tomentose. Cauline leaves alternate, sessile, fascicled and shorter on the older leafy stems, coriaceous, linear-lanceolate, strongly spine-tipped, the margins with few to several more slender spines, glabrous or the young leaves finely tomentulose, 1–2 cm. long; floral leaves oval to ovate, spiny-margined, persistent, 10–20 mm. long, 6–10 mm. wide, conspicuously reticulate-veined, becoming stramineous and papery; involucres 1-flowered, narrowly cylindric, the phyllaries in 4–5 series, appressed, cuspidate-acuminate, loosely woolly on the margin; achenes sparsely tomentose when young, glabrous in age, subcyclindric, somewhat narrower at base, the covering thinner in texture than the coroniform laciniate pappus.

Desert mountains, Upper Sonoran Zone; western Esmeralda and Mineral Counties to Clark County, Nevada, westward to the Death Valley region and the Inyo Mountains, California. Type locality: Candelaria, Esmeralda County, Nevada, altitude 6,000 feet. Said by Shockley to have been collected at Silver Peak, Esmeralda County, instead of the locality given by Gray in the original description. June–July.

Tribe 12. CICHORIEAE.

Achenes without pappus (with short crown in variety of *Phalacroseris*).
    Leaves basal, the cauline none or bract-like; native species.
        Desert annuals; flowers white or pinkish.   163. *Atrichoseris.*
        Perennials of the Sierra Nevada; flowers yellow.   154. *Phalacroseris.*
    Leaves both basal and cauline; introduced species.   176. *Lapsana.*
Achenes with pappus.
    Pappus partly or completely paleaceous, or plumose, or with plumose awns.
        Pappus paleaceous, often awned above the paleaceous base (capillary bristles in *Nothocalais alpestris,*
            Microseris borealis).
            Phyllaries enclosing the outer achenes; pappus of the outer achenes differing from that of the inner.
                    167. *Hedypnois.*
            Phyllaries not enclosing the outer achenes; pappus of the inner and outer achenes alike.
                Pappus of 2 or 3 series of unawned paleae; ligules bright blue.   150. *Cichorium.*
                Pappus-paleae in a single series, surmounted by a smooth, scabrous or barbellate awn (pappus
                  almost obsolete on some forms of *Microseris*); ligules shades of yellow.
                  Perennial with heads always erect; outer and inner phyllaries equal or nearly so.
                      151. *Nothocalais.*
                  Annual or if perennial with heads nodding in bud; outer phyllaries distinctly shorter than
                  the inner.   152. *Microseris.*
        Pappus-bristles conspicuously plumose and sometimes paleaceous-dilated at the extreme base or with
            plumose awns.
            Achenes with long and slender or short and stout beaks (outer achenes truncate in *Hypochoeris*
            glabra; outer pappus paleaceous in *Leontodon leysseri*).
                Achenes contracted into a short beak.   169. *Leontodon.*
                Achenes with a long slender beak.
                  Phyllaries in 2 series, unlike, the outer broadly ovate, subcordate; pubescence coarse-
                  hispid.   168. *Picris.*
                  Phyllaries in 1 series or loosely imbricated, alike; plants glabrous (except *Hypochoeris*).
                  Stems scapose.   166. *Hypochoeris.*
                  Stems leafy.
                    Leaves grass-like; involucre without calyculate bractlets. 164. *Tragopogon.*
                    Leaves pinnatifid; involucre with calyculate bractlets. 158. *Rafinesquia.*
    Achenes truncate.
        Involucres strongly imbricated, the phyllaries rounded and broadly scarious-margined.
                    162. *Anisocoma.*
        Involucres not or but little imbricated, subtended by calyculate bractlets, the phyllaries narrowly or not at all scarious-margined.
             Pappus of stout awns with several rigid bristles toward the base. 155. *Chaetadelpha.*
             Pappus of slender plumose bristles.
                  Ligules yellow; bristles in more than 1 series, unequal. 165. *Scorzonera.*
                  Ligules pale or dark pink; bristles in 1 series, equal. 157. *Stephanomeria.*
    Pappus-bristles capillary, either smooth or scabrous, rarely barbellate.
        Achenes flattened (obscurely so in *Sonchus*).

Achenes not beaked.   170. *Sonchus.*
Achenes beaked.   171. *Lactuca.*
Achenes not flattened, either angulate or terete.
  Ligules pink or purplish.
    Heads nodding before and after anthesis; plants of coastal Oregon and Washington.
                                 172. *Prenanthes.*
    Heads erect before and after anthesis; plants of the desert.   156. *Lygodesmia.*
  Ligules shades of yellow, sometimes white or cream-colored (pink in *Malacothrix blairii*).
    Leaves all basal; heads solitary on scapose peduncles.
      Achenes truncate; pappus-bristles barbellate.   151. *Nothocalais* and
                                 152. *Microseris,* in part.
      Achenes beaked or merely tapering at the summit in some species of *Agoseris;* pappus-bristles not barbellate.
        Achenes 4–5-ribbed, spinulose-muricate above.   175. *Taraxacum.*
        Achenes 10–15-ribbed or -nerved, not spinulose-muricate above.
                               153. *Agoseris.*
  Stems leafy; heads not on scapose peduncles.
    Achenes minutely to strongly rugulose or tuberculate between the angles.
      Depressed branching annuals with crustaceous-margined leaves; achenes abruptly beaked.   160. *Glyptopleura.*
      Erect, stipitate, glandular annuals without crustaceous-margined leaves; achenes tapering to the beak.   161. *Calycoseris.*
    Achenes striate between the angles.
      Pappus-bristles early deciduous, 1–8 bristles remaining in a few species.
                               159. *Malacothrix.*
      Pappus-bristles persistent (deciduous at maturity in some species of *Crepis*).
        Phyllaries not at all thickened; pappus sordid or brownish, rarely white.
                               173. *Hieracium.*
        Phyllaries somewhat thickened at base or on midrib; pappus soft, white.
                               174. *Crepis.*

150. **CICHÒRIUM** [Tourn.] L. Sp. Pl. 813. 1753.

Herbs with erect branching stems, alternate and basal leaves, and large heads of usually blue flowers, peduncled or in sessile clusters along the branches. Involucral bracts in 2 series, herbaceous, the inner erect and subtending or partly enclosing the outer achenes. Receptacle flat, naked or slightly fimbrillate. Ligules truncate and 5-toothed at the apex. Anthers sagittate at the base. Style-branches slender, obtuse. Achenes 5-angled or 5-ribbed, truncate, beakless. Pappus of 2–3 series of short blunt scales. [From the Arabic name.]

An Old World genus of about 8 species. Type species, *Cichorium intybus* L.

1. **Cichorium íntybus** L. Chicory. Fig. 5956.

Cichorium intybus L. Sp. Pl. 813. 1753.

Perennial from a long deep taproot, the stems sparsely hispid, stiff, branched, 3–10 dm. or more high. Basal leaves spreading on the ground, runcinate-pinnatifid, spatulate in outline, 8–20 cm. long, tapering into long petioles; upper leaves much smaller, lanceolate to oblong, entire or lobed, clasping or auricled at base; heads many, 2.5–4 cm. broad, 1–4 together in sessile clusters on the nearly naked or bracted branches; flowers bright blue or rarely white; involucres 9–15 mm. high; achenes 2–3 mm. long, 5-angled, truncate above, the pappus reduced to a minute fringed crown.

Roadsides and waste places, in all the Pacific States, and more or less generally throughout the United States and Canada. A substitute for coffee is sometimes made from the roots. March–July.

Scólymus hispánicus L. Sp. Pl. 813. 1753. A thistle-like herb 3–5 dm. high, more or less arachnoid-pubescent or sometimes glabrate; leaves alternate, 4–8 cm. long, sinuate-dentate or pinnatifid, the lobes strongly spinescent, midrib broadened below the confluent ribs from the lobes, and decurrent on the stem; heads in the axils of the smaller upper leaves; involucral bracts spine-tipped; flowers yellow; chaffy bracts more or less enclosing the beakless achenes. Native of the Mediterranean region and sparingly adventive at Los Gatos, Santa Clara County, California. Golden Thistle.

151 **NOTHOCÁLAIS*** (A. Gray) Greene, Bull. Calif. Acad. **2**: 54. 1886.

Scapose, acaulescent, perennial herbs with milky juice, the stout underground caudex bearing 1 to few long-lived, often rope-like taproots. Leaves entire (often pinnatifid in *N. alpestris*), glabrous or villous along the midrib, the margins frequently undulate and usually minutely ciliate. Scapes unbranched, naked or with a single leaf near the base, glabrous or white-villous under the many-flowered head. Involucre campanulate to hemispherical, the outer phyllaries only slightly shorter than the innermost, glabrous to ciliate or white-villous along the midline. Receptacle naked. Corollas all ligulate, yellow, well exceeding the involucre. Achenes columnar-fusiform, smooth or scabrous above, not beaked, the upper portion often vacant but not externally differentiated from the body of

---

* Text contributed by Kenton Lee Chambers.

the fruit. Pappus of 10 to many silvery bristles, some or all of which are gradually compressed and widened downward into a narrowly paleaceous basal portion (or the bristles essentially capillary in *N. alpestris*). [From the Greek word meaning spurious, plus *Calais*, a synonym of *Microseris*.]

A genus of 4 species native to central and western North America. Type species, *Microseris troximoides* A. Gray.

Nothocalais appears to be most closely allied to *Microseris* and to *Agoseris*, resembling each of these genera in one or another of its gross morphological features. The distinctive combination of characteristics displayed by its species, however, marks it as a natural group separate from either related genus.

Pappus of 30–50 capillary bristles; leaves usually toothed or pinnatifid.   1. *N. alpestris.*
Pappus of 10–30 elongate narrow paleae tapering into short awns; leaves entire or undulate-margined.
                                                    2. *N. troximoides.*

1. **Nothocalais alpéstris** (A. Gray) Chambers. Alpine Lake-agoseris. Fig. 5957.

Troximon alpestre A. Gray, Proc. Amer. Acad. **19**: 70. 1883.
Troximon barbellulatum Greene ex A. Gray, Syn. Fl. N. Amer. **1²**: 437. 1884.
Agoseris alpestris Greene, Pittonia **2**: 177. 1891.
Agoseris barbellulata Greene, loc. cit.
Microseris alpestris Q. Jones ex Cronquist, Vasc. Pl. Pacif. Northw. **5**: 267. 1955.
Nothocalais alpestris Chambers, Contr. Dudley Herb. **5**: 66. 1957.

Glabrous or puberulent, perennial herb with an elongate, blackish, simple or multicipital, underground caudex and deep taproot. Leaves all basal, 3–20 cm. long, linear to oblanceolate or spatulate, entire or more often remotely toothed or pinnatifid with linear to broadly deltoid lobes; scapes 1 to several from the basal rosette, erect, simple, 3–30 cm. tall, glabrous above, sometimes bracteate near the base; heads erect, 10- to many-flowered; involucres 10–20 mm. high, broadly or narrowly campanulate, strictly glabrous, the outer phyllaries lanceolate or ovate-lanceolate, attenuate, slightly shorter than the narrower inner ones, finely dotted with purple dorsally; achenes brown, about 10-ribbed, 5–10 mm. long, truncate or the upper 1–3 mm. somewhat narrower and lightly scabrous; pappus 6–10 mm. long, the 30–50 white capillary bristles barbellulate or rarely subplumose.

Open meadows and gravelly slopes, Boreal Zones; Cascade Mountains from Mount Rainier, Washington, to Siskiyou County, California, and in the Sierra Nevada near Lake Tahoe. Type locality: Mount Adams, Washington. June–Aug.

2. **Nothocalais troximoìdes** (A. Gray) Greene. False-agoseris. Fig. 5958.

Microseris troximoides A. Gray, Proc. Amer. Acad. **9**: 211. 1874.
Nothocalais troximoides Greene, Bull. Calif. Acad. **2**: 55. 1886.
Nothocalais suksdorfii Greene, op. cit. 54.
Scorzonella troximoides Jepson, Man. Fl. Pl. Calif. 994. 1925.

Glabrous or sparsely villous, perennial herb with a stout, blackish, underground caudex and long, often rope like taproot. Leaves in a basal cluster, rather stiff and elongated, linear to linear-lanceolate, attenuate, the margins entire or distinctly undulate, usually minutely ciliate; scapes 1 to several, erect, 5–35 cm. tall, simple or only bracteate below, glabrous or white-villous toward the apex; heads erect, 10- to many-flowered; involucres 14–25(30) mm. high, broadly or narrowly campanulate, the phyllaries narrowly or broadly lanceolate, attenuate, glabrous or ciliate and white-villous along the midline, usually dotted or lined with purple on the midline, the outer

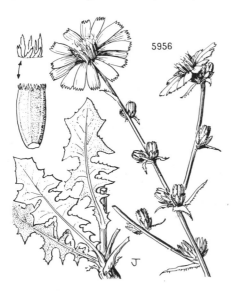

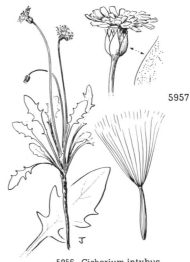

5956. Cichorium intybus
5957. Nothocalais alpestris

phyllaries more than three-fourths as long as the inner ones; achenes straw-colored or tan, 8–13 mm. long, about 10-ribbed, slender, columnar-fusiform, scabrous toward the apex; pappus-parts 10–30, silvery, 10–20 mm. long, rather unequal, each gradually tapering downward from a short terminal awn into a long, narrowly palaceous portion about 0.5 mm. wide near the base. $n = 9$.

Open slopes and flats in dry, often rocky soil, Arid Transition Zone; plains and foothills east of the Cascade Range from British Columbia to Sierra County, California, east through northern Nevada and the Snake River Plain to southwestern Montana and northern Utah. Type locality: "hills on the Clear Water River, Oregon (now Idaho Terr.)." April–June.

East of the Pacific States, especially in Idaho, *N. troximoides* tends to vary in the direction of *N. nigrescens* (Henders.) Heller, often acquiring the broader, plane, acute leaves and glabrous, diffusely speckled outer phyllaries of that species. *Nothocalais nigrescens* itself grows in wet meadows in the Rocky Mountains of Idaho, Montana, and Wyoming; it may occur in our area in the high mountains of northeastern Oregon.

152. MICRÓSERIS* D. Don, Phil. Mag. 11: 388. 1832.

Caulescent or acaulescent herbs with milky juice, the plants perennial, with 1 or more fleshy taproots, or annual. Leaves entire to variously toothed or pinnatifid, glabrous or scurfy-puberulent (villous in section *Calocalais*). Heads many-flowered, erect or nodding in bud, borne singly on naked or remotely bracteate, glabrous or scurfy, scape-like peduncles. Involucre cylindrical to campanulate or hemispherical, the innermost phyllaries lanceolate-attenuate, subequal, appressed black-tomentose (at least within), the outer phyllaries shorter, imbricate to calyculate. Receptacle naked or bearing caducous black hairs. Corollas all ligulate, elongate and showy or short and inconspicuous, yellow-orange, yellow, or white. Achenes columnar to truncate-fusiform, not beaked (or short-beaked in section *Calocalais*), about 10-ribbed, smooth or scabrous, the outermost often densely white-villous. Pappus of 2–30 awn-tipped paleae (or of 30–60 brownish capillary bristles in subgenus *Apargidium*), the paleae sometimes nearly obsolete, the awns usually elongate, denticulate to plumose. [Name Greek, meaning small chicory.]

A genius of 16 species, occurring principally in western North America but with one species in Chile and one in Australia and New Zealand. Type species, *Microseris pygmaea* D. Don. Chromosome numbers, determined or verified by the contributor, are given where known.

Plants perennial; florets conspicuous, well exceeding the involucre; pappus-parts 6–60.
    Pappus of 30–60 brownish capillary bristles; plants acaulescent, the scapes naked and unbranched. (subgenus *Apargidium*)   1. *M. borealis*.
    Pappus of 6–30 white or tawny, awn-tipped paleae; plants caulescent, sparingly branched and leafy above or the inflorescence sometimes simple and only bracteate. (subgenus *Scorzonella*)
        Pappus-parts 15–30, bright white, the awns soft, plumose.   2. *M. nutans*.
        Pappus-parts 6–12, the awns rather stiff, minutely spiculate to barbellulate or if subplumose then the pappus tawny.
            Pappus-paleae 5–10 mm. long, fimbriate at the apex about the base of the awn; awns subplumose, tawny.   6. *M. sylvatica*.
            Pappus-paleae usually less than 5 mm. long, tapering into the awn or emarginate; awns minutely spiculate to barbellulate, white or tawny.
                Outer phyllaries ovate-lanceolate to broadly ovate or circular, acute or cuspidate, usually glabrous dorsally, all over 2.5 mm. broad.   3. *M. laciniata laciniata*.
                Outer phyllaries linear to ovate-lanceolate, acute to attenuate, often scurfy-puberulent dorsally, the smallest on each head not over 2.5 mm. broad.
                    Involucre cylindrical or narrowly campanulate, 15–25-flowered, 8–18 mm. high; Siskiyou Mountains.   4. *M. howellii*.
                    Involucre narrowly or broadly campanulate, 25–75-flowered, 10–25 mm. high; Coastal Oregon and California.
                        Pappus-paleae 0.5–1.5 mm. long, deltoid or ovate; inner achenes brown; awns white or tawny.   3. *M. laciniata leptosepala*.
                        Pappus-paleae 2–4 mm. long, lanceolate; inner achenes pale straw-colored or dull white; awns tawny.   5. *M. paludosa*.
Plants annual; florets inconspicuous, equaling or barely exceeding the involucre; pappus-parts 2–5. (subgenus *Microseris*)
    Heads erect before anthesis; achenes attenuate into a short slender beak; pappus silvery, deciduous. (section *Calocalais*)   7. *M. linearifolia*.
    Heads inclined or nodding before anthesis; achenes truncate or slightly narrower toward the apex; pappus silvery or sordid, persistent.
        Pappus-paleae linear-lanceolate, bifid or minutely lacerate at the apex; plants frequently caulescent. (section *Brachycarpa*)
            Achenes columnar, flared at the apex, gray or straw-colored to brown or bluish violet or if blackish then equaling the paleae in length.   8. *M. heterocarpa*.
            Achenes columnar-fusiform, not flared at the apex, blackish, about twice the length of the paleae.   9. *M. decipiens*.
        Pappus-paleae various but if linear-lanceolate then entire at the apex, tapering evenly into the awn; plants acaulescent. (section *Microseris*)
            Achenes 3.0 mm. long or less.
                Achenes columnar to obconical, the flared apex as broad as or broader than the body of the fruit.
                    Paleae over 2.0 mm. long, the margins strongly incurved.   10. *M. douglasii platycarpha*.
                Paleae 2.0 mm. long or less, flat or the margins slightly incurved.   14. *M. elegans*.

---

* Text contributed by by Kenton Lee Chambers. (For complete synonymy of the annual species, see Contr. Dudley Herb. 4: 283–306. 1955.)

Achenes truncate-fusiform, the apex narrower than the body of the fruit.
    Paleae conspicuous, at least 1.0 mm. long.   13. *M. bigelovii.*
    Paleae essentially obsolete, scarcely 0.5 mm. long.   10. *M. douglasii tenella.*
Achenes over 3.0 mm. long.
    Paleae averaging 1.0 mm. long or longer.
        Paleae linear-lanceolate, the midrib very stout, tapering, forming one-third to one-
            fifth of the maximum palea width.   11. *M. acuminata.*
        Paleae circular to ovate or lanceolate, the stout or slender midrib forming less than
            one-fifth of the maximum palea width.
            Paleae arcuate only at the base, straight and flat above, the midrib linear, abruptly
                thickened at the base.   13. *M. bigelovii.*
            Paleae arcuate throughout, the margins incurved to convolute, the midrib linear
                or tapering from the base.
                Paleae uniformly 5, minutely scabrous, clear or lightly smoky, the margins
                    only slightly incurved.   12. *M. campestris.*
                Paleae 5 or fewer, villous to minutely scabrous, clear or chalky to tawny or
                    blackish, the margins strongly incurved to convolute.
                    Paleae nearly equaling to exceeding the length of the achene; achenes
                        4.5 mm. long or less.   10. *M. douglasii platycarpha.*
                    Paleae 1–6 mm. shorter than the achene or if equaling or exceeding it
                        then the latter more than 4.5 mm. long.
                                                                 10. *M. douglasii douglasii.*
    Paleae mostly less than 1.0 mm. long.   10. *M. douglasii tenella.*

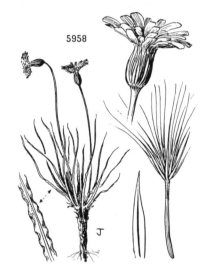

5958

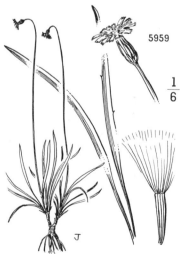

5959

$\dfrac{1}{6}$

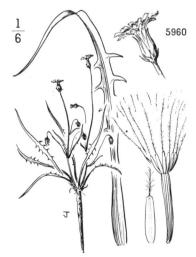

$\dfrac{1}{6}$

5960

5958. Nothocalais troximoides
5959. Microseris borealis
5960. Microseris nutans

1. **Microseris boreàlis** (Bong.) Sch.-Bip. Apargidium. Fig. 5959.

Apargia borealis Bong. Mém. Acad. St. Pétersb. VI. Math. Phys. Nat. **2**: 146. 1832.
Leontodon boreale DC. Prod. **7**: 102. 1838.
Apargidium boreale Torr. & Gray, Fl. N. Amer. **2**: 474. 1843.
Microseris borealis Sch.-Bip. Pollichia **22–24**: 310. 1866.
Scorzonella borealis Greene, Pittonia **2**: 19. 1889.

Glabrous, acaulescent, perennial herb with numerous fleshy roots from a short rootstock. Leaves up to 30 cm. long, linear-oblanceolate, acute to acuminate, entire or remotely denticulate; scapes slender, naked and unbranched, erect or curved at the base, 15–60 cm. tall, sometimes lightly scurfy above; heads 20–50-flowered, nodding before anthesis; involucres 10–18 mm. high, campanulate, the inner phyllaries lanceolate-attenuate, the outer phyllaries shorter, loosely imbricate, linear or lanceolate-attenuate, glabrous or sometimes lightly black-tomentose; florets yellow; achenes brown, 4–8 mm. long, columnar, attenuate toward the base; pappus 5–10 mm. long, of 30–60 brownish barbellulate awns only slightly thickened at the base.

Marshy meadows and sphagnum bogs, Humid Transition and Boreal Zones; along the coast from southern Alaska to British Columbia and south at scattered localities in the Cascade Mountains of Washington and Oregon to Humboldt County, California. Type locality: Sitka, Alaska. Collected by Mertens. June–Aug.

2. **Microseris nùtans** (Hook.) Sch.-Bip. Nodding Scorzonella. Fig. 5960.

Scorzonella nutans Hook. Lond. Journ. Bot. **6**: 253. 1847.
Crepis nutans Geyer ex Hook. loc. cit., as a synonym.
Ptilophora nutans A. Gray, Mem. Amer. Acad. II. **4**: 113. 1849.
Ptilophora major A. Gray, loc. cit.
Ptilophora major β *laciniata* A. Gray, loc. cit.
Calais nutans A. Gray in Torr. Pacif. R. Rep. **4**: 113. 1857.
Calais major A. Gray in Torr. loc. cit.
Microseris nutans Sch.-Bip. Pollichia **22–24**: 309. 1866.
Calais nutans var. *latifolia* D. C. Eaton, Bot. King Expl. 197. 1871.
Stephanomeria (?) intermedia Kellogg, Proc. Calif. Acad. **5**: 39. 1873.
Calais gracililoba Kellogg, op. cit. 48.
Ptilocalais nutans Greene, Bull. Calif. Acad. **2**: 54. 1886.
Ptilocalais major Greene, loc. cit.
Ptilocalais gracililoba Greene, loc. cit.
Ptilocalais tenuifolia Osterhout, Muhlenbergia **1**: 142. 1906.
Microseris nutans var. *major* Nels. & Macbr. Bot. Gaz. **61**: 47. 1916.
Scorzonella nutans var. *laciniata* Jepson, Man. Fl. Pl. Calif. 993. 1925.

Glabrous or scurfy-puberulent, perennial herb with 1 to several fleshy taproots, the stems few, erect or curved at the base, slender, 10–70 cm. tall, usually branched and leafy above. Leaves up to 30 cm. long, extremely variable, linear or lanceolate-attenuate to spatulate, entire to toothed or laciniate-pinnatifid with slender lobes; heads 10–75-flowered, nodding before anthesis; involucres 8–22 mm. high, cylindrical to campanulate, the inner phyllaries lanceolate-attenuate, the outer phyllaries unequal, shorter and nearly calyculate, glabrous dorsally or often scurfy-puberulent, linear to ovate, acuminate, the smallest ones on each head 2 mm. broad or less; florets yellow; achenes gray to brown, 3.5–8.0 mm. long, columnar or narrowly fusiform; pappus 5–10 mm. long, bright white, the paleae 15–30, oblong or lanceolate, 1–3 mm. long, obtuse or somewhat fimbriate above, the awns slender, distinctly plumose.

Open sites in various habitats ranging from grass and sagebrush communities to coniferous forests, Arid Transition and Boreal Zones; British Columbia to southern Oregon mostly east of the summit of the Cascade Range and south in the Sierra Nevada and North Coast Range of California east to Wyoming and Colorado. Type locality: "Dry sunny loamy declivities of Spokan and Coeur d'Aleine mountains." April–Aug.

3. **Microseris laciniàta** (Hook.) Sch.-Bip. Cut-leaved Scorzonella. Fig. 5961.

Hymenonema ? glaucum Hook. Fl. Bor. Amer. **1**: 300. 1833.
Hymenonema ? laciniatum Hook. op. cit. 301.
Scorzonella laciniata Nutt. Trans. Amer. Phil. Soc. II. **7**: 426. 1841.
Scorzonella glauca Nutt. loc. cit.
Calais laciniata A. Gray in Torr. Pacif. R. Rep. **4**: 113. 1857.
Microseris laciniata Sch.-Bip. Pollichia **22–24**: 309. 1866.
Calais glauca A. Gray, Proc. Amer. Acad. **7**: 364. 1868.
Calais glauca var. *procera* A. Gray, op. cit. 365.
Microseris laciniata var. *procera* A. Gray, op. cit. **9**: 209. 1874.
Microseris procera A. Gray, op. cit. **19**: 64. 1883.
Scorzonella megacephala Greene, Bull. Calif. Acad. **2**: 50. 1886.
Scorzonella procera Greene, loc. cit.
Scorzonella pratensis Greene, op. cit. 51.
Scorzonella arguta Drew, Bull. Torrey Club **16**: 152. 1889.
Scorzonella maxima Bioletti, Erythea **1**: 69. 1893.
Scorzonella procera var. *pratensis* Jepson, Man. Fl. Pl. Calif. 994. 1925.

Glabrous or lightly scurfy-puberulent, perennial herb, usually with a single fleshy fusiform taproot, the stems few, erect, slender to very stout, 15–120 cm. tall, rather sparingly leafy or bracteate above, branched above or sometimes only near the base, or unbranched in small individuals. Leaves chiefly basal, up to 50 cm. long, linear to broadly oblanceolate, attenuate to obtuse, entire to few-toothed or laciniate-pinnatifid; heads 30–250-flowered, usually nodding before anthesis; involucres 14–30 mm. high, campanulate to hemispherical, the inner phyllaries subequal, lanceolate, attenuate, the outer phyllaries unequal, shorter, more or less imbricate in 1 or more series, glabrous dorsally or very rarely scurfy-puberulent, broadly ovate to ovate-lanceolate, acute to cuspidate, the smallest ones on each head more than 2.5 mm. broad and rarely less than one-third as long as the innermost phyllaries; florets yellow; achenes brown, 3.5–6.5 mm. long, columnar, attenuate toward the base; pappus 7–15 mm. long, the paleae 6–10, deltoid to lanceolate, 0.5–4.0 (rarely –5.0) mm. long, each tapering into a white or tawny, minutely spiculate to barbellulate awn.

Moist meadows and grassy hillsides, Upper Sonoran and Transition Zones; Pacific Slope mostly west of the Cascade Mountains from the vicinity of Puget Sound, Washington, south to Sonoma County, California, and east through the Klamath region to Lake County, Oregon, and northern Lassen County, California; usually not on the immediate coast. Type locality: "Dry plains of the Columbia, from the Rocky Mountains to the ocean." Collected by Douglas. April–July.

Toward the eastern limit of the range, individuals of this subspecies may possess smaller heads and involucral phyllaries superficially resembling those found in the generally coastal subspecies *leptosepala*, below.

Microseris laciniata subsp. **leptosépala** (Nutt.) K. Chambers, Contr. Dudley Herb. **5**: 61. 1957. (*Scorzonella leptosepala* Nutt. Trans. Amer. Phil. Soc. II. **7**: 426. 1841; *Calais bolanderi* A. Gray, Proc. Amer. Acad. **7**: 365. 1868; *Microseris leptosepala* A. Gray, op. cit. **9**: 209. 1874; *M. bolanderi* A. Gray, op. cit. **19**: 64. 1883; *Scorzonella bolanderi* Greene, Bull. Calif. Acad. **2**: 52. 1886; *S. laciniata* var. *bolanderi* Jepson, Man. Fl. Pl. Calif. 994. 1925; *S. leachiana* M. E. Peck, Proc. Biol. Soc. Wash. **47**: 188. 1934.) Involucre narrowly to broadly campanulate, 10–25 mm. high, 25–75-flowered, the outer phyllaries frequently scurfy-puberulent, linear to ovate, the smallest one on each head 0.5–2.5 mm. broad, often less than one-third as long as the innermost phyllaries; pappus-paleae deltoid to ovate, 0.5–1.5 mm. long, tapering into the awn or obtuse. Moist meadows and swampy areas, Humid Transition Zone; plains of the lower Willamette Valley, Oregon, and south along the coast to Mendocino County, California, extending inland on the lower Klamath and Trinity Rivers and probably elsewhere on the western drainage of the Coast Ranges. Type locality: "near the outlet of the Wahlamet," that is, the mouth of the Willamette River.

4. **Microseris howéllii** A. Gray. Howell's Scorzonella. Fig. 5962.

Microseris howellii A. Gray, Proc. Amer. Acad. **20**: 300. 1885, emend. Syn. Fl. N. Amer. **1**[2]: suppl. 454. 1886. *Scorzonella howellii* Greene, Bull. Calif. Acad. **2**: 52. 1886.

Glabrous or lightly scurfy-puberulent,. perennial herb with a fleshy fusiform taproot, the stem slender, erect, usually single, 15–50 cm. tall, sparingly leafy, branched below the middle or sometimes unbranched and only bracteate below. Leaves chiefly basal, up to 30 cm. long, linear to narrowly lanceolate, attenuate, entire or laciniate-pinnatifid with slender lobes; heads 15–25-flowered, nodding before anthesis; involucres 8–18 mm. high, cylindrical to narrowly campanulate, the principal phyllaries lanceolate, attenuate, the outer phyllaries shorter, unequal, nearly calyculate, glabrous dorsally or more often scurfy-puberulent, deltoid to lanceolate, the smallest ones on each head only 0.5–1.5 mm. broad and less than one-third as long as the innermost phyllaries; florets pale yellow; achenes brown, 3.5–5.0 mm. long, columnar, attenuate toward the base; pappus 6–12 mm. long, the paleae 6–10, lanceolate, 3–6 mm. long (or infrequently only 1 mm. long), each tapering into a white, minutely spiculate awn.

Moist rocky flats, frequently in serpentine soils, Transition Zones; Siskiyou Mountains of southern Josephine and Jackson Counties, Oregon, and adjacent California. Type locality: Waldo, Josephine County, Oregon. April–July.

This taxon is morphologically rather similar to *M. laciniata* subsp. *leptosepala*, but its habitat and typically much larger pappus-paleae appear to be distinctive.

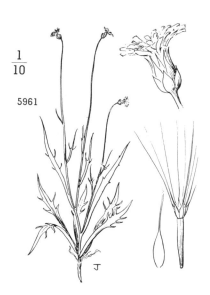

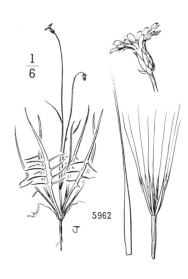

5961. Microseris laciniata   5962. Microseris howellii

5. Microseris paludòsa (Greene) J. T. Howell. Marsh Scorzonella. Fig. 5963.

Microseris sylvatica var. *stillmanii* A. Gray, Proc. Amer. Acad. **9**: 208. 1874.
Scorzonella paludosa Greene, Bull. Calif. Acad. **2**: 52. 1886.
Scorzonella sylvatica var. *stillmanii* Jepson, Man. Fl. Pl. Calif. 994. 1925.
Scorzonella paludosa var. *integrifolia* Jepson, loc. cit.
Microseris paludosa J. T. Howell, Leaflets West Bot. **5**: 108. 1948.

Glabrous or scurfy-puberulent, perennial herb with a fleshy, slender-fusiform taproot, the stems few, erect or curved at the base, slender, 15–70 cm. tall, leafy and branched toward the base or sometimes unbranched and only bracteate above. Leaves chiefly basal, up to 35 cm. long, linear to oblanceolate, attenuate above, entire to toothed or laciniate-pinnatifid with slender lobes; heads 35–75-flowered, nodding before anthesis; involucres 10–20 mm. high, broadly campanulate, the inner phyllaries lanceolate-attenuate, the outer phyllaries unequal, shorter and loosely imbricate, glabrous dorsally or often scurfy, lanceolate- or ovate-attenuate, the smallest ones on each head 2 mm. broad or less; florets yellow; achenes (except the outermost) pale straw-colored or dull white, 4–7 mm. long, columnar, attenuate toward the base; pappus 8–12 mm. long, tawny, the paleae 6–10, lanceolate, 2–4 mm. long, each tapering into a slender barbellulate awn.

Moist grassy meadows and open woods, Upper Sonoran and Transition Zones; in California from central Sonoma County to San Francisco and south along the coast to the Monterey Peninsula. Type locality: "On Mark West's creek," Sonoma County, California. Collected by Bigelow. April–June.

6. Microseris sylvática (Benth.) Sch.-Bip. Sylvan Scorzonella. Fig. 5964.

Scorzonella sylvatica Benth. Pl. Hartw. 320. 1849.
Calais sylvatica A. Gray in Torr. Pacif. R. Rep. **4**: 114. 1857.
Microseris sylvatica Sch.-Bip. Pollichia **22–24**: 309. 1866.
Scorzonella montana Greene, Bull. Calif. Acad. **2**: 53. 1886.
Scorzonella lepidota Heller, Muhlenbergia **2**: 146. 1906.
Microseris montana H. M. Hall, Univ. Calif. Pub. Bot. **3**: 252. 1907.

Glabrous or scurfy-puberulent, perennial herb with a slender elongate taproot, the stems few, erect or curved at the base, slender or stout, 15–75 cm. tall, leafy and branched near the base or sometimes unbranched and only bracteate above. Leaves chiefly basal, up to 35 cm. long, narrowly elliptic to broadly oblanceolate, attenuate above, denticulate to laciniately toothed or pinnatifid, undulate; heads 30–100-flowered, nodding before anthesis; involucres 12–25 mm. high, campanulate to hemispherical, the inner phyllaries lanceolate-attenuate, the outer phyllaries unequal, shorter and loosely imbricate, glabrous or scurfy-puberulent, linear to lanceolate- or ovate-attenuate, the shortest ones on each head often over 2 mm. broad; florets yellow; achenes pale straw-colored or dull white, 5–11 mm. long, columnar; pappus 12–18 mm. long, tawny, the paleae 6–12, narrowly lanceolate, 5–10 mm. long, minutely fimbriate at the apex about the base of the subplumose, rather stout awn.

Grassy flats, hill slopes, and open woods, Upper Sonoran Zone; Inner Coast Ranges from Glenn County to San Benito County, California, and Sierra Nevada foothills from Butte County to the Greenhorn and Tehachapi Mountains, and northern Los Angeles County. Type locality: "In sylvis vallis Sacramento." March–May.

7. Microseris linearifòlia (Nutt.) Sch.-Bip. Uropappus. Fig. 5965.

Calais lindleyi DC. Prod. **7**: 85. 1838.
Calais linearifolia DC. loc. cit. Illegitimate name.
Uropappus lindleyi Nutt. Trans. Amer. Phil. Soc. II. **7**: 425. 1841. Confused name.
Uropappus linearifolius Nutt. loc. cit.
Uropappus grandiflorus Nutt. loc. cit.
Calais macrochaeta A. Gray, Mem. Amer. Acad. II. **4**: 112. 1849.
Microseris linearifolia Sch.-Bip. Pollichia **22–24**: 308. 1866.
Microseris lindleyi A. Gray, Proc. Amer. Acad. **9**: 210. 1874, except for description, which applies to *M. hetero-carpa*. Confused name.

Slender, puberulent or nearly glabrous, annual herb, 10–70 cm. tall, nearly acaulescent to rather long-stemmed, bearing solitary head on naked, scape-like peduncles. Leaves often reddish, up to 3 dm. long, linear, long-attenuate at the tip, entire or pinnatifid with linear retrorse lobes, tapering below into a broad, somewhat clasping, often ciliate and sparsely villous petiole; peduncles erect, scurfy-puberulent or glabrescent, stout and fistulous at the apex; head glabrous, 5–150-flowered, erect at all times, conical to broadly fusiform in fruit, the inner phyllaries lanceolate-attenuate, subequal, the lanceolate outer ones shorter, very unequal, few and somewhat imbricate but not truly calyculate; florets yellow; achenes blackish (rarely gray), 7–17 mm. long, slender-fusiform, the upper portion attenuate and beak-like (or stout and truncate in some desert forms); pappus 10–20 mm. long, silvery, deciduous after maturity, consisting of 5 linear-lanceolate paleae, each terminating in a short, hair-like awn that arises from a distinct notch in the palea apex. $n = 9$.

Dry scrub and grasslands, in well-drained soil, often on open slopes, Sonoran and Transition Zones; Whitman County, Washington, south to Lower California, east to south-central Idaho, and through southern Utah and Arizona to southwestern New Mexico. Type locality: California. Collected by Douglas. April–June.

8. Microseris heterocárpa (Nutt.) Chambers. Derived Microseris. Fig. 5966.

Uropappus heterocarpus Nutt. Trans. Amer. Phil. Soc. II. **7**: 425. 1841.
Calais parryi A. Gray, Pacif. R. Rep. **4**: 112. 1857.
Microseris parryi Sch.-Bip. Pollichia **22–24**: 309. 1866.
Calais kelloggii Greene, Bull. Calif. Acad. **2**: 49. 1886.

Calais clevelandii Greene, op. cit. 153.
Uropappus leucocarpus Greene, Erythea 1: 260. 1893.
Uropappus lindleyi var. *clevelandii* Jepson, Fl. W. Mid. Calif. 494. 1901.
Microseris lindleyi var. *clevelandii* Hall, Univ. Calif. Pub. Bot. **3**: 251. 1907.
Uropappus lindleyi var. *leucocarpus* Jepson, Man. Fl. Pl. Calif. 993. 1925.
Microseris heterocarpa Chambers, Contr. Dudley Herb. **4**: 286. 1955.

Glabrous or puberulent, annual herb, 10–60 cm. tall, essentially acaulescent or the leafy stem sometimes elongated, the head borne singly on long naked peduncles. Leaves scurfy, 5–35 cm. long, linear or narrowly elliptic, attenuate, entire or pinnatifid with narrow spreading lobes; peduncles erect, glabrescent, slender and scarcely fistulous above; head 5–125-flowered, broadly fusiform, inclined or nodding before anthesis, erect later, glabrous dorsally, loosely black-villous within, the lanceolate inner phyllaries subequal, the few outer ones shorter, lanceolate or ovate, unequal, nearly calyculate; florets white or yellow; achenes 4.5–12.0 mm. long, gray to brown or bluish or reddish violet, often flecked with purple, columnar or slightly narrower above, more or less flared at the apex, the outer sometimes densely villous; pappus 8–19 mm. long, pale or tawny, persistent, the 5 lanceolate paleae 4–11 mm. long, each with a stout brown midrib and erose or notched apex bearing a slender spiculate awn. $n = 18$.

Open grassy flats and hillsides and on dry slopes, Sonoran Zones; California in the Coast Ranges and Sierra Nevada foothills from Lake and Butte Counties to Santa Barbara and Kern Counties and the islands and cismontane region of southern California to Lower California. Type locality: San Diego, California. April–June. This species is an amphiploid derivative of *M. lindleyi* and *M. douglasii.*

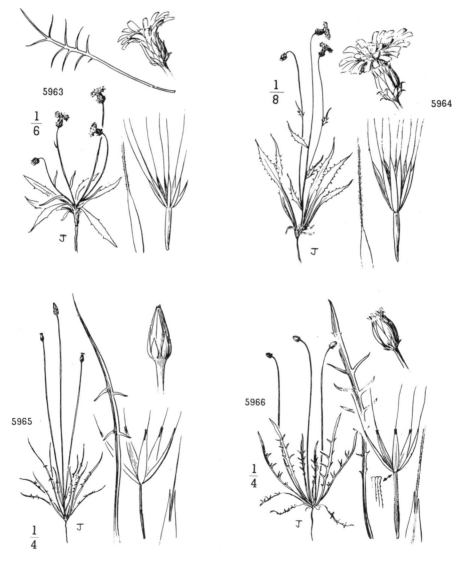

5963. Microseris paludosa
5964. Microseris sylvatica

5965. Microseris linearifolia
5966. Microseris heterocarpa

9. **Microseris decípiens** Chambers. Santa Cruz Microseris. Fig. 5967.

Microseris decipiens Chambers, Contr. Dudley Herb. **4**: 290. 1955.

Scurfy, somewhat glabrescent, annual herb. 10–40 cm. tall, acaulescent or with a short basal stem, the head solitary on slender scapose peduncles. Leaves 5–15 cm. long, linear or narrowly elliptic, attenuate, entire or with linear spreading lobes; head 10–40-flowered, broadly fusiform in fruit, inclined before anthesis, erect near maturity, glabrous externally, black-villous within, the few outer phyllaries lanceolate, shorter than the inner ones, unequal and nearly calyculate; florets yellow; achenes 6–8 mm. long, brown or purplish brown to nearly black (the paler outer fruits often with purple flecks), columnar-fusiform, not flaring outward at the apex; pappus 7–10 mm. long, persistent but fragile, the 5 lanceolate paleae 2.5–5.0 mm. long, silvery or dull, with a yellowish midrib and erose apex bearing a slender spiculate awn. $n = 18$.

Open places in loose soil of shale or stabilized sand and on serpentine, Humid Transition Zone; central California on the seaward slope of the Outer Coast Range from Marin County to Santa Cruz County and perhaps near Monterey. Type locality: west slope of Santa Cruz Mountains between Scott Creek and Mill Creek, Santa Cruz County, California. April–June. This species resembles *M. heterocarpa*, but it is quite restricted in range and represents an amphiploid derivative of *M. lindleyi* and *M. bigelovii*.

10. **Microseris douglásii** (DC.) Sch.-Bip. Douglas' Microseris. Fig. 5968.

Calais douglasii DC. Prod. **7**: 85. 1838.
Calais cyclocarpha A. Gray, Pacif. R. Rep. **4**: 113, *p. 18, figs. 1–6.* 1857.
Calais eriocarpha A. Gray, Proc. Amer. Acad. **6**: 552. 1865.
Microseris douglasii Sch.-Bip. Pollichia **22–24**: 308. 1866.
Microseris cyclocarpha Sch.-Bip. loc. cit.
Microseris cyclocarpha var. *eriocarpha* A. Gray, Proc. Amer. Acad. **9**: 210. 1874.
Microseris attenuata Greene, Bull. Torrey Club **9**: 111. 1882.
Microseris parishii Greene, Bull. Calif. Acad. **2**: 46. 1886, as to type, not of authors.
Microseris platycarpha var. *parishii* H. M. Hall, Univ. Calif. Pub. Bot. **3**: 249. 1907.

Scurfy-puberulent, acaulescent, annual herb bearing solitary head on slender scapose peduncles curved at the base and 5–60 cm. tall. Leaves in a basal cluster, 3–25 cm. long, linear to oblanceolate, attenuate or acute, entire or coarsely pinnatifid with mostly slender tapering lobes; head 50–100-flowered, distinctly nodding before anthesis, nodding or erect near maturity, ovoid or broadly fusiform, the principal phyllaries 5–18, equal, glabrous externally, black-villous on the inner surfaces, somewhat carinate, often with reddish midline, the small outer phyllaries calyculate; florets yellow or white; achenes highly variable, 4–10 mm. long, gray to brown or blackish, often spotted, turbinate to columnar-fusiform or attenuate into a stout, vacant upper portion, the longitudinal ribs usually distinctly scabrous and flared outward at the apex, the outer fruits often densely villous; pappus-parts 5 or fewer, 5–13 mm. long, the paleae lanceolate to circular, arcuate, 1.0–6.5 mm. long, chalky-white to straw-colored or blackish, with incurved or convolute margins and conspicuous midrib, mostly scabrous or hairy dorsally, gradually or abruptly tipped by an equally long or longer barbellulate awn. $n = 9$.

Grassy flats and hillsides in heavy, often hard-packed soil, Sonoran Zones; Jackson County, Oregon, south through the interior valleys and foothills of central California and cismontane southern California to northern San Diego County. Type locality: California. Collected by Douglas. April–June.

Microseris douglasii subsp. **tenélla** (A. Gray) Chambers, Contr. Dudley Herb. **4**: 294. 1955. (*Calais tenella* A. Gray, Pacif. R. Rep. **4**: 114. *pl. 17, figs. 6–10.* 1857; *C. aphantocarpha* A. Gray, Proc. Amer. Acad. **6**: 552. 1865; *Microseris tenella* Sch.-Bip. Pollichia **22–24**: 308. 1866; *M. aphantocarpha* Sch.-Bip. loc. cit.; *M. aphantocarpha* var. *tenella* A. Gray, Proc. Amer. Acad. **9**: 209. 1874; *M. indivisa* Greene, Erythea **1**: 7. 1893; *M. aphantocarpha* var. *indivisa* Jepson, Fl. W. Mid. Calif. 495. 1901; *M. tenella* var. *aphantocarpha* Blake, Journ. Wash. Acad. **25**: 325. 1935.) Heads ovoid to nearly globose near maturity, up to 300-flowered, the principal phyllaries 5–30; achenes 3.0–6.5 mm. long, brown, truncate-fusiform or clavate, usually not flared at the apex; pappus fragile, 3.5–9.0 mm. long, the paleae deltoid, about 1.0 mm. long or less, often nearly obsolete. $n = 9$. Open grassy places in heavy soil, Sonoran Zones; central California in the foothills of the Coast Ranges and near the coast, Yolo County to Los Angeles County, rarely in the Central Valley. Type locality: Napa Valley, Napa County, California. This taxon. intergrades extensively with subsp. *douglasii* in the Coast Ranges of central California and also occasionally with *M. bigelovii* in coastal areas.

Microseris douglasii subsp. **platycárpha** (A. Gray) Chambers, Contr. Dudley Herb. **4**: 296. 1955. *Calais platycarpha* A. Gray, Pacif. R. Rep. **4**: 113. 1857; *Microseris platycarpha* Sch.-Bip. Pollichia **22–24**: 308. 1866.) Heads ovoid or fusiform, 5–50-flowered, the principal phyllaries 5–15; achenes 3.0–4.5 mm. long, straw-colored to brown or blackish, turbinate to columnar, flared at the apex; pappus stout, 4–10 mm. long, the lanceolate to circular paleae 2.5–6.5 mm. long, involute at maturity, tipped with a rather short stout awn. $n = 9$. Grassy slopes and mesas, Sonoran Zones; coastal southern California from Los Angeles County to northern Lower California and on Santa Catalina and San Clemente Islands. Type locality: San Luis Rey, San Diego County, California.

11. **Microseris acumináta** Greene. Sierra Foothills Microseris. Fig. 5969.

Microseris acuminata Greene, Bull. Torrey Club **10**: 88. 1883.

Strictly acaulescent, puberulent, annual herb, the erect scapes 5–35 cm. tall, curved or decumbent at the base. Leaves basal, 3–20 cm. long, deeply pinnatifid, the entire slender lobes spreading from a linear-attenuate rachis; heads 5–50-flowered, nodding until mature, broadly to narrowly fusiform, large in proportion to the plant, the inner phyllaries 11–19 mm. long in nearly mature fruit, black-villous ventrally, carinate, the much shorter outer ones unequal, calyculate; florets yellow; achenes 4.5–7.0 mm. long, uniformly brown, columnar, flared at the apex, scabrous, the outer fruits never villous; pappus 9–18 mm. long, persistent, the 5 linear-lanceolate paleae 3.5–11.0 mm. long, shining or dull, straw-colored, lightly scabrous or villous dorsally, scarcely involute at the margins, tapering evenly above into a barbellulate awn formed as a continuation of the stout flat midrib. $n = 18$.

Grassy flats and hog-wallows, and on flats or pans on foothill terraces, mostly Upper Sonoran Zone; Jackson County, Oregon, south in the Inner Coast Range, Sacramento Valley, and Sierra Nevada foothills to Fresno County, California. Type locality: Elmira, Solano County, California. April–June.

12. **Microseris campéstris** Greene. San Joaquin Microseris. Fig. 5970.

Microseris campestris Greene, Pittonia 5: 15. 1902.

Acaulescent, scurfy-puberulent, annual herb, the solitary head borne on erect curving scapes up to 50 cm. tall. Leaves all basal, 3–20 cm. long, linear to oblanceolate, acute or attenuate, entire or more often pinnatifid with slender spreading lobes; head 5–125-flowered, nodding until mature, ovoid or fusiform, the inner phyllaries 5–20, carinate, often red on the midrib, black-villous on inner surfaces, the small outer phyllaries calyculate; florets yellow or white; achenes 3.25–5.25 mm. long, gray or pale brown, often darkly spotted, turbinate or columnar, not attenuate above, more or less flared at the apex, only minutely scabrous, the outer fruits often paler but rarely white-villous; pappus-parts uniformly 5, persistent, 5.0–8.5 mm. long, the paleae ovate or lanceolate, 1.0–4.5 mm. long, silvery or smoky, with only slightly involute margins and slender midib, very minutely scabrous dorsally, tapering into a slender barbellulate awn. $n = 18$.

Grassy flats and slopes in adobe soils, Sonoran Zones; California in the Central Valley and Inner Coast Range from Colusa County to Kern County. Type locality: Byron, Contra Costa County, California. April–June.

13. **Microseris bigelòvii** (A. Gray) Sch.-Bip. Coast Microseris. Fig. 5971.

Calais bigelovii A. Gray, Pacif. R. Rep. **4**: 113. *pl. 17, figs. 1–5*. 1857.
Microseris bigelovii Sch.-Bip. Pollichia **22–24**: 308. 1866.

Acaulescent, often somewhat fleshy, annual herb with curving, decumbent or erect scapes. Leaves in a basal tuft, 3–25 cm. long, glabrous or scurfy, linear to oblanceolate or spatulate, obtuse to attenuate, entire or coarsely pinnatifid with slender or broad, tapering lobes or teeth; heads

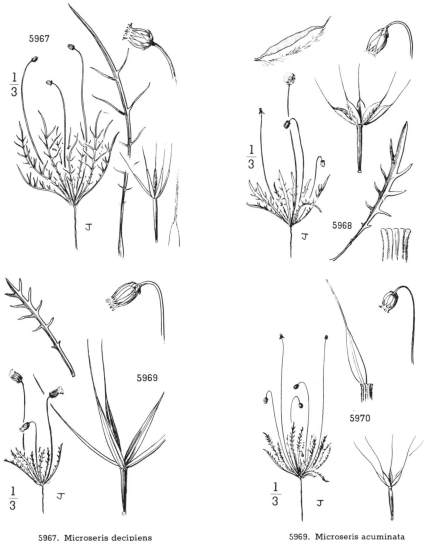

5967. Microseris decipiens
5968. Microseris douglasii

5969. Microseris acuminata
5970. Microseris campestris

ovoid or fusiform, nodding until mature, the principal phyllaries carinate, often reddish dorsally, black-villous within, the outer phyllaries short, essentially calyculate; florets yellow-orange or yellow; achenes 2.50–5.25 mm. long, brown or bronze, sometimes darkly spotted, truncate-fusiform to turbinate or clavate, not flared at the apex, very minutely scabrous, the outer fruits often densely pubescent; pappus-parts always 5, 4.5–10.5 mm. long, the paleae lanceolate, 1–4 mm. long, straight and flat above the middle, silvery to smoky or bronze, tapering or lacerate above, the linear, band-like midrib projected as a slender spiculate awn. $n = 9$.

Coastal bluffs, hillsides, and sandy flats, Humid Transition Zone; Vancouver Island, British Columbia, San Juan County, Washington, and south along the coast from Lincoln County, Oregon, to San Luis Obispo County, California. Type locality: Corte Madera, Marin County, California. April–July.

14. Microseris élegans Greene ex A. Gray. Elegant Microseris. Fig. 5972.

Microseris elegans Greene ex A. Gray, Syn. Fl. N. Amer. 1²: 419. 1884.
Microseris aphantocarpha var. *elegans* Jepson, Man. Fl. Pl. Calif. 991. 1925.
Microseris aphantocarpha var. *mariposana* Jepson, loc. cit.

Delicate, acaulescent, lightly puberulent, annual herb, the slender scapes curved at the base, erect, 5–35 cm. tall. Leaves basal, 2–20 cm. long, linear or narrowly oblanceolate, acute or attenuate, toothed or deeply pinnatifid with slender tapering lobes or entire; heads 10–100-flowered, ovoid or globose, nodding until almost mature, the principal phyllaries carinate, often lined with red, black-villous on inner surfaces, the short outer phyllaries calyculate; florets yellow-orange or yellow; achenes 1.5–3.5 mm. long, smooth, gray-brown to brown or blackish, not darkly spotted, slenderly obconical, the slightly flaring apex as broad as or broader than the body of the achene, the marginal fruits often densely villous; pappus always of 5 parts, 3.5–7.0 mm. long, the paleae minute, delicate, ovate or deltoid, scarious, somewhat lacerate, with slightly involute margins and a brown, finely linear midrib projected as a hair-like terminal awn. $n = 9$.

Open flats and hillsides, on bare spots in adobe soils, Sonoran Zones; Shasta County, California, south through the interior valleys and foothills to the islands and cismontane region of southern California and northern Lower California. Type locality: Byron Springs, Contra Costa County, California. April–June.

153. AGÓSERIS* Raf. Fl. Ludov. 58. 1817.

Scapose perennials, biennials, or annuals, the latter sometimes short-caulescent, from a taproot surmounted by a more or less well-developed caudex. Leaves in a basal rosette, linear to narrowly ovate or oblanceolate, entire to laciniate, glabrous and glaucous to tomentose. Scapes 1 to several, usually longer than the leaves, glabrous or sparsely floccose-pubescent, frequently villous at the summit. Heads solitary, erect, from 1–6 cm. in diameter, narrowly campanulate to hemispheric; phyllaries imbricated in 2–4 series. Florets yellow or burnt orange, frequently drying purplish; ligules shorter than to well exceeding the phyllaries. Achenes fusiform or linear, terete or angular, usually more or less conspicuously 10-nerved, distinctly beaked (with rare exceptions in *A. glauca*) when mature. Pappus about twice the length of the achene, of numerous barbellulate, capillary bristles, silky-white to sordid. [Name Greek, *aix*, meaning goat, and *seris*, chicory.]

A genus of 7 species in North America, primarily western, and one in southern South America. Type species, *Troximon glaucum* Pursh.

Plants perennial or biennial.
    Beak of mature achenes less than twice as long as the body; nonmaritime plants.
        Flowers yellow, sometimes drying pinkish.
            Beaks commonly shorter than the body; ligules 7–19 mm. long; widespread species.
                        1. *A. glauca.*
            Beaks usually much longer than the body; ligules 5–8 mm. long; rare plants of the central Sierra Nevada and central Cascade Mountains at intermediate to low elevations. 2. *A. elata.*
        Flowers burnt orange or pinkish, drying purplish.
            Trichomes of head eglandular; beaks mostly 6–8 mm. long; pappus mostly 10–12 mm. long; common, widespread.
                        3. *A. aurantiaca.*
            Some trichomes of head gland-tipped; beaks commonly 8–10 mm. long; pappus 12–14 mm. long; this color form occasional in the central Sierra Nevada. 2. *A. elata.*
    Beak of mature achenes more than twice as long as the body, or sometimes shorter in maritime plants.
        Heads relatively large, rarely less than 25 mm. high in fruit; beaks 11–25 (commonly 14–21) mm. long; pappus 8–18 (mostly 10–17) mm. long; coastal hills and inland.
            Achenes abruptly beaked from a truncate apex; leaves uniformly retrorsely lobed; ligules 6–14 mm. long; pappus 14–18 mm. long. 4. *A. retrorsa.*
            Achenes tapering gradually into the beak; leaves irregularly lobed; ligules 3–6 mm. long (rarely longer); pappus 8–12 (rarely –14) mm. long. 5. *A. grandiflora.*
        Heads relatively small, less than 25 mm. high in fruit; beaks 4–9 (rarely –12) mm. long; pappus 5–9 mm. long; plants of maritime and coastal hills. 6. *A. apargioides.*
Plants annual; purplish-segmented and gland-tipped trichomes common on involucre. 7. *A. heterophylla.*

1. Agoseris glaùca (Pursh) Raf. Short-beaked Agoseris. Fig. 5973.

Troximon glaucum Pursh, Fl. Amer. Sept. 2: 505. 1814.
Agoseris glauca Raf. Herb. Raf. 39. 1833.
Macrorhynchus glaucus D. C. Eaton, Bot. King Expl. 204. 1871.

---

* Text contributed by Quentin Jones.

Agoseris longissima Greene, Leaflets Bot. Obs. **2**: 122. 1911.
Agoseris isomeris Greene, op. cit. 123.
Agoseris longula Greene, op. cit. 125.

Slender perennials, glabrate, mostly 20–45 cm. tall, from a long heavy taproot. Leaves numerous, linear-lanceolate to narrowly oblanceolate, acuminate, entire or rarely somewhat denticulate, glaucous and glabrous or sparsely puberulent on the margins and midrib below, ascending and falcate to erect, usually much shorter than the 1 to several scapes; scapes slender, glabrous; heads usually narrowly campanulate, florets yellow, in fruit 21–22 mm. high, 15–20(–25) mm. wide; phyllaries lanceolate, tapering gradually from the base, frequently purple-spotted, glabrous, the outer series noticeably shorter; ligules mostly 12–15 mm. long; anther-tubes 3.5–5.2 (mostly 4.0–4.8) mm. long; achenes mostly 6–7 mm. long, the beak 1.5–6.0 mm. long; pappus commonly 10–12 mm. long.

Prairies and lower mountain meadows, Arid Transition Zone; east of the Cascade Mountains in Washington and Oregon, in northeastern California, and eastward beyond our range. Type locality: "On the banks of the Missouri." July–Aug.

Agoseris glauca var. **dasycéphala** (Torr. & Gray) Jepson, Man. Fl. Pl. Calif. 1005. 1925. (*Ammogeton scorzoneraefolius* Schrad. Ind. Sem. Hort. Gotting. 1. 1833; *Troximon glaucum β dasycephalum* Torr. & Gray, Fl. N. Amer. 2: 490. 1843; *Macrorhynchus glaucus* var. *dasycephalus* D. C. Eaton, Bot. King Expl. 205. 1871; *Agoseris scorzoneraefolia* Greene, Pittonia 2: 177. 1891; *A. leontodon aspera* Rydb. Mem. N.Y. Bot. Gard. 1: 457. 1900; *A. villosa* Rydb. op. cit. 458; *Troximon glaucum asperum* Piper, Mazama 2: 96. 1901; *Agoseris glauca scorzoneraefolia* Piper, Contr. U.S. Nat. Herb. 11: 542. 1906; *A. glauca aspera* Piper, loc. cit.; *Troximon villosum* A. Nels. in Coult. & Nels. New Man. Bot. Rocky Mts. 598. 1909; *Agoseris aspera* Rydb. Fl. Rocky Mts. 1030. 1917; *A. scorzoneraefolia aspera* Blake, Contr. U.S. Nat. Herb. 25: 630. 1925; *A. glauca* var. *villosa* Wittr. Publ. Puget Sound Biol. Sta. 6: 253. 1928; *A. glauca* var. *aspera* Cronquist, Leaflets West. Bot. 6: 41. 1950.) Mostly 10–20 cm. tall; leaves narrowly oblanceolate to ovate or spatulate, tips commonly obtuse, usually entire, sometimes weakly laciniate below, rarely glabrous, usually evenly short-pubescent to occasionally villous, ascending; scapes evenly puberulent, becoming villous beneath the head; head broadly campanulate to hemispheric, about as broad as long in fruit; phyllaries in 3 or 4 evenly graduated series, villous and ciliate with translucent and frequently glandular trichomes; ligules mostly 13–15 mm. long; anther-tubes 4.5–5.7 mm. long; achenes mostly 7–9 mm. long; beak 1–3 mm. long; pappus mostly 14–15 mm. long. Near and above timberline in the Olympic and Cascade Mountains of Washington and in the Rocky Mountains from Colorado into Alberta and British Columbia. Type locality: "Saskatchewan and prairies of the Rocky Mountains to the Arctic Coast." Collected by Richardson. July–Sept.

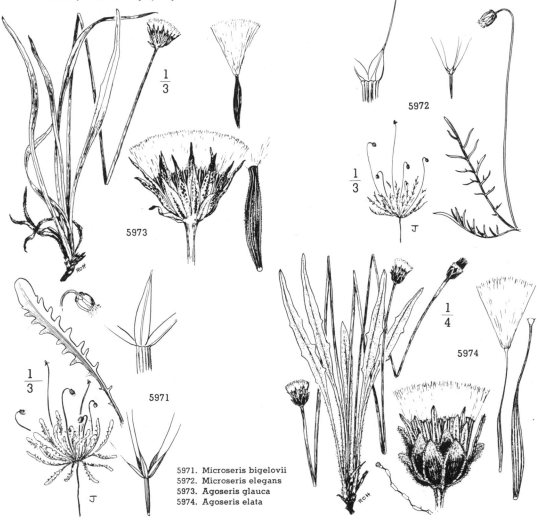

5971. Microseris bigelovii
5972. Microseris elegans
5973. Agoseris glauca
5974. Agoseris elata

Agoseris glauca var. agréstis (Osterh.) Q. Jones ex Cronquist, Vasc. Pl. Pacif. Northw. 5: 26. 1955. (*Agoseris agrestis* Osterh. Bull. Torrey Club 28: 645. 1901.) Robust, mostly 25–50 cm. tall, from a relatively small caudex; leaves lanceolate to narrowly oblanceolate, acute, laciniate or merely dentate or occasionally entire, commonly glaucous and glabrous, rarely sparsely ciliate, ascending to erect; scapes about twice the leaf length, scattered-pubescent, becoming villous beneath the head; heads large, broadly campanulate, 25–30 mm. high, 20–26 mm. wide in fruit; phyllaries usually in 3 series, the outer noticeably shorter, usually with a darker mid-stripe that fades marginally to a light pink or lavender, commonly only ciliate, occasionally slightly villous; florets drying yellow except for the purplish veins; ligules mostly 13–16 mm. long; anther-tubes 5–6 mm. long; achenes 7.0–8.5 mm. long; beaks 2.5–4.0 mm. long; pappus mostly 14–16 mm. long. Brush and woodlands or frequently in weedy situations, at low to intermediate elevations in eastern Washington eastward beyond our range from southern Colorado to Alberta and Minnesota. Type locality: Estes Park, Larimer County, Colorado. Collected by Osterhout. July–Sept.

Agoseris glauca var. laciniàta (D. C. Eaton) Smiley, Univ. Calif. Pub. Bot. 9: 404. 1921. (*Troximon taraxacifolium* Nutt. Trans. Amer. Phil. Soc. II. 7: 434. 1841; *T. parviflorum* Nutt. loc. cit.; *Agoseris parviflora* D. Dietr. Syn. Pl. 4: 1932. 1847; *A. taraxacifolia* D. Dietr. loc. cit.; *Macrorrhynchus glaucus* var. *laciniatus* D. C. Eaton, Bot. King Expl. 204. 1871; *Troximon glaucum* var. *taraxacifolium* A. Gray, Bot. Calif. 1: 437. 1876; *T. glaucum* var. *laciniatum* A. Gray, loc. cit.; *T. glaucum* var. *parviflorum* A. Gray, Proc. Amer. Acad. 19: 71. 1883; *Agoseris dens leonis* Greene, Erythea 3: 23. 1895; *A. glauca parviflora* Rydb. Contr. U.S. Nat. Herb. 3: 511. 1896; *A. tomentosa* Howell, Fl. N.W. Amer. 1: 401. 1901.) Low, mostly 10–25 cm. tall, from a weakly developed caudex; leaves lanceolate, nearly always laciniate except for the long-acuminate, entire tip, the lobes usually retrorse; petioles and leaf-bases arachnoid, scattered-pubescent or occasionally tomentose distally, midrib whitish, conspicuous; commonly a single scape, well exceeding the leaves, scattered-pubescent to more or less villous beneath the head; heads large for the plant, narrowly campanulate, 27–35 mm. high, 10–18 mm. wide in fruit; phyllaries in 3 series, the outer somewhat shorter, the two inner subequal, outer phyllaries lanceolate, usually rather abruptly acuminate, ciliate below, rarely for most of the length, trichomes always white-opaque, never glaudular, all of the phyllaries with a dark median line, fading to the hyaline, straw-colored margins; florets drying yellow with purplish veins or sometimes quite purplish; ligules commonly 11–14 mm. long; anther-tubes mostly 4–5 mm. long; achenes mostly 7.5–10.0 mm. long; beaks 4.5–9.0 (mostly 5–7) mm. long; pappus mostly 14–17 mm. long. Widespread in the mountains bordering the Great Basin region, at lower elevations. Type locality: Rocky Mountains, Colorado. Collected by Hall and Harbour. May–July.

Agoseris glauca var. montícola (Greene) Q. Jones ex Cronquist, Vasc. Pl. Pacif. Northw. 5: 26. 1955. (*Agoseris monticola* Greene, Pittonia 4: 37. 1899; *A. covillei* Greene, Leaflets Bot. Obs. 2: 130. 1911; *A. decumbens* Greene, loc. cit.) Low, mostly 7–22 cm. tall, from a long taproot and well-developed caudex; leaves narrowly lanceolate, oblanceolate, or elliptic, entire to laciniate (high altitude forms), short-acuminate or obtuse, glabrous and glaucous or sometimes evenly puberulent, never arachnoid below; scapes lanate at base, glandular-puberulent beneath the head; head broadly campanulate, variable in size, 18–30 mm. high, 10–25 mm. wide in fruit; phyllaries in 3 length-classes, the outer series conspicuously shorter, triangular to ovate, not abruptly tapered; florets drying yellow with purplish veins; ligules mostly 7–9 mm. long; anther-tubes 4–5 mm. long; achenes 6.0–9.5 mm. long; beak up to 4.5 mm. long or sometimes essentially absent; pappus 10–15 (mostly 10–11) mm. long. At high elevations in the Sierra Nevada and the Warner and Klamath Mountains, extending northward and eastward through the southern Cascade, Steen, and Blue Mountains of Oregon and on Mount Adams, Washington. Type locality: Mount Shasta, Siskiyou County, California. Collected by Merriam. July–Sept.

2. Agoseris elàta (Nutt.) Greene. Tall Agoseris. Fig. 5974.

Stylopappus elatus Nutt. Trans. Amer. Phil. Soc. II. 7: 433. 1841.
Macrorynchus elatus Torr. & Gray, Fl. N. Amer. 2: 492. 1843.
Troximon nuttallii A. Gray, Proc. Amer. Acad. 9: 216. 1874.
Agoseris elata Greene, Pittonia 2: 177. 1891.
Troximon elatum A. Nels. in Coult. & Nels. New Man. Bot. Rocky Mts. 599. 1909.

Slender to stout perennials, mostly 3.0–6.5 dm. tall, from a relatively slender taproot. Leaves narrowly lanceolate to broadly oblanceolate, entire to pinnatifid, glabrous or rarely sparsely ciliate; scapes usually slender, glabrous to floccose-pubescent, becoming lanulose beneath the head; heads broadly campanulate, 3.1–3.8 cm. high in fruit; phyllaries imbricated in 3–4 length-classes, the outer pubescent on the abaxial surface and ciliate with translucent, gland-tipped trichomes; florets yellow or burnt orange (Sierra Nevada), drying yellow or pinkish; ligules 5–8 mm. long; anther-tubes 2.5–4.5 mm. long; achenes mostly 7.5–9.0 mm. long, usually somewhat arcuate, tapering gradually into the 7.5–15.0 mm. long beak; pappus mostly 12–14 mm. long.

At intermediate elevations, Transition and Canadian Zones; lower elevations in the Cascade Mountains of southern Washington and northern Oregon, spreading to low-lying prairies in the Willamette and Puget Sound basins, and the Sierra Nevada, California, from Placer County to Tulare County. Type locality: "Plains of the Wahlamet, near its estuary." Collected by Nuttall. June–July.

3. Agoseris aurantìaca (Hook.) Greene. Orange-flowered Agoseris. Fig. 5975.

Troximon aurantiacum Hook. Fl. Bor. Amer. 1: 300. *pl. 104*. 1833.
Macrorhynchus aurantiacus Fisch. & Mey. Ind. Sem. Hort. Petrop. 3: 40. 1837.
Macrorhynchus troximoides Torr. & Gray, Fl. N. Amer. 2: 491. 1843.
Troximon gracilens A. Gray, Proc. Amer. Acad. 19: 71. 1883.
Troximon gracilens var. *greenei* A. Gray, loc. cit.
Agoseris gracilenta Greene, Pittonia 2: 177. 1891.
Agoseris gracilenta var. *greenei* Greene, loc. cit.
Agoseris aurantiaca Greene, loc. cit.
Agoseris gracilens Kuntze, Rev. Gen. Pl. 1: 304. 1891.
Agoseris greenei Rydb. Mem. N.Y. Bot. Gard. 1: 459. 1900.
Agoseris angustissima Greene, Leaflets Bot. Obs. 2: 129. 1911.
Agoseris vulcanica Greene, loc. cit.
Agoseris prionophylla Greene, op. cit. 131.
Agoseris howellii Greene, loc. cit.
Agoseris gracilens greenei Blake, Contr. U.S. Nat. Herb. 25: 629. 1925.
Agoseris gracilens var. *greenei* Jepson, Man. Fl. Pl. Calif. 1007. 1925.
Agoseris subalpina G. N. Jones, Univ. Wash. Publ. Biol. 5: 262. 1936.

Perennials from a usually well-developed, branching caudex, mostly 1.8–4.0 dm. tall. Leaves extremely variable in shape and cutting, narrowly linear-lanceolate to broadly oblanceolate, entire to laciniate, most commonly glabrate with a sparse pubescence along the midrib below, margins frequently ciliolate toward the base, the petioles usually anthocyanous; scapes floccose-pubescent, becoming densely lanate beneath the head, rarely sparsely so; heads turbinate to narrowly campanulate; phyllaries in about 3 series, slightly or not at all imbricate, narrow, acuminate, the outer series ciliate, glabrous (usually) to villous on the surface; trichomes eglandular, usually white-opaque, sometimes translucent; florets burnt orange to pinkish, drying light to dark purplish; ligules mostly 6–9 mm. long; anther-tubes 1.5–4.0 mm. long; achenes mostly 6–9 mm. long, commonly hispid distally, tapering gradually into the beak or tumid on one side distally and abruptly tapering; beak mostly 6–8(3–11) mm. long; pappus silky-white or sordid, 9–14 mm. long.

At medium to high elevations in the mountains of western North America, mostly Boreal Zone; Skagway, Alaska, south to Tulare County, California, and southern Colorado, with disjuncts in Quebec. Type locality: "Alpine Prairies of the Rocky Mountains." Collected by Drummond. June–Aug. Only the typical variety, described above, occurs in our range.

4. Agoseris retrórsa (Benth.) Greene. Spear-leaved Agoseris. Fig. 5976.

Macrorhynchus retrorsum Benth. Pl. Hartw. 320. 1849.
Macrorhynchus angustifolium Kell. Proc. Calif. Acad. **5**: 47. 1873.
Troximon retrorsum A. Gray, Proc. Amer. Acad. **9**: 216. 1874.
Agoseris retrorsa Greene, Pittonia **2**: 178. 1891.

Stout perennials, mostly 2.5–4.5 dm. tall, from a well-developed taproot. Leaves numerous, 1.2–3.5 dm. long, lanceolate, uniformly laciniate except for the long-acuminate, entire tip, the lobes more or less retrorse, all callous-tipped, usually canescent toward the base; 1–2 numerous stout scapes, glabrous or more commonly floccose-pubescent, canescent at base of involucre; heads uniformly campanulate, 3–6 cm. high in fruit; phyllaries in about 3 series, the outer strikingly shorter in fruit, the inner about equaling the pappus; florets yellow, becoming orange after anthesis, frequently drying pinkish; ligules usually 6–10 mm. long; anther-tubes 2.5–4.0 mm. long, dark yellow; achenes 5–7 mm. long, more or less truncate at the apex; beak smooth, 15–25 (mostly 18–21) mm. long; pappus 14–18 mm. long.

At low altitudes, upper Sonoran Zone; the Cascade and Blue Mountains of Oregon and Washington, south to the Coast Ranges and the Sierra Nevada, California, to the Angeles Ranges of southern California. Type locality: "In montibus Sacramento," California. Collected by Hartweg. April–July.

5. Agoseris grandiflòra (Nutt.) Greene. Large-flowered Agoseris. Fig. 5977.

Stylopappus grandiflorus Nutt. Trans. Amer. Phil. Soc. II. **7**: 432. 1841.
Macrorhynchus grandiflorus Torr. & Gray, Fl. N. Amer. **2**: 492. 1843.
Troximon grandiflorum A. Gray, Proc. Amer. Acad. **9**: 216. 1874.
Troximon marshallii Greene, Pittonia **1**: 174. 1888.
Troximon plebeium Greene, op. cit. **2**: 79. 1890.
Agoseris grandiflora Greene, op. cit. 178. 1891.
Agoseris marshallii Greene, loc. cit.
Agoseris plebeia Greene, loc. cit.
Agoseris intermedia Greene, Erythea **1**: 175. 1893.
Troximon grandiflorum var. *obtusifolium* Suksd. Deutsch. Bot. Monatss. **18**: 98. 1900.
Agoseris grandiflora var. *intermedia* Jepson, Fl. W. Mid. Calif. 500. 1901.
Agoseris cinerea Greene, Leaflets Bot. Obs. **2**: 132. 1911.
Agoseris obtusifolia Rydb. Bull. Torrey Club **38**: 20. 1911.
Agoseris grandiflora var. *plebeia* Wittr. Publ. Puget Sound Biol. Sta. **6**: 253. 1928.

Perennials, mostly 2.5–6.0 dm. tall, from a multicipital caudex. Leaves polymorphic, ranging from linear and subentire to broadly spatulate and laciniate, the lobes irregularly disposed, glabrous to canescent, usually at least pubescent on the midribs; scapes coarse, usually conspicuously

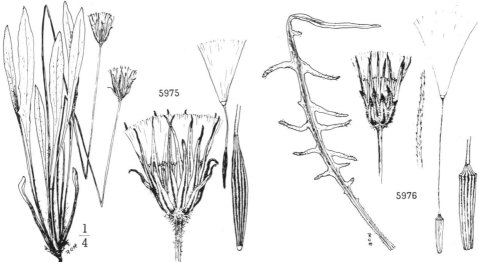

$\frac{1}{4}$

5975

5976

5975. Agoseris aurantiaca

5976. Agoseris retrorsa

ribbed distally, tomentose below, glabrous or scattered-pubescent above, usually canescent beneath the head; heads large, broadly campanulate to hemispheric, 2.2–4.0 cm. high, 2.3–6.0 cm. wide in fruit; phyllaries in 4–5 series, the outer conspicuously ciliate with white-opaque, never glandular or translucent, trichomes; florets yellow, drying purplish, shorter than the involucre; ligules 3–6 mm. long; anther-tubes 1.0–2.5 mm. long; achenes spindle-shaped, tapering gradually into the long filiform beak; pappus 8–12 (rarely –14) mm. long.

Grassy areas and open timber in the lowlands and lower elevations in the mountains, Upper Sonoran and Transition Zones; Vancouver Island, British Columbia, to southern California. Type locality: "High plains of the Wahlamet." Collected by Nuttall. May–July.

6. Agoseris apargioìdes (Less.) Greene. Seaside Agoseris. Fig. 5978.

Troximon apargioides Less. Linnaea **6**: 501. 1831.
Leontodon hirsutum Hook. Fl. Bor. Amer. **1**: 296. 1833.
Borkhausia lessingii Hook. & Arn. Bot. Beechey 145. 1833.
Macrorhynchus lessingii Hook. & Arn. op. cit. 361. 1840.
Taraxacum hirsutum Torr. & Gray, Fl. N. Amer. **2**: 494. 1843.
Macrorhynchus humilis Benth. Pl. Hartw. 320. 1849.
Macrorhynchus harfordii Kell. Proc. Calif. Acad. **5**: 47. 1873.
Troximon humile A. Gray, Proc. Amer. Acad. **19**: 72. 1883.
Agoseris apargioides Greene, Pittonia **2**: 177. 1891.
Agoseris hirsuta Greene, loc. cit.
Agoseris humilis Kuntze, Rev. Gen. Pl. **1**: 304. 1891.

Perennials, mostly 1.2–3.5 dm. tall, with a long slender taproot. Leaves mostly more than 9 cm. long, ascending, polymorphic, linear, lanceolate or narrowly oblanceolate, usually laciniate, the lobes uniform, rarely entire or merely dentate, glabrous or sparsely puberulent to tomentose below and on the clasping petioles; scapes usually several, well exceeding the leaves, glabrous or scattered-pubescent, usually villous beneath the head; heads broadly campanulate to hemispheric, in fruit 1.3–2.3 cm. high, 1.1–2.0 cm. broad; phyllaries in 3 or 4 series, those of the outer 2 or 3 series usually obtuse, commonly ciliate and rather densely villous on the outer surface, trichomes in part glandular and translucent or frequently with scattered, glandular or white-opaque trichomes; florets yellow, frequently drying purplish; ligules 3.5–6.0 mm. long; anther-tubes 1.5–2.5 mm. long; achenes mostly 3.5–5.0 mm. long; beak fine, glabrous or hispidulous, mostly 4–9 mm. long; pappus plainly barbellulate, 5–9 mm. long.

Grassy hillsides, Humid Transition Zone; near the coast from Humboldt County south to Santa Barbara County, California. Type locality: California. Collected by Chamisso. May–June.

Agoseris apargioides subsp. **marítima** (Shelton) Q. Jones ex Cronquist, Vasc. Pl. Pacif. Northw. **5**: 24. 1955. (*Agoseris maritima* Sheldon, Bull. Torrey Club **30**: 310. 1903.) Leaves mostly 4–10 cm. long, usually reclined, broadly oblanceolate or spatulate, the tips obtuse, irregularly and remotely lobed, or more commonly merely dentate below; trichomes of head eglandular and white-opaque; anther-tubes 1.5–2.5 mm. long; achenes mostly over 4.5 mm. long. Coastal sand dunes, beaches, and maritime bluffs, from Clallam County, Washington, south to Monterey County, California. Type locality: Clatsop Beach, Clatsop County, Oregon. Collected by Sheldon. May–Aug.

Agoseris apargioides var. **eastwoódiae** (Fedde) Munz, Aliso **4**: 100. 1958. (*Agoseris eastwoodiae* Fedde, Bot. Jahresb. 31[1]: 808. 1904.) A related form which differs from the preceding in having translucent trichomes on the heads; outer ligules 8–16 mm. long; anther-tubes 3.5–4.5 mm. long; achenes mostly less than 4.5 mm. long. It occurs commonly from Humboldt County, California, to Monterey County.

7. Agoseris heterophýlla (Nutt.) Greene. Annual Agoseris. Fig. 5979.

Macrorhynchus heterophylla Nutt. Trans. Amer. Phil. Sec. II. **7**: 430. 1841.
Kymapleura heterophylla Nutt. loc. cit., in errata.
Cryptopleura californica Nutt. op. cit. 431.
Macrorhynchus californicus Torr. & Gray, Fl. N. Amer. **2**: 493. 1843.
Troximon heterophyllum Greene, Bull. Torrey Club **10**: 88. 1883.
Troximon heterophyllum var. *cryptopleura* Greene, loc. cit.
Troximon heterophyllum var. *kymapleura* Greene, loc. cit.
Agoseris heterophylla Greene, Pittonia **2**: 178. 1891.
Agoseris heterophylla var. *cryptopleura* Greene, op. cit. 179.
Agoseris heterophylla var. *kymapleura* Greene, loc. cit.
Agoseris major Jepson ex Greene, loc. cit.
Troximon heterophyllum var. *cryptopleuroides* Suksd. Deutsch. Bot. Monatss. **18**: 98. 1900.
Troximon heterophyllum var. *glabratum* Suksd. loc. cit.
Agoseris heterophylla var. *glabra* Howell, Fl. N.W. Amer. **1**: 402. 1901.
Agoseris heterophylla subsp. *normalis* Piper, Contr. U.S. Nat. Herb. **11**: 544. 1906.
Agoseris heterophylla californica Piper, loc. cit.
Agoseris heterophylla glabrata Piper, loc. cit.
Troximon heterophyllum f. *kymapleurum* H. M. Hall, Univ. Calif. Pub. Bot. **3**: 278. 1907.
Troximon heterophyllum f. *cryptopleurum* H. M. Hall, loc. cit.
Troximon heterophyllum var. *californicum* f. *crenulatum* H. M. Hall, op. cit. 279.
Troximon heterophyllum var. *californicum* f. *turgidum* H. M. Hall, loc. cit.
Agoseris heterophylla var. *californica* Jepson, Man. Fl. Pl. Calif. 1007. 1925.
Agoseris heterophylla var. *crenulata* Jepson, loc. cit.
Agoseris heterophylla var. *turgida* Jepson, loc. cit.

Slender erect annuals, 3–45 (mostly 6–20) cm. tall, sometimes weakly caulescent. Leaves ascending, usually oblanceolate, denticulate, occasionally linear or lanceolate, entire or laciniate; scapes frequently several, glabrous or with a sparse pubescence, becoming glandular-pubescent beneath the head; heads narrowly campanulate to hemispheric, in fruit mostly 1.4–1.7 cm. high,

1.7–2.8 cm. in diameter; phyllaries commonly in 2 length-classes, more or less villous with septate twisted trichomes, some of these purplish and gland-tipped; florets yellow, frequently pinkish on drying; ligules 2.5–6.0 mm. or (plants from central California) 9–15 mm. long; anther-tubes 1.0–1.7 mm. or (plants from central California) 2.0–3.9 mm. long; outer achenes polymorphic, usually spindle-shaped, acutely 10-ribbed, tawny, sometimes the ribs abnormally developed, straight or sinuate, glabrous or hispid, whitish to purplish, or the pericarp inflated to the obliteration of the ribs; beak capillary, mostly 6–10 mm. long; pappus 4–7 mm. long.

Dry, open, frequently weedy situations, Upper Sonoran Zone; mostly at lower elevations from British Columbia to northern Lower California. Type locality: "The plains of Oregon." Collected by Nuttall. April–July. Only the typical subspecies, described above, occurs in our range.

154. **PHALACRÓSERIS** A. Gray, Proc. Amer. Acad. 7: 364. 1868.

Glabrous perennial herbs with a tuft of basal leaves and 1 to several 1-headed scapes. Involucre campanulate, of 12–16 equal, narrowly lanceolate phyllaries. Flowers white. Achenes short-oblong, obscurely 4-angled, slightly incurved. Pappus none. [Name from the Greek word meaning bald-headed, and *seris*, the Greek name for chicory.]

A monotypic genus.

5977. Agoseris grandiflora
5978. Agoseris apargioides
5979. Agoseris heterophylla
5980. Phalacroseris bolanderi

1. Phalacroseris bolánderi A. Gray. Bolander's Dandelion. Fig. 5980.

Phalacroseris bolanderi A. Gray, Proc. Amer. Acad. **7**: 364. 1868.

Acaulescent with a stout perennial root, glabrous throughout. Leaves narrowly lanceolate, linear-lanceolate, or some linear-oblanceolate, borne on the simple or branched, woody root-crown, slightly succulent, 8–20 cm. long; scape 10–25 cm. long, 1-flowered; involucre about 1 cm. high, 5–6 mm. in diameter; phyllaries linear-lanceolate, 6–7 mm. long; ligules yellow, well exserted; achenes somewhat 4-angled, truncate at both ends, about 3 mm. long.

Wet meadows, Arid Transition and Canadian Zones; western slopes of the Sierra Nevada in Mariposa and Madera Counties. Type locality: "Westfall's Meadows, above Yosemite Valley, alt. 8,000 feet." June–Aug.

Phalacroseris bolanderi var. **coronàta** H. M. Hall, Bot. Gaz. **31**: 393. 1901. Closely resembling the typical species in general habit but achenes with a short crown. Western slopes of the Sierra Nevada from Mariposa County to Fresno County, California. Type locality: Pine Ridge, Sierra Nevada, alt. 5,400 feet, Fresno County.

155. CHAETADÉLPHA A. Gray, Proc. Amer. Acad. **9**: 218. 1874.

Perennial herbs with a deep-seated rootstock, many-branched from the base and dichotomously branching above. Leaves alternate, linear or linear-lanceolate. Heads 5-flowered, solitary at the ends of the branchlets. Phyllaries 5, keeled at the base, membranous, subtended by 5 calyculate bractlets. Receptacle naked. Achenes 5-angled. Pappus persistent, the bristles 5, rigid, branched from the base and also about one-half their length into slender divisions, the ultimate portion aristate and surpassing the involucre. [From the Greek words meaning bristle and sister.]

A monotypic genus of the western part of the Great Basin.

1. Chaetadelpha weèleri A. Gray. Chaetadelpha. Fig. 5981.

Chaetadelpha wheeleri A. Gray, Proc. Amer. Acad. **9**: 218. 1874.

Glabrous perennials, freely branching above, 1–3 dm. high. Leaves glaucous, 2–5 cm. long, thickish; phyllaries 5–15 mm. high, the subtending calyculus of 5 bractlets 2–3 mm. long; flowers sordid-white or pink; achenes 8–10 mm. long, the 5 pappus-bristles brownish, as long as or longer than the achene.

Desert valley bottoms, Upper Sonoran Zone; southern Malheur County, Oregon, south through western Nevada and adjacent Inyo County, California. Type locality: "Southern Nevada." Collected by Wheeler. June–July.

156. LYGODÉSMIA D. Don, Edinb. New Phil. Journ. **1829**: 311. 1829

Perennial or annual, glabrous or pubescent herbs, branched, and in 1 species spinescent. Leaves linear and entire or the lower and basal ones broader and sometimes pinnatifid, the uppermost sometimes reduced to scales. Flowers pink or purple, solitary, and erect at the ends of the branches or sometimes racemose. Involucre cylindric, the phyllaries 5–8, equal, linear, scarious-margined, slightly united at base and subtended by several calyculate bractlets. Receptacle flat, naked. Ligules truncate and 5-toothed at the apex. Anthers sagittate at the base. Style-branches slender. Achenes beakless, few-ribbed, cylindric or nearly so or angled and linear, the pappus of capillary bristles. [Name from Greek words meaning pliant twig and bundle, from the numerous branches.]

A genus of about 6 species, natives of western and southern North America. Type species, *Prenanthes juncea* Pursh.

Perennial, with rigid spine-tipped branches; pappus-bristles tawny, 6–10 mm. long.   1. *L. spinosa.*
Annual, the very slender branches not spine-tipped; pappus-bristles white, 2–2.5 mm. long.   2. *L. exigua.*

1. Lygodesmia spinòsa Nutt. Thorny Skeleton-plant or Spiny Lygodesmia. Fig. 5982.

Lygodesmia spinosa Nutt. Trans. Amer. Phil. Soc. II. **7**: 444. 1841.

Stems several from a branching woody root-crown, with conspicuous tufts of wool at base, 2–5 dm. high, intricately branched into rigid, spine-tipped, pale green branchlets. Leaves few, the lower broadly linear, often 3–4 cm. long, the upper becoming reduced and bract-like; heads subsessile or often terminating short rigid branchlets; involucres 7–9 mm. high, 3–5-flowered; ligules rose-colored, well exserted; achenes 4–5 mm. long, 4–5-ribbed and -angled, slightly narrowing toward the apex; pappus tawny, 6 mm. long, short-plumose.

Dry plains and arid desert slopes, mainly Upper Sonoran Zone; Montana and British Columbia south to northern Arizona; occurring in the Pacific States in Lake, Harney, and Malheur Counties, Oregon, and Modoc County, California, south to east of the Sierra Nevada to the San Gabriel (Swartout Canyon) and San Bernardino (Baldwin Lake) Mountains. Type locality: "In the Rocky Mountain plains towards California." June–Oct.

2. Lygodesmia exígua A. Gray. Annual Lygodesmia or Egbertia. Fig. 5983.

Prenanthes exigua A. Gray, Smiths. Contr. **5**[6]: 105. 1853.
Lygodesmia exigua A. Gray, Proc. Amer. Acad. **9**: 217. 1874.
Stephanomeria minima M. E. Jones, Contr. West. Bot. No. 17: 31. 1930.

Slender, diffusely much-branched annual, glabrous or the stems sparsely glandular-puberulent below, the branchlets slender and divergent. Lower leaves oblanceolate or spatulate, 5–20 mm. long, toothed or lobed, those of the branchlets reduced to often minute bracts; heads terminating very slender branchlets, subtended by a minute, bract-like leaflet; phyllaries usually 4, erect, 3 mm. long; flowers 3–4; ligules about 5 mm. long, rather deeply lobed, pale pink or whitish; achenes 3–4 mm. long, pale, 5-angled, the pappus-bristles rather rigid, bright white.

Rocky ledges and desert flats, Sonoran Zones; Mojave and Colorado Deserts, southern California east to Colorado and Texas. Type locality: El Paso, Texas. April–May.

Lygodesmia júncea (Pursh) D. Don, Edinb. New Phil. Journ. **1829**: 311. 1829. (*Prenanthes juncea* Pursh, Fl. Amer. Sept. 498. 1814.) A fastigiately much-branched perennial from a deep-seated, creeping root and bearing no basal tufts of wool. This may be expected in eastern Washington, as it occurs in British Columbia and Idaho and Montana, but no specimens have been seen.

157. **STEPHANOMÈRIA** Nutt. Trans. Amer. Phil. Soc. II. **7** : 427. 1841. Nomen conservandum.

Annual or perennial, mostly glabrous and often glaucous herbs, sometimes woody at the base, with erect, simple or branched stems. Leaves alternate or basal, entire or runcinate-pinnatifid, those of the branches often small and scale-like. Heads subsessile or pedunculate in panicles or solitary at the ends of the branches. Involucre cylindric or oblong, the phyl-

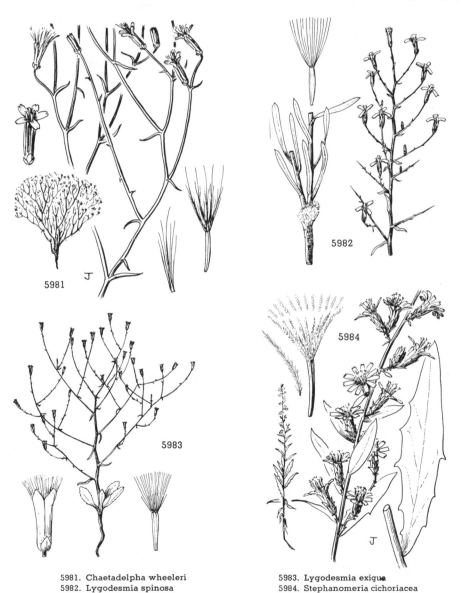

5981. Chaetadelpha wheeleri
5982. Lygodesmia spinosa

5983. Lygodesmia exigua
5984. Stephanomeria cichoriacea

laries few, mostly in 1 series, equal, scarious-margined, slightly united at base, subtended by several calyculate bractlets, thin and withering in age. Flowers pink, flesh-colored or whitish, opening in the morning. Receptacle flat, naked or hirsutulous. Anthers sagittate at base. Style-branches slender. Achenes oblong or columnar, sometimes slightly curved, truncate at the apex, 5-ribbed or -angled, the intervals between the ribs more or less grooved and often tuberculate. Pappus of 1 series of early-deciduous plumose bristles, these often expanded below into a paleaceous base and connate in groups. [From the Greek words meaning wreath and division, perhaps referring to the virgate branches.]

A western North American genus of about 15 species. Type species, *Stephanomeria minor* (Hook.) Nutt.

Phyllaries 10–15 mm. long; leaves (3.5–11 cm. long) present on midstems at flowering time.
 Receptacle deeply pitted, hirsute; heads sessile or subsessile on elongated naked branches of the inflorescence.
                            1. *S. cichoriacea.*
 Receptacle naked; heads terminating the branchlets, corymbosely or racemosely arranged.
  Leaves thin, entire or saliently toothed.   2. *S. lactucina.*
  Leaves thick and usually callous-margined, runcinately pinnatifid. 3. *S. parryi.*
Phyllaries 6–9 (rarely 10) mm. long; leaves mostly bract-like on midstems at flowering time or absent.
 Perennials; achenes striate or minutely rugulose.
  Pappus-bristles plumose to the base or nearly so; plants mostly herbaceous at the base, the stems with many ascending branches.   4. *S. tenuifolia.*
  Pappus-bristles hirsutulous for about one-fourth or more of their length and long-plumose above; plants woody below and divaricately branching above.
   Plants glabrous.   5. *S. pauciflora.*
   Plants densely tomentulose.   6. *S. cinerea.*
 Annuals or biennials; mature fertile achenes more or less tuberculate or rugose between the longitudinal ribs.
  Pappus-bristles connate below in groups of 3 or 4, plumose to point of attachment or merely hirsutulous for the lower one-half or one-third; heads pedunculate (shortly so in robust forms of *S. paniculata* and the intermediates between *S. exigua* and *S. virgata*).
   Stems stout; peduncles 0.5–1.5 cm. long; achenes oblong-clavate, a little one-sided, minutely hispidulous, angled but not grooved between the angles.   7. *S. paniculata.*
   Stems slender; peduncles 1–4 cm. long; achenes linear, glabrous, angled, and grooves evident.
                        8. *S. exigua.*
  Pappus-bristles free to the base and deciduous, or breaking just above the base leaving a minute crown, plumose throughout; heads typically subsessile.   9. *S. virgata.*

1. Stephanomeria cichoriàcea A. Gray. Chicory-leaved Stephanomeria. Fig. 5984.

Stephanomeria cichoriacea A. Gray, Proc. Amer. Acad. **6**: 552. 1865.

Perennial from a long deep taproot and somewhat woody, the stems erect, simple or virgately branched, 6–15 dm. high, stout, glaucous, glabrous or commonly more or less woolly-pubescent especially when young. Basal leaves spreading on the ground, runcinate-pinnatifid, spatulate in outline, 8–16 mm. long, narrowed into petioles; stem-leaves sessile and clasping or auricled, the upper smaller, lanceolate or oblong, entire or commonly irregularly and saliently toothed; heads numerous, mostly on short bracteate peduncles along the stems, mostly 10–12-flowered; corolla pink, rarely white; achenes smooth, striate; pappus sordid, plumose throughout, 10–12 mm. long.

Dry washes and slopes, Upper Sonoran and Transition Zones; southern Monterey and San Benito Counties and Tejon Pass south to the southern slopes of the San Bernardino Mountains and the Santa Ana Mountains, and Santa Cruz Island, California. Type locality: Fort Tejon, Tehachapi Mountains. Aug.–Dec. Tejon Milk-aster.

2. Stephanomeria lactucìna A. Gray. Large-flowered Stephanomeria. Fig. 5985.

Stephanomeria lactucina A. Gray, Proc. Amer. Acad. **6**: 552. 1865.
Ptiloria lactucina Greene, Pittonia **2**: 133. 1890.

Perennial with slender, deep-seated, creeping rootstock, the stems arising singly, branching near the base with the slender branches spreading, or strictly erect with ascending branches, 6–40 cm. high, glabrous or sparsely puberulent. Leaves linear, 2.5–8 cm. long, 2–7 mm. wide, more or less attenuate at apex, entire or more or less sparingly runcinate-denticulate; heads terminating the naked or bracteate, slender branches and someyhat corymbose or sometimes somewhat racemose; involucres narrowly campanulate, the inner phyllaries linear-attenuate, about 10–12 mm. long, the few outer about 5 mm. long, about twice the length of the calyculate bractlets, pale green or rose-tinged; flowers usually 7 or 8, the ligules well exserted, rose-colored or lilac-purple; achenes 5–6 mm. long, ribbed, glabrous, light brown; pappus-bristles 9–10 mm. long, about 20, white, plumose throughout, slightly broadened and somewhat adhering in groups of 5 or 6.

Sandy flats and open pine forests, Arid Transition and Canadian Zones; eastern slope of the Cascade Mountains from southwestern Jackson County and Deschutes County, Oregon, southward through the Cascade Mountains and Sierra Nevada to Mariposa County, California; also in the Siskiyou Mountains and Mount Sanhedrin in the North Coast Ranges, California. Type locality: "Dry hill near Big-tree Road in the Sierra Nevada, alt. 6,000 feet." July–Aug.

3. Stephanomeria párryi A. Gray. Parry's Stephanomeria. Fig. 5986.

Stephanomeria parryi A. Gray, Proc. Amer. Acad. **19**: 61. 1883.
Ptiloria parryi Coville, Contr. U.S. Nat. Herb. **4**: 144. 1893.

Perennial, pale glaucous-green and glabrous throughout, the stems rather stout, erect, 15–30 cm. high, widely branching from near the base, the branches ascending or more commonly divaricately spreading. Leaves linear-lanceolate, rather thick and firm, deeply runcinately pinnatifid, toothed, the margin somewhat calloused, glabrous or sparsely pubescent beneath, 2.5–4 cm. long; heads terminating the branches and the short, lateral, bracteate branchlets, 10–14-flowered; inner phyllaries 10–12 mm. long, scarious-margined; corolla pale pink to white, well exserted; pappus-bristles 7–8 mm. long, long-plumose above, barbellate at base and widened, deciduous in groups of 2 or 3; achenes glabrous, with narrow rugulose ribs.

Sandy and gravelly slopes, Upper Sonoran Zone; Inyo and Panamint Mountains, Inyo County, and western Mojave Desert, California, east to northwestern Arizona and southern Utah. Type locality: "near St. George, S. Utah." May–July. Parry's Rock-pink.

4. **Stephanomeria tenuifòlia** (Torr.) H. M. Hall. Narrow-leaved Stephanomeria. Fig. 5987.

Prenanthes ? tenuifolia Torr. Ann. Lyc. N.Y. **2**: 210. 1827.
Ptiloria tenuifolia Raf. Atl. Journ. 145. 1832.
Lygodesmia minor Hook. Fl. Bor. Amer. **1**: 295. *pl. 103, fig. A.* 1833.
Stephanomeria minor Nutt. Trans. Amer. Phil. Soc. II. **7**: 427. 1841.
Ptiloria filifolia Greene, Pittonia **3**: 311. 1898.
Stephanomeria tenuifolia H. M. Hall. Univ. Calif. Pub. Bot. **3**: 256. 1907.

Stems herbaceous, pale green and glabrous, 2–5 dm. long, erect from a perennial root, with several slender, flexuose, ascending branches. Lower leaves narrow, runcinate-pinnatifid, about 5 cm. long, the upper leaves linear, ascending, somewhat shorter; heads terminal, mostly 5-flow-

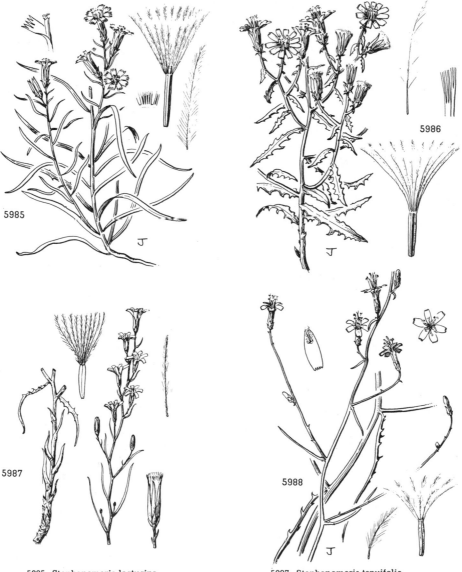

5985. Stephanomeria lactucina
5986. Stephanomeria parryi

5987. Stephanomeria tenuifolia
5988. Stephanomeria pauciflora

ered; involucre 8–10 mm. long, the phyllaries usually 5; achenes 5-angled and grooved, the pappus-bristles 15–25, white or sordid, plumose throughout.

Plains and lower mountain slopes, Upper Sonoran and Transition Zones; southeastern British Columbia south to Modoc County, California; also Montana south to Colorado and northern Arizona. Type locality: "near the Rocky Mountains." Collected by E. P. James. June–Aug.
A form differing from the name-bearing taxon in having divergent branches and shorter stature has been collected in the Lake Tahoe region in Nevada and California and also at Sonora Pass, Tuolumne County, and in Inyo County. The name *Ptiloria divergens* was applied to this entity by Greene in C. F. Baker, West Amer. Pl. [1]: 19. 1902.

Stephanomeria tenuifolia var. **myrioclàda** (D. C. Eaton) Cronquist, Leaflets West Bot. **6**: 48. 1950. (*Stephanomeria myrioclada* D. C. Eaton. Bot. King Expl. **5**: 198. *pl. 20, figs. 1–4.* 1871; *Ptiloria tenuifolia myrioclada* Blake, Contr. U.S. Nat. Herb. **25**: 623. 1925; *Stephanomeria pauciflora* var. *myrioclada* Munz, Man. S. Calif. 589, 601. 1935.) Plants 1–3 dm. high of densely crowded, slender stems arising from a woody root; leaves linear, filiform; involucre 5–6 mm. high; phyllaries 3–5; achenes as in the species. Arid Transition and Canadian Zones in the desert ranges of Nevada and Wyoming westward to the eastern slopes of the Cascade Mountains in Oregon south to the southern Sierra Nevada, California. Type locality: "Thousand Spring and Goose Creek Valleys, [Elko County] Nevada; 6–6,500 feet elevation."

5. Stephanomeria pauciflòra (Torr.) A. Nels. Few-flowered Stephanomeria or Wire Lettuce. Fig. 5988.

Prenanthes (?) pauciflora Torr. Ann. Lyc. N.Y. **2**: 210. 1827.
Ptiloria pauciflora Raf. Atl. Journ. 145. 1832.
Ptiloria divaricata Greene, Erythea **1**: 224. 1893.
Stephanomeria pauciflora A. Nels. in Coult. & Nels. New Man. Bot. Rocky Mts. 588. 1909.
Stephanomeria haleyi Eastw. Leaflets West. Bot. **2**: 55. 1937.

Perennial, woody at the base, the several stems divaricately and intricately branched and more or less rigid, forming rounded, glaucous and glabrous bushes 3–5 dm. high. Lower leaves runcinate-pinnatifid with narrow segments, the upper entire and spreading, with tufts of wool at the base, or reduced to scales, these also sometimes woolly; heads solitary, terminal on the branches and also scattered along the branchlets on short bracteate peduncles, 5-flowered; involucre 7–8(10) mm. high, the flowers 3–5; achenes grooved between the 5 ribs and rather inconspicuously transversely rugulose; pappus somewhat tawny, the bristles rather firm, hispidulous from the base for about one-fourth their length and long-plumose above, deciduous in groups.

Sandy or gravelly desert slopes, Sonoran Zones; western Kern County and the Mojave and Colorado Deserts, California, eastward through Nevada to western Kansas and south to northern Sonora and Texas. Type locality: "near the Rocky Mountains." Collected by E. P. James. April–Oct. Desert Straw.

6. Stephanomeria cinèrea (Blake) Blake. Gray Stephanomeria. Fig. 5989.

Ptiloria cinerea Blake, Proc. Biol. Soc. Wash. **35**: 177. 1922.
Stephanomeria runcinata var. *parishii* Jepson, Man. Fl. Pl. Calif. 998. 1925.
Stephanomeria cinerea Blake, Journ. Wash. Acad. **33**: 272. 1943.
Stephanomeria pauciflora var. *parishii* Munz, Aliso **4**: 100. 1958.

Perennial, woody at base and divaricately much-branched above, densely cinereous-tomentose throughout, 3–4 dm. high. Leaves of the midstem linear-lanceolate, acuminate, runcinate-toothed and reduced to scales above; heads solitary at the tips of the branchlets or on short peduncles, about 5-flowered; phyllaries 5, 7–8 mm. high, subtended by calyculate bractlets; achenes 5-angled, whitish, finely transverse-rugulose when mature; pappus-bristles about 14, somewhat tawny, hispidulous at the extreme base, plumose above and deciduous in a ring.

Gravelly soil, Lower Sonoran Zone; Death Valley region in Inyo County, California, and adjacent Nevada south to Lancaster, Los Angeles County, California, in the Mojave Desert. Type locality: "Pahrump Valley, Nevada, altitude 610–915 meters." June–July.
Much resembling the western desert form of *S. pauciflora* and differing principally in the cinerous tomentum of the stems.

7. Stephanomeria paniculàta Nutt. Stiff-branched Stephanomeria. Fig. 5990.

Stephanomeria paniculata Nutt. Trans. Amer. Phil. Soc. II. **7**: 428. 1841.
Ptiloria paniculata Greene, Pittonia **2**: 132. 1890.
Stephanomeria oregonensis Gandoger, Bull. Soc. Bot. Fr. **65**: 53. 1918.
Stephanomeria suksdorfii Gandoger, loc. cit.

Annual, stems usually simple, rather stout, paniculately branching above, 3–6 dm. high, glabrous throughout or the branches of the inflorescence sometimes sparsely short-pubescent. Lower leaves narrowly oblanceolate, entire or denticulate, the upper narrowly linear, much reduced and bract-like in the inflorescence; heads on short divergent bractlets forming narrow or sometimes more widely branching panicles, 5–7-flowered; phyllaries 5, linear, 5–7 mm. long, subtended by calyculate bractlets; ligules bright pink; achenes sharply 5-angled, the intervals tuberculate; pappus-bristles 15–20, united in groups of 2 or 3, white or more commonly tawny, rather short-plumose above and becoming diminished near the base.

Dry hillsides and plains, especially in open sagebrush, Arid Transition and Upper Sonoran Zones; Washington east of the Cascade Mountains and Idaho southward to eastern Oregon and northern California. Type locality: "On the Rocky Mountain plains, towards the Colorado." July–Aug.

8. Stephanomeria exígua Nutt. Small Stephanomeria. Fig. 5991.

Stephanomeria exigua Nutt. Trans. Amer. Phil. Soc. II. **7**: 428. 1841.
Ptiloria exigua Greene, Pittonia **2**: 132. 1890.

Annual or rarely biennial, 1–5 dm. high, commonly fastigiately branched from the base in older or more vigorous plants, occasionally nearly simple with paniculately ascending branches; stems usually slender, pale bluish green, glabrous or with a few minute glandular hairs. Lower and basal leaves narrowly oblong, coarsely toothed or runcinate-pinnatifid, 5 cm. or less long, the upper leaves small and bract-like; heads at the tips of the branches, more rarely clustered on short divaricate peduncles; involucres 8–10 mm. high, bearing calyculate bractlets at base; phyllaries 4–6, linear; flowers 3–7, bright pink; achenes 5-angled, longitudinally grooved between the angles, the intervals with 2 longitudinal rows of tubercles; pappus-bristles 8–18, white, plumose above, the lower third naked, breaking off toward the base, more or less adhering in groups of 4 or 5 by the widened bases, these often with minute marginal bristles.

Brushy slopes and in forests, Upper Sonoran and Transition Zones; Wyoming, Idaho, and eastern Oregon south to New Mexico and California on the eastern face of the Sierra Nevada and the mountains of southern California; also on the western face of the Sierra Nevada in Kern County and near the coast in southern California and Lower California where several diverse forms are to be found, some of which appear to be intermediates between *S. exigua, S. exigua* var. *deanei,* and *S. virgata.* In some of these forms the pappus-bristles are more copious and tend to shatter readily, leaving an irregular crown on the achenes formed by the broadened and partially united, nonplumose bases of the bristles. These forms have been interpreted by authors as *S. exigua* var. *coronaria* (Greene) Jepson (Man. Fl. Pl. Calif. 998. 1925.) but are different from *S. coronaria* Greene, a plant of the central Coast Ranges of California which is a part of the *S. virgata* complex. Type locality: "On the Rocky Mountain plains, towards the Colorado." July–Oct.

Stephanomeria exigua var. **pentachaèta** (D. C. Eaton) H. M. Hall, Univ. Calif. Pub. Bot. **3**: 260. 1907. (*Stephanomeria pentachaeta* D. C. Eaton, Bot. King Expl. 199. *pl. 20, figs. 8–10.* 1871; *Ptiloria pentachaeta* Greene, Pittonia **2**: 133. 1890.) Rather stout annual plants conspicuously glaucous and sometimes glandular above, stems whitish, divaricately branched above from the axis of the main stem or branched from the base with fewer ascending branches than the species, the flowering branchlets ascending or divaricate, conspicuously bracteate with 6–9 much-reduced bracteate leaves; pappus of 5–7(8) bristles, these coarse, nearly distinct to the base but deciduous in groups, plumose only on the upper half, somewhat dilated at the base and usually with a few minute teeth. Dry desert regions, Harney and Malheur Counties, southeastern Oregon, southward east of the

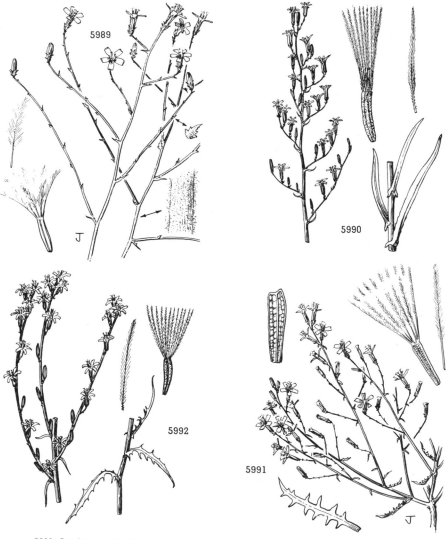

5989. Stephanomeria cinerea
5990. Stephanomeria paniculata

5991. Stephanomeria exigua
5992. Stetphanomeria virgata

Cascade Mountains and the Sierra Nevada to Inyo County and the deserts of southern California and adjacent Lower California; also Nevada, southwestern Utah. and northwestern Arizona. Occurring with *S. exigua* var *exigua*. Type locality: "Truckee and Humboldt Valleys," Nevada.

  Stephanomeria exigua var. **dèanei** J. F. Macbride, Contr. Gray Herb. No. 53: 22. 1918. (*Ptiloria exigua* var. *deanei* J. F. Macbride ex Davids. & Moxley, Fl. S. Calif. 355. 1923.) Slender annual, divaricate-paniculate with bracteate peduncles, or persisting more than one season and becoming intricately branched; more or less glandular especially in the inflorescence; heads small as in the species; intervals of the achenes strongly grooved and little or not at all tuberculate; pappus-bristles as in the species, often breaking off irregularly above the base. Hill slopes, San Diego County, California, and mountains of adjacent Lower California; intergrading to the north with the slender nonglandular form of *S. exigua* in which the achenes are slightly rugose and strongly grooved. Type locality: Sweetwater Valley, San Diego County.

9. **Stephanomeria virgàta** Benth. Virgate or Tall Stephanomeria. Fig. 5992.

Stephanomeria virgata Benth. Bot. Sulph. 32. 1844.
Stephanomeria elata Nutt. Journ. Acad. Phila. II. 1: 173. 1848.
Stephanomeria coronaria Greene, Bull. Calif. Acad. 1: 194. 1885.
Stephanomeria tomentosa Greene, op. cit. 2: 152. 1886.
Ptiloria canescens Greene, Pittonia 2: 131. 1890.
Ptiloria pleurocarpa Greene, loc. cit.
Stephanomeria virgata var. *tomentosa* Munz, Aliso 4: 100. 1958.

  Annual, simple, erect with rigid stems or virgately branching above, 3–10 dm. high or rarely 20 dm.; herbage usually glabrous throughout, in some forms tomentose. Lower leaves oblong or spatulate, often sinuate or shallowly pinnatifid, soon deciduous, the upper ones linear and entire; heads solitary and subsessile or fascicled on short peduncles along the naked branches, more rarely with the peduncles more elongate; heads 6–16-flowered, sometimes 20-flowered on robust plants; involucres 7–8 mm. high, the phyllaries 5–6, subtended by calyculate bractlets; ligules pink, occasionally white on the upper surface; achenes 4–5 mm. long, oblong to somewhat clavate or sometimes fusiform, dark or light brown and sometimes mottled, 4–5-ribbed, the interspaces with or without an evident groove and strongly rugose to nearly smooth; pappus-bristles white, 4–5 mm. long, copious, about 20 in number, fragile, the plumose hairs usually a little shorter toward the base, completely deciduous or the bases of the bristles in some forms remaining on the achene as an extremely short, even crown.

  Dry hills and valleys, Transition and Sonoran Zones; southwestern Oregon southward to northern Lower California. Type locality: San Pedro, Los Angeles County, California. July–Oct.

  Future field studies may prove that some of the synonyms listed under this variable taxon may have some taxonomic status. Different strains are sometimes locally recognizable but intergrading forms are constantly to be found even in the same region.

  The photograph of the type of *S. virgata* var. *virgata* from San Pedro shows a plant with a virgate-paniculate inflorescence in which the heads are subsessile and fascicled along the stems, a form which is found rather commonly northward along the coast. The achenes are angled, slightly incurved, and rugose-tuberculate, and usually dark or mottled. Apparently they are typically not grooved on the faces between the angles as this character is not mentioned in the original description.

  The type of *S. elata* from Santa Barbara has not been seen. As interpreted from the original description it belongs to the *S. virgata* complex but differs from the typical form in the "small terminal panicle" in which each of the "flower branches . . . bear three or four flowers." The blue color assigned to the ligules in the original description was evidently not obtained from field notes but was doubtfully applied to the dried specimens by Nuttall. Some specimens are to be found in collections from the Santa Barbara region in which the heads are definitely pedunculate in an evident panicle rather than clustered along the branches of a virgate-paniculate inflorescence. The achenes of some of these specimens are pale brown, 5-grooved, and somewhat rugose, all characters which are assigned to *S. elata* in the original dscription.

  Stephanomeria coronaria, described originally from the Santa Lucia Mountains, California, and found more commonly away from the coast in this range and in the Mount Hamilton Range, is characterized by having a minute even crown on the achene formed from the bases of the copious, white, deciduous bristles. This is quite different, as pointed out by Greene (Pittonia 2: 132. 1890), from the pappus of *S. exigua* in which the bristles are attached at the base in groups of 3 or 4, are not plumose to the base, and break off rather irregularly. The other character, absence of tubercles between the angles, is met with sporadically throughout the species. Also some specimens having the short crown on the achene have grooves and low tubercles rather than smooth faces on the achenes.

  Stephanomeria tomentosa, described by Greene from Santa Cruz Island, Santa Barbara County, is described as a stout plant having the inflorescence and achene characters of *S. virgata* var. *virgata*, but as "white tomentose throughout when young, the inflorescence becoming glabrate." This type of pubescence made up of soft, short, multi-cellular, unbranched hairs is found in varying abundance in other specimens from the islands and adjacent mainland. It can even be demonstrated on specimens having the "*elata*" type of inflorescence which are found in the same area.

  Ptiloria canescens, described from Napa Valley, and *P. pleurocarpa* from Redding, Shasta County, as well as other forms of the complex, occur throughout northern and central California and the Sierra Nevada. *Ptiloria canescens* is best distinguished by the "sparse. not virgately disposed heads" but its character of canescent pubescence appears quite as often on plants having many heads on virgately disposed branches. *Ptiloria pleurocarpa* is distinguished principally from *Stephanomeria virgata* var. *virgata* by pale brown achenes.

158. **RAFINÉSQUIA** Nutt. Trans. Amer. Phil. Soc. II. 7: 429. 1841.

  Glabrous and slightly succulent, branching annuals with fistulose stems and pinnatifid leaves. Heads rather large, with white or rose-tinged flowers. Involucre conic or cylindric, with 7–15 linear acuminate phyllaries, somewhat fleshy at base, with a few loose, calyculate, outer ones. Achenes terete, somewhat fusiform, obscurely few-ribbed, attenuate into a beak. Pappus white or sordid, of 8–15 slender bristles, these softly long-plumose from the base to near the tip. [Name in honor of Constantine Rafinesque, an American naturalist and traveler.]

  A genus of 2 species, natives of southwestern United States and adjacent Mexico. Type species, *Rafinesquia californica* Nutt.

Rays about 5–8 mm. long; beak of the achene as long as the body; pappus capillary and plumose with straight hairs.   1. *R. californica*.
Rays about 15 mm. long; beak of the achene shorter than the body; pappus flattened at base, plumose with arachnoid hairs, these sometimes lacking at the attenuate tip.   2. *R. neo-mexicana*.

1. **Rafinesquia califórnica** Nutt. California Chicory. Fig. 5993.

Rafinesquia californica Nutt. Trans. Amer. Phil. Soc. II. 7: 429. 1841.
Nemoseris californica Greene, Pittonia 2: 193. 1891.

Stems stout, 6–12 dm. high, glabrous, simple or sometimes branched above. Basal leaves coarsely toothed, soon withering; stem-leaves oblong or lanceolate, auriculate-clasping, dentate to runcinate-pinnatifid, the divisions rather broad and not cut to the midrib, the upper leaves much reduced; heads several in a paniculate-corymbose inflorescence, 15–20 mm. high; ligules 5–8 mm. long, white; outer achenes pubescent and somewhat tuberculate, beak slender, equaling the body; pappus dull white or sordid, plumose with straight hairs to the top of the bristles.

Foothills and valleys, Upper Sonoran Zone; Humboldt and Mariposa Counties southward to coastal southern California and northern Lower California; occasionally eastward at higher elevations to the Mojave and Colorado Deserts and southwestern Arizona. Type locality: "Near the sea-coast in the vicinity of St. Diego, Upper California." April–July.

2. **Rafinesquia neo-mexicàna** A. Gray. Desert Chicory. Fig. 5994.

Rafinesquia neo-mexicana A. Gray, Smiths. Contr. 5[6]: 103. 1853.
Nemoseris neo-mexicana Greene, Pittonia 2: 193. 1891.

Stems simple or branched from the base, 1.5–5 dm. high, often purplish, rather weak and usually supported by desert shrubs. Basal leaves 3–9 cm. long, mostly runcinate-pinnatifid, acute at the apex, usually persisting during anthesis, the stem-leaves sessile and auriculate-clasping, much reduced above and becoming bracteate in the inflorescence, deeply runcinate-pinnatifid, the nonincised portion of the leaf narrowly lanceolate to linear; flowers solitary at the ends of the branchlets; phyllaries about 2 cm. long, linear-lanceolate, scarious-margined, the outer calyculate bractlets much shorter, ligules 15–20 mm. long, laciniate-toothed at the apex, white within, tinged with rose or purple without; achenes 12–15 mm. long, obscurely angled, tapering abruptly into a beak shorter than the body of the achene, the surface appearing grayish-mottled with a papillate puberulence, the pappus-bristles bright white, the base somewhat flattened, plumose with soft arachnoid hairs, these often lacking at the apex of the bristle.

Mesas and canyons, usually in the shade of shrubs, Sonoran Zones; Inyo County, California, south through the Mojave and Colorado Deserts to central Lower California and eastward to southern Utah and through Arizona to western Texas, and in adjacent Sonora. Type locality: El Paso, Texas. March–May.

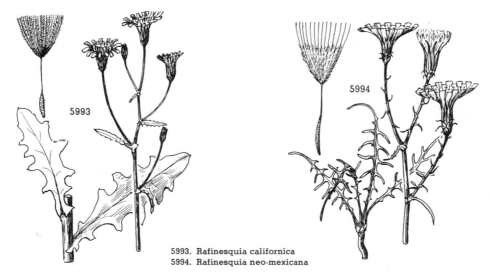

5993. Rafinesquia californica
5994. Rafinesquia neo-mexicana

159. **MALACÓTHRIX** DC. Prod. 7: 192. 1838.

Annual or perennial herbs. Leaves alternate or basal, mostly pinnatifid. Heads long-peduncled, panicled, or solitary; flowers yellow· or white or sometimes purplish-tinged. Involucre campanulate, its principal phyllaries reflexed in age, the phyllaries nearly equal in 1–2 series, with few short outer ones or these larger, several, and much imbricated, the margins narrowly or broadly scarious. Receptacle flat, naked or bristly. Ligules truncate and 5-toothed at apex. Achenes oblong or linear, glabrous, 10–15-ribbed, truncate or margined at the summit, sometimes minutely toothed. Pappus-bristles in 1 series, naked or minutely serrulate, often softly ciliolate below, slender, coherent at the base and deciduous in an easily fragmented ring, or a few persisting on the achene. [Name Greek, meaning soft and hair, in reference to the soft woolly pubescence on the leaves of the original species.]

A western North American genus of about 15–20 species. Type species, *Malacothrix californica* DC.

Annuals, 0.5–5 dm. high.
  Outer phyllaries well imbricated in 3–4 series, orbicular to narrowly or broadly ovate with conspicuous
    scarious margins.
    Pappus-bristles 1–4, persisting on the achene; outer phyllaries orbicular, 4–5 mm. broad; plants of desert
      or arid areas.   1. *M. coulteri.*
    Pappus-bristles not persisting on the achene; outer phyllaries ovate-acute, 2 mm. broad; plants of
      Channel Islands.   2. *M. insularis squalida.*
  Outer phyllaries in 1–2 series, the outer much shorter than the inner, with narrow scarious margins.
    Stems leafy to the inflorescence; leaves not much reduced above.
      Plants erect; leaves laciniate-pinnatifid.   3. *M. foliosa.*
      Plants depressed (or erect); lobes of the leaves obtuse.   4. *M. indecora.*
    Leaves mostly basal, reduced and sparse above.
      Basal leaves pinnatifid with long, linear, filiform lobes; flower-heads large, the ligules 12–16 mm.
        long.   5. *M. californica.*
      Basal leaves pinnatifid with short, broad, mostly toothed lobes; flower-heads medium to small, the
        ligules 2–10 mm. long.
        Involucres 4–7 mm. high; achenes finely 10–15 striate with the 5 ribs not markedly more promi-
          nent.
          Ligules about 2–3 mm. long; achenes with a setose-denticulate crown and 1–2 persistent
            bristles; leaf-margins without tufts of wool.   6. *M. clevelandii.*
          Ligules 3–5 mm. long; achenes with an entire crown and no persistent bristles; leaf-margins
            with tufts of wool.   7. *M. floccifera.*
        Involucres 8–13 mm. high; achenes 15-striate, every third rib prominent.
          Achenes 2–3 mm. long; pappus-bristles all deciduous; inflorescence essentially glabrous.
                                               8. *M. sonchoides.*
          Achenes 3–4 mm. long; 1–8 pappus-bristles persistent; inflorescence with scattered, gland-
            tipped hairs.   9. *M. torreyi.*
Perennials; shrubs, stout herbaceous plants or the more or less decumbent forms from a woody root-crown.
  Ligules yellow; dune plants prostrate or ascending from a stout root-crown.   10. *M. incana.*
  Ligules pink, purplish, or white; shrubs or erect herbaceous plants.
    Heads many-flowered (40–100); herbaceous or shrubby only at base (except *M. saxatilis* var. *implicata*);
      leaves not obovate or oblong-obovate.
      Stems herbaceous, arising singly from a deep-seated, branched root, the entire leaves of midstem,
        if present, acute or acuminate at the apex.   11. *M. saxatilis* vars.
      Stems several from a woody root, usually suffrutescent at base; entire leaves of midstem obtuse or
        broadly acute at the apex (except in vars. *implicata* and *tenuifolia*).
                                             11. *M. saxatilis.*
    Heads 9–12-flowered; shrubs; leaves obovate or oblong-obovate.   12. *M. blairii.*

1. Malacothrix coùlteri Harv. & Gray. Snakes-head or Coulter's Malacothrix. Fig. 5995.

Malacothrix coulteri Harv. & Gray in A. Gray, Mem. Amer. Acad. II. **4**: 113. 1849.
Malacolepis coulteri Heller, Muhlenbergia **2**: 147. 1906.

Stems branching above and sometimes also at base, erect or sometimes more or less decumbent, 1–5 dm. high, glabrous and pale. Leaves sinuate-dentate or somewhat pinnatifid, the basal
oblong or spatulate, 5–10 cm. long, the cauline clasping, ovate to lanceolate in outline, 2–4 cm.
long; heads terminating the short branches, short-peduncled; involucres hemispherical, 10–15
mm. high; phyllaries well imbricated, suborbicular to ovate, with broad scarious margins and a
broad, purplish or green, central band; ligules pale, yellow to nearly white, the outer ones sometimes lined with reddish purple, 5–18 mm. long; achenes light greenish brown, 2–2.75 mm. long,
4–5-angled, with 2 striae between the sharply acute angles, these extended above the body of the
achene in short points; 1–4 of pappus-bristles persistent.

    Flats and low hills, mainly Upper Sonoran Zone; Alameda, San Joaquin, and Mariposa Counties, central
California, south to northern Lower California and east in the desert regions to southwestern Utah and Arizona.
Type locality: California. Collected by Coulter. March–May.
    Malacothrix coulteri var. **cognàta** Jepson, Man. Fl. Pl. Calif. 1001. 1925. Heads and phyllaries as in
M. coulteri var. *coulteri*; stem-leaves clasping but parted nearly to the midrib into linear divisions; achenes and
pappus-bristles as in *M. coulteri* var. *coulteri*. Rarely collected; Santa Cruz and Santa Rosa Islands in Santa
Barbara County, the San Pedro hills in Los Angeles County (*Abrams 3133*), and Tia Juana, San Diego County
(*Jones*), California. Type locality: Santa Cruz Island. Collected by Brandegee.

2. Malacothrix insulàris var. squálida (Greene) Ferris. Island Malacothrix. Fig. 5996.

Malacothrix squalida Greene, Bull. Calif. Acad. **2**: 152. 1886.
Malacothrix foliosa var. *squalida* E. Williams, Amer. Midl. Nat. **58**: 507. 1957.
Malacothrix insularis var. *squalida* Ferris, Contr. Dudley Herb. **5**: 102. 1958.

Glabrous annuals, the stout stems decumbent or sometimes erect, 1.5–4 dm. high, quite densely
leafy throughout, branched at base and above. Leaves sessile, 2–7 cm. long, broadly lanceolate to
ovate in outline, with laciniately pinnatifid, acute lobes, the shorter upper leaves also deeply laciniate
and clasping at the base; heads many, on short, leafy, bracteate peduncles; involucres 9–11 mm.
high, campanulate, 4–5-seriate, the outer 2 or 3 rows of phyllaries ovate, about 2 mm. broad at
base, acute at apex, with a wide scarious margin and a dark central midsection; ligules bright
yellow, 4–5 mm. long; achenes 2 mm. long, dark brown, 5-angled or -ridged, the angles extended
above the apex of the achene in minute blunt points and having usually 2 less prominent striae
between the angles; pappus-bristles all deciduous.

    Hillsides and terraces, Upper Sonoran Zone; Santa Cruz, Santa Rosa, Santa Barbara, and Anacapa Islands
off the coast of southern California. Type locality: Santa Cruz Island. Collected by Greene. April–July.
    Some specimens from Santa Barbara Island appear to be quite close to *M. insularis* var. *insularis*.

3. **Malacothrix foliòsa** A. Gray. Leafy Malacothrix. Fig. 5997.

Malacothrix foliosa A. Gray, Syn. Fl. N. Amer. ed. 2. 1²: 455. 1886.

Glabrous annuals, the rather slender stems often purplish, erect, 1.5–3.5 dm. high, leafy throughout, usually simple below and paniculately branched above, rarely with 2 or 3 branches from the base. Leaves sessile, 1.5–8 cm. long, lanceolate to linear-lanceolate, irregularly laciniate-pinnatifid, with slender-attenuate lobes; heads several to many, on short bracteate peduncles, 2 or 3 on dwarfed plants; involucres campanulate, 7–9 mm. high; inner phyllaries linear or linear-lanceolate, with narrow scarious margins, the outer reduced to short, obtuse, purple-tipped, calyculate bracts, 2 or 3 of these sometimes about one-half the length of the inner phyllaries; ligules bright yellow, surpassing the inner phyllaries by 2.5–3 mm.; achenes 1.2–1.3 mm. long, obscurely angular, 12–15-ribbed or striate, the ribs nearly of equal width, those on the angles not noticeably larger; pappus-bristles all deciduous.

Hillsides and terraces, Upper Sonoran Zone; Santa Barbara, San Clemente, and Coronados Islands off the coast of southern California and northern Lcwer California. Type locality: San Clemente Island, Los Angeles County, California. Collected by Nevin and Lyon. April–July.

4. **Malacothrix indécora** Greene. Fleshy Malacothrix. Fig. 5998.

Malacothrix indecora Greene, Bull. Calif. Acad. **2**: 152. 1886.
Malacothrix foliosa var. *indecora* E. Williams, Amer. Midl. Nat. **58**: 507. 1957.

Low, leafy annuals 2–4 cm. high, at first single-stemmed, becoming much branched from the base, usually depressed. Leaves glabrous and succulent, oblong or often oblanceolate, more or less pinnatifid with broad obtuse lobes, the lower 2–3 cm. long, the upper about half as long; heads

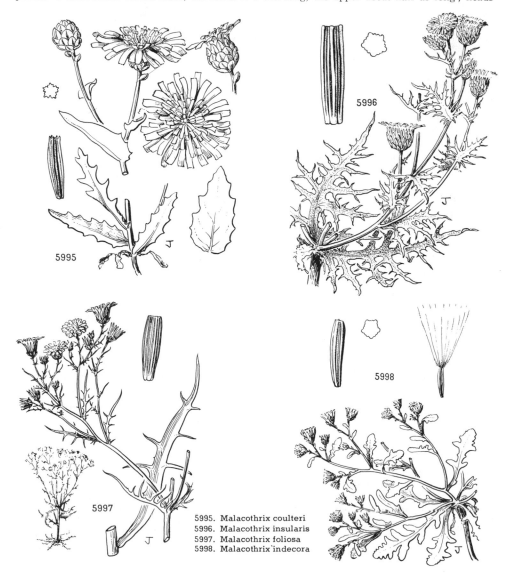

5995. Malacothrix coulteri
5996. Malacothrix insularis
5997. Malacothrix foliosa
5998. Malacothrix indecora

on short bracteate peduncles; involucre 6–7 mm. high; inner phyllaries linear-lanceolate, usually greenish, the outer shorter and purple-tinged; ligules short, greenish yellow; achenes about 1 mm. long, 5-angled and 2–3-striate between the obtuse angles; pappus deciduous in a ring with no outer persistent bristles, the bristles retrorsely ciliate at base, barbellate above; receptacle with minute paleae.

An insular species, occurring on Santa Cruz and San Miguel Islands, Santa Barbara County, and San Nicolas Island, Ventura County, off the coast of southern California. Type locality: Santa Cruz Island. April–Sept.

5. **Malacothrix califórnica** DC. California Malacothrix. Fig. 5999.

Malacothrix californica DC. Prod. 7: 192. 1838.

Acaulescent scapose annual. Leaves basal, forming a dense rosette, pinnatifid, 6–12 cm. long, the lobes narrowly linear or almost filiform, more or less densely woolly-tomentose when young, partially glabrate in age; scapes usually several, 10–35 cm. high, erect or somewhat decumbent at base, glabrous, simple and bractless, bearing a solitary terminal head with pale yellow ligules; involucres broadly campanulate, 8–15 mm. high, densely woolly at base; outer phyllaries narrowly subulate, the inner much longer and linear-lanceolate, with broadly scarious margins; receptacle-bristles usually present but delicate; achenes about 2.5 mm. long, slender and narrowed toward the base, 4–5-ribbed, with 2 slender striae between the ribs; pappus white, about 8 mm. long, the 2 outer bristles persistent, the inner ones united at base and falling away together.

Sandy or gravelly soils, Sonoran Zones; inland and along the coast, Contra Costa County, and Fresno and Tulare Counties south to Lower California; also on the western edges of the Mojave and Colorado Deserts. Type locality: California. Collected by Douglas. March–May.

Malacothrix californica var. **glabràta** A. Gray in D. C. Eaton. Bot. King Expl. 201. 1871. (*Malacothrix glabrata* A. Gray, Syn. Fl. N. Amer. 1²: 422. 1884.) Growth habit and height of *M. californica* var. *californica*; herbage glabrous throughout, rarely a little woolly on young basal leaves; flowering stems bearing 2 or 3 large heads (rarely 1); 1 or 2 cauline leaves sometimes present; achenes and pappus as in *M. californica* var. *californica*. Sandy soil, southwestern Idaho and eastern Oregon and Nevada through the deserts of California and Arizona to Lower California; occurring rarely with the species in the San Joaquin Valley in Kern County, California, and eastern part of the Cuyama Valley, Santa Barbara County. Type locality: "Carson City (Anderson!) Foothills of the Trinity Mountains, Nevada." March–July. Desert Dandelion.

6. **Malacothrix clevelándii** A. Gray. Cleveland's Malacothrix. Fig. 6000.

Malacothrix clevelandii A. Gray, Bot. Calif. 1: 433. 1876.

Annual, the stems usually several from the base, paniculately branched above, the branches ascending, 1–4 dm. high, glabrous, often reddish. Leaves mostly basal, more or less pinnatifid, 2–10 cm. long, the lobes sometimes reduced to teeth, 1–3 mm. wide; heads many, terminating the slender paniculate branches; involucre narrowly campanulate, 6–7 mm. high, the inner phyllaries linear, green, often purple-tipped and with a narrow scarious margin; ligules yellow, about 2 mm. long; achenes linear, minutely striate-costate; pappus mostly deciduous leaving only 1 (rarely 2) persistent bristle (this usually tardily deciduous) and a circle of minute white teeth.

Chaparral and open hillsides, especially on burns or disturbed places, mainly Upper Sonoran Zone; occasional in the foothills of the Sierra Nevada, California, from Tuolumne County southward and the Coast Ranges from Glenn County to southern California, where it is more common, and northern Lower California; also on Santa Cruz Island (*Brandegee*) and occasional in the Colorado and Mojave Deserts to southwestern Nevada and Arizona, and Sonora. Type locality: near San Diego. April–June.

7. **Malacothrix floccífera** (DC.) Blake. Woolly Malacothrix. Fig. 6001.

Senecio flocciferus DC. Prod. 6: 426. 1837.
Malacothrix obtusa Benth. Pl. Hartw. 321. 1849.
Malacothrix parviflora Benth. loc. cit.
Malacothrix floccifera Blake, Contr. U.S. Nat. Herb. 22: 656. 1924.

Annual, stems simple at base and erect or often several with the outer somewhat decumbent, paniculately branched above, 1–4 dm. high. Leaves nearly all broad, oblong or oblong-spatulate, 1.5–10 cm. long, pinnatifid or the smaller often merely dentate, often bearing tufts of wool on the lower surface, especially on the lobes; involucre narrowly campanulate, 3–5 mm. broad; inner phyllaries 4–5 mm. high, linear, acute or short-acuminate, with broad scarious margins, often purplish at least at the tip; ligules white or sometimes pale yellow, often tinged with pink, 5–10 mm. long; achenes oblong-obovoid, entire at summit; pappus-bristles all deciduous.

Dry slopes and rocky banks, Upper Sonoran and Transition Zones; Coast Ranges from Siskiyou County to Mount Pinos, Ventura County, California, and in the Sierra Nevada from Lassen County to Mariposa County; east to western Nevada. Type locality: California. Collected by Douglas. April–Oct.

8. **Malacothrix sonchoìdes** (Nutt.) Torr. & Gray. Sow-thistle Malacothrix. Fig. 6002.

Leptoseris sonchoides Nutt. Trans. Amer. Phil. Soc. II. 7: 439. 1841.
Malacothrix sonchoides Torr. & Gray, Fl. N. Amer. 2: 486. 1843.
Malacothrix runcinata A. Nels. Bull. Torrey Club 26: 485. 1899.

Annual with a single erect stem or with several stems and somewhat decumbent at base, branching above, 1–5 dm. high, glabrous or essentially so. Leaves in a basal tuft, rather thick, 3–10 cm. long, oblong in outline, regularly pinnatifid, the lobes and teeth callous-tipped, lower stem-leaves few, similar to the basal, the uppermost reduced to small subulate bracts; heads few to numerous, short-pedunculate; involucres campanulate, 7–8 mm. high, the inner phyllaries linear-acuminate;

ligules bright yellow, about 1 cm. long; achenes with 5 prominent ribs and 2 finer ones in the intervals; pappus-bristles all deciduous.

Sandy or gravelly slopes and washes, Sonoran Zones; desert regions of Inyo and San Bernardino Counties, California, east through the Great Basin region to western Nebraska. Type locality: "The plains of the Platte." April–June.

9. **Malacothrix tórreyi** A. Gray. Torrey's Malacothrix. Fig. 6003.

Malacothrix torreyi A. Gray, Proc. Amer. Acad. **9**: 213. 1874.
Malacothrix sonchoides var. *torreyi* E. Williams, Amer. Midl. Nat. **58**: 503. 1957.

Annual, the stems simple below and erect or often branched at base and somewhat decumbent, 10–30 cm. high, herbage glabrous or usually sparsely puberulent with gland-tipped hairs. Leaves 2–6 cm. long, pinnatifid, the lobes divergent or runcinate, the lower wing-petioled, the cauline several to many, sessile or nearly so, with the lobes entire or toothed; heads usually many on short slender peduncles; involcre 7–11 mm. high, the phyllaries acuminate, 16–20; outer ligules 6–8 mm. long; achenes about 4 mm. long, slender, prominently keeled on the angles, striate between; persistent pappus-bristles 2–5, white, 6–7 mm. long.

Sandy or gravelly places, mainly Arid Transition Zone; southeastern Oregon in Harney and Malheur Counties, east to Idaho and Utah, and south to Nevada. Type locality: Salt Lake City, Utah. May–July.

10. **Malacothrix incàna** (Nutt.) Torr. & Gray. Dune Malacothrix. Fig. 6004.

Malacomeris incanus Nutt. Trans. Amer. Phil. Soc. II. **7**: 435. 1841.
Malacothrix incana Torr. & Gray, Fl. N. Amer. **2**: 486. 1843.

Perennial with a stout, rather deep-seated root, the whole plant more or less densely white-woolly especially when young, the stems branching from the root-crown, short and stout, forming tufts or mats, or branches elongate, 2–3 dm. long, and decumbent. Leaves spatulate, entire or some-

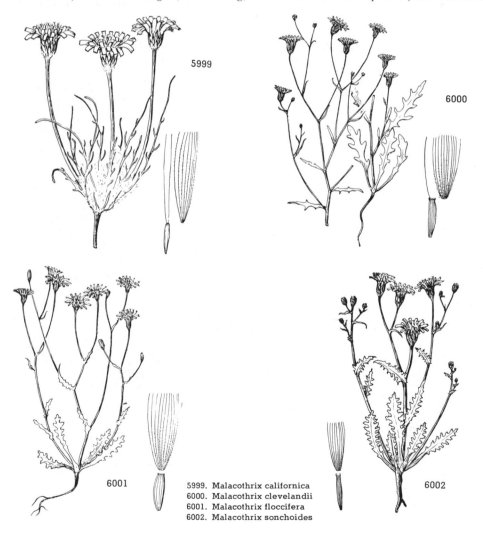

5999. Malacothrix californica
6000. Malacothrix clevelandii
6001. Malacothrix floccifera
6002. Malacothrix sonchoides

what pinnatifid with a few broad lobes or teeth, 3–6 cm. long including the winged petiole, 5–20 mm. wide, those of the stems reduced; peduncles several to many, bracteate; heads 15–30 mm. broad; involucres 10–13 mm. high, attenuate at base; phyllaries in 4 or 5 series, the upper linear-oblong, acute, those on the attenuate base much reduced; ligules lemon yellow; achenes oblong, 15-striate; pappus-bristles all deciduous.

On or near sand dunes along the coast, Upper Sonoran Zone; Pismo, San Luis Obispo County, California, to Carpinteria and the Channel Islands off the coast of Santa Barbara County. Type locality: Island in the bay at San Diego, California, but not since collected in that region. April–Aug. Some collections by Hoffman from Santa Rosa Island approach var. *succulenta*.

Malacothrix incana var. **succulénta** (Elmer) E. Williams, Amer. Midl. Nat. 58: 506. 1957. (*Malacothrix succulenta* Elmer, Bot. Gaz. 39: 44. 1905.) Growth habit of the species but completely glabrous or the young leaves sometimes woolly beneath. On dunes along the coast, southern San Luis Obispo County (Oso Flaco Lake) to Surf, Santa Barbara County, California, the type locality.

11. **Malacothrix saxátilis** (Nutt.) Torr. & Gray. Cliff Malacothrix. Fig. 6005.

Leucoseris saxatilis Nutt. Trans. Amer. Phil. Soc. II. 7: 440. 1841.
Malacothrix saxatilis Torr. & Gray, Fl. N. Amer. 2: 486. 1843.

Stems stout and fistulose, erect or decumbent from a woody base, 3–7 dm. high, simple and very leafy below, branched and leafy above, glabrous or somewhat floccose at the leaf-axils. Leaves thickish and somewhat succulent, narrowly oblong to linear-spatulate or more rarely broadly acute, slightly clasping at the base, 5–12 cm. long, 0.5–1 cm. wide, entire, the lowest leaves occasionally irregularly toothed or lobed; inflorescence cymose, the heads large; involucres 12–15 mm. high; phyllaries in 3–4 series, narrowly scarious-margined, the inner linear-lanceolate, attenuate, the outer much reduced, almost subulate and often spreading; ligules white with a pink or purplish midsection; achenes 10–15-ribbed, 5 of the ribs more prominent; pappus-bristles all deciduous, leaving the crown of the achenes obscurely denticulate.

Ocean bluffs, Upper Sonoran Zone; Point Conception, Santa Barbara County, California, south to Santa Barbara and also along the coast in northwestern Ventura County. Type locality: Santa Barbara. Collected by Nuttall. May–Sept. Cliff-aster.

Stems usually several, arising from the root at the surface of the ground, suffrutescent at base or herbaceous.
  Leaves leathery, obtuse, oblanceolate, or broadly linear, entire or the lowest sometimes irregularly lobed; plants of ocean bluffs, Point Conception to Santa Barbara. var. *saxatilis*.
  Leaves not leathery, pinnatifid or pinnately or bipinnately divided into linear divisions, if entire then linear-filiform.
    Leaves bipinnately divided; stems woody, densely leafy to the inflorescence; plants of the Channel Islands. var. *implicata*.
    Leaves pinnately parted; stems not densely leafy to the inflorescence; mostly herbaceous; plants of the coastal mountains of southern California. var. *tenuifolia*.
Stems partly subterranean, arising singly from branches of a deep-seated root, essentially herbaceous.
  Lower stem-leaves thin, deeply laciniately lobed to near the midrib, the lobes irregular, few on each side, acute, the upper part of the leaf entire or toothed, acuminate; plants of the Tehachapi Mountains, Mount Pinos region, the higher elevations in the mountains of Santa Barbara County, and the Santa Monica Mountains. var. *altissima*.
  Lower stem-leaves firm, entire or short-toothed, the teeth of irregular lengths, midstem-leaves entire or merely denticulate.
    Permanently arachnoid-tomentose throughout, 0.5–2 m. tall; plants of Carmel Valley, Monterey County. var. *arachnoidea*.
    Completely glabrous or sparsely arachnoid-tomentose on the leaf-bases or stems; plants of Salinas Valley and eastern face of Santa Lucia Mountains, and adjacent Santa Barbara County. var. *commutata*.

Malacothrix saxatilis var. **implicàta** (Eastw.) H. M. Hall, Univ. Calif. Pub. Bot. 3: 269. 1907. (*Malacothrix implicata* Eastw. Proc. Calif. Acad. III. 1: 113. 1898.) Stems woody below, branched, densely leafy up to the inflorescence; leaves irregularly bipinnately divided into narrowly linear or filiform segments; ligules purplish. San Nicolas, San Miguel, Santa Rosa, and Santa Cruz Islands off the coast of Santa Barbara, California. Type locality: San Nicolas Island, Ventura County.

Malacothrix saxatilis var. **tenuifòlia** (Nutt.) A. Gray, Syn. Fl. N. Amer. 1²: 423. 1884. (*Leucoseris tenuifolia* Nutt. Trans. Amer. Phil. Soc. II. 7: 440. 1841; *Malacothrix tenuifolia* Torr. & Gray, Fl. N. Amer. 2: 487. 1843; *M. saxatilis* var. *tenuissima* Munz, Man. S. Calif. 591, 601. 1935.) Stems several from a woody root, 0.5–1 m. high, often branched; glabrous or somewhat floccose throughout; lower leaves pinnatifid, upper stem-leaves reduced upward, entire, linear-filiform or pinnately parted into linear divisions. The most common form. Hillsides and canyons in the coastal mountains from the Santa Maria River, Santa Barbara County, to Orange County, California, and extending to the eastern border of Los Angeles County; also Santa Catalina Island. Type locality: "St. Barbara, on the mountains near the town."

Malacothrix saxatilis var. **altíssima** (Greene) Ferris. Contr. Dudley Herb. 5: 102. 1958. (*Malacothrix altissima* Greene, Bull. Calif. Acad. 1: 195. 1885.) Leafy-stemmed plants 1–2 m. high, typically glabrous throughout; leaves thin, 8–20 cm. long, irregularly and deeply laciniate-lobed and also denticulate, the uppermost leaves tending to be entire, acuminate at apex, tapering at the base. Tehachapi Mountains, Kern County, westward to the Mount Pinos region, Ventura County, and the higher slopes of the mountains in Santa Barbara County southward to the Santa Monica Mountains in Los Angeles County. Type locality: near Tehachapi Station, Kern County. Collected by Curran. Intergrading forms are to be found that show relationship with var. *tenuifolia* while others approach var. *commutata*.

Malacothrix saxatilis var. **arachnoìdea** (McGregor) E. Williams, Amer. Midl. Nat. 58: 509. 1957. (*Malacothrix arachnoidea* McGregor, Bull. Torrey Club 36: 605. *fig. 3*. 1909.) Densely leafy-stemmed plants 0.5–2 m. high; herbage persistently and densely arachnoid-tomentose throughout; leaves firm, 4–9 cm. long, narrowed to a petiolar base, usually entire or sometimes denticulate, acute at apex; crown of the achene minutely crenulate. Hill slopes; known only from Carmel Valley, Monterey County, California, the type locality.

Malacothrix saxatilis var. **commutàta** (Torr. & Gray) Ferris, in Munz, Aliso 4: 100. 1958. (*Hieracium californicum* DC. Prod. 7: 235. 1838, not *Malacothrix californicum* DC. op. cit. 192; *Sonchus californicus* Hook. & Arn. Bot. Beechey 361. 1840; *Leucoseris californicus* Nutt. Trans. Amer. Phil. Soc. II. 7: 441. 1841; *Malacothrix commutata* Torr. & Gray, Fl. N. Amer. 2: 487. 1843.) Densely leafy-stemmed plants 0.5–1 m. high, glabrous to sparsely arachnoid-tomentose about the leaf-bases, midveins, and lower stems; leaves firm, 4–9 cm. long, narrowed at the base, mostly denticulate, sometimes entire, if at all lobed these much shorter than the width of the leaf, acute at apex; crown of the achene minutely crenulate. Salinas Valley, eastern face of the Santa Lucia Mountains in Monterey and San Luis Obispo Counties, and in adjacent Santa Barbara County. The collections upon which the names listed above were based were made by Douglas presumably on his journey from Monterey to Santa Barbara, an area where *M. saxatilis* var. *commutata* commonly occurs. The specimen collected by Douglas deposited at Kew (mounted on a sheet with one of Coulter's collection) is of the inflorescence only, but corresponds to this form rather than *M. saxatilis* var. *tenuifolia*. This entity, with the related *M. saxatilis* var. *arachnoidea*, is quite distinct in its typical form from *M. saxatilis* var. *altissima*, but intergrading forms are to be found at the southern extension of its range.

12. **Malacothrix blàirii** (Munz & Jtn.) Munz. Blair's Malacothrix. Fig. 6006.

Stephanomeria blairi Munz & Jtn. Bull. Torrey Club **51**: 301. 1924.
Malacothrix blairi Munz, Man. S. Calif. 591, 601. 1935.

Coarse straggling shrub 1–2 m. high, branches thick, fleshy, thinly brownish-tomentose, glabrate, scarred with persistent leaf-bases subtending small, brownish-tomentose buds. Leaves approximate; petioles about 1 cm. long, narrowly winged, broadened at base; leaf-blades obovate or oblong-obovate, 6–13 cm. long, 3–6 cm. wide, rounded or obtuse, cuneate at base, irregularly sinuate-toothed with 3–4 pairs of very blunt, unequal teeth, thin, light green, at first thinly tomentose, quickly glabrate; panicles terminal, up to 20 cm. long and 10 cm. thick, many-headed, erectish-branched, thinly tomentose, glabrescent, densely stipitate-glandular above, with small squamiform bracts; peduncles mostly 2–6 mm. long; heads cylindric, 9–12-flowered; involucre 8–9 mm. high, the inner phyllaries linear, obtusish, obscurely granular-pulverulent, slightly hispidulous above, purplish toward the tip, the outer about 2-seriate, narrowly triangular to ovate, acute, glandular-ciliolate; corollas rose-colored, the ligules about 7 mm. long; achenes brownish white, subcylindric-pentagonal, 3 mm. long, minutely puberulous, not rugose, each face shallowly 1–2-grooved; pappus entirely deciduous, of about 25 white, minutely hispidulous bristles.

Known only from the type locality, rocky canyon wall near Lemon Tank, San Clemente Island, Los Angeles County, California. Late summer.

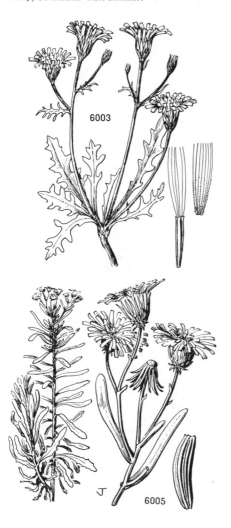

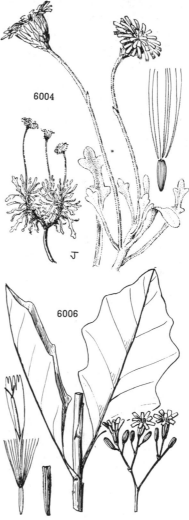

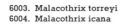

6003. Malacothrix torreyi
6004. Malacothrix icana

6005. Malacothrix saxatilis
6006. Malacothrix blairii

160. **GLYPTOPLEÙRA** D. C. Eaton, Bot. King Expl. 207. *pl. 20, figs. 11–18.* 1871.

Low, tufted or matted, glabrous and somewhat fleshy, winter annuals with a deep tap-root. Leaves with a broad midrib, pinnatifid, the margins white-crustaceous and prickly toothed, often equaling or surpassing the flowers. Involucres cylindric, the phyllaries 7–12, erect, linear-lanceolate, entire and scarious-margined; the calyculate bractlets spreading, spatulate, the apex foliaceous with a white-crustaceous, lacerate-toothed or -fringed margin. Ligules pale yellow, white or sometimes pinkish. Receptacle naked. Achenes oblong, 5-angled, straight or curved, the intervals between the angles with a row of pits separated by cross ridges; apex of achenes constricted into a short cupped beak. Pappus bristles copious, white, deciduous. [Name Greek, from *glyptos*, carved, and *pleura*, side, in reference to the marked sculpturing of the seed.]

A genus of 2 known species, of the arid desert regions of the southwestern United States. Type species, *Glyptopleura marginata* D. C. Eaton.

Ligules little exserted; margins of the calyculate bractlets lacerate-toothed at the apex, lacerate-fringed below.
1. *G. marginata.*

Ligules much surpassing the phyllaries; margins of calyculate bractlets lacerate-fringed at the dilated apex, entire or nearly so below.
2. *G. setulosa.*

1. **Glyptopleura marginàta** D. C. Eaton. Carved-seed. Fig. 6007.

Glyptopleura marginata D. C. Eaton, Bot. King Expl. 207. *pl. 20, figs. 11–18.* 1871.

Glabrous, fleshy, densely flowered annual, much branched and forming depressed tufts 5–15 cm. in diameter. Leaves 1–4 cm. long, pinnatifid, oblanceolate to obovate, narrowed to broad flattened petioles, the lobes distinctly white-margined and pectinately toothed; involucre broadly cylindric, on short peduncles and usually not surpassing the leaves; inner phylllaries linear and acuminate, 9–12 mm. long, broadly scarious-margined; outer calyculate bractlets loosely spreading, spatulate and pectinate at apex, the apex with broad pectinate teeth, the lower part deeply lacerate-fringed; ligules short, little exserted, white turning pink in age; achenes about 4 mm. long; pappus very soft and fine, longer than the achene, the inner bristles usually united at the base.

Dry sandy washes and dunes, Sonoran Zones; southeastern Oregon from Lake to Malheur Counties southward east of the Sierra Nevada to the Mojave Desert, California, and east through Nevada. Type locality: "Sandy Artemisia plain in Truckee Pass of the Virginia Mountains, in a cañon of the Trinity Mountains, and in Unionville Valley, Nevada." April–July.

2. **Glyptopleura setulòsa** A. Gray. Large-flowered Carved-seed or Keyesia. Fig. 6008.

Glyptopleura setulosa A. Gray, Proc. Amer. Acad. **9**: 211. 1874.
Glyptopleura marginata var. *setulosa* Jepson, Man. Fl. Pl. Calif. 1008. 1925.

Low, depressed, densely flowered annual closely resembling the preceding species in general habit. Leaves like those of the preceding species, the white-crustaceous margin usually narrower; phyllaries linear-lanceolate, scarious-margined; calyculate bractlets with the white margins at apex pectinately dissected into slender teeth as long as the width of the green, almost circular tip; ligules well exserted, 10–15 mm. long, twice as long as the involucre, cream-colored or yellowish, aging pink; achenes and pappus similar to preceding species.

Sandy desert flats, Lower Sonoran Zone; western and northern Mojave Desert in southern Kern County and in Inyo County, California, eastward to northwestern Arizona and southwestern Utah. Type locality: near St. George, Utah. April–June.

161. **CALYCÓSERIS** A. Gray, Smiths. Contr. **5**[6]: 104. *pl. 14.* 1853.

Annuals branching from the base, glabrous and glaucescent below, glandular with tack-shaped glands above. Leaves pinnately parted into linear divisions. Ligules showy, in rather large heads terminating the branches. Phyllaries linear, scarious-margined, much longer than the calyculate bractlets. Receptacle with capillary bristles. Achenes 5–6-ribbed, narrowed above into a short beak tipped by a shallow denticulate crown. Pappus-bristles white and copious, all united at base and falling away in a ring. [Name from Greek meaning cup, alluding to the shallow cup at the summit of the achene, and *seris*, chicory.]

A genus of 2 species, inhabiting southwestern United States and adjacent Mexico. Type species, *Calycoseris wrightii* A. Gray.

Ligules white, purplish-veined; ribs of achenes more or less tuberculate; tack-shaped glands pale yellowish throughout.
1. *C. wrightii.*

Ligules yellow; ribs of achenes smooth; tack-shaped glands blackish-purple-tipped.
2. *C. parryi.*

1. **Calycoseris wrìghtii** A. Gray. White Tack-stem. Fig. 6009.

Calycoseris wrightii A. Gray, Smiths. Contr. **5**[6]: 104. *pl. 14.* 1853.
Calycoseris wrightii var. *californica* Brandg. Zoe **5**: 155. 1903.

Stems several from the base or rarely solitary, erect or ascending, simple or usually branching,

1–5 dm. high, pale green and glaucous or sometimes rose-colored below, glandular above with light yellowish, tack-shaped glands. Leaves basal and cauline, the lower 6–15 cm. long, pinnately divided into remote, linear or almost filiform and divergent segments, the uppermost reduced and bract-like; involucres campanulate, 10–12 mm. high; phyllaries 12–18, ascending, linear-lanceolate, broadly scarious-margined, usually purplish at tip, and with scattered, yellowish, tack-shaped glands mostly along the midrib below the middle, the calyculate bractlets reduced, with their acuminate tips more or less recurved; ligules 2–3 cm. long, white with rose-colored veins on the back extending down the tube from the rather prominent, attenuate teeth; achenes dark brown, tuberculate on the ribs, 6 mm. long including the beak; pappus-bristles 7–8 mm. long.

Desert washes and hillsides, Lower Sonoran Zone; Shoshone and Death Valley regions, Inyo County south along the western edges of the Colorado Desert of southern California east to southern Utah and Texas, and in northern Sonora. Type locality: "Stony hills around El Paso," Texas. March–May.

2. **Calycoseris párryi** A. Gray. Yellow Tack-stem. Fig. 6010.

Calycoseris parryi A. Gray in Torr. Bot. Mex. Bound. 106. 1859.

Stems several from the base, erect or ascending, simple or branching above, 6–30 cm. high, glandular above with tack-shaped, blackish-purple glands. Basal leaves more or less glaucous, 5–12 cm. long, pinnately divided into narrowly linear lobes of uneven length and widely divergent from the narrowly linear blade; involucre rather sparsely beset with tack-shaped, purple-tipped glands; phyllaries 10–13 mm. long, linear, rather abruptly attenuate at apex, broadly scarious-margined, greenish dorsally or more or less purplish at the tip; calyculate bractlets few, 2–4 mm. long, scarious; ligules showy, 1.5–2 cm. long, yellow; achenes light brown or gray with smooth or nearly smooth ribs.

Usually in sandy soils, Upper and (mainly) Lower Sonoran Zones; desert regions from Inyo County, California, and adjacent Nevada, south through the Mojave and Colorado Deserts to northern Lower California and east to southern Utah and Arizona. Type locality: Reported as "Mountains east of Monterey, California"; but probably not from that locality. Collected by Parry. March–June.

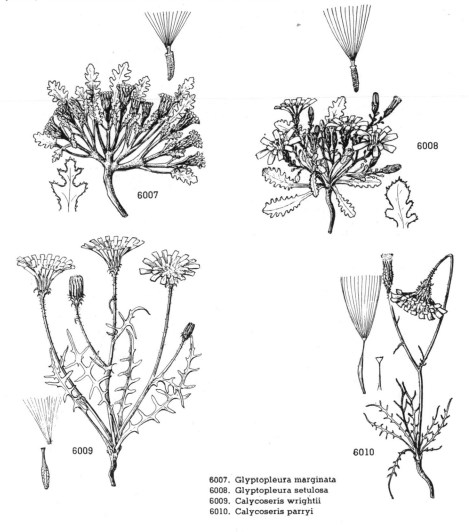

6007. Glyptopleura marginata
6008. Glyptopleura setulosa
6009. Calycoseris wrightii
6010. Calycoseris parryi

162. **ANISÓCOMA** Torr. & Gray, Bost. Journ. Nat. Hist. **5**: 111. 1845.

Annual herbs with basal leaves toothed or pinnately parted and several monocephalous scapes. Involucre cylindric, the phyllaries with broad scarious margins, the outer short, broad and obtuse, the inner linear and acutish. Receptacle flat, the chaffy bracts linear-filiform and scarious. Ligules pale yellow. Achenes terete, linear-turbinate, 10–15-nerved, the truncate apex with a narrow cup-like rim within which is inserted 10–12 long, white, plumose pappus-bristles. [Name Greek, meaning unequal and a tuft of hair, in reference to the pappus.]

A monotypic genus of the desert regions of southwestern United States.

1. **Anisocoma acaùlis** Torr. & Gray. Anisocoma or Scale-bud. Fig. 6011.

Anisocoma acaulis Torr. & Gray, Bost. Journ. Nat. Hist. **5**: 111. *pl. 13, figs. 7–11.* 1845.
Pterostephanus runcinatus Kell. Proc. Calif. Acad. **3**: 21. *fig. 4.* 1863.

Low winter annual with a rather deep-seated taproot. Leaves in a rather dense basal rosette, pinnatifid, 2–7 cm. long, 3–10 mm. wide, the later often runcinate and woolly-tomentose beneath, plants otherwise glabrous; scapes usually several to many, 5–25 cm. high, glabrous and bractless; involucre 1.5–3 cm. high, the phyllaries often margined with reddish brown, the outer short and orbicular, the innermost elongated and oblong-linear; ligules conspicuous and pale yellow.

Sandy desert washes, Sonoran Zones; Tehachapi Mountains and east of the Sierra Nevada in Inyo County and central Nevada south in the desert regions to northern Lower California and northwestern Arizona. Type locality: not indicated. April–July.

163. **ATRICHÓSERIS** A. Gray, Syn. Fl. N. Amer. **1²**: 410. 1884.

Annual herb with a branched stem, glabrous throughout. Leaves in a basal tuft, cuneate or obovate, spinulose-denticulate. Stem-leaves reduced to small foliaceous bracts. Stems solitary, slender-branched above. Heads in a cymose panicle on slender peduncles. Phyllaries scarious, lanceolate, about 15, with a few small ones at base. Receptacle scrobiculate. Ligules white or pinkish. Achenes oblong, corky-ribbed, without pappus. [Name Greek, meaning without hair, and *seris*, a name applied to chicory.]

A monotypic genus.

1. **Atrichoseris platyphýlla** A. Gray. Tobacco-weed or Parachute Plant. Fig. 6012.

Malacothrix platyphylla A. Gray, Proc. Amer. Acad. **9**: 214. 1874.
Atrichoseris platyphylla A. Gray, Syn. Fl. N. Amer. **1²**: 410. 1884.

Annual, glabrous throughout and somewhat glaucous, the stems solitary, erect, cymosely branched above, 3–7 dm. high. Leaves basal, wide-spreading and usually flat on the ground, oblong-obovate or oblong-oblanceolate, 3–10 cm. long, 2–5 cm. wide, narrowed at base to a usually short, broad, winged petiole, glaucous and often purplish or spotted with purple; lower part of stem often with a few scattered, clasping, foliaceous bracts, laciniate-toothed like the leaves; phyllaries about 6 mm. long, ovate-lanceolate, with broad scarious margins; corollas white or pinkish, the outer ligules 10–15 mm. long, prominently toothed at apex; achenes white or stramineous, 4 mm. long.

Sandy desert washes, Lower Sonoran Zone; Mojave and Colorado Deserts of California. Inyo County to Imperial County, California, and adjacent Nevada, east to southwestern Utah and western Arizona. Type locality: near Fort Mohave, Arizona. Collected by Cooper. Feb.–May.

164. **TRAGOPÒGON** [Tourn.] L. Sp. Pl. 789. 1753.

Glabrous biennial or rarely perennial herbs with long taproots. Leaves entire, grass-like, with conspicuous nerves, clasping at base. Peduncles terminating the stems long, stout, often thickened and fistulous; phyllaries in 1 series. Heads yellow- or purple-flowered, opening in early morning, usually closed by noon. Receptacles without bracts. Achenes fusiform, long-beaked or the outer beakless, 5–10-ribbed. Pappus a single row of long, plumose bristles cuneate at the base. [Name Greek, meaning goat's-beard.]

A genus of about 35 species, all natives of the Old World. Type species, *Tragopogon pratensis* L.

Peduncles scarcely inflated even in fruit; outer ligules equaling the phyllaries.   1. *T. pratensis.*
Peduncles strongly inflated; ligules shorter than the phyllaries.
     Ligules pale to deep violet-purple; phyllaries usually 8–9.   2. *T. porrifolius.*
     Ligules pale yellow; phyllaries usually 13.   3. *T. dubius.*

1. **Tragopogon praténsis** L. Meadow Salsify. Fig. 6013.

Tragopogon pratensis L. Sp. Pl. 789. 1753.

Stems from a taproot, usually branched, rather slender, 4–10 dm. high. Leaves keeled from the more or less clasping base to a long-acuminate, usually recurved tip, the margins concave and crisped below the heads even in fruit; phyllaries 8 or 9, rarely 13, broadly lanceolate, acuminate, shorter than or rarely slightly exceeding the chrome-yellow ligules; marginal achenes striate, smooth or roughened, usually dark brown, the inner paler, 20–25 mm. long, abruptly tapering to the beak; pappus whitish.

Fields and waste places, sparingly adventive in western Washington and Oregon to Shasta County, California, and more frequently east of the Cascade Mountains, especially in Union and Wallowa Counties, Oregon; also northern Arizona. Widely distributed in Canada and eastern United States. Native of Europe. June–Sept.

2. **Tragopogon porrifòlius** L. Salsify or Oyster Plant. Fig. 6014.

Tragopogon porrifolius L. Sp. Pl. 789. 1753.

Stems simple from a long taproot, 5–12 dm. high. Leaves tapering uniformly to the apex, dilated and clasping at base; peduncles much inflated and hollow from 2.5–4 cm. below the heads; phyllaries 8–9, rarely 12, linear-lanceolate, acuminate, usually much longer than the outer ligules; ligules purple; achenes about 3–4 mm. long, the outer ones covered with scale-like tubercles, especially on the ribs, the beak much longer than the body and abruptly narrowed; plumose branches of the pappus-bristles brownish, long and mat-like.

Naturalized as a weed in the Pacific States, also, but less frequently, eastward across the continent. Native of the Old World. April–Nov.

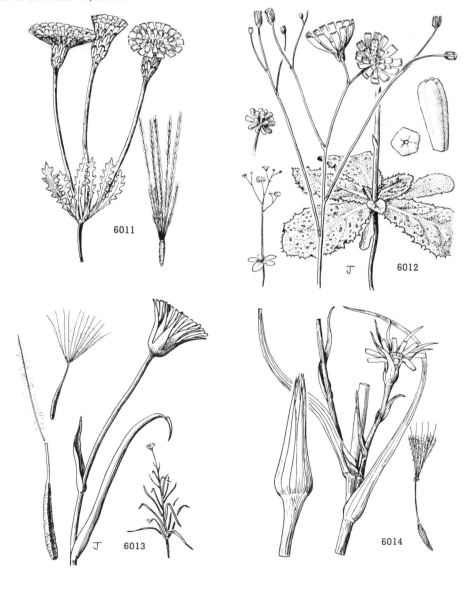

6011. Anisocoma acaulis
6012. Atrichoseris platyphylla
6013. Tragopogon pratensis
6014. Tragopogon porrifolius

3. Tragopogon dùbius Scop. Yellow Salsify. Fig. 6015.

Tragopogon dubius Scop. Fl. Carn. ed. 2. **2**: 95. 1772.
Tragopogon major Jacq. Fl. Austr. Select. Icones **1**: 19. *pl. 29.* 1773.

Stems from a taproot, 4–8 dm. high, rather bushy, the branches from near the base. Leaves linear-lanceolate, 12–15 cm. long, tapering evenly from base to apex, ratner floccose when young, becoming glabrous and glaucous; heads many-flowered, the peduncle much inflated below the involucre; phyllaries about 13, about 33 mm. long, lanceolate to linear-lanceolate, attenuate; ligules pale yellow, shorter than the phyllaries; achenes slender, 25–36 mm. long including the beak, the body gradually narrowed to the beak, the outer achenes pale brown, the inner paler and straw-colored; pappus whitish.

In fields and waste places in southeastern Washington and adjacent Idaho and occurring sporadically southward; also eastward to Arizona and Texas. It has been collected in Siskiyou, Santa Clara, and San Bernardino Counties, California. Native of Europe. May–Sept.
The recent work on natural hybridization and amphiploidy in *Tragopogon* (Amer. Journ. Bot. **37**: 487–499. 1950.) by Marion Ownbey demonstrates the occurrence of two new amphiploids which have arisen by natural hybridization in southeastern Washington and adjacent Idaho, where all three previously known species occur. The amphiploids are *Tragopogon mirus* Ownbey, Amer. Journ. Bot. **37**: 497. 1950 (*T. dubius* × *porrifolius*) and *T. miscellus* Ownbey, op. cit. 498, derived from *T. dubius* × *pratensis*.

165. SCORZONÈRA L. Sp. Pl. 790. 1753.

Perennial, rarely annual, herbs. Leaves alternate, usually entire, often grass-like, sometimes pinnately lobed or dissected. Flowering heads on long peduncles, yellow, rose, or lilac. Involucres several-seriate, the outer phyllaries much shorter than the inner. Receptacle naked, foveolate. Ligules truncate, 5–toothed at apex. Anthers sagittate at base, the style-branches slender. Achenes linear, subterete, many-nerved. Pappus-bristles in more than 1 series, unequal, serrulate or more or less plumose. [From the old French, meaning serpent.]

A genus of about 120 species, natives of the Old World. Type species, *Scorzonera humilis* L.

1. Scorzonera hispánica L. Black Salsify. Fig. 6016.

Scorzonera hispanica L. Sp. Pl. ed. 2. 1112. 1763.
Scorzonera glatifolia Willd. Sp. Pl. **3**: 1499. 1804.

Branching perennial 6–10 dm. high from a fleshy, black-skinned taproot, the herbage woolly-pubescent to nearly glabrous. Leaves oblong to lanceolate or linear, the margin entire, undulate, tapering below to a petioliform base, the stem-leaves few, much reduced above; heads solitary at the ends of the branches, the peduncles 15–30 cm. long; involucre urceolate in flower, 2–2.5 cm. long, the ligules yellow; phyllaries few, usually 3-seriate, ovate-acuminate, somewhat woolly on the margin, the outer much shorter than the inner; achenes 1.5 cm. long, whitish, slender, many-nerved, those of the outer flowers sparsely tuberculate; pappus-bristles of unequal length, 1 cm. long or less, plumose with soft hairs.

A native of southern and central Europe which has become established on roadsides and waste places in Napa and Mendocino Counties, California. June–July. Viper's Grass.

166. HYPOCHOÈRIS L. Sp. Pl. 810. 1753.

Herbs with a basal tuft of leaves and bracteolate, usually branched, scapose stems. Involucres solitary at the ends of the scapose branches, oblong-cylindric to campanulate, the phyllaries in several series, herbaceous. Receptacle flat, chaffy. Ligules yellow. Achenes oblong to linear, 10-ribbed, constricted above or the outer truncate. Pappus of a single row of plumose bristles, or sometimes the outer ones shorter and not plumose. [A name of Theophrastus for some member of this tribe.]

About 50 species, natives of Europe, the Mediterranean Region, Asia, and South America. Type species, *Hypochoeris glabra* L.

Annual; heads 4–6 mm. broad; outermost achenes truncate at apex, not beaked.   1. *H. glabra.*
Perennial; heads 20–40 mm. broad; achenes all beaked.   2. *H. radicata.*

1. Hypochoeris glàbra L. Smooth Cat's-ear. Fig. 6017.

Hypochoeris glabra L. Sp. Pl. 811. 1753.

Annual with a slender taproot; stems solitary or commonly several, erect or somewhat decumbent, simple or often corymbosely branched, 1–4 dm. high, glabrous, bracteate especially at the base of the branches. Leaves spreading on the ground, denticulate to pinnatifid, oblanceolate to oblong-oblanceolate, 2–15 cm. long, glabrous or sometimes ciliate on the margins; involucres campanulate, 10–16 mm. long; ligules scarcely exceeding the phyllaries; achenes dark brown, the outermost beakless, the others with slender beaks as long as the body; pappus 1 cm. long.

Common on roadsides, fields, and waste places, Sonoran and Transition Zones; Washington, Oregon, and California, also eastern United States and Canada. Native of the Old World. March–Sept.

2. Hypochoeris radicàta L. Hairy or Long-rooted Cat's-ear. Fig. 6018.

Hypochoeris radicata L. Sp. Pl. 811. 1753.

Perennial; stems several, slender, 3–8 dm. high, branched or rarely simple, bracteate. Leaves basal and spreading on the ground, oblanceolate to obovate, pinnatifid to dentate, 5–30 cm. long, hirsute; involucre oblong-cylindric, 2–2.5 cm. long; ligules usually well exceeding the phyllaries; achenes brown, all with slender beaks as long as the body; pappus plumose.

Roadsides, waste places, and pasture land, Sonoran and Transition Zones; frequent weed throughout the Pacific States and across the Continent. Native of the Old World. April–Nov.

167. HEDÝPNOIS [Tourn.] Mill. Gard. Dict. abr. ed. 4. 1754.

Annual herbs with branched stems and yellow flowers. Phyllaries in 1 series, narrow and incurved, enfolding the marginal achenes, spreading in age. Receptacle naked. Achenes 5–10-ribbed, the ribs barbellate. Pappus on the outer achenes in 1 series of denticulate or fimbriate scales, that of the inner achenes usually in 2 series; the outer series composed of short scales or sometimes absent, the inner of bristles dilated toward the base. [Name given by Pliny to a kind of endive.]

A genus of 3 species, natives of the Mediterranean Region and the Canary Islands. Type species, *Hedypnois annua* Mill. (*Hyoseris hedypnois* L. 1753).

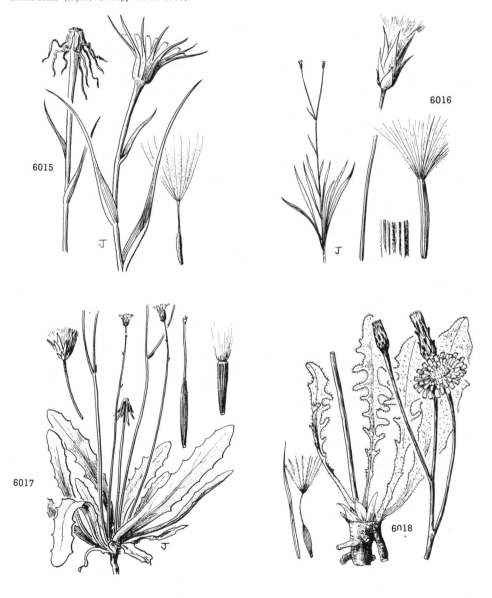

6015. Tragopogon dubius
6016. Scorzonera hispanica

6017. Hypochoeris glabra
6018. Hypochoeris radicata

1. **Hedypnois crètica** (L.) Willd. Crete Hedypnois. Fig. 6019.

Hyoseris cretica L. Sp. Pl. ed. 2. 1139. 1763.
Rhagadiolus creticus All. Fl. Ped. 1: 226. 1785.
Hedypnois cretica Willd. Sp. Pl. 3: 1617. 1804.

Stems simple or usually branched from the base, 1–3 dm. long, hispidulous, the hairs simple or minutely forked at the tip. Basal leaves oblanceolate, tapering to a winged petiole, entire or sparsely denticulate or sometimes lobed, 2–10 cm. long, 3–15 mm. wide, sparsely hispidulous; peduncles naked, solitary, or on 2 or more branched stems; involucre 8–10 mm. high, enlarged in fruit, hispidulous and sometimes ciliate on the margins; achenes appressed-scabrous on the ribs; pappus-bristles 5 mm. long.

Locally naturalized in dry interior valleys of the Inner Coast Range and the Sierra Nevada foothills west to the coast and south to San Diego, California. Native of Eurasia and originally described from Crete. March–May.

168. **PÍCRIS** L. Sp. Pl. 792. 1753.

Erect, hispid, leafy, and mostly branching herbs with mostly alternate leaves and rather large, corymbose or paniculate heads of yellow flowers. Involucre campanulate or cupulate, its inner phyllaries in 1 series, nearly equal and erect, its exterior ones in 2–3 series, small or large and spreading. Receptacle flat, short-fimbrillate. Rays truncate and 5-toothed at apex. Anthers sagittate at the base. Style-branches slender. Achenes linear or oblong, slightly curved, terete or angled, 5–10-ribbed and transversely wrinkled, narrowed at the base and apex, or beaked in some species. [Name Greek, meaning bitter.]

About 35 species, natives of the Old World, or one species possibly indigenous in Alaska. Type species, *Picris asplenioides* L.

1. **Picris echioìdes** L. Bristly Ox-tongue. Fig. 6020.

Picris echioides L. Sp. Pl. 792. 1753.

Perennial with branched hispid stems 5–12 dm. high. Leaves beset with coarse, barbed, usually pustulate-based bristles, the basal and lower leaves spatulate or oblong, obtuse, repand-dentate, 5–15 cm. long, narrowed into petioles; upper leaves smaller, sessile and clasping, oblong or lanceolate, the uppermost mainly acute and entire; heads many, usually rather crowded, short-peduncled, about 1.5 mm. broad; outer phyllaries of the involucre 4 or 5, foliaceous, subcordate, hispid-ciliate; inner phyllaries lanceolate, thin and tipped with a prickly awn.

A native of Europe, naturalized in waste places, especially along roadsides, widely distributed but most common in central and southern California; also Michigan east to the Atlantic seaboard. May–Nov.

169. **LEÓNTODON** L. Sp. Pl. 798. 1753.

Annual or perennial herbs with heads borne on simple or branched, naked or scaly-bracted scapes. Leaves in a basal rosette, entire, toothed, or pinnatifid. Heads many-flowered, yellow. Phyllaries of the inner series of the involucre subequal, the outer much shorter, calyculate. Flowers perfect, ligulate, the ligules yellow. Receptacle alveolate, the pits often with toothed or ciliate margins. Anthers not appendaged below, with blunt appendages apically. Style-branches long and slender. Achenes fusiform, with a short beak or merely narrowed above, with more or less muriculate ribs. Pappus persistent, of plumose bristles flattened at the base and sometimes with an outer row of short paleae, the pappus of the marginal flowers often of scarious paleae only. [From the Greek words meaning lion and tooth, in allusion to the toothed leaves.]

A genus of about 45 species, natives of Europe, western and central Asia, and northern Africa. Type species, *Leontodon hispidus* L.

1. **Leontodon lèysseri** (Wallr.) G. Beck. Hairy Hawkbit. Fig. 6021.

Crepis nudicaulis L. Sp. Pl. 805. 1753. (Nomen confusum)
Leontodon hirtum of authors, not L. Syst. ed. 10. 1194. 1759.
Hyoseris taraxacoides Vill. Prosp. Pl. Delph. 33. 1779.
Thrincia hispida Roth, Catal. Bot. 1: 99. 1797.
Thrincia leysseri Wallr. Sched. Crit. 441. 1822.
Leontodon nudicaulis Merat, Ann. Sci. Nat. 22: 109. 1831.
Leontodon rothii Ball, Journ. Linn. Soc. 16: 543. 1878.
Leontodon leysseri G. Beck, Fl. Nied-Oesterr. 2²: 1312. 1893.
Leontodon nudicaulis subsp. *rothii* Schinz & Thell. Bull. Herb. Boiss. II. 7: 389. 1907.
Leontodon nudicaulis subsp. *taraxacoides* Schinz & Thell. loc. cit.

Scapose perennial with fibrous roots and short caudex. Leaves many in a basal rosette, 5–15 cm. long, narrowly oblanceolate, tapering to a petioliform base, midvein prominent, sometimes nearly entire, more often shallowly toothed to runcinate-pinnatifid, with hirsute spreading pubescence; heads solitary, the slender scapes several to many, 10–35 cm. long, naked above, spreading-hirsute toward the base; involucre 6–10 mm. long, the inner phyllaries subequal, lanceolate, glabrous or hirsute, usually medially darkened, the outer reduced to small calyculate bractlets; achenes 3–4 mm. long, fusiform, more or less shortly beaked above, striate, the striae muricate, pappus of the marginal flowers usually reduced to short laciniate scales, that of the inner of plumose bristles, exceeding the achene in length and flattened below to a scarious base.

A European weed of lawns, waste places, and roadsides from British Columbia west of the Cascade Mountains south to central California; also introduced in eastern United States.

Because of the confusion created by efforts to assign a precise determination of the basonym *Crepis nudicaulis*, authors using the next clearly available name are here followed.

Leontodon autumnàlis L. Sp. Pl. 798. 1753. Differs from *L. leysseri* in having the inflorescence of several heads, the peduncles scaly-bracted and enlarged below the head, and the pappus of both inner and outer flowers a single row of plumose bristles. A European weed rather commonly distributed in the east but known in the Pacific States from a single collection in Bellingham, Washington (Blake, Leaflets West, Bot. 7: 285–86. 1955) and also from Alaska.

170. SÓNCHUS [Tourn.] L. Sp. Pl. 793. 1753.

Annual, biennial or perennial, rather fleshy coarse herbs with alternate, mostly auriculate-clasping, entire or pinnatifid, prickly-margined leaves. Inflorescence corymbose, paniculate, or subumbellate, the heads yellow-flowered. Involucre in ours becoming thickened and indurated at base in age, the phyllaries somewhat imbricated, the thin tips reflexed in age, subtended by calyculate bractlets. Receptacle flattened, naked. Achenes obcompressed, ribbed or striate, often rugose, truncate at apex. Pappus much surpassing the achene in length, of copious, soft, capillary bristles adhering at the base and falling away in a ring. [The Greek name for the sow-thistle.]

An Old World genus of about 45 species. Type species, *Sonchus oleraceus* L.

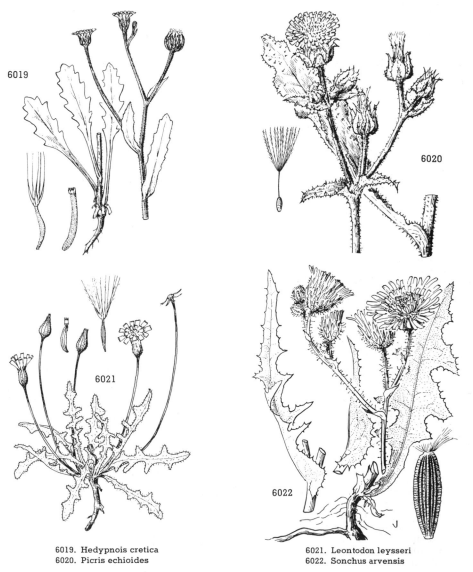

6019. Hedypnois cretica
6020. Picris echioides

6021. Leontodon leysseri
6022. Sonchus arvensis

Perennial with deep vertical root and horizontal creeping root-branches; flowering-heads 4–5 cm. broad.
                                                                                    1. *S. arvensis.*
Annual or biennial; flowering-heads less than 4–5 cm. broad.
    Auricles of the leaves acute; achenes striate, transversely wrinkled.
        Leaf-blades serrate or with lobes 0.5–3 cm. wide; achenes about as broad as thick.   *2. S. oleraceus.*
        Leaf-blades pinnately parted with narrow lobes mostly less than 0.5 cm. wide; achenes only slightly wider
          than thick.   3. *S. tenerrimus.*
    Auricles of the leaves rounded; achenes usually conspicuously 3-nerved, smooth or nearly so, and thin-
      margined.   4. *S. asper.*

1. Sonchus arvénsis L. Perennial Sow-thistle. Fig. 6022.

Sonchus arvensis L. Sp. Pl. 793. 1753.

Perennial with creeping rootstocks; stems 6–10 dm. high, glabrous below, usually glandular-bristly above, especially on the peduncles. Leaves 3–10 cm. long, runcinate-divided or sometimes merely denticulate, the margins spinose-denticulate, the lower narrowed to a winged petiole, the upper sessile, auriculate-clasping at base and rounded; inflorescence terminal, corymbose or even subumbelliferous; involucres 14–20 mm. high, the phyllaries attenuate at apex, glandular-bristly as are also the peduncles; ligules bright orange-yellow, the heads up to 3–5 cm. broad at anthesis; achenes oblong, flattened, 2.5–3.5 mm. long, usually 5-ribbed on each face, transversely rugose; pappus 8–10 mm. long.

Naturalized from Europe but not common in the Pacific States. Puget Sound area and Whitman County, Washington; eastern Oregon and west of the Cascade Mountains in Willamette Valley; occurring sporadically in California where it has been collected at a few scattered localities throughout the state. April–July.

Sonchus arvensis var. **glabréscens** Guenth. Graeb. & Wimmer, Enum. Stirp. Phan. Siles. 127. 1824. (*Sonchus uliginosus* Bieb. Fl. Taur. Cauc. 2: 238. 1808.) Resembling *S. arvensis* var. *arvensis* in habit and variability of leaves but lacking the glandular bristles on involucre and peduncles of that taxon. The phyllaries are often more pale. A European weed more common in north central and northeastern United States, and occurring sporadically in the Pacific States as a weed but less common than *S. arvensis* var. *arvensis*.

2. Sonchus oleràceus L. Common Sow-thistle. Fig. 6023.

Sonchus oleraceus L. Sp. Pl. 794. 1753.

Plants usually stoutish, 5–10 dm. high, sparingly leafy, glabrous or with a few glandular hairs on the peduncles and involucres, often glaucescent. Leaves variable in shape, ovoid or narrower, simple or runcinate-pinnatifid, toothed but not prickly-margined, amplexicaul, the auricles straight and the lobes acute; ligules yellow; achenes about 2 mm. long, striate-nerved, strongly transversely rugose-scabrous on nerves and intervals, not strongly flattened on the margins.

A common garden and roadside weed throughout the Pacific States and across the continent. Native of Europe Spring–Autumn.

3. Sonchus tenérrimus L. Slender Sow-thistle. Fig. 6024.

Sonchus tenerrimus L. Sp. Pl. 794. 1753.
Sonchus tenuifolius Nutt. Trans. Amer. Phil. Soc. II. 7: 438. 1841.

Plants rather slender, leafy-stemmed to the inflorescence, glabrous, 3–8 dm. high. Leaves oblong in outline, deeply divided into several pairs of linear spreading lobes, these entire or denticulate, the auricles narrow, lanceolate to ovate, sometimes attenuate; inflorescence a short corymb, mostly with few heads; phyllaries linear-lanceolate, 8–12 mm. long, thickened below in age; ligules yellow; achenes narrowly ovoid, 2.5 mm. long, but little compressed, longitudinally striate and finely and distinctly rugose on both striae and intervals.

Introduced from Europe, sandy and rocky soil; Tulare County, the Channel Islands, Santa Barbara County, south to western San Diego County, California, south to Lower California. March–June.

4. Sonchus ásper (L.) Hill. Prickly Sow-thistle. Fig. 6025.

Sonchus oleraceus var. *asper* L. Sp. Pl. 794. 1753.
Sonchus asper Hill, Herbarium Brit. 1: 47. 1769.

Similar to *Sonchus oleraceus* but stems usually stouter, distinctly angled, and more leafy. Leaves undivided, lobed or sometimes pinnatifid, the margins spinulose-dentate or spinulose-denticulate, the lower and basal obovate to spatulate and petioled, the upper oblong or lanceolate and clasping by an auriculate base, the auricles helicoid and appressed to the stem, the lobes rounded; ligules yellow; achenes 2.5–3 mm. long, strongly usually 3-ribbed on each face, not at all transversely rugose, the thin, wing-like margin serrulate.

A fairly common field and roadside weed in the Pacific States and also across the continent. Naturalized from Europe. June–Nov.

171. LACTÙCA [Tourn.] L. Sp. Pl. 795. 1753.

Leafy-stemmed annual or perennial herbs, glabrous or nearly so. Leaves linear and entire to broader and pinnatifid. Heads 5–56-flowered; ligules yellow or blue, in small or medium-sized, paniculate heads. Involucre cylindrical, the phyllaries usually imbricated in several series. Receptacle flat, naked. Ligules truncate, 5-toothed at the apex. Achenes oval, oblong, or linear, flattened, ribbed on each face, abruptly or gradually beaked. Pappus copious, of soft capillary bristles. [The ancient Latin name, from *lac*, milk, referring to the milky juice.]

A genus of about 50 or more species, natives of the northern hemisphere. Type species, *Lactuca sativa* L.

The common garden lettuce is occasionally collected in waste places but apparently does not become established.

Annual or biennial; ligules not long-exserted, shades of yellow (bluish in *L. biennis* and forms of *L. ludoviciana*).
    Achenes prominently 1-nerved on each side.
        Fruiting involucres 10–15 mm. high; pappus 5–7 mm. long.   1. *L. canadensis.*
        Fruiting involucres 15–22 mm. high; pappus 7–12 mm. long.   2. *L. ludoviciana.*
    Achenes evidently several-nerved on each side.
        Achenes with a long filiform beak as long as or longer than the achene.
            Achenes black, conspicuously wing-margined.   3. *L. virosa.*
            Achenes gray or brownish, not wing-margined.
                Inflorescence broadly paniculate with spreading branches; stem-leaves oblong to elliptic in outline.
                        4. *L. serriola.*
                Inflorescence spiciform-paniculate; stem-leaves mostly linear-lanceolate.
                        5. *L. saligna.*
        Achenes with a short beak much shorter than the achene or beakless.
            Pappus tawny; ligules 15–35, white or bluish (in ours).   6. *L. biennis.*
            Pappus white; ligules 5, yellow.   7. *L. muralis.*
Perennial with taproot and deep-seated crown; ligules long-exserted and showy, blue or bluish purple.
                        8. *L. tartarica pulchella.*

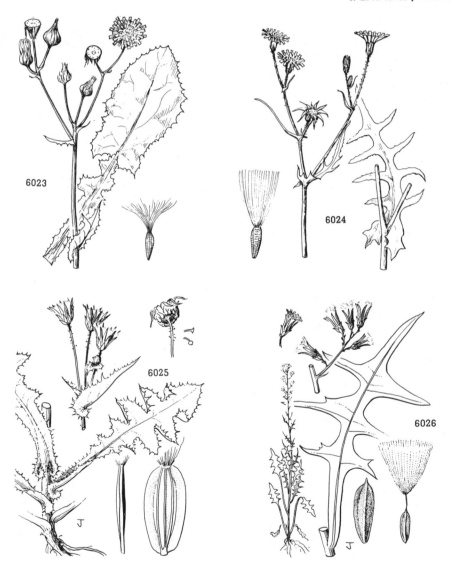

6023. Sonchus oleraceus   6025. Sonchus asper
6024. Sonchus tenerrimus   6026. Lactuca canadensis

1. Lactuca canadénsis L. Trumpet Fireweed. Fig. 6026.

Lactuca canadensis L. Sp. Pl. 796. 1753.
Lactuca elongata Muhl. ex. Willd. Sp. Pl. **3**: 1525. 1804.
Lactuca polyphylla Rydb. Bull. Torrey Club **38**: 23. 1911.

Stout, weedy, simple-stemmed annual or perennial 3–25 dm. high, glabrous or sometimes with coarse hirsute hairs. Leaves entire, toothed, or deeply pinnately lobed, sagittate and often narrowed at base, 10–35 cm. long; panicle elongate, spreading, the peduncles bracteate; heads 13–22-flowered with yellow ligules; fruiting involucres usually 10–15 mm. high, the phyllaries in 4 or 5 series; achenes black, obovate, flat, transversely rugose with 1 prominent nerve on each face, the beak about one-half as long as to equal the body of the achene; pappus white, 7–10 mm. long.

Fields and woodland, introduced from the eastern United States, occurring locally in Washington, in Wallowa County, Oregon, and in Siskiyou, Shasta, Plumas, and Amador Counties, California; also Idaho. Type locality: Canada. July–Sept.

A taxon quite variable as to leaf-shape and many named forms have been described. Both the entire-leaved and pinnatifid-leaved forms are found in the Pacific States.

2. Lactuca ludoviciàna (Nutt.) DC. Western Lettuce. Fig. 6027.

Sonchus ludovicianus Nutt. Gen. **2**: 125. 1818.
Lactuca ludoviciana DC. Prod. **7**: 141. 1838.
Lactuca campestris Greene, Pittonia **4**: 37. 1899.

Biennial, the stems leafy up to the inflorescence, paniculately branched above, 6–20 dm. high. Leaves auriculate-clasping, spinulose-denticulate, sinuate-lobed or pinnatifid with spinulose segments, setose-hispid on the midrib, otherwise glabrous throughout; heads numerous in a large open panicle, their peduncles bracteate; fruiting involucre 15–22 mm. high, cyclindric or ovoid-cylindric, glabrous, the phyllaries in 4 or 5 series and successively shorter and broader; ligules yellow; achenes oval to obovate, flat, mottled with gray and black, transversely rugose, 1-ribbed on each face, beak about equaling the body, dilated below; pappus white.

Sparingly introduced in the Pacific States, Klickitat County, Washington; Wallowa County, Oregon; Shasta County, California, and vicinity of San Bernardino, southern California. Type locality: "Fort Mandan on the Missouri." June.

3. Lactuca viròsa L. Wild Lettuce. Fig. 6028.

Lactuca virosa L. Sp. Pl. 795. 1753.

Biennial, the stems stout, often 6–8 dm. high. Leaves ample, broadly obovate, 6–15 cm. long, strongly and sharply dentate or sometimes divided, glabrous, the midvein often prickly on the dorsal side; panicle ample, often 30 cm. long and 15–20 cm. broad; heads 6–12-flowered, yellow; involucres 10–12 mm. high; achenes 3.5–4 mm. long, narrowly obovate, black, 4–5-ribbed within the thickened winged margin, inconspicuously spinulose at apex; beak slender, about equaling the body of the achene; pappus white.

Usually on banks or partly shaded slopes; a European weed rather sparingly introduced in the San Francisco Bay region, California. June–Sept.

4. Lactuca serriòla L. Prickly Lettuce. Fig. 6029.

Lactuca serriola L. Cent. Pl. **2**: 29. 1756; Amoen. Acad. **4**: 328. 1759.
Lactuca scariola L. Sp. Pl. ed. 2. 1119. 1763.

Biennial, the stems leafy, paniculately branched above, 5–20 dm. high, glabrous throughout or prickly-hispid below, pale green or straw-colored. Leaves oblong or oblanceolate, sagittate-clasping, lobed or pinnatifid, the lower 5–20 cm. long, the upper much smaller, spinulose-hispid on the midrib and margins, otherwise glabrous or more or less hispid; heads numerous in a usually large, ovoid panicle, 6–12-flowered, ligules yellow; involucre subcylindric or narrowly conical, 12–16 mm. high, the phyllaries irregularly imbricated in about 4 series, linear-lanceolate; achenes obovate-oblong, about equaled by the slender beak; pappus white.

Fields and waste places; a common weed in the Pacific States, also generously distributed in the United States. Native of Europe. June–Sept.

The plant with the leaves simple instead of lobed or divided has been recognized as a form (*L. serriola* var. *integrata* Gren. & Godr. Fl. France **2**: 320. 1850).

5. Lactuca salígna L. Willow Lettuce. Fig. 6030.

Lactuca saligna L. Sp. Pl. 796. 1753.

Annual, glabrous throughout, the stems usually several from a rather stout taproot, decumbent at base, 3–7 dm. high. Leaves pale green, narrowly linear-lanceolate, attenuate at apex, sagittate-clasping at base, entire or sometimes toothed or few-lobed; heads 8–10-flowered, yellow, in narrow spicate panicles with short ascending branches; involucres narrowly cylindric, 12–15 mm. long; achenes light brown, 3 mm. long, oblong-oblanceolate, rather abruptly attenuate at apex, the ribs 5–7 on each, minutely spiculate near the apex, the beak slender, 5–6 mm. long; pappus-bristles white, about as long as the beak.

This European weed has become well established in many places throughout California and has also been found in Douglas and Marion Counties, Oregon; also eastward to the Atlantic seaboard. Type locality: Europe. Aug.–Oct.

6. Lactuca biénnis (Moench) Fernald. Tall Blue Lettuce. Fig. 6031.

Sonchus biennis Moench, Meth. 545. 1794.
Sonchus racemosus Lam. Encycl. **3**: 400. 1789. Not *Lactuca racemosa* Willd. 1804.

Sonchus leucophaeus Willd. Sp. Pl. **3**: 1520. 1804.
Lactuca spicata Hitchcock in Britt. & Brown, Ill. Fl. **3**: 276. 1898, as to plant described; not as to basic type
    Sonchus spicatus Lam.
Lactuca biennis Fernald, Rhodora **42**: 300. 1940.

Biennial, glabrous throughout, the stems simple, stout, erect, 1–3.5 m. high, leafy. Leaves pin-natifid or runcinate, irregularly and sharply dentate, the upper auriculate-clasping at base, 15–30 cm. long; heads in an elongated, many-flowered panicle, ligules little-exserted, cream-colored to bluish; involucres 9–14 mm. high, somewhat constricted above, the phyllaries often purplish at tip; achenes 4 mm. long, 1.5 mm. broad, narrowly winged on the margins, 3-ribbed on each side, mottled with different shades of brown, narrowed at apex to a very short beak; pappus tawny, or some-times tinged with purple.

Wet grassy places in rich soils, Transition and Canadian Zones; Alaska, British Columbia, and Olympic Moun-tains, Washington, south to Humboldt County, California, and east across Washington and Oregon to the eastern United States and Canada. Type locality: not indicated. July–Aug.

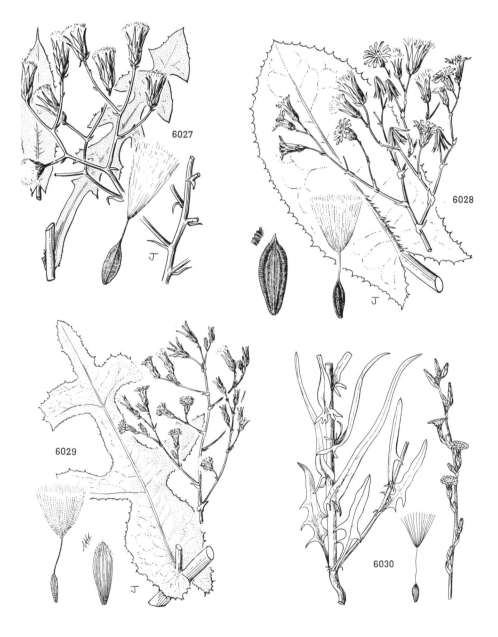

6027. Lactuca ludoviciana
6028. Lactuca virosa

6029. Lactuca serriola
6030. Lactuca saligna

7. Lactuca muràlis (L.) Fresen. Wall Lettuce. Fig. 6032.

Prenanthes muralis L. Sp. Pl. 797. 1753.
Cicerbita muralis Wallr. Sched. Crit. 436. 1822.
Mycelis muralis Reichb. Fl. Germ. Excurs. 272. 1830–32.
Lactuca muralis Fresen. Taschenb. 484. 1832.

Slender glabrous annual or perennial, stems simple, 4–9 dm. high. Lower stem-leaves thin and often purplish, lyrate- or runcinate-pinnatifid, narrowed below and auriculate-clasping, the terminal segment broad and angularly lobed, up to 18 cm. long, the upper stem-leaves few, much reduced, sessile and auriculate-clasping; inflorescence a divaricately branched panicle; involucres narrowly cylindric, 9–11 mm. high, 5-flowered, the ligules yellow; phyllaries in 2 series, the inner long, the outer calyculate; achenes about 4 mm. long, dark brownish or reddish, flattened, several-nerved on each face, the beak pale, much shorter than the achene.

A European species, found in moist woods, that has become established on Vancouver Island, British Columbia, and in Island and Clallam Counties, Washington; also New York and Quebec. June–Aug.

8. Lactuca tartárica subsp. pulchélla (Pursh) Stebbins. Blue Lettuce. Fig. 6033.

Sonchus pulchellus Pursh, Fl. Amer. Sept. 502. 1814.
Lactuca pulchella DC. Prod. 7 : 134. 1838.
Lactuca tartarica subsp. *pulchella* Stebbins, Madroño 5 : 123. 1939.

Plants with a deep perennial rootstock, glabrous throughout and somewhat glaucous, the stems leafy up to the inflorescence, 5–12 dm. high. Leaves variable, linear to lanceolate or oblong, entire, dentate, lobed, or pinnatifid, the lower sometimes petioled, the others sessile or somewhat clasping, 5–20 cm. long; heads in an elongated panicle, often leafy below; involucre cylindric, 10–12 mm. high; phyllaries well imbricated, the outer successively shorter, ovate-lanceolate, often purplish; ligules blue or violet, well exserted; achenes oblong-lanceolate, flat, short-beaked; pappus white.

Moist ground, Arid Transition Zone; British Columbia southward east of the Cascade Mountains through eastern Washington and Oregon to Mono County, California, and as a weed in the San Joaquin Valley; east to western Ontario and Kansas. Type locality: "On the banks of the Missouri." June–Oct.

172. PRENÀNTHES [Vaill.] L. Sp. Pl. 797. 1753.

Leafy-stemmed, perennial herbs. Leaves alternate, petioled below and sessile or auriculate-clasping above. Inflorescence thyrsoid or paniculate, the flowers drooping. Involucres cylindric and usually narrow, 7–15-flowered, the phyllaries in 1 or 2 series. Receptacle naked. Ligules white, shades of pink or purple, or yellowish. Style-branches slender. Achenes columnar, 4–5-angled or terete, striate. Pappus copious, white or reddish brown, capillary. [From the Greek, meaning drooping and flower.]

A genus of 27 species or less, natives of northern Europe and Asia, North America, and the Mediterranean Basin. Type species, *Prenanthes purpurea* L.

1. Prenanthes alàta (Hook.) D. Dietr. Western Rattlesnake-root. Fig. 6034.

Sonchus hastatus Less. Linnaea 6 : 99. 1831.
Nabalus alatus Hook. Fl. Bor. Amer. 1 : 294. *pl. 102.* 1833.
Prenanthes alata D. Dietr. Syn. Pl. 1309. 1847.
Prenanthes hastata M. E. Jones, Bull. Univ. Mont. Biol. Ser. No. 15 : 47. 1910. Not Thunb. 1784.
Prenanthes lessingii Hultén, Fl. Aleut. Isl. 335. 1937.

Perennial herbs from running rootstocks, stems simple or somewhat branched above, 3–6 dm. high, glabrous or nearly so below the inflorescence. Lower and middle leaves 5–12 cm. long, paler below than above, deltoid to deltoid-hastate, attenuate at apex, sharply and irregularly dentate, narrowed below to a petiolar base, the upper scarcely narrowed below and shorter than the lower; heads 10–15-flowered, the ligules purplish, loosely and somewhat paniculately corymbose, the peduncles tomentose; involucre narrowly campanulate, about 12 mm. high, the phyllaries linear, becoming glabrate, the calyculate bractlets very small; achenes about 5 mm. long, finely striate between the ribs; pappus-bristles capillary, copious, brownish.

In rich soil in woods, Humid Transition and Canadian Zones; Alaska southward along the Pacific slope to Clatsop and Hood River Counties, Oregon. Type locality: Fort Vancouver, Washington. Collected by Scouler. June–Aug.

173. HIERÀCIUM [Tourn.] L. Sp. Pl. 799. 1753.

Perennial herbs, the stems scapiform or leafy, arising from a branched caudex, stout rhizome, or from slender stolons. Leaves entire or toothed in most of ours, more abundant at or near the base of the stems, persisting or early deciduous, reduced above. Inflorescence various, the heads rarely solitary. Heads small or medium, few- to many-flowered, with yellow, white, or orange-red ligules. Involucres cylindric or campanulate; phyllaries in 2 or 3 series, often obscurely imbricate. Receptacle flat, mostly naked. Achenes truncate, striate; pappus a single row of white or tawny, capillary bristles. [From the Greek, meaning hawk.]

A variable genus, in which nearly 1000 species have been described, native of both hemispheres. Type species, *Hieracium murorum* L.

In all our native species, perhaps to a less degree in *H. horridum*, extensive field and experimental studies should be made to arrive at an understanding of true relationships. Then some of the names here listed as synonyms may be found to represent genuine biological entities at least at the subspecific level. Probably in the *albertinum-scouleri-cynoglossoides* complex, a situation exists comparable to that which has been demonstrated to occur in the polyploid apomictic species-complexes of the genus in Europe.

Plants not stoloniferous; native species.
  Basal and subbasal leaves rather small and early deciduous, those of midstem many and mostly larger.
    Stems with long setose hairs; leaves two to five times as long as wide.
      Leaves fringed with very long setose hairs; plants of the Columbia River Gorge.
                                                               4. *H. longiberbe.*
      Leaves essentially glabrous; plants widely distributed in Canada and northern United States.
                                                            11. *H. canadense.*
    Stems without long setose hairs; leaves four to twelve times as long as wide.   12. *H. umbellatum.*
  Basal and subbasal leaves the larger, persistent, those of midstem few or none, progressively and sometimes markedly reduced upward.
    Stellate pubescence lacking.
      Flowers white, 18–30 per head; plants 3–12 dm. high; widely distributed in the Pacific States.
                                                       6. *H. albiflorum.*
      Flowers yellow, 5–15 per head; plants 1–3 dm. high; restricted to the Siskiyou region.
                                                   8. *H. bolanderi.*
    Stellate pubescence present (sometimes only on the inflorescence), sparse or dense.
      Plants canescent throughout with a dense stellate pubescence (sometimes also with long setose hairs on the lower leaves.)
                                                   7. *H. greenei.*
      Stellate pubescence sparse and often restricted to leaves or inflorescence, obscure.
        Nearly all leaves remotely but conspicuously sinuate-dentate; plants of west-central and southern California.
                                              5. *H. argutum.*
        Leaves all entire or a few obscurely denticulate; plants not of west-central and southern California.
          Basal leaves glabrous or if at all pubescent lacking setose hairs; slender, naked-stemmed plants, mostly of moist mountainous situations.   9. *H. gracile.*
          Basal and stem-leaves when present densely or sparsely covered with long setose hairs; plants principally of dry, open or forested areas.
            Heads (5)10-12-flowered, the involucres narrow; plants crinate, slender, 1–3 dm. high.
                                      10. *H. horridum.*
            Heads 15–50-flowered, the involucres broad; plants if crinate rather stout, 3.5–7 dm. high.
              Herbage sparsely to moderately long-setose above (occasional on involucres) and sometimes glaucous; plants of Cascade, Siskiyou, and Sierra Nevada ranges.
                                    1. *H. scouleri.*
              Herbage evidently long-setose above and below, not glaucous; plants east of the Cascade, Siskiyou, and northern Sierra Nevada ranges.
                Involucres and stems rather sparsely long-setose, noticeably glandular.
                                  2. *H. cynoglossoides.*
                Involucres and sometimes the stems crinate, scarcely if at all glandular.
                                  3. *H. albertinum.*
Plants stoloniferous; introduced species.
  Inflorescence of 5–10 heads, orange- to orange-red-flowered.   13. *H. aurantiacum.*
  Heads solitary, yellow-flowered.   14. *H. pilosella.*

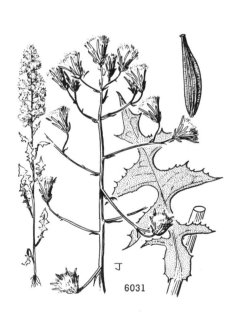

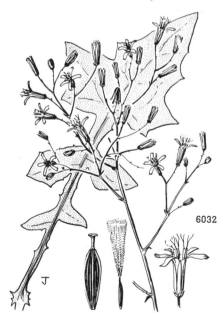

6031. Lactuca biennis   6032. Lactuca muralis

1. **Hieracium scoùleri** Hook. Scouler's Hawkweed. Fig. 6035.

Hieracium scouleri Hook. Fl. Bor. Amer. **1**: 298. 1833.
Hieracium amplum Greene, Erythea **3**: 101. 1895.
Hieracium cinereum Howell, Fl. N.W. Amer. 396. 1901. Not Tausch. 1819.
Hieracium cineritium Nels. & Macbr. Bot. Gaz. **61**: 47. 1916.
Hieracium idahoense Gandoger, Bull. Soc. Bot. Fr. **65**: 48. 1918.
Hieracium paddoense Gandoger, op. cit. 51.
Hieracium washingtonense Gandoger, loc. cit.
Hieracium chelannense Zahn, Pflanzenreich 4²⁸⁰: 1127. 1922.
?Hieracium parryi Zahn, op. cit. 1128.
Hieracium chapacanum Zahn, op. cit. 1129.
Hieracium hemipoliodes Zahn, op. cit. 1130.

Perennial from a short rootstock, the stems simple, rather slender, 2.5–7 dm. high, subglabrous and often glaucous above, long-setose below and more or less stellate. Lower leaves several at or near the base of the stem, 8–25 cm. long, 1–3 cm. wide, narrowed to a petioliform base, oblanceolate to lanceolate, essentially entire, the stem-leaves few, progressively reduced in size, sessile, setose-hairy and inconspicuously stellate; inflorescence rather open, the branching tending to be paniculate, the heads few to many and about 15–20-flowered; involucre 7–11 mm. high, the phyllaries not conspicuously imbricate; phyllaries linear-lanceolate, finely stellate-pubescent with some gland-tipped bristles and some setose hairs often present; achenes 3–3.5 mm. long, exceeded by the sordid pappus.

Open woods, Humid Transition and Canadian Zones; commonly occurring west of and through the Cascade Mountains from British Columbia to southern Oregon and possibly adjacent northern California; also occurring less frequently eastward to Montana and western Wyoming. Type locality: mouth of the Columbia River. Collected by Scouler. June–Aug.

A highly variable species merging to the south with var. *nudicaule* and with *H. cynoglossoides* and *H. albertinum* in its more easterly range.

Hieracium scouleri var. **nudicaùle** (A. Gray) Cronquist, Vasc. Pl. Pacif. Northw. **5**: 238. 1955. (*Hieracium cynoglossoides* var. *nudicaule* A. Gray, Proc. Amer. Acad. **19**: 68. 1883; *H. nudicaule* Heller, Muhlenbergia **2**: 149. 1906; *H. cascadorum* Zahn, Pflanzenreich 4²⁸⁰: 1127. 1922; *H. babcockii* Zahn, loc. cit.; *H. babcockii* var. *setosiceps* Zahn, loc. cit.; *H. cusickianum* Zahn, op. cit. 1129.) Plants 1.5–3.5(4) dm. high, the stems several from a woody root; basal leaves as in *H. scouleri* var. *scouleri* but stem-leaves completely lacking or reduced to 1 or 2 narrow bracts; heads 2–11 in an open inflorescence, the phyllaries with conspicuous setose hairs as in *H. cynoglossoides* and also glandular. Dry open woods or gravelly flats in the southern Cascade and Siskiyou Mountains, Oregon, south through northern California to Placer County in the Sierra Nevada and to Trinity County in the Coast Ranges. Type locality: Sierra County, California. Collected by Lemmon.

2. **Hieracium cynoglossoìdes** Arv.-Touv. Houndstongue Hawkweed. Fig. 6036.

Hieracium cynoglossoides Arv.-Touv. Spicil. Hier. 20. 1881.
Hieracium griseum Rydb. Mem. N.Y. Bot. Gard. **1**: 464. 1901. Not Form. 1896.
Hieracium cusickii Gandoger, Bull. Soc. Bot. Fr. **65**: 48. 1918.
Hieracium rydbergii Zahn, Pflanzenreich 4²⁸⁰: 1130. 1922.
?Hieracium flettii St. John & Warren, Proc. Biol. Soc. Wash. **41**: 108. 1928.

Habit much resembling *H. scouleri* var. *scouleri* but setose and not subglabrose or glaucous. Leaves and inflorescence as in the preceding taxon; involucres finely stellate, also with short, blackish, gland-tipped bristles and some long setae; achenes and pappus as in *H. scouleri* and *H. albertinum*.

Dry open places at low elevations, Sonoran and Arid Transition Zones; British Columbia and Alberta south to eastern Washington, eastern Oregon, and Modoc County, California, east to Utah and Wyoming. Type locality: northern Wyoming. June–Aug. Intermediate between *H. scouleri* and *H. albertinum*.

3. **Hieracium albertìnum** Farr. Western Hawkweed. Fig. 6037.

Hieracium albertinum Farr, Ottawa Nat. **20**: 109. 1906.
Hieracium absonum Macbride & Payson, Contr. Gray Herb. No. **49**: 71. 1917.

Rather coarse perennial 4–12 dm. high, the stems 1 or more from a stout rootstock, leafy especially below, with the uppermost leaves few and somewhat reduced in size, abundantly pubescent throughout with long setose hairs and rather sparingly stellate. Leaves 8–25 cm. long, 1–3 cm. wide, lanceolate or oblanceolate, entire, narrowing to a petioliform base, midrib prominent, whitish, the upper leaves sessile; inflorescence usually with many heads, corymbiform to paniculate-corymbiform, the peduncles 5–20 mm. long; involucres 9–12 mm. high, many-flowered (15–20); phyllaries loosely imbricate, linear, with dark midsection, stellate-pubescent and copiously long-setose usually with black-based hairs, these sometimes dense enough to conceal the phyllaries; achenes 3 mm. or more long, the pappus whitish.

Dry open places in prairies and open pine forests, Arid Transition Zone; southern British Columbia and Alberta south to eastern Washington and northeastern Oregon and also to Idaho and western Montana. Type locality: between Lake Louise and Moraine Lake, Alberta. July–Aug.

4. **Hieracium longibérbe** Howell. Long-bearded Hawkweed. Fig. 6038.

Hieracium longiberbe Howell, Fl. N.W. Amer. 395. 1901.
Hieracium piperi St. John & Warren, Proc. Biol. Soc. Wash. **41**: 109. 1928.

Perennial, 3–6 dm. high, the stems erect, 1 or more from a short woody caudex, rather slender, erect, reddish below and rather sparsely beset with long setose hairs 5–10 mm. long, nearly glabrous above. Basal and lowest stem-leaves soon withering, the midstem-leaves 6–15 cm. long, the uppermost somewhat smaller, all leaves lanceolate, oblanceolate, or elliptic, narrowed to a petiolar base, becoming sessile above, entire or rarely denticulate, thin, green but pubescent with setose hairs up to 10 mm. long, conspicuously so on midrib and margins, otherwise glabrous; inflorescence open, the heads mostly few on peduncles 3–5 cm. long, the flowers yellow; involucre 7.5–9 mm. high, the phyllaries with dark midrib, long, often black-based, setose hairs and sparsely stellate-pubescent, not glandular; achenes 3.5 mm. long, exceeded by the white or sordid-white pappus.

Cliffs and rocky bluffs, Humid Transition Zone; Columbia River Gorge, Skamania County, Washington, and Multnomah County, Oregon. Type locality: Columbia River near The Cascades. July.

A distinct local form recognizable by its unusually long setose hairs, suggesting relationship with the *scouleri-cynoglossoides* complex on one hand and with the *canadense-umbellatum* complex on the other. The habit and the less reduced and more numerous stem-leaves indicate relationship to the latter group.

5. **Hieracium argùtum** Nutt. Southern California Hawkweed. Fig. 6039.

Hieracium argutum Nutt. Trans. Amer. Phil. Soc. II. **7**: 447. 1841.

Perennial, 3–7 dm. high from a branched woody caudex, the stems beset with long setose hairs below, these lacking or nearly so on the upper stems. Lower leaves often densely clustered below,

6033. Lactuca tartarica
6034. Prenanthes alata

6035. Hieracium scouleri
6036. Hieracium cynoglossoides

10–20 cm. long, 1–4 cm. wide, oblong or oblong-lanceolate, narrowed to a petiole or a petioliform base, the margins remotely but deeply sinuate-dentate or -denticulate, setose-hairy; middle and upper leaves sessile, often much reduced, linear, essentially entire, stellate-pubescent to nearly naked; inflorescence paniculate, the branches glabrous to stipitate-glandular with black hairs, the stellate pubescence sparse; heads 8–9 mm. high, yellow-flowered; phyllaries loosely imbricate, linear, dark, densely to rather sparsely black-stipitate-glandular with some yellow-stipitate-glandular hairs, the stellate pubescence sparse; achenes 2.5–3 mm. long, dark brown, the pappus whitish.

Rocky and wooded slopes, Upper Sonoran and Transition Zones; Santa Rosa and Santa Cruz Islands, Santa Barbara County, and on the mainland in Santa Barbara County, where it merges with forms of the following variety. Type locality: Santa Barbara. Collected by Nuttall. June–Aug.

Hieracium argutum var. paríshii (A. Gray) Jepson, Man. Fl. Pl. Calif. 1009. 1925. (*Hieracium parishii* A. Gray, Proc. Amer. Acad. **19**: 67. 1883; *H. brandegeei* Greene, Bull. Calif. Acad. **1**: 195. 1885; *H. grinnellii* Eastw. Bull. Torrey Club **32**: 217. 1905.) Leaves as in *H. argutum* var. *argutum*, the northern forms more often denticulate and not dentate; inflorescence usually more openly branched; pubescence of the phyllaries merely stellate-pubescent (*H. parishii*) or more commonly also with yellow-stipitate-glandular hairs, the black-stipitate-glandular hairs completely lacking or sparse. Dry wooded canyons, Santa Lucia Mountains, Monterey County, California, south to mountains of Los Angeles and San Bernardino Counties. Type locality: rock crevices, Waterman Canyon, San Bernardino Mountains.

6. **Hieracium albiflòrum** Hook. White-flowered Hawkweed. Fig. 6040.

Hieracium albiflorum Hook. Fl. Bor. Amer. **1**: 298. 1833.
?*Hieracium vancouverianum* Arv.-Touv. Spicil. Hier. 10. 1881.
Hieracium helleri Gandoger, Bull. Soc. Bot. Fr. **65**: 51. 1918.
Hieracium leptopodanthum Gandoger, loc. cit.
Hieracium candelabrum Gandoger, op. cit. 52.
Hieracium albiflorum subvar. *rosendahlii* Zahn, Pflanzenreich **4²⁸⁰**: 1123. 1922.
Hieracium albiflorum f. *lyallii* Zahn, loc. cit.
Hieracium albiflorum subvar. *sanhedrinense* Zahn, loc. cit.
?*Hieracium pacificum* Zahn, loc. cit.

Perennial, 3–12 dm. high, the stems usually solitary, arising from a woody root and commonly unbranched short caudex, leafless or with few stem-leaves, with long setose hairs below and mostly naked above. Leaves mostly basal, 4–18 cm. long, oblanceolate, narrowed to a petiole except in the stem-leaves, moderately long-setose, entire or repand-denticulate; inflorescence corymbiform, the rather small heads several to many, the peduncles slender, 2–7 cm. long, the flowers white or pale yellow (?); involucres narrow, usually 18–30-flowered; phyllaries scarcely imbricate, the longer inner ones 7–9 mm. long, linear-lanceolate, greenish or blackish, glabrous or often sparsely glandular with pale or black hairs, sometimes with a few setose hairs; achenes about 2.5–3 mm. long, the pappus white or sordid, longer than the achene.

Open woods in moist or dry situations, Upper Sonoran Zone to Hudsonian Zone; Alaska and northwestern Canada south to southern California and Colorado. Type locality: "mouth of the Columbia." Collected by Scouler. June–Aug.

The most widely distributed species in the Pacific States both altitudinally and geographically. Quite variable as to type of margin of the lower leaves and in vesture of the phyllaries but having in common growth habit, linear-lanceolate phyllaries, absence of stellate pubescence, and white or cream-colored flowering heads.

7. **Hieracium greènei** A. Gray. Greene's Hawkweed. Fig. 6041.

Hieracium greenei A. Gray, Proc. Amer. Acad. **19**: 69. 1883.
Hieracium howellii A. Gray, Bot. Gaz. **13**: 73. 1888.
?*Hieracium barbigerum* Greene, Pittonia **3**: 228. 1897. Not Stenstr. 1896.
?*Hieracium oregonicum* Zahn, Pflanzenreich **4²⁸⁰**: 1130. 1922.

Perennial, 1.5–3.5 dm. high, the stems few to several from a woody root and stout branched caudex, stoutish, characteristically densely stellate-tomentose throughout, the lower leaves also long-setose. Basal and lower stem-leaves 4–10(15) cm. long, rather thick, oblanceolate to spatulate, narrowed to a petiolar base, those above the base shorter and becoming sessile, the margins entire or irregularly repand-denticulate; inflorescence paniculate with divaricate branchlets, the heads several to many, the stoutish peduncles 0.5–2 cm. long, the flowers yellow; involucres narrow, 10–15-flowered; phyllaries 3-seriate, the inner 9–11 mm. long, usually 1.5–1.75 mm. wide, broadly linear, obtusish at apex, pinkish-tinged and more or less stellate-pubescent; achenes about 5 mm. long, the pappus longer than the achenes, white or brownish.

In rocky places, often in serpentine, Arid Transition and Canadian Zones; Klamath County in the southern Cascade Mountains and Josephine County in the Siskiyou Mountains, Oregon, southward to inner Humboldt County, California, and Shasta County and also occurring as far south as Sierra County. Type locality: pine woods, Scott Mountain, Siskiyou County, California. July–Aug.

In some of the Californian specimens, the stellate pubescence is quite sparse but the plants possess the characteristic stiffly divaricate inflorescence of *H. greenei* and are here considered to be aberrant forms of that species.

8. **Hieracium bolánderi** A. Gray. Bolander's Hawkweed. Fig. 6042.

Hieracium bolanderi A. Gray, Proc. Amer. Acad. **7**: 365. 1868.
Hieracium siskiyouense M. E. Peck, Proc. Biol. Soc. Wash. **47**: 188. 1934.

Rather slender perennial 1–3 dm. high, the glabrous and glaucous stems solitary or few from a simple or branched, woody caudex, unbranched below the inflorescence, (occasionally with a slender flowering branchlet at midstem), glabrous and naked except for occasional small, linear, setose bracts below and in the inflorescence. Basal leaves 1.5–7 cm. long, spatulate to oblanceolate, sessile (the longest leaves usually with petioliform bases), thin, entire or obscurely denticulate, conspicuously but not densely long-setose; inflorescence open with ascending branches, the heads 3–10, borne on slender, glabrous or very sparsely glandular peduncles (2)4–8 cm. long, about 5–10-flowered, pale yellow, rarely white; involucres narrow, the phyllaries few, scarcely imbricate, the inner up to 1 cm. long, 1–1.3 mm. wide, acute at apex, often blackish-tinged and glaucous, entirely glabrous, or with a very few short, gland-tipped hairs and an occasional long seta; achenes about 3 mm. long, the pappus exceeding the achenes, white or whitish.

Forested or open areas, Upper Arid Transition and Canadian Zones; mountains of Curry, Josephine, and Jackson Counties, Oregon, south to Siskiyou and Humboldt Counties, California. Type locality: Red Mountain, Humboldt County. June–July.

9. **Hieracium grácile** Hook. Alpine Hawkweed. Fig. 6043.

Hieracium gracile Hook. Fl. Bor. Amer. 1 : 298. 1833.
Hieracium hookeri Steud. Nom. ed. 2. 1 : 763. 1840.
Hieracium triste var. *detonsum* A. Gray, Bot. Calif. 1 : 441. 1876.
Hieracium gracile var. *detonsum* A. Gray, Proc. Amer. Acad. 19 : 67. 1883.
Hieracium gracile subsp. *gracile* var. *detonsum* subvar. *densifloccum* Zahn, Pflanzenreich 4[280] : 1133. 1922.

Perennials, 1–3 dm. high, the subscapose slender stems clustered, few to several from a horizontal rootstock and short caudex, simple or sometimes branched, stellate-puberulent, especially

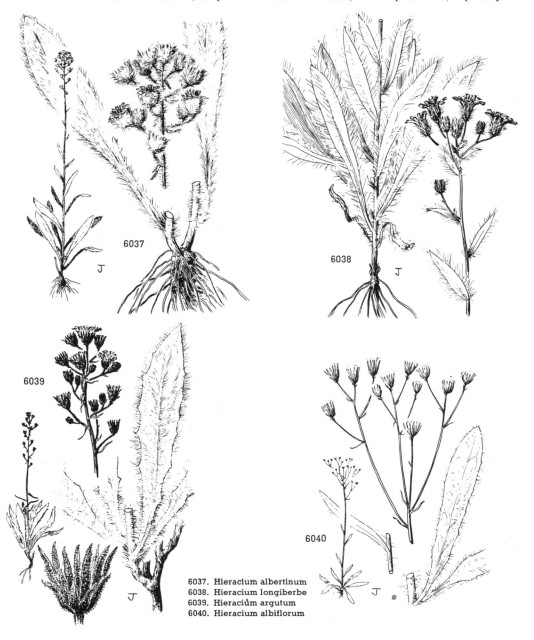

6037. Hieracium albertinum
6038. Hieracium longiberbe
6039. Hieracium argutum
6040. Hieracium albiflorum

above, to glabrous. Leaves 3–10 cm. long, many-clustered at the base, the occasional stem-leaves small, bracteate; spatulate to oblanceolate, narrowed below to a petiole 1–5 cm. long, essentially glabrous; heads yellow-flowered, solitary or 2–15 in racemes, pedunculate; involucres 6–8 mm. high, the phyllaries with black, usually gland-tipped setae or these whitish or lacking, also often stellate-pubescent; achenes 2 mm. long, surpassed by the white pappus.

Meadows and open places, Boreal Zone; Alaska and Canada south in the Pacific States to the Siskiyou region and the southern Sierra Nevada in California and the Rocky Mountains region south to northern New Mexico. Type locality: Rocky Mountains. Collected by Drummond. July–Sept.

10. Hieracium hórridum Fries. Shaggy Hawkweed. Fig. 6044.

Hieracium horridum Fries, Epicris. Hier. 154. 1862.
Pilosella relicina Sz. Sz. Flora 439. 1862.
Hieracium breweri A. Gray, Proc. Amer. Acad. **6**: 553. 1865.

Rather slender perennial 1–3 dm. high, the stems clustered, simple or with few slender lower branches, arising from a woody rootstock and stout, simple or branched caudex, stellate to nearly glabrate and also with long setose hairs or these lacking especially on the upper stems. Basal and lower stem-leaves many, 6–15 cm. long, spatulate to oblanceolate, entire, narrowed to a petioliform base, shaggy with long, spreading, setose, white or brownish hairs, the leaves below the bracteate inflorescence sessile, similar, longer than the internodes and not greatly reduced; inflorescence openly branched, corymbiform, with few to many heads, the heads rather small, yellow, 10–12-flowered, the peduncles mostly 1–1.5 cm. long; involucre 7.5–10 mm. high, the phyllaries scarcely imbricate, the outer series very short, stellate to nearly glabrate, the usually darkened midsection with long, pale, black-based setae; achenes about 2 mm. long, the pappus tawny, longer than the achenes.

Common among rocks, mostly Boreal Zone; eastern Lane County, Oregon, in the Cascade Mountains, to the mountains of northern California and southward in the Sierra Nevada to Tulare County; also the higher mountains of Southern California. Type locality: California. Collected by Bridges. July–Sept.

11. Hieracium canadénse Michx. Canada Hawkweed. Fig. 6045.

Hieracium canadense Michx. Fl. Bor. Amer. **2**: 86. 1803.
Hieracium columbianum Rydb. Bull. Torrey Club **28**: 513. 1901.

Perennial, 4–12 dm. high, the stems 1 or more from a short woody caudex, rather stout, pubescent below with long spreading hairs, glabrous to sparingly stellate-pubescent above. Basal and lowest stem-leaves rather small, soon withering and deciduous, the midstem-leaves about 6–12 cm. long, 1–4 cm. wide, many, longer than the internodes, mostly lanceolate to oblanceolate, entire or sparingly denticulate, sessile and subclasping at base, rather thin, the margins and often veins and undersurface with spreading setose hairs, rarely sparingly stellate-pubescent; inflorescence corymbiform-paniculate with spreading peduncles, the heads 6–20 or more, yellow-flowered; involucres 6–10 mm. high, the phyllaries imbricate, dark, glabrous or nearly so; achenes about 3 mm. long, purplish black, the pappus tawny, a little longer than the achene.

Woods and moist open places, Transition and lower Boreal Zones; Canada south to Washington, the Rocky Mountains region, and eastward to New Jersey. Type locality: Canada. July–Sept. Much resembling *H. umbellatum*.

12. Hieracium umbellàtum L. Umbellate Hawkweed. Fig. 6046.

Hieracium umbellatum L. Sp. Pl. 804. 1753.
Hieracium scabriusculum Schwein. Long. Exp. **2**: 394. 1824. Not *H. umbellatum* var. *scabriusculum* Farw. 1927.
Hieracium macranthum Nutt. Trans. Amer. Phil. Soc. II. **7**: 446. 1841.
Hieracium suksdorfii Gandoger, Bull. Soc. Bot. Fr. **65**: 51. 1918.

Perennial, 4–12 dm. high, the stems 1 or more from a short woody rootstock, stout, glabrous or nearly so below, usually stellate-puberulent above. Basal and lowest stem-leaves rather small, soon withering and deciduous, the midstem-leaves 6–10 cm. long, 1–2 cm. wide, sessile, not clasping and sometimes narrowed at base, many, much longer than the internodes, entire or sparingly denticulate, lanceolate to narrowly oblong, rather firm, stellate and more or less scabridulous on margins and veins with broad-based hairs; inflorescence corymbiform to subumbellate, stellate-pubescent, the heads 3–10 or more, yellow-flowered; involucres 6–13 mm. high, the phyllaries imbricate, glabrous or nearly so; achenes 3–3.5 mm. long, reddish, the pappus tawny, about equaling the achenes.

Moist thickets and woods, Transition and lower Boreal Zones; circumboreal southward in North America to Michigan and Wisconsin and in the west to Colorado and northwestern Oregon. Type locality: Europe. July–Sept.

13. Hieracium aurantìacum L. Orange Hawkweed. Fig. 6047.

Hieracium aurantiacum L. Sp. Pl. 801. 1753.

Resembling *H. pilosella* in habit, the stems 2–5 dm. high, subscapose, bearing 1 or 2 reduced leaves, long-setose, and also stellate and glandular above. Leaves 4–20 cm. long, oblanceolate to elliptic, narrowed to a petiolar base, usually long-setose above and below; inflorescence corymbiform, the heads 5–25 on short peduncles, the flowers red-orange; involucre 6–8 mm. high, the phyllaries long-setose and also bearing black, gland-tipped hairs, often slightly tomentose; achenes and pappus as in *H. pilosella*.

A European weed sparingly established in Bremerton, Kitsap County, Washington, and in Multnomah and Deschutes Counties, Oregon; also introduced in eastern United States. June–Sept. King-devil.

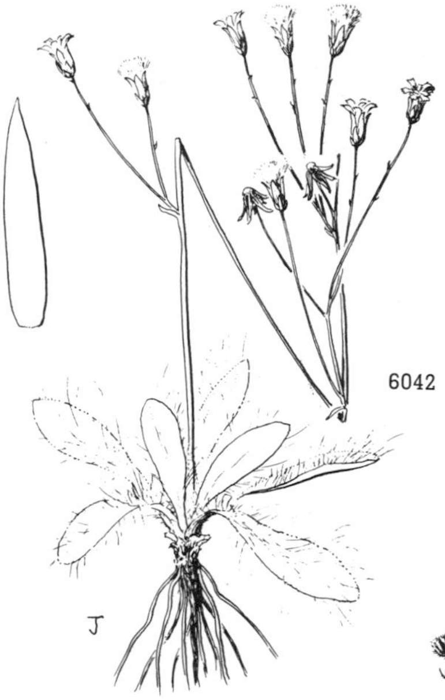

6041

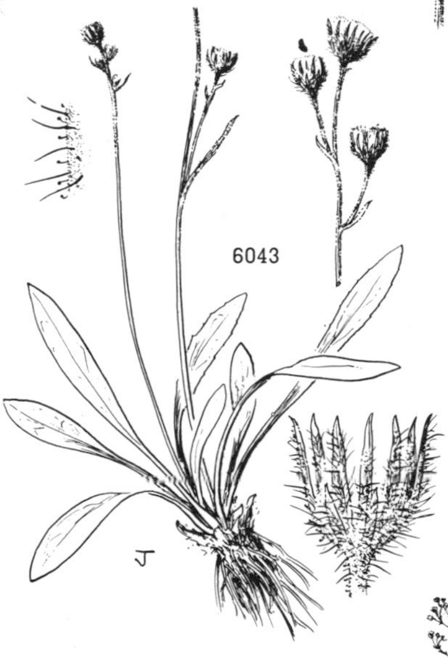

6042

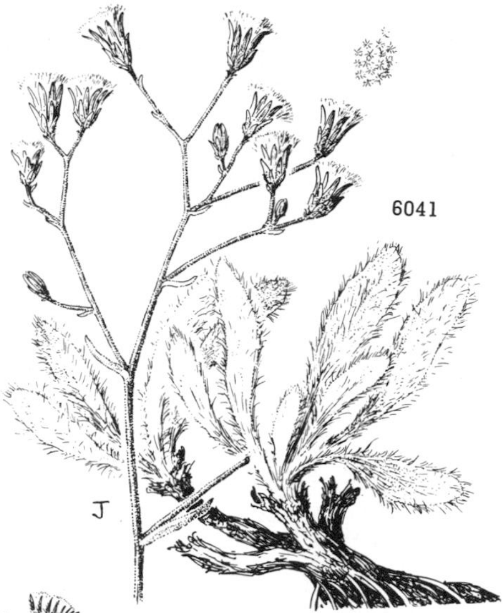

J

J

6043

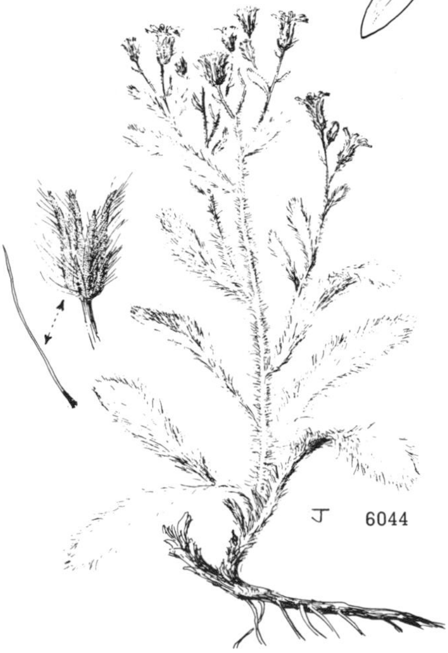

6044

J

6045

J

6041. Hieracium greenei
6042. Hieracium bolanderi
6043. Hieracium gracile
6044. Hieracium horridum
6045. Hieracium canadense

14. **Hieracium pilosélla** L. Mouse-ear Hawkweed. Fig. 6048.

Hieracium pilosella L. Sp. Pl. 800. 1753.

Stoloniferous perennial, the slender stems 5–20 cm. high arising from a rather slender root-stock, scapose, rarely with a single leaf, more or less spreading-pubescent with gland-tipped, often black hairs. Leaves 3–12 cm. long (those of the stolons small), oblanceolate to spatulate, tapering to a petiolar base, entire, pale beneath with a dense stellate pubescence and with some long setose hairs, green above and glabrous except for long setose hairs; heads solitary, yellow-flowered; involucres 7–11 mm. high, the phyllaries stellate-pubescent and also with spreading, often gland-tipped, black hairs, sometimes with long setose hairs; achenes about 1.5–2 mm. long, the pappus exceeding the achene, sordid.

A European weed sparingly established in lawns and waste places, Willamette Valley and Multnomah County, Oregon; also introduced in eastern United States. May–Sept.

174. **CRÈPIS*** L. Sp. Pl. 805. 1753.

Perennial, biennial, or annual herbs; stems scapiform or branched. Leaves chiefly basal, the cauline alternate, many, few, or none, entire or pinnatifid. Heads small to large, few- to many-flowered. Involucres usually 3-seriate, the inner phyllaries much longer than the outer. Receptacle plane or convex, naked or somewhat hairy or paleaceous; ligules yellow, white, or pink, sometimes reddish-tinged, usually 5-toothed. Anthers appendaged. Style-branches filiform, attenuate or truncate at apex. Achenes columnar or fusiform, ribbed or striate, beaked or beakless; pappus-bristles in more than 1 series, white to tawny, persistent or deciduous. [From the Greek, meaning sandal, the ancient name of some plant.]

A genus of 196 species, natives of Eurasia, Africa, and North America. Type species, *Crepis biennis* L.

Leaves dentate or parted, not spatulate or obovate; not dwarf alpine plants.
  Involucres 5–12 mm. high; introduced annual or biennial species (perennial in *C. bursifolia*).
    Achenes beakless, not attenuate above.   2. *C. capillaris.*
    Achenes beaked, the beak slender or filiform.
      Perennial; beak filiform, about twice the length of the achene.   1. *C. bursifolia.*
      Annual or biennial; beak not filiform, about the length of the achene or slightly longer.
        Involucres and stems strongly setose with yellow bristles.   3. *C. setosa.*
        Involucres and stems tomentulose and pubescent or setulose with short black bristles.
           4. *C. vesicaria taraxacifolia.*
  Involucres usually 9–22 mm. high; native perennials.
    Stems and leaves mostly glabrous and glaucous; cauline leaves absent or essentially so; plants of moist situations.   6. *C. runcinata.*
    Stems and leaves more or less tomentose in addition to other types of pubescence; 1–3 cauline leaves present (except *C. pleurocarpa*); plants of dry situations.
      Herbage and involucres shaggy-hirsute with gland-tipped hairs; outer phyllaries linear-lanceolate.
         7. *C. monticola.*
      Herbage and involucres variously pubescent, if glandular of short hairs only; outer phyllaries lanceolate or ovate-lanceolate.
        Involucres or lower stem or both conspicuously setose but not glandular.
          Inflorescence of 1–9 heads; inner phyllaries 10–15.   8. *C. modocensis.*
          Inflorescence of 6–70 heads; inner phyllaries 5–10.   9. *C. barbigera.*
        Involucres and stems with setae sparse or lacking.
          Heads 5–10-flowered; inner phyllaries 5–7.
            Cauline leaves 1–3; phyllaries glabrous or evenly and lightly tomentose.
               10. *C. acuminata.*
            Cauline leaves reduced or wanting; phyllaries glabrous in midportion, conspicuously tomentose on the margins.   11. *C. pleurocarpa.*
          Heads 10–40-flowered; inner phyllaries 8–10.
            Divisions of leaves linear or narrowly lanceolate, mostly entire; achenes greenish.
               12. *C. atribarba.*
            Divisions of leaves lanceolate or deltoid, some usually toothed; achenes yellowish or brownish.
              Plants mostly 3 dm. or less high; involucres broadly cylindric; heads 12–30-flowered.
                Leaves grayish-tomentulose, not glandular; peduncles not dilated above.
                   13. *C. occidentalis.*
                Leaves green and glandular; peduncles fistulose and dilated above.
                   14. *C. bakeri.*
              Plants mostly 3–6 dm. high; involucres narrowly cylindric; heads 8–10-flowered.
                 15. *C. intermedia.*
Leaves all entire or few-toothed, spatulate or obovate; dwarf alpine plants.   5. *C. nana.*

1. **Crepis bursifòlia** L. Italian Hawksbeard. Fig. 6049.

Crepis bursifolia L. Sp. Pl. 805. 1753.

Perennial, 0.5–3.5 dm. high, stems several, decumbent or arcuate, arising from a woody caudex and vertical root, slender, tomentulose, cymosely branched above, bearing 2–14 heads.

---

* Key and text adapted from Babcock and Stebbins (*Carnegie Inst. Pub.* No. 504, 1938) and Babcock (*Univ. Calif. Pub. Bot.* **22**: 1947), where complete synonymy is also to be found.

Leaves mostly basal, the cauline few, linear and bract-like, the basal leaves 5–25 cm. long, oblanceolate in outline, lyrately pinnatifid, the obtuse terminal portion dentate or denticulate; heads 30–60-flowered, on peduncles 0.5–6 cm. long; involucres 9–11 mm. long, cylindric, canescent-farinose and sometimes somewhat yellow-setulose; outer phyllaries unequal, one-fourth to one-third as long as the inner, these 8–10; achenes 6–7.5 mm. long, 10-ribbed, pale brown, fusiform and attenuate into a filamentous beak about twice the length of the body of the achene; pappus 3–4 mm. long.

A native of Italy found growing in waste places in the San Francisco Bay region, California (Berkeley; Stanford). May–June.

2. Crepis capillàris (L.) Wallr. Smooth Hawksbeard. Fig. 6050.

Lapsana capillaris L. Sp. Pl. 812. 1753.
Crepis virens L. op. cit. ed. 2. 1134. 1763.
Crepis capillaris Wallr. Linnaea 14: 657. 1840.
Malacothrix crepoides A. Gray in J. G. Cooper, Pacif. R. Rep. 12²: 53. 1860.

Annual or biennial, 0.2–9 dm. high, with several stems from a taproot, these erect, rather slender, hispidulous at the base and sometimes above. Basal leaves 3–30 cm. long, lanceolate in outline, runcinate-pinnatifid or lyrately pinnate, petiolate, glabrous or hispidulous beneath on the

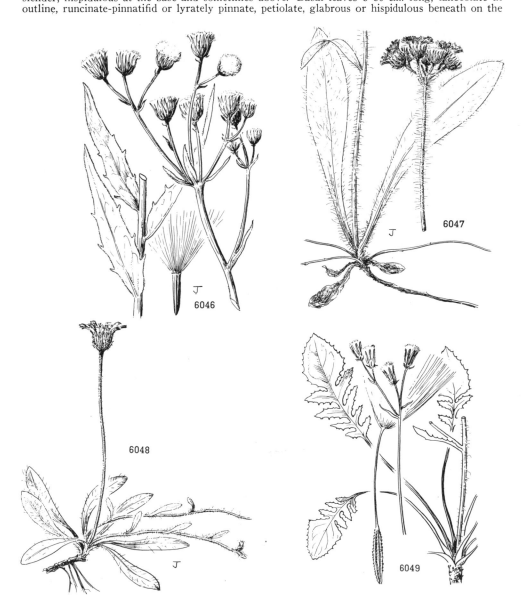

6046. Hieracium umbellatum
6047. Hieracium aurantiacum
6048. Hieracium pilosella
6049. Crepis bursifolia

midrib with short yellow hairs, sometimes sparsely hispidulous above, the cauline leaves sessile, becoming reduced above; inflorescence corymbiform, the heads several to many, 20–60-flowered, on peduncles 0.5–3.5 cm. long; involucres 5–8 mm. high, turbinate in fruit; outer phyllaries one-third to one-half the length of the inner, glabrous or tomentose, the inner phyllaries 8–16, lanceolate, glabrous, tomentose or sparingly glandular, sometimes sparsely setulose with black glandular hairs; receptacle glabrous; achenes 1.5–2.5 mm. long, 10-ribbed, brownish yellow to dark brown, fusiform and beakless; pappus 3–4 mm. long.

A native of Europe, growing in meadows, lawns, and waste places, widely introduced in the Pacific States from western Washington to central California; also occurring in Canada and eastern United States. May–Aug.

Crepis nicaeénsis Balbis ex Pers. Syn. Pl. **2**: 376. 1807. Annual or biennial, 2.5–11 dm. high, related to and resembling *C. capillaris*; differing in having more pubescent herbage; involucres 8–10 mm. high and becoming campanulate in fruit; receptacle ciliate, achenes 2.5–4 mm. long, golden-brown. A native of southern Europe and sparingly introduced in eastern United States, and known in the Pacific States from one collection at Marysville, Snohomish County, Washington.

Crepis púlchra L. Sp. Pl. 806. 1753. Annuals, 3–7 dm. high from a slender taproot, pubescent below, glabrous above; basal leaves 3–15 cm. long, oblanceolate, acute or obtuse, runcinate-pinnatifid, narrowed to a winged petiole, pubescent; stem-leaves somewhat reduced, sessile and amplexicaul and less divided than the basal; involucres cylindric, glabrous; achenes 4–4.5 mm. long, the outer compressed and often without pappus, the inner terete, bearing copious pappus. A native of Europe, locally established in southern Jackson County, Oregon.

Crepis rùbra L. Sp. Pl. 806. 1753. A garden annual, native of Italy and the Balkans, mostly 1–2.5 dm. high with basal, dentate or runcinate, pinnatifid leaves and scapiform stems, sometimes branched below, terminated by pink- or white-flowered heads, nodding before anthesis; achenes 9–20 mm. long, with a coarse or slender beak as long or twice as long as the body. Rarely escaping from cultivation; in the Pacific States it has been collected at Belvedere, Marin County, California.

Crepis tectòrum L. Sp. Pl. 807. 1753. Glabrous or puberulent annual, 1–10 dm. high, with runcinate-pinnatifid basal leaves and sessile, linear cauline leaves; inflorescence usually with few heads; involucres cylindric-campanulate, 9 mm. high, the inner phyllaries pubescent on the inner faces; achenes usually 3–4 mm. long, 10-ribbed, dark purplish brown, fusiform and attenuate above but scarcely beaked. A native of Europe, sparingly introduced in Canada and the United States; on the Pacific Slope, introduced in British Columbia and to be expected in northern Washington.

3. **Crepis setòsa** Hall. f. Rough Hawksbeard. Fig. 6051.

Crepis setosa Hall. f. Roem. Arch. Bot. 1²: 1. 1797.

Annual, 1.8–8 dm. high, the erect stem from a taproot, simple or branched from near the base and also above, hispid. Basal leaves mostly few, 6–30 cm. long, oblanceolate in outline, narrowed to a petioliform base, runcinate-pinnatifid, the terminal portion merely toothed, acute at apex, finely hispid, the cauline leaves sessile, auriculate, lanceolate, toothed at the base, entire and acuminate above; heads rather few on the branches, many-flowered, on peduncles 0.5–6.5 mm. long, hispid and setose with yellow bristles; involucre 8–10 mm. long, somewhat turbinate in fruit, the outer phyllaries not more than one-half the length of the inner, the inner linear-lanceolate, thickened at the base, setose with yellow bristles, the yellow ligules often reddish on outer face; receptacle pubescent; achenes about 3.5–5 mm. long, 10-ribbed, fusiform and tapering into a beak 1–2.5 mm. long; pappus 2.5–5 mm. long.

A native of southeastern Europe, widely introduced, but not common in the Pacific States; in the Willamette Valley, Oregon, and Humboldt County, California. June–Aug.

4. **Crepis vesicària** subsp. **taraxacifòlia** (Thuill.) Thell. Weedy Hawksbeard. Fig. 6052.

Crepis taraxacifolia Thuill. Fl. Par. 409. 1799.
Crepis vesicaria subsp. *taraxacifolia* Thell. ex Schinz & Keller, Krit. Fl. Schweiz ed. 3. 361. 1914.

Annual or biennial (rarely perennial), 0.3–8 dm. high, stems 1 to several from a thickened taproot, often branched near the base as well as above, more or less purplish, striate, more or less tomentose. Leaves mostly basal, the cauline leaves sessile and amplexicaul, nearly entire to pinnately parted, the basal leaves 10–20 cm. long or somewhat longer, mostly narrowly oblanceolate in outline, mostly runcinate-pinnatifid or lyrately pinnate with a large terminal lobe, petiolate, pubescent on both surfaces with pale glandless hairs; inflorescence corymbiform, with several heads, on peduncles 1–13 cm. long; involucre 8–12 mm. high; outer phyllaries short, lanceolate or ovate-lanceolate, spreading, the inner phyllaries 9–13, lanceolate, obtuse, tomentose, often glandular-pubescent and usually with black glandless setae; receptacle ciliate; achenes 4.5–9 mm. long, 10-ribbed, pale brown, fusiform, attenuate into a slender beak as long as or longer than the body of the achene; pappus 4–6 mm. long.

A variable subspecies, native of Europe and widely introduced elsewhere, becoming more abundant in waste places and borders of woods in California; Mendocino, Marin, Alameda, San Mateo, and Santa Clara Counties, and also adventive in Los Angeles County. April–June.

5. **Crepis nàna** Richards. Dwarf Hawksbeard. Fig. 6053.

Prenanthes pygmaea Ledeb. Mém. Acad. St.-Pétersb. V. **5**: 553. 1815.
Crepis nana Richards. in Frankl. 1st Journ. Bot. App. 746. 1823.
Prenanthes polymorpha Ledeb. vars. a et b Fl. Alt. **4**: 144. 1833.
Crepis humilis Fisch. ex Herder, Bull. Soc. Impér. Nat. Mosc. **43**: 190. 1870.

Dwarfed tufted perennials, the vertical or creeping, underground stems arising from a taproot, the leafless upper stems usually about 5–10 cm. high, about as long as or a little shorter than the leaves. Leaves glaucous, often somewhat purplish, many-clustered basally, up to 7 cm. long, ovate-acute to suborbicular, the blade tapering abruptly into the petiole, margin entire or a little lyrate-pinnatifid; involucres 24 on the congested branches; inner phyllaries about 10 in 1 series, 10–13 mm. long, subtended by the calyculate outer phyllaries; ligules very short, yellow, tinged without with purple; achenes subterete, striate, with 10–13 smooth or rugulose ribs, somewhat narrowed and attenuate above; pappus white, 4–6 mm. long.

Mostly in loose gravel and rock, Arctic-Alpine Zone; Alaska east to the Atlantic Ocean and south to Utah and Nevada, and in the Pacific States from Washington to the higher peaks of southern California; also eastern Asia. Type locality: Arctic sea coast "On the Copper-mine River." July–Aug.

Crepis nana subsp. **ramòsa** Babcock, Univ. Calif. Pub. Bot. **22**: 542. 1947. Differing principally from *C. nana* subsp. *nana* by the more elongated leafy branches, with heads borne well above the basal leaves. British Columbia and Idaho south to Tulare County in the southern Sierra Nevada, California. Type locality: above Lake Constance, Jefferson County, Washington.

6. **Crepis montícola** Coville. Mountain Hawksbeard. Fig. 6054.

Crepis occidentalis var. *crinata* A. Gray, Bot. Calif. 1: 435. 1876, in part.
Crepis monticola Coville, Contr. U.S. Nat. Herb. **3**: 562. *pl. 22.* 1896.

Perennial, stems 1.3–3.5 dm. high, from a simple or 1-forked caudex and a stout woody root, branching near the base, sparingly tomentulose and densely hirsute with long glandular hairs. Basal leaves 10–20 cm. long, pinnatifid with rather broad, angular, dentate lobes, attenuate into a broadly winged petiole, the stem-leaves several, gradually reduced above and becoming less divided, sessile, broad-based and somewhat clasping, the pubescence of the leaves like that of the stems; inflorescence of 4–20 heads, the heads 16–20-flowered; involucre campanulate, densely hirsute with long glandular hairs; phyllaries 18–24 mm. long, the inner lanceolate, attenuate at the apex, the outer one-half to three-fourths the length of the inner, narrowly lanceolate to linear; achenes 5.5–9 mm. long, rather strongly ribbed, reddish brown, shorter than the pappus.

Open woods, Arid Transition Zone; Josephine, Jackson, and Lake Counties, Oregon, south in the Coast Ranges to Lake County, California, and Mount Hamilton, Santa Clara County, and from Modoc County south to Sierra County. Type locality: Yreka, Siskiyou County, California. May–July.

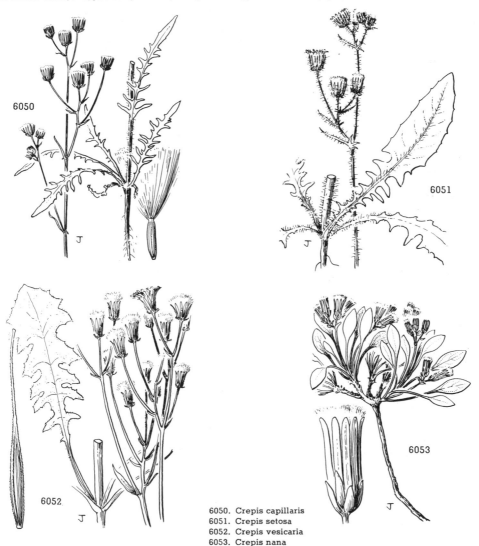

6050. Crepis capillaris
6051. Crepis setosa
6052. Crepis vesicaria
6053. Crepis nana

7. Crepis occidentàlis Nutt. Western Hawksbeard. Fig. 6055.

Crepis occidentalis Nutt. Journ. Acad. Phila. 7 : 29. 1834.

Perennial, stems 1–3, mostly 1–4 dm. high from a somewhat swollen caudex and long slender root, the stems and leaves with a close dense tomentum. Basal leaves 8–35 cm. long, variable, laciniate-dentate to pinnatifid with toothed lobes, the stem-leaves somewhat reduced and sessile; inflorescence of 10–30 heads, the heads 12–30-flowered, the peduncles usually glandular-pubescent; involucre cylindric-campanulate, 12–19 mm. long, with some glandular pubescence, the inner phyllaries 8–13, lanceolate, the outer phyllaries 6–8, about one-third shorter than the inner; achenes 6–10 mm. long, 10–18-ribbed, slightly attenuate at the apex, mostly medium brown, somewhat shorter than the pappus.

Dry rocky slopes, Arid Transition Zone; southeastern Washington south through Nevada and northeastern California along the east face of the Sierra Nevada to the Mount Piños region and Bear Valley in the San Bernardino Mountains; also from Idaho and western Wyoming south to New Mexico. Type locality: "on the borders and in the vicinity of the river Columbia." Collected by Wyeth. May–July.

Crepis occidentalis subsp. **costàta** (A. Gray) Babc. & Stebbins, Carnegie Inst. Wash. Pub. No. 504: 124. 1938. (*Crepis occidentalis* var. *costata* A. Gray, Bot. Calif. 1: 435. 1876; *C. grandifolia* Greene, Pittonia 3: 107. 1896.) Stems 0.8–4 dm. high; involucres, peduncles, and usually the upper cauline leaves with conspicuous, dark (rarely pale), gland-tipped setae; inflorescence rather narrow, of 15–30 heads, the heads 10–14-flowered, usually with 7–8 inner phyllaries. Gravelly or rocky, open slopes and flats; common from British Columbia southward, east of the Cascade Mountains, to northern California and east to South Dakota. Type locality: Stansbury Island, Great Salt Lake, Utah.

Crepis occidentalis subsp. **pùmila** (Rydb.) Babc. & Stebbins, Carnegie Inst. Wash. Pub. No. 504: 128. 1938. (*Crepis pumila* Rydb. Mem. N.Y. Bot. Gard. 1: 462. 1900.) Stems 1–4 dm. high, branched above, glandular pubescence completely lacking; heads few to many, 12–20-flowered, the inner phyllaries 8. Rocky open slopes and flats, occurring locally in the mountains of Montana, Idaho, and eastern Washington but more common in Nevada, eastern Oregon, and California as far south as Kern and Ventura Counties. Type locality: Bridger Mountains, Montana.

Crepis occidentalis subsp. **conjúncta** (Jepson) Babc. & Stebbins, Carnegie Inst. Wash. Pub. No. 504: 134. 1938. (*Crepis occidentalis* var. *conjuncta* Jepson ex Babc. & Stebbins, loc. cit.; *C. occidentalis* var. *nevadensis* Kell. Proc. Calif. Acad. 5: 50. 1873, in part.) Stems 0.5–2 dm. high, branched from near the base, the branches arcuate; segments of the pinnatifid leaves remote; inflorescence with 2–9 heads, the heads 12–30-flowered, phyllaries tomentose but not glandular or setose. Rocky slopes at rather high altitudes in Jackson and Josephine Counties, Oregon, and the mountains of adjacent California south to Mendocino County, and south in the Sierra Nevada to Placer County; also southeastern Washington and northwestern Wyoming. Type locality: Cisco, Placer County, California.

8. Crepis bàkeri Greene. Baker's Hawksbeard. Fig. 6056.

Crepis bakeri Greene, Erythea 3 : 73. 1895.

Perennial, stems 0.8–3 dm. high from a slightly thickened caudex and slender elongated root, stout, divaricately branching from near the base, reddish, glabrate or glandular-hispid. Basal leaves 8–20 cm. long, deeply pinnatifid with lanceolate or elliptic, usually dentate lobes, the midrib often reddish, narrowing below into a winged petiole, more or less canescent-tomentose and also glandular, the stem-leaves few, reduced above, sessile; inflorescences with 2–13 heads, the heads large, 11–40-flowered, on stout peduncles; involucres 11–20 mm. high, broadly cylindric to cyathiform, dark green; inner phyllaries 10–14, lanceolate, sparsely tomentose and conspicuously glandular-hispid, the outer phyllaries one-half to two-thirds the length of the inner; achenes 8–10.5 mm. long, slightly contracted at the apex, about 13-ribbed, yellowish to dark brown, the pappus about equalling to a little longer than the achenes.

Rocky slopes and flats, Arid Transition Zone; southern Washington in Klickitat and Kittitas Counties; southern Oregon in Crook, Jackson, and Josephine Counties; in California south in the Coast Ranges to Mendocino County and in eastern California from Modoc County to Placer County. Type locality: Egg Lake, Modoc County. May–June.

Crepis bakeri subsp. **cusíckii** (Eastw.) Babc. & Stebbins, Carnegie Inst. Wash. Pub. No. 504: 140. 1938. (*Crepis cusickii* Eastw. Bull. Torrey Club 30: 502. 1903.) Differing from *C. bakeri* subsp. *bakeri* in having smaller flowers, involucres 10–14 mm. long in flower, 13–17 mm. long in fruit, the pappus 6–9 mm. long, and achenes attenuate at the apex rather than being scarcely contracted. Lake and Jackson Counties, Oregon, south to Siskiyou and Lassen Counties, California. Type locality: 15 miles east of Ashland, Jackson County, Oregon.

9. Crepis modocénsis Greene. Low Hawksbeard. Fig. 6057.

Crepis modocensis Greene, Erythea 3 : 48. 1895.
Crepis scopulorum Coville, Contr. U.S. Nat. Herb. 3 : 563. *pl. 24.* 1896.

Perennials, stems 1 to several, rather slender, 0.5–4 dm. high, from a simple or 2–4-forked caudex and slender woody root, sparsely leafy, branching above, glabrate to tomentulose and bearing stiff yellowish setae. Basal leaves narrowly elliptic in outline and deeply pinnatifid or bipinnatifid, the segments acute; stem-leaves sessile, less divided, glabrate or tomentulose but bearing some stiff yellowish setae on the rachis; inflorescence of 1–8 heads, the heads 10–60-flowered; involucre cylindric-campanulate, the phyllaries 13–16 mm. long, rather densely beset throughout with stiff black setae; achenes 7–12 mm. long, somewhat striate to nearly smooth, dark, greenish black to reddish brown, mostly exceeding the pappus in length.

Sagebrush slopes and open forest, Arid Transition and Canadian Zones; Montana to British Columbia southward, east of the Cascade Mountains in Oregon and Washington, to northeastern California and east to Colorado. Type locality: lava beds, Modoc County, California. May–June.

Crepis modocensis subsp. **subacaùlis** (Kell.) Babc. & Stebbins, Carnegie Inst. Wash. Pub. No. 504. 148. 1938. (*Crepis occidentalis* var. *subacaulis* Kell. Proc. Calif. Acad. 5: 50. 1873; *C. occidentalis* var. *nevadensis* Kell. loc. cit.) Stout stems 6–20 cm. high branching from near the base; setae of stem and petiole straight and yellowish; involucres sparsely beset with black setae, the pappus-bristles 9–13.5 mm. long, longer than or equaling the achenes, which are dark, more or less brownish, and not at all beaked.

Rocky places, Warner Mountains. Lake County, Oregon, south to Placer County, California, and also in the San Bernardino Mountains. Type locality: Cisco, Placer County, California.

Crepis modocensis subsp. **rostràta** (Cov.) Babc. & Stebbins, Carnegie Inst. Wash. Pub. No. 504: 152. 1938. (*Crepis rostrata* Coville, Contr. U.S. Nat. Herb. 3 : 564. *pl. 25.* 1896; *C. occidentalis* var. *crinita* A. Gray, Bot.

Calif. 1: 435. 1876, in part.) Stems 1.5–3 dm. high, rather stout, densely covered with crisped whitish setae or tomentulose; involucres 14–17 mm. high; inner phyllaries with whitish crisped setae and also tomentulose; pappus-bristles 7–10 mm. long, shorter than or equaling the coarse-beaked achene. South-central British Columbia southward to Klickitat County, Washington. Type locality: near Crab Creek, Grant County, Washington.

Crepis modocensis subsp **glareòsa** (Piper) Babc. & Stebbins, Carnegie Inst. Wash. Pub. No. 504: 154. 1938. (*Crepis glareosa* Piper, Bull. Torrey Club **28**: 42. 1901.) Stems 0.6–1.3 dm. high, tomentulose, the whitish crisped setae lacking or nearly so; involucres 11–13 mm. high, the phyllaries hirsute with whitish crisped setae; pappus-bristles shorter than the greenish or yellowish achenes which are attenuate at the apex. Known only from the type locality at Ellensburg, Kittitas County, Washington.

10. **Crepis acumináta** Nutt. Long-leaved Hawksbeard. Fig. 6058.

Crepis acuminata Nutt. Trans. Amer. Phil. Soc. II. 7: 437. 1841.

Perennial, 2–7 dm. high, stems 1–3 from a swollen, 2–3-forked caudex and woody, deep-seated root, stout, tomentose at least at the base. Basal leaves 12–40 cm. long, pinnatifid with rather even lanceolate lobes, attenuate into a stout, narrow-winged petiole, densely to sparsely canescent-tomentose, the stem-leaves few, sessile, reduced above; inflorescences with 30–100 heads, the heads small, 5–12-flowered, the peduncles short and slender; involucres 9–15 mm. long, cylindric-campanulate; inner phyllaries 5–8, lanceolate, glabrous or tomentulose, the outer phyllaries 5–7, lanceolate-deltoid, mostly 1–3 mm. long; achenes 5.5–9 mm. long, about 12-ribbed, pale yellow or brownish, equaling the pappus.

Dry open places, Arid Transition Zone; eastern Washington southward to the mountains of southern California east to Montana and south to northern Arizona and New Mexico. Type locality: "Plains of the Platte." May–Aug.

6054. Crepis monticola
6055. Crepis occidentalis
6056. Crepis bakeri
6057. Crepis modocensis

11. Crepis pleurocárpa A. Gray. Naked-stemmed Hawksbeard. Fig. 6059.

Crepis pleurocarpa A. Gray, Proc. Amer. Acad. 17: 221. 1882.
Crepis intermedia var. *pleurocarpa* A. Gray, Syn. Fl. N. Amer. 1²: 432. 1884.
Crepis acuminata var. *pleurocarpa* Jepson, Man. Fl. Pl. Calif. 1011. 1925.

Perennial, stems mostly 2–4 dm. high, from a simple or 1-forked, thickened caudex and a slender elongated root, branching above the middle, slender or stoutish, greenish, tomentulose or glabrate and sometimes glandular. Basal leaves 7–28 cm. long, denticulate, dentate, or more or less runcinate-pinnatifid, with deltoid or lanceolate, acuminate lobes, attenuate into a broad petioli-form base, the stem-leaves very few, sessile, much reduced above; inflorescences mostly of 15–30 heads, the heads small, 4–12-flowered; involucre cylindric-campanulate, 8–16 mm. long, the phyl-laries densely floccose-tomentulose, the midportion of the phyllaries glabrate; inner phyllaries mostly 5(6–8), lanceolate, with scarious margins, the outer 5–6, 1.5–4 mm. long; achenes 5–8 mm. long, 10-ribbed, dark brown, shorter than the pappus.

In forests or on rocky slopes, mostly Arid Transition Zone; Curry County to Douglas County, Oregon, and the mountains in adjacent California south to Lake County; also Wenatchee Mountains in central Washington. Type locality: headwaters of the Sacramento River. Collected by Pringle. June–Aug.

12. Crepis atribárba Heller. Slender Hawksbeard. Fig. 6060.

Crepis occidentalis var. *gracilis* D. C. Eaton, Bot. King Expl. 203. 1871, in part.
Crepis atribarba Heller, Bull. Torrey Club 26: 314. 1899.
Crepis gracilis Rydb. Mem. N.Y. Bot. Gard. 1: 461. 1900. Not Hook f. & Thoms. 1876.
Crepis exilis Osterh. Muhlenbergia 1: 142. 1906, in part.
Crepis atribarba subsp. *typica* Babc. & Stebbins, Carnegie Inst. Wash. Pub. No. 504; Suppl. 1939.

Perennial, stems 1.5–3.5 dm. high, rather slender, 1 or 2 from a simple or 1-forked caudex and slender woody root, usually branched at or above the middle, grayish-tomentulose, becoming glabrate. Basal leaves 10–35 cm. long, pinnatifid, with lanceolate attenuate lobes; stem-leaves similar, becoming reduced in size, pubescence like the stems; inflorescences of 3–18 heads, the heads up to 65-flowered; involucre cylindric-campanulate, canescent-tomentulose to glabrate; phyllaries 8–14 mm. long, lanceolate, the inner and sometimes the outer with black glandless setae; achenes 3–10 mm. long, rather strongly ribbed, mostly greenish, about equaling the pappus in length.

Grassy open places and in yellow pine forests, Arid Transition Zone; common from British Columbia to Montana south to Colorado, Utah, Nevada, and central Oregon. Type locality: Lake Waha, Nez Perces County, Idaho. May–July.

Crepis atribarba subsp. *originàlis* Babc. & Stebbins, Carnegie Inst. Wash. Pub. No. 504; Suppl. 1939. (*Crepis occidentalis* var. *gracilis* D. C. Eaton, Bot. King Expl. 203. 1871, in part; *C. gracilis* Rydb. Mem. N.Y. Bot. Gard. 1: 461. 1900. Not Hook. f. & Thoms. 1876; *C. exilis* subsp. *originalis* Babc. & Stebbins, Carnegie Inst. Wash. Pub. No. 504: 162. 1938.) Stems slender, 3.5–7 dm. high; leaves pinnatifid with linear falcate divisions and usually glabrate; heads 10–40 in an inflorescence, few-flowered; phyllaries nearly or completely without setae. Grassy slopes, mostly at lower elevations than the preceding taxon and less common in the Rocky Mountains; British Columbia to Montana south to Utah, Nevada, and central Oregon. Type locality: near Hedley, British Columbia.

13. Crepis intermèdia A. Gray. Intermediate Hawksbeard. Fig. 6061.

Crepis intermedia A. Gray, Syn. Fl. N. Amer. 1²: 432. 1884.
Crepis acuminata var. *intermedia* Jepson, Man. Fl. Pl. Calif. 1011. 1925.

Perennial, 3–7 dm. high, stems 1–2 from a swollen, simple or 1-forked caudex and woody, deep-seated root, rather stout, branching at or above the middle, more or less canescent-tomentose. Basal leaves 15–40 cm. long, pinnatifid, the lanceolate lobes entire or dentate, narrowed below into a slender or narrowly winged petiole, the stem-leaves few, sessile, reduced above; inflorescences with 10–60 heads, the heads of medium size, mostly 7–12-flowered, the peduncles rather slender; involucre cylindric-campanulate, 10–16 mm. long, canescent, glabrous, sometimes with a few gland-less setae; inner phyllaries 7–8, lanceolate, outer phyllaries 6–8, lanceolate-deltoid, one-fifth to one-third as long as the inner; achenes 5.5–9 mm. long, 10–12-ribbed, yellowish to brown, longer than the pappus.

In forests and open places, Canadian Zone to Arid Transition Zone; eastern Washington south to Inyo County, California; also southern Alberta to Colorado, northern Arizona, and New Mexico. Type locality: indicated by Babcock (Univ. Calif. Pub. Bot. 22: 601. 1947) as Yosemite Valley, based upon a collection of Bolander. June–Aug.

14. Crepis barbígera Leiberg ex Coville. Bearded Hawksbeard. Fig. 6062.

Crepis barbigera Leiberg ex Coville, Contr. U.S. Nat. Herb. 3: 565. 1896.

Perennial, stems 1 to several, 2–8 dm. high, from a swollen, simple or 1–2-forked caudex and rather slender, woody root, mostly branched above the middle, the herbage glabrate to sparsely tomentulose, stout, often with yellow or greenish, nonglandular setae. Basal leaves 1–4 dm. long, pinnately or bipinnately toothed or parted, the lobes acute or acuminate, tapering below to a long, narrow, winged petiole, the cauline leaves few, similar, reduced in length, the lowermost petiolate; inflorescences of about 7–30 heads, the heads 8–25-flowered; involucre cylindric, with spreading yellowish setae and tomentose; inner phyllaries 9–17 mm. long, lanceolate, acute, the outer less than one-third the length of the inner, similar in outline; achenes 6.5–10 mm. long, strongly ribbed, oblong with a broad summit, olive-green or yellowish, slightly longer than the pappus.

Dry places, Arid Transition Zone; eastern Washington east to northern Idaho and south to Harney County, eastern Oregon. Type locality: Alkali Lake, Douglas County, Washington. June–July.

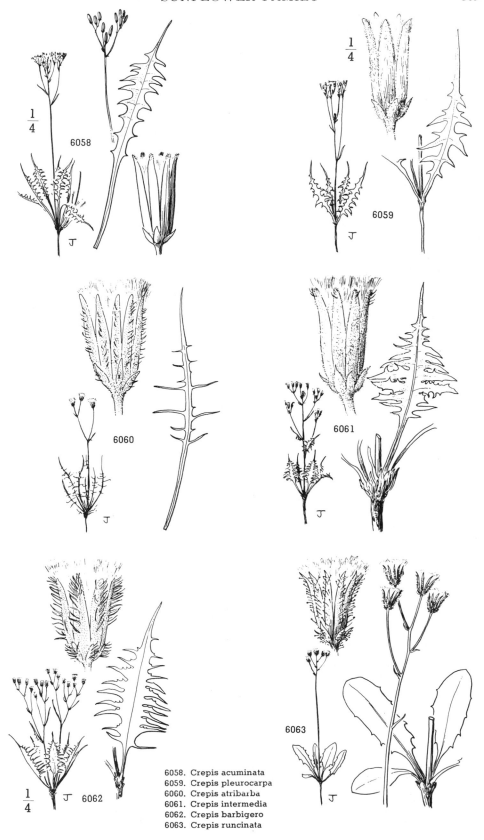

6058. Crepis acuminata
6059. Crepis pleurocarpa
6060. Crepis atribarba
6061. Crepis intermedia
6062. Crepis barbigero
6063. Crepis runcinata

15. Crepis runcinàta subsp. hispidulòsa (Howell) Babc. & Stebbins.
Meadow Hawksbeard. Fig. 6063.

Crepis platyphylla Greene, Pittonia **3**: 27. 1896.
Crepis runcinata var. *hispidulosa* Howell ex Rydb., Mem. N.Y. Bot. Gard. **1**: 461. 1900.
Crepis aculeolata Greene, Leaflets Bot. Obs. **2**: 86. 1910.
Crepis pallens Greene, loc. cit.
Crepis runcinata subsp. *hispidulosa* (Howell) Babc. & Stebbins, Carnegie Inst. Wash. Pub. No. 504: 96. 1938.

Perennial, stems 1–3 from a swollen, simple or 1-forked, thickened caudex and a fleshy root, usually 2.5–5 dm. high, glabrous or glandular-hispid. Basal leaves obovate, rounded at the apex, 6–25 cm. long, 3–8 cm. wide, the margin toothed toward the base, dentate above; cauline leaves reduced to bracts; inflorescences of 10–30 heads, the heads 20–50-flowered; involucre 8–12 mm. high, campanulate; phyllaries linear-lanceolate, acute, more or less glandular-hispid, the outer few and short, much shorter than the inner; achenes 3.5–5 mm. long, light to dark brown, mostly 10-ribbed, narrowed but not beaked, the pappus 4–8 mm. long.

Mostly moist alkaline meadows, Arid Transition Zone; eastern Washington and Oregon in the Pacific States; also Montana, Idaho, Colorado, and northern Utah. Type locality: base of Steen Mountains, Harney County, Oregon. June–July.

Crepis runcinata subsp. imbricàta Babc. & Stebbins, Carnegie Inst. Wash. Pub. No. 504: 102. 1938. Stems bearing 3–7 heads; basal leaves oblanceolate, 5–11 cm. long, 1.5–3 cm. broad, closely dentate, with the teeth tipped by white coreaceous mucros; phyllaries imbricate, abruptly contracted to an obtuse or broadly acute tip, the inner 10–12 mm. high; achenes 4.5–5 mm. long, tapering above but not beaked. Alkaline meadows in southeastern Oregon south to Elko and Washoe Counties, Nevada. Type locality: Alvord Valley, Harney County, Oregon.

Crepis runcinata subsp. andersònii (A. Gray) Babc. & Stebbins, Carnegie Inst. Wash. Pub. No. 504: 104. 1938. (*Crepis andersonii* A. Gray, Proc. Amer. Acad. **6**: 553. 1865.) Stems bearing 6–20 heads; basal leaves as in subsp. *imbricata*; phyllaries attenuate at the apex, the inner 13–21 mm. high; achenes 6–8 mm. long, more or less strongly beaked. Alkaline meadows, Sierra County, California, and Washoe, Ormsby, and Esmeralda Counties, Nevada. Type locality: Carson City, Ormsby County, Nevada.

Crepis runcinata subsp. hállii Babc. & Stebbins, Carnegie Inst. Wash. Pub. No. 504: 104. 1938. Stems bearing 5–14 heads on ascending peduncles; basal leaves 6.5–27 cm. long, 1.5–3 cm. broad, oblanceolate, coarsely toothed to subpinnatifid, narrowed to a long, broadly winged, petiolar base; achenes 4.5–6.5 mm. long, chestnut brown, tapering to a short broad beak. Alkaline meadows, Mono County, California, east to central Nevada. Type locality: Benton, Mono County, California.

175. TARÁXACUM Wiggers, Prim. Fl. Holst. 56. 1780.

Perennial acaulescent herbs from a taproot, with basal, sinuate-dentate or pinnatifid leaves and large heads of yellow flowers solitary at the ends of naked hollow scapes. Involucre oblong or campanulate, its inner phyllaries in 1 series, slightly united at base and often corniculate below the apex, the outer in several series of shorter, somewhat erect, reflexed or spreading ones. Receptacle flat, naked; ligules yellow (in ours), truncate, 5-toothed at the summit. Anthers sagittate at the base; style-branches slender, obtuse. Achenes oblong or linear-fusiform, 4–5-angled, 5–10-nerved, roughened or spinulose at least above, tapering into a slender beak. Pappus of many unequal, filiform, persistent bristles. [Name from the Greek, meaning to stir up, alluding to medical virtues.]

A genetically complex genus of perhaps 50 or 60 species, though many more have been described; occurring in the Old and New Worlds and in both the northern and southern hemispheres. Type species, *Taraxacum officinale* Weber.

Only limited synonymy is shown for the taxa given below. For a more complete listing of the taxa described see: *Monographie der Gattung Taraxacum*, Handel-Mazzetti, 1907; *North American Species of Taraxacum*, Sherff, Bot. Gaz. **70**: 329–359, 1920; and various papers by Haglund and Dahlstedt.

Introduced, widespread, weedy species; outer phyllaries reflexed to spreading (often less so in *T. laevigatum*).
    Leaves usually not deeply dissected, the terminal lobe larger than the lateral ones; inner phyllaries scarcely or not at all corniculate.   1. *T. officinale*.
    Leaves usually deeply dissected, the terminal lobe scarcely or not larger than the lateral ones; inner phyllaries commonly corniculate.   2. *T. laevigatum*.
Native species of high elevations; outer phyllaries appressed or somewhat spreading.
    Achenes brown, olive-brown, or stramineous when ripe, obscurely quadrangular.
        Inner phyllaries mostly corniculate; plants circumboreal in mountains of Washington, Oregon, and probably northern California.   3. *T. ceratophorum*.
        Inner phyllaries seldom if at all corniculate; plants of southern California mountains.
                                                                                              4. *T. californicum*.
    Achenes red or reddish brown when ripe, mostly sharply quadrangular.   5. *T. eriophorum*.

1. Taraxacum officinàle Weber. Common Dandelion. Fig. 6064.

Leontodon taraxacum L. Sp. Pl. 798. 1753.
Leontodon vulgare Lam. Fl. Franc. **2**: 113. 1778.
Taraxacum officinale Weber in Wiggers, Prim. Fl. Holst. 56. 1780.
Taraxacum dens-leonis Desf. Fl. Atlantica **2**: 228. 1799.
Taraxacum palustre var. *vulgare* Fernald, Rhodora **35**: 380. 1933.

Root thick and deep. Leaves oblong to spatulate in outline, pinnatifid, sinuate-dentate or rarely nearly entire, usually with a large terminal lobe, 1–3 dm. long, 2–5 cm. wide, glabrous or pubescent; scapes usually several, 5–40 cm. high; heads solitary, 2.5–5 cm. broad, with numerous yellow flowers; inner phyllaries erect, linear to linear-lanceolate, scarcely corniculate; outer phyllaries similar but shorter and reflexed; achenes grayish or greenish brown, fusiform, spinulose above, narrowed into a filiform beak two to three times as long as the body; pappus white, 6–8 mm. long.

Fields, roadsides, and lawns; naturalized from Europe; Alaska to southern California, well established in the Pacific States as well as in the eastern states. Native of the Old World. May–Oct.

2. **Taraxacum laevigàtum** (Willd.) DC. Red-seeded Dandelion. Fig. 6065.

Leontodon laevigatus Willd. Sp. Pl. **3** : 1546. 1804.
Taraxacum laevigatum DC. Cat. Hort. Monsp. 149. 1813.
Taraxacum erythrospermum Andrz. ex Bess. Enum. Pl. Volh. 75. 1822.

Root and general habit similar to the preceding species. Leaves deeply runcinate-pinnatifid or pinnately divided into narrow, trianguiar-lanceolate, and usually long-pointed segments, the terminal lobe not much larger than the lateral ones; scapes glabrous or pubescent above; heads many-flowered, 2–3 mm. broad; involucral phyllaries glaucous, the outer spreading or somewhat reflexed, lanceolate, attenuate, the inner linear, 7–10 mm. long, erect, each usually corniculate just below the tip; flowers sulphur-yellow or the outer usually purplish dorsally; achenes red or reddish brown, rather sharply ribbed and conspicuously spinulose above, the beak slender, not over twice the length of the body; pappus 4–7 mm. long, rather dull white.

Fields and dooryards; naturalized from Europe but less common in the Pacific States than *T. officinale*; common throughout the United States. May–Oct.

3. **Taraxacum ceratóphorum** (Ledeb.) DC. Horned Dandelion. Fig. 6066.

Leontodon ceratophorus Ledeb. Ic. Pl. Ross. **1** : 9. *pl. 34.* 1829.
Taraxacum ceratophorum DC. Prod. **7** : 146. 1838.
Taraxacum paucisquamosum M. E. Peck, Madroño **5** : 247. 1940.

Root thick, blackish on the surface. Leaves procumbent, ascending, or erect, lanceolate or oblanceolate, dentate or often retrorsely toothed or lobed, glabrous; scapes about as long as the leaves (shorter in flower), well exceeding them in fruit; outer phyllaries erect, broadly lanceolate to ovate, glabrous or villous; inner phyllaries linear-lanceolate, 10–12 mm. long, typically distinctly corniculate at apex, dark purplish, scarious-margined; rays well exserted, purple-veined; achenes stramineous or brown, prominently muricate above the middle, the beak as long as or exceeding the body; pappus white, 5–6 mm. long.

Banks and meadows, Boreal Zone; circumboreal southward in the mountains from Alaska to northern California; also Rocky Mountains and eastern United States. Type locality: Kamchatka. June–Oct.

Taraxacum lyràtum (Ledeb.) DC. Prod. **7** : 148. 1838. (*Leontodon lyratus* Ledeb. Fl. Alt. **4** : 152. 1833.) A dwarf alpine herb, differing chiefly from *T. ceratophorum* by the mature blackish achenes, is found in North America in the alpine regions in rocky localities in Colorado, Nevada, and Arizona. This taxon may be expected to occur in the White Mountains on the eastern border of California. Type locality: Siberia.

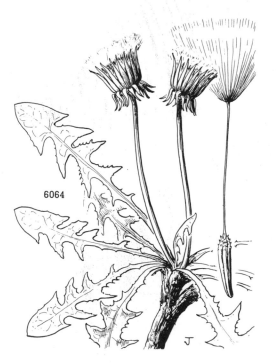

6064. Taraxacum officinale   6065. Taraxacum laevigatum

4. **Taraxacum califórnicum** Munz & Jtn. California Dandelion. Fig. 6067.

Taraxacum californicum Munz. & Jtn. Bull. Torrey Club **52**: 227. 1925.
Taraxacum ceratophorum var. *bernardinum* Jepson, Man. Fl. Pl. Calif. 1004. 1925.

Root thick and dark, the crown simple or sometimes divided into 2 short crowns, the herbage glabrous. Leaves ascending or widely spreading, oblanceolate, 5–12 cm. long, 1–3 cm. wide, obtuse to acutish, subentire to sinuate, dentate, rarely runcinate-incised, glabrous and pale green; scape solitary to several, ascending, longer than the leaves in fruit, glabrous; involucres broadly cylindric, 10–15 mm. high, nearly truncate at base; inner phyllaries erect, not corniculate or only obscurely so; outer phyllaries ovate-lanceolate, barely half as long as the inner, glabrous; flowers yellow, erect or nearly so, 2–3 mm. longer than the involucre; achenes light brown, about 3 mm. long, rugose below, narrowly tuberculate above, abruptly attenuate into the beak, this 7–9 mm. long; pappus white, about 5 mm. long.

Moist places, Arid Transition and Canadian Zones; San Bernardino Mountains, San Bernardino County, southern California. Type locality: Bear Valley, San Bernardino Mountains. June–Aug.

5. **Taraxacum erióphorum** Rydb. Rocky Mountain Dandelion. Fig. 6068.

Taraxacum eriophorum Rydb. Mem. N.Y. Bot. Gard. **1**: 454. 1900.
Taraxacum olympicum G. N. Jones, Univ. Wash. Pub. Bot. **5**: 263. 1936.

Perennial, much resembling *T. ceratophorum* in growth, habit, and leaves; inner phyllaries usually not corniculate; outer phyllaries and summit of scape with some woolly hairs; achenes red to reddish brown at maturity, sharply quadrangular.

Moist meadows, Boreal Zone; Alaska, the Olympic Mountains in Washington, Montana, and Wyoming. Type locality: Sheridan, Madison County, Montana. May–Sept.

176. **LÁPSANA** L. Sp. Pl. 811. 1753.

Annual herbs with branching stems and alternate leaves. Heads on slender peduncles, paniculate. Involucre cylindric, the bracts erect in 1 series with a few minute outer ones at base. Corolla yellow. Achenes oblong-obovoid, rounded at apex, 20–30-nerved. Pappus none. Receptacle naked. [Ancient Greek name of obscure derivation.]

An Old World genus of 9 species. Type species, *Lapsana communis* L.

1. **Lapsana commùnis** L. Nipplewort. Fig. 6069.

Lapsana communis L. Sp. Pl. 811. 1753.

Stems simple below, 1 to several from the annual root, branching paniculately above, 3–10 dm. high, glabrous above, more or less villous-hirsute below and sometimes glandular. Lower leaves ovate or deltoid-ovate, 3–5 cm. long, thin, pubescent or glabrate; petioles often as long as or longer

6066

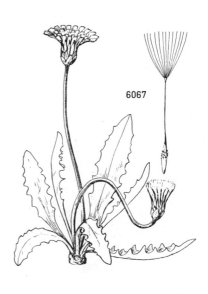

6067

6066. Taraxacum ceratophorum   6067. Taraxacum californicum

than the blade, often with 2 to several lateral lobes; upper leaves often deltoid or deltoid-lanceolate, with shorter winged petioles or nearly sessile; heads on slender peduncles, usually numerous; involucre oblong-cylindric, 4–6 mm. high, with about 8 principal phyllaries and a few small outer ones at base, these erect in fruit.

Mostly in moist shady situations, principally Humid Transition Zone; British Columbia southward, mostly west of the Cascade Mountains, to Oregon and south along the coast in California to San Mateo County; also in the eastern United States. Naturalized from Eurasia. June–Aug.

Lapsana apogonoïdes Maxim. Bull. Acad. St. Pétersb. **18**: 288. 1873. A glabrous flaccid-stemmed annual with phyllaries of the mature heads divaricately spreading. A native of Japan, growing as a ground cover on Sauvies Island, Multnomah County, Oregon.

6068

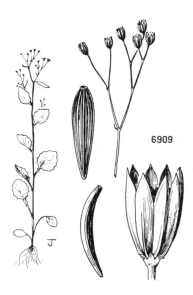

6909

6068. Taraxacum eriophorum

6069. Lapsana communis

APPENDIX

KEY TO FAMILIES DESCRIBED IN VOLUMES I–IV

(Characters used apply to genera and species of the Pacific States)

Plants without true flowers or seeds, reproducing by spores. Division Pteridophyta.
Plants with true flowers and seeds. Division Spermatophyta.
   Ovules and seeds naked on surface of scale; stigmas none. Subdivision Gymnospermae.
   Ovules and seeds enclosed in a cavity (ovary); stigmas present. Subdivision Angiospermae.
     Floral parts usually 3 or 6 (sometimes greatly modified or reduced); leaf-veins typically parallel longitudinally (except *Trillium* and *Smilax*); vascular bundles scattered. Class Monocotyledoneae.
     Floral parts numerous or usually in multiples of 4 or 5 (sometimes greatly reduced or modified); leaf-veins typically divided palmately or pinnately (except *Plantago* spp. and others); vascular bundles not scattered. Class Dicotyledoneae.
       Petals distinct or lacking, or calyx as well as corolla lacking. Subclass Choripetalae* (Archichlamydeae).
       Petals fused and deciduous as a unit. Subclass Sympetalae† (Metachlamydeae).

Pteridophyta

Stems not jointed, not with siliceous areas; leaves not obscure and not fused around the stem.
   Spores enclosed in a stalked sporocarp including both megaspores and microspores.
     Semiaquatics, becoming terrestrial, with either filiform or clover-like leaves; sporocarp-covering firm.
       3. Marsileaceae 1: 33.
     Minute floating aquatics with scale-like leaves; sporocarp-covering thin. 4. Salviniaceae 1: 35.
   Spores in sporangia in leaf-axils, in cone-like clusters, or borne marginally or on the underside of the fronds; spores alike or dissimilar.
     Plants with several tubular elongate leaves from a corm-like base; sporangia borne in cavities at the bases of the leaves. 5. Isoetaceae 1: 35.
     Plants otherwise.
       Leaves large in relation to the stem (rhizome), much divided or at least pinnatisect (entire in *Ophioglossum*); sporangia marginal or on undersurface of fronds.
         Expanding leaves circinate; rhizome with scales or hairs, often branched.
           2. Polypodiaceae 1: 5.
         Expanding leaves erect or bent; rhizome short, glabrous. 1. Ophioglossaceae 1: 1.
       Leaves small or scale-like, closely covering the usually branched stems; sporangia in the axils of the leaves or in cone-like clusters.
         All spores alike. 7. Lycopodiaceae 1: 43.
         Spores of two sizes (microspores and megaspores). 8. Selaginellaceae 1: 46.
Stems jointed, the internodes with siliceous tubercles or crossbands; leaves obscure and fused around the stems.
       6. Equisetaceae 1: 38.

Spermatophyta

Gymnospermae

Stems not jointed; leaves green, linear or needle-like or, when scale-like, imbricate and persistent.
   Seeds borne singly, drupe-like or subtended by a red aril; plants dioecious. 1. Taxaceae 1: 50.
   Seeds in a cone of several woody scales or the scales coalescent and berry-like in *Juniperus* and *Sabina*; plants monoecious.
     Cone-scales spirally arranged; leaves needle-like or linear (decurrent and awl-shaped in *Sequoia gigantea*), solitary and spirally arranged on the stems or in clusters of 2–5.
       Pollen-sacs and ovules 2 on each scale; seeds with a distinct wing. 2. Pinaceae 1: 52.
       Pollen-sacs and ovules several on each scale; seeds wingless or merely wing-margined.
         3. Taxodiaceae 1: 68.
     Leaves and cone-scales opposite or in whorls of 3, the leaves scale-like. 4. Cupressaceae 1: 70.
Stems jointed; leaves paired or ternate, of thin dry scales; plants of desert areas. 5. Ephedraceae 1: 77.

Angiospermae

Monocotyledoneae

Plants strictly aquatic, either floating or submersed, not merely persisting in mud as the water recedes.
   Minute, floating, thallus-like plants not differentiated into stems and leaves.
     14. Lemnaceae 1: 346.
Plants not minute, differentiated into leaves and stems.
   Perianth absent or greatly modified (sepal-like appendages on staminate flowers of *Potamogeton*).
     Flowers in spikes, heads, or umbels, the inflorescence more or less pedunculate or enclosed in a spathe.

---

* Petals fused at the base or some distally in *Fumariaceae, Fabaceae, Krameriaceae, Oxalidaceae, Polygalaceae, Balsaminaceae, Malvaceae.*

† Calyx lacking in *Compositae*; corolla in *Glaux (Primulaceae)* and *Allotropa (Monotropaceae)*; corolla with free or nearly free segments in some genera of *Monotropaceae, Pyrolaceae, Plumbaginaceae, Oleaceae, Gentianaceae, Marah* spp. in *Cucurbitaceae.*

Spike-like inflorescences, flattened and 1-sided; plants of tide pools or estuaries.
　　　　5. Zosteraceae 1: 93.
Inflorescences of heads, umbels, or if spikes these not flattened and 1-sided; plants of fresh or brackish water.
　　Style 2-cleft, sometimes simple (achenes with 1 or more dry bracts); plants monoecious (most species emersed and persisiting in mud). 2. Sparganiaceae 1: 80.
　　Stigmas sessile, disk-like; flowers perfect. 　　3. Potamogetonaceae 1: 83
Flowers in the leaf-axils.
　　Leaves serrate or serrulate; nutlet ellipsoid, enclosed in a loose membranous coat.
　　　　4. Naiadaceae 1: 92.
　　Leaves entire; nutlet obliquely oblong with a prominent beak, not enclosed in a membranous coat (*Zannichellia*). 　　3. Potamogetonaceae 1: 83.
Perianth in 2 series, the inner petaloid.
　　Ovary inferior; inner perianth white. 　　9. Vallisneriaceae 1: 103.
　　Ovary superior; inner perianth yellow or blue. 　　15. Ponterderiaceae 1: 349.
Plants terrestrial or semiaquatic with the stems partially emersed and the plants persisting in mud as the water recedes.
　Trees up to 9–12 m. high.
　　Leaves fan-shaped and lobed; inflorescence in a large spathe; trunks columnar.
　　　　12. Phoenicaceae (Palmae) 1: 344.
　　Leaves linear-lanceolate; inflorescence not in a large spathe; trunks normally branched (*Yucca brevifolia* and *Y. mohavensis*). 　　18. Liliaceae 1: 379.
　Plants herbaceous or shrub-like, or acaulescent from a woody caudex, or woody vines.
　　Plants woody vines; northern California and southern Oregon. 　　20. Smilacaceae 1: 458.
　　Plants not woody vines.
　　　Perianth none or rudimentary (of bristles or inconspicuous scales).
　　　　Plants semiaquatic.
　　　　　Acaulescent, more or less delicate annuals sometimes completely submerged; flowers few, perfect, staminate or pistillate. 　　6. Lilaeaceae 1: 95.
　　　　　Caulescent herbs, usually rather stout; monoecious, many-flowered, the staminate flowers borne above the pistillate.
　　　　　　Flowers in dense spikes; perianth of bristles. 　　1. Typhaceae 1: 79.
　　　　　　Flowers in globose heads; perianth of scales. 　　2. Sparganiaceae 1: 80.
　　　　Plants terrestrial, sometimes in wet ground.
　　　　　Flowers subtended by 1 or 2 chaffy bracts; basal portion of leaves sheathing the stems (grasses and sedges).
　　　　　　Stems normally hollow between the nodes; leaf-sheaths split on 1 side; flowers subtended by 2 scarious bracts. 　　10. Poaceae (Gramineae) 1: 103.
　　　　　　Stems normally solid; leaf-sheaths not split; flowers subtended by 1 scarious bract or by several bristles. 　　11. Cyperaceae 1: 255.
　　　　　Flowers crowded on a stout spadix subtended by a pale yellow spathe.
　　　　　　13. Araceae 1: 345.
　　　Perianth present, 6-parted or -lobed (3-parted in *Triglochin stricta*), regular or irregular.
　　　　Plants grass-like or sedge-like; perianth inconspicuous, greenish or greenish yellow (inner segments white in *Scheuchzeria*) and herbaceous, or straw-colored or brown and dry and firm.
　　　　　Fruit of follicles, distinct or separating from the base upward when ripe.
　　　　　　7. Scheuchzeriaceae 1: 96.
　　　　　Fruit a loculicidal capsule. 　　16. Juncaceae 1: 350.
　　　　Plants not grass-like or sedge-like; perianth petaloid or the segments fused, conspicuous and usually white or brightly colored (greenish in some *Convallariaceae, Orchidaceae*).
　　　　　Fruit of many achenes arranged in whorls or heads; plants semiaquatic.
　　　　　　8. Alismaceae 1: 97.
　　　　　Fruit capsular or berry-like, never of distinct achenes; plants terrestrial.
　　　　　　Ovary superior.
　　　　　　　Fruit a capsule.
　　　　　　　　Styles 3-parted, completely distinct or nearly so; plants from an erect or creeping rootstock (from bulbs in *Zygadenus, Stenanthium*).
　　　　　　　　　17. Melanthaceae 1: 371.
　　　　　　　　Style 1, the parts fused; plants commonly from bulbs (rootstock in *Leucocrinum*; woody caudex in *Yucca* spp., *Nolina*).
　　　　　　　　　18. Liliaceae 1: 379.
　　　　　　　Fruit a berry (capsular in *Scoliopus*). 　　19. Convallariaceae 1: 448.
　　　　　　Ovary inferior.
　　　　　　　Plants essentially acaulescent with a stout woody crown; deserts.
　　　　　　　　21. Amaryllidaceae 1: 459.
　　　　　　　Plants herbaceous, not acaulescent, from rootstocks, corms, or tubers.
　　　　　　　　Perianth regular, the 2 series of segments similar or dissimilar.
　　　　　　　　　22. Iridaceae 1: 461.
　　　　　　　　Perianth irregular. 　　23. Orchidaceae 1: 469.

Dicotyledoneae

Choripetalae

Calyx and corolla both wanting, or if a perianth present this vestigial or obscure (flowers sometimes subtended by bracts simulating a calyx).
　Trees or shrubs; flowers unisexual (see also pistillate plants of *Atriplex, Grayia, Forestiera,* and pistillate flowers in some genera of *Ambrosiinae*).
　　Flowers, at least the staminate, in erect or drooping aments.
　　　Fruit a capsule or a dry berry; seeds many or at least 2 (sometimes 1 by abortion).
　　　　Leaves alternate; seeds minute, many, comose (bearing a tuft of hair).
　　　　　25. Salicaceae 1: 486.

Leaves opposite; seeds not minute, few, not comose.   111. Garryaceae **3**: 284.
Fruit not capsular, either a nut, a winged nutlet, or a waxy drupe.
    Leaves compound; nut enclosed in a fibrous husk (vestigial corolla of pistillate flowers).
        27. Juglandaceae **1**: 509.
    Leaves simple.
        Ovary superior; fruit a waxy drupe.   26. Myricaceae **1**: 508.
        Ovary inferior; fruit otherwise.
          Seeds winged nutlets in a woody or thin-scaled cone, or a single nut in a tubular involucre.   28. Betulaceae **1**: 510.
          Fruit a single nut (acorn) in a scaly cup or more than one in a bur.
          29. Fagaceae **1**: 514.
Both staminate and pistillate flowers in heads (minute petals present in our species).
    62. Platanaceae **2**: 403.
Herbs or subshrubs; flowers perfect or unisexual.
    Plants parasitic.
        Plants branched, some woody; fruit a 1-seeded berry.   33. Loranthaceae **1**: 528.
        Plants minute, sessile on stems of *Dalea*; fruit a many-seeded capsule.
        Rafflesiaceae.*
    Plants not parasitic.
        Aquatic plants; floating or growing on mud as the water recedes.
          Fruit a nutlet with a calyx-like involucre; leaves dissected.
          45. Ceratophyllaceae **2**: 174.
          Fruit compressed, 4-celled, without involucre; leaves entire.
          80. Callitrichaceae **3**: 42.
        Terrestrial plants (calyx lacking also in pistillate flowers of *Eremocarpus* and *Stillingia* in *Euphorbiaceae*).
          Inflorescence of small, staminate and pistillate flowers set within a modified involucre which simulates a complete flower (*Euphorbia*).   79. Euphorbiaceae **3**: 23.
          Inflorescence otherwise.
            Succulent-leaved subshrubs; salt marshes of southern California.
            40. Batidaceae **2**: 114.
            Perennial herbs, the leaves not succulent; not plants of salt marshes.
              Flowers in spikes subtended by white petaloid bracts; leaves entire.
              24. Saururaceae **1**: 485.
              Flowers in spikes not subtended by petaloid bracts; leaves trifoliolate (*Achlys*).
              47. Berberidaceae **2**: 216.
Calyx only present, or both calyx and corolla present.
    Flowers apetalous.
        Calyx conspicuously colored or white and corolla-like, or large (2–8 cm. long except in some spp. of *Anemone*), or with both characters. (See also *Mesembryanthemum* in *Aizoaceae*, with petaloid staminodes).
          Calyx-base fused to the solitary achene (anthocarp), the upper portion brightly colored.
          38. Nyctaginaceae **2**: 100.
          Calyx-base and fruit otherwise.
             Ovary inferior; flowers brownish purple and strongly tinged with green.
             34. Aristolochiaceae **1**: 534.
            Ovary superior; flowers variously colored, not strongly tinged with green.
              Fruit a drupe; glabrous shrub with pliant stems and soft wood.
              103. Thymelaeaceae **3**: 163.
              Fruit either capsular or the achenes in a head; if shrubs the stems not with soft wood.
                Vines or herbs; achenes in heads; flowers (in ours) never shades of yellow (*Anemone, Clematis, Pulsatilla*).   46. Ranunculaceae **2**: 174.
                Shrubs or small trees; fruit capsular; flowers bright yellow (*Fremontia*).
                93. Sterculiaceae **3**: 112.
        Calyx small (1 cm. long or less), green or colored.
          Trees, shrubs, or subshrubs.
            Fruit a samara (winged nutlet); leaves compound (sometimes simple in *Fraxinus anomala*).
              Styles 2; fruit a double samara (*Acer negundo californica*).
              87. Aceraceae **3**: 56.
              Style 1; fruit a single samara (*Fraxinus* spp.).   119. Oleaceae† **3**: 346.
            Fruit otherwise; leaves simple, entire or toothed.
               Style and stigma 1 (style shortened and stigma 2-lobed in *Ulmaceae*); ovary 1-celled, 1-seeded.
                Flowers perfect.
                  Fruit a drupe; stamens 9, opening with flap-like valves; trees with aromatic evergreen foliage.   49. Lauraceae **2**: 222.
                  Fruit an achene seated within a hypanthium; stamens 15–40, opening by slits; shrubs (*Coleogyne, Cercocarpus*).   64. Rosaceae **2**: 407.
               Flowers monoecious or dioecious; leaves deciduous.
                Fruit a drupe, completely superior.
                  Tree; pistillate flowers solitary; leaves alternate, serrate-margined.
                  30. Ulmaceae **1**: 523.
                  Shrub; pistillate flowers clustered: leaves opposite, entire (*Forestiera*).
                  119. Oleaceae† **3**: 346.
               Fruit drupe-like, the fleshy calyx-base enclosing the achene; shrub with silvery-scurfy foliage.   104. Eleagnaceae **3**: 163.

---

* Not included in Volume I; represented in our area in the southern Colorado Desert, California, by *Pilostyles thurberi* A. Gray (Mem. Amer. Acad. II. **5**: 326. 1854). The plants are composed of tiny brown flowers and a few subtending bracts.

† In *Sympetalae*.

Styles and/or stigmas 2–3 or more.
  Flowers perfect (see also *Allenrolfia* in *Chenopodiaceae*); perianth white or colored (*Eriogonum*).   35. Polygonaceae **2**: 1.
  Flowers monoecious or dioecious; perianth not white or colored.
    Ovary superior.
      Fruit a capsule or the seed solitary in a loose pericarp (utricle).
        Fruit a 3–4-celled capsule; herbage not fleshy or scurfy.
          Capsule not seated in a persistent calyx; carpels separating from a central column (*Tetracoccus, Acalypha, Bernardia, Ricinus*). 79. Euphorbiaceae **3**: 23.
          Capsule seated in a persistent calyx; acorn-like capsule with 1 seed developing.   81. Buxaceae **3**: 45.
        Fruit a utricle; herbage mostly fleshy or scurfy (*Atriplex, Grayia, Eurotia, Sarcobatus*).   36. Chenopodiaceae **2**: 66.
      Fruit a fleshy drupe or berry.
        Creeping, heath-like shrubs; leaves crowded and linear (petals occasionally present).   82. Empetraceae **3**: 45.
        Shrubs 1–3 m. high; leaves if small not linear (*Rhamnus* spp., *Condalia* sp.).   90. Rhamnaceae **3**: 59.
    Ovary inferior; subshrubs (*Galium* spp.). 143. Rubiaceae* **4**: 22.
Annual or perennial herbs.
  Stamens perigynous, attached to the hypanthium (see also some *Aizoaceae*).
    Stem-leaves lacking or essentially so; stipules lacking; leaves simple and shallowly lobed (*Heuchera* spp., *Ozomelis* sp.).   59. Saxifragaceae **2**: 349.
    Stem-leaves present; stipules present; leaves compound (*Acaena, Sanguisorba*, but not in *Alchemilla*).   64. Rosaceae **2**: 407.
  Stamens hypogynous or epigynous.
    Ovary inferior.
      Leaves opposite or verticillate.
        Aquatic plants.
          Fruit 1-seeded and drupe-like (*Hippuris*). 107. Haloragidaceae **3**: 212.
          Fruit a many-seeded, dry capsule (*Ludwigia* sp.). 106. Onagraceae **3**: 167.
        Plants not aquatic (*Galium* spp.).   143. Rubiaceae* **4**: 22.
      Leaves alternate.
        Plants erect, not succulent.
          Leaves entire; fruit a 1-seeded drupe. 32. Santalaceae **1**: 527.
          Leaves unequally pinnatifid; fruit a dry, many-seeded capsule. 101. Datiscaceae **3**: 142.
        Plants fleshy; decumbent or creeping; leaves entire (*Tetragonia, Mesembryanthemum*).   41. Aizoaceae **2**: 115.
    Ovary superior.
      Fruit a capsule.
        Capsule opening by valves (*Loeflingia* sp., *Arenaria* spp., *Sagina* sp.). 43. Caryophyllaceae **2**: 137.
        Capsule circumscissile or loculicidally dehiscent. 41. Aizoaceae **2**: 115.
      Fruit otherwise.
        Sepals falling at anthesis.   46. Ranunculaceae **2**: 174.
        Sepals persistent or tardily deciduous
          Fruit an achene or a utricle.
            Fruit a solitary achene.
              Flowers perfect, the perianth colored in most genera; achenes 3–4-angled or sometimes lenticular. 35. Polygonaceae **2**: 1.
              Flowers monoecious or dioecious, sometimes polygamous, the perianth never colored; achenes flattened or ovoid. 31. Urticaceae **1**: 523.
          Fruit a utricle.
            Stipules present except in *Scleranthus*. 39. Illecebraceae **2**: 111.
            Stipules lacking.
              Calyx herbaceous.   36. Chenopodiaceae **2**: 66.
              Calyx scarious or chartaceous (somewhat herbaceous in *Tidestromia*).   37. Amaranthaceae **2**: 98.
          Fruit a schizocarp or a 2-valved capsule (*Tragia, Eremocarpus, Stillingia*). 79. Euphorbiaceae **3**: 23.
Flowers with both calyx and corolla present.
  Leaves insectivorous; plants of boggy areas.
    Leaves large and hood-like.   55. Sarraceniaceae **2**: 329.
    Leaves small, with sensitive glandular hairs.   56. Droseraceae **2**: 329.
  Leaves not insectivorous; plants of various habitats.
    Corolla or calyx or both irregular.
      Calyx with the sepals showy and colored as well as the petals and often much modified in shape.
        Pistils 2–5 (1 in *Delphinium ajacis*); fruit of clustered follicles (*Delphinium, Aquilegia, Aconitum*).   46. Ranunculaceae **2**: 174.
        Pistil 1; fruit not of follicles.
          Fruit 1-celled and 1-seeded, indehiscent and armed with prickles. 70. Krameriaceae **2**: 627.

---

* In *Sympetalae.*

Fruit a 2–5-celled capsule.

Capsule 2-celled, compressed contrary to the partition; anthers opening by sub-terminal pores. 78. Polygalaceae **3**: 21.

Capsule 5-celled, elastically dehiscent; anthers opening by longitudinal slits. 89. Balsaminaceae **3**: 59.

Calyx with sepals relatively small and regular and, if colored, inconspicuous; petals of markedly different shape or spurred or otherwise modified.

Flowers papilionaceous (petal 1 in *Amorpha*); stamens 10. 69. Fabaceae **2**: 480.

Flowers not papilionaceous; stamens less than 10.

Capsule 2-valved or an indehiscent nutlet; leaves pinnately decompound. 51. Fumariaceae **2**: 233.

Capsule 3-valved; leaves entire or palmately cleft or palmately decompound. 99. Violaceae **3**: 123.

Corolla regular, the petals alike or essentially so.

Ovary completely inferior or if half-inferior the ovary not fused the entire length of the calyx (see also *Ceanothus* spp., *Colubrina*, *Adolphia* in *Rhamnaceae* with capsule half-immersed in disk).

Ovary completely inferior.

Plants leafless with thick, succulent, spiny stems; perianth-segments numerous, grading from sepals to petals. 102. Cactaceae **3**: 143.

Plants not as in the preceding; perianth-segments few, the sepals and petals well differentiated.

Fruit a berry, drupe or pome, juicy or somewhat fleshy.

Leaves mostly opposite, simple, the margins entire. 110. Cornaceae **3**: 283.

Leaves alternate, compound, or if simple, the margin serrulate, toothed, or lobed.

Inflorescence not umbellate.

Hypanthium produced above the ovary; fruit a berry; shrubs often with spiny stems. 61. Grossulariaceae **2**: 388.

Hypanthium not produced above the ovary; fruit a pome; stems without spines. 66. Malaceae **2**: 469.

Inflorescence of umbels. 108. Araliaceae **3**: 213.

Fruit not a berry, drupe, or pome, dry.

Fruit a capsule with many seeds; petals in most species showy and rather large; inflorescence relatively few-flowered.

Herbage markedly rough-pubescent; sepals 5; petals 5 or more. 100. Loasaceae **3**: 133.

Herbage glabrous or soft-hairy; sepals and petals normally 4. 106. Onagraceae **3**: 167.

Fruit at maturity splitting into 1-seeded divisions; petals small; inflorescences many-flowered in *Umbelliferae*.

Fruit of 4 divisions; aquatics, the submerged leaves with capillary divisions (*Myriophyllum*). 107. Haloragaceae **3**: 212.

Fruit of 2 divisions; if aquatics the submerged leaves not as in the preceding; herbage odorous. 109. Umbelliferae **3**: 215.

Ovary half-inferior.

Capsule circumscissile; fleshy herbs (*Portulaca*). 42. Portulacaceae **2**: 119.

Fruit capsular and not circumscissile or of separate follicles; herbage not fleshy.

Leaves principally basal; stem-leaves when present very few (except *Saxifraga* spp.); herbaceous perennials or rarely annuals. 59. Saxifragaceae **2**: 349.

Leaves opposite; shrubs or trailing vines. 60. Hydrangeaceae **2**: 384.

Ovary superior, either obviously so or set within the floral tube and free from it.

Stamens normally more than twice as many as the petals (in some species 10 or less).

Stamens united in a tube around the pistil. 92. Malvaceae **3**: 82.

Stamens separate or in clustered bundles, or if all united not as in the preceding.

Aquatics with large, entire, peltate or cordate, floating leaves. 44. Nymphaeaceae **2**: 172.

Plants not aquatics, except *Ranunculus* spp. with leaves not as in the preceding.

Sepals and petals alike, red; shrubs with opposite leaves. 48. Calycanthaceae **2**: 221.

Sepals and petals differentiated.

Sepals caducous at anthesis (except *Paeonia* in *Ranunculaceae*).

Petals the same number as the sepals, rarely more.

Fruit a cylindrical pod 3–5 cm. long; plant with a disagreeable odor (*Polanisia*). 53. Capparidaceae **2**: 322.

Fruit not as in the preceding; plants not with a disagreeable odor. 46. Ranunculaceae **2**: 174.

Petals twice the number of sepals (fused into a cap in *Eschscholzia*). 50. Papaveraceae **2**: 223.

Sepals not caducous at anthesis.

Trees or shrubs.

Carpels 1; fruit a legume (stamens 10 in *Prosopis*). 67. Mimosaceae **2**: 474.

Carpels 3 to many; fruit various.

Seeds arillate; follicles 10–15 mm. long; stipules lacking. 63. Crossosomataceae **2**: 406.

Seeds not arillate; fruit various, if follicular the follicles mostly 3–6 mm. long; stipules present in most genera.

Fruit never a fleshy drupe with a woody or bony stone. 64. Rosaceae **2**: 407.

Fruit a fleshy drupe with a woody or bony stone.
65. Amygdalaceae **2**: 465.
Annual herbs or perennial herbs which may be cespitose cushion
plants or slightly woody at base.
Sepals 2, except in *Lewisia rediviva* (*Lewisia* spp., *Talinum*,
Calandrinia spp.).
42. Portulacaceae **2**: 119.
Sepals as many as the petals or more.
Fruit a lobed capsule gaping at the apex before the seeds are
mature; petals toothed or cleft.
54. Resedaceae **2**: 327.
Fruit if capsular not as in the preceding; petals not at all
lobed.
Stamens perigynous.
64. Rosaceae **2**: 407.
Stamens hypogynous.
Leaves all opposite; stamens usually basally united.
94. Hypericaceae **3**: 115.
Leaves (in ours) alternate; stamens free; plants
from a woody base.
98. Cistaceae **3**: 122.
Stamens few, usually 5 or 10, not more than twice as many as the petals.
Trees, shrubs, or woody vines.
Woody vines with juicy berries (see also climbing forms of *Rhus diversiloba* with
dry fruit).   91. Vitaceae **3**: 81.
Shrubs or trees (see also *Frankenia* sp. and *Lepidium* sp.).
Fruit a samara.
Leaves opposite; fruit single or double samaras with a terminal wing.
Leaves pinnately compound; samara with a single wing (*Fraxinus
dipetala*).   119. Oleaceae **3**: 346.
Leaves simple, palmately lobed (trifoliolate in variety of *Acer doug-
lasii*); samara with a double wing.
87. Aceraceae **3**: 56.
Leaves alternate; samara with a circular or ellipsoid wing.
Leaves trifoliolate, glandular-dotted (*Ptelea*).
75. Rutaceae **3**: 17.
Leaves pinnate, not glandular-dotted (*Ailanthus*).
76. Simaroubaceae **3**: 19.
Fruit not a samara.
Leaves simple or reduced to scales.
Flowers papilionaceaus; leaves round or round-reniform; fruit a
legume (*Cercis*). 68. Caesalpinaceae **2**: 478.
Flowers, leaves, and fruit not as in the preceding.
Fruit of several dry, stellately spreading, drupe-like carpels;
leafless, spinescent, desert shrubs (*Holocantha*).
76. Simaroubaceae **3**: 19.
Fruit not as in the preceding; shrubs not leafless and spinescent
(except *Adolphia* in *Rhamnaceae*; *Glossopetalon* in *Celas-
traceae*).
Stamens opposite the petals, 4 or 5; petals mostly hooded.
90. Rhamnaceae **3**: 59.
Stamens alternate with the petals, 10 or less; petals not
hooded.
Fruit a small drupe, often glandular; leaves persistent
(*Rhus* spp.).
84. Anacardiaceae **3**: 50.
Fruit a capsule or a 1–2-seeded berry.
Herbage strongly scented, conspicuously dotted
with punctate glands (*Thamnosma, Cneori-
dium*)
75. Rutaceae **3**: 17.
Herbage not strongly scented and glandular-punctate.
Seeds with long plumose hairs; flowers extremely numerous in dense spikes and
panicles.
97. Tamaricaceae **3**: 120.
Seeds glabrous or arillate; flowers few or, if
more abundant, then in leafy thyrsoid
cymes.
85. Celastraceae **3**: 53.
Leaves compound.
Leaves opposite; fruit a bladdery, 3-lobed capsule.
86. Staphyleaceae **3**: 55.
Leaves alternate (sometimes uppermost leaves opposite); fruit not a
bladdery, 3-lobed capsule.
Leaves palmately compound; seed 1 only, 2–3 cm. in diameter.
88. Aesculaceae **3**: 58.
Leaves pinnately or bipinnately compound, or trifoliolate (leaflets only 2 in *Larrea*); seeds much less than 1 cm. in
diameter.
Leaves deciduous.
Fruit a legume (*Cassia* sp., *Hoffmannseggia* sp., *Cer-
cidium*).
68. Caesalpinaceae **2**: 478.
Fruit a drupe, the covering deciduous, smooth or pubescent.

Leaflets many, 7–8 mm. long.
77. Burseraceae **3**: 20.

Leaflets, if more than 3, 2.5–10 cm. long (*Rhus* spp.).
84. Anacardiaceae **3**: 50.

Leaves persistent.

Fruit an inflated capsule with a stipe about as long as the capsule; leaves mostly 3-foliolate (*Isomeris*).
53. Capparidaceae **2**: 322.

Fruit sessile, either a berry or a schizocarp.

Fruit a berry; not desert shrubs of low elevations (*Berberis*).
47. Berberidaceae **2**: 216.

Schizocarp separating into 5 hirsute nutlets; common desert shrub of low elevations (*Larrea*).
74. Zygophyllaceae **3**: 14.

Herbs, annual or perennial.

Flowers monoecious (*Ditaxis*, *Croton*, with petals usually lacking in pistillate flowers of *Croton*).
79. Euphorbiaceae **3**: 23.

Flowers perfect.

Leaves compound, pinnate, bipinnate, trifoliolate, or pinnately or palmately dissected (see also *Cassia* sp., *Hoffmannseggia* sp. in *Caesalpinaceae*).

Sepals and petals reflexed at anthesis; fruit a 2-valved follicle (*Vancouveria*).
47. Berberidaceae **2**: 216.

Sepals and petals not reflexed at anthesis; fruit not as in the preceding.

Petals and sepals 5 (3 in *Floerkea*); stamens often 10.

Fruit seemingly of separate nutlets or a schizocarp separating into variously modified, indehiscent, normally 1-seeded parts; sap not acid.

Style persistent on the mature carpel-body and much surpassing it, either curved or coiled.
71. Geraniaceae **3**: 1.

Style not persisting on the mature carpel-body.

Leaves pinnate; styles terminal.
74. Zygophyllaceae **3**; 14.

Leaves pinnately dissected; styles arising from the base of the ovaries.
83. Limnanthaceae **3**: 46.

Fruit a capsule; sap with acid taste.
72. Oxalidaceae **3**: 8.

Petals and sepals 4; stamens 6.

Leaves pinnately or bipinnately divided.
52. Brassicaceae (Cruciferae) **2**: 237.

Leaves trifoliolate or palmately divided (reduced to 1 leaflet in *Wislizenia palmeri*).
53. Capparidaceae **2**: 322.

Leaves simple, entire, or the margins sometimes serrulate or crenulate (see also *Erodium* sp. in *Geraniaceae*).

Petals with a filamentous claw and hooded limb; stamens united in a column; low, woody-based, desert plant (*Ayenia*).
93. Sterculiaceae **3**: 112.

Petals, if more or less clawed, lacking a hooded limb; stamens free or united only at the base.

Fruit usually a silique or a silicle, the false partition remaining after the valves have fallen (indehiscent in *Raphanus*, *Thysanocarpus*); sepals and petals 4; stamens 6.
52. Brassicaceae (Cruciferae) **2**: 237.

Fruit not as in the preceding; stamens as many or twice as many as the petals.

Carpels free or joined basally, the fruit follicular; succulent or fleshy plants mostly of rocky ledges and crevices.
57. Crassulaceae **2**: 330.

Carpels united; fruit a capsule; plants not noticeably succulent (somewhat so in *Portulacaceae*).

Scapose herbs; flowers solitary; lobed or glandular-fringed staminodia always present.
58. Parnassiaceae **2**: 347.

Plants not scapose; staminodia if present not as in the preceding.

Stamens and petals borne on the tubular calyx.
105. Lythraceae **3**: 164.

Stamens and petals free from the calyx.

Seeds borne on the walls of the capsule; small-leaved plants of salt marshes and alkaline habitats (1 sp. a low shrub).
96. Frankeniaceae **3**: 119.

Seeds borne basally or on the central axis of the capsule; plants not of saline or alkaline habitats (e x c e p t *Calandrinia ambigua*, *Calyptridium* sp. in *Portulacaceae*; *Spergularia* spp. in *Caryophyllaceae*).

Stamens distinctly united into a ring at the base.
73. Linaceae **3**: 9.

Stamens free or nearly so.

Capsule completely 1-celled or incompletely 2–5-celled basally.

Sepals 2; stems never with swollen nodes; seeds usually few.

42. Portulacaceae **2**: 119.

>Sepals typically 5; stems with swollen nodes; seeds usually many.
>43. Caryophyllaceae **2**: 137.
>Capsule 3-celled; semiaquatic plants persisting in mud, except *Bergia*.
>95. Elatinaceae **3**: 118.

Sympetalae

Plants parasitic or saprophytic, either obviously attached to the host or lacking chlorophyll (or having both characters), or terrestrial and lacking chlorophyll.

Plants attached to the host above the ground level; plants twining.   127. Cuscutaceae **3**: 390.

Plants terrestrial, saprophytes or root-parasites.

Corolla regular, the petals free or fused (lacking in *Allotropa*).

Filaments free or united only at the base; plants of coniferous forests.
113. Monotropaceae **3**: 292.

Filaments adnate to the corolla for most of their length; plants of sandy deserts or the sandy seashore of southern California.   130. Lennoaceae **3**: 475.

Corolla irregular, bilabiate.   139. Orobanchaceae **4**: 3.

Plants not parasitic, or if partially so having chlorophyll (some genera of *Scrophulariaceae* partially dependent on surrounding hosts).

Ovary superior.

Corolla regular (lobes of equal length or nearly so; tube not gibbose or spurred).

Stamens 5 or less.

Fruit of paired or single follicles much exceeding the flower in length; seeds comose (except *Amsonia, Vinca*); juice milky in most genera.

Stamens connivent above the style; pollen not cohering in masses (pollinia) in each anther-cell.   123. Apocynaceae **3**: 367.

Stamens adnate to the style-column; pollinia of each anther-cell linked to that of a cell of the next anther.   124. Asclepiadaceae **3**: 372.

Fruit a capsule, a berry, or of 2–4 distinct nutlets; seeds not comose; plants without milky juice.

Stamens as many as the corolla-lobes and opposite them (staminodes sometimes present); some genera with corolla deeply parted seemingly into distinct petals.

Seeds few to many; calyx herbaceous.   116. Primulaceae **3**: 331.

Seed 1; calyx scarious.   117. Plumbaginaceae **3**: 344.

Stamens alternate with the corolla-lobes, of equal number or fewer; corollas not deeply parted.

Capsule circumscissile.

Capsule prominently 2-lobed; corolla not scarious; desert shrubs or shrubby perennials (*Menodora*).   119. Oleaceae **3**: 346.

Capsule not 2-lobed; corolla scarious; herbs.
142. Plantaginaceae **4**: 14.

Fruit if capsular not circumscissile.

Fruit of 2–4 nutlets (sometimes 1 by abortion).

Leaves alternate (rarely whorled or opposite); inflorescences mostly scorpioid.
132. Boraginaceae **3**: 532.

Leaves opposite; inflorescences of spikes or heads (*Verbena*).
133. Verbenaceae **3**: 609.

Fruit not of nutlets.

Anthers opening by terminal pores (*Rhododendron*).
114. Ericaceae **3**: 297.

Anthers opening by longitudinal slits.

Fruit a capsule.

Style 3-parted; ovary 3-celled; capsule opening by 3 valves.
128. Polemoniaceae **3**: 396.

Style, ovary, and capsule otherwise.

Rhizome thick and long, covered with membranous leaf-bases; perennials of lakes and bogs of high elevations.
122. Menyanthaceae **3**: 365.

Rhizome if present not thick; plants of various habitats, if semiaquatic not as in *Menyanthaceae*.

Ovary and capsule 2-lobed (see also *Veronica* spp., *Synthyris* spp. in *Scrophulariaceae*, with capsule more or less 2-lobed).
125. Dichondraceae **3**: 380.

Ovary and capsule not 2-lobed.

Densely tomentose desert shrubs of southeastern California; flowers congested in globose cymules.
120. Loganiaceae **3**: 350.

Herbs or, if shrubs (*Eriodyctyon, Turricula*), not as in the preceding.

Stem-leaves when present opposite (except *Verbascum*).

Capsule 1-celled; stamens 4 or 5.
121. Gentianaceae **3**: 350.

Capsule 2-celled; stamens 2 or 4 (5 in *Verbascum*).
136. Scrophulariaceae **3**: 686.

Stem-leaves alternate.

Calyx divided to the base, the sepals distinct.

Flowers mostly solitary, sessile or pedicellate in the leaf-axils, margin of
corollas entire or nearly so (except *Cressa*).
126. Convolvulaceae **3**: 380.
Inflorescence cymose, the branches usually coiled; corollas lobed.
131. Hydrophyllaceae **3**: 476.
Calyx toothed or shortly lobed, deeply so in
Petunia, Oryctes.
135. Solanaceae **3**: 662.
Fruit a many-seeded berry.   135. Solanaceae **3**: 662.
Stamens at least twice as many as the corolla-lobes (petals united only at base in some genera).*
Anthers opening by longitudinal slits; spiny shrubs of California deserts.
129. Fouquieriaceae **3**: 474.
Anthers opening by pores or tubes; plants not of the desert.
Herbs; corolla-lobes united only at the base.   112. Pyrolaceae **3**: 287.
Mostly shrubs or trees; corollas campanulate or urceolate (lobes nearly free in *Clado-
thamnus, Ledum*).   114. Ericaceae **3**: 297.
Corolla irregular, bilabiate or spurred, with lobes obviously unequal.†
Fruit 10–25 cm. long.
Deciduous desert trees; capsule linear, membranous.   137. Bignoniaceae **4**: 1.
Herbs, introduced, or indigenous to the desert; capsule woody beneath the deciduous exocarp,
the persistent style splitting into 2 curved horns. 138. Martyniaceae **4**: 1.
Fruit not more than 2 cm. long.
Plants aquatic with dissected leaves bearing minute bladders (except *Pinguicula* of bogs, with
entire basal leaves).   140. Lentibulariaceae **4**: 10.
Plants not aquatic.
Fruit a capsule, abruptly narrowed basally and bivalvate, the valves elastically dehiscing;
red-flowered desert shrubs.   141. Acanthaceae **4**: 13.
Fruit if a capsule not dehiscing elastically nor narrowed below.
Fruit of 2–4 nutlets; most genera with aromatic herbage.
Ovary 2-celled with 2 nutlets; style terminal (*Phyla, Aloysia*).
133. Verbenaceae **3**: 609.
Ovary 4-celled and -lobed with 4 nutlets; style arising from depression between
the ovary lobes.   34. Menthaceae (Labiatae) **3**: 614.
Fruit a several- to many-seeded capsule, opening by valves or pores.
136. Scrophulariaceae **3**: 686.
Ovary inferior (partially so in *Styracaceae*).
Stamens 10; deciduous shrubs with nut-like, 1-seeded fruits.   118. Styracaceae **3**: 345.
Stamens 2–5; if shrubs the fruit otherwise.
Stamens free; leaves opposite (alternate in *Campanuloideae*).
Stipules always present (leaf-like in *Galium*, the leaves thus appearing whorled).
143. Rubiaceae **4**: 22.
Stipules absent (except *Sambucus* in *Caprifoliaceae*).
Fruit a berry, drupe, or often modified achene (dry, 3-celled, 1-seeded fruit in *Linnaea*).
Shrubs or woody vines (*Linnaea* a creeping evergreen herb).
144. Caprifoliaceae **4**: 42.
Perennial or annual herbs.
Flowers irregular, the tube usually spurred or gibbous.
145. Valerianaceae **4**: 56.
Flowers regular; our species all introduced weeds.
146. Dipsacaceae **4**: 64.
Fruit a capsule, dehiscing by valve-like openings or splitting between the ribs (*Campanu-
loideae*).   148. Campanulaceae **4**: 72.
Stamens with either filaments or anthers or both parts completely or partially fused; leaves alternate
(sometimes opposite or completely basal in *Compositae*).
Individual flowers small, aggregated into a head on a receptacle and subtended by a calyx-like
involucre; the whole appearing flower-like; flowers asepalous.
149. Compositae‡ **4**: 98.
Flowers if in heads not aggregated as in the preceding; sepals present.
Fruit a juicy berry; stamens opening by pores (see also *Gaultheria*, with fleshy calyx).
115. Vacciniaceae **3**: 326.
Fruit capsular or a dry berry (pepo) either indehiscent or rupturing irregularly; stamens
opening by slits.
Flowers monoecious or dioecious, regular (deeply parted in *Marah*); vines with
tendrils.   147. Cucurbitaceae **4**: 65.
Flowers perfect, irregular; herbs (*Lobelioideae*).
148. Campanulaceae **4**: 72.

---

* See *Choripetalae* for genera with petals cohering at the base in *Crassulaceae, Oxalidaceae, Mimosaceae,
Malvaceae*.

† See *Choripetalae* for genera with petals partly united in *Fumariaceae, Fabaceae, Polygalaceae, Krameriaceae*.

‡ Marginal flowers of the heads often with colored rays, as in sunflower and daisy, or all flowers of the
heads bearing such appendages, as in dandelion, or all flowers without such appendages, as in thistle; corolla
vestigial in some *Ambrosiinae*).

INDEX OF COMMON NAMES

[Families in SMALL CAPITALS; genera in Roman; Roman numerals refer to volume numbers.]

Abrojo
  California, III, 61
  Gray, III, 61
  Spiny, III, 61
Abronia
  Alpine, II, 111
  Beach, II, 110
  Coville's Dwarf,
    II, 110
  Desert, II, 108
  Honey-scented, II, 108
  Mojave, II, 111
  Red, II, 110
  Transmontane, II, 111
  Yellow, II, 110
Absinthe, IV 410
Abutilon, Dwarf,
  III, 83
Acacia, Gregg's, II, 475
Acaena, California,
  II, 446
Acalypha, California,
  III, 28
ACANTHUS FAMILY,
  IV, 13
ADDER'S-TONGUE
  FAMILY, I, 1
Adder's Tongue, I, 2
Fetid
  California, I, 449
  Oregon, I, 450
  Western, I, 2
Adolphia, California,
  III, 81
Agastache
  Cusick's, III, 626
  Small-leaved, III, 626
Agave
  Desert, I, 460
  Shaw's, I, 461
  Utah, I, 460
Agoseris
  Annual, IV, 566
  Large-flowered,
    IV, 565
  Orange-flowered,
    IV, 564
  Seaside, IV, 566
  Short-beaked, IV 562
  Spear-leaved, IV, 565
  Tall, IV, 564
Agrimony, Tall Hairy,
  II, 446
Ajamete, III, 378
Alder
  Red, I, 513
  Sitka, I, 513
  Thin-leaved, I, 514
  Wavy-leaved, I, 513
  White, I, 514
Alexanders, Heart-
  leaved, III, 241
Alfalfa, II, 520;
  III, 395, 396
Alkanet, III, 541
All-heal, IV, 56
Allionia, II, 104
Allocarya

Adobe, III, 561
  Artist's, III, 563
  Austin's, III, 562
  Bearded, III, 561
  Bracted, III, 570
  California, III, 571
  Calistoga, III, 561
  Coast, III, 563
  Cognate, III, 567
  Cooper's, III, 566
  Cusick's, III, 567
  Diffuse, III, 570
  Downy, III, 559
  Dwarf, III, 559
  Fragrant, III, 565
  Glabrous, III, 563
  Greene's, III, 561
  Harsh, III, 570
  Oregon, III, 566
  Rough, III, 565
  Rough-fruited,
    III, 570
  Salty, III, 566
  Scouler's, III, 566
  Scribe's, III, 559
  Sculptured,
    III, 562, 565
  Slender, III, 567
  Smooth-stemmed,
    III, 562
  Stipitate, III, 563
All-scale, II, 86
Aloysia, Wright's,
  III, 613
Alum Root, II, 381
Alyssum
  Sweet, II, 319
  Yellow, II, 319
AMARANTH FAMILY,
  II, 98
Amaranth
  California, II, 99
  Fringed, II, 99
  Green, II, 98
  Low, II, 99
  Palmer's, II, 99
  Prostrate, II, 99
  Spleen, II, 98
AMARYLLIS FAMILY,
  I, 459
Amblyopappus, IV, 248
Ammannia, Long-
  leaved, III, 165
Ammi, Toothpick,
  III, 230
Amole
  Common, I, 413
  Narrow-leaved, I, 414
  Purple, I, 414
  Small-flowered, I, 414
Amsinckia
  Douglas', III, 604
  Seaside, III, 606
Amsonia
  Short-leaved, III, 367
  Woolly, III, 368
Anchusa, Italian,
  III, 541

Androsace
  California, III, 337
  Northern, III, 337
  Slender, III, 337
  Western, III, 337
Androstephium,
  Small-flowered, I, 410
Anemone
  Desert, II, 191
  Drummond's, II, 191
  Globose, II, 191
  Northern, II, 191
  Small-flowered, II, 191
  Western Wood,
    II, 191
Angelica
  Brewer's, III, 272
  California, III, 269
  Canby's, III, 272
  Henderson's, III, 269
  King's, III, 269
  Kneeling, III, 272
  Lyall's, III, 272
  Seacoast, III, 269
  Sea-watch, III, 269
  Sierra, III, 269
Anisocoma, IV, 584
Anoda, Crested,
  III, 110
Antelope Bush
  Mojave, II, 451
  Northern, II, 451
Antelope Horns,
  III, 378
Apache Plume, II, 450
Apargidium, IV, 556
Aphanisma, II, 68
APPLE FAMILY, II, 469
Aristida
  California, I, 125
  Divaricate, I, 125
  Fendler's, I, 128
  Few-flowered, I, 126
  Parish's, I, 126
  Purple, I, 127
  Reverchon's, I, 127
  Schiede's, I, 125
  Wright's, I, 127
Arnica
  Cordilleran, IV, 420
  Heart-leaved, IV, 424
  Hillside, IV, 422
  Klamath, IV, 426
  Lawless, IV, 421
  Meadow, IV, 419
  Mountain, IV, 423
  Nodding, IV, 422
  Rayless, IV, 426
  Recondite, IV, 424
  Seep-spring, IV, 420
  Serpentine, IV, 424
  Shasta, IV, 426
  Sierra, IV, 424
  Streambank, IV, 420
  Twin, IV, 422
  Veiny, IV, 426
ARROW-GRASS FAMILY,
  I, 96

Arrow-grass
  Marsh, I, 96
  Seaside, I, 97
  Three-ribbed, I, 96
Arrow-head
  Arum-leaved, I, 101
  Broad-leaved, I, 101
  Gregg's, I, 102
  Montevideo, I, 102
  Sanford's, I, 102
Arrow-leaf, IV, 498
Arrow-weed, IV, 466
Artichoke
  Garden, IV, 513
  Globe, IV, 513
  Jerusalem, IV, 116
  Thistle, IV, 513
ARUM FAMILY, I, 345
Ash
  Arizona, III, 347
  California Flowering,
    III, 347
  Dwarf, III, 347
  Fringe-flowered,
    III, 347
  Foothill, III, 347
  Oregon, III, 346
Asparagus, I, 452
Aspen, American, I, 486
Astephanus, Utah,
  III, 372
Aster
  Alkali
    Marsh, IV, 330
    Rayless, IV, 332
    Short-rayed, IV, 330
    Shrubby, IV, 330
  Alpine, IV, 328
    Plumas, IV, 328
  Arctic, IV, 314
  Ash-colored, IV, 333
  Brickellbush, IV, 323
  Cascade, IV, 324
  Cliff-, IV, 580
  Common California,
    IV, 320
  Cudweed, IV, 341
  Desert, IV, 336
  Douglas', IV, 318
  Eaton's, IV, 319
  Elegant, IV, 325
  Engelmann's, IV, 326
  Entire-leaved, IV, 313
  Geyer's, IV, 316
  Gorman's, IV, 326
  Great Northern,
    IV, 314
  Greata's, IV, 316
  Hall's, IV, 319
  Heath-like, IV, 322
  Hoary, IV, 334
  Klickitat, IV, 326
  Laguna, IV, 333
  Lava, IV, 328
  Leafy, IV, 318
  Little Gray, IV, 322
  Long-leaved, IV, 320
  Marsh, IV, 316

INDEX OF SCIENTIFIC NAMES

[Families and tribes in SMALL CAPITALS ; genera and species in Roman ; synonyms and references in *italic*.]

653